Food, Pharmaceutical and Bioengineering Division

Core Programming Topic at the 2012 AIChE Annual Meeting

Pittsburgh, Pennsylvania, USA
28 October - 2 November 2012

Volume 1 of 2

ISBN: 978-1-62276-728-1

Printed from e-media with permission by:

Curran Associates, Inc.
57 Morehouse Lane
Red Hook, NY 12571

proceedings.com

Some format issues inherent in the e-media version may also appear in this print version.

Printed by Curran Associates, Inc. (2013)

AIChE
3 Park Avenue
New York, NY 10016-5991

Phone: (203) 702-7660
Fax: (203) 775-5177

www.aiche.org

Additional copies of this publication are available from:

Curran Associates, Inc.
57 Morehouse Lane
Red Hook, NY 12571 USA
Phone: 845-758-0400
Fax: 845-758-2634
Email: curran@proceedings.com
Web: www.proceedings.com

TABLE OF CONTENTS

Volume 1

Volume 2

(6a) Topical Plenary: Nanotechnology and Bioengineering in an Evolving Chemical Engineering World: The Next Generation of Recognitive, Intelligent Medical Microdevices

Nicholas A. Peppas, Dept. of Chemical Engineering, The University of Texas at Austin, Austin, TX

We will discuss first how the field of chemical engineering has contributed to major developments in bioengineering and nanotechnology, offering a historical prospective of the field and its leading scientists. Then, we will discuss how engineering the molecular design of intelligent biomaterials by controlling recognition and specificity is the first step in coordinating and duplicating complex biological and physiological processes. We will address design and synthesis characteristics of artificial molecular structures capable of specific molecular recognition of biological molecules. Recent developments in protein delivery have been directed towards the preparation of targeted formulations for protein delivery to specific sites, use of environmentally-responsive polymers to achieve pH- or temperature-triggered delivery, usually in modulated mode, and improvement of the behavior of their mucoadhesive behavior and cell recognition. Molecular imprinting and microimprinting techniques, which create stereo-specific three-dimensional binding cavities based on a biological compound of interest can lead to preparation of biomimetic materials for intelligent drug delivery, drug targeting, and tissue engineering. We have been successful in synthesizing novel glucose-binding molecules based on non-covalent directed interactions formed via molecular imprinting techniques within aqueous media.

(11a) Modular Biosynthesis for the Production of Advanced Fuels and Chemicals

Ramon Gonzalez, Departments of Chemical & Biomolecular Engineering and Bioengineering, Rice University, Houston, TX

Advanced (long-chain) fuels and chemicals are generated from short-chain metabolic intermediates through pathways that require carbon-chain elongation. The condensation reactions mediating this carbon-carbon bond formation can be catalyzed by enzymes from the thiolase superfamily, including beta-ketoacyl-acyl-carrier protein (ACP) synthases, polyketide synthases, 3 hydroxy-3-methylglutaryl-CoA synthases, and biosynthetic thiolases. Pathways involving these enzymes have been exploited for fuel and chemical production, with fatty acid biosynthesis (β ketoacyl ACP synthases) attracting the most attention in recent years. Degradative thiolases, which are part of the thiolase superfamily and naturally function in the b oxidation of fatty acids, can also operate in the synthetic direction and thus enable carbon-chain elongation. We have demonstrated that a functional reversal of the β-oxidation cycle can be engineered in *Escherichia coli* and used for the synthesis of alcohols and carboxylic acids of various chain lengths. The engineered reversal of the β oxidation cycle is composed of three core modules: i) priming, ii) elongation, and iii) termination. This modular nature allows the use of the engineered pathway as a general platform for the synthesis of a wide array of functional molecules with applications to fuel and chemical production.

(11b) Production of Bioethanol From Saccharomyces Pastorianus and Escherichia Coli Using a Two-Stage Fermentation Process

Yogender Kumar Gowtham, Bioengineering, Clemson University, Clemson, SC, Kristen P. Miller, Microbiology, Clemson University, Clemson, SC, J. Michael Henson, Biological Sciences, Clemson University, Clemson, SC and Sarah W. Harcum, Department of Bioengineering, Clemson University, Clemson, SC

Non-renewable energy sources are being consumed at an alarming rate and are estimated to be depleted over the next 25 to 30 years. Hence there is a great need for an alternative source of energy. Bioethanol is a biofuel that is produced from lignocellulosic biomass, which primarily consists of plant wastes, forestry residues, paper waste etc. Lignocellulosic biomass is rich in cellulose and hemicellulose, the majority of which are converted into glucose and xylose via pretreatment and hydrolysis processes. Yeast can easily convert glucose into ethanol; however, converting xylose to ethanol is a major hurdle in producing economic levels of bioethanol. This major hurdle is primarily due to the lack of appropriate xylose metabolizing pathways in yeast, of the genus *Saccharomyces*. Previously we showed that xylose isomerase could improve ethanol productivity. Additionally other microorganisms can metabolize xylose; including *Escherichia coli*. In this study, a novel two-stage fermentation process was proposed to improve the bioethanol productivity. In the first stage, glucose in the hydrolysate was converted to ethanol by *S. pastorianus*. After the fermentation was completed, the mixture of ethanol, biomass and unconverted xylose was separated. In the second stage, the xylose separated was used as a feed for *E. coli* to grow. The *E. coli* cells obtained were killed by heating and recycled back to first stage yeast fermentation. Overall the combination of xylose isomerase and killed *E. coli* extract had significantly higher ethanol productivity.

(11c) Engineering Acetogenic Clostridia for Ethanol and n-Butanol Production From CO2

Chih-Chin Chen, Department of Chemical and Biomolecular Engineering, The Ohio State University , Columbus, OH and Shang-Tian Yang, Department of Chemical and Biomolecular Engineering, The Ohio State University, Columbus, OH

Some *Clostridia*, including *C. ljungdahlii* and *C. carboxidivorans*, are capable of producing ethanol and butanol from CO_2 and H_2. However, the alcohol productivity, yield, and titer from these native strains are low and uneconomical for industrial applications. In contrast, several acetogenic *Clostridia* can convert CO_2 and H_2 to acetate with high yield, titer, and productivity comparable to those from glucose as the substrate. A high metabolic flux from CO_2 to acetyl-CoA, the immediate precursor for both acetate and ethanol, is the prerequisite for developing a highly productive ethanol and butanol producers. In this work, we engineered homoacetogens for ethanol and n-butanol production by introducing aldehyde/alcohol dehydrogenase and genes in the acetyl-CoA to butyryl-CoA pathway using modular clostridia plasmids. The fermentation kinetics of these engineered mutants and their ability to produce ethanol and butanol from CO_2 and H_2were studied and will be presented in this paper.

AIChE 2012 Annual Meeting, Pittsburgh

(11d) Bacteria Engineered to Produce the Chiral Building Block (S)-Styrene Oxide From Renewable Feedstocks

Rebekah McKenna[1], Shawn Pugh[1], Matthew Sawtelle[2] and **David R. Nielsen**[1], (1)Chemical Engineering, Arizona State University, Tempe, AZ, (2)Chemical Engineering, Arizona State University

Aromatic compounds represent a diverse class of fine and commodity chemicals with important commercial applications ranging from their use as molecular building blocks, flavor agents, and monomers. Today, conventional chemocatalytic synthesis routes for all aromatic compounds rely upon petroleum-derived BTEX (benzene, toluene, ethylbenzene, and xylenes) compounds as feedstock. However, through *de novo* pathway engineering, our lab has been exploring the 'bottom up' development of microbial biocatalysts to produce a number of useful aromatics renewable feedstocks. One such example is the chiral building block (*S*)-styrene oxide whose conventional applications include as a reactive plasticizer and as a chemical intermediate for cosmetics, surface coatings, and agricultural and biological chemicals. Enantiopure (*S*)-Styrene oxide production from glucose was accomplished by the co-expression of styrene monooxygenase (SMO) in a previously engineered styrene producing strain of *Escherichia coli*. SMO from 3 strains of styrene-degrading Pseudomonas was assayed for function and performance. Ultimately, styAB from *P. putida* S12 was found to display the greatest inherent activity. Expressing *styAB* with *PAL2* from *Arabidopsis thaliana* and *FDC1* from *Saccharomyces cerevisiae* in a phenylalanine over-producing strain of *E. coli*, initial titers of more than (*S*)-styrene oxide titers exceeded 1 g/L in shake flask cultures. Metabolite flux has been improved and balanced through modulation of pathway enzyme expression, including PAL2 the initial pathway bottleneck. This was achieved through the systematic tuning of each of gene copy number, promoter strength, and RBS strength. As a result, current titers now approach the toxic threshold of (*S*)-styrene oxide, determined as ~1.6 g/L. The (*S*)-styrene oxide pathway has most recently been translated into the aromatic tolerant host *P. putida* S12 upon its engineering to: i) over-produce phenylalanine and ii) eliminate its natural ability to degrade (*S*)-styrene oxide. The resultant strain has been characterized with regards to its ability to produce (*S*)-styrene oxide from different renewable substrates and survive in presence of elevated product titers.

(11f) A Synthetic E. Coli Consortium for Efficient Conversion of Hexose and Pentose Monomers and Oligomers to Isobutanol

Alissa Kerner, Department of Biomedical Engineering, University of Michigan, Ann Arbor, MI, Jeremy J. Minty, Department of Chemical Engineering, University of Michigan, Ann Arbor, MI, Bruce E. Dale, Chemical Engineering and Material Science, Great Lakes Bioenergy Center, Michigan State University, Lansing, MI, Venkatesh Balan, Chemical Engineering and Material Science, Michigan State University, East Lansing, MI and Xiaoxia (Nina) Lin, Chemical Engineering, University of Michigan, Ann Arbor, MI

Complete and efficient co-fermentation of both hexose and pentose mono- and oligo-saccharides is one of the largest obstacles to cost-effective and scalable lignocellulosic biofuel production. To address this issue we have engineered a consortium of two isobutanol-producing *E. coli* strains which utilizes both hexose and pentose monosaccharides. Wild-type *E. coli* can serve as the hexose specialist due to its preference for hexose sugars, whereas the pentose specialist was created by deleting genes involved in hexose transport and assimilation. We then further engineered these strains to utilize oligomers, including cellobiose and xylodextrin, for maximum conversion of sugar mixtures generated from the enzymatic hydrolysis of cellulosic materials. Next, we investigate the growth characteristics and isobutanol production of this engineered consortium with minimal media containing various combinations of the sugars. In addition, we test the synthetic consortium on AFEX-pretreated corn stover hydrolysates in an effort to optimize isobutanol productivity on an industrially relevant feedstock.

(11g) High-Yield and High-Titer n-Butanol Production From Lignocellulosic Biomass by Engineered Clostridium Tyrobutyricum

Yinming Du, Chemical and Biomolecular Engineering, The Ohio State University, Columbus, OH and Shang-Tian Yang, Department of Chemical and Biomolecular Engineering, Ohio State University, Columbus, OH

The cost of substrates and nutrients, including organic nitrogen source and carbon materials, is one of the key factors that could determine the development and application of biofuels as an alternative transportation fuel. In order to make the biofuels more cost-effective and competitive over petrochemical fuels, we investigated the utilization of corn steep liquor (CSL) and lignocellulosic biomass for n-butanol, a superior biofuel over bioethanol, production by a metabolically engineered *Clostridium tyrobutyricum*. In general, a high n-butanol titer of ~17 g/L with a yield of 0.33 g/g glucose, 80% of the theoretical yield, was achieved in batch fermentation using corn steep liquor as the nitrogen source and glucose as the carbon source. Further improvements in n-butanol titer (~100 g/L) and yield (~0.35 g/g glucose) were achieved by applying integrated gas-stripping process for *in situ* butanol recovery and cell immobilization in a fibrous-bed bioreactor (FBB). Next, the feasibility of using various lignocellulosic biomass, including Jerusalem artichoke, cotton stalk, sugarcane bagasse, and corn fiber, as alternative feedstocks for fermentative n-butanol production was studied. The results showed that *C. tyrobutyricum* was able to efficiently convert lignocellulosic biomass hydrolysates containing glucose and xylose to n-butanol, achieving a high titer of over 15 g/L and a high yield of over 0.30 g/g substrate. This study demonstrated an economically competitive process for n-butanol production from abundant and renewable lignocellulosic biomass.

(11h) Engineering the Cyanobacterium Synechococcus Sp. PCC 7002 to Produce Commodity Chemicals

Matthew B. Begemann, Microbiology Doctoral Training PROGRAM, UNIVERSITY OF WISCONSIN-MADISON, Madison, WI, Daniel Mendez-Perez, Department of Chemical and BIOLOGICAL Engineering, UNIVERSITY OF WISCONSIN-MADISON, Madison, WI and **Brian F. Pfleger**, Chemical and Biological Engineering, University of Wisconsin Madison, Madison, WI

Metabolic engineering of cyanobacteria allows for the direct conversion of CO_2 and sun light into biofuels and petrochemical replacements without the need for plant derived sugars. In addition to the photosynthetic abilities of these organisms, their inherent physiology and metabolism may make them more effective hosts for certain pathways than heterotrophic bacteria. The goal of this research is to demonstrate a pathway for the production of 3-hydroxypropionate in which cyanobacteria are a more effective host than *E. coli*. 3-hydroxypropionate (3HP) is a compound of interest because it can be easily converted to a variety of commodity chemicals including acrylic acid. The cyanobacterium *Synechococcus* sp. PCC 7002 was chosen as a potential host because of its fast growth rate, tractable genetics systems, and wide tolerance range to salinity and light intensity. The first hurdle in engineering this system was addressing end product toxicity. A loss of function mutation was identified that increases the minimum inhibitory concentration of 3HP 50-fold. This mutation was in the gene *acsA*, which was annotated as any acetyl-CoA ligase. Knocking out *acsA* resulted in increased tolerance to 3HP, acrylic acid, and propionate. An enzyme assay of the purified protein demonstrated that acsA has activity towards these compounds. The *acsA* knock out strain was used as a starting platform for 3HP production. The pathway chosen for 3HP production involves the heterologous expression of a malonyl-CoA reductase from *Chloroflexus aurantiacus*. This enzyme converts malonyl-CoA to 3HP. This reaction requires the cofactor NADPH, which is at a high concentration in cyanobacteria relative to *E. coli*. Expression of this gene from a native multi-copy plasmid resulted in the accumulation of 3HP. In order to increase the production of 3HP from malonyl-CoA, additional acetyl-CoA carboxylase genes will be expressed to increase the intracellular concentration of malonyl-CoA. The engineered strains from these experiments will be grown in a photobioreactor to determine how 3HP yields compare to *E. coli* strains expressing the same pathway.

(12a) Heterologous Expression of Three Tetracycline Biosynthetic Pathways and Identification of New Tailoring Enzymes

Peng Wang, UCLA, Los Angeles, CA and Yi Tang, Chemical and Biomolecular Engineering, University of California, Los Angeles, Los Angeles, CA

Tetracyclines are members of the polyketide family produced by microbial sources. The chemical complexity of tetracyclines leads to a variety of biological activities including antibacterial, anticancer and activities against tetracycline resistant strains. Therefore, understanding the biosynthesis of tetracyclines not only provides genetic and biochemical insights but also sets the stage to engineer biosynthesis for generating new tetracycline analogs. To this end, we here report the heterologous expression of three structurally diverse tetracyclines and the identification of a number of new tailoring enzymes that can modify the tetracycline scaffold.

(12b) Synthetic Biology Enabled Refactoring of Cryptic Fungal Gene Clusters for Natural Product Discovery

Ryan E. Cobb, Chemical and Biomolecular Engineering, University of Illinois at Urbana-Champaign, Urbana, IL and Huimin Zhao, Chemical and Biomolecular Engineering, University of Illinois at Urbana Champaign, Urbana, IL

In recent years, the exponential increase in genome sequence data has revealed a myriad of secondary metabolite gene clusters across all domains of life for which the cognate product is unknown. The presence of these so-called "cryptic" gene clusters suggests that many organisms possess the potential to produce a far greater diversity of natural products than have ever been detected in a laboratory setting. Many methods have been developed to unlock this potential in the native host, including screening of varied culture conditions, co-culture with different competitive species, over-expression of putative pathway-specific regulators, and addition of epigenetic factors. However, the results of these approaches vary on a case-by-case basis, and are difficult or even impossible to predict *a priori*. As a result, it is our goal to develop a robust method for natural product discovery based on complete refactoring of the target cryptic pathway and heterologous expression in a well-characterized, highly tractable host. Our current focus is on the refactoring of fungal polyketide synthase (PKS) clusters and their expression in *Saccharomyces cerevisiae*. One such cluster has been identified from *Penicillium marneffei*, a pathogenic fungus containing 25 PKS genes but only one known polyketide metabolite. The target cluster contains two PKS genes with ~40 % amino acid identity to those of multiple known resorcylic acid lactone (RAL) gene clusters. RALs are a class of macrocyclic compounds that possess a wide range of bioactivities, from antibacterials and antifungals to kinase inhibitors. While multiple fungal RALs have previously been identified, the three tailoring genes found in the target cluster do not match those of any other known cluster, suggesting that the target cluster encodes a novel member of the RAL family. Using the DNA assembler method, we have rapidly removed the introns from each gene in the cluster and constructed an expression vector with each gene placed under a characterized constitutive promoter. We have also constructed a set of control vectors with individual genes disrupted. Transcription of the refactored heterologous genes in *S. cerevisiae* has been confirmed by qRT-PCR. A novel product was detected by HPLC-MS/MS and its structural characterization is currently in progress.

(12c) Development of a Synthetic Biology Approach to Demystify the Target Cryptic Pathway for Novel Natural Product Discovery

Yunzi Luo[1], Zengyi Shao[1] and Huimin Zhao[2], (1)Chemical & Biomolecular Engineering, University of Illinois at Urbana-Champaign, Urbana, IL, (2)Department of Chemical and Biomolecular Engineering, Department of Chemistry and Biochemistry, Institute for Genomic Biology, University of Illinois at Urbana-Champaign, Urbana, IL

The increasing availability of whole genome sequences has revised our view of the metabolic capabilities of microorganisms. Analyses of more than 500 microbial genome sequences currently in the publicly-accessible databases have revealed numerous examples of gene clusters encoding enzymes similar to those known to be involved in the biosynthesis of many important natural products. We recently developed a highly efficient one-step method, so-called "DNA assembler", for rapid construction of a multi-gene biochemical pathway via *in vivo* homologous recombination in *Saccharomyces cerevisiae*. Here we combined this synthetic biology approach with metabolic engineering to study cryptic biosynthetic pathways. As proof of concept, one pathway from *Streptomyces griseus*, SGR810-815, containing a PKS/NRPS hybrid gene and several genes encoding potential tailing enzymes, was chosen as a model system. To decipher this target gene cluster, a library of promoters with different strengths was discovered and characterized and the promoter-gene cassettes were assembled by using DNA assembler. Verified constructs were selected and transformed into *Streptomyces lividans* for heterologous expression of the target gene cluster. qRT-PCR was used to track the transcription of each gene. It was found that the original gene cluster showed poor expression of every gene and the native promoter(s) were inactive. In contrast, our refactored gene cluster showed enhanced levels of gene expression for each gene and the levels were comparable to the native producers'. LC-MS analysis of the cell extracts from the constructs indicated that a peak with molecular weight of 510 may be the potential product of the gene cluster. The compound was purified and the structure determination by NMR is in progress. Meanwhile, gene deletion was performed to elucidate the biosynthetic mechanism of the target pathway and the minimal set of genes involved in the target pathway.

(12d) Towards Understanding the Kinetic Enhancements in Spatially Organized Multi-Enzyme Structures

Jyun-Liang Lin and Ian Wheeldon, Chemical and Environmental Engineering, UC Riverside, Riverside, CA

In nature, multi-enzyme pathways are often spatially organized into well-defined nanostructures with precise inter-enzyme distances and quaternary structures. Such organization creates the chemical and physical conditions that allows for complex reactions to occur with high specificity and yield in highly heterogeneous environments. Here, we use DNA nanostructures as scaffolds with controlled inter-enzyme distance and shape to study kinetic enhancements in multi-enzyme complexes. The commonly used enzyme horseradish peroxidase (HRP) is used to demonstrate the beneficial effects that DNA scaffolds can impart on single reactions. Additionally, DNA nanostructures and chemical crosslinkers are used to link enzymes in cascade reactions. The glucose oxidase (GOx) and HRP enzyme pair are used to demonstrate the kinetic benefit of spatially organized cascade reactions. The inter-enzyme distances and shapes of multi-enzyme/DNA nanostructures are characterized by Förester resonance energy transfer (FRET) and atomic force microscopy (AFM), respectively. Kinetic enhancements are evaluated by determining the apparent Michaelis-Menton binding constants of the enzyme/DNA nanostructures and by determining the fold enhancement of the initial reaction rates in single and cascade reactions, respectively. In this work we begin to develop a fundamental understanding of the underlining principles that govern the kinetic benefits of spatial organization that are essential to the engineering of multi-enzyme reaction systems.

(12e) Solvent-Enhanced Biotransformations of Chemicals by Beauveria Bassiana As Biocatalyst

Richard González and Tonya L. Peeples, Chemical and Biochemical Engineering, University of Iowa, Iowa City, IA

This research evaluates the oxidative capacity of n-alkane-induced fungal *Beauveria bassiana;* a versatile whole cell biocatalyst used in the biotransformation of steroids, and in the production of valuable organic compounds. n-alkane solvents like dodecane, and hexadecane serve as carbon sources during growth of *B. bassiana,* inducing the expression of oxidative enzymes, and affecting the growth rate. n-Piperidinylacetophenone is a classic substrate used for the hydroxylation of an unfunctionalized carbon; an attractive reaction that results in valuable intermediates for specialty chemical synthesis. As well, steroid 7-alpha-hydroxy-4-cholesten-3-one (NS) is hydrolylated to 7-alpha-12 alpha-dihydroxy-4-cholesten-3-one, to provide metabolites of interest for medical purposes. Our research targets the hydroxylated products of these chemicals. Reactions with n-alkane adapted strains (ATCC 90517, ATCC 90518, ATCC 7159) on a resting cell environment; were grown in media containing different nitrogen sources (Corn Step Liquor, Soybean flour, peptone, and ammonium nitrate). 1H-NMR, and HPLC were used to characterize, and isolate the biotransformation products. Results include biotransformation selectivity, reaction yield, n-alkane effect analysis in biocatalyst performance, and correlations between biomass and reaction yields.

(12f) Nitroreductases: A Biocatalytic Alternative to Produce Aromatic Amines

Jonathan T. Park[1], Yanto Yanto[1], Kyle Ferguson[2] and Andreas S. Bommarius[1], (1)School of Chemical and Biomolecular Engineering, Georgia Institute of Technology, Atlanta, GA, (2)Chemistry and Biochemistry, Georgia Institute of Technology, Atlanta, GA

Nitroreductases are enzymes that reduce nitro compounds to corresponding nitroso, hydroxylamino, and amino compounds. In the past 20 years, they have gathered interest in several applications such as degradation of xenobiotic nitroaromatics, pharmaceutical production of amines, and activation of prodrugs in medical applications.

In previous studies, nitroreductases have shown to produce the hydroxylamine equivalent of the substrate. The instability and carcinogenicity of this product has been a barrier in using this enzyme. Also, the lack of a method to detect aromatic amines over hydroxylamines was a major barrier in improving the selectivity of nitroreductases

We show the development of a novel high-throughput colorimetric assay that can identify the aromatic amine. This assay is used to screen multiple libraries that consist of mutations in the active site. Variants that were identified to have improved production of amine were characterized and compared to the wild-type enzyme.

In addition, by studying nitroreductases from *Salmonella typhimurium* (NRSal), *Enterobacter cloacae* (EntNR) and *Mycobacterium smegmatis* (NfnB) we have found a Hammett correlation in the reduction rate of various substituted nitro aromatic compounds. This in turn helps us identify specific targets for large-scale synthesis.

(12g) Enzymatic Oligomerization of Resveratrol to Generate Novel Anti-Microbial Analogs

Namita Bhan[1], Mauricio Mora-Pale[1], Julia Wood[1], Jonathan S. Dordick[2] and Mattheos A.G. Koffas[3], (1)Chemical and Biological Engineering, Rensselear Polytechnic Institute, Troy, NY, (2)Dept of Chemical & Biological Engineering, Rensselaer Polytechnic Institute, Troy, NY, (3)Chemical and Biological Engineering, Rensselaer Polytechnic Institute, Troy, NY

Enzymatic oligomerization of Resveratrol to generate novel anti-microbial analogs

Plant polyphenols such as flavonoids and stilbenoids have been associated with several health benefits and thus have been widely studied in recent years. Despite the current controversy regarding their potential for pharmacological applications, recent studies suggest that natural phenols can be converted into active metabolites *in vivo*, and more attention is required on the *in vitro* synthesis of analogs to enhance their efficacy. Also, a constant growth of anti-microbial resistance and rapid microbial evolution; underscores the importance of identification of alternatives to generate new and novel antimicrobial compounds. We propose that the antimicrobial activity can be significantly enhanced by generating oligomers of these compounds. In particular, we produced, isolated and characterized a dimer from resveratrol (d-viniferin, the tri-dehydrodimer of resveratrol) from an *in vitro* enzymatic oligomerization reaction mediated by soybean peroxidase (SBP) in aqueous buffer. Dose response studies in gram- positive and gram- negative bacteria show great antimicrobial potential of d-viniferin as compared to resveratrol. For example, d-viniferin showed MIC and IC_{50} values of 125 μM and 59 μM respectively against *E. coli* (in the presence of efflux pump inhibitor Phe-Arg-ß-naphthylamide, PABN, 25 μg/mL), while resveratrol shows MIC and IC_{50} values of 2000 μM and 250 μM respectively. Significant improvement was observed against *B. cereus* with d-viniferin giving MIC and IC_{50} values of 5 μM and 15 μM, while for resveratrol MIC and IC_{50} values were >150 μM, all in presence of the efflux pump inhibitor 1-(1-Naphthylmethyl) piperazine (NMP).

The significant increase in anti-microbial activity in presence of efflux pump inhibitor PABN in the case of *E. coli* indicated that efflux pumps are responsible for the extrusion of resveratrol and its analogs from the cell. Thus, d-viniferin was tested against an *E. coli* strain with a deletion of *tolC* gene (part of AcrAB-tolC efflux pump) in the presence and absence of PABN. In both cases, significant but similar growth inhibition was observed ([+]PABN, MIC and IC_{50} values were 1.2 and 3 μM respectively; [-]PABN, MIC and IC_{50} values were 2.5 and 7 μM respectively). Moreover, in the absence of PABN both the dimer and monomer had no effect on *E. coli* at these concentrations. The effect of efflux pump inhibitor on *B. cereus* was less significant; MIC and IC_{50} values in the absence of NMP were 12 and 30 μM respectively.

Visual evidence discarded the possibility of d-viniferin causing cellular aggregation leading to a lower cell density or count. Also, morphological changes in *B. cereus* were observed before and after incubation with d-viniferin; there was a decrease in the cell size while cell wall was found to be unaltered and intact.

In summary, it was demonstrated that novel and active antimicrobials from plant polyphenols can be generated *in vitro*. In particular, d-viniferin has shown great antimicrobial activity and its low toxicity against mammalian cells provides a new insight in the development of antimicrobial with potential therapeutic applications.

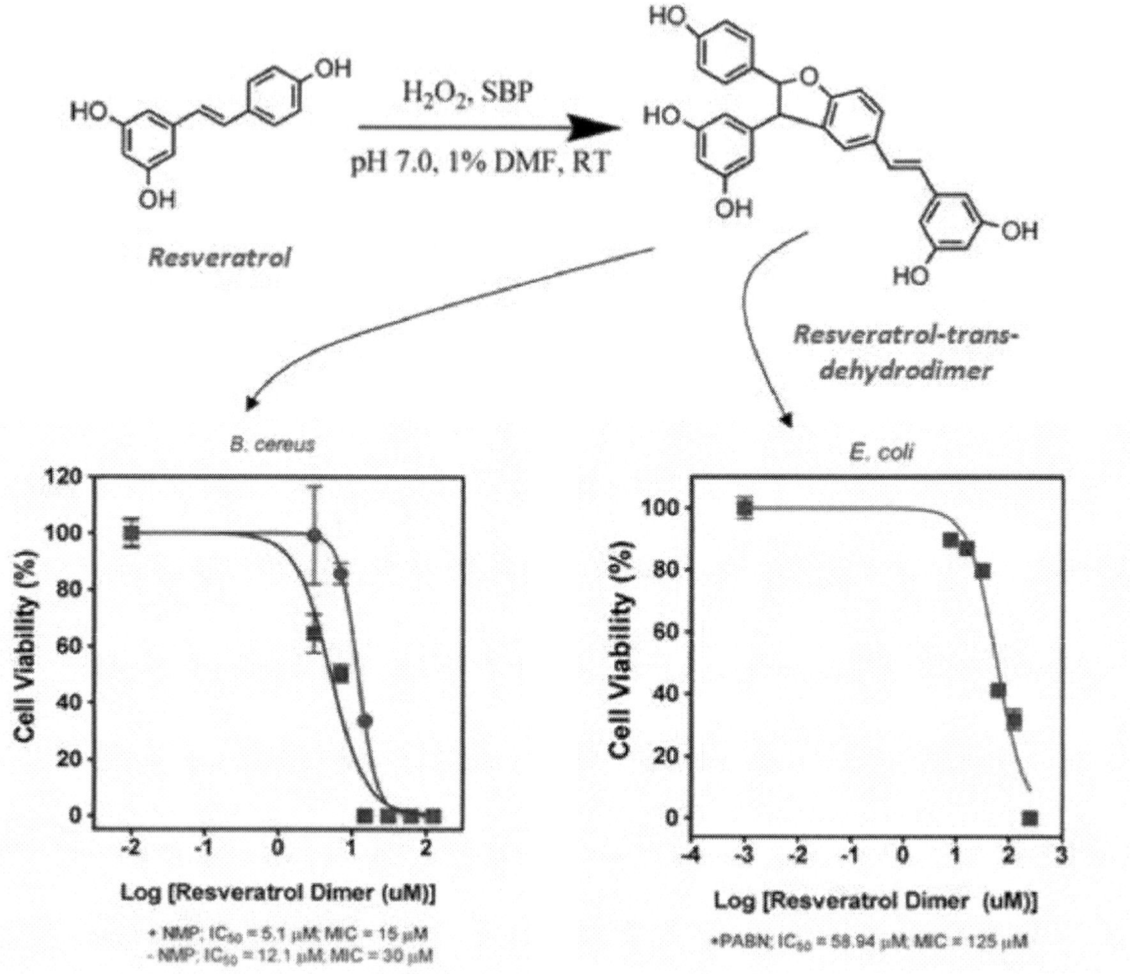

H2O2, SBP
pH 7.0, 1% DMF, RT
Resveratrol
Resveratrol-trans-dehydrodimer
B. cereus
E. coli
Cell Viability (%)
Log [Resveratrol Dimer (uM)]
+ NMP; IC50 = 5.1 μM; MIC = 15 μM
- NMP; IC50 = 12.1 μM; MIC = 30 μM
+PABN; IC50 = 58.94 μM; MIC = 125 μM

(12h) Aryl-Aldehydes in Fungal Polyketides: Discovery and Characterization of Novel Biosynthesis Pathways

Meng Wang, Chemical & Biomolecular Engineering, University of Illinois at Urbana-Champaign, urbana, IL and Huimin Zhao, Department of Chemical and Biomolecular Engineering, Department of Chemistry and Biochemistry, Institute for Genomic Biology, University of Illinois at Urbana-Champaign, Urbana, IL

Fungal polyketides are an important class of natural products, which leads to the development of numerous pharmaceuticals such as blockbusting drug lovastatin. Genome sequencing of various fungal strains reveals that each strain contains many more polyketide and/or non-ribosomal peptide (NRP) biosynthesis pathways than what has been reported. These unknown or "cryptic" pathways present both great opportunities and challenges. An aryl-aldehyde is a common feature in fungal polyketides. It can be retained in the final product such as 3-(2,4-dihydroxy-6-methylbenzyl)-orsellinaldehyde, or serves as an active intermediate for further modifications like in asperfuranone biosynthesis. On the genetic and biochemical level, all previously characterized aryl-aldehydes in fungal polyketides are generated via a reducing domain (R domain), which is attached to the end of a non-reducing polyketide synthase (NR-PKS). However, we discovered and characterized a second mechanism for generating aryl-aldehydes via cloning and heterologous expression of NR-PKS and NRPS-like genes in *Saccharomyces cerevisiae*. We first cloned and expressed NR-PKS (ATEG_03629) and NRPS-like protein (ATEG_03630) from *Aspergillus terreus*. We discovered that ATEG_03629 alone can produce 2,4-dihydroxy-5,6-dimethyl benzoic acid and the ATEG_03629_03630 two genes cassette can produce 2,4-dihydroxy-5,6-dimethyl benzaldehyde. Because NRPS-like gene contains an activation domain (A domain), an ACP domain and a R domain, we proposed that 2,4-dihydroxy-5,6-dimethyl benzoic acid produced from NR-PKS is first activated and attached to the ACP domain via adenylation, and then reduced by the R domain to generate aryl-aldehydes. Further genome mining of all sequenced fungal strains shows that such NR-PKS and NRPS-like pairs can be found in over 20 different fungal strains, which indicates such novel mechanism is common and highly conserved in all fungi. Therefore, we expanded our cloning and expression effort towards three additional similar gene clusters from *Aspergillus nidulans*, *Aspergillus terreus* and *Penicillium marneffei*. Our results show that the same mechanism was also used in these gene clusters, although products with different methylation patterns on the aromatic ring are produced from different pathways.

(15a) Chemical Grafting of Chromatographic Resins for the Selective Adsorption of Pegylated Proteins

Aguilar Oscar and Hernández-Martínez Agustín, Center of Biotechnology-FEMSA, Tec de Monterrey, Monterrey, Mexico

Chemical grafting of chromatographic resins for the selective adsorption of PEGylated proteins.

Hernandez-Martínez Agustín; Aguilar Oscar*

Centro de Biotecnología-FEMSA. Tecnológico de Monterrey, Campus Monterrey.

Av. Eugenio Garza Sada 2501 Sur, Col. Tecnológico Monterrey, México C.P. 64849

*corresponding author: alex.aguilar@itesm.mx

PEGylation reaction, or covalent attachment of a polyethylene glycol (PEG) molecule to a therapeutic protein is a recent strategy still in development aiming to increase the pharmacologic effect and decrease the degradation rate of the protein during its circulation in the body. One of the main challenges in the production of PEGylated proteins is related with the downstream processing, due to the complex mixture of reaction products, resulting in a population comprising mono- and di-PEGylated species as well as several positional isomers. The resolution and characterization of these positional isomers gains higher relevance due to the fact that in most of the cases only one of the isomers exhibits the desired therapeutic effect.

Recent reports of our group suggest that PEGyilated proteing exhibit a marked prefference for the PEG-rich phase of a PEG/salt aqueous two-phase system (ATPS)[1] suggesting a posible affinity partitioning that could be applied to a PEGylated surface such as the chromatografic resin.

Sepharose 6B was subjected to chemical modification according to the protocol reported by Sundberg and Porath[2] for the covalent attachment of PEG 550, PEG 2000 and PEG 5000. The amount of polymer covalently linked to the resin was determined to be from 138 to 180 mg PEG/g resin. An increase in the static binding capacity in HIC conditions was observed for the three modified resins (3.0, 3.6 and 2.6 mg BSA/mL resin respectively), as compared to the unmodified Sepharose 6B (1.2 mg BSA/mL), suggesting changes in hydrophobicity of the resins due to the PEG molecule.

Dynamic binding capacities were determined for the three modified supports in 1 mL Tricorn columns using a Äkta Prime system (GE Healthcare) under the conditions reported by Cisneros-Ruiz *et al.* (2009)[3] (Figure 1), showing a decrease in DBC fro BSA between the unmodified Sepharose compared to modified resins. Different capacities were observed between the molecular weight of the polymer used for grafting the chromatographic curface, mainly due to the difference in the chain length, having a direct impact on the hydrophobicity of the resin. Additionally a change in the conformation of the polymer attached to the resin creates a different environment for protein adsorption.

The potential

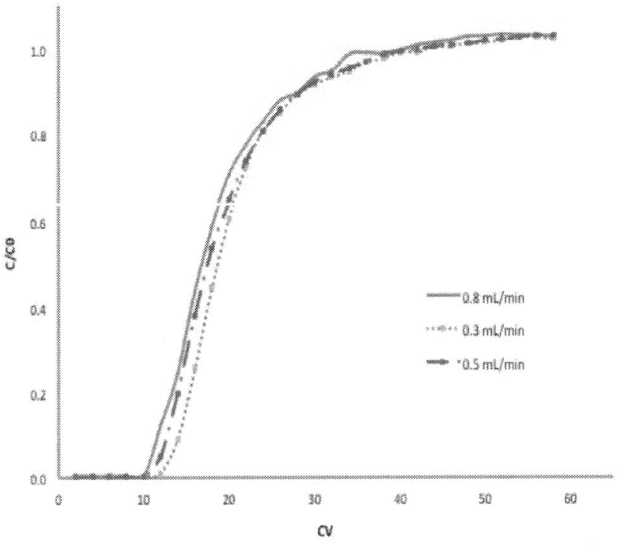

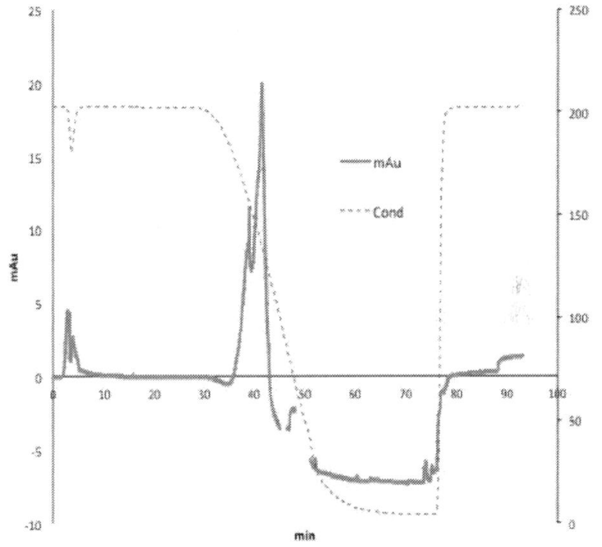

application of these modified resins for the isolation and purification of PEGylated products is under study. Preliminaty results indicate the potential use of these resins for the isolation of PEGylated products from the reaction mixture, without the need for additional SEC or IEX previous steps (Figure 2). An optimization of the separation conditions is needed, but the potential for peak resolution would open the way for an integrated process for primary extraction and purification of PEGylated proteins from reaction mixtures.

1- González-Valdez J, Cueto L, Benavides J, Rito-Palomares M. Potential application of aqueous two-phase systems for the fractionation of RNase A and a-Lactalbumin from their PEGylated conjugates. *J Chem Technol Biotechnol.* 86 (2011) 26–33.

2- **Sundberg** L, Porath J. Preparation of adsorbents for biospecific affinity chromatography: I. Attachment of group-containing ligands to insoluble polymers by means of bifunctional oxiranes. *J. Chromatogr.* 90 (1974) 87-96

3- Cisneros-Ruiz M, Mayolo-Deloisa K, Przybycien TM, Rito-Palomares, M. Separation of PEGylated from unmodified ribonuclease A using sepharose media. Sep Pur Technol 65 (2009) 105–109

(15b) Control of Intein-Mediated Recombinant Protein Purification in a Chinese Hamster Ovary (CHO) Cell Expression System

Tzu-Chiang Han, Chemical and Biomolecular Engineering, The Ohio State University, COLUMBUS, OH and David W. Wood, Chemical and Biomolecular Engineering, The Ohio State University, Columbus, OH

Control of Intein-mediated Recombinant Protein Purification

in a Chinese Hamster Ovary (CHO) Cell Expression System

Tzu-Chiang Han

Dr. David W. Wood

Inteins are genetic elements that can self-excise from various host proteins in a post-translational protein splicing process. This activity can be modified by simple mutation to yield isolated, controllable cleaving activity at either the N- or C-terminus of the intein. By combining self-cleaving inteins with conventional affinity tags, self-cleaving affinity tags have been developed. Once the tagged target protein is purified, the affinity tag and intein components are removed through the controlled self-cleavage of the intein, in a single step without adding any protease. Although this efficient tag-removal method has been successfully applied in prokaryotic cell systems in our previous studies, premature cleaving of the tag *in vivo* has been a serious drawback for this method in eukaryotic cell systems. Here we present our most recent work in developing a next-generation intein for use in mammalian cells. In particular, we have engineered a new intein with enhanced control, where the control mechanism is compatible with CHO cell culture. This intein has been tested with several target proteins in CHO cell culture, and results of these experiments will be presented. By controlling *in vivo* intein cleavage in a CHO cell culture context, we have generated a potential platform technology for the manufacture of non-antibody glycoprotein therapeutics at large scale. Additionally, this intein has been modified for high-throughput applications in drug and target discovery and development.

(15c) Excipients Removal by Capto S Chromatography and UF/DF Process Development for High-Concentration Drug Substance Production

Gregory Waszak, Purification Process Development, Pfizer Inc, Chesterfield , MO, Larry Determan Jr., Bioprocess R&D, Pfizer Inc., Chesterfield and Joseph Martin Jr., Purification Process Development, Pfizer Inc, Chesterfield

The scope of this work is two-fold: 1) excipients removal from formulated mAb drug substance by Capto S chromatography and 2) UF/DF process development to make high-concentration drug substance (DS) for subcutaneous injection. During process scale-up, the project team decided to make a high-concentration mAb drug substance for subcutaneous injection and change the formulation. Since the pilot plants were not available for large-scale campaigns, a creative alternative was needed to produce 2 kg of antibody from formulated DS for these studies. Fortunately, a sufficient supply of formulated DS was available for reprocessing. A Capto S column was chosen over Protein A chromatography to remove excipients from formulated drug substance because of its higher binding capacity. In addition, Capto S has lower resin costs, takes less time to process, and uses milder elution conditions. Analysis showed that the Capto S run removed the excipients with yields of ≥ 96%. During the UF/DF process, the antibody was initially concentrated to 90 g/L , diafiltered, and concentrated to ≥ 180 g/L, then the retentate was collected. After the retentate was collected, a minimal volume of buffer was added for the UF rinse. Most of the rinse was combined with the retentate to hit the target protein concentration. UF/DF yields of 92% have been obtained. This method enabled > 90 % overall recovery of unformulated DS at ≥ 150 g/L.

(15d) Designing in Process Robustness: Minimizing Proteolysis Through Analytical and Chromatographic Tools

Janelle Konietzko[1], Andrew Englehart[2], Thomas Svab[2], Adam Kristopeit[2], Marc Wenger[3], Michael E. Laska[4] and Aaron R. Goerke[2], (1)Vaccine Process Development, Merck & Co., West Point, PA, (2)Vaccine Process Development, Merck & Co., (3)Bioprocess and Bioanalytical Research, Merck & Co., Inc., West Point, PA, (4)BioPurification Development, Merck & Co., West Point, PA

Host cell protein removal is a primary goal during purification process development for vaccines and therapeutic proteins, however the subset of protease impurities is a particular threat to product yield, quality, and product/process consistency. The presence of such enzymes in the process can lead to proteolytic clipping of the product, resulting in a heterogeneous mixture of degraded product fragments and isoforms. Resolution of the desired product from these closely-related degradates represents a significant separation challenge and can result in project delays and elevated developmental cost (and cost of goods) for vaccines and therapeutics. Due to the complexity and constraints of genetically removing proteases from the host cell system, the purification process is often tasked with solving these problems by minimizing degradation and eliminating the offending protease activity. The exact solution to the problem, however, is highly dependent on the particular host cell, product, and process conditions. In this work we outline a strategic methodology that has been consistently utilized to rapidly identify and minimize protease activity. Examples comprising unique experimental designs and statistics will be used to illustrate chromatographic and filtration development aimed at minimizing residual proteases while maintaining high product recovery. The resolving power of mixed-mode resins combined with tailored buffer/wash matrixes was instrumental in identifying conditions yielding the lower proteolytic activity. Rapid development methods, such as high-throughput screening, were exploited to shorten the development life cycle and map the chromatography design space for optimal removal of protease activity and product degradates. Novel fluorescent linkers enabling a low level of detection and real time analytics were used to detect protease and inform development decisions. By understanding the operating conditions that minimize protease activity, we have been able to design multiple tiers of robustness into several high value biological processes.

(15e) Proteomics Based Multivariate Random Forest Method for Prediction of Protein Separation Behavior During Downstream Purification

Ryan K. Swanson[1], Ruo Xu[2], Dan Nettleton[2] and Charles E. Glatz[1], (1)Chemical and Biological Engineering, Iowa State University, Ames, IA, (2)Statistics, Iowa State University, Ames, IA

Proteomics based multivariate random forest method for prediction of protein separation behavior during downstream purification

Ryan K. Swanson[1], Ruo Xu[2], Dan Nettleton[2] and Charles E. Glatz[1]

[1]Department of Chemical and Biological Engineering, Iowa State University, Ames, IA 50011

[2]Department of Statistics, Iowa State University, Ames, IA 50011

Optimizing the downstream purification process remains a significant challenge when using biological expression hosts for the production of recombinant proteins. An important reason why is due to the lack of knowledge of the host cell proteins' (HCP) separation behavior during the selected purification methods. The work to be presented addresses this issue by modeling the separation behavior of a HCP extract from an arbitrary expression host (corn) during commonly used chromatographic and non-chromatographic methods. Aqueous two phase system (ATPS) partitioning followed by two dimensional electrophoresis (2DE) provided data on the three physicochemical properties most commonly exploited during downstream purification for each HCP; isoelectric point, molecular weight and surface hydrophobicity. A multivariate random forest (MVRF) statistical methodology was then applied to this database of characterized proteins creating an accurate tool for predicting the separation behavior of a mixture of proteins for three commonly selected purification methods; cation exchange chromatography (CEX), hydrophobic interaction chromatography (HIC) and ammonium sulfate precipitation (ASP). The next logical goal will be to link these three methods in series (i.e. ASP followed by HIC followed by CEX) and test the prediction accuracy of the MVRF methodology during this simulated downstream purification process. Also discussed will be the future capabilities of this work including the ability to optimize a downstream purification process without requiring any initial product or time spent screening potential methods in the lab regardless of expression host.

(15f) Virus Purification by Aqueous Two Phase Extraction System

K. Saagar Vijayaragavan, Chemical Engineering, Michigan Technological University, Houghton, MI and Caryn L. Heldt, Department of Chemical Engineering, Michigan Technological University, Houghton, MI

AIChE 2012 Annual Meeting, Pittsburg, PA October 28-November 2, 2012

Division: Bioengineering

Group: Bioseparations and Downstream Processing

Virus Purification by Aqueous Two Phase Extraction System

K. SAAGAR VIJAYARAGAVAN (kskhrupa@mtu.edu) & CARYN L. HELDT (heldt@mtu.edu)

Department of Chemical Engineering, MTU

Rising social concerns over potential viral contamination and infection from blood plasma derivatives in the last few decades have raised the need of biotheraputic safety measures for downstream processes in the biopharmaceutical industry. One strong technique that is often applied to protein purification, but rarely to virus removal is aqueous two phase systems (ATPS). ATPS was first developed in Sweden by Huddleston in 1950's and consists of a liquid-liquid biphasic system widely recognized for separation and purification of biomolecules, like proteins and cell organelles. Due to its ecofriendly, economic, and easy scale up, it is possible to implement this separation system to different biotechnological applications. The partitioning of protein molecules into different aqueous phases is a highly complex phenomenon due to hydrogen bonding, charge interactions, Van der Waal's forces and surface hydrophobicity. Porcine Parvovirus (PPV), a non-enveloped DNA virus of 18-26 nm diameter, is a model virus used to test virus removal. It is a model of the human B19 parvovirus, which has been reported to contaminate biotheraputic products. PPV has been studied and recently validated to be more hydrophobic than typical blood and host cell proteins, by our laboratory. Based on this experimental data, we are proposing ATPS to effectively separate PPV to one of the liquid-liquid immiscible phases due largely on hydrophobicity. In our ATPS experiments we use both a polymer-polymer and polymer-salt system. We are currently focusing our efforts on a polymer-salt system comprising of PEG and three different salts, namely phosphate, sulphate and citrate. We have successfully identified the phase separation based on binodial phase curve lines and are using Alexa fluor-tagged PPV to determine the separation of the virus in individual phases of the biphasic system. We have successfully separated PPV with this system. Some of the factors affecting partition coefficient (K) are pH, neutral salts addition, salt concentration, PEG molecular weight and PEG concentration. The study will provide scientific understanding into viral clearance strategies and improve biotheraputic safety in manufacturing processes.

(15g) The Effect of Single-Use Retentate Recycle Bag Design On Diafiltration Efficiency

David M. Bohonak and Elizabeth Goodrich, Biomanufacturing Science Network, EMD Millipore Corporation, Billerica, MA

Performance of ultrafiltration/diafiltration (UF/DF) systems depends, among other factors, on highly effective mixing in the retentate recycle tank. Retentate mixing requirements are unique as compared to mixing requirements for other bioprocessing unit operations such as buffer preparation, because UF/DF processes must achieve a high level of tank homogeneity and buffer exchange (e.g. > 99.9%) under continuously changing level, protein concentration, and/or buffer concentration conditions. In ultrafiltration processes, poor retentate mixing may result in a higher protein concentration in the feed stream than intended, leading to reduced flux and issues with protein quality or solubility. In diafiltration processes, poor retentate mixing can lead to reduced exchange of buffer components and inability to meet a final excipient specification. Consideration of effective retentate mixing is of particular importance for processes employing single-use flow paths since component and system availability is more limited and less established for rigorous downstream separation capability than more mature stainless steel system designs.

The design of two different styles of commercially available, single-use retentate recycle bags was experimentally investigated, specifically to assess the effect on diafiltration performance. The use of active versus passive agitation, different inlet/outlet configurations, and a range of process parameters such as tank level, recirculation rate, and fluid viscosity were evaluated using a simulated diafiltration process to characterize the effect on solute clearance. A mechanistic model of solute clearance during diafiltration is introduced which explicitly accounts for unmixed regions or deadlegs. The model is used to explain the experimental results and illustrates shortcomings of common practices used to evaluate tank mixing in UF/DF processes.

(15h) Flux Enhancement in Recycling Cellulase From Enzymatic Hydrolyzate of Acid Treated Wheat Straw by Electroultrafiltration

Guoqiang Chen, National Key Laboratory of Biochemical Engineering, Institute of Process Engineering, Chinese Academy of Sciences, Beijing, China and Yinhua Wan, National Key Laboratory of Biochemical Engineering, Institute of Process Engineering, Chinese Academy Of Sciences, Beijing, China

Ultrafiltration has been applied extensively in concentration and separation of biomacromolecules. It also shows great potential in recovering both cellulase and β-glucosidase from lignocellulosic hydrolyzates by one step without addition of any chemical. In addition, ultrafiltration membrane reactor coupled with microfiltration or not, could recycle cellulase while removing end-products continuously during hydrolysis. In spite of these advantages, fouling and concentration polarization, which cause flux decline limiting application of membrane processes significantly, are two severe problems remaining to be resolved. In the present work, to enhance the membrane filtration performance in recycling cellulase from lignocellulosic hydrolyzate by ultrafiltration, an electric field was applied across the membrane. The feasibility of electroultrafiltration (EUF) to recycle cellulase was first determined by measuring the filter paper activity of cellulase after EUF of cellulase solution, and the effect of electric field under different operating conditions (molar concentration of buffer, acid treated wheat straw concentration, current and temperature) on permeate flux during EUF was examined indetail. The results showed that EUF was an effective method to recycle cellulase due to its high cellulase activity recovery and high flux. Under appropriate operating conditions, the fllux for 2% (substrate concentration, w/v) lignocellulosic hydrolyzate increased by a factor of 5.0 when an electric strength of 920 V/m was applied, compared to that without electric field. The improvement of flux was independent on substrate concentration. Therefore, EUF could be a useful approach to recycle cellulase for reducing hydrolysis cost.

(18a) A Nuclear Rheostat That Couples Microenvironment Rigidity to Cell Lineage

Dennis E. Discher, Chemical and Biomolecular Engineering, University of Pennsylvania, Philadelphia, PA

A solid tissue can be soft like fat or brain, stiff like striated muscle and heart, or rigid like bone. Proteomic profiling of tissue nuclei shows that Lamin-A/C expression increases more than 30-fold and in near-proportion to micro-elasticity of adult tissue, while other nuclear envelope components such as Lamin-B exhibit small variations. Lamin-A/C has been implicated in aging syndromes that affect muscle and fat but not brain, and we find nuclei in brain-derived cells are indeed dominated by Lamin-B and are much softer than nuclei derived from muscle cells with predominantly Lamin-A/C. *In vitro*, matrix elasticity can affect expression of nuclear envelope components in adult stem cells and major changes in Lamin-A/C direct lineage: lower levels favor soft tissue and higher levels promote rigid tissue lineage. Stress response elements in the Lamin-A/C promoter are just beginning to be elaborated, but tagging of cryptic sites while physically stressing isolated nuclei reveals stress-driven, mass spectrometry-mapped changes in various nuclear proteins including Lamin-A/C, consistent with cell and tissue evidence that the nucleus transduces physical stress.

(18b) Nuclear Mechanics of VEGF-Stimulated Endothelial Cells

Stephen T. Spagnol[1], James S. Weltz[1] and Kris Noel Dahl[2], (1)Department of Chemical Engineering, Carnegie Mellon University, Pittsburgh, PA, (2)Chemical Engineering, Carnegie Mellon University, Pittsburgh, PA

The packaging of the genome into higher order structures of chromatin and chromosomes has been well established. However, the complementary knowledge of this hierarchical organization within the nucleus has proven elusive. Recent evidence indicates that genome organization plays a role in gene regulation and, further, that gene position within the nucleus may correlate with expression. In response to a stimulus, chromatin must decondense and reorganize for protein binding and transcription. The signal transduction cascades leading to stimulated gene expression have been a major focus of research, but stimulated cytoskeletal rearrangement frequently occurs in parallel to the biochemical signaling leading to mechanical force on the nucleus that has been implicated in chromatin agitation. We hypothesize that this aids in gene expression through motor protein-induced cytoskeletal stresses on the nucleus that act globally and nonspecifically to move gene domains (strain) and alter their probability of expression. Our work aims to correlate these cytoskeletal stresses with the underlying nuclear strain and known changes in gene expression through particle-tracking microrheology.

As a model system, we focus on vascular endothelial growth factor (VEGF) stimulated angiogenesis in human umbilical vein endothelial cells (HUVECs), which is a well-characterized and direct pathway linked to known changes in gene expression and cytoskeletal rearrangement. We transfect cells with GFP-tagged sub-nuclear markers and track them in live cells during treatment with up to 50 ng/mL VEGF. Particle trajectories are obtained (using previously published custom software), from which we compute the mean square displacement (MSD) versus time. The MSD is fit to a power-law model ($MSD(\tau)=D_{eff}*\tau^{\beta}$), as consistent with previous work. In response to VEGF stimulation, the nuclear interior undergoes an effective softening evinced by an increase in MSD magnitudes over all time points ($p < 0.001$ compared to unstimulated controls). This effective softening is consistent with global chromatin decondensation necessary for nuclear reorganization of the genome and the transition in gene expression. It also results from an increased driving force from the motor protein-driven cytoskeletal stress necessary to enhance gene and protein motion as well as the frequency of binding events to increase the probability of transcription. Thus, our results suggest global nuclear reorganization is necessary for stimulated gene expression despite the fact that chemical stimulation only leads to the upregulation of specific genes. This lends credibility to the dynamic view of nuclear organization as a means of regulating the genome, with implications for cytoskeletal stress in driving stimulated genome rearrangement.

(18c) Control of Dorsal Ruffle Dynamics in Cells Through Substrate Stiffness

Yukai Zeng[1], Tanny Lai[2], Philip R. LeDuc[1] and Keng Hwee Chiam[2], (1)Mechanical Engineering, Carnegie Mellon University, Pittsburgh, PA, (2)Biophysics, A*STAR, Institute of High Performance Computing, Singapore, Singapore

Control of Dorsal Ruffle Dynamics in Cells through Substrate Stiffness

Yukai Zeng[1,2], Tanny Lai[2], Philip R. LeDuc[1] and K.-H Chiam[2]

[1] Mechanical Engineering, Carnegie Mellon University, Pittsburgh, PA, 15213, USA

[2] A*STAR, Institute of High Performance Computing, Singapore

Circular dorsal ruffles (CDRs) are structures which are rich in actin, formed in many mammalian cells after stimulation with various growth factors, such as platelet-derived growth factor (PDGF). These structures and their chemical environment are important in many areas such as PDGF playing a role in cell proliferation, angiogenesis and cell migration. CDRs are hypothesized to aid in cell migration by initiating large scale reorganization of the actin cytoskeleton in cells [1] and form minutes after PDGF stimulation, before persisting and disappearing within tens of minutes in fibroblasts [1,2]. In addition, cells which exhibit CDRs are thought to show an increase in macropinocytosis activity [2], suggesting this as a possible function of CDRs. From a mechanics standpoint, cells seeded on varying substrate stiffnesses exhibit differences in cell migration speed [3]. We used this knowledge to investigate the migration speed of cells which exhibit CDRs and compare them to those which do not. The size of the CDRs formed in cells were also quantified, since when cells are seeded on 2 dimensional substrates and observed under a fluorescence microscope after staining for F-actin, the CDRs appear as circular rings of F-actin and thus provide a basis for stimulation based comparisons.

Our experiments show that although changing substrate stiffness does not change the physical properties such as the size of CDRs, their kinetics is changed [4]. This is based on the increase in persistence time of these CDRs when substrate stiffness is increased. We hypothesize that the CDRs transverse at a reduced speed below the cell membrane with increasing substrate stiffness. In addition, the percentage of cells which exhibit CDRs in a population, when stimulated with PDGF, decreases with increasing substrate stiffness. The ability of the CDR formation model, based on Rac-Rho antagonism, to replicate our experimental results suggest that the system of reaction-diffusion reactions based on the PDGF stimulated cell signaling cascade and cell compartmentalization are able to explain the experimental observations of the spatiotemporal response. This indicates that while the geometric alterations of the CDRs may not be directly related to substrate stiffness, the kinetics of the response may be very important.

rEferences

1. Krueger, E. W., Orth, J. D., Cao, H. & McNiven, M. A. A dynamin–cortactin–Arp2/3 complex mediates actin reorganization in growth factor-stimulated cells. *Mol. Biol. Cell* 14, 1085–1096 (2003).

2. Buccione, R., Orth, J.D. & McNiven, M.A. Foot and mouth: podosomes, invadopodia and circular dorsal ruffles. *Nat Rev Mol Cell Bio.* 5, 647–657 (2004).

3. Lo, C.M, Wang, H.B., Dembo, M & Wang, Y.L. Cell Movement Is Guided by the Rigidity of the Substrate. *Biophysical Journal* 79, 144-152 (2000).

4. Zeng, Y.*, Lai, T.*, Koh, C. G., LeDuc, P.R., & Chiam, K.-H. Investigating Circular Dorsal Ruffles through Varying Substrate Stiffness and Mathematical Modeling. *Biophysical Journal* 101, 2122–2130 (2011).

*Equal contribution from both authors.

(18d) Cell Shape Regulates Myofibroblast Activation

Joseph O'Connor, Chemical Engineering, The Pennsylvania State University, University Park, PA and **Esther W. Gomez**, Chemical Engineering and Bioengineering, The Pennsylvania State University, University Park, PA

Myofibroblasts, specialized cells within the body that aid in normal wound healing processes, exert large contractile forces on their microenvironment with the primary job of these cells being to assist in the closure of wounds. Aberrant and chronic activation of myofibroblasts can lead to the development of pathological conditions including fibrosis and cancer. Mechanical tension is thought to be crucial for the activation of myofibroblasts from other cell types, however, a mechanistic understanding of how biophysical cues regulate myofibroblast phenotype is not clear. Here, we employ a microfabrication approach to investigate the role of cellular shape and tension in the development of myofibroblasts. We find that cell shape controls the subcellular localization of key mechanoresponsive molecules and subsequent expression of cytoskeletal proteins that contribute to the increased contractile features characteristic of activated myofibroblasts. Results provide insight into how biophysical cues contribute to the development of myofibroblasts and may suggest ways to treat or prevent pathologies including fibrosis and cancer.

(18e) 3D Traction Force Microscopy in Fibrin Gels

Arjun S. Adhikari[1], Natascha Leijnse[1,2] and Alexander R. Dunn[1], (1)Chemical Engineering, Stanford University, Stanford, CA, (2)Niels Bohr Institute, Copenhagen, Denmark

The mechanical forces exerted and detected by living cells play integral roles in diverse biological phenomena, including growth and development, wound healing, and cancer metastasis. In the past decade, techniques such as traction force microscopy and micropost arrays have proven to be powerful tools for measuring the forces generated by cells. In particular, traction force microscopy has recently been extended to three-dimensional cell culture environments by embedding tracer beads in either a synthetic polyethyleneglycol hydrogel (PEG; Legant *et. al*, Nat Met. 2010) or in collagen gels (Koch *et. al*, Plos One, 2012). The embedded beads move in response to cell-generated distortions of the matrix, allowing cell-generated forces to be calculated. We sought to develop an experimental system that would exhibit the excellent mechanical properties of the PEG hydrogel while using a naturally occurring biological matrix. Fibrin gels fulfill both of these requirements: fibrin is elastic up to ~50% strain (Brown *et. al*, Science, 2009) and is also widely used for 3D cell culture. Here we describe the use of fluorescently labeled fibrin gels to measure the forces generated by cells in 3D culture. We observe dramatic but elastic deformations of the fibrin matrix surrounding cells as they grow, divide, and migrate. Further, we find that the forces generated by the cell can be measured using the deformations of the matrix itself, obviating the need for tracer particles, and providing a direct observation of how the cell modifies its surroundings. We discuss the use of this new technique in studying a variety of cellular processes, for example matrix remodeling and cell migration.

(18f) A Microfluidic Device to Measure Traction Forces During Confined Cancer Cell Migration towards Chemoattractant

Phrabha Raman, Colin Paul, Kimberly Stroka and Konstantinos Konstantopoulos, Chemical and Biomolecular Engineering, Johns Hopkins University, Baltimore, MD

Cell migration represents a key facet of cancer metastasis, since cells depend on motility to navigate through interstitial tissues, intravasate to the circulatory or lymph systems, and seed new metastatic foci. Cell motility mechanisms are influenced by the physical and chemical cues of the microenvironment. Mechanical forces are constantly being exerted by migrating cells on the surrounding cells as well as the extracellular matrix. Simultaneously, forces are being continuously exerted on the cells themselves from the extracellular environment. To fully comprehend the combinatorial effects of mechanical forces and chemotaxis, we engineered a novel two-layer microfluidic-based migration device to measure traction forces exerted by cells migrating in microchannels in response to a chemotactic stimulus.

The fully formed migration device consists of confining microchannels with a "floor" consisting of deflectable micropillars. Cells are induced to migrate through these channels by a chemotactic stimulus. Molds for the top and bottom layers are fabricated using multi-step photolithography. The device layers themselves are formed by replica molding with PDMS. The top layer of the device is composed of parallel microchannels (separated by 50 μm) of height 4 μm and length 200 μm. Channels vary in width from 10 to 50 μm. The microchannels are aligned in a ladder-like configuration orthogonal to and connecting two larger main channels which serve as a cell seeding source and a chemokine reservoir, respectively. The bottom portion of the microfluidic migration chamber consists of micropillars (diameter - 3 μm; height - 9 μm; center-to-center spacing - 6 μm) that are located within channels of different widths, mirroring the top portion of the device. Using a micromanipulator and a stereoscopic microscope, the top portion of the microfluidic device is aligned precisely over the bottom portion, thereby defining the channels.

To demonstrate the general utility of our migration device for traction force measurement, we performed experiments with metastatic human breast adenocarcinoma cells (MDA-MB-231). MDA-MB-231 cells were seeded and allowed to migrate within the microchannels over micropillars coated with type I collagen towards the chemoattractrant. Once the cells were on top of the micropillars, time lapse video microscopy was performed to record the real-time pillar deflections due to cellular traction forces. The pillar deflections were converted into forces by multiplying by the spring constant of the micropillar, which was calculated using beam bending theory. Our preliminary data revealed that colchicine treatment leading to the disruption of microtubules results in higher pillar deflections relative to untreated controls for cells migrating through a 50 μm-wide channel. This increase can be explained by an increase in traction that can be attributed to the loss of compression-supporting capacity of microtubules, resulting in a shift of load and energy from the microtubules to the substrate. We wish to further characterize the cell traction forces in response to increasing confinement, manifested in our design by decreasing channel width. Thus, our novel microfluidic design, which incorporates micropillars within microchannels to quantify cellular tractions forces during directed migration under confinement, is a powerful tool for the study of cellular chemotaxis and is suitable for a wide range of biological and biomedical applications.

(18g) Differential Effect of Shear Stress On Endothelial Cells Response to TNF-Alpha and IL1-Beta Stimulation Is Linked to Cell Shear History

Ryan B. Huang and **Omolola Eniola-Adefeso**, Chemical Engineering, University of Michigan, Ann Arbor, MI

Introduction: The ability to discriminate between healthy and diseased endothelial cells (ECs) via cell adhesion molecule (CAM) expression will have significant impact on the design of vascular-targeted imaging and drug delivery vehicles for treatment of several diseases. However, EC behavior under chronic inflammation has yet to be fully understood due to the use of complex *in vitro* models that fail to simulate actual physiological inflammation conditions – wherein the combined effects of chemical and mechanical stimulus are simultaneously present. Work done by this group showed previously unreported patterns of E-selectin CAM expression on EC exposed to shear-interleukin-1β (IL-1β) co-stimulation; in this report, human umbilical vein EC (HUVEC) exposure to combined fluid shear stress and tumor necrosis factor α (TNFα) is investigated.

Materials and Methods: HUVEC sources were harvested from fresh cords and subcultured for no more than two passages. HUVEC monolayers were prepared and subjected to 0 or 12 hrs of shear stress imparted by fluid flow in a parallel plate flow chamber prior to simultaneous exposure to shear-cytokine stimulation. This pre-shearing period preconditions the mechanically-sensitive cells to a phenotypic state mimicking physiological profiles. IL-1β and TNFα were used in the subsequent stimulations (individual and combined) to simulate EC exposure to inflammatory chemicals. To determine HUVEC response, E-selectin (E-sel) – the hallmark CAM expressed during inflammation – was measured via immunofluorescence microscopy.

Results and Discussion: We found that like activation with IL-1β, the response of HUVECs to co-stimulation with fluid shear and TNFα is dependent on the shear history and the time frame of the following shear-cytokine activation. The presented data (Fig. 1) suggests that fluid shear differently affects IL-1β and TNFα signaling –prolonged exposure to high laminar shear prior (preconditioning) to cytokine-shear stimulation elicits muted E-sel response for TNFα activation (clear bars in C versus A) whereas response to IL-1β (filled bars) is enhanced (Fig. 1B) or muted (Fig. 1C) depending on the length of shear-cytokine stimulation post-preconditioning.

The observed data for dual-stimulation of HUVECs with both TNFα/IL-1β, seems to correlate with E-sel expression for shear-IL-1β activated monolayers (Fig. 1C). This indicates that the observed E-sel expression trend for shear-IL-1β-only activated monolayers likely remains relevant even under more complex activation conditions.

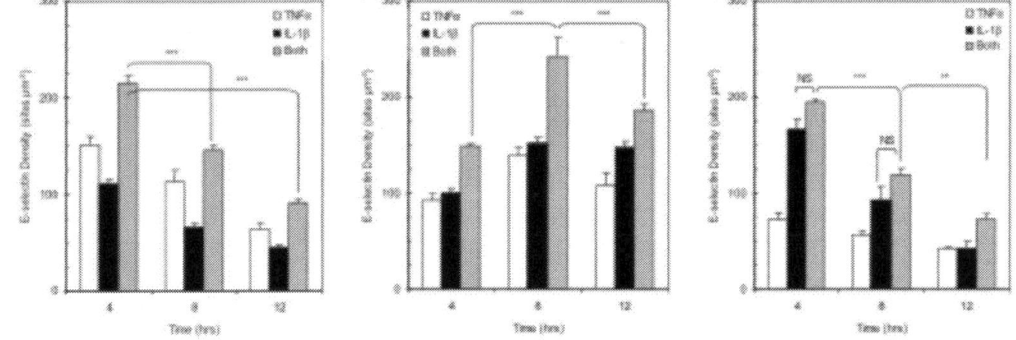

Figure 1. E-sel site density of combined and individual TNFα or IL-1β activation. HUVEC monolayers were treated with 10 ng/mL TNFα (clear), 0.1 ng/mL IL-1β (filled), or simultaneously with both cytokines at the same concentrations (grey) for up to 12 hrs under (A) static-cytokine and shear-cytokine (10 dyn/cm2) activation of (B) na•ve cells and (C) 12 hrs preconditioned monolayers.

Conclusions: This study explores longer term activation of HUVECs relevant to time frames encountered during physiological chronic inflammation and observes previously unreported patterns of shear-TNFα-induced E-sel expression. However, we also report that shear-cytokine stimulation of HUVECs is also cytokine-specific: TNFα activated monolayers elicit an increasingly muted E-sel expression response relative to IL-1β activated cells with increasing exposure to fluid shear stimulus. This indicates that fluid shear plays a significant role in modulating the pathways for TNFα and IL-1β signaling cascades and should be further explored to determine additional specific endothelial responses for potential application in developing targeted therapeutic treatments.

(18h) Migratory Single Cell Force Measurements Using Deformable Suspended Nanofiber Networks

Kevin Sheets[1], Puja Sharma[2], Ji Wang[3], Brian Koons[4], Tim O'Brien[4], Bahareh Behkam[5] and Amrinder S. Nain[5], (1)School of Biomedical Engineering and Sciences, Virginia Tech, Blacksburg, VA, (2)Biomedical Engineering, Virginia Tech, Blacksburg, VA, (3)Engineering Science and Mechanics, Virginia Polytechnic Institute and State University, Blacksburg, VA, (4)Virginia Tech, Blacksburg, VA, (5)Mechanical Engineering, Virginia Tech, Blacksburg, VA

Introduction

During migration, cells exert forces in their immediate microenvironment known as the extracellular matrix (ECM), a nanofibrous mesh which provides geometrical and chemical cues that directly affect cell behavior[1]. While the nature and magnitude of these forces are dependent on the ECM itself, they can also reveal important information about the mechanical state of the cell[2]. For instance, recent works have demonstrated that traction forces can be directly related to cellular differentiation, and cancer progression has also been shown to have an effect on the cell's mechanical state, thereby altering contractility[3,4]. Many recent efforts have been devoted to measuring cell traction and migration forces using micropost arrays and deformable gel surfaces, which are unable to capture entire 3D effects on force generation[3,5,6]. Therefore, there is clearly both great interest and need for the ability to quantify the forces a cell exerts on the ECM using a system which allows cells to migrate along 3D geometries mimicking ECM mechanical environments. In this work, we present a method to measure real-time migration forces of single cells using a system of suspended nanofiber beams whose stiffness (N/m) is readily adjustable. As cells attach and begin moving along these scaffolds, individual fibers with clamped boundary conditions deflect in the elastic limit, allowing cell migration forces to be calculated via classical beam theory. With this technology, cells may be monitored in real-time for the forces they exert on their ECM-mimicking environment. Implications of this work include understanding internal cell forces during differentiation and therefore the future design of accurate tissue engineering scaffolds, drug testing platforms, and identification of diseased cells including cancer.

Materials and Methods

Scaffolds were fabricated using the previously reported STEP (Spinneret-based Tunable Engineered Parameters) platform, a non-electrospinning fiber manufacturing technique[7–10]. Briefly, polymer solution is injected through a microneedle and a rotating substrate is brought into contact with the resulting droplet. As a controlled stage translates the hollow scaffold, aligned fibers are deposited on hollow substrates in criss-cross layers which were fused at intersections forming nanonet structures.

After creating the scaffolds, C2C12 mouse myoblasts were seeded onto the STEP scaffolds and attached within 4-6 hours. Cells then began migrating along the structures. .Migration was visualized via time-lapse videography using an incubating microscope with a digital three-axis stage (Zeiss). Images were taken every 10 minutes for 12 hours, which captured cells deflecting STEP fibers. Built-in software allowed measurements to be taken relevant to calculating forces including fiber length, positioning of the load, and deflection distance.

Cells were assumed to cause deformation to the nanofibers according to classical beam theory with clamped-clamped boundary conditions in the elastic limit, described by the following equation:

$$P = \left(\frac{3EIL^3}{a^3 b^3} \right) \delta$$

where 'P' is the applied load to fiber length 'L' at distance from the boundary 'a'. The elastic modulus 'E' is a material dependent parameter, and the moment of inertia 'I' is shape dependent. Therefore, as a deflection 'd' is noted, the applied load may be calculated. Forces were assumed to be applied as a single load perpendicular to the point of maximum deflection, an assumption that will be further investigated in the future.

Results and Discussion

After spinning two layers of fibers on the STEP scaffolds, fiber intersections were fused together (Fig. 1a). As shown in Fig. 1b, the resulting grid-like nanostructure can be considered as a system of springs with individual fiber stiffnesses (N/m) being dependent on the fiber material, diameter, and length as well as positioning along the fiber. Designing a scaffold in which the longitudinal fibers are stiffer than the latitudinal fibers results in a system in which the less stiff fibers deflect while the larger fibers can be considered immobile.

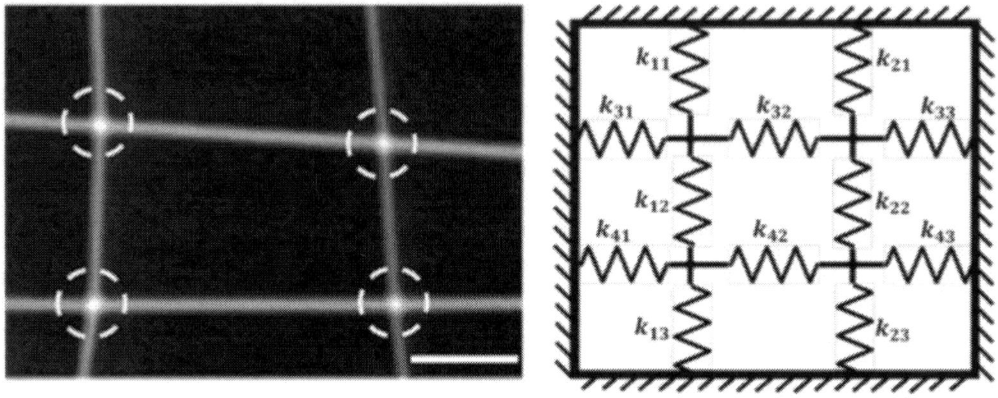

Figure 1: a) SEM images of PS nanofibers with fused intersections shown in circles. b) Spring model of the fiber system when the intersections between criss-cross fibers are fused together.

Cells were found to attach to the scaffolds and began migrating along the suspended fibers. As cells attached to both the stiffer and less stiff fibers simultaneously, the less stiff fibers deflected under forces exerted by the cells (Fig. 2). These deflections were measured and then calculated using the aforementioned equation to obtain migratory cell forces in a natural 3D fibrous microenvironment.

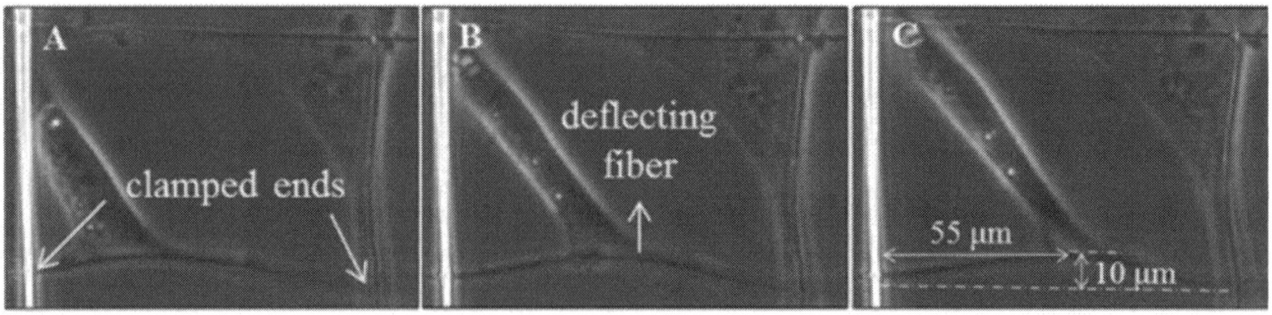

Figure 2: Timelapse images showing a C2C12 pulling against a nanofiber. Using these deflections, force are calculated according to clamped-clamped boundary conditions. In this example, the calculated force i frame C is estimated to be 18 nN

Figure 2 shows an example timelapse of a C2C12 mouse myoblast migrating and deflecting a horizontal fiber. In this 30 minute sequence, the cell progresses along both fiber axes, causing the cell to stretch further and further apart.

Using frame C as an example calculation, a PS fiber of d=500 nm diameter spanning L=100 μm, pulled a=55μm from the boundary would require an 18 nN force to deflect a distance of d=10μm. This is in good agreement with other force measurement systems, which often predict migratory forces ranging from tens to hundreds of nN[11]. While this does not necessarily represent the maximum force that a C2C12 is able to exert, it does show what the cell was experiencing during that stretching phase.

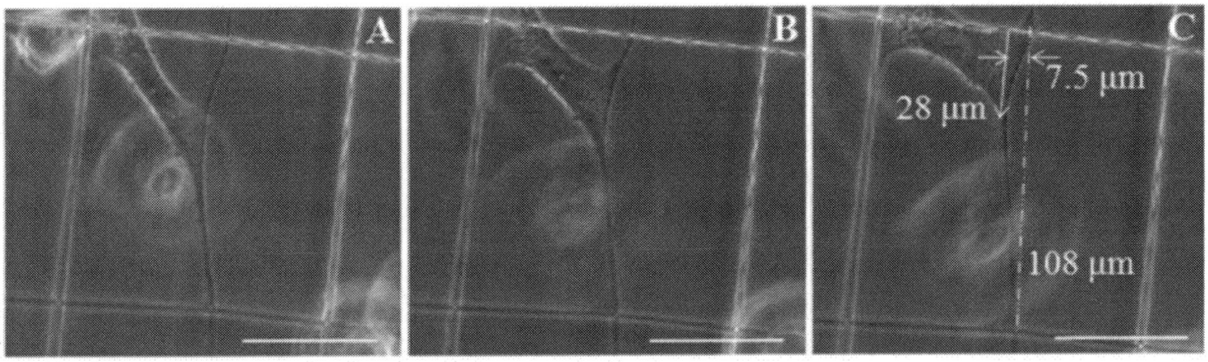

Figure 3: Timelapse images of a cancerous DBTRG pulling against a nanofiber. In this example, th calculated force in frame C is estimated to be 23 nN.

In addition to testing normal cell lines, DBTRG (Denver Brain Tumor Research Group) cells were also examined. These cancerous cells, which originate from an aggressive brain cancer called glioblastoma multiforme, were also seeded onto the STEP scaffolds for force evaluation (Fig. 3). Again, as an example calculation using the above timelapse in frame C, a PS fiber d=500 nm in diameter spanning L=108 μm, pulled a=28 μm from the boundary would require a 23 nN force to deflect a distance of d=7.5 μm. As more cells are evaluated in each condition, we will be able to distinguish weather or not the cancerous state of a cell changes its migratory forces.

Conclusions

We have presented a method which for the first time measures real-time migratory forces of single cells using fused, suspended nanofibers. This system is advantageous in that it accounts for 3D topographies that would be experienced by cells *in vivo*. With this method, migration forces have been measured in the 10-100 nN range, but more or less sensitive structures may be developed simply by adjusting the stiffness of the fibers which makeup the scaffold. Decoupling force vectors from attached cells remains a challenge and will continue to be investigated in the future. In addition to gathering additional data for cancerous and non-cancerous cell lines to determine a baseline migratory force value for each case, this system will be used to simultaneously evaluate migratory forces and associated differentiation of stem cells in future work.

References

(1) Lock, J. G.; Wehrle-Haller, B.; Strömblad, S. *Seminars in cancer biology* **2008**, *18*, 65-76.

(2) Crow, A.; Webster, K. D.; Hohlfeld, E.; Ng, W. P.; Geissler, P.; Fletcher, D. A. *Biophysical Journal* **2012**, *102*, 443-451.

(3) Fu, J.; Wang, Y.-K.; Yang, M. T.; Desai, R. a; Yu, X.; Liu, Z.; Chen, C. S. *Nature Methods* **2010**, *7*, 733-736.

(4) Ketene, A. N.; Schmelz, E. M.; Roberts, P. C.; Agah, M. *Nanomedicine : nanotechnology, biology, and medicine* **2011**, *8*, 93-102.

(5) Ricart, B. G.; Yang, M. T.; Hunter, C. a; Chen, C. S.; Hammer, D. a *Biophysical journal* **2011**, *101*, 2620-8.

(6) Rape, A. D.; Guo, W.-H.; Wang, Y.-L. *Biomaterials* **2011**, *32*, 2043-51.

(7) Ker, E. D. F.; Nain, A. S.; Weiss, L. E.; Wang, J.; Suhan, J.; Amon, C. H.; Campbell, P. G. *Biomaterials* **2011**, *32*, 8097-107.

(8) Nain, A. S.; Phillippi, J. a; Sitti, M.; Mackrell, J.; Campbell, P. G.; Amon, C. *Small (Weinheim an der Bergstrasse, Germany)* **2008**, *4*, 1153-9.

(9) Bakhru, S.; Nain, A. S.; Highley, C.; Wang, J.; Campbell, P.; Amon, C.; Zappe, S. *Integrative biology : quantitative biosciences from nano to macro* **2011**, *3*, 1207-14.

(10) Nain, A. S.; Sitti, M.; Jacobson, A.; Kowalewski, T.; Amon, C. *Macromolecular rapid communications* **2009**, *30*, 1406-12.

(11) Ferrell, N.; Woodard, J.; Hansford, D. J. *Sensors and Actuators A: Physical* **2011**, *170*, 84-89.

(23a) Microstructural Characterization and Dissolution Behavior of Drug/Semicrystalline Polymer Systems On Substrates

Hsin-Yun Hsu, Chemical Engineering, Purdue University, West Lafayette, IN and Michael T. Harris, School of Chemical Engineering, Purdue University, West Lafayette, IN

Solid dispersions have been used to improve the bioavailability of poorly water-soluble drugs. However, controlling the stability of the products can still be challenging. Recently, solidification on certain substrates can result in the control of polymorph and crystallization kinetics. The presence of the substrate surface provides heterogeneous nucleation sites that enables the control and scale-up of solidification processes.

In this study, the Drop on Demand technique is used to dispense a solution of Naproxen (NAP) and polyvinylpyrrolidone (PVP) on Chitosan or HPMC films to investigate the role of the substrate surface on the crystallization of the solid dispersion. XRD curves show that NAP has higher crystallinity when crystallized on Chitosan film compared to the HPMC film. The SEM and SHG images also show that NAP forms larger crystal sizes on the Chitosan film. AFM was used to characterize the films and it shows that the Chitosan film has a rougher surface that may provide more sites for NAP to nucleate. The dissolution experiments of those samples on different substrates were also conducted and the results are correlated to their crystallinity and microstructure on the films.

(23b) Slurry Crystallization of Water-Insoluble Drug Substance – Overcoming Challenges in Solubility and Miscibility Requirements for Solvents and Anti-Solvents

Alexander Chu-Kung, Engineering, Abbott Laboratories, North Chicago, IN and Hsien-Hsin Tung, Global Pharmaceutical Research & Development, Abbott Laboratories, North Chicago, IL

In the crystallization development of water-insoluble drug substances, a common issue encountered is the availability of proper solvents and anti-solvents. Ideally, a proper solvent would possess a sufficient solubility at the beginning to dissolve the drug substance, and then its solubility can be sufficiently reduced by adding a solvent-miscible anti-solvent, to crystallize out the drug substance. The dual requirements of solubility and miscibility for both solvents and anti-solvents can be challenging from time to time.

In our case, we present a unique slurry crystallization approach as one solution to bypass such constraints. In the 1st part of the slurry crystallization process, the drug substance is slurried in the solvent which has only limited drug substance solubility. The slurry is seeded with the drug substance of the right crystal form, which subsequently converts all un-dissolved drug substance to the correct crystal form. By this approach, the solubility constraint to dissolve the entire drug substance is successfully bypassed.

In the 2nd part of the slurry crystallization process, the slurry is charged to an anti-solvent, which has a very low solubility, in order to minimize the loss in the mother liquor. As mentioned, the anti-solvent may not be miscible with the solvent used in the 1st part of the process. To overcome this, the solvent is distilled off during the transfer so that its level is kept at a level below the miscibility boundary of solvent and anti-solvent. As a result, the miscibility requirement of solvent and anti-solvent can be successfully bypassed.

In this presentation, we demonstrated a case of slurry crystallization to illustrate the this features, using ethyl-acetate as solvent and water as anti-solvent.

(23c) A Microfluidic Platform for *In Meso* Crystallization and *In Situ* Crystal X-Ray Diffraction of Membrane Proteins

Daria S. Khvostichenko, Jeremy M. Schieferstein, Ashtamurthy Pawate and Paul J.A. Kenis, Department of Chemical & Biomolecular Engineering, University of Illinois at Urbana-Champaign, Urbana, IL

Membrane proteins are key actors in a number of biological processes, which is highlighted by the fact that they account for ~60% of all available drug targets. To better understand functional mechanisms, high-resolution crystal structures of proteins are required. The success rate of membrane protein crystallization trials, however, is notoriously low, and growing diffraction-quality crystals of membrane proteins is a major bottleneck in structural biology.

In meso crystallization exploits properties of certain lipids that self-assemble into continuous bilayer systems, known as lipidic mesophases, when mixed with water or aqueous solutions. During *in meso* crystallization trials membrane protein molecules are incorporated into the bilayers of the mesophase and are thus maintained in a native-like environment. In spite of a number of recent impressive successes, the *in meso* method is not used widely in the structural biology community, presumably due to extra steps required in the crystallization protocol compared to traditional crystallization from solutions, and due to the difficulties of handling highly viscous lipidic mesophases. Similarly, harvesting fragile protein crystals from gel-like lipidic mesophases as necessary for crystal X-ray diffraction data collection is challenging and may compromise the quality of crystals.

We present a microfluidic platform for membrane protein crystallization in lipidic mesophases and *in situ* X-ray analysis. The multi-well platform requires 60 nL of the protein solution and 25 nL of lipid per well, with a potential for further reduction in the sample amount, and automates simultaneous formulation of multiple samples. The platform design is optimized for diffusional mixing of the protein solution with the lipid to formulate protein-loaded mesophases. For fluid metering and routing the platform relies on the multilayer polydimethylsiloxane (PDMS) architecture with monolithic pneumatic valves. For X-ray transparency thin layers of PDMS patterned with features for metering and routing are sandwiched between layers of cyclic olefin copolymer (COC), which provides rigidity and servers as an evaporation barrier. We validate our approach by crystallizing membrane protein photosynthetic reaction center (*R. Sphaeroides*) and collecting protein crystal X-ray diffraction data on-chip.

(23d) Modeling and Simulation of Coaxial Crystallizers by Dynamically Coupled Population Balance, Macromixing, and Micromixing Models

J. Carl Pirkle Jr.[1], Lucas Foguth[1], Steven J. Brenek[2], Kevin Girard[2] and Richard D. Braatz[3], (1)Massachusetts Institute of Technology, Cambridge, MA, (2)Chemical R&D, Pfizer, Inc, Groton, CT, (3)Department of Chemical Engineering, Massachusetts Institute of Technology, Cambridge, MA

A leading method for the crystallization of pharmaceutical compounds is to rapidly mix an antisolvent with a solvent saturated with the desired drug. Many crystallizer designs have been explored to generate high supersaturation in such mixtures as an approach for generating consistent crystal nuclei that are subsequently grown to a desired size [1-5]. Compared to cross-flow mixers, coaxial jet mixers have negligible buildup of crystalline material on their surfaces and are less likely to plug. Coaxial mixers can be designed to deliver rapid turbulent mixing using short sections of pipe. As the energy required for mixing is provided by the inlet streams, with no moving metal parts and no bearings, these devices have simple maintenance and operation. Some experimental and modeling studies of coaxial crystallizers have been published to gain deeper understanding and to facilitate more efficient development and optimization of the coaxial mixer crystallization process [6-9]. Important design parameters for crystallization in coaxial mixers are the length of pipe downstream of the injection point, the velocity and temperature of the inlet streams, and the inner and outer pipe diameters.

This presentation describes an effort whose goal is to speed up the design of the coaxial crystallizers to tailor the crystal size distribution according to the bioavailability and drug administration requirements. Dynamic simulations of a confirmed coaxial crystallizer were carried out that simultaneously solve partial differential equations for macromixing, micromixing, and a population balance for the crystals. The computational model [10-11] was used, which replaced a quadrature-method-of-moments model used to simulate the time evolution of the particle size distribution by Rodney Fox [12] with a full spatially varying population balance model implemented using a high resolution finite-volume method. This presentation employs an extension of the model [10-11] to include the temperature effects on the crystallization. Our simulations were used to perform a parameter sensitivity analysis (see [13] for background on such analyses) to identify the key model parameters and to simulate variations on their values on the full crystal size distribution (CSD) in the antisolvent crystallization of Lovastatin, using kinetics reported in the literature [14]. The effects of inlet concentrations and stream flow rates on CSD were numerically investigated and compared with CSDs obtained in a dual-impinging jet crystallizer [11]. As observed in simulations of dual impinging jets, the mean crystal size and the width of the distribution were found to decrease with an increase in inlet stream velocity. The simulation results showed different degrees of inhomogeneity in the supersaturation and the nucleation and growth rates for different inlet stream flow rates.

To the authors' knowledge, this is the most detailed simulation study on coaxial crystallizers reported to date. The simulation results show the feasibility of tailoring a specific crystal size distribution by adjusting the operating conditions (such as inlet stream velocities) of the coaxial crystallizer. Such computational tools can provide a technological advancement for the process development in the pharmaceutical industry by providing a more in-depth understanding of the crystallization process, and by reducing the number of experiments required to determine the optimum operating conditions, to reduce the quantify of active pharmaceutical ingredient (API) needed for the experiments. Consequently, the crystallizer process design can be performed much earlier during the drug development process, where a limited quantity of API is available.

References

1. Midler, M., E.L. Paul, E.F. Whittington, M. Futran, P.D. Liu, J. Hsu, and S.-H. Pan, U.S. Patent 5,314,506, 1994.
2. Dauer, R., J.E. Mokrauer, and W.J. McKeel, U.S. Patent 5,578,279, 1996.

3. Lindrud, M.D., S. Kim, and C. Wei, U.S. Patent 6,302,958, 2001.
4. am Ende, D.J., T.C. Crawford, and N.P. Weston, U.S. Patent 6,558,435, 2003.
5. Wang, X., J.M. Gillian, and D.J. Kirwan, Quasi-emulsion precipitation of pharmaceuticals. 1. Conditions for formation and crystal nucleation and growth behavior, Crystal Growth & Design, 2006, 6, 2214-2227.
6. Wei, H., and J. Garside, Application of CFD modelling to precipitation systems, Chem. Eng. Res. Des., 1995, 75(2), 219-227.
7. Baldyga, J., and W. Orciuch, Barium sulphate precipitation in a pipe: An experimental study and CFD modeling, Chem. Eng. Sci., 2001, 56, 2435-2444.
8. Marchisio, D.L., A.A. Barresi, and M. Garbero, Nucleation, growth, and agglomeration in barium sulfate turbulent precipitation. AIChE J., 2001, 48, 2039-2050.
9. Öncül, A.A., K. Sundmacher, A. Seidel-Morgenstern, and D. Thévenin, Numerical and analytical investigation of barium sulphate crystallization, Chem. Eng. Sci., 2006, 61, 652-664.
10. Woo, X.Y., R.B.H. Tan, P.S. Chow, and R.D. Braatz, Simulation of mixing effects in antisolvent crystallization using a coupled CFD-PDF-PBE approach, Crystal Growth & Design, 2006, 6,1291-1303.OWTH
11. Woo, X.Y., R.B.H. Tan, and R.D. Braatz, Modeling and computational fluid dynamics-population balance equation-micromixing simulation of impinging jet crystallizers, Crystal Growth & Design, 2009, 9(1), 156-164.
12. Fox, R.O., Computational Models for Turbulent Reactive Flows, 2003, Cambridge: Cambridge University Press.
13. Varma, A., M. Morbidelli, and H. Wu, Parametric Sensitivity in Chemical Systems, 1999, Cambridge: Cambridge University Press.
14. Mahajan, A.J., and D.J. Kirwan, Nucleation and growth-kinetics of biochemicals measured at high supersaturations. J. of Crystal Growth, 1994, 144, 281-290.

(23e) Impact of a Reversible Solid State Form Change On Particle Morphology and Bulk Powder Properties of a Pharmaceutical Compound

Joshua D. Engstrom[1], Qi Gao[2], Daniel Roberts[3], Chenkou Wei[2], Chiajen Lai[4] and Jonathan Brown[1], (1)Drug Product Science and Technology, Bristol-Myers Squibb Company, New Brunswick, NJ, (2)Drug Product Science & Technology, Bristol-Myers Squibb Co., New Brunswick, NJ, (3)Drug Product Science & Technology, Bristol-Myers Squibb Co., New Brunswick, NJ, (4)Drug Product Science and Technology, Research and Development, Bristol-Myers Squibb Company, New Brunswick, NJ

For efficient formulation processing of a pharmaceutical drug product, it is desirable to have an active pharmaceutical ingredient (API) that handles well in terms of flow and bulk density, especially for products with high drug loadings. To improve bulk API powder property behavior, particle engineering techniques can be employed during the crystallization step to alter the morphology of the API crystals from needles, which normally have poor flowing powders, to bricks or rods. Although particle engineering can have a significant impact on particle morphology, the susceptibility of the API to crystalline form change from environmental factors such as relative humidity must also be carefully studied and controlled.

A case study is presented for Compound A whose thermodynamically desired crystalline form has an undesired morphology but can be modified to bricks or rods through a particle engineering approach. Compound A is known to crystallize as iso-structural solvates (Form I) in all solvent systems that have been studied. Upon isolation and drying, Form I desolvates to Form II which can reversibly convert to Form III in the solid state at a critical relative humidity. In this presentation, the crystal structures of Forms I, II and III will be discussed and used to explain how the form changes impact the integrity and morphology of the API crystal as well as the bulk powder properties. The importance of this work is to gain a fundamental understanding of how the API crystal morphology improvement made during crystallization can be retained in the post crystallization steps of drying, storage and handling. The thorough understanding of the crystal structures can provide valuable insights for handling bulk powders to maintain desired API powder properties for drug product processing.

(23f) The Morphological-Population Balance Model (M-PBM) Generator. Application to Additives Controlled Crystallization of KAP

Meenesh R. Singh and Doraiswami Ramkrishna, School of Chemical Engineering, Purdue University, West Lafayette, IN

This talk will present a comprehensive framework to develop M-PBM for crystal morphology distributions evolving in a supersaturated solution. The model considers faces on seed crystals and energetically favorable F-faces to assess the polyhedral symmetry and identify internal coordinates. A set of all possible morphologies is generated using a combinatorial approach. These morphologies are further decomposed to yield polyhedral cones representing morphology domains in crystal state space. A bidirected graph is constructed where the nodes and edges correspond to morphology domains and their transitions, respectively. The population balance equations are identified for each node accounting for morphology transformations in dynamic environments. The solution of M-PBM can be readily obtained by the method of characteristics which takes considerably less computation time as compared to other numerical methods. The application of M-PBM to additives controlled crystallization of Potassium Acid Phthalate (KAP) is considered. The optimal profiles of additives concentration are computed for maximizing purity, productivity and sphericity of KAP crystals.

(43a) Looking to the Future in Continuous Pharmaceutical Manufacturing

Richard D. Braatz, Chemical Engineering, Massachusetts Institute of Technology, Cambridge, MA

This presentation describes the motivations, challenges, and opportunities in continuous pharmaceutical manufacturing. Some of the motivations are particularly intriguing, such as semicontinuous and continuous-flow unit operations that have the potential to have order-of-magnitude higher controllability and/or are much easier to scale up than batch and semibatch processes. Theoretical, computational, and experimental results are provided as evidence of the potential opportunities of continuous-flow pharmaceutical manufacturing operation. One of the conclusions of reviewing such unit operations is that they can look and operate very different from the batch and semibatch unit operations currently used in the pharmaceutical industry.

Example unit operations are drawn from the published literature, which include both journal papers and patents from various academic and industrial research groups. The presentation will also review and comment on the application of Quality-by-Design methodologies to pharmaceutical manufacturing. Of particular interest is the generation of design spaces, and the pros and cons of different approaches. In particular, it is shown that many of the methods for the construction of design spaces that have been proposed in the literature are useless from a control or operations point of view.

(43b) New Challenges and Opportunities for Pharmaceutical Manufacturing Science

Vincent Vilker, Office of Testing and Research, FDA, Chevy Chase, MD and Mansoor A. Khan, Cder/Ops/Otr/Dpqr, Food and Drug Administratin, Silver Spring, MD

New Challenges and Opportunities for Pharmaceutical Manufacturing Science

Vincent L Vilker and Mansoor A Khan, CDER/FDA

The increasing sophistication of modern drug products is leading to unprecedented challenges and opportunities for pharmaceutical manufacturing. At the same time, scientific and technological advancements in analytical instrumentation, coupled with implementation of the principles embedded in quality by design, enable the improved process monitoring and control that lead to better and more consistent product quality. ICH Q8(R2), the first of three International Conference on Harmonization (ICH) Guidance for Industry that spell out the concepts of Quality by Design, states that critical formulation attributes and process parameters should be identified through an assessment of the extent to which their variation can have impact on the quality of the drug product. In this presentation, examples will be given where modern instrumentation and manufacturing practices are required to maintain the critical formulation attributes of several drug products that have unique physicochemical and/or biological properties. Examples may include properties such as very low solubility, polymorphic solid states, poorly defined macromolecular size distributions, or where there is a tendency to become unstable in the formulated product.

() Recent Advances in the Molecular Modeling Foundation of Systems-Based Pharmaceutics

Constantinos C. Pantelides, Department of Chemical Engineering, Imperial College London,Center for Process Systems Engineering, London, United Kingdom

There are potentially significant benefits from the adoption of a systems-based approach to pharmaceutical science and engineering based on detailed mathematical first-principles models. However, the success of such an approach is predicated on the accurate description of the thermodynamic, kinetic and mechanical properties of the materials involved.

In this presentation, we review some recent advances in the description of the solid and fluid phases arising from work by the Molecular Systems Engineering group at Imperial College London. In particular, we review the state of the art for the *ab initio* prediction of crystal structure and polymorphism, and the prediction of vapour/liquid and liquid/solid phase equilibria using the SAFT-gamma group-contribution based equation of state. Significant progress has been achieved in all these areas over the last few years, but there are also significant challenges that still need to be addressed.

(43d) Experience with Regulatory Issues for Biomedical Devices and Vision for the Use of Advanced Control In Pharmaceutical Operations

Francis J. Doyle III, Chemical Engineering, UCSB, Santa Barbara, CA, Eyal Dassau, UCSB, Santa Barbara, CA and Howard Zisser, Sansum Diabetes Research Institute, Santa Barbara, CA

The landscape for clinical testing of biomedical devices has evolved rapidly over the last 5-10 years. In particular, the use of advanced control algorithms, such as model predictive control, are now being reviewed by the FDA as components of closed-loop devices. In our work with the Artificial Pancreas (AP), we have developed master files and other supporting materials to support IDE (Investigational Device Exemption) studies. In this talk, I will share some of the challenges in developing algorithms for regulatory review, including the role of simulation studies, communication architectures, algorithmic design, and IDE studies. I will extrapolate from these experiences to project the opportunities for advanced control in pharmaceutical operations.

Chemical Engineering & Biopharmaceutics: Linking the Process with the Ultimate Quality Attribute

Ravi M. Shanker

The discovery and development of heterocyclic, small molecule new drug entities is a highly sophisticated, multi-disciplinary, scientifically advanced, collaborative, highly competitive, extremely expensive and multi-year endeavor. The heavily R&D dependent pharmaceutical industry utilizes cutting edge scientific and technological tools and techniques encompassing the broad disciplines of chemistry, biology and medicine. To provide significantly improved therapeutic treatments for various acute and chronic diseases and to be able to meet the requirements and expectations of the regulatory agencies, the patients and the society is a constant challenge for the innovative pharmaceutical industry. The challenges forge deeper dives in science among the wide array of disciplines that constitute the pharmaceutical industry and consequently have provided tremendous opportunities for the advancement of computational and predictive tools within each of these disciplines and sub-disciplines. A key segment of R&D and industrial scale manufacture is pharmaceutical sciences. It is an amalgamation of applied disciplines of chemical engineering, physical and physical organic chemistry, synthetic organic chemistry, process chemistry, analytical chemistry, polymer chemistry, material sciences, crystallization science, drug product design and manufacture and evaluation of performance of drug product in the patients. Pharmaceutical sciences have also embraced the development and use of predictive computational tools among each of its areas of specialization and disciplines to ensure the highest quality of drug product being delivered to the patient. The multi-disciplinary components of pharmaceutical sciences are operationally integrated to provide the highest quality of drug product to meet the regulatory requirements in this highly regulated industry. However, there is a lack of integration of such predictive tools across disciplines to enable the assessment of local perturbations in quality during any segment of work flow on the ultimate quality and performance of the drug product. The ultimate quality attribute of a drug product is it's performance in patients, however, linking that performance to the upstream attributes of drug substance and drug product via engineering control during manufacture with emphasis on the principles of chemical engineering continues to be unrealized and unfulfilled possibility. The vision to provide such integrated approach of existing computational and predictive tools within disciplines and sub-disciplines in pharmaceutical sciences and to provide opportunities for further development of advanced integrated tools as a holistic system is the underlying concept for the creation of Systems Based Pharmaceutics.

(47a) Design of Potent Antibody Inhibitors of Amyloid Fibrillization

Ali Reza A. Ladiwala, Moumita Bhattacharya, Joseph M. Perchiacca and Peter M. Tessier, Dept of Chemical & Biological Engineering, Rensselaer Polytechnic Institute, Troy, NY

Antibodies hold significant potential for specifically inhibiting toxic protein aggregation associated with conformational disorders such as Alzheimer's and prion disease. Unfortunately, conventional antibodies are poor inhibitors of protein aggregation. We aim to harness the homotypic interactions between hydrophobic peptide segments that mediate amyloid fibrillization to design antibodies with potent anti-aggregation activity. We have discovered that grafting small hydrophobic peptides from amyloidogenic polypeptides into the complementarity determining regions of single-domain antibodies yields antibody variants that potently inhibit amyloid formation. Grafted AMyloid-Motif AntiBODIES (gammabodies) presenting hydrophobic peptide segments from amyloidogenic proteins inhibit aggregation of the corresponding polypeptide at substoichiometric concentrations. In contrast, sequence- and conformation-specific antibodies obtained via immunization are unable to prevent aggregation at substoichiometric concentrations. We expect that our antibody design approach – which eliminates the need for immunization or screening to identify sequence-specific antibodies – can be readily extended to generate potent antibody inhibitors for other amyloidogenic polypeptides linked to human diseases.

(47b) Development of Computational Methods to Support De Novo Antibody Design

Robert J. Pantazes, Deparment of Chemical Engineering, The Pennsylvania State University, University Park, PA and Costas D. Maranas, Department of Chemical Engineering, The Pennsylvania State University, University Park, PA

The *de novo* design of a novel protein with a particular function remains a formidable challenge with only isolated and hard-to-repeat successes to date. Due to their many structurally conserved features, antibodies are a family of proteins more amenable to predictable rational design. Previously, we have developed the Optimal Complementarity Determining Regions (OptCDR) method to design antigen binding sites targeted against any specified epitope. OptCDR functions by selecting optimal combinations of backbone structures that are most likely to have favorable interactions with an antigen and then filling in their amino acids during a computational affinity maturation step.

Here we will present how we are moving beyond designing just the antigen recognition sites to the design of fully-human antibody variable domains. In this effort, antibody sequences, genes, structures, and standards are imported from the IMGT database [http://www.imgt.org]. The immune system naturally generates billions of antibodies from only hundreds of genes using VDJ recombination, where random variable (V), diversity (D), and joining (J) genes, each of which correspond to a specific region of antibody variable domain structure, are combined to create an antibody. Using a clustering procedure we mapped the observed structural diversity onto less than 1,000 structures that span three regions, analogous to the V, D, and J genes. The recombination of this collection of structures spans a potential diversity of over 6*10^9 unique human antibodies. We find that using this structural description we can predict the experimentally resolved structures with an average rmsd of 1.28 Å. Subsequently, we analyzed (i) the broadly neutralizing anti-influenza antibody CH65 and (ii) anti-HIV antibody 4E10 to provide starting points for the design of affinity-matured antibody libraries and examine the frequency and distribution of mutations during the affinity maturation process. Ultimately, the goal is to use the developed database within a computational method for the de novo design fully human antibody variable domains.

(47c) Computationally Driven Therapeutic Protein Deimmunization

Regina S. Salvat[1], Andrew S. Parker[2], Daniel C. Osipovitch[3], Chris Bailey-Kellogg[4] and **Karl E. Griswold**[1], (1)Thayer School of Engineering, Dartmouth, Hanover, NH, (2)Department of Computer Science, Dartmouth, Hanover, NH, (3)Program in Experimental and Molecular Medicine, Dartmouth, Hanover, NH, (4)Department of Computer Science, Dartmouth College, Hanover, NH

The unparalleled specificity and activity of therapeutic proteins has reshaped many aspects of modern clinical practice, and aggressive development of new protein drugs promises a continued revolution in disease therapy. As a result of their biological origins, therapeutic proteins present unique design challenges for the biomolecular engineer. One distinguishing risk factor is the prospect of eliciting an immune response upon administration to human patients. An anti-drug immune response can compromise therapeutic efficacy and even threaten patient safety. In the case of therapeutic antibodies, codified humanization technologies can mitigate much of the risk associated with immune reactions. These grafting-based deimmunization strategies are not, however, broadly applicable to other protein classes. In particular, they require a detailed understanding of structure-function relationships and the availability of homologous human protein scaffolds. Looking to the future, our capacity to effectively tap the full diversity of non-immunoglobulin therapeutic candidates will require more broadly applicable protein deimmunization technologies. To help meet this challenge, we have recently developed optimization algorithms that seamlessly integrate computational prediction of T cell epitopes and bioinformatics-based assessment of the structural and functional consequences of epitope-deleting mutations. The algorithms have been extensively tested and refined using retrospective studies of therapeutic candidates from the literature. More recently, we have experimentally validated the algorithms with the Enterobacter cloacae P99 β-lactamase enzyme, a component of Antibody Directed Enzyme Prodrug Therapies for cancer. This talk will highlight recent results in which we have explored the dual-objective protein design space and mapped the Pareto optimal frontier, i.e. testing undominated enzyme designs whose stability and immunogenicity scores are not simultaneously bested by any other protein. We conclude that our deimmunization algorithms, which can be applied to virtually any protein target, guide the protein engineer towards promising immunoevasive therapeutic candidates.

(47d) Improved Protein Switches for Cancer-Activated Enzyme Prodrug Therapy

R. Clay Wright[1], Arjun Khakhar[2], Chapman M. Wright[1], James R. Eshleman[3] and Marc Ostermeier[1], (1)Chemical and Biomolecular Engineering, Johns Hopkins University, Baltimore, MD, (2)Biomedical Engineering, Johns Hopkins University, Baltimore, MD, (3)Pathology and Oncology, Johns Hopkins University School of Medicine, Baltimore, MD

Prevailing approaches for developing cancer protein therapeutics focus on creating proteins that therapeutically modulate a cancer marker's function. Such an approach limits the therapeutic mechanism to those that naturally arise from modulation of the cancer marker and precludes the use of cancer markers for which therapeutic modulation is not feasible. Furthermore, many potential protein therapies lack the desired cancer targeting. The ability to link recognition of any cancer marker with activation of any desired therapeutic function would enormously expand the number of possible protein therapeutics.

We have previously engineered a switchable prodrug-activating enzyme that selectively kills human cancer cells that accumulate the cancer marker hypoxia-inducible factor 1α (HIF-1α). This HIF-1α-activated enzyme switch (Haps59) was created by fusing the prodrug-converting enzyme yeast cytosine deaminase (yCD) and the CH1 domain of the p300 protein, which binds HIF-1α. Haps59 autonomously increases its ability to convert the prodrug 5-fluorcytosine (5FC) into the chemotherapeutic 5-fluorouracil (5FU) in a HIF-1α-dependent manner, rendering colon and breast cancer cells more susceptible to the prodrug. However, the difference in 5FC sensitivity between the presence and absence of HIF-1α was not as large as desired. Using circular permutation as well as random and cassette mutagenesis followed by a two-tiered genetic selection for improved switches, we have identified new HIF-1α-activated enzymes that confer increased 5FC sensitivity in the presence of HIF-1α and reduced 5FC sensitivity in the absence of HIF-1α.

Our strategy offers a platform for the development of inherently selective protein therapeutics for cancer and other diseases. By introducing protein-level regulation into cancer prodrug therapies, our approach skirts the problematic selective transduction requirement currently limiting enzyme-prodrug therapies. In addition, our approach is complementary to both transcriptional and transductional targeting and might be combined with these approaches to afford a double or triple layer of specificity: at the gene delivery level, at the transcription level and at the protein level.

(47e) Engineered Hepatocyte Growth Factor Mutants: New Tools for Tissue Regeneration and Vascularization

Cassie Liu[1], Douglas S. Jones[2], Ping-Chuan Tsai[2] and Jennifer R. Cochran[2], (1)Chemical Engineering, Stanford, STANFORD, CA, (2)Bioengineering, Stanford University, Stanford, CA

Hepatocyte growth factor (HGF), through activation of its tyrosine kinase receptor c-MET, has shown great promise in regenerative medicine; however, the instability and poor expression yield of recombinant HGF have been a severe bottleneck to its clinical translation. Using rational and combinatorial methods, we engineered a protein fragment composed of the N terminal domain and first Kringle domain of HGF, termed NK1. Compared to full-length HGF (~105 kDa), the engineered NK1 fragment (termed eNK1) is much smaller (~20 kDa), has increased thermal stability by more than 15°C, and can be recombinantly expressed at an order of magnitude higher yield in yeast than native NK1. Dimerization of HGF's NK1 domains is a prerequisite to c-MET dimerization and activation. Thus, to develop a more potent agonist, we created a dimeric eNK1 protein via covalent crosslinking through an N-terminal disulfide bond. Biophysical and biological characterization of the dimeric eNK1 protein will be discussed, including affinity to c-MET, downstream phosphorylation, endothelial migration, and vascularization. Remarkably, we found that the eNK1 dimers elicited significantly greater agonistic activity than the eNK1 monomer, approaching activity levels of full-length HGF. Our ultimate goal is to develop these engineered proteins as novel therapeutics for promoting tissue regeneration. Funded by NIH NCI R01CA151706.

(47f) Enzyme Prodrug Therapy Targeted to Breast Tumor Vasculature Using a Recombinant Fusion Protein

John Krais[1], Vassilios I. Sikavitsas[1], Carla Kurkjian[2] and Roger G. Harrison[1], (1)Chemical, Biological, and Materials Engineering, University of Oklahoma, Norman, OK, (2)Hematology-Oncology, University of Oklahoma Health Sciences Center, Oklahoma City, OK

Invasive breast cancer will appear in almost a quarter of a million people in the United States this year, with a current mortality rate of almost 25%. Limited efficacy and severe side effects of current breast cancer treatments have created a demand for improved and innovative therapeutic strategies. Targeted therapeutics is a promising approach for the development of new cancer treatments that seek to address these challenges. A targeted enzyme prodrug therapy has been developed to localize cytotoxicity to the site of the tumor using a protein fusion of annexin V (AV) and bacterial purine nucleoside phosphorylase (PNP). The high affinity, calcium-dependent binding of annexin V acts as the tumor targeting component of the fusion, as it has been shown to bind to phosphatidylserine (PS) expressed externally almost exclusively on cancer cells and the endothelial cells of the tumor vasculature. The enzymatic component of the fusion, PNP, converts the prodrug fludarabine into a more cytotoxic form, 2-fluoroadenine. This conversion allows for increased impact at the site of the tumor, while reducing side effects on normal tissue.

Recombinant production and purification of the fusion protein has yielded an active enzyme and promising *in vitro* results. Binding strength and stability were determined for breast cancer cell lines MDA-MB-231 and MCF-7, as well as non-confluent endothelial cells HAAE-1 that mimic tumor vasculature with externalized phosphatidylserine. These studies reveal that the dissociation constant for each cell line is less than 100 pM and the fusion protein remains bound for a minimum of three days. Cytotoxicity tests indicated susceptibility of both cancer cell lines to prodrug and prodrug with fusion protein at concentrations as low as 1.0 μM fludarabine. Non-confluent endothelial cells showed a dose dependent cytotoxic effect of fludarabine with PNP-AV with a reduced impact of fludarabine alone.

This novel enzyme-prodrug therapy warrants *in vivo* studies, which will be performed with SCID mice and MDA-MB-231/GFP breast cancer xenografts. PEGylation of the fusion protein will allow for future studies in immunocompetent mice. Combinatory treatments with immunotherapeutics are also exciting possible future directions for vasculature targeted strategies such as this enzyme prodrug therapy.

(47g) Engineering Polymerized Hemoglobins for Use In Transfusion Medicine

Andre Palmer, Chemical and Biomolecular Engineering, The Ohio State University, Columbus, OH

Universal O_2-carrying solutions that can replace the O_2 storage and transport functions of red blood cells will greatly improve clinical outcomes both for trauma victims and for patients undergoing high-blood-loss surgical procedures. These O_2 carriers also will prevent the many serious complications associated with blood transfusions. My talk will address a simple approach for designing hemoglobin-based O_2 carriers (HBOCs). Our design strategy is based on the observation that transfusion of hemoglobin results in vasoconstriction at the microcirculatory level and the development of systemic hypertension. The root cause of this effect stems from the ability of hemoglobin to extravasate through pores in the wall of blood vessels and scavenge nitric oxide (NO) from the surrounding vasculature. This in turn leads to blood vessel constriction and results in the observed hypertensive effect. Therefore, our design strategy will focus on increasing the size of HBOCs so that these molecules are unable to traverse across the wall of blood vessels. In this talk, I will discuss the design, biophysical properties and *in vivo* response of polymerized hemoglobins. This work is significant in that it will lead to the development of novel O_2 carriers that are safe and efficacious for therapeutic use.

(47h) Cell-Based Biosensors for Dynamically Imaging Immune Function in Vivo

Nichole Daringer, Rachel M. Dudek and Joshua N. Leonard, Chemical and Biological Engineering, Northwestern University, Evanston, IL

The ability to dynamically monitor or visualize local immune states *in vivo* would transform our ability to understand processes such as the establishment and development of dysfunctional immune states at tumor sites and the response of these multicellular networks to potential therapeutic interventions. To date, however, we are limited to systemic measures (such as profiles of serum cytokines or circulating immune cells) or terminal assays (requiring tissue biopsy or, more commonly, sacrifice of the experimental animal). We have established a novel mammalian synthetic biology technology enabling us to engineer cells into living "biosensors" responsive to environmental stimuli that are exclusively extracellular, such as the cytokines that characterize specific immunological states. These cell-based biosensors detect a relevant analyte of interest and then produce an optical signal that can be imaged in whole, living animals (e.g., luminescence, near-infrared fluorescence). This approach is fundamentally distinct from a reporter gene approach, in which one can only detect whether a given gene is expressed in the engineered cell, not whether the engineered cell is exposed to the analyte. As an initial proof of principle, we have engineered tumor cells to function as biosensors, such that following adoptive transfer of engineered tumor cells to a syngeneic mouse host, the tumor monitors and reports upon its own immunological milieu. We are initially applying this technology to help understand how the immune response shifts from anti-tumor (immunostimulatory) to pro-tumor (immunosuppressive) during the course of tumor progression. This foundational technology could eventually be extended to generate cell-based therapies by tying biosensing to the expression of therapeutically-relevant gene products.

(48a) Development of a Bacterial System for the Expression of Full-Length Monoclonal Antibodies

Michael-Paul Robinson, Chemical and Biomolecular Engineering, Cornell University, Ithaca , NY and Matthew P. DeLisa, Chemical and Biomolecular Engineering, Cornell University, Ithaca, NY

Monoclonal antibodies (mAbs) have proven to be an excellent basis for developing diverse and powerful clinical therapeutics. Owing to their complex multimeric structure that requires oxidation of multiple disulfide bonds, mAbs are typically produced in mammalian cell systems. However, full-length mAb expression has been reported in *Escherichia coli* and required targeting to the periplasmic space, where disulfide bond formation/oxidation is favored. This strategy, while successful, relies on membrane co-localization and assembly of heavy and light chains in the periplasm, which due to its small volume is unlikely to be optimal for high-level expression of large, multi-subunit proteins like mAbs. Moreover, periplasmic expression of full-length mAbs also requires the energy intensive inner membrane translocation of both heavy and light chains, which likely presents another barrier to optimal mAb production.

Expression in the cytoplasm of *E. coli* can potentially provide a means to circumvent these limitations. However, the glutaredoxin and thioredoxin pathways provide a major obstacle to cytoplasmic expression and assembly of full-length mAbs. In wild type *E. coli,* these pathways maintain thiol groups on cysteines in their reduced, non-disulfide bonded state. To address this complication, *E. coli* strain SHuffle™T7 Express was engineered for the cytoplasmic expression of proteins requiring multiple disulfide bonds for proper folding. The SHuffle™T7 Express strain carries mutations in the oxidoreductase genes of the thioredoxin and glutaredoxin pathways (trxB, gor), a growth suppressor mutation (ahpC*), as well as a chromosomal copy of the periplasmic disulfide bond isomerase DsbC lacking its native signal sequence. Collectively, these mutations render the cytoplasm a favorable environment for the correct folding of proteins requiring disulfide bonds, such as mAbs. Here, we report a new expression platform called MAXmAb (Membrane-Autonomous eXpression of monoclonal Antibodies) that utilizes *E. coli* SHuffle™T7 Express cells for cytoplasmic production of mAbs. Using MAXmAb, full-length immunoglobulin G (IgG) antibodies in three formats - fully mouse, mouse/human chimeric, and humanized - specific for *E. coli* maltose binding protein (MBP) were expressed and assembled in the cytoplasm. The cytoplasmically-expressed IgGs, which we call cytoclonal antibodies (cytAbs), displayed strong binding activity for MBP that was comparable to that for IgGs expressed from hydridoma cells. Additionally, we reprogrammed the humanized anti-MBP IgG using a variable domain sequence substitution strategy. This resulted in a panel of new cytAbs that exhibited altered antigen specificity, in particular, for *Bacillus anthracis* protective antigen (PA63), the cardiac glycoside digoxin (Dig), the leucine zipper domain of the yeast transcriptional activator Gcn4 (Gcn4LZ), a peptide derived from the influenza hemagglutinin protein (HAG), and human mitogen activated protein kinase 9 (JNK2). Each of the newly generated cytAbs exhibited high in vitro binding affinity for their respective antigens and comparable expression to the original anti-MBP cytAb. We also show the potential advantage of the cytoplasmic localization and assembly of light and heavy IgG chains system over the existing periplasmic strategy. Finally, building upon a previous study, we demonstrate the recovery of immunological activity, specifically binding activity for FcγRI/CD64, by site-directed mutagenesis of the CH3 domain of cytAb heavy chains. Overall, our study opens the possibility of developing *E.coli* as a robust and rapid platform for the development, discovery, and engineering of mAbs for therapeutic and diagnostic applications.

(48b) Protein Endocytosis and Degradation: Impact On Secreted Protein Titers in Yeast

Keith E.J. Tyo[1], Zihe Liu[2], Dina Petranovic[2] and Jens Nielsen[3], (1)Chemical and Biological Engineering, Northwestern University, Evanston, IL, (2)Department of Chemical and Biological Engineering, Chalmers University of Technology, Gothenburg, Sweden, (3)Department of Chemical and Biological Engineering, Chalmers University of Technology, Göteborg, Sweden

Recombinant protein production is an important activity in the production of drugs, fuels, and chemicals. Secreting protein into minimal media simplifies downstream purification, therefore reducing cost of manufacture. To date, considerable effort has been placed on increasing protein titers through improving recombinant protein *synthesis* and *secretion*. In the present work, we unexpectedly identified protein *uptake* and *degradation* as a significant process that decreases protein titers in *Saccharomyces cerevisiae*. Experiments showing (1) competitive endocytosis of insulin, bovine serum albumin (BSA), and yeast extract protein and (2) growth on BSA as a sole carbon source was used establish endocytosis as the mechanism of protein loss. Transcriptomics showed decreased amino acid synthesis, increased amino acid degradation/utilization, and metabolomics showed increased concentrations of amino acids, indicating large scale uptake of protein for carbon and energy. We identify key mechanisms that can be further engineered to reduce the re-uptake of recombinant proteins. This work has revealed an unanticipated mechanism for product loss that impact many industrial and pharmaceutical bioprocesses.

(48c) Engineering G-Protein Coupled Receptor Signaling in Yeast

Patrick M. McNeely, University of Delaware, Newark, DE and Anne S. Robinson, Department of Chemical and Biomolecular Engineering, Tulane University, New Orleans, LA; Department of Chemical and Biomolecular Engineering, University of Delaware, Newark, DE

G Protein-Coupled Receptors (GPCRs) are a large family of integral membrane proteins that transmit extracellular stimuli across the plasma membrane of cells in the human body. This signaling activity has been implicated in numerous diseases and is thus a major target for the development of novel therapeutics [1]. However, drug discovery is challenging because of the paucity of structures and the complex mechanisms for signal transduction in living cells. The ability to express GPCRs in large quantities and study the effect of ligand binding on signaling in a well-controlled manner would enable improved understanding of the biological binding and selectivity mechanisms and improving drug discovery efforts.

The use of a model cell systems to express and purify recombinant GPCRs is often able to address the quantity of receptor available for study in isolation. To date, however, few systems have been used to directly observe changes in signaling behavior of engineered receptors in these systems. In this work, the yeast *Saccharomyces cerevisiae* has been engineered to fluorescently report on the signaling activity of the GPCRs expressed in the cell, in addition to expressing recombinant receptor molecules. Using this strain, it is possible to elucidate biomolecular mechanisms of signaling activity *in vivo* and in membrane-mimetic environments.

[1] RM Millar and CL Newton, *Mol. Endocrinology* (2010)

(48d) Soluble Human Lysozyme Production in Escherichia Coli Via an Engineered Anti-Toxin Switch

Jonathan Guerrette, John W. Lamppa, Samuel Tanyos and Karl E. Griswold, Thayer School of Engineering, Dartmouth College, Hanover, NH

The accelerating spread of drug-resistance among bacterial pathogens underlies a critical need for novel antibacterial therapies capable of treating drug-resistant infections while slowing the development of new resistance phenotypes. Natural bactericidal peptides and proteins represent potential leads in the search for new drugs, and human lysozyme (hLYS) constitutes one particularly promising molecular scaffold. hLYS exerts both catalytic and non-catalytic antibacterial activities, and due in part to this multimodal action hLYS is relatively broad-spectrum, killing both Gram-positive and Gram-negative bacteria. Moreover, the redesign of hLYS through biomolecular engineering has the potential to yield a diverse panel of antibacterial biocatalysts that are tailored to meet specific clinical needs. However, efficient development of such therapeutic candidates requires laboratory access to gram-plus scale quantities of material. While hundreds of kilograms of wild-type hLYS can be produced in recombinant rice, this plant-based system is not readily scaled down to bench top production. To address this issue, we have recently developed the first expression system capable of producing folded, soluble hLYS in *E. coli* cells. Our system harnesses protein folding chaperones and a natural anti-lysozyme protective protein to generate >1000-fold increases in soluble hLYS relative to prior reports in *E. coli*. Our original design exploited *E. coli*'s ease of culture, short doubling time, and facile genetics to yield an expression platform that was moderately scalable and benefited from rapid turnaround. Critically, the system could be implemented in even the most minimally equipped biotechnology laboratory. Unfortunately, the overly robust nature of the hLYS-inhibitor interaction reduced the efficiency of purification, which was originally envisioned as a one-step IMAC process. We have therefore redesigned the natural inhibitor to create a switchable entity that protects the *E. coli* host during induction but efficiently releases the desired hLYS product upon a mild shift in environmental conditions. We anticipate that our redesigned expression and purification platform will facilitate further development of engineered lysozymes having utility in disease treatment and other practical applications.

(48e) Glycans-by-Design: Bottom-up Engineering of a Eukaryotic Protein Glycosylation Pathway In Escherichia Coli

Matthew P. DeLisa, Chemical and Biomolecular Engineering, Cornell University, Ithaca, NY

Until recently, asparagine-linked (*N*-linked) protein glycosylation was thought to be the exclusive domain of eukaryotes. However, studies over the past decade have firmly established that certain bacteria also perform this important post-translational modification. However, bacterial *N*-glycans are structurally distinct from their eukaryotic counterparts, severely limiting the use of the bacterial *N*-glycosylation system for the production of therapeutic *N*-glycoproteins. Here we report the bottom-up engineering of a eukaryotic protein glycosylation pathway in the bacterium *Escherichia coli*. This involved the heterologous expression of four eukaryotic glycosyltransferases, including yeast uridine diphosphate-*N*-acetylglucosamine transferases (Alg13, Alg14) and mannosyltransferases (Alg1, Alg2), and a bacterial oligosaccharyltransferase (PglB). The resulting synthetic pathway enabled *E. coli* to produce secretory glycoproteins bearing *N*-linked eukaryotic trimannosyl core glycans (mannose$_3$-*N*-acetylglucosamine$_2$). Overall, the engineering of defined glycosylation pathways in *E. coli* sets the stage for further engineering of this host for the production of vaccines and therapeutics with even more structurally complex human-like glycans. Moreover, glycoengineered *E. coli* has the potential to serve as a model genetic system for deciphering the "glycosylation code" which governs the non-template driven synthesis of diverse glycans and their specific attachment to proteins.

(48f) Engineering Synthetic Organelles: Encapsulating Heterologous Proteins Into Bacterial Microcompartments

Edward Y. Kim, Chemical and Biomolecular Engineering, University of California, Berkeley, CA and Danielle Tullman-Ercek, Chemical and Biomolecular Engineering, University of California, Berkeley, Berkeley, CA

Subcellular bacterial microcompartments are protein scaffolds of approximately 100 nm diameter that serve as a promising starting point for the creation of organelles to spatially organize biochemical reactions within the bacterial cell. Targeting enzymes or pathways of enzymes to such organelles would enable a wide variety of biotechnological applications by enhancing synthetic metabolic pathway yields, shielding the cell from toxic intermediates and products, and providing a distinct local environment for biochemical reactions that are incompatible with the cytosol. Here, we demonstrate the ability to encapsulate heterologous proteins into the lumen of bacterial microcompartments. We then investigate the role of N-terminal signal sequence peptides in localizing proteins to microcompartments. Through single amino acid substitution mutations, we find that a motif with specific spacing between charged and hydrophobic amino acids play an important role in making proteins competent for microcompartment localization. These results will aid in the creation of a consensus sequence, and allow for the eventual colocalization of multiple enzymes to bacterial microcompartments to serve as nanobioreactors for biochemical pathways.

(48g) Expression and Characterization of Three Trichoderma Reesei Cellulose Hydrolases in Kluyveromyces Lactis

Michael J. Brodeur-Campbell and David R. Shonnard, Chemical Engineering, Michigan Technological University, Houghton, MI

The recalcitrance of cellulose to hydrolysis is a major technological barrier to commercial implementation of products based on cellulose degradation. Enzymatic hydrolysis is seen as an environmentally friendly way of breaking down cellulose into its constituent glucose molecules for fermentation to fuels and other fine chemicals. However, the enzymes which are currently available to hydrolyze cellulose have low specific activities and high production costs. Expression in yeast expression hosts is a promising method for doing fundamental research on protein activity, with the goal of increasing enzyme specific activity and lowering production costs. The first step in heterologous enzyme production is characterization of the enzyme when produced in a foreign host.

In this research genes for two exoglucanases and one endoglucanase were isolated from *Trichoderma reesei* strain QM9414 (ATCC #26921) and inserted into *Kluyveromyces lactis* GG799 (New England Biolabs) for expression and characterization. Messenger RNA (mRNA) was isolated using the RNeasy kit, and cDNA synthesized with the Omniscript Reverse Transcription kit (Qiagen) using a 15 residue oligo-dT primer. Genes for Cel7A (CBH I), Cel6A (CBH II), and Cel7B (EG I) were amplified from the cDNA by polymerase chain reaction (PCR) using primers designed to eliminate the native *T. reesei* secretory signal sequence and fuse to the α-mating factor secretory signal sequence of *K. lactis* on the pKLAC1 vector plasmid (NEB) under control of the inducible $P_{LAC4-PBI}$ promoter. After transformation and selection, *K. lactis* transformants were grown on rich media containing 10 g/L yeast extract, 20 g/L peptone, and 20 g/L galactose (YPGal) to induce heterologous protein production.

Cell culture supernatant from *K. lactis* strains containing each of the three genes was spotted onto Petri dishes containing 50 mM citrate buffer pH 4.5, 0.5% carboxymethyl cellulose (CMC), 0.7% agar, and allowed to incubate overnight at room temperature. Plates were stained with 0.3% Congo Red and destained with 1 M NaCl. Clearing zones around the areas spotted with recombinant Cel7B indicated active endoglucanase enzyme.

Next, cell culture supernatant containing rCel7B was desalted by size exclusion chromatography (SEC) over Toyopearl HW-40F resin (Tosoh Biosciences) with a size exclusion limit of 10 kDa. Protein concentration was performed by anion exchange chromatography (AEX) using Toyopearl DEAE-650M (Tosoh Biosciences) and eluted with 0.5 M NaCl. Sodium dodecyl sulfate-polyacrylamide gel electrophoresis (SDS-PAGE) with Coomassie Brilliant Blue staining of the rCel7B showed a tight but continuous distribution of molecular weights centered at an approximate value of 67 kDa. This is slightly higher than the value of 54 kDa reported literature for the native protein[1]. Protein concentration was determined by the Bradford method, and activity assays of the rCel7B were performed using 1% CMC as the substrate in 50 mM citrate buffer, pH 4.5, at 40°C, and the dinitro-salicylic acid (DNS) method was used to detect reducing end formation. Specific activity of the rCel7B was determined to be approximately 0.5 M glucose/(g enzyme-hr), compared to a glucose standard curve, or about one-fifth of reported values for the native enzyme[2].

SDS-PAGE of cell culture supernatant containing rCel6A showed three distinct bands of differing molecular weights at approximately 50, 60, and 65 kDa. This is comparable to the reported value of 58 kDa for the native Cel6A[3]. SDS-PAGE following SEC and AEX chromatography of rCel6A in the same manner as for rCel7B discovered that the largest molecular weight form of rCel6A was not bound to the AEX resin, and was therefore not concentrated. Exoglucanase activity assays were performed on the concentrated rCel6A fraction by incubating 0.1 mg/mL enzyme with 1 mg/mL Avicel PH-101 powdered cellulose (Sigma-Aldrich) in 50 mM citrate, pH 4.5, at 50°C with 250 RPM orbital shaking for three days. Samples were taken every 24 hours and analyzed by HPLC for glucose production. No significant exoglucanase activity was detected. Ultrafiltration was then performed using a Vivacell 70 protein concentrator (Sartorius Stedim) with a 10 kDa cut-off membrane, on the crude rCel6A cell culture supernatant, and on the flowthrough from SEC and AEX chromatography. Exoglucanase activity assays detected glucose production both in the concentrated supernatant, and in the concentrated flowthrough, indicating that rCel6A is produced in both active and inactive forms by *K. lactis*, and activity is found only in the largest 65 kDa form of the enzyme. Furthermore, glucose production from rCel6A on Avicel was found to increase with the addition of 0.01 mg/mL of rCel7B, even though no glucose production was detected from rCel7B acting on Avicel alone. This result indicates that the recombinant forms of the enzymes retain their specificity for endo- or exo- activity, and can function together synergistically to increase cellulose hydrolysis rate.

SDS-PAGE of rCel7A showed a similar continuous distribution of molecular weights, much like rCel7B. Concentration was performed by ultrafiltration as for rCel6A. No detectable exo- or endo-glucanase activity was detected for rCel7A either alone or in combination with other enzymes. Post-processing, likely glycosylation, is apparently incompatible with producing active recombinant forms of this enzyme in the *K. lactis* expression system. DNA sequencing is being performed to verify that complete copies of the gene sequences were correctly inserted into the yeast genome.

References

1. Shoemaker, S.; Watt, K.; Tsitovsky, G.; Cox, R., Characterization and properties of cellulases purified from *Trichoderma reesei* strain L27. *Nature Biotechnology* **1983,** 1, 687-690.

2. Karlsson, J.; Saloheimo, M.; Siika-aho, M.; Tenkanen, M.; Penttilä, M.; Tjerneld, F., Homologous expression and characterization of Cel61A (EG IV) of *Trichoderma reesei*. *European Journal of Biochemistry* **2001,** 268, 6498-6507.

3. Tomme, P.; Van Tilbeurgh, H.; Pettersson, G.; Van Damme, J.; Vandekerckhove, J., Studies of the cellulolytic system of *Trichoderma reesei* QM9414. *European Journal of Biochemistry* **1988,** 170, 575-581.

(48h) Functional Self-Assembly of Artificial Cellulosomes for Efficient Cellulose Hydrolysis

Qing Sun, Shen-Long Tsai, Bhawna Madan and Wilfred Chen, Chemical and Biomolecular Engineering, University of Delaware, Newark, DE

Efficient hydrolysis of cellulose to glucose is gaining momentum because of its significant role in biofuel production. Cellulosome is a naturally occurring multi-enzyme system with a substantially enhanced ability for cellulose hydrolysis due to the effect of substrate targeting and enzyme clustering. The major component of cellulosome is a structural scaffoldin that consists of at least one cellulose binding module (CBM) and repeating cohesin domains, which are docked individually with an enzyme tagged with a corresponding dockerin domain. Although artificial cellulosomes containing at most four enzymes and a CBM have been constructed with up to 4-fold improvement in hydrolysis efficiency, efforts to engineer more complex cellulosome structures have been lacking. In this talk, we will highlight our recent strategies in engineering more complex artificial cellulosome structures using either protein or DNA as the template. The effect of enzyme loading and ordering on the overall enhancement in cellulose hydrolysis will be discussed.

(52a) The Interaction of a Bacterial Toxin (Aggregatibacter actinomycetemcomitans leukotoxin) with Its Receptor Depends On Both Lipid-Protein and Protein-Protein Interactions

Angela C. Brown[1], Patrik Nygren[2], Kathleen Boesze-Battaglia[3] and Edward T. Lally[1], (1)Department of Pathology, University of Pennsylvania, School of Dental Medicine, Philadelphia, PA, (2)University of Pennsylvania, Philadelphia, PA, (3)Department of Biochemistry, University of Pennsylvania, School of Dental Medicine, Philadelphia, PA

The Interaction of a Bacterial Toxin (*Aggregatibacter actinomycetemcomitans* leukotoxin) With Its Receptor Depends on Both Lipid-Protein and Protein-Protein Interactions

Angela C. Brown, Patrik Nygren, Kathleen Boesze-Battaglia, Edward T. Lally

The gram negative bacterium *Aggregatibacter actinomycetemcomitans* is associated with human periodontal disease as well as systemic diseases such as infective endocarditis. As part of its toxic arsenal, *A. actinomycetemcomitans* secretes a 114-kDa protein called leukotoxin (LtxA). This toxin kills human white blood cells, thereby allowing the bacterium to flourish in the host. The ability of LtxA to kill cells has been shown to require both cholesterol and the expression of lymphocyte-function associated antigen 1 (LFA-1), an alpha-L(CD11a)/beta-2(CD18) integrin. Binding of LtxA to LFA-1 initiates a multi-step process that results in the cleavage of talin from the cytoskeleton and clustering of the LtxA/LFA-1 complex in large, cholesterol-rich rafts. In this work, we sought to identify the role of both cholesterol and LFA-1 in this process, using confocal microscopy, surface plasmon resonance (SPR) and fluorescence resonance energy transfer (FRET) microscopy.

We first visualized LtxA internalization using confocal microscopy and demonstrated that internalization occurs only in the presence of LFA-1, and the toxin is located close to the membrane after internalization. The functional significance of the interaction of LtxA and LFA-was explored using a K562 cell line (Science **301**:1720, 2003) which expresses a CFP-tagged cytosolic alpha-L domain and a YFP-tagged beta-2 domain. In these FRET microscopy experiments, we found that activation of LFA-1 with PMA causes transient cytosolic domain separation (measured by a decrease in FRET), as expected. However, binding of LtxA to LFA-1 results in an increase in FRET, which is consistent with a mechanism in which LtxA binds to and brings the cytosolic domains closer together than they are in the inactive state. Unlike activation, this ligating effect is not transient, lasting for at least 30 mins.

The binding of LtxA to cytoplasmic domains of alpha-L (residues 1112-1170) and beta-2 (residues 724-769) was measured using SPR. A peptide corresponding to the cytoplasmic domain of the beta-3 cytoplasmic domain (residues 746-786) served as a control for the experiments. We determined that LtxA has a strong affinity for the cytosolic domains of both the alpha-L ($K_D = 1.5 \times 10^{-8}$ M) and beta-2 ($K_D = 4.2 \times 10^{-9}$ M) subunits and a lower affinity for the beta-3 cytoplasmic domain ($K_D = 2.3 \times 10^{-7}$ M). A series of shorter peptides were then synthesized to locate the regions of the cytoplasmic domains that interact with LtxA. These results demonstrated that LtxA bound to the membrane-proximal region of the alpha-L domain and an intermediate region of the beta-2 domain. Overall, these SPR results indicate that LtxA binds the cytoplasmic domains of LFA-1 strongly, suggesting that LtxA may displace cytoskeletal-linking proteins, such as talin, from LFA-1. This is consistent with our previous observation that the LtxA/LFA-1 clustering requires cleavage of LFA-1 from its cytoskeletal tether.

In addition, the LtxA/LFA-1 clustering requires the presence of cholesterol, suggesting that either LtxA or LFA-1 binds to cholesterol, in addition to binding one another. Because LFA-1 is not known to have an affinity for cholesterol, we measured the affinity of LtxA for membranes containing varying amounts of cholesterol. We found that the affinity of LtxA for membranes in the absence of cholesterol (1.9×10^{-8} M) is comparable to the affinity of LtxA for the alpha-L cytosolic domain of LFA-1. In the presence of cholesterol, however, the affinity of LtxA for membranes increases substantially (4.6×10^{-12} M), due to a drastic difference in the dissociation rate. These results suggest that LtxA may bind to the membrane and LFA-1 simultaneously and that the binding of LtxA to cholesterol may act to hold the toxin and LFA-1 in a cholesterol-rich environment, thereby driving the cholesterol-dependent clustering of LtxA and LFA-1.

This work provides additional insight into the complex interaction of LtxA with its receptor and demonstrates that the process is driven by both protein-protein and lipid-protein interactions.

This work was supported by the NIH: R01DE09517 and F32DE020950.

(52b) The Pivotal Role of Motifs: Elucidating Mechanisms of Heterologous GPCR Expression and Trafficking in Yeast Through Chimeric Receptors

Carissa L. Young[1], Emily C. McCusker[2], Zachary T. Britton[1], Shannon Modla[3], Jeffrey Caplan[3], Kirk J. Czymmek[3] and Anne S. Robinson[4], (1)Department of Chemical and Biomolecular Engineering, University of Delaware, Newark, DE, (2)Department of Biochemistry, University of California, San Francisco, San Francisco, CA, (3)UD Bio-Imaging Center, Delaware Biotechnology Institute, Newark, DE, (4)Department of Chemical and Biomolecular Engineering, Tulane University, New Orleans, LA

G protein-coupled receptors (GPCRs) constitute a superfamily of cell surface receptors responsible for the transduction of extracellular stimuli into intracellular signals. One sub-family of GPCRs is the adenosine family, identified as A_1R, A_2aR, A_2bR, and A_3R, which mediates the physiological effects of extracellular adenosine. Structurally analogous to other GPCRs, each adenosine receptor consists seven transmembrane-spanning a-helices, an extracellular N-terminal domain, three extracellular and intracellular loops, and a cytoplasmic C-terminal domain. Despite this common architecture, adenosine receptor subtypes are subject to regulation by distinct molecular mechanisms, couple promiscuously to multiple signaling pathways, and maintain unique ligand binding properties. Furthermore, the synthesis and export of polytopic membrane proteins is a complex process involving multiple steps of regulatory control to ensure proper folding, assembly, selective retention, and transport. As newly synthesized proteins transverse the secretory pathway from the endoplasmic reticulum (ER) to their final destination, quality control processes ensure that the ER-folding capacity is sufficient and adjusted appropriately, in order to maintain ER homeostasis. Compared to extensive studies of ER quality control and endogenous trafficking mechanisms in *S. cerevisiae*, molecular mechanisms underlying the transport processes of heterologous membrane proteins from the ER to cell surface are less well understood. Notably, the regulation of receptor signaling at the plasma membrane in non-native hosts has remained elusive.

In this study, specific recognition signals that govern quality control and secretion-dependent interactions provide a basic framework for determining trafficking mechanisms of recombinantly expressed GPCRs. We have investigated processes governing the heterologous expression of two human adenosine receptors, hA_2aR and hA_3R, and engineered chimeras of these two receptors in yeast, *Saccharomyces cerevisiae*. To improve functional production (i.e. ligand-binding yields indicative of active receptors) of GPCRs, we have assessed protein structure/function through rational engineered chimeras, determined *in vivo* receptor activity using fluorescent ligands, confirmed the activity of purified GPCRs by XAC ligand chromatography, and quantified total yields of GPCR production using novel standards. Multiple tags for identification and purification were incorporated (as reviewed [1]). By implementing DNA recombination strategies combined with high-resolution imaging techniques, we have determined intracellular the intracellular localization of heterologous GPCRs in yeast compared to endogenous organelle markers at the appropriate spatiotemporal resolution using Structured Illumination Microscopy (SIM) and evaluated alterations in organelle morphology, such as autophagy due to prolonged stress, via conventional TEM. In pursuit of a thorough analysis of protein distribution at the subcellular level, multiple yeast expression cassettes [2] have been created to test the effects of codon-optimized fluorescent variants, small epitope tags, polylinker length for N- and C- terminal tags, and the inclusion of essential retrieval sequences for ER luminal chaperones and foldases [3]. *S. cerevisiae* strains were further engineered to express fluorescent proteins targeted to various organelles [4]. To investigate discrete subpopulations of tagged proteins using live-cell imaging methods and super-resolution techniques (e.g. Fluorescence-Photoactivation Localization Microscopy, F-PALM), a photoconvertible GFP variant (i.e. mEos2) and six-residue tetracysteine motif required for FlAsH (fluorescein arsenical helix binder)-based technology were implemented.

Utilizing time course analyses, quantitative PCR, co-immunoprecipitation of select proteins, and novel tags with state-of-the-art high-resolution imaging techniques, we have shown differences in GPCR trafficking and quality control initiation, including the UPR, autophagy, and ER associated degradation (ERAD) pathways. We have evaluated GPCR expression profiles, optimized conditions to minimize UPR induction, determined colocalization with organelles and sub-compartments, identified select protein interactions with ER folding factors, and confirmed the activity of human adenosine receptors (i.e. hA_2aR, hA_1R, hA_2bR, hA_3R) [5,6]. Furthermore, the rational design of chimeric receptors altered localization and functional production. We continue to analyze the expression, trafficking, and activity of additional GPCR families, including the neurokinin receptors (hNK_1R) in order to determine the generalizability of our observations [7]. Collectively, our analyses provide mechanistic insight for the successful production of recombinant membrane proteins in the model eukaryote, *S. cerevisiae*.

Publications

1. C. L. Young, Z. T. Britton, A. S. Robinson *Recombinant Protein Expression and Purification: A Comprehensive Review of Affinity Tags and Microbial Applications*, Biotechnology Journal, 7(4), Jan 10 2012 doi:10.1002/biot.201100155. [Epub ahead of print]

2. C. L. Young, D. Raden, J. Caplan, K. Czymmek, A. S. Robinson *Optimized Cassettes for Live-Cell Imaging of Proteins and High Resolution Techniques in Yeast*, Yeast, 2012 doi:10.1002/yea.2895. [Epub 2012 Apr 4]

3. C. L. Young, D. L. Raden, A. S. Robinson, *Analysis of Endoplasmic Reticulum Resident Proteins in S. cerevisiae: Implementation of H/KDEL Retrieval Sequences,* 2012 (submitted).

4. C. Young, Z. Britton, J. Caplan, K. Czymmek, A. Robinson *Exploiting S. cerevisiae: Cellular Systems & Techniques Aimed at Identifying the Localization of Targeted Proteins,* 2012 (in preparation).

5. C. Young, E. McCusker, Z. Britton, S. Modla, J. Caplan, B. Chhun, K. Czymmek, A. Robinson *Conserved Trafficking Motifs Regulate Adenosine Receptors in Yeast: Rational Design of Chimeric hA2a/hA3 Receptors,* 2012 (in preparation).

6. C. Young, E. McCusker, Z. Britton, S. Modla, B. Chhun, J. Caplan, K. Czymmek, A. Robinson *Subcellular Localization and Trafficking Mechanisms of Functional Adenosine A_1 Receptor Expressed in Saccharomyces cerevisiae: Implications for Pharmacological Assays,* 2012 (in preparation)

7. C. Young, E. McCusker, B. Chhun, J. Caplan, K. Czymmek, A. Robinson *Regulation of G protein-coupled receptor, NK_1R: Mechanisms of ER Export and Plasma Membrane Localization in S. cerevisiae,* 2012 (in preparation)

(52c) Silencing α(1,3)Fucosyltransferases in Human Leukocytes Reveals Differences in E-Selectin Ligand Synthesis Between Humans and Mice

Alexander Buffone Jr.[1], Nandini Mondal[1], Kyle P. McHugh[1], Joseph T.Y. Lau[2] and Sriram Neelamegham[1], (1)Chemical and Biological Engineering, State University of New York at Buffalo, Buffalo, NY, (2)Molecular and Cellular Biology, Roswell Park Cancer Institute, Buffalo, NY

Introduction: Glyosyltransferases are a family of about 200 enzymes that constitute ~1% of the human genome. By catalyzing the attachment of complex carbohydrate chains on protein scaffolds, members of this enzyme family post-translationally modify a majority of mammalian proteins. The binding of leukocyte borne, glycosylated selectin-ligands to E-, P- and L-selectin expressed on vascular cells plays a critical role in regulating inflammation. The α1,3 fucosyltransferase (FUT) family of enzymes has three members called FUT4, 7, and 9 that potentially contribute to the formation of sLe^X type glycans on various selectin-ligands, including P-selectin glycoprotein ligand-1 (PSGL-1) in humans. While much of the data regarding FUT function in selectin binding has been found through use of transgenic mice knockouts, specifically the $Fut4^{-/-}Fut7^{-/-}$ dual knockout which completely removes binding to all selectins, important species specific exist in selectin function in humans and mice, particularly in E-selectin.

Materials and Methods: For the over-expression studies, full length FUT fusion proteins with C-terminal EGFP tag were expressed in the mammalian Golgi of HEK293T cells. Their function was verified using enzyme activity assays and immunocytochemistry. The PSGL-1 scaffold was co-expressed in these cells and its effect on leukocyte adhesion function was determined using microfluidic flow cells with substrates bearing L-/E-/P-selectin.

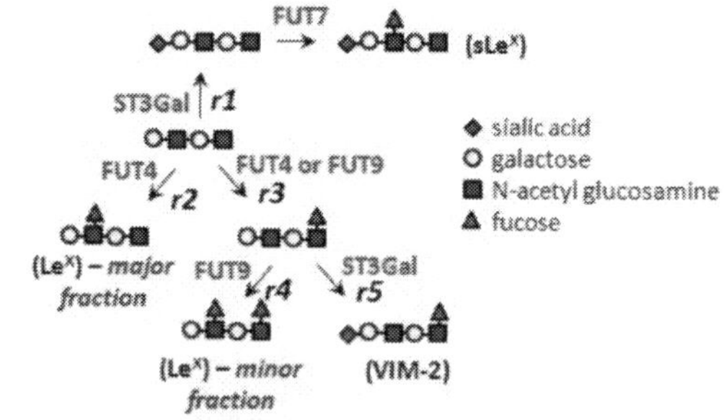

Figure 1. Proposed model for fucosylation in HL-60. Interpreting the effect of gene silencing requires consideration of the competition between reactions for a common substrate. This is illustrated here assuming the di-lactosamine substrate to be the starting material with biochemical reactions *r1-r5* acting on it. The reactions depicted in this model are based on the following: i) α(2,3)sialylation by ST3Gal followed α(1,3)fucosylation by FUT7 results in sLe^X. ii) FUT4 can synthesize Le^X-type antigens (*r2*). iii) Both FUT4 and FUT9 can fucosylate the inner GlcNAc residue (*r3*). Sialylation of this intermediate structure results in the VIM-2 epitope.

For the RNAi assays, shRNA sequences that reduce by >85% each of the α1,3 FUT activities were determined. This was done by transducing FUT specific shRNA and monitoring reduction in EGFP signal associated with the corresponding FUT-EGFP cell lines. Results were validated using quantitative RT-PCR. Once identified, shRNA were co-expressed with either DsRed or Cerulean and packaged into lentivirus. Suitable promoters were then determined to drive high levels of gene expression and RNA interference in HL-60 cells, a human leukocytic cell line. Stable HL-60 cells lacking either one, two or all three of the FUTs were then constructed. The ability of these FUT knockdowns to modify selectin-dependent cell adhesion function was assayed using flow cytometry and microfluidic assays and compared directly in our system to neutrophils from $Fut4^{-/-}Fut7^{-/-}$ mice.

Results and Discussion: α1,3 FUT over-expression studies show that FUT7 confer selectin binding for all three selectins (E/P/L). FUT9 alone also enabled E-selectin binding suggesting a previously unnoted importance for this enzyme. Knocking-down single enzymes in HL-60 caused specific alterations to the cell surface glycan structures with $FUT4^-$ HL-60 reducing cell-surface Le^X (CD15) expression, $FUT7^-$ HL-60 reducing sLe^X (CD15s) and $FUT9^-$ HL-60 up-regulating VIM-2 (CD65s) antigen expression by 2-3 fold.

Microfluidic based rolling studies compared the rolling and attachment to E-/P-/L-selectin for bothe the HL-60 FUT^-

cells and $Fut4^{-/-}Fut7^{-/-}$ mouse neutrophils run in parallel. In complete agreement with the $Fut4^{-/-}Fut7^{-/-}$ neutrophils, FUT7 and FUT4 co-mediated attachment, rolling and adhesion on cell lines bearing P- or L-selectin in HL-60. This process was fully dependent on the KPL-1 binding epitope of PSGL-1. E-selectin was the most promiscuous selectin ligand with cell rolling studies on L-E Cells and IL-1β stimulated HUVECs demonstrating a role for all three FUTs (FUT9>FUT7> FUT4). KPL-1 blocking or $PSGL\text{-}1^{-}$ HL-60 only reduced rolling by 40%, suggesting at least one other E-selectin ligand on these cells was present. FUT9 played major role in E-selectin mediated adhesion as the $FUT9^{-}$ HL-60 reduced the number of rolling and adherent cells on E-selectin by >50% alone, >75 percent in tandem with FUT7, and >90% in the $FUT4^{-}7^{-}9^{-}$ HL-60. The role for FUT9 in human E-selectin binding highlights stark species-specific differences as rolling and adhesion was decreased >95% on E-selectin in the $Fut4^{-/-}Fut7^{-/-}$ mouse neutrophils.

Conclusions: This study introduces novel molecular engineering tools to study both gain and loss of glycosyltransferase function in leukocytes and related cell adhesion function. To this end, data generated using these methods highlight a critical and predominant role for FUT9, in addition to FUT4 and FUT7, in the synthesis of ligands for E-selectin in human leukocytes. This is in contrast to the observation that all selectin binding function (including E-selectin) is lost in $Fut4^{-/-}Fut7^{-/-}$ mouse neutrophils. Thus, we observe notable species-specific differences in FUT and E-selectin ligand function that may be considered during the design of anti-inflammatory therapies that target the selectins, PSGL-1, or fucosyltransferases.

(52d) Analysis of Time Scales Involved in Ligand-Mediated Endocytosis of the EGF Receptor

Calixte S. Monast and Matthew J. Lazzara, Chemical and Biomolecular Engineering, University of Pennsylvania, Philadelphia, PA

Ligand-mediated endocytosis is an important cellular mechanism for regulating signaling initiated by the epidermal growth factor receptor (EGFR), a ubiquitously expressed receptor tyrosine kinase which plays critical roles in physiology and is frequently over-expressed in cancer. The classical understanding of EGFR endocytosis holds that ligand binding promotes receptor kinase activity, phosphorylation of receptor cytoplasmic tyrosines, and recruitment of proteins involved in the formation of endocytic vesicles. Recent work from our group has identified the presence of certain hidden time scales in this process which arise due to the reversibility of binding of proteins to phosphorylated EGFR tyrosines and the activity of protein tyrosine phosphatases at the plasma membrane. In this presentation, we will describe our work developing a kinetic model of EGFR endocytosis which incorporates the processes of ligand binding, receptor dimerization and phosphorylation, receptor dephosphorylation by protein tyrosine phosphatases, binding of adapter proteins, recruitment into coated pits, and translocation of receptors from the plasma membrane to the cell interior. Our model was constructed as a set of coupled ordinary differential equations describing the kinetics of these rate processes. The model was regressed against quantitative experimental measurements of the kinetics of receptor phosphorylation, dephosphorylation, endocytosis, and binding of relevant intracellular adapter proteins including Grb2 and epsin. All measurements were taken in cultured wild-type HeLa cells. A variety of perturbations to the wild-type cellular environment were also explored in a HeLa cell background, including knockdown of specific receptor-like protein tyrosine phosphatases, expression of specific adapter proteins, and chemical inhibition of phosphatases and endocytosis. After regression against experimental data, our model was able to explain the seemingly contradictory experimental findings that protein tyrosine phosphatases regulate EGFR phosphorylation at the plasma membrane but exert no control over EGFR endocytosis. Given the known requirement of EGFR phosphorylation for EGFR endocytosis, a feature which is encoded in our model, the model was only able to explain these experimental observations through identification of a time scale for EGFR pit recruitment which was very small compared to that for receptor dephosphorylation. This model prediction was validated by experimental assessment of the kinetics of binding of pit-associated adapter proteins to EGFR in response to EGF. Thus, our study suggests that phosphatases act upon EGFR at the plasma membrane but only after EGFR recruitment to coated pits. Our model further resolved that the rate limiting step in the ultimate translocation of EGFR to the cell interior is EGFR movement out of coated pits. Our study represents a novel experimental and computational approach for understanding the complex process of ligand-mediated EGFR endocytosis within a framework which incorporates multiple competing rate processes occurring across multiple time scales. The results of our analysis suggest interesting new possibilities for regulating EGFR endocytosis as a means to control EGFR-mediated signaling.

(52e) TLR4 Signals Via a Novel Allosteric Switch-Like Mechanism

Nichole Daringer and Joshua N. Leonard, Chemical and Biological Engineering, Northwestern University, Evanston, IL

Toll-like Receptor 4 (TLR4) signaling is important for activation of the innate immune system in response to pathogenic bacteria. TLR4 can respond to very low levels of endotoxin (lipopolysaccharide, LPS) providing early warning of an infection. However, improper or excessive activation of TLR4 is associated with diseases such as Chron's disease, some types of allergy, and in acute scenarios, sepsis and death. In Chron's disease and allergy, TLR4 is activated by self-derived ligands, leading to inflammation that can damage tissues and organs. In sepsis, systemic infection results in over-activation of TLR4 causing severe damage, organ failure, and in severe cases death. For all these reasons, understanding how TLR4 signals at a mechanistic level is essential for developing therapeutics that modulate its activity in a safe and effective fashion. However, several key aspects of TLR4 signaling remain poorly understood.

The current dominant viewpoint is that signaling through TLR4 is initiated via ligand-induced dimerization. However, TLR4 posses a unique hydrophobic region adjacent to the transmembrane domain in the cytosol, which results in oligomerization in the absence of ligand, challenging this viewpoint. We have interrogated this receptor mechanism using a combination of protein engineering, synthetic biology, and computational modeling techniques. Based on these findings, we now propose that TLR4 signaling is induced via a novel "mousetrap"-like mechanism, whereby in the absence of ligand, TLR4 is dimerized but held in a conformation that precludes signaling. Upon ligand binding, this constraint is alleviated and the receptor is able to induce signaling. This type of signaling mechanism could play an important role in the ability of TLR4 to respond quickly to very low concentrations of LPS. Understanding this mechanism could also help to engineer sensitive protein-based switches for applications in which detection of the presence of a ligand at any concentration (i.e., in a binary fashion) is most important. Moreover, elucidating this novel relationships between receptor structure and function will enable targeted development of therapeutics that modulate TLR4 signaling.

(52f) Feedback Control of Gene Expression in Tissue Patterning

Sophia Carrell, Chemical and Biomolecular Engineering, North Carolina State University, Raleigh, NC

Receptor-mediated signals are crucial for tissue patterning in development, stem cell maintenance, and adult homeostasis. Many diseases, most notably cancer, occur when cells misinterpret or ignore signals regarding proliferation, migration, and/or apoptosis, implying these signaling pathways must be tightly regulated. Since these same signals are highly conserved between species, we study the regulation of signaling pathways in the context of patterning the dorsal-ventral (DV) axis in the early *Drosophila melanogaster* embryo.

The transcription factor Dorsal, homologous to NFκB, is responsible for expression of the genes that generate the DV pattern in the *Drosophila*embryo. Dorsal is present in the nuclei in a gradient, with the highest concentration at the ventral midline and a steady decay to about 40% of the embryo's circumference. This gradient is established on the ventral side of the embryo by signaling through the Toll receptor, which phosphorylates the inhibitor protein Cactus (homologous to I-κB), marking it for degradation. In the absence of Cactus, Dorsal is free to enter the nuclei and direct expression of target genes in a concentration-dependent manner.

Because the Dorsal gradient is essential to correct patterning of the early embryo, there must be multiple sets of regulatory loops to ensure proper development in the face of perturbed conditions. Preliminary evidence indicates the existence of such feedback loops through Cactus, Toll, or other receptor-mediated signaling networks. Here we present work on a negative feedback loop that acts to stabilize the Dorsal gradient and minimize patterning defects in the face of large changes in input parameters.

(52g) Functional Single-Cell Analysis of T-Cell Activation by Supported Lipid Bilayer Tethered Ligands On Arrays of Nanowells

Alexis J. Torres[1], Rita Lucia Contento[1], Susana Gordo[2], Kai W. Wucherpfennig[2] and J. Christopher Love[1], (1)Department of Chemical Engineering, The David H. Koch Institute for Integrative Cancer Research, Massachusetts Institute of Technology, Cambridge, MA, (2)Department of Cancer Immunology and AIDS, Dana-Farber Cancer Institute, Boston, MA

The use of ligands immobilized on supported lipid bilayers (SLBs) has been an invaluable tool for characterizing T cell receptor (TCR)-mediated activation and re-distribution of associated membrane-expressed proteins during the formation of the immunological synapse. The use of bilayers on planar supports and fluorescent visualization of protein dynamics is limited in its ability to determine other activation-dependent cellular responses such as cytokine secretion that occur on timescales longer than the typical periods of observation for molecular dynamics. We have developed a method to measure the functional responses of individual T cells when engaged by lipid bilayers supported on an elastomeric array of subnanoliter wells (nanowells). In this approach, each nanowell artificially mimics an antigen-presenting cell (APC), thus allowing the formation of a T cell synapse upon contact with the surface. After cell activation by membrane-bound ligands, secretion of cytokines from single T cells were captured and quantified by microengraving in both antigen-independent and dependent manners. This method thus allows a flexible system for relating molecular events in TCR signaling to complex functional responses such as cytokine secretion and cell proliferation.

(52h) Cell-Cell Contact Regulates Myogenic Fate Differentiation of Mesenchymal Stem Cell Through OB-Cadherin

Stella Alimperti, Chemical and Biological Engineering, State University of New York at Buffalo, Amherst, NY and Stelios T. Andreadis, Department of Chemical and Biological Engineering, University at Buffalo, the State University of New York, Amherst, NY

Mesenchymal stem cells (MSCs) are multipotent cells that can be differentiated into fat, bone, cartilage and muscle. Although TGFβ is a soluble factor that promotes myogenic differentiation, the role of cell-cell adhesion in this process is not well understood. In this study we examined how cell-cell communication determines the fate of MSC towards smooth muscle cells (SMC). In particular, we hypothesized that cadherin mediated cell-cell contact might promote MSC differentiation into contractile SMCs. To address this hypothesis, we examined the effect of cell-cell adhesion on myogenic differentiation of MSC, in the absence of soluble factors promoting MSC differentiation. First, we observed that increased cell density increased adherens junction (AJ) formation, as shown by translocation of b-catenin to the cell-cell contact sites. Simultaneously, high cell density increased expression of alpha smooth muscle actin (aSMA) as well as N-cadherin and OB-cadherin even in the presence of bFGF, which is known to suppress MSC differentiation. Gene knockdown studies revealed that N cadherin and OB-cadherin mediated myogenic differentiation of MSC. Specifically, knockdown of OB-cadherin diminished expression of all myogenic genes including aSMA, calponin, caldesmon and SM22. We also found that two pathways mediated cell-cell contact-induced differentiation: a TGF-b1-dependent and a TGF-b1 independent pathway. Interestingly, the TGF-b1 independent pathway was mediated by the activity of ROCK as the chemical inhibitor, Y27632 or shRNA diminished cell-cell contact induced differentiation. These results are novel as they identify cell-cell contact as an inducer of myogenic differentiation through cadherins and their intracellular mediators. Our data demonstrated that OB-cadherin mediated ROCK activation and was also necessary for the contractile SMC phenotype. Although activation of the TGF-b1-dependent pathway enhanced this process, it was not necessary. Our results demonstrated the mechanisms of MSC differentiation and may have implications for development of strategies to control stem cell fate.

(56a) Synthetic Autoinducer-2 Triggered Expression for Quorum Sensing Surveillance

Jessica Terrell[1], Hsuan Chen Wu[2], Chenyu Tsao[3], Matthew Servinsky[4] and William Bentley[2], (1)Bioengineering, University of Maryland, College Park, MD, (2)Fischell Department of Bioengineering, University of Maryland, College Park, College Park, MD, (3)University of Maryland, College Park, MD, (4)U.S. Army Research Laboratory, Adelphi, MD

Bacterial multicellularity is primarily coordinated by a communicative process known as quorum sensing (QS). Its signal-based and "switch-like" regulation of phenotypic changes provides an inspiring framework for interaction-driven expression among distinct populations within a complex bacterial community. Accordingly, we seek to emulate the QS "switch" for useful purposes including controlled protein expression.[1] In the current work, we have built on this theme to use bacteria to report on their nearest neighbors and on their immediate environment. We demonstrate expression of streptavidin-binding peptide as a response to the presence of autoinducer-2, a QS signaling molecule. The engineered cell surveys its environment for AI-2, and upon recognition of the signal, the cell then expresses a surface-displayed streptavidin binding peptide (SBP), rendering it readily susceptible for concentration, separation, and amplified detection. This strategy expands upon previous work that demonstrated the ability of an operon under AI-2 regulation to be rewired within a dual plasmid system for the fluorescent reporting of AI-2. Here, we have associated SBP expression with fluorescence- both by labeling the cell surface with an affinity tag and by intrinsically expressing a fluorescent protein. In this way, the bacteria synthetically feature a surveillance mechanism to interact amongst a quorum sensing population yet remain fluorescently distinguishable, to sense and respond to the event of AI-2 exchange, and to be recovered. Our data characterizes each of these features. Furthermore, we have engineered the sensitivity of the surveyor cells by tuning the QS "switch" at specific thresholds of autoinducer signaling. Our surveillance mechanism uniquely capitalizes on the event of QS to achieve selective separation from a mixed, larger community. In this way, cell-based surveillance could be applicable as an alternative to current detection methods for bacterial contamination, especially that which is pathogenic. Benefit is provided by the sensitive molecular recognition capability per single surveyor cell. Additionally, concentration of fluorescent output via cellular affinity to streptavidin enables an amplification of the original signal and further enhances detection ability. Overall, this work is representative of a fertile opportunity for engineers to interrogate and manipulate cell behavior by eavesdropping on cell-to-cell communication at the level of individual cells.

[1]Tsao et al., *Metab. Eng.*, 12, 291 (2010)

(56b) Microelectrode Analysis of an Artificial Phototrophic Biofilm Consortia Reveals a Positive Feedback Basis of Syntrophic Interactions

Hans C. Bernstein[1], Alissa Bleem[2] and Ross P. Carlson[1], (1)Chemical and Biological Engineering, Montana State University, Bozeman, MT, (2)Montana State University, Bozeman, MT

Consortia can be engineered to optimize multiple functions simultaneously providing a robust platform for bioprocesses. Photoautotrophic consortia have the added benefit of using inexpensive sunlight and CO_2 as energy and carbon sources. Naturally occurring phototrophic organisms often couple with heterotrophic counterparts in an ancient and globally distributed ecological partnership based on syntrophic exchange of mass and energy via chemical species. This naturally occurring ecological template has been reconstructed using an artificial binary community comprised of photoautotrophic *Synechococcus* and heterotrophic *Escherichia coli*. Both *Synecochoccus* and *E. coli* are industrially relevant microbial strains often engineered for commercial products such as biofuels, bioplastics, amino acids and organic acids. Coculturing permits a mutually beneficial metabolite exchange between the strains; *Synechococcus* photosynthates support *E. coli* growth while *E. coli* consumption of potentially inhibitory organic acids and photosynthetically derived O_2 increases *Synecochoccus* biomass productivity. Cocultured *Synechococcus* biomass productivity was 50% higher than monocultures. A detailed micron-scale spatial analysis of oxygen gradients within the binary biofilms provided a basis for estimating photosynthetic rates in monocultures and cocultures. The binary system exhibited decreased local oxygen concentration with respect to position. This result corresponds with increased rates of respiration and reduced product inhibition of *Synechococcus* with respects to O_2 and photorespiration byproducts. This artificial binary biofilm culture represents a novel and adaptable microbial catalyst platform which exploits renewable carbon and energy inputs.

(56c) Design and Construction of Synthetic Fungi-Bacteria Consortia for Direct Production of Isobutanol From Cellulosic Feedstocks

Jeremy J. Minty[1], Marc E. Singer[1], Chang Hoon Bae[1], Jungho Ahn[1], Cliff Foster[2], James C. Liao[3] and Xiaoxia Lin[1], (1)Department of Chemical Engineering, University of Michigan, Ann Arbor, MI, (2)Great Lakes Bioenergy Research Center, Michigan State University, East Lansing, MI, (3)Chemical and Biomolecular Engineering, University of California Los Angeles, Los Angeles, CA

Present approaches to microbial cellulosic biofuel production are largely based on the "superbug" paradigm of aiming to incorporate all required functionalities (optimized biofuel production pathways, cellulase production, etc.) into a single organism. Inspired by the ubiquity of synergistic microbial communities in nature, we explore an alternative approach for biofuel production: the design, construction, and characterization of a synthetic microbial consortium consisting of multiple species which cooperate to directly convert lignocellulosic biomass to isobutanol, a promising next-generation biofuel. Three microbial specialists are utilized: the fungus *Trichoderma reesei* RUTC30, which secretes cellulase enzymes to hydrolyze lignocellulosic biomass into component hexose and pentose sugars, and engineered *Escherichia coli* hexose and pentose specialists which ferment hexose and pentose sugars to isobutanol, respectively. We develop and experimentally validate a comprehensive modeling framework for cellulolytic fungi-bacteria consortia, allowing us to elucidate key ecological interactions. Using stability and sensitivity analyses, we identify possible population equilibria and key parameters controlling the partition of carbon flow between *T. reesei* and *E. coli*. We demonstrate direct conversion of microcrystalline cellulose and AFEX pre-treated biomass to isobutanol with the *T. reesei* / *E. coli* consortium, with titers up to 1.7 g/L and 41% of theoretical yield. While we offer isobutanol production as a proof-of-concept application, the modularity of our system will allow it to be readily adapted to the large portfolio of existing *E. coli* strains metabolically engineered to produce biofuels or commodity chemicals.

(56d) Peptide-Based Communication Platform for Interspecies Communication

Nicholas Marchand and **Cynthia H. Collins**, Chemical and Biological Engineering, Rensselaer Polytechnic Institute, Troy, NY

Traditionally, industrial processes requiring bacteria have used a single organism. However, future processes are likely to take advantage of the many advantages available that come from utilize microbial consortia. In order to control interactions and complex dynamics required for these cells to work together, it is important for engineers and synthetic biologists to create robust communication pathways between biotechnologically relevant species. *Escherichia coli* and *Bacillus megaterium* are two such bacterial species that have proven themselves useful in industry in the past. Here we present a synthetic communication pathway between the two using a heterologous quorum sensing (QS) mechanism taken from *Staphylococcus aureus*. The QS system of *S. aureus* is composed of four genes making up the accessory gene regulator (agr) locus. The *agrB* and *agrD* genes, along with a Type I Signal Peptidase, produce and secrete the signaling molecule termed Autoinducing Peptide (AIP). The *agrC* and *agrA* genes encode a two component signaling cascade that leads to up-regulation of all four QS genes. We have engineered an *E. coli* strain that is able to produce and secrete mature AIP into the cell supernatant. Furthermore, introducing this supernatant to our engineered *B. megaterium* cultures effectively induces target gene expression. We anticipate that our communication system, when combined with successful co-culture systems, will be a key technology for the implementation of synthetic consortia for bioprocessing applications.

(56e) The Growth Promoting Effect of Symbiont Bacteria On Chlorella Vulgaris

Yen Wah Tong, Department of Chemical and Biomolecular Engineering, National University of Singapore, Singapore, Singapore and Zhi Guo, Department of Chemical and Biomolecular Engineering, National University of Singapore, Faculty of Engineering, Singapore

Symbioses between microalgae and bacteria are common in nature but are usually a complex phenomena. Other works have shown that algal-bacterial system have many interactions such as the microalgae producing O_2 which is necessary for bacterial aerobic respiration while at the same time, CO_2 from bacteria can be utilized by microalgae to perform photosynthesis. Some bacteria excrete phytohormone like indole-3-acetic acid which can induce algal cell division while microalgae also secrete extracellular compounds which can similarly enhance the bacterial activity. However, not all the interactions between algae and bacteria are synergistic. Microalgae can produce and accumulate bactericides, such as chlorellin, which can inhibit bacterial growth while toxic algaecide released from bacteria can affect the growth of algae. Recent studies confirmed that some small molecule modulators from algae can control the algal-bacterial symbiosis.

Here, we report the growth promoting effect of a symbiont bacteria isolated from a *chlorella vulgaris* culture. The algae culture ran for about one year and were found to be growing with other microorganisms to form the algae-bacterial system in our photobioreactor. Several bacteria strains were then isolated based on the agar plate culture method. The symbiont bacteria were co-cultured with axenic *chlorella vulgaris* and growth promoting effects on the microalgae were found. The number of bacteria increased over several days before remaining constant during the culture period. The findings that synergistic interaction between *chlorella vulgaris* and the symbiont bacteria could aid our understanding of algal-bacterial symbiosis in nature.

(56f) Effects of Quorum-Signaling Molecules On Human Epithelial Cells: Implications for Interkingdom Response and Communication

Amin Zargar, Center for Biosystems Research, College Park, MD; Bioengineering, University of Maryland, College Park, College Park, MD and William Bentley, Fischell Department of Bioengineering, University of Maryland, College Park, College Park, MD; Institute for Bioscience and Biotechnology Research, College Park, MD

Effect of quorum-signaling molecules on human epithelial cells: Implications for interkingdom response and communication

Introduction: It has been established that bacteria use quorum sensing molecules to coordinate gene expression in a cell density-dependent manner. This group of cell communication molecules is displayed throughout the bacterial hierarchy influence characteristics such as swarming motility, biofilm formation, virulence, among others. A distinct class of signaling molecules, autoinducer-2, is capable of interspecies communication, and is so prevalent throughout the bacterial domain that it is thought to be a "universal" signaling molecule. While the mechanisms behind autoinducer-2 have been clearly elucidated, the interkingdom relationship between quorum signaling molecules and human cells are not yet understood. This relationship is of particular importance considering the many species of bacteria that constitute the microflora inside the intestine. In this study, we explore the effects of autoinducer-2 on gene expression of human epithelial cells using the next-generation RNA-seq.

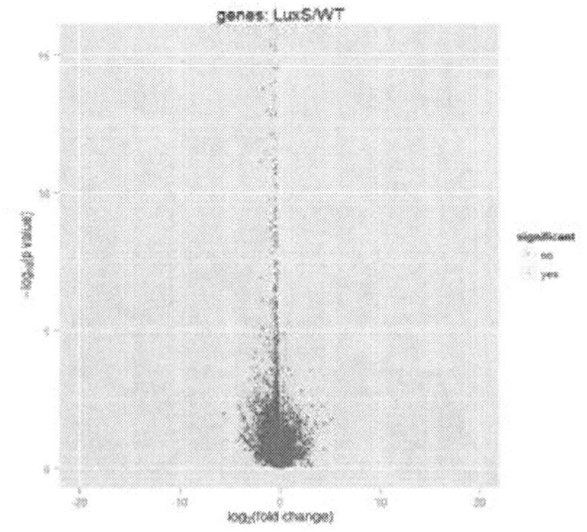

Human epithelial cells, HCT-8, are exposed to BL21, a wild-type bacteria that produce AI-2, and a BL21 LuxS knockout that does not produce AI-2. The RNA is extracted and is sequenced in triplicate on two lanes of the Illumina HiSeq1000 Sequencing System. Over 100 million reads for each condition per biological replicate were analyzed using the Tuxedo Suite (Bowtie, Tophat & Cufflinks). With 90% alignment, a total of 318 differentially expressed genes and 8773 differentially expressed isoforms were found. Gene clustering analysis using DAVID (Database for Annotation, Visualization and Integrated Discovery) led to significant 28 gene clusters.

With the advancements of next generation RNA-Seq technology, sequencing of the eukaryotic transcriptome is now possible with newfound depth and breadth, to determine eukaryotic responses. In this study, transcriptional and post-transcriptional differential processing was explored and biological network paths were inferred to indicate response. Reinforced with this understanding, two-way interkingdom communication can potentially be mediated with genetically engineered bacteria that can act on eukaryotic response mechanisms.

(56g) Synthetic Microbes Engineered to Fight Human Pathogens

Mui Hua Tan[1], Choon Kit Wong[2], Nazanin Saeidi[2], Tat-Ming Lo[2], In Young Hwang[1], Chueh Loo Poh[2] and Matthew Wook Chang[2], (1)School of Chemical and Biomedical Engineering, Nanyang Technological University, Singapore, Singapore, (2)School of Chemical and Biomedical Engineering, Nanyang Technological University, Singaproe, Singapore

Given the stalled development of new antibiotics and the increasing emergence of multi-drug resistant pathogens, there is an urgent need for designing a new treatment regimen for infectious diseases. In our previous study, we engineered *Escherichia coli* to effectively sense and kill *Pseudomonas aeruginosa,* as a proof of concept of a novel microbe-based antimicrobial strategy. In this study, we have further designed and constructed additional genetic circuits that equip *E. coli* with other clinically relevant functionalities. In particular, we have shown that the implementation of genetic circuits that enable E. coli to move towards *P. aeruginosa* and produce biofilm-degrading enzymes in response to *P. aeruginosa* biofilm with a precise time-delay element improved killing efficiency. We have also demonstrated these killing efficiencies in animal models. This study suggests that synthetic microbes may provide a novel synthetic biology-driven antimicrobial strategy that could potentially be applied to fighting P. aeruginosa and other infectious pathogens.

(56h) Inducible Cell Communication Amplifies Salmonella Gene Expression In Tumor Tissue

Neil S. Forbes, Chemical Engineering, University of Massachusetts, Amherst, MA, Yumei Dai, UMass Amherst, Amherst, MA, Charles Swofford, Chemical Engineering, University of Massachusetts Amherst, Amherst, MA and Bhushan J. Toley, Chemical Engineering, University of Massachusetts, Amherst, Amherst, MA

Bacterial anticancer therapies have the potential to overcome resistances that limit the efficacy of chemotherapy. Triggered production of anticancer agents by bacteria, which have penetrated into tumor tissue, would kill more cancer cells while damaging less normal tissue. However, the use of chemical triggers is hampered by uptake, clearance and diffusion, which reduce the number of activated bacteria. We have solved this problem by enabling bacteria to communicate. The engineered cell-communication system integrates the P_{BAD} promoter and components from the quorum-sensing machinery of *V. fisheri*. Cell communication enables activated bacteria to induce neighbors by producing a signaling molecule that can be 'heard' by responsive bacteria. The system was tested by connecting it to *ZsGreen*, a fluorescent reporter gene. Gene expression and sensitivity were measured by comparison with non-communicating controls that induce gene expression with direct induction of P_{BAD}. Function in three-dimensional tissue was tested in an *in vitro* tumor-on-a-chip device specifically designed to quantify the interaction of bacteria and cancer cells. Bacterial communication increased gene expression 40-fold and increased sensitivity to inducer molecules more than 10,000-fold. Tests with sensing bacteria showed that the AI-1 autoinducer was produced, secreted into the environment, and taken into bacterial cells. The system enabled individual bacteria to activate at least 100 neighbors and significantly increased the time-scale of protein production. Gene expression with this system was controllable and tightly regulated. In tissue, induction of protein production was possible with communicating *Salmonella* but not with control bacteria. At the optimal inducing signal, communicating bacteria produced 350 times more protein than non-communicating bacteria. The cell-communication system created in this study has numerous uses beyond cancer therapy, including protein manufacturing, bioremediation and biosensing. It would enable amplified induction of gene expression in any environment that limits availability of inducer molecules. Ultimately, because inducible cellular communication enables gene expression in tissue, it will be a critical component of every successful bacteria-based anticancer therapy.

(60a) Nanoparticle Cancer Therapeutics: Concept to Clinic

Mark E. Davis, Chemical Engineering, California Institute of Technology, Pasadena, CA

Nanoparticle-based therapeutics containing either small molecule anti-cancer agents (CRLX101 (formerly called IT-101); Phase II) or siRNA (CALAA-01; Phase I) are described. The key component in each nanoparticle is a cyclodextrin-containing polymer that endows it with multifunctional behavior. Relationships between nanoparticle design and function in animal models are illustrated. Results from preclinical animal models and human clinical trials are compared to illustrate similarities and differences between behaviors in animals and humans. For example, with CRLX101, features designed into the nanoparticles such as the ability to have long circulation times and long drug release in tumors cells has been observed in mice and humans. With CALAA-01, short circulation times observed in mice and non-human primates is also seen in humans. Additionally, the ability of CALAA-01 to release siRNA and have it perform RNAi inside of cancer cells within the tumors of animals and humans is verified using RACE analyses. While many of the features of the nanoparticles do translate from animals to humans, there are issues that do arise when treating humans. Some of these issues will be discussed.

(60b) Multi-Reservoir Drug Therapies...From Academic Lab to Startup Company to Clinical Demonstration

John T. Santini Jr., On Demand Therapeutics, Inc., Tyngsboro, MA

As pharmaceutical drugs become more potent, targeted, and complex, the importance of systems designed to maintain drug stability and precisely control the delivery of those drugs to the body increases. Therefore, a real need exists for novel materials and devices that can: (1) store one or more drugs, (2) protect them from the body until they are needed, and (3) controllably release the drugs at the desired time. Advances in microfabrication and hermetic sealing technologies have enabled the creation of multi-reservoir drug delivery systems that address these needs. Precise control over the release of drug from each reservoir is enabled by novel formulation techniques, the application of outside energy sources such as lasers, and/or the integration of pre-programmed microprocessors, wireless telemetry, or biosensors.

Startup companies have proven to be an effective way to quickly translate scientific and technological discoveries from academic labs into new medical products that benefit patients. This talk will include a discussion of: (1) the development and first clinical demonstration of an implantable, wireless, multi-reservoir drug therapy for osteoporosis, and (2) the various issues impacting the successful translation of academic science to the clinic.

(60d) Transdermal Drug Delivery: Translation From Chemical Engineering Laboratories Into the Clinic

Samir Mitragotri, Chemical Engineering, University of California, Santa Barbara, Santa Barbara, CA

Transdermal drug delivery provides an advantageous alternative to injections for delivering drugs into the body. It avoids the pain and needle—phobia associated with injections while providing sustained release of drugs. However, delivery of drugs into skin is limited by the high barrier properties of skin owing largely to its topmost layer, the stratum corneum. Accordingly, applications of transdermal delivery are limited to a handful of small drugs. Several technologies have been developed to enhance skip permeability to drugs after topical application. These technologies utilize physical and chemical means to reversibly permeabilize stratum corneum. These technologies have yielded devices as well as formulations that can be applied on the skin for local and systemic delivery of drugs. My presentation will present an overview of these technologies with an emphasis on their transition into the clinic. Applications of these technologies in the clinical setting and clinical problems addressed by these technologies will be discussed.

(60e) New Approaches to Treating Brain Tumors

Gary Gallia, Neurosurgery, Johns Hopkins Univ. SOM, Baltimore, MD, Betty Tyler, Neurosurgery, Johns Hopkins University SOM, Baltimore, MD and Henry Brem, Neurosurgery, Oncology and Biomedical Engineering, Johns Hopkins University, Baltimore, MD

New Approaches to Treating Brain Tumors

Gary Gallia, Betty Tyler, and Henry Brem

Departments of Neurosurgery, Oncology, and Biomedical Engineering

Johns Hopkins University, Baltimore, Maryland

The technology of local delivery is being utilized to target multiple pathways for brain tumor therapy. We will review the challenges of developing Gliadel, the first polymer drug delivery system for brain tumors. Through combination of surgical resection, Gliadel, radiation therapy and oral temozolomide (TMZ), median patient survival for brain tumor patients has increased from 9 to 21 months. Recent clinical trials with resistance modifiers and trials with Gliadel for brain metastases have also led to improved survival. To further increase survival we have been exploring the local delivery of temozolomide and have shown that in animal glioma models we can achieve 37.5% long term survival. We have also incorporated paclitaxel, a mitotic inhibitor, into a thermo-sensitive gel depot which provides local delivery, enhances efficacy, and limits systemic toxicity. In combination with TMZ and radiation therapy this depot significantly increases long term survivors and improves median survival compared to either therapy alone. This data led to a Phase I clinical trial. Rapamycin, an MTOR inhibitor, anti-angiogenic agent and anti-proliferative agent, currently used for coronary stents, has been efficiently incorporated and intracranially delivered using biodegradable controlled release beads. We demonstrated its safety for CNS delivery as well as it significantly prolonged survival in our 9L rodent glioma model. Other anti-angiogenic agents are being explored as well, including Fc-endostatin, a recombinant human endostatin conjugated to the Fc domain of IgG, which has shown significantly prolonged survival. Minocycline, a semi-synthetic tetracycline, has been shown to be an effective adjuvant therapy when combined with either radiation therapy or oral temozolomide, two common methods of treating glioblastoma. Direct intracranial delivery using microchips has been a focus of our laboratory in collaboration with colleagues at MIT. Both biodegradable "passive" chips as well as microelectromechanical systems, or "active" chips, have shown efficacy in delivering chemotherapeutic agents for the treatment of both glioma as well as metastatic cancer models. Intracranial microdialysis has proven to be a very useful technique to determine real time intracranial concentrations of systemically delivered drugs. This helps to determine if effective concentrations are being reached at the site of the brain tumor. To date we have contributed to our overall objective of developing better technologies for treating brain tumor patients using locally delivered therapies.

(69a) Engineering Enzymes Using Nonnatural Amino Acids

Shun Zheng, Sung In Lim, H. Edward Wong and **Inchan Kwon**, Department of Chemical Engineering, University of Virginia, Charlottesville, VA

Enzymes have become a very important player for modern biotechnlogoy. Since wild-type enzymes do not always have the properties needed for various applications, enzymes are often engineered to obtain desirable properties using protein engineering techniques. In the past decade, complementary to the widely used rational protein design and directed evolution techniques, nonnatural amino acid incorporation (NAAI) has become a new and effective protein engineering technique. As an emerging application, we have explored whether NAAI can be used to improve enzyme properties. As a test protein, we chose murine dihydroxyfolate reductase (mDHFR) and incorporated several nonnatural amino acids in a site-specific manner. Site-specific incorporation of nonnatural amino acids into mDHFR did not perturb the correct folding of mDHFR. Properties of the mDHFR variants containing nonnatural amino acids (nonnatural mDHFR variants) including caatalytic activities, thermal stability, and resistance to inhibitors are being evaluated in order to obtain knowledges on how nonnatural amino acids modulate enzyme structure and properties. These findings would facilitate the validation of NAAI as a tool to engineer enzymes.

(69b) Enzyme Biocatalysis Orientation Control During Immobilization by Unnatural Amino Acid Incorporation

Bradley C. Bundy[1], Jeffery C. Wu[2], Mark T. Smith[1] and Chad T. Varner[1], (1)Chemical Engineering, Brigham Young University, Provo, UT, (2)Microbiology and Molecular Biology, Brigham Young University, Provo, UT

The \$3+ Trillion global chemical industry is built predominantly upon catalytic and increasingly biocatalytic processing. The enzyme biocatalysis market has increased by 15% annually to the current \$5+ billion market and has many advantages including high chemo-, regio-, and stereospecificity and sustainability. However, biocatalysis technology is often limited by factors such as enzyme instability and difficulties recovering and reusing the enzyme. Enzyme immobilization has been shown in many cases to overcome these limitations. While a number of immobilization methods have been developed (entrapment, cross-linking, absorption, covalent attachment, etc.), covalent attachment is particularly promising due to enzyme leaching and transport limitations of other methods. Unfortunately, enzymes are composed of only 20 amino acids which complicate our ability to control the exact site where the enzyme is covalently attached to the surface without significant mutagenesis. Here we present the controlled site-specific incorporation of the unnatural amino acid propargyloxyphenylalanine to directly attach proteins and enzymes to surfaces at a specified orientation for optimal performance. The immobilization method requires only a single insertion or substitution of the unnatural amino acid which can be positioned at virtually any location on the enzyme. The biocompatible copper-catalyzed azide alkyne cycloaddition is used to attach our unnatural amino acid containing protein to azide groups exposed on the solid support's surface. With this method we have demonstrated orientation controlled covalent immobilization at high attachment efficiency and retained activity.

(69c) Engineering Novel Tandem Catalytic Reactions Using Organometallic Catalysts and Metalloenzymes

Carl A. Denard[1], Lars Martin Jarenmark[2], Ramesh Giri[3], Yichen Tan[4], Levi Stanley[5], John F. Hartwig[3] and Huimin Zhao[1,6,7,8], (1)Department of Chemical and Biomolecular Engineering, University of Illinois at Urbana-Champaign, Urbana, IL, (2)Chemistry, University of Illinois at Urbana-Champaign, Urbana, IL, (3)Chemistry, University of California-Berkeley, Berkeley, CA, (4)University of California-Berkeley, Berkeley, CA, (5)Chemistry, Iowa State University, Ames, IA, (6)Department of Chemistry, University of Illinois at Urbana-Champaign, Urbana, IL, (7)Department of Biochemistry, University of Illinois at Urbana-Champaign, Urbana, IL, (8)Institute for Genomic Biology, University of Illinois at Urbana-Champaign, Urbana, IL

Concurrent asymmetric tandem reactions, so called "multicast promoted tandem reactions," still remain a synthetic challenge in terms of matching catalysts' reaction rates, fine-tuning substrate and product selectivities, and reaction condition compatibility. Nonetheless, many wonderful examples of multicast promoted one-pot syntheses using chemical catalysts from various disciplines have found success. In contrast, multienzymatic systems in nature provide ample evidence of tandem catalysis in an aqueous environment. In addition, protein engineering towards industrial biocatalysis has emerged as a powerful tool for the biosynthesis of fine chemicals with high regio- and enantioselectivity. The combination of catalysts from these two very different disciplines (chemical catalysts and enzymes) for the synthesis of high-value compounds is an attractive target and has recently seen increased interest. In the Zhao and Hartwig labs, we are focused on developing novel catalytic asymmetric tandem reactions using organometallic catalysts and metalloenzymes.

We have developed a tandem process to achieve successive olefin isomerization-enantioselective epoxidation, a reaction difficult to achieve occurring with a single catalytic system. This process can potentially provide a practical and efficient strategy for the enantioselective preparation of linear terminal epoxides in high *ee* from a mixture of simple substrates. By combining an isomerization catalyst with a purified P450 BM3 mutant in a biphasic reaction, we report for the first time that the two reactions work in tandem. Using our tandem reaction, we have achieved the successive terminal to internal olefin isomerization followed by epoxidation to yield the internal epoxides, as well as the thermodynamically challenging internal to terminal olefin isomerization followed by the terminal olefin epoxidation. We have further expanded the substrate scope of the reaction system to straight and branched aliphatic olefins, unsaturated methyl esters, acids and ketones. Overall yields of 35% can be obtained, as well as interesting substrate-dependent reaction regioselectivities.

In parallel, using the robust and established chemistry of ruthenium carbene metathesis catalysts, we are investigating ways to couple them with metalloenzymes to develop novel tandem reaction systems. Several ruthenium metathesis catalysts were found to be active, stable in the presence of air and water, at room temperature and in several solvent: buffer systems to afford cross metatheses with moderate to high selectivities. We obtained preliminary data of a novel biphasic tandem cross metathesis-epoxidation/hydroxylation approach involving a Hoveyda-Grubbs 2nd generation complex and the versatile cytochrome P450 from *Bacillus megaterium*, for the selective hydroxylation and epoxidation of 5-dodecenoic acids in high yields. We are extending this platform to other challenging transformations, such as the epoxidation of 1-butenylbenzene generated by cross metathesis of *cis*-stilbene and *trans*-3-hexene as well as the enantioselective epoxidation of long chain alkenes generated by alkene chain elongation through cross- or self-metathesis of short chain alkenes.

(69d) A Self-Assembling Protein Hydrogel for Enzyme Incorporation Onto Electrodes in Biofuel Cells

Dongli Guan, Miguel Ramirez, Ivan B.F. Rincon, Daniel Jacobson and Zhilei Chen, Chemical Engineering, Texas A&M University, College Station, TX

We developed a self-assembling protein hydrogel for enzyme immobilization on electrodes for enzymatic biofuel cell applications. The building blocks of this hydrogel are two protein block copolymers (PC1 and PC2). PC1 contains a triangular trimer protein fused to a recombinant PDZ protein, and PC2 contains the same trimer protein and the peptide ligand of PDZ, as well as a repetitive non-structural S peptide $[((AG)_3PEG)_{10}]$ and a docking station peptide $SH3_{lig}$. The S peptide facilitates the retention of water in the hydrogel and $SH3_{lig}$ enables stable non-covalent immobilization of the SH3 (Src homology 3 domain from the adaptor protein Crk)-modified target protein. Individually, PC1 and PC2 exist as viscous solutions composed of trimeric proteins. Mixing solutions of PC1 and PC2 triggers the physical association of PDZ with its peptide ligand, joining individual PC1 and PC2 trimers and giving rise to the formation of a highly crosslinked protein network (hydrogel). Rheological measurements confirmed the formation of a protein hydrogel and shear-thinning properties reflective of a physical hydrogel. This hydrogel is stable in solution with a very low erosion rate. To demonstrate the utility of this hydrogel for enzyme immobilization, two recombinant proteins, PTDH (phosphite dehydrogenase) and Slac (laccase), were purified as fusions to the SH3 protein and immobilized individually in the hydrogel. A very low leaching rate of the target protein from the hydrogel in solution was observed. The Slac-containing hydrogel was subsequently rendered conductive for application in a direct electron transfer enzymatic electrode. The potential of this hydrogel for power generation will be discussed.

(69e) Efficient Stereoselective Synthesis of Ethyl (R)-2-Hydroxy-4-Phenylbutyrate by a Bacterial Reductase Coupled with Cofactor Regeneration

Ye Ni[1], Yuning Su[2] and Zhihao Sun[2], (1)The Key Laboratory of Industrial Biotechnology, Ministry of Education, School of Biotechnology, Jiangnan University, Wuxi, China, (2)Jiangnan University

(*R*)-2-Hydroxy-4-phenylbutyrate is an important chiral intermediate in the synthesis of ACE inhibitors. Asymmetric reduction of OPBE to (*R*)-HPBE using recombinant strain could provide high enantioselectivity. However, cofactor regeneration is a critical issue in the application of recombinant strain. In our study, a carbonyl reductase gene (*iolS*) and a glucose dehydrogenase gene (*gdh*) from *Bacillus subtilis* were cloned, and three coexpression strategies were attempted for the coexpression vector construction. Recombinant *E. coli* strain containing polycistronic plasmid pET-G-T7-I showed excellent carbonyl reductase activity, and the specific activity of both IolS and GDH in crude cell extract reached 1.5 U/mg. In the asymmetric reducing of OPBE to (*R*)-HPBE by recombinant *E. coli* cells in aqueous system, the yield and optical purity of (*R*)-HPBE reached 100% and 99.5% enantiomeric excess (*ee*) at 10 g/L of OPBE. In water/octanol biphasic system, higher product titer could be achieved with 99.5% *ee*. This study provides a practical approach for further the asymmetric synthesis of (*R*)-HPBE and other chiral alcohol enantiomers.

ENANTIOSELECTIVE REDUCTION OF A-BROMO AROMATIC KETONES USING CARROT CELLS

Yi Wang, Jimaio Li, Xiang Liu, Jianhe Xu*

School of Chemical and Material Engineering, Jiangnan University, Wuxi, China

Abstract

Chiral 1-aryl-2-bromo alcohols with special functional groups on the aromatic ring are potentially valuable synthetic intermediates for the synthesis of drugs, agrochemicals, flavors and pigments. These chiral alcohols can be obtained from the enantioselective reduction of prochiral aromatic ketones by biological methods. Among these biological methods, microbes and plant cell cultures are more useful than isolated enzymes due to the existence of many enzymes that catalyze various reactions. In addition, the use of microbes and plant cell cultures is particularly advantageous for carrying out the desired reduction since they do not require the addition of cofactors for their regeneration. By comparison, less attention has been given to the use of plant cell cultures as a potential enzyme source for catalyzing selected organic reactions. In this paper, the enantioselective reduction of α-bromo aromatic ketones was carried out by using carrot cell cultures and the corresponding chiral 1-aryl-2-bromo alcohols with special functional groups were produced. After the reaction conditions were optimized, these chiral 1-aryl-2-bromo alcohols were obtained with good yields and excellent enantiomeric excesses under the optimal conditions. Meanwhile, the steric factors and electronic effects of the substituents on the aromatic ring were shown to have an interesting influence on both yield and enantioselectivity in our experiment.

Keywords: Enantioselective reduction, α-Bromo aromatic ketone, Chiral alcohol, Carrot cell

Introduction

The asymmetric reduction of ketones is one of the most important, fundamental and practical reactions, the chiral alcohol product which can be transformed into various functionalities, without racemization, to synthesize industrially important chemicals such as pharmaceuticals[1], agrochemicals and natural products. The catalysts for the asymmetric reduction of ketones can be classified into two categories: chemical and biological methodologies[2,3]. Biocatalyst including either isolated oxidoreductases[4] or living organisms [5] is considered as a promising catalyst due to its advantages, such as specificity, enantioselectivity, mild reaction condition, environment friendly and renewable capability. Whole cell or tissue is an excellent alternative to the isolated enzyme, since the oxidoreductase, cofactor and the cofactor regeneration system are all within the cell. Many reactions with excellent stereoselectivity and high yield are reported, but the previous reports mainly focus on using microbe as the biocatalyst. It is valuable to explore the asymmetric reduction reaction catalyzed by vegetable or plant tissue in detail. Plant cell is another potential biocatalyst [6,7], since diverse oxidoreductases and the cofactor regeneration system exist in the plant cell. Unfortunately, there are few reports about the biotransformation catalyzed by plant cell. But the previous reported works is mainly concentrated on natural substrates. In this paper, 2-bromine-1-phenylethanol derivatives were chosen as the model substrates catalyzed by carrot (*Daucus carota L.*) was studied (Scheme 1).

E-mail address: liuxiang@jiangnan.edu.cn (X. Liu)

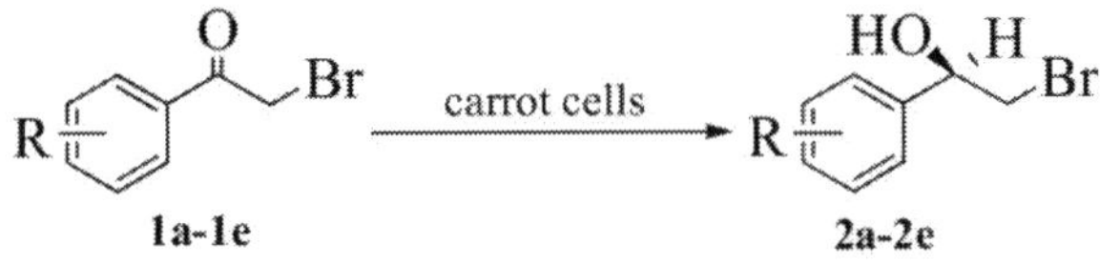

1a,2a:R=H
1b,2b:R=*m*-Cl
1c,2c:R=*p*-Cl
1d,2d:R=*p*-CH_3
1e,2e:R=*p*-OCH_3

Scheme 1. Asymmetric reduction of **1a–1e** catalyzed by carrot cells

Materials and methods

Materials and Chemicals

Fresh and healthy carrot roots were purchased from a local market. The external layer of these roots was removed and the rest was carefully cut into small thin pieces (approximately 5 mm × 5 mm ×2 mm). Alpha-bromine aromatic ketones **1a–1e** were synthesized through the transformation of aromatic ketones. The racemic alcohol standards were prepared by chemical reduction of the **1a–1e** with sodium borohydride in ethanol [8]. (*S*)-**2a** and (*R*)-**2a** were purchased from Aldrich. All other chemicals and reagents were purchased in analytic grade from local suppliers.

Biotransformation and HPLC Chromatographic analysis

General procedures for the asymmetric reduction of α–bromine aromatic ketones catalyzed by carrot cells are as follows: the small thin pieces of fresh carrot roots (10.0 g) were placed in flasks with 50 mL of 0.05 mol/L phosphate buffer or distilled water. To this suspension, 10–30 mg of **1a** was added and shaken at 28–36℃ for 20–36 h in an orbital shaker (120 r/min). The suspension was filtered and the filtrate was extracted with ethyl acetate (3 × 20 mL). The organic phases were centrifuged for 20 min at 4000 r/m to remove water, and dried by Na_2SO_4. The dried organic phases concentrated to a certain volume in vacuum. The concentrations of α–bromine acetophenone, (*R*)-**2a** and (*S*)-**2a** was established by HPLC equipped with Chiralcel OD-H column(4.6 mm × 250 mm; Daicel Chemical Industry, Ltd., Japan), and the others were established by HPLC equipped with Chiralcel OB-H column (4.6 mm × 250 mm; Daicel Chemical Industry, Ltd., Japan).

Results and discussion

Effect of substrate concentration on yield and ee of (R)-1-phenyl-2-bromoethanol

The effect of substrate concentration on the yield and enantiomeric excess (*e.e.*) of **2a** was investigated. As shown in Fig.1a, the yield of **2a** was heavily influenced by the substrate concentration. The probable reason for this was that the intracellular activity of carrot cells gradually decreased until being inactivated because of the toxicity of substrate to the cells. On the other hand, the *e.e.* of **2a** gave no noticeable difference at varied concentrations ranging from 0.2 g/L to 0.6 g/L, keeping above 87% in all the cases tested. Therefore, a substrate concentration at 0.4g/L was appropriate when considering both yield and *e.e.* in this work.

Effect of temperature on yield and ee of (R)-1-phenyl-2-bromoethanol

Fig.1b demonstrated that the effect of temperature on the yield of **2a**, but little impact on *e.e.*. As the reaction temperature was increased, yield rose firstly and then diminished gradually, reached their peak values at 32℃. The *e.e.* of **2a** value maintained at 87% to 88%. The result can be regarded as that the reaction rate increased as the temperature climbed up, which enhanced the yield in a certain period of reaction time. But on the other hand, a too high temperature would inactivate enzymes and then reducd the reaction rate. So the optimum temperature for the tested cases was 32℃. At this temperature, the highest yield and relatively high enantioselectivity were obtained.

Effect of reaction medium on yield and ee of (R)-1-phenyl-2-bromoethanol

As can be seen in Fig.1c, the pH of reaction media exhibited noticeable effects on yield and enantiopurity of **2a**. Both yield and *e.e.* value markedly descended as the pH of reaction media increased. The interesting results are that both yield and enantiopurity of **2a** reached the peak value in distilled water not in aqueous buffer system. **2a** was obtained with 87.7% *e.e.* and 95.6 % yield because of high activity of the carrot cells. The pH of reaction media affected reaction mainly as a result of influencing the enzymatic activity, it can damage the spatial structure of enzyme protein if too high or too low, causing changes in enzyme conformation and thus affecting the enantioselectivity and activity. Furthermore, the pH affects the ionization of substrates which influence the interaction between substrates and enzyme, or the products can not be generated. In our work, the distilled water of pH 7.0 was chosen as the reaction medium.

Effect of reaction time on yield and ee of (R)-1-phenyl-2-bromoethanol

Fig.1d shows the yield and *e.e.* changed depending reaction time. On the one hand, the yield didn't instantly went-up with the reaction time's lengthening until at 32 h. On the other hand, the *e.e.* of the product gradually started to go down until at 28 h. The reason for this may be assumed that some adverse reactions happened in such conditions and product inhibition happened as the over much reaction time. The phenomenon indicated that the reaction time must be moderate. The 28 h reaction time was selected in this experiment.

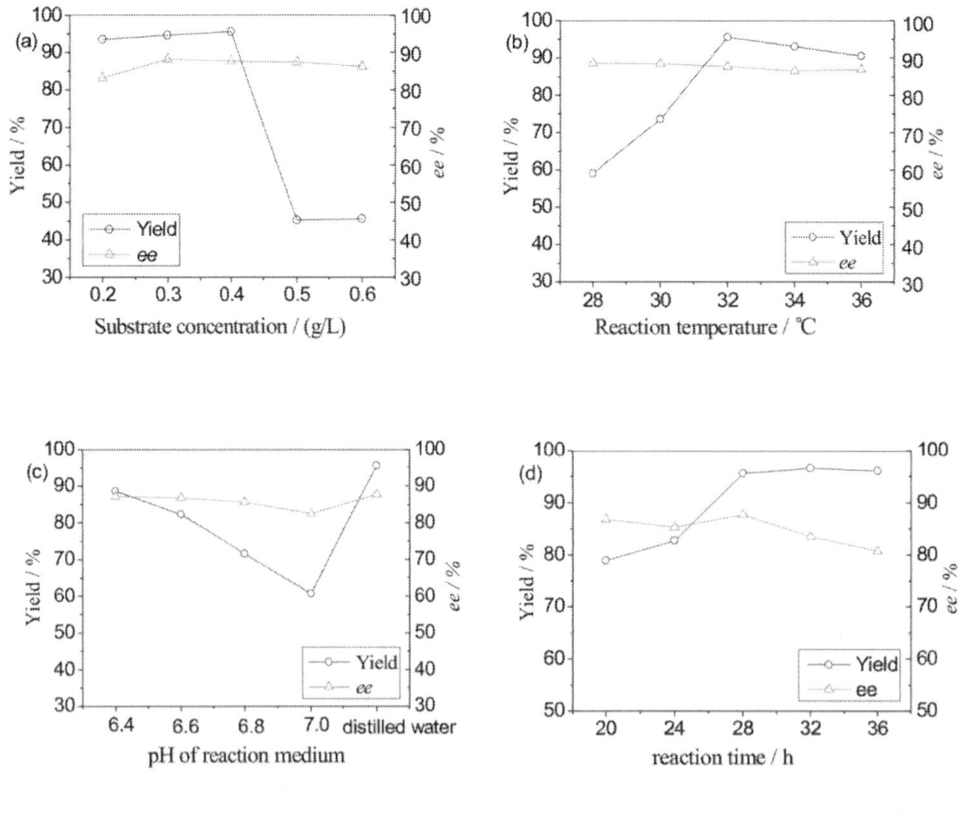

Fig. 1. The effect of reaction conditions on the yield and e.e. of **2a**. Reduction conditions: (a) 32℃, distilled water, 28 h; (b) 0.4 g/L **1a**, distilled water, 28 h; (c) 0.4 g/L **1a**, 32℃, 28 h; and (d) 0.4 g/L

1a,32℃,distilled water.

Effect of substituted groups on yield and ee of Chiral 1-aryl-2-bromo alcohols

The optimal reaction conditions were applied to the asymmetric reduction of a series of α-hydroxy aromatic ketones **1b**-**1e**, and the results are shown in Table 1. As can be seen in Table 1, a series of substrates was successfully reduced under the appropriate conditions, giving the corresponding alcohol **2a**-**2e** in good yield and excellent *e.e.*. As the best model substrate, α-bromoacetophenone (**1a**) can be effectively reduced to (*R*)-1-phenyl-2-bromoethanol (**2a**), the yield and *e.e.* can reach to 95.6 % and 87.7 %, respectively, which indicated that steric factors of substituted obstructed interaction between substrates and enzyme, leading yield declined. However, the results of reduction α-bromine-3-chloroacetophenone (**1d**) and α-hydroxy-4-chloroacetophenone(**1e**) were more attractive in term of the yield than the rest, which inferred substrates with electron-withdrawing substituted groups are more acceptable to plant tissue. Besides, the substrates with electron-donating groups afforded their products (**2b**, **2c**) with higher *e.e.* than those of the others. It seems that the electron-donating groups are benefit to enantiopurity. Therefore, the electronic effects, caused by the substituted group, makes the carbonyl reductases present in plant tissue more efficient.

Table 1. The variation trends of yields and *e.e.* of **2a**–**2e**.

Entry	Substituent on the aromatic ring	Yield (%)	e.e. (%)
2a	H	95.6	87.7
2b	m-Cl	47.3	88.2
2c	p-Cl	48.5	86.8
2d	p-CH3	64.7	85.3
2e	p-OCH3	67.1	84.5

Reaction conditions: 0.4 g/L substrates, 32 ℃, distilled water, 28 h..

Conclusions

Bio-reduction catalyzed by dehydrogenases enzymes present in carrot cells provides an attractive approach in reducing various types of α–bromine aromatic ketones with good yields and enantioselectivities. The enantioselectivity was the same as Prelog's rule as expected. The steric factors and electronic effects of substituted groups had a clear effect on the yield and enantioselectivity of the reaction. This provides a very clean, simple and convenient method for preparation chiral alpha bromide aromatic alcohol which are important chiral building blo often involving in the production of a wide range of fine chemicals and pharmaceutical molecules.

Acknowledgments

This research was financially supported by the National Special Fund for State Key Laboratory of Bioreactor Engineering (No. 2060204) and the National Natural Science Foundation of China (No. 20672037).

References

1. Jin Li, Kevin K, C Liu(2004), "Synthetic Approaches to 2002 New Drug," *Mini-Rev. Med. Chem.*, 4,pp.207-233.

2. Tracy D Nixon, Michael K Whittlesey, Jonathan M. J. Williams(2011), "Ruthenium-Catalysed transfer hydrogenation reactions with dimethylamine borane," *Tetrahedron Lett.,* 52,pp.6652 – 6654.
3. Kaoru Nakamura, Rio Yamanaka. Tomoko Matsuda, Tadao Harada(2003), "Recent developments in asymmetric reduction of ketones with biocatalysts," *Tetrahedron: Asymmetry.,* 14,pp.2659–2681.
4. Kathrin Holsch, Dirk Weuster-Botz(2011), "Kinetic mechanism of 3-ketoacyl-(acyl-Carrier -protein) reductase from Synechococcus sp. strain PCC 7942: A useful enzyme for the production of chiral alcohols," *J. Mol. Catal. B: Enzym.,* 69,pp.89–94.
5. Hiroyuki Nagaoka(2004), "Ability of Different Biomaterials to Enantioselectively Catalyze Oxidation and Reduction Reactions," *Biotechnol. Progr.,* 20, pp.128-133.
6. Zhong-Hua Yang, Rong Zeng, Gai Yang(2008), "Asymmetric reduction of prochiral ketones to chiral alcohols catalyzed by plants tissue. *J Ind Microbiol Biotechnol.,* 35,pp.1047–1051.
7. Takamitsu Utsukihara, Satoshi Watanabe, Atsushi Tomiyama(2006), "Stereoselective reduction of ketones by various vegetables," *J. Mol. Catal. B: Enzym.,* 41,pp.103–109.
8. Mariappan Periasamy, Muniappan Thirumalaikumar(2000), "Methods of enhancement of reactivity and selectivity of sodium borohydride for applications in organic synthesis," *J. Organomet. Chem.,* 609,pp.137–151.

(69g) Engineering Biosynthesis and Applications of Medium-Chain Carboxylic Acids

Kechun Zhang, Chemical Engineering and Materials Science, University of Minnesota, Minneapolis, MN

Depleting oil reserves and growing environmental concerns have necessitated the development of new approaches to fuels and chemicals from renewable carbon sources. This talk will summarize my lab's recent progress in metabolic engineering of E. coli for the sustainable production of C4-C6 carboxylic acids from glucose. Since these compounds are not native metabolites, we have designed an oxidation metabolic platform and engineered novel enzymes to enable the bio-based production. Furthermore, after several rounds of optimization, we have achieved high production titers. In particular, our engineered strains could biosynthesize C4 isobutyrate with a final concentration of 90 g/L. The concentration of C5 isovalerate could reach 32 g/L as well in a bioreactor. The success in fermentative manufacturing of C4-C6 acids provides a renewable approach to the synthesis of a variety of high-volume industrial chemicals. For example, isobutyrate is used to manufacture poly(methyl methacrylate); isovalerate is used to manufacture methyl isobutyl ketone.

(69h) *In Situ* Real-Time Monitoring of Biotransformations

Sean Cusack and Vlachos Vlachos, Mettler Toledo AutoChem, Inc., Columbia, MD

Enzymes, as biocatalysts, are a highly suitable, environmentally friendly alternative to heavy metal industrial catalysts. The "Green Chemistry" nature of enzymes as catalysts is based on a number of characteristics. Enzymes are biodegradable and typically produced through fermentation of renewable feedstocks or easily mass produced through recombinant technologies. Enzymes also catalyze chemical reactions under mild temperature (20°C-40°C) and pH conditions (pH 5-8), and perform well in aqueous environments. The chemical, petroleum, agriculture, polymer, electronics and pharmaceutical industries rely on catalysis to optimize their chemical processes. It is estimated that at least 90% of chemicals produced rely to some extent on catalysis. In recent years researchers have increasingly focused their attention on green chemistry, including catalysis. Examples of enzyme-catalyzed industrial-scale organic reactions include hydrolysis, oxidation, reduction, addition – elimination, halogenation and dehalogenation, and transesterification. This presentation reviews case studies from pharma, academia and the military, showing how *in situ* mid-IR was used to monitor the enzyme-catalyzed reactions in real time, to enable researchers to understand reaction mechanisms, and determine kinetic parameters and reduce costs.

(72a) Amphiphilic Macromolecules to Manage Atherosclerosis: Quantitative Structure Activity Relationships

Daniel R. Lewis[1], Vladyslav Kholodovych[2], Li Gu[3], Dawanne Poree[3], Kathryn E. Uhrich[3] and Prabhas V. Moghe[4], (1)Chemical and Biochemical Engineering, Rutgers University, Piscataway, NJ, (2)IST/High Performance and Research Computing and Dept of Pharmacology , University of Medicine and Dentistry of New Jersey, Newark, NJ, (3)Chemistry and Chemical Biology, Rutgers University, Piscataway, NJ, (4)Biomedical Engineering, Rutgers University, Piscataway, NJ

Introduction: Atherosclerosis is the primary component of pathologies underlying cardiovascular disease (CVD), which is the leading cause of death in western countries. Atherosclerosis is characterized as an inflammatory disease involving macrophage scavenger receptor (SR) interactions with oxidized low-density lipoproteins (oxLDL), leading to plaque growth. Conventional therapies are often plagued by off-target effects and the inability to address localized inflammation from preexisting lipid deposits. Therefore, it is necessary to develop a therapeutic that can address localized lipid uptake and inflammation. We have designed a new library of bioactive Amphiphilic Macromolecules (AM) to competitively occupy SRs and inhibit oxLDL uptake, foam cell formation and inflammatory signaling in human monocyte derived macrophages (MDM). Modulating the AM architecture resulted in varying degrees of oxLDL uptake and inflammation. Molecular modeling was used to study the correlation between AM structural features and their corresponding biological behavior.

Materials and Methods: AMs were synthesized by acylating the hydroxyl groups of sugar acids with dodecanoyl chloride in the presence of zinc chloride, then 5000 Da monohydroxyl-substituted PEG is coupled using 1,3-dicyclohexylcarbodiimide and 4-(dimethylamino)pyridinium p-toluene-sulfonate as catalyst before precipitation in CH_2Cl_2 and diethyl ether. In vitro efficacy of AM was investigated in MDM differentiated from peripheral blood mononuclear cells. MDM were treated with oxLDL (5μg/mL DiO labeled for uptake, 50μg/mL unlabeled for foam cell formation, 100μg/mL unlabeled for inflammatory response) in the presence of 10^{-4}-10^{-6}M AM for 6-24 h in RPMI media. OxLDL uptake and foam cell formation were determined by cell associated fluorescence (DiO and post treatment Oil Red O staining, respectively) via epifluorescent microscopy and quantified with ImageJ. The inflammatory response was determined using qRT-PCR ($\Delta\Delta C_t$), for which RNA was extracted and reverse transcribed to cDNA before cycling on a LightCycler480 with SYBR green and primers specific for ACTB, IL-1β, IL-6 and IL-8. ELISA's were used to validate IL-8 cytokine secretion in post-treatment cell culture supernatant. For modeling studies, all AM were constructed and minimized in Molecular Operating Environment (MOE). Molecular descriptors (only molecular properties, atom types and chemical group counts) were calculated with Dragon software v.5.4- 2006. Partial least squares regression method implemented in MOE was used to model the experimental data.

Results and Discussion: Seven different AM structures were synthesized, varying the overall charge, hydrophobicity, sugar structure (linear vs. cyclic) and stereochemistry by using the sugar acids: mucic (1cM), neutral mucic (0cM), saccharic (1cS), tartaric (1cT), galacturonic(0cG), glucuronic (0cGL), di-mucic (1cMM). Cells treated with oxLDL alone displayed significant amounts of oxLDL uptake and lipid accumulation (foam cell phenotype) in addition to upregulation of IL-1β, IL-6 and IL-8. While some AMs were effective at reducing these endpoints, small changes in AM sugar structures resulted in large differences in potency. This effect was most drastic with the substitution of saccharic acid for mucic acid (the only difference between the structures being one stereocenter), in which 1cM was the most effective at reducing oxLDL uptake (80% inhibited relative to oxLDL alone), whereas 1cS increased uptake. Without an external pro-inflammatory stimulus, linear monosugar AM (1cM, 0cM, 1cS and 1cT) displayed no upregulation of cytokines IL-1β, IL-6 and IL-8, whereas AM with cyclic sugars (0cG and 0cGL) or two mucic acids (1cMM) upregulated the tested genes to a similar level as oxLDL. Linear monosugar 1cM and 1cT were both effective at reducing oxLDL induced inflammatory gene upregulation. Levels of IL-8 cytokine secretion paralleled IL-8 gene upregulation for all tested conditions. Modeling studies found that a good correlation between predicted and observed results could be achieved for oxLDL uptake with only three descriptors (r^2=0.97, q^2=0.73) while only two descriptors are needed to model foam cell formation (r^2=0.97, q^2=0.86).

Conclusions: Previous work with 1st generation AM structures qualitatively studied placement and quantity of charge, but lacked the sophistication to accurately identify key features. Using quantitative structure activity relationships will be able to guide future development of an optimal AM structure for mitigating adverse athero-relevant endpoints.

(72b) A Peptide Probe for Rapid Detection of Various Amyloid Oligomers

Yang Hu, Baihao Su, Michael Hernandez, He Qiu Zheng and **Jin Ryoun Kim**, Chemical and Biological Engineering, Polytechnic Institute of New York University, Brooklyn, NY

Aggregation of amyloid is implicated in the pathology of various neurodegenerative diseases. A considerable amount of data has identified soluble amyloid oligomers as potentially significant toxic agents. Rapid, specific and quantitative detection is preferred for accurate profiling of structurally unstable oligomers as well as for implementation of high-throughput assays for pharmaceutical applications. PG46 is an engineered beta-amyloid (Abeta) variant, constructed by integrating Abeta self-recognition sequences with the conformation-sensitive biarsenical fluorescent dye, FlAsH. PG46 was found to be an effective peptide probe for rapid, specific and quantitative detection of Abeta oligomers. However, PG46 was highly aggregation-prone and displayed a limited repertoire of detectable Abeta oligomers. Here, we report the creation of a novel molecular probe, PG44, by C-terminal truncation of PG46. PG44 exhibited a reduced self-aggregation propensity and a different conformation when compared to PG46, and generated specific FlAsH fluorescence signals as a result of binding to various Abeta oligomers, including those not readily detectable by PG46. We also show that sensitivity of PG44 for detection of certain Abeta oligomers may be increased by lowering PG44 concentration and thus decreasing the extent of self-aggregation of PG44. Our results suggest that PG44 can serve as an important molecular probe with a broadened repertoire of detectable Abeta oligomeric aggregates. We believe that detection of Abeta oligomers using our peptide probe would potentially contribute toward a better understanding of the molecular basis of Abeta oligomerization and the development of Abeta oligomer-based early diagnostics as well as therapeutic drugs targeting Abeta oligomers. We will also discuss our efforts to create a molecular probe for rapid, specific and quantitative detection of oligomers formed by other amyloid proteins.

(72c) Development and Characterization of Degron-Based Substrates Capable of E3 Ligase-Mediated Ubiquitination

Adam Melvin[1], Marcey Waters[1] and Nancy Allbritton[1,2,3], (1)Chemistry, Univeristy of North Carolina, Chapel Hill, NC, (2)Biomedical Engineering, University of North Carolina, Chapel Hill, NC, (3)Biomedical Engineering, North Carolina State University, Raleigh, NC

The ubiquitin proteasome system (UPS) is responsible for the recognition and degradation of misfolded, damaged, or tightly regulated proteins in the cell. Deregulation of this critical process is the hallmark of many different forms of cancer leading to failed tumor suppression, loss of cell cycle control, and an increased metastatic response. The specificity of this pathway comes from the >600 E3 ubiquitin ligases, each of which recognizes a specific sequence (termed a 'degron') on a specific protein to undergo ubiquitination. To gain a better understanding of protein ubiquitination and E3 ligase-degron kinetics, we sought to generate and test a library of substrates based on known degrons. In this study, we demonstrate the utility of the degron-based substrates, the variability in E3 ubiquitin ligase performance, and, ultimately identify substrates that could serve as potential reporters for E3 ubiquitin ligase-targeted chemotherapeutics.

Using solid phase peptide synthesis (SPPS), we generated small (~25-30 amino acids) substrates consisting of a degron, a proximal lysine, and a fluorescein tag. The degrons were selected across a wide array of cellular functions from proteins associated with various cancers including β-Catenin, iNOS, TAZ, HIF-1α, p53, and others. Substrate ubiquitination was evaluated by an *in vitro* HeLa S100 cytosolic-based ubiquitin pull down followed by gel electrophoresis. We discovered that each of the nine degron-based substrates was ubiquitinated in our *in vitro* system, although the extent of ubiquitination varied across the library, suggesting a hierarchy of degrons could be established. Additionally, using a methylated ubiquitin (effectively preventing ubiquitin chain formation), we found that our substrates could be mono-, di-, and tri-ubiquitinated. To understand this observation and quantify substrate performance, we constructed a mathematical model, based on first order reaction kinetics, to evaluate kinetic rate constants for mono-, di-, and tri- substrate ubiquitination. Our model indicates a lack of fidelity by E3 ubiquitin ligases when it comes to lysine preference proximal to the degron. The kinetic rates constants also identified two substrates, β-Catenin and Bonger, which exhibited rapid ubiquitination when compared with the rest of the field. The robustness of these substrates was then tested in lysates from multiple myeloma (OPM-2 and MM1.S) and leukemia (U937, THP-1, HL-60, and K562) cell lines. Our degron-based substrates provide a robust, easy to synthesize, and highly useful tool in the study of the ubiquitin proteasome system and demonstrate a promising new technology that could be used as a reporting tool for chemotherapeutics designed to specifically target the UPS.

(72d) The Roles of Palmitate On IRE1alpha Enzymatic Activities

Hyunju Cho[1], Liang Fang[2], Pratheeba Palasuberniam[2], Michael Feig[2] and Christina Chan[1], (1)Chemical Engineering and Materials Science, Michigan State University, East Lansing, MI, (2)Biochemistry and Molecular Biology, Michigan State University, East Lansing, MI

IRE1α (inositol-requiring enzyme 1α) is the most highly conserved signaling node of the unfolded protein response (UPR). IRE1α consists of an N-terminal luminal domain and C-terminal cytosolic domain containing kinase/endoribonuclease activity, and are connected by a signal helix transmembrane domain. Upon ER stress, the signal in the luminal domain is transduced to the cytosol by sequential dimerization/multimerization, trans-autophosphorylation and activation of its endoribonuclease. The activated IRE1α splices Xbp1(X-box binding protein 1) mRNA generating an active Xbp1s transcriptional activator that upregulates expression of UPR target genes. It is well known that palmitate, C16-saturated fatty acid, induces ER stress by stimulating IRE1α activities in the cells. However, the molecular mechanism by which palmitate activates IRE1α is not well understood. Our group previously showed that palmitate directly interacts with the cytosolic domain of the IRE1α. We address whether palmitate regulates the enzymatic activities of IRE1α-CD protein using *in-vitro* kinase and RNase assays. Computational molecular dynamics simulations suggested several potential palmitate-binding sites (R611, R635, R687, K716, R722, R727, and R728) which are mainly located in the activation loop, dimerization interfaces, and αC-helix. We tested whether mutations of these sites are able to alter the palmitate-binding as well as the enzymatic activities of IRE1α-CD protein. The mutation analyses suggested that palmitate plays a role as an allosteric activator of IRE1α. We further introduced the mutant genes into the $IRE1^{-/-}$-MEF (Mouse Embryonic Fibroblast) cells, confirming that mutation blocks the palmitate-mediated activation of IRE1α. Therefore, these findings suggest that palmitate is an endogenous cytosplamic ligand that may activate the signals from the ER lumen to modulate IRE1α activities. In addition, the results provide insights into the development of drugs that modify IRE1α activities, and the signaling of the UPR.

(72e) Discovery and Characterization of a Novel Actin-Binding Protein From Stenotrophomonas Maltophilia

Logan MacDonald[1], Sean O'Keefe[1], Emily L. Wong[2] and Bryan W. Berger[3], (1)Bioengineering, Lehigh University, Bethlehem, PA, (2)Division of Infectious Diseases, Lehigh Valley Hospital, Allentown, PA, (3)Chemical Engineering and Program in Bioengineering, Lehigh University, Bethlehem, PA

Stenotrophomonas maltophilia is a ubiquitous, Gram-negative bacteria that is of increasing concern to healthcare professionals due to intrinsic multidrug resistance and prevalence as a cause of infection in immunocompromised patients. Major challenges in preventing *S. maltophilia* exist, such as its high degree of genetic diversity and ability to rapidly acquire new mechanisms of drug resistance. However, relatively little is known about specific, secreted effectors from *S. maltophilia* that may contribute to its virulence. Through a bioinformatics-based approach, we identified a series of proteins that resembled potential T4SS effectors and investigated their potential as virulence factors. Co-localization experiments using RFP-labeled actin indicated that one of the predicted proteins (smlt3054) disrupted the organization of the actin cytoskeleton in transfected HEK293 and A549 cell lines. *In vitro* actin polymerization experiments and *in vivo* dose dependence experiments were performed using recombinantly expressed and purified smlt3054, which confirmed the actin-binding ability of smlt3054. Expression of labeled smlt3054 in clinical strains of *S. maltophilia* indicate the protein is expressed at the cell surface, consistent with the prediction it is a secreted effector. We also find despite the wide genetic variability of *S. maltophilia*, several clinical strains isolated from actual infections contain smlt3054 as well. Thus, smlt3054 represents a potential secreted effector from *S. maltophilia* that we believe may play a key role in its virulence.

(72f) Engineering Probiotic Bacteria to Help Us Fight the Emergence of Antibiotic Resistant Bugs

Katherine G. Volzing, Juan Borrero del Pino and Yiannis N. Kaznessis, Chemical Engineering and Materials Science, University of Minnesota, Minneapolis, MN

Gastrointestinal (GI) infections currently take a significant toll on human health and constitute a serious concern worldwide (1). The Centers for Disease Control estimates that more than 650 million people globally are affected each year by bacterial diarrhea (2). This number is rising as antibiotic use in agriculture is hastening the emergence of drug-resistant bacteria. The effectiveness of many antibiotics has been lost as the number of drug resistant bacteria strains increases. There is now a substantial overlap in between the classes of antibiotics listed as critically important for human health by the World Health Organization and the antibiotics listed as critically important for agriculture by the World Organization for Animal Health (3-5). Currently, it is estimated that over 70% of antibiotics produced in the United States are given to cattle, pigs, and poultry to improve feed efficiency at sub-therapeutic levels and in the absence of infection (6). This use favors the emergence of antibiotic resistant bacterial strains and ignites a significant threat to human health worldwide (5-7).

Herein, we present a new approach to reducing the widespread use of antibiotics in agriculture. This work has aimed to engineer probiotic bacteria strains with inducible antimicrobial peptide (AMP) expression systems. AMPs are small proteins with remarkable bactericidal character (8, 9) while probiotics are live bacteria that can be safely delivered to animal GI tracks as supplements in food or water (10, 11). To date, we have built inducible AMP expression systems in probiotic bacteria strains. We have characterized both their antimicrobial activity against specific pathogens of interest as well as their toxicity to host cells. We are currently working towards characterizing our systems in animal models. More specifically, we are seeking to quantify the changes in gut microflora driven by the administration of our engineered probiotics and evaluate their effectiveness and impact on animal health in the presence of pathogen challenges.

Our engineered probiotics will be optimized to a) improve overall animal health, and b) fight off bacterial infections by expressing AMPs only in the presence of pathogens or upon induction with a small molecule.

1. Turner SM, Scott-Tucker A, Cooper LM, & Henderson IR (2006) *FEMS Microbiol Lett* **263,** 10-20.
2. Sanders JW, Putnam SD, Frankart C, Frenck RW, Monteville MR, Riddle MS, Rockabrand DM, Sharp TW, & Tribble DR (2005) *Am J Trop Med Hyg* **73,** 713-719.
3. Barie P (1998) *World Journal of Surgery* **22,** 118-126.
4. Domin M (1998) *Br J Theatre Nurs* **8,** 14-18.
5. (2007) *Joint FAO/WHO/OIE Expert Meeting on Critically Important Antimicrobials* **Rome, Italy**.
6. White W (1998) *Science.*
7. Mellon CB, KL Benbrook et al (2001) *Union of Concerned Scientists.*
8. Bolintineanu D, Hazrati E, Davis HT, Lehrer RI, & Kaznessis YN *Peptides* **31,** 1-8.
9. Hancock RE (2001) *Lancet Infect Dis* **1,** 156-164.
10. Pagnini C, Saeed R, Bamias G, Arseneau KO, Pizarro TT, & Cominelli F *Proc Natl Acad Sci U S A* **107,** 454-459.
11. Van Niel CW, Feudtner C, Garrison MM, & Christakis DA (2002) *Pediatrics* **109,** 678-684.

(72g) Discovery of Novel Plantaricin-423 Analogs From Synthetic Oligonucleotide Libraries

Saadet Albayrak Guralp, Katie J. Miskovich, Jean-Marie Rouillard and Erdogan Gulari, Chemical Engineering, University of Michigan, Ann Arbor, MI

Excessive use of antibiotics in human and animal health care for the treatment and prevention of bacterial infections led to a rapid increase in antibiotic-resistant pathogenic bacteria. Antimicrobial peptides (AMPs) belong to a class of natural microbicidal molecules with lower propensity for inducing drug resistance; hence, they present significant potential to be used as alternative drugs to conventional antibiotics. By generating AMP libraries, one can study a large number of candidates for their activities simultaneously in a timely manner. However, current methodologies for the construction of AMP libraries present many limitations when it comes to sequence design, peptide length, or the library size. We have previously reported a novel technology for the construction of custom peptide libraries, where fully defined collections of *in silico*-designed peptides are encoded into DNA oligonucleotides and expressed in a cellular host for rapid screening of active molecules. We now present an application of this method to discovery of novel Plantaricin-423 analogs. Plantaricin-423 (Pln-423) is a 37-amino acid Class II-a bacteriocin and it displays bactericidal activities against many foodborne pathogens and spoilage bacteria. A custom mutant library coding for 12,000 unique Pln-423 analogs was constructed in *Escherichia coli* and screened against *Listeria innocua* and several other gram positive bacteria. Analysis of the selected clones by both Sanger-sequencing and high-throughput 454 pyrosequencing resulted in identification of more than 20 novel sequences with greater activities. Further investigation of the sequencing data revealed the specific mutations that are responsible for the change in antimicrobial activity of the wild-type peptide. The data generated in this study can be used for the construction of secondary libraries to optimize the anti-listerial activity of Pln-423 mutants or broaden their range of activity against other gram positive bacteria. Therefore, this study successfully demonstrates the great potential of our method for AMP discovery and development.

(72h) Targeted DNA Methylation Using a Bisected M.HhaI Fused to Zinc Fingers

Brian Chaikind[1], Krishna Praneeth Kilambi[2], Jeffrey J Gray[2] and Marc Ostermeier[2], (1)Chemical Biology Interface Program, Johns Hopkins University, Baltimore, MD, (2)Chemical and Biomolecular Engineering, Johns Hopkins University, Baltimore, MD

CpG methylation of promoters is an important form of epigenetic control in eukaryotic cells, causing transcriptional repression. It is therefore involved in many important cellular processes, such as cellular differentiation, X-chromosomal inactivation, and chromatin remodeling. As such, aberrant methylation patterns have been implicated in many disease states such as cancer, psychological disorders, and a range of diseases caused by the deregulation of epigenetic markers. The ability to target methylation toward a specific site within the genome would have broad applicability as a tool to study the effect of single CpG methylation events on transcription and as a potential therapeutic to target diseases resulting from the hypomethylation of particular promoters.

We have engineered a targeted methyltransferase by splitting a prokaryotic, monomeric methyltransferases into a heterodimeric methyltransferase and fusing each heterodimeric fragment to a sequence specific DNA binding protein. The recognition sequences for the DNA binding protein flank a desired CpG site. When bound, the DNA binding proteins increase the local concentration of the methyltransferase fragments over the CpG site, forcing functional reassembly of the enzyme only over the target site. Thus, the enzyme activity at a desired site will be contingent upon DNA binding recognition of the adjacent sites. This strategy works because the fragments are engineered to have reduced affinity for another, and do not efficiently assemble away from the target DNA sequence. Using this strategy we created a target methyltransferase that methylates ~60% of a desired site with undetectable methylation at non-targeted sites. *In vivo* experimentation and *in silico* modeling of these methyltransferases has elucidated some of the design parameters and current limitations of these targeted methyltransferases.

(74a) Osmotic Regulation of Cell Migration and Volume in Microfluidic Channels

Kimberly M. Stroka[1], Hongyuan Jiang[2], ZiQiu Tong[1], Sean X. Sun[3] and Konstantinos Konstantopoulos[1], (1)Chemical and Biomolecular Engineering, Johns Hopkins University, Baltimore, MD, (2)Mechanical Engineering, Johns Hopkins University , Baltimore, MD, (3)Departments of Mechanical Engineering and Biomedical Engineering, Johns Hopkins University, Baltimore, MD

Introduction: Cell homeostasis and diverse processes, including migration, are tightly regulated by cell volume. During migration through tissues, metastatic cancer cells experience varying degrees of physical confinement. We hypothesized that cell volume regulation by osmolarity is especially important for migration in confined microenvironments, due to recent work from our lab suggesting that actomyosin-based motility and adhesion are dispensable for confined migration, in contrast to 2D planar migration. We engineered a microfluidic-based chemotactic device with 3 µm-wide extracellular matrix-coated channels to maintain distinct extracellular osmolarities at the leading and trailing edges of the cell. Using phase contrast timelapse microscopy, we imaged live cancer cell migration in isotonic medium everywhere, or with hypo/hypertonic medium in the top, bottom, or top and bottom of the chamber.

Results: Upon switching from isotonic to hypotonic medium at only the top of the chamber, we observed a rapid reversal of cell migration direction (away from the hypotonic medium and chemoattractant). The speed of migration in the opposite direction decreased as the osmolarity of the medium approached isotonic. Meanwhile, switching to hypotonic medium at only the bottom of the chamber allowed cells to continue migrating towards the chemoattractant. Finally, switching to hypotonic medium in both the top and bottom of the chamber caused cells to migrate twice as fast towards the FBS. Similar responses were observed in wider channels (6, 10, 20, and 50 µm) and also in multiple cancer cell lines (AB3 mouse sarcoma, HTB-94 human sarcoma, and MDA-MB-231 breast cancer), though the threshold osmolarity necessary to produce a response varied between cell types. We also evaluated the effects of hypertonic shock at the top, bottom, or top and bottom of the chamber.

Changes in cell migration velocity were accompanied by a decrease in length (and therefore volume), the kinetics and final value of which were dependent on the polarity of the osmotic shock (top, bottom, or top and bottom of the chamber), osmolarity, and cell type. To confirm the decreases in volume and verify that cells occupied the entire channel, we used reduced-height channels (6 µm) and confocal imaging of F-actin-stained cells. Importantly, inhibition of actin polymerization (via latrunculin-A) stalled cell reversal after switching medium to hypotonic at the top of the chamber, even though latrunculin-A-treated cells migrated normally during confined migration in isotonic medium. In future work, we expect that microtubule dynamics will be important for cell polarity change, while ion channels and aquaporins will regulate cell volume changes that occur in these confined spaces upon osmotic shock.

Conclusions: Our results indicate that cancer cell migration in confined microenvironments is sensitive to fluctuations in extracellular osmolarity, and the location of osmotic shock (leading vs. trailing edge of the cell) is extremely important in determining the cellular response. Importantly, actin polymerization is a key process driving cell migration away from a hypotonic medium, even though it is dispensable for normal migration in isotonic medium in confined spaces.

(74b) Molecular Mechanisms in Durotaxis

Dennis E. Discher, Chemical and Biomolecular Engineering, University of Pennsylvania, Philadelphia, PA

On rigid substrates, the cytoskeleton is polarized during cell migration, but extracellular matrix is normally soft. Here we show that Myosin-IIB (MIIB) is unpolarized in cells on 2D and 3D soft matrix with rear/front polarization emerging as cells migrate onto stiff matrix, including gradients in elasticity relevant to wound-healing. Within Mesenchymal Stem Cells (MSCs) that are known to traffic to stiff scars, MIIB is random on soft matrix but becomes polarized as cells migrate from soft to stiff matrix. This 'durotaxis' is more sensitive to MIIB levels than the more abundant, unpolarized Myosin-IIA (MIIA), but Mass Spectrometry suggested that stiffness-dependent changes in phosphorylation of MIIA might play a key role. Indeed, phospho-Ser1943 in MIIA proves higher on soft matrix versus stiff matrix; in addition, phosphomimetic-MIIA proves highly mobile in cells, suppresses both MIIA-filamentation and MIIB-polarization on stiff matrix, and ultimately inhibits durotaxis. Such perturbations and more show that both durotaxis and polarization are maximal at wild-type levels of MIIB expression as coupled to phospho-dynamics of MIIA. Cytoskeletal polarization thus acts as a highly regulated, mechanosensitive compass in directed migration.

(74c) JNK Regulates Rigidity-Dependent Adherence Junction Formation of Epithelia in Vivo and in Vitro

Hui You, Aishwarya Ranganathan and Stelios T. Andreadis, Department of Chemical and Biological Engineering, University at Buffalo, the State University of New York, Amherst, NY

Introduction: In previous work we showed JNK phosphorylated b-catenin and regulated adherent junction (AJ) formation in human primary keratinocytes (hKC) and ME180 epithelial cancer cells. Inhibition of JNK activity caused translocation of E-cadherin/b-catenin complex to cell-cell contact sites leading to formation of AJ. It is well known that there is cross talk between cell-cell and cell-substrate adhesion as manifested by inhibition of cadherin function upon integrin engagement. This delicate balance between integrin and cadherin signaling was found to be regulated by substrate rigidity, which also regulates cell spreading, migration, proliferation and stem cell differentiation and tissue maintenance. Extracellular matrix stiffening has been shown to increase integrin expression and drive malignant behavior of tumor cells through Rho-mediated cytoskeletal tension. Tumor invasiveness was also associated with reduced E-cadherin expression and dissolution of AJ. Here we hypothesized that JNK may regulate cell-cell adhesion in a manner that depends on substrate mechanics.

Materials and Methods: Isolation and culture of primary keratinocytes and preparation of skin equivalent were performed as described previously[1]. shRNA Targeting the *jnk1, jnk2* and *jnk1/2* mRNAs was cloned downstream of the H1 promoter of pLVTHM lentiviral vector. A fusion protein of JNK and *its upstream activator MKK7 was cloned in p-TRIP-Z vector.* Polydimethylsiloxane (PDMS) and a crosslinked siloxane polymer were used as 9:1 (w/w, base to crosslinker) for hard substrates (1MPa) and 9:0.2 for soft substrates (16 kPa).

<>Results and Discussion: *In vitro* 2D studies show that the substrate rigidity regulates JNK activity of hKC. The levels of p-JNK were significantly reduced in the hKC on soft substrates. E-cadherin and b-catenin, two components of AJ complex, co-localized at the cell-cell contact. However, the expressions of p-FAK and p-Paxillin, two of the focal adhesion (FA) markers, at *cell protrusions* were significantly reduced. hKC with constitutive active JNK had well developed FA but could not form AJ on the soft substrate. However, hKC knocked down JNK1, JNK2, or JNK1/2 formed AJ even on the rigid substrates. Notably, our observations extended *in vivo*, where we observed a negative correlation between AJ formation and JNK activity in human foreskin epidermis and bioengineered skin. Keratinocytes closer to basal layer (stiff substrate) expressed more p-JNK and p-cJun and lacked AJ. However, keratinocytes on upper suprabasal layers of the epidermis (soft substrate) formed tight AJ but expressed low levels of p-JNK and p-cJun (Fig.1A-B). To test if the stiffness gradient may in turn determine the gradients of p-JNK and AJ formation observed across the epidermis, we increased the stiffness of the dermal matrix using a natural crosslinker genipin. Interestingly, expression of p-JNK and p-cJun extended to the upper suprabasal layers of the epidermis grown on stiffer dermis. Accordingly, the levels of E-cadherin and b-catenin at the cell-cell contact sites were significantly reduced even in the upper suprabasal layers (Fig.1C). To verify that the disruption of AJ in the suprabasal cells was due to JNK activation, we generated epidermis with hKC overexpressing the constitutively active JNK1 (MMK7-JNK1 fusion protein). Immunostaining of tissue sections for E-cadherin and b-catenin showed that MKK7-JNK1 bioengineered epidermis exhibited weak and disorganized *AJ* (Fig.1D). In agreement, staining for p-JNK and p-cJun was extended in the suprabasal layers of MKK7-JNK1 bioengineered epidermis. On the other hand, in either JNK1 and JNK2 deficient mouse skin or bioengineered skin, robust AJ were formed throughout the epidermis even in the basal layer (Fig.1E-F), which is normally devoid of AJ.

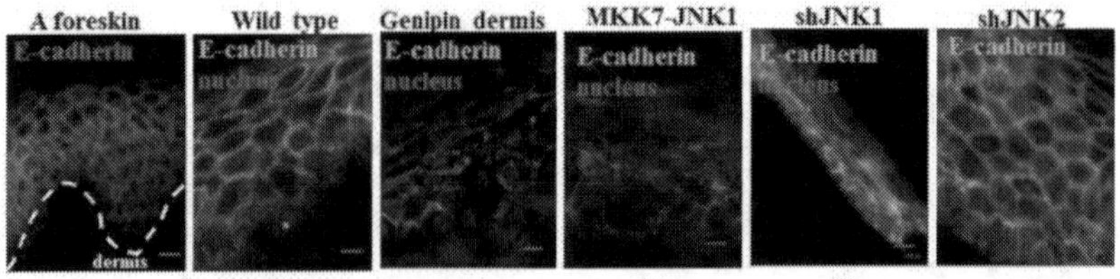

A B C D E F

Fig.1 Immunofluorescent staining of E-cadherin in (A) human foreskin, (B) wild type bioengineered skin **(C)** bioengineered skin **on genipin-treated dermis, (D) MKK7-JNK1 hKC** bioengineered skin, **(E) shRNA JNK1** bioengineered skin, **(F) shRNA JNK2** bioengineered skin**. View: x63 scale =10μm**

Conclusions: Our results clearly suggest that formation of AJ is regulated by the activity of JNK, which in turn may be regulated by substrate stiffness, ultimately resulting in molecular gradients that may affect tissue development, wound healing and cancer progression.

Reference: 1. Andreadis, S.T., *et al.* 2001. *FASEB J* **15**, 898-906

(74d) Modulating Megakaryocyte Adhesion for Improved Proplatelet Formation

Alaina C. Schlinker, David C. Whitehead and William M. Miller, Department of Chemical and Biological Engineering, Northwestern University, Evanston, IL

Megakaryocytes generate long, cytoplasmic extensions, known as proplatelets, to package platelets and release them into the circulation. In vivo, this process occurs in the bone marrow vascular niche. Mature megakaryocytes express receptors for many of the ligands present in this niche, suggesting that cell adhesion may play an important role in proplatelet formation. We aim to improve proplatelet formation for the purpose of in vitro platelet production. We found that the CHRF megakaryoblastic cell line and megakaryocytes derived from CD34+ mobilized peripheral blood stem and progenitor cells formed proplatelets on a tissue-culture treated surface, but not on a neutral, hydrophilic polymer that inhibits protein adsorption and cell adhesion. Importantly, CHRF cells cultured on the nonadhesive surface attained much higher mean ploidy by day 11 (8.6±0.5N vs. 5.8±0.6N, $P<0.005$) and maintained higher viability. Subsequent transfer to tissue-culture plastic yielded faster and more extensive proplatelet formation compared to cells initially cultured on tissue-culture plastic, with a maximal increase for day-5 transfer. Interestingly, acetylated alpha tubulin was abundant in the proplatelet extensions of CHRF cells on tissue-culture plastic, but was present at a minimal level in CHRF cells on the nonadhesive surface. We are currently exploring the role of tubulin acetylation, a hallmark of microtubule stability, in polyploidization and proplatelet formation.

(74e) The α2,3 Sialyltransferase ST3Gal-IV Regulates Human Leukocyte Binding to All Three Selectins: Distinction Between Mice and Men

Nandini Mondal[1], Alexander Buffone Jr.[2] and Sriram Neelamegham[1], (1)Chemical and Biological Engineering, State University of New York at buffalo, Buffalo, NY, (2)Chemical and Biological Engineering, State University of New York at Buffalo, Buffalo, NY

Introduction: Leukocyte adhesion to the inflamed vascular endothelium is initiated by the capture and rolling of cells on E- and P-selectin expressed on the endothelial cells. Here, sialo-fucosylated glycans, such as sialyl Lewis-X (sLe^X), expressed on leukocytes act as the selectin ligand. Such glycans are primarily synthesized in the Golgi by the action of glycosyltransferases, a family of enzymes that constitute ~1% of the human genome. The current paper applies RNAi strategy to determine the human α2,3 sialyltransferases responsible for the synthesis of P-, L- and E-selectin ligands on leukocytes. The data reveal that, in humans, ST3Gal-IV is the dominant enzyme responsible for the synthesis of selectin-ligands against all members of the selectin family.

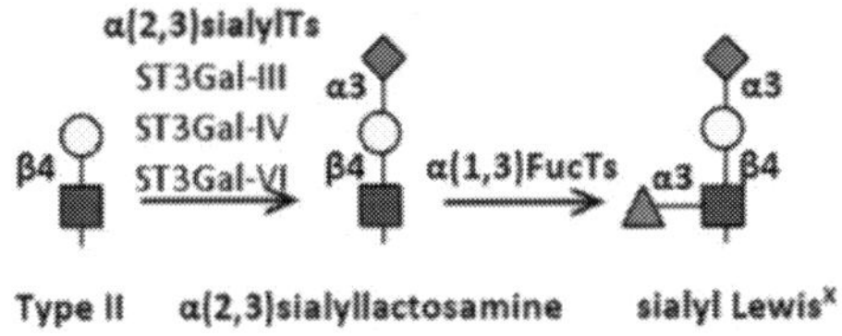

Figure 1. Sialyl Lewis X biosynthesis: Three α(2,3) sialyltransferases, ST3Gal-III, -IV and VI can act on Type II N-acetyllactosamine substrate. Action of α(1,3) fucosyltransferase on sialyllactosamine results in the sialyl Lewis-X antigen.

Materials and Methods: We studied all three leukocyte-specific, human α2,3 sialyltransferases, ST3Gal-III, -IV and -VI, that attach sialic acid residues to the N-Acetyl Lactosamine or LacNAc substrate (Galβ1,4GlcNAc) (**Figure 1**). ShRNA that suppress these enzyme expression by >85% were determined. Methods to introduce these constructs into hard-to-transfect leukocytes (HL-60 promyelocytic cells) were developed using lentivirus. RT-PCR confirmed gene silencing in leukocytes. A variety of fluorescent reporters and drug selection markers were introduced to ensure stable gene knock-down in these cells. Stable single (ST3Gal-III$^-$ HL-60, ST3Gal-IV$^-$ HL-60, ST3Gal-VI$^-$ HL-60) and dual (ST3Gal-III$^-$-IV$^-$ HL-60, and ST3Gal-IV$^-$-VI$^-$ HL-60) HL-60 knockdown cells were created. Glycan structure analysis on these cells was performed using antibody staining and flow cytometry. Microfluidic cell adhesion assays measured selectin dependent cell rolling function. Results were compared to primary neutrophils isolated from ST3Gal-IV$^{-/-}$ knock-out mice.

Results and Discussion: The ST3Gal knock-down cell lines displayed distinct cell surface glycans and rolling pheonotypes. Silencing ST3Gal-IV substantially reduced cell surface CLA/HECA-452 (cutaneous lymphocyte antigen) and sLe^X expression. ST3Gal-III has a modest role in reducing the CLA epitope while ST3Gal-VI upregulated this epitope. In this regard, while the CLA epitope is primarily found on leukocyte PSGL-1 (P-selectin glycoprotein ligand-1), sLe^X is more ubiquitously expressed on other molecular entities also. Thus, all enzymes ST3Gal-III, -IV and -VI appear to regulate the glycan structure on PSGL-1 and also other molecular entities. When these cells were perfused over either IL-1β stimulated HUVEC monolayers or cells bearing E-selectin (selectin associated with inflammation) or P-selectin (involved in platelet-neutrophil interactions) or L-selectin (involved in lymphocyte homing and leukocyte-leukocyte interaction), different enzymes were observed to contribute to different selectin-ligand interactions. E-selectin mediated binding was exclusively dependent on ST3Gal-IV. Silencing this gene abolished leukocyte rolling and adhesion on stimulated HUVEC and E-selectin bearing substrates. This observation is in marked contrast to ST3Gal-IV$^{-/-}$ mice where ST3Gal-IV only partially reduces leukocyte rolling. Human ST3Gal-IV also played a dominant role in P-selectin dependent adhesion, unlike mice where this gene had a partial effect. Human ST3Gal-IV and ST3Gal-VI in tandem regulated leukocyte interaction with substrates bearing L-selectin.

Conclusions: Knocking down ST3Gal-IV alone is sufficient to abolish E and P-Selectin mediated leukocyte recruitment in human systems. This is markedly different from murine systems where granulocytes from ST3GalIV$^{-/-}$ mice only display partial reduction in selectin-ligand binding activity. Overall, human ST3Gal-IV may be a viable target for anti-inflammatory therapy since it alters leukocyte interaction with endothelial cells and platelets.

Shear-Induced Resistance to Neutrophil Activation via the Formyl Peptide Receptor

Michael J. Mitchell[1] and Michael R. King[1]

[1]Department of Biomedical Engineering, Cornell University, Ithaca, NY 14853 USA

I. Introduction

The adhesion of leukocytes to the luminal surface of the microvasculature plays an important role in the inflammatory response and lymphocyte homing to lymphatic tissues (1). The initial step of the leukocyte adhesion cascade involves the capture and rolling of leukocytes on the receptor bearing endothelial cell layer, with L-selectin acting as an important mediator on the leukocyte surface. L-selectin is constitutively expressed on microvilli tips of neutrophils, and is rapidly cleaved from the neutrophil surface due to an inflammatory stimulus. Firm adhesion to the endothelium is mediated via ICAM-1 on endothelial cells binding to β_2 integrins on neutrophils. Stimulus by fMLP, which binds to the formyl peptide receptor (FPR) on neutrophils, induces $\alpha_M\beta_2$ integrin conformational changes and downregulation of L-selectin, which are key indicators of neutrophil activation (2).

FPR, a chemoattractant G-protein coupled receptor (GPCR), exhibits high constitutive activity in addition to activity due to agonist binding (3). Under static conditions, neutrophils spread their cytoplasm and migrate on glass substrates, in part due to constitutive GPCR activity. In the presence of fluid shear stress, neutrophils rapidly retract lamellipodia, assume a round resting state, and decrease GPCR constitutive activity (4). Neutrophils treated with pertussis toxin significantly attenuated the fluid shear stress-induced pseudopod retraction response, demonstrating the role of GPCR activity changes due to fluid shear stress. Transfection of cDNA for FPR into HL60 cells with low levels of FPR and low pseudopod activity led to the projection of pseudopods, which then retracted following exposure to fluid shear stress. FPR depletion via siRNA delivery in differentiated HL60 cells significantly reduced fluid shear stress-induced pseudopod retraction.

While the ability of activated leukocytes to retract pseudopods in response to fluid shear stress has been documented, a variety of different responses have been observed. Leukocytes treated with dexamethasone or after centrifugation have shown to reverse their shear stress response, and can project pseudopods when exposed to fluid shear stress (5). The effect of fluid shear stress on earlier indicators of neutrophil activation, such as L-selectin shedding and $\alpha_M\beta_2$ integrin activation induced by fMLP, has not yet been addressed. Here, we examined the quantitative dynamics of the shear stress-dependent response of fMLP-induced L-selectin shedding and $\alpha_M\beta_2$ integrin activation in neutrophils.

II. Results

A. Fluid shear stress reduces fMLP-induced L-selectin shedding and $\alpha_M\beta_2$ integrin activation in neutrophils

We initially studied the fluid shear stress response of neutrophils to fMLP-induced activation. Neutrophils were exposed to static conditions or 4.0 dyn/cm^2 of fluid shear stress in a cone-and-plate viscometer for 2 h at 23°C, followed by stimulation with 0.5 nM fMLP for 10 min. Neutrophils exposed to static conditions (Fig. 1A) or shear stress (Fig. 1D) in the absence of fMLP did not show appreciable differences in activation, as expected. However, neutrophils exposed to fMLP after fluid shear stress (Fig. 1E) showed a measurable reduction in L-selectin shedding and $\alpha_M\beta_2$ integrin activation compared to neutrophils exposed to static conditions (Fig. 1B) followed by fMLP. A significant reduction in fMLP-induced L-selectin shedding and $\alpha_M\beta_2$ integrin activation was found in sheared neutrophils compared to neutrophils under static conditions.

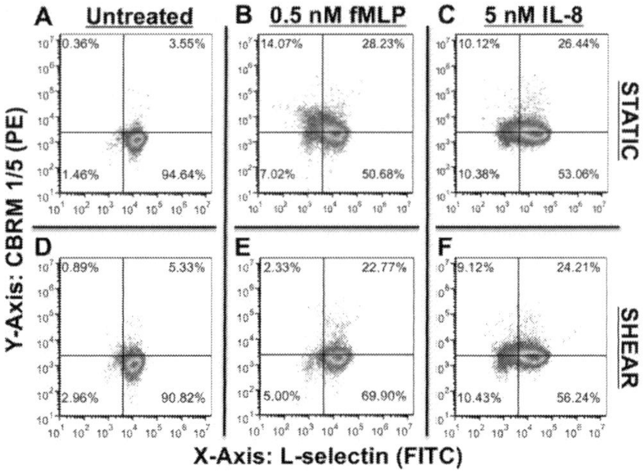

Figure 1: Neutrophils exposed to static conditions (A) and 4 dyn/cm^2 of shear (B) for 2 h. Neutrophils exposed to static conditions and 4 dyn/cm^2 for 2 h then exposed to 0.5 nM fMLP (C, D) or 5 nM IL-8 for 10 min (E, F). Upper two quadrants of plots represent CBRM1/5 staining. Two righthand quadrants represent L-selectin staining. Quadrants determined using isotype antibodies labeling nonspecific binding sites on neutrophils. PE: R-Phycoerythrin. FITC: fluorescein isothiocyanate.

B. Fluid shear stress does not affect IL-8-induced L-selectin shedding and $\alpha_M\beta_2$ integrin activation

To assess whether fluid shear stress alters activation via other major GPCRs, neutrophils were exposed to static conditions or fluid shear stress, followed by stimulation with 5 nM of IL-8, which binds to GPCRs CXCR1 and CXCR2. Neutrophils exposed to static conditions (Fig. 1C) or fluid shear stress (Fig. 1F) followed by IL-8 did not show significant differences in L-selectin shedding or $\alpha_M\beta_2$ integrin activation.

C. fMLP-induced L-selectin shedding and $\alpha_M\beta_2$ integrin activation is fluid shear stress dose-dependent

To study the effect of shear stress magnitude on fMLP-induced activation, neutrophils were exposed to shear stresses of 0.1-4.0 dyn/cm^2 over 2 h. At shear stresses of 0.10 and 0.25 dyn/cm^2, no significant differences in L-selectin shedding (Fig. 2A) or $\alpha_M\beta_2$ integrin activation (Fig. 2C) were observed between cells exposed to shear and static conditions followed by fMLP. However, a shear stress of 0.75 dyn/cm^2 yielded a significant reduction in L-selectin shedding (Fig. 2A) and $\alpha_M\beta_2$ integrin activation (Fig. 2C) after fMLP exposure. Shear stresses of 2.5 and 4.0 dyn/cm^2 showed an even greater reduction of neutrophil activation.

D. fMLP-induced L-selectin shedding and $\alpha_M\beta_2$ integrin activation is shear stress exposure time-dependent

To assess the kinetics of the resistance response, neutrophils were exposed to a shear stress of 4.0 dyn/cm^2 while increasing exposure time from 1-120 min. No significant difference in fMLP-induced L-selectin shedding in neutrophils was found over a 1-10 min shear stress exposure time (Fig. 2B). A significant decrease in fMLP-induced L-selectin shedding was found at a threshold exposure time of 20 min, and was significantly less than neutrophils in static conditions over a 20-120 min range. fMLP-induced $\alpha_M\beta_2$ integrin activation in showed no significant difference in neutrophils exposed to shear and static conditions over 1-30 min (Fig. 2D). A threshold value of 60 min was required to produce a significant difference in $\alpha_M\beta_2$ integrin activation in neutrophils.

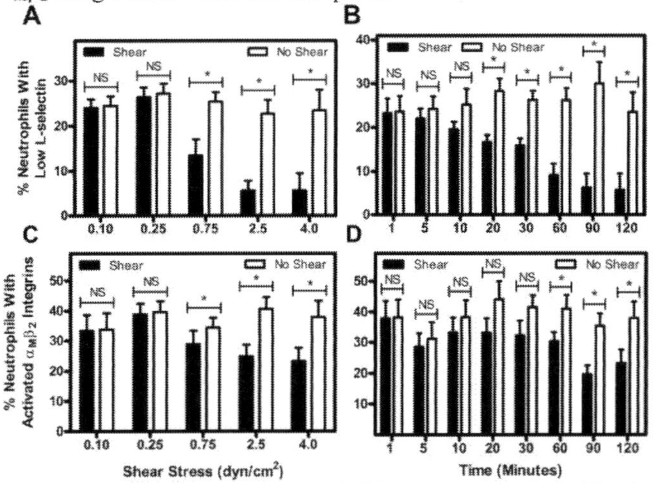

Figure 2: Increasing shear stress reduces fMLP-induced L-selectin shedding (*A*) and $\alpha_M\beta_2$ integrin activation (*C*). Shear stress varied in experiments from 0.1-4.0 dyn/cm² for 120 min, followed by 0.5 nM fMLP stimulation. Time dependence of resistance to fMLP-induced L-selectin shedding (*B*) and $\alpha_M\beta_2$ integrin activation (*D*) determined by increasing exposure time from 1-120 min at 4.0 dyn/cm². n = 5 donors. *P < 0.05.

E. Fluid shear stress reduces FPR surface expression

To investigate shear stress effects on FPR surface expression, neutrophils were exposed to shear stress and static conditions and immediately labeled with anti-FPR antibodies for flow cytometry analysis. Sheared neutrophils displayed a reduction in FPR expression (Fig. 3A) compared to nonsheared samples. A significant difference in FPR receptor count was revealed, as sheared neutrophils averaged 14,600 receptors/cell, while nonsheared samples averaged 20,100 receptors/cell (Fig. 3B). IL-8 receptor CXCR1 and CXCR2 receptor counts showed no significant differences due to fluid shear stress (Fig. 3A,B).

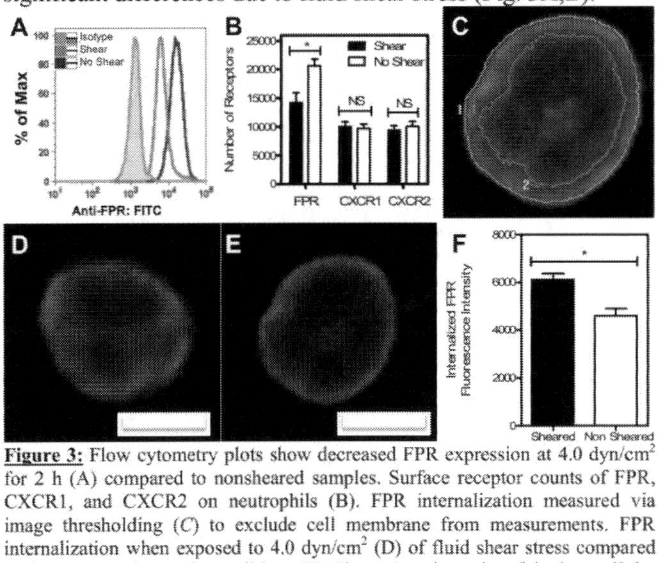

Figure 3: Flow cytometry plots show decreased FPR expression at 4.0 dyn/cm² for 2 h (A) compared to nonsheared samples. Surface receptor counts of FPR, CXCR1, and CXCR2 on neutrophils (B). FPR internalization measured via image thresholding (*C*) to exclude cell membrane from measurements. FPR internalization when exposed to 4.0 dyn/cm² (D) of fluid shear stress compared to those exposed to static conditions (E). Fluorescent intensity of the intracellular region (*F*) was quantified. Scale bars = 5 µm. n = 3 donors. *P < 0.05.

F. Neutrophils undergo FPR internalization under fluid shear

To examine FPR internalization as a cause for surface expression decrease following shear stress exposure, neutrophils were exposed to fluid shear stress or static conditions for 2 h, permeabilized and labeled with FPR antibodies for examination via confocal microscopy. Images were thresholded (Fig. 3C) to exclude the cell membrane from fluorescence measurements. Immunostaining revealed FPR to be clearly localized within the cell in sheared samples (Fig. 3D), while minimal FPR was shown within neutrophils exposed to static conditions (Fig. 3E). Average pixel intensities of the intracellular region showed a significant increase in fluorescence intensity in sheared neutrophils, compared to neutrophils exposed to static conditions (Fig. 3F).

III. Conclusion

Results from this study suggest that fluid shear stress has a significant effect on the activation of circulating neutrophils. Neutrophils acquired a fluid shear stress-induced resistance to activation via FPR. Resistance was shown to be dependent on shear stress magnitude, as the response increased with increasing shear stress. The mechanical response was also dependent on shear stress duration, as neutrophils increased resistance with increased shear exposure time. A decrease in FPR surface expression was observed upon exposure shear stress, and high-resolution confocal microscopy revealed FPR internalized within cells. While other studies on mechanotransduction in neutrophils mostly focused on morphological changes, the present study focused on earlier indicators of activation, specifically fMLP-induced L-selectin shedding and $\alpha_M\beta_2$ integrin activation. The complete signaling pathways of these receptors deserve further study, along with molecules that mediate GPCR internalization. Other receptors that display high constitutive activity should be investigated to understand their contributions to the mechanosensing response of cells within the vascular microenvironment.

References

1. Springer, T.A. 1994. *Cell.* 76: 301-314.
2. Fan, H. et al. 1999. *EMBO J.* 18: 6962-6972.
3. Wenzel-Seifert, K. et al. 1998. *J. Biol. Chem.* 273: 24181-24189.
4. Moazzam, F. et al. 1997. *PNAS.* 94: 5338-5343.
5. Fukuda, S., et al. 2002. *J. Leukoc. Biol.* 72: 133-139.

(74g) Dynamic Adhesion Assays for Understanding the Significance of CD151 in the Metastatic Cascade

Jennifer Fischer[1], Adrianne Shearer[1], Xiuwei Yang[2], Christine Trinkle[3], Richard Eitel[1] and Kimberly W. Anderson[1], (1)Chemical and Materials Engineering, University of Kentucky, Lexington, KY, (2)Molecular and Biomedical Pharmacology, University of Kentucky, Lexington, KY, (3)Mechanical Engineering, University of Kentucky, Lexington, KY

Cancer is the second leading cause of death in the United States, with metastatic lesions resulting from circulating tumor cells (CTCs) most often causing the mortality. The basic steps of the metastatic cascade, from initial tumor formation to growth at a secondary site, are a linear sequence of steps composed of a network of mechanisms that are not well understood. One of these steps, the adhesion of cancer cells to endothelial cells during dissemination, is the focus of this research. Specifically, the influence of CD151, a gene that encodes a cell surface glycoprotein that is known to complex with integrins and enhance cancer cell motility, invasion and metastasis was studied. To study the influence of CD151, a prostate cancer cell line (PC3) and a breast cancer cell line (sum149) were employed in dynamic adhesion assays in which the expression of CD151 was knocked down on the cancer cells using shRNA gene silencing. Both the knocked down and unmodified lines were utilized to quantify adhesive strength and frequency of adherence to endothelial monolayers originating from different microvessels. The microvessels compared were veins and arteries, which were simulated using Human Umbilical Vein Endothelial Cell (HUVEC) monolayers and Human Umbilical Arterial Endothelial Cell (HUAEC) monolayers. A parallel plate flow chamber system was utilized to investigate these properties under flow at various shear stresses ranging from 0.25 dynes/cm^2 to 15 dynes/cm^2 to cover the range of most physiologically relevant shears. Preliminary results indicate that the presence of CD151 leads to enhanced initial adhesion of prostate cancer cells to HUVECs, but retention rates were unaffected by CD151 expression once those cells became firmly adhered. Results from PC3 adhesion to HUAECs as well as from the sum149 adhesion assays will also be presented.

(74h) Distinct Glycosylations On the Hcellv Isoform of CD44 Mediate Breast Cancer Cell Adhesion to E-Selectin

Tiantian Liu, Venktesh S. Shirure and **Monica M. Burdick**, Department of Chemical and Biomolecular Engineering, Ohio University, Athens, OH

Metastasis is the culmination of an elaborate cascade of events in which cancer cells separate from the primary mass, intravasate through blood vessel walls into the vasculature, traverse throughout the body, and finally seed new colonies in distant organs. E-selectin expressed by vascular endothelium plays a significant role in the cascade, mediating the tethering and rolling of circulating tumor cells (CTCs) that initiate the recruitment process in the secondary organ. We have previously shown that sialylated glycolipids on breast cancer cells are E-selectin ligands. Glycoprotein E-selectin ligands are also expressed by breast cancer cells, but it is not known whether O-linked and/or N-linked glycans confer ligand activity. Therefore, we sought to define the type(s) of glycosylation(s) leading to E-selectin ligand function of cell surface proteins expressed by breast cancer cells. When BT-20 breast cancer cells were perfused over IL-1β-stimulated human umbilical vein endothelial cells (HUVECs) at 0.8 dyn/cm^2 in the parallel plate flow chamber adhesion assay, the breast cancer cells attached in an endothelial E-selectin-dependent manner. Pretreatment of the BT-20 cells with Vibrio cholerae neuraminidase, which enzymatically cleaves terminal sialic acid residues, essentially abrogated attachment to HUVECs, consistent with the sialylation requirement of E-selectin ligands. To more specifically characterize the glycans expressed by glycoprotein E-selectin ligands, BT-20 cells were treated with deoxymannojirimycin (DMJ) to prevent N-linked glycosylation of glycoproteins, whereas cells were treated with benzyl-N-acetyl-α-galactosaminide (Bzl-GalNAc) to prevent O-linked glycosylation of glycoproteins. Immunoprecipitation, Western blotting, flow cytometry, and parallel plate flow chamber adhesion assays revealed that an N-glycosylated high molecular weight variant of CD44 (~150 kDa; CD44v) was a major contributor to E-selectin-mediated adhesion. O-glycans also contributed to E-selectin ligand activity of CD44v (~150 kDa), albeit to a lesser extent than N-glycans. However, no E-selectin ligand activity was attributed to CD44s (~100 kDa). The breast cancer glyco-isoforms of CD44 possessing E-selectin ligand activity, termed HCELLv, thereby demonstrate similarities to both N-glycosylated HCELLs and HCELLv predominantly expressed by normal hematopoietic stem cells and leukemic cells, as well as O-glycosylated HCELLv expressed by colon cancer cells. With increased understanding of the role that glycosylation plays in E-selectin ligand function of breast cancer cells, vital improvements in the diagnosis, prognosis, and treatment of breast cancer can be made.

(75a) Using Atmospheric Carbon Dioxide Environments to Improve Control of Stem Cell Expansions

Sarah W. Harcum, Department of Bioengineering, Clemson University, Clemson, SC and Arthur Nathan Brodsky, Bioengineering, Clemson University, Simpsonville, SC

Mesenchymal stem cells (MSCs) are found in many tissues in adult mammals, including but not limited to bone marrow. MSCs replenish lost cells of many different lineages throughout the life of an adult organism as well as possess the ability to modulate the immune system. Among other applications, MSCs have already been implemented to prevent graft-versus-host-disease in humans and in mice have improved myocardial compliance after infarction. As the potential applications of MSCs increase, improving expansion practices will be crucial. Already, fed-batch bioreactors allow volumetric scale-up and enable better control of mammalian cell cultures to improve batch-to-batch reproducibility; however, all stem cell medias currently utilize the bicarbonate pH buffering system, which requires elevated carbon dioxide levels that must be supplied from the gas phase along with crucial oxygen. While the ability to dynamically alter culture parameters is beneficial, the challenge of ensuring sufficient mixing of all these components can result in cyclic fluctuations in pH and dissolved oxygen that can temporarily upset culture conditions. The cells can be stressed in the time it takes for the system to rebalance itself, which can be especially detrimental to stem cells, which rely on precise culture conditions to ensure generation and expansion of a desired phenotype. Eliminating carbon dioxide as a requirement in pH control can simplify bioreactor control and minimize these disruptive shocks and improve the consistency with which stem cells can be expanded for both therapeutic and research applications.

This current research developed the first media for MSCs in atmospheric carbon dioxide conditions. A MOPS-HEPES buffered media enabled MSCs to achieve equivalent growth rates to MSCs grown in the standard, bicarbonate-buffered media in 5% carbon dioxide. All cell densities and growth rates between the two conditions were determined to be not significantly different after performing a t-test ($p \leq 0.05$). After accomplishing equivalent growth kinetics, MSCs grown in each of the two conditions were differentiated along the osteogenic, adipogenic, and chondrogenic lineages. Imaging helped visualize differentiation and through staining of lineage-specific markers, we confirmed that MSCs grown in each condition successfully differentiated and also that undifferentiated MSCs were negative for differentiated markers. Real time qrt-PCR analysis confirmed gene expression levels for both the undifferentiated and differentiated MSCs were not significantly different between the custom and standard conditions ($p \leq 0.05$). This study demonstrated the proof of principle that MSCs can be expanded effectively in atmospheric carbon dioxide. These results are encouraging, but further study is needed to confirm these results for suspension and large-scale stem cell expansions. In addition, serum-free medias that do not rely on the bicarbonate-buffering system will have to be developed before this method can be used to expand MSCs for use in human clinical treatments. By eliminating the requirement of carbon dioxide sparging, bioreactor control can be simplified and allow more consistent culture conditions to be maintained. This will grant the potential to improve yields and batch-to-batch reproducibility of stem cells, which rely on the ability to expand stem cells and ensure emergence of a precise phenotype, as well as other biotechnology-relevant cell lines such as Chinese hamster ovary (CHO) and NS0 cells.

Presentation Title: Characterizing formation and mitigation of precipitates in mammalian cell culture media during heat treatment for viral inactivation

Presenter: Prince Bhebe

Conference: American Institute of Chemical Engineers (AIChE) Annual Meeting

Venue: Pittsburgh Convention Center, Pittsburgh, PA

Dates: 10/28 – 11/2/2012

Abstract

In mammalian cell culture production processes it is essential to maintain a sterile envelop around the bioreactor system against adventitious agents. Heat treatment methods for inactivating viruses in cell culture media have been widely adopted in the biotechnology industry. The presence of salts of inverse solubility in media presents a precipitation challenge during heat inactivation. In this study we characterized the formation of inverse solubility phosphate salts during heating by differential scanning calorimetry (DSC) and the real time, physical behavior of the particles using focused beam reflectance measurements (FBRM) and particle vision microscopy (PVM). We also identified a temperature zone of 85 - 90°C where the onset of precipitation occurs. We also defined low pH conditions of ≤ 6.65 which suppress the formation of phosphate salts in chemically defined platform media. Removal of cationic or anionic species of the precipitating salts in media has also been demonstrated to be effective tools in mitigating precipitation.

(75c) Investigation of Process Parameters and Their Effect On Cell Metabolism and N-Linked Glycosylation

Marija Sarenac, Miroslav Soos and Massimo Morbidelli, Institute of Chemical and Bioengineering, ETH, Zurich, Switzerland

The efficacy of monoclonal antibodies arises from their specificity to the target antigen and their ability to trigger immune responses by activating effector functions. This is critically dependent on appropriate posttranslational modification of the antibodies, e.g. glycosylation. It has been shown that glycosylation can dramatically influence the efficacy of mAbs by either reducing the serum half-life or by altering the triggered effector mechanism.

In this study we systematically investigate operating conditions in bioreactor systems (pH, osmolarity, dissolved gases, hydrodynamic forces) in a range that covers the heterogeneous conditions observed in large-scale reactors. By performing parameter shifting experiments we investigated the effect of those external variables on cell metabolism and the glycosylation profile of the mAb. Our goal is to further understand the intracellular trigger of such alternations in order to be able to predict metabolic switches during cultivation as well as changes in product quality. In particular, we observe altered lactate and amino acid metabolism induced by process operating conditions. To further understand the effect of external factors on product quality we investigate the glycosylation profile of our model antibody at various cultivation conditions. Especially for pH and osmolarity changes, an influence on glycosylation towards less sialylated forms could be observed.

Finally, the knowledge gained will be used to develop a multi-scale model that is able to describe the effect of operating parameters on metabolism and glycosylation. On the way to targeted glycosylation such a model will help to better understand the connection of product quality attributes to process parameters.

(75d) Application of Multivariate Analysis Tools and Design of Experiments (DoE) to Model the Design Space for Characterization of a Mammalian Cell Culture Process

Vijayakumar Janakiraman[1], Benjamin Orbon[1], Yao-Ming Huang[1], Barry Wolf[2], Maartje Verschuur[2], Amit Varma[3], Marty Sinacore[4] and Thomas Ryll[4], (1)Cell Culture Development, Biogen Idec, Research Triangle Park, NC, (2)Cell Culture Development, Abbott Biotherapeutics Corp., Redwood City, CA, (3)Abbott Biotherapeutics Corp., Redwood City, CA, (4)Cell Culture Development, Biogen Idec, Cambridge, MA

This work will cover the design space characterization of a mammalian cell culture process used for the production of a humanized monoclonal antibody. In any late stage process characterization, the first step is to develop a valid scale-down model, which will ensure the results from small scale testing are representative of manufacturing scale. The first part of this presentation will focus on developing and qualifying a scale-down model using statistical multivariate analysis tools. The selection of an appropriate scale-down parameter, such as k_La or power per unit volume, is important and scenarios will be presented. Furthermore, experimental strategies to model the interrelationship between culture pH, lactate and dissolved CO_2 will be discussed. In the second half, the application of design of experiments (DoE) strategy to fully characterize the cell culture design space will be presented. A fractional factorial central composite design was used to minimize the number of experiments within an aggressive timeline. A response surface model describing the impact of pH, dissolved oxygen, temperature and seed density on growth, titer and critical quality attributes (CQAs) was developed. The process optimum was identified and verified, and the model had good predictability over the entire design space. Based on the outcome of these studies, critical controlled parameters, and critical in-process tests and controls were identified, as well as appropriate action limits set. Collectively, this formed our process control strategy to ensure consistent production of a product that would meet its pre-determined specifications and quality attributes. The modeling and experimental strategies discussed here will have potential application for any stage of process development.

(75e) Facing Heterogeneities in Bioprocesses, a Downscale Approach Tailored to Suit Large Scale Conditions

Benjamin Neunstoecklin, Miroslav Soos and Massimo Morbidelli, Institute of Chemical and Bioengineering, ETH, Zurich, Switzerland

Nowadays, a majority of mammalian cell cultivations is performed in stirred bioreactors of various sizes. While heterogeneity of the hydrodynamic stress, covering several orders of magnitude, is well documented with increasing bioreactor size, additional heterogeneities such as DO and CO2 are becoming more important. Furthermore, perturbations in pH can be observed when pH is controlled via acid/base addition. Commonly in the process development stage, scale dependent parameters are usually neglected, often resulting in a different behavior of the culture when transferred to large scale production bioreactors with volumes of several thousand liters.

The aim of this work is to incorporate the variation of scale dependent variables, e.g. shear stress, DO and pH, into a typical 3L development scale bioreactor. A trajectory analysis performed in a validated CFD model was used to evaluate the frequencies, maximum values and exposure times of hydrodynamic stress, DO and pH to which cells are exposed along their journey through a large scale bioreactor. Obtained values were used to design an external loop, driven by a centrifugal pump, attached to a 3L bioreactor, enabling us to independently vary shear frequency, shear value and shear exposure time. After proper testing and characterization, further development was done by considering the heterogeneity of DO and pH perturbations commonly seen in large scale bioreactors. Comparison of cell cultivations in such scale-down model showed better agreement with large scale data compared to those obtained from conventional 3L bioreactors.

(75f) Microfluidic Scale-Down of Upstream Biopharmaceutical Production

Shireen Goh[1], Michelangelo Canzoneri[2], Horst Blum[2], Rajeev J. Ram[3] and Anthony J. Sinskey[4], (1)Department of Materials Science and Engineering, Massachusetts Institute of Technology , Cambridge, MA, (2)Sanofi-Aventis Deutschland GmbH, (3)Electrical Engineering & Computer Science, Massachusetts Institute of Technology, Cambridge, MA, (4)Biology and Health Sciences and Technology, MIT, Cambridge, MA

The production of therapeutic proteins is the fastest growing segment of the pharmaceutical industry. The post-translational addition of carbohydrate groups to the protein product has been shown to alter the efficacy, immunogenicity, clearance time in the body, and solubility of therapeutic proteins. Chinese Hamster Ovary (CHO) cells represent the preferred expression system for biopharmaceutical production because of the cell's inherent ability to perform post-translational modification within the secrationary pathway as the protein is transported in the endoplasmic reticulum (ER). As a result, approximately 70% of biopharmaceuticals in the market are produced in CHO. Glycosylation of proteins is sensitive to pH, temperature, dissolved oxygen (DO), dissolved CO_2, shear, and specific metabolite concentrations. Shear sensitivity (due to morphological deformation of the ER) is what distinguishes CHO cell culture from microbial expression systems. Shear stresses as low as 0.10 Pa have been shown to suppress productivity by 34.2% and smaller variations of shear have been shown to alter the glycosylation of proteins at an energy dissipation rate (EDR) of 6.0×10^4 W/m^3. Microscale mammalian culture has been explored previously, however either small volumes or laborious sampling have greatly limited in-line and off-line measurement techniques. In Amanullah et.al., 700uL volumes are insufficient for both in-line characterization of the process (e.g. metabolite profiling [RX Daytona]) or product characterization (e.g.glycosylation profiling [HPAEC-PAD]).

The device shown in Figure 1 contains a 2mL micro-bioreactor for batch and fed-batch culture of CHO cells with online sensors to monitor pH, dissolved oxygen (DO), dissolved carbon dioxide (DCO_2), optical density (OD) and temperature in real time. The pH is controlled by varying the partial pressure of carbon dioxide in the gas headspace during the early stages of the culture and by injecting liquid base during the latter part of the culture. The maximum shear stress experienced by cells has been modeled and the effects of shear stress on the growth and productivity of the CHO cells are experimentally measured for different mixing rates. The gas transfer rate, k_La, is measured for oxygen (O_2) and carbon dioxide (CO_2) gas, which can be independently controlled and adjusted till the k_La for both O_2 and CO_2 are similar to that of large scale bioreactors. A batch culture of CHO (Invitrogen CHO-S) cells is grown in the micro-bioreactor and the growth rate and productivity of cells grown in the micro-bioreactor are compared with shake flask cultures. Periodic samplings for offline measurements are performed to validate the online sensors and track the osmolarity increase of the culture medium. The cell viability and growth rate are monitored throughout the culture.

ENHANCEMENT OF CELL ADHESION AND PROLIFERATION THROUGH SURFACE ROUGHNESS MODIFICATION IN CHO CELL CULTURES IN PMMA-PDMS MICRODEVICES

Lucía D. Garza-García[1], Eduardo J. Tapia-Mejía[2], Erika García-López[3], Sergio Camacho-León, Ciro Ángel Rodríguez-González [3], Leydi-Maribel Carrillo-Cocom and Mario Moisés Álvarez[1,]*

[1]Centro de Biotecnología-FEMSA, [2]Departamento de Ingeniería Eléctrica y Computacional, and [3]Centro de Innovación en Diseño y Tecnología, Tecnológico de Monterrey at Monterrey, Monterrey, México

ABSTRACT: Recently, the production of therapeutic proteins, particularly monoclonal antibodies, has grown significantly in the biopharmaceutical market. Chinese Hamster Ovary cells (CHO cells) are considered to be the standard mammalian cell host for the production of recombinant proteins. At the lab scale, this cell line is usually grown in bioreactors and Erlenmeyer flasks, where the culture medium has to be changed periodically. Culture of CHO cells in micro-devices could be a cost-effective strategy for screening and eventually commercial production of therapeutic proteins.

We document the continuous perfusion culture of a normally non-adherent CHO cell line, producer of a monoclonal antibody (mAb), in zig-zag shaped PMMA-PDMS micro-devices. Cell morphology, cell production, and mAb productivity of this CHO cell line were evaluated in PMMA-PDMS micro-devices. Surface Micro-devices surfaces were textured with 15 μm diameter pores. A textured surface was key to promote cell anchorage, since a non-adherent cell line was employed. Cell adhesion, cell proliferation, and mAb production over time were evaluated under different operative conditions: (a) two different flow rates (0.1 $\mu l\ min^{-1}$ or 3 $\mu l\ min^{-1}$), and (b) with or without fibronectin coating in the micro-device surface. Computational Fluid Dynamics (CFD) was used to characterize the velocity field within the different micro-device designs.

For cell production and mAb productivity, the best results were achieved when applying fibronectin coating at low flow rate. In the absence of fibronectin coating, a considerable amount of cells were detached from surface during perfusion operation. Remarkably, under the best set of operative conditions, mAb productivity was comparable to that measured in stirred tank suspended culture (1.34 $mg\ L^{-1}\ h^{-1}$). Results demonstrated that non-adherent cell lines can be successfully culture on textured surfaces coated with fibronectin under a continuous perfusion system. Our findings suggest that continuous perfusion systems with micro-devices can be used for the production of high added value biopharmaceuticals, as well as for screening purposes.

KEYWORDS: CHO cells, microdevices, PMMA, PDMS, monoclonal antibody, perfusion

* Ave. Eugenio Garza Sada 2501 Sur Col. Tecnológico C.P. 64849, Monterrey, Nuevo León, México, +52 (81) 8358-2000, +52 (81) 8328-4136, mario.alvarez@itesm.mx.

1 INTRODUCTION

Today, the trend in technology is toward miniaturization: use of microscale systems or microdevices. Microdevices offer several benefits, including shorter response time; high surface/volume ratio (with important implications for transfer phenomena); portability; ease of storage and use; reduced sample and reagent consumption, etc. In the area of diagnostics and biosensing, these "lab on a chip" technologies have revolutionized research and will soon massively impact the market. However, in the area of biopharmaceutical research, development, and production, only an incipient body of publications exists pertaining to the use of microdevices. In general, reports of cell culture experiments in microdevices, particularly, continuous flow microdevices, are still scarce.

In this work, we explore the specific application of microdevices built on PMMA (polymethyl methacrylate) and PDMS (polydimethyl siloxane) for the production of high value recombinant therapeutic proteins (i.e., monoclonal antibodies) in Chinese Hamster Ovary (CHO) cell cultures. In the last decade, therapeutic protein production has grown significantly in the pharmaceutical market, and CHO cells are considered the standard mammalian cell host for the production of recombinant proteins of pharmaceutical interest. In 2004, the value of biopharmaceutical market was estimated at $33 billion dollars and is projected to be at $70 billion dollars by 2010 [1].

When designing microdevices for cell culture applications, the choice of construction materials is an important consideration. In the present study, we developed microdevices built from PDMS and PMMA, which are two of the most commonly used materials for microdevices. Both PDMS and PMMA are non-toxic to cells, and are less expensive and more manageable materials for microdevice construction. Cellular systems grown on PDMS/PMMA surfaces exhibit high viability and other relevant cell growth characteristics that are comparable to those observed in their counterparts grown on polystyrene [2].

Another important consideration in the fabrication of microdevices for cell growth is the presence of microscale patterns and features on the surface, which promote cell adhesion and growth or that facilitate proper cell or protein patterning, among other objectives [3]. For example, Folch and co-workers generated a microfluidic perfusion system, where cell-adhesive and cell-repellent micropatterns were created in a microchannel network to control the proliferation and transformation of C2C12 myoblasts into myotubes [4]. Takayama and Dixon used a corona discharge treatment as an inexpensive and simple method for creating a linear protein pattern in a PDMS microchannel and found that C2C12 myoblasts adhered only to the patterned region during 21 days of culture [5].

Some applications related to the analysis of cell behaviour (i.e., biosensing) require a more specific control and finer manipulation of cell morphology and spatial location. Recent studies demonstrate that cell morphology is significantly influenced by the microscale landscape of cell culture system. In 2010, Hansford and his research group created a vacuum-assisted cell seeding system for patterning adherent (NIH 3T3), non.adherent (THP-1), and aggregate forming cells (PANC-1) into circular and rectangular geometries in microwells. They found that cell morphology was defined by the microwell geometry [6]. In another investigation, Moghe and Ranucci demonstrated that HepG2 cells cultured in a microtextured PGLA ((poly)-glycolic-co-lactic acid) exhibited stronger adhesion (200% higher) than did cells seeded on a non-textured surface [7]. Consistently, when microtextured surface was coated with collagen, adhesion strength increased in 420% [7]. These reports demonstrate the advantages of microtextured surfaces for enhanced cell anchorage, even when employing non-adherent cells. In the present paper, we evaluated cell attachment, cell proliferation, and mAb productivity of a non-adherent CHO cell line using PMMA-PDMS microdevices consisting of a "zigzag" microchannel. Cell culture performance was compared in

* Ave. Eugenio Garza Sada 2501 Sur Col. Tecnológico C.P. 64849, Monterrey, Nuevo León, México, +52 (81) 8358-2000, +52 (81) 8328-4136, mario.alvarez@itesm.mx.

microdevices with (a) a smooth or textured surface, and (b) in the presence or absence of an extracellular ligand (attachment factor).

2 MATERIALS AND METHODS

2.1 General experimental Strategy

The attachment, cell proliferation, and production of an anti-TNF mAb by CHO cells were evaluated in microdevice systems made of PMMA and PDMS. Two different factors were assayed to evaluate cell attachment: the presence or absence of texture in the surface of microchannels (specifically, the roughness caused by the presence of 15 μm diameter pores); and the presence or absence of a surface treatment with Fibronectin (FN). Consequently, cell attachment under four different conditions was evaluated: (a) in smooth surface channels without FN treatment; (b) in smooth surface channels with FN treatment; (c) in rough surfaced channels without FN treatment; and (d) in rough surfaced channels with FN treatment.

2.2 Microfluidic device fabrication

Microfluidic devices with a "zigzag" channel design were fabricated using PMMA and PDMS. Geometric details are presented in Figure 1a. The lower portion of each microdevice (i.e., the bottom), where the channel was imprinted, was fabricated from a rectangular PMMA wafer 6 mm wide, using a micromilling manufacturing center machine MAKINO F3 with a cutting diameter of 400 μm. Devices were constructed with either smooth or rough (i.e., with 15 μm diameter pores) surfaces. The surface roughness was mainly defined during the pre-cooling process with carbon dioxide. Each "zigzag" microchannel was 44 mm long, 0.5 mm wide, and 0.1 mm deep, with a 1 mm diameter reservoir (Figure 1b,c).

The upper section of each microdevice (i.e., the lid) was built from PDMS (Sylgard, 184, Dow Corning). PDMS and curing agent were mixed to obtain a prepolymer with a 10:1 weight ratio. The resulting mixture was degassed under vacuum conditions for 20 min and poured in a mask. The PDMS was cured at 120°C for 20 min and then peeled from the mask to obtain a 5 mm PDMS wafer. Reservoirs were drilled into the PDMS surface with a sharpened blunt-tip needle.

PDMS and PMMA surfaces were treated with a handheld corona treater (Electro-Technic-Products Inc., model BD20A) for 15 sec to achieve an efficient bonding between both surfaces and to convert their nature from hydrophilic to hydrophobic. In addition, PDMS and PMMA layers were sealed together using liquid PDMS around the borders and then thermally bonded by incubation in an oven at 120°C for 20 min.

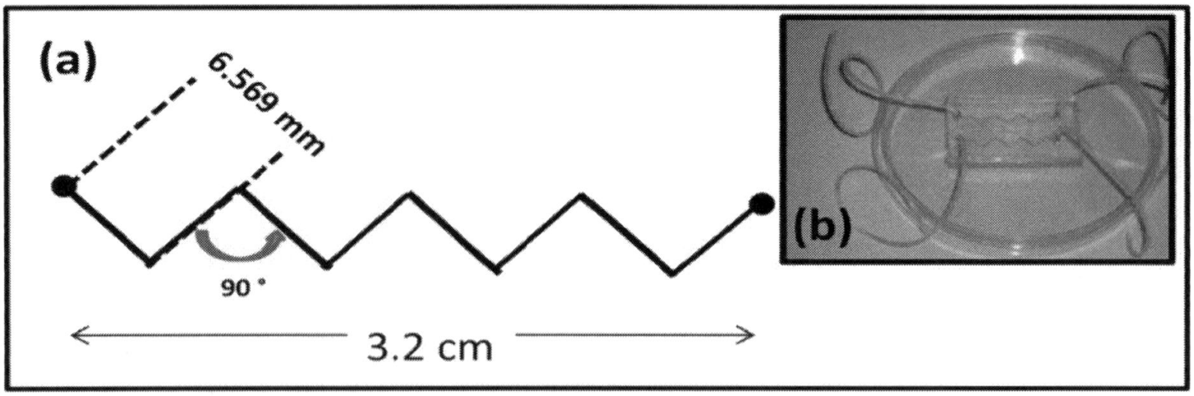

Figure 1: a) Schematic representation of the zig-zag microchannel, and b) Actual microdevice image.

Subsequently, tubes were inserted into the reservoirs and glued in place with liquid PDMS. As a final manufacturing step, microdevices with their corresponding tubes were also thermally bonded at least 24 hours before use.

2.3 Cell culture

The CHO cells, producers of a recombinant monoclonal antibody (mAb), were cultured in a medium mixture containing CD OPTICHO (Invitrogen, USA) and CHO CD EfficientFeed™ B (Invitrogen, USA) supplemented with 200 mM L-glutamine (Invitrogen, USA). Non-adherent CHO cultures were maintained in a 50 ml plastic Erlenmeyer shake flask (Corning, USA) in an incubator with a humidified environment at 33°C in a 8% CO_2 atmosphere. Cells were passaged when 95% confluent, and for this investigation, passages 3 and 4 were used. Cells were counted using a hemocytometer.

2.4 Experimental set-up

Microdevices were sterilized for 30 min using 70% ethanol and UV light. Each microdevice was inoculated with a viable density of 1×10^6 cells mL^{-1}. Cell inoculation was performed inside an air purifier. Briefly, a disposable plastic syringe was filled with phosphate-buffered solution (PBS). A 300 µl volume of PBS was introduced into the microchannel with a Harvard 33 twin syringe pump (Harvard Instruments, Inc.). Subsequently, In the experiments where FN was used, 100 µl of FN (Invitrogen, Carlsbad, CA, USA) was injected into the microchannel and incubated at 37°C in a humidified atmosphere for 1 h. To remove excess FN from the microchannel, 300 µl of PBS were injected into the microdevice. Prior to cell inoculation CD OPTICHO medium was infused into all microdevices. Lastly, 1×10^6 cells mL^{-1} were introduced in the microchannel at a flow rate of 5 µl min^{-1}.

During the first stage of culture, microdevices were incubated at static conditions for 96 h, inside an incubator with a humidified environment at 33°C in an 8% CO_2 atmosphere. The extent of cell proliferation was monitored photographically. Pictures were taken every day with a Carl Zeiss 200 inverted fluorescence video microscope equipped with a 20X microscope objective.

After static culture for 96 h, and to test proper cell attachment and mAb production under perfusion, microdevices were submitted to continuous flow conditions. Medium mixture (CD OPTICHO and CHO CD EfficientFeed™ B) was injected continuously for 4 h at a flow rate of 5 µl min^{-1} using a syringe pump system installed inside an incubator. Samples (1.0 ml) were collected on day three (during static culture) and on day 5 (during continuous culture) for determination of mAb concentrations using an enzyme-linked immunosorbent assay (ELISA).

3 RESULTS AND DISCUSSION

3.1 Rationale for microdevice design

Preliminary experiments using different channel geometries (not reported here) indicated that, after inoculation, cells preferentially attached to corners and wall surfaces. The "zigzag" design used here as a model presents multiple "corners" along a relatively short distance, thereby maximizing the number of locations where cells can locate during inoculation.

We determined the roughness pattern to use in this study by considering the typical shape and size of a CHO cell. When suspended in culture, CHO cells typically exhibit a spherical shape with an average diameter of approximately 15 µm (range from 5 to 20 µm). Therefore, a surface with 15 µm diameter micropores will offer sites for deposition/attachment of individual cells. Conceptually, these single cells will later develop colonies to achieve a confluent culture. Given the flow rate used in our experiments, and the effective volume in the microchannel, the residence time within the microdevice is approximately 0.44 min.

3.2 Cell adhesion and proliferation

The CHO cells used in the experiments reported here were adapted to grow in suspension. Effective operation of a perfusion system requires that proper cell attachment be maximized. In this study, cell attachment resulting from four different attachment strategies was evaluated: (a) use of smooth surface channels without FN treatment (negative control); (b) use of smooth surface channels with FN treatment; (c) use of rough surfaces (with 15 μm diameter micropores) without FN treatment; and (d) use of rough surfaces with FN treatment.

As expected, negligible cell attachment was observed in systems with smooth surfaces without FN treatment. When perfusion was initiated (after 96 h of static conditions) almost all of the cells within the channel were washed out by continuous flow (Figure 2a). Similarly, in systems with a smooth surface but coated with FN, cell attachment was poor and most of the cells were washed out once perfusion was activated (Figure 2b).

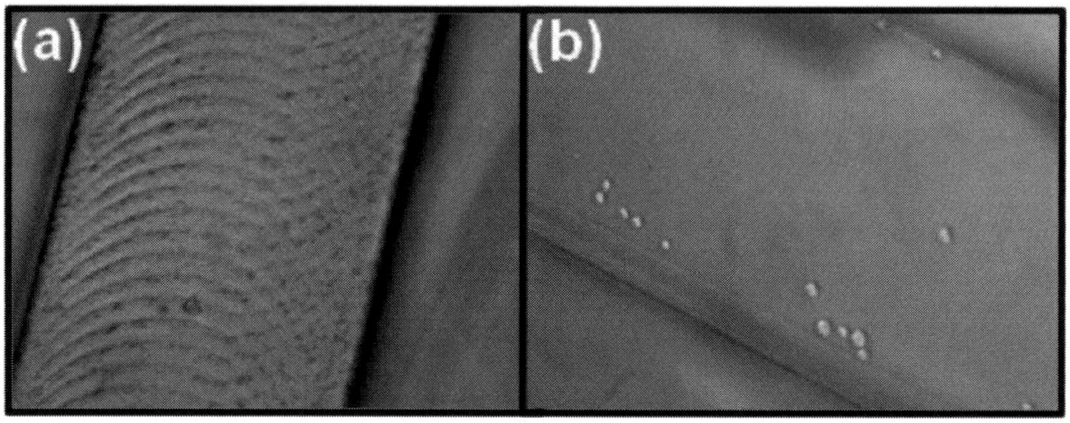

Figure 2: Cell attachment/proliferation resulting from different attachment enhancement treatments in "zigzag" channel microdevices. Negligible cell attachment was observed in channels with **a)** a smooth surface and no FN coating (images taken after perfusion operation; and **b)** a smooth surface with FN coating (images taken after perfusion).

Our results suggest that surface roughness has a dramatic effect on cell attachment (Figure 3). The textured surface systems with no FN treatment showed a similar cell density both before and after a period of 4 h of continuous perfusion (Figure 3a and 3b). This suggests that cells attached to the channel surface with enough strength to withstand continuous flow at 5 μl min^{-1}. However, note that even after 96 hours of incubation/proliferation at static conditions, the cell culture was not completely confluent at the moment of activation of perfusion. Cell attachment and proliferation was further enhanced by treatment of textured surfaces with FN. In channels treated this way, the cell culture appeared confluent after incubation at static conditions, and after perfusion for 48 h (Figure 3c and 3d). An increase in attached cell density was evident after 48 h of continuous flow at 5 μl min^{-1}. In general, these observations were valid for different locations along the microchannels. Cell morphology was also different in smooth and textured microchannels. In smooth surface microchannels, cell morphology maintained a circular shape at the end of the perfusion experiment (similar to that adopted by CHO cells in suspended culture). In contrast, cells cultured in textured microchannels, particularly in those with FN coating, elongated along the microchannel and exhibited a more irregular border shape that was suggestive of better attachment to the surface.

3.3 mAb productivity

Recombinant mAb production was consistent with the observed cell growth and proliferation. Cell cultures on smooth surface microchannels (with or without FN coating), where no cell proliferation was observed, did not exhibit any mAb production during perfusion. In In textured micro-channels without FN coating, the mAb concentration at the micro-reactor outlet fluctuated between o0.095 mg L^{-1} and 0.629 mg L^{-1} during 60 hours of continuous operation. In textured surfaces with FN coating, cell attachment and proliferation seemed to have improved significantly. During 48 h of continuous perfusion, mAb concentrations at the outlet of the micro-device progressively and linearly increased from 1.66 to 2.04 mg L^{-1}.

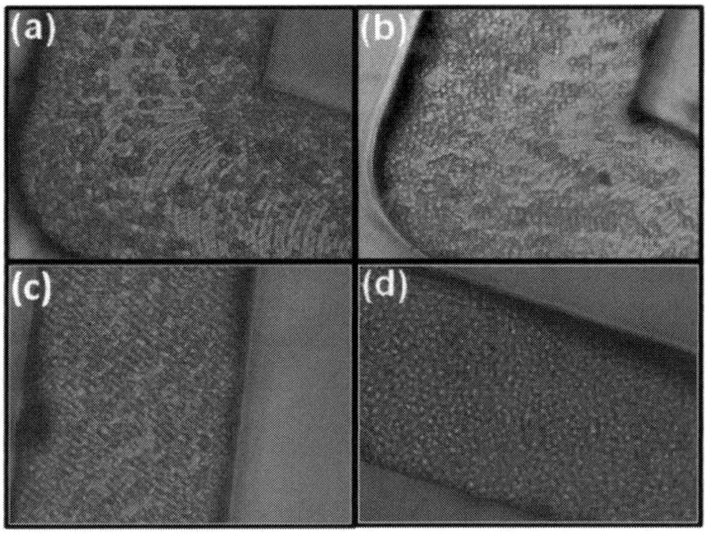

Figure 3: Cell attachment improves on rough surfaces. Cell attachment in microchannels with a rough surface without AF coating: **a)** image taken before perfusion; and **b)** after perfusion. Further cell attachment is observed on rough surfaces treated with AF: **c)** image taken before perfusion; and **d)** image taken after perfusion.

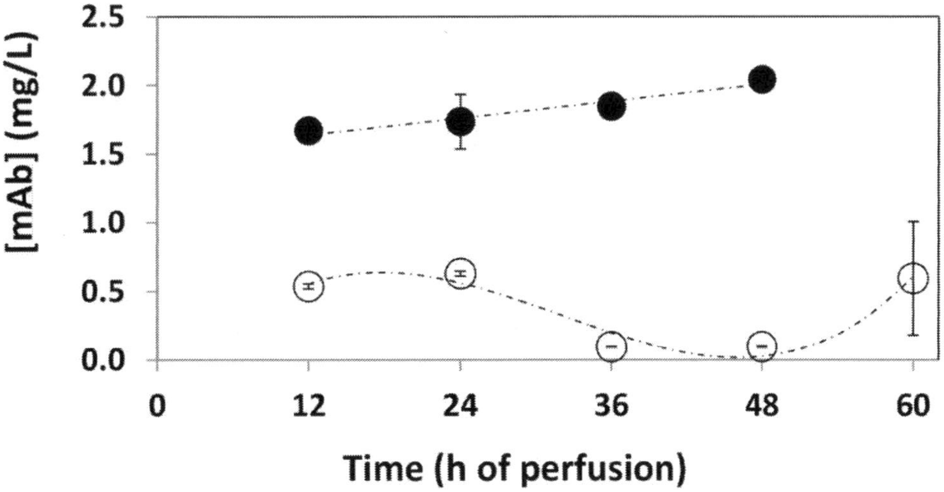

Figure 4. Production of a monoclonal anti-body under continuous flow. MAb concentration (as measured by ELISA) of active mAb concentration at different times during a period of continuous feed (single pass) of culture media at 5 μL min^{-1} in textured and FN coated micro-channels (blue dots) and textured but not FN coated micro-channels. Images show details of the cell density associated with the stage of continuous operation in (b) FN coated and (c) not FN coated micro-channels.

Assuming that mAb production is exactly proportional to cell density, and that the productivity of each cell remains constant regardless of FN surface treatment, the number of cells properly attached to the rough FN coated surfaces can be inferred to be 4fold higher than for rough surfaces without FN.

Given the flow rate (5 μl min^{-1}) and the dimensions of the microchannel used here (V = 2.2 μL), the average residence time in the device was 0.44 min. Therefore, during a perfusion period of 4 h, more than 500 reactor volumes passed through the device. Consequently, the cumulative mAb concentrations observed after 4 h of perfusion are truly representative of production under continuous operation. In textured and FN coated microdevices, the average mAb productivity for the perfusion period can be calculated as 285.75 mg L^{-1} h^{-1}. This value is remarkably higher to that we have observed previously in fed-batch suspension cell cultures using the same cells and substrates (i.e., a range from 1.52 to 4.16 mg L^{-1} h^{-1}). Our experiments suggest that mAb productivity under perfusion conditions is at least three orders of magnitude higher than that observed under static conditions in the same microdevice.

4 CONCLUSIONS

This paper reports the culture of a non-adherent monoclonal antibody-producing CHO cell line within PMMA-PDMS microdevices. Cell adhesion within microdevices enables a continuous perfusion culture at a microscale and requires only a small amount of medium to sustain cell growth and recombinant protein production. Cell production and mAb productivity of CHO cell cultures were evaluated in a PMMA-PDMS microdevice with a "zigzag" channel design. Different factors were tested to maximize cell proliferation and adhesion: smooth or textured microdevice surface; and presence or absence of attachment factor (FN) coating. For non-adherent cells, continuous perfusion culture is feasible through the use of microdevices with textured surfaces. Cell adhesion and proliferation, and mAb productivity was enhanced by treatment with FN. The mAb productivity of CHO cells in continuous perfusion microdevices was two orders of magnitude higher to that observed in 1 L fed-batch bioreactor systems.

ACKNOWLEDGEMENT

The authors acknowledge the financial support provided by the Mexican government through CONACYT scholarships for LDGG. The authors are grateful for the financial support provided by Tecnológico de Monterrey at Monterrey through Cátedra de Investigación (CAT 122).

REFERENCES

[1] Coco-Martin J. M., Harmsen M. M.: A review of therapeutic protein expression by mammalian cells. *BioProcess International*, 4:28-33, 2008.

[2] Wlodkowic D., Faley S., Skommer J., McGuinness D., Cooper J.: Biological implications of polymeric microdevices for live cell assays. *Analytical Chemistry*, 81(23): 9828-9833, 2009.

[3] Xia Y., Whitesides G. M.: Soft Lithography. *Annu. Rev. Mater. Sci.*, 28, 153-184, 1998.

[4] Tourovskaia A., Figueroa-Masot X., Folch A.: Differentiation-on-a-chip: A microfluidic platform for long-term cell culture studies. *Lab on a C*hip, 5: 14-19, 2005.

[5] Dixon A., Takayama Sh.: Guided corona generates wettability patterns that selectively direct cell attachment inside closed microchannels. *Biomed Microdevices*, 12: 769-775, 2010.

[6] Ferrell N., Gallego-Perez D., Higuita-Castro N., Butler R. T., Reen R. K., Gooch K. J., Hansford D. J.: Vacuum-Assisted Cell Seeding in a Microwell Cell Culture System. *Anal. Chem.*, 82:2380-2386, 2010.

[7] Ranucci C. S., Moghe P. V.: Substrate microtopography can enhance cell adhesive and migratory responsiveness to matrix ligand density. *John Wiley & Sons, Inc.*, 149-161, 2000.

(83a) Robust Particle Size Control During API Crystallization

Nathan Domagalski[1], Brendan C. Mack[2], Amanda Rogers[1], Chenchi Wang[2], Jose E. Tabora[3] and Lindsay Hobson[2], (1)Chemical Development, Bristol-Myers Squibb Co., New Brunswick, NJ, (2)Bristol-Myers Squibb Company, New Brunswick, NJ, (3)Chemical Development, Bristol Myers Squibb, New Brunswick, NJ

A robust process offers flexibility to large scale manufacturing by permitting execution within broad parameter ranges without impact to product quality. In this paper the development of a controlled API batch crystallization will be described which delivers both a highly reproducible and predictable particle size specification. The process includes simultaneous cooling and wet milling that favors secondary nucleation over growth under low supersaturation conditions. Investigation includes a fractional factorial design of experiments (DoE) approach studied on a greater than 100 kg pilot plant scale. This DoE enabled identification of key process parameters using an equipment train nearly equivalent to commercial manufacturing. Collected data served as the basis for an empirical model used to estimate particle size and establish a demonstrable design space for the crystallization process. Comparison with similarly executed batches showed that the model may be used to accurately predict future performance. The described crystallization procedure was superior to conventional wet milling in producing unimodal particle size distribution in the isolated product.

(83b) Effects of an Impurity On the Morphology of an API: Using ATR-FTIR, FBRM and PVM to Confirm a Hypothesis

Christopher S. Polster[1], Michael A. Lovette[1], Lori R. Hilden[1] and Christopher L. Burcham[2], (1)Chemical Product Research and Development, Eli Lilly and Company, Indianapolis, IN, (2)Chemical Product Research and Development; Particle Design Laboratory, Eli Lilly and Company, Indianapolis, IN

During the crystallization of multiple 1-2 kg batches of API, a substantial change in morphology from needles (aspect ratio 10-20) to plates (aspect ratio 1-2) was observed in some of the lots. An unknown impurity found at trace levels was found to correlate with the change in morphology. This impurity was isolated chromatographically and identified as a dimeric species of the parent compound. Several milligrams of the impurity were isolated and used to spike into subsequent crystallizations in order to confirm the hypothesis that the impurity alone was responsible for the morphology change. These crystallizations were characterized by using ATR-FTIR to monitor solute concentration changes with time, FBRM to monitor particle chord length (along with aspect ratio) changes with time, and PVM was used as visual confirmation of the aspect ratio changes. Crystallizations (over a range of impurity levels) show an evolution from expected needle morphology to plate morphology. The results find that the aspect ratio is altered by slowing the growth of the expected axial face.

(83c) Microfluidic Platforms for Time Resolved Laue Crystallography

Sarah L. Perry[1], Sudipto Guha[2], Ashtamurthy Pawate[2], Zhong Ren[3] and Paul J.A. Kenis[4], (1)Institute for Molecular Engineering, University of Chicago, Chicago, IL, (2)Department of Chemical & Biomolecular Engineering, University of Illinois at Urbana-Champaign, Urbana, IL, (3)Department of Biochemistry and Molecular Biology and the Consortium for Advanced Radiation Sources, University of Chicago, Chicago, IL, (4)Department of Chemical & Biomolecular Engineering, University of Illinois at Urbana Champaign, Urbana, IL

While efforts in microfluidics for protein crystallization have developed dramatically in the past ten years [1-3], the idea of coupling microfluidic technology with advanced protein crystallography techniques for *in situ* analysis is an area where significant advances can be made. Microfluidic devices are particularly applicable for Laue methods [4], where polychromatic X-rays allow for significantly faster data collection that enables the analysis of tiny or fragile crystals that are susceptible to radiation damage as well as kinetic experiments [5,6]. This high rate of data collection enables time-resolved crystallography experiments by matching the X-ray exposure time to the timescale of chemical and structural changes in the protein. Thus Laue crystallography provides a very elegant platform for performing kinetic experiments that directly probe changes that occur during enzymatic function.

Microfluidic platforms have the benefit of not only enabling experiments at very small volumes, but also by creating an environment free of inertial or convective effects and allowing for exquisite control over local conditions and gradients. We have developed microfluidic platforms for protein crystallization which utilize this high degree of control for both the growth of high quality protein crystals, and also as a triggering mechanism for use in time resolved crystallography experiments. Here we report on microfluidic crystallization platforms that have been optimized for the growth of a large number of crystals for *in situ* time resolved Laue analysis. We used photoactive yellow protein (PYP) as a model system for the first time resolved crystallography experiments. The enzymatic reaction is first triggered by a laser light pulse and then a series of X-ray diffraction images are taken at varying time delays. This large set of diffraction data is then analyzed to yield a time resolved picture of the various structural changes within the crystal. We are working on extending the capabilities of our microfluidic systems to include triggering of enzyme reactions by flowing a solution over the crystals. This capability will enable the time resolved structural analysis of a large range of protein systems, which had previously been inaccessible for study.

1. Hansen, C. and S.R. Quake, *Microfluidics in structural biology: smaller, faster... better.* Current Opinion in Structural Biology, 2003. **13**(5): p. 538-544.

2. van der Woerd, M., D. Ferree, and M. Pusey, *The promise of macromolecular crystallization in microfluidic chips.* Journal of Structural Biology, 2003. **142**(1): p. 180-187.

3. Perry, S.L., et al., *Microfluidic Generation of Lipidic Mesophases for Membrane Protein Crystallization.* Crystal Growth & Design, 2009. **9**(6): p. 2566-2569.

4. Ren, Z., et al., *Laue crystallography: coming of age.* Journal of Synchrotron Radiation, 1999. **6**: p. 891-917.

5. Cornaby, S., et al., *Feasibility of one-shot-per-crystal structure determination using Laue diffraction.* Acta Crystallographica Section D, 2010. **66**(1): p. 2-11.

6. Bourgeois, D., et al., *Time-resolved methods in biophysics. 6. Time-resolved Laue crystallography as a tool to investigate photo-activated protein dynamics.* Photochemical & Photobiological Sciences, 2007. **6**(10): p. 1047-1056.

(83d) Crystalline Particle Formation and Polymorphism From Uniform Droplet Evaporation

Kelly M. Carver and **Ryan C. Snyder**, Chemical Engineering, Bucknell University, Lewisburg, PA

Particle size, shape and structure are all important product qualities with direct implications for the pharmaceutical, food product and dye industries. Crystal size directly impacts dissolution rate. Crystal shape affects both downstream processes such as filtering, washing and drying as well as material properties such as mechanical strength, bioavailability and surface functionality. Finally, different crystal structures, or polymorphs, impact the entire scope of the product from melting point and solubility to patent considerations. Thus, methods to manipulate and control size, shape and structure are continuously being sought. Droplet evaporation, commonly used industrially through spray drying, offers some advantages such as a reduction in production steps when compared to traditional crystallizations; however, it is notoriously difficult to control and model the product properties because of the varied droplet sizes and flow patterns of the droplets. Monodisperse droplet generation offers a useful tool for better controlling droplet evaporation by eliminating the challenges associated with droplet size distributions in spray dryers. Monodisperse droplet generators have previously been used to produce uniform particle sizes (Nandiyanto and Okuyama 2011, Vehring 2008); however, crystalline properties have not been the primary focus of previous work.

In this talk, we will highlight our recent work on producing crystalline particles through the evaporation of monodisperse droplets produced using a vibrating orifice aerosol generator (VOAG). In this study, we use succinic acid as a test case molecule. The experiments show that we not only can produce monodisperse particle sizes, but that our particles are also fully crystalline, containing no amorphous characteristics. As with other crystallization methods, solvent choice is a key variable in determining the morphology of the resulting particles, and we demonstrate the variability in morphology across several solvents (e.g. acetone, water, isopropanol). We will show that solvent choice indeed impacts the morphologies of these particles and demonstrate that the morphologies fall into a new class of particle shapes that previously have not been obtained from droplet evaporation. Additionally, we will show that the polymorphs obtained through this method are unique when compared to the crystallization of succinic acid from bulk solution. Finally, we will discuss the potential for this type of droplet evaporation as a tool for studying polymorphism of organic molecules in general and as a good system to develop fundamental models that could later be applied to industrial scale processes such as spray drying.

References

Nandiyanto, A. B. D.; Okuyama, K. *Adv. Powder Technol.* **2011**, *22*, 1-19.

Vehring, R. *Pharm. Res.* **2008**, *25*, 999-1022.

(83e) Influence of Additives On Crystal Morphology of Organic Molecular Crystals

Zubin B. Kuvadia, Department of Chemical Engineering, University of California, Santa Barbara, Santa Barbara, CA and Michael F. Doherty, Chemical Engineering, UC Santa Barbara, Santa Barbara, CA

Crystal morphology has a great impact on the end-use efficacy of solid products as well as on the downstream performance of the entire manufacturing and formulation process. Foreign molecules like additives or impurities influence the crystal morphology to a significant extent by inhibiting the growth mechanism of certain crystal faces. The additives can be byproducts of the preceding step of reaction or they could be specially designed to obtain specific crystal morphologies for focused applications, for example, using crystal growth inhibitors to reduce the size of kidney stones[1].

Because of the specific chemical structure of structurally similar additives, they can incorporate or lock into the original lattice only in certain configurations, hence only some of the crystal faces can recognize these additive molecules. Once attached, they interfere with the first turn of a rotating growth spiral and hence slow down the growth rate of the face [2]. In this work, we develop a generic probabilistic scheme for quantitatively estimating imposter recognition on each crystal face using a combination of mean field theory and energy minimization. Based on this recognition, the mechanistic effect of impurity on the first turn of the spiral and hence modified growth rates and modified crystal habits are computed. These concepts are generalized for all molecular crystals, including non-centrosymmetric molecules. Finally, we present work on specific systems of immense industrial and scientific interest such as paracetamol/PAA[3] and L-cystine/L-CDME[1] and compare our results with experiments from the literature and our laboratory.

References:

[1]: Rimer J . D, et al, Science **330**, 337(2010)

[2]: Sizemore J.P. and Doherty M.F, Cryst. Growth Des., **9**,6 (2009)

[3]: Chow et al, Intnl. Journal of Pharmaceutics, **24** (2-3) (1985)

(83f) Impact of Pre-Heating On Nucleation Kinetics and Polymorph

Srividya Ramakrishnan[1], Vasu Dev Rudraraju[2], Sundaralakshmi Kanniah[1] and Rashmi Agrawal[1], (1)Dr. Reddy's Laboratories Ltd., Hyderabad, India, (2)Aurigene Discovery Technologies Ltd., Hyderabad, India

Polymorphs play an important role in pharmaceutical development as they can impact the stability and bioavailability of the active pharmaceutical ingredient (API). However, isolating and stabilizing the desired form can often be a challenging task as several factors can affect the crystallization of a particular polymorph. In this study, the impact of thermal history of the under-saturated solution on the nucleation kinetics was studied for an API. It was found that pre-heating a solution of the API at a temperature higher than the dissolution temperature could significantly increase the induction time or suppress the nucleation temperature during cooling crystallization. In addition, pre-heating was also found to have an effect on the polymorph that nucleated. While typically the metastable form would nucleate during cooling, pre-heating resulted in concomitant nucleation of the metastable as well as stable form, with a correlation between the duration of pre-heating and the proportion of the stable form in the polymorphic mixture. This suppression of nucleation by pre-heating may due to a conformational transition at higher temperatures. Conversion to the wrong conformer(s) would not only reduce the effective supersaturation but might also inhibit nucleation. Spectroscopic studies are under way to study the conformers and pre-nucleation aggregates in solution.

Communication number: IPDO IPM-00269

(83g) Integration of Biocatalysis and Crystallization towards the Manufacture of Enantiomerically Pure Compounds

Luis G. Encarnación-Gómez, School of Chemical and Biomolecular Engineering, Georgia Institute of Technology, Atlanta, GA, Andreas S. Bommarius, Chemical & Biomolecular Engineering, Georgia Institute of Technology, Atlanta, GA and Ronald W. Rousseau, School of Chemical & Biomolecular Engineering, Georgia Institute of Technology, Atlanta, GA

Crystallization of enantiomerically pure compounds from racemic mixtures that form racemic compounds requires an enantiomeric excess greater than the eutectic point between the compound and the pure enantiomer. It is estimated that roughly 90% of racemic mixtures fall into this category; hence, development of a process that utilizes such an approach could be of significant value.

To achieve a necessary enantiomeric excess, we are investigating the chemoenzymatic reaction of racemic mixtures of amino acids. We have demonstrated the possibility of performing such a reaction at high concentrations, even at those that require substrate dissolution. Preliminary results show that even in supersaturated solutions, enantiomeric excesses greater than 99% can be achieved. The chemoenzymatic reaction of (RS)-phenylalanine to obtain (S)-phenylalanine has been performed in undersaturated, saturated, and supersaturated solutions. The reaction proceeds through the selective oxidation of the (R)-amino acid utilizing D-amino acid oxidase (D-AAO), which produces an achiral imino acid that is reduced into a racemic amino acid utilizing ammonia-borane. Moreover, extending such reaction to saturated or supersaturated solutions allows us to operate crystallization processes such that enantiomerically pure compounds are continuously recovered in a solid phase. Therefore, the implementation of this reactive separation process is beneficial because of the following factors: (1) this process allows higher productivity, (2) the rate of reaction is maximized, (3) the process has potential to be run continuously, and (4) organic solvents or low-MW chiral acid/bases (i.e., the formation of diastereomers) are not needed during separation.

(96a) Engineering Escherichia Coli for Adipic Acid Production

Yuanqing Wu, Wujiu Chen, Yuanyuan Zhang, Li Han, Yanfeng Peng, Ran Tu, **Qinhong Wang** and Yanhe Ma, Tianjin Institute of Industrial Biotechnology, Chinese Academy of Sciences, Tianjin, China

Adipic Acid, six-carbon straight-chain dicarboxylic acid, is linked almost 90% to nylon production by the polycondensation with hexamethylenediamine and further processed into fibers for applications in carpeting, automobile tire cord and clothing. Almost all of the commercial adipic acid is produced from cyclohexane through two sequent oxidation processes. The process is regarded as poor because the yield is low and the chemicals used are not environmentally friendly. Consequently, extensive research has been directed towards developing an alternative, clean process based on microbial production. Here, we reported that *Escherichia coli* was engineered to produce adipic acid. Three genes of aroZ, aroY and catA from different bacteria with or without codon optimization, which encoded dehdroshikimate dehydratase, protocatechuate decarboxylase and catechol 1, 2-dioxygenase, respectively, were combinatorially assembled into the heterogeneous synthetic pathway and introduced into the cell of *E. coli* to complete the whole biosynthetic pathway for conversing glucose to cis, cis-muconic aicd, the precursor of adipic aicd, and then cis, cis-muconic acid was hydrogenated into adipic acid. With blocking the pathway to aromatic acid biosynthsis by mutating shikimate dehydrogenase (AroE) and combinatirally regulating the heterogeneous synthetic pathway with different constitute promoter with different strength in *E. coli*, the production of cis, cis-muconic acid was improved. In the shaking-flask test, 11.2g/L adipic acid was produced from 30g/L glucose. The genetic modification in combination with metabolic evolution was ongoing to further improve adipic acid production in engineered *E. coli*. Our study showed the potential to provide a rationale for the engineering and commercialization of microorganisms capable of producing adipic acid.

(96b) Metabolic Engineering of Escherichia Coli K5 for Production of Heparosan, a Precursor to the Anticoagulant Pharmaceutical Heparin

Brady Cress[1], Mattheos A.G. Koffas[1] and Robert J. Linhardt[2], (1)Chemical and Biological Engineering, Rensselaer Polytechnic Institute, Troy, NY, (2)Department of Chemistry and Chemical Biology, Rensselaer Polytechnic Institute, Troy, NY

Heparosan is a key precursor of the anticoagulant heparin (a drug with a market size greater than $10 billion annually) that is currently derived from porcine intestinal extract. Animal-sourced heparin production has several disadvantages and imposes quality concerns on the product. This issue was brought to international attention in 2008 when a heparin contamination crisis resulted in hundreds of deaths. To overcome the drawbacks of the current heparin production process, we are utilizing stoichiometric-based modeling to improve the production of heparosan in *Escherichia coli* K5. We have developed a multilevel optimization algorithm capable of predicting combinations of genetic manipulations for increased production of target metabolites. The algorithm builds upon the success of Cipher of Evolutionary Design (CiED), a genetic algorithm developed in our lab for predicting gene knockouts, to allow for prediction of sets of gene overexpressions and deletions consistent with superior cellular phenotype. Here we present experimental validation of computational predictions leading to an engineered *E. coli* K5 strain that achieves improved heparosan titers. In conjunction with computational predictions, traditional fermentation optimization strategies make heparosan production in *E. coli* K5 a safe and viable heparin production route.

(96c) Reusable Pathway Optimization with Quantitative Sequence-Expression-Flux Maps

Iman Farasat, Chemical Engineering, Pennsylvania State University, State College, PA, Jason Collens, Chemical Engineering, Pennsylvania State University, University Park, PA and Howard Salis, Chemical Engineering / Biological Engineering, Pennsylvania State University, University Park, PA

Metabolic pathway optimization commonly relies on trial-and-error mutagenesis to vary enzyme expression levels and maximize pathway flux. By combining biophysical models of translation initiation and advanced optimization algorithms, we present a new approach to pathway optimization that efficiently characterizes the relationship between enzyme expression level and pathway flux and identifies the optimal stoichiometries in the pathway, while using a minimal number of characterization experiments. We utilize this data to generate a Quantitative Sequence-Expression-Flux Map (QSEF Map) that enables one to rationally control the flux through a pathway for diverse applications without requiring pathway optimization to be performed again "from scratch". We experimentally demonstrate this new approach, called QSEF Mapping, on a 3-enzyme carotenoid biosynthesis pathway. In the process, we efficiently explore the three-dimensional enzyme expression level space across a 500,000-fold range, rationally control pathway flux over a 600-fold range, and achieve a maximal carotenoid titer of at least 20 mg/g cell dry weight in 7 hours.

(96d) Ctfab and adhE2 Can Induce Acetone Production and Increase Butanol Production Yield From Glucose in Clostridium Tyrobutyricum

Le Yu, The Ohio State University, Columbus, OH, Jingbo Zhao, William G. Lowrie Department of Chemical and Biomolecular Engineering, The Ohio State University, Columbus, OH and Shang-Tian Yang, Department of Chemical and Biomolecular Engineering, Ohio State University, Columbus, OH

C. tyrobutyricum ATCC 25755 is an acidogen with butyric acid and acetic acid as two main fermentation products from glucose and xylose. In addition to the high butyric acid yield, *C. tyrobutyricum* also has a high butyric acid tolerance, making it an ideal candidate for industrial production of biofuels such as n-butanol. In our previous study, we introduced *adhE2* gene encoding a bifunctional aldehyde/alcohol dehydrogenase from *C. acetobutylicum* into an *ack* knock-out strain of *C. tyrobutyricum*, and the mutant produced ~11 g/L butanol from glucose with butyric and acetic acids as two byproducts. To further improve butanol production, it is desirable to re-assimilate the butyric acid and acetic acid and convert them to corresponding alcohols. In this study, we constructed three different plasmids containing *ctfAB* genes which encode for a CoA-transferase together with *adhE2* gene in *C. tyrobutyricum*. As expected, *ctfAB* overexpression increased butanol production to 12-14 g/L, with an improved butanol yield and productivity of ~0.24 g/g glucose and ~0.38 g/L·h, and both butyric acid and acetic acid production decreased significantly. Interestingly, *ctfAB* overexpression alone also induced acetone production without decarboxylation process. Batch fermentation kinetics in pH-controlled fermentors with these mutants was further investigated at different pH values, and the results will be presented in this paper.

(96e) Engineering Auto-Regulatory Genetic Circuits for Production of Hydroxycinnamic Acid-Derived Chemicals

Tat-Ming Lo, Chueh Loo Poh and Matthew Wook Chang, School of Chemical and Biomedical Engineering, Nanyang Technological University, Singaproe, Singapore

In the biological production of chemicals, inducers such as arabinose and IPTG are often used. Artificial induction in many cases is less favorable due to high economic cost of inducers, inducer toxicity, and incompatibilities with industrial scale-up. Hence, in this study, we have attempted to engineer *E. coli* that is capable of auto-inducing and self-regulating its enzyme expressions by applying control theory through the introduction of a genetic bio-sensor with a time-delay element. Implementation of control theory in enzyme expression not only eliminates the need for artificial inducers, it also allows optimal performance in dynamic biological systems and achieving system stability through corrective actions. As a proof of concept, we forward engineered *E. coli* to detect ferulic acid and convert it to vanillin, a key compound for vanilla aroma which is used in food, cosmetic and pharmaceutical industries. To ensure that the host cells would not be induced at the early stage of growth, we installed a time-delay element to the ferulic acid sensor. Together, our engineered *E. coli* attained a much improved growth rate, and reached maximum vanillin production in a highly efficient manner. We envision that our auto-regulation approach can be readily extended to improving production of a range of chemicals derived from hydroxycinnamic acid.

(96f) Metabolic Engineering of Escherichia Coli for Production of Mcl-PHA From Unrelated Carbon Sources

Daniel E. Agnew[1], J. Tyler Youngquist[2] and **Brian F. Pfleger**[2], (1)Chemical and Biological Engineering, University of Wisconsin - Madison, Madison, WI, (2)Chemical and Biological Engineering, University of Wisconsin Madison, Madison, WI

Polyhydroxyalkanoates (PHA) are a class of naturally occurring microbial polyesters whose metabolism and material properties have been well characterized owing to their renewable nature and potential to obviate petroleum derived plastics such as polypropylene. Polyesters with monomers consisting of ≥6 carbons are known as medium chain-length (mcl)-PHA and are industrially relevant, allowing for greater variation in the material properties of the polymer with longer chain-lengths resulting in lower glass transition and melting temperatures. Production of mcl-PHA is often achieved through feeding of exogenous fatty acids (FA) and related compounds and generally results in a polymer with a heterogeneous monomer composition due to the iterative nature of fatty acid metabolism. However, this method of production is limited by the cost of the FA feedstock and in many cases, the lack of control over the composition of the final product. To overcome these obstacles, we have engineered a strain of *Escherichia coli* that is capable of growth on a non-FA feedstock (*i.e.*, glucose) for production of mcl-PHA with a pre-determined composition. Endogenous FA production on glucose was achieved by overexpression of a plant thioesterase (BTE) and the resulting free FAs routed through β-oxidation toward mcl-PHA via overexpression of PHA biosynthesis genes from *Pseudomonas*. In addition, experimental analysis of a β-oxidation knockout library led to insight into the role of *fad* genes in the presence of exogenous and endogenous free FAs.

(96g) Expanding the Synthetic Biology Toolbox for Engineering Metabolic Pathways in *E. Coli*

Peng Xu and Mattheos A.G. Koffas, Chemical and Biological Engineering, Rensselaer Polytechnic Institute, Troy, NY

Expanding the Synthetic Biology Toolbox for Engineering Metabolic Pathways in *E. coli*

Harnessing cell factories for producing biofuel and pharmaceutical molecules has stimulated efforts to develop novel synthetic biology tools customized for modular pathway engineering and optimization. Here we report the development of a set of vectors compatible with BioBrick^{TM} standards and its application in metabolic engineering. The engineered ePathBrick vectors comprise four compatible restriction enzyme sites allocated on strategic positions so that different regulatory control signals can be reused and manipulation of expression cassette be streamlined. Specifically, these vectors allow for fine-tuning gene expression by integrating multiple transcriptional activation or repression signals into the operator region. At the same time, ePathBrick vectors support the modular assembly of pathway components and combinatorial generation of pathway diversities with three distinct configurations. We also demonstrated the functionality of a seven gene pathway (~9Kb) assembled on one single ePathBrick vector. The ePathBrick vectors presented here provide a versatile platform for rapid design and optimization of metabolic pathways in *E. coli*.

Keywords: metabolic engineering; transcriptional fine-tuning; gene assembly; pathway configuration; T7 promoter activity

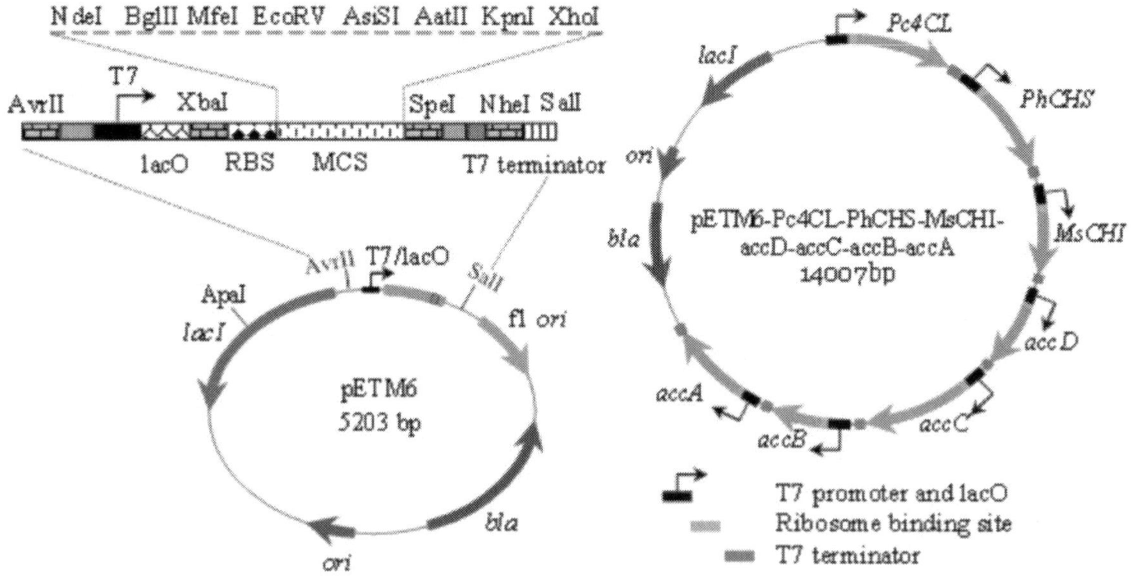

Reference

Xu P, Vansiri A, Bhan N, and Koffas M.A.G. (2012) ePathBrick: A synthetic biology platform for engineering metabolic pathways in *E. coli*. *ACS Synthetic Biology*, **DOI:** 10.1021/sb300016b, In Press.

(96h) Metabolic Engineering of Escherichia Coli for (2S)-Pinocembrin Production From Glucose

Jingwen Zhou, Junjun Wu, Guocheng Du and Jian Chen, School of Biotechnology, Jiangnan University, Wuxi, China

Flavonoids are valuable natural products widely used in human health and nutrition. Recent advances in synthetic biology and metabolic engineering have yielded improved strain titers and yields. However, current fermentation strategies often require supplementation of expensive phenylpropanoic precursors in the media and separate evaluation of each strategy in turn as part of the flavonoid pathway, implicitly assuming the modifications are additive. In this study, an *Escherichia coli* fermentation system was developed to bypass both of these problems. An eight-step pathway, consisting of 3-deoxy-D-arabinoheptulosonate 7-phosphate synthase (DAHPS), chorismate mutase/prephenate dehydratase (CM/PDT), phenylalanine ammonia lyase (PAL), 4-coumarate:CoA ligase (4CL), chalcone synthase (CHS), chalcone isomerase (CHI), malonate synthetase, and malonate carrier protein, was assembled on four vectors in order to produce the flavonoid precursor (2S)-pinocembrin directly from glucose. Furthermore, a modular metabolic strategy was employed to identify conditions that optimally balance the four pathway modules. Once this metabolic balance was achieved, such strains were capable of producing 40.02 mg/L (2S)-pinocembrin directly from glucose. These results were attained by culturing engineered cells in minimal medium without additional precursor supplementation. The fermentation platform described here paves the way for the development of an economical process for microbial production of flavonoids directly from glucose.

(100a) A Systems-Level Analysis Approach for Identifying Genetic Targets to Treat Biofilm-Forming Pathogens: An Application to Pseudomonas Aeruginosa

Zhaobin Xu, Department of Chemical Engineering , Villanova University, Villanova, PA, Xin Fang, Department of Systems Biology, Henry M. Jackson Foundation, Bethesda, MD, Thomas K. Wood, Department of Chemical Engineering & Biochemistry and Molecular Biology, Pennsylvania State University, University Park, PA and Zuyi (Jacky) Huang, Department of Chemical Engineering & The Center for Nonlinear Dynamics and Control, Villanova University, Villanova, PA

The investigation of biofilm formation is an urgent research theme in systems biology, as biofilms are heavily involved in bacterial chronic inflammatory and infectious diseases (Sauer et al., 2007). It is reported that nearly 65% of all nosocomial infections in the USA are associated with biofilms. Prevention of the initiation of biofilm formation is the most important step for combating biofilm-forming pathogens, as the ability of pathogens to resist antibiotics is enhanced 10 to 1000 times once they form biofilms (Thien and O'toole, 2001). Both experimental approaches (Gerdes et al., 2003) and model-based approaches (Edwards and Palsson, 2000) have been developed to identify genes essential for bacterial growth in the planktonic state. However, inhibition of some essential planktonic-growth genes might prevent the growth of microorganisms in the planktonic state but may not prevent the formation of biofilms. In these cases, microorganisms can still switch from planktonic growth to the biofilm lifestyle before antibiotics completely inhibit biological functions associated with these genes. For example, Zhang et al., 2007 has reported that biofilm formation is a general strategy utilized by pathogens to survive stress such as that imposed by antibiotic treatment. On the other hand, a large amount of genes are involved in biofilm formation. Even though we inhibit some of these biofilm-associated genes, we cannot eliminate pathogenicity as these genes might not be essential for microorganism metabolism. Therefore, the ideal gene targets to treat biofilm-forming pathogens should be those that prevent the formation of biofilms and also eliminate pathogens in the planktonic state.

In this work, we develop a systems-level analysis approach to determine target genes that are essential for bacterial metabolism but which do not induce biofilm formation when they are inactivated. The proposed approach is based on flux balance analysis (FBA) (Edwards and Palsson, 2000), artificial center hint and run sampling (ACHR) (Becker et al., 2007), and hierarchical clustering. Specifically, fluxes of reactions positively associated with biofilm formation are used as sensors to monitor the trend of gene-inhibition mutants to form biofilms. Mechanisms for inhibition mutants of essential genes to form biofilms are investigated by identifying those biofilm-associated reactions whose fluxes surge for most gene-inhibition mutants. Target genes to treat biofilm-forming pathogens are then defined as those essential planktonic-growth genes whose mutants cannot induce the formation of biofilms.

The proposed approach was applied to identify genetic targets to treat Pseudomonas aeruginosa infections, as it is one of the leading causes of nosocomial infections in hospitalized patients and P. aeruginosa displays resistance to a wide array of antibiotics (Cao et al., 2004). The model presented in Oberhardt et al., 2008, was used to predict the trend for P. aeruginosa to switch from the planktonic state to the biofilm growth mode when essential planktonic-growth genes are inhibited. It is interesting to find that that inhibition of most essential planktonic-growth genes induces the formation of biofilms. Specifically, P. aeruginosa tends to survive the essential-gene inhibition treatment (i.e., a pseudo antibiotic-treatment) by mainly increasing the fluxes through 10 reactions that regulate amino acid metabolism, acetate metabolism, hydrogen peroxide, and ABC phosphate transporter. This phenomenon indicates the adaptability of P. aeruginosa toward unfavorable environments. This work presents the first computational approach to address the mechanism under this phenomenon. In addition, three essential planktonic-growth genes, i.e., PA1756 (cysH), PA3686 (adk), and PA2023 (galU), cannot induce biofilm formation when they are inhibited. They constitute gene targets to treat P. aeruginosa as inhibition of them can eliminate pathogens without inducing biofilm formation. All these three genes have been validated by existing data as potential drug targets for their crucial role in the survival or virulence of P. aeruginosa (Bonofiglio et al., 2005; Markaryan et al., 2001; Ren et al., 2005; Williams et al., 2002).

Becker SA, Feist AM, Mo ML, Hannum G, Palsson BO, Herrgard MJ. Quantitative prediction of cellular metabolism with constraint-based models: the COBRA Toolbox. Nat Protoc 2:727-738, 2007
Bonofiglio L, García E, Mollerach M. Biochemical characterization of the pneumococcal glucose 1-phosphate uridylyltransferase (GalU) essential for capsule biosynthesis. Curr Microbiol 51(4):217-221, 2005
Cao B, Wang H, Sun H, Zhu Y, Che M. Risk factors and clinical outcomes of nosocomial multi-drug resistant Pseudomonas aeruginosa infections. J Hosp Infect 57(2): 112-118, 2004
Edwards JS, Palsson BO. Metabolic flux balance analysis and the in silico analysis of Escherichia coli K-12 gene deletions. BMC Bioinformatics 1:1-1, 2000
Gerdes SY, Scholle MD, Campbell JW, et al. Experimental determination and system level analysis of essential genes in Escherichia coli MG1655. J Bacteriol 185(19): 5673-5684, 2003
Markaryan A, Zaborina O, Punj V, Chakrabarty AM. Adenylate kinase as a virulence factor of Pseudomonas aeruginosa. J Bacteriol 183(11):3345-3352, 2001
Oberhardt MA, Puchalka J, Fryer KE, Martins dos Santos VAP, Papin JA. Genome-scale metabolic network analysis of the opportunistic pathogen Pseudomonas aeruginosa PAO1. J Bacteriol 190(8): 2790–2803, 2008
Ren D, Zuo R, Gonzalez Barrios AF, Bedzyk LA, Eldridge GR, Pasmore ME, Wood TK. Differential Gene Expression for Investigation of Escherichia coli Biofilm Inhibition by Plant Extract Ursolic Acid. Appl Environ Microb 71(7): 4022–4034, 2005
Sauer K, Rickard AH, Davies DG. Biofilms and biocomplexity. Microbe 2:347–353, 2007
Thien FCM, O'Toole GA. Mechanisms of biofilm resistance to antimicrobial agents. Trends Microbiol 9(1): 34-39, 2001
Williams SJ, Senaratne RH, Mougous JD, Riley LW, Bertozzi CR. 5'-Adenosinephosphosulfate lies at a metabolic branch point in mycobacteria. J Biol Chem 277(36):32606-32615, 2002
Zhang X, Garcia-Contreras R, Wood TK. YcfR (BhsA) Influences Escherichia coli Biofilm Formation through Stress Response and Surface Hydrophobicity. J Bacteriol 189(8): 3051-3062, 2007

(100b) Engineering ROS Metabolism with Futile Cycles

Kristin Adolfsen, Princeton University, Princeton, NJ and Mark P. Brynildsen, Chemical and Biological Engineering, Princenton University, Princeton, NJ

Reactive oxygen species (ROS) are toxic molecules utilized by the innate immune system and antibiotics to kill bacteria. In addition to their role in host defense, a basal level of ROS is produced inadvertently in all aerobically growing organisms when reduced electron carriers collide with O_2 and transfer their electrons. For this reason, many bacteria have evolved defense capabilities against these toxic species, such as producing detoxification enzymes and antioxidant metabolites. When the level of ROS within a cell exceeds its capacity for detoxification, bacterial death occurs. In previous work, we have shown that increased ROS production potentiates the effects of oxidative stress. An important insight from that study was that ROS production increases as inefficient use or production of ATP increases. In this study, we sought to increase ATP inefficiency by engaging futile cycles and determine if such a perturbation increases sensitivity toward oxidative stress. Futile cycles are sets of reactions whose net activity results in ATP hydrolysis. Using mixed-integer linear optimization, we have identified all futile cycles in *E. coli* metabolism containing ≤ 3 reactions, as well as those that require the expression of only one enzyme in aerobic growth in minimal glucose media. We have experimentally validated the presence of futile cycling and measured an increase in sensitivity to H_2O_2 for strains that are actively futile cycling when compared to strains carrying an inactive form of the cycle-enabling enzyme. The knowledge gained from this project could lead to the development of new therapies to sensitize microbes to immunity and antibiotics.

(100c) Dynamic Modelling As a Tool for Increasing Single-Chain Antibody Fragment Specific Productivity in Pichia Pastoris

Kate Royle, Life Sciences, Imperial College London, London, United Kingdom, David Leak, Biology & Biochemistry , University of Bath, Bath, United Kingdom and Cleo Kontoravdi, Chemical Engineering and Chemical Technology, Imperial College London, London, United Kingdom

Pichia pastoris is a commonly used expression host for heterologous protein production, predominantly because it is amenable to genetic manipulation and can grow to high cell densities in cheap culture media. While considerable yields can be achieved in this way, the specific productivity is relatively low. Consequently, the full impact of this host on industrial biotechnology has not yet been realised.

Previous studies to increase productivity have focused on bioengineering, such as targeting gene and strain characteristics, and fermentation conditions. Whereas the latter has been subject to multiple systematic multivariate optimisations with computational modelling techniques, the former has not. The majority of bioengineering studies have targeted one factor in isolation, and despite comparable strategies have variable outcomes. Here, a dynamic computational modelling approach has been taken to understand how the factors interact and develop a global optimisation strategy.

Initially, a deterministic single-cell model was derived recapitulating the essential features of protein production: transcription and translation were modelled with mass action kinetics, and the folding of nascent protein into its three-dimensional conformation with Michaelis-Menten kinetics based on the foldase Pdi. Additionally an endoplasmic reticulum (ER) stress response, the unfolded protein response (UPR), was included as it is well documented that this is activated during heterologous protein production. In essence, too much unfolded protein in the ER is detected by the receptor Ire1 which triggers a cascade of reactions concluding in an upregulation of proteins to counteract stress. Although this pathway has been modelled in yeast once before, our work has expanded both the breadth and the kinetics. Specifically, stress occurs when the chaperone Kar2 dissociates from Ire1 to bind unfolded protein, allowing excess unfolded protein to bind Ire1. This activates the endonuclease domain of Ire1, which subsequently splices *HAC1* mRNA allowing translation of this transcription factor. The protein-protein interactions have been modelled with equilibrium kinetics, and the splicing of *HAC1* mRNA with Michaelis-Menten kinetics. The transcription factor Hac1 retrotranslocates to the nucleus and the upregulation of *KAR2*, *PDI*, *HAC1* and *E3* has been modelled with Michaelis-Menten kinetics. Finally, the ER associated degradation pathway (ERAD), which acts to remove misfolded protein from the ER and is intertwined with the UPR, has been incorporated - a unique feature making this model more comprehensive than those previously published. Specifically, the binding of Kar2 to misfolded protein has been modelled with equilibrium kinetics and the subsequent degradation with Michaelis-Menten kinetics based on E3, the rate limiting step in the ERAD pathway.

Whilst simulations of the model proved qualitatively accurate, a lack of literature values hindered quantitative accuracy. Consequently, we sought to construct and characterise strains of *P. pastoris* producing a heterologous protein for which single-chain antibody fragments (scFvs) were chosen. Although monoclonal antibodies are currently the dominant biopharmaceutical in manufacture, smaller antibody fragments such as scFvs are expected to feature more heavily in the future. While mammalian cells are most frequently used to produce full antibodies, the smaller fragments are better suited to a microbial host.

Therefore, novel *P. pastoris* strains have been constructed producing two scFvs: BC1, which targets new blood vessels of tumours; and MFE23, which targets gastrointestinal cancers. These genes were cloned into the commercial vector pPICZαA with C-terminal polyhistidine and myc tags, and subsequently integrated into the *AOX1* locus. Expression was driven by the *AOX1* promoter, and secretion conferred by the N-terminal α-Factor from *Saccharomyces cerevisiae*. As there is biological variation in specific productivity, different clones of the same strain can produce varying amounts of scFv. In order to include this variation in the model and help identify relevant factors, both low and high yield clones were isolated. As multiple copies of the heterologous gene can increase productivity, the low and high yield clones were subjected to Southern blot analysis to confirm single copy integrants, allowing the gene copy number in the model to be set to one.

A major source of quantitative inaccuracy in the model was the concentration of two proteins in the ER, the chaperone Kar2 and the foldase Pdi. These proteins were identified as having a large influence over yield; however, literature values varied over three-fold. We therefore sought to accurately quantify these and developed an LC-MS/MS method to do so. Initially, an *in silico* tryptic digestion of the proteins was conducted and the peptides ranked according to their enhanced signature peptide prediction score, a computational method which evaluates their physicochemical properties. MRM methods were written to target the superior peptides, and these were experimentally validated with analysis of a tryptic digest of *Escherichia coli* BL21 DE3 strains overexpressing the proteins. Isotopically labelled versions were subsequently expressed, purified and used to optimise the MRM in terms of collision energy and collision cell exit potential. As the proteins are difficult to express at high yields, however, isotopically labelled peptides were purchased from JPT Peptide Technologies for the experiment.

As the UPR upregulates Kar2 and Pdi, the model required experimental data for the baseline level in unstressed cells, the higher level in stressed cells and the time over which this occurs. As transcription of the *SCFV* genes are specifically induced by changing the carbon source, we have devised an experimental system to generate this profile. Specifically, uninduced strains at an OD_{600} of 1 were incubated at 30°C, 250 rpm for 24 hours. Samples were taken at 22 and 23 hours, at which point the culture is in stationary phase, allowing evaluation of the baseline Kar2 and Pdi values. At 24 hours, scFv expression was induced and further samples taken at 0, 2, 4 and 6 hours. These data provide a complete picture of the unstressed baseline, activation of stress and the final maximum the system can achieve.

Simultaneously, the samples were analysed for the magnitude of the UPR and ERAD pathways. Traditionally, activation of the UPR has been measured using reverse-transcriptase quantitative PCR targeting *KAR2*, *PDI* and also *HAC1* mRNA, and this strategy was implemented here. Indeed, for Kar2 and Pdi it also allowed for interesting comparisons between transcriptional and protein upregulation as it is widely appreciated that they do not always equate. Therefore, primers were designed and optimised to target *KAR2*, *PDI* and *HAC1* in conjunction with those against actin, an internal control for the experiment. Upregulation was calculated with the Pfaffl method, compared to the wild-type GS115. Activation of the ERAD was assessed in a similar way; however, this pathway is poorly described compared to the UPR. Primers were designed to target major contributors, specifically a ubiquitin-activating enzyme (*E1*), a ubiquitin-conjugating enzyme (*UBC7*), two ubiquitin protein ligases (*HRD1* and *DOA10*) and an ATPase (*CDC48*). Interestingly, some studies have suggested that the ubiquitin protein ligases may be substrate specific; consequently analysing both *HRD1* and *DOA10* allowed an insight into whether this is the case for scFvs in *P. pastoris*.

Whilst the above data refined model accuracy, validation was required. Many studies of the UPR chemically induce stress to test hypotheses. This, however, produces pleiotropic effects which are not ideal. Here, we constructed strains which overexpress Kar2 and Pdi both individually and together. The increase in availability of these proteins provides the necessary perturbation to the system, yet they are also expected to have beneficial effects on yield imparting added industrial relevance. Specifically, each gene was cloned into the pIB2 vector under the control of the *GAP* promoter, with an *AOX1* transcription terminator and homology to the *his4* locus for integration. This approach resulted in a constitutive upregulation of the proteins for incorporation into the model. Simultaneous upregulation required a more developed strategy. Ideally, the same promoter was required; however, ongoing research suggests that *P. pastoris* frequently recombines between large regions of homology preventing the use of two copies of the promoter. To circumvent this, both genes were expressed as a single bicistronic mRNA with an IRES sequence joining the two. IRES sequences allow cap-independent translation such that the second gene in the bicistronic mRNA can be translated in addition to the first. Interestingly, there is some evidence that *P. pastoris* may be subject to translational downregulation during ER stress. If this is the case, cap-independent translation will not be hindered. Therefore, *KAR2* was strategically cloned into pIB2 after *PDI* and under control of an IRES sequence from *Saccharomyces cerevisiae*, as it is better able to relieve ER stress. Further RT-QPCR and mass spectrometry analysis confirmed the perturbation and provided a more thorough analysis of the use of IRES sequences in *P. pastoris* – a technique used only once previously.

In conclusion, we have developed a deterministic model which mimics scFv production and the initiation of ER stress pathways. This model has been tested with constructs that over-express Kar2 and Pdi, both validating its predictive capabilities and generating industrially relevant constructs. Finally, it has been simulated to predict the optimal strategy for increasing specific productivity in *P. pastoris*, constituting an integrated modelling and experimental approach to solving a life sciences problem.

(100d) Modeling Intra- and Inter-Kingdom Signaling Through NF-Kb Pathway in Dendritic Cells

Shreya Maiti[1], Robert Alaniz[2], Juergen Hahn[3] and Arul Jayaraman[1], (1)Department of Chemical Engineering, Texas A&M University, College Station, TX, (2)Department of Microbial and Molecular Pathogenesis, Texas A&M Health Science Center, College Station, TX, (3)Biomedical Engineering and Chemical & Biological Engineering, Rensselaer Polytechnic Institute, Troy, NY

The human gastro-intestinal (GI) tract is colonized by ~10^{14} commensal (non-pathogenic) bacteria in close proximity to host epithelial cells [1]. Intra-kingdom (host-host) and inter-kingdom (host-bacteria) communication is integral to proper functioning of GI tract, and helps maintain a delicate balance between a pro-inflammatory response to invading pathogens and tolerance towards commensal bacteria (anti-inflammatory response) [1]. Disruption of this homeostatic balance can lead to indiscriminate phagocytosis of both bacterial and host cells resulting in diseases such as Crohn's disease and Ulcerative colitis [2]. This study focuses on investigating the inter- and intra-kingdom signaling mechanisms involved in maintaining homeostasis in the GI tract through mathematical modeling and experimental validation.

One of the anti-inflammatory metabolites produced by the commensal bacteria is indole. Indole is generated from dietary tryptophan by the commensal bacteria, using the enzyme tryptophanase A, and is released into the lumen of GI tract [3]. Prior work from our lab [3] has shown that microbiota-derived indole attenuates indicators of inflammation by decreasing production of IL-12 and IL-8 while increasing production of IL-10. More importantly, these studies also showed that indole decreased activation of the pro-inflammatory transcription factor NF-kB in intestinal epithelial cells. Since IL-10 is also shown to decrease NF-kB activation [4], these results suggest that NF-kB signaling is a putative mechanism through which inter-kingdom signaling effects are mediated.

In this work, we have developed a mathematical model that represents the signal transduction pathway in dendritic cells through which each host-derived pro-inflammatory signaling and microbiota-driven anti-inflammatory signaling converge at NF-kB. TNF-α is used as the prototypic pro-inflammatory cytokine while IL-10 and indole are used as the anti-inflammatory input to the model. The integrated model has been developed in MATLAB using ordinary differential equations and unknown parameters have been estimated from experimental data. A regulatory feedback loop for TNF-α has been included to account for the effect of *de novo* synthesized TNF-α on NF-kB activation. IL-10 and indole signaling models have been integrated into TNF-α model through common nodes among them. Quantitative experimental data on NF-kB activation has been obtained using dendritic cells transduced with a secreted luciferase reporter system.

The model developed in this study is expected to be helpful in determining the points of intersection of pro- and anti-inflammatory responses and the molecular mediators involved in maintaining homeostasis in the human GI tract, which would in turn help in designing therapeutic strategies for inflammatory disorders.

References:

1. Berg, R.D., *The indigenous gastrointestinal microflora.* Trends in Microbiology, 1996. **4**(11): p. 430-435.

2. Bouma, G. and W. Strober, *The immunological and genetic basis of inflammatory bowel disease.* Nat Rev Immunol, 2003. **3**(7): p. 521-533.

3. Bansal, T., et al., *The bacterial signal indole increases epithelial-cell tight-junction resistance and attenuates indicators of inflammation.* Proceedings of the National Academy of Sciences, 2010. **107**(1): p. 228-233.

4. Dhingra, S., et al., *IL-10 attenuates TNF-α-induced NFκB pathway activation and cardiomyocyte apoptosis.* Cardiovascular Research, 2009. **82**(1): p. 59-66.

(100e) Optimum Perfusion Duration for Machine-Perfused Rat Livers

Sinem Perk[1], Maria-Louisa Izamis[1], Herman Tolboom[2], Basak Uygun[1], Francois Berthiaume[3], Martin L. Yarmush[4] and Korkut Uygun[1], (1)Center for Engineering in Medicine, Massachusetts General Hospital, Harvard Medical School, Shriners Hospital for Children, Boston, MA, (2)Division of Cardiac and Vascular Surgery, University Hospital Zurich, Zurich, Switzerland, (3)Department of Biomedical Engineering, Rutgers, The State University of New Jersey, Piscataway, NJ, (4)Center for Engineering in Medicine, MGH/Harvard Medical School, Boston, MA

With the crisis in donor organ availability, recent studies on liver preservation focus on new, dynamic modalities such as machine perfusion to recover discarded donor organs to a functional state. Compared to static cold storage, machine perfusion provides oxygen and nutrition enabling recovery of the organ. Despite years of research however, little is known about key factors such as how to evaluate the degree of recovery during perfusion and what the ideal perfusion duration is to maximize the graft recovery and transplantation success. Development of such approaches are complicated by the fact that every donor liver is different and therefore it is very likely that the optimum perfusion time, as well as other operating parameters and specific treatments, may be different for every organ.

It is well known that the existing organ preservation systems are limited in duration that the success rate falls dramatically in time. While many studies demonstrate that hours of ex-vivo machine perfusion are necessary for resuscitation of a DCD organ, it is also well known that organ function deteriorates if it is kept in perfusion for too long. Accordingly, the organ viability and transplant success probability likely goes through an optimum, before which the recovery from ischemia is incomplete, and beyond which the organ begins to deteriorate due to damage by the perfusion system.

The aim of this work is the development of an approach that enables assessment of liver viability during perfusion, the prediction of clinical outcome post-transplantation, and the determination of the optimum duration of perfusion.

The ideal method in perfused graft viability assessment would be based on online measurements that can be performed accurately and quickly such as Blood Gas Analyses. The dynamic patterns of organ function during perfusion are evaluated and compared between successful and unsuccessful transplants in a manner that allows a statistical analysis of transplant success, providing an estimate for the likelihood of graft success to the surgeon as a decision support system for transplantation.

With the on-line data obtained during machine perfusion, a suitable methodology for such dynamic organ viability assessment is statistical process monitoring (SPM) based on partial least squares (PLS) methodology and its extension for dynamic processes: multi-way PLS (MPLS). Constructing a statistical model of liver dynamic response to perfusion enables in-silico predictions of the liver trajectory during perfusion. These predictions of the trajectory of the organ in the following hours can then be used to detect any emerging trends that indicate deterioration of organ viability, and therefore enable identifying the optimum perfusion duration, and potentially other operational variables, for each organ based on its function on-line during perfusion.

This work presents a novel approach and a two-step MPLS/PLS algorithm for the prediction of perfused liver survival, based on measurements taken during perfusions of rat livers for 6 and 12 hours, with recipients of the first group surviving and latter group mostly failing. Via an online MPLS analysis, blood gas data taken hourly is used to estimate the post-transplantation survival rates dynamically during perfusion, which are compared with the actual post-transplantation clinical outcomes for each animal. The results demonstrate that with the proposed methodology, graft viability can be monitored and transplant success rate can be predicted online during perfusion with cross-validated sensitivity and specificity over 91%.

(100f) Analyzing the Dynamics of Cell Cycle Transition in Differentiating Embryonic Stem Cells Through an Integrated Experimental and Modeling Approach

Keith Task, Chemical Engineering, University of Pittsburgh, Pittsburgh, PA and Ipsita Banerjee, Chemical and Petroleum Engg, University of Pittsburgh, Pittsburgh, PA

Embryonic stem cells (ESC) have the potential to be used in many therapeutic applications due to their ability to expand in an undifferentiated state while being able to differentiate to any cell type under specific conditions. One important characteristic vital to this behavior is the cell cycle. In somatic cells, the cell cycle is characterized by a lengthy G1 (growth) phase and a long doubling time. ESC are typically characterized by a shortened G1 phase, with an increased proportion of cells residing in the S phase (DNA replication). This leads a shorter time between mitotic events. These traits are crucial for ESC to facilitate quick self-renewal and proliferation. Although these unique cell cycle traits are central to ESC, attempts towards a thorough quantitative analysis of the cell cycle behavior in human ESC (hESC) and in reprogrammed induced pluripotent cells (iPS) is lacking thus far. Furthermore, little work has been done in the area of mechanistic information and dynamics of cell cycle transition from undifferentiated to germ layer. In this project we are using an integrated experimental and computational approach to investigate cell cycle traits with different hESC and iPS cell lines, both at the singular cell and population levels. Furthermore, we investigate the dynamics of cycle behavior transition when differentiation is induced.

Two different hESC lines (the use of which we have approval for) were used for the study, H1 and H9, as well as three iPS lines (Y1 and hFib2-iPS5, derived from dermal fibroblasts, and a type I diabetes specific line). Proportions of the cell cycle were quantified with the DNA stain propidium iodide (PI) and flow cytometry. Further characterization of the cell cycle was performed with Brdu, an S-phase proliferation maker, to extract information regarding the length of cell cycle phases. Quiescent cell state was determined by the lack of Ki67 stain. The undifferentiated cells were first analyzed to determine base level cycle phase characteristics. To study how this heterogeneous population is generated from a homogeneous starting point, the pluripotent cells were synchronized in the G2 phase by Nocodazole. Cells were released from cell synchronization, and the DNA content was analyzed every two hours to determine, temporally, how the cells revert to their asynchronous behavior. In the differentiation studies, endoderm was induced by supplementing culture media with Activin A. Four different induction pathways were compared: Activin A supplemented with either Wnt3a, Bmp4, Fgf2 or PI3K inhibition. PI, Brdu, and Ki67 stains were performed at different time points for each condition to determine the dynamics of the cell cycle phase shift. To determine the extent of differentiation, both quantitative polymerase chain reaction (qPCR) and intracellular flow cytometry were utilized. Furthermore a stochastic population based model was developed to extract mechanistic information from the experimental observations.

The simulated cell cycle data was compared to the experimental observations, and through this comparison probabilistic parameters and characteristics associated with ESC were determined. Only certain probability distributions and moments (e.g. mean and variance) were able to recapitulate experimental dynamics. It is shown that the simulated data can describe the experimental data well, being able to generate the PI flow cytometry histograms, both of the asynchronized pluripotent cells and the dynamics after release from synchronization. The doubling time predicted by the model was validated by time-lapse microscopy and Brdu analysis. Furthermore, by performing the analysis on differentiating cells, it is shown that different induction conditions show different behavior in terms of cell cycle, with different dynamics associated with the transition from "pluripotent cycle" (shorter doubling times and reduced G1 time) to "differentiated cycle." Our stochastic population model is able to draw mechanistic insight regarding both the nature of the cell cycle and of differentiation from experimental observations. Although applied to undifferentiated and germ layer cells, the model could be extended to later stages of differentiation.

(100g) Analysis of EGFR Flux Through Different Endocytic Pathways in Cancer Cells with Elevated EGFR Expression

Alice J. Macdonald, Bioengineering, University of Pennsylvania, Philadelphia, PA and **Matthew J. Lazzara**, Chemical and Biomolecular Engineering, University of Pennsylvania, Philadelphia, PA

In cancer cells characterized by elevated expression of the epidermal growth factor receptor (EGFR), perturbations to the kinetics of ligand-mediated EGFR endocytosis are frequently observed. Such perturbations can have important consequences for the ability of EGFR to activate downstream signaling pathways and may therefore affect cancer cell growth rates and response to therapeutics. Given the existence of endogenous mechanisms which both promote and impede EGFR endocytosis in response to EGFR activation, the question arises as to how these processes act in concert to determine rates of EGFR endocytosis in cancers with elevated EGFR expression. To quantitatively address this question in the context of non-small lung cancer (NSCLC) cells with elevated EGFR expression, we developed a novel computational model to parse the effects of Sprouty2, an antagonist of EGFR endocytosis through the CBL pathway, and MIG6, which promotes non-CBL-mediated EGFR endocytosis, on overall EGFR endocytosis kinetics. Our computational framework consisted of a set of coupled ordinary differential equations describing the kinetics of EGFR ligand binding, EGFR dimerization, and EGFR endocytosis through pathways involving CBL (which is sequestered by Sprouty2), MIG6, or neither protein. To determine the unknown rate constants describing the flux of EGFR through the different endocytosis pathways, we created a panel of H1666 and PC9 NSCLC cell lines with knockdown of Sprouty2, MIG6, or both proteins. Whereas H1666 cells express wild-type EGFR, PC9 cells express a constitutively active EGFR mutant which displays impaired ligand-mediated endocytosis. In addition to measuring rates of EGF-mediated EGFR endocytosis in these cell lines, we also measured EGFR expression levels and the fraction of endocytosed EGFR which was recycled to the cell surface. In both cell lines, SPRY2 knockdown promoted EGFR endocytosis and MIG6 knockdown impeded endocytosis. With SPRY2 knockdown alone or together with MIG6 knockdown, we also observed a marked reduction in EGFR expression in both cell lines. Based on these data, our model results suggested that the effects of Sprouty2 knockdown on EGFR endocytosis were almost completely attributable to changes in EGFR expression, a prediction which we validated by reconstituting EGFR expression in H1666 and PC9 cells with Sprouty2 knockdown. In addition, our model results suggested that the MIG6-mediated internalization pathway plays a surprisingly significant role in setting EGFR endocytosis kinetics of wild-type EGFR in H1666 cells. Not surprisingly, the CBL-mediated pathway was also found to play a significant role in EGFR endocytosis in H1666 cells. By contrast, the model results suggested that the CBL-mediated pathway plays essentially no role in mutant EGFR endocytosis in PC9 cells and that MIG6 may promote mutant EGFR endocytosis less efficiently than wild-type EGFR endocytosis. In addition to these specific conclusions regarding wild-type and mutant EGFR endocytosis in NSCLC cells, the results of this work demonstrate the utility of our computational and experimental approach for identifying the most relevant determinants of EGFR endocytosis in different cellular contexts. This approach may be of general use in identifying optimal ways to regulate EGFR endocytosis kinetics for therapeutic purposes.

(100h) A Stochastic Model of Glycosylation of Monoclonal Antibodies

Devesh Radhakrishnan[1], Andrew Bitner[1], Melissa St. Amand[1], Kevin Tran[1], Anne S. Robinson[1,2] and Babatunde A. Ogunnaike[1], (1)Department of Chemical and Biomolecular Engineering, University of Delaware, Newark, DE, (2)Department of Chemical and Biomolecular Engineering, Tulane University, New Orleans, LA

Most proteins expressed in eukaryotic cells undergo the post-translational modification known as glycosylation in which a glycan (sugar moiety) is attached to the core of the protein. The addition of the sugar can either be to the amine group on the protein (N-glycosylation) or to the hydroxyl group (O-glycosylation). Of these, N-glycosylation, in which the glycan is added to the core of the protein in the lumen of the endoplasmic reticulum of the cell, followed by subsequent modification in the Golgi apparatus, has the greater impact on protein structure and properties. (For instance, the different N-glycans attached to the protein affect immunogenicity, *in vivo* half-life, and the receptor recognition of the protein.) Several commercially relevant biopharmaceuticals such as monoclonal antibodies (mAbs), erythropoietin, etc., are glycosylated proteins whose final product quality characteristics depend on the distribution of the different glycoforms. In order to ensure consistent product quality, therefore, regulatory agencies are encouraging manufacturers to monitor and control the quality of the product during the course of manufacturing. This requires the implementation of an appropriate control strategy, and developing such a strategy for effective control of glycosylation requires a mathematical model that represents the process of glycosylation adequately. Several models exist for predicting the distribution of glycoforms in different cellular systems[1-4] on the basis of overall production of protein in the culture. However, the glycan addition to a protein is a molecular phenomenon occurring within the microscopic volumes of vesicles in the Golgi apparatus, essentially invalidating the continuum assumption for glycan concentrations. Under such circumstances, a stochastic model will provide a more appropriate representation of the mechanisms of glycosylation and glycoprotein formation.

In this work, we develop what is, to our knowledge, the first stochastic model of glycosylation, and discuss the key challenges in the implementation of such a model, along with our approach to handling these challenges. We present results of our stochastic model's prediction of the glycoform concentration profiles and compare them to corresponding deterministic model predictions.

References

1. Umaña, P.; Bailey, J. E. Biotechnology and Bioengineering 1997, 55, 890-908.

2. Krambeck, F. J.; Betenbaugh, M. J. Biotechnology and Bioengineering 2005, 92, 711-728.

3. Hossler, P.; Mulukutla, B. C.; Hu, W. S. Plos One 2007, 2.

4. Jimenez Del Val, I.; Nagy, J. M.; Kontoravdi, C. Biotechnology progress 2011, 27, 1730-1743.

(104a) Protein Analogous Micelles: Versatile, Modular Nanoparticles

Matthew Tirrell, Institute for Molecular Engineering, University of Chicago, Chicago, IL

Peptides are functional modules of protein macromolecules that can be displayed apart from the whole protein to create biofunctional surfaces and interfaces, or can be re-assembled in new ways to create synthetic mimics of protein structures. Each of these routes are being employed to gain new insight into protein folding and to develop new, functional, biomolecular materials. Examples of work from our laboratory in this area using peptide-lipid conjugate molecules (peptide amphiphiles) will be discussed relating to multi-functional surfaces, DNA-binding peptide assemblies, synthetic vaccines, and protein analogous micelles for cancer and cardiovascular therapeutics. Spherical micelles with characteristic dimensions of less than 20 nanometers are very effective in avoiding clearnace before reaching targeted pathlogical tissues. Particular applications to targeting vulnerable atherosclerotic plaque and solid tumors will be described. Aspects of structure-function relationships, stability and control of geometry of these assemblies will be discussed. Protein anaologous micelles represent the most flexible, potent and versatile class of nanoparticles for any applications.

(104b) Molecular Transport and Structure of Surface Modified Nanopores:

Igal Szleifer, Biomedical Engineering, Northwestern University, Evanston, IL

Natural and man-made pores and channels of nanoscale dimensions display unique ionic and protein transport behavior. Nanopores modified with supramolecular chemical species (such as polyelectrolyte brushes) have dimensions that are similar to the range of the electrostatic interactions, and also to the molecular size of the tethered macromolecules. In cells, Nuclear Pore Complexes (NPC) control the transport of species between the cytoplasm and the nucleus using disordered proteins as gate keepers. The competition between molecular and interaction length scales, as well as the geometry of the surfaces, creates interesting possibilities for the creation of stimuli responsive gates and ion channels and for the fundamental understanding of the interplay between molecular organization and charge transport in nanoconfined environments. In this talk we will discuss recent theoretical developments that enable us to study the structural properties of the modified pores and the coupling that exists between the protonation state of the polyelectrolytes and ionic transport through them. The predictions from the theory are in excellent agreement with experimental observation and they provide a direct understanding of the important coupling that exists between protonation state, molecular organization and transport. We will demonstrate how the molecular organization within the pore is a strong function of the pore geometry and how these effects translate into the transport behavior. We will show a novel non-equilibrium approach to treat transport of ions and discuss the intrinsic coupling that exists between macromolecular organization within the nanopore and ion transport. Finally, we will show preliminary calculations of the spatial distribution of amino acids when the full NPC disorder proteins is taken into account. The importance of considering explicit molecular details of the surface modifiers as well as the coupling between chemical equilibrium, physical interactions and molecular organization will be discussed in detail.

(104c) The Micromechanics and Physics of Cancerous Cells Revealed Using the Tools of Nanoscience

Bartosz Grzybowski, Chemical and Biological Engineering, Northwestern University, Evanston, IL

The mainstream research in cancer biology focuses on the genetic and molecular signatures underlying the disease. An alternative and perhaps more physics-oriented approach is to study cancer from a perspective of an entire cell treated as a complex system characterized by a set of micromechanical properties and cell-motility characteristics. In my lecture, I will talk about both of these aspects. In the context of micromechanics, I will first describe a new cell-tomography technique which allows for the reconstruction of 3D shapes of cells subject to well-defined, geometrically regular microconfinements. The reconstructed shapes allow for the derivation of an analytical micromechanics model according to which cell's mechanical properties are determined by the deformability of the cell membrane and cortical actin and by the limited compressibility of the nucleus. This model does not involve any arbitrary parameters and is universal in the sense that it reproduces the shapes, volumes and surface areas of cells of different types under different experimental conditions. Unfortunately, analysis of cancerous vs. non-cancerous cells indicates little – if any – differences in their micromechanical properties. Thus, in search for the features distinguishing malignant from non-malignant cells the second part of the talk will focus on the ways in which cells move – again, over well defined microenvironments allowing for quantitative analyses. As I will show, non-cancerous and non-metastatic cells migrate in ways fundamentally different from metastatic cells. While non-metastatic cells are simple, diffusive random walkers, their non-metastatic variants employ space-searching strategies used by animal predators. In this spirit, metastatic "cellular predators" can be viewed as navigating optimally through human body in search for suitable conditions to thrive (as in the "seed and soil"). Importantly, in doing so they desynchronize their cytoskeletal dynamics and "lose" a characteristic time scale present in non-metastatic cells. These observations prompt a hypothesis that cancer metastasis entails "desynchronization" of a cytoskeletal "clock".

(104d) Artificial Viruses

Nicholas Kotov, Chemical Engineering, University of Michigan, Ann Arbor, MI

Self-organization of nanoparticles (NPs) and nanoscale objects represents one of the most dynamic areas of materials science. Better understanding of these phenomena is important from both fundamental and practical perspectives because nanoparticle self-organization processes

(1) identify similarities between biological and non-biological nanoscale species;

(2) lead to unusual optical properties from different combinations of nano- and microscale features;

(3) can potentially simplify manufacturing of electronic, photonic, and sensing devices.

In this presentation, the analogies between self-organized nanoparticle assemblies and viruses will be discussed and analyzed. We shall start with their thermodynamics and continue with the preparation of different NP assemblies resembling the structure and geometry of viruses. The individual NPs in the virus-like superstructures will be treated as an analog of a protein. Dynamic reorganization of NP superstructures will also be considered. The presentation will be concluded with parallels in the interactions of NP assemblies with cells and their utilization for drug delivery.

Relevant References.

1. Z. Tang, N. A. Kotov, M. Giersig, *Science,* **2002**, *297,* 237-240.
2. Tang, Z. Zhang, Z.; Wang Y.; Glotzer, S. C. Kotov, N. A., *Science*, **2006**, 314 (5797) 274-278.
3. J. Lee, A. O. Govorov, N. A. Kotov, *Angew. Chem. Intern. Ed.* **2005**, *44,* 7439-7442.
4. J. Lee, A. O. Govorov, N. A. Kotov, *Nature Materials.* **2007**, 6(4), 291-295.
5. S. Srivastava, A. Santos, K. Critchley, K.-S. Kim, P. Podsiadlo, K. Sun, J. Lee, C. Xu, G. D. Lilly, S. C. Glotzer, and N. A. Kotov, *Science*, **2010**, 327, 1355-1359.
6. M. Yang, K. Sun, N. A. Kotov, *J. Am. Chem. Soc.*, **2010**, *132* (6), pp 1860–187.
7. Y. Zhou, M. Yang, K. Sun, Z. Tang and N. A. Kotov, *J. Am. Chem. Soc.*, **2010**, *132* (17), 6006–6013
8. N. A. Kotov, Inorganic Nanoparticles as Protein Mimics, *Science,* **2010**, 330 (6001), 188-189.
9. S. I.Yoo, N.A. Kotov, *Angewandte Chemie*, **2011,** 50(22), 5110–5115.
10. Y. Xia et al, *Nature Nanotechnology*, **2011**, 6, 580-587.

(104e) Confinement Protection Effects of Mesoporous Silica SBA-15 On Myoglobin, in an Environment Inspired by the Groel/ES Chaperonin System

Michele Lynch[1], Justin Siefker[2] and Marc-Olivier Coppens[1], (1)Chemical & Biological Engineering, Rensselaer Polytechnic Institute, Troy, NY, (2)Dept. Chemical & Biological Engineering, Rensselaer Polytechnic Institute, Troy, NY

With the increasing prevalence of protein therapy drugs, such as antibodies, it is important to stabilize and protect these proteins in the body. The structure of proteins is more delicate than that of small-molecule drugs; e.g., many proteins denature and become inactive when exposed to proteases or extreme pH [1]. Microscopic, mesoporous silica SBA-15 particles are interesting candidates for drug delivery and sensing applications. SBA-15 contains a hexagonally ordered array of cylindrical pores. These pores have certain commonalities with the GroEL/ES complex, a pair of chaperone proteins found in bacteria cells that assist in the folding of new proteins [3]. More in particular, the pores of SBA-15 have a hydrophilic core, similar to the GroEL/ES complex, and exhibit similar electrostatics with respect to protein adsorption. We report on the denaturation protection from exposure to proteases and extreme pH, conferred to proteins adsorbed on the interior pore surface of SBA-15.

To test for the protective effects of SBA-15, myoglobin is adsorbed on the internal pore surface of SBA-15 particles, and these composites are exposed to the protease pepsin. Both free myoglobin and myoglobin/SBA-15 composites are suspended in phosphate buffer with a pH of 4.6. This pH was chosen because pepsin is only active in environments with a pH lower than 6.5, and optimally at a pH of 2.0 [2]. The activity of free and confined myoglobin is tested via a peroxidase enzyme activity assay, which is measured by UV-Vis spectroscopy. Activity measurements are normalized to those of free myoglobin in buffer with a pH of 7.2 (myoglobin's isoelectric point), and to the measured amount of adsorbed myoglobin. The myoglobin loading of all composites is below 50 mg/g SBA-15, this is within the Henry adsorption regime.

Our previous work had shown that adsorbing myoglobin to SBA-15 at its isoelectric point (pH 7.2) leads to an extraordinary increase in activity, around 700% [1]. However, at pH 4.6, the activity of confined myoglobin is reduced with respect to that of free myoglobin at the same pH. When free myoglobin is digested by active pepsin, its activity is reduced substantially. In contrast, confined myoglobin shows a nearly imperceptible drop in activity when exposed to pepsin. This result is encouraging, as it suggests that confinement protects the myoglobin from proteolysis.

Changes in secondary structure are observed with Fourier transform infrared spectroscopy (FTIR). Also the effect on myoglobin catalytic activity of restoring the pH of all samples to 7.2 after a temporary decrease in pH will be reported. This experiment is inspired by the behavior of the GroEL/ES complex, which can also refold denatured proteins. The results of this study elucidate the effects of confinement in SBA-15 on myoglobin with respect to extreme pH exposure.

[1] L.C. Sang and M.-O. Coppens (2011), *Phys. Chem. Chem. Phys.* **13**, 6689.

(104f) Inhibition of Viral Infection by Molecularly Imprinted Nanoparticles - A Synthetic Approach to Antiviral Therapy

Yen Wah Tong, Department of Chemical and Biomolecular Engineering, National University of Singapore, Singapore, Singapore and Niranjani Sankarakumar, Chemical and Biomolecular Engineering, National University of Singapore, Singapore, Singapore

Despite the current advances in the field of medicine and pharmaceutics, virus transmitted diseases are still one of the major health issues today with some of the most deadly and perplexing viruses being Ebola, HIV and SARS. Unfortunately there are no cure nor viable vaccines for all types of viral infections while currently existing therapies lose their effectiveness with time or due to viral mutations. To cater to the mounting demand for new antiviral strategies, we herein propose the use of high affinity polymeric receptors prepared by molecular imprinting technique to "catch" viruses.

Molecular imprinting is an effective technique that can generate tailor-made artificial materials possessing comparable molecular recognition and binding capabilities as with natural biomolecules like antibodies. It involves the creation of three dimensional highly specific binding sites in a polymer matrix ('the lock') that are complementary to the size, shape and functional group orientation of the 'substrate' molecule ('the key'). Thus, molecularly imprinted polymers (MIPs) have been extensively used as substitutes for antibodies or enzymes in a wide range of technologies such as separation and purification, sensing, catalysis and drug delivery. In this research, our aim is to create surface imprints of viruses on polymeric nanoparticles and to employ the particles for "capture" and removal of any active, disease-causing viruses.

Virus imprinted polymeric nanoparticles (vMIPs) were fabricated using different functional monomers and surface imprinting strategies with fr bacteriophages employed as a model virus. The imprinted polymeric particles prepared were of sizes ranging from 40 nm to 570 nm using methyl methacrylate and ethylene glycol dimethacrylate as the functional and cross-linker monomers, respectively. A significant reduction in phage concentrations were observed after just 3 h of exposure to vMIPs. The imprinted particles further displayed remarkable positive anti-viral effect that significantly hindered phage infection of host bacteria as compared to the controls. To the best of our knowledge, this is the first report demonstrating the use of molecularly imprinted polymers as anti-viral agents targeted at infectious disease treatment.

The synthesis of selenium nanoparticles via a simple fast reaction and its antibacterial application

Qi Wang[1] *and Thomas J. Webster*[1,2]
[1]*Bioengineering Program and* [2]*Department of Chemical Engineering, College of Engineering, Northeastern University, Boston, MA 02115, U.S.A.*

Abstract

In this study, selenium nanoparticles were synthesized and coated on materials using a simple fast precipitation method. The amount, roughness, adhesion and other properties of the nano-sized coatings were characterized. By coating with selenium nanoparticles, the materials, specifically polycarbonate and paper towels, were introduced with antibacterial properties inhibiting bacterial growth. The objective of the present study was to coat selenium nanoparticles on materials and test their effectiveness preventing bacterial growth. The results showed that the selenium nanoparticles coated on the materials were well distributed semi spherical and 50-100nm in diameter. Most importantly, selenium coatings significantly inhibited *Staphylococcus aureus* growth to 27% and 10% after 72 hours on polycarbonate and paper towels, respectively, compared to the uncoated materials. Therefore, this study suggested that coating materials with selenium nanoparticles is a fast and effective way to reduce bacterial infections, which could potentially used for various materials.

Introduction

Bacterial infections are commonly found in hospital environment, such as forming biofilms on various medical devices, bacteria attachment on the wall paper of the clinic room resulting in cross contamination during flu seasons and contamination of hands caused by paper towels hanging in sink splash zones.[1-4] *S. aureus* (*Staphylococcus aureus*) is one of the bacterium commonly found in numerous infections. Each year, there are 11 million outpatient/emergency room visits and 464,000 hospital admissions in the U.S. alone due to *S. aureus* infections.[5] *S. aureus* biofilms have been found on a wide range of medical devices including prosthetic heart valves, central venous catheters, orthopedic prostheses, contact lenses, endocarditis, otitis media and osteomyelitis.[6] Therefore, it is significant to develop a method to prevent bacterial infections on materials.

Different from plenty of nano-sized metal-based antibacterial materials, such as zinc oxide nanoparticles,[7] silver nanoparticles,[8-10] copper and its alloys,[11-12] that may cause potential toxicity to mammalian cells,[13-15] selenium is a naturally occurring micronutrient, which is fundamental for 25 selenoproteins or enzymes in the human body with selenocysteine at their active center.[16] Therefore, in this study, selenium nanoparticles were coated on polycarbonate and paper towel surfaces through a quick precipitation method and the properties of selenium coated paper were characterized. In addition, such materials were tested for their effectiveness to kill bacteria and/or prevent bacteria from attaching in the bacterial assays. The results showed that the selenium coatings successfully introduced antibacterial properties to polycarbonate and paper towels, revealing a promising nano-selenium based method to prevent bacterial infections on various materials.

Materials and Methods

Selenium nanoparticles were synthesized through a simple reaction between glutathione and sodium selenite (4:1 molar mixture) and were precipitated on the surface of polycarbonate (McMASTER-CARR, 85585k82, 0.13mm thickness, cut into round films 7.01mm in diameter) and paper towels (Tork Advanced, MB550A Hand Towel, cut into round chips, 7.01mm in diameter) for either 30s or 60s. Sodium hydroxide was added to bring the pH of the solution to the alkaline regimen to start the reaction. As the size of selenium nanoparticles on the surface is influenced by the concentration of NaOH and the coating time, two coating conditions were used in this study. One was using a 0.5M NaOH solution for 30s and the other one was using a 1.0M NaOH solution for 60s. After coating, tape tests (D3359-09e2: Standard Test Methods for Measuring Adhesion by Tape Test) were used to test the strength of adhesion of the selenium nanoparticles on the substrate surfaces. AAS (Atomic Absorption Spectroscopy) was used to measure the amount of coated selenium on polycarbonate surfaces. The SEM (Scanning Electron Microscope, HITACHI 2700) images of the substrate surfaces were taken to determine the size, the distribution and coverage of selenium nanoparticles. AFM (Atomic Force Microscope, MFP3D, Asylum Research, Sharp tipped cantilever, K = 0.06N/M, Contact Mode) was used to demonstrate that the coated selenium nanoparticles increased the surface selenium exposure and surface roughness of polycarbonate.

In order to test the effectiveness of selenium coatings inhibiting biofilm formation, bacteria experiments were implemented. A bacteria cell line of *Staphylococcus aureus* was obtained in freeze-dried form from the American Type Culture Collection (catalog number 25923). Selenium coated samples were rinsed with 75% ethanol for 20 minutes for sterilization purposes and were left in the sterile petri dishes for 30 minutes to completely dry. Then, the samples were transferred to a 24-well plate, treated with the prepared bacterial solutions (10^6 bacteria/ml, *Staphylococcus aureus*) and cultured for either 24, 48 or 72 hours in an incubator (37°C, humidified, 5% CO_2). After the treatment, the samples were rinsed with a PBS (phosphate buffered saline) solution twice and placed into 1.5ml microfuge tubes with 1ml of PBS. These tubes were shaken at 3000 rmp for 15 minutes on a vortex mixer to release the bacteria attached on the surface into the solution. Solutions with bacteria were spread on agar plates and bacteria colonies were counted after 18 hours of incubation. Bacterial tests were conducted in triplicate and repeated three times. Data were collected and the significant differences were assessed with the probability associated with a one-tailed Student's t-test. Statistical analyses were performed using Microsoft Excel (Redmond, WA).

Results and Discussion

Most of the nanoparticles were approximately 50-100nm in diameter and well distributed on the polycarbonate (Figure 1) and paper towel (Figure 2) surfaces. On polycarbonate surfaces, selenium nanoparticles coated with a condition of 1.0M NaOH for 60s were larger than those with the condition of 0.5M NaOH for 30s. The concentration of nanoparticles on the polycarbonate surfaces were on average 19.34g/m^2 and 20.95g/m^2 for the 0.5M NaOH for 30s condition and for the 1.0M NaOH for 60s condition, respectively. While on paper towels, the concentration of of selenium nanoparticles on the coated paper towel surface was 69.00 g/m^2, which was 4 times larger than the amount on polycarbonate under the same coating conditions. One of the reason was that the fibrous structure of the paper towel allowed the samples to hold more selenium nanoparticles. Tape tests removed about 50% and 75% of the selenium nanoparticles from polycarbonate surfaces for the 0.5M NaOH for 30s condition and for the 1.0M NaOH for 60s condition, respectively, according to AAS results. This result indicated that selenium nanoparticles with larger sizes might have less adhesion to an underlying substrate than smaller selenium nanoparticles.

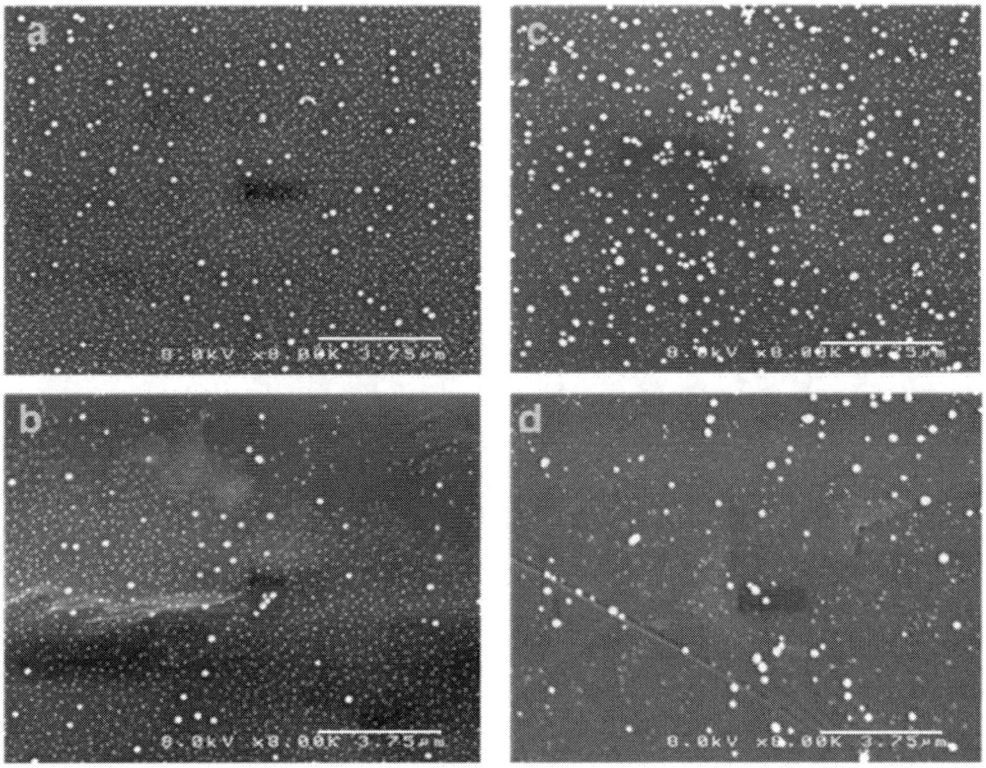

Figure 1. SEM images of selenium coated polycarbonate samples before and after a tape test. (a) 0.5M NaOH, coating for 30 seconds and before a tape test. (b) 0.5M NaOH, coating for 30 seconds and after a tape test. (c) 1.0M NaOH, coating for 60 seconds and before a tape test. (d) 1.0M NaOH, coating for 60 seconds and after a tape test. The coverage of selenium nanoparticles on the surface for image (a), (b), (c) , (d) are 11.74%, 6.94%, 11.38%, 5.88% respectively.

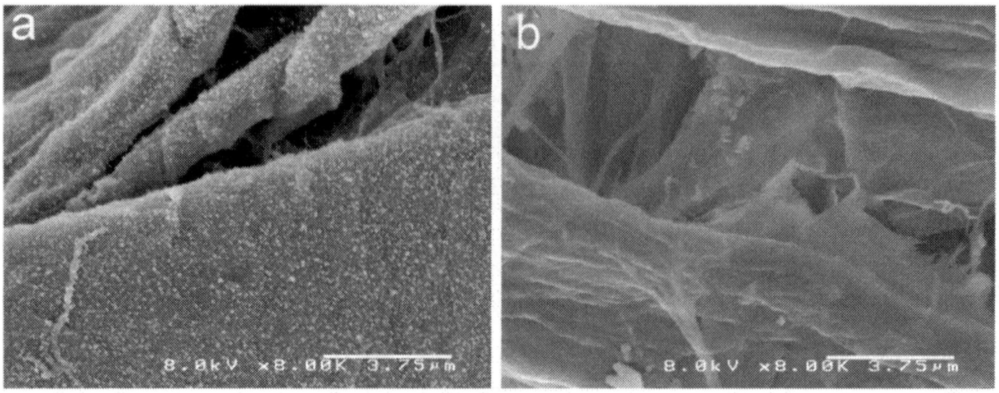

Figure 2. SEM images of selenium coated (image a) and uncoated paper (image b) towel samples. The coating condition for image a was 0.5M NaOH for 30 seconds. The concentrations of selenium on paper towels from AAS measurement were 69.00 g/m^2 for selenium coated paper towels and 0 g/m^2 for uncoated paper towels.

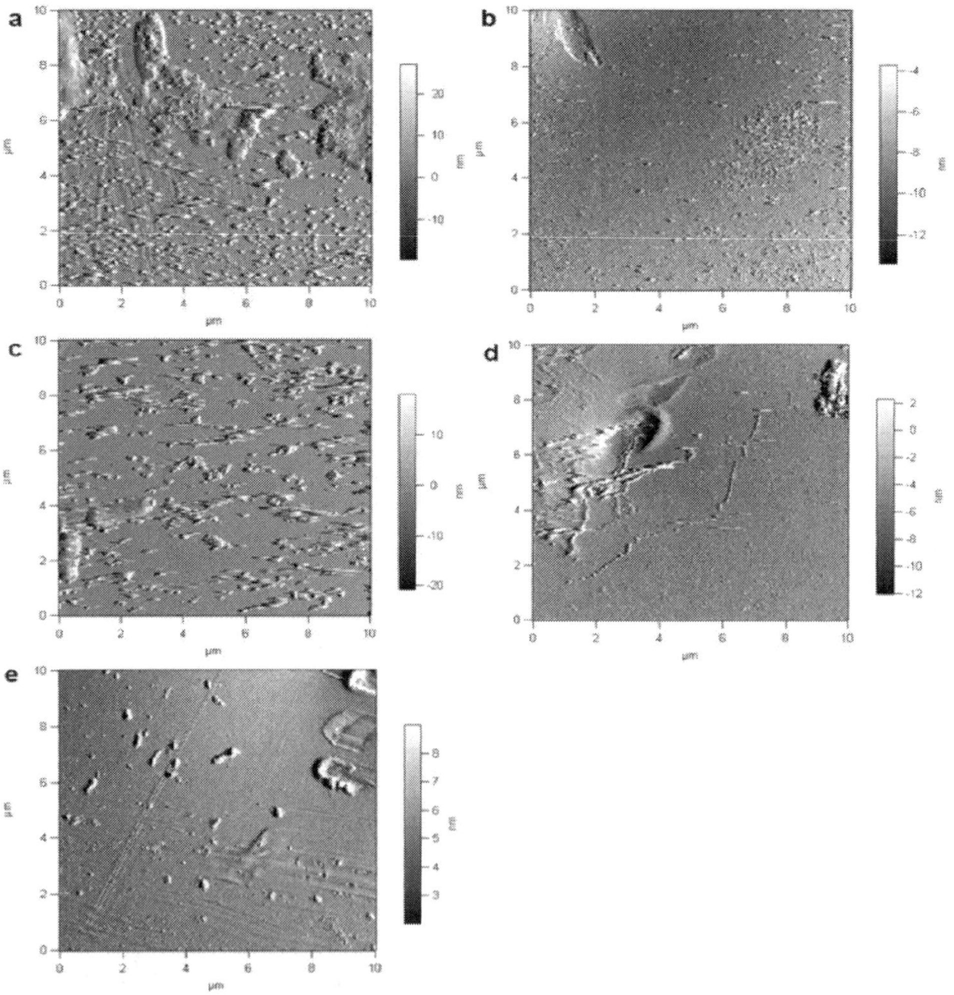

Figure 3. AFM images and RMS (root mean square, scan area = 10um×10um) roughness of

polycarbonate surface. (a) Selenium coated with 0.5M NaOH for 30s, RMS roughness = 45.997nm. (b) Selenium coated with 0.5M NaOH for 30s, after tape test, RMS roughness = 21.731nm. (c) Selenium coated with 1.0M NaOH for 60s, RMS roughness = 53.084nm. (d) Selenium coated with 1.0M NaOH for 60s, after tape test, RMS roughness = 34.925nm. (e)Uncoated sample, RMS roughness = 14.898nm.

From the AFM images (Figure 3), The RMS (root mean square) roughness for the selenium coated surface at a condition of 0.5M NaOH for 30s and at a condition of 1.0M NaOH for 60s were 45.997nm and 53.084nm, respectively. After the tape test, the RMS roughness decreased to 21.731nm and 34.925nm for the these two coating conditions. The RMS roughness of the uncoated surface was 14.898nm, which was much smaller than the selenium coated polycarbonate samples. So, there was a significant increase in roughness, therefore surface area, after coating with selenium nanoparticles.

Experiments with bacteria (specifically, *Staphylococcus aureus*) showed high effectiveness for the selenium coated materials at killing bacteria or preventing bacteria from attaching compared with the uncoated surfaces (Figure 4 and Figure 5). Compared to uncoated samples, more than 91%, 15% and 73% of the bacteria (compared with uncoated polycarbonate) died or had been removed after 24, 48 and 72 hours on the selenium coated polycarbonate. The selenium coated paper towels had 88.6%, 88.9% and 88.8% less bacteria attached than the uncoated paper towels after treatment with the bacteria solution for 24, 48 and 72 hours, respectively. Importantly, this was accomplished without using antibiotics.

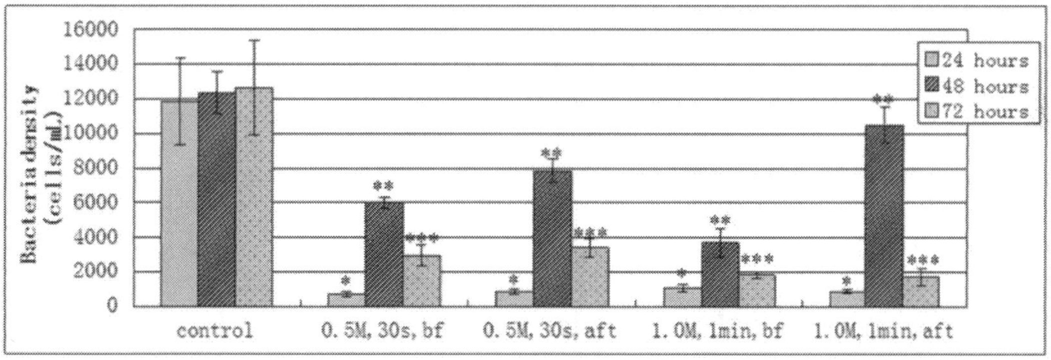

Figure 4. Bacteria (*S. aureus*) growth on the surface of polycabonate. Polycarbonate samples were treated with bacteria (*S. aureus*) in 0.03% TSB (Tryptic Soy Broth) and were incubated for 24, 48 or 72 hours. The media was changed with 0.03% TSB every 24 hours for those samples incubated at 48 hours or 72 hours. The control group is uncoated polycarbonate. bf = before tape test; aft = after tape test. Data=Mean ± standard deviation by mean, n=3; *$p<0.05$ compared with control group (uncoated polycarbonate) after 24 hours; **$p<0.004$ compared with control group after 48 hours; ***$p<0.02$ compared with control group after 72 hours.

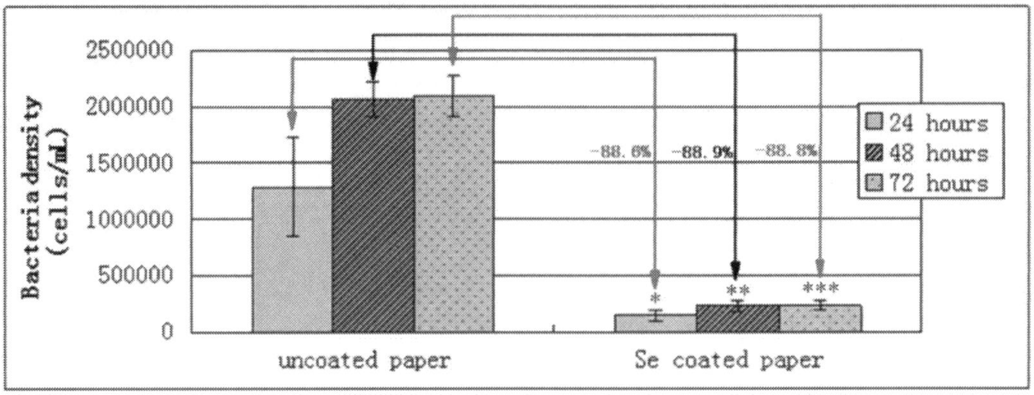

Figure 5. Bacteria (*S. aureus*) growth on the surface of uncoated selenium coated paper towels. Paper towel samples were treated with bacteria (*S. aureus*) in 0.03% TSB (Tryptic Soy Broth) and were incubated for 24, 48 or 72 hours. The control group is the uncoated paper towels. Data=Mean ± standard deviation, n=3; *p<0.02 compared with control group (uncoated paper) after 24 hours; **p<0.002 compared with the control group after 48 hours; ***p<0.002 compared with control group after 72 hours.

Conclusions

Selenium nanoparticles were synthesized and coated on polycarbonate and paper towels by a simple fast reaction occurring in 60s. The coating of selenium nanoparticles on polycarbonate and paper towels introduced significant and effective anti-bacterial properties to the material surfaces which deserves further investigation.

Acknowledgements

The authors thank Dr. Justin Seil for help with the bacteria experiments and Dr. Vera Fonseca for help with the AFM images. Funding from the Hermann Foundation is also acknowledged.

References

1. Costerton JW, Z. Lewandowski, Caldwell DE, Korber DR, Lappin-Scott HM. Microbial biofilms. *Annu. Rev. Microbiol.* 1995; 49:711–745.
2. Zottola EA., and Sasahara KC. Microbial biofilms in the food processing industry—should they be a concern. *Int. J. Food Microbiol.* 1994; 23: 125–148.
3. Darouiche RO, Treatment of infections associated with surgical implants. *N. Engl. J. Med.* 2004; 350:1422–1429.
4. Hattula JL, Steven PE. A descriptive study of the handwashing environment in a long-term

care facility. Clin Nurs Res. 1997;6:363-74.
5. Martinez LR, Han G, Chacko M, et al. Antimicrobial and healing efficacy of sustained release nitric oxide nanoparticles against Staphylococcus aureus skin infection. *J Invest Dermatol.* 2009;129(10):2463–2469.
6. Singh R, Ray P, Das A, Sharma M. Role of persisters and small-colony variants in antibiotic resistance of planktonic and biofilm-associated *Staphylococcus aureus*: an in vitro study. *J Med Microbiology.* 2009;58(8):1067-1073.
7. Chule K, Ling YC, et al. Preparation and characterization of ZnO nanoparticles coated paper and its antibacterial activity study. Green Chemistry. 2006;8(12):1034-1041.
8. Chiang WC, Schroll C, et al. Silver-palladium surfaces inhibit biofilm formation, *Appl. Environ. Microbiol.* 2009; 75(6):1674.
9. Kutsal Devrim Secinti, Hakan Ozalp, et al. Nanoparticle silver ion coatings inhibit biofilm formation on titanium implants. *Journal of Clinical Neuroscience.* 2011; 18: 391-395.
10. Yuan Liao, Yaqin Wang, et al. Antibacterial surfaces through dopamine functionalization and silver nanoparticle immobilization. *Materials Chemistry and Physics.* 2010; 121: 534-540.
11. Robert JC, Timothy JN Smith, et al. Antibacterial activity of multilayer silver-copper surface films on catheter material. *Canadian Journal of Microbiology.* 1993; 39(9): 895-899
12. Mulligan AM, Wilson M, Knowels JC. The effect of increasing copper content in phosphate-based glasses on biofilms of Streptococcus sanguis. *Biomaterials.* 2003; 24: 1797-1807.
13. Brewer GJ. Copper toxicity in the general population. *Clin Neurophysiol.* 2010; 121(4):459-460.
14. Foldbjerg R, Autrup H, et al. PVP-coated silver nanoparticles and silver ions induce reactive oxygen species, apoptosis and necrosis in THP-1 mono-cytes. *Toxicol. Lett.* 2009; 190:156-162.
15. Xia T, Nel AE, et al. Comparison of the Mechanism of Toxicity of Zinc Oxide and Cerium Oxide Nanoparticles Based on Dissolution and Oxidative Stress Properties. ACS Nano. 2008;2(10):2121-2134.
16. Kryukov GV, Castellano S, Novoselov SV, et al. Characterization of mammalian selenoproteomes. *Science* 2003; 300: 1439 – 43.

(107a) Strategies in Advancing the Pharmaceutical Pipeline - an Overview of API Pilot Plant

Jim Ratway, R452, Abbott, North Chicago, IL

This is a place holder that must be 122 words long entered by David Carl Attride of Eastman Chemical Corp. the Chair of the Area 12B section of the Process Development Division on Tuesday, May 1, 2012. Mary had a little lamb, little lamb, little lamb. Is that enough text to satisfy?

This is a place holder that must be 122 words long entered by David Carl Attride of Eastman Chemical Corp. the Chair of the Area 12B section of the Process Development Division on Tuesday, May 1, 2012. Mary had a little lamb, little lamb, little lamb. Is that enough text to satisfy?

This is a place holder that must be 122 words long entered by David Carl Attride of Eastman Chemical Corp. the Chair of the Area 12B section of the Process Development Division on Tuesday, May 1, 2012. Mary had a little lamb, little lamb, little lamb. Is that enough text to satisfy?

(107b) A Pharmaceutical Engineering Process/Product Life-Cycle Approach to Drug Product Development and Manufacturing

Jose C. Menezes[1,2], Stephan Sacher[2], Georg Scharrer[3], Stefan Leitgeb[4], Christine Voura[5], Simon D. Fraser[5] and Johannes G. Khinast[6], (1)Institute for Biotechnology & Bioengineering, I.S.T., Lisbon, Portugal, (2)Research Center for Pharmaceutical Engineering, Graz, Austria, (3)Advanced Simulation, Research Center for Pharmaceutical Engineering, Graz, Austria, (4)TU Graz, Institute of Biotechnology and Biochemical Engineering, Research Center Pharmaceutical Engineering GmbH, Graz, Austria, (5)Research Center Pharmaceutical Engineering (RCPE), Graz, Austria, (6)Research Center Pharmaceutical Engineering GmbH, Graz, Austria

A Pharmaceutical Engineering Process/Product Life-Cycle Approach to Drug Product Development and Manufacturing

Jose C. Menezes, Sephan Sacher, Georg Scharrer, Stefan Leitgeb, Christine Voura, Simon Fraser, Johannes Khinast

Research Center Pharmaceutical Engineering GmbH, A-8010 Graz, Inffeldgasse 21a/II, Austria

Drug product design, development and manufacturing involve different goals and use different tools throughout product and process engineering life-cycle. At the formulation design stage a materials science-based approach should dominate. During development and especially later during manufacturing (continuous or batch) a process-centered perspective prevails. At the end of processing a drug product of required purity, safety and efficacy should be obtained. A stronger link between the different goals throughout those life-cycle stages must be established to ensure modern and competitive pharmaceutical manufacturing.

PAT monitoring tools, however, when used as such (i.e., as process state estimation and not as in-process quality controls of specific parameters) are process-centered but can be used to link process to product (viz., process parameters and attributes to product critical quality attributes). Here we discuss several case-studies involving RCPE's recent activities, highlighting the use of advanced simulation and innovative PAT tools in formulation design, process development and different operational requirements for routine manufacturing within the QbD context. A balance summary perspective of where the area is evolving into after a decade of PAT and QbD will be given.

(107c) Leveraging HPLC Data in a QbD Approach, for Right-First Time Scale-up of API Reaction Steps

Wilfried Hoffmann[1], Andrew Bird[1] and Peter Clark[2], (1)Scale-up Systems, Dublin, Ireland, (2)DynoChem Inc, Yardley, PA

HPLC data that are routinely generated can be leveraged to deliver much additional value; when interpreted quantitatively, they can be used to develop mechanistic models. These models in turn can be used to probe and question the data in a way that leads to increased process understanding. There is also the opportunity to run experiments that are designed to produce impurity so that the mechanism of impurity formation can be understood. The above are core elements in developing a process that works right first time on scale-up, achieving Quality by Design.

In more detail:

Experimental data are the basis for adjusting rate parameters in kinetic models. The most powerful data sets are analytical profiles, as these data are able to show species levels lower than 1%, particularly important in the Pharmaceutical Industry. At the same time these are one of the most difficult to handle, as chromatographic detection is in the case of the most common UV-detectors linked to molecular absorbance, and this can vary by orders of magnitude.

The talk will present an approach to convert analytical data to absolute values by use of Relative Response Factors and a link to the mass balance implied in a kinetic model. The data and model may not agree and in this case either the data or the model may be wrong.

By changing one or the other, we can ultimately reach data consistent with the model's mass balance, learning much in the process about our analytical method and our reaction scheme; this consistency is a requirement to get a good match in fitting with kinetic models,. The other factor is the rate law assumption, which we also address in this talk.

In an enhancement of this approach, a method is proposed which allows the calculation of the Relative Response Factors under the conditions of overlapping mass balances, which is very often the case.

Extended Abstract for American Institute of Chemical Engineers (AIChE)
Presentation ID# 279310
Meeting: Fall Meeting 2012
Location: Pittsburgh, PA
Date: 28-Oct-2012 to 02-Nov-2012

Title: Continuous Distillation Operations and Application of the Wiped Film Evaporator (WFE) to Pharmaceutical Pilot Plant Processing

Primary Author/Presenter: Anne Mohan
Contributing Author(s): Beth Fisher, Rukyah Hennessey, Ozgur Karahan, Joe Kukura, Dave Lashen, Crystal Miranda, Glenn Spencer, Michael Ward, Ming Yue, Dimitrios Zarkadas, George Zhou

Objective: To provide a detailed analysis of wiped film evaporation, a type of continuous distillation technology, by utilizing a case study from a Merck pharmaceutical pilot plant. In addition, this presentation will provide a comprehensive technical explanation of the technology and its application to industrial chemical processing.

Background:

The utilization of continuous processing technology provides increased productivity and versatility to scaled processes. Continuous *distillation* technology, in addition to providing these general benefits, provides the opportunity to process streams at higher temperatures and lower pressure ranges than *batch* distillation operations. Wiped film evaporation (hereafter referred to as "WFE"), which is a type of continuous distillation technology, is ideal for heat sensitive and/or high-boiling products due to the short residence time (less than 60 sec) and the low vacuum capability (down to ~1 mm Hg) of the unit. In addition, the volume of batch required for WFE operation is significantly less that that required for batch distillation, making WFE appropriate for low-volume products. Wiped film evaporation (WFE) is a type of continuous distillation technology recently investigated, developed, and ultimately applied to a pharmaceutical project at the Merck Rahway site at the pilot plant scale. This technology was utilized in order to address the thermal instability and high boiling issues associated with a specific process. The implementation of wiped film evaporation to this process is an example of the appropriate application of continuous distillation technology at scale, and the benefits of such an application.

The objective of this presentation is to communicate and discuss the recent lab development and pilot plant work with wiped film evaporation technology, which is used for continuous distillation processing. The presentation will focus on the methodology used to scale-up the technology, the challenges faced in the pilot plant, how these challenges were handled, and the accomplishments of the campaign both in the lab and in the pilot plant.

Methods:

The 2" and 4" lab scale, and the 6" pilot plant scale WFE units were used for continuous processing technology development and for specific project-related process development during 2011 and 2012 at Merck & Co., Inc. Lab development was performed on the smaller WFE units to gain an understanding of WFE operating parameters and factors which affect the distillation operation. For 2" and 4" WFE operations, material was fed into the WFE from a Wheaton bottle or poly jug using a peristaltic pump. The feed rate was controlled under the set pressure and temperature conditions in order to obtain the desired split between the distillates and residue. These exiting streams were collected in round bottom flasks attached to the respective WFE ports, and the product concentration in both was determined by HPLC (high performance liquid chromatography) at the conclusion of the experimental runs.

The data and knowledge gained from the lab developmental work was applied to the operation of the pilot plant scale, 6", WFE unit. This larger unit was used during batch processing throughout the month of October 2011 at the Merck Rahway site. During 6" WFE operation, the batch was fed from drums to the WFE system via an in-line mass meter and diaphragm pump. Under specific temperature and pressure conditions, the batch was boiled in the column of the WFE, and the distillates and residue generated were collected into respective drums. Samples of the exiting streams were pulled at various points throughout the operations to either ascertain the split being generated, or to ensure the targeted split was achieved. In this manner, the WFE equipment was used for both technological development and process-specific operation.

The data gathered from these operations was analyzed in order to understand and characterize the heat transfer of the WFE unit and the batch performance in the WFE. Additionally, the data was used to develop a scaling procedure which can be applied to take a process from the lab to the pilot plant scale. The experiments used to accomplish these tasks explored the controllable variables of the WFE unit, such as feed rate, wiper speed, pressure, and temperature.

Results and Conclusions:

Through analysis of the data obtained from the operations of the WFE, the following conclusions were drawn:

1. The material of construction plays a large role in the heat transfer characterization of the WFE unit. At the lab scale, the limiting factor for heat transfer is the glass wall of the WFE body whereas at the pilot plant scale, the limiting factor for heat transfer was the solvent, or batch, itself.
2. The distillate and residue split for a given temperature and pressure condition can be controlled most tangibly by the feed rate.
3. Of the controllable factors explored, wiper speed has the least effect on the distillate and residue split during a WFE distillation.

The associated notable accomplishments of this endeavor were:

1. Increased WFE equipment knowledge regarding general operation and function, heat transfer capabilities, and overall productivity and efficiency
2. Attainment of data and knowledge regarding the calculations required to scale-up from the lab scale WFE units to the larger scale WFE units
3. Increased industrial processing knowledge in the continuous realm

This presentation will provide: (1) an in-depth overview of the wiped film evaporation operation, (2) a detailed description of the data collected and knowledge gained by all WFE units, with specific focus on the larger 6" unit, (3) a list of the advantages and disadvantages observed through the operation of all size WFE units, and (4) a report of the quantitative and qualitative results of the WFE usage for the batch processing. The technological and process-specific knowledge that was gained through the Merck WFE experiences can be used to promote the continuous platform.

The utilization of continuous processing technology provides increased productivity and versatility to scaled processes. Continuous *distillation* technology, in addition to providing these general benefits, provides the opportunity to process streams at higher temperatures and lower pressure ranges than *batch* distillation operations. Wiped film evaporation (hereafter referred to as "WFE"), which is a type of continuous distillation technology, is ideal for heat sensitive and/or high-boiling products due to the short residence time (less than 60 sec) and the low vacuum capability (down to ~1 mm Hg) of the unit. In addition, the volume of batch required for WFE operation is significantly less that that required for batch distillation, making WFE appropriate for low-volume products. Wiped film evaporation (WFE) is a type of continuous distillation technology recently investigated, developed, and ultimately applied to a pharmaceutical project at the Merck Rahway site at the pilot plant scale. This technology was utilized in order to address the thermal instability and high boiling issues associated with a specific process. The data gathered from this operation was analyzed in order to understand and characterize the heat transfer of the WFE unit and the batch performance in the WFE. Additionally, the data was used to develop a scaling procedure which can be applied to take a process from the lab to the pilot plant scale. The experiments used to accomplish these tasks explored the controllable variables of the WFE unit, such as feed rate, wiper speed, pressure, and temperature. The implementation of wiped film evaporation to this process is an example of the appropriate application of continuous distillation technology at scale, and the benefits of such an application.

The objective of this presentation is to communicate and discuss the recent lab development and pilot plant work with wiped film evaporation technology, which is used for continuous distillation processing. The presentation will focus on the methodology used to scale-up the technology, the challenges faced in the pilot plant, how these challenges were handled, and the accomplishments of the campaign both in the lab and in the pilot plant. The presentation will provide: (1) an in-depth overview of the wiped film evaporation operation, (2) a detailed description of the data collected and knowledge gained by all WFE units, with specific focus on the larger 6" unit, (3) a list of the advantages and disadvantages observed through the operation of all size WFE units, and (4) a report of the quantitative and qualitative results of the WFE usage for the batch processing. The technological and process-specific knowledge that was gained through the Merck WFE experiences can be used to promote the continuous platform.

(107e) Optimization of Amidecoupling Reaction Conditions Using Kinetic Modeling

Ozgur Karahan, Chemical Process Development and Commercialization, Merck & Co. Inc., Rahway, NJ, Camille O. Anderson, Global Pharmaceutical Commercialization, Merck & Co., Inc., Rahway, NJ and Dimitrios Zarkadas, Chemical Process Development & Commercialization, Merck & Co. Inc., Rahway, NJ

An amide coupling reaction between a carboxylic acid and an amine in a solvent in the presence of EDC-HCl and an amine base was studied to optimize the reaction conditions and develop a robust process for large scale production. Results of a fractional-factorial DOE in combination with reaction understanding suggested a mechanism with multiple elementary steps, and overall apparent reaction kinetics depended on the concentration of all reactants as well as temperature. The Dynochem software was successfully utilized to model this complex reaction system and identify optimal conditions. In addition, the DOE results were used to generate a model that predicts oiling or the formation of a gummy phase in the reaction mixture, phenomena that can affect reaction performance. This information was then coupled with a cost function to determine conditions that ensure robust commercial manufacturing and minimize manufacturing costs. The amide coupling reaction with optimized conditions was successfully demonstrated on manufacturing scale.

(107f) Scale-up of Superficial Fluidization Air Velocity for a Pharmaceutical Fluid Bed Granulation

Brendon G. Ricart, Pharmaceutical Commercialization Technology, Merck & Co., Lansdale, PA, Alan R. Silverman, Pharmaceutical Commercialization Technology, Merck & Co., Inc., West Point, PA, Kristin J. Ploeger, Pharmaceutical Commercialization Technology, Merck & Co, Lansdale, PA and Edward J. Smith, Merck & Co., Inc., West Point, PA

Fluid bed granulation is a common unit operation in pharmaceutical manufacturing. Although the technology is mature, the complexity of the solid-fluid system has left incomplete scale-up principles. In the current study, fluidizing air velocity scaling techniques are examined across equipment design, equipment scale, and batch size. The scale of the current study ranges from 10-20 kg pilot-scale studies to 500 kg commercial scale batches. We find that scaling superficial air velocity at the top of the fluidized bed (v_{TOB}) provides a more robust process scaling than at the distributor plate, which is commonly used. Additionally, a dimensional analysis is performed for general fluid bed granulations to identify 5 relevant dimensionless groups which are interpreted to describe scale-up of process moisture, fluidizing air velocity (2), bowl fill, and total spray time. We find conflicting guidance from 2 dimensionless groups for scaling fluidization air, but constant v_{TOB} represents a compromise between the two. Together, these analyses led to a unique experimental design for a 500 kg scale developmental drug product campaign.

(107g) Experimental Determination of Maximum Shear Stress in Multiphase Flow – From Micro Scale to Pilot Scale Reactors

Thomas K. Villiger, Miroslav Soos and Massimo Morbidelli, Institute for Chemical and Bioengineering, ETH Zurich, Zurich, Switzerland

Although a tremendous effort has been made to determine velocity, turbulence and shear level distribution including their maximum values in stirred tanks, resolving the small sizes of the dissipative scales applying either experimental techniques or computational approaches still remains a challenge, particularly for multiphase systems. Such flow fields occur frequently in bioreactors where mechanical stress caused by stirring or bubble burst, entrainment, coalescence and break-up has been reported to have detrimental effects on mammalian cell cultures and product quality.

In order to construct a rational process design space, we have developed an aggregate – breakup system which is able to determine the maximum shear stress present in any device with arbitrary geometry. The presented methodology is based on the measurement of the steady state aggregates size broken by the hydrodynamic stress generated by stirring, sparging or their combination. Different contractive nozzles characterized by well-defined fluid dynamic conditions are used to calibrate the aggregate size to the maximum hydrodynamic stress. This system has been applied to biotechnological equipment ranging from micro scale bioreactors (0.015 L) up to a pilot scale bioreactor (300 L). The results show that not only the size and geometry of the vessel but also the flow field has a significant impact on the highest shear stress level. A cell will therefore exhibit a profoundly different hydrodynamic stress from expansion to production which should be carefully taken into account during scale-up as well as during the entire bioprocess development.

(110a) A Platform for Antibody Engineering Against Membrane Protein Targets Using Detergent-Solubilized Cell Lysates As Antigen Sources

Yong Ku Cho[1], Ben J. Tillotson[1], Inigo de Larrinoa[2] and **Eric V. Shusta**[1], (1)Chemical and Biological Engineering, University of Wisconsin-Madison, Madison, WI, (2)Applied Chemistry, University of the Basque Country, San Sebastian, Spain

Antibodies that target plasma membrane proteins are often attractive reagents for research and therapeutic purposes. However, membrane proteins are challenging to work with in terms of antibody selection, engineering, and antigen identification as a result of their insolubility in aqueous solutions. We have therefore developed a platform for antibody engineering using detergent-solubilized cell lysates as antigen sources. Such approaches are compatible with membrane protein targets, subcellular selections, and the rapid assessment of antibody specificity. To demonstrate, we screened a nonimmune human single-chain antibody library directly against detergent-solubilized cell lysates to identify antibodies binding plasma membrane proteins. Because they were selected using nondenaturing detergents, these antibodies retained functional binding capacity under physiological conditions. Moreover, when combined with yeast display immunoprecipitation procedures, antigen analysis was facilitated. For example, the antibody-displaying yeast were used directly to immunoprecipitate cognate membrane protein ligands from cell lysates and this immunoprecipitation product was used to explore antigen size and diversity as well as for *de novo* antigen identification through tandem mass spectrometric protein sequencing. Finally, we show that detergent-solubilized mammalian cells constitute a crude antigen solution that can be used for antibody affinity maturation. Yeast surface display was employed as a platform to screen a mutagenic library of anti-human transferrin receptor (hTfR) antibodies using detergent-solubilized cell lysates. An off-rate selection yielded scFv having 2-8 fold improved binding affinity to hTfR. Taken together, the combined cell lysate and yeast antibody display platform facilitates antibody engineering against membrane proteins at many levels and could prove to be a powerful platform as the interest in membrane proteins continues to increase.

(110b) Directed Evolution Strategies for Escaping Local Fitness Maxima

Barrett Steinberg and Marc Ostermeier, Chemical and Biomolecular Engineering, Johns Hopkins University, Baltimore, MD

Directed evolution is a powerful tool for enhancing or creating protein function without the existing knowledge of its structure. However, proteins exhibit a high degree of sign epistasis, implying a rugged fitness landscape containing many local maxima in fitness. Most directed evolution studies exclusively use positive selection (i.e. selection for improved variants). Present approaches for bypassing local fitness maxima are limited. One can expand the initial variation in sequences to attempt a variety of evolutionary pathways, use neutral drift to expand the diversity in a manner enriched for function, or simply recombine evolved sequences. However, these techniques rely on chance to escape a fitness peak since they do not traverse the valleys of the fitness landscape but instead rely on single-step hopping from one peak to another. Here we compare the efficacy of four evolutionary strategies for bypassing local maxima: positive selection, neutral selection, negative selection, and oscillating positive and negative selection. All four strategies end with a final positive selection step. Specifically, we evolved variants of *β-lactamase*, an antibiotic resistance gene, and make use of a synthetic gene circuit that that kills cells possessing *β-lactamase* activity above or below a desired threshold to perform the neutral and negative selections. The implications of our results for applied directed evolution and for natural evolutionary mechanisms, particularly of antibiotic resistance, will be discussed.

(110c) Enhancing FLI-TRAP for Rapid Isolation of Solubility and Affinity Matured Antibodies

Dujduan Waraho and Matthew P. DeLisa, Chemical and Biomolecular Engineering, Cornell University, Ithaca, NY

We successfully developed FLI-TRAP (Functional Ligand-binding Identification by Tat-based Recognition of Associating Proteins), a simple and versatile selection method for isolating interacting proteins. The technique utilizes the unique ability of the *Escherichia coli* twin-arginine translocation (Tat) pathway to transport non-covalent complexes of two folded polypeptides, termed the 'hitchhiker' mechanism. In the FLI-TRAP assay, one polypeptide consists of a fusion between an N-terminal Tat signal peptide (ssTorA) and the protein to be screened for interaction, and the second of a fusion between a known or putative partner protein and mature TEM-1 β-lactamase (Bla). In solubility and/or affinity maturation processes, mutations that improve solubility often compromise function and vice versa. FLI-TRAP, on the other hand, couples selection for solubility and binding activity in a single step. Through selection of yeast GCN4 binders from a randomized CDR-H3 anti-GCN4 scFv library, we previously validated the power of FLI-TRAP in isolating interacting, soluble, non-aggregating, and protease-resistance proteins pairs. Here, we show that this technique can be easily extended to screen naïve libraries for isolation of intracellular antibodies against protein antigens of interest. By simply increasing the selection stringency, FLI-TRAP can be further extended as a tool for solubility and/or affinity maturation of selected clones. Finally, we have developed 'competitive FLI-TRAP' whereby selection is performed under competitive binding *in vivo* and thus favors the isolation of affinity-matured clones. Collectively, these advances reveal the versatility of the FLI-TRAP method and highlight its potential for routine selection of high-affinity binding proteins (e.g., antibody-antigen) without the need for purification or immobilization of the binding target.

(110d) Development of a High-Throughput Screen for Directed Evolution of Inteins

Michael J. Coolbaugh, Chemical and Biomolecular Engineering, The Ohio State University, Columbus, OH and David W. Wood, Department of Chemical and Biomolecular Engineering, the Ohio State University, Columbus, OH

Screen Development for Directed Evolution of Controllable Inteins Using Yeast Surface Display

Inteins are naturally occurring-protein elements that splice themselves out of precursor proteins. Useful inteins have been engineered to cleave at only one of their two termini, facilitating their use in self-cleaving protein-purification tags. The major issue with the use of intein-based self-cleaving purification tag technology is controlling the cleaving reaction. Currently, intein cleavage can be induced in one of two manners: addition of a thiol as reducing agent, or a shift in temperature/pH. The use of thiols such as DTT to induce cleavage precludes their use for purification of disulfide-bond-containing target proteins such as mAbs. The pH inducible inteins do not have this shortcoming, but premature cleavage at physiological conditions (pH 7.2 and 37°C) precludes their use in mammalian expression hosts. Here we aim to address this problem by developing a new pH-inducible intein variant that retains high cleavage rates at low pH, but which does not cleave at physiological conditions.

We have developed a screen for the directed evolution of inteins based on the yeast surface display system developed by K. Dane Wittrup. Briefly, the screen involves expressing an intein mutant library between two epitope tags and anchoring the library to the cell wall of *S. cerevisiae*. The epitope tags are then labeled with fluorescent antibodies and the cells analyzed using flow cytometry. Comparison of the relative fluorescence of the upstream versus downstream tags allows for the distinction between intein cleavage and non-cleavage events. This then provides the basis for screening mutant libraries with enhanced cleaving control using fluorescence activated cell sorting (FACS).

We report the optimization of the above mentioned screen, as well as its use in screening a mutant intein library. The intein library was constructed by introducing orthologous substitutions into Block F, a highly conserved region believed to play a role in controlling the rate-determining step of intein cleavage. We also report progress toward the development of a pH inducible intein that does not cleave readily at physiological conditions but retains high cleavage activity at reduced pH, allowing for the use of this intein in mammalian cell expression.

(110e) Mod Designer: A Robust Optimization Framework for the Design of Proteins with Post-Translational Modifications and Unnatural Amino Acids and Its Applications to Cancer

George A. Khoury, James Smadbeck and Christodoulos A. Floudas, Department of Chemical and Biological Engineering, Princeton University, Princeton, NJ

Most proteins being used for drug applications contain post-translational modifications (PTMs) [1]. It can become prohibitively expensive and time-consuming to experimentally modify and test the effects of large numbers of modifications. To enable academic and industrial researchers to minimize time and monetary cost searching for new functional designs, in this work we introduce a method to *in silico* design proteins and peptides with post-translational modifications and unnatural amino acids.

We develop a new integer linear optimization design formulation and complementary parallelized algorithm exploiting MPI to introduce the modified amino acids, drawing on our recently developed forcefields. Given a ligand or an ensemble of ligand conformations of a peptide bound in a receptor protein, the formulation maximizes the favorable (Van der Waals and hydrogen bond) contacts and minimizes steric clashes while maintaining or enhancing the binding free energy. To reduce the combinatorial complexity of the problem, logical restraints to preserve the residue-wise charge and hydrophobicity are applied to reduce the number of allowed modifications in each position. For each configuration, the initial contacts, clashes, and hydrogen bonds are populated. Next, for each configuration, each design position is modified, a local energy minimization is performed, and the new contacts, clashes, hydrogen bonds and interaction energy are populated. Finally, the ILP is solved to global optimality, generating a rank-ordered list of designs using integer cuts. The importance of using a flexible template in using these design metrics is demonstrated using an ensemble of conformers generated from a molecular dynamics simulation.

The algorithm is designed to serve as ***Stage 3: Modification Selection*** of our generalized *de novo* design framework [2], with ***Stage 1*** being an optimization driven ***Sequence Selection***, and ***Stage 2*** being ***Fold Specificity*** and statistical mechanically-based ***Approximate Binding Affinity*** calculations. We describe our efforts to use the method to design new variants of Compstatin to inhibit Complement C3c and new variants of peptide inhibitors targeting the methyltransferase EZH2. For Compstatin, we begin design using the extremely potent Compstatin analog, Variant E1. It was found as one of the top ten solutions in the rank-ordered list of designs [3]. In all efforts we begin with template sequences and structures containing only the 20 amino acids discovered to experimentally bind and inhibit their target receptors [4-6]. In this way, the modified amino acids serve to *fine-tune specificity* and potentially *enhance* the *affinity* of a given inhibitor.

References:

1. Walsh, C., Posttranslational modification of proteins: expanding nature's inventory2006, Englewood, Colo.: Roberts and Co. Publishers. xxi, 490 p.

2. Bellows, M.L., H.K. Fung, and C.A. Floudas, Recent Advances in De Novo Protein Design, in Process Systems Engineering2011, Wiley-VCH Verlag GmbH & Co. KGaA. p. 207-232.

3. Mallik, B., et al., Design and NMR Characterization of Active Analogues of Compstatin Containing Non-Natural Amino Acids. Journal of Medicinal Chemistry, 2004. 48(1): p. 274-286.

4. Bellows, M.L., et al., Discovery of Entry Inhibitors for HIV-1 via a New De Novo Protein Design Framework. Biophysical Journal, 2010. 99(10): p. 3445-3453.

5. Bellows, M.L., et al., New Compstatin Variants through Two De Novo Protein Design Frameworks. Biophysical Journal, 2010. 98(10): p. 2337-2346.

6. López de Victoria, A., et al., A New Generation of Potent Complement Inhibitors of the Compstatin Family. Chemical Biology & Drug Design, 2011. 77(6): p. 431-440.

(110f) A Novel Amine Dehydrogenase for the Selective Production of Chiral Amines

Michael J. Abrahamson[1], Eduardo Vazquez-Figueroa[1], Nicholas Woodall[2], Jeffery Moore[3] and Andreas S. Bommarius[1], (1)School of Chemical and Biomolecular Engineering, Georgia Institute of Technology, Atlanta, GA, (2)Georgia Institute of Technology, Atlanta, GA, (3)Department of Process Research, Merck and Company, Inc., Rahway, NJ

Biocatalysts are increasingly used in industry to create enantiomerically pure compounds; routes employing dehydrogenases and other redox enzymes have been especially successful.[1] Enantiomeric compounds have a broad range of uses including an utmost importance in pharmaceuticals. Contemporary methods of protein engineering, such as applying rational design guided by mechanistic and structural knowledge, have greatly increased the ability to create novel enzyme functionality.[2-3]

Application of degenerate codons and a high-throughput screening assay allowed for simplified, rapid evaluation of enzyme variants. Constraint of mutant library sizes was guided by mechanistic and structural knowledge, ultimately reducing the screening requirements while maintaining an increased chance of generating alternate substrate specificity. Expansion of its substrate specificity was achieved through several rounds of focused mutagenesis, allowing for the creation of chiral amines. Beneficial mutations of each round were carried over into subsquent variants.

Over the course of eleven libraries, we have sucessfully developed a novel amine dehydrogenase (AmDH) from an existing leucine dehydrogenase scaffold with broad substrate specificity. Reductive amination activity was not measureable with the original leucine dehydrogenase scaffold, and was increased to a k_{cat} of 0.46 s^{-1} with a K_M of 15.1 mM. This creates a new green route in the production of chiral amines utilizing ketone and NH_3 substrates. Novel activity was achieved toward a number of compounds while maintaining the enzymes' native enantioselectivity of 99.8% e.e.[4]

---

[1] M. Hall and A.S. Bommarius, *Chem. Rev.* **2011**, *111*, 4088-4110.

[2] A.S. Bommarius, J.K. Blum, M.J. Abrahamson, *Curr. Opin. Chem. Biol.* **2011**, *15*, 194-200.

[3] A.S. Bommarius, *Nature Chem. Biol.* **2010**, *6*, 793-794.

[4] M. Abrahamson, *et. al*, *Angewandte Chemie*, **2012**, *51*, 3969-3972.

(110g) A Platform for the Discovery of Proteasomal Degradation Activators

Wenting Zhao, Chemical and Biomolecular Engineering, Rice University, Houston, TX and **Laura Segatori**, Chemical and Biomolecular Engineering and Bioengineering, Rice University, Houston, TX

Protein misfolding and aggregation currently comprise one of the main challenges of the fields of bioengineering and biotechnology. Protein folding in the crowded cellular milieu leads to the formation of natively folded as well as off-pathway misfolded species. Misfolded proteins are intrinsically prone to aggregation and, since aggregates are toxic to cells, they must be immediately degraded. In eukaryotic cells, the degradation of misfolded proteins is catalyzed by the ubiquitin proteasome system (UPS). However, aberrant accumulation of misfolded proteins often overloads the UPS capacity and causes the formation of insoluble aggregates. Aggregation is a key impediment to the high expression of recombinant proteins not only in bacteria but also in eukaryotic cells and limits the high-yield production of proteins for research, diagnostic and industrial applications. Protein misfolding and aggregation are also hallmarks of pathogenesis in a number of human diseases, such as Parkinson's and Alzheimer's. Enhancement of proteasomal degradation is hypothesized to facilitate intracellular clearance of misfolded proteins, thus restoring protein homeostasis. Compounds that inhibit proteasomal degradation are widely used in research and have transitioned to clinical therapeutics. Although the mechanism of proteasomal inhibition is understood, we currently do not know how to enhance proteasomal degradation. Our laboratory developed a technology for detecting proteasomal degradation enhancement based on coupling the increase in degradation of a UPS model substrate to an easily detectable fluorescent output. We employed a synthetic biology approach to generate a cellular circuit in which the expression of a fluorescent protein is under the transcriptional repression of an UPS model substrate. This novel strategy overcomes limitations of current methods, which are based on detecting a decrease in concentration of a degradation-prone protein and thus are prone to artifactual results. We also generated a computational model to predict how fine-tuning this cellular circuit and, particularly, inducing self-activation of the UPS model substrate's expression affect the intensity of cell fluorescence. The results of this simulation led to assay optimization for high-throughput screening applications. We report here the development of this innovative technology and its use for the discovery of molecular probes that enhance the cell's innate ability to dispose of misfolded proteins.

(110h) Quantitative Functional Single-Cell Characterization of Surface Display of Recombinant Proteins Using A Bacterial Autotransporter

Balakrishnan Ramesh, Chemical and Biomolecular Engineering, University of Houston, Houston, TX, Patrick C. Cirino, Chemical & Biomolecular Engineering, University of Houston, Houston, TX and Navin Varadarajan, Dept. of Chemical & Biomolecular Engineering, University of Houston, Houston, TX

Quantitative functional single-cell characterization of surface display of recombinant proteins using a bacterial autotransporter

Balakrishnan Ramesh, Patrick C. Cirino, Navin Varadarajan

University of Houston, S222 Engineering Bldg 1, Houston, TX, 77204-4004, USA

Expressing functional recombinant proteins on the surface of bacteria has generated substantial interest due to its utility in biotechnological applications such as whole-cell biocatalysis, bioremediation, responsive biosensors and protein engineering. The mostly widespread protein family in nature that facilitates surface display of native proteins is the autotransporter (AT) pathway. ATs are ubiquitous in Gram negative bacteria and facilitate the translocation and secretion/ display of diverse passenger proteins with functions such as adhesion, proteolysis, and pore formation. ATs comprise of an extended N-terminal leader sequence cleaved at the inner membrane followed by an N-terminal passenger (20-400 kDa) typically associated with virulence functions and a conserved C-terminal ~30kDa β-barrel. Replacing native passenger domain with heterologous protein facilitates the surface display of large proteins at a high number of copies per cell. However, there exists considerable controversy in the ability of ATs to transport passengers, either native or recombinant, containing folded elements, especially those containing disulfide bonds.

We have systematically investigated the ability of the *E. coli* AT Antigen 43 (Ag43) to display two different recombinant reporter proteins, a single-chain antibody (scFv) that contains two disulfides and chymotrypsin that contains four disulfides. Using flow-cytometry to quantify surface display of functional protein at the single-cell level, we demonstrate that in spite of the known propensity of these passengers to fold in the periplasm, surface display of these proteins is rather efficient and can be achieved using only the C-terminal domain containing the α-helix and the β-barrel. Our flow-cytometric data is consistent with data obtained via either fluorescence microscopy or western blotting on fractionated cell-compartments. In contrast to previous reports, no genetic manipulation like the use of ompT- strains or the use of reducing agents like β-mercaptoethanol was necessary to accomplish efficient display. Our results indicate that display of recombinant proteins containing multiple disulfides can be achieved by employing the Ag43 system and that the vast majority of native AT including the autochaperone domain are not indispensable for heterologous protein display. We anticipate that this methodology will be especially useful for the recombinant display and screening of protease libraries for a broad variety of applications.

(119a) Effective Application of Established Treatment Technologies for Metals, Metalloids, and Oxyanions

James Patterson, Patterson Environmental Consultants, Chicago, IL

The underlying theories and fundamentals of application of treatment technologies for control of metals, metalloids and oxyanions are well established. However, the complexity of the wastewater matrix to be treated can introduce unanticipated responses to the applications of such technologies, including when discharge limits are unusually restrictive due for example to water quality-based effluent limitations, or in scale-up from treatability investigations to full-scale application. This Presentation will provide an overview of such theories and applications, and address pollutant speciation, wastewater matrix, and treatment technology aspects which can enhance or interfere with the effectiveness of treatability performance. Such aspects include target pollutant oxidation state and speciation, interfering (or enhancing) interactive species, wastewaters containing multiple (potentially non-compatible) target species, and other issues. Effective pretreatment can be requisite to the efficient application of the appropriate treatment technology. Pretreatment options may include extraction or chemical modification of complexing or other interfering species, electrochemical pre-oxidation or reduction of a target speci, and management of adsorptive surface area and electrical charge in applications of inherent primary, or applied secondary co-precipitation. Improved awareness and understanding of such issues support cost- and performance-effective application of treatment technology.

(119b) Impact of Water Quality and Assessment of Adsorption Technologies for Arsenic Removal

Kashinath Banerjee, Veolia Water Solution & Tech, Moon Township, PA, Gary L. Amy, Water Desalination and Reuse Research Center, 4700 King Abdullah University of Science and Technology, Thuwal, Saudi Arabia and Tapas K. Das, School of Engineering, Saint Martin's University, Lacey, WA

Abstract

The sources of arsenic in drinking water are both natural and anthropogenic. In order to protect public health, the U.S. EPA, World Health Organization (WHO), and the European Commission have decided to lower the MCL of arsenic in drinking water to < 10 µg/L. This stringent arsenic standard will inevitably require many utilities to upgrade their present systems or consider new treatment options. Extensive literature and vendor surveys revealed that a large number of commercial and experimental adsorbents are available for arsenic removal. Some of the commercially available materials are matured products (e.g., activated alumina) that have been widely tested, revealing both their attributes and limitations, while others are more recent products. Results from several investigators (DeMarco *et al.*, 2003; Greenleaf and SenGupta, 2005; Jekel et al., 1999; Pena *et al.*, 2005) indicate that some of these adsorbents are very promising.

Bench-scale studies were conducted to develop a comparative assessment of several commercially available adsorbents for arsenic removal, and to assess their performance in the presence of phosphate, vanadium, and silica. The primary objective of this work was to identify the impacts of water quality, and to define adsorbent properties that will minimize the effects of water quality and/or maximize arsenic adsorption capacity. Additionally, the effect of temperature on the adsorption kinetics and capacities were investigated. Iron modified activated alumina (AA-FS50), granular ferric hydroxide (GFH), granular ferric oxide (GFO), and a titanium-based media (MetSorb G) were used as the adsorbent. In order to better understand the adsorption mechanics of the selected adsorbents, each was characterized according to its important physical/chemical properties, including particle size, bulk density, pH of zero-point charge (pH_{zpc}), surface area, and mineralogy. The concentrations of dissolved arsenic and vanadium were analyzed using Inductively Coupled Plasma – Mass Spectroscopy (ICP-MS); Ion Chromatography (IC) was used to analyze phosphorus. The concentrations of molybdate-reactive (monomeric) silica in the samples were measured, and polymeric silica was estimated by the difference between the total and monomeric silica.

Results revealed that pH has a significant impact on the adsorption capacities. AA-FS50 showed a moderate decrease in adsorption capacity as pH increased, and phosphate had a small impact. GFO media was only slightly affected by increased pH, but was significantly impacted by phosphate, and monomeric silica had some minor effects. GFH performance decreased as pH increased, with phosphate and monomeric silica exhibiting some impact. The adsorption performance of MetSorb G also decreased as pH increased, and phosphate and monomeric silica had some impact, although the interfering effects were less at a lower pH for this media. The two iron oxide media and MetSorb G generally exhibited comparable capacities. For iron oxide based media, rate data are best described by a pseudo-first order kinetic model at each temperature and pH condition studied. At lower pH values, arsenic (V) exhibits greater removal rates than arsenic (III). An increase in temperature increases the overall adsorption reaction rate constant values for both arsenic (V) and arsenic (III). An examination of thermodynamic parameters reveals that adsorption of arsenic (III) as well as arsenic (V) by iron oxide media is an endothermic process and is spontaneous at the specific temperatures investigated.

References

DeMarco, M.J., SenGupta A.K., Greenleaf J.E., 2003. Arsenic removal using a polymeric/inorganic hybrid sorbent, *Water Res.* 37, 164-176.

Greenleaf, J. E. and SenGupta, A. K., 2005, *Investigating New Arsenic Removal Technologies*, Chapter 9.3.2, p. 524-

544; In Das, T. K., (ed.), *Toward Zero Discharge: Innovative Methodology and Technologies for Process Pollution Prevention,* John Wiley and Sons, Hoboken, NJ.

Jekel, M, and Seith, R., 1999. Comparison of conventional and new techniques for the removal of arsenic in a full-scale water treatment plant. Proceedings World Water Conference, Buenos Aires, Argentina.

Pena, M.E., Korfiatis, G.P., Patel, M., Lippincott, L., Meng, X., 2005. Adsorption of As (V) and As (III) by nanocrystalline titanium dioxide. *Water Res.* 39, 2327 – 2337.

(119c) Removal of Arsenic by a Nano-Particle Blend Hollow Fiber Membrane

Jinsong He, Civil and Env Engineering, National University of Singapore, Singapore, Singapore and J. Paul Chen, Department of Civil & Environmental Engineering, National University of Singapore, Singapore, Singapore

The objective of this research is to prepare a based nano-particle blend hollow fiber membrane to adsorb arsenate from groundwater. A new nano-sized metal oxide was first prepared, which was then used in fabrication of the membrane. The hydrophobicity, porosity, mechanical strength, water flux and arsenate adsorption capacity of the membrane were investigated. The additional zirconium based nano-particle increased the hydrophilicity, which resulted in a significant enhancement on the water flux of the blend membrane. The batch adsorption experiments indicated that the membrane could effectively remove arsenate in a wide pH ranging from 2 to 9. The adsorption equilibrium could be reached in 24 h, and the maximum adsorption capacity was above 120mg/g. Ionic strength did not obviously affect the adsorption. The co-anions such as F^-, HCO_3^-, $SiO_3^{2-}$and PO_3^{2-} had different effects on the adsorption. In addition, the membrane could remove arsenate efficiently though the continuous filtration. The mechanism for the removal of arsenate was proposed based on the FTIR and XPS studies.

COMPETITIVE ADSORPTION OF NICKEL, MANGANESE, CHROMIUM AND CADMIUM FROM AQUEOUS SOLUTIONS BY DIFFERENT ADSORBENTS

Malyuba Abu-Daabes [a,], Hani Abu Qdais [b], Hatem Alsyouri [c]*

[a] *Pharmaceutical-Chemical Engineering Department, School of Applied Medical Sciences, German-Jordanian University, Amman, Jordan*

[b] *Department of Civil Engineering, Faculty of Engineering, Jordan University of Science and Technology, Irbid, Jordan*

[c] *Department of Chemical Engineering, Faculty of Engineering and Technology, University of Jordan, Amman, Jordan*

**corresponding author:*

Telephone: +962-6-530-0666, Fax: +962-6-5300-672, Email address: malyuba.abudaabes@gju.edu.jo

Abstract

Landfilling is the primary solid waste disposal method in Jordan. Since Jordan is considered one of the poorest countries with water resources, it is crucial to protect its groundwater resources from any possible risk of contamination from landfill leachates. A survey on leachates from landfills of different ages and control systems in Jordan, performed by our group during the years 2009 and 2010, showed that heavy metals concentrations in leachates were high exceeding the allowable limits of both Jordan Industrial wastewater and US Environmental Protection Agency. The main heavy metals with high risk are identified to be: Cr (0.66-20.5 mg/L), Mn (0.41-10.37 mg/L), Cd (0.48-4.5 mg/L), and Ni (0.74-6.69 mg/L). In the present study, competitive adsorption of these four metals on different adsorbents including: natural untreated shrimp shells (SH), granular activated carbon (GC), acid-treated granular activated carbon (AGC) and acid-treated powder activated carbon (APC) were investigated under simulated conditions. All adsorption isotherms were generated using the bottle point method, following the ASTM standard procedure (ASTM D3860-89a). Untreated granular carbon (GC) was the best adsorbent for Ni removal from a simulated solution with 90% removal efficiency. AGC and APC both showed very high selectivity for Cr over other present metals with removal efficiency of 99%. For Mn, shrimp shells (SH) was the best adsorbent with 79% removal efficiency, while both GC and SH adsorbents have comparable efficiency for Cd removal from the simulated solution (92%). Order of metals selectivity for SH adsorbent was Cd > Mn > Ni > Cr, for GC the selectivity was Cd > Ni > Cr > Mn, for both AGC and APC the order was Cr > Cd > Ni > Mn.

Key words: competitive metal ions, Leachate, Nickel, Manganese, Chromium, Cadmium, activated carbon, shrimp shells, chitin.

Introduction

Landfilling remains the predominant management method for the disposal of Municipal Solid Waste which leads to the formation of leachates. Leachates may act as a vehicle for the introduction of heavy metals to groundwater resources. Contamination of water resources by heavy metals is a great environmental problem, mainly due to their high toxicity even at very low concentrations (Jaramillo et al., 2009). Heavy metals are not biodegradable and cannot be removed by natural processes from the ecosystems. A previous study by our group on leachates collected from three landfills with different ages and control systems showed that leachates from landfills possess high contamination potential to the limited water sources in Jordan. Heavy metals, mainly Cr, Mn, Cd, and Ni, were found to exceed the allowable limits set by Jordan Industrial Estates Corporation and US Environmental Protection Agency for industrial wastewater discharge. Pollution of heavy metals may arise from different sources. Cd, Cr, Mn and Ni are mainly introduced to the environment from spent rechargeable and household batteries, electroplating, (Xara et al. 2009; Karnchanawong and Limpiteeprakan 2009) and tannery industry (Gotvajn et al., 2009).

One of the promising methods for heavy metals removal from wastewater is the use of natural and surface-modified adsorbents for selective removal of metal ions. Activated carbon is widely used in many applications. In this study the effect of modifying activated carbon surface with acidic groups on competitive metal ions removal from aquatic solution is evaluated. Shrimp shells are also used as a low cost natural adsorbent without any surface modification.

Materials and Methods

1) Adsorbents

Activated carbon was used both in granular (GC) and powder form. All carbon was boiled in deionised water, dried in an oven at 110 °C for 24 hours. Further surface modification was performed by boiling part of them with 4M HNO_3 solution for 1 hour with total condensation. After cooling, the solution of the acid-treated granular activated carbon (AGC) and acid-treated powder activated carbon (APC) was decanted and the solids were washed with deionised water several times until a neutral, constant pH was achieved. Shrimp shells (SH) were washed thoroughly, grounded and sieved without further modification.

2) Characterization of Adsorbents

Surface area and pore size distribution of different adsorbents were measured following the Brunauer, Emmett, and Teller (BET) method (Quantachrome Instruments, Model Autosorb-1, USA).

Table 1. Adsorbents characterization

Adsorbent	BET	HK cumulative pore volume	HK pore wedth
	m^2/g	cc/g	nm
GC	635	2.99E-01	1.398
AGC	591	2.79E-01	1.388
APC	892	4.22E-01	1.388
SH	25.6	9.95E-03	1.058

3) Adsorption isotherms

Stock solutions of heavy metals of 100 mg/L concentration (Ni(II), Mn(II), Cr(VI)) and 50 mg/L Cd(II) were prepared separately by dissolving appropriate quantities of nickel (II) nitrate hexahydrate, cadmium nitrate in nitric acid 0.5 mol/L (Fluka, Switzerland), manganese (II) sulfate monohydrate and chromium (VI) oxide (Merck, Germany). Adsorption isotherms were generated using the bottle point method, following the ASTM standard procedure (ASTM D3860-89a). Adsorption isotherms were generated at 25 °C, using simulated solution of the main heavy metals found in the landfill leachates. The composition of the simulated solution was 100 mg Ni/L, 100 mg Cr/L, 100 mg Mn/L, and 50 mg Cd/L. The pH of the solution was adjusted to 6. Ten ml of the simulated solution was added to 15-ml bottles that contained preweighed amounts (25, 50, 100, 500 and 1000 mg) of the adsorbents. All samples were agitated for 3 days in a temperature controlled water bath to ensure that equilibrium was achieved. The samples were then centrifuged and filtered using 0.45 μm syringe filters before being analysed. The samples were analyzed for their heavy metals content using Inductive Coupled Plasma Technique (ICP, Perkin-Elmer, Optima DV2000, USA). Blank samples were used with every isotherm in order to account for any loss and adsorption on the walls of the tube.

The equilibrium adsorptive quantities of the metal ions on the adsorbents surfaces were calculated using the following relation:

$$q_e = \frac{C_0 - C_e}{M} \times V \qquad (1)$$

Where q_e the amount of metal ion adsorbed per unit mass of adsorbent (mg/g), C_0 and C_e are the original ion concentration and the equilibrium solution concentration (mg/L), respectively, M is the mass of adsorbent, V is the volume of solution (L).

The percent removal of the metal ion by the adsorbent was calculated using the following mass balance relation:

$$\%\ Removal = \frac{C_0 - C_e}{C_0} \times 100\% \qquad (2)$$

Results and Discussion

1) Properties of the adsorbents

The highest surface area is reported for the powdered activated carbon treated with acid (APC) (892 m^2/g) and the lowest one for shrimp shells (25.6 m^2/g), as shown in Table 1. All measured samples have microporous structure with pore sizes less than 2 nm based on Horvath-Kawazoe (HK) model for the determination of pore size distribution in microporous sorbents such as activated carbon and zeolites. Acid modification of granular activated carbon lowered its surface area from 635 m^2/g to 591 m^2/g for GC and AGC, respectively. Previous work showed that oxygenation of activated carbon with 4M HNO_3 significantly increased the concentration of acidic surface oxygen complexes on the surface (carboxylic and carbonyl groups), reduced the total number of basic groups on the surface and caused the surface to be more hydrophilic (Franz et al., 2000; Abu-Daabes and Pinto, 2004; Jaramillo et al., 2009). These acidic sites are bonded to the edges of the graphitic basal plane.

2) *Competitive adsorption of heavy metals*

Adsorption isotherms for all four metal ions, Ni^{+2}, Cr^{+6}, Mn^{+2}, and Cd^{+2}, are shown in Figure 1 (a-d). All isotherms don't show uniform curve which could be due to the competition between the ions for the adsorption sites as well as the repulsion and synergy between them. Similar behaviour for competitive adsorption of metal ions was observed by Covelo et al.,(2004). In general, the mechanism of heavy metal ions adsorption on the surface of activated carbon is by electrostatic attraction to the carbon surface, ion-exchange and surface complexation (Jaramillo et al., 2009; Radovic et al., 2001). Adsorption of heavy metals on the unmodified carbon surface was suggested to be through interactions of the metal ion with the π electrons of the carbon basal planes according to the reactions (Jaramillo et al., 2009):

$$-C\pi + H_2O \leftrightarrow -C\pi\text{-}H^+ + OH^- \qquad (3)$$

$$2(-C\pi\text{-}H^+) + M^{2+} \leftrightarrow (-C\pi)_2 - M^{2+} + 2H^+ \qquad (4)$$

In the presence of acidic functional groups on the surface of activated carbon the mechanism become surface complexation or ion-exchange depending on the pH. If the solution has low pH, the weakly acidic groups are more likely to be protonated and will undergo ion-exchange reactions with the metal ions, as described in reaction (5).

$$2(-COOH) + M^{2+} \leftrightarrow (-COO)_2Cu + 2H^+ \qquad (5)$$

At higher pH, the acidic groups will be nonprotonated and the metal ions will bind to them by surface complexation, releasing less than one hydrogen ion by each metal ion adsorb (Jaramillo et al., 2009).

The selectivity sequences for heavy metal ions depends on the characteristics of the ions such as ionic radius ($Cd^{2+} > Ni^{+2} = Mn^{+2} > Cr^{+6}$), ionic potential or charge-to-radius ratio ($Cr^{+6} > Ni^{+2} = Mn^{+2} > Cd^{+2}$) and electronegativity (Ni > Cd > Cr > Mn) (Sposito, 1989; McBride, 1994; Covelo et al., 2004). The selectivity of each adsorbent for heavy metals under simulated conditions is shown in Figure 2 for the 500 mg adsorbents isotherms. The results show that acid treatment of both granular carbon and powder carbon (AGC and APC, respectively) changed the order of metals selectivity compared to the basic granular carbon (GC). Furthermore, acid treatment of granular carbon (AGC) increased Cr removal efficiency by 30% while it decreased for Ni, Mn and Cd by 23%, 36% and 37%, respectively. This can be explained by the much higher ionic potential of Cr^{+6} compared to other ions which will lead to strong attraction to the acidic surface groups of activated carbon and less accessibility for the ions with lower ionic potential.

Shrimp shells compose of 20% chitin by mass of dry weight, 37% protein and 43% minerals (Einbu, 2007). Chitin is a natural aminopolysaccharide. The results of this study showed that untreated shrimp shells have very high selectivity for Cd^{2+} and Mn^{2+}. This could be due to the presence of amine groups on the chitin of SH. The amine functional groups will form metal complexation. The order of metals affinity for different adsorbents under simulated conditions is summarized in Table 2.

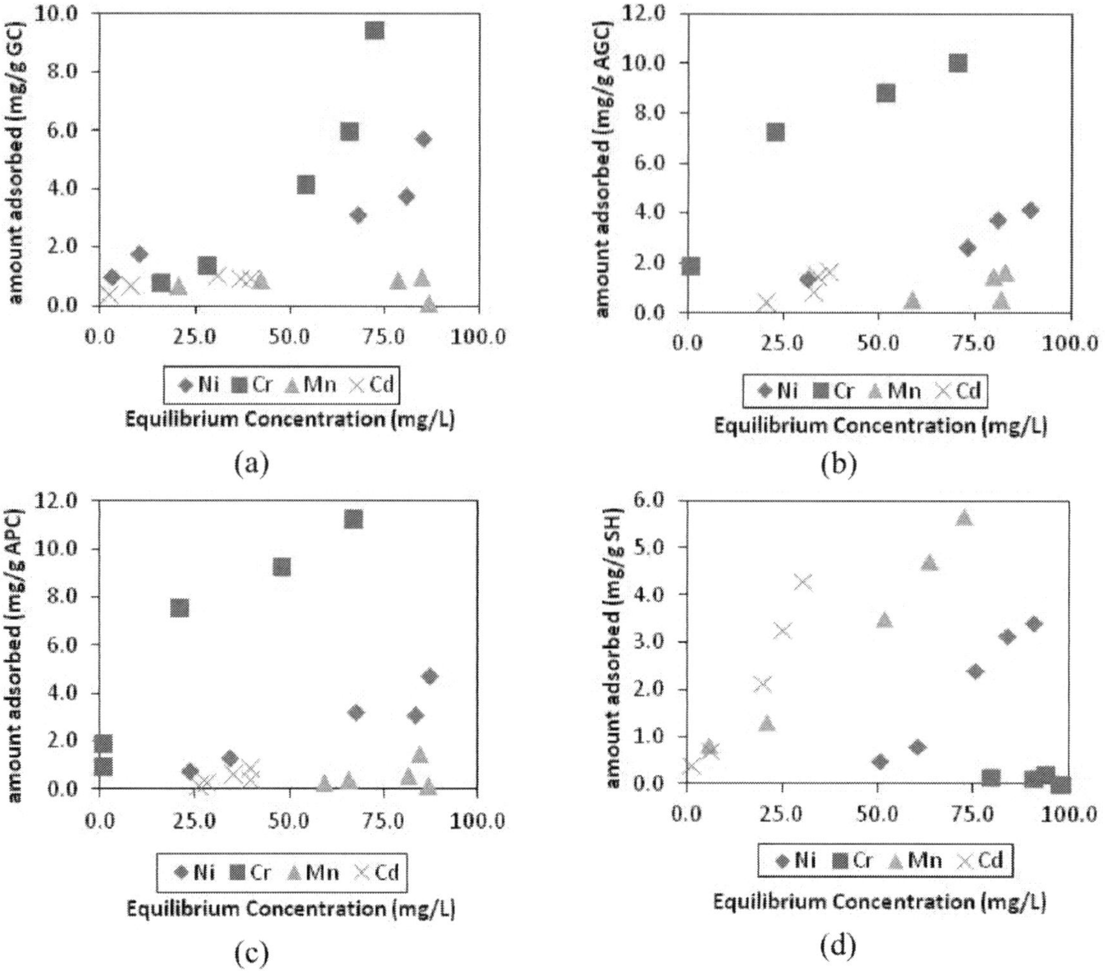

Figure 1. Adsorption Isotherms of Metal Ions on Different Adsorbents

Table 2. Order of metals selectivity for different adsorbents

Adsorbent	Metals order
GC	Cd > Ni > Cr > Mn
AGC	Cr > Cd > Ni > Mn
APC	Cr > Cd > Ni > Mn
SH	Cd > Mn > Ni > Cr

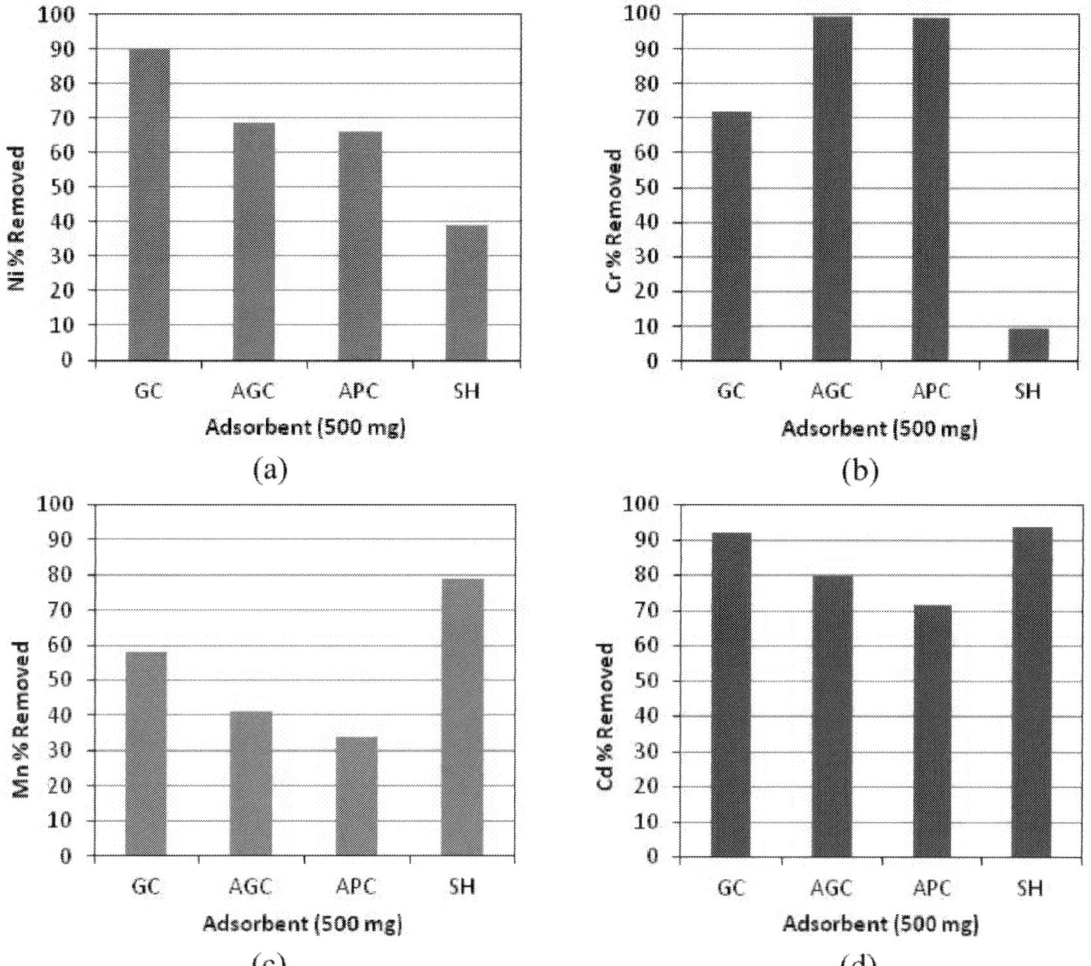

Figure 2. Percent removal of Metal Ions on Different Adsorbents

Conclusion

In this study activated carbon, surface modified activated carbon and unmodified natural shrimp shells were evaluated as potential adsorbents for competitive removal of Ni^{+2}, Cr^{+6}, Mn^{+2}, and Cd^{+2} from aqueous solutions. Untreated granular carbon was the best adsorbent for Ni^{+2} removal from a simulated solution with 90% removal. AGC and APC both showed very high selectivity for Cr^{+6} over other present metals with removal efficiency of 99%. This was due the high ionic potential of Cr^{+6} compared to other ions. For Mn^{+2}, shrimp shells (SH) was the best adsorbent with 79% removal efficiency, while both GC and SH adsorbents have comparable efficiency for Cd^{+2} removal from the simulated solution.

Acknowledgments

The authors wish to thank the German-Jordanian University, the Higher Council for Science and Technology in Jordan, and the European Union presented by Support to Research and Technological Development & Innovation Initiative & Strategies in Jordan (SRTD) Project for their financial support to this project.

References

Abu-Daabes, M., Pinto, N.G. (2004) "Effect of Surface Oxygen Complexes of Activated Carbon on Phenol Adsorption from Single and Mixed Non-Aqueous Solvents", *Separation Science and Technology*, 39, 2997-3009.

Covelo, E. F., Andrade Couce, M. L., Vega, F. A. (2004) "Competitive Adsorption and Desorption of Cadmium, Chromium, Copper, Nickel, Lead, and Zinc by Humic Umbrisols", *Communications in Soil Science and Plant Analysis*, 35, 2709-2729

Einbu, A. (2007), "Characterisation of Chitin and a Study of its Acid-Catalysed Hydrolysis", Ph.D. Thesis, Norwegian University ofScience and Technology.

Franz, M., Arafat, H.A., Pinto, N.G. (2000), "Effect of chemical surface heterogeneity on the adsorption mechanism of dissolved aromatics on activated carbon", *Carbon*, 38, 1807–1819

Gotvajn, Z., Tisler, T., Zagorc-Koncan, J. (2009), "Comparison of different treatment strategies for industrial landfill leachate". *Journal of Hazardous Materials*, 162, 1446-1456.

Jaramillo, J., Go´mez-Serrano, V., A´ lvarez P.M. (2009), "Enhanced adsorption of metal ions onto functionalized granular activated carbons prepared from cherry stones". *Journal of Hazardous Materials*, 161, 670–676

Karnchanawong, S., Limpiteeprakan, P. (2009), "Evaluation of heavy metal leaching from spent household batteries disposed in municipal solid waste". *Waste Management*, 29, 550–558.

McBride, M.B. Environmental Chemistry of Soils; Oxford University Press: New York, 1994.

Radovic, L.R., Moreno-Castilla, C., Rivera-Utrilla, J. (2001), "Carbon materials as adsorbents in aqueous solutions", in: L.R. Radovic (Ed.), Chemistry and Physics of Carbon, 27, Dekker, New York, pp. 227–405.

Sposito, G. (1989), "The Chemistry of Soils", Oxford University Press: New York,.

Strelko, V., Malik, D.J. (2002) "Characterization and metal sorptive properties of oxidized active carbon", *J. Colloid Interf. Sci.* 250 213–220

Xará, S., Delgado, J., Almeida, M., Costa, C. (2009), "Laboratory study on the leaching potential of spent alkaline batteries", *Waste Management*, 29, 2121–2131.

(119e) Arsenic Mobilization In Alluvial Soils of Punjab, North-West India Under Flooded Conditions

Kuldip Singh[1], Dhanwinder Singh[2], H. S. Hundal[1] and Raj Kumar[2], (1)Soils, Punjab Agricultural University, Ludhiana, India, (2)Soil Science, Punjab Agricultural University, Ludhiana, India

Elevated arsenic contaminant in drinking water above the WHO (1996) safe limits of 10 μgL^{-1} was reported at many locations in underground water of Punjab, North-West India. In rural areas, flooded irrigations and maintenance of standing water in paddy fields likely to create anoxic conditions in the soils over-lying above the shallow aquifers. The objective of the present investigation was to study the mechanism of arsenic mobilization in surface alluvial soils. Fifty gram of each of 18 soil samples were transferred to 200 ml polythene bottles, and 100 ml of deionized water was added to maintained flooded conditions. These soil-water systems were kept in BOD incubator at 25^0C for equilibration under anaerobic conditions of 24 and 240 hours period, respectively. After equilibration periods, pH, EC, soluble HCO_3 and Cl^- were determined for each soil-water equilibrated systems. Supernatant aliquots were withdrawn using a syringe, centrifuged, filtered and acidified with a drop of nitric acid prior to analysis for Fe, Mn, SO_4^{2-} and PO_4^{3-} on ICAP-AES. Arsenic in the filtrates solution was determined using hydride generation assembly on ICAP-AES.

Total dissolved As in 24 hour soil-water equilibrated systems varied from 3 to 16 μgL^{-1} with mean value of 9 μgL^{-1} (Figure 1), whereas its concentration range in different soil-water suspensions after 240 hour of equilibration period enhanced to 33 to 1761 μgL^{-1} with mean value of 392 μgL^{-1}. The median, 75th and 90th percentiles concentrations of As were 8, 12 and 14 μgL^{-1} after 24 hour and eventually elevated to 190, 490 and 1013 μgL^{-1}, respectively after 240 hours of anaerobic conditions. Arsenic concentration was enhanced with decreasing redox potential. The results showed antagonistic relationship of arsenic concentration with the magnitude of redox potential status of alluvial soil-water suspensions (Figure 2a). Dissolved Fe concentration versus redox status of soil-water systems determined after two equilibration periods are illustrated in Figure 2b. As the pe decreased, the Fe concentration declined in different alluvial soil water systems.

The effect of redox potential on soluble Mn concentrations determined at different flooding periods in alluvial soil-water suspensions is shown in Figure 2c. As the pe decreased, the soluble Mn concentration enhanced in soil water systems. The soluble concentrations of Mn was ranged between 0.034 to 0.407 mgL^{-1} with a mean value of 0.132 mgL^{-1} after 24 hour equilibration period, but were considerably enhanced to 0.449-44.44 with mean value of 22.398 mgL^{-1} after 240 hour of equilibration periods. The behavior of SO_4^{2-} in different alluvial soil-water systems under different redox status is shown in Figure 2d. With decrease in pe, SO_4^{2-} concentration declined in all soil water systems. Total dissolved concentration of SO_4^{2-} anion in soil-water suspensions also considerably lowered to range of 0.003-25.30 mgL^{-1} with mean value of 1.95 mgL^{-1} after 240 hour of equilibration periods from 4.56 to 81.6 mgL^{-1} with a mean value of 25 mgL^{-1} after 24 hours of equilibration periods. Under reducing conditions, sulfate concentrations decreased significantly.

Pourbaix (pe-pH) diagrams for arsenic and ferromagnesian oxides (Figure 3 and 4) were constructed by to identify the chemical species for each of these elements in different soil-water suspensions. Aqueous arsenic species, $HAsO_4^{2-}$ and $H_3AsO_3^0$ along with the possible solid phase, orpiment (amorphous As_2S_3) as were identified in soil-water systems (Figure 3). The mono valent arsenate oxy-anion ($HAsO_4$) specie was predominant in all the soil-water suspensions except soils 2, 4 and 8, where arsenite ($H_3AsO_3^0$) specie was identified after 24 hour of anaerobic condition. The predominance of neutral arsenite ($H_3AsO_3^0$) specie was observed after 240 hours of reduced incubated period in all soil water systems except soils 12, 15 and 16 in which orpiment (As_2S_3) as solid phase was identified. The neutral arsenite ($H_3AsO_3^0$) and arsenate oxy-anions ($H_2AsO_4^-$) has been found to be predominantly occurred in the alluvial aquifers of Punjab between the pH of 7.05 to 8.25 and under reduced pe value from 0.77 to -4.49, respectively. Thus, it is conclusively evident that two chemical aqueous species controlling As chemistry and its toxicity in soils are arsenate and arsenite. Arsenate is the thermodynamically stable form under aerobic conditions. Arsenite is the predominant species under anaerobic conditions and it is a neutral species at natural pH and is more soluble, mobile and phototoxic than arsenate. The orpiment (As_2S_3) solid phase domain predominates near the bottom of pe-pH diagram and lying over the H_2O / H_2g stability line in the Pourbaix diagram (Figure 3). After 240 hours of equilibrium period, arsenite ($H_3AsO_3^0$) specie was predominant in all alluvial soil-water suspensions. The increase in arsenic concentration was also primarily due to an increase in As (III) species in all the soil-water suspensions under more reduced conditions. In some soil water systems (soil number 12, 15 and 16) apparent precipitation of orpiment was identified on pe-pH diagram. It is evident that total arsenic concentration continuously enhanced in these soil-water systems with decreased magnitude of redox potential. The simultaneously drastic decline in concentration of both Fe and SO_4^{2-} in contrast to accumulation As in reduced soil-water suspensions of alluvial Holocene validated the formation of iron sulfide mineral in reduced alluvial soil-water systems of Punjab. Because of the differences in solubility of iron (DG= -117.11 Kcal $mole^{-1}$) versus arsenic (DG = -382.64 Kcal $mole^{-1}$) sulfides, precipitation of iron sulfide may remove sulfide from solution but not arsenic since the precipitation rates of iron sulfide was very rapid compared to amorphous arsenic sulfide.

The aqueous species of Fe^{2+}, Mn^{2+} and $MnHPO_4$ and solid phase of hematite (Fe_2O_3) were predominant in all the alluvial soil-water equilibrated solutions after 24 hours within the pe and pH ranges of -1.75 to 0.77 and 7.25-8.25, respectively (Figure 4). After 240 hours of incubation period, reduction potential status range in soil-water suspensions declined between -4.49 to -2.74 within pH range of 7.05 to 7.94. Aqueous species of Fe^{2+}, Mn^{2+} and $MnHPO_4$ and solid phase of hematite [Fe_2O_3] were predominant in soil-water suspensions of soils 1, 5, 6, 11 and 14. Pourbaix diagram (Figure 4) for iron illustrate that declined in iron was due to the predominantly precipitation of iron sulfide [FeS_2] and or as vivianite [$Fe_3(PO_4)_2 8H_2O$] minerals. The reduced solid phase vivianite [$Fe_3(PO_4)_2 8H_2O$] was found in soil 2, 3, 7 and 13, respectively. Pyrite [FeS_2] was found to be predominant in soil 8, 9, 10, 12, 16, 17 and 18, respectively. All the water composition falls in the stability field for aqueous Mn^{2+} ion (Figure 4) and its concentrations varied from 0.45 to 44.44 mgL^{-1}. Stability diagram for manganese showed that Mn^{2+} ion is thermodynamically more stable and predominant across a wide pe-pH range than for Fe^{2+} ion. Consequently, the Mn^{2+} ion remain in solution in aqueous environment in which Fe^{2+} ions is oxidized to Fe^{3+} and precipitated as Fe $(OH)_2.7Cl_{0.30}$ below pH 6.8 at pe > -1.8 and Fe_2O_3 above pH 6.8 at pe < -1.8 and thus eventually contributed to higher dissolved Mn^{2+} concentration under reduced environments in soils. Iron and manganese concentration in the anoxic soil-water systems were as high as 1.634 and 44.44 mgL^{-1}, respectively. The largest contributor of ferrous to the equilibrated soil-water interaction appears to be an iron oxide phase (probably maghemite, Gamma-Fe_3O_3, an alteration of magnetite). These results showed that dissolve Fe concentration were probably controlled by hematite (Fe_2O_3) under alkaline oxic condition and by iron sulfide/pyrite (FeS) or in some soils by vivianite [$Fe_3(PO_4)_2 8H_2O$] under reduced environment, where soluble phosphate concentration was high. Under reduced condition (developed after 240 hours of flooding), soil solution Fe and SO_4^{2-} are governed by newly formed precipitates of pyrite (FeS_2).

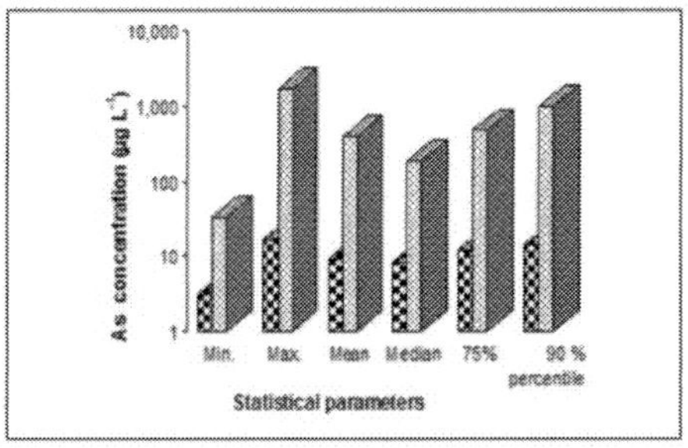

Fig.1 As concentrations measured in 18 alluvial soil-water suspensions of Punjab after equilibration of 24 ( ) and 240( ) hours

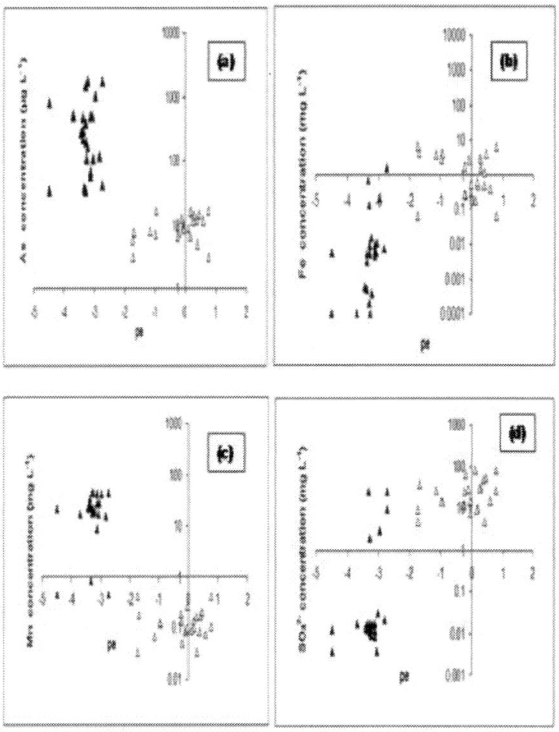

Fig. 2 Effect of redox potential on As (a), Fe (b), Mn (c) and SO_4^{2-} (d) concentrations measured in 18 alluvial soil-water suspensions after 24 (?) and 240 (?) hours of equilibration periods

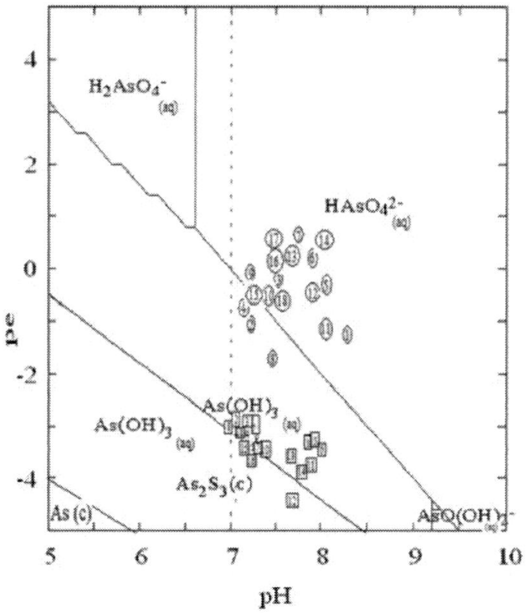

Fig. 3 Arsenic species identified in alluvial soil-water suspensions after equilibration of 24 (O) and 240 (?) hours at 25 °C. The concentration of [AsO_4^{3-}] =6.70µM, [PO_4^{3-}]=0.37mM, [SO_4^{2-}]= 0.25mM,[CO_3^{2-}]=0.20mM, [Cl^-]=0.13mMand ionic strength=0.020M were fixed to construct the Pourbaix diagram

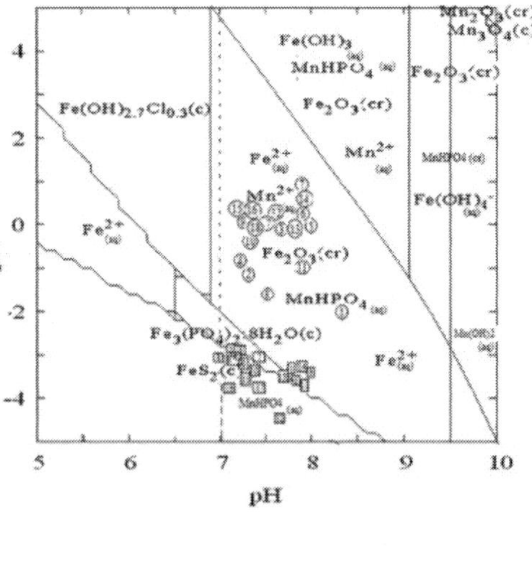

Fig.4 Iron and manganese species identified in alluvial soil-water suspensions after equilibration of 24 (O) and 240 (?) hours at 25 °C. The concentration of [Fe=15µM], [Mn=0.27mM], =6.70uM, [PO43-] =0.37mM, [SO_4^{2-}]= 0.25mM,[CO_3^{2-}]=0.20mM, [Cl^-]=0.13mM and ionic strength=0.020M were fixed to construct the Pourbaix diagram

ADSORPTION OF LEAD FROM AQUEOUS SOLUTIONS BY COTTON FABRICS MODIFIED WITH CITRIC ACID

Refugio Bernardo García Reyes, Saúl H. Martínez Treviño, Alcione García González and Ricardo Gómez González*

Universidad Autónoma de Nuevo León, Facultad de Ciencias Químicas, Av. Universidad s/n, Cd. Universitaria, C.P. 66400, San Nicolás de los Garza, N. L., México. FAX: +52(81)83294000 Ext. 6282, TEL: +52(81)83294000 Ext. 6288
*E-mail: bernardogarciareyes@yahoo.com.mx

ABSTRACT

The objective of this research was to modify cotton (i.e. cellulose) fabrics for the removal and recovery of lead (Pb(II)) from aqueous solutions. Cotton fabrics were modified with 0.6 M citric acid solutions and characterized by infrared spectroscopy (ATR-FTIR) and potentiometric titrations. Also, adsorption, desorption, and kinetic experiments in batch systems were conducted.

When modifying cotton fabrics with citric acid, density of functional groups increased up to 2.47 mmol/g in comparison with the non-modified fabric that had 0.61 mmol/g. In addition, estimated equilibrium constants for the modified fabrics can be associated in part to carboxylic groups of the citric acid. The appearance of carboxylic groups on modified fabrics was also confirmed by ATR-FTIR analyses. Maximum adsorption capacity of Pb(II) on modified fabrics, estimated with Langmuir isotherm, was 73.3 y 42.3 mg/g at pH 4 y 3, respectively, whereas non-modified cotton fabrics did not adsorb Pb(II) from water. Decrement in adsorption capacity suggests that Pb(II) and H^+ ions compete for the same adsorption sites. On the other hand, it was possible to desorb around 92% of Pb(II) previously adsorbed on the modified cotton with using 0.1 M HCl solutions since the chemical equilibrium is shifted. Finally, adsorption kinetics can be predicted with a pseudo-second order model.

In summary, cotton fabric treated with 0.6 M citric acid has an adsorption capacity of Pb(II) up to 3.5 times greater than the activated carbon, and the adsorbed metal can be recovered almost completely with 0.1 M HCl. These results suggest that modified cotton fabric is a highly competitive adsorbent for the treatment of metal-containing wastewater.

1. INTRODUCTION

Consumption of water polluted with lead (Pb(II)) is associated to human diseases such as anaemia, kidney damage, sterility, and it is suspected to be carcinogenic. For these reasons, several treatment technologies have been applied for the removal of this pollutant from water, for example: chemical precipitation, coagulation-flocculation, ion exchange, and adsorption with activated carbon, among others (Stumm y Morgan, 1996; Metcalf and Eddy, 2004). Adsorption process is one of the most commonly used technologies because it is useful as a polishing step in a water treatment plant. Nevertheless, it is important to explore the possible application of new adsorbents for removing lead from aqueous solutions. For instance, lignocelullosic materials can be used as new adsorbents because these are worldwide available and low-cost in comparison with polymeric resins and activated carbons. Moreover, it was reported that modification of lignocelullosic materials with citric acid enhanced the adsorption capacity of copper in comparison with non-treated adsorbents (Marshall *et al.*, 2007; Zhu *et al.*, 2008). In this way, the adsorption of metal cations in chemically modified lignocelullosic materials can be enhanced significantly. Nonetheless, to the best of our knowledge, at this time is not reported the density of functional groups added to the adsorbent and their equilibrium constants before and after chemical modification.

In this work, the effect of citric acid modification of cotton fabrics on lead adsorption from aqueous solutions is explored by means of ATR-FTIR analyses, potentiometric titrations, adsorption/desorption and kinetic experiments.

2. METHODOLOGY

2.1 Chemical modification. Thirty grams of cotton fabrics were added in a precipitated flask with 210 mL of 0.6 M citric acid solution. Suspension was heated at 60°C for 1 hour and, afterwards, water was evaporated at about 100°C until dryness. The remaining residues were dried at 120°C in an oven for 24 hours to carry out the esterification on cotton fabrics. Next, modified cotton fabrics were rinsed out with deionized water at 50°C and were dried at 100°C and stored in desiccators.

2.2 Adsorbent characterization. Non-modified and modified cotton fabrics were characterized by means of attenuated total reflection Fourier transform infrared spectroscopy (ATR-FTIR) and potentiometric titrations (Yun *et al.*, 2001; Garcia-Reyes *et al.*, 2009) to identify quantitative and qualitatively the functional groups incorporated to the adsorbent materials, respectively.

2.3 Adsorption experiments. Adsorption tests were conducted in batch systems at 30°C, 150 rpm and the selected pH (3 or 4) to keep lead in solution. A mass of 40 mg of cotton fabrics (non-modified and modified) were added to 40 mL of Pb(II) solutions with different concentration (5-100 mg/L). Solution pH was daily adjusted (with HNO_3 or NaOH) along the Pb(II) adsorption tests at the chosen value (pH 3 or 4) until the equilibrium was reached. Aliquots were taken at the beginning and the ending of adsorption experiments to determine the metal concentration by atomic absorption spectroscopy (ASS).

2.4 Desorption trials. Cotton fabrics, previously used in lead adsorption experiments, were regenerated with 40 mL of 0.1 N HCl solutions after being separated by filtration and rinsed with 30 mL of deionized water at pH 4. These suspensions were continuously stirred at 150 rpm during 24 hours. After that, an aliquot was taken to determine the Pb(II) concentration in solution by AAS.

2.6 Adsorption kinetics. To determine the time required to achieve the equilibrium and to estimate the parameters of kinetic models, adsorption rate experiments of Pb(II) on cotton fabrics were carried out in a packed column with recirculation at initial Pb(II) concentration of 0.256 mM, 30°C, initial pH of 4 and a feeding flow of 10 mL/min. Previous to the adsorption kinetic experiments, modified cotton fabrics were conditioned by stirring the adsorbent in a solution with pH 4 for 24 hours.

3. RESULTS AND DISCUSSION

Figure 1 shows the infrared spectra of the adsorbents used in this research. It is observed that chemical modification of cotton fabrics shown a peak about 1740 cm^{-1} (absent in non-modified cotton fabrics) that can be associated to carboxylic groups of the citric acid (see Figure 1) covalently bound to hydroxyl groups of cellulose structure. It was reported that carboxylic functional groups play a major in adsorption of metal cations from aqueous solutions (Tiemann *et al.*, 1999; Garcia-Reyes *et al.*, 2009).

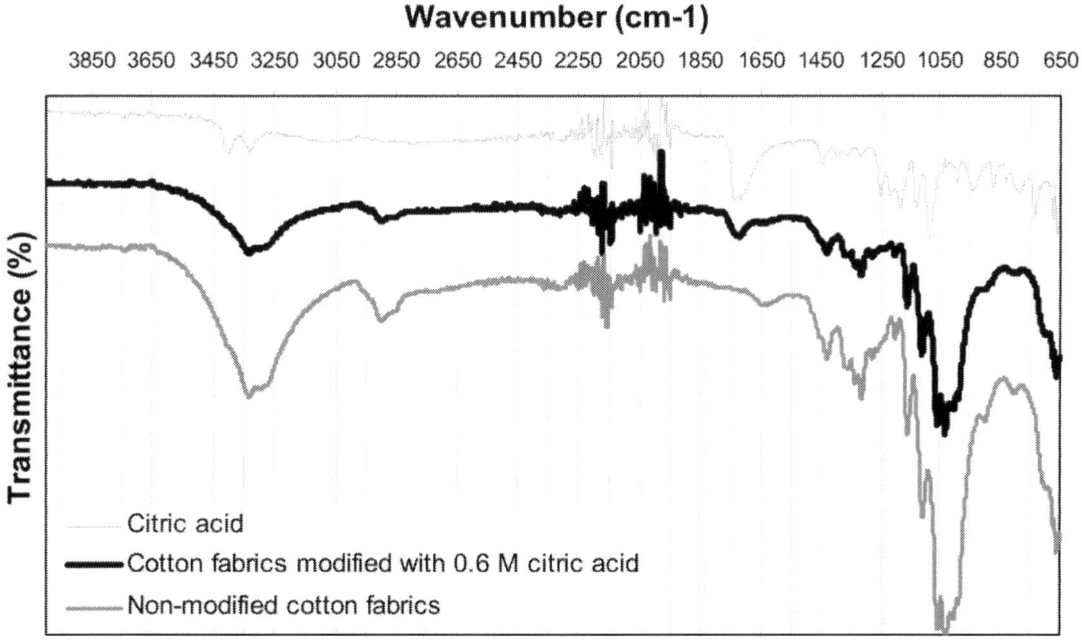

Figure 1. Infrared spectra (ATR-FTIR) of the studied adsorbents.

Table 1 shows the quantity of functional groups per unit mass of adsorbent (b_j) and their equilibrium constants (pk_{aj}) of non-modified and modified cotton fabrics. Non-modified cotton fabrics presented one functional group with an estimated quantity of 0.61 mmol/g and a pk_a value of 8.71 probably due to the hydroxyl groups of cellulose (peaks at 3350 and 1030 cm^{-1} in Figure 1) (Silverstein and Webster, 1998). In contrast, cotton fabrics modified with citric acid showed two functional groups (b_1=0.67 y b_2=1.80 mmol/g) with different pk_a values (4.06 and 8.08). These equilibrium constants are close to two of three pk_a values reported for citric acid (3.13, 4.8 y 6.40). It is important to mention that the functional group of modified cotton fabrics with pk_a=4.06 may correspond to a carboxylic group of citric acid and, therefore, these functional groups can be used as adsorption sites of positively-charged species in water such as Pb(II) species (Pb^{2+}, $PbNO_3^+$, $PbOH^+$, and Pb_2OH^{3+}) (Tiemann *et al.*, 1999; García-Reyes *et al.*, 2009) meanwhile pH of solution remains below 5.6 to avoid Pb(II) precipitation.

Table 1. Density of functional groups (b_j) and their equilibrium constants (pk_{ai}) of the studied adsorbents.

Adsorbent	b_1 (mmol/g)	pk_{a1}	b_2 (mmol/g)	pk_{a2}
Non-modified cotton fabrics	n.d.	n.d.	0.61	8.71
Modified cotton fabrics	0.67	4.06	1.80	8.08

n.d.: Not detected.

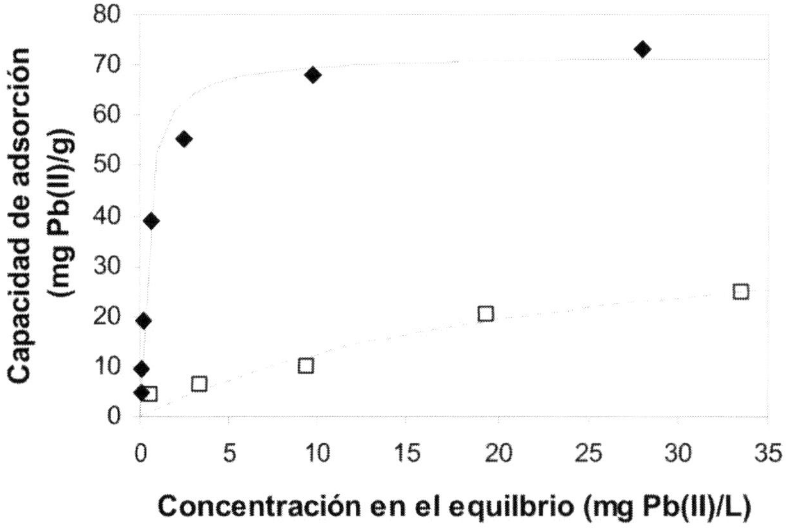

Figure 2. Adsorption isotherms of Pb(II) on cotton fabrics modified with 0.6 M citric acid at 30°C, pH 3 (□) y 4 (♦). Symbols represent experimental data and lines were predicted with the Langmuir isotherm.

Figure 3 shows the adsorption rate of Pb(II) on modified cotton fabrics without pH control. Kinetic experiments were conducted at initial pH 5 but final (at equilibrium) pH decreased to 3.36. This result seems to be reasonable because pH was not corrected along the kinetic experiments and Pb(II) species displaced H^+ ions from protonated carboxylic groups (covalently bound on the cotton fabrics). Adsorption kinetics was well predicted with the pseudo-second order model (Ho y

McKay, 1999). This result suggests that lead species interact with adsorption sites by physical forces such as electrostatic charges.

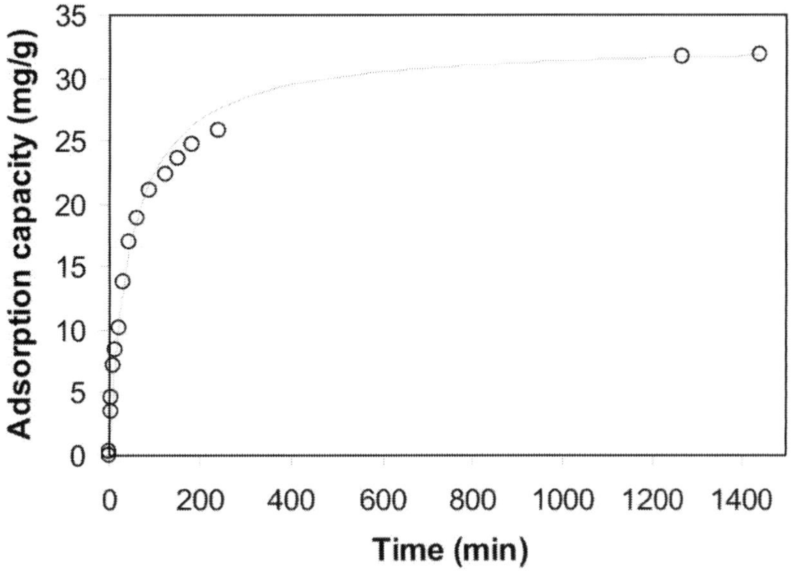

Figure 3. Adsorption kinetics of Pb(II) on modified cotton fabrics at 30°C, initial pH 5, initial concentration 53.18 mg/L, volumetric flow of 10 mL/min, and adsorbent dosage of 1 g/L. Experimental data are represented by symbols and line was predicted with the pseudo second order model.

4. CONCLUSIONS

- ✓ Chemical modification of cotton fabrics by citric acid incorporate functional groups (*i.e.* carboxyl groups) capable of binding metal cations such as Pb(II) from aqueous solutions.
- ✓ Adsorption capacity of Pb(II) on modified cotton fabrics at pH 4 was around 3.5 times greater than activated carbon.
- ✓ It is possible to recover up to 92% of the lead previously adsorbed on the modified cotton with using 0.1 M HCl solutions.
- ✓ Adsorption kinetics can be predicted by a pseudo-second order model and it suggests that lead species interact with adsorption sites by physical forces such as electrostatic charges.

5. ACKNOWLEDGMENTS

This research was financially supported in part by the Secretaría de Educación Pública (PROMEP/103.5/11/6627) y la Universidad Autónoma de Nuevo León (PAICYT IT577-10).

6. REFERENCES

Acharya, J., Sahu, J.N., Mohanty, C.R. and Meikap, B.C. (2009) Removal of lead(II) from wastewater by activated carbon developed from Tamarind wood by zinc chloride activation. Chemical Engineering Journal 149(1-3), 249-262.

Garcia-Reyes Refugio Bernardo, Rangel-Mendez Jose Rene, Alfaro-De la Torre Ma. Catalina (2009), Chromium (III) uptake by agro-waste biosorbents: chemical characterization, sorption-desorption studies, and mechanism, Journal of Hazardous Materials, 170(2-3) 845-854.

Ho, Y.S. and McKay, G. (1999) Pseudo-second order model for sorption processes. Process Biochemistry 34(5), 451-465.

Marshall, W.E., Akin, D.E., Wartelle, L.H. and Annis, P.A. (2007) Citric acid treatment of flax, cotton and blended nonwoven mats for copper ion absorption. Industrial Crops and Products 26(1), 8-13.

Metcalf and Eddy (2004), Waste Water Engineering Treatment and Reuse, 4th Ed., Ed. Mc Graw Hill, USA, 1819 + XXVIII.

Silverstein, R. M., and Webster, F. X. (1998), Spectrometric Identification of Organic Compounds. Wiley & Sons, New York.

Stumm, W. & J. J. Morgan, Aquatic Chemistry: Chemical Equilibria and Rates in Natural Waters, 3rd. ed., John Wiley & Sons, New York, 1996.

Tiemann, K. J., J. L. Gardea-Torresdey, G. Gamez, K. Dokken, S. Sias (1999), Use of X-ray absorption spectroscopy and esterification to investigate Cr(III) and Ni(II) ligands in alfalfa biomass. Environmetal Science and Technology, 33(1) 150-154.

Yun, Y. S., D. Park, J. M. Park, B. Volesky (2001), Biosorption of trivalent chromium on the brown seaweed biomass, Environmetal Science and Technology, 35(21) 4353-4358.

Zhu, B., T. X. Fan, D. Zhang (2008), Adsorption of copper ions from aqueous solution by citric acid modified soybean straw. Journal of Hazardous Materials, 153(1-2) 300-308.

(156a) Synthetic Antibacterial Probiotics

Yiannis Kaznessis, Department of Chemical Engineering and Materials Science, University of Minnesota, Minneapolis, MN

Foodborne bacterial gastrointestinal infections are significant causes of morbidity and mortality worldwide. Alarmingly, because of the extensive, non-therapeutic use of antibiotics in agriculture, foodborne pathogens are emerging that are resistant to our most potent drugs.

We pursue an integrated research approach to reduce the use of therapeutic antibiotics in animal feed and treat gastrointestinal infections in animals and humans. We employ synthetic and systems biology technologies to engineer probiotic bacteria that reside in animal gastrointestinal tracts, and then express and release antimicrobial peptides (AMPs). Probiotic bacteria are part of the gastrointestinal microbiota and have known benefits for humans and animals. AMPs are proteins and can be readily produced by bacteria. What is unique in our approach is the use of synthetic biological switches to precisely control the overexpression and delivery of AMP molecules.

AMPs are small molecules with remarkable bactericidal properties. Their use has been limited because they are quickly degraded by the host if administered orally or intravenously. We use probiotics as AMP-delivery vehicles. Probiotics are bile-resistant microorganisms that can be delivered safely in food or water. We engineer inducible AMP expression systems in probiotics. We examine the impact of controllable delivery of AMPs in swine and poultry challenged by infectious agents and explore AMP-carrying probiotics as alternatives to traditional antibiotics in agriculture. We also experiment with mice as models of human gastrointestinal tracts.

We use established protein expression systems in probiotic bacteria to express bacteriocins, AMPs that are naturally produced by bacteria. We also study pathogen-specific AMPs. We experiment with probiotic species *Lactococus lactis, Lactobacillus acidophilus,* known to be safe for consumption by animals and humans. These species are also well-annotated with known genomes and have established microbiological and genetic engineering techniques. Synthetic molecular devices are engineered to be robust. With established metagenomics, proteomics and bioinformatics techniques, we then analyze animal gut microbiomes and the relationships between probiotics, commensal microbes and health. We quantify the presence of engineered probiotics and of expressed AMPs in intestines, and quantify the disease outcome for animals challenged by pathogens.

(156b) Functionalization of Novel Polyanhydride Nanoparticle Adjuvants to Promote Pathogen-Mimicking Characteristics

Yashdeep Phanse[1], Brenda R. Carrillo-Conde[2], Amanda Ramer-Tait[1], Rajarshi Roychoudhuri[3], Scott Broderick[4], Nichola Pohl[3], Krishna Rajan[4], **Balaji Narasimhan**[5], Michael J. Wannemuehler[1] and Bryan H. Bellaire[1], (1)Department of Veterinary Microbiology & Preventive Medicine, Iowa State University, Ames, IA, (2)Department of Biomedical Engineering, University of Texas at Austin, Austin, TX, (3)Department of Chemistry, Iowa State University, Ames, IA, (4)Department of Materials Science and Engineering, Iowa State University, Ames, IA, (5)Department of Chemical and Biological Engineering, Iowa State University, Ames, IA

An imminent need exists to design improved adjuvants and vaccines to control emerging and re-emerging diseases. An ideal vaccine will mimic the way in which a naturally occurring infection induces a robust immune response yet avoids the undesirable effects of disease. To accomplish this, new adjuvants must engage pattern recognition receptors on antigen presenting cells (APCs). In this work, we employ novel strategies to enhance the adjuvanticity of polyanhydride nanoparticles composed of sebacic acid (SA) and 1,6-bis(p-carboxyphenoxy) hexane (CPH) by functionalizing their surface with glycolic acid linker or the pathogen associated molecular pattern di-mannose (diM) to confer "pathogen-like" properties. Co-incubation of linker functionalized nanoparticles with APCs elicited significant increases in surface expression of antigen presentation and co-stimulatory molecules and secretion of proinflammatory cytokines. An 800% increase in uptake of linker and diM-nanoparticles was also observed. Microscopic analysis revealed that majority of the linker particles trafficked to LAMP-1+ vesicles implicated in MHC II presentation pathway. A small proportion of linker nanoparticles trafficked to ER and a subsequent increase in MHC I MFI was observed suggesting that encapsulated antigen could be presented to $CD8^+$ T cells via MHC I in vivo. Together, our data show that APCs incubated with linker functionalized nanoparticles demonstrated similar patterns of uptake, intracellular fate, persistence and activation as did APCs exposed to live, pathogenic *Y. pestis* or *E. coli.* The innovative combination of polyanhydride nanoparticles and their surface modification provides mechanistic insight into the properties required for next-generation adjuvants to safely mimic the immune stimulating activities of bacterial pathogens.

(156d) An Investigation of the Pulmonary and Systemic Persistence of Polyanhydride Nanoparticle Vaccine Formulations

Kathleen A. Ross[1], Shannon Haughney[1], Lucas Huntimer[2], Latrisha Petersen[1], Amanda Ramer-Tait[2], Michael J. Wannemuehler[2] and Balaji Narasimhan[1], (1)Chemical and Biological Engineering, Iowa State University, Ames, IA, (2)Veterinary Microbiology and Preventive Medicine, Iowa State University, Ames, IA

Polyanhydrides have properties that make them ideal candidates for vaccine delivery including the ability to stabilize and protect fragile protein antigens, provide a sustained release of antigen, and modulate the immune response. Homopolymers and copolymers based on sebacic acid (SA), 1,6-bis-(*p*-carboxyphenoxy)hexane (CPH) and 1,8-bis-(*p*-carboxyphenoxy)-3,6-dioxaoctane (CPTEG) have shown efficacy in vaccine formulations. Encapsulation of F1-V protein (an antigen of *Yersinia pestis*) into nanoparticles made from CPTEG:CPH copolymers demonstrated the ability to provide protection against a lethal bacterial challenge. Additionally, polyanhydride nanoparticles made from CPH:SA copolymers have been shown to preserve the functionality of the protective antigen (PA) of *Bacillus anthracis* when stored at 37°C for up to one year.

In this work, we have integrated live animal imaging, confocal microscopy, and flow cytometry to analyze the deposition, uptake, and *in vivo* persistence of PA delivered via the polyanhydride nanoparticle platform over 12 weeks. Distribution and uptake of antigen-loaded polyanhydride nanoparticles administered intranasally is different than traditional adjuvants such as MPLA, reaching the deep recesses of the lungs, where they are internalized and persist for more than 42 days. Our previous work has demonstrated that protective polyanhydride vaccine formulations contain a combination of both soluble and encapsulated antigen by providing an immediate immune response stimulus followed by sustained antigen release to generate affinity matured antibodies and long-lasting immunity. This current study supports these observations, demonstrating the differential *in vivo* distribution kinetics of soluble versus nanoparticle encapsulated PA antigen through *ex vivo* organ imaging of fluorescently tagged PA. These studies lay the foundation for mechanistic insight into continued antigen persistence in the immunization site leading to high antibody titer with increased avidity.

(156e) Sputum Penetration and Enhanced Airway Gene Transfer by Mucus Penetrating Synthetic Gene Nanocarriers

Anthony J. Kim, Jung Soo Suk, Nicholas Boylan and Justin S. Hanes, The Center for Nanomedicine, Johns Hopkins University School of Medicine, Baltimore, MD

No gene delivery system, viral or non-viral, has shown evidence of efficiently penetrating the human mucus barrier; this reality may account in part for the limited success in clinical trials for CF gene therapy to date. Gene carriers that fail to penetrate airway mucus layers are ultimately cleared from the lungs via mucus clearance mechanisms. To potentially overcome this formidable barrier, we developed a synthetic gene delivery platform capable of penetrating human CF sputum while maintaining an ability to provide efficient airway gene transfer in mice and in human airway epithelial cells grown at air-liquid interface. Mucus penetrating gene nanocarriers markedly enhanced particle diffusion in CF sputum, leading to enhanced particle distribution, retention and gene transfer in the mouse airways. Enhanced airway gene transfer by mucus penetrating gene nanocarriers is most likely attributed to the improved mucus penetration, which leads to more uniform distribution and prolonged retention in the lung compared to conventional cationic gene carriers. Pulmonary administration of these nanocarriers exhibited safety and inflammation profile comparable to saline control, whereas conventional cationic gene carriers induced significant toxicity and inflammation.

(156f) Dendrimer Nanocarriers for Transport Modulation Across the Pulmonary Epithelium: Cellular Internalization, Transport and Formulation in Portable Inhalers

Balaji S. Bharatwaj[1,2], Radovan Dimovski[2], Denise S. Conti[2] and Sandro R. P. da Rocha[2], (1)Respiratory Product Development, Merck & Co., Summit, NJ, (2)Chemical Engineering and Materials Science, Wayne State University, Detroit, MI

The ability to modulate the transport of therapeutics across the pulmonary epithelium may provide new opportunities in terms of improving the treatment of regional lung diseases, and also in the development of new strategies for the systemic delivery of therapeutics through the lungs using the oral inhalation (OI) route. The lungs are an attractive pathway to the bloodstream as they have a large surface area, consist of a thin cellular barrier, and have lower proteolytic activity compared to those observed in traditional delivery routes.

In this work, the effect of PEGylation on the cellular internalization and transport of amine-terminated, polyamidoamine dendrimer nanocarriers across an *in vitro* model of the human airway epithelium (Calu-3 cells) was evaluated. Our results indicate that an increase in the density of PEG grafts onto the surface of dendrimer nanocarriers enhances their transport across polarized Calu-3 monolayers, and can also be used to tune their cellular uptake. We also demonstrate that these dendrimer nanocarriers can be formulated in portable inhalers (pressurized metered-dose inhalers), and that these suspensions possess enhanced aerosol characteristics, which are conducive to deep lung deposition.

(156g) Elastin Based Nano-Particles for Targeted Therapy of Lung Adenocarcinomas

Raul Iglesias, University of South Florida, Tampa, FL and Piyush Koria, Department of Chemical and Biomedical Engineering, University of South Florida, Tampa, FL

Introduction: Nanoparticles have become increasingly popular for disease treatment in the last decade. Their ability to be used as transport and delivery vehicles for drugs has made their synthesis very appealing. In this work, we propose the development of nanoparticles for targeting of lung adenocarcinomas; this type of cancer represents about 40% of all lung cancers. Our nanoparticles consist of chimeric fusion proteins comprising of recombinant keratinocyte growth factor (KGF) and elastin like peptides (ELP). KGF is the ligand for the keratinocyte growth factor receptor (KGFR); which is highly expressed in lung adenocarcinoma. This chimeric fusion retains the bioactivity of KGF and the characteristic inverse phase transitioning behavior of ELPs. The phase transition behavior of ELPs induces nanoparticle self-assembly with the hydrophobic ELPs forming the core and hydrophilic KGF displayed at the periphery. The size of the nanoparticles is a direct function of the unique ELP sequences chosen.

Materials and Methods: The chimeric fusion protein (KGF-ELP) was produced recombinantly in bacteria. The formed nanoparticles were characterized using dynamic light scattering and transmission electron microscopy. RT-PCR and western blots were used for quantifying the expression of the KGFR in different lung adenocarcinoma cells and fibroblasts (Fb). Lung adenocarcinoma cells or fibroblasts (500,000 / well) were serum starved overnight and then treated with Cy3 labeled KGF-ELP (0.5ug/ml). Twenty-four hours after treatment, the cells were analyzed by flow cytometry. The fluorescence geometric mean was calculated using FACSDiva version 6.1.3 software.

Results and Discussion: The KGF-ELP fusion protein self-assembled into nanoparticles with diameters ranging from 100-500 nm depending on the length of the chosen ELP cassette. The particles were fairly mono-dispersed and were stable over a range of temperatures. The particle internalization per cell was further quantified using flowcytometery for different lung cancer cell lines, which yielded significantly high particle internalization compared to fibroblasts.

Conclusions: Our results indicate specific targeting of the nanoparticles to lung cancer cells via the KGF receptor. The particles were selectively internalized by KGFR expressing lung cancer cells through receptor mediated internalization, while very little internalization was seen in fibroblasts that do not express KGFR. Therefore these nanoparticles represent a promising approach for developing targeted therapies and imaging modalities for lung adenocarcinoma. Moreover, the recombinant method of producing the chimeric fusion proteins enables one to include different targeting motifs in the nanoparticle with relative ease; thereby making this strategy extremely versatile for different applications in cancer and medicine.

(156h) Single-Walled Carbon Nanotubes Targeted to the Tumor Vasculature for the Treatment of Breast Cancer

Brent D. Van Rite, Chemical, Biological and Materials Engineering, University of Oklahoma , Norman , OK, Luis F. F. Neves, Chemcal, Biological and Materials Engineering, University of Oklahoma , Norman, OK, Whitney Prickett, Chemical, Biological and Materials Engineering, University of Oklahoma , Norman, OK, Antonietta Restuccia, Chemical, Biological and Mateials Engineering, University of Oklahoma, Norman, OK, Daniel Resasco, School of Chemical, Biological and Materials Engineering and Carbon Nanotube Technology Center, University of Oklahoma, Norman, OK and **Roger Harrison**, University of Oklahoma, Norman, OK

While nanoparticles have a large potential to be used to improve the treatment of cancer, a major barrier of further development is the navigation of the nanoparticle from its point of introduction into the body to its final destination. A new cancer treatment system is being developed with the goal of delivering single-walled carbon nanotubes (SWNTs) to the tumor rapidly and efficiently. This system uses SWNTs that are functionalized with the protein annexin V or the F3 peptide. Annexin V binds selectively to anionic phospholipids such as phosphatidylserine (PS). PS exposure on the outer cell membrane is almost completely exclusive to cancer cells and endothelial cells that line the tumor vasculature. F3, a fragment of the high mobility group protein, is a 31-amino acid peptide that has been shown to home to the nuclei of tumor vasculature cells and tumor cells and bind to the cell surface nucleolin receptor and become internalized. In a treatment of cancer by this system, SWNTs bound to the tumor vasculature are exposed to near-infrared light (NIR) at a wavelength that heats the SWNTS but not the surrounding normal tissue, resulting in the tumor being destroyed.

In this study, we have covalently coupled the human protein annexin V or the F3 peptide to a phospholipid-polyethylene glycol-maleimide linker that is strongly adsorbed to the SWNTs by the phospholipid in the linker. We avoided covalently coupling the protein to the SWNT sidewall to avoid loss of the NIR absorption by the SWNTs. Using non-confluent human endothelial cells *in vitro*, there was a significant loss of cell viability in cells that had been treated with the SWNT-annexin V or SWNT-F3 conjugate, washed, and then exposed to NIR light. We have found that non-confluent endothelial cells mimic these cells in the tumor vasculature in their expression of PS on the cell surface. Results of using this treatment in immune competent mice implanted with 4T1 murine mammary cancer cells will be reported. This treatment was also combined with simultaneous administration of the immunostimulant cycylophosphamide, which is known to deplete regulatory T cells. Regulatory T cells are believed to block antitumor immune responses mediated by tumor-specific T cells.

(169a) From Cancer to Hydrogen: Cell-free Biomolecular and Bioprocess Engineering

James R. Swartz, Department of Chemical Engineering, Stanford University, Stanford, CA

Normally, the design and production of cancer therapeutics would seem to have little in common with large-scale hydrogen production. Yet, the amazing capabilities of biological systems provide the components and principles for both pursuits. In particular, cell-free methods allow us to manipulate and harness even very complex biological reaction systems. In this presentation, I will describe progress toward: a) new cancer diagnostics and novel cancer vaccines, b) a potent and broadly protective influenza vaccine, and c) a potential highly productive process for sustainable hydrogen production from cellulosic biomass. Even though the applications are highly diverse, they can all be addressed by the power of fundamental biochemical engineering.

(169c) Biochemical Engineering in Half a Century: Tracing the Steps of Prof. Daniel IC Wang

Wei-Shou Hu, Chemical Engineering and Materials Science, University of Minnesota, Minneapolis, MN

For over five decades Prof. Wang has devoted himself to biochemical engineering education, research, and technology creation. The course of his career illustrates the tremendous advances that have taken place in the field. His work helped propel technological innovation and development as it reached critical junctures. This presentation will use Prof. DIC Wang's work to highlight the advances in biochemical engineering in the past half a century and invite the audience to reflect on its impact on the future of the profession.

(190a) Lung Targeting Hydrogel Microparticles with Embedded Nanoparticles for the Treatment of Lung Cancer

Nathalie M. Pinkerton[1], Bryan Benson[1], Dayuan Gao[2], Howard A. Stone[3], Patrick J. Sinko[4] and Robert K. Prud'homme[1], (1)Chemical and Biological Engineering, Princeton University, Princeton, NJ, (2)Rutgers University, New Brunswick, NJ, (3)Mechanical and Aerospace Engineering, Princeton University, Princeton, NJ, (4)Department of Pharmaceutics, Rutgers University, Piscataway, NJ

Although only the second most prevalent type of cancer, lung cancer is the most lethal. National Cancer Institute assesses that results of standard treatment are generally poor with only a 15 percent 5-year survival rate for combined cancer stages. Challenges facing the current chemotherapy drugs include, hydrophobicity of therapeutic agents, excessive systemic toxicity in healthy tissues and inability to specifically target lung tumors. The hydrogel-based delivery vehicle described herein aims to overcome these challenges by selectively targeting the lung to deliver hydrophobic, anti-cancer drugs.

The first layer of the delivery system is a poly(ethylene glycol) based gel microparticle (GMP) designed to take advantage of the venous lung filtration pathway and passively accumulate in the lungs after intravenous injection into the body. The size and modulus of the GMPs affect the targeting, accumulation and clearance from the lungs. The second layer of the delivery system consists of nanoparticles (NP) embedded within the GMP matrix. The NPs are loaded with a camptothecin prodrug designed with a hydrolytically unstable ester linker. As the ester is cleaved, camptothecin is released. This dual-delivery system affords the ability to sustain a high concentration of anti-cancer drugs in the lungs while minimizing the systemic exposure and accordingly reducing the side effects.

(190b) Overcoming Transport Barriers in Drug Delivery to Tumors

Jennifer Anne Pascal[1], Carlee E. Ashley[2], Zhihui Wang[1], C. Jeffrey Brinker[2] and Vittorio Cristini[1], (1)Pathology, University of New Mexico, Albuquerque, NM, (2)Department of Chemical Engineering, University of New Mexico, Albuquerque, NM

The natural defenses of the human body create many challenges for chemotherapeutics to reach their intended targets. In order to be effective, chemotherapy agents must travel through the blood vessels, extravasate out of the vessels into the interstitium and diffuse through tumor cells without being phagocytosed in the reticuloendothelial system, binding to proteins in serum or being metabolized by red blood cells. A largely overlooked limitation of traditional chemotherapy delivery is the transport within the tumor microenvironment. The unstructured and limited vasculature in tumors makes them a very heterogeneous microenvironment causing inconsistencies in the delivery and distribution of drugs. Because of these heterogeneities, in traditional chemotherapy administration, drugs are only transported to perfused regions. One approach to overcome these transport barriers is to utilize nanocarriers to selectively target and deliver drugs to malignant tumor cells.

Here, we present a mathematical model that describes *in vitro* cytotoxicity assay data for parental and multi-drug resistant human hepatocellular carcinoma cell lines using the common anticancer drug, doxorubicin. The mathematical model was solved analytically and fit to data for both free doxorubicin and doxorubicin-loaded protocells to obtain parameters describing cell death rate and delay time. The information from this model and experimental data can be used to predict the fraction of tumor volume that will be killed.

(190c) Transdermal Protein Delivery and Immunization Using Nano-Coating Drug Carrier

Masahiro Goto, Applied Chemistry & Center for Future Chemistry, Kyushu University, Fukuoka, Japan, Yoshiro Tahara, Kyushu University and Noriho Kamiya, Center for Future Chemistry, Kyushu University, Fukuoka, Japan

In the present study, we demonstrated that a solid-in-oil (S/O) nanodispersion, an oil-based nanodispersion of hydrophilic drugs, effectively enhanced the permeation and immunization of proteins through the skin. Two different model proteins, FITC-labeled insulin and HRP, permeated through the stratum corneum (SC) of Yucatan micropig skin *in vitro* by forming the S/O nanocarrier. The penetrated HRP exhibited catalytic activity, suggesting that the proteins can permeate into the skin in a functional form. It has been difficult to achieve therapeutic levels of relatively large drugs (over 500 Da) through intact skin to the systemic circulation by the chemical penetration enhancer alone. To overcome these situations, a surface nano-coating technique is proposed. In my presentation, the surfactant–protein complexes in which proteins are coated with hydrophobic surfactant molecules are prepared. The modified proteins can be highly dispersed in an oil of interest. The idea is that the dispersion of the surfactant–protein complex in the oil phase as a nano-order particle makes the proteins permeable into the skin. This concept is novel because to date, almost all studies on transdermal delivery of hydrophilic drugs have been based on aqueous vehicles. The results obtained in this study clearly suggest that the S/O nanodispersion technique has a high potential for the creation of a novel protein immunization system.

(190d) In Vivo Evaluation of Rod-Shaped Carriers for Vascular Imaging and Targeted Drug Delivery in Large Vessels in Mice

Alex Thompson[1], Katawut Namdee[2] and Omolola Eniola-Adefeso[1], (1)Chemical Engineering, University of Michigan, Ann Arbor, MI, (2)Biomedical Engineering, University of Michigan, Ann Arbor, MI

Introduction: The development of vascular-targeted carriers (VTC) for the delivery of therapeutics could greatly improve the treatment of many human diseases. Recently, particle shape has received attention as a parameter that can be used to improve the performance of VTCs. The purpose of this study is to examine the hemodynamics of rod-shaped particles of different aspect ratios relative to spheres of equal volume both *in vitro* in human blood flow under physiological conditions relevant in M/LBVs and *in vivo* in a mouse model of atherosclerosis.

Methods: For *in vitro* assays, rod-shaped and spherical particles were conjugated with targeting molecules to selectins (via Sialyl Lewis a) at a fixed surface density. The differently shaped particles were tested for their margination efficiency from human blood flow both via parallel plate flow chambers with human endothelial cells. The particles explored ranged in diameter from 0.5 – 2 μm (equivalent spherical diameter, ESD, for rods). For *in vivo* analysis, particles co-conjugated with sLe^A (ligands for selectin) and mouse VCAM-1 antibody (at a fixed surface density for all particle dimension) were injected via the tail vein into ApoE knockout mice with developed atherosclerosis. Particle binding efficiency was quantified along the aorta and biodistribution quantified in internal organs and blood as a function of particle size, particle shape and circulation time.

Results and Discussion: With *in vitro* assays, we find that rod shaped particles with ESD >2 μm have a higher binding efficiency to HUVEC monolayer at the wall from bulk human blood flow than spheres of equivalent volume and targeting ligand site density. There appears to be a minimum major axis length requirement for rods with ESD <2μm to display a significant increase in flow adhesion over their equivalent spheres. To further investigate this, nanoparticles (500 nm ESD) is fabricated which have relatively high major axis lengths (major axis≈6 μm). The adhesion of these rods from blood flow is compared to spheres to determine whether a long major axis can significantly improve nanoparticle binding

For *in vivo* assays, preliminary results suggest that particle size and shape play a significant role in particle binding efficiency in atherosclerotic mice. Specifically, 2 μm spherical particles showed higher adhesion to plaque surface in the aorta over 0.5 μm spheres in agreement with previous *in vitro* studies. Furthermore, ellipsoidal particles with aspect ratio (AR) 4 yielded higher adhesion in the aorta than 2 μm spheres, especially at the branch and bifurcation area where plaque were preferentially deposited. However, longer rods with AR 9 displayed minimal adhesion along the aorta while exhibiting significant accumulation in the lung and other internal organs possibly due to the physical entrapment in the capillaries. The biodistribution of particles after 30 min and 8 hr circulation times will also be discussed.

Conclusions: Our study shows that shape can be a very useful and tunable parameter in the design of vascular-targeted carriers for imaging and treatment of atherosclerosis and other diseases of medium and large blood vessels.

(190e) Engineering a New Microbial Production Platform for Fuels and Chemicals

Rachel Ruizhen Chen, Chemical&Biomolecular Engineering, Georgia Institute of Technology, Atlanta, GA

The prevailing biomass technology relies on a complete hydrolysis of cellulose to its monomer sugars before the fermentation step, entailing large amounts of enzymes. The cost of enzymes and slow hydrolysis step are widely recognized as important barriers for biofuel production. We have developed novel microbial catalysts capable of utilizing cellulose partial hydrolysis products, cellobiose or larger glucose oligomers (collectively known as cellodextrin). These microbial strains transport cellodextrin inside the cells, where they are subsequently metabolized and converted to biofuel. The new production scheme has several important advantages. 1.) Directly use of cellodextrin eliminates the need for complete hydrolysis. As a result, exogenous beta-glucosidase is no longer needed, generating significant cost savings. 2.) Once inside the cells, cellodextrin can be metabolized via a more energetically favorable mechanism, phosphorolysis. 3.) Transporting sugar oligomers inside the cells for intracellular metabolism bypasses the catabolite repression. Consequently, glucose oligomers and xylose can be simultaneously converted to biofuel molecules, reducing the fermentation time thereby enhancing productivities. 4.) The new technology reduces the burden of contamination as glucose concentration in the fermentation tank can be kept low.

In this presentation, we discuss the discovery of cellodextrin transporters from bacterial origin.
We show that the newly discovered transporters can be used to engineer new types of microbial catalysts that access partial hydrolysed cellulose and assimilate cellodextrin directly. We present a first successful E. coli biocatalyst that metabolizes cellodextrin through an alternative but more energy- efficient phosphorolytic mechanism. Finally, we show that the transport of cellodextrin and its subsequent intracellular metabolism masks the presence of glucose and allows to evade catabolite repression, thus cellodextrin and xylose were simultaneously utilized. In one example, simultaneous conversion of cellobiose and xylose led to a 100% increase in productivity, demonstrating the potential of this new microbial platform in biofuel production.

(190f) Synthesis and Characterization of an Acetalated Dextran Polymer and Microparticles with Ethanol As a Degradation Product

Kevin J. Kauffman[1], Clement Do[2], Sadhana Sharma[2], Matthew D. Gallovic[1], Eric M. Bachelder[2] and Kristy M. Ainslie[1,2], (1)Chemical and Biomolecular Engineering, The Ohio State University, Columbus, OH, (2)Division of Pharmaceutics, College of Pharmacy, The Ohio State University, Columbus, OH

Dextran is a polysaccharide biopolymer composed of linked glucose molecules. One of its several FDA-approved uses is as an antithrombotic agent, and it is highly biocompatible and biodegradable. Using dextran as the molecular foundation of our polymeric technology, our group has initiated and continues to expand a family of biobased polymers derived from dextran that have numerous potential drug delivery applications, including protein-based vaccine delivery, immunosuppressant delivery to phagocytic cells, and chemotherapeutic delivery to pulmonary cancer cells. Currently, poly(lactic-*co*-glycolic acid) (PLGA) is commonly used as a polymeric delivery vehicle. However, PLGA is pH-insensitive, its degradation profile is not widely tunable, and its degradation products create an acidic microenvironment. Thus, our group has created and continues to improve upon the biobased material acetalated dextran (Ac-DEX), a biopolymer synthesized by appending cyclic and acyclic methoxy acetal groups onto the hydroxyl groups of dextran. Ac-DEX possesses a tunable, acid-sensitive degradation profile and could be used in numerous *in vivo* applications where acidic environments are present (e.g. the delivery of vaccines via the endocytic pathway of immune cells). Additionally, Ac-DEX has three pH-neutral degradation products: acetone, dextran, and methanol. A drawback of Ac-DEX is the presence of its methanol degradation product. It is known that methanol is metabolized to formic acid, causing metabolic acidosis, which can lead to blindness or death. Although using lower doses of Ac-DEX microparticles would likely not pose a problem *in vivo*, other potential uses of Ac-DEX would require higher amounts of the biopolymer, such as multiple microparticle doses or microfibers in tissue engineering applications. Therefore, in this study our group has developed another biobased material, ethoxy-derivatized acetalated dextran (Ace-DEX). Ace-DEX has nearly the same structure as Ac-DEX; one structural difference is the presence of an extra methyl group on the acyclic acetal groups, creating acyclic ethoxy acetals. This additional methyl group eliminates the methanol byproduct seen with Ac-DEX degradation and replaces it with an ethanol byproduct. Ethanol is more desirable, because it is much more biologically tolerable. The presence of the ethanol has been confirmed using ^{1}H-NMR. Several material properties of Ac-DEX have been previously characterized: the effects of reaction time on cyclic and acyclic acetal coverage, the degradation profiles in extracellular (pH 7.4) and phagosomal (pH 5.0) conditions, and the cell viability of macrophages treated with Ac-DEX microparticles. Thus, in this study we have examined these same characteristics of Ace-DEX. As compared to Ac-DEX, using reaction times spanning from 5 minutes to 8 hours, Ace-DEX possesses a slightly lower degree of substitution of cyclic acetals and a slightly higher degree of substitution of acyclic acetals. This difference in the degrees of substitution indicates that the reaction rate of Ace-DEX is slower than that of Ac-DEX, because cyclic acetals replace acyclic acetals over time. It is hypothesized that the slower reaction rate is caused by a minor steric hindrance effect from the 2-ethoxypropene reactant (as opposed to the 2-methoxypropene reactant used in Ac-DEX synthesis). It is shown that the degradation profiles of Ace-DEX in acidic (pH 5.0) and extracellular (pH 7.4) conditions are easily tunable via simply altering the reaction time; extending the reaction time creates more cyclic acetals and causes a slower degradation to occur. As compared to Ac-DEX, Ace-DEX is shown to have less of a burst and a more sustained degradation (indicated by a longer half-life, the time at which 50% of the material has degraded). Thus, Ace-DEX could be used for a more sustained delivery of a therapeutic. Finally, the cell viabilities of macrophages treated with Ace-DEX microparticles are shown to be statistically similar to cells treated with Ac-DEX or PLGA microparticles at concentrations ranging from 0 to 1 mg/mL. It is very promising that the biobased Ace-DEX polymer shows very similar cell viabilities to that of FDA-approved PLGA. Overall, Ace-DEX represents the continuing development and growth of our family of biobased, acid-sensitive polymers that have tunable degradation profiles. Ace-DEX is very attractive for applications requiring higher amounts of polymer, because the methanol degradation product of Ac-DEX has been replaced with an ethanol byproduct. In the future, we intend to explore the use of Ace-DEX in studies requiring multiple microparticle doses or in tissue engineering applications that require microfibrous scaffolds.

(190g) Synthesis of Peg/PCL-Based Polyurethane Nanoparticles by Miniemulsion Polymerization for Pharmaceutical Applications

Alexsandra Valerio, Departamento de Engenharia Quimica e de Alimentos, Universidade Federal de Sanata Catarina, Florianopolis, Brazil; Chemical Engineering, Wayne State university, Detroit, MI, Claudia Sayer, Departamento de Engenharia Quimica e Alimentos, Universidade Federal de Santa Catarina, Santa Catarina, Brazil, Pedro H. H. Araujo, Departamento de Engenharia Quimica e de Alimentos, Universidade Federal de Santa Catarina, Florianoolis, Brazil and Sandro R. da Rocha, Wayne State University, Detroit, MI

In recent years polymeric nanoparticles have gained increasing popularity as promising vehicles for the delivery of drugs and others agents in a variety of biomedical and pharmaceutical applications. More recently, the development of colloidal delivery systems from biodegradable polyurethanes (PU) has attracted increasing interest as the highly variable synthetic chemistry of PU may be exploited to generate polymers having properties ranging from very soft elastomers to very rigid plastics.

In this presentation we will discuss the preparation of PU nanoparticles by miniemulsion interfacial polymerization using isophoronediisocyanate (IPDI), poly(-caprolactone) (PCL), and poly(ethylene glycol) (PEG). The effect of Tween 80 and sodium dodecyl sulfate as surfactants, and the effect of PEG molar mass on the stability, size and morphology of nanoparticles will be discussed. This is a particularly interesting and facile strategy for the formation of sub 100nm nanoparticles, and for the control of the chemistry of the particle shell, and for the grafting of PEG onto the particle.

(190h) Utilizing Peg-Derivative Hydrogels towards Engineering Compliant Vascular Grafts

Dhaval Patel, School of Chemical and Biomolecular Engineering, Georgia Institute of Technology, Atlanta, GA, Rohan Menon, Department of Biomedical Engeering, Georgia Institute of Technology, Atlanta and Lakeshia Taite, Chemical and Biomolecular Engineering, Georgia Institute of Technology, Atlanta, GA

Utilizing PEG-Derivative Hydrogels Towards Engineering Compliant Vascular Grafts

The formation of a suitable extracellular matrix (ECM) within a vascular graft is critical for graft survival post implantation over an extended period of time. Herein, we have utilized polyethylene glycol (PEG) based hydrogels in combination with a bio-inspired peptide towards developing a suitable material capable of promoting increased vascular ECM deposition. Three-dimensional (3D) hydrogel templates were engineered using a combination of PEG-DA, an elastin mimetic peptide, EM-23, and an enzymatically cleavable peptide for promoting mature vascular ECM production in the presence of smooth muscle cells (SMCs). The elastin mimetic peptide, EM-23 (AAKAAKVGVAPGRGDSAAKAAKK), was engineered to contain containing the hexapeptide sequence, VGVAPG, and the RGDS sequence which binds to the surface bound receptor elastin binding protein (EBP) and cell surface integrins, respectively. Also, a degradable sequence, (DS; AAAAAAAAAK), was also engineered to allow for ECM growth and remodeling within the hydrogel. Different hydrogels comprising of varying amounts of PEG-DA, EM-23, and DS were polymerized in the presence of SMCs and the amount of elastin deposition was quantified over a period of three weeks. Our results show that hydrogels comprised entirely of EM-23 were able to promote significant amounts of elastin over all the other hydrogel combinations after 7 and 21 days (Figure 1). Furthermore, after 21 days, a ten-fold increase in elastin production was observed within EM-23 hydrogels in comparison with previously reported values where elastin deposition was quantified within hydrogels comprised entirely of PEG-DA and RGDS[1]. The higher amounts of elastin production over a period of three weeks from EM-23 hydrogels indicate that this peptide contains all the beneficiaries of a suitable biomaterial capable of providing a 3D template for increased ECM deposition and can therefore be easily exploited to engineer compliant vascular grafts.

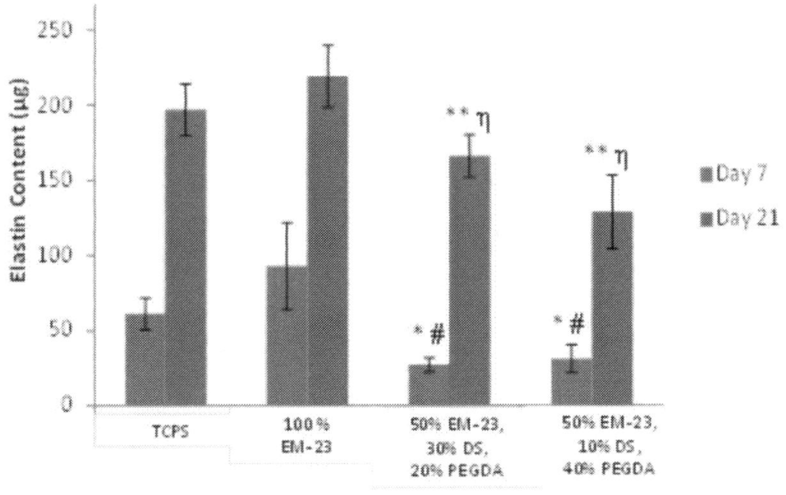

Figure 1. Elastin production over a period of 7 and 21 days. ($n = 3\text{-}4$, * $p < 0.05$ compared to TCPS at Day 7; ** $p < 0.05$ compared to TCPS at Day 21; $^{\#}$ $p < 0.05$ compared to 100% EM-23 at Day 7; $^{\eta}$ $p<0.05$ compared to 100% EM-23 at Day 21).

References:

1. Munoz-Pinto, D. J.; Bulick, A. S.; Hahn, M. S. J. Biomed. Mater. Res. Part A. 2009, 90A, 303-16.

(191a) Systems Biology Analysis of the Effect of Acetic Acid On S. Cerevisiae 424A(LNH-ST) and Resistant Strains During the Co-Fermentation of Glucose and Xylose

Nathan S. Mosier, Agricultural and Biological Engineering & Laboratory of Renewable Resources Engineering (LORRE), Purdue University, West Lafayette, IN, Miroslav Sedlak, Agricultural & Biological Engineering & Laboratory of Renewable Resources Engineering (LORRE), Purdue University, West Lafayette, IN and Nancy W. Y. Ho, ChemE/LORRE, Purdue University, West Lafayette, IN

Efficient conversion of hemicellulose-derived sugars to ethanol at high yields and titers are goals toward commercializing cellulosic ethanol production. *S. cerevisiae* 424A (LNH-ST) developed at Purdue University can efficiently ferment glucose and xylose. However, inhibitors present in cellulosic feedstocks (acetic acid) and the desired fermentation product (ethanol) reduce yeast growth rate and fermentation rates, especially during xylose fermentation.

Through selective culturing, we have developed new, evolved strains with improved xylose fermentation compared to the original strain. Under control fermentations (no acetic acid), the new strain consumes xylose (g/g cells/hr) 2.2 times faster than the original strain. When challenged with acetic acid (10 g/L), the rate of xylose consumption drops almost 90% in the control strain compared to a drop of only 47% in the evolved strain. We present the results from a system biology approach to analyzing differences between our original strain and newly developed strains. We report differences between the strains and the effect of acetic acid on both. Data from the concentrations of intracellular metabolites as well as gene expression for multiple points during batch fermentations are analyzed to give insights into changes in the evolved strain.

(191b) Enhanced Ethanol Tolerance in Yeast by Micronutrients

Felix Lam, Chemical engineering, Massachusetts Institute of Technology, Cambridge, MA, Gerald Fink, Whitehead Institute, Cambridge, MA and Greg Stephanopoulos, Chemical Engineering, Massachusetts Institute of Technology, Cambridge, MA

Ethanol will continue to be the dominant mass-market biofuel in the near to intermediate term. The yeast *Saccharomyces cerevisiae* is the catalyst of choice in the conversion of plant sugars to ethanol. One of the primary bottlenecks to higher ethanol production is the toxicity of ethanol itself as well as compounds present in the fermentation medium. Cell death occurs at ethanol concentrations greater than 15-20% (v/v), and hydrolysates of cellulosic biomass (widely considered to be the next generation feedstock for renewable fuels) contain a variety of organic molecules inhibitory to growth. Conditions that increase ethanol tolerance could improve both product yield and the viability of yeast for reuse in subsequent fermentations. However, the biological basis for ethanol tolerance remains elusive: whole genome studies have shown that tolerance is influenced by many genes, and thus far, none have elicited significant improvement when manipulated alone. An alternative approach is to alter extracellular conditions such that cells become less sensitized to high ethanol concentrations. To this end, we have screened and identified several micronutrients which, when added to a fermentation culture, can increase ethanol production in *S. cerevisiae* by ~50%. These modifications to the medium do not require any genetic manipulation of strains and, in fact, lead to increased ethanol output in both laboratory and industrial strains. Furthermore, we have found that the boost in ethanol production is attributable directly to an enhancement of population viability (i.e., tolerance), and occurs in both glucose- and xylose-fermenting strains alike. Our findings simultaneously hint at a novel, univariate determinant of ethanol tolerance and are a step forward in unlocking the economic potential of cellulosic biomass.

(191c) The Impacts of Glucose Repression on Invertase Expression in Various Yeast Strains During the Ethanol Fermentation of Sucrose-Containing Substrates

Lei Zhang, Institute of New Energy Technology, Tsinghua University, Beijing, China and Shi-zhong Li, Institute of Nuclear and New Energy Technology, Tsinghua University, Beijing, China

Recently, the sweet sorghum ethanol using advanced solid-state fermentation (SSF) developed by Tsinghua University in China has been applied in the largest solid-state fermenter (550 m^3) of the world, which makes sugar-based fuel ethanol production more cost-competitive, first demonstrating the SSF for fuel ethanol production is feasible in industrial scale. Although the relatively high ethanol productivity have been achieved, seeking the yeast strains with high fermentation rate is still an important means to further strengthen the cost competitiveness of the process, due to the inherently insufficient mass and heat transfer of SSF. Due to the high contents of sucrose in sweet sorghum and sugarcane, the hydrolysis rate of sucrose by yeast into fermentable sugars, including glucose and fructose, significantly determines the ethanol fermentation rate of sucrose-containing substrates. Invertase is the key enzyme for yeast involved in the cleavage of sucrose, which is subjected to the glucose repression that poses significantly adverse effects on the yeasts' fermentation of mix-sugar substrates by extension of fermentation time. The dynamic changes of glucose levels during the utilization of sucrose-containing substrates make the regulation of invertase expression in yeast more complicated. Thereby, in this study we investigated the impacts of invertase expression on the final fermentation performances of the various *Saccharomyces cerevisiae* strains through monitoring the dynamic variation of specific invertase acivities of yeasts during the fermentation of sucrose (120 g l^{-1} of sucrose) or mixed sugars (70 g l^{-1} of sucrose, 30 g l^{-1} of glucose, and 23 g l^{-1} of fructose). During the cell growth in YPD media with glucose as sole carbon source, the both commercial strains CA and AF showed much higher induction levels of invertase than both CL and AL did when the glucose levels in the media decreased to the threshold value of triggering the invertase induction. In both sucrose and mix sugars media, surprisingly, the both originally invertase-repressed CA and AF showed an obviously increasing trends of invertase activities after 7 h of fermentations even under the glucose repression conditions at the time. Moreover, the both CA and AF showed highest ethanol productivities in sucrose and mix sugars media. Meanwhile, AL showed a stable and relatively high background invertase activity during the whole process of ethanol fermentation even under the impact of glucose repression, which led the high sucrose hydrolysis rate of AL at the early stage of fermentation. On the other hand, ethanol inhibits the induction of invertase to a great extent, and the invertase induction of both CA and AF showed much stronger tolerance to ethanol. All these results suggested that the glucose repression might be not the only mechanism for the regulation of the invertase involved in the process of ethanol fermentation in *S. cerevisiae* but a net effect of multi-factors, such as glucose repression, stress response. Meanwhile, these results can provide us some useful indicators for isolating or developing the yeast strains with high ethanol productivities with sucrose-containing feedstocks as sustrates.

(191d) Effects of Dissolved Inorganic Carbon and Mixing On Growth and Lipid Formation of Chlorella Vulgaris

Jinsoo Kim, School of Energy, Environmental, Biological and Medical Engineering, University of Cincinnati, Cincinnati, OH, Zhouyang Liu, School of Energy, Environmental, Biological, and Medical Engineering , University of Cincinnati, Cincinnati, OH, Joo-Youp Lee, School of Energy, Environmental, Biological, and Medical Engineering, University of Cincinnati, Cincinnati, OH and Ting Lu, Wastewater Treatment Division, Metropolitan Sewer District of Greater Cincinnati, Cincinnati, OH

Microalgae have recently received great interests as a potential alternative biofuel source because they can generate a large amount of triacylglycerols (e.g. 20-50 % dry cell weight) for biodiesel production. *Chlorella vulgaris* has been extensively studied due to its relatively rapid growing rate and reasonable quantity of triacylglycerols (e.g. 14-22 % dry cell weight). Dissolved inorganic carbon including carbon dioxide (CO_2(aq)) and bicarbonate ion (HCO_3^-) is an important carbon source for the autotrophic growth of microalgae. However, the dissolved inorganic carbon concentration in culture mediums is not usually enough for the growth of microalgae in typical open ponds when CO_2(g) is added to the medium due to a narrow pH window suitable for the growth. In this study, sodium bicarbonate ($NaHCO_3$) was used to increase the inorganic carbon concentration. Then, different mixing speeds were employed to investigate the effect of dissolved inorganic carbon concentrations on the algal cell surface. During the entire growth phase, the lipid content in algal cell was also monitored using fluorescence spectroscopy *in vivo* to trace neutral lipid formation.

(191e) High Level Malic Acid Production From Polymalic Acid Fermentation by Aureobasidium Pullulans and Acid Hydrolysis

Xiang Zou Sr., College of Pharmaceutical Sciences, Southwest University, Chongqing, China, **Yipin Zhou**, William G. Lowrie Department of Chemical and Biomolecular Engineering, The Ohio State University, columbus, OH and S.T. Yang, Chemical Engineering, The Ohio State University, Columbus, OH

High level malic acid production from polymalic acid fermentation by *Aureobasidium pullulans* and acid hydrolysis

Xiang Zou[a], Yipin Zhou[b], Shang tian Yang[b*]

(a. *College of Pharmaceutical Sciences, Southwest University, Chongqing 400715, P.R China;*

(b. *William G. Lowrie Department of Chemical and Biomolecular Engineering, The Ohio State University, Columbus, OH 43210, USA*)

Abstract

Malic acid, a C4-dicarboxylic acid, is widely used in the food industry and a potential C4 platform chemical in biorefinery. Malic acid is commercially produced either through hydration of fumaric or maleic acids under high temperature and pressure yielding a racemic mixture or via the enzymatic transformation of fumarate producing L-malic acid by immobilized bacterial cells. However, these processes require expensive raw materials such as petroleum-derivative maleic anhydride or costly enzymes, which have limited their economical competitiveness. Compared with these methods, microbial fermentation using relatively inexpensive substrates is a promising way for economical production of malic acid. However, the low product yield from conventional malic acid fermentation limits further industrialization efforts due to product inhibition. In this work, we developed a novel process for malic acid production from fermentation produced polymalic acid (PMA) followed with acid hydrolysis. First, a high-yield strain *A. pullulans* ZX10 was isolated, which produced PMA at a high titer equivalent to 87.6 g/l malic acid and high productivity of 0.61 g/l h in free-cell fed-batch fermentation. Repeated fed-batch fermentations with cells immobilized in a fibrous-bed bioreactor achieved the highest product titer of 144.2 g/l and productivity of 0.74 g/l h. The fermentation produced PMA was separated by ethanol precipitation and then purified by adsorption with IRA-900 resin, achieving a ~100% purity. Pure malic acid was then produced from PMA by acid hydrolysis. This process provides an efficient and economical way for PMA and malic acid production, and is promising for industrial application.

(191f) Economical Production of Propionic Acid from Renewable Feedstocks by Propionibacteria in a Fibrous-bed Bioreactor

Zhongqiang Wang, Chemical & Biomolecular Engineering, The Ohio State University, Columbus, OH and Shang-Tian Yang, Department of Chemical and Biomolecular Engineering, Ohio State University, Columbus, OH

Propionic acid is an important chemical with various industrial applications. *Propionibacterium freudenreichii* subsp. *shermanii*, typically used as ripening culture for Swiss-type cheese, has been studied for its ability to produce propionic acid and vitamin B_{12}. A *P. shermanii* strain with high propionic acid yield was used for propionic acid production by cells immobilized in a fibrous-bed bioreactor (FBB). Batch and fed-batch fermentations with glycerol and glucose as carbon sources were investigated first. The results showed that the strain was able to convert both glucose and glycerol simultaneously to propionic acid at a high rate and with a high propionic acid yield. Then, the fermentation was studied using cassava bagasse hydrolysate and crude glycerol from biodiesel wastes to demonstrate the feasibility of using low-cost feedstocks to economically produce propionic acid.

(193a) Reduction of Virus Infectivity in the Presence of Protecting Osmolytes

Maria F. Tafur, Department of Chemical Engineering, Michigan Tech, Houghton, MI and Caryn L. Heldt, Department of Chemical Engineering, Michigan Technological University, Houghton, MI

Protecting osmolytes have the ability to reduce the infectivity by 4-logs of a model non-enveloped virus, porcine parvovirus (PPV) in cell culture. Osmolytes are natural, organic compounds that stabilize intracellular proteins against environmental stresses, such as severe temperature or high osmotic pressure by changing the water content of cells. Protecting osmolytes have the ability to fold proteins by structuring water around the protein surface, therefore stabilizing the protein. We are currently working with PPV, a non-enveloped, icosahedral, single-strand DNA virus, to demonstrate the ability of osmolytes to attenuate non-nenveloped viruses. We have discovered that both glycine and trimethylamine N-oxide (TMAO) reduce the initial 50% infectious dose of PPV from 8-logs to 4-logs. This was quantified by the MTT assay, a cell viability assay that measured the cytopathic effects of PPV virus in susceptible pig kidney cells (PK-13 cells). The reduction in infectivity decreased over time and approached 1 log reduction when the osmolyte was added 20 hours after initial virus infection. We hypothesize that the osmolytes are disrupting the PPV infection cycle. The PPV infection cycle starts with virus attachment to susceptible host cells, and then the virus penetrates the cells through the endosomal pathway to uncoat its genetic material. The virus then uses the host cell's replication machinery to produce viral DNA. After replication, the virus assembles a protein coat, enveloping the nucleic acids, and then breaks open the cell to release progeny PPV particles. It is possible that the osmolytes inhibit virus infectivity by disruption of either the virus interaction with receptors, the endosomal trafficking of virus through the cell or the transport of virus to the nucleus. It appears that protecting osmolytes stabilize intracellular proteins in the initial stages of the PPV infection cycle. We propose that the presence of protecting osmolytes can reduce virus infectivity of other non-enveloped viruses, such as the rhinovirus, which is known to cause the common cold in humans.

(193b) Alginate Encapsulated Activina Doped Chitosan Nanoparticles (CNP) to Promote Endodermal Differentiation in Human Embryonic Stem Cells

Joseph E. Candiello[1], Thomas Richardson[2], Prashant Kumta[1] and Ipsita Banerjee[3], (1)Bioengineering, University of Pittsburgh, Pittsburgh, PA, (2)Chemical Engineering, University of Pittsburgh, Pittsburgh, PA, (3)Chemical and Petroleum Engineering, University of Pittsburgh, Pittsburgh, PA

Introduction:

Human embryonic stem cells (hESCs) have the ability differentiate along pathways of all three primary germ layers: endoderm, ectoderm, and mesoderm. This ability renders them especially useful as a possible therapeutic solution to a variety of disease pathogenesis. hESCs can be directed along these differential pathways via both internal and external stimuli, such as genetic manipulation, chemical signaling, surface morphology, and the stiffness of the immediate microenvironment. A current challenge of significant interest is in driving hESCs to become a sustainable and viable source of beta-like insulin producing cells as a replacement for cellular function lost with type I diabetes, which currently affects over 1 million people in the United States alone. The initial step along the pathway of viable beta-like insulin producing cells is to first produce a definitive endoderm phenotype, which has been induced in embryonic stem cells in the presence of Activin A [1].

In addition to lineage fate signaling, in some therapeutic applications it may also be facilitative and even a necessity to protect a differentiating cell from the larger system while allowing the passage of waste, nutrients, and other chemical signals to and from the immediate cell microenvironment. This is of special importance in the transplantation of beta and beta-like insulin producing cells to combat diabetes. A larger goal application of such a system is deriving a source of functional insulin producing cells from hESCs while protecting them from the body's natural immune pathway. In this study we report a novel alginate/chitosan system to encapsulate the hESC while providing a source of Activin A chemical signaling for the purpose of inducing a definitive endoderm phenotype, an important first step towards attaining this larger goal.

Methods:

In our system we have used the ionic gelation method to create Activin A loaded chitosan nanoparticles (CNPs) by combining a solution of sodium tripolyphosphate (TPP) and Activin A with a solution of dissolved chitosan, a natural biocompatible polymer derived from sea shells. Chitosan nanoparticles have been utilized as a delivery method for both organic and inorganic particles. Activin A was chosen for the explicit purpose of inducing endodermal lineage in the hESCs. The size of the nanoparticles before and after Activin A loading was determined by both dynamic light scattering and atomic force microscopy. Further characterization of the loading efficiency and release dynamics of Activin A loaded in the CNPs were investigated by ELISA assay, targeted for Activin A only. Endoderm differentiation of the hESCs was induced by introducing the Activin loaded CNP to an otherwise signal and serum free media in 2D cell culture for 4 days. We used quantitative PCR to determine the changes in the hESC gene expression of FOXA2 and SOX17, markers indicative of endodermal gene expression.

After loading, the CNPs were encapsulated along with hESC colonies in a 3D alginate bead by the external gelation method. Alginate, also a well known natural polymer has been utilized as a non-biodegradable and biologically compatible encapsulation method for a variety of applications. Briefly, the CNPs and hESCs were suspended in a 1.1% medium viscosity sodium alginate solution and then extruded in droplets from a syringe into a calcium chloride bath. The size of the encapsulated beads was determined by light microscopy, and the release dynamics were determined for the alginate encapsulated CNPs by ELISA assay. Cell viability was qualitatively monitored after 4 days using a Calcein AM and ethidium homodimer-1 based live/dead assay. After 4 days of differentiation quantitative PCR analysis was used to determine the changes in FOXA2 and SOX17 endodermal gene expression of the hESCs in our 3D encapsulated system.

Results:

We successfully loaded Activin A into nanoparticles with a size range of 200-300 nm diameter and a high loading efficiency of ~87 to 90%. The size of the CNPs did not vary significantly before and after loading with Activin A. Release profiles determined by ELISA showed an initial "burst" release of 18% of the loaded Activin A after 1 day and ~30% release after 4 days in cell culture media. We also observed an increase in the endodermal gene markers SOX17 and FOXA2 in hESCs grown in 2D culture with the loaded CNPs when compared to hESCs spontaneously differentiated in a signal free environment. These results verified the feasibility of using Activin A doped chitosan nanoparticles to drive hESC differentiation to a definitive endoderm cell type.

The Activin A loaded CNPs were also successfully encapsulated along with hESC cells in alginate beads of a diameter in the range of 1 to 2 mm. A live/dead cell assay demonstrated cell survival in this environment after 4 days in culture. Activin A release in the media was determined to be reduced by the alginate encasement when compared to release in media only. Quantitative PCR analysis on the 3D system still demonstrated an increase in endodermal genetic markers SOX17 and FOXA2 after 4 days of culture in the alginate beads, demonstrating the feasibility of our system as a whole.

Conclusion:

This study verifies the feasibility of creating a system of Activin A loaded chitosan nanoparticles encapsulated along with human embryonic stem cells in an alginate microbead. We demonstrated both a high efficiency in loading of Activin A and a controlled release into the media. Gene expression consistent with an endodermal phenotype was found in both 2D cell culture and 3D alginate encapsulated cell culture when Activin A loaded CNPs were introduced. Based on the results produced in this study, we have demonstrated a closed system useful for both protecting and signaling human embryonic stem cells towards the definitive endodermal cell fate, and an important first step in producing insulin producing beta-like cells while containing in a closed system for protection and possible downstream implementation in vivo studies.

1. Kubo, A., et al., *Development of definitive endoderm from embryonic stem cells in culture.* Development, 2004. **131**(7): p. 1651-62.

(193c) Electroporation of Nano-Sized Quantum Dots to Track Cancer Cell Transport

Kyung A. Kang, Chemical Engineering, University of Louisville, Louisville, KY and Kwang-Sup Soh, Nano Primo Research Center, Seoul National University, Suwon, South Korea

Primo Vascular System (PVS) is a newly found organ and it is composed of nodes (Primo Nodes; PN) and thin vessels (Primo Vessels; PV) branching out of the PNs. PVs are mechanically fragile but more elastic than lymphatic or blood vessels. PVS is found in/on most of organs, including liver, brain, intestine, heart, inside some lymphatic and blood vessels, etc., forming an extensive network system in the entire body. The liquid inside the PVS (Primo Fluid) flows at an average velocity of 0.3 mm/s. The diameter of rabbit PVs is only in the range of 20-50 micrometers and the size of the PN is in the range of 100-1000 micrometers. Due to its small size and optically transparent nature, PVS is usually difficult to observe. There are, however, three unique properties that have been used for distinguishing PVS from blood or lymphatic vessel. These are: the outermost layer of PV is more porous than those of blood or lymphatic vessels, taking up dye solution preferentially, a PV is composed of a bundle of multiple sub-vessels running in parallel and the nuclei of the PVS endothelial cells are rod-shaped, rather than circular.

Because the existence of the PVS system was confirmed less than ten years ago most of its roles and functions in the body are still yet to be revealed. Nevertheless, several important PVS properties are already reported and some of these are; in the primo fluid, there are cells presenting stem cell markers CD133, Oct4 and Nanog, which may imply that this system has a role in regeneration; these cells are highly populated in PNs, suggesting PNs have a role of storing these cells; PVS is more densely populated in the adipose tissue, which may be important in the study of obesity; and the PVS in the brain runs a long PV through the spinal canal, implying it to be a potential communication pathway between the brain and other organs in the body.

Another very important finding is the potential relevance of PVS to cancer. According to the animal study results using xenografts of various cancer types (lung, ovarian, skin, gastric, breast cancers, leukemia, etc.), as the cancer is formed, PVS is formed at a high density in/around the tumors. For the purpose of tracing the movement of cancer cells, cancer cells were first electroporated with nanosized near infrared quantum dots (QDs). As the cells were proliferated the new cells also carried QDs. It was found that as the cancer cells form tumors they formed PVS and cancer cells were transported via the PVs in a more actively than via lymph vessels and in some cases, secondary tumors were formed at the end of PVS started from the primary tumors.

In the future we planned to use high resolution CT and MRI with gold and iron oxide nanoparticles, respectively, as contrast agent to image the PVS in three dimensions.

(193d) Simultaneous Capture and Characterization of Circulating Exosomes/Microvesicles Using Tethered Lipoplex Nanoparticles for Lung Cancer Diagnosis and Surveillance

Yun Wu[1], Kwang Joo Kwak[2], Yicheng Mao[3], Melissa Crawford[4], Serge P. Nana-Sinkam[4] and Ly James Lee[5], (1)Nanoscale Science and Engineering Center, The Ohio State University, Columbus, OH, (2)Chemical and Biomolecular Engineering, The Ohio State University, Columbus, OH, (3)Division of Pharmaceutics, College of Pharmacy, the Ohio State University, Columbus, (4)Division of Pulmonary, Allergy, Critical Care and Sleep Medicine, The Ohio State University, Columbus, OH, (5)William G. Lowrie Department of Chemical and Biomolecular Engineering, The Ohio State University, Columbus, OH

Lung cancer is the leading cause of cancer deaths worldwide with a disappointing 15% overall 5-year survival rate. A patient-friendly early detection and surveillance method would substantially reduce the mortality in this serious disease. Circulating exosomes/microvesicles based diagnostic platform, as a non-invasive and sensitive detection method, may achieve this goal. Many studies have shown that circulating exosomes/microvesicles are potential biosignatures because the tumor is the primary source of circulating exosomes/microvesicles in cancer patients. In lung cancer, it has been shown that total exosomes and exosomal microRNA/mRNA levels in lung cancer patients are significantly higher than healthy controls, and the circulating exosomal microRNA/mRNA patterns may be a potential predictor of the overall survival in lung cancer patients.

Current technologies, such as qRT-PCR and NanoString nCounter®, need to first isolate total RNA and then measure the target microRNA/mRNA, which require large quantity of samples and are labor intensive and time consuming. We have developed a novel and simple tethered lipoplex nanoparticles (tLN) device that is capable to simultaneously capture and characterize circulating exosomes/microvesicles, and only consumes as less as 10uL of samples. In this study, microRNA-21 (miR-21) and lung specific TTF-1 mRNA were selected as the biomarkers for lung cancer early detection. In the tLN device, intracellular molecular probes, such as molecular beacons (MBs) that detect miR-21 and TTF-1 mRNA, were encapsulated in cationic lipoplex nanoparticles and tethered on the glass substrate.

The tLN device was first demonstrated using non-small cell lung cancer cells (A549) and normal human bronchial epithelial cells (HBEC). Exosomes secreted by A549 and HBEC cells were collected using ultracentrifugation and applied on tLN device to detect circulating miR-21 and TTF-1 mRNA. Conventional qRT-PCR was also conducted to quantitatively correlate the expressions of miR-21 and TTF-1 mRNA with those measured by tLN device. Exosomes produced by A549 showed higher miR-21 and TTF-1 mRNA expression compared to the exosomes secreted by normal HBEC cells, which was consistent with qRT-PCR results. Preliminary results using lung cancer patient blood samples also demonstrated promising results. Serum and serum exosomes were isolated from the whole blood, and then applied to tLN device for miR-21 and TTF-1 mRNA detection. Compared with normal donors, patient samples showed higher miR-21 and TTF-1 mRNA expression, which agreed with qRT-PCR results. The biological atomic force microscopy (Bio-AFM) and the total internal reflection fluorescence (TIRF) microscopy were used to investigate the interactions between exosomes/microvesicles and tethered lipoplex nanoparticles.

Our tLN device represents an innovative approach to characterizing the peripheral blood signature in lung cancer. This facile technique holds great promise in our fight against lung cancer. It may also be applied to many other cancers and deadly diseases.

(193e) Improved Methods to Characterize and Preserve Exosomes

Yueting Wu and David J. Klinke, Chemical Engineering, West Virginia University, Morgantown, WV

Most mammalian cells release nanoscaled membrane vesicles called exosomes. Exosomes transport membrane proteins, cytosolic proteins and nucleotides derived from the parent cells. The biological function of exosomes in regulating immunity and promoting oncogenes remains unclear. To help characterizing the biological role of exosome, the objective of this study was to assess existing methods to isolate and characterize exosomes. Scanning electron microscope (SEM) was used as an alternative to transmission electron microscope (TEM) to examine the morphology of exosomes. Changes in exosome morphology inspected by SEM were used to assess different exosome production and storage conditions. In addition to morphology, the quality of RNA contained within the exosomes was characterized using Agilent Bioanalyzer. Collectively the results suggested revised recommendations for processing and storage of exosomes that preserve morphological and biochemical characteristics. Maintaining exosomes in near native state is essential to reproducibly investigate exosome functions. This study was about improved methods to characterize and preserve exosomes.

(193f) 'Active Stealth' Signaling with a Synthetic 'self' Peptide

Dennis E. Discher, Chemical and Biomolecular Engineering, University of Pennsylvania, Philadelphia, PA

Foreign particles and cells are rapidly cleared from the body by professional phagocytes that incessantly encounter and recognize 'self' cells. The membrane protein CD47 is reportedly a 'Marker of Self' in mouse that impedes phagocytosis of self cells, with signaling through the highly polymorphic receptor CD172a on phagocytes. Here, minimal 'Self' peptides were computationally designed from $CD47^{human}$, synthesized with anchoring groups, and attached to virus-size nanoparticles for intravenous injection into NOD.SCID (NSG) mice that express a unique $CD172a^{NSG}$ (or SIRPA) compatible with $CD47^{human}$. 'Self' peptides delay splenic macrophage clearance of particles with an exponential advantage in persistent circulation, which we use to enhance near-infrared imaging of human tumor xenografts by >10-fold. The affinity of $CD47^{human}$ for $CD172a^{NSG}$ proves weak but within the broad range (0.1 ~ 5 microMolar) measured for ten constructed variants of $CD172a^{human}$; several versions of 'Self' peptide are likewise shown to bind and potently inhibit nanoparticle uptake by an unexpected cytoskeletal mechanism. The reductionist approach reveals the importance and utility as well as some limits of a human 'Self' peptide.

(193h) Increasing Single Wall Carbon Nanotube Delivery to Macrophages by Independent Modifications of the Material and Cellular Activity

Patrick D. Boyer[1], Brian D. Holt[2], Mohammad F. Islam[3] and Kris Noel Dahl[1,2], (1)Chemical Engineering, Carnegie Mellon University, Pittsburgh, PA, (2)Biomedical Engineering, Carnegie Mellon University, Pittsburgh, PA, (3)Materials Science & Engineering, Carnegie Mellon University, Pittsburgh, PA

Single wall carbon nanotubes (SWCNTs) are a class of nanomaterials with unique physical and optical properties which make SWCNTs well-suited for biomedical applications. Recent studies have shown the potential for SWCNT use as drug and gene delivery vectors, imaging probes for cellular labeling and tracking, molecular sensors, and for thermal ablation of cells *in situ*. SWCNTs are routinely dispersed in aqueous solution using polyethylene glycol (PEG) copolymers, lipids, and proteins. Yet, there remains a limited mechanistic understanding of how different molecular SWCNT surface coatings interact with a complex, dynamic cellular state, such as those found in immune cells. Nanomaterial interactions with cells of the immune system, such as macrophages, are particularly important for determining mechanisms of toxicity and developing specific medical therapeutics for inflammation and cancer detection. We investigated the differential effects of SWNCT dispersion with bioinert Pluronic F-127 (PF127) polymer and bioactive bovine serum albumin (BSA) protein on SWCNT uptake and localization in J774A.1 macrophages and NIH-3T3 fibroblasts. We also examined stimulation of macrophages with lipopolysaccharide (LPS) to measure effects of cell state on SWCNT uptake.

We quantified cellular uptake of SWCNTs over three orders of magnitude (1-100 μg/ml) using Raman spectroscopy. Macrophages exhibited a steady dose dependent increase in cell associated SWCNTs with approximately double the effect for BSA-SWCNTs compared to PF127-SWCNTs. Cells treated with BSA-SWCNTs showed gross changes in size, morphology, proliferation capability and other phenotypic changes. In contrast, these cellular changes were not observed for macrophages with PF127-SWCNTs or fibroblasts with either PF127-SWCNTs or BSA-SWCNTs. Cellular stimulation with LPS increased uptake, and an additive effect was observed for pre-stimulated macrophages with increasing SWCNTs, but the uptake of BSA-SWCNTs plateaued at higher exposure concentrations. Taken together, these results indicate the sensitivity of macrophage uptake to bioactive SWCNT coatings and cellular stimulation, as well as the ability of macrophages to adapt for increased uptake of bioactive coated SWCNTs.

(199a) Analysis of Metabolic Flux Rewiring in CHO Cells During Fed-Batch Culture

Maciek R. Antoniewicz, Department of Chemical and Biomolecular Engineering, University of Delaware, Newark, DE and Woo Suk Ahn, Chemical Engineering, University of Delaware, Newark, DE

Despite significant investments in research and development with the goal of improving understanding of CHO cells, relatively little is still known about intracellular metabolic fluxes in fed-batch cell cultures of CHO cells. In this work, the metabolism of CHO cells was studied in detail during a fed-batch fermentation. Metabolism was characterized by measured extracellular uptake and excretion rates of metabolites combined with a detailed metabolic network model for CHO cells. To obtain additional information ^{13}C-tracers were applied and labeling of intracellular metabolites was measured using mass spectrometry (GC-MS). Fluxes were quantified using ^{13}C-metabolic flux analysis. The flux results revealed significant rewiring of intracellular metabolic fluxes in the transition from growth phase to stationary phase, including changes in energy metabolism and redox metabolism. The results that we present provide a solid foundation for future studies of CHO cell metabolism for applications such as cell line development and medium optimization for high-titer production of recombinant proteins.

References

[1] Ahn WS, Antoniewicz MR. Metabolic flux analysis of CHO cells at growth and non-growth phases using isotopic tracers and mass spectrometry. Metab Eng, 13(5): 598-609, 2011

[2] Ahn WS, Antoniewicz MR. Towards dynamic metabolic flux analysis in CHO cell cultures. Biotechnol J, 7(1): 61-74, 2012

(199b) Quantifying the Impact of Bcl-2Δ Overexpression Upon Central Metabolism Through 13C Metabolic Flux Analysis

Neil Templeton[1], Abasha Lewis[2], Haimanti Dorai[3], Kevin Smith[3], Steven Lang[3], Michael J. Betenbaugh[2] and **Jamey D. Young**[1], (1)Chemical and Biomolecular Engineering, Vanderbilt University, Nashville, TN, (2)Chemical and Biomolecular Engineering, Johns Hopkins University, Baltimore, MD, (3)Janssen Pharmaceuticals (J&J), Radnor, PA

Monoclonal antibodies are among the most expensive class of drugs to produce in biopharma today. While the primary goal is to maximize the overall rate of antibody production, most industrial cell lines instead generate an excess of lactate. Lactate production often corresponds with limited mitochondrial capacity. Interestingly enough, apoptotic pathways are often associated with a shift in typical mitochondrial function. We have determined in prior work with Chinese Hamster Ovary (CHO) cells that by altering the apoptotic pathways through the overexpression of Bcl-2Δ, metabolic pathways are altered as well. Not only can the onset of programmed cell death be delayed by overexpression of Bcl-2Δ, but the activation of wasteful catabolic pathways, specifically lactate production, can be minimized during exponential phase. During stationary phase, cells overexpressing Bcl-2Δ have an increased capacity to consume lactate that was formerly produced, allowing overall viable cell density (VCD) to decline at a decreased rate, despite limited glucose availability. We will present how this corresponds with increased TCA activity and to what level overall metabolic reprogramming is taking place to lead to these exhibited phenotypes. These two unique phases will be independently discussed based upon a complete stoichiometric analysis and C13 metabolic flux analysis.

(199c) An Integrated Modeling and Experimental Framework for Predicting the N-Linked Glycosylation of Monoclonal Antibodies

Ioscani Jimenez del Val and Cleo Kontoravdi, Chemical Engineering and Chemical Technology, Imperial College London, London, United Kingdom

An integrated modeling and experimental framework for predicting the N-linked glycosylation of monoclonal antibodies

Ioscani Jiménez del Val and Cleo Kontoravdi

Department of Chemical Engineering and Chemical Technology, Imperial College London, South Kensington Campus, SW7 2AZ

Development and approval of pharmaceutical products is an extremely costly and time-consuming process which has resulted in unsustainable expense for healthcare providers and patients. In order to address this critical issue, regulatory agencies and industry experts have jointly encouraged implementation of the Quality by Design (QbD) paradigm for the development of all new drug products in the pipeline, including therapeutic proteins. Adoption of QbD principles in pharmaceutical product development is expected to reduce not only regulatory approval time and costs, but also to encourage innovation by building processes around the mechanistic relationships between their inputs and end product quality. Monoclonal antibodies (mAbs) are currently the highest-selling products of the biopharmaceutical industry and are expected to make up half of the top -ten selling pharmaceutical products by 2016 [1]. All approved mAbs contain N-linked carbohydrates on their crystallizable fragment (Fc), the distribution of which greatly impacts the safety and efficacy of these therapeutic glycoproteins and thus defines it as a critical quality attribute (CQA) of mAbs under the QbD scope.

Based on QbD guidelines, the purpose of this work has been to generate an integrated modeling and experimental framework to mechanistically and quantitatively tie process conditions –namely, nutrient availability- with the glycosylation-associated quality of mAbs. The integrated framework consists of a dynamic mathematical model that mechanistically couples mAb N-linked glycosylation with cellular metabolism and an experimental methodology that measures the key variables required for parameter estimation and model validation. In order to mechanistically link cellular metabolism with mAb glycosylation, the pathway for nucleotide sugar donors (NSDs) –which are metabolic products directly derived from glucose and glutamine and intrinsic co-substrates for the glycosylation reactions- has been modeled. A Monte Carlo-based global sensitivity analysis method [2] was then used both to reduce the NSD model and to determine which of its parameters are crucial for accurate representation of the system. Experimentally, batch cultures were performed with hybridoma (CRL-1606 from ATCC), GS-NS0 and GS-CHO cell lines. Typical data was collected from these batch cultures and included viable cell density, glucose, glutamine, lactate, and ammonia concentrations, as well as antibody titre. Intracellular samples were produced by quenching with ice-cold 0.9% NaCl solution and extracting with perchloric acid. The intracellular samples were then assayed for glucose and glutamine using fluorescent kits and used for NSD quantification using anion exchange HPLC. The HPLC method was based on a previously reported one [3], but optimized to resolve eight nucleotide sugars and four nucleotides in 30 minutes. Glycan analysis was performed at all time-points using MALDI-TOF mass spectrometry. The generated data was used to determine the unknown parameters of the model for all three cell lines. Subsequently, fed-batch validation experiments, where glucose and glutamine were added throughout culture, were performed in order to assess model and parameter validity. For all three cell lines tested, the results show accurate representation of the dependence between extracellular glucose and glutamine availability, intracellular NSD concentrations and mAb Fc glycosylation profile.

Our results suggest that this combined modelling and experimental framework is an effective way of mechanistically linking readily-measurable process inputs with critical quality attributes of biopharmaceuticals as underlined in the QbD methodology. Furthermore, the framework outlined here has the potential of aiding the development of novel processes for the manufacture of therapeutic glycoproteins under the Quality by Design paradigm in the future.

1. EvaluatePharma, *World Preview 2016 "Beyond the Patent Cliff"* (2011).

2. Kucherenko S, et al. Reliab. Eng. Syst. Safe 94 (7) (2009) 1135-1148.

3. Tomiya N, et al. Analytical Biochemistry 293(1) (2001) 129-137.

(199d) A Systems Biology Approach for Mechanistic Understanding of Glycan Structure and Function

Gang Liu, Apurv Puri and Sriram Neelamegham, Chemical and Biological Engineering, SUNY at Buffalo, Buffalo, NY

Introduction: Glycosylation is an important post-translational modification that alters more than 50% of mammalian secreted and cell surface proteins. Glycosylated proteins play structural and functional roles in diverse biological processes including inflammation, cancer and development. The recent development of advanced analytical tools has heralded the emergence of Glycomics, a field where entire glycomes of cell systems are characterized [1]. It is often challenging to understand glycosylation mechanisms using statistical analysis method alone. Herein, we developed a systems biology approach to unravel the complexity of glycosylation reaction network underlying complex glycan structures and to predict the structure and composition of the glycans when the cellular systems are perturbed.

Methods: Our approach uses the object-oriented concepts and graph theory to model the cellular glycosylation networks including the glycans, enzymes, reactions, pathways and compartments. With the aid of such definitions, we defines a streamlined modeling procedure to couple the mathematical model with the glycan structures. The mathematical model developed are linked to informatics databases [2]. Thus, information from these large databases are used to inform the model, and the model built can query the databases in a mechanistic manner.

Master pathway construction is implemented as the first step to construct a "maximum-size" glycosylation pathway, followed by subset modeling, hierarchical clustering and principal component analysis method to refine the network. Lastly, sensitivity analysis is applied to identify key regulating enzymes and to generate experimental testable hypothesis. All these procedures can be readily implemented in a MATLAB based toolbox called Glycosylation Network Analysis Toolbox (GNAT). In addition, this toolbox enables the description of glycan structures in Systems Biology Markup Language (SBML) files and it allows high-quality visualization of the glycosylation reaction network.

Results and Conclusions: Our modeling approach has been tested with two examples that simulate: 1) N-glycan initiation and branching; 2) O-glycosylation networks. These tests demonstrate that our modeling approach can be used to construct the "feasible" glycosylation reaction network from the experimentally measured mass spectrometry data. In addition, the distribution of the glycans can be computed using our reconstructed models and this enables fitting and comparison with experimental data. In summary, we present a novel, structured approach for the generation of systems-based models that enhance our understanding of cellular glycosylation pathways.

References:

1.Neelamegham, S., et al. Glycobiology, 2011. 21(12): p. 1541-53.
2.Ranzinger, R., et al. Nucleic Acids Res, 2011. 39(Database issue): p. D373-6.

(199e) Development of a Stochastic Model of Heterogeneity of Human Pluripotent Stem Cell Populations Under Conditions Promoting Self-Renewal or Differentiation

Jincheng Wu and Emmanuel S. Tzanakakis, Chemical and Biological Engineering, State University of New York at Buffalo, Buffalo, NY

Heterogeneity is an intrinsic feature of stem cell populations giving rise to different cell fate potentials under conditions promoting differentiation. Better understanding of the dynamics of heterogeneous stem cell populations can be achieved through coupling of sophisticated experimental techniques with computational modeling. In this study, we constructed a population balance equation (PBE) system for human embryonic stem cells (hESCs) under self-renewal and differentiation. Model development was accompanied by experimental studies conducted with hESCs under self-renewing and directed differentiation conditions.

A system of PBEs was constructed with each PBE representing different populations of cells depending on their differentiation state. The state vector comprised attributes such as cell size and expression level of lineage markers specific to the differentiated progeny. The heterogeneity within the same population stemmed from gene expression noise and stochastic partitioning at cell division. The model results are in excellent agreement with the experimental data obtained in our laboratory via different methods including flow cytometry. For hESCs undergoing self-renewal, the model predicted that both gene expression noise and partitioning at cell division have comparable contributions to the heterogeneity of hESC populations. Under differentiation conditions, the cell transition kinetics were modeled with respect to cell fate determinants such as gene expression noise and growth factor concentration.

The PBE framework captures the dynamics of stem cell populations with cell-to-cell variability. For self-renewing hESCs, we provide for the first time a quantitative account of the effects gene expression noise and cell division on population heterogeneity. The PBE system presented here also describes the evolution of stem cells and their progeny under conditions promoting cell commitment.

Modeling of Fluorescent Protein-labeled Cell Populations to Analyze Transcriptional and Division Effects on Fluorescent Intensity Distributions

Loveleena Bansal[1], Shreya Maiti[1], Arul Jayaraman[1], Carl Laird[1], Juergen Hahn[2]
[1]Texas A&M University, College Station, TX, [2] Rensselaer Polytechnic Institute, Troy, NY

Fluorescent proteins have been widely used as markers of gene expression and transcriptional regulation due to several reasons [1], such as that their plasmid can easily be integrated into the DNA of the cells, they do not have a toxic effect on cell growth and they can also be conveniently detected by illuminating the sample with suitable light. Thus, the fluorescence obtained in fluorescent protein based reporter systems can be used as an indicator of transcription and translation. However, there are phenotypic variations in cell populations as the observed fluorescence is affected by noise in gene expression [2] as well as physiological factors such as unequal partitioning of cellular material resulting from cell division [3]. Since most of the experimental data of fluorescence is obtained from techniques such as flow cytometry or fluorescence microscopy that utilize cell populations, using single cell models to estimate cell physiological parameters or transcriptional dynamics may lead to erroneous conclusions [4]. Furthermore, it is non-trivial to track individual cells in a population during the course of an experiment to analyze how fluorescence in a single cell evolves over time [5]. In this regard, we have developed a multiscale population balance model (PBM) that describes the dynamics of distributions of fluorescence intensity in a cell population labeled with a fluorescent protein.

The developed PBM of a fluorescent protein expressing cell population is represented by an integro-partial differential equation (PDE) with time and fluorescence intensity as independent variables. It models the dynamic fluorescence intensity distribution of cells by taking into account processes such as cell birth and unequal partitioning due to cell division as well as the increase in fluorescence due to gene expression. The initial-boundary value problem of this PDE model has been solved using an implicit finite difference scheme which is second order accurate in time and fluorescent intensity space. The resulting distributions are illustrated in Figure 1. This work also includes using optimization techniques to estimate unknown physiological parameters such as the division rate, rate of fluorescence increase due to gene expression, cell partitioning distribution parameter, etc. in the model. For this purpose, the experimental data are obtained from a previously developed GFP (green fluorescent protein) reporter system for STAT3 [6]. The cells are stimulated with IL-6 to start the expression of GFP and timely samples are taken. The samples are fixed with paraformaldehyde and then analyzed using flow cytometry to obtain fluorescence intensity distributions. These data are then used for estimating the unknown parameters in the population balance model.

Using the presented technique, it is possible to quantify the effects of gene expression and cell division on the observed fluorescence intensity distributions and analyze the time scales at which division becomes dominant in mammalian cells. The experimental data are also obtained for cell populations at different initial densities and different IL-6 concentrations to study their effect on division and death rate of cells and on GFP expression. Summarizing, this study integrates computational and experimental techniques to characterize certain aspects of fluorescent protein expressing cell populations and estimates unknown physiological population level parameters.

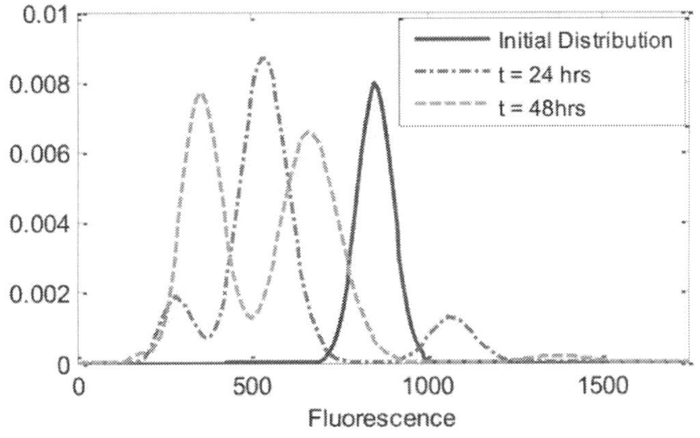

Figure 1 Fluorescence intensity distributions for a fluorescent protein-labeled population of cells obtained from the solution of the PBM using Gaussian initial distribution.

References

[1] M. Chalfie, Y. Tu, G. Euskirchen, W. W. Ward, and D. C. Prasher, "Green fluorescent protein as a marker for gene expression," *Science,* vol. 263, pp. 802-805, 1994.

[2] P. S. Swain, M. B. Elowitz, and E. D. Siggia, "Intrinsic and extrinsic contributions to stochasticity in gene expression " *PNAS,* vol. 99, pp. 12795-12800, 2002.

[3] M. A. Hjortsø, *Population balances in biomedical engineering: Segregation through the distribution of cell states*: McGraw-Hill, 2006.

[4] J. Hasenauer, S. Waldherr, M. Doszczak, N. Radde, P. Scheurich, and F. Allgöwer, "Identification of models of heterogeneous cell populations from population snapshot data," *BMC Bioinformatics,* vol. 12, pp. 1-15, 2011.

[5] Z. Huang, Y. Chu, and J. Hahn, "Computing transcription factor distribution profiles from green fluorescent protein reporter data," *Chemical Engineering Science,* vol. 68, pp. 340-354, 2012.

[6] C. Moya, Z. Huang, P. Cheng, A. Jayaraman, and J. Hahn, "Investigation of IL-6 and IL-10 signalling via mathematical modelling," *IET Systems Biology,* vol. 5, pp. 15-26, 2009.

(199g) Analysis of the Effects of Varying Carbon Dioxide Concentration in the Biomass Production and Metabolic Network of the Microalgae Chlamydomonas Reinhardtii

Flavia Vischi Winck[1], **Rubén Darío Lopez Parra**[1], Rossmary Jay Pang Moncada[1], Diego Mauricio Riaño-Pachón[2], Jorge M. Gomez[1], Nubia Milena Velasco Rodriguez[3] and Andrés Fernando González Barrios[1], (1)Chemical Engineering, Universidad de los Andes, Bogotá, Colombia, (2)Biological Sciences , Universidad de los Andes, Bogotá, Colombia, (3)Industrial Engineering, Universidad de los Andes, Bogotá, Colombia

Introduction

The increase of air emissions originated from the burning of fossil fuels and the expected decrease in fossil resources have been an issue of world impact and socio-economic importance. According to previous reports, the carbon emissions in Colombia increased from 66.4 to more than 120 Million tons per year in the period between 1996 and 2010. The accumulation of air emissions derived from the burning of fossil fuels and industrial activities may affect the biodiversity, geographical distribution of species, seasonal timing and overall climate change. It is then important to contribute in finding and optimizing the use of alternative energy sources with a lower net gas emission and to accelerate the development of environmentally friendly technologies. Algae-based technologies of bioremediation coupled to biomass production are alternative strategies for reducing levels of air contaminants and creating new sources of renewable biomass which can be used for energy production, e.g., biofuels, or for the accumulation of other by-products, e.g., pigments. However, there are still limitations in terms of the costs of biomass production and further conversion of this material into useful fuels; the increase in biomass per cell may help to reduce these production-associated drawbacks (Stephenson, 2011). A better understanding and improvement of the biomass production in algae may accelerate its use as alternative energy source in industrial scale. The process of biomass production in microalgae is influenced, among other factors, by the availability of carbon dioxide and cultivation conditions (Collins, 2006). Changes on these parameters have shown to result in more or less biomass yield (Kliphius,2011; Renberg, 2010). It seems feasible that the production of biomass can be modulated by key metabolic pathways through environmental changes (e.g. nutrient and CO_2 availability) or molecular approaches (e.g., gene overexpression or gene knock out). Nevertheless, how to optimize the CO_2 capture and increase the biomass and lipid production in algae in a cost-effective and sustainable way is a question that remains to be answered. Under conditions of air-level carbon dioxide (CO_2) concentration a Carbon Concentrating Mechanism (CCM) is induced to facilitate cellular carbon uptake. The CCM increases the availability of carbon dioxide at the site of cellular carbon fixation, increasing photosynthetic rates. However, how CCM is controlled at the transcriptional and metabolic level is not well understood. Furthermore, the effects of increased CO_2 concentration in the biomass accumulation have not been completely characterized either. In our present work, we analysed the effects of varying concentrations of carbon dioxide in the cellular responses of the photosynthetic unicellular green alga *Chlamydomonas reinhardtii*. Experimental data for the transcript abundance of selected enzymes were integrated into the previously reconstructed metabolic network of *C. reinhardtii* (Chang, 2011), together with data of the biomass composition. This approach revealed new target genes and metabolic pathways for further metabolic engineering of biomass accumulation process.

Material and methods

Cells and culturing conditions

The microalga *Chlamydomonas reinhardtii* CC-1690 was obtained from the culture collection at ChlamyCenter (http://www.chlamy.org/strains.html) (University of Minnesota). Cells were cultured in a Bioreactor Bioengineering (Bioengineering AG, Wald, Switzerland), which permits to cultivate the cells under controlled temperature (21°C) and photoautotrophic conditions (100 $\mu E.m^{-2}.s^{-1}$). Cells were cultivated under four different CO_2 concentrations, varying from 0.04% to 10% CO_2, and the experimental data obtained in triplicate.

Transcripts profiling

The abundance of mRNA transcripts of selected genes which code for enzymes of the primary metabolism was quantified using Reverse-Transcription real-time PCR (RT-qPCR) as previously described (Winck, 2011) for cells cultivated at 0.04% CO_2 and cultivated at CO_2 condition where the biomass was highest. Primers were designed and computationally validated using the tool QuantPrime (Arvidsson, 2008).

Biomass characterization

Biomass was determined as dry weight and its composition was further characterized. Protein, lipids, nucleic acids, starch, pigments and carbohydrate content were quantified as previously described (Kliphuis, 2011). Briefly, the determination of the biomass composition was conducted by colorimetric and fluorometric assays.

Metabolic modelling and simulation

The metabolic network of *Chlamydomonas reinhardtii* previously reconstructed, named iRC1080 (Chang, 2011), enables the phenotypic prediction under conditions fixed as in the experimental conditions cited above. We carried out a sensitivity analysis by perturbing the network under different CO_2 levels using the software Xpress IVE®. This allowed to detect the role of specific metabolic flows and this information could be matched with the abundance of mRNA transcripts in order to understand the underpinnings behind the responses of the microalga.

Results

The biomass of microalgae cultivated at different CO_2 concentrations was characterized and the variations on its composition were identified. Cells cultivated at high CO_2 concentration showed higher growth rates. Furthermore, the expression profiles for the enzyme coding genes were identified revealing changes on the cellular responses under varying CO_2 conditions. The experimental data from gene expression and biomass values were used for the metabolic network simulations. Our objective was to attain the maximum use of CO_2 with the maximum biomass production. For this, we have performed a sensitivity analysis, varying the limits of reactions, whereby we identified the reactions which play an important role in the uptake of CO_2, keeping an optimal biomass production. Consequently, it was possible to establish the required conditions to hold cellular viability within an optimal level concurrently with a maximum concentration of CO_2. This simulation were carried out taking advantage of the available resource of the network by means of which we can take different light sources to simulate the input photons to the metabolism, obtaining results in agreement with those reported previously. Finally, we fixed as light source White Led, which is the same used in our experiments, given that its applications are broader than those of other artificial light sources, because of their availability and economy. Our results revealed target genes and metabolic pathways which can be further investigated for the optimization of biomass accumulation using metabolic engineering approaches.

Conclusion

The integration of experimental data into metabolic networks permitted to perform simulation analysis of biomass accumulation in the microalgae *Chlamydomonas reinhardtii*. This approach revealed the identity of important metabolic pathways which can now be investigated on future strategies for the enhancement of biomass accumulation and constitutive high photosynthetic rates in microalgae species.

References

Arvidsson, S., Kwasniewski, M., Riaño-Pachón, D.M., and Mueller-Roeber, B. (2008) QuantPrime - a flexible tool for reliable high-throughput primer design for quantitative PCR. *BMC Bioinformatics*, *9*:465.

Chang, R., et al. (2011), "Metabolic Network Reconstruction of *Chlamydomonas* offers insight into light-driven algal metabolism", *Molecular Systems Biology 7:518*.

Collins, S., Sultemeyer, D., and Bell, G. (2006) Changes in C uptake in populations of Chlamydomonas reinhardtii selected at high CO2, *Plant, cell & environment 29*, 1812-1819.

Colombani, Y. and Heipcke, S. (2008), "Mosel: An Overview", FICO^{TM} Xpress Optimization Suite Whitepaper.

Kliphuis, A. M., Martens, D. E., Janssen, M., and Wijffels, R. H. (2011) Effect of O(2) : CO(2) ratio on the primary metabolism of Chlamydomonas reinhardtii, *Biotechnol Bioeng*. doi: 10.1002/bit.23194

Orth, J., Thiele, I. and Palsson, B. (2010), "What is Flux Balance Analysis?", *Nature Biotechnology 28*, 245-248.

Renberg, L., Johansson, A. I., Shutova, T., Stenlund, H., Aksmann, A., et al. (2010) A metabolomic approach to study major metabolite changes during acclimation to limiting CO2 in Chlamydomonas reinhardtii, *Plant physiology 154*, 187-196.

Stephenson, P. G., Moore, C. M., Terry, M. J., Zubkov, M. V., and Bibby, T. S. (2011) Improving photosynthesis for algal biofuels: toward a green revolution, *Trends Biotechnol 29*, 615-623.

Winck, F. V. (2011) Nuclear proteomics and transcription factor profiling in *Chlamydomonas reinhardtii*, In *Biology and Biochemistry* 1 ed., p 182, Universitaet Potsdam Potsdam.

(199h) Genome-Scale Flux Balance Analysis of Tree Metabolism

Ashish Misra[1], Margaret Simons[1], Matthew Conway[1], Gary D. Coleman[2] and Ganesh Sriram[1], (1)Chemical and Biomolecular Engineering, University of Maryland, College Park, MD, (2)Plant Science and Landscape Architecture, University of Maryland, College Park, College Park, MD

Photoautotrophic organisms are important sources for the production of renewable fuels and chemicals. Plants and trees form a subset of such organisms with unique properties and complex metabolism. Plant metabolism is much more complex than that of prokaryotes and other eukaroyotes with cellular compartmentation and replication of pathways in multiple compartments. Furthermore, trees exhibit interesting features like efficient usage of their nitrogen reserves via recycling to specialized storage proteins and generation of hydrocarbons to deal with heat stress. In plants, a few pioneering studies have applied flux balance analysis (FBA) for estimating fluxes using plant cell culture (Belau et al, Plant Physiol. 149: 585-598, 2009; Williams et al, Plant Physiol. 154: 311-323, 2010; Hay et al, The Plant Journal, 67(3): 513–525, 2011). The metabolism of trees has not been well studied compared to that of model plants such as Arabidopsis and maize.

This presentation will report an investigation of metabolism in a model tree using genome-scale FBA. We have reconstructed a genome scale model of a model tree using reactions from the BioCyc family of database. As part of the study, we will use FBA for estimating fluxes in the tree using previously determined biomass compositions, thus obtaining flux distributions on a genome scale. We will then estimate the flux distributions under a set of interesting and physiologically relevant conditions. We will investigate flux distributions for minimizing production of lignin to improve viability of the tree as a biofuel crop; maximizing production of nitrogen storage proteins to elucidate underlying fluxes during seasonal nitrogen cycling; and flux changes during biosynthesis of the hydrocarbon isoprene, whose production by trees impacts atmospheric chemistry. These studies will represent the first application of genome scale FBA to a model tree system, and additionally provide insight on the metabolic network in model tree organisms.

(210a) Characterization of Coherent Feedforward Motifs in Mammalian Cells Using Synthetic Gene Circuits

Richard Moore[1], Li Yi[1] and Leonidas Bleris[1,2], (1)Bioengineering, The University of Texas at Dallas, Richardson, TX, (2)Electrical Engineeering, The University of Texas at Dallas, Richardson, TX

Characterization of Coherent Feedforward Motifs in Mammalian Cells using Synthetic Gene Circuits

It is by now appreciated that the behavior of large biochemical networks can often be understood through the analysis of relatively small network motifs. Synthetic circuit engineering gives us the ability to use only a handful of components to build a range of small-scale networks. We report the construction and characterization of three (well-studied in bacteria) feedforward motifs, which we introduce in mammalian cells via transient transfections. Each motif consists of three nodes which code for transcription factors that interact to form the following specific architectures: a type 1 feedforward loop with OR decision at the final node, a type 1 feedforward loop with AND decision, and a cascade style architecture. We selected components that are known to have minimum interference with the host, and designed all necessary control experiments. Additionally, the edges to the second and final node are controlled by ligands, giving us the flexibility to study these networks under a range of different conditions.

We monitor each node via the use of the three different fluorescent reporters, which have negligible crosstalk (microscopy and flow cytometer measurements). Using control experiments, we have established that the AND and OR decisions indeed operate as expected at the final node for the respective architectures. In addition, we have established that the cascade shows a clear delay from the initial node to the secondary node and from the secondary node to the final node, under saturating levels of ligands. We will present time-lapse microscopy and flow cytometry data, for a range of induction levels and perturbations, and will discuss the observed properties of these feedforward motifs. Furthermore, we use population and single-cell data to build quantitative models of the architectures' characteristics. To conclude, the topologies of the circuits that we study are found in many types of natural mammalian networks, including neural and transcriptional networks. We will discuss the relevance to endogenous networks and the implications to understanding and mapping biological pathways.

(210b) High Throughput Monitoring of Pathway Activation Upon Ectopic Expression of Nanog in Human Mesenchymal Stem Cells Using Lentiviral Arrays

Janhavi Moharil[1], Panagiotis Mistriotis[1], Hui You[1], Pedro Lei[1], Jun Tian[1] and Stelios T. Andreadis[1,2], (1)Chemical and Biological Engineering, University at Buffalo, State University of New York, Amherst, NY, (2)Center of Excellence in Bioinformatics and Life Sciences, Buffalo, NY

Nanog has been shown to regulate the proliferation capacity of Mesenchymal Stem Cells (MSCs) and alter their differentiation potential as assessed by overexpression and knock-down studies. Overexpression of Nanog in MSCs is known to reduce their adipogenic differentiation potential while inducing osteogenesis. Our laboratory has shown that Nanog overexpression in MSCs enhances their myogenic differentiation potential as evidenced by improved contractility. However the molecular mechanisms by which Nanog plays a specific role in differentiation are not known. In this study we present a novel tool for high throughput screening of different pathways in Nanog expressing MSCs. We employed a novel dual-promoter lentiviral vector (LVDP) carrying human PGK promoter driving the expression of DsRed serving as an internal control and the promoter of interest driving the expression of ZsGreen to monitor gene expression upon induction of Nanog expression in human Hair Follicle derived MSCs (hHF-MSCs) and Bone Marrow derived MSCs (hBM-MSCs). We employed a Tet-On system for induced expression of Nanog in hHF-MSCs and hBM-MSCs in the presence of doxycycline. The MSCs were transduced with a lentiviral library of about 60 vectors containing promoters and transcriptional response elements (RE) targeting about 25 different pathways and lineage specific markers. The cells were subsequently induced to myogenic differentiation and the promoter/RE activity was monitored over time with automated fluorescence microscope. Fluorescence intensity was quantified by Cell Profiler and the green fluorescence intensity was normalized by red fluorescence intensity. First, we used the lentiviral library to monitor gene expression dynamics during myogenic differentiation of hHF-MSCs and hBM-MSCs over a period of 7 days. Smooth muscle cell markers such as α-SMA, Myocardin, RE-CArG, and components of TGF-β1 signaling pathway such as RE-SMAD2/3 and RE-SMAD7 showed significant upregulation. In the presence of TGF-β1 receptor inhibitor SB431542, RE-SMAD2/3 and RE-SMAD7 did not respond, indicating that the pathway was shutdown as expected. We then employed the lentiviral array to identify which pathways were activated upon ectopic expression of Nanog. Overexpression of Nanog alone induced a marked response in RE-CArG within 3 days showing an enhanced potential for myogenic differentiation **(Figure 1)**. Treatment of $Nanog^{+}$ cells with TGF-β1 and Heparin further increased the activity of RE-CArG, suggesting that effects of Nanog and TGF-β1 in promoting MSC myogenesis were additive. RE-ERK representing the MAPK signaling pathway showed an increased activity in $Nanog^{+}$ cells. However treatment of $Nanog^{+}$ cells with TGF-β1 blocked its response. RE-STAT3 (JAK/STAT signaling pathway), RE-KLF4, RE-GATA and RE-GATA2 also showed significant activation in $Nanog^{+}$ cells but treatment with TGF-β1 did not increase their response any further. We thus demonstrate the application of lentiviral arrays in monitoring gene expression during myogenic differentiation of hHF-MSCs and hBM-MSCs. We present a novel screening strategy to identify various pathways that are activated upon ectopic expression of Nanog in MSCs. Currently we are applying this approach to study the role of Nanog in osteogenic, chondrogenic and adipogenic differentiation.

Figure 1:.Heatmap representation of high throughput array. The figure shows promoter / response element activity in HF-MSCs and Nanog+ HF-MSCs in growth media and myogenic differentiation media after 3 days of treatment. The response in each condition was normalized to the response of HF-MSCs in growth media (control). The dendrogram was computed by hierarchical clustering.

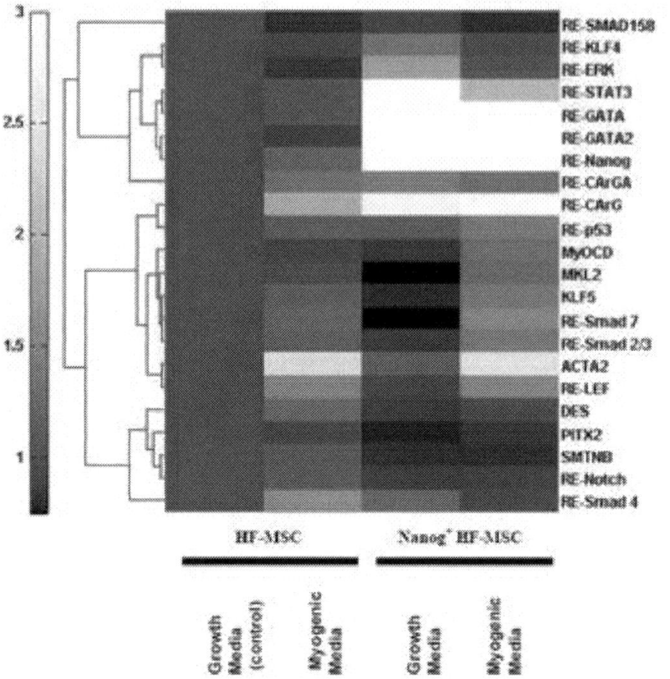

(210c) Integrated Single-Cell Analysis of Heterogeneous Secretory Profiles Exhibited by Human Primary Colorectal Tumor Cells

Viktor Adalsteinsson[1,2,3], Naren Tallapragada[1,4], Narmin Tahirova[1], Xiaosai Yao[1,5], Alessandro Angelini[1], K. Dane Wittrup[1,3,5] and J. Christopher Love[1,2,3], (1)The David H. Koch Institute for Integrative Cancer Research, Massachusetts Institute of Technology, Cambridge, MA, (2)The Broad Institute of Harvard and MIT, Cambridge, MA, (3)Department of Chemical Engineering, Massachusetts Institute of Technology, Cambridge, MA, (4)Department of Electrical Engineering and Computer Science, Massachusetts Institute of Technology, Cambridge, MA, (5)Department of Biological Engineering, Massachusetts Institute of Technology, Cambridge, MA

There is increasing evidence for genetic heterogeneity among individual tumor cells. In colon cancer, multilineage differentiation and epigenetic regulation have also been shown to induce transcriptional heterogeneity, which may affect other complex functional behaviors such as secretion of growth factors or chemokines. Since secreted factors produced by tumor and stromal cells can affect progression of disease, it is important to understand the functional heterogeneity that exists within the tumor microenvironment in order to begin mapping signaling networks. Bulk methods such as ELISAs or protein microarrays lack the ability to resolve functional heterogeneity among single cells. In this study, we employ quantitative microengraving, in combination with on-chip cytometry, to probe the secretomes of over 10,000 human primary colorectal tumor cells in parallel using arrays of subnanoliter wells (nanowells). Starting with a panel of 21 cytokines, chemokines, and growth factors relevant to cancer biology, we examined protein production from six human primary tumors and identified common responses among the samples, including chemokines such as IL-8. We show that individual cells from the tumors exhibit discrete secretory profiles, suggesting that functional heterogeneity is common among cells from a single tumor. We also find evidence for intercellular bypass signaling via production of redundant chemokines that agonize the same receptor.

(210d) Chance and Circumstance Govern Macrophage Functional Diversity

Yishan Chuang, Chemical and Biologial Engineering, Northwestern University, Evanston, IL and Joshua N. Leonard, Chemical and Biological Engineering, Chemistry of Life Processes Institute, R.H. Lurie Comprehensive Cancer Center, Northwestern University, Evanston, IL

Macrophages play a critical role in maintaining the balance between homeostasis and protective inflammation by adopting either immunostimulatory (M1) or immunosuppressive (M2) phenotypes. Tumors and pathogens manipulate this balance by inducing the production of immunosuppressive stimuli, which shifts macrophages into a M2 phenotype, supporting tumor growth and pathogen survival. In addition, M2 macrophages have been reported to convert to an M1 phenotype given certain stimuli, which is termed "plasticity". Despite the centrality of these phenomena to diverse disease processes, the mechanisms regulating macrophage "decision-making" vis-à-vis polarization remain poorly understood. Here, we describe several novel systems-level insights into how such dysfunctional immune network states are established, maintained, and potentially may be disrupted.

Since most macrophage polarization studies to date involve applying coherent stimuli (either pro-M1 or pro-M2 stimuli alone), and these stimuli rarely exist independently *in vivo*, we first investigated how macrophages integrate contradictory signals to "calculate" a response. Macrophages were exposed to combinatorial variations in dose of IL-10 and IL-12 (pro-M2 and pro-M1 stimuli, respectively), and macrophages were "activated" using bacterial endotoxin. Surprisingly, the presence of IL-10 blocked previously-described IL-12-mediated conversion of M2 cells to an M1 phenotype. Moreover, M2-type responses increased with IL-10 dose and were largely independent of IL-12 co-treatment. We also investigated a competition between IL-10 and the canonical pro-M1 stimulus, IFNγ. Interestingly, IFNγ specifically suppressed the production of additional IL-10 by M2 cells, potentially disrupting a key positive feedback loop involved with maintaining the M2 state. This suggests that IFNγ may enable the restoration of an immune network state that promotes M1 polarization and immunological clearance or control of cancer.

We also investigated whether incoherent stimuli induce heterogeneous responses among different macrophages within a single population. Using flow cytometric profiling, we observed that under some conditions, distinct M1 and M2 cells were induced within a single population. The probability of polarization towards an M2 state increased with IL-10 dose and was independent of IL-12 co-treatment. Other cells remained non-responsive to activation, and this probability of activation was unaffected by cytokine treatment. These data represent the first evidence to date that macrophage polarization may be a stochastic process and suggest that separate regulatory mechanisms govern activation and polarization. Subsequent investigations focus on determining whether plasticity is also heterogeneous, and if so, how this is dependent upon a cell's prior polarization history. This system-level analysis of macrophage function provides new insights into innate immune function and will help to formulate therapeutic strategies for treating the many chronic diseases characterized by local immune dysfunction.

(210e) Multiscale Prediction of Patient-Specific Platelet Function Under Flow

Matthew H. Flamm[1], Thomas Colace[1], Manash S. Chatterjee[1], Hiuyan Jing[2], Songtao Zhou[2], Daniel Jaeger[2], Lawrence F. Brass[3], Talid R. Sinno[1] and **Scott L. Diamond**[1], (1)Chemical and Biomolecular Engineering, University of Pennsylvania, Philadelphia, PA, (2)University of Pennsylvania, Philadelphia, PA, (3)School of Medicine, University of Pennsylvania, Philadelphia, PA

During thrombotic or hemostatic episodes, platelets bind collagen and release ADP and thromboxane A2, recruiting additional platelets to a growing deposit that distorts the flow field. Prediction of clotting function under hemodynamic conditions for an individual's platelet phenotype remains a challenge. A platelet signaling phenotype was obtained for 3 healthy donors using Pairwise Agonist Scanning (PAS), where calcium dyeloaded platelets were exposed to pairwise combinations of ADP, U46619, and convulxin to activate P2Y1/P2Y12, TP, and GPVI receptors, respectively, in the presence or absence of the IP receptor agonist, iloprost. A neural network model was trained on each donor's PAS experiment and then was embedded into a multiscale Monte Carlo simulation of donor-specific platelet deposition under flow. The simulations were directly compared to microfluidic experiments of whole blood flowing over collagen at 200 and 1000 s-1 wall shear rate. The simulations predicted the ranked order of drug sensitivity for indomethacin, aspirin, MRS-2179 (P2Y1 inhibitor), and iloprost. Consistent with measurement and simulation, one donor displayed larger clots, while another donor presented indomethacin-resistance (revealing a novel heterozygote TP-V241G mutation). *In silico* representations of an individual's platelet phenotype allowed prediction of blood function under flow, essential to identifying patient-specific cardiovascular risks, drug responses, and novel genotypes.

Figure 1

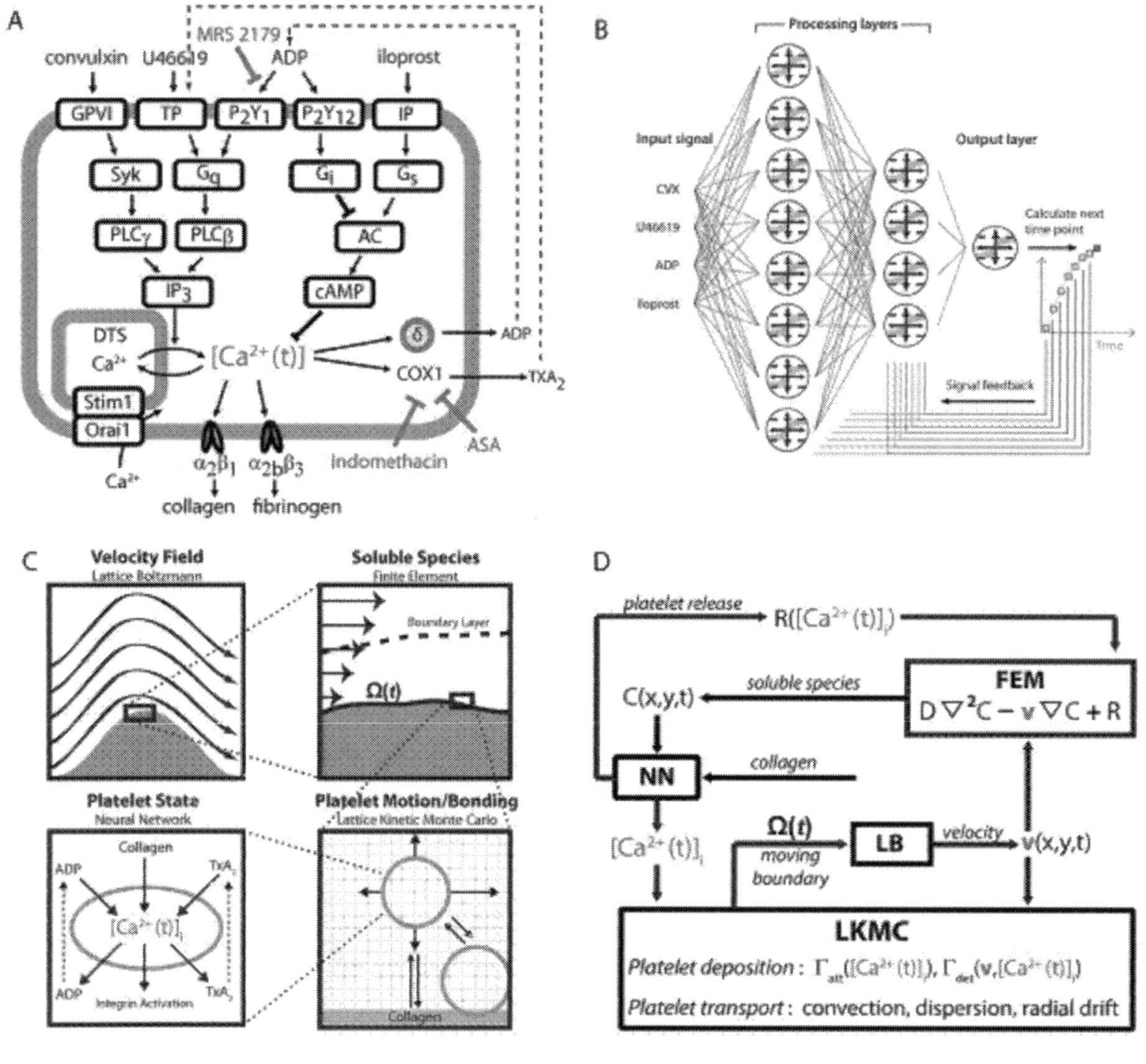

Fig. 1. Multiscale model of combinatorial platelet activation and thrombus formation under flow. Platelet agonists (*blue*) used individually or in pairs to activate GPVI or G-protein coupled receptors (thromboxane receptor, TP; purinergic receptors P2Y1 and P2Y12; and the prostacyclin receptor, IP) result in modulation of intracellular calcium (*green*) from intracellular stores distal of phospholipase C (PLC) activation or from store operated calcium entry via Stim1-Orai1 activation. Inhibitors (*red*) such as acetylsalicylic acid (ASA) or indomethacin inhibit cyclooxygenase-1 (COX-1). Autocrine pathways include release of TXA2 and ADP (**A**). A 2-layer, 8-node/4-node neural network (NN) with feedback is trained with 74 measured calcium traces to predict [Ca]i for each patient-specific platelet Pairwise Agonist Scan (**B**). The multiscale simulation of platelet deposition under flow requires simultaneous solution of the instantaneous velocity field over a complex and evolving platelet boundary Ω(t) by Lattice Boltzmann (LB), concentration fields of ADP and TXA2 by finite element method (FEM), individual intracellular platelet state ([Ca]i) and release reactions (R) for ADP and TXA2 by neural network, and all platelet positions and adhesion/detachment by lattice Kinetic Monte Carlo (LKMC) (**C,D**).

(210f) Prediction and Validation of Microbiota-Derived Tryptophan Metabolites with Anti-Inflammatory Properties

Kyungoh Choi[1], Gautham V. Sridharan[2], Robert Alaniz[3], Kyongbum Lee[2] and Arul Jayaraman[1], (1)Department of Chemical Engineering, Texas A&M University, College Station, TX, (2)Chemical and Biological Engineering, Tufts University, Medford, MA, (3)Department of Microbial and Molecular Pathogenesis, Texas A&M Health Science Center, College Station, TX

The human gastrointestinal (GI) tract is colonized by ~10^{14} bacteria belonging to ~ 500 species that are collectively termed the microbiota. Several clinical and *in vivo* studies have shown that alterations in the microbiota composition and/or function (i.e., its dysbiosis) are correlated to pathologies such as inflammatory bowel disease (IBD) and colitis. The importance of microbiota metabolites is also evident from a metabolomic study of patients with Crohn's disease where disease severity was correlated with alterations in the levels of tryptophan in feces. Recently, work from our lab has shown that indole, a metabolite derived by the microbiota from tryptophan, attenuates indicators of inflammation intestinal epithelial cell. These studies highlight the potentially important role for tryptophan metabolites in the regulation of inflammation in the GI tract. However, identification of metabolites that exist in the GI tract is very difficult because many of the gut microbiota species cannot be cultured *in vitro*. As metabolic pathways are highly conserved across different organisms and many of the microbiota metabolites can also be synthesized by host cells, the panel of metabolites that can be exclusively synthesized by the microbiota is largely unknown.

In this study, we developed a computational approach for predicting possible tryptophan-derived metabolites that can be uniquely ascribed to microbiota using probabilistic pathway analysis. The probabilistic pathway construction algorithm explores the space of possible biochemical transformations in the KEGG database from a starting metabolite (tryptophan) and predicts all possible product metabolites that can be formed from reactions catalyzed by enzymes that are expressed exclusively in bacteria but not in host cells (e.g. human or mouse). Each of these product metabolites, in turn, are used as substrates for the next round of possible biotransformations, and the algorithm continues until a product metabolite that cannot be uniquely ascribed to bacteria is reached. Starting from tryptophan and its immediate metabolites, the algorithm predicted ~10 possible metabolites that can be uniquely ascribed to microbiota reactions. A targeted metabolomics approach was then taken to identify and quantify four of the predicted metabolites – indole-3-acetate, indole-3-pyruvate, indole-3-acetamide, and tryptamine - that were present in murine intestinal tissue and fecal samples using an LC/MS-MS system operating in multiple reaction monitoring (MRM) mode. Preliminary data suggest that the identified metabolites attenuate indicators of inflammation in mucosal cells; thereby, demonstrating the potential and utility of our integrated bioinformatics and metabolomics approach. The identification of novel anti-inflammatory metabolites is expected to lead to impact the development of therapeutic molecules against inflammatory disorders such as IBD and colitis.

(210g) Understanding Nitrogen Metabolism in the Biofuel Crop Poplar by Isotope-Assisted Metabolic Flux Analysis

Shilpa Nargund[1], Ashish Misra[1], Xiaofeng Zhang[1], Gary D. Coleman[2] and Ganesh Sriram[1], (1)Chemical and Biomolecular Engineering, University of Maryland, College Park, MD, (2)Plant Science and Landscape Architecture, University of Maryland, College Park, College Park, MD

Nitrogen crucially determines the growth and productivity of photosynthetic organisms. However, it is often supplied by means of artificial fertilizers, whose production is energetically intensive and significantly relies on non-renewable fossil fuels. Therefore, the investigation of nitrogen use efficiency (NUE), particularly in plants that are potential biofuel crops, will have wide-ranging consequences for biofuel economy and sustainability. Poplar, a biofuel crop, uses an array of sophisticated metabolic and regulatory mechanisms to deposit, remobilize and thereby conserve nitrogen. The transport of nitrogen from senescing leaves to the bark storage protein (BSP) during winter is one such mechanism. This recycling of nitrogen is closely related to the status of carbon availability in plants as evidenced by previous studies which show that photoperiod perception, growth, presence of glutamine or calcium and protein phosphorylation affect accumulation of BSP (Zhu. B and Coleman. G.D., Plant Physiology 126:342-351, 2001). The involvement of multiple factors in regulation of BSP production suggests that significant metabolic rewiring occurs during nitrogen recycling.

To study this phenomenon we employed metabolic engineering tools such as steady state isotope-assisted metabolic flux analysis (MFA) to quantify metabolic fluxes in cells of the tree poplar (*Populus trichocarpa*). Metabolic fluxes are the rates of carbon flow through metabolic pathways (measured in mol $time^{-1}$ g $biomass^{-1}$) and are important indicators of cellular functions (Stephanopoulos, G. & Stafford, D.E., Chemical Engineering Science 57: 2595-2602, 2002). MFA involves conducting isotope labeling experiments (ILEs) wherein isotopically labeled (e.g. ^{13}C, ^{15}N, ^{17}O) substrates are fed to the plant cell culture or tissue and the labeling patterns of biomass components such as proteins, lipids, sugars and nucleotides, are measured using nuclear magnetic resonance (NMR) or mass spectrometry (MS). Fluxes are estimated by fitting the labeling patterns to a mathematical model of the metabolic network.

Poplar suspension cells, subjected to environmental or genetic perturbations that affect BSP metabolism, will be grown under 100% 1-^{13}C and 30% U-^{13}C glucose in separate ILEs. Comparative metabolic flux maps for poplar cells under the various treatments will be constructed to evaluate the changes in metabolism during BSP accumulation and N recycling. The differences in fluxes between conditions and their implications on cell physiology will be discussed. The results of our MFA experiments are anticipated to reveal the metabolic rewiring that occurs in poplar cells when they deposit and remobilize nitrogen, and eventually provide insights toward engineering photosynthetic organisms with high NUE.

(210h) Tandem Mass Spectrometry: The Next Frontier in 13C-Metabolic Flux Analysis

Maciek R. Antoniewicz, Department of Chemical and Biomolecular Engineering, University of Delaware, Newark, DE and Jungik Choi, Chemical Engineering, Univerisity of Delaware, Newark, DE

Tandem mass spectrometry is a new measurement technique for quantifying isotopic labeling for ^{13}C-metabolic flux analysis (^{13}C-MFA). Recently, we developed a general mathematical framework based on isotopomer simulations for incorporating tandem MS data into ^{13}C-MFA (Choi et al., Metab Eng 2011). Here, we present further advancements we have made towards establishing next-generation tandem MS based flux analysis, specifically: 1) measuring complete positional isotopomer distribution of aspartate by tandem MS; and 2) developing an elementary metabolite units (EMU) framework for modeling tandem mass isotopomers.

First, we developed a simple and accurate method for determining the complete positional isotopomer distribution of aspartate carbon atoms by gas chromatography tandem mass spectrometry (GC-MS/MS). To achieve this, we screened tandem MS spectra of the TBDMS-derivative of aspartate for daughter fragments with the necessary carbon atom fragmentations to fully determine all sixteen isotopomers of aspartate. We selected five fragments that provided a redundant set of labeling measurements to quantify all isotopomers by least-squares regression. We demonstrated the validity of our approach using ^{13}C-labeled aspartate standards and natural aspartate.

Next, we extended our modeling framework for tandem MS simulations. Here, we present for the first time our new framework based on EMUs, which are computationally more efficient to evaluate compared to previous methods based on isotopomers/cumomers. The new mathematical framework, which we called EMU^2, is based on two-dimensional representation of isotopic-labeling. Using this new formulation we take full advantage of the additional information that is obtained from tandem MS measurements for ^{13}C-MFA.

Production of Omega-3 Fatty Acids by Fermentation of Metabolically Engineered *Yarrowia lipolytica*

*Dongming Xie *, David Short, Quinn Zhu, Zhixiong Xue, Narendra Yadav, Pamela Sharpe, Raymond Hong, J Martin Odom, Boonchai Boonyaratanakornkit, Clementina Dellomonaco, Xiaochun Fan, J Martin Odom, Bjorn Tyreu, and Ethel Jackson*

Division of Biochemical Sciences and Engineering, Central R&D, DuPont Company, Wilmington, DE 19880, U.S.A.

Abstract

The omega-3 fatty acids, cis-5, 8,11,14,17-eicosapentaenoic acid (C20:5; EPA) and cis-4,7,10,13,16,19-docosahexaenoic acid (C22:6; DHA), have been shown to confer wide-ranging health benefits against e.g., cardiovascular disease, rheumatoid arthritis, Alzheimers disease, lung fibrosis and inflammatory bowel disease. Currently, the major sources for omega-3 fatty acids are fish oil, microalgae, and a limited number of species of plants. Due to high costs, uncertain availability, and many other disadvantages within the current sources, DuPont has developed a clean and sustainable alternative source of omega-3 fatty acids through fermentation using a metabolically engineered strain of *Yarrowia lipolytica*. Specifically, desaturase and elongase genes have been introduced into the oleaginous yeast to utilize certain carbon sources to synthesize omega-3 fatty acids under fermentation conditions. This metabolic engineering of *Yarrowia* also provides a platform technology to produce the tailored omega-3 (EPA, DHA) or omega-6 fatty acids (ARA, GLA) in the oil. The strain candidates were evaluated via culture tubes, shake flasks, micro-24 bioreactor, and lab-scale fermentation experiments. The typical fermentation process consists of a nitrogen-rich growth phase and a nitrogen-starved oleaginous phase. Medium components and process conditions were optimized to improve the omega-3 fatty acid production. Mathematical modeling was a useful tool to analyze, optimize, control, and scale up the fermentation process. The yeast triacylglyceride oil produced has a unique fatty acid profile with greater than 55% as EPA (corresponding to the highest omega-3 fatty acid reported so far in the oil), and less than 8% as saturated fatty acid. The developed fermentation process was successfully scaled up to commercial scale. The produced EPA oil (Newharvest™) has been used as human nutritional supplement; and the EPA-rich biomass has been used as salmon feed to raise brand salmon, Verlasso™ (www.verlasso.com).

(211b) High Yield Ethanol Production From Fermentation of C5 and C6 Biomass Sugars by Using Native Yeast

Bin Li[1], Sasidhar Varanasi[2] and Patricia Relue[1], (1)Bioengineering, The University of Toledo, Toledo, OH, (2)Chemical and Environmental Engineering, The University of Toledo, Toledo, OH

Bioconversion of lignocellulosic feedstocks to fuel-grade ethanol is a promising route towards reduced dependence on fossil fuels and the mitigation of greenhouse gas production. Baker's yeast (*Saccharomyces cerevisiae*) used for industrial-scale ethanol production is unable to utilize xylose, which accounts for about 30% of the total sugars released from biomass hydrolyzates to make ethanol. Current methods to overcome this limitation focus on the development of different types of genetically modified organisms (GMOs). However, several problems exist for these GMOs. First, the consumption rate of xylose is low, and the time required to produce ethanol is relatively long. In addition, the genetic stability of GMOs, costs for propagation, and robustness under industrial fermentation conditions are as yet unreported.

Alternatively, in this research, a non-GMO approach is proposed. Simultaneous-isomerization of xylose to xylulose and reactive-extraction of xylulose to an organic phase is used for high yield conversion and concentration of the xylulose. Recovery of xylulose to an aqueous phase and fermentation to ethanol is then efficiently accomplished by native yeast. The kinetics of glucose and xylulose utilization and yield of ethanol are presented for this fermentation system.

Optimization of Biodemulsifier production condition in order to crude oil demulsification

Shahab Amirabadi, M.R. Rahimpour[1], A. Jahanmiri, Babak Rafie nia

School of Chemical and Petroleum Engineering, Department of Chemical Engineering, Shiraz University, Shiraz, Iran

Abstract

Paenibacillus alvei ARN63 as a biodemulsifier-producing bacterium has represented incredible demulsifying ability. The produced lipopeptide biodemulsifier can be used in crude oil industry as an environmentally friendly (green) substance, containing no toxic component. Due to increase production of biodemulsifier, as a type of biosurfactant, the culture conditions for the strain were optimized. For this goal, the effect of different carbon sources and C/N ratios (carbon concentration) were examined in detail at constant temperature, pH, shaking speed (rpm) and nitrogen source. Biodemulsifier synthesis and cell growth were determined via measurement of surface tension and optical density (OD600nm), respectively. Subsequently, to investigate its demulsification performance, the bottle test was carried out. Normal paraffin compounds such as $C_{10}H_{22}$ and $C_{12}H_{26}$ and motor oil favored the synthesis of this biodemulsifier. The best results were obtained employing motor oil as carbon source with a concentration of 50 ml/l. The biosurfactant (biodemulsifier) production was probably growth associated as resulted from the biosurfactant production and growth kinetics. The surface tension was reduced (from 58 mN/m) to 24.7 mN/m in 72 h, and demulsification ratio reached to approximately 73%. Under optimal conditions, the biodemulsifier yield approached 2.1 g/l.

Introduction

An emulsion is a dispersion system, in which liquid droplets are dispersed throughout another immiscible liquid [1]. Large amount of crude oil processed worldwide exist in a state of emulsion. The most common emulsions in the oil industry are of the water-in-oil (w/o) type [2]. Petroleum-field emulsions may cause problems at several stages of the crude oil production, such as higher cost of transport and storage, corrosion and salt deposits in the refining equipment and poisoning of catalysts [3].

Therefore, the destabilization of the emulsions is a necessary step for the efficient operation of separating water from petroleum and to recover saleable oil. At commercial scale, emulsions are treated by a combination of physical and chemical treatments involving centrifugation, heat, electrical treatment or chemical substances containing fatty acids and long chain alcohols [4].

These chemical substances, which are called surfactants, are amphiphilic agents which, by accumulating at the interface between immiscible phases, can reduce surface and interfacial tensions [5]. In addition, chemically synthesized surfactants are not biodegradable and can be

---

[1] . Corresponding author. Tel.: +98 711 2303071; fax: +98 711 6287294.

E-mail address: rahimpor@shirazu.ac.ir (Prof. M. R. Rahimpour).

toxic to the environment [6, 7]. Therefore, in recent years much attention has been directed towards biosurfactants, which are produced by microorganisms, owing to their different advantages such as, lower toxicity, higher biodegradability, better environmental compatibility, higher foaming, high selectivity, specific activity at extreme temperatures, pH and salinity, and the ability to be synthesized from renewable and cheaper substrates. [8, 9]

The most important applications of these biological compounds in different fields are enhanced oil recovery, demulsification activity in the petroleum and bio-remediation in environment, emulsifiers and as a functional ingredient in the food industries, microbiological and pharmaceutical agents in the biological industries [5, 10-11]. Consequently, biodemulsifiers, as a type of biosurfactants, have attracted increasing attention in research.

On the other hand, studies on biodemulsifiers lag far behind research on other biosurfactants and remain in a preliminary stage [11-13].

In 2010, *Paenibacillus alvei* ARN63 was isolated from an Iranian oil field and it was showed that this bacterium was capable to produce lipopeptide biosurfactant which was able to lower the surface tension of the cultivation media to 35 mN/m [14].

The identification and optimization of the cultivation conditions that affect the surfactant production represent key points for the development of a cost-competitive process.

However, the optimization of cultivation conditions to increase biodemulsifier yield has not been studied for biodemulsifier-producing bacteria, other than for a few ones such as Liu, et al. [11, 13, 14]. It is necessary to say that temperature, pH, carbon source and carbon/nitrogen ratio (C/N) of the media play a great role in optimizing the cultures.

The objective of this work was to optimize the culture conditions for *Paenibacillus alvei* ARN63 growth and activity to produce biodemulsifier (as a kind of biourfactant) more effectively for its application in the destabilization of crude oil emulsions.

In this study, the effect of carbon source and carbon/nitrogen ratio (C/N) was investigated to determine optimal condition. However, it remains unknown how the production of biodemulsifiers will be affected by changes in carbon or nitrogen sources [11]. As the produced biourfactant could be effectively used to demulsify the water-in-oil emulsion, its demulsifying performance was also evaluated.

1. Materials and Methods

2.1. Organism

Paenibacillus alvei ARN63 which was isolated from an Iranian oil well was used in this work to produce biourfactant [14].

2.2. Culture media preparation

Brain Heart Infusion (BHI) agar was used for preparation of the inoculums. For the preparation of this culture, 37 gr/l BHI powder was dissolved in distilled water and autoclaving in 120 °C for 20 min, then biourfactant-producing bacterium was added to this media from a stoke which was held in -80 °C.

The cultures were grown in this media for 18-24 h at 37 °C and shaking condition of 180 rpm. This was used as inoculums at the 1% (v/v) level. For biourfactant production a mineral salt media (MSM) with the following composition was used (gr/l) [14]:

1 $(NH_4)_2SO_4$, 0.25 $MgSO_4$, 50 NaCl, 13.9 K_2HPO_4, 2.7 KH_2PO_4, 10 trace salt solution. The composition of the solution of trace elements was (g/l): 1 EDTA,3 $MnSO_4{\cdot}H_2O$, 0.1 $FeSO_4{\cdot}7H_2O$, 0.1 $CaCl_2{\cdot}2H_2O$, 0.1 $CoCl_2{\cdot}6H2O$, 0.1 $ZnSO_4{\cdot}7H_2O$, 0.01 $CuSO_4{\cdot}5H_2O$, 0.01

$AlK(SO_4)_2 \cdot 12H_2O$, 0.01 H_3BO_3, 0.01 $Na_2MoO_4 \cdot 2H_2O$. The pH of the medium was adjusted to 7 using 1N NaOH.

Carbon source filtering by 0.22 µm Filter was added to the autoclaved MSM by different concentrations. Cultivations were carried out in 250 ml Erlenmeyer flask containing 50 ml media and incubated on a shaker-incubator at temperature of 37 °C and be shaken at 180 rpm.

At the time intervals of 48 h and 72 h, the samples were collected and monitored for biosurfactant production and growth of bacterium. Therefore the optical density at 600 nm (OD600nm) and surface tension (ST) were measured.

2.3. Media optimization

In this study the effect of various carbon sources with different concentrations was performed in order to obtain the best carbon source and the appropriate carbon/nitrogen ratio or carbon concentration for higher productivity of the biosurfactant. The medium optimization was conducted in a series of experiments changing carbon source and its concentration and keeping nitrogen source ($(NH_4)_2SO_4$) fixed at 1g/l.

The concentration of carbon source which was chosen to add the MSM, were 5, 10, 20 and 30 (ml/l) for normal paraffin compounds and 10, 20, 30 and 50 (ml/l) for motor oil. Consequently, different C/N could be tested although, the nitrogen source and its concentration was fixed.

The normal paraffin such as $C_{10}H_{22}$, $C_{12}H_{26}$ (SIGMA-ALDRICH), were considered as carbon sources.

2.4. Optical density and surface tension measurement

The optical density measurements of the samples were determined at wavelength of 600 nm (OD600nm) in a Sp-3000plus spectrophotometer (OPTIMATOKYO, Japan). Moreover, the surface tension measurements of cell-free broth were carried out in a DSA100 (Krüss, Hamburg, Germany).

2.5. Biodemulsifier purification

After cultivation the cells on the MSM at optimum condition for 72 h, the bacterial cells were removed by centrifugation at 1000 g for 15 min and. The cell-free lipopeptide was precipitated by adding 3 N HCl to a final pH of 2.0 and stored for 30 min at 4 °C. The precipitates were collected by centrifugation at 10000 g for 20 min at 4 °C, and dissolved in a chloroform/methanol (2:1, v/v) solvent system. The solution was dried with a rotary vacuum evaporator RE200 (BIBBY STERILIN LTD, UK) [15].

2.6. Demulsification performance investigation

Fresh crude oil from Cheshmekhosh oilfield in Iran containing no water –water was separated- was used for the bottle test. In the bottle test, fresh crude oil emulsion was prepared by mixing 100 ml oil and 10 ml distilled water. This water-in-oil emulsion was stable, showing no evidence of water separation within 24 h. In order to perform demulsification test, 1 ml of fermented broth was added to prepared oil emulsion and then, mixed in a 250 ml Erlenmeyer flask with a mixer at 800 rpm for 20 min. After mixing, the mixture was transferred in a 100 ml glass bottle and put in an oven (Memmert, Germany) at 80 °C.

The change in volume of the water phase was recorded at certain intervals, and consequently demulsifying performance was evaluated by calculating the demulsification ratio [11]:

$$Demulsification\ Ratio\ \% = \frac{V}{V_0} \times 100$$

Where V is the water volume on the bottom and V_0 is water volume in original emulsion with added culture volume.

3. Results and discussion

3.1. Effect of carbon source and C/N ratio

Paenibacillus alvei was used to produce biodemulsifier using n-paraffin substrate as immiscible carbon source which was added to MSM.

Normal decane ($C_{10}H_{22}$), n-dodecane ($C_{12}H_{26}$) had different performances in biodemulsifier production, although both of them were able to be used as a suitable substrate, because surface tension of the media was reduced when they were present in the culture (Fig.1, 2).
As illustrated in Fig. 1, the production is achieved at low $C_{10}H_{22}$ concentrations (C/N) and the maximum surface tension decrease is associated to the C/N of 16.94, where surface tension reduction occurred from 55.4 to 33.16 mN/m. It is important to note that, the C/N values greater than 67.79 should not be applied for this source.
For $C_{12}H_{26}$, although the surface tension was reduced, its decline was not as well as the previous carbon sources (Fig. 2).
As seen from Figs. 1, 2, n-decane was more suitable for biosurfactant synthesis in comparison to n-dodecane.
Motor oil was similarly used as carbon source. It revealed a considerable strength as carbon source in biodemulsifier production. The Fig. 3 indicates that among these three substances, motor oil was rated as the best substrate. The surface tension reduction for motor oil was from 58 to 24.7 (mN/m) after 72 hours at concentration of 50 ml/l.
In comparison to n-paraffin compounds, not only this substrate is more effective in surface tension reduction and biodemulsifier synthesis, but also is more economical and cheaper, owing to this fact that the large amount of them can be prepared from industrial wastes. To sum up, according to above facts and strength of motor oil in high productivity of biodemulsifier, it can be considered as the best substrates for this purpose.
The strain growth on motor oil did not differ sharply from n-paraffin compounds. Moreover, time of 72 hours was the best time for cultivation except for some substrates for which 48 hours could be considered as optimum time. It should be noted that the maximum biodemulsifier yield (approximately 2.1 g/l) was obtained for optimum cultivation conditions (motor oil as carbon source at 50 ml/l concentration and 72 h).
Finally, the selection of appropriate C/N ratio play a great role in biodemulsifier production because from the figures, it can be observed that the performance of production varied dramatically by C/N ratio for all carbon sources.

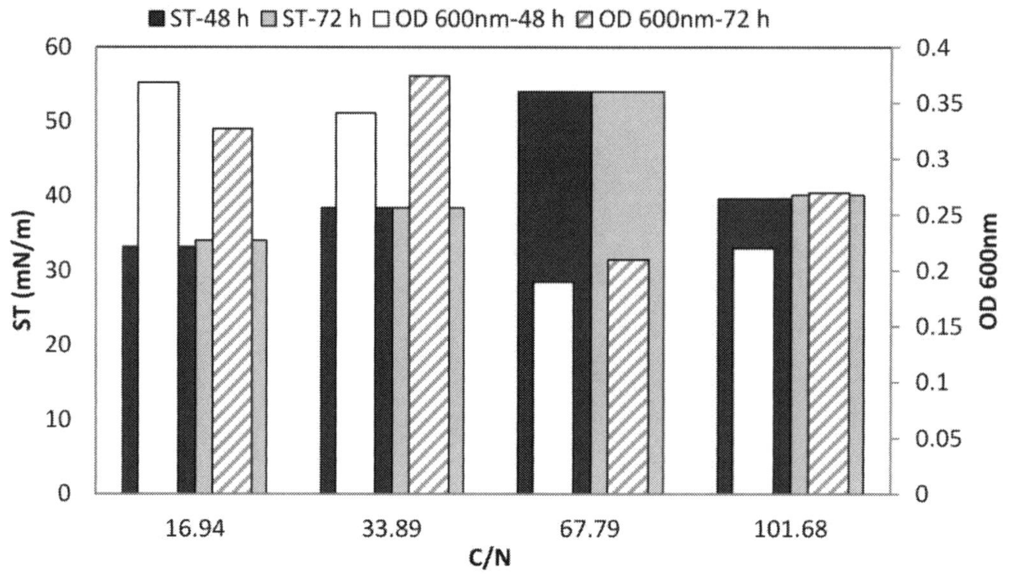

Figure 1

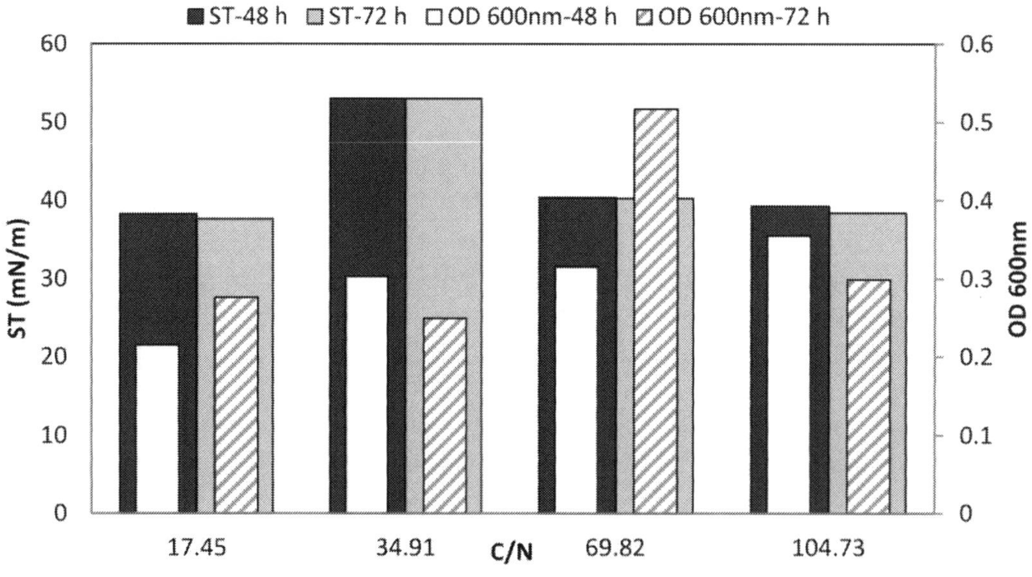

Figure 2

3.2. Kinetics of biodemulsifier synthesis

After the establishment of the best medium, the growth and biodemulsifier production kinetics were described for *P. alvei* cultivated in 250 ml Erlenmeyer flask containing 50 ml optimal medium at 37 °C during a 96-hour period of time under agitation speed of 180 rpm. The kinetic curves (Fig. 4) show that there was an almost parallel relationship between surface tension reduction (as a consequence of biodemulsifier production) and cell growth. Surface tension

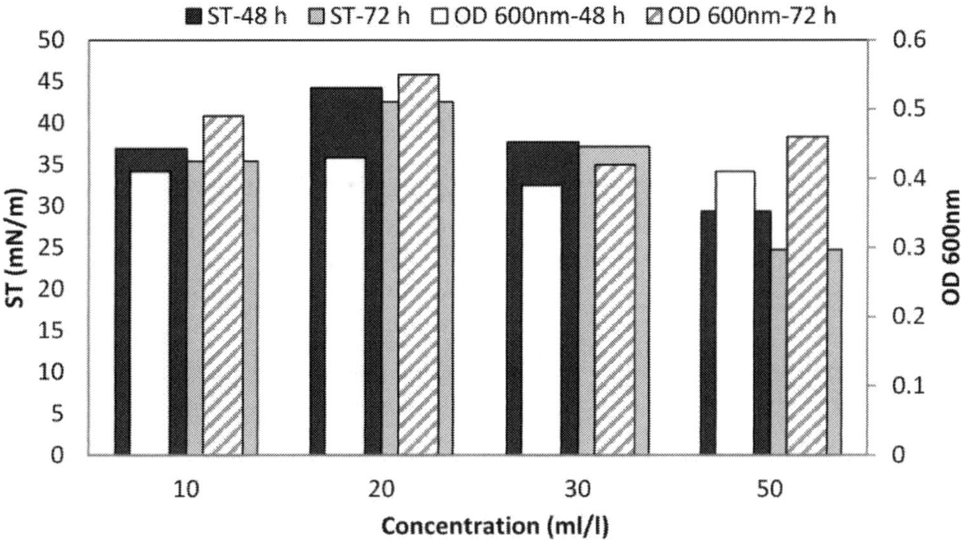

Figure 3

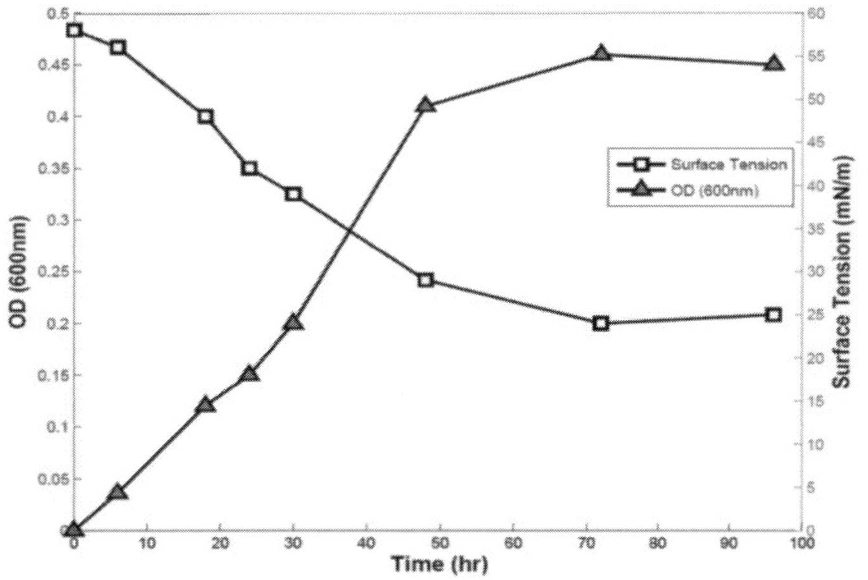

Figure 4

reduction occurred in logarithm phase owing to biodemulsifier production and the lowest surface tension was 24.7mN/m at the beginning of stationary phase after 72 hours.

In addition, surface tension stabilization after 72 hours illustrate that biodemulsifier production stopped and its maximum yield was 2.1 g/l. These observations (Fig. 4) indicate that as a result of biodemulsifier synthesis during logarithm growth phase, the biodemulsifier is probably produced as primary metabolite accompanying cellular biomass formation (growth-associated kinetics).

3.4. Demulsification capability of the produced biodemulsifier

The produced biodemulsifier obtaining from optimal cultivation was tested due to break water-in-oil emulsion and the demulsification ratio was evaluated in different intervals. It is shown in Fig.10 that the demulsification ratio rapidly raised for 120 minute and reach approximately 73%. However its performance did not vary after this time and the best performance was obtained for t=120 min. Although its performance was not better than chemical demulsifiers, its environmental compatibility lead to being significantly practical for industrial application, owing to this fact that it is not necessary to be separated during the process. Therefore, they can be more economical than chemical ones.

4. Conclusion

The produced biodemulsifier not only was able to demulsify water-in-oil emulsion effectively, but also because of its environmental compatibility, its demand might probably increase, especially in petroleum industry. It should be noted that the expenses related to separation after processing reduced and the processes did not become complex, because they did not contain any toxic components and are biodegradable. Furthermore, this biosynthesis is far cheaper than chemical demulsifiers, due to this fact that chemical carbon sources could be replaced by industrial wastes as economical substrates. As a result, effect of various cultivation factors on the production of biodemulsifiers were investigated to determine the best condition for production and consequently, higher production. The results showed that n-heptane could prepare the best condition for bacterial cells to produce biodemulsifier. It was observed that carbon source concentration or C/N ratio played a great role in biosynthesis. Finally, under the optimal conditions, the yield of biodemulsifier reached to 2.1 g/l. This biosynthesis is predicted to be promising for future applications.

References:

[1] Y. Wen, H. Cheng, L.J. Lu, J. Liu, Y. Feng, W. Guan, Q. Zhou, X. F. Huang, Analysis of biological demulsification process of water-in-oil emulsion by Alcaligense sp. S-XJ-1, Bioresour. Technol. 101 (2010) 8315-8322 .

[2] R. Del Colle, E.Longo, S.R. Fontes, Demulsification of water/sunflower oil emulsions by a tangential filtration process using chemically impregnated ceramic tubes, J. Membr. Sci. 289 (2007) 58-66.

[3] D. C.M. Filho, J. B.V.S. Ramalho, L. S. Spinelli, E. F.Lucas, Aging of water-in-oil emulsions: effect on water content, droplet size distribution, dynamic viscosity and stability, Colloids Surf., A 10.1016/j.colsurfa.2011.12.076.

[4] N. Nadarajah, A. Singh, O. P. Ward, De-emulsification of petroleum oil emulsion by a mixed bacterialculture, Process Biochem. 37 (2002) 1135–1141.

[5] H. Ghojavand, F. Vahabzadeh, E. Roayaei, A. Khodabandeh Shahraki, Production and properties of a biosurfactant obtained from a member of the Bacillus subtilis group (PTCC 1696), J. Colloid Interface Sci. 324 (2008) 172-176.

[6] A.J. Desai, K.M. Patel, J.D. Desai, Emulsifier production by Pseudomonas fluorescens during the growth on hydrocarbon, Curr. Sci. 57 (1998) 500–501.

[7] N.K. Bordoloi, B.K. Konwar, Microbial surfactant-enhanced mineral oil recovery under laboratory conditions, Colloids Surf., B 63 (2008) 73–82.

[8] J.D. Desai, I.M. Banat, Microbial production of surfactants and their commercial potential, Microbial. Mol. Biol. Rev. 61 (1997) 47–64.

[9] S. Joshi, C. Bharucha, S. Jha, S. Yadav, A. Nerurkar, A.J. Desai, Biosurfactant production using molasses and whey under thermophilic conditions, Bioresour. Technol. 99 (2008) 195–199.

[10] I.M. Banat, R.S. Makkar, S.S. Cameotra, Potential commercial applications of microbial surfactants, Appl. Microbiol. Biotechnol. 53 (2000) 495-508.

[11] J. Liu, X.F. Huang, L.J. Lu, J.C. Xu, Y. Wen, D.H. Yang, Q. Zhou, Optimization of biodemulsifier production from Alcaligense sp. S-XJ-1 and its application in breaking crude oil emulsion, J. Hazard. Mater. 183 (2010) 466-473.

[12] A.L. Stewart, N.C.C. Gray, W.L. Cairns, N. Kosaric, Bacteria-induced de-emulsification of water-in-oil petroleum emulsions, Biotechnol. 5 (1983) 725-730.

[13] J. Liu, K. Peng, X.f. Huang, L. Lu, H. Cheng, D. Yang, Q. Zhou, H. Deng, Application of waste frying oils in the biosynthesis of biodemulsifier by a demulsifying strain Alcaligense sp. SXJ-1, J. Environ. Sci., 23 (2011) 1020-1026.

[14] A.R. Najafi, M.R. Rahimpour, A.H. Jahanmiri, R. Roostaazad, D. Arabian, M. Soleimani, Z. Jamshidnejad, Interactive optimization of biosurfactant production by *Paenibacillus alvei* ARN63 isolated from an Iranian oil well, Colloids Surf., B 82 (2011) 33–39.

[15] P.I. Kim, H. Bai, D. Bai, H. Chae, S. Chung, Y. Kim, R. Park and Y.-T. Chi, Purification and characterization of a lipopeptide produced by Bacillus thuringiensis CMB26, Journal of Applied Microbiology 2004, 97, 942–949.

(211h) Strategies for Enhanced Production of Recombinant Staphylokinase by *Escherichia Coli*

Arshad Jawed, K.L. Dikshit and Debendra K. Sahoo, CSIR - Institute of Microbial Technology, Chandigarh, India

Thromboembolism, a significant medical problem, is the formation of clot in blood vessel and this disorder has emerged as one main cause of mortality around the globe. Plasmin, the active fibrinolytic component of circulating system, dissolves clots through limited proteolysis. The 15.5 kDa bacterial protein staphylokinase (SAK) isolated from *Staphylococcus aureus*, a non-enzymatic, forms a stochiometric protein-protein complex with plasmin(ogen). Cleavage of the peptide bond Lys10-Lys11 of SAK in this complex is the trigger for the formation of an enzyme species capable of activating plasminogen. The SAK-plasmin(ogen) complex shows a high degree of specificity for cleavage of blood clots with reduced side effects as the SAK-plasmin(ogen) complex is strongly inhibited by α_2-antiplasmin circulating in the blood. Owing to the market potential, we aimed for development of a large-scale process for production of staphylokinase at reduced cost.

The application of recombinant DNA technology to large-scale processes has enabled the production of therapeutic proteins in quantities that might otherwise have been difficult, if not impossible, to obtain from natural sources. The extensive employment in recombinant protein production and vast knowledge base on genetics and physiology of gram-negative *Escherichia coli* along with its ease of handling, inexpensive growth requirements and the accumulation of the product to higher levels in the cell cytoplasm has made it to be called as the "microbial factory". Improvements in fermentation processes is known to be an important avenue for increasing yield of a recombinant protein in *E. coli* and the efficiency of recombinant protein expression in *Escherichia coli* depends on the parameters affecting both cell growth and protein expression. The preferred method for increasing the volumetric yield of heterologous recombinant proteins, which is proportional to both cell density (unit cell mass per litre) and specific yield (amount of product per unit cell mass), is the fed-batch strategy.

We formulated strategies for large-scale production of staphylokinase, using *E. coli* cells transformed with plasmid pET-9b containing the gene for staphylokinase. The protein was expressed intracellularly in soluble form. Fed-batch cultivations were carried out that involved initial cell cultivation in batch mode followed by subsequent addition of nutrients using different feeding strategies to build up high density of *E. coli* cells for induction. We strategized to increase volumetric yield and productivity of target protein without compromising on its specific yield, i.e., protein yield per unit cell mass (SAK expression at shake flask level with LB medium was taken as control), thereby reducing the load of intracellular contaminants per unit of SAK produced at fermentation stage that was perceived to reduce purification cost and result in low processing cost of the target protein. Optimization of feeding was carried out to increase cell concentrations and SAK yield. The effects of media constituents such as sources of carbon and organic nitrogen, C/N ratio, inducer concentration and process parameters such as dissolved oxygen concentration (DOC) on cell growth and SAK production were investigated. A medium mainly consisting of sugar (glucose), simple salts and some low cost organic nitrogen sources was developed to produce staphylokinase at a concentration of 3500 mg/L, highest reported up to date providing added advantages of processing lower quantity of broth per unit of staphylokinase. Staphylokinase so obtained exhibited biological properties e.g., clot dissolution. The process for production of Staphylokinase was scaled up to 15 L fermenter. The results in detail will be presented.

(211d) A Novel Microbial Consortium for Direct Biofuels Production From Cellulosic Biomass with in-Situ Product Removal

Edyta Szewczyk, Department of Biological and Agricultural Engineering, University of California, Davis, CA, Takao Kasuga, USDA-ARS; Department of Plant Pathology, University of California, Davis, CA and **Zhiliang (Julia) Fan**, Department of Biological and Agricultural Engineering, University of California, Davis, Davis, CA

The conventional process for cellulosic biofuels production involves six distinct steps: pretreatment, cellulase production, enzymatic hydrolysis, hexose fermentation, pentose fermentation, and product recovery. Herein, a novel configuration for cellulosic biofuels production- Integrated bioconversion and separation (IBS) , which combines five of the six unit operations (all except pretreatment) into one step, was proposed. The IBS configuration is achieved by a mixed culture composed of a hydrolysis enzyme producer such as a lignocellulolytic fungus and a fermentative microorganism such as a bacterium. The lignocellulolytic fungus produces hydrolysis enzymes under ATP-plentiful conditions in an aerobic environment. The hydrolysis enzymes produced by the fungus will hydrolyze the cellulose and hemicellulose to resulting hydrolysis products. The bacterium will ferment the hydrolysis products to biofuels. A novel strain engineering strategy was adopted engineer the fungus and the bacterium strains to divert most of the hydrolysis product to the fermentative microorganism for biofuel production while maintaining a robust fungus and bacterium co-culture. The air supplied to the system for aeration can strip out the volatile biofuel products so that they will not accumulate to high concentrations that are inhibitory to the fungus or to the fermentative microorganism. Therefore, IBS is able to achieve cellulase production, enzymatic hydrolysis, fermentation, and product separation in one single reactor. Our preliminary results showed that such a configuration is able to achieve the production of biofuels directly from cellulose at high yiled without the addition of cellulase.

(211e) Rhamnolipid Production by Long-Term Pseudomonas Aeruginosa Fermentation

Maysam Sodagari and Lu-Kwang Ju, Department of Chemical and Biomolecular Engineering, The University of Akron, Akron, OH

Biosurfactans are considered as potential alternative to synthetic surfactants, since they are biodegradable and biocompatible. Rhamnolipids are one of the most promising biosurfactants due to their wide applications in bioremediation, enhanced oil recovery, anti-fouling, wound healing, etc. However, at this point rhamnolipids are not competitive with synthetic surfactants economically. The highly foaming nature of microbial cultures, low productivity and yield are some of the challenges leading to both very low amounts of rhamnlipid production and high production cost. In this study, rhamnolipid production by *P. aeruginosa* fermentation under various levels of oxygen has been studied. The effects of two renewable carbon sources including glycerol and soybean oil on rhamnolipid production have been investigated as well. By using soybean oil, the ramnolipid productivity was doubled compared to the productivity obtained from glycerol as the carbon source. A stagnation of rhamnolipid production was observed after few hundred hours of fermentation process. Several possible hypotheses such as rhamnolipid inhibition, accumulation of other inhibitory compounds, nutrients depletion, and accumulation of fatty acids and/or glycerol were investigated. According to these findings, a long-term rhamnolipid fermentation with high productivity has been developed.

(211f) Cell-Free Enzymatic Hydrogen: Conversion of Biomass Sugars to Biocommodity

Joseph A. Rollin[1,2], Julia S. Martin del Campo[1,3] and Y.-H. Percival Zhang[1,2,4], (1)Biological Systems Engineering, Virginia Polytechnic Institute and State University (Virginia Tech), Blacksburg, VA, (2)Gate Fuels Inc., Blacksburg, VA, (3)Applied Physics Department, Cinvestav-Mérida, Mérida, Mexico, (4)Institute for Critical Technology and Applied Science, Virginia Polytechnic Institute and State University

Biohydrogen has tremendous potential to minimize the greenhouse gas emissions and other environmental issues associated with current hydrogen production methods, which use natural gas. The economic potential of this alternative is tremendous as well. By harnessing the power of thermostable enzymes assembled into the proper pathway, sugars derived from cellulosic biomass can be used to create biohydrogen at a cost competitive with current fossil production methods. Synthetic Pathway Biotransformation (SyPaB) is one such scheme. In this method, twelve heterologously expressed, thermostable enzymes are assembled in one pot, to form a synthetic enzymatic pathway that fully oxidizes glucose to hydrogen and carbon dioxide. Industrial examples exist of enzymes with weight-based total turnover numbers (kg product produced per kg enzyme) of more than 1,000,000. Once this stability is achieved for the SyPaB enzymes, cost of the biocatalyst becomes a very small fraction of the biocommodity cost. Between this and other advantages over whole-cell catalysis, such as no cell wall transport required and better ability to monitor reaction conditions, SyPaB conversion has the potential to produce hydrogen from biomass sugars at a faster rate and higher yield than any other method. Previous studies by our group have demonstrated the feasibility of SyPaB conversion of starch and soluble cellodextrins to hydrogen. Here, we report a broadened substrate range and increases in reaction rate and yield. Our yield is now the highest of any biohydrogen production method, and our reaction rate is now greater than dark fermentation. Substrates now include glucose, fructose, xylose, and insoluble cellulose, the last of which is made possible by combining SyPaB and with cellulose hydrolysis. Rate and yield improvements were achieved using a metabolic model informed by enzyme kinetics and metabolomics assays of the reaction mixture throughout the time course of the reaction. These important results highlight the excellent potential of cell-free conversions for future biocommodity production, and demonstrate the power of metabolomics and enzyme kinetic modeling to greatly improve the performance of these systems.

(214a) Effect of Engineered Transthyretin Mutants On Beta-Amyloid Aggregation and Toxicity

Dennis T. Yang, Patricia C. Cho and Regina M. Murphy, Chemical and Biological Engineering, University of Wisconsin-Madison, Madison, WI

Alzheimer's disease (AD) is the most prevalent age-associated neurodegenerative disease. Some of the characteristic features associated with AD are extracellular senile plaques, intraneuronal neurofibrillary tangles, and extensive neuronal cell death. The plaques consist mostly of the 4kDa peptide β-amyloid (Aβ). Aβ spontaneously self-assembles through a multistep process into soluble oligomers and fibrillar aggregates. Numerous *in vitro* and transgenic animal studies have causally linked Aβ aggregation to neurotoxicity. The current paradigm is that the soluble oligomeric intermediates are the most toxic form of Aβ.

Transthyretin (TTR) is a homotetrameric transport protein present in both blood and cerebrospinal fluid (CSF). Upregulation of TTR expression in Tg2576 mice has been linked to protection against the toxic effects of Aβ deposition. This and other observations suggest that TTR serves as a natural anti-AD compouond, and motivate our detailed study of TTR-Aβ association.

Through enzyme-linked immunoassays, we demonstrated that Aβ binds directly to TTR; more Aβ binds if it is aggregated, or if TTR is in its monomeric form. To further identify the interaction region of TTR, we used crosslinking plus tandem mass spectrometry, and peptide arrays. These data show that there are two domains on TTR, strand G and EF helix/loop, that interact with Aβ. Scanning alanine mutagenesis was used to further pinpoint residues important for binding. Alanine mutants L82A (on the EF loop) and L110A (on strand G) bound significantly less Aβ compared to wildtype TTR; thus we conclude that Aβ binding is mediated through these bulky hydrophobic residues. Two other alanine mutants, S85A and L17A, showed increased binding to Aβ. S85A was less stable as a tetramer than wildtype; increased Aβ binding is postulated to be a direct consequence of reduced quaternary stability. Enhancement of binding in L17A is postulated to arise from reduced steric restriction to the interior L110 site.

The next set of experiments are aimed at connecting TTR binding to TTR's effect on Aβ aggregation and toxicity. Surface plasmon resonance was used to determine the kinetics and equilibrium of Aβ binding to TTR (wildtype and mutants). Via dynamic light scattering, we characterized the effect of TTR on Aβ aggregate growth. S85A and L17A were the most effective, completely inhibiting growth of Aβ aggregates and thus showing a causal relationship between binding and inhibition of aggregation. In in vitro assays, TTR inhibited Aβ toxicity in a dose-dependent manner; efficacy of mutants was linked to the strength of binding. Importantly, inhibition was observed at substoichiometric TTR: Aβ ratios. Our data can be explained by postulating that TTR sequesters Aβ toxic oligomers. These results open new avenues for AD prevention by bolstering or supplementing TTR's natural protective action.

(214b) Modulation of Amyloid-Beta Aggregation and Cytotoxicity by Halogenated Small Molecules

H. Edward Wong, Jacob Irwin and **Inchan Kwon**, Department of Chemical Engineering, University of Virginia, Charlottesville, VA

Numerous small molecules have been identified as amyloid-beta aggregation modulators. However, considering that no amyloid-beta aggregation modulator has been approved by the Food and Drug Administration (FDA), there still remains a need to develop a new type of amyloid-beta aggregation modulators. In order to search for novel small molecule modulators with good biocompatibility, we screened various FDA-approved food coloring dyes and their close analogs. We recently reported that brilliant blue G (BBG), brilliant blue R (BBR), and red food dye erythrosine B (ERB) are novel modulators of amyloid-beta aggregation in vitro and amyloid-beta cytotoxicity in cell-based assays. Furthermore, we hypothesized that ERB can be a new type of amyloid-beta aggregation/cytotoxicity modulators due to its halogen atoms. A significant number of drugs and drug candidates are halogenated. Halogen groups are often inserted during hit-to-lead or lead-to-drug conversions for several reasons, including enhanced antagonistic/agnostic effects due to increased volume and improved oral absorption/BBB permeability. However, halogenated amyloid-beta aggregation modulators were rarely reported. Comparative studies of ERB and its close structural analogs revealed that halogen atoms attached to either xanthene group or benzoate group play an important role in modulation of amyloid-beta aggregation and cytotoxicity. Different amyloid-beta aggregation modulatory mechanisms of ERB analogs were also compared. These findings would enhance our understanding of amyloid-beta aggregation mechanisms and facilitate discovery of lead compounds for future therapeutic development for treatment of Alzheimer's disease.

(214c) Conformational Differences Between Beta-Amyloid Oligomers of Similar Size and Dissimilar Toxicity

Ali Reza A. Ladiwala and Peter M. Tessier, Dept of Chemical & Biological Engineering, Rensselaer Polytechnic Institute, Troy, NY

Several protein conformational disorders (Parkinson's and prion diseases) are linked to aberrant folding of proteins into prefibrillar oligomers and amyloid fibrils. Although prefibrillar oligomers are more toxic than their fibrillar counterparts, it is difficult to decouple the origin of their dissimilar toxicity because oligomers and fibrils differ both in terms of structure and size. We have discovered that the 42-residue beta-amyloid peptide associated with Alzheimer's disease spontaneously assembles into two prefibrillar oligomers with similar size and dissimilar toxicity (Ladiwala et al., *JBC*, 2012). In this presentation, we will discuss our characterization of beta-amyloid oligomers using conventional and novel site-specific structural methods. We find that hydrophobic regions within beta-amyloid oligomers are more solvent exposed in the more toxic oligomer. Moreover, the more toxic oligomer disrupts lipid bilayers, while the less toxic oligomer is inactive. Our results suggest that the ability of beta-amyloid oligomers to interact with and disrupt cellular membranes is linked to the degree of solvent exposure of their hydrophobic residues.

(214d) Inhibition of Alzheimer's-Associated Aβ Aggregation by Gold Nanoparticles

Kelly A. Moore[1], Deborah Soto-Ortega[1], Mihyun Lim[2], Kayla Pate[3], Kaliah Jackson[4], Sam Lohse[5], Rahina Mahtab[4], Catherine Murphy[5] and Melissa Moss[1,3], (1)Biomedical Engineering Program, University of South Carolina, Columbia, SC, (2)Department of Biological Sciences, University of South Carolina, Columbia, SC, (3)Chemical Engineering, University of South Carolina, Columbia, SC, (4)Department of Chemistry, SC State University, Orangeburg, SC, (5)Department of Chemistry, University of Illinois at Urbana-Champaign, Champaign, IL

Introduction: Alzheimer's disease (AD) is currently the most common type of dementia and the 6th leading cause of death in the United States. One pathological hallmark of AD is amyloid plaques, which deposit around the neurons in the brain. These plaques are composed primarily of amyloid-beta protein (Aβ), which self-assembles to create aggregated structures in a process that is hypothesized to be closely tied to disease progression. During self-assembly, monomeric Aβ forms nuclei, which progress to form soluble aggregates and ultimately insoluble fibrils. Inhibition of Aβ aggregation is one therapeutic strategy for AD.

One potential therapeutic agent is gold nanoparticles (NP). Nanoparticles have emerged recently as effective therapeutics and diagnostic tools with applications in cancer, photo-thermal ablation, and enhanced medical imaging. In particular, gold NPs are highly attractive therapeutic candidates due to their stability, low toxicity, and ease of synthesis. In this study, we have examined gold nanospheres with varying electrostatic surface properties and particle diameters for their ability to attenuate Aβ aggregation.

Methods and Materials: 8nm, 18nm, or 40nm gold NPs were synthesized with citrate, cetrimonium bromide (CTAB), poly acrylic acid (PAA), and poly allylamine hydrochloride (PAH) surface properties. All NPs were tested for toxicity on SH-SY5Y cells using an XTT assay. NPs were further evaluated in a monomer aggregation assay. Briefly, 40μM Aβ monomer was agitated alone or in the presence of 200pM-20pM NPs. Reactions were monitored using thioflavin T (ThT), a fluorescent dye that specifically recognizes β-sheet structures of amyloid aggregates. Using this technique, the aggregation process, which includes a lag time, a period of growth, and an equilibrium plateau, can be observed. Reactions were periodically diluted into 10μM ThT and fluorescence was assessed. The effect of NP type on extension of the lag time and reduction of the equilibrium plateau was evaluated using a one-way analysis of variance with Dunnett's post test. $p<0.5$ was considered significant. After the completion of each reaction, samples were gridded for transmission electron microscopy (TEM) analysis.

Results and Discussion: SH-SY5Y cells treated with 100pM or 200pM CTAB NPs exhibited significant toxicity; however citrate, PAA and PAH NPs were non toxic. None of the NPs tested elicited any effect on monomer nucleation, which is evidenced by no extension of the lag time. 200pM citrate NPs were observed to have a small reduction of 19±8% on the equilibrium plateau, which is illustrative of their ability to attenuate the total quantity of amyloid material formed during monomer aggregation. PAH NPs had a stronger effect on abrogating aggregation with a 59±9% and 45±4% reduction of the equilibrium plateau at 200pM and 20pM, respectively. The greatest inhibition of monomer aggregation was observed in the presence of PAA NPs. At concentrations as low as 20pM, 96±2% inhibition was noted. This is equivalent to a substoichiometric ratio of NPs to $A\beta_{1-40}$ of 1:2,000,000. Inhibition was also monitored using TEM. Results from TEM images paralleled those of the ThT assay. In particular, no aggregates were observed in PAA reactions. To further study the positive inhibitory capabilities of PAA NPs, 8nm and 40nm particles were evaluated to determine whether NP size effects aggregation. 8nm PAA NPs proved to be equally effective at inhibiting monomer aggregation, with a 95±3% reduction in equilibrium plateau at 20pM. However, when particle diameter was increased to 40nm inhibitory capabilities decreased.

Conclusion: These results demonstrate that 8nm and 18nm PAA NPs are effective inhibitors of Aβ aggregation at concentrations as low as 20pM. It has also been evidenced that as particle diameter increases to 40nm, effectiveness decreases. Together, these results demonstrate that gold nanoparticles serve as effective inhibitors of Aβ self-assembly and modify aggregation at substoichiometric concentrations. Furthermore, inhibition of Aβ aggregation by gold nanoparticles is dependent upon the nanoparticle coating and particle diameter. Determining the mechanisms by which these various nanoparticles inhibit Aβ self-assembly could enable the development of an effective therapeutic for AD.

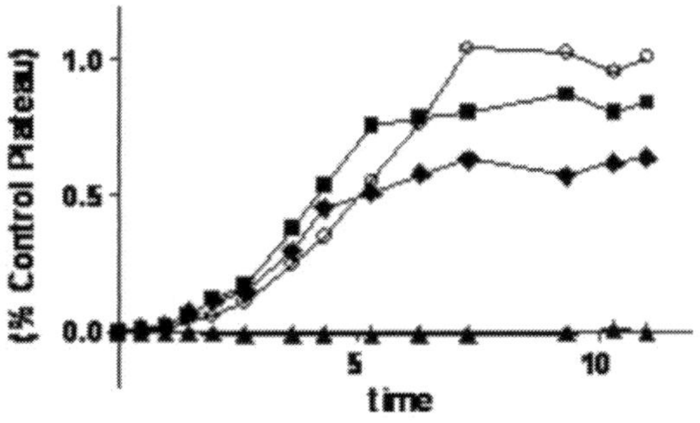

Figure 1: Effect of 18nm NPs with varying surface chemistries on $A\beta_{1-40}$ monomer aggregation. $A\beta_{1-40}$ monomer diluted to 40μM was incubated alone (control,○), or in the presence of 200pM citrate NPs (■), 200pM PAA NPs (▲), or 200pM PAH NPs (◆). Results represent three independent experiments.

Figure SEQ Figure * ARABIC 1: Effect of 18nm NPs with varying surface chemistries on $A\beta_{1-40}$ monomer aggregation. $A\beta_{1-40}$ monomer diluted to 40μM was incubated alone (control,Â), or in the presence of 200pM citrate NPs (Â¢), 200pM PAA NPs (p), or 200pM PAH NPs (z). Results represent three independent experiments.

(214e) Effect of Co-Solutes On the Aggregation of the Model Amyloid Protein Insulin

Brian Murray[1], **Mirco Sorci**[1], Zhongli Zheng[2], Y. Elaine Zhu[2] and Georges Belfort[1], (1)Rensselaer Polytechnic Institute, Troy, NY, (2)University of Notre Dame, Notre Dame, IN

In vitro experiments have been widely used to characterize the misfolding/unfolding pathway characteristic of amylodogenic proteins. Conversion from natively folded amyloidogenic proteins to oligomers via nucleation is the accepted path to fibril formation. In order to simulate the crowded environment of cells (protein concentration ~ 200 mg/ml), we investigated the aggregation kinetics of insulin, a well-established amyloid model protein, in the presence of different co-solutes including: macromolecules, such as polyvinylpyrrolidone (PVP), MW 40 kDa, or micromolecules such as glycerol (Gly), MW 92 Da. Using absorbance spectroscopy to measure turbidity of the solution, the rate of aggregation was measured to determine the effects co-solutes on insulin aggregation kinetics. AFM imaging was used to investigate the morphology of the aggregates and rheometry to measure the solution viscosity for different concentrations of co-solutes. The results show that (a) addition of co-solutes substantially delayed formation of fibrils, and (b) the delay was more pronounced for PVP than Gly. The results were confirmed using fluorescence correlation spectroscopy. Either viscosity and reduced solute diffusion or specific interaction effects of the co-solutes on the reaction rates of oligomers prior to the formation of fibrils are used to explain these results.

(214f) Hofmeister Effects and Amyloid Structure of Sup35

Jonathan Rubin[1], Sven H. Behrens[1], Yury O. Chernoff[2] and Andreas S. Bommarius[3], (1)Chemical & Biomolecular Engineering, Georgia Institute of Technology, Atlanta, GA, (2)Biology, Georgia Institute of Technology, Atlanta, GA, (3)School of Chemical and Biomolecular Engineering, Georgia Institute of Technology, Atlanta, GA

Prion proteins are capable of converting from their soluble, biologically-functional forms into highly ordered, fibrous aggregates, called amyloids. This conversion is associated with certain neurodegenerative conditions, such as Alzheimer's disease. Protein misfolding is strongly influenced by ion-specific solvent effects, known as Hofmeister effects. Using a yeast prion protein (Sup35), we tested the influence of a host of sodium salts on aggregation kinetics and investigated the aggregates that were formed. Aggregation kinetics was starkly different between the two main groups of salts (chaotropes and kosmotropes). The differences in aggregation kinetics profoundly effected the structures of the resulting aggregates. These structural differences were investigated by studying the amyloids' physical stability, thermostability, fiber length distributions, infectivity and TEM images. The fastest forming aggregates were found to be the most fragment prone, least thermostable, shortest, most infectious, and least orders when imaged. The opposite was true of slower forming aggregates. These findings are in line with previous works on temperature effects on Sup35 and what one might expect for a non-specific colloidal system. This work shows the profound effect of solvents and ion-specific/hydration effects on aggregation and disease.

(214g) Modulation of Amyloid Aggregation by Engineering the Sequence Connecting Beta-Strand Forming Domains

Yang Hu, Michael Hernandez, Ahra Ko and **Jin Ryoun Kim**, Chemical and Biological Engineering, Polytechnic Institute of New York University, Brooklyn, NY

Aggregation of amyloidogenic proteins, implicated in the pathology of various neurodegenerative diseases, such Alzheimer's disease and Parkinson's disease, produces fibrils where layers of parallel, in-register beta sheet-loop-beta sheets are formed. The effects of sequence variation in the loop forming region (referred to as the linker region) on amyloid aggregation have yet to be systematically studied. In the study presented here, we created and characterized variants of alpha synuclein (alphaS) and beta amyloid (Abeta) containing mutations in linker regions.

Our results with alphaS variants indicate that while physicochemical properties of the linker region evaluated based on an intrinsic property of a single amino acid still play a significant role in aggregation, additional factors may also determine aggregation of alphaS linker mutants. Our analyses suggest that these factors may include a pairwise potential for parallel in-register beta sheet formation. A linker variant displaying significantly reduced self-aggregation interfered with alphaS aggregation by inhibiting conversion of alphaS soluble species to alphaS insoluble fibrils.

Our results with Abeta variants indicate that a propensity to form oligomers may be changed by local sequence variation in the Abeta linker region without mutating the charged residues. Strikingly, one Abeta linker variant rapidly formed protofibrillar oligomers, which did not convert to fibrillar aggregates in contrast to Abeta aggregating to fibrils under similar incubation conditions. Moreover, our results suggest that molecular forces critical in oligomerization and fibrillization may differ at least for those involved in the linker region. When co-incubated with Abeta, some Abeta linker variants were found to induce accumulation of Abeta oligomers.

Taken together, our results suggest that engineering of the linker region may represent a novel approach to control amyloid aggregation and create amyloid aggregation modulators.

(214h) Elucidating the Role of N-Terminal Huntingtin Fragment in Polyglutamine Aggregation

Diwakar Shukla, Chemical Engineering, Massachusetts Institute of Technology, Cambridge, MA; Bioengineering, Stanford University, Stanford , CA and Vijay Pande, Chemistry, Stanford University, Stanford, CA

Huntington's disease is cause by the aggregation of poly-glutamine (35-40 residues) repeats forming toxic fibrils in the neuronal cells leading to the cell death. These repeats are attached at the N-terminal of the Huntingtin protein. It has been shown that the in the presence of the 17 residues N-terminal fragment, N17 of the Huntingtin protein, the poly-glutamine repeats aggregate faster. Several mechanisms have been proposed to explain the role of N17 fragment including poly-glutamine induced unfolding of N17 (Williamson et. al., JMB, 2010) or a conformational change in N17 to expose hydrophobic residues (Dlugosz et. al., JPCB, 2010) leading to the N17 aggregation. The aggregation of N17 fragment brings the poly-glutamine repeats closer and induces the fibril formation. In this talk, we report results from an extensive molecular dynamics simulations of the N17 fragment of Huntingtin protein along with pathogenic (35 residues) and non-pathogenic poly-glutamine repeats (15 residues) to understand the influence of polyglutamine repeats on the conformational preferences of the N17 fragment and viceversa.

(216a) Biobutanol Production Using Clostridium Beijerinckii From Ultrafiltered Corn Fiber Hydrolysate and Recycled Spent Broth

Michelle C. Almendrala, Philippine Sugar Research Institute Foundation, Makati City, Philippines and Shang-Tian Yang, William G. Lowrie Department of Chemical and Biomolecular Engineering, Ohio State University, Columbus, OH

In these studies, a mutant strain of Clostridium beijerinckii was used to produce butanol (acetone–butanol–ethanol, or ABE) from corn fiber in an immobilized fibrous-bed bioreactor. Inhibitors present in fermentation of renewable lignocelluloses requires complicated pretreatments which includes repeated adsorption and overliming of the hydrolysate. Thus, ultrafiltration, which is a cost-effective and environmentally benign process, was considered in the purification of the acid hydrolysate prior to fermentation. Sterilization of the medium was carried out using microfiltration to prevent degradation of sugars and formation of more products that may inhibit microbial growth. It has been demonstrated that the use of 80%(v/v) corn fiber hydrolysate as fermentation medium can produce ~3 to 4 g biobutanol per liter of fermentation broth with productivity of ~0.06 g/L-h. The highest titer of biobutanol was achieved within 72 hours of fermentation. The ratio of acetone-butanol-ethanol (ABE) mixture remained at 3:6:1 using the acid corn fiber hydrolysate, however, the concentration of acetic acid and butyric acid produced was more than two times higher than ABE concentration. Results showed that higher titer of biobutanol up to 16 g/L and productivity of ~0.11 g/L-h can be achieved when fermentation time was prolonged for ~6 days. Another significant result of the studies showed that reuse of fermentation broth was possible for the first recycle. Second recycle of spent broth produced very low biobutanol titer due to the accumulated inhibitors in the broth which cannot be reduced by ultrafiltration. Therefore, further purification of spent broth is needed to be able to reuse the broth in the next fermentation batch. It is recommended that further studies must be conducted for prolonged fermentation of corn fiber hydrolysate. The utilization of sugars in the corn fiber hydrolysate must be studied during the prolonged fermentation as well as the factors affecting fermentation process.

(216b) Re-Commercialization of the Butanol Fermentation

Edward Green and Edward T. Davies, Green Biologics, Abingdon, United Kingdom

Biobutanol is an important commodity chemical with a global market worth approximately $5 billion. Biobutanol is also a superior next generation biofuel with a higher energy density and better performance than ethanol. It can also be made from more sustainable feedstocks than bio-diesel. Biobutanol has the potential to substitute for both ethanol and biodiesel in the biofuel market estimated to be worth $247 billion by 2020.

The butanol fermentation process was first developed in the UK in 1912 and commercial production quickly spread around the globe during the first and second world wars; first to produce acetone for ammunitions and then later to produce butanol for paint lacquers. The fermentation process fell out of favour in the US and Europe in the 1950's when it could no longer compete with synthetic production from oil. Some fermentation remained in China, Russia and South Africa until the early 1980's.

China leads the effort to re-commercialize the butanol fermentation process. Over $200 million has been invested to install over 0.3 million t pa of solvent capacity. We report on this activity and also on some of the key technical advances required to drive down production cost. In particular, we will focus on the utilisation of low cost and sustainable cellulosic feedstocks and on our re-commercialization activities.

(216c) Construction of a Xylose and Arabinose Utilizing Zymomonas Mobilis Strain for Commercial Use in Fuel Ethanol Production

Brian G. Lefebvre[1], George C. Fox[1], William D. Hitz[1], Maria C. Leana[1], Min Qi[2], Dongmei Qian[1], Sarah E. Rush[1], Luan Tao[1], Paul V. Viitanen[1] and Jianjun Yang[2], (1)Industrial Biosciences, DuPont, Wilmington, DE, (2)Central Research and Development, DuPont, Wilmington, DE

Zymomonas mobilis is an obligate and rapid producer of ethanol but its normal substrate range is limited to glucose, fructose and sucrose. The addition of exogenous genes encoding enzymes to allow the use of xylose and arabinose to *Z. mobilis* was shown to be effective over 10 years ago so the main tasks in strain development have been to reach rates of 5 carbon sugar metabolism that are compatible with those required to be economically favorable in fuel ethanol production and to do that in strains that are compatible with the nutrient and inhibitor set that is present in the cellulosic substrate to ethanol process. The combination of iterative expression level changes in genes in the 5 carbon pathway, knock out of toxin producing genes, adaptation to growth on the 5 carbon substrates, and adaptation to growth in cellulosic process hydrolysate has produced several viable strains. Strains and process conditions that produce desired volumetric productivity have been created and tested at the pilot plant scale.

NOVEL ENZYMES BOOST PERFORMANCE IN HIGH PH, HIGH TEMPERATURE HYDRAULIC FRACTURING

C. David Armstrong and Prasad D.K. Dhulipala

Research and Development, Applied Liquid Technology,
Baker Hughes Incorporated, Tomball, TX 77375, USA

Introduction

Hydraulic fracturing is used to create subterranean fractures that extend from the borehole into rock formations in order to increase the rate at which fluids can be produced by the formation. Generally, a high-viscosity fracturing fluid is pumped into the well at sufficient pressure to fracture the formation. In order to maintain the increased formation exposure, a solid proppant is added to the fracturing fluid which is carried into the fracture by the high pressure applied to the fluid. Once the proppant is in place, chemical breakers are used to reduce the fluid's viscosity. This allows the proppant to settle into the fracture and thereby increase the formation's exposure to the well. Breakers work by reducing the molecular weight of the polymers, thus "breaking" the polymer and allowing the fracture to become a high permeability conduit for the production of fluids and gas to the well-bore.

More than 65% of conventional fracturing fluids are made of guar gum (galactomannans) or guar gum derivatives such as hydroxypropyl guar (HPG), carboxymethyl guar (CMG) or carboxymethylhydroxypropyl guar (CMHPG). These polymers can be crosslinked together to increase their viscosities and increase their proppant transport capabilities.

Chemical oxidizers and enzymes are most commonly used as breakers. The oxidizer produces a radical which then degrades the polymer. This reaction is limited by the fact that oxidizers are stoichiometric—they will attack not only the polymer but any molecule that is prone to oxidation. Enzymes, on the other hand, are catalytic and substrate specific and will catalyze the hydrolysis of specific bonds on the polymer. An enzyme will degrade many polymer bonds in the course of its useful lifetime. Unfortunately, enzymes typically operate under narrow pH and temperature ranges.

Enzymes isolated from extremophilic sources, however, have shown great promise as breakers and can be used under most fracturing conditions. Additionally, extremophilic enzymes are generally more robust than their mesophilic counterparts. They are much more able to withstand the various conditions of subterranean formations and the surfactant-based additives often used in fracturing fluid formulations.

This work details the use of several novel enzymes used in hydraulic fracturing. Since the genetic sequences for these enzymes are from thermo-tolerant and extremeophilic sources, the translated enzymes are more robust than conventional products. They are capable of degrading hydraulic fracturing biopolymers at elevated pH ranges (7-12), elevated temperatures (≥ 80°C), and do not generally denature in the presence of common fracturing fluid additives.

Experimental

Enzyme production:

Gene coding for the enzymes discussed here were isolated from thermo-tolerant and/or extremophilic sources and codon-optomized for expression in *E. coli*[1-4]. *E. coli* was transformed with a plasmid containing the gene for the enzyme breaker and incubated in LB-Miller for 18 hours at 100°F and 200 rpm agitation. Overexpression of enzyme was confirmed by SDS-PAGE and a cell-free lysate used as the enzyme solution in all tests. Rheology was used to determine enzyme activity and functionality against field-specific hydraulic fracturing fluid formulations.

Enzyme Testing:

Rheology: The reduction in viscosity of crosslinked guar polymer by the enzyme was measured across a range of temperatures (40 to 150°F using 17, 25 and 30 ppt crosslinked guar at pH 10.5. Rheology measurements were carried out in a Chandler model 5550 rheometer at a shear rate of 100 sec^{-1} and a pressure of 500 psi.

Effect of Allosteric Effectors: The effects of different anions and cations on the enzyme were determined by rheology. For the purpose of this paper, the mannanohydrolase enzyme was tested with and without a small concentration of allosteric effector at 150°F and pH 9.5. The effector did not change the pH of the fluid and did not, by itself, break the viscosity of the crosslinked fluid. Results are shown in Figure 3.

Assay of Oxidase-Type Enzyme: A solution of 6.25 ug/mL of oxidase-type enzyme was prepared in 100 mL of 25 ppt crosslinked guar fracturing fluid. Samples were prepared with either no glucose or with 0.3 mM glucose. Fluid viscosity was measured at 140°F on a Chandler rheometer. Results are displayed in Figure 4.

Results and Discussion

High-Temperature, High-pH Enzyme Breaker:

A mannanohydrolase enzyme from a thermophilic source was tested for its ability to break the viscosity of a crosslinked guar polymer at 150°F and pH 9.5 (Figure 1). The enzyme functioned remarkably well under these conditions. Further testing was performed with this enzyme under temperatures ranging from 75°F to 250°F and pH ranges from 9.5-10.5 (data not shown). Under all conditions, the enzyme performed as expected with lower activity at 75°F and approaching its maximum at 150°F. The activity of the enzyme would begin to decline as the temperature approached 250°F. However, lowering the enzyme dilution to 1:50 or increasing the loading in the fluid formulation would provide the necessary activity for breaking the viscosity of the guar polymer. In contrast to previous enzyme breaker packages, this new enzyme does not need an associated pH modifier nor does it need the assistance of an oxidative breaker to perform at its peak limits.

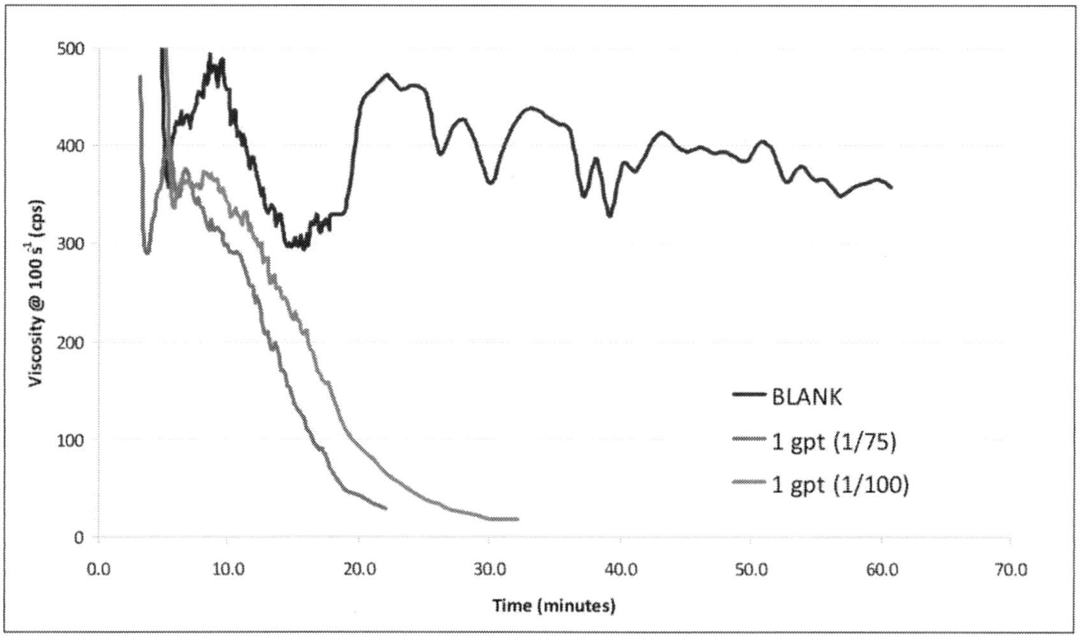

Figure 1. Activity of the enzyme breaker at 150°F and pH 9.5. The enzyme effectively breaks the viscosity of the crosslinked 17 ppt guar polymer. The figure shows the activity of two dilutions of enzyme stock, 1:75 and 1:100.

High-pH, Low-Temperature Enzyme Breaker:

A β-mannanase enzyme from an alkaliphilic source was examined for its ability to break the viscosity of guar at elevated pH ranges that are often found in fracturing conditions (Figure 2). The enzyme functioned amazingly well at pH 11 and 12. At pH 13, the activity of the enzyme was noticeably lessened. However, after 18 hours, the viscosity of the fluid had been greatly reduced signifying that the enzyme retained at least some activity at this point. The upper temperature limit for this enzyme was found to be approximately 150°F (data not shown). Increasing the enzyme concentration in the fluid formulation would extend the activity past this temperature at the cost of severely reducing the initial viscosity of the formulation, possibly resulting in a screen-out. However, the enzyme functions well at lower temperatures and elevated pH ranges.

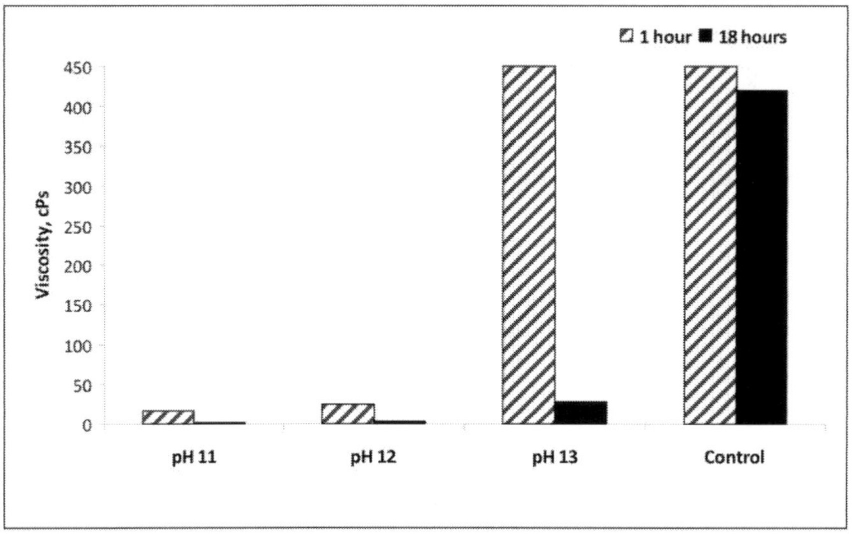

Figure 2. β-mannanase activity at pH 11, 12 and 13. The enzyme was incubated in an 18 ppt crosslinked guar fracturing fluid for 1 hour (dashed lines) or 18 hours (solid lines). The enzyme still retained activity at pH 13. The pH of the control sample was 11. Samples were incubated at room temperature.

Allosteric Effectors Improve Catalytic Rates:

It has been found that certain anions and cations can improve the reaction rates and/or stabilities of the β-mannanases and mannanohydrolases used in this study[5]. As seen in Figure 3, a small amount of allosteric effector greatly improves the catalytic activity of the enzyme. The identity of the effector will depend on which enzyme is used, with different effectors influencing the enzyme in different ways

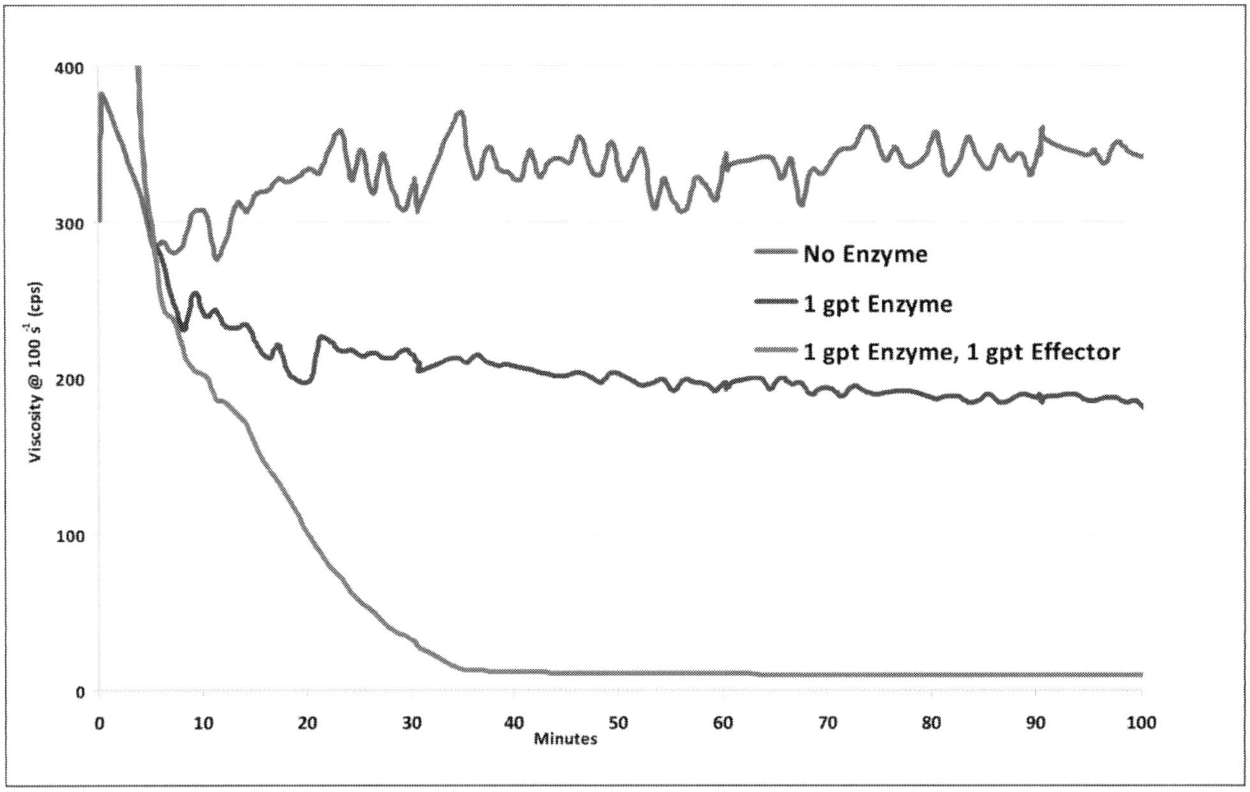

Figure 3. Ability of an allosteric effector to improve the catalytic ability of the enzyme against hydraulic fracturing fluid. The enzyme and effector was tested against 17 pound crosslinked guar fluid at pH 9.5 and 150 °F.

Universal Enzyme Breakers:

An oxidase-type enzyme was tested for its ability to degrade crosslinked guar. On its own, the enzyme had no activity against the guar polymer. However, when the reaction was "seeded" with a monosaccharide (glucose, mannose or galactose) the reaction was initiated and the enzyme effectively "broke" the viscosity of the crosslinked polymer (Figure 4). The enzyme's ability to reduce the fracturing fluid's viscosity occurs by two pathways. The first is production of a carboxylic acid which lowers the fluid pH. This results in a reduction of the efficacy of the crosslinking reaction. Second, the enzyme produces the oxidizer, H_2O_2, which is extremely effective in degrading the polymer's molecular weight. Because the oxidizer is produced *in situ*, the hazard to operators and the environment at the surface of the well is greatly reduced, if not eliminated.

Once the enzyme reaction is seeded with a monosaccharide, the guar is broken down by the oxidizer, which releases additional monosaccharides enabling the process to continue without further seeding by the operator.

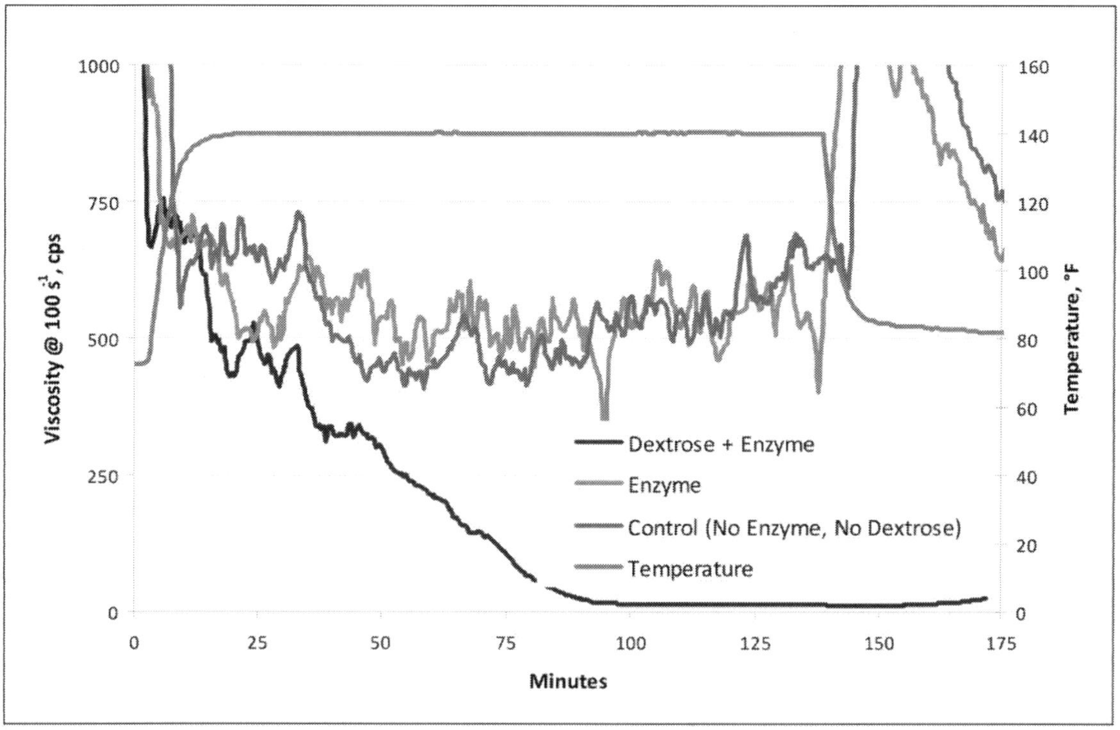

Figure 4. Rheology profile of the oxidase-type enzyme. When the enzyme is "seeded" with a monosaccharide, the viscosity of the polymer is broken.

In summary, enzymes are effective in hydraulic fracturing applications. The wide variety of enzymes available and the ability to modify their activity with different effectors makes it possible to customize a treatment fluid for a particular situation. They are also a more environmentally friendly alternative to traditional oxidative breaker chemicals and provide a safer working environment for the field operator.

Acknowledgments

The authors would like to thank Baker Hughes management for the opportunity to write and present this research.

References

1. Ma, Y., *et al.*: *Extremophiles*, 2004, 8, 447-454.
2. Armstrong, C.D., Patent No. US 8,096,360 B2, 2012.
3. Armstrong, C.D., Patent No. US 8,058,212.
4. Armstrong, C.D. Patent Application US 2012/0031618.
5. Armstrong, C.D. et al. Patent Application P210-1840-US.

(216e) How to Improve Biopharmaceutical Production Efficiency Using Integrated Bioprocessing?

Xiaoguang Liu, The University of Alabama, Tuscaloosa, AL

In the last decade of mammalian bioprocess development with novel medium and nutrient feed design, antibody titers in excess of 1 g/L have been achieved. However, it is still a challenge to consistently and rapidly construct a stable biopharmaceutical producing cell line and develop an efficient production bioprocess capable of supporting both high therapeutic protein yield and acceptable antibody glycosylation profile. To overcome these bottlenecks, it is highly desired to develop a novel bioprocessing strategy using advanced technologies, for example, integrating host cell engineering, systems biology and bioreactor process optimization to improve the bioprocess efficiency of therapeutic proteins. CHOnomics technology, including genomics, metabolomics and transcriptomics, can be applied to globally understand CHO metabolic profiling and genome transcription, and to identify the key metabolic activity and regulation genes or elements. The potential application of multiple cell engineering strategies in engineering CHO cell to achieve novel phenotypes will be discussed. The integrated bioprocessing can efficiently improve the biopharmaceuticals production by combining metabolite nutrient development, high producing cell line construction and bioreactor process optimization. Using the integrated bioprocessing technology, it is expected to achieve stable antibody production from gene with titer of 1-3g/L within 6 months. Finally, the development of a robust and scalable cell culture production process is critical in biopharmaceutical manufacturing to maintain both high protein productivity and high protein quality.

(216f) Technologies and Approaches for Reliable Scale-up of Complex Secondary Metabolite Fermentations

Chris Stowers, Karan Bansal and Nigel Mouncey, Bioprocess R&D, Dow AgroSciences, Indianapolis, IN

Reliable scaled-down fermentation platforms that are predictive of large-scale performance and simultaneously detect incremental improvements in yield can be challenging for mycelial (viscous, shear-thinning) fermentations, especially those utilizing industrial grade medium. Inherent differences across scales often have a physiological impact and make linear scaling impractical. Small scale micro titer plates and shake flasks used for mutant selection are often subject to mass transfer limitations that can mask higher yielding strains or give false positives. Amongst these challenges, there is a consistent trend across the Industry to increase experimental throughput and accelerate the scale-up process. Scale-down models must be predictive, but also should have low variance to maximize throughput and statistical power. Model variance and scale-up criteria must be well defined in order to develop robust experimental designs that maximize predictability and reduce scale-up time. This presentation will describe not only methods that have been employed at Dow AgroSciences to detect and overcome scale-up limitations but will also examine experimental considerations to improve statistical significance and experimental throughput.

(230a) Engineering Single Wall Carbon Nanotubes for Sub-Cellular Delivery

Brian D. Holt[1], Patrick D. Boyer[2], Kris Noel Dahl[1,2] and **Mohammad F. Islam**[3], (1)Biomedical Engineering, Carnegie Mellon University, Pittsburgh, PA, (2)Chemical Engineering, Carnegie Mellon University, Pittsburgh, PA, (3)Materials Science & Engineering, Carnegie Mellon University, Pittsburgh, PA

As part of nanomaterial applications in biology and medicine, we are interested in developing single wall carbon nanotubes (SWCNTs) as intracellular biomaterials. Our research in this area has been directed to characterizing, developing, and utilizing SWCNT functionalization to control SWCNT sub-cellular localization and probe fundamental cellular processes utilizing SWCNTs. To produce biologically relevant solutions of individualized SWCNTs and overcome aggregation of SWCNTs, we use a surfactant dispersant. Specifically, we are focused on developing bio-inert, biological, and bioactive SWCNTs dispersions to understand and control how the SWCNT dispersions enter cells, where they go and what they do once inside cells, and do they leave cells.

The first step in targeted cell delivery is to determine how SWCNTs get into cells. To determine if SWCNTs dispersed in Pluronic F127 (PF127) could physically penetrate a cell membrane, we performed complementary studies with *in vitro* synthetic membranes, plasma membrane vesicles and live cell experiments. Our results demonstrated that, while P-F127 dispersed SWCNTs are membrane active and associate with cell membranes, they lack sufficient energy to physically penetrate a cell membrane bilayer. Further, our experiments suggested that SWCNTs are directly present in endosomes. Therefore, we experimentally validated that SWCNTs enter cells *via* endocytosis and not membrane penetration.

Once inside the cell, we were interested in the sub-cellular localization and cellular effects of SWCNTs. We showed that highly purified, length-selected, and SWCNTs well-dispersed with PF127 (bio-inert) are not acutely cytotoxic but reduce proliferation in a dose-dependent manner. Due to the similar mechanical and anisotropic properties of SWCNTs and F-actin, we investigated the interaction and effects of SWCNTs on F-actin. These SWCNTs reorganized F-actin structures, leading to more F-actin intensity in the apical cellular regions. Fluorescence lifetime imaging microscopy further suggested that SWCNTs interact with F-actin structures both in the cells and in purified actin filaments reconstituted *in vitro*. Functionally, SWCNTs reduce forces that can be generated by NIH-3T3.

Endocytosed SWCNTs coated with PF127 were able to enter cellular compartments and interact with actin, likely from their membrane activity which could destabilize endosomes. However, dispersion with bovine serum albumin (BSA) caused markedly different cellular effects. SWCNTs-BSA were not acutely cytotoxic and did not significantly reduce proliferation. Confocal Raman imaging co-registered to phase contrast showed different patterns of sub-cellular localization for these dispersions within perinuclear region of the cell. We also observed a significant increase in cellular uptake of SWCNTs-BSA into cells with 10's of millions of SWCNTs entering human mesenchymal stem (hMSC) cells and HeLa and NIH-3T3 cells. Surprisingly, steady-state uptake per cell occurs in less than 1 minute, and there is a concentration threshold between 1 and 30 mg/mL as cell uptake machinery is saturated. Confocal Raman mapping of SWCNT radial breathing modes, fluorescence, and G-band demonstrated that SWCNTs primarily remain individually dispersed.

Finally, we determined that cells could recover from SWCNT exposure. Most cell lines exposed to SWCNTs-BSA were able to expel SWCNTs and recover cell phenotype, and after ~80 hr essentially no SWCNTs remained in cells. Conversely, cells exposed to SWCNTs-PF127 did not recover, maintaining substantial SWCNTs-PF127 throughout the cells, even after 80 hr. This differential ability to recover suggests different subcellular compartmentalization of SWCNTs as well as the possibility for cellular delivery of SWCNTs within the body.

In conclusion, we have dispersed SWCNTs – an important nanostructured material for biological applications – with different classes of biological dispersing agents to control uptake, sub-cellular processing, and recovery. Therefore, from these studies, we have shown that we can engineer and control SWCNT-cell interactions.

References

1. B. D. Holt, K. N. Dahl, and M. F. Islam, "Cells Uptake and Recover from Protein Stabilized Single Wall Carbon Nanotubes with Two Distinct Rates", *ACS Nano* **6**, 3481 (2012).
2. P. N. Yaron, B. D. Holt, P. A. Short, M. Lösche, M. F. Islam, and K. N. Dahl, "Single wall carbon nanotubes enter cells by endocytosis and not membrane penetration", *Journal of Nanobiotechnology* **9**, 45 (2011).
3. B. D. Holt, K. N. Dahl, and M. F. Islam, "Quantification of Uptake and Localization of Bovine Serum Albumin-Stabilized Single-Wall Carbon Nanotubes in Different Human Cell Types", *Small* **7**, 2348 (2011).
4. B. D. Holt, P. A. Short, A. D. Rape, Y. Wang, M. F. Islam, and K. N. Dahl, "Carbon Nanotube Reorganize Actin Structures in Cells and *ex vivo*", *ACS Nano* **4**, 4872 (2010).

(230b) Nanocapsules for Protein Delivery

Yunfeng Lu, Chemical and Biomolecular Engineering, University of California at Los Angeles, Los angeles, CA

An average cell contains thousands of proteins participating in its normal cellular functions. Most diseases, however, are somehow related to the malfunction of specific proteins. Protein therapy, endowed with more defined action mechanism than small molecule drugs and safer administration methods than gene therapy, is considered to be a treatment modality of choice for many disorders today and in the future. Advances in biotechnology have accelerated the economical, large-scale production of proteins, vaccines, and hormones, making them readily available for therapeutic applications in medical practices and clinical studies. Up to now, over 140 FDA-approved protein drugs have been placed on the market, and more are coming. Despite its fast growth in the recent years, protein therapy still faces critical challenges, such as poor *in-vitro* and *in-vivo* stability, fast degradation and clearance, and the incapability to delivery proteins across biological barriers. Developing efficient protein delivery strategy is still in high demand. Various factors affect the efficiency of protein delivery agents, including surface properties, size, and shape. To develop protein delivery agents for various therapeutic applications, we designed a general strategy based on protein nano-encapsulation, a novel protein-polymer conjugation technique achieved by in-situ polymerization on protein surface. This approach yielded protein nanocapsules with retained activity, enhanced stability and the ability to transport through biological barriers (such as cell membranes). This strategy is applicable over a variety of proteins, for a wide spectrum of therapeutic applications. Our in-vitro and in-vivo experiments demonstrated the potential of the nano-encapsulation strategy for the treatment of gout, phenylketonuria, alcohol intoxication, and other therapeutic applications.

(230c) Biomimetic Long-Circulating Nanoparticles for Combinatorial Drug Delivery

Liangfang Zhang, Nanoengineering Department, University of California, San Diego, La Jolla, CA

Efforts to extend nanoparticle residence time in vivo have inspired many strategies in particle surface modifications to bypass macrophage uptake and systemic clearance. Herein I report a top-down biomimetic approach in particle functionalization by coating biodegradable polymeric nanoparticles with natural red blood cell (RBC) membranes including both membrane lipids and associated membrane proteins for long-circulating cargo delivery. This approach aims to camouflage the nanoparticle surface with the erythrocyte exterior for long circulation while retaining the applicability of the polymeric core. In vivo results revealed superior pharmacokinetics and biodistribution by the RBC-mimicking nanoparticles as compared to control particles coated with the state-of-the-art synthetic stealth materials. I will also report several new strategies for nanoparticle-based combination chemotherapy by concurrently incorporating two different types of drugs into a single polymeric nanoparticle with ratiometric control over the loading of the two drugs. The cytotoxicity of these dual-drug carrying nanoparticles was compared with their cocktail mixtures of single-drug loaded nanoparticles and showed superior therapeutic effect.

(230d) Corking Carbon Nanotube Cups with Gold Nanoparticles for Drug Delivery Applications

Alexander Star, Department of Chemistry, University of Pittsburgh, Pittsburgh, PA

In this talk, I will describe an effective methodology of obtaining nitrogen-doped carbon nanotube cups (NCNCs), which have a hollow multiwall graphitic structure with one end sealed and the other open. We have previously shown that nitrogen doping is essential to determine the unique morphology and properties of NCNCs.[1,2] Other studies have demonstrated that nitrogen doped carbon nanotubes have better biocompatibility and mitigated cytotoxicity as compared to undoped carbon nanotubes.[3] These findings are important when considering NCNCs as potential carriers in drug delivery applications. Most recently, we have demonstrated[4] that by applying intense ultrasonication, stacked NCNC fibers, which were prepared by chemical vapor deposition (CVD) technique, can be greatly shortened to afford individual NCNCs with an efficient yield. Microscopic and spectroscopic characterization confirmed the cup-shaped hollow structure of isolated NCNCs and showed their robust chemical reactivity due to amine groups that are preferentially distributed at the open rims of the cups. We were further able to cork isolated nanocups with gold nanoparticles (GNPs) by utilizing the localized amine groups as reactive sites. These nanocup-nanoparticle hybrid structures may find future applications as nanoscale containers and drug delivery carriers.

1. Allen, B. L.; Kichambare, P. D.; Star, A. Synthesis, characterization, and manipulation of nitrogen-doped carbon nanotube cups. *ACS Nano* **2008**, *2*, 1914–1920.
2. Tang, Y.; Allen, B. L.; Kauffman, D. R.; Star, A. Electrocatalytic activity of nitrogen-doped carbon nanotube cups. *J. Am. Chem. Soc.* **2009**, *131*, 13200–13201.
3. Castle, A. B.; Gracia-Espino, E.; Nieto-Delgado, C.; Terrones, H.; Terrones, M.; Hussain, S. Hydroxyl-functionalized and N-doped multiwalled carbon nanotubes decorated with silver nanoparticles preserve cellular function. *ACS Nano* **2011**, *5*, 2458–2466.
4. Zhao, Y.; Tang, Y.; Chen, Y.; Star A. Corking carbon nanotube cups with gold nanoparticles. *Submitted.*

(230e) Low Frictional Mesoporous Titanium Dioxide Film by Geometrical Roughness-Induced Heterogeneous Nanostructure From Titanate

Rong An, Xiaohua Lu, Ximing Wu, Changsong Wang, Chang Liu and Shuangqin Fu, chemical engineering, Nanjing university of technology, Nanjing, China

It is of great importance to construct a low frictional heterogeneous titanium dioxide (TiO_2) surface by physical strategy. Native dense TiO_2 layers prefer to forming on bulk titanium (Ti) substrates. However, the poor frictional TiO_2 on Ti surface is not favorable to fluid flow, resulting in harmful effects including but not limited to thrombus, nonspecific adsorption of proteins and cells, osteoarthritis. Thus surface modification with chemical functional groups was utilized to provide a heterogeneous stable and bacteria-resistant TiO_2 surface. But the inducing poor biocompatibility is of great concern and the functional groups are susceptible to oxidation, hydrolysis or thermal degradation.

An approach to low surface friction by constructing geometrical roughness-induced nano-patterned heterogeneous surfaces could confer adsorbates resistance and maintain the surface biocompatibility. Atomic force microscopy (AFM), with a nanoscale sensitivity, offers an unprecedented opportunity to explore the molecular-level factors that govern friction on various surfaces.

Herein, we reported a simple physical strategy based on titanate by constructing a roughness-induced nano-patterned heterogeneous mesoporous TiO_2 film to induce long-range hydrophobic interactions to further low friction. With AFM measurement, the friction coefficient (0.0026, measured in open system: air environment, at 25°C, relative humidity: ~50%) of mesoporous TiO_2 film is lower than 1/25 of dense one (0.067). Due to the existence of TiO_2(B) (59.7% in weight), mesoporous TiO_2 film keeps the structure thermal stability, further leading to the secondary nano-patterned pores (<15nm). By AFM adhesion analysis, the distribution of adhesion with Si_3N_4 tip on mesoporous TiO_2 surface shows heterogeneous property with a bimodal peak (3nN, 12nN) caused by the presence of the aforesaid secondary nano-patterned pores and TiO_2 particles. While the adhesion distributions on dense TiO_2 surface is single(12nN), showing homogeneous, attributing to TiO_2 particles. In AFM phase results, the secondary nano-patterned pores on mesoporous TiO_2 film induce long-range hydrophobic interactions, leading to a relatively hydrophobic boundary region around each hydrophobic pore patch. The boundary length extension can effectively increase the coverage of the hydrophobic phase (nano-patterned pores) to further reduce friction. The strategy to low mesoporous TiO_2 film friction by geometrical roughness-induced heterogeneous nanostructure from titanate benefits the fluid flow to further block the propensity of various adsorbates to avidly stay on the interfaces and it can maintain TiO_2 native biocompatibility.

(230f) SAM-Modified Microdisc Electrode Arrays (MDEAs) with Functionalized Carbon Nanotubes for Amperometric Biosensors

Anthony Guiseppi-Elie, Center for Bioelectronics, Biosensors and Biochips, Clemson University, Anderson, SC

A novel biosensor for glucose, one that makes use of covalent conjugation of glucose oxidase (GOx) to acid-chopped, chemically modified single-walled carbon nanotubes (SWNT) that were themselves covalently immobilized via alkane thiol SAMs onto gold microdisc electrode arrays (MDEAs) has been constructed and evaluated. Amperometric biotransducers fabricated by chemisorbing self-assembled monolayers (SAM) of cysteamine (CA) and 11-amino-1-undecanethiol (11-AUT) onto 100 mm diameter gold microdisc arrays, resulting in MDEA-Au|CA|SWNT-COOH and MDEA-Au|11-AUT|SWNT-COOH electrodes, were characterized electrochemically using cyclic voltammetry and amperometry in the presence of ferrocene. Subsequent covalent conjugation of GOx to free -COOH groups using EDC NHS hetero bifunctional coupling, produced MDEA-Au|CA|SWNT-CONH-GOx and MDEA-Au|11-AUT|SWNT-CONH-GOx electrodes that were similarly characterized in changing concentrations of glucose. The combination of SWNT at the CA-modified MDEA surface compared to the other electrode surfaces gave a well-resolved voltammogram with highest peak currents ($I_{pa} = 5.54 \times 10^{-4}$ A, $I_{pc} = 2.34 \times 10^{-4}$ A at 100 mV/s). The resulting amperometric glucose biosensor displayed linear responses between 0.5 and 3.0 mM glucose with a response time of 40s and was superior to constructs using 11-AUT.

(232a) Elastin Based Nano-Particles for Treatment of Chronic Wounds

Yuan Yuan, Chemical and Biomedical Engineering, University of South Florida, Tampa, FL and Piyush Koria, Department of Chemical and Biomedical Engineering, University of South Florida, Tampa, FL

Introduction: Chronic wounds affect 5.7 million patients and cost an estimated 20 billion dollars annually. Recently, we developed a growth factor delivery platform based on elastin like peptides (ELPs) for growth factor delivery to chronic wounds. Elastin like peptides are repeats of the pentapeptide (VPGXG, X=guest residue can be anything but Proline) derived from natural elastin. They are attractive delivery vehicles due to their unique phase transition properties and are genetically encodable and biocompatible. The unique phase transition property of ELPs is a direct function of the chain length and guest residue, which allows us to design biomaterial delivery systems with desirable physical properties. Interestingly, we also observed that ELPs were not only useful for delivery of growth factors but also enhanced granulation in the wound through modulation of fibroblast growth and function. In this study, we aim to elucidate the effect of chain length and guest residues on the ELP induced modulation of fibroblast function.

Materials and Methods: We created several ELPs with different chain lengths (number of VPGVG repeats) and having different cysteines in the guest residue position. These ELPs were recombinantly produced in bacteria using standard molecular biology and protein expression techniques. The protein was purified to high homogeneity using inverse temperature cycling which exploits the phase transitioning property of ELPs. The purity was tested using SDS-PAGE and the physical properties of ELPs were characterized using dynamic light scattering and transmission electron microscopy. Normal human dermal fibroblasts were treated with different ELPs and proliferation was evaluated using Hoechst assay. We used RT-PCR to assess the effect of ELPs on several genes associated with normal fibroblast functions such as collagen production, matrix metalloproteinase activity and fibrillin expression.

Results: We observed that ELPs self-assembled into nanoparticles with different hydrodynamic diameters depending on their chain length and proportion of cysteines. We further found that ELPs increased fibroblast proliferation significantly. Interestingly, this increase was dependent on the chain length and proportion of cysteines. This increase of proliferation was successfully blocked using lactose suggesting that ELPs bind to the elastin receptor to induce proliferation.

Conclusions: Our study suggests that ELPs induce fibroblast proliferation. This biological activity of ELPs is dependent on chain length as well as guest residues which also determine the physical phase transition properties of ELPs. Furthermore, our data suggest that ELPs induce their biological effect on fibroblasts through interaction with the elastin receptor complex present on the fibroblast surface. An understanding of the exact interaction of ELPs and identifying the determinants of this interaction (i.e. chain length, guest residue) will help us in the design of unique ELP based biomaterials having the desired physical and biological properties that are conducive to the healing of chronic wounds. Thus they will not only be useful as drug delivery vehicles but also may induce healing synergistically through their intrinsic biological activities.

(232b) Kinetic Assembly of Macromolecular Therapeutics for the Potential Management of Atherosclerosis

Adam W. York[1], Kyle R Zablocki[2], Daniel Lewis[3], Li Gu[4], Kathryn E. Uhrich[4], Robert K. Prud'homme[5] and Prabhas V. Moghe[2], (1)New Jersey Center for Biomaterials, Piscataway, NJ, (2)Biomedical Engineering, Rutgers University, Piscataway, NJ, (3)Chemical and Biochemical Engineering, Rutgers University, Piscataway, NJ, (4)Chemistry and Chemical Biology, Rutgers University, Piscataway, NJ, (5)Chemical and Biological Engineering, Princeton University, Princeton, NJ

Nano-sized assemblies are attractive for developing novel therapies as they are small enough (10 to 200 nm) to interact with cellular surfaces but large enough to transport and protect molecular actives upon administration.[1] Several technologies exist (i.e. ligand exchange with inorganic nanoparticles, self-assembly of polymeric amphiphiles, mini emulsions, and rapid solvent exchange techniques) for the fabrication of molecular nanoassemblies used in either diagnostic or therapeutic applications. In particular, polymer amphiphiles assembled through kinetic (solvent exchange) rather than thermodynamic (self-assembly) means can provide a more colloidal stable nanoassembly that provides greater resistance to disruption in biological fluids.[2]

Amphiphilic macromolecules (AMs) synthetically prepared by our labs and derived from sugar backbones, have shown the capacity to manage atherosclerosis by competitively inhibiting the uptake of oxidized low-density lipoproteins (oxLDL) in both murine and human macrophages.[3] Recently, we demonstrated that kinetic assemblies of these bioactive AMs, via Flash NanoPrecipitation, into organic nanoparticles (NPs) have superior physiological stability and bioactivity relative to analogous micellar assemblies which lose structural integrity in the presence of serum proteins.[2c] Since most oxLDL uptake occurs through class A and B scavenger receptors,[4] we have chosen to investigate interactions between novel formulations of kinetically assembled NPs and macrophage scavenger receptors. Here, we have considered the interaction of a series of distinct nanoparticles with scavenger receptor A (SR-A) through surface plasmon resonance measurements. To verify these receptor-particle interactions, in vitro studies were carried out in human embryonic kidney cells engineered to express SR-A and the ability of the nanoparticle to inhibit modified LDL uptake was examined in human macrophages. The physiochemical properties of the formulated NPs and how these attributes influence the inhibition of modified LDL uptake in macrophages will be discussed. Furthermore, the use of Flash NanoPrecipitation to formulate NPs comprised of both bioactive AMs and small molecule actives at the NP core for the potential management of atherosclerosis will be highlighted.

1. Mailänder, V.; Landfester, K., Interaction of Nanoparticles with Cells. *Biomacromolecules* **2009,** *10* (9), 2379-2400.

2. (a) Ansell, S. M.; Johnstone, S. A.; Tardi, P. G.; Lo, L.; Xie, S.; Shu, Y.; Harasym, T. O.; Harasym, N. L.; Williams, L.; Bermudes, D.; Liboiron, B. D.; Saad, W.; Prud'homme, R. K.; Mayer, L. D., Modulating the Therapeutic Activity of Nanoparticle Delivered Paclitaxel by Manipulating the Hydrophobicity of Prodrug Conjugates. *J. Med. Chem.* **2008,** *51* (11), 3288-3296; (b) Johnson, B. K.; Prud'homme, R. K., Flash NanoPrecipitation of organic actives and block copolymers using a confined impinging jets mixer. *Aust. J. Chem.* **2003,** *56* (10), 1021-1024; (c) York, A. W.; Zablocki, K. R.; Lewis, D. R.; Gu, L.; Uhrich, K. E.; Prud'homme, R. K.; Moghe, P. V., Kinetically Assembled Nanoparticles of Bioactive Macromolecules Exhibit Enhanced Stability and Cell-Targeted Biological Efficacy. *Adv. Mater.* **2012,** *24* (6), 733-739.

3. (a) Iverson, N. M.; Sparks, S. M.; Demirdirek, B.; Uhrich, K. E.; Moghe, P. V., Controllable inhibition of cellular uptake of oxidized low-density lipoprotein: structure-function relationships for nanoscale amphiphilic polymers. *Acta Biomater.* **2010,** *6* (8), 3081-3091; (b) Tian, L.; Yam, L.; Zhou, N.; Tat, H.; Uhrich, K. E., Amphiphilic Scorpion-like Macromolecules:Design, Synthesis, and Characterization. *Macromolecules* **2004,** *37* (2), 538-543.

4. Shashkin, P.; Dragulev, B.; Ley, K., Macrophage Differentiation to Foam Cells. *Curr. Pharm. Des.* **2005,** *11*, 3061-3072.

(232c) Multi-Component Mass Transfer From a Polymeric Micelle Nanocontainer for Cancer Drug Delivery

David D. Shaw[1], Philip Smith[1] and Leonard F. Pease III[2], (1)Chemical Engineering, University of Utah, Salt Lake City, UT, (2)Chemical Engineering and Pharmaceutics & Pharmaceutical Chemistry, University of Utah, Salt Lake City, UT

Most cancer drug candidates are not water soluble but require carriers such as poly(ethylene glycol)-block-poly(D,L-lactic acid) (PEG-b-PLA) micelles to enhance their delivery. Experiments by Shin et al. [1] using PEG-b-PLA micelles as vectors for combinations of three hydrophobic cancer drugs (paclitaxel, 17-allylamino-17-demethoxygeldanamycin (17-AAG), and rapamycin) found the delivery half life to be a function of the oil-water partition coefficient in contrast to their assertion that the drug cocktail release rate is diffusion controlled. Here we explain this unexpected, but important, result using a multi-component mass transfer model to estimate the flux of drug delivery. We use linearized theory and a general Fickian mass transfer approach to determine the concentration over time and distance from the micelle center. Different diffusivity models were assessed to determine the optimal diffusion coefficients required to fit the experimental data. Our results are in reasonable agreement with the experimental data, validating the applied methodology. This work provides a sophisticated model of combination drug delivery using polymer micelles and a platform for design of carriers for hydrophobic drug cocktails that will be increasingly important as most new drug candidates for diseases from heart disease to Alzheimer's are large hydrophobic molecules.

1) Ho-Chul Shin, Adam W. G. Alani, Hyunah Cho, Younsoo Bae, Jill M. Kolesar, and Glen S. Kwon. A 3-in-1 polymeric micelle nanocontainer for poorly water-soluble drugs. Molecular Pharmaceutics, 8(4):1257–1265, 2011.

(232d) A Nanoemulsion Platform for Targeted Dendritic Cell Delivery

BiYun Zeng[1], Anton P. J. Middelberg[1], Yap Pang Chuan[1], Brendan O'Sullivan[2], Mireille Lahoud[3] and Ranjeny Thomas[2], (1)Australian Institute for Bioengineering and Nanotechnology, The University of Queensland, Brisbane QLD 4072, Australia, (2)Diamantina Institute, The University of Queensland, Brisbane QLD 4072, Australia, (3)Centre for Immunology, Burnet Institute, Melbourne VIC 3004, Australia

Nanoemulsions hold great potential as drug delivery vehicles as they can be tailored to deliver unique drug combinations needed for personalized medical interventions, in a cell-specific fashion. Here we report a tailorable nanocarrier emulsion (TNE) for the targeted co-delivery of therapeutics. A designed surface active peptide (SAP) with defined interfacial properties is used to stabilize an oil-in-water (O/W) emulsion. Incorporation of active components into the oil phase at the time of emulsion formation enables facile packaging, with decoration of the outer surface using simple additive processing steps that present functional antibody or PEG. *In vitro* results showed that TNE prepared in this way were internalized and well-tolerated by model APCs, with effective suppression of an internal protein target. Initial results also showed receptor-specific dendritic cell (DC) targeting both *in vitro* and *in vivo*. This work demonstrates a new type of biocompatible designer nanocarrier which can be manufactured and tailored for targeted cell therapy using simple additive preparation steps.

(232e) Nanoemulsions As Drug Delivery Systems for Poorly Water-Soluble Drugs: Formulation, Transport of Praziquantel Across Caco-2 Cell Monolayers and Cytotoxicity

Vania E.B. de Campos[1,2], Claudia E. Mansur[1], Eduardo Jr. Ricci[3] and Sandro R. da Rocha[2], (1)Institute of Macromolecules, Federal University of Rio de Janeiro, Rio de Janeiro, Brazil, (2)Chemical Engineering and Materials Science, Wayne State University, Detroit, MI, (3)Department of Medicines, Federal University of Rio de Janeiro, Rio de Janeiro, Brazil

Nanoemulsions (NEs) are oil-in-water (o/w) dispersions with droplets in the 20–200 nm range. Lipid based formulations have shown their utility to enhance the absorption of poorly absorbable drugs. Praziquantel (PZQ) is a schistosomicidal agent, which is only administrated via the oral route. High dosages of the drug are used to ensure efficacy to overcome its low bioavailability due to low aqueous solubility. NEs represent a powerful alternative carrier system to enhance the bioavailability of PZQ because of their high solubilization capacity.

In this presentation we will discuss the compatibility of the NEs with the Caco-2 cell monolayers. We investigated the cytotoxicity of the NE and the effects of the chemistry of the NE (surfactant, oil phase) in the integrity of polarized Caco-2 monolayers. Evaluation of the NEs in enhancing intestinal transport will be also discussed and compared with the free drug and a solid nanoparticle formulations, and correlated with the opening of the tight junctions of polarizedCaco-2 cells.

(232f) Ultrasound Induced Calcein Release From Eliposomes

William G. Pitt, Department of Chemical Engineering, Brigham Young University, Provo, UT, James R. Lattin, Chemical Engineering Dept, Brigham Young University, Provo, UT, Ghaleb A. Husseini, Chemical Engineering Department, American University of Sharjah, Sharjah, United Arab Emirates and David Belnap, Department of chemistry and Biochemistry, Brigham Young University, Provo, UT

Ultrasound Induced Calcein Release from eLiposomes

J.R. Lattin[1], W.G. Pitt, Ph.D.[1], D.M. Belnap, Ph.D.[2], G.A. Husseini, Ph.D.[3]

[1] Chemical Engineering Department, Brigham Young University, [2] Chemistry and Biochemistry Department, Brigham Young University, [3] Chemical Engineering Department, American University of Sharjah, Sharjah,

Liposomes - bilayer lipid vesicles - have proven to be effective and versatile drug carriers. Targeted drug delivery from liposomes could be used to further increase their delivery efficiency by more specifically controlling the site of drug delivery. One such method of targeting involves using ultrasound to release sequestered drugs at a specific target site from carriers such as microbubbles. However, liposomes are not inherently responsive to ultrasound. The goal of the current work is to develop a sub-micron sized liposomal drug carrier that is responsive to low intensity ultrasound. The proposed drug carrier is called an eLiposome, defined as a liposome containing vaporizable emulsion droplets. The nano-sized emulsion droplets are formed from perfluorocarbon liquids with high vapor pressures. During the low pressure phase of an ultrasound wave, the local pressure may drop below this high vapor pressure, allowing the formation and expansion of a vapor phase [1]. As this vapor phase forms and expands, it stretches and disrupts the liposomal membrane, causing local drug release. Emulsions were formed from perfluorohexane (PFC6) and perfluoropentane (PFC5) because of their relatively high vapor pressures and non-toxicity. Additionally, both of these perfluorocarbons have very low solubility in water, enabling nanodroplets to persist in an aqueous environment.

Emulsions were formed by inserting a 20 kHz ultrasound probe into a cuvette containing 0.2 g of PFC and 5 mg of dipalmitoyl phosphatidyl choline (DPPC) in 1.5 mL of water. Large (400 nm) and small (100 nm) emulsion droplets were prepared by varying sonication time and intensity and/or by extrusion. Sizes were verified by dynamic light scattering. eLiposomes were formed using a lipid sheet refolding technique [2]. Interdigitated DPPC sheets were formed from small DPPC vesicles by adding ethanol. After sheet formation, excess ethanol was removed from the sheet suspension by two cycles of centrifugation and washing. 0.2 mL of emulsion was added to 10 mg of interdigitated sheets along with 0.2 mL of water (for TEM imaging) or a 30 mM calcein solution (for release experiments). The solution was heated to 50°C and stirred with a magnetic stir bar for 30 minutes, allowing the sheets to fold back into vesicles and trapping nanoemulsion droplets inside [3]. The size distribution of the resulting eLiposomes was controlled by extrusion through an 800-nm polycarbonate filter at 50°C. Empty eLiposomes, defined as liposomes refolded from interdigitated sheets in the absence of emulsion, were prepared as a negative control.

The eLiposomes were imaged using cryogenic transmission electron microscopy (cryoTEM). The microscope stage was rotated and the sample was viewed at -45°, 0°, and +45° to verify encapsulation of the emulsion droplets [3]. The cryoTEM images verify that PFC6 emulsion droplets have been encapsulated inside the lipid vesicles using the sheet refolding technique and that this method was able to encapsulate both large (400 nm) and small (100 nm) emulsion droplets. Figure 1A shows an example of an 800-nm eLiposome with 3 distinct 100-150 nm emulsion droplets. Figure 1B shows an example of an 800-nm eLiposome with one 475-nm emulsion droplet as well as two smaller droplets.

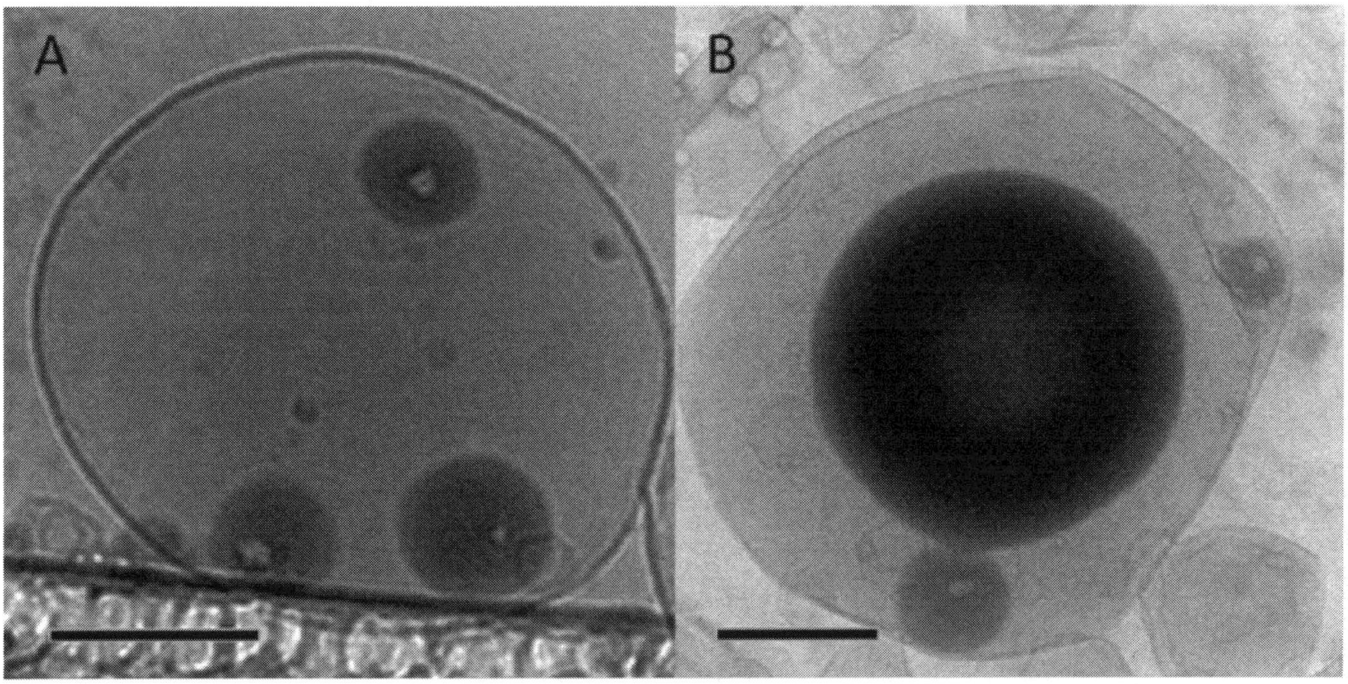

Figure 1. CryoTEM images of 800-nm eLiposomes containing A) three 100-150 nm PFC6 emulsion droplets or B) one 475-nm droplet and two smaller droplets. Scale bars represent 200 nm. The straight (B, top left) and mottled (A, bottom) structures are supporting entities of the holey carbon support film

Calcein containing samples were prepared for quantification of release. During vesicle formation, calcein was encapsulated inside of the eLiposomes at a concentration of approximately 15 mM. At this concentration, the calcein was self-quenched. The external calcein concentration was reduced by allowing the sample to settle at the bottom of the microcentrifuge tube for a few hours to form a gel-like pellet. The top phase was removed and replaced with an NaCl solution. Samples were further diluted by adding 20 μL of sample to a cuvette along with 2 mL of NaCl solution in order to leave an external calcein concentration of 1 to 5 μM so as to be within the linear region of the concentration curve for calcein. Sample fluorescence was measured using a QuantaMaster fluorometer (Photon Technology International, Birmingham, New Jersey) with excitation and emission wavelengths of 488 nm and 525 nm, respectively. Baseline fluorescence data was collected for 10 seconds at 4 data points per seconds. The cuvette was then removed from the fluorometer and 20-kHz ultrasound was applied using a 3 mm probe (Sonics and Materials, Newton, CT) inserted directly into the cuvette. As concentrated (self-quenched) calcein was released from the interior of the eLiposomes into the surrounding solution, it was diluted below its self-quenching concentration. Fluorescence was again measured after sonication. Finally 25 μL of 5% Triton X-100 was added to lyse any remaining liposomes and fluorescence was measured

Figure 2 shows the results as the time of ultrasound exposure was varied from 100 ms to 10 seconds at 1 W/cm^2. The eLiposome samples demonstrated increased calcein release compared to conventional liposomes (containing only calcein) and compared to conventional liposomes with external emulsions, suggesting that the encapsulated emulsion droplets add an ultrasound sensitive element to the eLiposomes. The eLiposomes were less sensitive to ultrasound when small emulsion droplets were encapsulated. This is most likely due to the greater Laplace pressure imposed on the droplets as their size decreases [4]. The increased pressure on the droplet adds to the ultrasound intensity required to vaporize the droplet. Because Laplace pressure is inversely proportional to radius, the amount of pressure on the droplet increases as droplets diameter decreases. In this study, the effect of droplet size can be best observed with PFC5 eLiposomes: eLiposomes with large emulsion droplets released approximately twice as much calcein as those with small droplets after 10 seconds of ultrasound exposure. There was also a noticeable difference between eLiposomes formed with PFC5 and PFC6. This difference is most likely due to the difference in vapor pressures; it is easier to promote ultrasonically induced gas formation with the higher vapor pressure of PFC5.

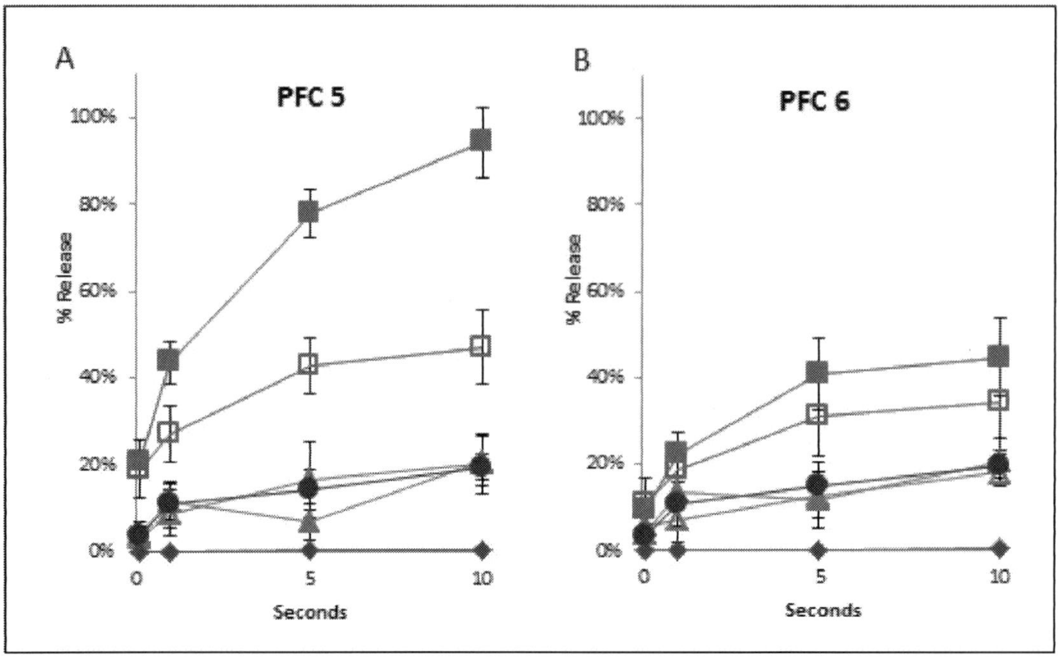

Figure 2. Calcein release from PFC5 eLiposomes (A) and PFC6 eLiposomes (B) when exposed to 20-kHz ultrasound at 1 W/cm^2 for varying times. Data is presented for eLiposomes with large (n) and small (o) emulsion droplets, empty control vesicles (l), and empty vesicles with Large (p) or small (r) emulsion droplets added to the exterior solution. Sham experiments were also performed using eLiposomes with large droplets (u). Error bars represent ± 1 standard deviation.

eLiposomes formed with PFC5 and PFC6 both released significantly more calcein than controls, suggesting that the internal emulsions do indeed add an ultrasound sensitive component to liposomes.

References

1. Rapoport, N.Y., et al., *Controlled and targeted tumor chemotherapy by ultrasound-activated nanoemulsions/microbubbles.* Journal of Controlled Release, 2009. **138**(3): p. 268-276.

2. Kisak, E.T., et al., *The vesosome - A multicompartment drug delivery vehicle.* Current Medicinal Chemistry, 2004. **11**(2): p. 199-219.

3. Lattin, J.R., D.M. Belnap, and W.G. Pitt, *Formation of eLiposomes as a drug delivery vehicle.* Colloids and Surfaces B-Biointerfaces, 2012. **89**: p. 93-100.

4. Sheeran, P.S., et al., *Formulation and Acoustic Studies of a New Phase-Shift Agent for Diagnostic and Therapeutic Ultrasound.* Langmuir, 2011. **27**(17): p. 10412-10420.

(232g) Engineering the Release of Protein Antigens by pH-Responsive Blend Polymer Particles

Xi Zhan, Kenny K. Tran and Hong Shen, Department of Chemical Engineering, University of Washington, Seattle, WA

Synthetic vaccines can provide more precise control of the quality and quantity of immune responses. In order to generate a broad spectrum of immune responses against a given infections, multiple antigens are required. Different antigens exhibit different size, charge and solubility, all of which affect their release from polymeric delivery systems. The release of antigens in extracellular space and intracellular compartments is critical for obtaining desired immune responses. The current study focuses on modulating the release of proteins with varied isoelectric points. pH-responsive blend polymer particles made from three polymers, poly(lactic-co-glycolic acid) (PLGA) and two random pH-sensitive copolymer, have been developed for the delivery of antigens. Three model proteins with the same sizes and different isoelectric points were effectively incorporated into these particles. By tuning the composition of the polymer blend, we were able to modulate the release of the three proteins at different pH environments. Our results demonstrate a promising approach to control the release of multiple antigens and thus corresponding immune responses.

(232h) pH-Responsive, Polycationic Nanoparticles Designed for Intracellular siRNA Delivery

Diane Forbes, Chemical Engineering, The University of Texas at Austin, Austin, TX and Nicholas Peppas, Dept. of Chemical Engineering, The University of Texas at Austin, Austin, TX; Dept. of Biomedical Engineering, The University of Texas at Austin; Division of Pharmaceutics, The University of Texas at Austin

INTRODUCTION: Harnessing the specificity and potency of siRNA to treat disease by correcting aberrant gene expression is limited by delivery challenges. The proposed pH-responsive, polycationic nanoparticles are designed to improve delivery by promoting endosomal escape and release of the siRNA into the cytosol (the site of action) following cellular uptake.

MATERIALS AND METHODS: The polycationic nanoparticles are composed of 2-(diethylamino) ethyl methacrylate (DEAEMA) for pH-responsiveness, poly(ethylene glycol) methyl ether methacrylate (PEGMA) for colloidal stability and biocompatibility, a crosslinking agent for enhanced nanoparticle stability even at low concentrations, and hydrophobic methacrylate monomers to tune the swelling and membrane disruptive characteristics. The emulsion-based ARGET ATRP scheme for nanoparticle synthesis was adapted from a free radical photoemulsion polymerization developed by Fisher et al.,[1]. This new methacrylate polymerization protocol occurs in water at ambient temperature without UV light.

RESULTS AND DISCUSSION: A panel of polycationic nanoparticles was synthesized and characterized. Dynamic light scattering indicated pH-responsive swelling near physiological pH which could be tuned as a function of composition. Hemolysis assays at pH values characteristic of the extracellular matrix (7.4) and early endosomes (6.5) were used as an indicator of endosomolytic activity. Incorporating a hydrophobic monomer resulted in the desired membrane disruption outcome: low membrane disruption at pH 7.4 (2-4% of the positive control) and strong membrane disruption at pH 6.5 (10-100% of the positive control).

Initial *in vitro* characterization indicates that the nanoparticle cytotoxicity is equivalent to that of particles synthesized using the traditional free radical polymerization scheme developed by Fisher et al.,[1], and Fisher and Peppas[2]. Additional characterization using a model intestinal epithelial cell culture model will enhance understanding of the polymer carrier cytotoxicity as well as the silencing efficiency of siRNA-loaded carriers.

ACKNOWLEDGMENTS: This work was supported by the U.S. National Science Foundation (CBET 10-33746) and a National Science Foundation Graduate Research Fellowship to DCF (DGE-1110007).

REFERENCES:

1. Fisher, O.; Kim, T.; Dietz, S.; Peppas, N., Enhanced Core Hydrophobicity, Functionalization and Cell Penetration of Polybasic Nanomatrices. Pharmaceutical Research 2009, 26 (1), 51-60.

2. Fisher, O. Z.; Peppas, N. A., Polybasic Nanomatrices Prepared by UV-Initiated Photopolymerization. Macromolecules 2009, 42 (9), 3391-3398.

(232i) Nanocrystalline Insulin: Molecular Mechanism of Poly Vinyl Alcohol (PVA) Induced Crystallinity

Sanjay Rawat, C. Raman Suri and Debendra K. Sahoo, CSIR - Institute of Microbial Technology, Chandigarh, India

Diabetes treatment (insulin dependent diabetes mellitus – IDDM) requires administration of multiple dose of insulin, a 51 amino acid dual chain hormone which is normally stored in pancreatic β-cells. Though protein therapeutics can be administered by oral, subcutaneous, intravenous and aerosol but daily subcutaneous injection is most preferred route of insulin administration. However, inadequate control of blood glucose and poor patient compliance are associated drawbacks with this treatment. Furthermore, most protein drugs are unstable in biological fluids and have short half lives and are not fully absorbed from their route of administration because of their relatively high molecular weights. Protein nanocrystals are expected to be useable for alternative, more attractive delivery forms such as sustained release formulation. However, formulation of biological macromolecules like insulin, that has an inherent tendency to undergo degradation processes, such as aggregation, oxidation, hydrolysis, and deamidation, for sustained release as stable and structurally integrated active protein constitutes a challenge Various polymers with different physicochemical properties have been explored for the crystallization behavior. The formulation of excipients favoring crystallization of a protein in encapsulated biopharmaceuticals is still a challenge. In the present study the Poly-(D, L lactide-co-glycolide) (PLGA) based nanoparticles formulation of bovine insulin has been prepared. The use of poly vinyl alcohol (PVA) as a surfactant in inner aqueous phase has been explored for imparting the crystallinity to the prepared formulation. The prepared nanoparticles were analyzed for size and morphology by dynamic light scattering (DLS), scanning electron microscopy (SEM) and transmission electron microscopy (TEM). TEM studies of prepared formulation, analyzed by electron diffraction, showed the crystalline behavior. The powder X-ray diffraction (PXRD) studies of lyophilized formulation further supported the crystallinity of the formulation. Differential scanning calorimetry (DSC) of lyophilized formulation showed the increase in enthalpy of fusion on increasing the PVA concentration in inner aqueous phase. On analyzing the released insulin by TEM and HRTEM (high resolution) the nanocrystalline insulin was observed. In *in vitro* release studies the released crystalline insulin was found to be conformationally and thermally stable using circular dichroism (CD) and differential scanning calorimetry (DSC). The results in detail will be presented.

(249a) Chemical Engineers and Tissue Engineering

W. Mark Saltzman, Biomedical Engineering, Yale University, New Haven, CT

Tissue engineering is a new field of inquiry, defined about 25 years ago, but it is already impacting medical care. Chemical engineers have been central contributors in this effort. Tissue engineering takes methods and materials from chemical engineering and deploys them for new purposes: bioreactors are used for tissue assembly, polymer materials provide scaffolds for tissue growth, and molecular delivery systems control the release of factors essential for cell survival and differentiation. Accordingly, the basic principles for tissue engineering align closely with chemical engineering science. Here, those basic principles are illustrated in examples of tissue engineering in the skin, the cardiovascular system, and the nervous system. These examples will be used to highlight the role of chemical engineers in the design of novel biomaterials, the delivery of agents that control cell function, and the optimization of vascular self-assembly.

Chemical Engineering Approaches to the Design and Fabrication of Extrusion-Based Functionally Graded Scaffolds Targeting Tissue Engineering Applications

Dilhan M. Kalyon

Stevens Institute of Technology

Introduction

The designing and fabrication of bioresorbable polymeric scaffolds for tissue engineering are challenged by the complex elegance of the structural and compositional gradations found in human tissues [1, 2]. For example, it would be desirable for the bone graft substitutes used in the repair of critical sized defects in long bones like femur and tibia to accommodate their changing porosities and moduli along their transverse and axial directions [3,4]. Therefore, mimicking of such complex gradations found in native tissues can require correspondingly complex gradations in tissue engineering scaffolds. Such gradations could be tailored as three-dimensional distributions in composition, structure and properties. Generation of tissue engineering scaffolds with such gradations are constrained by the available conventional scaffolding methodologies. Examples to fabrication of graded structures involve the generation of multilayered scaffolds that could be fabricated via electrospinning of separate meshes and pressing them under hydraulic pressure [5], layer-by-layer casting, freeze-drying, phase separation, and rapid prototyping techniques, including fused deposition modelling, 3D printing, selective laser sintering, and stereolithography [1, 2, 6-12]. However, additional methods that would allow the reproducible and industrially-scalable grading of bone graft substitutes and tissue engineering scaffolds for a wide range of compositions, porosities and mechanical properties are still needed.

In the fabrication of functionally-graded tissue engineering scaffolds extrusion technologies can play significant roles. Various variants of the extrusion processes, including single screw extrusion, twin screw extrusion using co-rotating or counter-rotating twin screw extruders, coupled to various dies and finishing equipment, offer significant advantages since they can be used to carry out multiple elemental steps of polymer processing [13]. These include the conveying of polymeric particles, melting, pressurization of the melt, devolatilization, shaping of the scaffolds. Furthermore, such extrusion equipment can be integrated with gravimetric and loss-in-weight feeders for solids and infusion and other pumps for liquids which allow precise and time-dependent control of the feeding rates of ingredients. Principally three extrusion–based processes were developed to enable the fabrication of functionally-graded scaffolds:

1. Hybrid process consisting of twin screw extrusion and electrospinning [14-17].
2. Twin screw extrusion and spiral winding [18-20].
3. Co-extrusion [21-25].

These three methods [14-25] are shown in Fig. 1 and were applied to the fabrication of scaffolds which target bone tissue engineering applications, and their biocompatibility was demonstrated employing osteoblasts or adipose or bone marrow derived stem cells. However, the three methods can be readily adopted and used for other tissue engineering applications. In the following the basic chemical engineering tools that can be utilized for the development and optimization of the three methods are outlined along with the descriptions of the capabilities of the three methods to fabricate functionally-graded scaffolds. In all cases poly(caprolactone) with or without solvents was employed as the bioresorbable polymeric matrix.

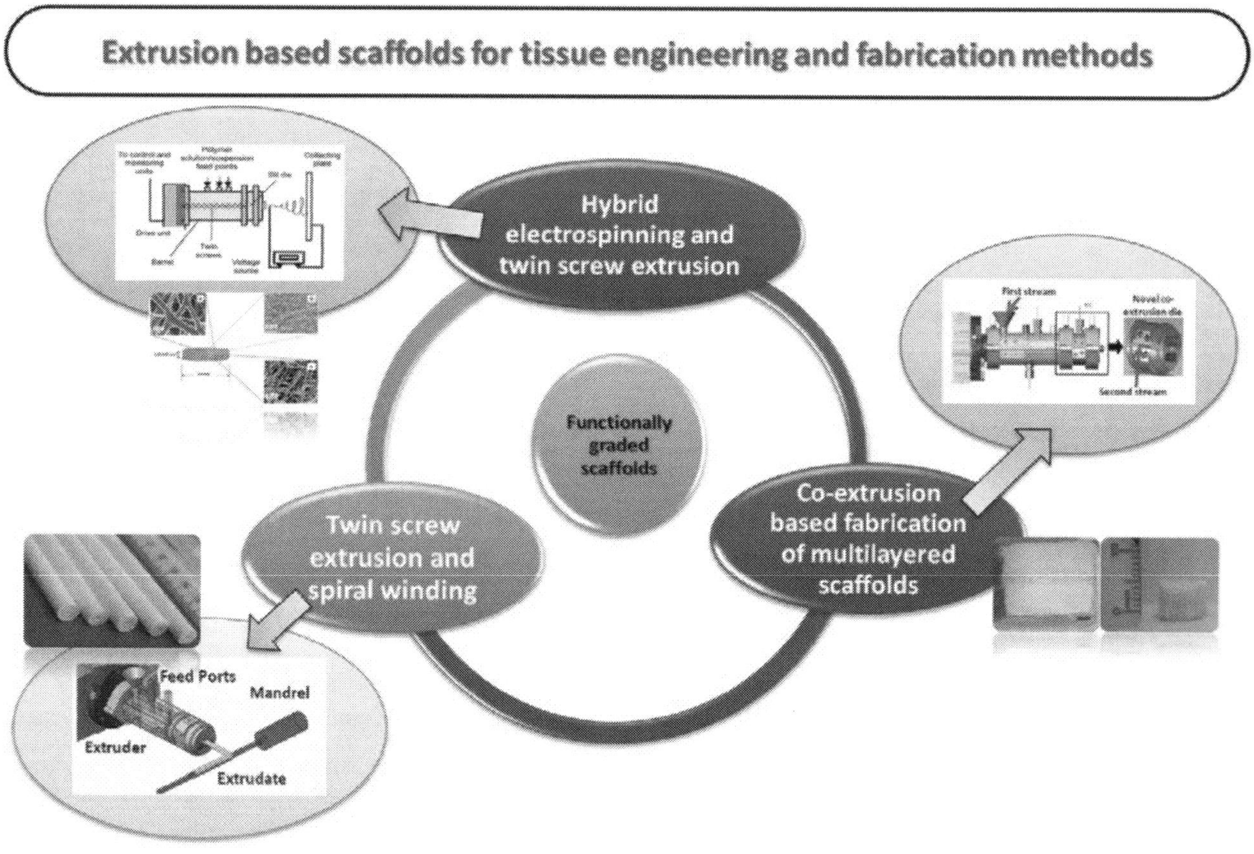

Fig. 1 Three types of functionally-graded scaffolds for tissue engineering applications via co-extrusion, hybrid method of twin screw extrusion and electrospinning and twin screw extrusion and spiral winding.

As summarized in Fig. 1, the twin screw extrusion, TSE, process can be applied concomitantly with electrospinning to generate nanofibrous meshes which can be graded with porosity, pore size and bioactive concentration distributions [14-17]. For example, concentration gradients of two bioactives, ß-gloycerophosphate and insulin could be generated using TSE-electropinning. The regions that are rich in insulin were determined to give rise to chondrocytic differentiation of adipose-derived stem cells, whereas the regions that were rich in ß-gloycerophosphate were determined to furnish greater degree of mineralization [16]. Other examples of extrusion based methods were also developed and applied to generate functionally-graded bone graft substitutes and scaffolds (Fig. 1). In twin screw extrusion and spiral winding the extrudates, the composition and porosity of which can be altered as a function of time, are wound on a simultaneously translating and rotating mandrel to generate radially and axially-graded scaffolds [18-20] and were incorporated with various bioactives with systematic changes in concentration and porosity. Furthermore, axially and radially graded scaffolds of PCL could be fabricated via the co-extrusion method to generate distributions of porosity, pore size and concentration distributions of osteoconductive particles, tricalcium phosphate, TCP, and hydroxyapatite, HA, in both the axial and radial directions (cage/core structures) [21-25] as shown in Fig. 2.

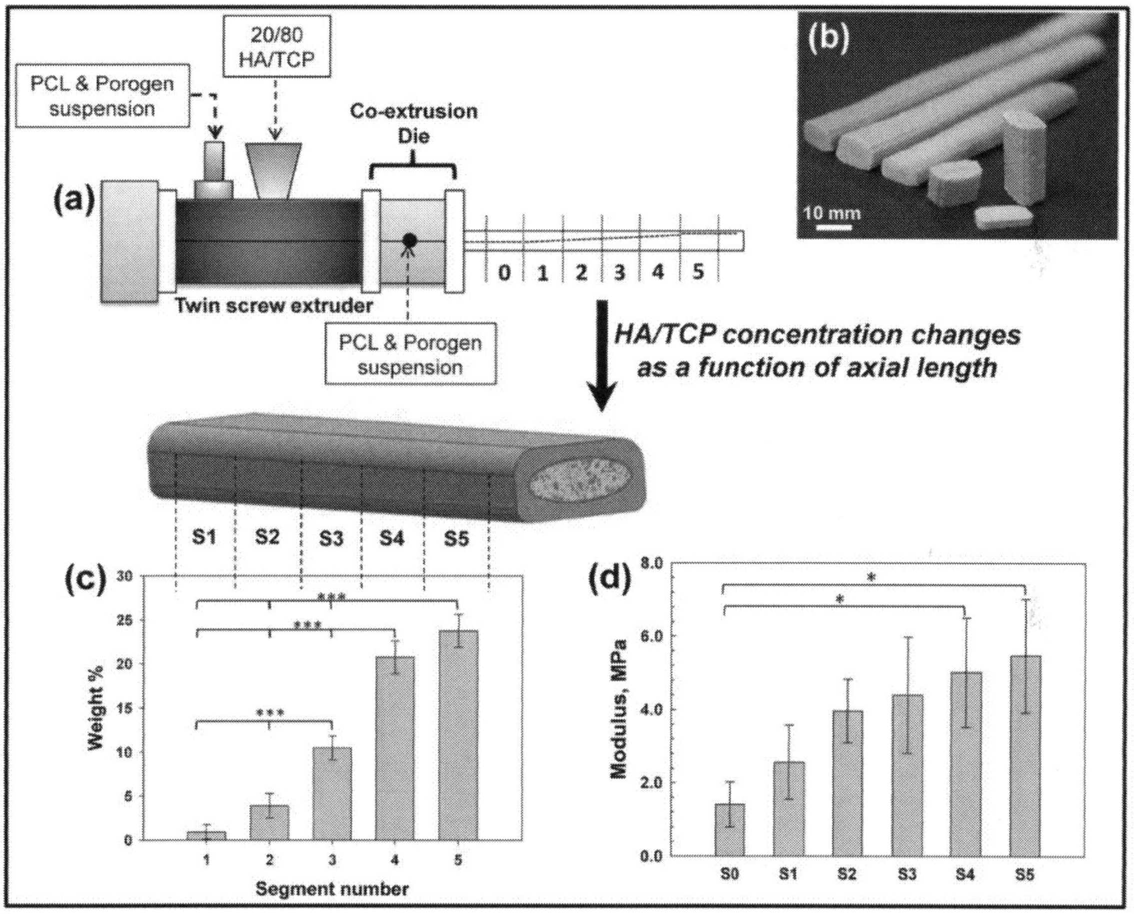

Fig. 2 Functionally-graded scaffolds for bone tissue engineering via co-extrusion with changes in HA/TCP particle concentrations in the axial direction with corresponding changes in the modulus of different segments in the axial direction and porosity gradation in the axial direction [25].

The modulus of the co-extruded scaffolds could be manipulated systematically in both the radial and axial directions. When tissue constructs were generated by seeding bone marrow derived stem cells onto the co-extruded graded scaffolds and samples were harvested as a function of time during cell proliferation and differentiation in media, the gene expression levels and μ-CT analysis revealed the osteogenenic differentiation, bone ECM formation and mineralization as affected by the gradations of the bioactive ingredient, i.e., HA and TCP concentrations [25].

Mathematical modeling of the extrusion flows

In all three methods the common denominator is the use of a twin screw extruder (fully-intermeshing co-rotating twin screw extruder). This is a very versatile extruder which is especially suitable for both distributive and dispersive mixing (thus ideal for mixing of porogens and osteoconductive particles). The basic mixing elements are the kneading disks which can be staggered at various angles in the forward and reverse directions. The following (Fig. 3) shows a schematic of the twin screw extrusion with a slit die in conjunction with two sets of kneading discs staggered at a stagger angle of 90°.

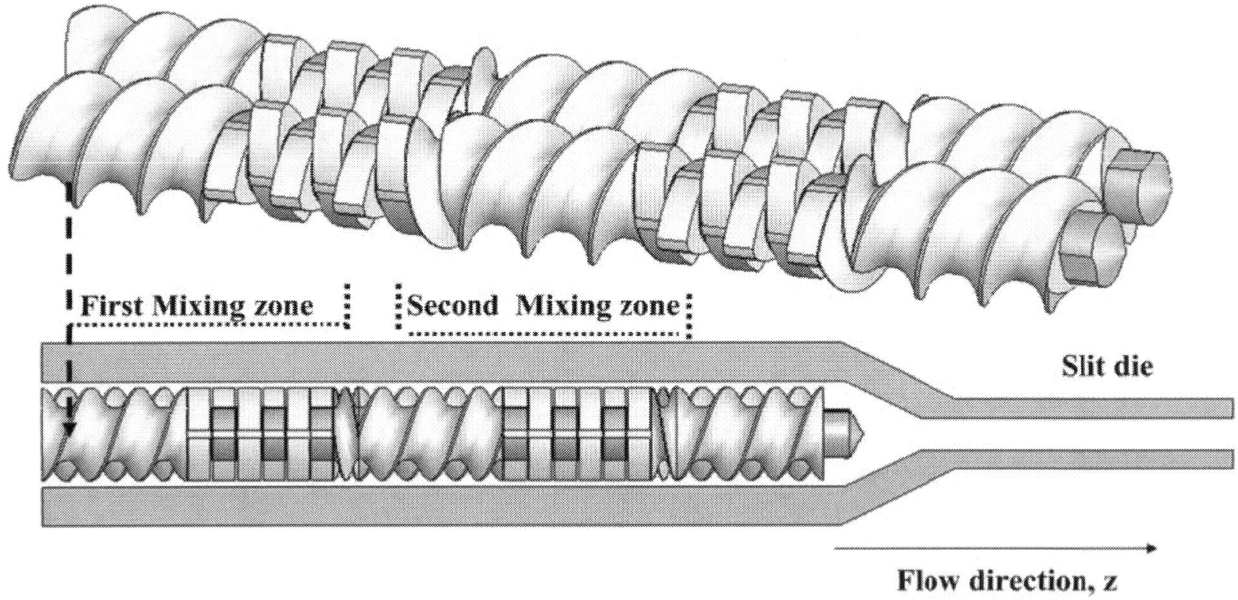

Fig.3 Schematics of the fully-intermeshing co-rotating twin screw extrusion showing a screw configuration involving two sets of kneading disks, regular flighted conveying screw sections and a slit die [34].

The optimum utilization of the extrusion based technologies requires the solution of the conservation equations for mass, momentum and energy in conjunction with appropriate rheological constitutive equations and equation of state. Especially, when salt particles are used as porogens at relatively high concentrations (typically over 60% by volume to generate adequate porosity with interconnected pores) a viscoplastic type constitutive equation can be utilized to represent the flow and deformation behavior of the concentrated suspension to be processed in the extruder (single or twin, co-rotating or counter-rotating, fully-intermeshing or tangential screws). For such filled systems the correct boundary condition is the wall slip condition. Mathematical models based on analytical and numerical solutions of the transport equations for processing of viscoplastic fluids subject to wall slip in various types of extrusion hardware and shaping using various types of extrusion dies are available [26-37]. The characterization of viscoplasticity and wall slip behavior also requires special rheological characterization apparati and methods [38-43].

Figure 4 shows some typical pressure distributions as a function of mass flow rate in the mixing zone of the twin screw extruder obtained via the 3-D Finite Element Method [34]:

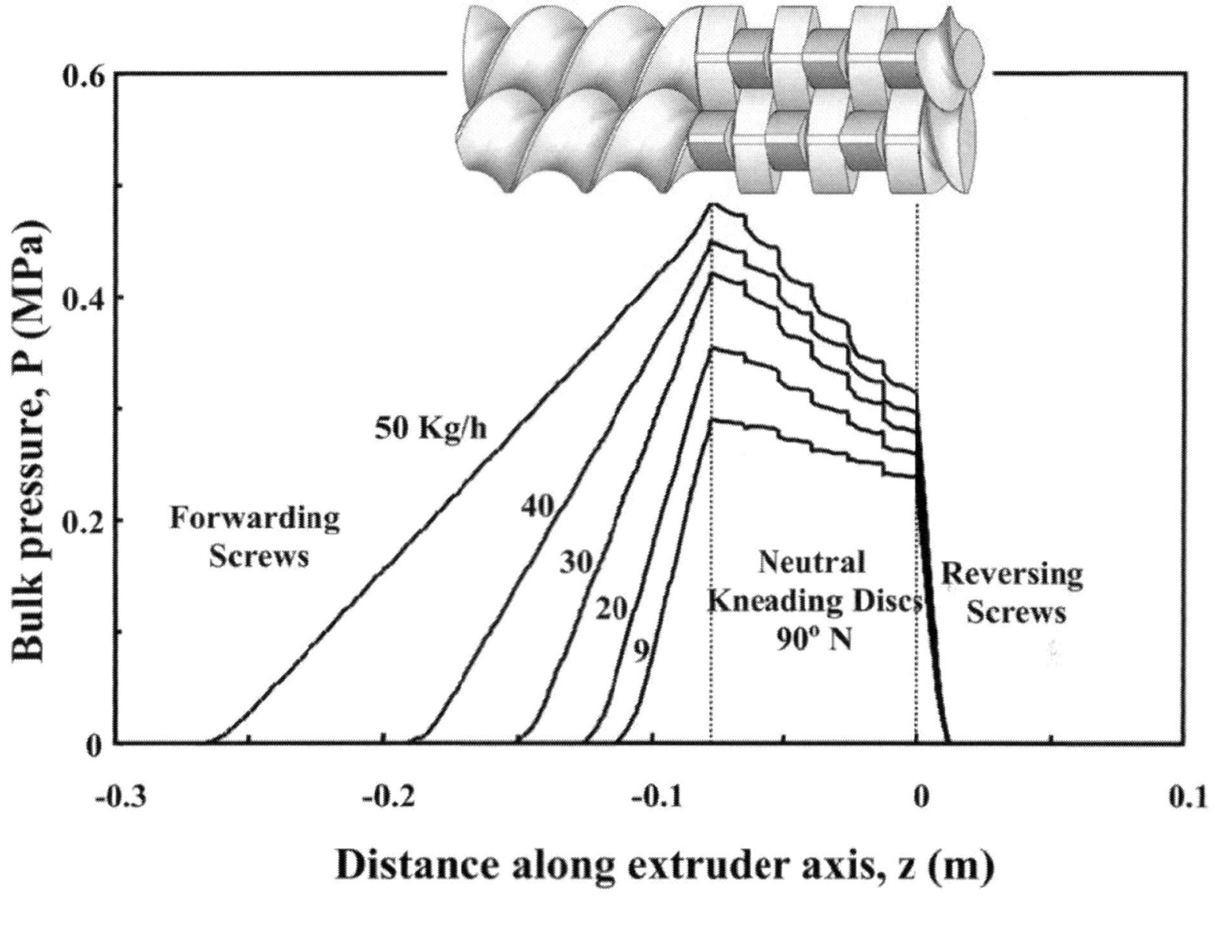

Fig.4 Pressure distributions and degree of fill as a function of flow rate in a mixing section of the fully-intermeshing co-rotating twin screw extruder as a function of mass flow rate [34].

Conclusion

During the last three decades the chemical engineering profession has gained specialized capabilities in rheology, mathematical modeling and simulation on various types of extrusion and fiber spinning methodologies, microstructural analysis and tailoring, and development of mechanical properties to allow the development of extrusion based technologies for the development of bone graft substitutes and tissue engineering scaffolds. The role of chemical engineering is expected to increase with greater utilization of these industrially-relevant methods upon commercial scale implementation to advance patient care by providing better graft substitutes and scaffolds which better mimic the complex elegance of native tissues.

References

1. Leong, K.F., Chua, C.K., Sudarmadji, N., and Yeong, W.Y. Engineering functionally graded tissue engineering scaffolds. J Mech Behav Biomed Mat 1, 140, 2008.
2. Pompe, W., Worch, H., Epple, M., Friess, W., Gelinsky, M., Greil, P., Hempel, U., Scharnweber, D., and Schulte, K. Functionally graded materials for biomedical applications. Mater Sci Eng A A362, 40, 2003.
3. Cuppone, M., Seedhom, B.B., Berry, E., and Ostell, A.E. The longitudinal young's modulus of cortical bone in the midshaft of human femur and its correlation with CT scanning data. Calcif Tissue Int 74, 302, 2004.
4. Bretcanu, O., Samaille, C., and Boccaccini, A.R. Simple methods to fabricate Bioglass-derived glass-ceramic scaffolds exhibiting porosity gradient. J Mater Sci 43, 4127, 2008.
5. Catledge, S.A., Clem, W.C., Shrikishen, N., Chowdhury, S., Stanishevsky, A.V., Koopman, M., and Vohra, Y.K. An electrospun triphasic nanofibrous scaffold for bone tissue engineering. Biomed Mater 2, 142, 2007.
6. Thomas, V., Zhang, X., Catledge, S.A., and Vohra, Y.K. Functionally graded electrospun scaffolds with tunable mechanical properties for vascular tissue regeneration, Biomed Mater 2, 224, 2007.
7. Salmoria, G.V., Ahrens, C.H., Klauss, P., Paggi, R.A., Oliveira, R.G., and Logo, A. Rapid manufacturing of polyethylene parts with controlled pore size gradients using selective sintering. Mat Res 10, 211, 2007.
8. Schaefer, D., Martin, I., Shastri, P., Padera, R.F., Langer, R., Freed, L.E., and Vunjak-Novakovic, G. In vitro generation of osteochondral composites. Biomaterials 21, 2599, 2000.
9. Gao, J., Dennis, J.E., Solchaga, L.A., Awadallah, A.S., Goldberg, V.M., and Caplan, A.I. Tissue engineered fabrication of an osteochondral composite graft using rat bone marrow-derived mesenchymal stem cells. Tissue Eng 7, 363, 2001.
10. Liao, S., Wang, W., Uo, M., Ohkawa, S., Akasaka, T., Tamura, K., Cui, F., and Watari, F. A three-layered nano-carbonated hydroxyapatite/collagen/PLGA composite membrane for guided tissue regeneration. Biomaterials 26, 7564, 2005.
11. Ghosh, S., Viana, J.C., Reis, R.L., and Mano, J.F. Bi-layered constructs based on poly(L-lactic acid) and starch for tissue engineering of osteochondral defects. Mater Sci Eng C 28, 80, 2008.

12. Sudarmadji, N., Tan, J.Y., Leong, K.F., Chua, C.K., and Loh, Y.T. Investigation of the mechanical properties and porosity relationships in selective laser-sintered polyhedral for functionally graded scaffolds. Acta Biomater 7, 530, 2011.
13. Tadmor Z. and C. Gogos, "Principles of polymer processing", Wiley, 2nd edition (2011).
14. Erisken, C., Kalyon, D.M., and Wang, H. A hybrid twin screw extrusion/electrospinning method to process nanoparticle-incorporated electrospun nanofibers. Nanotechnology 18, 1, 2008.
15. Erisken, C., Kalyon, D.M., and Wang, H. Functionally graded electrospun polycaprolactone and β-tricalcium phosphate nanocomposites for tissue engineering applications. Biomaterials 29, 4065, 2008.
16. Erisken, C., Kalyon, D.M., Wang, H., Ornek-Ballanco, C., and Xu, J. Osteochondral tissue formation through adipose-derived stromal cell differentiation on biomimetic polycaprolactone nanofibrous scaffolds with graded insulin and beta-glycerophosphate concentrations. Tissue Eng Part A 17, 1239, 2011.
17. Erisken, C., D. Kalyon and H. Wang, "Viscoelastic and biomechanical properties of tissue engineered constructs for osteochondral tissue regeneration", J. Biomechanical Engineering, 132, 091013-9 (2010).
18. Ozkan, S., D. Kalyon and X. Yu, "Multifunctional protein-encapsulated polycaprolactone scaffolds", NY Academy of Sciences, Chemical Engineering Approaches to Challenges in Energy and Biomedicine Meeting, March 30, 2012, New York, NY.
19. Ozkan, S., Kalyon, D.M., and Yu, X. Functionally graded β-TCP/PCL nanocomposite scaffolds for bone tissue engineering: In vitro evaluation with human fetal osteoblast cells. J Biomed Mater Res A 92, 1007, 2010.
20. S. Ozkan, D. M. Kalyon, X. Yu, C. McKelvey and M. Lowinger, "Multifunctional protein-encapsulated polycaprolactone scaffolds: Fabrication and in vitro assessment for tissue engineering", Biomaterials, 30, 4336-4347 (2009).
21. Ergun, A., Yu, X., Valdevit, A., Ritter, A., and Kalyon, D.M. In vitro analysis and mechanical properties of twin screw extruded single-layered and coextruded multilayered poly(caprolactone) scaffolds seeded with human fetal osteoblasts for bone tissue engineering. J Biomed Mater Res Part A 99A, 354 (2011).
22. Ergun A., D. Kalyon, A. Valdevit, A. Ritter , X. Yu, "Twin screw extrusion and co-extrusion based fabrication of uniform and multilayered scaffolds for bone tissue engineering", Orthopaedic Research Society, Transactions Vol. 36, 0285 (2011).
23. Ergun A., D. Kalyon, A. Valdevit and A. Ritter, "Compressive fatigue behavior of osteoblast seeded tissue constructs of poly(caprolactone) multilayered scaffolds for bone graft substitute applications", Orthopaedic Research Society, Transactions Vol. 36, 1850 (2011).
24. Ergun, A., Chung, R., Ward, D., Valdevit, A., Ritter, A. and Kalyon, D. M. Unitary bioresorbable cage/core bone graft substitutes for spinal arthrodesis coextruded from polycaprolactone biocomposites. Annals of Biomedical Engineering, 40, 1073, (2012).
25. Ergun, A., X. Yu, A. Valdevit, A. Ritter and D. M. Kalyon, "Radially and axially-graded multi-zonal scaffolds targeting critical-sized bone defects from polycaprolactone/hydroxyapatite/tricalcium phosphate", Tissue Engineering A (2012) doi:10.1089/ten.tea.2011.0625.
26. D. Gotsis, A. Ji and D. M. Kalyon, "3-D Analysis of the Flow in Co-rotating Twin Screw Extruders," Society of Plastics Engineers, ANTEC Technical Papers, 36, 139-142(1990).

27. Lawal A. and D. M. Kalyon, "Non-isothermal Model of Single Screw Extrusion of Generalized Newtonian Fluids," Numerical Heat Transfer, 26 (1), 103-121 (1994).
28. Lawal, A. and Kalyon D. M., "Non-Isothermal Extrusion Flow of Viscoplastic Fluids with Wall Slip", International Journal of Heat and Mass Transfer, 40 (1997) 3883-3897.
29. Lawal, A. and Kalyon D. M., "Viscous Heating in Nonisothermal Die Flows of Viscoplastic Fluids with Wall Slip," Chem. Eng. Sci., 52 (1997) 8, 1323-1337.
30. Kalyon D. M., A. Lawal, R. Yazici, P. Yaras and S. Railkar, "Mathematical Modeling and Experimental Studies of Twin Screw Extrusion of Filled Polymers", Polym. Eng. Sci., 39, 6, 1139-1151 (1999).
31. Lawal, A. and D. M. Kalyon, "Analysis of Nonisothermal Screw Extrusion Processing of Viscoplastic Fluids with Significant Backflow", Chem. Eng. Sci., 54, 999-1013 (1999).
32. Malik M. and Kalyon, D.M., "Three-dimensional Finite Element Simulation of Processing of Generalized Newtonian Fluids in Counter-rotating and Tangential Twin Screw Extruder and Die Combination", Int. Polym. Processing, 20 (2005) 398- 409.
33. Kalyon D. M., Apparent slip and viscoplasticity of concentrated suspensions, J. Rheol. 49 (2005) 621-640.
34. Kalyon D. M. and Malik M., "An integrated approach for numerical analysis of coupled flow and heat transfer in co-rotating twin screw extruders", International Polymer Processing, 22 (2007) 293-302.
35. Tang H. and Kalyon D. M., "Unsteady circular tube flow of compressible polymeric liquids subject to pressure-dependent wall slip", J. Rheology, 52 (2008) 507-526.
36. Tang H. S. and Kalyon D. M., "Time-dependent tube flow of compressible suspensions subject to pressure dependent wall slip: Ramifications on development of flow instabilities", Journal of Rheology 52 (2008) 5, 1069-1090.
37. Kalyon D.M. and Malik M., "Axial laminar flow of viscoplastic fluids in a concentric annulus subject to wall slip", Rheologica Acta (2012) 51:805–820.
38. Yilmazer U. and Kalyon D. M., Slip effects in capillary and parallel disk torsional flows of highly filled suspensions, J. Rheol. 33 (1989) 1197-1212.
39. Kalyon D. M. and Gokturk H. S., "Adjustable Gap Rheometer," U.S. Patent and Trademark Office # 5,277,058, Jan. 11, 1994.
40. Aral B. and Kalyon D. M., Effects of temperature and surface roughness on time-dependent development of wall slip in steady torsional flow of concentrated suspensions, J. Rheol. 38 (1994) 957-972.
41. Lawal, A., Kalyon D. M., Squeeze flow of viscoplastic fluids subject to wall slip, Polymer Eng. Sci. 38 (1998) 1793-1804.
42. Tang H. and Kalyon D. M., "Estimation of the Parameters of Herschel-Bulkley Fluid under Wall Slip Using a Combination of Capillary and Squeeze Flow Viscometers", Rheologica Acta, 43 (2004) 1 80-88.
43. Kalyon D. M. and Tang H., "Inverse problem solution of squeeze flow for parameters of generalized Newtonian fluid and wall slip", Journal of Non-Newtonian Fluid Mechanics, 143 (2007) 133-140.

(249c) Investigating Angiogenesis in Living Vascular Networks in Vitro

John P. Morgan, Chemical and Biomolecular Engineering, Cornell University, Ithaca, NY

Investigating Angiogenesis in Living Vascular Networks *In Vitro*

Microvascular networks support metabolic activity and define the microenvironmental conditions within tissues during healthy and pathological processes, and provide the foundation for embryonic development. Following its *de novo* synthesis, the microvasculature develops and remodels via the process of angiogenesis, the formation of new capillaries from preexisting blood vessels by the sprouting or splitting of preexisting vessels into a hierarchically branching vascular network. In this presentation, we describe how we are studying angiogenesis by engineering living vascular networks *in vitro* in a microfluidic culture, using lithographic techniques to form endothelialized microvessels within a native collagen matrix. The role of physical and chemical forces in angiogenesis is investigated experimentally and modeled mathematically.

Angiogenesis is a tightly controlled sequence of events involving endothelial cell sprouting, migration and proliferation; degradation of tissue; new capillary vessel formation and loop formation (anastomosis). However, none of these events occur until the onset of flow, with sufficient viscosity and hence shear stress, through the network. We hypothesize that mechanical stresses provide the key triggering event to initiate biochemical signaling pathways that guide subsequent growth. Further, these mechanically induced changes evolve both spatially and temporally in response to complimentary and competing effects of chemotaxis and perfusion-related hemodynamic forces. Post-development, angiogenesis also occurs during wound healing, menstruation, in response to hypoxia (low oxygen levels), and in pathologies such as cancer. Elucidation of the basic mechanism of angiogenesis and related triggering stimuli could, for example, lead to the clarification of mechanobiological signaling pathways, aid in the refinement of anti-angiogenic cancer therapies and form the basis of strategies to engineer vascularized tissues for regenerative medicine.

We will present our work toward the formation of functional microvascular structures *in vitro*. We will describe our strategies for the formation of endothelialized vessels within cell-remodelable matrices and our experimental methods for defining physiologically relevant perfusion stresses and biochemical gradients during extended cultures. We will then turn to a discussion of our observations of the response of these vessels, as evaluated via live microscopy and at immunohistochemically stained end points. We will focus on characterization of morphology, and long-term stability of endothelia, as a function of individual physical forces (e.g., shear stress and wall tension) and different flow regimes (e.g., pulsatile, laminar, and oscillatory). We will further discuss the behavior of co-cultures with perivascular cells in the bulk of the matrix. We will conclude with a discussion of how our observations relate to observations in *in vivo* studies and how they inform mathematical models of angiogenic maturation.

USE OF LATE EMBRYOGENESIS PROTEINS TO ENGINEER DESICCATION TOLERANCE IN MAMMALIAN CELLS

Nilay Chakraborty[1,2*], *Shumin Li*[3*], *Apurva Borcar*[3], *Steven C. Hand*[3], *Mehmet Toner*[2]

[1]*University of Michigan-Dearborn, Dearborn, MI, USA*
[2]*Harvard Medical School and Massachusetts General Hospital, Boston, MA, USA*
[3]*Louisiana State University, Baton Rouge, LA, USA*

Introduction

The mechanisms by which different anhydrobiotic organisms protect their cellular structure and function during desiccation can be an important strategy for cell stabilization to support cell-based therapies. While it is known that sugars such as trehalose can help in stabilizing biological structures during desiccation (1, 2), recent studies underscore the importance of various protective proteins during drying that include Late Embryogenesis Abundant (LEA) proteins (3, 4). Late embryogenesis abundant (LEA) proteins are associated with desiccation tolerance in several anhydrobiotic organisms including insects, tardigrades and rotifers (3, 4).

The objective of this study was to use the biomimetic strategy of using LEA proteins to engineer desiccation tolerance in mammalian cells for lyopreservation. We have identified and sequenced two LEA genes (*Afrlea2* and *Afrlea3m*) in the embryo of desiccation tolerant brine shrimps (*Artemia franciscana*). AfrLEA3m was shown to be localized in mitochondria while AfrLEA2 is cytoplasmic (REF). In this study, human hepatocellular carcinoma (HepG2) cells with AfrLEA2 and AfrLEA3m was used to evaluate the possibility of improved survivorship during drying. The technique of spin-drying was used to rapidly and uniformly desiccates cells to a moisture content of 0.2 gH2O/gdw at 22^{0}C. We report dramatic improvement of desiccation tolerance of HepG2 cells using AfrLEA3m proteins. Desiccated cells showed significantly higher survival after drying and rehydration.

Materials and Methods

Human hepatocellular carcinoma (HepG2) cells were stably transfected to express AfrLEA2 and AfrLEA3m using a tetracycline-inducible expression system. Trehalose was loaded into these cells using a TRET1 trehalose transporter isolated from anhydrobiotic larvae of Polypedilum vanderplanki (REF kikawada). The expression vector pcDNA 6.2/V5-DEST (Gateway Vector, Invitrogen) was used for constitutive expression of the TRET1 trehalose transporter (6). Transfection of HepG2 Tet-On cell lines (with or without the capacity for AfrLEA2/AfrLEA3m expression) was accomplished with Geneporter transfection reagent (Genlantis, San Diego, CA, USA). Stable cell lines were selected using blasticidine.

Monolayers of HepG2 cells expressing various combinations of AfrLEA2, AfrLEA3m and TRET1 were allowed to attach to 22 mm collagen coated round glass coverslips (BD BioCoat, San Jose, CA) by incubating 18 h in cell culture medium. For all cells expressing TRET1, the medium also contained 50 mM trehalose. To prevent cells from attaching to the edge of the coverslips, 20 mm diameter Press-to-Seal^{TM} silicone isolators (Invitrogen, Carlsbad, CA) were used. Immediately before spin-drying, the cell culture media was removed completely using a Pasteur pipette and replaced with the spin-drying solution (1.8 M trehalose, 10 mM KCl, 5 mM glucose, 20 mM HEPES, and 120 mM choline chloride at pH 7.4). This trehalose-containing solution was essential for successful cellular drying outcomes regardless of the intracellular variables tested to avoid extensive cellular injury due to ionic imbalance. Spin-drying was performed using a commercially available spin-coater (Brewer Science Inc. St. Louis, MO). The basic spin-drying process is described in the Figure 1. Cells attached to glass coverslips were spin dried at 1000 rpm for 60 sec in a trehalose buffer at room temperature. Following spin-drying, the samples were immediately rehydrated and the membrane integrity of the cells following desiccation was assessed using a Syto-13/Ethidium bromide fluorescence assay. Desiccated cells were cultured under normal tissue culture conditions following rehydration for 7 days to evaluate long-term growth characteristics.

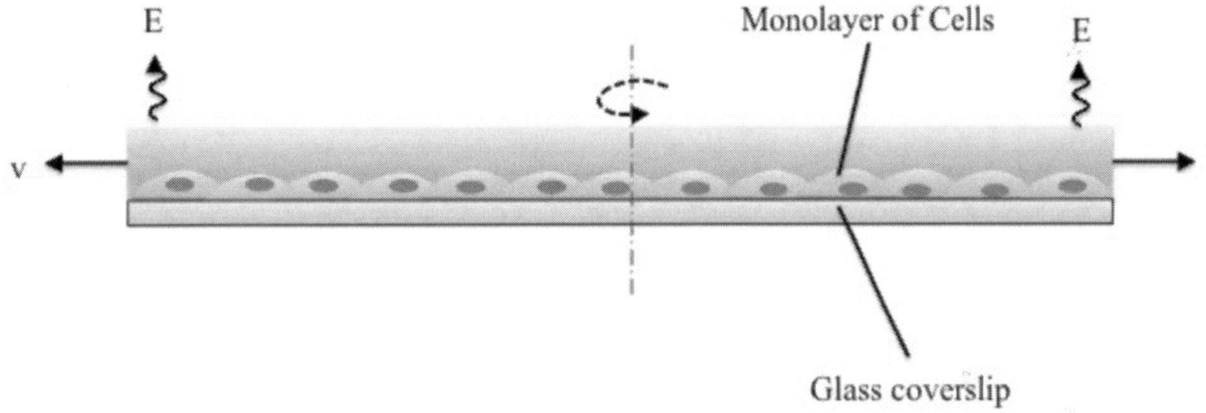

Figure 1: Schematic explaining the technique of spin-drying of cells. The arrows indicate different mechanism of moisture removal during spin-drying. At the beginning of the spin-drying process the excess liquid is removed from the sample by radial flow (v). After the excess liquid is removed by radial flow, the evaporation (E) into an environment continuously purged with dry nitrogen gas causes further drying of the samples.

Results and Discussion

Control HepG2 cells (without LEA protein or intracellular trehalose) exhibited 0 % membrane integrity (n = 9). HepG2-TRET1 cells preloaded with trehalose showed 44.5 ± 22.2 % integrity (mean ± SD, n = 3); HepG2-AfrLEA2 cells without trehalose, 57.2 ± 13.0 % (n = 9) (Figure 2); HepG2-AfrLEA2 cells preloaded with trehalose, 98.3 ± 2.2 % (n = 9); HepG2-AfrLEA3 cells without trehalose, 93.6 ± 4.6 % (n = 9); and HepG2-AfrLEA3 preloaded with trehalose, 97.7 ± 3.8 % (n = 9). Growth studies across subsequent days after rehydration revealed higher proliferation for cells containing LEA protein than those without. Between AfrLEA2 (cytoplasmic) and AfrLEA3m (mitochondrial) proteins, AfrLEA3m was able to provide better protection against desiccation stresses as evidenced by superior proliferation rates. Growth studies also demonstrate that the LEA proteins act synergistically with trehalose in providing protection against desiccation stresses and adding intracellular trehalose along with LEA proteins enhances the long-term growth of cells, as measured across 7 days post-rehydration, to proliferation rates that approach those of control cells never dried.

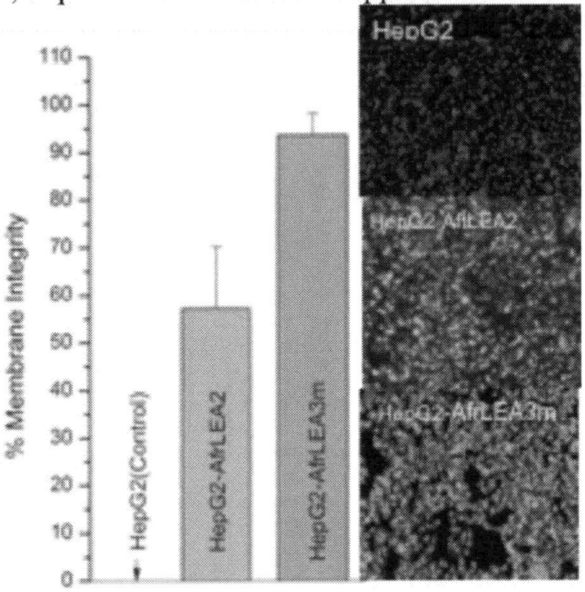

Figure 2. Membrane integrity of the HepG2 cells expressing Late Embryogenesis Proteins. Fluorescent micrographs of the rehydrated HepG2 cells indicate the protective of the LEA proteins.

Conclusion

This study outlines a biomimetic approach of using LEA proteins to dramatically enhance desiccation tolerance in mammalian cells. LEA proteins offer an excellent opportunity for engineering cellular biostability in the dried state.

References

1. Crowe JH., Crowe LM., Wolkers WF., Oliver AE., Ma X, Auh J-H, Tang M, Zhu S, Norris J, and Tablin F. (2005), "Stabilization of dry mammalian cells: Lessons from nature," *Integr Comp Biol,* 45(5):810-820.
2. Xie G, Timasheff SN (1997), "The thermodynamic mechanism of protein stabilization by trehalose," *Biophys Chem* 64(1-3):25-43.
3. Tunnacliffe A and Wise MJ (2007), "The continuing conundrum of the LEA proteins," *Naturwissenschaften* 94(10):791-812.
4. Hand SC, Menze MA, Toner M, Boswell L and Moore D (2011), "LEA proteins during water stress: Not just for plants anymore," *Annu Rev Physiol* 73(1):115-134.
5. T. Kikawada, A. Saito, Y. Kanamori, Y. Nakahara, K. Iwata, D. Tanaka, M. Watanabe, T. Okuda (2007), "Trehalose transporter 1, a facilitated and high-capacity trehalose transporter, allows exogenous trehalose uptake into cells," *Proc. Natl. Acad. Sci. USA* 104:11585–11590.
6. Chakraborty N, Menze M, Hand SC and Toner M (2012) Trehalose Transporter from African Chironomid Larvae Improves Desiccation Tolerance of Chinese Hamster Ovary Cells. *Cryobiology* 64:91-96.

(249f) Hemoglobin Regulates the Migration of Glioma Cells Along Poly(ε-caprolactone)-Aligned Nanofibers

Alexander Roth[1], Ruipeng Xue[2], Tyler Nelson[3], Jed Johnson[4], Jacob Elmer[1], Joseph Huntley[1], John J. Lannutti[5], Mariano S. Viapiano[6] and Andre Palmer[1], (1)Department of Chemical and Biomolecular Engineering, The Ohio State University, Columbus, OH, (2)Materials Science and Engineering, The Ohio State University, Columbus, OH, (3)Department of Biomedical Engineering, The Ohio State University, Columbus , OH, (4)Material Science and Engineering, Ohio State University, Columbus, OH, (5)Department of Material Science and Engineering, The Ohio State University, Columbus, OH, (6)Neurological Surgery, The Ohio State University, Columbus, OH

This work seeks to integrate two disparate technologies and examine their combined effect on U251 glioma cell migration. Aligned fibers have been shown to facilitate cell migration in the direction of fiber alignment while oxygen-carrying solutions improve the metabolism of cells in hypoxic culture. Therefore in this work, U251 aggregate migration on poly(ε-caprolactone) (PCL) aligned fibers was studied in cell culture media supplemented with the O_2 storage and transport protein hemoglobin (Hb) obtained from bovine, earthworm and human sources at concentrations ranging from 0-5 g/L within a cell culture incubator and oxygen tensions ranging from 1-19% O_2. In one set of experiments conducted on tissue culture polystyrene – representing a surface conducive to random cell movement – cell viability and the total number of live cells were compared for individually cultured U251 cells under different culture conditions. To determine the effect of Hb concentration and incubator O_2 tension on Hb oxidation, Hb auto-oxidation rate constants were measured 24 hours after Hb incubation at 37°C. Individual cell migration was also quantified using a wound healing assay up to 24 hours after cell seeding. In another set of experiments, U251 cell aggregates were developed and aggregate dispersion/cell migration quantified for cells cultured on PCL-aligned fibers up to 72 hours after aggregate seeding. The results of this work show that the presence of bovine or earthworm Hb improved individual cell viability at 1% O_2 while human Hb adversely affected cell viability at increasing Hb concentrations and decreasing oxygen levels. It was also observed that Hb auto-oxidation rate constants increased with decreasing O_2 levels, increasing Hb concentrations, and was greatest for bovine Hb and the least for earthworm Hb under identical culture conditions. The presence of cells decreased the rate of Hb auto-oxidation. The wound healing assay showed no differences in individual cell migration under most conditions, except for an increase in cell migration at 0.5 g/L bovine Hb at 5% O_2, and a reduction in cell migration at 1% O_2 in the presence of 5 g/L earthworm Hb. The control data also suggests that decreasing the O_2 tension in the incubator from 5% O_2 to 1% O_2 decreased aggregate dispersion on the PCL aligned fibers. However, the addition of bovine Hb at 5% O_2 significantly improved aggregate dispersion. At 19% O_2 Hb did not impact aggregate dispersion. Also at 1% O_2, aggregate dispersion appeared to increase in the presence of earthworm Hb, but only at the latter time points. Taken together, these results show that Hb-based O_2 carriers can be utilized to improve O_2 availability and the migration of glioma spheroids *in vitro*.

(254a) Integrated Processing of Microalgae for Valuable Food Ingredients, Chemicals and Fuel Production

Kaige Wang[1], Yi Liang[2], Zhiyou Wen[2] and Robert Brown[1], (1)Center for Sustainable Environmental Technologies, Iowa State University, Ames, IA, (2)Food Science and Human Nutrition, Iowa State University, Ames, IA

Lipid-based biofuel production by microalgae has captured interest of researchers due to microalgae's high biomass yield and ability to synthesize and accumulate large quantities of lipid. However, high cost of algal cultivation and processing has limited the commercialization of algal fuel production. Developing a microalgae biorefinery strategy, i.e., using every components of algal biomass for fuels and chemicals production is a feasible approach for commercially utilization of the algal biomass.

This study reports a hybrid processing for treating the microalgae *Schizochytrium limacinum*. Crude glycerol, a major byproduct in biodiesel production plants, was used as substrate for producing high content of algal oil containing DHA (22:6, n-3), a high value omega-3 polyunsaturated fatty acid with beneficial effects on human health. The residues after oil-extraction from *S. limacinum* were then catalytically pyrolyzed on a micro-furnace pyrolyzer with zeolite catalyst being used. Aromatics and ammonium was generated from this catalytic pyrolysis process. Among the aromatics, benzene, toluene and xylene, which are valuable basic petrochemicals,were the dominant ones. Most of the nitrogen in the residues was released in ammonium form, which may be in turn used as nutrient for algal culture. In summary, all the components in the biomass of the *S. limacinum* were utilized for producing chemicals, fuels and value-added products which can be future enhanced the economic for algal industry.

(254b) Pyrolytic Fractionation of Oleaginous Feedstocks

Balakrishna Maddi, Sridhar Viamajala and Sasidhar Varanasi, Chemical & Environmental Engineering, University of Toledo, Toledo, OH

Traditional pyrolysis of oleaginous biomass results in bio-oils that contain a mixture of compounds derived from protein, starch and lipids – constitutive biopolymers of these feedstocks. As a result, such biooils have to undergo energy-intensive denitrogenation and deoxygenation to render them suitable as fuels. We have recently developed a "*pyrolytic fractionation*" technique based on the observation that the protein, starch and lipid portions undergo thermal degradation over distinct non-overlapping temperature intervals. Proteins are most thermally labile followed by starch and lipids. Hence, conducting the pyrolysis in three temperature stages provides an opportunity to recover homogenous and biopolymer-specific biooils. Bio-oils from protein and starch can find other applications, in addition being lower grade fuels. For instance, N-rich biooil from protein can serve as a source of high value N-compounds and levoglucosans from starch pyrolysis could serve as precursors for value-added polymers. Similarly, bio-oil from lipid fraction, predominantly containing free fatty acids, can be converted to fuels (such as biodiesel or hydrocarbons) or other higher value fatty acid derivatives. Finally, the bio-char can be a soil amendment.

In this presentation, results showing component-specific biooil compositions and yields from pyrolytic fractionation of algal biomass and oil seeds will be presented. We will also present a comparison of energy needs for pyrolytic fractionation relative to other themochemical methods for conversion of oleaginous feedstocks.

(254c) Hydrothermal Carbonization and Supercritical Ethanol in Situ Transesterification for the Production of Algal Biodiesel

Robert Levine, Chemical Engineering, University of Michigan, Ann Arbor, MI and Phillip E. Savage, Chemical Engineering, University of Michigan, Ann Arbor, Ann Arbor, MI

The production of transportation fuels from microalgae is complicated by the fact that algae are typically grown suspended in a dilute aqueous media and have a high demand for nitrogen and phosphorous. In this work, we explore a new and environmentally preferable strategy for algal biofuel production. There are three major components of the bio-refinery model we have developed: 1) cultivation of oil-rich algae using recycled water and nutrients; 2) treatment of the harvested algae through hydrothermal carbonization to produce a biochar; and 3) conversion of the carbonized algal biomass into transportation fuels using in situ transesterification with supercritical ethanol.

Lipid-rich *Chlorella* was grown in bioreactors until it contained approximately 50% oil, then carbonized for 15 to 90 min at 180 to 235 °C. Biochar obtained by filtration retained >90% of lipids of the original biomass and was converted into biodiesel by reacting the solids with ethanol at 270-325 °C for 15-180 min. Fatty acid ethyl yields in excess of 90% were obtained while employing 10-100 times less ethanol than previously reported for comparable transesterification reactions. The nutrients liberated during hydrothermal carbonization were used to produce additional algal biomass, thereby recycling N and P. Overall, our work suggests that it is possible to produce net-energy positive algal biofuels with significantly less environmental impact than the petroleum fuels they replace.

(254d) Integrating Photo-Bioreactor and Fermentor to Produce Biofuels and Bioelectricity

Alim Dewan, Marci Kerls and M. Nazmul Karim, Chemical Engineering, Texas Tech University, Lubbock, TX

We have investigated simultaneous production of biofuels and bioelectricity from an integrated bioreactor that combines photo-synthesis and fermentation. In the integrated bioreactor, a fermentor and a photo-bioreactor were combined in a way that the carbon dioxide produced in the fermentation can be used to grow algae in the photo-bioreactor. Also, the reactors were attached to a proton exchange membrane in between so that they form a biological fuel cell. To synchronize the carbon dioxide production in the fermentation and the carbon dioxide utilization in photo-synthesis, the fermentor was run continuously but the photo-bioreactor was run in batch mode. An autotrophic microalgae, *Chlorella vulgaris,* and a bear yeast, *Saccharomyces bayanus,* were grown in the photo-bioreactor and in the fermentor, respectively. The ethanol production by the yeast, oil production by the algae, and electricity generation were monitored over time. The electricity generation was characterized by potentiodynamic polarization. In a two-liter size integrated bioreactor, we observed an ethanol productivity of 0.49 ± 0.37 g l^{-1} hr^{-1}, algae oil productivity 2.5 mg l^{-1} hr^{-1}, power density of 70.61 ± 12.41 mW/m^2 of the electrode, and current density of 155.02 ± 31.23 mA/m^2. Preliminary scale-up studies show a great potential for this integrated system.

(254e) Determination of Lipid Hydrolysis Kinetics in Soybean Oil and Algal Systems

Joshua C. Wissinger[1], Amber L. Bosley[1] and Constance Schall[2], (1)Chemical and Environmental Engineering, University of Toledo, Toledo, OH, (2)Chemical & Environmental Engineering, University of Toledo, Toledo, OH

Determination of lipid hydrolysis kinetics in soybean oil and algal systems

Joshua C. Wissinger, Amber L. Bosley, Constance A. Schall

The third generation of biofuels, algae, is of growing interest as a renewable feedstock due to its high yield of energy per unit area, use of carbon dioxide and other available resources for growth, and minimal competition with the food industry. Biodiesel and jet fuel can be produced from lipid-rich algae. However, extracting lipids from algae is complicated due to the high water content of harvested and mechanically dewatered algae (60 to 90% water). Thus, there is a need for processes that convert wet algal biomass into fuels or fuel precursors directly in the aqueous phase. Such a process alternative is to use the water in the harvested and partially dewatered algae for hydrolysis of lipids to free fatty acids (FFA), forming aqueous and oil rich phases for ease of extraction and separation of the lipid component. Lipid processing techniques include refining FFA by catalytic cracking and hydrotreating for jet fuel production, or the UOP/Eni Ecofining™ process to obtain clean green diesel. In our first step, subcritical water is reacted with lipids producing FFA by lipid hydrolysis eliminating the need for expensive and water sensitive catalysts. In the initial development of an empirical kinetic model, soybean oil, a system much less complex than alga, was used in hydrolysis studies at subcritical water (SCW) conditions. This model was applied to algal lipid hydrolysis. The robustness of our model has been tested by verifying both the autocatalytic behavior of lipid hydrolysis reactions, as well as hydrolysis reactions over a range of temperatures.

(254f) Microwave-Assisted Subcritical Water Extraction of Lipids From Wet Algae

Harvind Kumar Reddy[1], Yingqiang Sun[1], Yin Li[2], Tapaswy Muppaneni[1], Sundaravadivelnathan Ponnusamy[1], Shuguang Deng[1], Tanner Schaub[3], Barry Dungan[4], Francisco Holguin[4], Peter Lammers[5], Wayne Voorhies[6] and Peter Cooke[7], (1)Chemical Engineering, New Mexico State University, Las Cruces, NM, (2)Chemical and Biological Engineering, Zhejiang University, Hangzhou, China, (3)College of Agricultural, Consumer and Environmental Sciences, New Mexico State University, Las Cruces, NM, (4)College of Agricultural, Consumer and Environmental Sciences, New Mexico State university, Las Cruces, NM, (5)Energy Research Laboratory, New Mexico State University, Las Cruces, NM, (6)Molecular Biology, New Mexico State university, Las Cruces, NM, (7)Core University Research Resource Laboratory, New Mexico State University, Las Cruces, NM

Microwave-assisted extraction of algal oil at subcritical water conditions was demonstrated with wet algae samples containing up to 90% of water. The extraction efficiency of this method was compared to those of the conventional solvent extraction methods, Folch and Bligh & Dyer. The extraction parameters including biomass loading, temperature and extraction time are optimized after executing a set of designed experiments. Algal oil was produced after purification of bio crude with activated charcoal. The residual algal biomass was also analyzed for its fatty acid and energy content to validate potential byproduct development as part of algal biorefinery. Energy calculations were performed for microwave assisted subcritical water extraction and compared with those of the conventional methods. Analytical instruments including GC-MS, FT-ICR were used to analyze the biomass and biocrude. Thermogravimetric analysis (TGA), TEM and SEM equipped with EDS were used to characterize the biomass in different stages of the extraction process.

(257a) Metabolic Network Modeling of Redox Balancing and Ethanol Production in Scheffersomyces Stipitis

Meng Liang[1], Min Hea Kim[1], Q. Peter He[2], Thomas Jeffries[3] and Jin Wang[4], (1)Department of Chemical Engineering, Auburn University, Auburn, AL, (2)Chemical Engineering, Tuskegee University, Tuskegee, AL, (3)University of Wisconsin-Madison, Madison, WI, (4)Chemical Engineering, Auburn University, Auburn, AL

Keywords: *Scheffersomyces stipitis*; *Pichia stipitis*; stoichiometric model; xylose fermentation; ethanol production; xylitol production; redox balance; xylose reductase; xylitol dehydrogenase

In view of rising concerns over energy sustainability, global warming and feed stock availability, lignocellulosic ethanol has been identified as one of the most promising long-term renewable energy sources [1]. However, many barriers exist for industrializing lignocellulosic ethanol processes. The effective conversion of xylose, as the second abundant mono-saccharide and representative pentose, to ethanol is one of the major barriers of commercial application of lignocellulosic ethanol processes.

Scheffersomyces stipitis (previous named *Pichia stipitis*), as the most promising native strain for xylose fermentation [2], has shown good overall performance on hydrolysate [3]. Understanding its metabolism, especially the central carbon metabolism, is very important to improve the strain, or to use it as a gene provider for other strains, or to provide metabolism adjustment for other strains. Redox couples NAD+/NADH and NADP+/NADPH have been reported to play an important role in energy metabolism and product formation yeasts [4]. Specifically for *S. stipitis*, it has been reported that its product distribution is very sensitive to oxygen transfer rate due to redox imbalance. Therefore, understanding the mechanism of redox balance and tuning up the redox conditions have significant impacts on ethanol production for *S. stipitis*, particularly for xylose fermentation.

In spite of the abundant experimental evidence regarding the role of redox balance in xylose utilization [5], there is a lack of understanding on the mechanism and cellular details of how redox imbalance affects the distribution of different metabolic products, especially considering that the cofactor specificities of the enzymes change according to culture conditions [6]. In this work, we reconstructed the central carbon metabolic network model of *S. stipitis*, and used the model to gain better understanding of its metabolic capabilities, as well as studied the influence of redox balance to products distribution.

First, the topology of central carbon metabolism of *S. stipitis* was identified and the metabolic network was reconstructed by integrating genomic (Pichia stipitis v2.0, KEGG), biochemical (ChEBI, KEGG) and physiological information available for this microorganism and other related yeast. The stoichiometry of the metabolic reactions was used in combination with biosynthetic requirements for growth and pseudo-steady state mass balances over intracellular metabolites for the quantification of metabolic fluxes using flux balance analysis (FBA). The model was employed to perform *in silico* characterization of the phenotypic behavior of *S. stipitis* grown on different carbon sources (glucose and xylose) with various oxygen supplies. The model predictions are in general agreement with published and our own experimental results. The carbon flux through pentose phosphate pathway was consistent with published data[7] when glucose was used as carbon source and aerobic culture was applied. The effects on growth of single reaction deletions were assessed and essential biochemical reactions were identified for different carbon sources. Phenotype phase plane analysis was applied to the model to study the interactive influence of sugar (glucose or xylose) and oxygen. Several phase planes have been identified, all of which are both sugar and oxygen limited except one is solely oxygen-limited.

Following the development of the stoichiometric model, the influence of redox imbalance on products distribution was studied stoichiometrically. Specifically for xylose metabolism, the cofactor specificity of XR and XDH play very important roles for redox balance. The influence of alternating the cofactor specificities of the two enzymes has recently been studied by protein-engineering the two enzymes to reversing or adding or enhancing the affinity to the other cofactor. It has been shown that alternating the cofactor specificities of the enzymes could not always improve the strain performance as desired [8]. With the model properly constrained, the influences of changing cofactor specificities of XR and XDH have been investigated. The results provide important insight on how the alternations of the specificities of the enzymes change the redox balance and product distribution. The increase of affinity of XR to NADH will promote the ethanol production and depress the xylitol production while the increase of affinity of XDH to NADPH does not always improve the ethanol production, especially if XR already has a high affinity to NADH. A comparison of flux distribution involved in redox balance with different carbon sources and oxygenation conditions also provides insights on the role of redox balance in the metabolism of glucose/xylose and by-products such as xylitol, glycerol and acetic acid. By performing multivariate statistical analysis tools, we were able to illustrate how the cofactor specificities of the enzymes influenced the redox balance and carbon flux distribution.

References

1. Cherubini, F., The biorefinery concept: Using biomass instead of oil for producing energy and chemicals. *Energy Conversion and Management* **2010,** *51*, 1412-1421.

2. Jeffries, T. W.; Van Vleet, J. R. H., Pichia stipitis genomics, transcriptomics, and gene clusters. *FEMS yeast research* **2009,** *9*, 793-807.

3. Rumbold, K.; van Buijsen, H. J.; Gray, V. M.; van Groenestijn, J. W.; Overkamp, K. M.; Slomp, R. S.; van der Werf, M. J.; Punt, P. J., Microbial renewable feedstock utilization: A substrate-oriented approach. *Bioengineered bugs* **2010,** *1*, 359-66.

4. Bakker, B. M.; Overkamp, K. M.; van Maris, A. J.; Kötter, P.; Luttik, M. a.; van Dijken, J. P.; Pronk, J. T., Stoichiometry and compartmentation of NADH metabolism in Saccharomyces cerevisiae. *FEMS microbiology reviews* **2001,** *25* (1), 15-37.

5. Jeffries, T. W.; Grigoriev, I. V.; Grimwood, J.; Laplaza, J. M.; Aerts, A.; Salamov, A.; Schmutz, J.; Lindquist, E.; Dehal, P.; Shapiro, H.; Jin, Y.-S.; Passoth, V.; Richardson, P. M., Genome sequence of the lignocellulose-bioconverting and xylose-fermenting yeast Pichia stipitis. *Nature biotechnology* **2007,** *25*, 319-26.

6. Yablochkova, E. N.; Bolotnikova, O. I.; Mikhailova, N. P.; Nemova, N. N.; Ginak, A. I., The Activity of Key Enzymes in Xylose-Assimilating Yeasts at Different Rates of Oxygen Transfer to the Fermentation Medium. *Microbiology* **2004,** *73* (2), 129-133.

7. Fiaux, J.; Cakar, Z. P.; Sonderegger, M.; Wüthrich, K.; Szyperski, T.; Sauer, U., Metabolic-flux profiling of the yeasts Saccharomyces cerevisiae and Pichia stipitis. *Eukaryotic cell* **2003,** *2*, 170-80.

8. Krahulec, S.; Klimacek, M.; Nidetzky, B., Analysis and prediction of the physiological effects of altered coenzyme specificity in xylose reductase and xylitol dehydrogenase during xylose fermentation by Saccharomyces cerevisiae. *Journal of Biotechnology* **2012,** *158* (4), 192-202.

(257b) Engineering Exogenous Sugar Catabolism in Saccharomyces Cerevisiae for Improved Biofuels Production

Eric M. Young[1], Austin D. Comer[1] and Hal Alper[2], (1)Chemical Engineering, University of Texas at Austin, Austin, TX, (2)Chemical Engineering, The University of Texas at Austin, Austin, TX

Metabolic engineering is a powerful platform for controlling cellular production of biomolecules. However, the field is still limited by few effective methods to alter the carbon source preference of an organism. For the case of lignocellulosic biomass conversion, recombinant strains of *Saccharomyces cerevisiae* are limited in their catabolic rates of pentose sugars. Specifically, both sugar transport and assimilation steps limit pentose utilization in these strains. While much progress has been made in altering the pentose assimilation steps, transporter engineering has only recently been shown to be effective. As an example, we have recently demonstrated how this approach can improve growth rate by 70% by simply altering transporter function. It is clear that both transporter modifications and catabolic enzyme selection and/or evolution will be necessary for optimizing pentose consumption in *S. cerevisiae*. In this talk, we will describe recent progress towards this goal. Advances in molecular transporter engineering will be discussed alongside downstream pathway optimization strategies. Synergies between these two approaches will be discussed. These studies provide insight into the possibility of efficient industrial production of value-added chemicals from lignocellulosic biomass.

(257c) Directed Evolution of a Cellobiose Utilization Pathway for Biofuel Production

Dawn Eriksen, Department of Chemical and Biomolecular Engineering, University of Illinois, Urbana, IL and Huimin Zhao, Department of Chemical and Biomolecular Engineering, Department of Chemistry and Biochemistry, Institute for Genomic Biology, University of Illinois at Urbana-Champaign, Urbana, IL

Lignocellulosic biofuel production in S. *cerevisae* can be made more economical by pentose sugar utilization. However, slow growth occurs if pentose sugars are the sole carbon source. Cofermentation with glucose inhibits effective utilization of the pentose sugars, called glucose repression. One strategy to address glucose repression is to utilize cellobiose rather than glucose. The cellobiose utilization pathway consists of a cellodextrin transporter (CDT-1) coupled with an intracellular β-glucosidase (GH1-1). The process was optimized through the application of a new directed evolution technique. The pathway was constructed with error-prone libraries of the CDT-1 and GH1-1 via the DNA Assembler. Mutants with improved activities were identified by higher growth rate and then subsequently screened for increased ethanol yield. The results of this optimization produced a non-intuitive mutation in GH1-1, improving cellobiose utilization by over 30%. Further rounds of directed evolution identified mutants with reduced cellodextrin accumulation, a major setback in cellobiose utilization. This is one of the first examples of engineering multiple proteins simultaneously on the pathway level *in vivo* instead of independently engineering single proteins.

(257d) Adaptive Evolution of Saccharomyces Cerevisiae to Improve Ethanol Production From Corn Cob Hydrolyzate Under High Temperature

Xianni Qi, Baowei Wang, Ran Tu, **Qinhong Wang** and Yanhe Ma, Tianjin Institute of Industrial Biotechnology, Chinese Academy of Sciences, Tianjin, China

To improve cellulosic ethanol yield and develop the cost-effective simultaneous saccharification fermentation for ethanol production, the strain should be developed with improved tolerance to high temperature and inhibitory compounds from lignocellusic biomass. Here, an industrial strain of *S. cerevisiae* Ethanol Red E491 (Ethanol Performance Group, NJ, USA) was evolved using corn cob hydrolyzate at step-increased temperature (from 37-42°C) and the resulting daughter strain, TIB-S.C Y01, produced ethanol much more rapidly than its parent in fermentations of corn cob hydrolyzate at 42°C. Adaptation improved fermentation performance of the evolved strain. Based on transcriptome analysis of parent and evolved strains via RNA sequencing, we find evidence of dramatic up/down-regulation in Bsc1p, Hxk2p, Pfk27p, Mer1p, Gnd1p, and Tkl1p genes. Consistent with the complex, multigenic nature of multiple stresses, we observe adaptations in a diversity of cellular processes, implying a rugged fitness landscape for the tolerance of high temperature and inhibitory compounds in yeast.

(257e) Metabolic Flux Analysis of *Shewanella Oneidensis* MR-1 Producing Electricity in a Microbial Fuel Cell

Thomas Wasylenko, Brian Pereira, Hamid Rismani-Yazdi and Gregory Stephanopoulos, Department of Chemical Engineering, Massachusetts Institute of Technology, Cambridge, MA

Microbial fuel cells (MFCs) are bioelectrochemical systems in which microbes catalyze the production of electricity from the oxidation of organic substrates. In addition to having several applications, including wastewater treatment, MFCs present an interesting metabolic system: microbes convert chemical energy to external electrical energy by utilizing a solid electrode as the terminal electron acceptor. We have employed ^{13}C metabolic flux analysis, in which isotopically-labeled tracers are applied to elucidate fluxes, to investigate the metabolism of a microbe within an MFC. *Shewanella oneidensis* MR-1 was selected as the model organism because it is known to be electroactive and its metabolic network has been previously described. We present the fluxes through gluconeogenesis, the TCA cycle, the pentose phosphate pathway, and serine metabolism for *S. oneidensis* MR-1 in an MFC and compare them with the fluxes in systems utilizing other terminal electron acceptors. These results provide insight into engineering microbes for electricity production in MFCs.

(257f) Evaluating the Potential of Using the Thermophilic Bacterium Thermus Thermophilus for Biofuels Production

Maciek R. Antoniewicz[1], Jing Lu[2], Aditi Swarup[3] and Kathleen Casey DeWoody[3], (1)Department of Chemical and Biomolecular Engineering, University of Delaware, Newark, DE, (2)University of Delaware, Newark, DE, (3)Chemical Engineering, University of Delaware, Newark, DE

Thermus thermophilus is a thermophilic eubacterium that thrives at temperatures above 80°C. In this presentation, we will demonstrate the potential of using *T. thermophilus* for the production of volatile biofuels. *T. thermophilus* has a high growth rate (i.e. 0.55 1/hr on defined medium, and 0.95 1/hr on complex medium), high biomass yield (similar to *E. coli*), and shows high transformation competence making it amendable to genetic manipulations. A current bottleneck for biotechnological applications of *T. thermophilus* is the limited knowledge regarding its metabolism and growth characteristics. In this work, we reconstructed and validated the metabolic network model of *T. thermophilus* HB8 using ^{13}C tracer experiments and characterized its metabolism using ^{13}C-metabolic flux analysis (MFA). *T. thermophilus* HB8 was successfully grown in custom-designed mini-bioreactors (10 mL) on optimized minimal medium without yeast extract. A large scale metabolic network model was first extracted from KEGG and then validated using parallel labeling experiments with ^{13}C-glucose tracers. For each tracer experiment we measured the labeling of amino acids (using GC-MS) and the labeling of CO_2 (using an on-line mass spectrometer). Metabolic fluxes were determined by fitting the measured isotopomer data to the reconstructed metabolic model. After validating the model, we performed tolerance studies to evaluate the potential of using *T. thermophilus* for the production of biofuels, including short chain alcohols such as ethanol, propanol, n-butanol, and isobutanol. Our results that strongly support the biotechnological potential of this organism for efficient biofuels production at high temperatures.

(257g) Redirecting Carbon Flux in Clostridium Thermocellum to Increase Ethanol Yield

Yu Deng, Daniel Olson and Lee R. Lynd, Thayer School of Engineering, Dartmouth College, Hanover, NH

In *Clostridium thermocellum*, a thermophilic anaerobic bacterium able to rapidly ferment cellulose to ethanol and organic acids, pyruvate kinase (EC 2.7.1.40) appears to be absent based on both the genome sequence as well as enzymatic assays. Lamed and Zeikus hypothesized that a pyridine nucleotide transhydrogenation pathway through oxaloacetate and malate. We further hypothesize that there is a conversion of phosphoenolpyruvate (PEP) to pyruvate may occur via a three step pathway involving conversion of PEP to oxaloacetate(OAA) by phosphoenolpyruvate carboxykinase, OAA to malate by malate dehydrogenase, and conversion of malate to pyruvate by malic enzyme. In this work we examined the impact of targeted modification of enzymes associated with this pathway, termed the "malate shunt", including exogenous expression of the pyruvate kinase gene from *Thermoanaerobacterium saccharolyticum*, mutation of the phosphoenolpyruvate carboxykinase and deletion of malic enzyme gene. Cellobiose-grown cultures of strain YD01, in which phosphoenolpyruvate carboxykinase expression was diminished by modifying the start codon from ATG to GTG, exhibited an ethanol yield of 1.33 moles ethanol/mole-glucose equivalent, 3.25-fold higher than the wild-type strain. A second strain, YD02, in which the gene for malic enzyme and part of malate dehydrogenase were deleted, exhibited an ethanol yield of 1.28 mole-ethanol/mole-glucose equivalent, also over 3-fold higher than the wild-type strain. In addition to increased ethanol yields, other consequences of routing flux away from the malate shunt were observed including increased recovery of substrate carbon in documented fermentation products, decreased production of amino acids, and increased tolerance to added ethanol. We hypothesize that these may be consequences of relieving NADPH oversupply when flux is directed away from the malate shunt, an interpretation supported by the measured cofactor specificities of malate dehydrogenase and malic enzyme.

(257h) Metabolic Engineering of Propionibacteria for n-Propanol Production

Ehab M Ammar and Shang-Tian Yang, William G. Lowrie Department of Chemical and Biomolecular Engineering, Ohio State University, Columbus, OH

Propionibacteria are widely used in industry for manufacturing of Swiss cheese, vitamin B12 and propionic acid. Little is known about their genetics and only a few reports are available on their metabolic engineering. Most previous metabolic engineering work on propionibacteria aimed at improving the fermentative production of either vitamin B12 or propionic acid. n-Propanol is a common solvent, intermediate in many industrial applications and a promising biofuel. No known microorganism can produce n-propanol naturally in sufficient quantities for manufacturing purposes. In this research, a bifunctional alcohol aldehyde deydrogenase (*aad*) was cloned from *E. coli* and overexpressed in propionibacteria. The mutant expressing the *aad* gene produced n-propanol from propionyl-CoA, which is the precursor for propionic acid biosynthesis. The production of n-propanol was limited by available NADH, which was improved significantly by using a mixture of different carbon sources. To our best knowledge, this is the first report on producing n-propanol by metabolically engineered propionibacteria, which offers a novel route to produce n-propanol from biomass.

(260a) Characterization of Carbohydrate Accessibility and Enzyme Adsorption Capacity for Diverse Cell Wall Phenotypes Subjected to Alkaline Hydrogen Peroxide Pretreatment

Muyang Li[1], Sivakumar Pattathil[2], Michael G. Hahn[2] and David Hodge[1,3], (1)Biosystems and Agricultural Engineering, Michigan State University, East Lansing, MI, (2)Complex Carbohydrate Research Center, University of Georgia, Athens, GA, (3)Chemical Engineering and Material Science, Michigan State University, East Lansing, MI

This study characterizes the changes to polysaccharides in terms of their abundance, accessibility, and affinity for binding by a range of cellulolytic enzymes and CBMs in the cell walls of grasses and dicots that were alkaline hydrogen peroxide (AHP) pretreated at varied severities or hydrothermally pretreated followed by AHP post-treatment. These bioenergy feedstocks represent relatively diverse cell wall phenotypes, including the monocots switchgrass (*Panicum virgatum* cv. Cave-In-Rock), corn stovers (a commercial hybrid and inbred brown midrib lines bm1 and bm3), and *Miscanthus* spp. and herbaceous and woody dicots including goldenrod (*Solidago* spp.) and hybrid poplar (*Populus* spp.). It was shown that increasing H_2O_2 loading results in increasing digestibility only for monocot grasses, while the digestibility of dicots is saturated at low H_2O_2 loadings. Polysaccharide accessibility in the residual cell wall is characterized through "glycome profile" which involves screening successive cell wall extracts with an array of 156 monoclonal antibodies with affinities for the entire spectrum of non-cellulosic cell wall polysaccharides revealing changes in their accessibility and extractability as a function of pretreatment. Enzyme binding is assessed by characterizing binding isotherms for the proteins contained commercial cellulase enzyme cocktails, purified cellulase components, and a library of GFP-linked CBMs.

(260b) Lignin Structural Changes Associated with Oxidative Pretreatment Catalyzed by Cu-Diimine Complexes

Zhenglun Li, Department of Chemical Engineering and Material Science and DOE Great Lakes Bioenergy Research Center, Michigan State University, East Lansing, MI and David Hodge, Chemical Engineering and Material Science, Michigan State University, East Lansing, MI

A wide range of alkaline-oxidative treatments of lignocellulose are known from the literature and practice of chemical pulping and bleaching and from more recent cellulosic biofuels applications. Alkaline hydrogen peroxide (AHP) has been developed as a pretreatment or post-treatment for increasing biomass digestibility by decreasing the recalcitrance of high-lignin biomass. Traditional AHP pretreatment uses the high pH of 11.5 for deprotonation of hydrogen peroxide which leads to the formation of oxidative radicals (·OH and $\cdot O_2^-$), whereas the low selectivity of the radical reactions and the non-productive decomposition of hydrogen peroxide to oxygen at alkaline pH limit the economic competitiveness of AHP pretreatment. Significantly, oxidative delignification catalyzed by coordinated transition metal ligands within laccases and peroxidases is the route taken for lignin degradation in nature. Catalyzed oxidative pretreatments offer the promise of mild temperatures, high reaction specificities, and high oxidizing efficiency that can simultaneously generate a material that is susceptible to enzymatic carbohydrate depolymerization, be performed at decreased pH with correspondingly less inorganics load, and generate modified low MW aromatics that may be potentially upgraded to petroleum-displacing fuels and chemicals. In this work, the effect of Cu-diimine complexes during catalyzed oxidative pretreatment on model lignins is characterized. Model lignins include a Kraft lignin and dioxane milled-wood and enzymatic lignins from select hardwoods and grasses, while characterization is performed using ^{31}P NMR, ^{1}H-^{13}C HSQC NMR and SEC which show distinct structural changes during pretreatment including, among others, clear evidence of lignin modification and depolymerization and cellulose depolymerization. Additionally, significant improved cellulolytic enzyme digestibility is found for select catalyzed pretreatments in woody dicots.

(260c) Mechanism of Exo-Exo Synergism Between CBH1 and CBH2 During the Hydrolysis of Cellulose

Makoto Ikeo, Shohei Okino, Yoshiki Ueno and Daisuke Taneda, Technology Development Department, JGC Corporation, Higashiibaraki-gun, Ibaraki Prefecture, Japan

In past studies, JGC corporation has reported that the efficiency of enzymatic hydrolysis of cellulose using cellulase with low enzyme loadings can be significantly improved by placing the reaction under static condition (2010 AIChE Annual Meeting, 617c Breaking the Barrier to Obtain High Yields On Enzymatic Hydrolysis Using Low Enzyme Loadings). The hydrolysis of 100ml 10% filter paper slurry at the enzyme loading of 6 mg protein/g-dry substrate was conducted using Cellulase SS (Nagase Chemtex). The agitated condition obtained a hydrolyzate with a final glucose concentration of 60 g/L. In contrast, a hydrolyzate with a final glucose concentration of 90 g/L was obtained under the static condition.

Further investigations found that this improvement occurs by the preservation of CBH2 and the exo-exo synergism between it and CBH1 under the static condition. Cellulase SS solution was agitated without any substrates under 50 °C, and SDS-PAGE analysis was conducted to measure the concentration of individual components. Of all the cellulase components, only the band of CBH2 disappeared within 6 hours of agitation, indicating its deactivation. Further examination using fractionated enzyme solutions, one of which lacked CBH1 and the other lacking CBH2, revealed that both CBH1 and CBH2 are required for the efficient hydrolysis of cellulose.

In this study, the mechanism of the exo-exo synergism between CBH1 and CBH2 was investigated by comparing the characteristics of enzyme adsorption and desorption to the cellulose under agitated and static conditions. The hydrolysis of 10% filter paper slurry with the enzyme loading of 12 mg-protein/ g-dry-substrate under both static and agitated conditions were conducted, and the changes in concentration of each cellulase component in the hydrolyzate was monitored using SDS-PAGE and a densitograph. The glucose concentrations of the hydrolyzates obtained were 88 g/L and 60 g/L under the static and agitated conditions respectively. Under both conditions, the CBH1 and CBH2 concentrations in the hydrolyzate dropped sharply, showing concentration decreases of 100% and 80% respectively, at the initial stage of the reaction, indicating their adsorption to the substrate. Under the static condition in which CBH2 is active, the CBH1 and CBH2 concentrations gradually rose as the reaction progressed, indicating desorption from the substrate. The final concentrations of CBH1 and CBH2 after the reaction were 80% and 90% of the original concentration respectively. Under the agitated condition in which CBH2 is deactivated, no increase in the concentration of CBH1 or CBH2 was observed after desorption. CBH2 is likely to be deactivated by shear stress from agitation. However, CBH1 is stable even under the agitated condition, indicating that, under agitated condition, CBH1 remains adsorbed to the substrate and the hydrolysis remains incomplete.

From this result, the exo-exo synergism was hypothesized as enhancing the hydrolysis by preventing the unproductive binding of CBH1 during the reaction. For further investigation, the adsorption and desorption characteristics of enzymes during the sequential addition of CBH1 and CBH2 fractions was observed. The hydrolysis of 10 wt% filter paper using CBH1 fraction yielded hydrolyzate with glucose concentration of 60 g/L at the end of the reaction, and no desorption of CBH1 from substrate was observed. The addition of CBH2 fraction at the later stage of this reaction reaccelerated the reaction, increasing the final glucose concentration to 90 g/L. Also, CBH1 desorption was observed after the addition of the CBH2 fraction. The results obtained suggested that CBH1 desorbs from the substrate and completes the hydrolysis under the presence of active CBH2, but remains absorbed on the substrate in an inactive form when CBH2 is absent.

(260d) Cellulose Thin Film Hydrolysis by Cellulase Enzymes in the Presence of Inhibitors – a QCM Study

Utshab Chakravorty[1], Stephen E. Rankin[1], Sue E. Nokes[2] and Barbara L. Knutson[1], (1)Chemical and Materials Engineering, University of Kentucky, Lexington, KY, (2)Department of Biosystems and Agricultural Engineering, University of Kentucky, Lexington, KY

Cellulose thin film hydrolysis by cellulase enzymes in the presence of inhibitors – a QCM study

Utshab Chakravorty, Barbara Knutson, Stephen E. Rankin, Sue Nokes

Abstract

Conversion of biomass to chemicals and biofuels using the products of cellulose hydrolysis is an important area of research for the production of the next generation of biofuels. However, the saccharification and biochemical conversion of cellulose from lignocellulosic biomass is potentially limited by the release of inhibitors during the processing steps which are used to make the cellulose accessible. Understanding the role and sensitivity of these inhibitors in altering the activity of cellulase enzymes is necessary to the design of pretreatment and separation approaches that maximize cellulose conversion. Bulk studies of cellulose degradation do not provide sufficiently detailed time-dependent and concentration-dependent information to delineate the effect of inhibitors on the adsorption of cellulases to cellulose surface and the subsequent hydrolytic activity of the cellulases. This study uses the frequency response of model cellulose thin films cast onto Quartz Crystal Microbalance (QCM) sensors to model the binding and hydrolysis of cellulose by cellulases (from *Trichoderma reesei*) in the presence of furfural and acetic acid. The sensitivity of the frequency response during cellulose hydrolysis to the specific inhibitor and inhibitor concentration is demonstrated. The concentration-dependent inhibition frequency response data are modeled using a reaction network that accounts for enzyme adsorption, hydrolysis, and several modes of enzyme inhibition and deactivation. The model results can be used to interpret the mode of inhibition. The rate constants of the binding and hydrolysis reactions determined by regression to sets of QCM data with varying inhibitor concentration indicate a strong binding affinity of the enzyme to the substrate and consistent rate coefficient for hydrolysis. The QCM data are used to understand the possible roles of various modes of active site inhibition and inhibition of binding to cellulose on changes in overall conversion in the presence of the inhibitors.

(260e) Investigation of Hemicellulase Inhibition in the Production of Bioethanol

Erik Kuhn[1], Xiaowen Chen[2], Clare J. Dibble[2], Joseph Shekiro[3], Nicholas J. Nagle[1] and Richard T. Elander[4], (1)Bioprocessing R&D, National Renewable Energy Laboratory, Golden, CO, (2)National Bioenergy Center, National Renewable Energy Laboratory, Golden, CO, (3)National Renewable Energy Laboratory, Golden, CO, (4)National Bioenergy Center, National Renewable Energy Lab, Golden, CO

Many of the lignocellulosic bioconversion routes utilize xylose and/or glucose produced by operations such as pretreatment and enzymatic hydrolysis. The insoluble solids (IS) from pretreatment enter enzymatic hydrolysis and exist either in the presence of the solubilized pretreatment fraction (i.e. whole slurry) or absent of the solubilized fraction (i.e. washed). The difference between washed IS and whole slurry enzymatic hydrolysis is the background concentration of sugars, organic acids, and other compounds prior to enzymatic hydrolysis. Whole slurry enzymatic hydrolysis is preferred as it decreases water and utility use, but higher glucose and xylose yields have been observed utilizing washed solids. Using commercially available hemicellulase and cellulase preparations we have observed little to no xylose production from high solids, whole slurry enzymatic hydrolysis on select dilute acid pretreatment slurries while greater than 80% xylan to xylose yields have been observed from high solids, washed IS enzymatic hydrolysis. Previous work has shown end product inhibition of Aspergillus niger beta-D-xylosidase by D-xylose, and inhibition from phenolic acid compounds on the endoxylanase, XynA, from Clostridium cellulovorans. Organisms that utilize C-5 sugars in fermentation decrease the xylose concentration of the fermentation broth, which may allow for further xylose production during fermentation. In the present work a range of potential xylanase inhibitors observed in dilute-acid pretreatment slurries and fermentation broths are dosed into washed IS and beechwood xylan at varied concentrations prior to enzymatic hydrolysis, and the effect on xylose production is presented.

(260f) Increased Enzyme Binding to Substrates Does Not Always Increase Catalytic Activity

Shishir Chundawat[1], Dahai Gao[1], Anurag Sethi[2], Stephanie Crews[1], Leonardo da costa Sousa[1], Nirmal Uppugundla[1], Venkatesh Balan[1], Sandrasegaram Gnanakaran[2] and Bruce E. Dale[1], (1)Chemical Engineering and Material Science, Great Lakes Bioenergy Center, Michigan State University, Lansing, MI, (2)Theoretical Division, Los Alamos National Laboratory, Los Alamos, NM

Substrate binding affinity is typically thought to be one of the primary rate-limiting steps preceding enzyme catalytic action during most homogeneous reactions. However, interfacial-based enzyme catalysis on insoluble crystalline substrates, like cellulose, has an additional step comprising of individual biopolymer chain decrystallization from the substrate interface. This additional decrystallization step has ramifications on the role of enzyme-substrate binding and its relationship to overall catalytic efficiency. We find that altering the crystalline structure of cellulose from its native allomorphic state called cellulose I to cellulose III results in reduced overall cellulase binding but enhanced hydrolytic activity on the latter substrate. We have constructed a comprehensive kinetic model for processive enzymes acting on insoluble substrates to explain this anomalous finding. Our study indicates that this phenomenon can be explained by the enhanced decrystallization kinetics of individual cellulose chains from the surface of cellulose III that despite reduced overall binding affinity results in enhanced catalytic activity. These findings have implications on our understanding of interfacial enzyme catalysis for biopolymers that incorporate analogous decrystallization events.

(283a) Biodistribution and Cellular Uptake Upon Intranasal Administration of Monodisperse Biodegradable Particles

Timothy Brenza[1], Latrisha Petersen[1], Yanjie Zhang[1], Lucas Huntimer[2], Amanda Ramer-Tait[2], Michael J. Wannemuehler[2] and Balaji Narasimhan[1], (1)Chemical and Biological Engineering, Iowa State University, Ames, IA, (2)Veterinary Microbiology and Preventive Medicine, Iowa State University, Ames, IA

Size and chemistry are known to play an important role in particle biodistribution and cellular uptake. The synthesis of monodisperse biodegradable polyanhydride particles of multiple sizes have enabled us to systematically evaluate the role of initial particle size and chemistry on biodistribution and trafficking. In this work we used a murine model to evaluate the biodistribution of intranasally administered 200 nm, 400 nm, and 1.5 μm (nominal size) poly(sebacic acid) particles and commercial polystyrene particles of the same size. The biodistribution of these particles was evaluated using fluorescent imaging of excised organs one day and seven days post administration. Cellular uptake was evaluated for the lung and lymph nodes by labeling tissue homogenates for specific cell types. While all three particle sizes are present in the lungs one day after administration, the quantity of 200 and 400 nm particles present, as determined through mean fluorescence intensity, was greater than the 1.5 μm particles. The percentage of neutrophils, macrophages, dendritic cells and epithelial cells with particles was also found to be dependent upon the size and chemistry of the particle. The insights gained from these studies will enable the rational design of nanoparticle-based adjuvants and delivery systems for vaccines against respiratory pathogens.

Title: Applications of nanoclays in controlled release of API-polymer matrix prepared by hot-melt mixing

Authors: Nonjaros Chomcharn and Marino Xanthos

Affiliates: Otto H. York Department of Chemical, Biological and Pharmaceutical Engineering, New Jersey Institute of Technology, Newark, NJ 07102

Abstract

Polymers have been widely used in oral pharmaceutical products to control the release of active pharmaceutical ingredients (APIs) from a dosage. Adding suitable nanoclays to the polymer matrix can enhance the strength of the matrix resulting in further extended release profile of APIs. This work investigates the effect of anionic clays (hydrotalcite, HT) on the bioavailability of a pH-dependent polymer (Eudragit® L100-55) containing anionic APIs (aspirin, Asp). Eudragit® L100-55 is an anionic copolymer based on methacrylic acid and ethyl acrylate and is dissolved at pH 5.5 and above. Triethyl citrate (TEC) was employed as a plasticizer at the ratio of 1:4 (TEC:Eudragit® L100-55) to reduce the processing temperature. The samples were prepared by hot melt mixing using batch mixer (Brabender®) at 50 rpm and 100°C for 5 minutes. After melt mixing, the samples were pressed into thin discs for further characterization.

In the formulation containing Asp and nanoclays HT (Asp/HT), Asp enhanced the collapse of nanoclay aggregates from multilayer structure to platelets dispersed throughout the polymer matrix, resulting in an increase of tortuosity and melt viscosity. XRD confirmed the collapse of HT due to the lower peak at the 2Θ angle of 11.2° and 22.4°. Thermogravimetric analysis (TGA) revealed greater stability of the aspirin-TEC/Eudragit® L100-55 matrix containing 10% HT. The release rate of aspirin from aspirin-TEC/Eudragit® L100-55 matrix containing HT was significantly prolonged presumably as a result of increased viscosity and possible crosslinking in the presence of Asp/HT combinations.

Dissolution study under simulated intestinal fluid (pH 7.4) at 37°C of the formulation containing Asp/HT revealed that approximately 80% and 90% of aspirin was released over the period of 12 hours and 24 hours respectively whereas an identical formulation without HT released nearly 100% of aspirin within 4 hours. Delayed-release dissolution showed that less than 10% of Asp was released in acid stage (2 hours) in both formulations containing only Asp and Asp/HT. After the pH was adjusted to pH 7.4 (buffer stage), all formulations have similar dissolution profiles as shown under the condition of pH 7.4 medium without exposed to acid stage.

(283c) Rationally Engineered Nanoparticles for Overcoming Drug Resistance in Multiple Myeloma

Tanyel Kiziltepe[1], Jonathan D. Ashley[2], **Jared F. Stefanick**[2], Yu Qi[2], Nathan J. Alves[2], Michael W. Handlogten[2], Mark A. Suckow[3], Rudolph M. Navari[4] and Basar Bilgicer[2], (1)Advanced Diagnostics and Therapeutics / Chemical and Biomolecular Engineering, University of Notre Dame, Notre Dame, IN, (2)Chemical and Biomolecular Engineering, University of Notre Dame, Notre Dame, IN, (3)Biological Sciences, University of Notre Dame, Notre Dame, IN, (4)Indiana University School South Bend of Medicine, South Bend, IN

Multiple myeloma (MM), a B-cell malignancy characterized by proliferation of monoclonal plasma cells in the bone marrow (BM), is the second most common type of blood cancer in the U.S. Despite the recent advances in treatment strategies and the emergence of novel therapies, it still remains incurable due to the development of drug resistance with a median survival of 4-5 years. A major factor that contributes to development of drug resistance in MM is the interaction of MM cells with the BM microenvironment. It has been demonstrated that the adhesion of MM cells to the BM stroma via the $\alpha_4\beta_1$ integrin (a.k.a. Very Late Antigen-4, VLA-4) leads to cell adhesion mediated drug resistance (CAM-DR), which enables MM cells to gain resistance to drugs such as doxorubicin (Dox)–a 1st line chemotherapeutic in the treatment of MM.

In our design, we used micellar nanoparticles as dynamic self-assembling scaffolds to present VLA-4 antagonist peptides and Dox conjugates, simultaneously, to selectively target MM cells and to overcome CAM-DR. VLA-4-antagonist peptides were conjugated via a multifaceted synthetic procedure for generating precisely controlled number of targeting functionalities, while Dox was conjugated to the nanoparticles through a hydrazone bond. When the nanoparticles are delivered to the MM cells, as a first step they interact with the cell surface VLA-4 integrins and inhibit MM cell adhesion to the stroma, thereby preventing development of CAM-DR. In the second step, Dox exerts its cytotoxic effects after cellular uptake, as the nanoparticles are designed to require a low pH environment, such as the endocytic vesicles, to release active Dox.

Our studies have shown that the nanoparticles were efficiently internalized by MM cells and induced cytotoxicity. Mechanistic studies revealed that nanoparticles induced DNA double-strand breaks and apoptosis in MM cells. Importantly, multifunctional nanoparticles overcame CAM-DR and were more efficacious than Dox when MM cells were cultured on fibronectin-coated plates. Finally, in a MM xenograft model, nanoparticles preferentially homed to MM tumors with ~10 fold more drug accumulation and demonstrated dramatic tumor growth inhibition with a reduced overall systemic toxicity. Altogether, we demonstrate the disease driven engineering of a nanoparticle-based drug delivery system, enabling the model of an integrative approach in the treatment of MM.

(283d) Potential Pharmaceutical Applications of Uniform-Sized Chitosan Micro/Nanospheres with Autofluorescent Property

Wei Wei and Guanghui Ma, National Key Laboratory of Biochemical Engineering, Institute of Process Engineering, Chinese Academy of Sciences, Peking, China

Chitosan has been adopted to have potential application in pharmaceutical science due to its outstanding properties of nontoxicity, biocompatibility, biodegradability, mucus-adhesion, and low cost. To cater to specific clinical requirements, we developed a series of novel chitosan-based mciro/nanospheres and testified their potenials in pharmaceutical applications.

First, uniform-sized chitosan microspheres were prepared by porous membrane emulsification technique. These microspheres were found to be autofluorescent due to the crosslinking reaction. Their uniform size and autofluorescent property significantly facilitated our following studies.

When microspheres were used as a drug carrier, their particle size highly correlated to subsequent transportation and metabolism in vivo. These microspheres with different size (2.1 μm, 7.2 μm, and 12.5 μm) exhibited distinguishing fates in bioadhesion, absorption and distribution, while the biodegradation also varied due to the different RES (reticuloendothelial system) locations.

The demands on drug release behavior vary from case to case in clinical therapy. To solve this problem, we developed a series of novel chitosan microspheres, which could exhibit different release profiles. The tunable ability of structural properties, such as surface charge, cavity size, and wall porosity, enabled the modification of these microspheres to cater to specific requirements in clinic.

Having revealed the influences of particle size and structure properties, we next used hollow-porous chitosan microspheres for oral administration of insulin. The hollow structure could conserve the bioactivity of encapsulated insulin, the quaternized group in matrix could decrease drug leakage by steric hindrance and enhance bioadhesive ability by electrostatic interaction, and the wall porosity could promote the insulin release. As expected, in vivo evaluation saw an optimal reduction in blood glucose level and powerful therapeutic effects after treatment with as-designed microspheres.

We also used premix membrane emulsification technique to prepare chitosan (CS) and its derivative N-trimethyl chitosan chloride (TMC) nanoparticles (NPs) and testified their feasibility as carrier for insoluble anticancer drug paclitaxel (PTX). Both NPs displayed a narrow size distribution (150-200 nm), a high encapsulation efficiency (>80%), and a relatively high loading content (>30%). PTX could be well distributed in the NPs through an in-situ crystallization process. PTX-TMC NPs could be more readily taken up by LLC (Lewis Lung Cancer) cells due to the larger amount of positive charge on the surface of TMC NPs, thus leading to a better cytotoxicity than PTX-CS NPs and free drug (Taxol®). Moreover, PTX-TMC NPs could be absorbed more effectively by the gastrointestinal after orally administrated to LLC tumor-bearing C57/BL6 mice, accumulated in tumor tissue to a larger extent, exhibited greater tumor growth inhibition effect, and caused lower side effects than PTX-CS NPs and Taxol®.

All these results strongly supported that these uniform-sized autofluorescent chitosan micro/nanospheres hold great potentials for oral delivery of protein drugs and anticancer chemotherapy.

(283e) Preparation of Polymeric Micro- Nano- Spheres Using a Microfluidic Approach for Delivery of Nutraeuticals

Raziye Samimi[1,2], Mehrnaz Salarian[2,3], Kazi Farida Akhter[1,2], **Paul A. Charpentier**[1,2,3] and Edmond Lui[2,4], (1)Department of Chemical and Biochemical Engineering, the University of Western Ontario, Canada, London, ON, Canada, (2)The Ontario Ginseng Innovation & Research Consortium, London, ON, Canada, (3)Biomedical Engineering Graduate Program, the University of Western Ontario, Canada, London, ON, Canada, (4)Department of Physiology and Pharmacology, the University of Western Ontario, London, ON, Canada

North American ginseng is a medicinal plant used in traditional herbal medicine whose root extracts are being used for commercial over the counter anti-cold formulations such as Cold-FX™. The main components of ginseng are ginsenosides and polysaccharides which provide biological activities such as anti- and pro-inflammatory effects. In this work, development of appropriate delivery systems for oral administration and bone cements was investigated, with release of bioactive and bioassay results described. Micro/nanospheres of both ginseng polysaccharide extracts and biodegradable poly (propylene fumarate) (PPF), and poly (DL-lactic-co-glycolic acid) (PLGA) containing ginsenocide extracts were prepared using a microfluidic device and compared to other conventional anti-solvent techniques. The morphology and size of the carriers was characterized by scanning electron microscopy (SEM), Transmission electron microscopy (TEM), ZEISS light microscopy, and dynamic light scattering (DLS). Depending on experimental conditions, microfluidics was found to provide unimodal spheres in the 10-50 um region or down to 20 nm with very narrow particle size distributions. Nanosized extracts were then encapsulated with gelatin, to give 180 nm spheres suitable for integration into capsules for subsequent oral drug delivery. Microsphere formation in the microfluidic system was modeled using computational fluid dynamics (CFD) while release profiles were studied and quantified by UV-Vis spectrophotometry. Results showed that the release of ginseng components was controllable with lower initial bursts than those observed with conventional microparticles. Bioassays including stimulation of cultured murine macrophages were investigated showing significant immunoactivty of extracts showing the utility of these materials for next-generation biomedical agents as controlled release devices for capsules and bone cements.

(283f) A Novel Approach for Transformation of Nanosuspensions Into Polymer Films Containing Nanoparticles

Lucas Sievens-Figueroa[1], Anagha Bhakay[2], Ramana Susarla[3], Jackeline jerez-Rozo[4], Natasha Pandya[5], Rodolfo Romañach[6], Bozena Michniak-Kohn[7], Zafar Iqbal[8], Ecevit Bilgili[9] and **Rajesh N. Dave**[9], (1)New Jersey Center for Engineered Particulates, New Jersey Institute of Technology, Newark, NJ, (2)Chemical engineering, New Jersey Institute of Technology, Newark, NJ, (3)Chemical, Biological & Pharmaceutical Engineering, New Jersey Institute of Technology, Newark, NJ, (4)Chemistry, University of Puerto Rico, Mayagüez Campus, Mayagüez, PR, (5)Chemical, Biochemical, and Pharmaceutical Engineering, , New Jersey Institute of Technology, Newark, NJ, (6)Chemistry, University of Puerto Rico at Mayaguez, Mayaguez, PR, (7)Rutgers, The State University of New Jersey, Piscataway, NJ, (8)Department of Chemistry and Environmental Sciences, New Jersey Institute of Tecgnology, Newark, NJ, (9)Chemical, Biological and Pharmaceutical Engineering, New Jersey Institute of Technology, Newark, NJ

In this work, the feasibility of a simple, continuous process of incorporating stable nanoparticles into edible polymer films was examined. The final goal of this work was to enhance the dissolution rate of poorly water soluble drugs by preserving the particle size after film disintegration. Nanosuspensions produced from wet stirred media milling (WSMM) were transformed into polymer films containing drug nanoparticles by mixing with a low molecular weight hydroxylpropyl methyl cellulose (HPMC E15LV) solution containing glycerin followed by film casting and drying in an oven; both with or without the use of convective drying. Three different BCS Class II drugs, naproxen (NPX), fenofibrate (FNB) and griseofulvin (GF) were studied. Drug particle size was obtained by using laser diffraction. Drug crystallinity was studied by XRD and Raman spectroscopy. Particle dispersion in polymer films was studied by using SEM and NIR chemical imaging. Film mechanical properties were studied by using a texture analyzer. Drug content was studied by HPLC. Dissolution was also studied by using a USP IV apparatus. When convective drying was employed, dried films were obtained within 60 minutes or less depending on the drying conditions. The drying rates increased with an increase in temperature and airflow rate. A two stage drying process was observed, a linear behavior (stage I) followed by a nonlinear behavior of drying (stage II). It was shown that film processing methodology employed has no effect on the drug crystallinity. NPX exhibited the strongest aggregation compared to the other drugs. The aggregation had a direct effect on drug content uniformity in the film. Due to strong hydrogen bonding with the polymer, NPX exhibited an increase in Young's Modulus (YM) of approximately 200%, among other mechanical properties, compared to GF films. A synergistic effect between surfactant/polymer and drug/polymer interactions in the FNB film resulted in an increase of more than 600 % in YM compared to the GF film. NPX also exhibit the stronger aggregation and less uniform drug content in films. A faster dissolution rate was also observed for films containing nanoparticles compared to films containing microparticles and other solid dosage forms with similar formulation. These results set the foundation for continuously operating process development of films containing nanoparticles, specifically, for poorly water-soluble drugs, for various drug delivery applications. As compared to various routes to making solid dosage forms from drug nanosuspensions such as freeze drying, lyophilization, and spray drying, film formation may offer a more economical and scalable option.

(283g) Magnetic Resonance Imaging As a Powerful Tool for Visualizing Controlled Release From Biodegradable Microparticles

Tianzhou Wu[1,2], Lilian Ngobi[1,2], Sam N. Rothstein[1,2], Steven R. Yutzy[3], Eric Wiener[3], Robert S. Parker[1,2,4] and Steven R. Little[1,2,5,6], (1)Department of Chemical and Petroleum Engineering, University of Pittsburgh, Pittsburgh, PA, (2)McGowan Institute for Regenerative Medicine, Pittsburgh, PA, (3)Department of Radiology, University of Pittsburgh, Pittsburgh, PA, (4)University of Pittsburgh Cancer Institute, Pittsburgh, PA, (5)Bioengineering, University of Pittsburgh, Pittsburgh, PA, (6)Immunology, University of Pittsburgh, Pittsburgh, PA

Controlled release technology has the potential to enhance drug safety and efficacy if proper control of the resulting in vivo drug distribution can be exhibited. While characterization studies and predictions of drug release *in vitro* have been performed extensively, the difficulty translating results to the clinic implies that these results do not correspond to the complex *in vivo* environment. Non-invasive visualization of the drug distribution in time and/or space originating from a controlled release system could aid in the characterization of factors that control local drug distribution and inhibit the achievement of undesired release behavior. We hypothesized that magnetic resonance imaging (MRI) could be used to noninvasively quantify concentration profiles of controllably released gadolinium contrast agent, which could ultimately be used to track controlled release in vivo non invasively.

MRI contrast agent gadolinium tetraaza-cyclododecanetetraacetic acid (Gd-DOTA) was encapsulated in biodegradable poly(lactic-co-glycolic acid (PLGA) microparticles using water-in-oil-in-water (w/o/w) double emulsion technique, controlling for the inherent osmotic gradient that results from Gd-DOTA solutions. Particle characterization was performed using SEM to examine morphology and size. Gadolinium loading and encapsulation efficiency were determined via inductive coupled plasma mass spectrometry (ICP-MS). Degradable microparticles were placed under a hydrogel composed of 5% carboxylmethyl cellulose (CMC) (mimicking a depot injected into extracellular matrix). The releasing microparticle depot surrounded by hydrogel was analyzed using a 7-T MRI scanner repeatedly over a period of 22 days. Spin-echo sequence was applied and imaging was based on T1-weighted measurement. Quantification of Gd-DOTA was obtained by nonlinear least square error curve fitting. Concentrations of agent were plotted with respect to the distance from the release depot over various time points.

The w/o/w double emulsion fabrication consistently yielded spherical particles with smooth surfaces, with size ranging from 10mm to 40mm. ICP-MS analysis determined that Gd-DOTA loading was between 5.6-34.4 mg/mg polymer, corresponding to an encapsulation efficiency of 30-70%. Through high-resolution scanning, Gd-DOTA appeared to be highly concentrated in each individual particle, which is significantly different from blank particles as control. Similar behavior was observed in the microparticle depot in which Gd-DOTA loaded microparticles showed in black because of high concentration of loading. Analysis of signal intensity from MRI yielded spatio-temporal concentration profiles of Gd-DOTA in the surrounding hydrogel. These profiles were established at time 0 by a depot of Gd-DOTA loaded microparticles and reached equilibrium within 8 days. This result was consistent for all four Gd-DOTA loading levels tested which spanned an order of magnitude.

Noninvasive visualization and quantitative monitoring of drug distribution can provide physiological insight into the local drug delivery system, providing insight into the spatial distribution of drug radially outward from the release system. Our study has revealed that degradable, PLGA microparticles can act as an effective carrier for imaging agent that can subsequently be visualized via MRI. The hydrogel "phantom" system has demonstrated to be a suitable model to monitor release from a depot. Future work will compare this data to models of diffusion and distribution as well as using these models to make predictions on live imaging data collected from depots injected into mice.

References:

1. S.N. Rothstein. J Mater. Chem. 2011;21:29-39.

2. A.L. Doiron. PNAS. 2008;105:17232-17237.

3. J. Siepmann. Adv. Drug Deliv. Rev. 2001;48:139-157.

4. A.L. Doiron. Pharm. Res. 2009;26:674-682.

(284a) Microalgae-Based CO2 Re-Utilization for Biofuels Production and Biorefinery

Jo-Shu Chang, Department of Chemical Engineering, National Cheng Kung University, Tainan, Taiwan

Considering environmental sustainability and economical feasibility, the new generation of biofuels has to be produced from non-food and low-cost feedstock. Microalgal biomass is considered the feedstock of third generation biofuels with the benefit of efficient mitigating CO_2 emissions while producing the raw materials for biofuels and other bio-based chemicals. Taiwan is located in a perfect location for microalgae cultivation with a high diversity in microalgae species. Therefore, developing microalgae industry is very promising. However, there are still challenges on developing microalgal biofuels and biorefinery technology, including low photosynthesis efficiency, low biomass concentration, difficulty in mass production and harvest. This presentation will focus on how to apply biochemical engineering tools to develop core technologies leading to efficient microalgal CO_2 fixation and biofuels/chemicals production. This talk will also discuss our bioprocess engineering concepts and strategies that are effective in improving microalgal growth, CO_2 capture, and formation of target products (e.g., lipids, carbohydrates, pigments, and so on). Recent progress on photobioreactor design for the production of target microalgal components as well as process integration and scale up for microalgae cultivation will be described. Recent national and industrial projects related to microalgae biofuels and biorefineries will also be briefly introduced. Finally, the perspective and vision of microalgal biofuels and biorefinery will be proposed.

Keywords: Microalgae, biofuels, biorefinery, CO_2 mitigation, photobioreactor

(284b) Recent Advances in Bioprocess Development for Mass Production of Medicinal Polysaccharides and Beneficial Microbes for Health and Agricultural Industries in Malaysia'

Hesham A. El Enshasy, Institute of Bioproduct Development , Universiti Teknologi Malaysia , Johor Bahru, Malaysia

Biotechnology industries are classified under the category of new technology based firms (n-TBF) and considered as main components of knowledge based economy. Based on this fact, different Asian countries paid more interest in the development and growth of this business sector. The Development of biotechnology incubators to act as techo-industrial platform for RTD and large scale manufacturing of biomass and bioactive metabolites is one of the main drivers to promote industrial biotechnology sector in national level. This presentation will focuses on finding the best answer for the question, How can process based and biofactory based facilities in research organization be integrated to form techno-industrial based platform for technology transfer and industrialization of biotechnology products?. In this context, two successful industrialized products will be given as examples. The first is the development of industrial platform for large scale production of immunomodular exopolysaccharides by medicinal mushrooms. This research was designed to achieve three main targets. First, optimization of cultivation process in shake flask level to increase the polysaccharide yield in short production time. Second, bioprocess development in semi industrial scale 16-L stirred tank bioreactor followed by scaling up of the process to 300-L bioreactor. Third, development of an efficient solvent free downstream process for exopolysaccharide extraction, purification and precipitation. The second example in this presentation will be the development of industrial platform and large scale production of nitrogen fixing bacteria for agriculture application. In this process, medium composition and cultivation conditions were developed to face the main challenge in this process in terms of high polysaccharide production during cultivation of nitrogen fixing bacteria. Under normal batch culture cultivations, low cell mass was produced and oxygen became the limited substrate in this process. In the process developed in this work, we achieved high cell density cultivation of nitrogen fixing bacteria up to 45 g/L biomass with minimal polysaccharide production using an optimized fed-batch cultivation strategy. This production process was scaled up successfully to 150L stirred tank bioreactor. The produced cells were used as one of the major component of biofertilizers developed by one of the major companies in Malaysia. These two examples show clearly how universities and research centers can work effectively in cooperate with private and public companies to harmonize activities in industrial biotechnology sector and work as industrialization hub for bioprocess industries in Malaysia.

(284c) When Gene Therapy Meets Cell Therapy: Applications In Bone Tissue Engineering

Yu-Chen Hu, Institute of Biomedical Engineering, National Tsing Hua University, Hsinchu, Taiwan; Department of Chemical Engineering, National Tsing Hua University, Hsinchu, Taiwan

Presentation type: (A) Oral presentation

When Gene Therapy Meets Cell Therapy: Applications in Bone Tissue Engineering

Yu-Chen Hu[1, 2]

[1] Department of Chemical Engineering, National Tsing Hua University, Hsinchu, Taiwan

[2] Institute of Biomedical Engineering

*E-mail: ychu@mx.nthu.edu.tw

Baculovirus is an effective vector for gene delivery into various mammalian cells and has been employed for a variety of applications. We have demonstrated that baculovirus also efficiently transduce mesenchymal stem cells derived from bone marrow and adipose tissues. Transduction of stem cells with recombinant baculovirus expressing growth factors can modulate cellular differentiation state and stimulate the extracellular matrix synthesis and tissue formation. Implantation of genetically modified stem cells into the massive segmental bone defects successfully improved the angiogenesis and accelerated the segmental bone repair at the femora and calvaria of New Zealand White rabbits, as demonstrated by X-ray radiography, positron emission tomography, micro computed tomography, immunohistochemical staining and biomechanical testing. This presentation focuses on how baculovirus-mediated gene delivery aids in the bone regeneration and recent progresses.

(284d) Bioplastic: Current Trends and Future Prospects in Thailand

Nuttha Thongchul, Institute of Biotechnology and Genetic Engineering, Chulalongkorn University, Bangkok, Thailand

Bioplastic is the plastic derived from plant – based materials or other renewable resources through fermentation process. This process requires low energy input and produces the plastic with the ability to degrade when being composted, resulting in carbon dioxide and water as the end products. In fact, bioplastic has become one of the most innovative materials which answer to an increasing demand for more environmentally friendly solutions. Around the world, there is a great awareness in searching for new raw materials or polymers to improve the quality and functionality of bioplastic so that they can replace the conventional plastics that are produced by petrochemical industries.

Thailand is an agricultural country with various biomass materials and agricultural products including rice, sugarcane, cassava, corn, cellulose, and palm available. These products contain abundant composition of either starch (carbohydrates), sugar (glucose), or fibers (cellulose). All of which can be used as feedstock for the production of bioplastic monomers such as lactic acid and succinic acid. Favored with the price competitiveness and strength in the cultivation techniques, Thailand is clearly standing at the prime spot. In addition, Thailand's plastic industry is a full cycle mature industry sector with a world class potential because Thailand has an active molding machinery production sub-sector, including die and mold machinery manufacture. Nonetheless, to date, there is no complete line bioplastic industry existing in Thailand. Only Purac Biochem B.V. produces lactic acid, a monomer for polylactic acid, at the annual capacity of 100,000 metric ton. To expedite bioplastic industry in Thailand, the national roadmap for the development of bioplastic industry has been implemented since 2008. Since then many researchers from the universities through the private sectors have been encouraged by this policy to develop the upstream technology including feedstock pretreatment and fermentation in order to increase the production of monomer for bioplastic from the biomass locally available. The first bioplastic production facility is expected by 2013. This presentation will review the current technological trends and the future prospects in bioplastic production in Thailand.

(284e) Renewable Bioenergy Development in Singapore and Recent Discoveries On Biofuel Generation

Jianzhong He, CEE, National Univeristy of Singapore, Singapore, Singapore

As a modern urbanized city, Singapore faces two great challenges (1) how to manage wastes properly and (2) how to obtain new clean energy. In Singapore, more than 2 million tonnes of organic wastes (comprising mainly of food, plastics, glass, paper, and wood/ horticulture wastes) were generated each year. This staggering amount of waste generation is expected to increase along with the rapid economic and population growth. Being a small and densely populated country with a hot and humid climate (land area: 710.2 km^2; population of 4.8 million), Singapore has to face the challenge of linking wastes to energy, thus research is centering on converting organic waste biomass into sustainable form of clean energy—biofuels. This is of tremendous significance as Singapore lacks natural resources as well as free land. This talk will cover an overview on recent developments in biotechnology for biofuels generation and funding investments into sustainable energy in Singapore/Southeast Asia. Additionally, our interesting research discoveries on lignocellulosic biomass and food wastes to biofuels will also be discussed. Singapore is aiming to be recognized globally as an eco-friendly city with sustainable prosperity.

Advances in Aerobic Granule Formation and Granule Stability

Duu-Jong Lee
Department of Chemical Engineering, National Taiwan University of Science and Technology, Taipei, 106, Taiwan
Department of Chemical Engineering, National Taiwan University, Taipei, 106, Taiwan

Abstract

Aerobic granulation is drawing increasing global interest in a quest for an efficient and innovative technology in wastewater treatment. Developed less than two decades ago, extensive research work on aerobic granulation has been reported. Instability of the granule, which is one of the main problems that hinder practical application of aerobic granulation technology, is still to be resolved. This paper presents a review of the literature in aerobic granulation focusing on factors that influence granule formation, granule development and their stability in the context of sludge granulation. The review attempts to shed light on the potential of developing granules with adequate structural stability for practical applications. Possibilities and perspective of using stored granule as inoculums for rapid startup, and as microbial supplement to enhance treatment of bioreactor systems are also discussed.

Keywords: aerobic granulation; stability; operation; storage; extracellular polymeric substances; continuous-flow reactor

(294a) Experimental and Computational Comparison of the Surface Hydrophobicity of Viruses and Model Proteins

Caryn L. Heldt, Amna Zahid and K. Saagar Vijayaragavan, Chemical Engineering, Michigan Technological University, Houghton, MI

The physical characteristics of viruses needs to be understood in order to fully understand the interaction of viruses with host cells, as well as to create specific molecular recognition techniques to detect, purify, and remove viruses. Viruses are generally believed to be positively charged at physiological pH, but there are few other defining characteristics. Here, we have experimentally and computationally demonstrated that a nonenveloped virus is more hydrophobic than a subset of well-characterized proteins. This adds an additional dimension to currently known virus characteristics. In this study we determined the relative hydrophobicity of several proteins and compared them to a model nonenveloped virus. The percentage of acetonitrile required to elute the proteins and virus from a C18 chromatography column was interpreted as the hydrophobic strength of each species. We obtained surface hydrophobicity data by coloring the surface of the protein and virus according to published hydrophobicity scales. The amino acid sequence hydrophobicity was also determined. The results obtained indicate a strong correlation between surface hydrophobicity and the percent of acetonitrile needed to elute the proteins from a reverse phase column. This demonstrates that hydrophobicity may be calculated, instead of experimentally determined. We also determined that the nonenveloped virus we are modeling, porcine parvovirus (PPV) is highly hydrophobic. Differences in hydrophobicity of proteins and viruses can be used to develop novel separation techniques to specifically adsorb viruses for detection, removal and purification, as well as interpret the interaction of viruses with cellular membranes.

(294b) Ligand Modulated Conformational Landscape of G-Protein Coupled Receptors

Diwakar Shukla[1], Morgan Lawrenz[2] and Vijay Pande[2], (1)Chemical Engineering, Massachusetts Institute of Technology, Cambridge, MA, (2)Chemistry, Stanford University, Stanford, CA

G-protein coupled receptors (GPCRs) are membrane proteins that modulate many important biological signal cascades, and are the targets of approximately 40% of all commercially available drugs. Besides binding G proteins, activated β_2AR can be phosphorylated and bind arrestin, which redirects signaling to other pathways. The crystal structures of GPCR's reveal the structural differences between the active and inactive states but do not give insights into the activation mechanism or pathway relating the active and inactive states. An understanding of GPCR activation mechanism may facilitate the development of more selective drugs with improved therapeutic activity and minimal undesirable side effects. Recent simulation studies (Dror et. al., PNAS, 108, 46, 18684, 2011) have reported a deactivation pathway for β_2AR. However, a more detailed and complete picture of the GPCR conformational landscape is desired. Furthermore, the conformational landscape of the apo-receptor changes due to binding of ligands (confirmed by the NMR spectroscopy results (Liu et. al., Science, 335, 1055, 2012)). Here, we report results from extensive molecular dynamics (MD) simulations of the apo and ligand bound β_2 adrenergic receptor(β_2AR) to elucidate the activation and deactivation mechanism of GPCRs and to identify the ligand induced conformational changes in the landscape.

(294c) Sequence Specific Recognition of Cancer Drug-DNA Adducts by HMGB1a Repair Protein

Robert M. Elder and Arthi Jayaraman, Chemical and Biological Engineering, University of Colorado at Boulder, Boulder, CO

Platinum-based anti-cancer drugs are designed to covalently bind to DNA to form adducts in order to inhibit continuous cell growth. Differential recognition of various drug-DNA adducts by repair proteins, like HMGB1a, has been linked to differential tumor resistance. We use molecular simulations to show the molecular reasons behind why HMGB1a differentially binds to adducts of cisplatin and oxaliplatin, two drugs widely used for treatment of cancer. We also explain why the binding affinity of HMGB1a for these two drug-DNA adducts is dependent on sequence context, i.e. the bases flanking the drug-DNA adduct. Our work has shown that the structure of drug-DNA *bound to* HMGB1a protein cannot explain differential recognition [1]. Instead the differences in conformational dynamics of the Cp-DNA and Ox-DNA in various sequence contexts *before the protein binds* and the free energy of deforming (e.g. bending and minor groove opening) the drug-DNA *during protein binding* explain the differential protein binding affinity for the two drugs in the various sequence contexts [1, 2].

[1] R. M. Elder and A. Jayaraman. "Role of structure and dynamics of DNA with cisplatin and oxaliplatin adducts in various sequence contexts on binding of HMGB1a." *Mol. Sim.* (in press), 2012.

[2] R. M. Elder and A. Jayaraman. "Sequence specific recognition of cancer drug-DNA adducts by HMGB1a repair protein." *Biophys. J.* (in press), 2012.

(294d) Exploring the Functional Conformational Transitions in Proteins Using Atomistic Simulations and Elastic Network Models

Harish Vashisth, Chemistry and Biophysics, University of Michigan, Ann Arbor, MI and Charles L. Brooks III, Department of Chemistry and Biophysics Program, University of Michigan, Ann Arbor, MI

Proteins often undergo ligand-triggered large-scale conformational changes that are involved in many cellular processes such as transport, signal transduction, and catalysis, etc. High-resolution crystallographic studies have indeed provided structural bases for the stable open and closed conformations of various proteins, but mechanistic details of the observed structural transitions are only marginally understood. Detailed experimental characterization of these functional transitions is significantly challenging because many intermediate conformations of a protein along the "activation-pathway" are only transiently populated. Molecular dynamics (MD) simulations can in principle provide atomistically-resolved details of the conformational ensembles of various proteins, but due to difficulties in observing large-scale conformational changes on reasonable time-scales, new promising simulation approaches have been proposed. These successful approaches often exploit the dimensionality-reduction as a tool to explore the free-energy landscape of proteins in large yet finite collective variables (CVs). Traditionally, collective motions of proteins have been analyzed successfully using normal mode analysis (NMA) of equilibrium structures because normal modes solely arising from the structure are intrinsically accessible to proteins and collective protein dynamics can be usually described by the low-frequency end of the mode spectrum. Particularly, normal mode analyses of coarse-grained (usually C_α-based) elastic network models (ENMs) of proteins with a single-parameter potential have been remarkably successful in describing large-scale conformational transitions in many biological complexes such as the ribosome. In this work, we aim to understand large-scale conformational changes in a class of integral membrane-proteins known as the ATP-binding cassette (ABC) transporters that are responsible for nutrient uptake (importers) or toxins export (exporters) using a combination of atomistic simulations and elastic network normal mode analysis. Specifically, we generate multiple independent temperature-accelerated molecular dynamics (TAMD) simulations of the components (maltose-binding protein, MBP, and nucleotide-binding domains, NBD) of a maltose transporter. We find that a few low-frequency normal modes of the open states of these proteins can account for the functional transitions observed in atomistic TAMD simulations. Therefore, we extend the NMA to understand the interplay of conformational changes in the transporter components and the transmembrane domains, a dynamic coupling of which leads to successful transport events. We find reasonably high projections of the modes of the intracellular ATP-domains on the functional displacement of the entire transporter, and suggest that NBDs are likely the drivers of this class of transporters.

(294e) Recovery of Chromosome Structural Ensembles From Contact Probabilities

Dario Meluzzi and Gaurav Arya, Department of Nanoengineering, University of California, San Diego, La Jolla, CA

Chromatin is a complex filamentous assembly of DNA and histone proteins that fills the nucleus of eukaryotic cells. The higher-order folding of chromatin into chromosomes is thought to play important roles in regulating vital genomic processes, such as transcription, replication, recombination, and DNA repair. However, the specific spatial organization of chromatin within the nucleus and the effects of this organization on the above processes are not well understood. Recently, an elaborate set of experimental techniques known as Chromosome Conformation Capture (3C) have allowed researchers to measure the frequency of interactions between different fragments of the chromatin fiber *in vivo*, providing the first glimpse into the complex looping interactions present inside chromosomes. Here we present a computational approach for inferring an ensemble of chromatin configurations from such interaction frequencies. In particular, we perform simulations of the chromatin fiber to measure probabilities of contact between different parts of the fiber and we restrain the fiber to match a given set of known contact probabilities. Using this approach we are able to recover known spatial measures for a number of artificial systems of increasing complexity. We are next applying this method to deduce ensembles of chromatin structure consistent with 3C experimental data.

(294f) Identification of Novel Intein Inhibitors with Relevance to Tuberculosis

C. Seth Pearson[1], Brian Callahan[2], Georges Belfort[1] and Marlene Belfort[2], (1)Howard P. Isermann Dept of Chemical & Biological Engineering and The Center for Biotechnology & Interdisciplinary Studies, Rensselaer Polytechnic Institute, Troy, NY, (2)University at Albany, Albany, NY

Inteins are naturally occurring protein elements that autocatalytically excise themselves from a non-functional precursor and ligate the flanking protein segments with a peptide bond, resulting in a functional protein. Inteins interrupt three proteins important for the viability of *Mycobacterium tuberculosis*. Preventing intein splicing, and thus the formation of functional post-processed proteins, suggests that intein inhibition may be used as a novel anti-mycobacterial strategy (M. Belfort, US Patent, 5,795,731). Due to the growing problem of multiple drug-resistant tuberculosis infections, such alternatives to traditional antibiotic regimens are especially appealing. It has been shown that cisplatin, an FDA approved anti-cancer drug, is a potent inhibitor of intein splicing, both *in vitro* and *in vivo* (Zhang et al. (2011) *JBC*, 286, 1277). Due to its high toxicity, however, cisplatin has limited clinical value as an anti-mycobacterial. Structural and mechanistic insights gained from studies with the Mtu RecA intein were applied in a virtual screen using the Autodock suite of docking software (http://autodock.scripps.edu/), designed to identify compounds that retained potent inhibition activity while minimizing the toxicity displayed by cisplatin. The lead compounds from the virtual screen were then tested for inhibition activity *in vitro* using a fluorescent splicing reporter. This study is laying the groundwork for potential *de novo* design of novel anti-microbials.

(294g) Predicting and Evaluating Protein Binding Motifs in the Chlorosomes of Green Sulfur Bacteria

Sándor Á. Kovács and Cynthia S. Lo, Energy, Environmental and Chemical Engineering, Washington University in St. Louis, St. Louis, MO

We present the first model of the chlorosome baseplate in the green sulfur bacteria *Chlorobaculum tepidum* binding to the Fenna-Matthews-Olson protein (FMO). Our model of the baseplate includes the photosynthetic pigment Bacteriochlorophyll *a* (BChl *a*) bound to chlorosome protein A (CsmA). Until now, atomistic structural models have been difficult to obtain for this system due to limitations in experimentally characterizing the large and complicated photosynthetic antenna complex. Our multiscale predictive modeling approach uses a combination of protein-ligand docking and semi-empirical quantum calculations to predict and evaluate potential BChl *a* binding sites on CsmA and protein-protein docking and molecular dynamics free energy methods to predict and evaluate the potential CsmA binding sites on FMO. Our computational approaches refine existing models and agree with available experimental data, including newly-identified CsmA binding sites on FMO. Our model consists of a Bchl *a* molecule, with the magnesium center of the bacteriochlorin ring located within 3 Å of the imidazole nitrogen of CsmA's HIS25 residue, and the phytyl tail situated on the hydrophobic side of the CsmA α-helix so that it is near the non-polar residues (TRP26, VAL29, LEU32, and PHE33). Our predicted structure reproduces the experimentally observed absorption redshift when comparing the absorption of the bare BChl *a* pigment in an organic solvent environment to our computed model of the pigment in its native protein environment. The rigorous understanding of the interaction of a single BChl *a* pigment with a single CsmA protein is necessary for the construction of CsmA multimers. In turn, these CsmA multimers bind to FMO, forming the baseplate of the photosynthetic antenna complex present in green sulfur bacteria. The stability of this baseplate-FMO complex in its native environment, along with possible mechanisms of assembly, will be discussed. This structure prediction is a necessary first step in characterizing and understanding more complicated photosynthetic systems composed of multiple pigment-protein aggregates.

(298a) Design and Characterization of Micro-Porous Hyaluronic Acid Hydrogels for Non-Viral DNA Delivery

Talar Tokatlian[1], Cynthia Cam[2], Shayne Siegman[1], Yuguo Lei[1] and Tatiana Segura[3], (1)Chemical and Biomolecular Engineering, University of California, Los Angeles, Los Angeles, CA, (2)Biomedical Engineering, University of California, Los Angeles, Los Angeles, CA, (3)Chemical and Biomolecular Engineering, University of California Los Angeles, Los Angeles, CA

Vascularization within tissue-engineered constructs remains the primary cause of construct failure following implantation. We are currently investigating two hypotheses to enhance hydrogel scaffold vascularization, both long-term mechanical support and cell-demanded release of non-viral DNA nanoparticles. Our preliminary in vivo studies show that after subcutaneous implantation for 3 weeks enzymatically degradable hyaluronic acid (HA) hydrogels have cellular infiltration only at the periphery of the hydrogel, while hydrogels with micron sized interconnected pores (micro-pore) are extensively infiltrated. Significant positive staining for endothelial markers (PECAM) was also found for micro-pore implants and not for nano-pore implants, even in the absence of pro-angiogenic factors. We believe that an open pore structure will increase the rate of vascularization through enhanced cellular infiltration and that the added delivery of DNA encoding for angiogenic growth factors will result in long lasting angiogenic signals. We have thusfar designed and characterized various pore size (30, 60, and 100 micron) micro-pore HA hydrogels loaded with a high concentration of non-viral DNA/PEI polyplexes, using a previously developed caged nanoparticle encapsulation (CnE) technique. These hydrogels allowed for long-term sustained transfection and transgene expression of incorporated mMSCs in vitro. For all investigated pore sizes, encapsulated DNA polyplexes were released steadily starting by day 4 for up to 10 days. Likewise, transgene expression was sustained over this period, although significant differences between different pore sizes were not observed. Cell viability was also shown to remain high over time, even in the presence of high concentrations of DNA polyplexes. Presently, we are using the knowledge acquired through this in vitro model to design and better predict scaffold-mediated gene delivery for local gene therapy in both subcutaneous implant and wound healing mouse models. We believe the proposed hydrogel system has applications for the controlled release of various DNA particles and other gene delivery vectors for in vivo tissue engineering and blood vessel formation.

(298b) Airway Gene Transfer and Intracellular Trafficking of Highly Compacted DNA Nanoparticles

Anthony J. Kim, Nicholas Boylan, Jung Soo Suk and Justin S. Hanes, The Center for Nanomedicine, Johns Hopkins University School of Medicine, Baltimore, MD

Despite over two decades of cystic fibrosis (CF) gene therapy research, development and clinical testing, no patient has yet been cured. Poor gene transfer to the airway epithelium has been attributed to a number of cellular barriers, including poor cellular uptake across the apical membrane, unproductive intracellular trafficking, and inefficient nuclear import. Highly compacted DNA nanoparticles, composed of single molecules of plasmid DNA compacted with block copolymers of poly-L-lysine and 10 kDa polyethylene glycol ($CK_{30}PEG_{10k}$), have shown considerable promise in human gene therapy clinical trials in the nares, but efficacy has not been established in the CF lung. Starting with the clinically tested DNA nanoparticle system, we have systematically varied the PEG MW (from 10kDa to 2kDa) and poly-L-lysine architecture and studied their effects on particle morphology, cellular uptake and intracellular trafficking, and gene transfer to lung airways in BALB/c mice. In addition, we will discuss a correlation between our intracellular trafficking data and *in vivo* gene transfer results. Together, these results represent an important step toward the rational development of an efficient gene delivery platform for the lungs based on highly compacted DNA nanoparticles.

(298c) Telecommunications Model of Lipoplex-Mediated Gene Delivery

Timothy M. Martin[1], Tadeusz Wysocki[2], Beata Wysocki[2] and Angela K. Pannier[1], (1)Biological Systems Engineering, University of Nebraska-Lincoln, Lincoln, NE, (2)Computer & Electronics Engineering, UNL, Omaha, NE

Gene delivery systems transport genetic information encoded as DNA to cells or biological systems, which then transcribe and translate that information into functional proteins. This process can be thought of as a type of nano-scale communication with the potential to directly alter endogenous gene expression and behavior with applications in functional genomics, tissue engineering, medical devices, and gene therapy. Although viral gene delivery systems are efficient at gene transfer, nonviral systems are preferred because of their low immunogenicity, inexpensive synthesis, and easy modification. In nonviral systems, the DNA is typically complexed with cationic lipids or polymers to facilitate gene transfer as the complex traverses putative intracellular barriers, the common culprits to low efficacy of these systems. Current lack of understanding of the biological mediators of gene delivery limits the design potential of new systems. Kinetic models can offer perspective and interpretation of unknown or complex cellular mechanisms involved in gene transfer. However, current models are inadequate and provide only a segmented view of gene delivery kinetics and treat complex routing as averaged or rigid events that are not biologically accurate. There is a need for new types of models that describe the complete gene delivery process with biological fitness to enable the design of more efficient systems.

In this work, we propose taking a novel approach to model the nonviral gene delivery process using telecommunications modeling. Delivery of genetic information to the cell nucleus can be considered in the same way as delivery of a packet of information to the destination computer within a packet switched computer network. In such a network, a packet containing the information is routed through several intermediate nodes before it reaches the destination, sometimes undergoing processing or transcoding within those nodes. This process is similar to gene transfer in which the complex must undergo various steps to achieve transgene expression: internalization, endosome escape, nuclear import, transcription, and translation. These steps are synonymous with intermediate nodes modeled using queuing theory with several input and output servers. Since biological processes within a cell can be considered as highly parallel, the queuing network model of gene delivery is solely dependent on kinetic constants. Through various particle tracking studies, kinetic constants are available in literature which represent the average time for complexes to travel through the various steps in the gene delivery process.

Here we used a layered communication protocol to model lipoplex delivery to an A431 epithelial cell line taking complex arrival and service time as Poisson processes. Each cell in the system was considered to consist of a queuing network as simulated lipoplexes were fed into the system and routed to servers. Stochastic averaging was used to simulate the transport process for a fixed number of individual cells. Simulation results show a high degree of agreement with experimental data, including internalization, nuclear accumulation, and transfection efficiency. Interestingly, a kinetic constant for nuclear localized plasmid transcriptional activity had to be incorporated, indicating an unaccounted for process or biological mechanism involved in gene transfer. Our novel application of queuing theory for modeling nonviral gene delivery provides insight into the pathways and mechanisms of nonviral gene transfer, so that more efficient systems can be designed.

(298d) Microrna-29b Delivery Via EpCAM Targeted Cationic Lipoplexes in Lung Cancer Treatment

Yun Wu[1], Yicheng Mao[2], Melissa Crawford[3], L. James Lee[4] and Serge P. Nana-Sinkam[3], (1)Nanoscale Science and Engineering Center, The Ohio State University, Columbus, OH, (2)Division of Pharmaceutics, College of Pharmacy, the Ohio State University, Columbus, (3)Division of Pulmonary, Allergy, Critical Care and Sleep Medicine, The Ohio State University, Columbus, OH, (4)Chemical and Biomolecular, the Ohio State University, Columbus

Lung cancer is the leading cause of cancer deaths in western countries and carries a poor overall five year survival rate. MicroRNAs are non-coding RNAs which regulate fundamental cell processes and have the potential to target multiple genes simultaneously, i.e. "one for all". MicroRNAs have been implicated in lung cancer development by serving as tumor suppressors or oncogenes. However, miRNA-based therapeutics are faced with several challenges including lack of tissue specificity, lack of optimal delivery systems, poor cellular uptake and risk of systemic toxicity. Here, we report an EpCAM (epithelial cell adhesion molecule)-targeted cationic lipoplexes based miRNA delivery system to address some of these challenges.

Among many lung cancer related miRNAs, miR-29b, a tumor suppressor, was selected as the representative therapeutic target because it has been associated with patient survival rate and sensitivity to chemotherapy. The miR-29b delivery efficiency of EpCAM targeted lipoplexes was first evaluated in A549 non-small cell lung cancer (NSCLC) cells and compared with siPORT™ NeoFX™ transfection agent (Life Technology Inc.). The expression of mature miR-29b and its 3 target genes (CDK6, DNMT3b, MCL-1) was measured using qRT-PCR and western blotting. The A549 cell growth and sensitivity to chemotherapeutic agents were characterized 24h after miR-29b overexpression. Subsequently, both the biodistribution and cytotoxicity *in vivo*were evaluated in ICR mice.

Compared to NeoFX, EpCAM targeted lipoplexes delivered miR-29b in a more efficient manner with ~10.5 fold increase in mature miR-29b expression in A549 cells. Results of flow cytometry and confocal microscopy demonstrated that the higher mature miR-29b expression was led by higher cellular uptake of EpCAM targeted lipoplexes. MiR-29b delivered by EpCAM targeted lipoplexes successfully down-regulated the expression of 3 target genes, CDK6, DNMT3b, MCL-1, by 61%, 40% and 21% respectively, but only by 20%, 22% and 17% with NeoFX. miR-29b delivered by lipoplexes significantly inhibited A549 cell growth by 62.3%, compared with 18.4% by NeoFX. Overexpression of miR-29b increased the sensitivity of A549 cells to chemotherapeutic agents. The IC50 value of cisplatin for A549 cells was 7.90uM 24h after miR-29b delivery by lipoplexes, compared to 12.94 uM by NeoFX. *In vivo*, EpCAM targeted lipoplexes achieved ~30% accumulation in lung tissue, which was ~50 fold higher than NeoFX. Lipoplexes mainly accumulated in pneumocytes, a region for lung cancer initiation. No short term cytotoxicity was observed *in vivo*. Mice systemically treated by miR-29b containing lipoplexes had a 23.6-fold induction in mature miR-29b expression in the lung compared to untreated mice.

Our results have demonstrated that the novel EpCAM targeted cationic lipoplexes may have potential as carriers for miRNA based therapeutics in NSCLC treatment. In the future, further *in vivo* miRNA-based delivery studies will be done to evaluate long-term local and systemic toxicity and target gene effects prior to clinical application.

(298e) Cellular Trafficking of Dextran Functionalized Silica Nanoparticles for Effective siRNA Delivery

Amanda P. Malefyt[1], Georgina Comiskey[2], Elizabeth Hinds[1], Stephen Lindeman[1], Gregory L. Baker[2], Christina Chan[1] and S. Patrick Walton[1], (1)Chemical Engineering and Materials Science, Michigan State University, East Lansing, MI, (2)Chemistry, Michigan State University, East Lansing, MI

Clinical applications of genetic therapies, including delivery of short, interfering RNAs (siRNAs) for RNA interference (RNAi), have been hampered by difficulties in delivering nucleic acids to cells in sufficient quantities to achieve silencing. On the cellular level, delivery efficiency is hindered by our lack of knowledge on how siRNA-vehicle complexes traffic across the plasma membrane and into the cytoplasm. Although some information is emerging, additional studies on the relationship between chemical and physical characteristics of the vehicles and their mode of uptake or trafficking within the cells are required. To address this challenge, we have developed novel dextran functionalized silica solid-core nanoparticles (NPs) with the ability to effectively deliver siRNA to human lung cancer (H1299) cells expressing an enhanced green fluorescent protein (EGFP). Through modifications to the particle size and amine content, we are able to relate the chemical and physical characteristics to changes in cellular uptake, endocytotic trafficking, and ultimately silencing efficiency. We will discuss the effect of endocytotic inhibitors on these cellular processes with respect to time and NP characteristics. Finally, we will explore the differences in silencing kinetics between lipid, polymer, and solid-core delivery vehicles.

(298f) Co-Delivery of cDNA and siRNA Using Viral/Nonviral Chimeric Nanoparticles for Synergistic Cancer Gene Therapy

Soo Kyung Cho, Chemical Engineering and Materials Science, University of California, Irvine, Irvine, CA, Shirley Wong, Pharmacology, Unversity of California, Irvine, Irvine, CA and Young Jik Kwon, Pharmaceutical Sciences, Chemical Engineering and Material Science, Biomedical Engineering, University of California, Irvine, Irvine, CA

Development of safe and efficient delivery vectors has been a pivotal demand in gene therapy. Viral vectors are popularly used in clinical trials, due to their high transduction efficiency, but their utilities have been limited by immunogenicity, pathogenicity, oncogenicity, limited gene packaging capacity, and difficulties in large-scale preparation and purification. Importantly, engineering viral vectors for versatile and multi-functional gene delivery is a cumbersome task and often infeasible. Nonviral vectors, on the other hand, offer relatively low immunogenicity, unlimited size in gene cargos, and high structural and functional flexibility for modifications, although their low transfection efficiency compared to viral counterparts is still a major drawback. Therefore, the advantages of viral vectors (i.e., high transduction efficiency) and nonviral vectors (i.e., low immunogenicity, and versatility in engineering structure and functions) were combined in viral/nonviral chimeric nanoparticles (ChNPs) with adeno-associated virus (AAV) as the viral core with acid-degradable polyketal (PK) shell as nonviral polymeric shell. The AAV core, which is one of the smallest viral vectors used in clinical gene therapy with no known linkages to human diseases, was shelled with PK shell that is designed to protect the AAV core from immune responses, promote endosomal escape, and enable the conjugation of targeting ligands on the surface. Most importantly, siRNA was incorporated in the PK shell during surface-initiated photopolymerization, in order to achieve dual modal gene delivery (i.e., simultaneous expression and silencing of multiple genes).

The core-shell structure of the resulting ChNPs, and the ordered release of siRNA and the AAV core in the presence of multiple intracellular stimuli were confirmed. Confocal microscopy demonstrated that ChNPs selectively released siRNA into the cytoplasm from the PK shell and facilitated the transnuclear localization of the AAV core. In vitro study with human lymphoma B cells showed significantly enhanced transduction by ChNPs, compared to free AAVs, with simultaneous gene silencing of additional target gene. It was also shown that the PK shell of ChNPs effectively shielded their AAV core from AAV-neutralizing antibodies and sialic acid-conjugated ChNPs selectively transduced CD22+ B-cells. Efficient and synergistic eradication of Philadelphia chromosome-positive (Ph+) leukemia was explored using ChNPs with Bim-encoding AAV core and Mcl-1 siRNA-encapsulating PK shell. Ph+ leukemia cells over-express a pro-survival Mcl-1 while down-regulating pro-apoptotic Bim. Cell apoptosis assay as well as cell growth kinetics indicated synergistic gene therapy for leukemia using ChNPs. This talk will highlight the design, synthesis, and characterization of novel gene carriers, along with a clinically relevant application.

(298g) Development of an Adenovirus Gene Therapy Vector with Improved Transduction Efficacy and Reduced Innate and Adaptive Immune Response

Adane Nigatu and Joshua Ramsey, Chemical Engineering, Oklahoma State University, Stillwater, OK

Adenoviruses-based vectors, which are currently used in more than 25 % of ongoing gene therapy clinical trials, are a promising and relatively safe class of gene therapy vectors. Adenovirus, however, has limitations that hinder its use. For example, adenovirus may initiate an innate inflammatory immune response or may be cleared rapidly in patients with preexisting immunity. Even in patients without preexisting immunity, the adaptive immune response greatly limits repeat administration of the virus. In addition, the dependency of the virus on the coxsackie-adenovirus receptor (CAR) compromises the ability of the virus to deliver genes to CAR-negative cells.

We developed a nanoparticle complex composed of adenovirus and a cell penetrating peptide (CPP)/polyethylene glycol (PEG) conjugate to reduce these drawback and produce an improved hybrid vector. The CPPs were used to enable the virus to transform cells independent of CAR, and PEG was used to prevent immune recognition, thereby reducing the innate and adaptive immune responses. The immune response towards the complexes and transduction efficiencies of the particles were optimized by varying the type of CPP (e.g., Tat, Penetratin, polyarginine and Pep1), degree of PEGylation, PEG molecular weight, and ratio of CPPs to adenovirus. In this study we found that CPP-PEG-Ad showed a 90-fold enhancement in gene transduction compared to the unmodified adenovirus when used to transduce CAR-negative cells. Moreover, the complexes were able to evade neutralizing antibodies and retained 90 % of their activity. Further, the complexes reduced the amount of proinflammatory cytokine (IL-6) produced by mouse macrophage cells. The results of the study demonstrate that modification of adenovirus with PEG and CPP can reduce immunogenic limitations associated with the virus while promoting transduction of cells not normally infected by adenovirus.

(306a) Evaluation of Process Capability and Robustness for a Lyophilized Product

Michael M. Choi, GBPRD, Teva, West Chester, PA

Evaluation of Process Capability and Robustness for a Lyophilized Product

Michael Choi, Daniel Kim, Marazban Sarkari, Ronald Simko, David Unger, Stelios Tsinontides

Teva Global Branded Products Research and Development

Abstract

Process capability for the manufacturing of lyophilized commercial batches at a new contract manufacturing organization (CMO) site was evaluated against the current CMO site using statistical process control (SPC) calculations. Key process parameters, material attributes, and quality attributes were examined using univariate and multivariate control charts, Box plots, probability plots, and capability ratios. Within[i] and between[ii] variations were evaluated to arrive at the type of control charts used in the analysis. Outliers were verified using Box and probability plots in conjunction with control charts. Simultaneous control charts were used to assess the impact of the critical process parameters (CPP) on critical quality attributes (CQA). Overlaid control charts were used to compare CMO performances side by side. The analysis was used to determine the adequacy of the controls and monitors at the new site, success of tech transfer, and impact of statistical outliers on critical quality attributes.

---

[i] variation within group of samples

[ii] variation between groups of samples

(306b) Understanding Thermodynamics of Aqueous Film Coating Process Using Data Logger

Ganeshkumar A. Subramanian, Drug Product Science and Technology, Bristol Myers Squibb, New Brunswick, NJ

Understanding Thermodynamics of Aqueous Film Coating Process Using Data Logger

Authors: G. Subramanian, P. Pandey, S. Gour, D. Bindra, S. Badawy

Abstract:

Pharmaceutical film coating process is typically monitored by inlet and exhaust condition of temperature and humidity with measurement devices located outside the coating pan. Current configurations of coating pans do not allow continuous measurement of the temperature and humidity within the tablet bed. Vast differences in heat and mass transfer efficiencies across different coating pan sizes can result in differences in drying and film forming dynamics across various coating pans at the same outlet temperature and humidity. Similarly, different local temperature and humidity conditions (and hence film forming dynamics) can be expected for coating operations carried out at same exhaust temperature, if such temperature is achieved by different means such as changing inlet air volume, air temperature, suspension spray rate or suspension solids concentration. Considerable insight into the dynamics of the coating process can be gained by using data loggers that can be incorporated in the tablet bed and placed such that they can collect temperature and humidity conditions in the tablet bed continuously. A design of experiments (DOE) with coating process performed at various macro-environmental conditions was carried out to determine their effects on the local-environmental thermodynamics in the tablet bed. The data collected from the data loggers enabled correlation between the changes to the macro-environment with changes in the local environment temperature and humidity in the tablet bed. Data from the data logger was used to develop a simple thermodynamic model based on heat and mass balance at different coating conditions across different pan sizes.

(306c) Process Robustness Aided by Retrospective Multivariate Data Analysis of Factory Data - a Case Study

Sankar Raghavan, Technical Operations, Merck & Company, Riverside, PA, Jeffrey Foster, Manuf. Tech. Support, Merck & Company , Riverside, PA, Dale Kopas, American Quality Associates and Ed Warner, Merck & Company

Process robustness aided by retrospective multivariate data analysis of factory data – a case study

Sankar Raghavan, Cherokee Pharmaceuticals L.L.C. a wholly owned subsidiary of Merck & Company, Sankar.Raghavan@merck.com , 570-271-2106

Jeff Foster, Cherokee Pharmaceuticals L.L.C. a wholly owned subsidiary of Merck & Company, Jeffrey.Foster@merck.com, 570-271-2173

Dale Kopas, dakopas@gmail.com, 609-417-1707

Ed Warner, Merck & Company, edwarner4@gmail.com, 908-670-1902

Presenter: Sankar Raghavan

Keywords: Multistep process, Noisy data, Multivariate analysis, Continuous improvement

Purpose: Illustrate the power of combining process knowledge and advanced statistical analysis in addressing intricate process issues

Abstract

Quality by Design is a powerful tool that can be applied to continuously improve performance and robustness of processes. However, with complex processes, it can be effort intensive and costly to do particularly when the process has multiple steps and the scale-up and the interaction effects are not fully understood. Since most pharmaceutical processes are batch, factory data with its normal variability provides a vehicle to generate such data without incurring excessive cost. Such a case study is presented – The example involves a manufacturing process for an antibiotic drug substance that involved four reactions and several down stream purification steps. Multiple factors influenced each of these steps. Identification of key factors that affected the process proved difficult as univariate and bivariate approaches such as SQC/SPC control charts and simple linear regression models explained only a small fraction of the process variability. Multiple interactions added to the complexity, and the presence of a high degree of co-linearity between process variables in the X-matrix lead to problems with matrix inversion in traditional multivariate regression models such as stepwise regression. To correct the situation, data mining was attempted by advanced multivariate methods, specifically Random Forest and Partial Least Squares (PLS) to supplement process knowledge.

The paper describes the analysis of inherently noisy factory data for an entire campaign consisting of over 200 individual batches and more than 100 variables by a combination of advanced statistical methods and physical reasoning based on expert knowledge of the process. The method was effective in explaining the broad factory trends. Key outcomes were:

1. The methodology helped rank factors in terms of their importance. This made it possible to focus on a subset of few factors.
2. Implementation of lessons derived from the studies has resulted in a more robust process with improvements in purity, yield and productivity.

3. The work pointed to areas of opportunity to improve the process.

The method aligns well with the F&DA initiative to continuously improve robustness of processes through the life cycle of a process. The methodology should be broadly applicable for the continuous improvement of complex processes.

(306d) Application of Weber Number-Based Regime Map for Scaling and Control of Pharmaceutical Coating Sprays

Sip C Liew[1], **Ariel Muliadi**[1] and Paul Sojka[2], (1)Chemical Engineering, Purdue University, West Lafayette, IN, (2)Mechanical Engineering, Purdue University, West Lafayette, IN

A Weber-number based regime map for describing the patternation of twin-fluid pharmaceutical coating sprays is proposed. The map was developed according to the dynamics of the spray evolution at the point where the two shaping air flows meet (the 'junction' point)—depending on the spray drop size, drop velocity, shaping air velocity, and surface tension at the junction point, the shaping air can either redistribute the drops, cause additional (secondary) drop breakup, or pinch the spray. A Gaussian spray results when the spray evolution process is driven mainly by drop redistribution; a homogeneous pattern is observed when there exists balance between the redistribution and secondary breakup effects; a dumbbell-shaped spray results when the pinching effect dominates spray evolution. The We-based regime map was constructed by plotting the junction point We against a dimensionless spray parameter calculated from the product of two flow ratios: the shaping-to-atomizing air flow rates ratio and the liquid supply rate-to-maximum nozzle liquid flow rate capacity ratio. Experimental results show a linear relationship between the junction point We and this dimensionless spray parameter. More importantly, different regions of the We-dimensionless spray parameter plot correspond to the three different spray patternations. Several scale-up scenarios we have explored demonstrate the We-based regime map reliability in achieving homogenous spray patterns. These examples also support the We-based regime map applicability in providing quantitative solutions for the design and optimization of sprays for pharmaceutical coating applications.

(306e) Characterizing Heterogeneity of Environmental Conditions in Various Bioreactor Scales Used for Cell Cultivation

Miroslav Soos, Benjamin Neunstoecklin and Massimo Morbidelli, Institute of Chemical and Bioengineering, ETH, Zurich, Switzerland

Large scale mammalian cell culture up to 20000L is developing more and more to an industrial standard. Process up scaling to those volumes is a fundamental necessity to guarantee process performance, product amount and product quality. Therefore, it is of outmost importance to be aware of scale dependent variables influencing the process. Reactor geometry, shear stress distribution, dissolved O2 and CO2 distribution and gas mass transfer rates can be very different in between scales, resulting in non-desired variations of process performance.

We studied scale dependent parameters (e.g. shear stress, kLa, DO, pCO2, and mixing time) from small to largest scale bioreactors, covering a volume range from 3L up to 15000L, using CFD simulation. At each scale, CFD simulations were validated against relevant scale dependent parameters (e.g. kLa, DO distribution, mixing time) measured experimentally. Comparisons of the computational and experimental values were in good agreement supporting the used approach. Such CFD simulations were consequently used to characterize heterogeneity of the environmental conditions (i.e. shear stress, DO, mixing time etc.) and to understand limitations of actual lab scale bioreactors currently used for process development. We will discuss those limitations and propose more suitable downscale approaches to fit large scale conditions already during process development.

(307a) All-Protein Vehicles for siRNA Delivery

Nicole J. Yang, Chemical Engineering, Massachusetts Institute of Technology, Cambridge, MA, Daniel G. Anderson, Chemical Engineering, Health Sciences & Technology, Massachusetts Institute of Technology, Cambridge, MA, Robert S. Langer, Massachusetts Institute of Technology, Cambridge, MA and K. Dane Wittrup, Chemical Engineering, Biological Engineering, Massachusetts Institute of Technology, Cambridge, MA

RNA interference (RNAi) is a naturally occurring gene-silencing mechanism which can be exploited to reduce the expression of disease-causing genes. Short-interfering RNA (siRNA) can artificially trigger RNAi, but cannot reach the cytoplasm of target cells efficiently on its own. As such, a protein-based delivery system for siRNA, composed of two separate agents, has been developed. The first construct binds to siRNA through a double-stranded RNA binding domain (dsRBD) and mediates internalization into cells through its targeting moiety against the epidermal growth factor receptor (EGFR). The second construct, a pore-forming protein taken from bacteria, allows endosomal release of siRNA. Administered together, this delivery system achieved efficient and specific silencing *in vitro*.

Towards adapting this delivery system *in vivo*, we have optimized both the siRNA carrier and the pore-forming protein. First, we identified that the rapid dissociation of siRNA from the dsRBD is a major limitation for efficient tumor targeting, and improved the binding affinity of the dsRBD via yeast-surface display and directed evolution. The resulting dsRBD could bind to both unmodified and chemically-modified siRNA with higher affinity, independent of sequence. The accrued mutations displayed an overall increase in charge of the amino acid side chains, suggesting that interactions with the negatively-charged siRNA backbone were increased. Second, a reversible protein mask was engineered for the pore-former, resulting in a significant decrease in cytotoxicity without compromising efficacy. Consequently, the therapeutic window was increased by an order of magnitude *in vitro*. Ultimately, our objective is to achieve efficient gene silencing in an animal model to validate therapeutic applications of the current all-protein siRNA delivery system.

(307b) Quantitative Assessment of in Vivo HIV Protease Activity Using Genetically Engineered QD-Based FRET Probes

Lakshmi N Cella, Chemical and Biomolecular Engineering, University of Delaware, Newark , DE, Marylynn V. Yates, Environmental Sciences, University of California, Riverside, Riverside, CA, Ashok Mulchandani, University of California, Riverside, Riverside, CA and Wilfred Chen, Chemical and Biomolecular Engineering, University of Delaware, Newark, DE

Inhibitors targeting the HIV protease are a major component of anti-HIV therapy. Genetic drift in the virus due to constantly accumulating mutations leads to resistance and failure of therapy. The resistance in case of protease inhibitors is a combined effect of mutations accumulating in the protease and its cleavage sites. We have developed several quantum dot (QD)-based FRET probes sensitive to HIV protease. These FRET probes consisted of known HIV protease cleavage site sequences and employed QD as the donor and alexa as the acceptor moiety. Upon introduction of this probe into the HIV infected cells, protease cleaves the probe and disrupts FRET. The changes in the FRET can be assessed both qualitatively and quantitatively. The FRET changes quantified in terms of probe cleaved in vivo over time helps us capture the kinetics of HIV protease. We have used different protease wild type substrates, reported to have varied processing rate and successfully demonstrated the kinetic differences in their processing with our assay. Apart from FRET probes with wild type cleavage sequences, we have constructed probes with mutated cleavage sequences found in drug resistant HIV and captured their reported differences in their cleavage rates. We will extend this study to include commonly reported drug resistant protease mutations along with the cleavage site mutations. The complete system displaying both protease mutations and cleavage site mutations, reflective of drug resistant HIV will be used as a screen to identify newer inhibitors.

(307c) Construction of Neurexin Biosensor and Screening of Potential Ligands

Jeevan Baretto[1], Jingjing Li[2] and David W. Wood[1], (1)Chemical and Biomolecular Engineering, The Ohio State University, Columbus, OH, (2)Department of Chemical and Biomolecular Engineering, the Ohio State University, Columbus, OH

Neurexins are neuronal presynaptic proteins that bind to the postsynaptic protein neuroligin at the synaptic junction. Both neurexins and neuroligins are involved in transmitting signals across the neural synapse. Many neurodevelopmental disorders involve abnormal synaptic function. The Neurexin-neuroligin interaction appears to play a key role in autism and other mental disorders. We have constructed a protein-based biosensor that can detect chemical ligands that bind to a given receptor protein. Our engineered bacterial biosensor fusion protein is expressed from a bacterial plasmid DNA, and consists of multiple protein domains including a solubility/stability platform (Maltose Binding protein and an engineered intein domain) and a reporter gene (phage T4 thymidylate synthase - TS) fused to a ligand-binding domain (LBD). The TS assay involves expression of this fusion protein in thymine-knocked out *E. coli* cells (D1210 ΔthyA) in a thymine-less media. Binding of the ligand in turn activates TS which is required for DNA synthesis in this *E. coli* strain. By measuring the growth phenotype of these cells at different ligand concentrations, we can calculate the EC50 for each of the ligands binding to Neurexin 2b. The biosensor system successfully developed in our lab has been previously used with estrogen receptors, thyroid receptors and Peroxisome Proliferator-Activated Receptor γ (PPARG), and has been published elsewhere. Here we describe the construction of a biosensor variant in which Neurexin 2b serves as an LBD within the intein domain, such that binding of ligands is transduced into a growth phenotype via TS. Using this neurexin-LBD biosensor strain, we found that Rosiglitazone, an anti-diabetic drug belonging to the thiazolidinedione class of drugs, causes a growth phenotype at low ligand concentrations, suggesting strong binding. This general approach could help identify ligands of Neurexin that interfere with the signaling across the synapse. Further, this might lead to a better understanding of the mechanism behind autism and other neural disorders.

(307d) Engineered Multi-Input Protein Switches As Versatile Biosensors with Target-Specific Controls

Jay H. Choi and Marc Ostermeier, Chemical and Biomolecular Engineering, Johns Hopkins University, Baltimore, MD

Engineered multi-input protein switches as versatile biosensors with target-specific controls

Jay H. Choi and Marc Ostermeier

Department of Chemical and Biomolecular Engineering, Johns Hopkins University, Baltimore, Maryland

Protein switches are engineered proteins that are composed of an input domain that recognizes and responds to an input signal and an output domain whose function is regulated by the state of the input domain. Engineered protein switches have a number of exemplary properties for sensing applications including a large dynamic range, high specificity for the activating ligand, and a modular architecture that will facilitate fine-tuning of the desired properties. Thus, protein switches can provide a unique platform for the next generation biosensor systems. The modular regulation mechanism of protein switches allows the possibility of developing versatile protein switches, in which different input domains can be coupled to the same output domain. However, switches engineered to date are limited to have natural ligands or similar biomolecules as input signals. For the development of versatile biosensors, it would be desirable to have input signals that can be easily controlled or to have multiple input signals that can be manipulated independently. Especially in cellular environment, multi-regulated switches is highly desirable for the target specific control. We have developed protein switches that can be simultaneously regulated by multiple input signals including small molecules, temperature, pH, and redox potential. These protein switches are designed and constructed based on a platform of a previously developed small molecule activated protein switch. We have taken advantage of its modular regulation mechanism that controls the activation of enzyme in the output domain by manipulating active conformational states. These protein switches function via the dual regulation mechanisms such that they are pre-activated by changes in temperature, redox potential, or pH, and then subsequently activated by small molecules. Thus, the engineered multi-regulated protein switches require dual input signals for the complete activation of enzyme. This design principle has also been applied to develop reversely regulated protein switches.

(307e) Endogenous Molecular Biosensors From Engineered Regulatory Proteins AraC and TetR

Christopher S. Frei[1], Joseph A. Gredell[1], Shuang-Yan Tang[2] and Patrick C. Cirino[1], (1)Chemical & Biomolecular Engineering, University of Houston, Houston, TX, (2)Penn State University, University Park, PA

Endogenous molecular biosensors for specific, sensitive, and high-throughput detection/monitoring of small molecules are emerging as invaluable tools in metabolic engineering and synthetic biology. Regulatory proteins controlled by "effector" molecules naturally couple molecular recognition to changes in gene expression, providing a platform for endogenous molecular reporters by linking *in vivo* molecular synthesis to a readily detectable phenotype (e.g. GFP expression). Numerous advances have been made in the detection of endogenous small molecules through whole-cell regulatory protein-based biosensors [1], but limitations arise due to the narrow range of natural effectors recognized by regulatory proteins. We are developing customized molecular biosensors by engineering regulatory protein effector recognition through directed evolution techniques. In general, these customized molecular biosensors are applicable to high-throughput screening of combinatorial enzyme libraries to improve enzymatic or microbial production of metabolites.

For these purposes, our initial focus was the well-studied L-arabinose-responsive AraC dual regulatory protein from *E. coli*. Saturation mutagenesis libraries, targeting up to 5 residues (~10^7 mutants), were generated based on analysis of the AraC ligand binding pocket. Using a pBAD-GFP construct expressed in *E. coli* and FACS, AraC variants responsive to D-arabinose, mevalonate, and triacetic acid lactone (TAL) were isolated [2-4]. The mevalonate and TAL reporter systems were subsequently used to screen for improved production of these compounds from *E. coli* expressed pathway-specific mutant libraries. Following the success of the previously engineered AraC-based biosensors, further AraC libraries were screened for a response to p-coumaric acid, trans-cinnamic acid, and vanillin. Continued engineering of AraC effector specificity is now aimed at exploring the range of ligands accessible with this motif as well as improving the understanding of the molecular determinants of ligand binding and regulatory switching.

We next sought to engineer reporter systems for considerably larger and more complex natural products such as polyketides. The regulatory protein TetR from the Tn10 transposon was selected due to its well-characterized functions and ability to naturally recognize large polycyclic molecules (i.e. tetracycline). Combinatorial mutant libraries of TetR were created through restricted saturation mutagenesis. Due to the sophisticated hydrogen bonding network found within the ligand binding domain of TetR and the limitations of FACS, our selection of specific target residues was critical in our library design. Restricted saturation mutagenesis allowed us to scan more residues per library while maintaining potential cooperative interactions and a workable library size (~10^7). The TetR mutants were expressed in *E. coli* cells containing a TetR-regulated GFP reporter construct. The libraries are being screened for specificity toward various polyketide compounds. Improved understanding of effector recognition will help optimize the mutational composition of various libraries generated, resulting in more rapid screening and isolation of mutants responsive to compounds of interest. These customized reporters will prove valuable in engineering heterologous polyketide biosynthesis pathways.

1. Gredell, J.A., C.S. Frei, and P.C. Cirino, *Protein and RNA engineering to customize microbial molecular reporting.* Biotechnology Journal, 2012. **7**(4): p. 477-499.

2. Tang, S.Y., H. Fazelinia, and P.C. Cirino, *AraC regulatory protein mutants with altered effector specificity.* Journal of the American Chemical Society, 2008. **130**(15): p. 5267-5271.

3. Tang, S.-Y., O. Akinterinwa, and P.C. Cirino, *A novel endogenous reporter of triacetic acid lactone (TAL) enables screening for improved TAL production by E. coli.* submitted.

4. Tang, S.-Y. and P.C. Cirino, *Design and Application of a Mevalonate-Responsive Regulatory Protein.* Angewandte Chemie International Edition, 2011. **50**(5): p. 1084-1086.

(307f) De Novo Protein Design of Multimeric Proteins with Flexible Templates for Application to Designing Aggregating Peptides

James Smadbeck, George A. Khoury, Meghan B. Peterson and Christodoulos A. Floudas, Department of Chemical and Biological Engineering, Princeton University, Princeton, NJ

As protein design methods develop and computing power improves it becomes increasingly possible to shift focus from the tertiary structure of the proteins to the quaternary structure of protein complexes. In such structures any number of identical (homo) or non-identical (hetero) proteins can come together to form a specified macromolecular structure. Such multimeric protein systems are prevalent in biological systems and can be either essential (e.g. enzymatic complexes) or disrupting (e.g. amyloid fibrils) to correct cell function. Taking into account all parts of such multimeric systems is important for accurate design. In recent years there has been much effort in producing design frameworks for the design of multimeric systems and protein-protein interfaces [1-3]. Here we introduce a novel de novo design method capable of designing for general multimeric peptides/proteins. Such a framework is important for the study of amyloid aggregation, as well as the design of peptides capable of forming self-assembling superstructures or disrupting such formation [4]. The framework is applied to the design of novel aggregating tripeptides and hexapeptides and validated experimentally.

The novel protein design method capable of improving binding affinity (aggregation) of a multimeric system involves three stages. The first stage is a Mixed Integer Linear Programming model [5-7] implemented using a framework enhanced for use in the peptide aggregation design problem by allowing for the possibility of identical amino acid positions in homo-multimer systems. A flexible template structure is initially determined through Molecular Dynamics simulations and a centroid-centroid 8-bin force field is employed to assess the potential energy of the system. This potential energy is minimized to generate a rank-ordered list of designed peptide sequences. The second stage calculates the Fold Specificity of each designed sequence for the flexible template structure.

The third stage is a novel framework that employs Molecular Dynamics simulations in CHARMM [8] to predict the assembly of multiple peptides into a macromolecular structure. These dynamic runs are used to generate an ensemble of peptide structures in solution, as well as ensembles of aggregate structures of varying size. From these ensembles we calculate approximate molecular partition functions of the structures, which are used to calculate an approximate binding affinity [9]. The more accurate the partition functions are, the more precise the binding affinity will be, and so this process is iterated, sampling more and more conformations, until the partition functions converge.

The work detailed in the presentation shows the development of a general de novo protein design method for multimer systems and its application to the design of aggregating tripeptide and hexapeptide systems. The first stage of the method optimizes protein-protein interactions that take place in an aggregate system based on a template generated through molecular dynamics. The rank ordered list of sequences are then run through a molecular dynamics framework to generate an ensemble of structures for approximate binding affinity calculations. The top sequences have been experimentally tested for aggregation through the detection of hydrogel formation.

[1] Kortemme, T. and D. Baker. Computational design of protein–protein interactions. *Current Opinion in Chemical Biology*, 8:91–97, 2004.

[2] Bellows, M. L., H. K. Fung, C. A. Floudas, A. Lopez de Victoria, and D. Morikis. New Compstatin Variants Through Two De Novo Protein Design Frameworks. *Biophysical Journal*,98:2337–2346, 2010.

[3] Bellows, M. L., M. S. Taylor, P. A. Cole, L. Shen, R. F. Siliciano, H. K. Fung, and C. A. Floudas. Discovery of entry inhibitors for HIV-1 via a new de novo protein design framework. *Biophysical Journal*, 99:3445–3453, 2010.

[4] Colombo, G., P. Soto, and E. Gazit. Peptide self-assembly at the nanoscale: a challenging target for computational and experimental biotechnology. *Trends in Biotechnology*, 25(5):211-218, 2007.

[5] Fung, H. K., M. S. Taylor, and C. A. Floudas. Novel formulations for the sequence selection problem in de novo protein design with flexible templates. *Optimization Methods and Software*, 22:51-71, 2007.

[6] Fung, H. K., C. A. Floudas, M. S. Taylor, L. Zhang, and D. Morikis. Toward Full-Sequence De Novo Protein Design with Flexible Templates for Human Beta-Defensin-2. *Biophysical Journal*94:584-599, 2008.

[7] Fung, H. K., W. J. Welsh, and C. A. Floudas. Computational De Novo Peptide and Protein Design: Rigid Templates versus Flexible Templates. *Industrial & Engineering Chemistry Research*, 47:993-1001, 2008.

[8] Brooks, B. R., et al. CHARMM: the biomolecular simulation program. *Journal of Computational Chemistry*, 30(10):1545-1614, 2009.

[9] Lilien, R. H., B. W. Stevens, A. C. Anderson, and B. R. Donald. A novel ensemble-based scoring and search algorithm for protein redesign and its application to modify the substrate specificity of the Gramicidin Synthetase A Phenylalanine Adenylation enzyme. *Journal of Computational Biology*, 12:740-761, 2005.

(307g) Engineering the Beta Roll Domain for Use in Stimulus Responsive Systems

Kevin Dooley[1], Oren Shur[1], Raymond Tu[2] and **Scott Banta**[3], (1)Chemical Engineering, Columbia University, New York, NY, (2)Department of Chemical Engineering, The City College of City University of New York, New York, NY, (3)Department of Chemical Engineering, Columbia University, New York, NY

We have extensively characterized a stimulus responsive repeats-in-toxin (RTX) domain and evaluated its potential as a unique scaffold for protein engineering studies. The RTX domain is intrinsically disordered in the absence of calcium and folds into a beta roll secondary structure consisting of two parallel beta sheet faces in calcium rich environments. Each beta strand contains two solvent exposed residues that are highly variable in naturally occurring beta roll domains and therefore amenable to mutation. In one line of work, we have rationally engineered one face to contain leucine residues which facilitate dimerization in the presence of calcium. This leucine mutant construct can be used as an allosterically regulated cross-linking domain for stimulus responsive protein hydrogel formation. The conformational change in response to calcium of the leucine mutant has been compared to the wild type through several biophysical techniques including CD spectroscopy, bis-ANS binding, and terbium binding. The mechanical properties of the mutant beta roll have been characterized through video particle tracking and microrheology. Secondly, we have used the beta roll domain as a scaffold to evolve novel stimulus responsive binding peptides by using the calcium dependent conformational response to trigger binding and release. We have created a library of beta roll domains by randomizing eight positions on one beta sheet face. This library has been expressed monovalently as fusions to M13 phage particles and screened against streptavidin as a proof of concept work. In the future, we hope to evolve stimulus responsive binding peptides for more useful targets for applications in clinical diagnostics and biosensing.

(307h) Design and Prediction of Peptides That Control Biomineralization

David Masica[1], Sarah Schrier[2], Marianna Sayeg[2] and **Jeffrey J. Gray**[1,2], (1)Program in Molecular Biophysics, Johns Hopkins University, Baltimore, MD, (2)Chemical and Biomolecular Engineering, Johns Hopkins University, Baltimore, MD

While x-ray crystallography and NMR approaches have determined many protein structures yielding great insight into the behavior of proteins in solution, there are no solved structures of proteins interacting with a solid surface. Yet, protein interactions with solid surfaces are critical for formation of hard tissue such as bone or shell, and the design of biomaterials or nanobiotechnology components will require engineering of protein interactions with surfaces. We have used computational protein structure prediction and design approaches to investigate and control the behavior of proteins on solid surfaces. In this talk, I will briefly review our predictions of the structure of statherin on hydroxyapatite incorporating solid-state NMR measurements, and then I will discuss recent efforts to affect calcite growth by the addition of computationally designed peptides. Mutant proteins are compared both by computational analysis and by experimental explorations of the resulting mineral structure by SEM and SAED. Energetic determinants of peptide interactions *in silico* largely correspond to morphological observations *in vitro*, but significant work remains to be able to create custom materials through computationally designed proteins. I will summarize the opportunities for discovery through biophysical approaches in this relatively unexplored but critically important field.

(308a) Elongation Kinetics of Fibrillar Polyglutamine Aggregates

Regina M. Murphy[1], Robert H. Walters[1], Kurt Jacobson[2] and Joel Pedersen[3], (1)Chemical and Biological Engineering, University of Wisconsin-Madison, Madison, WI, (2)Civil and Environmental Engineering, University of Wisconsin, Madison, WI, (3)Soil Science, University of Wisconsin, Madison, WI

Repeat-containing domains are remarkably common in eukaryotes. These domains are frequently natively disordered and play important roles in cellular processes such as transcription, signaling, and formation of multi-protein networks. The most common residues in repeat-containing proteins include glutamine, alanine and (in invertebrates), asparagine. In some cases, expansion of the repeat domain beyond normal length occurs and is associated with a number of disorders; the best known example of this is polyglutamine expansion in Huntington's disease. Polyglutamine expansion leads to abnormal self-assembly; aggregate formation is causally linked to the neuronal degeneration characteristic of Huntington's. This link between polyglutamine expansion, aggregation, and disease motivates our detailed examination of the mechanism and kinetics of polyglutamine-mediated aggregation.

Here we present results in which extension of pre-existing polyglutamine fibrils was explored using two techniques: quartz crystal microbalance with dissipation (QCM-D) and online waveguide spectroscopy (OWLS). Polyglutamine fibrils were immobilized on sensor surfaces, and the adsorption of monomeric polyglutamine-containing peptides was monitored over time. We observed a strong length dependence of elongation: peptides containing only 8 glutamine (Q8) associated only weakly and reversibly, while Q20 and Q24 associated to a much larger extent, and association was only partially reversible. We tested the effect of di-amino acid insertions between two Q10 blocks: one insertion (PG) serves as a beta-turn template whereas another insertion (PP) enforces local rigidity. Association rates, and particularly reversibility, were strongly affected by monomer conformation. Specifically, the PG domain increased the association rate and reduced dissociation, whereas PP had the opposite effect. Dissipation measurements provide additional insight into the elongation process via characterization of the viscoelastic properties of the adsorbed layer.

QCM-D measurement of the adsorbed mass includes not only the peptide itself but also any associated water. For comparison, we measured adsorption rates of monomer to immobilized polyglutamine fibrils using OWLS, which measures only peptide mass. The difference between QCM-D and OWLS mass measurements is a direct measure of associated water. We observed that when Q20 added to the fibrils, it brought with it a significant amount of water that remained associated with the fibrils over time. Peptides containing the PG beta-turn template carried less water. Interestingly, after peptide addition was stopped, we observed a further loss of water from the elongating fibrils. We speculate that this is direct observation of fibril maturation, as the beta-sheet propagates through the fibril end and water is expelled. Taken together, our data support an 'association-conformational conversion' mechanism of fibril elongation.

(308b) Antioxidative Oligomeric Procyanidins Prevent Insulin Fibrillation Via the Formation of Unstructured, off-Pathway Aggregates

Rui Liu, Rongxin Su, Mengfan Wang, Wei Qi and Zhimin He, School of Chemical Engineering and Technology, Tianjin University, Tianjin, China

β-sheet–rich amyloid fibrils or aggregates accumulation has been implicated in a number of human diseases [Ross and Poirier, 2004; Stoppini et al, 2004]. Numerous studies demonstrate that natural polyphenols decrease the risk of degenerative diseases and inhibit in vitro amyloid formation. However, the molecular mechanism for the anti-amyloidogenesis of polyphenols is still unclear. We have investigated the effects of oligomeric procyanidins (OPCs), resveratrol, and trehalose on the amyloidogenicity of insulin using thioflavin-T fluorescence, dynamic light scattering, circular dichrosim, transmission electronic microscopy, and cyclic voltammograms. The main ideas and conclusions are summarized as follows.

(1) The order of inhibitory effect was OPCs > resveratrol > trehalose, suggesting that polyphenolic structure is essential for fibril deposition, as observed by ThT fluorescence. DLS results showed that the presence of OPCs significantly inhibited the aggregation process, at which only smaller particles (d_H = 15 nm and 200 nm) were found after incubation for 105 h. Meanwhile, resveratrol and trehalose failed to show an apparent inhibitory effect on protofibrils, with large particles observed at a similar size to that of insulin alone. TEM results indicated that OPCs show potent inhibitory effects at all stages of insulin fibrillation and redirect insulin aggregation pathway via the formation of unstructured, off-pathway aggregates.

(2) Electrochemical measurements indicate that anti-amyloidogenic effect also correlates with the antioxidative activity of amyloid inhibitor. OPCs were found to be more reducing than resveratrol and trehalose, with evident redox peaks at 0.441 and 0.385 V, respectively. The potential difference (ΔE) between the anodic and cathodic peaks was approximately 56 mV. The anodic peak current was nearly equal to the cathodic peak current, indicating a reversible process that involves one proton and one electron [Liu R et al, 2011]. Meanwhile, the OPCs solution showed a gradual decrease in current intensity ($E = -0.624$ V) with the increase in incubation time, and the reduction peak potential slightly shifted from −0.624 to −0.61V; whereas in the resveratrol and trehalose solutions, only shifts in the reduction peak potential were observed. These results suggest that the antioxidant capabilities of the compounds weakened with increased incubation time.

(3) This study showed, for the first time, how the aggregation pathway of bovine insulin can be redirected with OPCs. The interactions among aggregation-prone domains of bovine insulin were suppressed via the formation of OPC-containing, unstructured off-pathway aggregates, whereas this phenomenon was not observed in the presence of resveratrol, suggesting that the properties of proteins, such as hydrophobicity and charge, play a crucial role in regulating the aggregation pathway, and that OPCs are more accessible than resveratrol for insulin intermolecular interactions.

In summary, the present work demonstrated that OPCs in grape seeds have potent inhibitory and disruptive effects on amyloid fibrillar structures in vitro, inducing the transformation of insulin monomers into unstructured, off-pathway aggregates. The structural properties of compounds, including aromatic structures, and the number of hydroxyl groups must be taken into account in disrupting the aggregation pathway of proteins. Electrochemical measurements indicated that the anti-amyloidogenic effect was also correlated with the antioxidative activity of the amyloid inhibitor. These findings contribute to the development of novel anti-amyloidogenic products from naturally occurring materials.

This work was supported by the Natural Science Foundation of China (No. 20806057 and 31071509), the Ministry of Science and Technology of China (Nos. 2012BAD29B05 and 2012AA06A303), and the Ministry of Education (No.

NCET-11-0372).

References

1. Ross C. A. and Poirier M. A. Protein aggregation and neurodegenerative disease, *Nat. Med.*, 10, S10–S17, 2004
2. Stoppini M.; Andreola A.; Foresti G.; Bellotti V. Neurodegenerative diseases caused by protein aggregation: a phenomenon at the borderline between molecular evolution and ageing. *Pharmacol. Res.*, 50, 419–431, 2004
3. Liu R.; Su R.; Qi W.; He Z. Photo-induced inhibition of insulin amyloid fibrillation on online laser measurement. *Biochem. Bioph. Res. Co.*, 409, 229–234, 2011

(308c) Design and Control of Protein-Protein Interactions to Enhance Stability and Solubility

Marco A. Blanco[1], Christopher J. O'Brien[1], Joseph Costanzo[2], Anne S. Robinson[1], Erik J. Fernandez[2] and **Christopher J. Roberts**[1], (1)Department of Chemical and Biomolecular Engineering, University of Delaware, Newark, DE, (2)Department of Chemical Engineering, University of Virginia, Charlottesville, VA

Biopharmaceuticals are arguably the fastest growing segment of the innovator pharmaceutical pipeline, with a market value in 2010 of $140 billion for just the biotechnology sector alone, and represent a large and growing investment in pharmaceutical research and development. Nonnative protein aggregation is a ubiquitous concern during biopharmaceutical product formulation and process development. The presence of even relatively small quantities of soluble or insoluble non-native aggregates may significantly increase product development time and expenses, as well as cause safety issues in the clinic. There is evidence that the size, structure, and/or morphology of aggregates are important factors in the biological response(s) to aggregates in vivo. Aggregation and phase separation of native or folded proteins is also a long-standing research area for pharmaceutical proteins, with great impact on both bioseparations and drug delivery. This presentation focuses on combined experimental, modeling, and protein engineering approaches to control or predict the effects of mutations and solvent conditions on aggregation and self-assembly of therapeutic and model proteins, as well as new approaches to predict high-concentration behavior from low-concentration light scattering experiments. Our complementary experimental and modeling approaches offer a unique perspective on these problems, and the results highlight an important interplay between thermodynamics and kinetics of folding, self-assembly, and amyloid formation. In addition, they underscore potential roles for coarse-grained molecular modeling to shed insight on engineering design strategies for multi-domain proteins, as well as predictive approaches for high-concentration protein systems of interest for delivery and manufacture of therapeutic proteins. In particular, we find that seemingly minor changes in surface charge (via mutation or solvent conditions) can have dramatic affects on processes such as native dimerization, phase transitions, and the growth processes and phase separation for non-native aggregates. Overall, the results illustrate opportunities to gain better control over native and non-native protein self-assembly via predictive engineering approaches.

(308d) Prediction and Accelerated Test of Antibody Aggregation

Jonathan Rubin[1], Lars Linden[2], Wayne M. Coco[3], Andreas S. Bommarius[4] and Sven H. Behrens[1], (1)Chemical & Biomolecular Engineering, Georgia Institute of Technology, Atlanta, GA, (2)Bayer Pharmaceuticals, Wuppertal, Germany, (3)Bayer Healthcare AG, Cologne, Germany, (4)School of Chemical and Biomolecular Engineering, Georgia Institute of Technology, Atlanta, GA

Physical stability is critical for any therapeutic protein's efficacy and economic viability. No reliable theory exists to predict stability *de novo,* and modeling aggregation is challenging as this phenomenon involves unfolding and rearrangement of non-covalent bonds inter- and intra-molecularly in a complex sequence of poorly understood events. Despite this complexity, the simple observation of protein concentration-dependent diffusivity in *stable*, low ionic-strength solutions can provide valuable information about a protein's propensity to aggregate at much higher salt concentrations and over longer time scales. We recently verified this notion using two model proteins, and other studies have shown that this strategy may be applicable to antibodies as well. In this paper, we expand our previous study to a pharmaceutically relevant hIgG1 antibody and discuss the merits and limitations of stability assessments based on the diffusional virial coefficient k_D. We find this interaction parameter to be a good predictor of relative protein stability in solutions of different chaotropic salts, and a telling heuristic for the effect of kosmotropes. Both temperature and glycosylation are seen to have a strong influence on k_D, and we examine how these factors affect stability assessments. Further, we monitor protein unfolding using a fluorescence assay to assist in interpreting the observed rates of salt-induced aggregation.

(308e) A Non-Invasive, Quantitative Method to Monitor Misfolding/Aggregation of Protein in Mammalian Cells

Simpson Gregoire, Joseph Costanzo, Shaojie Zhang, Erik J. Fernandez and Inchan Kwon, Department of Chemical Engineering, University of Virginia, Charlottesville, VA

Protein misfolding and aggregation are often causative to the onset of numerous neurodegenerative diseases such as Alzheimer's disease, Parkinson's disease, Huntington's disease, and Amyotrophic lateral sclerosis (ALS). Therefore, quantitative analysis of protein misfolding and aggregation in mammalian cells will facilitate enhanced understanding of the molecular mechanisms for various human diseases. We have developed a non-invasive assay that quantifies the extent of deviation from the correctly folded structure of proteins expressed in mammalian cells. In order to validate the assay, we chose wild-type human copper/zinc superoxide dismutase (SOD1) and its mutants (A4V, A4V/C57S, and A4V/C111S) because of its ability to form intracellular aggregates in a short period of time, its variation in aggregation propensity and misfolding and its implications in familial amyotrophic lateral sclerosis (fALS). ALS is a progressive neurodegenerative disorder that inevitably leads to the death of motor neurons. In particular, the A4V mutation is the most prevalent mutation in fALS patients in the United States and leads to death within a year after diagnosis. It is believed that the development of SOD1_A4V aggregates in the spinal cord and brain of fALS patients is pathologically linked to fALS. We have begun screening variants of SOD1_A4V in order to identify less aggregation-prone SOD1_A4V using the aforementioned monitoring system. Biophysical properties of the less aggregation-prone mutants identified will be characterized in vitro.

(308f) Design of Antibodies Specific for Unfolded Proteins

Ali Reza A. Ladiwala, Joseph M. Perchiacca, Moumita Bhattacharya and Peter M. Tessier, Dept of Chemical & Biological Engineering, Rensselaer Polytechnic Institute, Troy, NY

Protein unfolding leads to aggregation and loss of function. No antibodies have been reported that recognize generic conformational epitopes within unfolded proteins despite the importance of detecting protein unfolding at extremely low concentrations and in the presence of background (folded) proteins. We posit that antibodies specific for solvent-exposed aromatic residues (which are normally solvent-shielded within folded proteins) will selectively recognize unfolded conformations of diverse proteins. Based on this hypothesis, we have engineered the complementarity determining regions of small antibodies with aromatic repeat sequences to enable them to bind to unfolded proteins via intermolecular pi-stacking interactions. We find that such antibodies recognize diverse proteins in their unfolded state and bind only to solvent-exposed aromatic residues. In contrast, the same aromatic repeat peptides (apart from the antibody) bind weakly to unfolded proteins, revealing that constraint of aromatic peptides within antibody loops significantly increases their binding affinity. We will discuss how these novel antibodies can be used to monitor protein unfolding in a highly sensitive and site-specific manner.

(308g) Interaction of Tau Protein with Model Lipid Membranes Induces Tau Structural Compaction

Emmalee M. Jones[1], Manish Dubey[2], Saurabh Singh[3], Ann Junghans[3], Phillip J. Camp[4], Briana C. Vernon[1], Jacek Biernat[5], Eckhard Mandelkow[5], Jaroslaw Majewski[2] and Eva Y. Chi[6], (1)University of New Mexico, Albuquerque, NM, (2)Los Alamos Neutron Science Center, Los Alamos National Laboratory, Los Alamos, NM, (3)Los Alamos Neutron Science Center, Los Alamos National Laboratory, (4)Department of Chemical and Nuclear Engineering, University of New Mexico, Albuquerque, NM, (5)Max Planck Unit for Structural Biology, Hamburg, Germany, (6)Chemical and Nuclear Engineering, University of New Mexico, Albuquerque, NM

The misfolding and aggregation of proteins into b-sheet rich fibrillar aggregates *in vivo* are linked to the pathogenesis of over twenty neurodegenerative diseases, including Alzheimer's disease. Alzheimer's disease is a progressive neurodegenerative disease characterized in part by neurofibrillary tangles (NFTs) found in the brains of affected patients. NFTs are composed of misfolded aggregates of the intrinsically disordered microtubule associated protein tau. In order to aggregate into fibrils, the intrinsically disordered, highly soluble, and highly stable tau needs to undergo conformational changes that render the protein aggregation-competent or "pro-aggregant". The "pro-aggregant" form is currently unknown, but the formation of β-sheet enriched fibrils from tau likely proceeds through the formation of partially folded, or structurally compact, conformations with increased aggregation propensities. *In vitro*, aggregation can be induced by polyanionic cofactors such as heparin and anionic micelles or vesicles, which compensate for tau's positive charges. This suggests that aggregation proceeds through a nucleation controlled polymerization pathway, but the molecular basis of the early aggregation events, such as the structural fluctuations that trigger the aberrant accumulation of tau into NFTs rich in β-sheets *in vivo*, remains unknown.This prompted our investigation to assess tau's propensity to interact with membranes and to elucidate the structural perturbations those interactions induce in the tau protein. We show that although highly charged and soluble, tau is also highly surface active. Tau strongly interacts with anionic membranes and does not exhibit any favorable interaction with neutrally charged membranes. To resolve molecular-scale structural details of protein associated with anionic lipid membranes, we utilized X-ray scattering techniques. X-ray reflectivity indicated tau's presence underneath an anionic DMPG monolayer at the air/water interface and penetration into the lipid headgroups and tailgroups. More significantly, both air/water and DMPG lipid membrane interfaces induce the disordered, molten globule-like, tau to partially adopt a more compact conformation with a density similar to that of a folded protein. To investigate the effect of tau-membrane interactions on tau fibril formation, we also incubate the protein with lipid vesicles containing either neutrally charged or anionic lipids. Thioflavin-S binding assay is used to detect fibril formation and transmission electron microscopy is used to image the morphology of aggregates formed. Our results show that tau is highly surface active and strongly interacts with anionic lipid membranes, leading to tau structural compaction which may render the protein "pro-aggregant" or aggregation-competent. This suggests possible membrane-based mechanisms of tau aggregation in neurodegenerative diseases.

(308h) Studies of IRE1alpha Transmembrane Domain Dimerization in Lipid Bilayers

Hyunju Cho[1], Ryan LaMarca[2] and Christina Chan[1], (1)Chemical Engineering and Materials Science, Michigan State University, East Lansing, MI, (2)Community, Agriculture, Recreation, and Resource Studies, Michigan State University, East Lansing

IRE1α (inositol-requiring enzyme 1α) is an integral membrane protein with an N-terminal luminal domain and a C-terminal cytosolic domain, which are connected by a single transmembrane (TM) helix. As one of the three transmembrane sensor proteins, IRE1α propagates the signal from the ER (endoplasmic reticulum) lumen to the cytosol. It has been known that the ER luminal domain of IRE1α is maintained in its inactive monomeric state by interacting with ER chaperone, Bip. However, upon ER stress, Bip is able to release from the luminal domain, allowing the luminal domain to dimerise or oligomerize. This promotes the dimerization of the TM domains and cytosolic domains, which results in activation of the cytosolic domain and induction of several signaling pathways to relieve ER stress. However, the molecular mechanism of signal transduction across the membrane remains unknown. In addition, although the TM domains are expected to play an important role in the dimerization process, the contribution of IRE1α TM domain to dimer stability has not been investigated. We address whether the TM domains are able to form dimers/oligomers in a membrane-like environment. The sodium dodecyl sulfate polyacrylamide gel electrophoresis (SDS-PAGE) and silver staining analyses show that three states (monomers, dimers, and high-oligomers) of TM domain peptides exist in SDS micelles, even in the absence of both the luminal and cytosolic domains. Circular Dichroism (CD) spectra suggested that IRE1α TM domain peptide is helical in SDS micelles and in liposome. We performed FRET (F rster resonance energy transfer) using FITC-labeled IRE1α TM peptide (donor) and Rhodamine-labeled IRE1α TM peptide (acceptor). FRET results suggest that IRE1α TM domain forms dimers in lipid bilayers. Since it is well known that palmitic acid (C16-saturated fatty acid) induces ER stress by activating IRE1α protein, using FRET energy analyses, we further determine whether palmitic acid is able to promote dimerization of IRE1α TM peptide in liposomes. Therefore, the results provide the first evidence that the IRE1α TM domains are capable of significant protein–protein interactions to thereby contribute to ER stress-signaling propagation.

(309a) Modeling the Diffusive Behavior of 3D Stem Cell Migration

Joshua D. Cohen[1], Tyler Vlass[2], Dayong Chen[3], Ryan C. Hayward[4] and **Shelly R. Peyton**[2], (1)Chemical Engineering, MIT, Cambridge, MA, (2)Chemical Engineering, University of Massachussetts, Amherst, Amherst, MA, (3)Polymer Science and Engineering, University of Massachussetts, Amherst, Amherst, MA, (4)Polymer Science and Engineering, University of Massachusetts, Amherst, Amherst, MA

Design of 3D scaffolds that can facilitate proper survival, proliferation, and differentiation of progenitor cells is a challenge for clinical applications involving large connective tissue defects, with cell migration within scaffolds a critical process governing tissue integration. In previous work, we have quantified how physical properties of 3D scaffolds, such as pore diameter, modulus, and integrin-binding peptides, as independently tunable parameters, govern the motility and infiltration of marrow-derived stem cells (MSCs). The relationships between migration speed, displacement, and total path length were found to be strongly dependent on pore size, most likely resulting from the non-intuitive convolution of pore diameter and void chamber diameter yielding different geometric environments experienced by the cells within.

A looming challenge in the quantification of cell motility in these complex microenvironments is whether or not we will be able to *predict* the extent of cell infiltration given a set of scaffold physical parameters. We are building computational models, based on anomalous diffusion, to determine whether or not MSC migration on 2D polyacrylamide substrates, with tunable stiffness and integrin-binding peptides, is predictive of MSC migration in more complex 2.5D and 3D microenvironments. An anomalous diffusion model of cell motility exhibits superior fitting to the canonical persistent random walk (PRW) model since it is able to capture both the superdiffusive and subdiffusive behavior of MSC migration within 3D microenvironments. This model relates the mean squared displacement (MSD) to time by a power law and is characterized by two parameters, a diffusion coefficient and a scaling index. To develop a predictive quantitative model of cell motility, we are determining diffusion coefficients and scaling indices using migration paths from single cell tracking experiments and correlating these model parameters to specific physicochemical cues from the microenvironment. Experimentally, we have observed that MSC migration speed has a biphasic dependence on the density of surface-bound RGD, but only at intermediate substrate stiffness. This biphasic relationship was observed in 3D gels, but only when the pores were much larger than the average cell diameter, and therefore likely a quasi-2D environment. In smaller pore diameters, where migration was truly 3D, this biphasic dependence disappeared. We are currently designing novel "2.5D" substrates, wherein cells cannot infiltrate the scaffold, but still experience topographical cues and ridges that approximate 3D macroporous scaffolds. We hypothesize that this hybrid microenvironment will bridge our computational efforts in 2D and 3D, and allow us to predict 3D migration given the scaffolds pore size, modulus, and density of integrin-binding peptides. This information on MSC motility could be very powerful for future intelligent scaffold design for MSC-directed bone regeneration *in vivo*.

(309b) Regulatory Role of D'D3 Domain in VWF-A1 Mediated Platelet Thrombus Formation: Application towards Understanding Von Willebrand Disease

Sri R. Madabhushi[1], Kannayakanahalli Dayananda[1], Chengwei Shang[1], Jun Qu[2] and Sriram Neelamegham[1], (1)Chemical and Biological Engineering, State University of New York at Buffalo, Buffalo, NY, (2)Department of Pharmaceutical Sciences, State University of New York at Buffalo, Buffalo, NY

Introduction: von Willebrand factor (VWF) is a multimeric blood glycoprotein that plays a crucial role during blood coagulation by recruiting platelets to vascular injury site [1]. Improper regulation of VWF-platelet interaction results in various cardiovascular disorders like stroke, heart attack and other bleeding disorders called von Willebrand disease. The 3-D structural organization of VWF is believed to regulate the platelet binding function of VWF-A1 domain [2]. However, the precise mechanism by which domain level organization controls VWF function is not known. Here, we tried to address this by applying a "systems approach".

Materials and Methods: Dimeric VWF (ΔPro-VWF) was constructed by deleting the propeptide section of full-length VWF cDNA. ΔD'D3-VWF was identical to ΔPro-VWF only it lacked the VWF-D'D3 domain. Recombinant GpIbα was synthesized as a fusion protein with human IgG tail. All proteins were purified from mammalian expression systems. ΔPro-VWF was crosslinked using BS^3, a reagent that links proximal amines. This was proteolytically cleaved and subjected to high resolution tandem mass spectrometry (LTQ Orbitrap). C++ programs were written to identify cross-linked peptide pairs from the MS data. Experiments that validated the MS studies examined the effect of D'D3 domain deletion on the binding of VWF-A1 to platelet receptor GpIbα in static ELISA studies and flow chamber based shear assays. Additionally, anti-D'D3 mAbs were generated in mice and the effect of these reagents on shear induced platelet aggregation (SIPA) was studied.

Results and Discussion: Analysis of the cross-linked VWF tandem mass spectrometry data revealed three cross-linked peptide pairs linking VWF D'D3 and A1 domain. One of these A1 peptides formed a major interaction interface with platelet GpIbα. Based on this, studies were designed to assay the role of D'D3 in regulating VWF-A1 function. To this end, the deletion of the D'D3 domain (i.e. ΔD'D3-VWF) resulted in a dimeric protein that efficiently bound GpIbα even under static conditions and in the absence of ristocetin. ΔPro-VWF, on the other hand, only bound GpIbα when ristocetin was added. Under low shear conditions (1dyn/cm^2) robust platelet translocation was observed on ΔD'D3-VWF, but not ΔPro-VWF, coated surfaces. At higher shear conditions (10-20 dyn/cm^2), while platelet accumulation was comparable on the two surfaces, platelet rolling velocity on ΔD'D3-VWF was ~ 50% lower compared to ΔPro-VWF. Finally, one of the anti-D'D3 mAb generated (clone DD3.1) blocked SIPA by ~50% in the absence but not presence of ristocetin.

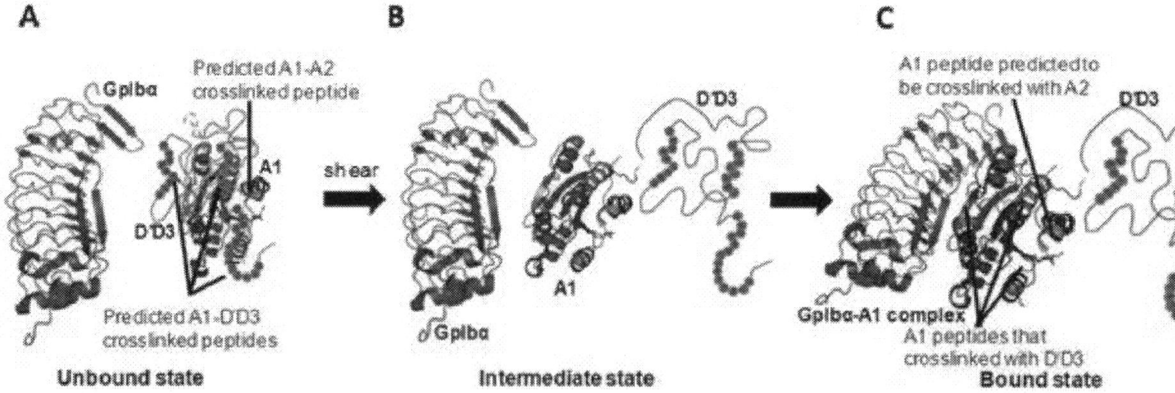

Model for VWF- GpIbα interaction under shear. A. D'D3 domain shields VWF-A1 binding to GpIbα. **B.** Upon shear, D'D3 releases A1. **C.** This enables VWF-A1 binding to GpIbα.

Conclusions: The data demonstrate that, in the native state, VWF-D'D3 domain sterically inhibits the VWF-A1 domain and this reduces the rate of VWF binding to platelets. Shear stress or agonist (e.g. ristocetin) releases this steric inhibition. This then allows VWF to bind platelet-Gp1bα efficiently. Such regulation of D'D3-A1 interaction may be a critical feature that controls platelet adhesion rates at sites of vascular injury. While von Willebrand disease (VWD) mutations have been identified, the molecular mechanism of mutation manifesting into the disease state is not well understood. Based on these results, it is likely that some of the VWD 2B or 2M mutations affecting VWF-GpIbα interaction affect the domain level regulation of VWF instead of directly affecting A1-GpIbα affinity.

References: [1] Madabhushi, S.R., et al., Blood, 2012, March 27 (*Epub ahead of print*).

[2] Singh, I et al., Biophys J., 2009;96(6):2313-20.

(309c) Thrombus Growth and Embolism On Tissue Factor-Bearing Collagen Surfaces Under Flow Role of Thrombin with and without Fibrin

Thomas Colace[1], Ryan Muthard[2] and Scott L. Diamond[1], (1)Chemical and Biomolecular Engineering, University of Pennsylvania, Philadelphia, PA, (2)Department of Chemical Engineering, Penn State University, University Park, University Park, PA

At sites of vascular injury thrombin is an important mediator in thrombus growth and stability. Using microfluidic flow devices as well as patterned surfaces of collagen and tissue factor we sought to determine the role that fibrin plays in clot stability without interfering with the production of thrombin. We deployed an 8-channel microfluidic device to study coagulation during corn trypsin inhibitor-treated (XIIa-inhibited) whole blood perfusion over lipidated tissue factor (TF) linked to a fibrillar collagen type 1 surface. Clot growth and embolization were measured at initial inlet venous (200 s^{-1}) or arterial (1000 s^{-1}) wall shear rates under constant flow rate or pressure relief mode in the presence or absence of Gly-Pro-Arg-Pro (GPRP) to block fibrin polymerization. Numerical calculations for each mode defined hemodynamic forces on the growing thrombi. In either mode at inlet venous flow, increasing amounts of TF on the surface led to a modest dose-dependent increase (up to 2-fold) in platelet deposition, but resulted in massive fibrin accumulation (> 50-fold) only when exceeding a critical TF threshold. At a venous inlet flow, GPRP led to a slight 20% increase in platelet accumulation ($p < 0.01$) in pressure relief mode with thrombi resisting ~1500 s^{-1} prior to full channel occlusion. GPRP-treated thrombi were unstable under constant flow rate, where shear forces caused embolization at a maximum shear rate of ~2300 s^{-1} (69 dynes/cm^2). In constant flow rate mode, the nonocclusive platelet-fibrin deposits (no GPRP) withstood maximum shear rates of ~29,000 s^{-1} (870 dyne/cm^2) or ~95% of full channel occlusion. For arterial inlet shear rate, embolization was marked for either mode with GPRP present when shear forces reached 87 dynes/cm^2 (~2900 s^{-1}). Under constant flow rate, platelet-fibrin deposits (no GPRP) withstood maximums of 2400 dynes/cm^2 (80,000 s^{-1}) or ~90% of full channel occlusion before embolization. Fibrin increased clot strength by 12 to 28-fold. Under pressure relief mode, ~2-fold more fibrin was produced under venous flow ($p < 0.001$). These studies define embolization criteria for clots formed with surface TF-triggered thrombin production (± fibrin) under venous and arterial flows.

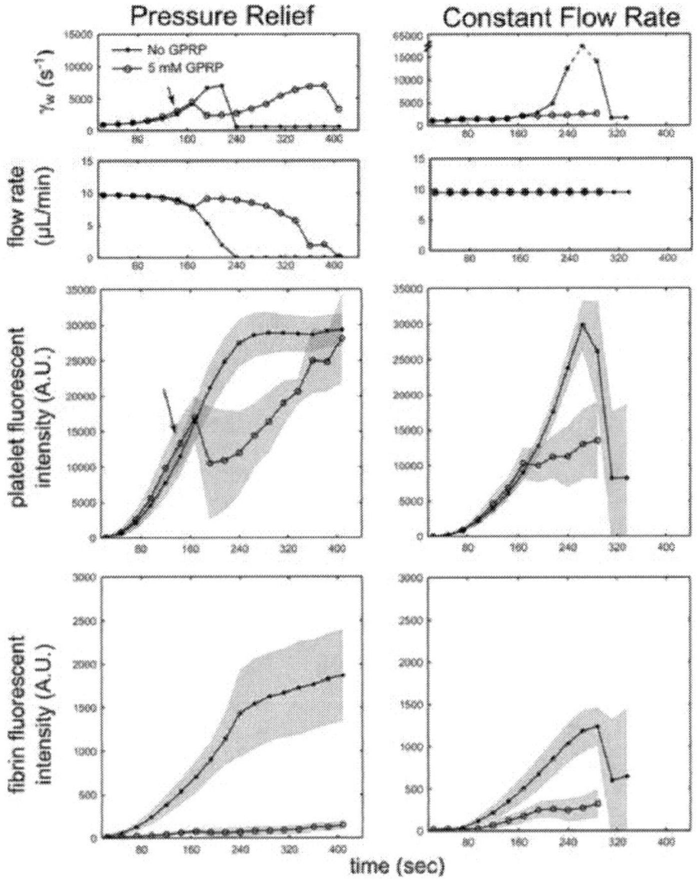

Figure 1 Whole blood (CTI 40 μg/mL) in the presence or absence of 5 mM GPRP was perfused over a fibrillar collagen surface at an initial arterial wall shear rate of 1000 s^{-1}, with (left, n=4) or without (right, n=8) pressure relief. Platelet accumulation data was used to calculate the hemodynamic conditions at the thrombus boundary based on simulation data.

A Control Engineering Perspective to Modeling Calcium Regulation and Related Pathologies

Christopher R. Christie[§], Luke E.K Achenie[§], Babatunde A. Ogunnaike[Ψ]
[§]Department of Chemical Engineering, Virginia Polytechnic Institute and State University, Blacksburg, Virginia 24060
[Ψ]Department of Chemical Engineering, University of Delaware, Newark, Delaware 19716

Calcium (Ca) homeostasis, the maintenance of a stable plasma Ca concentration in the body, is essential to normal physiological function where the total plasma Ca concentration in a healthy human body must be maintained within a very narrow range (2.2-2.4mM) [1] in the face of significant variations in Ca ingestion and/or excretion. Meeting such stringent requirements is the task of an effective, high performance physiological regulatory system that employs parathyroid hormone (PTH) and calcitriol to regulate Ca flux between the plasma and kidneys, intestines and bones. Disorders in the regulatory organs cause abnormal hormonal secretion and activity and contribute to chronic imbalances in plasma Ca levels. Furthermore, these changes in hormonal activity often lead to long-term physiological problems: for example, osteoporosis (increased loss of bone mineral density) arises as a consequence of primary hyperparathyroidism (PHPT) – a condition characterized by hyper secretion of PTH [2].

Our primary objective is to develop a quantitative understanding of how normocalcemia (the maintenance of Ca concentration in the appropriate narrow range) is achieved, and to employ such understanding to elucidate the mechanisms by which associated pathological conditions such as PHPT, hypoparathyroidism, humoral hypercalcemia of malignancy (HHM) and vitamin D deficiency affect Ca homeostasis.

In this presentation, we discuss our representation of the physiological process of Ca regulation in the form of an engineering control system block diagram with explicitly identified sensor, controller, actuators and process. Mathematical models are generated for each component sub-process by applying standard conservation of mass principles augmented with available mechanistic information. The result is a series of ordinary differential equations (ODE) that describe the dynamics of: (i) PTH in the controller; (ii) calcitriol from the kidneys in the actuator; (iii) bone cells (osteoblasts and osteoclasts) from the bone in the actuator; and (iv) plasma calcium (the "process"). The overall model is then validated with published clinical data. Pathological conditions are simulated as defects in appropriate control system components (sensor, controller and/or actuator) and the simulation results validated against published clinical data of the Ca-related disorders in question.

We demonstrate how our modeling approach provides information about important quantities that are typically not measured and are hence not available in published data sets, and how such information can be used to understand Ca homeostasis and the onset and propagation of Ca-related disorders.

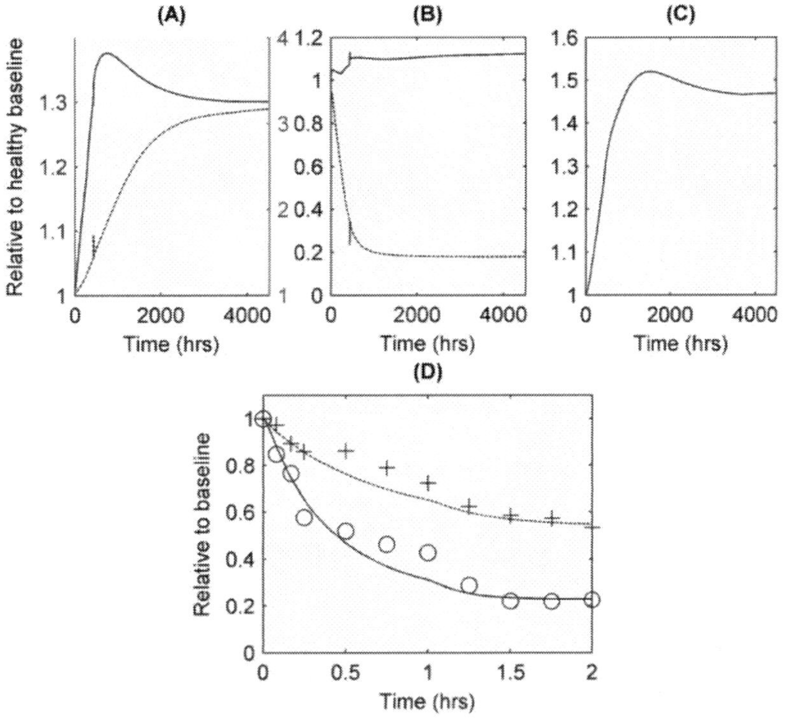

Figure 1. Controller defects: increase in PTH secretory rate (PHPT) – (A): Ca (—), PTH (--); (B) CTL (—), Cau (--); (C) bone cell ratio; (D) Comparison of clinical PTH data: Healthy: (—,o), PHPT (--- , +);

PHPT is represented in our model (Fig. 1A-C) by a defect in the controller, resulting in an increase in the PTH basal levels leading to additional increases in CTL, bone cell proliferation and reduced renal Ca excretion, consistent with clinical observations [3, 4]. This underscores the utility of our model as we are able to provide quantitative data that is usually unavailable from clinical tests. Additionally, we compare the response to hypercalcemic induction in both the healthy and PHPT models (Fig. 1D). From the modifications used to achieve the fit, we can postulate the mechanism of progression of the disease and thus provide a useful research tool in investigating, and ultimately, predicting the stage of PHPT in patients.

References

1. Raposo, J.F., L.G. Sobrinho, and H.G. Ferreira, *A minimal mathematical model of calcium homeostasis.* J Clin Endocrinol Metab, 2002. **87**(9): p. 4330-40.
2. Fraser, W.D., *Hyperparathyroidism.* Lancet, 2009. **374**(9684): p. 145-58.
3. Khosla, S., et al., *Calcium infusion suggests a "set-point" abnormality of parathyroid gland function in familial benign hypercalcemia and more complex disturbances in primary hyperparathyroidism.* J Clin Endocrinol Metab, 1993. **76**(3): p. 715-20.
4. Rubin, M.R., et al., *The natural history of primary hyperparathyroidism with or without parathyroid surgery after 15 years.* J Clin Endocrinol Metab, 2008. **93**(9): p. 3462-70.

(309e) A Quantitative Systems Approach to Identify Paracrine Mechanisms That Locally Suppress Immune Response in Melanoma

David J. Klinke, Chemical Engineering, West Virginia University, Morgantown, WV; Microbiology, Immunology and Cell Biology, West Virginia University, Morgantown, WV

Therapeutic cytokines enhance anti-tumor immunity when delivered to the tumor microenvironment. However, local immunoregulatory elements dampen the efficacy of many of these cytokines. The identity of these local mechanisms used by tumors to suppress immunosurveillance represents a key knowledge gap for improving tumor immunotherapy. From a systems perspective, local suppression of anti-tumor immunity is a closed-loop system - where system response is determined by an unknown combination of external inputs and local cellular cross-talk. Here, we hypothesized that the B16 cell line, a transplantable mouse model for spontaneous metastatic melanoma, locally suppresses the action of Interleukin-12 (IL-12) , a key cytokine that links innate to adaptive immunity. To test this hypothesis, we recreated this closed-loop system in vitro and combined quantitative high content assays, in silico model-based inference, and a proteomic workflow to identify the biochemical cues responsible for immunosuppression. Following an induction period, the B16 melanoma cell model inhibited the response of a T helper cell model to IL12. This paracrine effect was not explained by induction of apoptosis or creation of a cytokine sink, despite both mechanisms present within the co-culture assay. Tumor-derived Wnt-inducible signaling protein-1 (WISP-1) was identified to exert paracrine action on immune cells by inhibiting their response to IL12. Moreover, WISP-1 was expressed in vivo following intradermal challenge with B16F10 cells and was inferred to be expressed at the tumor periphery. This spatial pattern of WISP-1 expression is also observed in human skin melanoma tissue. Collectively, the data suggest that (1) biochemical cues associated with epithelial-to-mesenchymal transition can shape anti-tumor immunity through paracrine action and (2) remnants of the immunoselective pressure associated with evolution in cancer include both sculpting of tumor antigens and expression of proteins that proactively shape anti-tumor immunity.

(309f) A Multiscale Model of Acute Insulin Resistance in Critical Illness

Ari Pritchard-Bell, University of Pittsburgh, Pittsburgh, PA, Gilles Clermont, Critical Care Medicine, University of Pittsburgh School of Medicine, Pittsburgh, PA, Balaji Yegneswaran, Critical Care Medicine, University of Pittsburgh, Pittsburgh, PA and Robert S. Parker, Chemical and Petroleum Engineering, University of Pittsburgh, Pittsburgh, PA

Robust blood glucose regulation is an evolutionarily conserved mechanism present in many living organisms and vital to the survival of a species. These mechanisms have developed aggressive measures to prevent blood glucose from falling below threshold (hypoglycemia), while loosely regulating overshoot, which results in hyperglycemia that has potentially harmful, but less immediate, consequences. The immune response plays an important role in this regulation by overriding insulin-triggered glucose uptake to provide resources needed to fight against infection and starvation[1]. The immune activation hijacking of the body's metabolism, while evolutionarily rational for survival, induces a potentially harmful increase in glucose levels via insulin resistance. Elevated blood glucose following injury, so called "stress hyperglycemia" or "diabetes of injury", has been correlated with increased mortality rates in the ICU[2]. We aim to develop a model-based decision support system (glucose sensor, control algorithm, insulin pump, user-friendly interface) to maintain glucose homeostasis without hypoglycemia in critically ill patients. Key to the control algorithm component is a multiscale system model linking cellular response to systemic response. This does not presently exist.

We synthesize a dynamic model of glucose metabolism in relation to the acute immune response by focusing on three submodels. Our multi-scale model consists of: (i) a cellular level model of the insulin-action pathway[3]; (ii) a compartmental model describing tissue-level interaction[4]; and (iii) a novel data-driven model describing the immune response to injury. At the cellular level the antagonist(s) of insulin sensitivity deactivate the tyrosine phosphorylation cascade, following insulin binding, through serine phosphorylation[5]. We use the detailed mechanistic model proposed by Sedaghat and colleagues as a functioning insulin-responsive cell to which we add a state corresponding to the serine kinase population that acts to inhibit insulin action. The cellular model receives input from the injury model in the form of increased serine kinase levels, as well as extravascular insulin from the tissue model. The cellular insulin-responsive glucose uptake is fed back into the tissue-level model as a variable glucose clearance rate. The tissue-level model from Roy[4]also accounts for exercise and fatty acids which can be used as inputs for the injury model.

The injury model characterizes the immune response to trauma using biometrics, cytokines and stress-hormones as inputs that are believed to influence serine kinase activity. Data from APACHE III scores (a recognized metric describing severity of illness[6]) are used as inputs that have been correlated to insulin sensitivity[7]. Fatty acids from the tissue model are classified by their inflammatory characteristics and accordingly propagate the immune response. Stress-induced hormones such as catecholamines and glucocorticoids[8]also influence insulin resistance, and these are described in the injury model as additional inputs to the cellular model to attenuate insulin-dependent glucose uptake.

The wide range of patient conditions and biomolecules contributing to stress hyperglycemia makes accurate characterization a serious challenge. The composite model allows us to depict the overall dynamics of glucose metabolism while detailing important molecular events. Analysis at the cellular level allows us to identify dynamic micro-scale factors influencing biological regulation of systemic glucose and insulin in patients recovering from trauma. Using multiple scales we are able to describe key phenomena relevant to glucose control, as applied to critical care patients. Molecular-level insight into stress hyperglycemia could reveal key aspects of patient specific acute insulin resistance which could be used in a model-based decision support system for controlling glucose in the intensive care unit.

1. Fernandez-Real, J.-M., et al. (1999). Insulin Resistance and Inflammation in an Evolutionary Perspective: the Contribution of Cytokine Genotype/Phenotype to Thriftiness. *Diabetologia,*42: 1367-1374

2. Van den Berghe, G. et al., (2001). Intensive Insulin Therapy in Critically Ill Patients. *The New England Journal of Medicine*. 345:1359-1367.

3. Sedaghat, A. R. et al. (2002). A Mathematical Model of Metabolic Insulin Signaling Pathways. *American Journal of Physiology Endocrinology and Metabolism*, 283: E1084-E11101

4. Roy, Anirban (2008). Dynamic Modeling of Free Fatty Acid, Glucose, and Insulin During Rest and Exercise in Insulin Dependent Diabetes Mellitus Patients. Ph.D dissertation, University of Pittsburgh.

5. Kitano, H. et al. (2004). Metabolic Syndrome and Robustness Tradeoffs. *Diabetes,*53 Suppl. 3: S6-S15.

6. Knaus, W. et al., (1991). The APACHE III Prognostic System. *CHEST*. 100(6): 1619-1636.

7. Zauner, A. et al. (2007). Severity of Insulin Resistance in Critically Ill Medical Patients. *Metabolism*. 56(1): 1-5.

8. Li, L. et al. (2009). Acute Insulin Resistance Following Injury. *Trends in Endocrinology and Metabolism*. 20(9): 429-435.

(309g) Systems Biology of Host-Pathogen Interactions

Anu Raghunathan, Chemical Engineering, National Chemical Laboratory, Pune, India, Sookil Shin, Mount Sinai School of Medicine, NY, NY and Simon Daefler, Division of Infectious Diseases, Mt Sinai School of Medicine, New York, NY

Francisella tularensis is an ill-characterized pathogen and a potential biowarfare agent necessitating a better understanding of its metabolism and virulence mechanisms. Genome scale constraints based metabolic models provide a systems framework to study metabolism, infection and pathogenesis. We have developed a metabolic network reconstruction for *F. tularensis* using the current genome sequence annotation and biochemical legacy data. Several mathematical techniques like Flux balance analysis (FBA), Flux Variability Analysis (FVA), Robustness Analysis (RA) were used to explore the metabolic capabilities of the *in silico* pathogen under *in vitro* and *in vivo* conditions. Gene expression, growth and physiological data were obtained experimentally and used as additional constraints in the model to compute and predict functional states of the pathogen. Integrating this data into the model as constraints allow us to reduce the solution space representing cell behavior and predict cellular function with more accuracy. *Francisella* was capable of growth with different specific growth rates on glucose and alternate carbon sources including glycerol, xylose, ribose, fructose and arabinose in a chemically defined medium with differential and preferential consumption of amino acids thus validating our model prediction that they can make up the bulk of the biomass. Gene expression of selected metabolic genes at specific time points (12hr and 24hr pi) of *francisella* during infection provide a snapshot and implicate certain metabolic pathways as critical to growth and survival in the host cell. Several gene transcripts detected in this data set were present in the optimal reactome calculated using FVA. Such an analysis provides insights into the lifestyle of the bacterium and the metabolic dependence on the host. Iterative model building can further help refine our understanding of the intracellular milieu encountered by the bacterium inside the host-cell and point towards genes essential in growth, survival and pathogenesis. Collectively, the results presented herein suggest an effective systems biology strategy of combining *in silico* modeling with experimental technologies to enhance biological discovery and discover novel drug targets.

(309h) Modulating Lipid Fate Controls Lipotoxicity in Palmitate-Treated Hepatic Cells

Alexandra K. Leamy, Chemical & Biomolecular Engineering, Vanderbilt University, Nashville, TN and Jamey D. Young, Chemical and Biomolecular Engineering, Vanderbilt University, Nashville, TN

As the American waistline continues to expand so too does the burgeoning health crisis associated with obesity, diabetes and the all-encompassing metabolic syndrome. The pathogenesis of these now common conditions involves abnormal accumulation of lipids (steatosis) in non-adipose tissues due to elevated free fatty acid export from adipose tissue. In the liver, chronic steatosis and saturated fatty acids (SFA) overexposure can result in dysregulation of ER function and oxidative stress. However, the lipid metabolic pathways and mechanisms responsible for these deleterious alterations remain largely undefined. Our central hypothesis is that overexposure to SFAs (e.g., palmitate) leads to increased saturation of ER membrane phospholipids, which alters membrane stability and induces ER stress.

In order to better characterize the phenotypic response of hepatic cells to SFA overexposure, we have sought to systematically perturb the three main pathways of fatty acid utilization and disposal: (1) phospholipid (PL) synthesis and remodeling, (2) energy generation by beta-oxidation, and (3) triglyceride (TG) synthesis. Experimentally, we simulated hepatic lipotoxicity using palmitate-treated H4IIEC3 rat hepatocytes as an *in vitro*model. This system demonstrated significantly higher levels of reactive oxygen species (ROS) accumulation, endoplasmic reticulum (ER) stress associated protein release (CHOP), and cell death as well as increased saturation of membrane phospholipid species. Using treatments designed to modulate specific lipid metabolic pathways, we have found that blocking phospholipid breakdown and turnover (using phospholipase inhibitors) or enhancing fatty acid disposal (using AICAR to stimulate beta-oxidation) can suppress the lipotoxic effects of palmitate.

A detailed investigation is currently underway to elucidate the mechanisms by which these treatments are able to reverse the lipotoxic effects of SFA overexposure. We are applying isotopically labeled ^{13}C-palmitate to trace the fate of exogenous fatty acids using metabolic flux analysis (MFA) and lipidomic profiling. This will enable us to identify pathways that are responsible for the observed metabolic phenotypes and develop a better understanding of the mechanisms leading to ROS accumulation and lipoapotosis in the presence of elevated SFAs. This understanding is expected to provide more effective and targeted therapies for treatment of fatty liver disease and related complications associated with diabetes and obesity.

(312a) Analytical Model of Local Distribution of Chemicals in Tissues with First Order Rate Metabolism Kinetics

Alexander Golberg, UC Berkeley, Berkeley, CA; CEM/MGH/Harvard Medical School, Boston, MA

Analytical model of local distribution of chemicals in tissues with first order rate metabolism kinetics.

Alexander Golberg,

Department of Mechanical Engineering, *Etcheverry* Hall, 6124, University of California at Berkeley, Berkeley, CA 94720 USA

Email: agolberg@gmail.com

Tel: +1 -510-2216557

Abstract

The profile of regional distribution of chemicals in tissue is important for the fundamental analyses of metabolism, morphogenetic growth control and the development of novel regional therapies. One of the challenges in tissue and whole organ research is predicting local chemical distribution. Here we introduce an analytical approach for the temporal and spatial regional distribution mapping of chemicals in metabolically active, perfused tissues. The new aspect of this model is in local chemical concentration dependence on of tissue perfusion rate, in addition to diffusion and metabolism rates. Using Duhamel theorem we performed dimensionless and dimensional analysis for local chemical distribution and report on steady-state and transient solutions for tissues with the first rate kinetic order of clearance by perfusion and metabolism. The predictions of our model are in good correlation with clinically observed data on percutanious ethanol ablation volumes of liver. Our study shows the importance to incorporate tissue perfusion rate to the fundamental understanding of *in vivo* intercellular signaling and treatment planning of regional therapies, which include local drug injection, which include electrochemotherapy, percutanious ethanol ablation and regional anesthesia.

Table 1. Model inputs and values investigated in the parametric study.

Variable	Physical Property	Units	Investigated range
t	Time from the compound introduction	s	1E-3-1E+3
x	Distance from the source	m	1E-3-1E-1
B	Boundary concentration	M	1

D	Diffusion coefficient of a chemical compound in the tissue	$m^2 \cdot s^{-1}$	5E-6
$w_p + k_b$	Average tissue perfusion rate and first order metabolic rate.	s^{-1}	1E-4 – 7E-2
ε	Minimal fraction of the chemical compound that causes to a desired effect		4E-1-9.5E-1

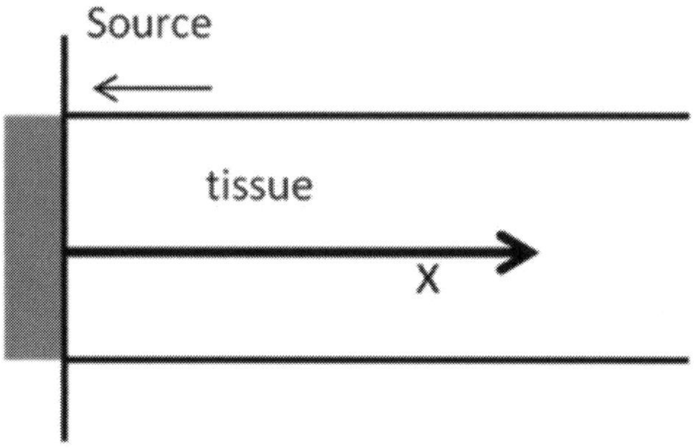

Fig.1 Analyzed physical system.

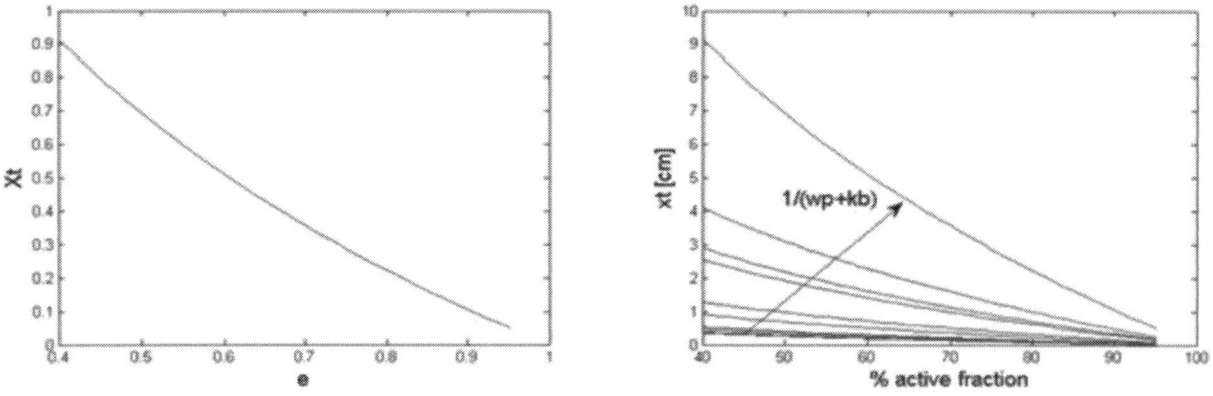

a b

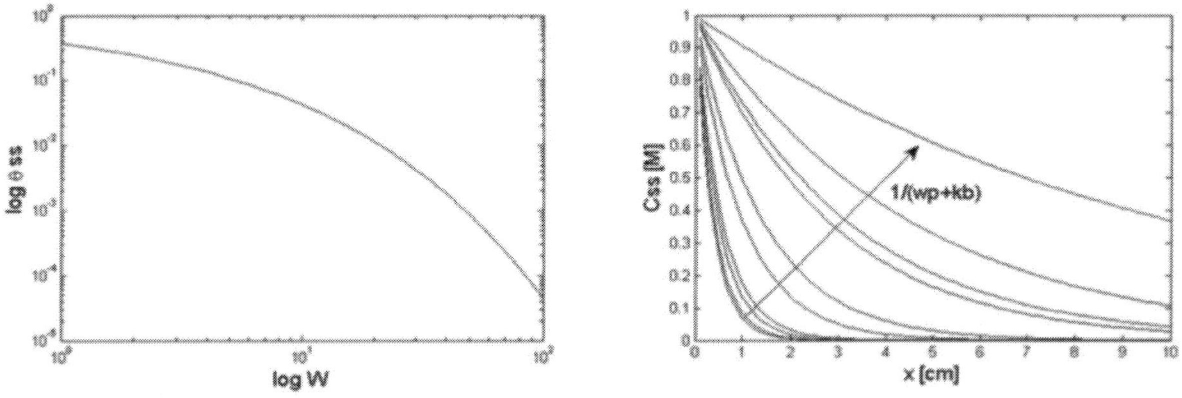

c d

Fig.2 Parametric study results. **a.** Dimensionless drug penetration depth (X_t) as a function of minimum active concentration(ε). **b.** Drug penetration depth as a function of a drug fraction in %. **c.** Dimensionless steady state concentration as a function of dimensionless metabolic rate (W). **d.** Drug steady-state concentration (C_{ss}) for various perfusion and absorption rates.

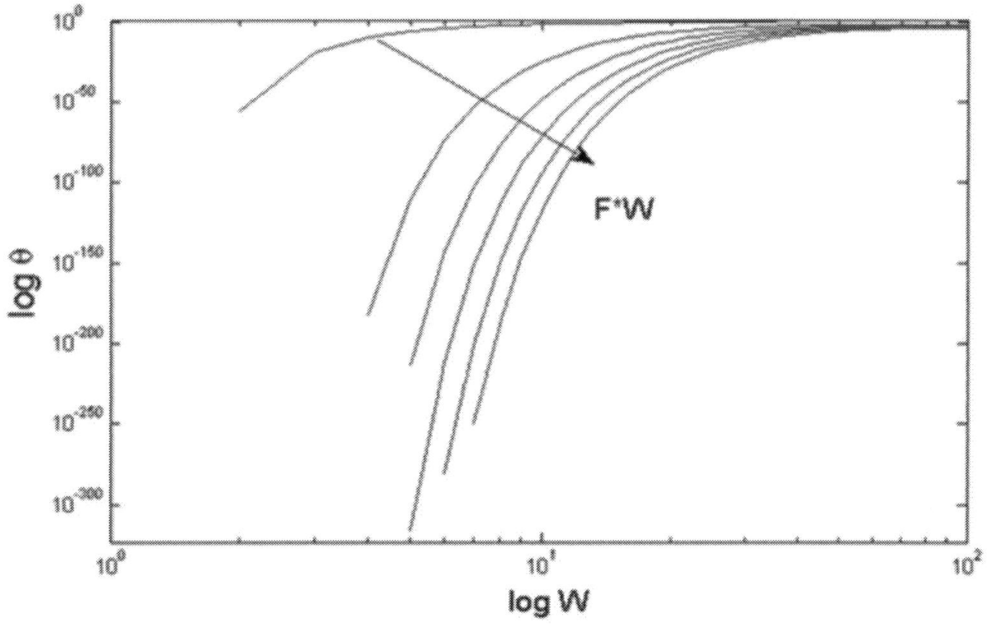

Fig.3 Parametric study results. Transient dimensionless concentration as a function of dimensionless perfusion rate (W) and time (F).

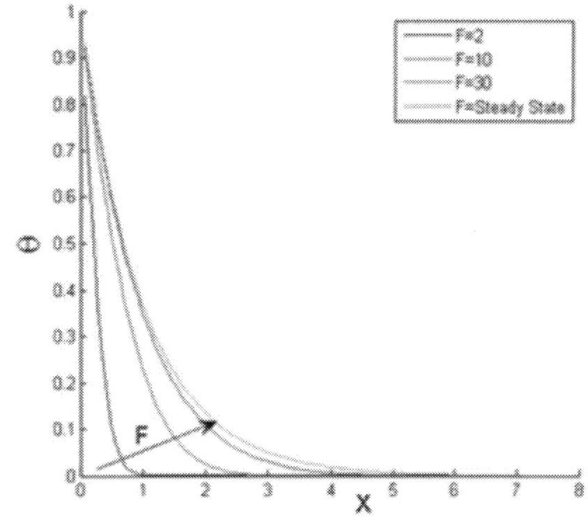

a

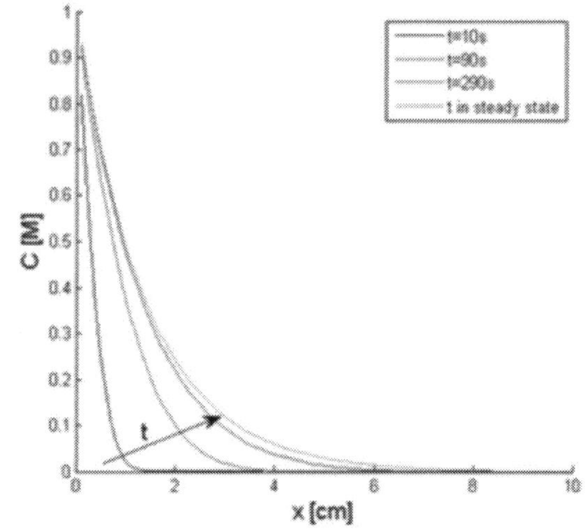

b

Fig.4 Parametric study results. a. Transient dimensionless concentration as a function of dimensionless time and location for dimensionless perfusion rate W=50 **b.** Transient concentration as a function of time and location for perfusion rate $w_p+k_b=5*10^{-3}\ s^{-1}$.

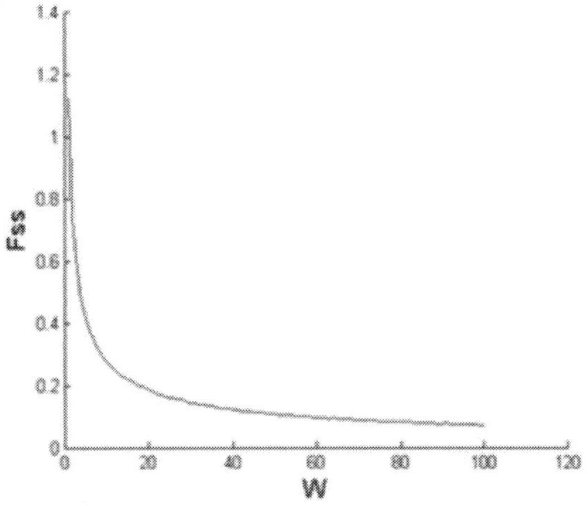

a

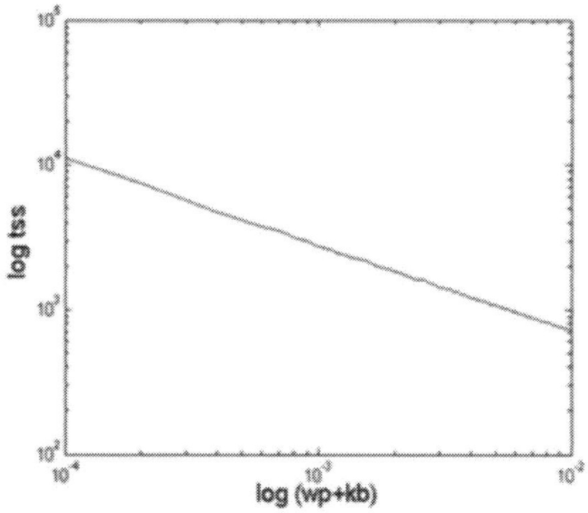

b

Fig.5 Parametric study results. a. Dimensionless time to steady-state (F_{ss}) as a function of dimensionless metabolic rate (W) **b**. Time to reach steady state (t_{ss}) as function of perfusion (w_p) and metabolism rates (k_b).

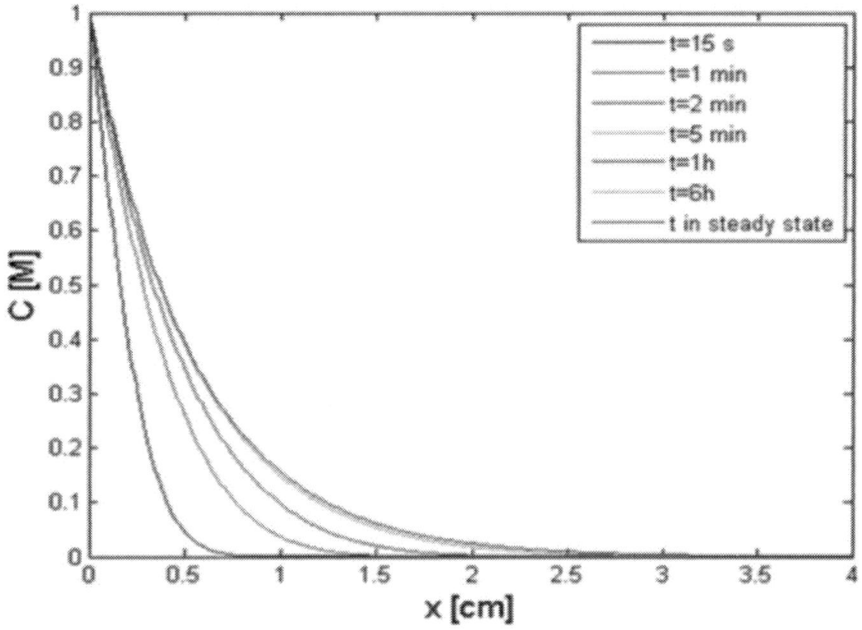

Fig.6 Practical example. Ethanol concentration in a liver as a function of location and time during percutaneous ethanol injection therapy. The system parameters appear in Table 2. 98% ethanol was injected in the liver (Boundary condition).

Table 2. Diffusion and metabolic parameters for percutaneous ethanol injection of liver model.

Variable	Physical meaning	Values
D	Diffusivity	2.2E-3 cm s^{-1}
k_b	First order kinetics of ethanol absorption in liver	8.1E-4 s^{-1}
w_p	Average perfusion rate of small rat liver tumor	45.5 ml $100gr^{-1}min^{-1}$

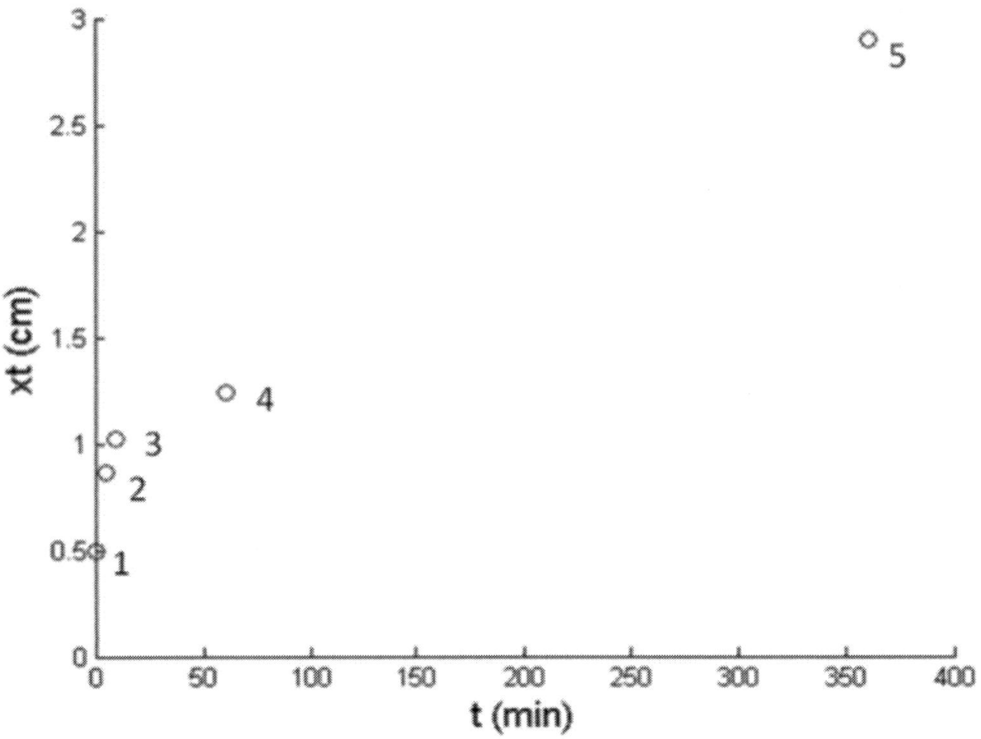

Fig.7 Practical example. Calculated distance from the injection point at which ethanol causes liver tumor ablation by necroses (points 1-4) and apoptosis (point 5) as described in Table 3:

Table 3. Toxic ethanol concentration for various exposure durations.

Time [s]	Critical concentration %	ε	Point N on Fig. 6
15	40	0.4	1
300	20	0.2	2
600	15	0.15	3
3600	10	0.1	4
21600	0.046	0.0046	5

(312b) Modeling the Superovulation Stage in in-Vitro Fertilization (IVF)

Kirti Maheshkumar Yenkie, Bioengineering, University of Illinois, Chicago, Chicago, IL, Urmila Diwekar, Vishwamitra Research Institute, Center for Uncertain Systems: Tools for Optimization and Management, Clarendon Hills, IL and Vibha Bhalerao, Jijamata Hospital, Nanded, India, Nanded, India

Abstract

In-vitro fertilization (IVF) is the most common technique in assisted reproductive technology and in most cases the last resort for infertility treatment. It has four basic stages: superovulation, egg retrieval, insemination/fertilization and embryo transfer. Superovulation is a drug induced method to enable multiple ovulation per menstrual cycle. The success of IVF majorly depends upon successful superovulation, defined by the number and similar quality of eggs retrieved in a cycle.

The model is adapted from the theory of batch crystallization. The aim of crystallization is to get maximum crystals of similar size and purity, while superovulation aims at eggs of similar quality and size. The rate of crystallization and superovulation are both dependent on the process conditions and varies with time. Thus, model formulation for multiple ovulation is on parallel lines to crystal formation in a batch process and is modeled as such. The kinetics of follicle growth is modeled as a function of injected hormones and the follicle properties are represented in terms of the moments. Hence, modeling of this stage in terms of distribution of eggs (oocytes) obtained per cycle involving the chemical interactions of drugs used and the conditions imposed on the patient during the process provides a basis for predicting the possible outcome. The results from the model prediction were verified with the known data from Jijamata Hospital Nanded, India. The predictions were found to be in well matched with the actual observations.

Thus, a phenomenon currently based on trial and error will get a strong base in terms of a predictive model. It will help the patient to decide whether to undergo superovulation or start the IVF from donor eggs, which in turn would save the patient from financial loss as well as emotional distress.

1. Introduction

Around 80 million people in the world are suffering from infertility issues. The rate of fertility is constantly declining in the developed nations due to late marriages, postponed childbearing and primary infertility. On the contrary, in the developing world the reasons for infertility involve prevalence of sexually transmitted diseases, infections which increase the rate of secondary infertility. Childlessness is often stigmatized and leads to profound social suffering for women in the developing nations (de Melo Martin, 1998).

1.1. In-vitro fertilization:

It is a process by which oocytes or egg cells are fertilized by a sperm outside the body in a laboratory simulating similar conditions in the body and then the fertilized eggs are implanted back in the uterus for full term completion of pregnancy.

Four stages in IVF:

i. Superovulation: It is method to retrieve multiple eggs using drug induced stimulation. In normal female body only one egg is ovulated per menstrual cycle, but with the use of fertility drugs and hormones, more number of eggs can be ovulated per cycle.

ii. Egg collection (retrieval): On the maturation of the multiple eggs produced in the previous stage, the eggs are retrieved through special techniques like ultrasonically guided transvaginal oocyte retrieval.

iii. In vitro fertilization (inseminaton/fertilization): Fertilization is done in the incubator using the retrieved oocytes and sperms. The conditions are maintained so as to mimic the invivo environment.

iv. Embryo transfer: The fertilized embryos are implanted into the uterus via a non-surgical technique using ultrasound guidance.

The major cost of IVF is associated with the superovulation stage where expensive drugs are used and almost daily monitoring is required. Success of this stage in terms of number and quality of eggs affects the outcome of IVF. In this work, we concentrated on modeling this stage and the approach is presented in the next section.

1.2. Superovulation:

In this work, we follow the analogy between batch crystallization and superovulation to model the process.

Analogy between Superovulation and Batch crystallization:

The moment model for follicle number and size is adapted from the concept of batch crystallization (Q. Hu. et. al., 2005) based on the analogy between batch crystallization and superovulation presented in Table 1.

Table 1: Analogy between batch crystallization and IVF superovulation stage

Batch Crystallization	Superovulation (IVF stage I)
Production of multiple crystals	Production of multiple oocytes or eggs
Crystal quality is determined in terms of size distribution and purity	Oocyte quality is determined in terms of no abnormalities, similar size.
The rate of crystallization or crystal growth varies with time and process conditions	The rate of ovulation or oocyte growth varies with time and drug interactions
Process is affected by external variables like agitation, and process operating variables like temperature, pressure, etc.	Process is affected by externally administered drugs and body conditions of the patient undergoing the process

The superovulation follicle growth model in general resembles greatly to the growth of seeded batch crystals. Thus, the moment model for both the processes remain the same the growth term which is a function of process variables like temperature and supersaturation in batch seeded crystallization will become a function of medicinal dosage in case of superovulation process.

2. Model details

Due to ovarian stimulation using externally injected hormones the number of follicles activated to enter into the ovulation stage are more in number as compared to a single follicle in a normal menstrual cycle. From the current data on successful superovulation for patient 1, organized in Table 2; it can be observed that during FSH dosage regime, as the time progresses the size of the eggs increase.

Table 2: Variation of Follicle size (diameter) with time and FSH dose

Size range (mm) dia Sr. No	↓ → Days	Day 1	Day 2	Day 5	Day 7	Day 9
1.	0-4	15	8	6	0	0
2.	4-8	3	6	4	2	0
3.	8-12	0	4	10	14	4
4.	12-16	0	0	0	2	11
5.	16-20	0	0	0	0	3
6.	20-24	0	0	0	0	0
	FSH dose	0	300	300	300	225

2.1. Model Assumptions:

The rate expression for follicle growth is dependent on FSH administered. Thus, we can write the growth term as;

G=kCfshα $G = kC_{fsh}^{\alpha}$ (1)

Assuming moment model, we consider only the first seven moments; zeroth moment corresponding to follicle number, the first moment corresponding to follicle size and the other 5 moments. 1st to 6th moments are being used since they help in recovering the size distributions more precisely as against lower number of moments.

2.2. Model equations (Q. Hu. et. al., 2005):

μ0=constant $\mu_0 = constant$ (2)

dμ1dt=Gtμ0t $\frac{d\mu_1}{dt} = G(t)\mu_0(t)$ (3)

dμ2dt=2Gtμ1t $\frac{d\mu_2}{dt} = 2G(t)\mu_1(t)$ (4)

dμ3dt=3Gtμ2t $\frac{d\mu_3}{dt} = 3G(t)\mu_2(t)$ (5)

dμ4dt=4Gtμ3t $\frac{d\mu_4}{dt} = 4G(t)\mu_3(t)$ (6)

dμ5dt=5Gtμ4t $\frac{d\mu_5}{dt} = 5G(t)\mu_4(t)$ (7)

dμ6dt=6Gtμ5t $\frac{d\mu_6}{dt} = 6G(t)\mu_5(t)$ (8)

Conversion of the data available on follicle number and size to moment using the expression given in literature by Flood, 2002:

μi=nir,trinDri $\mu_i = \sum n_i(r,t) r_i^n \Delta r_i$ (9)

Here, μ_i = i^{th} moment

$n_i(r,t)$ = number of follicles in bin of mean radius '*r*' at time '*t*'.

r_i = mean radius of ith bin

Δr = range of radii variation in each bin

3. Results:

We integrate the equations 2-8 for predicting the kinetic constants in the follicle growth expression. Later we use non-linear optimization algorithm to predict the values of these kinetic constants along with the integration constants obtained after integrating the set of moment equations. In real practice, the model will be calibrated with the first two days of data and then used for prediction of the complete cycle.

3.1. Model Validation:

The current moment model predicts the moment values, however our final output desired is the follicle size distribution, thus in model validation the approach to obtain follicle size distribution from moment values are shown. The method is adapted from the literature by Flood 2002; where he shows the method to recover particle size distribution from moments in batch crystallization. Using the model predicted moment values we evaluate n(r,t) and compare with the actual data to check the model accuracy. We plot the follicle size distribution for four patients for various days. The experimental size distribution is shown by symbols while the continuous curve shows the model predicted values after using the inversion method.

4. Conclusion:

The moment model developed for IVF superovulation predicts the follicle size distribution which is in well agreement with the actual size distribution seen in the IVF cycle data. The model can be used to predict the outcome. This will reduce the almost daily requirement of testing. The model can also provide a basis for predicting the optimum dosage for the desired outcome from the superovulation stage. The model used here is a very basic model and the complexities present in the patient are not considered. Later, we aim to include these complexities and model the system uncertainties, using more data for analysis, modeling and validation.

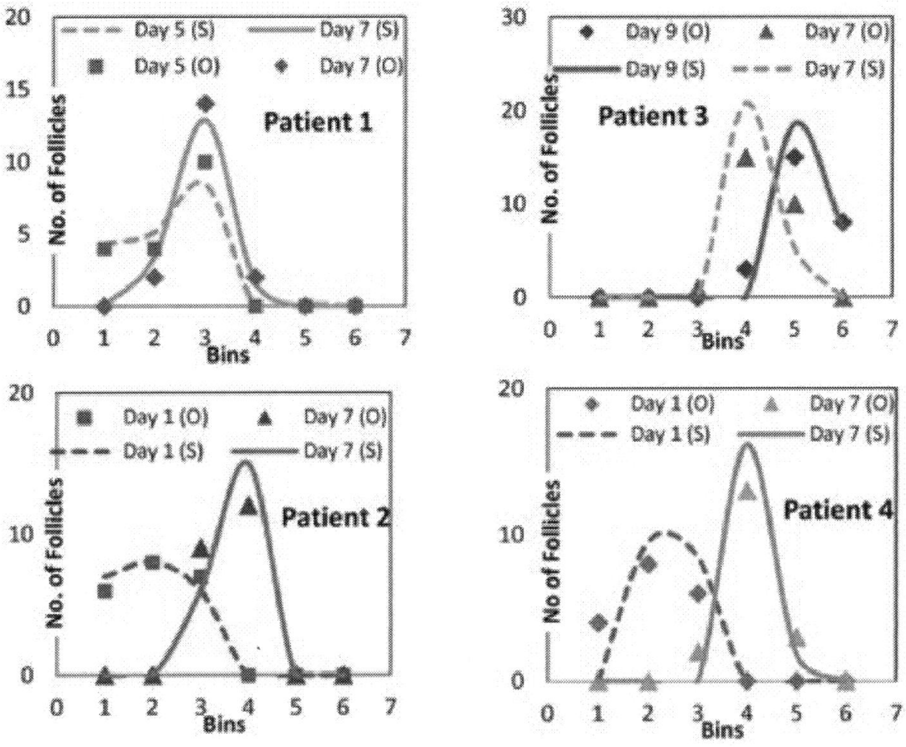

Fig 1. Follicle size distribution for four patients

References

de Melo-Martin, I., 1998. "Ethics and Uncertainty: In Vitro Fertilization and Risks to Women's Health" Risk: Health, Safety & Environment 201.

Flood, A. E., 2002. "Thoughts on recovering particle size distributions from the moment form of the population balance." *Dev. Chem. Eng. Mineral Process,* Vol. 10, No. 5/6, pp. 501-519.

Q. Hu, S. Rohani, A. Jutan, 2005. "Modelling and optimization of seeded batch crystallizers" *Computers and Chemical Engg.* Vol. 29, pp. 911-918.

(312c) Interactions of PEO-PPO-PEO Block Copolymers with Lipid Membranes: A Computational and Experimental Study Linking Membrane Lysis with Polymer Structure

Paola Carbone, School of Chemical Engineering and Analytical Science, The University of Manchester, Manchester, United Kingdom and Selina Nawaz, Chemical Engineering and Analytical Science, University of Manchester, Manchester, United Kingdom

Poly(ethylene oxide) (PEO), and poly(propylene oxide) (PPO) block co-polymers (PEO-PPO-PEO, sold as Pluronics, Poloxamers, Tetronics) are a widely-used class of amphiphilic materials for different biological applications. In fact for certain members of the Pluronics series, the interactions of block segments with living cells alter the lipid membrane properties and facilitate the permeation of drugs. A fuller understanding the molecular mechanisms underpinning these interactions is essential for ensuring their safety and efficacy in biomedical applications and to inform the design of new amphiphilic copolymers for potential use in a clinical setting. In this paper, by means of atomistic molecular dynamics simulations and membrane lysis assays, we investigate the relationship between the molecular conformations of a subset of the Pluronic copolymers (L31, L61, L62 and L64) and their haemolytic activity. Our computational studies suggest that the hydrophilic blocks in these copolymers interact with the polar head groups of lipid molecules, resulting in a predicted modification of the structure of the membranes. Parallel membrane lysis assays in human erythrocytes indicate differences in the rates of haemolysis, as a result of incubation with these polymers, that correlate well with the predicted interactions from the atomistic simulations. The computational data thus provide a putative mechanism to rationalize the available experimental data on membrane lysis by these copolymers and quantitatively agree with haemoglobin release endpoints measured when copolymers with the same molecular weight and structure of those modelled are incubated with erythyocytes. The data further suggest some new structure-function relationships at the nanoscale that are likely to be of importance in determining the biological activity of these otherwise inert copolymers.

(312d) Computational Study of Drug Transport in Realistic Models of Solid Tumour

Wenbo Zhan[1], Wladyslaw Gedroyc[2] and Xiao Yun Xu[1], (1)Department of Chemical Engineering and Chemical Technology, Imperial College London, London, United Kingdom, (2)Department of Medicine, Imperial College London, London, United Kingdom

Background: The efficacy of anticancer therapy is often offset by serious side effects. This is largely due to toxicity of the chemotherapy drug and its undesirable uptake in *vivo*. Various attempts have been made to retain the anticancer effect while reducing toxicities, including continuous infusion and liposomal delivery of chemotherapy drugs. Since the outcome of an anticancer therapy depends strongly on the pharmacokinetics of the drug as well as the properties of tumour and its surrounding tissue, rational criteria for choosing an optimal delivery mode and dose for different types of tumours are not yet available.

Methods: In this study, a computational model is developed which incorporates real tumour geometry reconstructed from magnetic resonance (MR) images, and the key physical and biochemical processes involved. These include time-dependent plasma clearance, drug transport through the blood and lymphatic vessels, extracellular drug transport (convection and diffusion), binding with proteins, lymphatic drainage, interactions with the surrounding normal tissue and drug uptake by tumour cells. Anticancer efficacy is evaluated based on the faction of survival tumour cells by directly solving the pharmacodynamics equation using the predicted intracellular doxorubicin concentration. Comparisons are made of the predicted efficacies of different modes of drug administration and doses in a clinical relevant range. As a typical anti-tumour drug, doxorubicin is employed in the computational model.

Results and Conclusion: Our results show that liposome-mediated drug delivery performs best in reducing drug concentration in normal tissues, which will help lower the risks of associated side effects. Bolus injection leads to the highest extracellular concentration in both tumour and normal tissues, giving the worst efficacy. Continuous infusion can effectively reduce the extracellular concentration while increasing the intracellular concentration (shown in Fig. 1), which mau help reduce the side effects caused by high concentration in normal tissues. As shown in Fig. 2, rapid continuous infusions lead to quick cell killing in the early phase of drug administration, but slow infusion can kill more tumour cells over a longer time period. Moreover, continuous infusion over 2-hour duration seems to offer the highest intracellular concentration. In addition, the effects of tumour size and shape have been investigated based on patient-specific MR images.

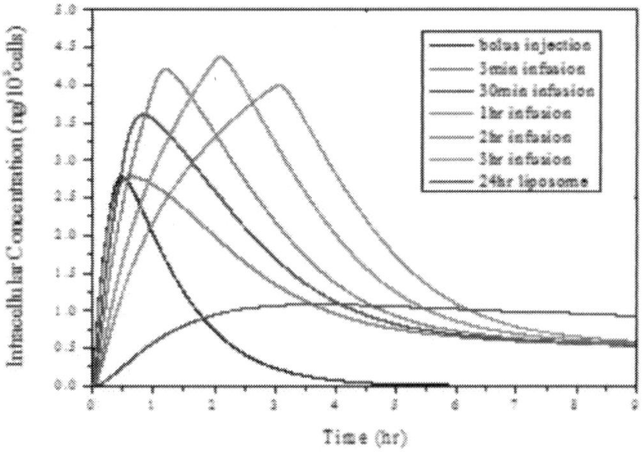

Fig. 1 Comparison of intracellular concentration among different modes of drug administration (bolus injection, continuous infusion over durations of 3 minutes, 30 minutes, 1 hour, 2 hours, and 3 hours, respectively, and liposomal release over 24 hours)

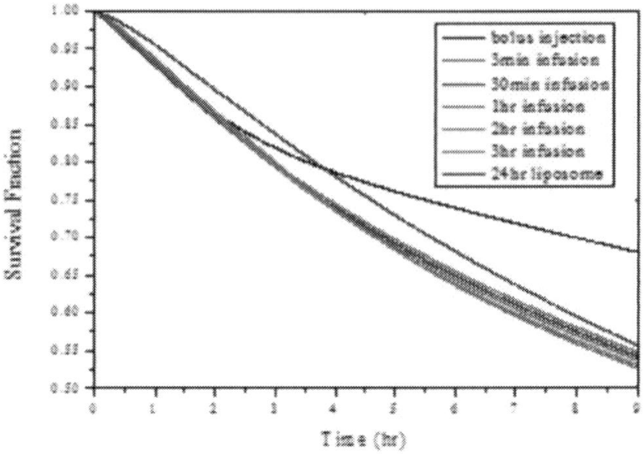

Fig. 2 Predicted anticancer efficacy for different modes of drug administration

(312e) Unraveling the Mechanism of a DNA Nanotechnology: The 10-23 Dnazyme

Margaret C. Linak and Kevin D. Dorfman, Chemical Engineering and Materials Science, University of Minnesota, Minneapolis, MN

Many emerging DNA nanotechnologies are difficult or even impossible to crystallize, frustrating attempts to connect their structure and function. Using a straightforward multi-scale simulation approach, we developed an atomic scale model of one such nanotechnology that has resisted attempts at crystallization, the widely used 10-23 DNAzyme. Deoxyribozymes (DNAzymes) are a new class of small catalytic oligodeoxynucleotides composed entirely of DNA. The 10-23 DNAzyme consists of a conserved 15-base catalytic core that is flanked by binding arms typically 7-10 bases in length with sequence complementary to a target single-stranded RNA molecule. The 10-23 DNAzyme is capable of cleaving at a purine-pyrimidine junction with both high turnover rate and substrate affinity. The ability of the 10-23 DNAzyme to cleave an RNA substrate has lead to both *in vitro* and *in vivo* applications although the evidence for its biochemical activity remains indirect. Our simulations suggest a plausible mechanism for the catalysis of RNA cleavage by this single-stranded DNA molecule. Two critical structural features were found: (i) the unstacking and rotation of the RNA nucleotide at the cleavage site, and (ii) the formation of electrostatic traps for cofactor metal ions.

(312f) Individualized Physiologically Based Modeling and Model Predictive Control of Volatile Anesthesia

Alexandra Krieger, Chemical Engineering, Imperial College London, London, United Kingdom, Nicki Panoskaltsis, Department of Haematology, Northwick Park Hospital, London, United Kingdom, Athanasios Mantalaris, Chemical Engineering and Chemical Technology, Imperial College London, London, United Kingdom, Michael C. Georgiadis, Centre for Process Systems Engineering, Chemical Engineering, Imperial College London, London, United Kingdom and Efstratios N. Pistikopoulos, Chemical Engineering, Imperial College London, Centre for Process Systems Engineering, London, United Kingdom

The modeling and control of anesthesia is believed to benefit the safety of the patient undergoing surgery and provide anesthetists and researchers with valuable insights. It is expected that mathematical modeling and optimized control could (i) pave the way for personalized health care, taking into account the individual patient characteristics for optimal and flexible drug infusion rates, (ii) provide the anesthetist with more time to focus on critical issues, and (iii) guarantee the safety of the patient minimizing side-effects.

This work presents a physiologically based, patient-specific, compartmental model for volatile anesthesia, consisting of a multiple blood and tissue compartmental model, adjusted to the weight, height, sex and age of the patient (Krieger et al. 2011). The predicted hypnotic effect and cardiovascular and respiratory depression is directly correlated to the arterial anesthetic concentration via an effect site concentration and the Hill-equation and (Gentilini et al. 2001). The individual patient characteristics in the pharmacodynamic equations are included via an initial estimate based on the physiologic variables of the patient and updated during the course of anesthesia as a function of the obtained measurements, such as the mean arterial blood pressure.

The derived model is used as the underlying model for model predictive control (MPC) of anesthesia. MPC is the method of choice for the biomedical system of anesthesia, as it can handle multiple-input multiple-output (MIMO) systems and safety constraints on the drug infusion rates, measured physiological variables and drug concentrations. Via applying multi-parametric programming for MPC, a fast and patient specific calculation of the optimal control inputs (Dua & Pistikopoulos 2005), e.g. the drug infusion rates and the inspired concentration of the volatile anesthetic agent is assured.

References

Dua, P. & Pistikopoulos, E.N., 2005. Modelling and control of drug delivery systems. *Computers and Chemical Engineering*, 29, p.2290-2296.

Gentilini, A.L. et al., 2001. Modeling and closed-loop control of hypnosis by means of bispectral index (BIS) with isoflurane. *IEEE Transactions on Biomedical Engineering*, 48(8), p.874-889.

Krieger, A. et al., 2011. A novel physiologically based compartmental model for volatile anaesthesia, *21st European Symposium on Computer Aided Process Engineering*. Elsevier, p. 1495 – 1499

(312g) Computational Model for Nanocarrier Adhesion to Cell Surfaces Validated Using in Vivo, in Vitro, and Atomic Force Microscopy Experiments

Jin Liu, University of Pennsylvania, Philadelphia, PA, Portonovo S. Ayyaswamy, Mechanical Engineering and Applied Mechanics, University of Pennsylvania, Philadelphia, PA, David M. Eckmann, School of Medicine, University of Pennsylvania, Philadelphia, PA, Vladimir Muzykantov, Institute for Environmental Medicine, Department of Pharmacology, University of Pennsylvania School of Medicine, Philadelphia, PA and Ravi Radhakrishnan, Department of Bioengineering, University of Pennsylvania, Philadelphia, PA

Our objective is to develop a computational platform for targeting of functionalized nanocarriers to optimize experimental design protocols for drug delivery. A computational methodology based on Monte Carlo and the weighted histogram analysis method has been developed to calculate the absolute binding free energy between functionalized nanocarriers (NC) and endothelial cell (EC) surfaces. The calculated NC binding free energy landscapes yield binding affinities that agree quantitatively when directly compared against analogous measurements of specific antibody-coated NCs (100 nm in diameter) to intracellular adhesion molecule-1 (ICAM-1) expressing EC surface in in vitro cell-culture experiments. The effect of antibody surface coverage (σs) of NC on binding simulations reveals a threshold σs value below which the NC binding affinities reduce drastically and drop lower than that of single anti-ICAM-1 molecule to ICAM-1. The model suggests that the dominant effect of changing σs around the threshold is through a change in multivalent interactions; however, the loss in translational and rotational entropies are also important. We also discuss the role of membrane deformation in mediating the NC adhesion. We validate the simulations against three distinct classes of experiments: invivo targeting in mice, cellular targeting in cultured cells under flow, and binding assays using AFM.

References:

1. A Computational Model for Nanocarrier Binding to Endothelium Validated Using In Vivo, In Vitro, and Atomic Force Microscopy Experiments, J. Liu, G. E. R. Weller, B. Zern, P. S. Ayyaswamy, D. M. Eckmann, V. Muzykantov, R. Radhakrishnan, Proceedings of the National Academy of Sciences, 107(38), 16530-16535, 2010. Pubmed ID: 20823256.

2. Dynamic factors controlling carrier anchoring on vascular cells, T. N. Swaminathan, J. Liu, B. Uma, P. S. Ayyaswamy, R. Radhakrishnan, D. M. Eckmann, IUBMB Life, 2011, 63(8):640-647. Pubmed ID: 21721099

3. Dynamic factors controlling targeting nanocarriers to vascular endothelium, D. M. Eckmann, R. Radhakrishnan, V. Muzykantov, R. Radhakrishnan, D. M. Eckmann, Current Drug Metabolism, 2012, 113, 70. Pubmed ID: 22292809.

4. Multiscale Modeling of Functionalized Nanocarriers in Targeted Drug Delivery, J. Liu, R. P. Bradley, D. M. Eckmann, P. S. Ayyaswamy, R. Radhakrishnan, Current Nanoscience, 2011, 7(5): 727-735. Pubmed ID: 22116782.

5. Multivalent binding of nanocarrier to endothelial cells under shear flow, J. Liu, N. J. Agrawal, A. Calderon, P. S. Ayyaswamy, D. M. Eckmann, R. Radhakrishnan, Biophysical J., 2011, 101(2):319-326. Pubmed ID: 21767483

An Automated Tissue Digester for Pancreatic Islet Production

Zhongliang Lu[a], Paul W. Todd[b], and Thomas R. Hanley[a]
[a]Department of Chemical Engineering, Auburn University, Auburn, AL USA
[b]Techshot, Inc., 7200 Highway 150, Greenville, IN USA

Introduction

Large quantities of pure viable islets are necessary to treat diabetes. Isolation of the islets is the primary step in preparing the islets for the transplantation. Moskalewski[1] in 1965 used collagenase digestion of the pancreas and was the first to obtain large number of islets. Lacy and Kostianovsky[2] in 1967 improved islet yield by cannulation of the pancreatic duct and injection of Hanks solution, which mechanically distended and disrupted the gland before digestion. In 1981, Horaguchi and Merrel[3] described a new technique for the isolation of islets from dog pancreas, but it could not be applied to other larger animals or the human pancreas. Noel et al.[4] in 1982 modified the technique of Horaguchi by injecting collagenase in two separate sections, adding a calcium chelator (EGTA) into the duct. In 1984 Gray et al.[5] modified the previous technique and applied it to the human and monkey pancreas. Warnock et al.[6], from the same group in Edmonton, compared the method of perfusion with single intraductal injection of collagenase. Finally, Ricordi et al.[7] further developed a semi-automated method for the isolation of human pancreatic islets, a definite improvement on existing methods.

In the Ricordi digestion, the pancreas and glass balls are placed into the chamber, creating a closed-loop recirculation and dissociation system operated by intermittent manual shaking of the vessel. However, the yield of islets was still variable (Iwashita, 1996)[8]. One of the most important reasons is the collagenase, essential for islets release from the surrounding tissues, can cause damage to islets by both warm ischemic injury and enzymatic disruption. The problem is further compounded by the variable activation degree of native proteolytic enzymes within acinar cells. Therefore, an automated system that is capable of introducing rigorous standards into the preparation of islets must be established to eliminate the uncertainty. By utilizing an established enzyme digestion system and by creating a reproducible, controllable tissue digester, two of the major variables in islet production are eliminated. The Dynacult reactor has wide application as documentation in micro gravity research (Ivan, 2006)[9]. One advantage is that the fluid flow within the vessel can be carefully controlled, enabling the shear stress experienced by the islets to be optimized when the islets rotate with the fluid flow simultaneously. All above features lay a good foundation for the new digester design. The modification of this reactor was accomplished by introducing wall vanes to enhance turbulence. The details of digester design and subsequent results of FLUENT simulations will be illustrated by diagrams.

Materials and Methods

The design of new digester. The modified Dynacult reactor (MDR) incorporates hemispherical vanes to create turbulence similar to that created by moving glass balls inside the Ricordi chamber. Contrary to Ricordi chamber, the MDR provides a controllable shear rate that allows for a controllable and more favorable digestion environment. In addition, the digestion is automated as the cylinder walls are controlled by motors in contract the manual operation of the Ricordi chamber.

When the bioreactor is operating, the rotation of the core is independent of the exterior wall. The operator may choose to rotate the core at the same rate and in the same direction as the wall, hold it stationary; or rotate it in opposite directions. With the progress of tissue dissociation, the vanes reduce the tissue size and allow maximum islet release and production. As rotational speed increases and the spacing between the walls decreases, the size of the tissue fragments is reduced.

Computational Fluid Dynamics (CFD) Simulation. FLUENT simulates fluid flow by using a discrete-phase Reynolds stress model. The results for both the MDR and the Ricordi chamber are compared. Velocity inlet boundary conditions assume velocities normal to inlet surface, the outlet pressure is set at 0 Pa gauge, and the no-slip condition is assumed at all walls. In addition, FLUENT's preprocessor "Mechanical" is used to generate flow-field meshes. An unstructured mesh consisting of tetrahedral volumetric elements in three dimensions is used in the entire domain with discrete control volumes. The operation mode is optimized by investigating rotating speed, inlet flow rate and vessel dimensions, each separately considered as a variable while others are held constant. The distribution of hydrodynamic numbers and relevant parameters is then displayed by contours generated by CFD-post. Therefore, the optimal operation condition can be determined by comparing all the parameter value and distribution diagrams under different examined conditions.

Experimental Measurements Using Cultisphers. Cultisphers (gelatin bead, size range from 200 to 380 microns) are soaked for two hours in phosphate buffered saline (PBS) prior to use. The particles are rinsed several times and mixed with water to prepare a suspension with known concentration. The motor is operated under four different voltages: 6, 8, 10 and 12 volts. The experiment has two parts. In the input dispersion process, the particle suspension is pumped into the water-filled digester at 60 ml/min while rotating the double walls in opposite directions. The particle motion and collisions are observed. The MDR is operated in a closed circulation at maximum rotation speed. The time interval (dispersion time) is measured between when the particles are first seen and when particles completely fill the vessel. In the output dispersion process, inlet is switched to pure water and the particle elution time is estimated. Particle motion is recorded by eScope digital microscope.

Results and Discussion

MDR configuration and operation modes. The configuration of the MDR is illustrated in Figure 1. The dimensions are listed in Table 1. In order to imitate the turbulent environment caused by glass balls in the Ricordi chamber, the "half-ball baffle" vane was employed to dissociate the pancreatic tissue. In earlier testing all vanes were cemented onto the annular wall evenly distributing four in one circle. The base system, including a stationary base that is mounted to the experiment enclosure and a mobile base that can be repositioned between experiments in order to vary the length of the bioreactor, has been designed to allow the shell, core, and rotational speeds to be changed (Paul, 2004)[10].

Distributions of Hydrodynamic Parameters for the Ricordi chamber. Manual shaking mode is divided into two motions: (1) rolling motion with the rotational axis parallel to central line of the Ricordi chamber at a maximum rotation speed of 66 rpm; (2) simple harmonic motion; velocity formula $V_z = -0.921534 * \sin(16.755161 * t)$ (m/s). Figure 2 shows the distributions of four important hydrodynamic parameters, and Figure 3 displays velocity magnitude contours.

Figure 1 - The Modified Dynacult Digester Design

Table 1 - Dimensions of Tissue Digester

No.	Shell diameter (inch)	Core diameter (inch)	Shell length (inch)	Core length (inch)	Shell shaft (inch)	Core shaft (inch)	Wall distance (inch)	Ball radius (inch)	No. of balls on shell	No. of balls on core	Specific Area (in^{-1})
1	OD=2, ID=1.75	1.5	2.5	1.87	0.362	0.268	0.125	0.083333333	7	5	12.64938126
2	OD=2, ID=1.75	1.25	2	1.875	0	0.125	0.25	0.166666667	3	2	10.79504838
3	OD=2, ID=1.75	1	2	1.3	0.4	0.3	0.375	0.21	2	1	6.827977316
4	OD=2, ID=1.75	1	3	2.37	0.362	0.268	0.375	0.25	3	2	7.460899369
5	OD=2, ID=1.75	1	3	2.969	0	0.031	0.375	0.27	3	2	8.536585366
6	OD=2, ID=1.75	1	3.5	2.85	0.372	0.278	0.375	0.29	3	2	7.865438369
7	OD=2, ID=1.75	1	3.953	3.375	0.237	0.341	0.375	0.3	3	2	7.826278815
8	OD=2, ID=1.75	0.5	2	1.3	0.4	0.3	0.625	0.125	4	2	4.536291251

- **Distributions of Cell Reynolds Number:**

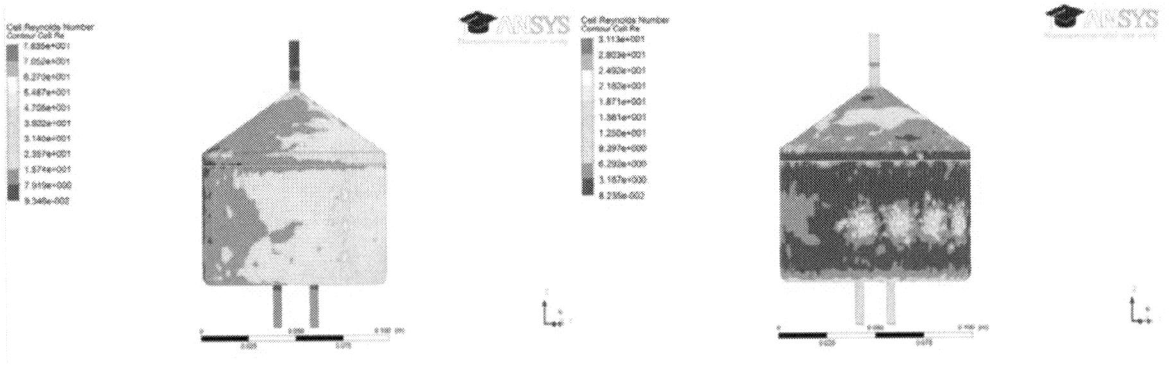

- **Distributions of Shear Rate (s^{-1}):**

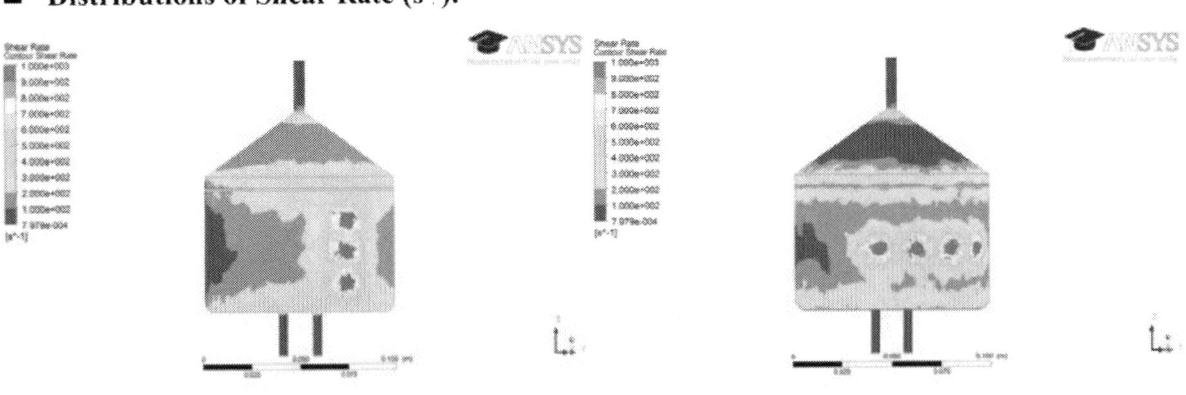

- **Distributions of Energy Dissipation Rate per Volume (W/m^3):**

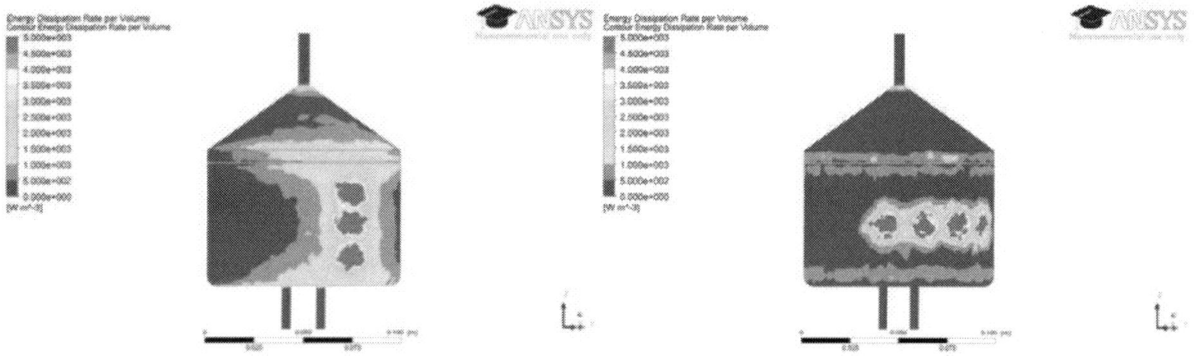

- **Distributions of Kolmogoroff Length(m):**

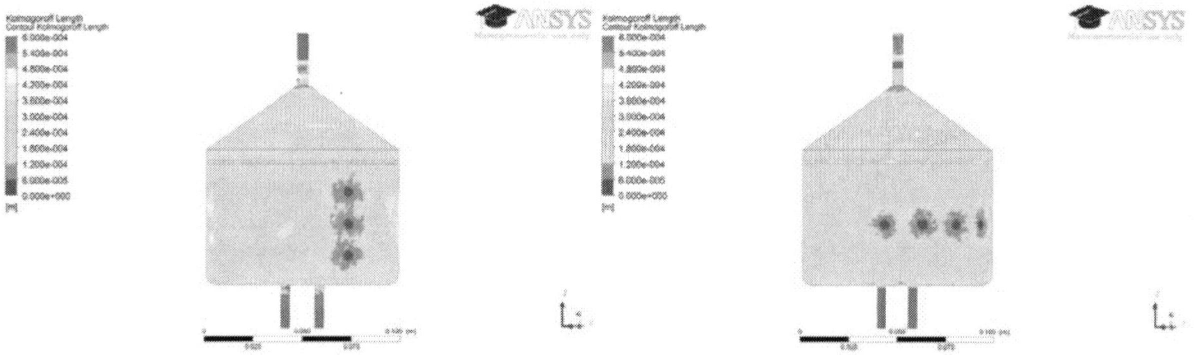

Figure 2 - The contour of hydrodynamic parameters for the chamber. (left column - rolling motion; right column - up and down motion)

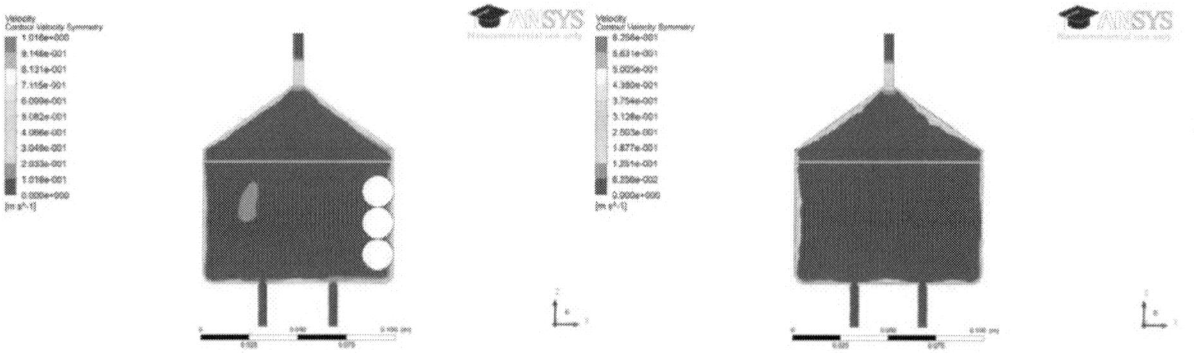

Figure 3 - The velocity magnitude contours on symmetric plane. (left column - rolling motion; right column - up and down motion)

Distributions of hydrodynamic parameters for MDR. The rotation speed, inlet flow rate and vessel dimensions were investigated independently; the double wall rotation position is assumed to have impact on shear environment. Based on FLUENT simulation results, the preferred 100 rpm rotation speed is firstly determined through comparing the shear rate and energy dissipation rate per volume at the same inlet flow rate and dimension. It is also found that the cylinder wall rotation has little impact on flow parameters so that all the contours can be displayed at a single position with minimum vane spacing. Subsequently, the optimal inlet flow rate 120 ml/min and vessel dimension 3 (see Table 1) are obtained by similar comparison under corresponding variable gradients. Totally,

35 combinations of triple variables are verified by FLUENT simulation and displayed by CFD-post. The distributions of hydrodynamic parameters, velocity magnitude contours and particle pathline under optimal operation condition are shown in Figure 4 and 5, respectively.

- **Distributions of Cell Reynolds Number:** **Distributions of Shear Rate (s^{-1}):**

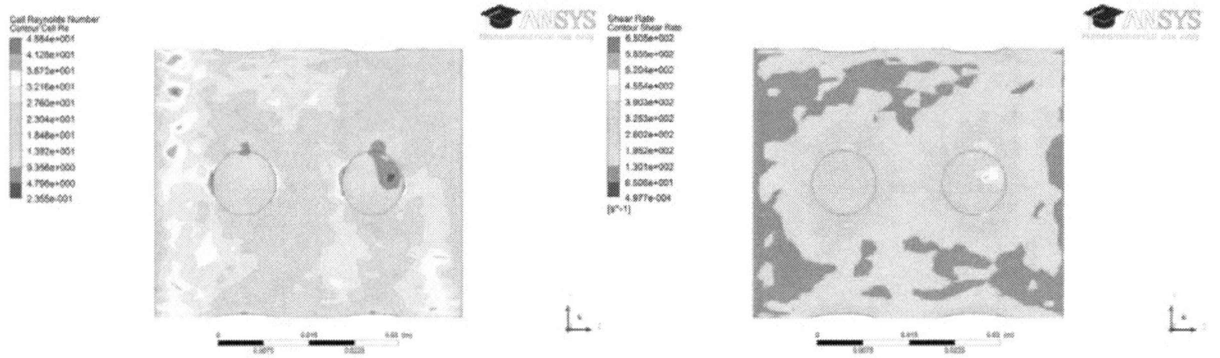

- **Distributions of Energy Dissipation Rate per Volume(W/m^3) (Left column below)**
- **Distributions of Kolmogoroff Length(m) (Right column below)**

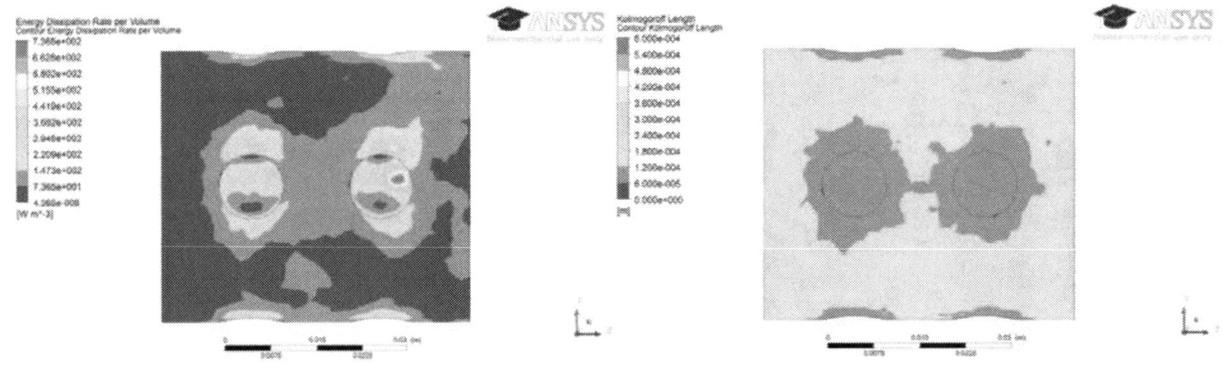

Figure 4 - The contour of hydrodynamic parameters for the MDR

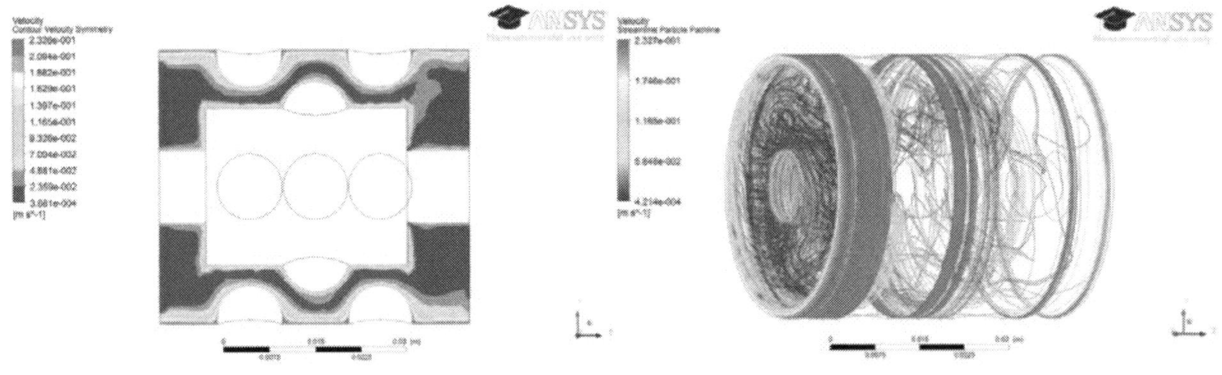

Figure 5 - The velocity magnitude contours on symmetric plane (Left)
The particle pathline (unit: m/s) shown in the MDR (Right)

Comparison of results between Ricordi chamber and MDR. Three flow parameters, including shear rate, shear stress, and energy dissipation rate per volume, are summarized as volume average, minimum and maximum value in Table 2, respectively.

Table 2 - Numerical simulation results for Ricordi chamber and MDR

Flow parameters \ Digester type		Ricordi chamber		MDR	
		Rolling motion	Up & down motion	A position	B position
Shear Rate (s^{-1})	VolAve	28.28	17.61	49.5	48.08
	Min	0.000797941	0.00219062	0.000497681	5.781E-05
	Max	4360.54	3363.51	650.549	620.068
Shear Stress (pa)	VolAve	0.4874	0.131	2.279E-01	2.141E-01
	Min	2.402E-05	1.420E-05	3.959E-06	3.969E-06
	Max	14.3727	12.956	2.23194	1.26812
P/V (W/m^3)	VolAve	46.3	20.52	18.9	17.41
	Min	1.92E-08	3.61E-07	4.36E-08	2.061E-08
	Max	57879.7	30878.6	736.461	754.62

P/V - energy dissipation rate per volume; VolAve - volume average value of corresponding parameters. Spacing between two adjacent vanes is minimum at A , maximum at B

The comparison of above parameters indicate that volume average shear stress and P/V for the MDR are lower than that for Ricordi chamber, with lower maximum values as well. Moreover, the maximum of shear rate has the same results as that of shear stress and P/V. However, volume average of shear rate shows the opposite results.

The General Description of Particle Motion and the Mixing Process in the MDR. An experimental analysis of the relative particle-to-fluid motion under the gradient of rotation speed shows that the particle will move along a certain path controlled by the action of different hydrodynamic forces. The particle dispersion/ residence time and relevant phenomenon are summarized in Table 3. The particle distribution is snapshotted using the eScope digital microscope under 12.5 magnification, both in stationary and rotating state (Figure 6). The effect of opposite rotation of the cylinder walls on the particles tracks is illustrated. Increasing the rotating speed of the inner wall will cause particles to migrate toward the outer wall. The greater the rotating speeds of

Table 3 - The Dispersion Process for Cultisphers

Voltage (V)	RPM	dispersion mixing time (s)	Input phenomenon	Residence time (min)	Output phenomenon
6, shell only	7	22	Most particles are sedimentary.	null	null
8, shell only	10	19	Sedimentary and suspended particles are nearly equal.	null	null
10, shell only	13	15	More particles are suspended.	null	null
12, shell only	17	13	More particles are suspended with less and scattered sedimentary ones.	null	null
12, in opposite directions	17&21 (shell & core)	11	No sedimentary particles are left	10~11	Particles are in radial symmetry distribution; turbulent flow

the outer wall, the less eccentric rate of the curve and the greater migrating rate of the particle in radial direction. Also, the final trajectory of the particle is closer to the inner cylinder. Fully developed turbulent flow can be seen through the transparent shell wall. The complete mixing process occurs in the intermediate region between adjacent vanes and double walls without any sedimentary particles while the particles near the walls will move with the shell/core in the same direction. The particles collide with each other in two directions due to radial and axial motion.

Figure 6 - Image from eScope microscope using Cultisphers as tissue fragment surrogates. Left: particles suspending in stationary state; Right: particles moving with walls rotating

The Potential Problems and Alternative Strategy for MDR Design. Existing technologies of pancreatic digestion have fundamental limitations. The design of the MDR, including the spacing between vanes and the rotational rates of the outer and inner modules, may need to be studied in greater details if we are unable to either suspend tissue during the digestion or if we are unable to separate islets from acinar tissue. The complex interaction between the tissue fragments and the MDR walls/vanes may require additional analysis resulting in a redesign. Problems with this design may require additional CFD analysis in order to estimate the shear stress and the convection forces in the MDR. The preliminary design of interacting hemispherical vanes may be altered. We also may only allow one of these modules to implement in order to simplify the simulation.

Conclusions

The new design of the MDR digester succeeding partial feature of Ricordi chamber overcomes the deficiency of manual operation. This digester configuration is designed to enhance the turbulent effect and maximize contact area with more probability of collision and mixing between tissue fragments and vanes. FLUENT simulations confirm a more uniform and lower shear environment in the MDR that should produce more viable pancreatic islets. The values of the shear and energy dissipation rates at the ball contact area in the Ricordi chamber are significantly larger than values in the MDR, resulting in an expected loss islet viability. The volume average value of shear rate for the Ricordi chamber is slightly lower than that for the MDR. Ricordi chamber fluid away from walls is near stationary. Comparing MDR results between rotation position A and B indicates the cylindrical wall motion has little effect on the shear and energy dissipation rates.

References

1. Moskalewski, S. (1965), "Isolation and Culture of the Islets of Langerhans of the Guinea Pig," *Gen. & Comp. Endocrinology*, **5**: 342-353.
2. Lacy, P. E., and M. Kostianovsky (1967), "Method for the Isolation of Intact Islets of Langerhans from the Rat Pancreas," *Diabetes*, **16**: 35-39.
3. Horaguchi, A. and R. C. Merrell (1981), "Preparation of Viable Islet Cells from Dogs by a New Method," *Diabetes*, **30**: 455-458.
4. Noel, J., A. Rabinovitch, A, L. Olson et al. (1982), "A Method for Large Scale High-Yield Isolation of Canine Pancreatic Islets of Langerhans,' *Metabolism*, **31**: 184-187.
5. Gray, D. W. R., P. McShane, A. Grant and P. J. Morris (1984), "A Method for Isolation of Islets of Langerhans from the Human Pancreas, *Diabetes*, **33**, 1055-1061.
6. Warnock, G. L., D. Ellis, R. V. Rajotte et al. (1988), « Studies of the Isolation and Viability of Human Islets of Langerhans, *Transplantation*, **45**, 957-963.
7. Ricordi C, P. E. Lacy, E. H. Finke et al. (1988), « Automated Method for Isolation of Human Pancreatic Islets, *Diabetes*, **37**, 413-420.
8. Iwashita, C., T. Asano, T. Kenmochi, K. Jingu, T. Uematsu, T. Nakagohri, M. Hasegawa, M. Maruyama, H. Miyauchi, and K. Isono (1996), "Combined Method of Mechanical Chopper and Automated Two-Step Digestion Technique for Islet Isolation from Canine Pancreas, *Transplantation Proceedings*, **28(1)**, 337.
9. Rivera-Solorio, I., and, S. J. Kleis (2006), "Model of the Mass Transport to the Surface of Animal Cells Cultured in a Rotating Bioreactor Operated in Micro gravity,. *Biotech. & Bioengr.*, **94(3)**, 496.
10. Todd, P. W., L. Hodapp, A. Jones, D. Kennedy, T. Popson and , E. Taylor (2004), "Robotic Bioreactor for Cell Culture in Space Applications (Dynacult). Final Progress Report,". 1-119.

(312i) An Integrated Computational Model of Powder Release, Dispersion, and Deposition in a Dry Powder Inhaler

Jovana Milenkovic, CERTH/CPERI, Thessaloniki, Greece, Aleck Alexopoulos, Chemical Process Engineering Research Institute, Thessaloniki, Greece and Costas Kiparissides, Department of Chemical Engineering, Aristotle University of Thessaloniki & Chemical Process Engineering Research Institute, Thessaloniki, Greece

Dry Powder Inhalers, DPIs, are the principle means of delivering pharmaceuticals due to their ease of use and cost-effectiveness. The main function of a DPI device is the adequate dispersion and delivery of particles. Initially the particles are in the forms of a loose powder which, under the action of airflow, is broken up and dispersed as particle aggregates which are then further broken up into fine particles. Powder properties, e.g., cohesion, charge, size, and size distribution, influence powder dispersion and the breakage of particle agglomerates. One of the problems with DPIs is the loss of powders due to deposition within the device which ranges from 20-40% of the total drug. In order to provide the maximum drug dose per inhalation and to ensure proper function of the DPI device for each dose (i.e., dose-to-dose variation) it is desirable to minimize the drug loses due to internal deposition. It is also desired to have good control over the dispersibility of the powder, release of drug (if attached to powder particles), and breakup of agglomerates in order to achieve the desired specification of particle/agglomerate size distributions at the exit flow of the DPI. The Turbuhaler (AstraZeneca) is a multidose dry powder inhaler that is widely used to deliver a number of drugs (typically for asthma) to the upper respiratory tract. Each dose is initially in the form of loosely packed drug aggregates, ~10-20μm in size, which are released into a mixing/dispersion chamber, where they are broken up into particles, ~1-2μm in size, which are then directed to the inhalation channel of the device (Tsima et al., 1994; Wetterlin, 1988).

In this work the airflow in a Turbuhaler DPI is determined by computational fluid dynamics, CFD, simulations. The DPI geometry is first constructed in a CAD/CAM environment (i.e., CATIA V5 R19) and then imported into GAMBIT (v2.1) where a tetrahedral grid is constructed. The computational grids consisted of 1-2 10^6 tetrahedral cells with a maximum skewness of 0.85. The equations for flow were solved using commercial CFD software (i.e., FLUENT v6.3). The SST k-ω model for transient turbulent flows was employed. Different steady state airflows were simulated by imposing a wide range of pressure drops at the mouthpiece outflow of -200 to -10000Pa corresponding to airflow rates of 2.7 and 18.6 l/min, respectively. Dynamic airflow simulations were performed by imposing a transient pressure boundary condition at the mouthpiece channel. Eulerian/Lagrnagian simulations of particle motion and deposition were conducted for particles between 1-20μm in size corresponding to the single particle and particle agglomerate size ranges of the powder. Particles and particle aggregates are assumed to be released instantaneously and uniformly from a surface located upstream from the powder storage site. For dynamic simulations the particles are released either instantaneously or at a constant rate over a period of time < the inhalation time.

The computed airflow in the DPI revealed a strongly non-uniform flow field. Particle depositions occurred mostly in the dispersion chamber and the helical region of the DPI where strong tangential motion was observed. The computed pressure drops were in agreement to available experimental data. In order to compute realistic particle depositions in the device a nonideal sticking efficiency, determined from particle/wall adhesion models, was employed. The predicted total particle depositions were in agreement to available literature data. This work represents the first coupling of particle adhesion models with fluid dynamics simulations for inhalation devices and permits the investigation of the effect of particle formulation properties on the DPI outflow and internal loses.

M.P. Tsima, G.P. Martin, C. Marriott, D. Gardenton and M. Yianneskis, 1994. Drug delivery to the respiratory tract using dry powder inhaler, International Journal of Pharmaceutics, 101 () 1-13.

K. Wetterlin, 1988 Turbuhaler:A New Powder Inhaler for Administration of Drugs to the Airways. Pharmaceutical Research, Vol, 5, No.8.

(322a) Magnetic Separation of Algal for Biofuel Production

Jeffrey J. Chalmers, William G. Lowrie Department of Chemical and Biomolecular Engineering, The Ohio State University, Columbus, OH, Wei Xue, Chemical and Biomedical Engineering, The Ohio State University, Columbus, Jie Xu, Chemical and Biomolecular Engineering, The Ohio State University, Columbus, OH, Maciej Zborowski, Department of Biomedical Engineering, Cleveland Clinic, Cleveland, OH and Brad Postier, Phycal, St. Louis, MO

Phycal is an algal biofuels company focusing on a hybrid photoautotrophic/heterotrophic open pond system using freshwater algal species. To **potentially reduce costs associated with algal biofuels production and/or improve the quality of intermediate steps through improved efficiencies downstream, Phycal is d**eveloping strains that can transport and store significantly more iron than wild-type strains. The potential reward is that they can out-compete contaminating species (*e.g.*, algae, bacteria, protozoa, and *etc.*) for the available iron. Once stored, the iron is in the form of magnetite inside ferritin complexes that are paramagnetic. Thus, in addition to the advantage of iron scavenging, these strains can be manipulated magnetically with the potential of magnetic separation.

In this presentation we will present progress on our Phase II SBIR work to 1) characterize the magnetic susceptibility of genetically engineered and intrinsic magnetic susceptibility as a result of culture conditions, and 2) pilot scale separation systems to "dewater" algae cultures.

(322b) A Novel Taylor-Couette Photobioreactor for Energy Efficient Micro Algae Cultivation

Bo Kong, Department of Chemical and Biological Engineering, Iowa State University, Ames, IA and **R. Dennis Vigil**, Iowa State University, Ames, IA

Numerous studies have demonstrated that algal growth rate and light utilization efficiency is dramatically impacted by the cycling in the intensity of light exposure as cells move between well-illuminated photic regions near a photobioreactor external surface and the darker regions of the reactor. In particular, when the characteristic frequency of the light/dark cycles is greater than approximately 1 Hz, the "flashing light effect" can boost cell growth rates 50~100% compared to growth rates obtained without light/dark cycling. However, high frequency light/dark cycles are not easy to achieve in large-scale reactors and may be one of the reasons for low yields. In commonly used flat panel bubble column and airlift reactors, the random nature of the agitation provided by the gas bubbles produces a wide spectrum of random light/dark frequencies. In the work presented here, we make use of robust and coherent hydrodynamic structures known as Taylor vortices that arise in the annular region of a Taylor-Couette flow cell (concentric cylinders with the inner cylinder rotating) in order to rapidly and periodically shuttle algal cells between photic and non-photic regions of the reactor. Because the fluid agitation in a Taylor vortex algal reactor is not dependent on bubble-induced turbulence, much lower carbon dioxide flow rates can be used and the carbon capture efficiency is significantly higher than in conventional algal reactors. Here, we present experimental results (using Chlorella vulgaris) and analyses based upon computational fluid dynamics simulations to demonstrate that even at relatively low light-dark cycling frequencies, the Taylor vortex algal reactor offers higher biomass yield than can be achieved in corresponding bubble column experiments. Furthermore, we consider important factors for design and scale up of Taylor vortex algal reactors.

(322c) Effect of Growing Conditions On Algal Carbohydrate to Butanol Production

Alice C. Jernigan, Chemical Engineering, University of Arkansas, Fayetteville, AR and Christa N. Hestekin, Ralph E. Martin Department of Chemical Engineering, University of Arkansas, Fayetteville, AR

There are several environmentally sound reasons to use algae as a feedstock for biofuels. The fact that they are readily abundant, can absorb unwanted elements out of waste water, and can use carbon dioxide as a carbon source are just a few of the reasons that algae are being examined as one of the most likely medium to replace petroleum fuels. An issue of some concern is that microalgae have a high rate of growth in the summer and lower rates in the winter, which can cause a depletion of feedstock for a bioreactor. We are using a species of *Clostridium* bacteria to convert algal sugars to butanol in a bioreactor. In order to facilitate localized bioreactors that can operate year-round, we examined dried and stored stocks of algae taken during a six month harvesting period and examined the effects of climate, location, and processing on the amount of carbohydrates extracted from the algae and converted into butanol.

(322d) Photoautotrophic Growth and Lipid Production Kinetics of the Microalgae Scenedesmus Dimorphus

Joanne Belovich, Jacob Schwenk, Christopher Hardulak and John Van Blargan, Cleveland State University, Cleveland, OH

The design and optimization of culture systems for microalgae for biofuel production are complicated by the conflicting requirements for high biomass productivity and high oil content. Reactor models that are needed for this optimization rely on models of the algae growth and lipid productivity kinetics. Unlike heterotrophic growth of most microorganisms, photoautotrophic growth of algae can have multiple limiting nutrients, further complicated by potential light limitation. Factorial experimental designs were conducted to identify the nutrients, and their concentration ranges, that are potentially limiting for growth and oil production of the microalgae *Scendesmus dimorphus*. The individual and interactive effects of six groupings of the components of 3N-BB medium on growth rate, lipid content, and total biomass yield of *S. dimorphus* were determined. Trace metal and vitamin concentrations were reduced to 1/6 the level of 3N-BB medium without adversely affecting growth rates and biomass concentration, while concentrations of $CaCl_2$ and K_2HPO_4/KH_2PO_4 were reduced to 1/10 that of 3N-BB without adversely affecting biomass and lipid concentrations. Lipid productivity was maximized at the lowest $NaNO_3$ concentration (1/10 that in 3N-BB) and independent of $MgSO_4$ concentration, while the interaction of these two chemicals enhanced biomass concentration. Initial substrate experiments were then conducted to determine in greater detail the effects of magnesium sulfate, sodium nitrate, and potassium phosphate on biomass growth. Kinetic models for growth rate and substrate consumption were developed from the data and validated. These fundamental kinetic models can then be used in a variety of reactor styles and configurations in order to identify the system and operating parameters

(322e) Use of Sodium Bicarbonate for Efficient Carbon and Water Management for Autotrophic Microalgae Cultivation in Open Pond System

Jinsoo Kim, School of Energy, Environmental, Biological and Medical Engineering, University of Cincinnati, Cincinnati, OH, Joo-Youp Lee, School of Energy, Environmental, Biological, and Medical Engineering, University of Cincinnati, Cincinnati, OH and Ting Lu, Wastewater Treatment Division, Metropolitan Sewer District of Greater Cincinnati, Cincinnati, OH

Dissolved inorganic carbon including carbon dioxide(aq) and bicarbonate ion (HCO_3^-) is an essential carbon source for the autotrophic growth of microalgae. However, the use of CO_2 gas only cannot provide a sufficient amount of dissolved inorganic carbon concentrations in a narrow pH window suitable for the growth. In this study, sodium bicarbonate was used to increase dissolved inorganic carbon concentrations in conjunction with CO_2 gas. Added CO_2 gas was able to suppress the increased pH values as a result of bicarbonate ion uptake and to supply additional dissolved inorganic carbon. The fates of inorganic carbon and sodium ion were studied by taking mass balances for the two components. The effects of sodium ion on CO_2 gas absorption and algal productivity will be presented for sustainable water reuse and autotrophic microalgal culture in a typical open pond system. A preliminary economic analysis of the use of sodium bicarbonate based on the experimental results will also be presented.

(322f) Transgenic Expression of a Bacterial Exo-Acting Intracellular α-Amylase in the Chlamydomonas Reinhardtii Chloroplast

Xiaoqing Wang[1], Barbara Sears[2], Yan Susie) Liu[3] and Wei Liao[1], (1)Biosystems and Agricultural Engineering, Michigan State University, East Lansing, MI, (2)Plant Biology, Michigan State University, East Lansing, MI, (3)Department of Biosystems and Agricultural Engineering, Michigan State University, East Lansing, MI

Microalgae with their relatively high starch and lipid contents are one of the most promising non-food-crop-based feedstocks for biofuels production. To date, most studies have focused on accumulating algal lipids for biodiesel production and some have focused on hydrogen generation. Very little attention has been paid to simultaneously converting algal carbohydrate and lipids into biofuels. Although some algal species are able to accumulate significant amounts of carbohydrates (mainly starch) and lipid, their rigid cell wall retards the extraction of those carbon-rich compounds. In order to overcome this barrier, this study constructed an amylase-producing strain of the unicellular green alga *Chlamydomonas reinhardtii*, with the goal of converting algal starch into soluble sugars *in vivo*, which would facilitate their recovery. An exo-acting intracellular thermophilic α-amylase AmyB from the hyperthermophilic bacterium *Thermotoga neapolitana* was used as the amylase gene of choice. The AmyB gene was synthesized with codon usage optimized for the *C. reinhardtii* chloroplast; it was cloned into chloroplast expression vector p666 to yield plasmid p666AmyB. The vector p666AmyB was inserted into the site between the *psaB* and *atpA* genes on the chloroplast genome of *C. reinhardtii* by micro-particle bombardment, without interruption of chloroplast gene function. PCR analysis indicated that *AmyB* was integrated into the chloroplast genome of the transformants, and western blotting demonstrated the AmyB protein accumulation. The activity of AmyB can reach 17.4 U/g dry biomass. Thus, α-amylase was successfully expressed within the chloroplast, accomplishing an important step towards improving the ability to harness and harvest the carbon-compounds synthesized by micoalgae.

(342a) Engineering of Microfuidic Systems for Rapid Detection of Food Pathogens

Michael R. Ladisch, Laboratory of Renewable Resources Engineering Department, Purdue University, West Lafayette, IN, Eduardo Ximenes, Agricultural and Biological Engineering & Laboratory of Renewable Resources Engineering (LORRE), Purdue University, West Lafayette, IN and Arden Bement, Purdue University, West Lafayette, IN

Microbial contamination of fresh food causes severe illness, compromises food security and has major economic and international trade consequences. Food safety and security, and disruption of intra-national and international trade in fresh foods have obvious impacts on the sustainability of global food supplies. Protocols to enable rapid detection of pathogens in foods that are both exported and imported into the U.S. are of importance for enhancing food safety, as well as informing global trade policy and stakeholders responsible for ensuring the security of the US food supply. The technology is still being developed, but at the same time we are beginning to analyze potential impacts.

The concentration of cells from food derived samples that utilizes hollow fibers in an integrated microfluidic system provides a window on future technology for rapid pathogen detection. The lessons learned have led to a number of prototypes that are currently being validated for automated detection of pathogens in food derived samples. In these systems, a large volume of extract is concentrated 500 to 1000 fold. The resulting small volume has a sufficient concentration of bacterial cells, so that they may be effectively probed for presence of pathogens on a "first alert" basis. A positive result, obtained in a time span of about an hour, would signal the need for subsequent analysis using PCR, antibody based capture and identification, analysis in a microfluidic system, and/or plating coupled with light scattering identification.

The results of rapid methods, currently under development, must be coupled with statistical techniques to define the best way to selectively sample the huge quantities of vegetable produce that are shipped into and across the US. Rapid detection is important since it could help to support both source and border interventions to detect biotoxins and other pathogens in foods and guard against their distribution, in a manner not previously possible. This paper presents the current status of microfluidic technology for rapid cell concentration and its possible impacts on both national and international intervention alternatives and their consequences, both positive and negative.

(342b) Hitchhiking of Nanoparticles On Red Blood Cells for Targeted Delivery to Lungs

Samir Mitragotri, Aaron C. Anselmo and Vivek Gupta, Chemical Engineering, University of California, Santa Barbara, CA

Nanoparticulate drug delivery systems are one of the most widely investigated approaches for developing novel therapies for a variety of diseases. However, there are two major roadblocks for clinical interventions of these exciting therapeutic approaches: (i) rapid clearance of foreign materials by the immune system, and (ii) limited ability to target delivery of therapeutics to sites of interest. Here, we describe an approach to harness the flexibility, circulation, and vascular mobility of red blood cells (RBCs) to overcome both of these roadblocks (Cellular Hitchhiking). Anchoring nanoparticles onto the surface of red blood cells has been proven to reduce clearance by the immune system and extend the *in vivo* circulation time when compared to non-anchored counterparts [1]. In this work, we investigate the potential of "Cellular Hitchhiking" for targeted delivery of therapeutic carriers to the respiratory system via surface attachment of nanoparticles to RBCs. This novel delivery system was evaluated *in vitro* by characterizing (i) attachment efficiency, and (ii) cellular morphological changes following particle attachment to red blood cells. Time-dependent biodistribution of RBC-anchored nanoparticles was determined *in vivo*, which exhibited several fold higher lung accumulation compared to that of unattached nanoparticles. Parameters defining mechanisms for enhanced lung accumulation, and its therapeutic potential will be discussed. These results highlight the efficacy of cellular hitchhiking in providing targeted delivery of therapeutic agents for treatment of various cardiopulmonary and respiratory ailments.

1. Chambers, E., & Mitragotri, S., *Prolonged circulation of large polymeric nanoparticles by non-covalent adsorption on erythrocytes*. Journal of controlled release, 2004, **100**(1): pg. 111-119.

(342c) Engineering Natural Product Biosynthesis and Biocatalysis

Yi Tang, Chemical and Biomolecular Engineering, University of California, Los Angeles, Los Angeles, CA

Microbial natural products that have potent biological activities are important sources of pharmaceuticals. In nature, these complex small molecules are assembled from simple primary metabolites using elaborate biosynthetic pathways and enzymes. Understanding the biosynthesis of these compounds is therefore an important goal towards rationally engineering new drug candidates. Our lab is interested in the biosynthesis of compounds that have antibiotic, anticancer and cholesterol lowering activities. In this presentation, I will discuss the strategies towards manipulating natural product pathways in microorganisms, especially towards the mining of new compounds with novel activities. Recent sequencing efforts have revealed that an overwhelming number of natural products gene clusters are silent under normal laboratory culturing conditions. Hence activation of these pathways is potentially a new source for drug discovery. Several examples in both soil-borne actinomycetes and filamentous fungi will be presented.

(342d) Engineering Ovarian Follicle Maturation

Lonnie Shea, Chemical and Biological Engineering, Northwestern University, Evanston, IL

In vitro culture and maturation of ovarian follicles is a promising method for treating infertility when IVF is not suitable, such as chemotherapy induced sterility or disorders such as polycystic ovarian syndrome (PCOS). Immature follicles, consisting of a centrally located oocyte surrounded by 1-2 layers of granulosa cells, can be isolated from the ovary and cryopreserved. The development of methods to promote follicle development and oocyte maturation in vitro, or to transplant early stage follicles with high engraftment and controlled development could provide reproductive options. We have pioneered the development of controllable environments for the in vitro culture and in vivo transplantation of immature ovarian follicles.

Our three-dimensional culture systems promote growth of early stage follicles that can produce fertilizable oocytes and live births. The 3D culture maintains the follicular architecture and cell-cell connectivity, while allowing for growth. Maintaining the architecture is particularly critical for small follicles and has enabled the individual culture of primary follicles, which was not possible with previous culture systems. Our culture systems has been extended to the culture of canine, rhesus macaques, baboons, and human follicles. We identified that the mechanical properties of the gel are critical regulators of follicle development, and the changes associated with the mechanical environment were recently correlated with patients that have PCOS in a genome wide association study. Ongoing research is applying systems biology tools to analyze folliculogenesis, and is identifying mechanisms governing follicle growth and oocyte quality. Modifications to the media composition derived from our analysis has substantially enhanced growth and survival of primary follicles. Identifying a TF regulatory network that underlies the highly coordinated process of folliculogenesis can be employed to engineer systems that promote follicle development *in vitro* and *in vivo*.

Ovarian tissue transplantation is an alternative to follicle culture that has restored fertility for women; however, survival, engraftment, and duration of function of the transplanted tissue remain challenging. Early stage follicles can be transplanted, however these follicles may not persist due to low survival following transplantation, or recruitment into the growing pool by the local environment. Our biomaterials facilitated host-graft integration for graft function, which was enhanced by delivery of angiogenic factors locally. Live births conceived naturally resulted from transplantation. Taken together, results from these biomaterial systems will contribute to our understanding of folliculogenesis and the impact of the local environment, which may ultimately lead to novel fertility options for women.

(342e) Building the Biotechnology Toolbox: At the Intersection of Biofabrication and Synthetic Biology

William E. Bentley, Fischell Department of Bioengineering, University of Maryland, College Park, MD

Synthetic biology provides a means for articulating concepts into new products and products. Its toolbox is extensive, including the ability to create synthetic genomes and tailor their regulation. Early successes augmented the cell's biosynthetic capacity and rewired its regulation, transforming our ability to produce an array of small molecules and fully functional therapeutic proteins at high yield. Also, the theoretical formalisms of metabolic engineering provide a basis for optimally routing its biochemical flux. These activities focused largely on the cell's intracellular biochemical network and rely less on molecular cues from a cell's immediate surroundings. Bacterial quorum sensing (QS) has enabled a reexamination of metabolic flux and regulation by hardwiring population-scale biological function to extracellular cues. Regulatory and fabrication modules are feasible, owing to a few relatively simple QS signal transduction cascades. QS represents a model for connecting environmental cues for the regulation of biosynthesis; a new context for entirely new products and processes that consider the individual or small populations of cells. Indeed, the intersection of QS and synthetic biology is just beginning to yield new commercialized processes and products - and its concepts are no longer limited to the cell. On an analogous track, biofabrication is an assembly process that relies less on the cell and more on enzymatic, self assembly, and stimuli responsive properties of the various building blocks. Also, it enables connectivity to microfabricated electronics. There exists, therefore, a potential bridge between electronics and synthetic biology that is thus far largely untapped. We will describe our efforts to bride this gap – using quorum sensing as a model pathway.

(351a) System Identification and Frequency Response Techniques for the Design of Controlled Release Drug Delivery Systems

Timothy Knab[1], Sam N. Rothstein[2], Steven R. Little[2] and Robert S. Parker[1], (1)Chemical and Petroleum Engineering, University of Pittsburgh, Pittsburgh, PA, (2)Department of Chemical and Petroleum Engineering, University of Pittsburgh, Pittsburgh, PA

System Identification and Frequency Response Techniques for the Design of Controlled Release Drug Delivery Systems

Timothy Knab, Sam Rothstein, Steven R. Little, and Robert S. Parker

Department of Chemical and Petroleum Engineering

University of Pittsburgh, Pittsburgh, PA 15237

There is great interest in developing truly programmable controlled release systems and, to that end, numerous models have been developed. However most models are not broadly predictive and fewer still allow for direct correlation of release behavior to the controllable physical properties of the device. To address this issue, our group has recently produced a model for well-defined, programmable controlled release in vitro[1], which to our knowledge, can be applied to more drugs and polymers than systems described to date. However, the current model is limited to an idealized in vitro environment and further work is required to account for the non-ideal in vivo processes that may impact release behavior. Our goal is, for the first time, to extend the current release model to have predictive capabilities in an in vivo environment and systematically determine the most important parameters governing release in vivo. We hypothesize that tissue environment-specific effects, affecting release and transport rates, can be detected in systemic circulation and characterized using system identification techniques, namely frequency response analysis. Using the predictive power of the aforementioned model, we can synthesize particles that yield specific release profiles designed to inform the data-driven analysis.

As a basis for this study, a proof of concept diffusion experiment has been developed. Changes in membrane diffusivities in a simulation of this experiment lead to changes in reservoir concentrations over time, and these differences produce identifiable changes in the frequency response characteristics as evidenced from Bode plots. This type of analysis requires a well-controlled input function -- taken from the mathematical model of controlled release - and an output signals that is easily measurable. For this we use flux, which is easily attainable from an actual experimental measurement of concentration via a derivative relationship. Over a two order of magnitude span in diffusivities these simulations show these plots changing from flat, indicating no lag, (rapid diffusion) to a rapid drop off in the magnitude and phase indicating slow diffusion and equilibration of the second tank.

Experimental studies on the physical realization of the simulated diffusion cell are being compared against the simulated data as a test/validation of the experimental system and the numerical tools. Although the dynamics of diffusion cells are well studied, this system can be extended to be more analogous to in vivo physiology through experimental modifications, such as using an ECM seeded with cells as a barrier to transport or the introduction of a diffusive barrier mimicking the effects of fibroblast encapsulation of foreign microparticles. If successful, this system would provide a design framework for controlled release where exquisite control of the release system can be used to characterize -- and overcome -- potential barriers to drug release. The result is the ability to "pre-program" a microparticle for a desired release profile that results in a specified concentration profile at a location remote to the particle.

The system identification techniques we propose to use require reliable, high resolution tracking of spatially disparate drug or tracker molecule concentrations. To that end, Gadolinium-tetraazacyclododecanetetraacetic acid (Gd-DOTA) is being investigated as a potential tracking agent. Gd-DOTA concentrations at various locations within a system can be directly related to magnetic resonance imaging (MRI) T1 relaxation times. The availability of a non-invasive non-destructive imaging platform facilitates the translation of our tools to spatio-temporal analysis of experimental and in vivo systems. Initial work has focused on the design and release of Gd-DOTA from microparticles and the characterization of the diffusion profile in a polymer gel matrix. Gd-DOTA seems to be capable of coordinating with the microparticle polymer matrix, which, in effect, acts as a ligand and results in extremely delayed release compared to the model-simulated release that neglects this drug-polymer coordination. This is thought to be due to the interplay between intra-microparticle pH and Gd-DOTA charge -- an observation consistent with protein release studies also taking place in our lab. For our techniques to be successful, the effects of pH and charge needed to be included in the current model. We are currently working on a mechanistic description of these effects that will ultimately lead to an even more broadly predictive model for controlled release that can also be used in the development of a model of pre-programmable, controlled release, in vivo.

[1] Rothstein, S. N., Federspiel, W. J., & Little, S. R. (2008). A simple model framework for the prediction of controlled release from bulk eroding polymer matrices. *Journal of Materials Chemistry*, *18*(16), 1873. doi:10.1039/b718277e

(351b) Quantification of Intracellular Distribution of Agents by pH-Responsive Blend Polymer Particles

Xi Zhan, Kenny K. Tran and Hong Shen, Department of Chemical Engineering, University of Washington, Seattle, WA

The ability to deliver biological agents into defined intracellular compartments offers a promising strategy for targeting multiple signaling pathways in cells. In this study, pH-responsive polymer particles made from a blend of multiple polymers, poly(lactic-co-glycolic acid) (PLGA) and random pH-sensitive copolymer, have been developed for the intracellular delivery of agents. The polymeric blend particles exhibited different pH sensitivity and conformation depending on the ratios of two polymers. These blend particles were able to induce endosomal/lysosomal leakage in antigen-presenting cells (APCs). Multiple methods were used to quantify the intracellular distribution of cargos with varying sizes in APCs. The results showed that the distribution of agents in two intracellular compartments, lysosomes and cytosol, depended on the composition of the copolymer, the ratio of copolymer in blend particles as well as the size of cargos. This study presents a mechanistic understanding of routing agents into different intracellular compartments by pH-responsive carriers for intracellular delivery.

(351c) Pulsed Release Through Layered Polymers

Swapnil Gandhi and Eric Nuxoll, Chemical and Biochemical Engineering, University of Iowa, Iowa City, IA

In recent years, pulsed release of drugs has been of interest to achieve improved pharmaceutical therapies. To date, most pulsed release systems have relied on user intervention or complex devices (timers, pumps) limiting the application in which they are feasible. We have developed a barrier-mediated pulsed release (BMPR) system in which individual doses of drug are loaded within stimuli-sensitive hydrogels (depots) which release solute only upon contact by a stimulant. These depots are protected by reactive barrier layers which scavenge the stimulant for a controllable period of time, tuning the delay time between pulses over a large range. We have demonstrated this tuned release from a delaminating BMPR system using citric acid as model stimulant to produce multiple pulses of model drug (methylene blue) over a different range of timescales. We are now adapting this approach to a non-delaminating system whose performance cannot be predicted analytically. We have therefore built a computational model to predict the release profiles of these systems. The model framework assumes on Fickian diffusion of mobile species, binary hydrogel states and transport limited reaction kinetics. This presentation will discuss both the experimental and computational BMPR systems, analyze their agreement, and introduce efforts to further generalize this delivery strategy by incorporation of glucose as the stimulant.

(351d) The Impact of Polymer Blends and Solid Dispersion Technologies On Drug Release Rates

Adeyinka Adegoke, Department of Chemical Engineering, University of Rhode Island, Kingston, RI and David Worthen, Department of Biomedical and Pharmaceutical Sciences, University of Rhode Island, Kingston, RI

The objective of this research was to investigate the impact of polymer blends and solid dispersion formulation technologies on drug forms and release rates. An hydrophobic carrier, poly(methyl methacrylate PMMA), an hydrophilic carrier Eudragit EPO, together with an hydrophobic drug indomethacin (INDO), were used to produce various dispersion formulations through hot melt mixing (HMM) and solvent evaporation methods. Characterization of the various solid dispersion formulations was performed using standard analytical methods such as differential scanning calorimetry (DSC), polarized light microscopy (PLM), environmental scanning electron microscopy (ESEM), powder X-ray diffraction (PXRD), and Fourier transform infrared spectroscopy (FTIR). The results obtained from the various characterizations suggest that the drug was molecularly dispersed in the polymer matrix. In-vitro dissolution studies were carried out in simple model gastric fluid (pH 1.2) and also in phosphate buffer dissolution medium (pH 7.2). Approximately, 19.79 % of the INDO contained in the HMM formulation comprising (PMMA 70 EPO 10 INDO 20 by percent weight was released in approximately 30 minutes while about 54% of the solvent evaporation formulation (PMMA 70 EPO 10 INDO 20) was released after 22 hours. Dissolution results demonstrated that a combination of hydrophobic and hydrophilic polymers will effectively sustain the drug release for a prolonged period of time depending on the formulation composition and the solid dispersion technology employed to produce it.

(351e) A Predictive Model for Coupled Polymer Degradation, Erosion, and Drug Release in PLGA Biodegradable Stent Coatings

Xiaoxiang Zhu and Richard D. Braatz, Chemical Engineering, Massachusetts Institute of Technology, Cambridge, MA

Drug release from Poly(D,L-lactic-co-glycolic acid) (PLGA) stent coatings typically has multiple stages and is an intricate process controlled by polymer degradation (change of polymer molecular weight), erosion (change of coating mass), and drug transport [1]. Designing such controlled release systems still heavily rely on trial-and-error experimental procedures. While models describing drug delivery from biodurable stent coatings are relatively abundant (e.g., [2, 3]), very limited models are currently available for bulk-eroding PLGA systems that take into account the impact of degradation and erosion on drug release, especially for PLGA stent coatings (e.g., [1, 4]).

This work focuses on the development of a predictive model for the coupled degradation, erosion, and drug release in PLGA stent coatings. Analytical expressions of PLGA coating degradation and erosion are derived, which greatly reduces the complexity of using a computational intensive model such as in [5]. The analytical model demonstrate good predictions of polymer molecular weight and coating erosion rates in different experiments reported in the literature [6-8]. Then an integrated model of drug release from a PLGA coating is presented. Simultaneous drug diffusion transport through the polymer matrix and pores is modeled for the first time for a stent coating and the model is validated for *in vitro*sirolimus release with good agreement with the experimental data in [6]. Such a contribution of drug release from diffusion through the porous structure is often overlooked in published models that assume the drug diffusivity is constant or solely dependent on polymer molecular weight change. The model developed here is potentially useful for optimizing the design of PLGA coatings to produce target drug release profiles.

References

[1] S. Fredenberg, M. Wahlgren, M. Reslow, and A. Axelsson, "The mechanisms of drug release in poly(lactic-co-glycolic acid)-based drug delivery systems-A review," *International Journal of Pharmaceutics,* vol. 415, p. 34-52, 2011.

[2] S. Venkatraman and F. Boey, "Release profiles in drug-eluting stents: Issues and uncertainties," *Journal of Controlled Release,* vol. 120, p. 149-160, 2007.

[3] X. Zhu, D. W. Pack, and R. D. Braatz, "Modelling intravascular delivery from drug-eluting stents with biodurable coating: investigation of anisotropic vascular drug diffusivity and arterial drug distribution," *Computer Methods in Biomechanics and Biomedical Engineering,* p. 1-12, 2012.

[4] A. N. Ford, R. D. Braatz, and D. W. Pack, "Multi-Scale Modeling of PLGA Microparticle Drug Delivery Systems," in *Proceedings of the 21st European Symposium on Computer Aided Process Engineering (ESCAPE-21)*, Chalkidiki, Greece, 2011, p. 1475-1479.

[5] X. Zhu and R. D. Braatz, "Mathematical Modeling of Intravascular Drug Delivery In Drug-Eluting Stents with Biodegradable Coating," in *AIChE Annual Meeting*, Minneapolis, MN, 2011.

[6] X. T. Wang, S. S. Venkatraman, F. Y. C. Boey, J. S. C. Loo, and L. P. Tan, "Controlled release of sirolimus from a multilayered PLGA stent matrix," *Biomaterials,* vol. 27, p. 5588-5595, 2006.

[7] R. P. Batycky, J. Hanes, R. Langer, and D. A. Edwards, "A theoretical model of erosion and macromolecular drug release from biodegrading microspheres," *Journal of Pharmaceutical Sciences,* vol. 86, p. 1464-1477, 1997.

[8] T. Xi, R. Gao, B. Xu, L. Chen, T. Luo, J. Liu, Y. Wei, and S. Zhong, "In vitro and in vivo changes to PLGA/sirolimus coating on drug eluting stents," *Biomaterials,* vol. 31, p. 5151-5158, 2010.

(351f) Impingement of Printed Droplets Into Porous Medium: Rapid Manufacturing of Personalized Oral Dosages

Marlena Brown, Chemical and Biochemical Engineering/Food Science, Rutgers University, Ocean, NJ and Paul Takhistov, Food Science, Cook College, Rutgers, The State University of New Jersey, New Brunswick, NJ

Deposition of drug solution onto tablets utilizing Drop-on-Demand can be used to predict and manipulate final drug structure resulting in control of the disintegration of the drug delivery dosage form, API dissolution, and bioavailability in vivo. Spreading and sorption of inkjet-printed droplets of ibuprofen and griseofulvin has been studied. Small droplets of an aqueous solution (8.8 % w/w ibuprofen, ethanol, and ethylene glycol monoethyl ether) were deposited onto porous methylcellulose tablets compressed at 10kN, 15kN, and 20kN pressures. The wetting dynamics, spreading, and infiltration of the droplets were characterized to determine the effect on API dissolution profiles. Mass transfer of the droplet was also examined to determine the coupling between the processes of the spreading, infiltration, and evaporation dynamics. We observed that tablets compressed at higher pressures have greater sorption and decreased dissolution rates as compared to tablets compressed at lower pressures. The sorption of the liquid into the tablet can be modeled by the evolution of the contact radius, R(t) and the penetration of the wetting front, h(t) .

(351g) Rapid Vaccination Via Acetalated Dextran Microparticulate Subunit Vaccine for Protection Against Bacillus Anthracis Challenge

Kevin Schully[1], John Pesce[1], Sadhana Sharma[2], Margaret Elberson[1], Kevin Peine[2], Eric M. Bachelder[3], Andrea Keane-Myers[1] and Kristy M. Ainslie[3,4], (1)Biological Defense Research Directorate, Naval Medical Research Center, Fort Detrick, MD, (2)Ohio State, Columbus, OH, (3)Pharmaceutics, Ohio State, Columbus, OH, (4)Chemical Engineering, Ohio State, Columbus, OH

For many diseases, the time to death is greater than the time it takes to confer a protective immunity. Anthrax, or infection from *Bacillus anthracis* (*B. anthracis*), is one such disease. The Centers for Disease Control classify *B. anthracis* as a category A agent due to its ability to diffuse across large areas easily with high lethality. The first reported case of bioterrorism using Anthrax dates back to 1916, with the most recent during a domestic bioterrorism attack in 2001. For pulmonary infection, death can result within one week, whereas the current vaccine (Anthrax Vaccine Adsorbed (AVA)) requires 6 doses and 18 months to achieve protection. AVA has an increased rate of incidence reported in Vaccine Adverse Events Reporting System, indicating safety concerns are also present with the vaccine. For these reasons, we have formulated a vaccine carrier comprised of microparticulate acetalated dextran (Ac-DEX), which encapsulates recombinant protective antigen (rPA) and the FDA approved adjuvant resiquimod. Ac-DEX is a unique polymer that has been shown to have enhanced presentation of subunit antigen to T cells, compared to other common biomaterials. The enhanced presentation of Ac-DEX is due in part due to its acid-sensitivity and broad range of degradation rates, which can range from hours to months. Using standard emulsion chemistry with homogenization, we encapsulated rPA at an efficiency of 88.2% for an overall loading of 1.33 mg of rPA/100 mg Ac-DEX. Resiquimod was also loaded into Ac-DEX particles using emulsion chemistry with sonication, yielding an encapsulation efficiency of 8.3% or 0.25 mg resiquimod per 100 mg Ac-DEX. SEM micrographs of both particle sets indicate the particles are spherical in nature and sub-micron in range for the resiquimod particles and approximately a micron in size for rPA particles. To evaluate the efficacy of the particles, mice were vaccinated on day 0 and 7, and collected blood samples were analyzed for antibody titer and toxin neutralization. Significantly higher levels of antibodies were generated by day 14 with encapsulated rPA and encapsulated resiquimod co-delivered, as well as encapsulated resiquimod with rPA and alum, compared to the conventional rPA and alum vaccination. In comparison of these same groups, the levels of IgG2a antibodies were also significantly higher, although the levels of IgG1 were not. An in vitro toxin neutralization assay (TNA) was performed with the isolated blood antibodies. The results indicated that neutralizing levels of antibodies were significantly higher in the rPA + alum, rPA + encapsulated resiquimod, and rPA +alum + encapsulated resiquimod groups, with the last group having significantly higher neutralizing equivalence units on day 14, 30 and 45, compared to all other experimental groups. Next the vaccinated mice were evaluated for survival with aggressive challenge. After two subcutaneous vaccinations (0, 7 days) a week apart, A/J mice were intratracheally challenged with *B. anthracis* spores (50xLD_{50}) three times, seven days apart (Days 14, 21, 28). Survival was not observed with the blank particle and PBS controls, but was observed with other groups, regardless of toxin neutralization results. Additionally, sterile cultures of swabs taken from mice hearts were found in several groups including the encapsulated rPA and encapsulated resiquimod co-delivered group. Overall, we report the use of a novel particulate carrier to generate a rapid protective immunity against anthrax.

(351h) Transdermal Delivery of Biopharmaceuticals Using Dissolving Microneedles Patch

Jeong Woo Lee, Chemical and Biomolecular Engineering, Georgia Institute of Technology, Atlanta, GA, Seong-O. Choi, Chemical & Biomolecular Engineering, Georgia Institute of Technology; Electrical and Computer Engineering, Georgia Institute of Technology, Atlanta, GA, Eric Felner, Emory University and Mark R. Prausnitz, Chemical and Biomolecular Engineering, Biomedical Engineering, Georgia Institute of Technology, Atlanta, GA

SEQ CHAPTER \h \r 1**TRANSDERMAL DELIVERY OF BIOPHARMACEUTICALS USING DISSOLVING MICRONEEDLES PATCH**

Jeong Woo Lee[1], Seong-O Choi[1], Eric I. Felner[1,2], Mark R. Prausnitz[1]

[1]School of Chemical and Biomolecular Engineering, Georgia Institute of Technology, Atlanta, GA 30332

[2]Division of Pediatric Endocrinology, Hughes Spalding Children's Hospital, Emory University School of Medicine, Atlanta, GA 30332

Introduction

The delivery of biopharmaceuticals via the oral route is not practically possible due to enzymatic degradation in the gastrointestinal tract, low absorption through mucous membranes of intestine, and hepatic degradation due to the first-pass effect in the systemic circulation. Hypodermic injection can ensure higher bioavailability; however, it has low patient compliance because of fear and pain caused by hypodermic needles and it often requires patients to visit a clinic for correct injection, disposal of used needles and maintaining the integrity of biopharmaceuticals in the cold chain. As an alternative, transdermal delivery with a dissolving microneedles patch can address these problems associated with oral and injection delivery. Here, we present a dissolving microneedles patch loaded with human growth hormone (hGH) by examining the functional activity of hGH and in vivo pharmacokinetics of hGH to assess transdermal delivery of biopharmaceuticals with a dissolving microneedles patch and ultimately for the self-administration of biopharmaceuticals.

Methods and Materials

Dissolving microneedles patches as shown in Figure 1A were prepared using microfabrication technologies, molding techniques, and modified solvent-casting methods.[1] First, microneedle master structures were fabricated using UV photolithography processes and an inverse mold was made by casting master structures in polydimethylsiloxane (PDMS). To fabricate dissolving microneedles patches, carboxymethylcellulose (CMC) was dissolved in deionized water and then dehydrated to form a viscous hydrogel. In the case of CMC/trehalose microneedles, CMC and trehalose were dissolved in deionized water at a ratio of 1:1. Then, recombinant hGH (Genotropin, Pfizer) was added to the concentrated hydrogel, which contains hGH and matrix at a mass ratio of 1:9 on a dry basis. Finally, the hydrogel containing hGH was cast onto the mold and dried under centrifugation used to fill the microneedle cavities.

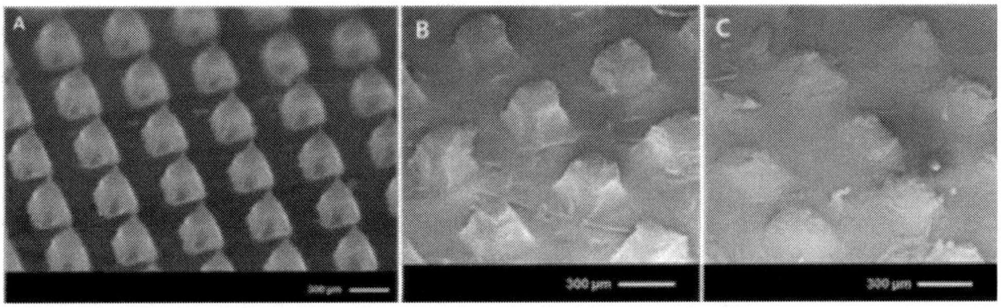

Figure 1. Brightfield micrograph of (A) CMC/trehalose dissolving microneedles patch encapsulating hGH. Scanning electron micrographs of (B) CMC microneedles patch and (C) CMC/trehalose microneedles patch after 24 h application to rat skin.

hGH functional activity was determined by measurement of hGH-stimulated growth of Nb2 rat lymphoma cells.[2] Nb2 cells in stationary culture medium will resume replication when stimulated by hGH. The viable Nb2 cell population at 3 days after the addition of hGH was measured using a cell viability analyzer (Vi-CELL, Beckman Coulter). Five different hGH solutions were added to the Nb2 cell culture; (i) CMC placebo solution (negative control), (ii) hGH solution (Genotropin, positive control), (iii) hGH solution mixed with CMC reconstituted from dissolved placebo microneedles, (iv) a reconstituted CMC microneedles patch encapsulating hGH, and (v) a reconstituted CMC microneedles patch encapsulating hGH after 3 months storage at ambient conditions (23±2°C and 38±5% relative humidity).

The pharmacokinetic study was performed with wild-type male hairless rats with approval by the Institutional Animal Care and Use Committee of Georgia Tech. Four groups were involved; Group 1 - subcutaneous injection of hGH from the manufacturer (positive control), Group 2 - CMC microneedles patch encapsulating hGH, Group 3 - CMC/trehalose microneedles patch encapsulating hGH, and Group 4 - subcutaneous injection of a reconstituted CMC/trehalose microneedle patch encapsulating hGH Rat blood was drawn from the tail vein at various time points up to 24 h after administration and collected in a CAPIJECT tube (T-MG, Terumo Medical). The collected blood was spun to isolate serum, which was stored at -70°C until hGH concentration was determined by enzyme-linked immunosorbent assay (ELISA). Areas under the serum hGH concentration curve were computed to calculate bioavailability.

Results and Discussion

We assessed the stability of hGH in dissolving microneedles patches by determining the level of cell proliferation of each group at the saturation level of hGH stimulation. As shown in Figure 2A, unprocessed hGH (positive control) stimulated cell growth resulting in almost 750% of initial population. Addition of CMC to unprocessed hGH stimulated cell growth to the statistically same level as the positive control, and reconstituted CMC microneedles (containing no hGH, placebo) did not stimulate cell growth, suggesting that CMC had no effect on hGH activity or cell proliferation, further indicating that CMC was inert. The cell population increase by reconstituted hGH in a CMC microneedles patch was not statistically different from the positive control, showing that the fabrication and reconstitution of hGH microneedles patches were not detrimental to hGH stability. Moreover, storage of hGH microneedles patches for 3 months in air at ambient temperature (23°) and relative humidity (~40%), caused only a 15% loss of hGH activity. This hGH functional activity study demonstrates the suitability of the formulation and process to fabricate hGH dissolving microneedles patches.

We next administered hGH into hairless rats and measured plasma concentration of hGH to assess bioavailability of hGH from a dissolving microneedles patch. As shown in Figure 2B, hGH administered by any group quickly had a peak hGH plasma concentration (C_{max}) at $t_{max} \approx 0.7$ h (ANOVA, $p > 0.05$) and then decreased over the next 6 hours. Subcutaneous injection of hGH was used as the positive control corresponding to 100% bioavailability. A reconstituted microneedles patch injected subcutaneously had the statistically same pharmacokinetic profile and bioavailability in comparison with the positive control, implying that the fabrication process of hGH microneedles patch did not degrade hGH integrity.

We next studied hGH administration using microneedle patches prepared with two different formulations. The first formulation contained only CMC as the microneedle material. In this case, the hGH serum concentration profile was significantly lower than the positive control, showing a C_{max} value that was 20% of the positive control (Student's t-test, $p < 0.05$) and a bioavailability of 13%. We hypothesized that this low bioavailability can be attributed to slow and incomplete dissolution of the CMC microneedle matrix in the limited amount of interstitial fluid in the skin. To examine this, we prepared microneedles formulated with a mixture of CMC and trehalose, which can dissolve more quickly with less fluid than CMC. Consistent with this hypothesis, CMC/trehalose microneedles achieved four times higher C_{max} and five times higher bioavailability than CMC microneedles (Student's t-test, $p < 0.05$).

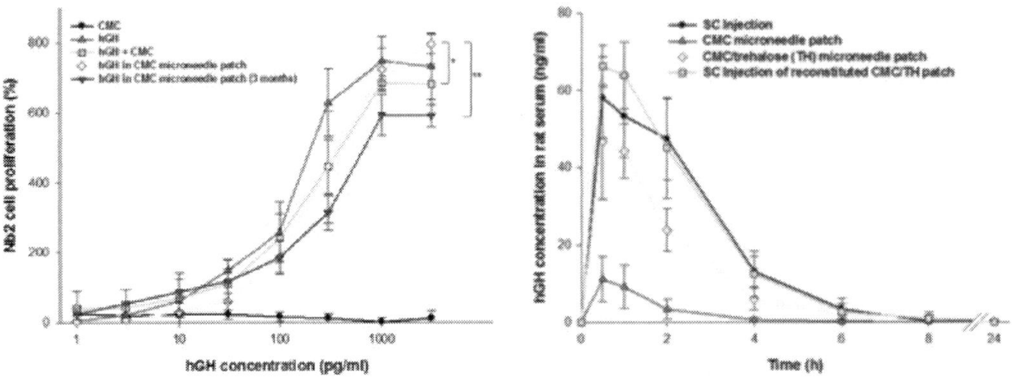

(A) (B)

Figure 2. (A) Stability of hGH in a dissolving microneedles patch. Asterisk indicates statistical comparison with positive control. (two-way ANOVA, P>0.05 for * and P<0.05 for **) (B) Pharmacokinetic profile of hGH in rat serum after administration.

Scanning electron micrographs and the measurement of remaining hGH in the patch could support the difference between the two formulations. As shown in Figure 1B, CMC microneedles were partially dissolved, leaving much of the microneedle shaft. In contrast, CMC/trehalose microneedles (Figure 1C) were mostly dissolved with just short stubs remaining behind. These microneedle patches were then dissolved in vitro after application and analyzed to determine residual hGH content, which was measured as 69±3% of the initial dose in CMC microneedles patch and 17±3% of the initial dose in CMC/trehalose microneedles patch. In future studies, microneedle design needs to be modified to improve insertion more deeply into the skin, to incorporate more soluble matrix materials and/or to localize active therapeutics more toward the tip.

Conclusion

This study addresses transdermal delivery of hGH with a dissolving microneedles patch designed for safe and simple self-administration by patients. Microneedle patches were fabricated to encapsulate and deliver hGH without loss of functional activity. The dissolving microneedles patches made of CMC and trehalose demonstrated 71% bioavailability of hGH; the loss of bioavailability was mostly accounted for by incomplete dissolution of microneedles. Overall, this study demonstrates the feasibility of a dissolving microneedles patch for transdermal delivery of hGH and other biopharmaceuticals.

Reference

[1] J. W. Lee, J. H. Park, M. R. Prausnitz, *Biomaterials* **2008**, *29*, 2113-2124.

[2] P. W. Gout, C. T. Beer, R. L. Noble, *Cancer Research* **1980**, *40*, 2433-2436.

(366a) Coupling of Crystallizers for Efficient Enantioseparation - Comparison of Two Different Process Strategies

Matthias Johannes Eicke, Physical and Chemical Foundations of Process Engineering, Max Planck Institute for Dynamics of Complex Technical Systems, Magdeburg, Germany, Guillaume Levilain, Max Planck Institute for Dynamics of Complex Technical Systems, Magdeburg, Germany, Martin Peter Elsner, Georg-Simon-Ohm University of Applied Sciences, Nuremberg, Germany and Andreas Seidel-Morgenstern, Physical and Chemical Foundations of Process Engineering, Max-Planck Institute for Dynamics of Complex Technical Systems, Magdeburg, Germany

Chiral molecules can be a particular challenge when used in drugs because of the existence of their stereoisomers. Although both enantiomers are merely mirror images, they can show drastically different effects in living organisms, which are chiral themselves and thus able to distinguish between the two molecules. Owing to this, it is essential to obtain enantiomers in their pure form. A wide array of methods has been studied and brought to application which on the one hand is concerned with the direct synthesis and on the other hand focuses on a more engineering biased approach that is, the separation of the racemate using well-known as well as novel techniques.

This contribution takes on the engineering route and puts the application of preferential crystallization (PC) as a separation technique at its center. It is well known that a racemic solution containing both enantiomers at equal proportions can be subcooled and still remain in a state where only the liquid phase exists. This phenomenon, called metastability, applies to other solutions as well but it is the key for applying PC. Within this region of kinetic inhibition it is possible to grow seed crystals of only one enantiomer. At some point however, the unseeded species will undergo nucleation and contaminates the product. In order to prevent this, the process must be stopped and the product has to be removed from the mother liquor. Certainly, this leads to great limitations with respect to yield because a considerable amount of product is still left in the dissolved state.

An improvement is possible for systems that crystallize as conglomerates such as DL-threonine/H_2O. It has been shown that two crystallizers coupled via a constant exchange of crystal free solution and seeded with opposite enantiomers can overcome these limitations [1, 2].

Recently more work has been done by the authors to further improve coupled PC. Different polythermal modes of operation have been investigated theoretically indicating the possibility for significant enhancements with respect to yield and productivity. Furthermore, a variant of coupled PC originally described in [3] has been investigated in detail both theoretically and experimentally for the model system DL-threonine/H_2O. The scheme of the process is depicted in Figure 1.

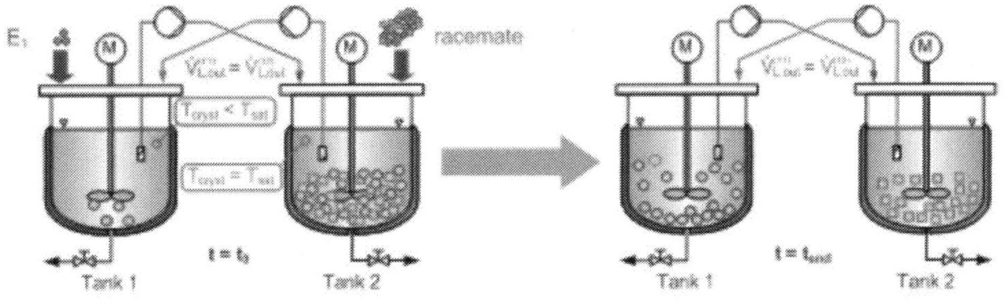

Figure 1: Operating principle for the connection of selective crystallization and dissolution.

Contrary to the first strategy, where both tanks are seeded with the respective pure enantiomers, this variant requires only one type of homochiral seeds. One crystallizer (Tank 1) holds a supersaturated racemic solution to which the preferred enantiomer E_1 is added in solid form. The other crystallizer (Tank 2) contains a saturated racemic solution with the same concentration to which solid racemate is added. Since the temperature in Tank 1 is below saturation level the seeds will grow reducing the level of supersaturation. Due to the exchange of clear solution a transient undersaturation with respect to E_1 will appear in Tank 2, leading to a selective dissolution of this enantiomer from the solid racemate. In this case preferential crystallization is happening in one tank while a selective dissolution in the other leads to a purification of the supplied racemate with respect to the second enantiomer E_2. The process thus yields two pure molecules while only one is needed as an investment.

This contribution will focus on how coupling crystallizers can help to make enantioseparation more efficient. The core will be a comparison of both process strategies with respect to their productivity but also their strengths and weaknesses. Further insight will be given by complementing experimental results with simulation studies directed at the influence of different process parameters.

1. Elsner, M.P., G. Ziomek, and A. Seidel-Morgenstern, *Efficient Separation of Enantiomers by Preferential Crystallization in Two Coupled Vessels*. AIChE Journal, 2009. **55**(3): p. 640-649.

2. Elsner, M.P., G. Ziomek, and A. Seidel-Morgenstern, *Simultaneous preferential crystallization in a coupled batch operation mode. Part II: Experimental study and model refinement*. Chemical Engineering Science, 2011. **66**(6): p. 1269-1284.

3. Merck & Co. Inc., *Resolution of Racemic Mixtures of Optically Active Enantiomorphs*. Patent DE1543238, 1965.

(366b) PBE-Aided Design of Slurry Milling Process to Reduce Particle Size of Intermediate and Active Pharmaceutical Ingredients

Carla Luciani, CPR&D, Eli Lilly & Co, Indianapolis, IN, Kevin D. Seibert, Chermical Product Research and Development, Eli Lilly & Co., Indianapolis, IN, Daniel Jarmer, CPRD, Eli Lilly & Co., Indianapolis, IN and Eric Moher, CPR&D, Eli Lilly & Co., Indianapolis, IN

A variety of milling processes are used in pharmaceutical industry to reduce the particle size of intermediate and active pharmaceutical ingredients. Slurry milling processes are of particular interest because they significantly reduce the transfer of heat to the product compared to dry mill equipment and they can eliminate a unit operation when crystallization-milling sequence are combined prior to the isolation. In spite of its extensive use in the pharmaceutical industry, the design of slurry milling processes are not trivial due to the multivariate nature of the process. One of the most intuitive ways to describe milling processes is the use of population balances equations (PBEs). However, relating PBE parameters (i.e., breakage rate and breakage kernel) to actual milling conditions (i.e., rotor tip speed or shear rate, slurry density, mechanical, morphological, and rheological properties of the milled material milled and the suspension medium, etc.) is challenging.

In this work, a PBE approach is applied to describe the slurry milling process of a pharmaceutical intermediate. Correlations between milling conditions and model parameters are proposed. Finally, the model is inverted to determine set of operating conditions that produce desired final particle sizes.

(366c) Development of a High Shear Crystallization Process for a Pharmaceutical Intermediate

Brenda Remy, Amit Joshi, Junying Fan, Jason Sweeney, Kenneth Natalie, Aghogho Pedro and Shawn Pack, Chemical Development, Bristol-Myers Squibb Co., New Brunswick, NJ

Batch crystallization processes for pharmaceutical intermediates are generally design to maximize the removal of impurities and to minimize yield losses. Crystallization performance is dependent on the ability to control several mass transport mechanisms such as growth, nucleation, impurity deposition etc. These mechanisms, in turn, can be impacted by controlling the amount of shear during the crystallization. Wet milling via rotor-stator devices is commonly employed in batch crystallization to control shear. While wet milling is mostly used as a particle size reduction technique, it can be used to control the purity of the isolated solids and to improve yield.

This talk will highlight the use of wet milling in a pharmaceutical intermediate crystallization to improve quality, purity and cycle time. The crystallization involves an initial mass nucleation followed by a solid-solid form conversion. Upon initial scale-up the process suffered from inconsistent yields and impurity rejection. These inconsistencies appeared to correlate with final product particle size. Wet milling achieved consistently smaller crystals and enabled faster desaturation leading to consistently higher yields. Additionally, wet milling improved the purity of the crystallized material by mitigating impurity inclusion. The use of a DoE study to determine the main parameters affecting desaturation and impurity removal will be highlighted. Scale up results from a pliot plant campaign will also be presented.

(366d) Developing a Selective Crystallization Process for a Complex Reaction System

Chenchi Wang, Truc Vu and Lucius Rossano, Research and Development, Bristol-Myers Squibb Company, New Brunswick, NJ

Developing a Selective Crystallization Process for a Complex Reaction System

Authors: Chenchi Wang, Truc Vu, Lucius Rossano.

Research and Development, Bristol-Myers Squibb Company, New Brunswick, New Jersey

A reactive crystallization process often involves multiple complex steps, such as chemical reaction, solid precipitation/crystallization, and particle size control if required. In this case study, an active pharmaceutical ingredient (API) is prepared in a telescoped process integrating deprotection, free basing and salt formation. Apart from the chemistry issues encountered that are not the focus of this presentation, the challenges of the crystallization lie in (1) maintaining chemical stability of the free base prior to crystallization, (2) quantitating the free base available for salt formation, (3) maximizing yield of the free base under process conditions, (4) enhancing purity by trace metal removal, and (v) achieving desired particle size for drug product development. In order to address these important process issues and improve process robustness, a two-step, selective crystallization process was developed. As a result of this development, a robust API synthesis step, incorporating this well-controlled reactive crystallization, was demonstrated at various scales and consistently generated material with desired purity, yield, and particle size. The designing and outcomes of this process will be discussed in details.

(366e) Crystal Nucleation Control Using Microscopic Online Imaging

Tod Canty, J.M. Canty Inc., Buffalo, NY

An innovative method of inline microscopic imaging has been developed to allow crystal viewing and measurement of size, shape, color and concentration in batch or continuous lab reactors ,pilot plant crystallizers and full production vessel at pressure(10,000psi) and temperature (-450 to 800 deg F). The high resolution images presented will show how this technology succesfully over comes the limitations of other systems where aprticle resoltions where limit to 30 micron and above due to front lighting even though the optical lenses where capable of higher pixel resoltuion. The particle imaging system captures high resolution 2 megapixel images down to .7 micron at speeds up to 1/100,000 of a second using back lighting. The system has a fiber optic lighting system that allows viewing through a wide variety of fluids (crude oil, polymer ,inks ,paint etc.). It does this at speeds through 10ft/sec in liquids. The software then analyzes the size and shape of particles compiling the data to determine particle distribution by length, width or volume along with ppm concentration. The Lab scale reactors then allow the user to optimize the seeding process to provide control of Nucleation to optimize yield and filterability.

(366f) Precipitation Kinetics of Aluminum Solids Formation During the Caustic Side Solvent Extraction (CSSX) Process

Rebecca K. Toghiani[1], Punith P. Naik[2], Jeffrey S. Lindner[2], Laura T. Smith[2] and Larry Pearson[2], (1)Dave C. Swalm School of Chemical Engineering, Mississippi State University, MS State, MS, (2)Institute for Clean Energy Technology, Mississippi State University, Starkville, MS

The Caustic Side Solvent Extraction Process (CSSX) process is used for cesium separation at the Savannah River Site (SRS). The CSSX process is comprised of 4 steps: 1) extracting cesium into the solvent; 2) scrubbed of the loaded solvent to remove co-extracted sodium and potassium; 3) stripping to transfer the cesium out of the solvent; and 4) washing of the stripped solvent to remove solvent degradation products if present. During the scrubbing process, aluminum precipitation has been observed. This is a concern as solids formation may have undesirable consequences during operation. The work reported was undertaken to identify conditions under which solids formation may occur and to identify an operating region where solids formation is avoided or minimized.

Experiments were conducted at room temperature using simulants based on different dissolution fractions of SRS Tank 25F contents. Combination of a given amount of simulant with a 0.05 M nitric acid solution mimicked the carryover during the CSSX process. Aluminum concentration and solution pH were measured over time for the prepared samples. Simulation of these experiments were also completed using the Environmental Simulation Program (ESP, OLI Systems, Inc.) and comparisons made. Solids formation was observed for a select subset of samples both experimentally and through ESP predictions. For these samples, a second set of samples was prepared. The temporal variation of aluminum concentration was measured, until solids were visually observed in each sample. The resulting data were analyzed using a first order reversible rate expression, and forward and backward reaction rate constants determined. Solids formed were isolated, dried and identified using polarized light microscopy.

(371a) Application of Split Intein in an ELP-Intein System

Changhua Shi, Chemical engineering, Ohio State University, Columbus, OH; Institute of bilological science and biotechnology, Donghua University, Shanghai, China, Qing Meng, Institute of biological science and biotechnology, Donghua University, Shanghai, China and David W. Wood, Department of Chemical and Biomolecular Engineering, the Ohio State University, Columbus, OH

Application of Split intein in an ELP-intein system

Changhua Shi, Qing Meng, David Wood

Inteins are internal protein sequences that can auto-excise from precursor proteins and splice together the flanking sequences. By substitution of amino acids, inteins can be transformed into self-cleavage elements, where cleavage is controlled by pH or thio-reagents. In our previous work, we have successfully combined a pH sensitive cleaving intein with a non-chromatographic purification tag (ELP, elastin like protein) to create a self-cleaving aggregation tag. The target protein can be easily purified by several steps of centrifugation and without expensive chromatography steps. However, the engineered intein usually has some level of pre-cleavage in vivo, depending on the target protein and expression temperature. This affects the final yield of the protein and the cost of production, especially for some highly expensive pharmaceutical proteins. This is also a problem that hampers the use of this technology at large scale. Here we introduce a split intein into our ELP-intein system to tackle this problem. The N terminal fragment of intein is removed from the ELP-intein-target protein precursor, and fused to a second ELP tag to be used as a switch for the cleavage activity of the tagged target precursor. The activity of truncated intein is tightly controlled by addition of the complementary N terminal fragment. The trans-cleaving efficiency of our split intein is similar to the full-sized intein. Our study also partially showed that the recognition and association of the intein fragments is strong even though there is a large tag, ELP, between them. Further, mutation of the first of amino acid of the target protein into the native conserved residues in this site, such as serine and cysteine can increase the cleavage efficiency. This dual ELP-intein technology can also be applied to do cost-effective protein purification in many other systems, such as yeast and mammalian cells.

(371b) Use of Flocculation to Improve Clarification Efficiency in a Microbial Feedstream

Jessica R. Molek, Jason M. Reck and Kent E. Goklen, Downstream Process Development, GlaxoSmithKline, King of Prussia, PA

E. coli is one of the fastest growing, least expensive, and highest product-to-volume systems available for the production of therapeutic proteins. Despite these advantages, efficient clarification of the product stream can be challenging due to its high solids content, propensity for cell lysis, as well as high concentrations of impurities such as DNA and host cell protein. While centrifugation can be used to remove larger particles, such as whole cells, many small particles, such as cell debris and protein DNA complexes, are difficult to sediment. Depth filtration can be used to remove fine particulates but capacities of the depth filters can be limited, especially by high DNA concentration. In this presentation, polycation and divalent cation precipitation are explored as methods to enhance the removal of cells, cell debris and DNA during centrifugation processes. Screening is performed using a combination of high throughput methods and bench scale experimentation. We examined the impact of several polycations and divalent salts, as well as changes to the pH and conductivity of the fermentation broth on the overall clarification process. Subsequent scale-up work has demonstrated a minimum four-fold increase in depth filter capacity using optimized flocculation conditions.

(371c) Fumaric Acid Recovery From Fermentation Broth by Activated Carbon Adsorption Followed by Desorption and Crystallization with Acetone

Kun Zhang, Chemical and Biomolecular Engineering, The Ohio State University, Columbus, OH, Lijie Zhang, Dept. of Pharmacy and Bioengineering, Chongqing University of Technology, Chongqing, China and Shang-Tian Yang, Department of Chemical and Biomolecular Engineering, Ohio State University, Columbus, OH

Fumaric acid, a dicarboxylic acid extensively used as a food acidulant and in the manufacturing of synthetic resins, can be produced from glucose in aerobic fermentation by the filamentous fungus *Rhizopus oryzae*. For fumaric acid recovery, acidification at a high temperature is usually required because of the low solubility of fumaric acid. Then, fumaric acid crystals can be obtained by cooling. However, the produced crystals may contain various impurities and need re-crystallization. In order to enhance the purity of fumaric acid, adsorption with activated carbon was investigated as a more economical separation method. The results from model fermentation broth (15 g/L fumaric acid, 5 g/L glucose, 5 g/L malic acid) showed that activated carbon had a high selectivity and high adsorption capacity (~250 mg/g) for fumaric acid, which generally increased with increasing the fumaric acid concentration in the solution. Desorption of fumaric acid from the activated carbon was then studied using acetone, which can effectively strip fumaric acid from activated carbon. After removing acetone by evaporation at 50 ^{o}C, fumaric acid crystals were obtained with a near 100% purity. The furmaric acid recovery yield from this process was affected by the amount of acetone and temperature used in desorption. After optimization, over 95% fumaric acid was recovered from activated carbon. Both activated carbon and acetone can be recovered and reused in the adsorption process. This process is effective and economical for fumaric acid recovery and purification, and should have great potential in industrial production of fumaric acid.

(371d) Membrane Fouling Mechanism in Ultrafiltration of Succinic Acid Fermentation Broth

Jianmin Xing, National Key Lab of Biochemical Engineering, Institute of Process Engineering, Chinese Academy of Sciences, Beijing, China, Caixia Wang, Institute of Process Engineering, Chinese Academy of Sciences, Beijing, China and Yinhua Wan, National Key Laboratory of Biochemical Engineering, Institute of Process Engineering, Chinese Academy Of Sciences, Beijing, China

Succinic acid (1, 4-butanedioic acid) is identified as an important building-block chemical which can be used for the synthesis of high value-added derivatives, such as 1, 4-butanediol (BDO), g-butyrolactone (GBL), tetrahydrofuran (THF), N-methyl-2-pyrrolidone (NMP), 2-pyrrolidone (2-Pyrr), succinimide, succinic esters, etc. Great efforts have been made to develop biotechnology methods to produce succinic acid from renewable resources since last decade.

However, fermentation process is still not economically feasible, mainly due to the low succinic acid concentration in the fermentation broth, and the high cost of succinic acid separation and purification. Considerable efforts have been made to improve the succinic acid concentration in the fermentation broth and several possible operation units such as adsorption, extraction and electrolysis have been investigated for the recovery of succinic acid.

In this study, fouling mechanisms of different membranes were fist studied in treating succinic acid fermentation broth during ultrafiltration process. The membrane fouling mechanism was studied during dead-end ultrafiltration. Different membranes were used and two models were applied to analyze the fouling mechanism. Resistance-in-series model was applied to determine the main factor that caused the operation resistance. Results indicated that most membranes tended to be fouled by cake layer or concentration polarization. Hermia's model, which is composed of four individual sub-models, was used to analyze the predominant fouling mechanism. Results showed that the fouling of RC 10kDa and PES 30kDa was controlled by the complete blocking mechanism, while PES 100kDa was controlled by the intermediate blocking and PES 10kDa was controlled by cake layer. This conclusion was also proved by SEM photos. Membrane characteristics were monitored before and after ultrafiltration by AFM and goniometer. Both contact angle and roughness of most membranes increased after ultrafiltration.

(371e) Separation of n-Butanol From Dilute Solution by Pervaporation Using Composite PDMS Membrane

Fangfang Liu, The Ohio State University, Columbus, OH, Kaijun Xiao, College of Light Industry and Food Science, South China University of Technology, Guangzhou, China and Shang-Tian Yang, Department of Chemical and Biomolecular Engineering, Ohio State University, Columbus, OH

Butanol has attracted increasing attention as an alternative to the fossil fuel. Before we can use butanol as a liquid fuel, it is necessary to refine the low-concentration butanol produced in fermentation. Pervaporation, as one of the membrane separation technologies, has high selectivity and low energy consumption compared to other separation techniques. Pervaporation can be used to separate azeotropic mixtures and other kind of mixtures, which are usually difficult to separate by conventional techniques like distillation. Poly(dimethyl siloxane) (PDMS) membrane is the most effective and widely used membrane for separating butanol from an aqueous solution. However, butanol flux and selectivity of PDMS membranes vary with fabrication methods and process conditions, and need further improvements for economical applications. In this work, different membrane fabrication techniques were explored, including plasma treatment, dipping, and electrospinning. The effects of membrane supporting materials, including poly(vinylidene fluoride) (PVDF), polysulfone (PS), and poly(tetrafluoroethylene) (PTFE), on the pervaporation performance of PDMS membrane were also studied. Further improvements in butanol selectivity and flux by the addition of hydrophobic fillers to the PDMS membrane were also investigated. Among the membranes studied, the PDMS membrane with a thickness of ~70 μm, supported by plasma treated PTFE and filled with 20 wt% CBV28014, gave the best separation performance, with a separation factor of 27 and a butanol flux of 48 $g/m^2 \cdot h$. The PTFE-supported PDMS membrane can concentrate a 15 g/L butanol solution to a 290 g/L butanol solution, and is promising for use in pervaporation for butanol recovery from dilute fermentation broth.

A TECHNIQUE FOR MASS BALANCE OF PROCESS DATA USING LINEAR COMBINATION OF LSQ AND LRSQ

A. I. A. Salama, W. Friesen, and T. Dabros
CanmetENERGY-Devon, Natural Resources Canada, Devon, AB, Canada

Introduction

In many oil sands, mineral, and chemical processing plants, plant audits are conducted, routinely, to evaluate the plant/process performance using mass recoveries and other separation criteria. The plant audit starts with sampling of plant streams and samples are collected using well-designed sampling scheme (Smith and Frew, 1983; Frew, 1983) which was improved, later, by Salama (1999a). The stream samples are subjected to analytical techniques to determine stream components (SCs). Furthermore, stream mass splits (SMS) / rates may be measured using online instrumentation. In most cases, the collected samples and SMS contain errors (e.g., sampling, measurement, and analytical) and must be balanced (or adjusted). The MB problem can be stated as: given a plant flowsheet and an associated set of observed (raw) data, find the optimal balanced set of SCs and SMSs such that an objective function is optimized. Simultaneously, a set of mass conservation constraints is satisfied.

However, in real time on-line optimization provides set points to the processes distributed control system (DCS) and therefore maintains the process near its optimum operating conditions (Ozyrut and Pike, 2004). Therefore, the process flowsheet mass balance and the real time on-line optimization are addressing two different objectives.

In almost all MB techniques, the stream consistency conditions (i.e., the sum of SCs in each stream adds up to 100%) are imposed and the SCs and SMSs are balanced at each node of the flowsheet. The only difference between the different techniques is the objective function of the optimization problem (which depends on the observed data error distributions or error model). In the:

- Conventional least error squares (LSQ) case, the objective function is defined as the sum of error (difference between the raw and estimate SCs) squares referenced to the standard deviations of raw SC errors. In this case, the stream data error distributions are assumed normal (Gaussian) (Kaput et al., 1993, Hodouin and Vaz Coelho, 1987, and Klimpel, 1979).
- Maximum likelihood technique, the data errors are assumed uniform or exponential to cover cases where many outliers are present in the raw data. For each error distribution a corresponding objective function is derived (Klimpel, 1979).
- Maximum entropy technique, the objective function is derived based on multinomial error distribution (Klimpel, 1979 and Kullback, 1959). The property of this technique is that no bias towards any particular data point (Jefferson and Scott, 1986, Kullback, 1959, Jaynes, 1957).
- Least relative error squares (LRSQ) case which is the subject of the present paper, the objective function is defined as the sum of, data error referenced to the raw data, squared.

Note that the focus of the present paper is to investigate the mass balance problem rather than modeling of the process. Using an observed data set try to search for a corresponding optimal data set such that a linear combination of the error and relative error squares is minimized. The advantages of the technique include:

- Cases where raw data error model is not available.
- Economic constraints may prevent the use of complete sampling campaign, especially, in early stages of new or proposed flowsheet developments.

- Conducting several preliminary tests to evaluate the feasibility and optimization of new or proposed or expanded flowsheet.

The process

One empirical method for obtaining data required in the engineering design of oil sands deposition process is to perform a simulation using a flume. A known amount of slurry is discharged continuously at one end of an open channel and flows toward the exit at the other end. A picture of the CanmetENERGY-Devon flume unit is shown in Figure 1. The mechanical dimensions of the unit are length (7.25 m), depth (0.6 m), and width (0.6 m). It should be emphasized that while the following developments are focused on a flume process, the results can be modified to handle any other general process (see the results of Salama, 2001 and 1999b).

Mathematical formulation

Let the raw data of feed, stages, and *ro* SCs be designated: p_{ij}, i=1,2,…., N_c, j=1,2,…., N_{st}+2. Where N_c, N_{st}, and $N_{st}+2$ are the number of stream components (SC)s, stages, and flume streams, respectively. Furthermore, the observed SMSs: (feed) $\boldsymbol{P}_1=P^f$, (stages) $\boldsymbol{P}_j$, j=2,3,….,N_{st}+1, and (*ro*) $\boldsymbol{P}_{Ns+2}=\boldsymbol{P}^{ro}$, respectively. In most oil sands applications, the SCs are bitumen (hydrocarbons), minerals (solids), and water contents and are labeled B, M, and W, respectively. This order will be kept during the presentation. In this case the number of SCs N_c =3. The raw data may be arranged in an array $\boldsymbol{P}$

$$\boldsymbol{P}=\begin{bmatrix} p_{11} & p_{12} & \cdot & \cdot & \cdot & \cdot & p_{1N_{St}+1} & p_{1N_{St}+2} \\ p_{21} & p_{22} & \cdot & \cdot & \cdot & \cdot & p_{2N_{St}+1} & p_{2N_{St}+2} \\ p_{31} & p_{32} & \cdot & \cdot & \cdot & \cdot & p_{3N_{St}+1} & p_{3N_{St}+2} \\ \boldsymbol{P}_1 & \boldsymbol{P}_2 & \cdot & \cdot & \cdot & \cdot & \boldsymbol{P}_{N_{St}+1} & \boldsymbol{P}_{N_{St}+2} \end{bmatrix} \tag{1}$$

Let the optimal estimates or mass balanced data of SCs be designated as: x_{ij}, i=1,2,….,N_c, j=1,2,…., N_{st}+2. And the associated estimated SMSs: (feed) $\boldsymbol{X}_1=X^f$, (stages) $\boldsymbol{X}_j$, j=2,3,….,N_{st}+1, (*ro*) and $\boldsymbol{X}_{Nst+2}=\boldsymbol{X}^{ro}$. The mass balanced data may be arranged in another array $\boldsymbol{X}$

$$\boldsymbol{X}=\begin{bmatrix} x_{11} & x_{12} & \cdot & \cdot & \cdot & \cdot & x_{1N_{St}+1} & x_{1N_{St}+2} \\ x_{21} & x_{22} & \cdot & \cdot & \cdot & \cdot & x_{2N_{St}+1} & x_{2N_{St}+2} \\ x_{31} & x_{32} & \cdot & \cdot & \cdot & \cdot & x_{3N_{St}+1} & x_{3N_{St}+2} \\ \boldsymbol{X}_1 & \boldsymbol{X}_2 & \cdot & \cdot & \cdot & \cdot & \boldsymbol{X}_{N_{St}+1} & \boldsymbol{X}_{N_{St}+2} \end{bmatrix} \tag{2}$$

Note that any other format different from the array representation is also applicable. In the present development the flowsheet estimated SMSs must satisfy the condition

$$\boldsymbol{X}^f=\boldsymbol{X}_1=\sum_{j=2}^{N_{st}+2}\boldsymbol{X}_j \tag{3}$$

i.e., the SMS of the feed equals the sum of stages and *ro* SMSs. Commonly, the SMS of the feed, $\boldsymbol{X}^f=\boldsymbol{X}_1$, is used as a reference and set equals 100% or unity; however, the label $\boldsymbol{X}_1$ will be kept in the subsequent equations. Using Eq. (3), the $\boldsymbol{X}_{Nst+2}$ (*ro* SMS, $\boldsymbol{X}^{ro}$) can be expressed as

$$\boldsymbol{X}^{ro}=\boldsymbol{X}_1-\sum_{j=2}^{N_{st}}\boldsymbol{X}_j \tag{4}$$

In Eq. (3), the circuit SMSs can be normalized relative to $\boldsymbol{X}_1$ as

$$1=\sum_{j=2}^{N_{st}+2}\frac{\boldsymbol{X}_j}{\boldsymbol{X}_1} \tag{5}$$

Furthermore, the flume feed, stages, and *ro* SCs must satisfy the SC consistency conditions or

$$\sum_{i=1}^{N_c} x_{ij}=1, \quad j=1,2,\ldots,N_{st}+2 \tag{6}$$

Moreover, the SCs mass balance at the flume node of the flume circuit must satisfy the conditions

$$\boldsymbol{X}_1 x_{i1}-\sum_{j=2}^{N_{st}+2}\boldsymbol{X}_j x_{ij}=0, \quad i=1,2,\ldots,N_c-1 \tag{7}$$

It should be emphasized that we considered only N_c-1 component mass balances (see Salama, 1999b). The reason is that the conditions in Eqs. (3-6) can be shown to imply the condition in Eq. (7) at $i=N_c$. This is important because the use of more constraints than what is needed may complicate the search for the optimal solution. One more note, if N_c=1, only one component is considered in Eq. (7).

Mass balance

Optimization Problem

Given the observed SCs: p_{ij}, i=1,2,….,N_c, j=1,2, ….,N_{st}+2 and the associated SMSs: $\boldsymbol{P}_j$, j=1,2,…., N_{st}+1, and $\boldsymbol{P}_{Nst+2}=\boldsymbol{P}^{ro}$, find the optimal SCs: x_{ij}, i=1,2,…., N_c, j=1,2,….,N_{st}+2 and the associated SMS: $\boldsymbol{X}_j$, j=1,2, ….,N_{st}+1, and $\boldsymbol{X}_{Nst+2}=\boldsymbol{X}^{ro}$, such that the objective function

$$\phi=\sum_{i=1}^{N_c}\sum_{j=1}^{N_{st}+2}\left(\frac{x_{ij}-p_{ij}}{\left[\alpha p_{ij}+(1-\alpha)\sigma_{ij}\right]}\right)^2+\sum_{j=1}^{N_{st}+2}\left(\frac{\boldsymbol{X}_j-\boldsymbol{P}_j}{\left[\alpha \boldsymbol{P}_j+(1-\alpha)\sigma_j\right]}\right)^2 \tag{8}$$

is minimized and simultaneously the constraints in Eqs. (3-7) are satisfied, where σ_{ij}'s and σ_j's are the standard deviations of SC ij and SMS j, respectively. Note that in Eq. (8), selecting α=1 corresponds to LRSQ case, selecting α=0 corresponds to the conventional LSQ case, and for any general α, $\alpha\in[0,1]$, corresponds to linear combination of LSQ and LRSQ cases.

Necessary Conditions

The solution of the present optimization problem can be determined using the necessary conditions of the stationary point of the optimization problem Lagrangian defined as (Fletcher, 1987)

$$\begin{aligned}\mathscr{L}=&\sum_{i=1}^{N_c}\sum_{j=1}^{N_{st}+2}\left(\frac{x_{ij}-p_{ij}}{\left[\alpha p_{ij}+(1-\alpha)\sigma_{ij}\right]}\right)^2+\sum_{j=1}^{N_{st}+2}\left(\frac{\boldsymbol{X}_j-\boldsymbol{P}_j}{\left[\alpha \boldsymbol{P}_j+(1-\alpha)\sigma_j\right]}\right)^2+\\&+\sum_{j=1}^{N_{st}+2}\lambda_j\left(1-\sum_{i=1}^{N_c}x_{ij}\right)+\sum_{i=1}^{N_c-1}\psi_i\left(\boldsymbol{X}_1x_{i1}-\sum_{j=2}^{N_{st}+2}\boldsymbol{X}_jx_{ij}\right)\end{aligned} \tag{9}$$

where λ_j, j=1,2,….,N_{st}+1, $\lambda_{Nst+2}=\lambda^{ro}$, and ψ_i, i=1,2,….,N_c-1 are constant multipliers. The parameter $\alpha\in[0,1]$ is introduced to determine general optimization results. The cases of available raw data error model and relative error squares case are special cases of the general results.

The necessary conditions are:

for i=1,2,….,N_c-1

$$\frac{\partial\mathscr{L}}{\partial x_{i1}}=2\left(\frac{x_{i1}-p_{i1}}{\left[\alpha p_{i1}+(1-\alpha)\sigma_{i1}\right]}\right)\left(\frac{1}{\left[\alpha p_{i1}+(1-\alpha)\sigma_{i1}\right]}\right)-\lambda_1+\psi_i\boldsymbol{X}_1=0 \tag{10}$$

$$\frac{\partial \mathcal{L}}{\partial x_{ij}} = 2\left(\frac{x_{ij} - p_{ij}}{\left[\alpha p_{ij} + (1-\alpha)\sigma_{ij}\right]}\right)\left(\frac{1}{\left[\alpha p_{ij} + (1-\alpha)\sigma_{ij}\right]}\right) - \lambda_j - \psi_i \boldsymbol{X}_j = 0, \quad j=2,\dots,N_{st}+2 \tag{11}$$

and for $i=N_c$

$$\frac{\partial \mathcal{L}}{\partial x_{N_c j}} = 2\left(\frac{x_{N_c j} - p_{N_c j}}{\left[\alpha p_{N_c j} + (1-\alpha)\sigma_{N_c j}\right]}\right)\left(\frac{1}{\left[\alpha p_{N_c j} + (1-\alpha)\sigma_{N_c j}\right]}\right) - \lambda_j = 0, \quad j=1,2,\dots,N_{st}+2 \tag{12}$$

and for $j=2,\dots,N_{st}+1$

$$\begin{aligned}\frac{\partial \mathcal{L}}{\partial \boldsymbol{X}_j} &= 2\left(\frac{\boldsymbol{X}_j - \boldsymbol{P}_j}{\left[\alpha \boldsymbol{P}_j + (1-\alpha)\sigma_{ij}\right]}\right)\left(\frac{1}{\left[\alpha \boldsymbol{P}_j + (1-\alpha)\sigma_{ij}\right]}\right) + \\ &+ 2\left(\frac{\boldsymbol{X}^{ro} - \boldsymbol{P}^{ro}}{\left[\alpha \boldsymbol{P}^{ro} + (1-\alpha)\sigma^{ro}\right]}\right)\left(\frac{1}{\left[\alpha \boldsymbol{P}^{ro} + (1-\alpha)\sigma^{ro}\right]}\right)(-1) - \sum_{i=1}^{N_c-1}\psi_i\left(x_{ij} - x_i^{ro}\right) = 0\end{aligned} \tag{13}$$

and Eqs. (6 and 7). Using Eqs. (10-12) and after some manipulations we get

$$x_{i1} = p_{i1} + 0.5\left[\alpha p_{i1} + (1-\alpha)\sigma_{i1}\right]^2\left[\lambda_1 - \psi_i \boldsymbol{X}_1\right] \tag{14}$$

for $i=1,2,\dots, N_c-1$

$$x_{ij} = p_{ij} + 0.5\left[\alpha p_{ij} + (1-\alpha)\sigma_{ij}\right]^2\left[\lambda_j + \psi_i \boldsymbol{X}_j\right], \quad j=2,3,\dots,N_{st}+1 \tag{15}$$

where $\lambda_{Nst+2} = \lambda^{ro}$, for $i=N_c$

$$x_{N_c j} = p_{N_c j} + 0.5\left[\alpha p_{N_c j} + (1-\alpha)\sigma_{N_c j}\right]^2\left[\lambda_j\right], \quad j=1,2,\dots,N_{st}+2 \tag{16}$$

To facilitate the presentation, the following condition is assumed;

$$\sum_{i=1}^{N_c} p_{ij} = 1, \quad j=1,2,\dots,N_{st} \tag{17}$$

and using Eqs. (14-16) and after some manipulations we get

$$\lambda_1 = \boldsymbol{X}_1 \frac{\sum_{i=1}^{N_c-1}\psi_i\left[\alpha p_{i1} + (1-\alpha)\sigma_1\right]^2}{\sum_{i=1}^{N_c}\left[\alpha p_{i1} + (1-\alpha)\sigma_1\right]^2} \tag{18}$$

$$\lambda_j = -\boldsymbol{X}_j \frac{\sum_{i=1}^{N_c-1}\psi_i\left[\alpha p_{ij} + (1-\alpha)\sigma_{ij}\right]^2}{\sum_{i=1}^{N_c}\left[\alpha p_{ij} + (1-\alpha)\sigma_{ij}\right]^2}, \quad j=2,\dots,N_{st}+1 \tag{19}$$

where $\lambda_{Nst+2} = \lambda^{ro}$. It should be noted that, if the conditions in Eq. (17) are not satisfied, they don't cause any loss of generality because Eqs. (18 and 19) can be easily modified.

Demonstration

To visualize how the LSQ and LRSQ adjust raw data let us consider a two-dimensional case. The raw SCs are p_1 and p_2 and the corresponding estimates are x_1 and x_2 as shown in Figure 2. The constraint, $x_1+x_2=1$ is imposed. In the LSQ case the objective function is defined as $\phi=(x_1-p_1)^2+(x_2-p_2)^2$. The optimal solution is located when a concentric circle becomes tangent to the $x_1+x_2=1$ (X_L*). Close observation of the LSQ solution indicates that: a) for all raw points on the lines parallel above and below the line $x_1+x_2=1$ generates the same adjustments and b) if either one of the SCs p_1 or p_2 becomes greater than one the optimal solutions have negative values. Both results are disadvantages. The LRSQ solution is a directed line from the raw data point (p_1, p_2) to the $x_1+x_2=1$ is labeled X_R* (see Figure 2). The X_R* solution indicates that: a) the corrections between X_R* and P are proportional to the ratio of p_1 and p_2 squared, and b) if one of SCs, p_1 or p_2, becomes zero the corresponding optimal SCs, x_1* or x_2*, are close to zero. In this case both results are advantageous. It should be emphasized that Figure 2 is only given as demonstration but no further conclusions can be drawn.

A raw data set collected around the CanmetENERGY-Devon flume unit is used to generate the data presented in Table 1. The normalization of the raw data of Table 1 is given in Table 2. The proposed technique is implemented in spread sheet format using Microsoft Excel^{TM}. Furthermore, the solver module embedded within Excel is used to solve the optimization problem stated in the previous subsection. Using the data in Table 2 and implementing the proposed technique for $\alpha = 1$, i.e., LRSQ case, the results obtained are given in Table 3. The corresponding SC's distributions along the flume are shown in Figure 3. Close examination of Figure 3 indicates that; a) the estimated and the observed data are very close, which reflects the reliability of the raw data, b) naturally, most of the hydrocarbons and solids in the feed slurry are deposited longitudinally along the flume sections. The case $\alpha = 0.8$, i.e., linear combination of LRSQ and LSQ cases. In this case a set of σ_{ij}s is considered and is given in Table 4. The σ_{ij}s of *f* and *ro* streams are two's and other σ_{ij}s are ones. The reason for this selection is that the *f* and *ro* streams produce good representative samples and their analyses are more reliable. Implementation of the proposed technique produced the results given in Table 5. The corresponding SC's are shown in Figure 4. Using the SCs given in Tables 3 and 5, the corresponding cumulative SC recoveries are given in Figures 5 and 6. For $\alpha = 1$ and close examination of Figure 5 shows that ~ 96% of B and M are deposited longitudinally along the flume bed and the remaining ~ 4% are carried over in the *ro* stream. And ~ 20% of water is retained over the flume bed and ~ 80% are discharged from the flume bed. For $\alpha = 0.8$ and close examination of Figure 6 indicates that ~ 13% M and 2% B are discharged and the remaining material are deposited along the flume bed.

Conclusions

Based on the results presented, the following conclusions may be drawn:

- The proposed technique is recommended to handle cases where the raw data statistics (i.e., errors model) are not available.
- The dimension of the optimization problem is reduced to $N_{st}+N_c$ manipulating variables (N_{st} SMSs and N_c constant multipliers) and the $N_{st}+N_c$ equality constraints.
- The resulting balanced data are used to calculate performance measures such as performance (separation) measures.
- Deposition of solids and liquid on inclined surfaces can be modeled using the proposed technique results.
- Note that the mechanical dimensions of CanmetENERGY-Devon flume unit along with mass rate and SCs data can be used in the scale up to industrial oil sand flumes.
- Data can assist the process engineer to analyze/modify his process flowsheet.
- It should be emphasized that the development in this paper is given to demonstrate how the estimated data are related to the observed data in the mass balance of LSQ, LRSQ, and linear combination of LSQ and LRSQ cases.

Finally, it should be emphasized that the present technique can be extended to handle any other general process (using the details given in Salama, 2001 and 1999b).

Acknowledgement

The author would like to acknowledge support from the Government of Canada's interdepartmental Program of Energy Research and Development, PERD POL.

References

1. Draper, N.R. and Smith, H. (1966), "Applied Regression Analysis," John Wiley and Sons Ltd., New York.
2. Fletcher, R. (1987), "*Practical Methods of Optimization*,". John Wiley and Sons Ltd., New York.
3. Frew, J.A. (1983), "Computer-aided design of sampling schemes," *Int. J. Mineral Process.*, 11: pp. 255-265.
4. Hodouin, D. and Vaz Coelho, S. (1987), "Mass balance calculations around mineral processing unit using composition analyses within particle-size classes," *Int. J. Mineral Process.*, 21: pp. 65-82.
5. Jaynes, E.T. (1957), "Information theory and statistical mechanics," *Phys. Review*, 106: pp. 620-630.
6. Jefferson, T.R. and Scott, C.H. (1986), "The entropy approach to material balancing in mineral processing," *Int. J. Mineral Process.*, 18: pp. 251-261.
7. Kapur, P.C., Agrawal, M., and Fuerstenau, D.W. (1993), A two-tier approach to the mass balance of multi-level mineral processing data. Par 1: performance index and local balance around a node. *Int. J. Mineral Process.*, 39: pp. 209-223.
6. Kapur, P.C., Agrawal, M., and Fuerstenau, D.W. (1993), A two-tier approach to the mass balance of multi-level Mineral processing data. Par 2: splicing of stream flows and global balance over a circuit. *Int. J. Mineral Process.*, 39: pp. 225-237.
7. Klimpel, R. (1979), "Estimation of weight ratios given component make-up analyses of streams," *SME/ AIME Trans.*, 266: pp. 1882-1886.
8. Kullback, S. (1959), "Information Theory and Statistics," Dover Publications, New York.
9. Ozyrut, D.B. and R.W. Pike (2004), "Theory and Practice of simultaneous Data reconciliation and Gross Error detection for chemical processes," *Computers and Chemical Engineering.*, 28, pp. 381–402.
10. Salama, A.I.A. (1999a), "Sampling schemes for mass balance of flowsheets with multi-component streams." *Int. J. Mineral Process.*, 55: pp. 219-229.
11. Salama, A.I.A. (1999b), "The maximum entropy approach and comparison with the least-squares approach for flowsheet mass balance," *Int. J. Mineral Process.*, 57: pp. 219-241.
12. Salama, A.I.A. (2001), "Flowsheet Mass Balance with Partial Observation of Stream Components, *Int. J. Mineral Process.*, 63, pp. 59–77.
13. Smith, H.W. and Frew, J.A., 1983. Design and analysis of sampling experiment - a sensitivity approach. *Int. J. Mineral Process.*, 11: pp. 267-284.

Table 1 – A data set generated based on a CanmetENERGY–Devon observed data set

OBSERVED DATA (%)									
S / C	*f*	1	2	3	4	5	6	7	*ro*
Hydrocarbon, B	4.6000	19.6000	19.4000	16.5000	13.4000	11.4000	12.6000	11.6000	0.2800
Minerals, M	6.5000	17.1000	14.1000	17.3000	16.2000	15.6000	15.5000	14.5000	0.3200
Water, W	88.9000	63.3000	66.5000	66.2000	70.4000	73.0000	71.9000	73.9000	99.4000
Sum	100.00	100.00	100.00	100.00	100.00	100.00	100.00	100.00	100.00

Table 2 – Normalization of the data set of Table 1

OBSERVED DATA (FRACTIONAL)									
S / C	*f*	1	2	3	4	5	6	7	*ro*
Hydrocarbon, B	0.046	0.196	0.194	0.165	0.134	0.114	0.126	0.116	0.003
Minerals, M	0.065	0.171	0.141	0.173	0.162	0.156	0.155	0.145	0.003
Water, W	0.889	0.633	0.665	0.662	0.704	0.730	0.719	0.739	0.994
Sum	1.000	1.000	1.000	1.000	1.000	1.000	1.000	1.000	1.000
Mass rate	1200.000	97.660	78.791	66.354	46.934	29.975	14.221	3.079	862.986
Pi's	1.000	0.081	0.066	0.055	0.039	0.025	0.012	0.003	0.719

Table 3– Proposed technique results using the data in Table 2, α=1

ESTIMATED DATA - PROPOSED MASS BALANCE APPROACH										
Y_i's	1.000	0.0835	0.0665	0.0564	0.0397	0.0252	0.0119	0.0026	0.714	-7.201	ψ's
λ's	-7.172	0.586	0.476	0.392	0.276	0.176	0.083	0.018	5.144	-10.539	
Hydrocarbon, B	0.0496	0.190	0.190	0.162	0.133	0.113	0.126	0.116	0.003	0.000	
Minerals, M	0.0498	0.180	0.146	0.179	0.166	0.158	0.156	0.145	0.003	0.000	
Water, W	0.9006	0.630	0.664	0.659	0.702	0.728	0.718	0.739	0.994		
Sum	1.0000	1.000	1.000	1.000	1.000	1.000	1.000	1.000	1.000		
$\partial\mathcal{L}/\partial X$=0		0.000	0.000	0.000	0.000	0.000	0.000	0.000		5.1E-29	ϕ
Stream #	*f*	1	2	3	4	5	6	7	*ro*	Bal 1^2 + Bal 2^2	

Table 4 – Proposed σ_{ij} using LSQ

σ_{ij}									
S / C	*f*	1	2	3	4	5	6	7	*ro*
Hydrocarbon, B	1.000	1.000	1.000	1.000	1.000	1.000	1.000	1.000	1.000
Minerals, M	1.000	1.000	1.000	1.000	1.000	1.000	1.000	1.000	1.000
Water, W	1.000	1.000	1.000	1.000	1.000	1.000	1.000	1.000	1.000
Pi's	1.000	1.000	1.000	1.000	1.000	1.000	1.000	1.000	1.000

Table 5 – Proposed technique results using the data in Tables 2 and Table 4, α=0.8

ESTIMATED DATA - PROPOSED MASS BALANCE APPROACH										
Y_i's	1.000	0.0819	0.0660	0.0557	0.0395	0.0253	0.0121	0.0028	0.717	-0.132	ψ's
λ's	-0.117	0.009	0.008	0.006	0.005	0.003	0.001	0.000	0.087	-0.148	
Hydrocarbon, B	0.0489	0.196	0.194	0.165	0.134	0.114	0.126	0.116	0.001	0.000	
Minerals, M	0.0530	0.172	0.141	0.173	0.162	0.156	0.155	0.145	0.010	0.000	
Water, W	0.8981	0.633	0.665	0.662	0.704	0.730	0.719	0.739	0.988		
Sum	1.0000	1.000	1.000	1.000	1.000	1.000	1.000	1.000	1.000		
$\partial\mathcal{L}/\partial X$=0		0.000	0.000	0.000	0.000	0.000	0.000	0.000		9.7E-31	ϕ
Stream #	*f*	1	2	3	4	5	6	7	*ro*	Bal 1^2 + Bal 2^2	

Figure 1 – CanmetENERGY –Devon Flume

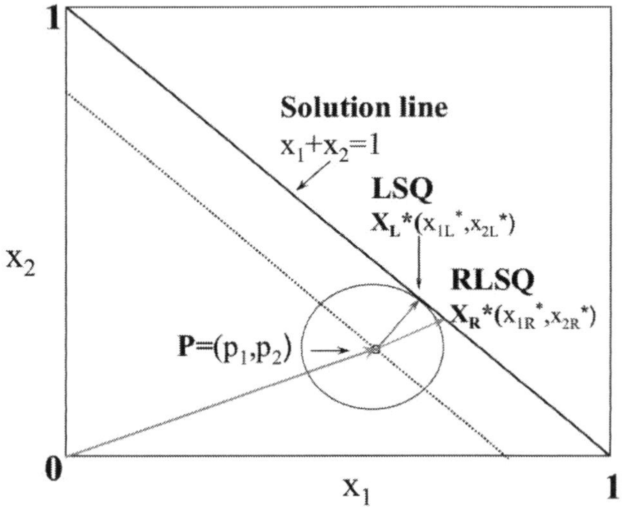

Figure 2 – A two-dimensional demonstration of LSQ and LRSQ formulations

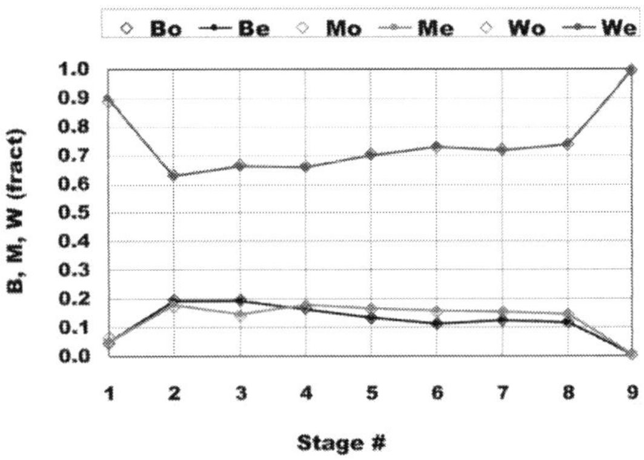

Figure 3 – Estimated SC distribution using the data of Table 2, α=1.0

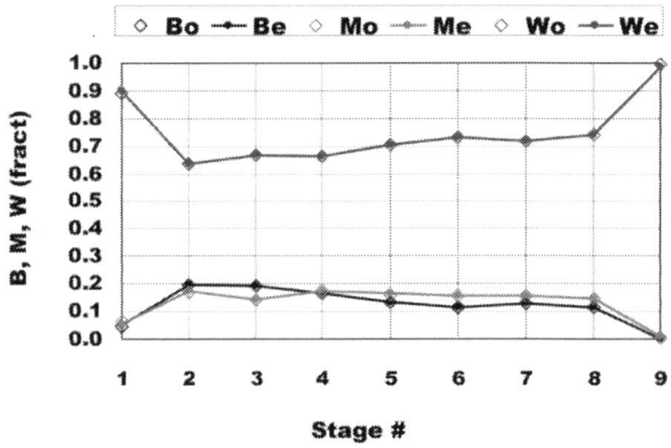

Figure 4 – Estimated SC distribution using the data of Table 2, α=0.8

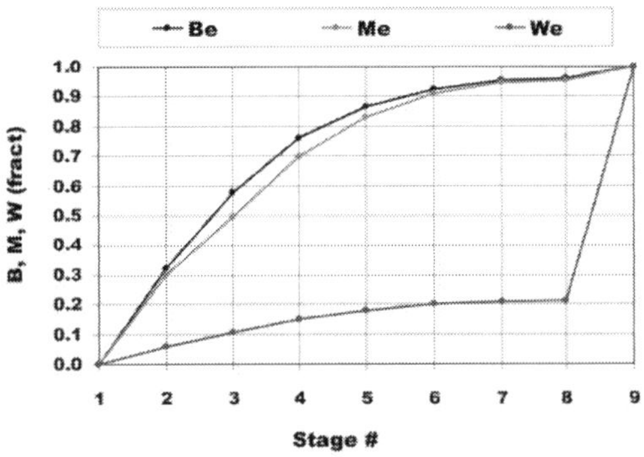

Figure 5 – Estimated SC cumulative recovery distributions α=1

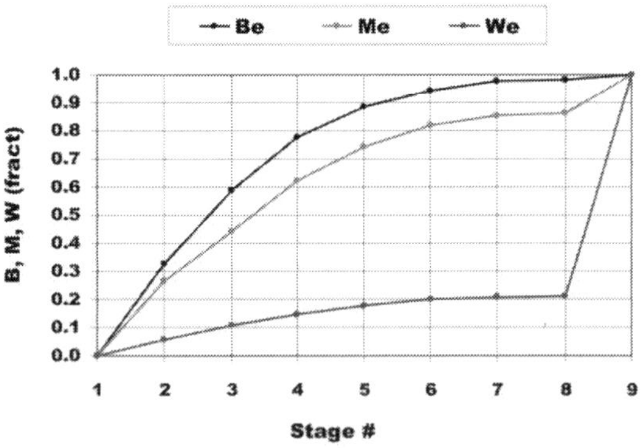

Figure 6 – Estimated SC cumulative recovery distributions α=0.8

(402a) Continuous Pharmaceutical Hot-Melt Extrusion and Hot-Die Face Pelletizing

Daniel Treffer[1], Andreas Eitzlmayr[1], Clemens-Michael Smola[2], Gerold Koscher[2], Eva Roblegg[3] and Johannes G. Khinast[1,2], (1)Institute for Process and Particle Engineering, Graz University of Technology, Graz, Austria, (2)Research Center Pharmaceutical Engineering GmbH, Graz, Austria, (3)Institute of Pharmaceutical Sciences, University of Graz, Graz, Austria

The pharmaceutical industry goes through a paradigm change. So far batch manufacturing was nearly the only way to produce pharmaceuticals. A new approach, however, is continuous manufacturing. Hot melt extrusion (HME) is such a continuous process with significant potential for the manufacturing of different dosage forms with an improved control of the drug release profile[1,2]. HME produces homogeneous strands of molten material, which are further processed in a downstream process. A hot die-face pelletizer is one possible downstream option. It cuts the strands directly at the die plate into pellets. Some authors reported that spherical pellets can be obtained directly with a hot die-face pelletizer[3]. The benefit of spherical pellets is better flowability compared to non-spherical pellets, which is an important requirement for further particle processes. In this work the application of a hot die-face pelletizer as a continuous downstream process is demonstrated.

The experimental setup consists of a co-rotating twin screw extruder ZSK 18 (Coperion, Germany) and a hot die-face pelletizer (Automatik Plastics Machinery GmbH, Germany). Powders or pellets consisting of active pharmaceutical ingredient (API) and excipients are fed continuously to the intake zone of the extruder with loss-in-weight feeders (K-Tron, Switzerland). The extruder itself combines several unit operations, such as conveying, mixing, melting, and degassing within one apparatus. At the extruder outlet the molten material is pushed through an extrusion die and homogeneous strands emerge from the die plate. The strands are cut continuously into pellets with rotating knifes directly at the die plate. An airstream cools the strands before the cutting as wells as the particles after the cutting process. Furthermore, it conveys the particles to the next processing step. In order to ensure good particle qualities the particle size and shape is monitored in-line with the Eyecon (Innopharma Labs, Ireland). Most process parameters are collected with SIPAT (Siemens, Belgium) which enables an efficient process development and monitoring.

Several pharmaceutical polymers and formulations have been investigated for hot die face pelletizing. Not all formulations are suitable for this downstream process. Rheological investigations of the formulations have been performed and approaches to predict applicability of the hot die-face pelletizer are derived.

References:

1. Breitenbach J.: Melt extrusion: from process to drug delivery technology, *European Journal of Pharmaceutics and Biopharmaceutics* 2002, 54, 107-117.
2. Roblegg E, et. al.: Development of sustained-release lipophilic calcium stearate pellets via hot melt extrusion, *European Journal of Pharmaceutics and Biopharmaceutics 2011, 79, 635-645.*
3. Bailleck S; Rein H, Preparation of starch-based pellets by hot-melt extrusion, *European Journal of Pharmaceutics and Biopharmaceutics, 2011, 79, 440-448*

(402b) Predictive Modeling of Hot Melt Extruders

Andreas Eitzlmayr[1], Daniel Treffer[1], Gudrun Hoerl[2], Sarah Windhab[2], Gerold Koscher[2] and Johannes G. Khinast[1], (1)Institute for Process and Particle Engineering, Graz University of Technology, Graz, Austria, (2)Research Center Pharmaceutical Engineering GmbH, Graz, Austria

Introduction

During recent years, extrusion processes increasingly attracted attention in the area of pharmaceutical manufacturing. As a continuous manufacturing operation, extrusion provides the potential of increased efficiency and reduced operation costs together with a steady product quality. Depending on the used materials and conditions, different types of pharmaceutical extrusion processes can be distinguished, as hot melt extrusion, wet extrusion and solid lipid extrusion. In contrast to wet extrusion and solid lipid extrusion, hot melt extrusion is characterized by a temperature level above the melting point of the material. Since the production of solid dispersions and solid solutions can be achieved, hot melt extrusion is a potential pathway to increase the bioavailability of poorly soluble drugs.

The most common extrusion devices are co-rotating twin screws. Compared to different types as counter-rotating twin screws and single screws the co-rotating twin screws are often preferred due to their mixing performance and self cleaning screw profiles.

Although melt extrusion processes are known from plastics industry for many years and spatially resolved CFD (Computational Fluid Dynamics) simulations of extruders screws are already achieved today [1], there is still a lack of predictive tools supporting design, scale-up and optimization of extrusion processes and a high amount of empirical methods and experimental effort is required. Finite element and finite volume CFD methods can be used to solve the hydrodynamics of completely filled screws. Partially filled screw sections cannot be addressed by conventional methods, since their facilities to recover free surface flows are very limited. The interactions of the involved physical phenomena are highly complex and first-principle design methods are not available for extrusion processes.

Methodology

In our work, we are following two different approaches. The first is a one-dimensional (1D) screw model, which describes physical process variables as pressure, material temperature and screw filling ratio along the screw, and follows the idea of a continuous stirred tank reactor cascade with backmixing, as shown by Choulak et al. [2]. This model targets a global, approximate description of the material along the twin-screw. Since a 1D model cannot resolve the three-dimensional (3D) geometry of the screws, it depends on empirical screw parameters which close the gap between 3D geometry and 1D flow description. These parameters can be determined either by experiments or by 3D simulations. The latter is a target of our second approach, which is a spatially resolved screw model based on a relatively new CFD method providing increased potential for problems with complex, moving boundaries as extruder screws.

Results

Results of the 1D model are profiles of filling ratio, pressure, material temperature and power input along the screw. A residence time distribution gives a measure for back-mixing in screw direction. The calculated profiles are tightly related to the geometrical configuration of the screw. Experimental work to validate the results is ongoing.

References

[1] K. Kohlgrüber, *Der gleichläufige Doppelschneckenextruder*, Carl Hanser Verlag, München, 1st edn., 2007, ch. 7, pp. 140 – 142.

[2] S. Choulak, F. Couenne, Y. Le Gorrec, C. Jallut, P. Cassagnau, A. Michel, Generic Dynamic Model for Simulation and Control of Reactive Extrusion, *Ind. Eng. Chem. Res.*, 2004, 43, 7373-7382.

(402c) Foam Granulation Extrusion: A Novel Method to Continuous Wet Granulation of Powder Drug Formulations

Michael R. Thompson[1], Sharleen Weatherley[1], Rohan Pukadyil[1] and Paul Sheskey[2], (1)Dept. of Chemical Engineering, McMaster University, Hamilton, ON, Canada, (2)The Dow Chemical Company, Midland, MI

The present work seeks to demonstrate a new continuous method for wet granulation using an aqueous foam binder to produce stable continuous operations in a twin screw extruder. Furthermore, we endeavor to understand the influence that foam properties have on the performance of a binder during foam granulation.

EXPERIMENTAL: The study was conducted using α-lactose monohydrate (Pharma FlowLac® 100, Meggle) as a model excipient to be granulated in a co-rotating twin screw extruder (ZSE 27HP, Leistritz) with side stuffer at 10-40 kg/h and 220-320 RPM. The extruder barrel was controlled to 25°C using a cooling system. An aqueous binder solution, composed of 4% (w/w) hypromellose (either METHOCEL™ A15PLV, E3PLV, E6PLV or E15PLV; Dow Chemical, Midland, MI), was foamed to 75-95% foam quality (FQ) using a mechanical foam generator and added at 6% or 11% (w/w) to the extruder. Foam properties were evaluated using two shear stability tests. Evaluation of granule properties was done by particle size analysis and determination of the characteristic fracture strength.

RESULTS: The initial studies to realize the method were done with A15 binder. Using a side stuffer (auxiliary feeder affixed to the extruder) it was found that the foamed binder could be accurately metered into the extruder at all operating conditions. A visualization window showed spreading of the foam over the surface of the lactose bed along an axial distance more than 4 L/D. To determine the value of the new method, wet granulation was compared by introducing the same aqueous binder solution at 6% or 11% (w/w) either as a foam (FA) or by direct addition using a pulseless metering pump (LI). The FA process demonstrated excellent stability over different flow rates (up to 40 kg/h) and screw speeds with no surging, attributed to uniform wetting of the powder; no segregated dry regions were noted, unlike by the LI process. Exiting temperatures of granules varied from 23-40°C (FA) versus 23-80°C (LI). Final granule properties were shown to be comparable between the two methods of binder addition though not at similar operating conditions and fewer operating conditions produced good yields with the LI approach. Processing was only found possible for the drier formulation with 6% binder solution using the FA approach.

A second set of studies looked more directly at the influence of the foamed binder on the granulation process to better understand the mechanism. In this case, granulation was done using binders of comparable surface tension but different molecular weights selected from the E-series of METHOCEL products. The half-life drainage time for the low MW binder, E3, was 9±3 min at 75% FQ (70 Pa-s), increasing to 23±2 min at 95% FQ (109 Pa-s); the value in brackets was the apparent viscosity of the foam to indicate its stiffness. For the highest MW binder, E15, its foam half-life was 82±6 min at 75% FQ (103 Pa-s) and 165±12 min at 95% FQ (232 Pa-s). Under shear in a cylinder viscometer, the average collapse rate of foam (based on foam stiffness) was 60% greater using E15 than E3. In comparison, the average collapse rate was 168% higher with foams of 95% FQ versus 75% FQ. The total fraction of particles larger than 1180μm was a good differentiating response variable to determine effectiveness of granulation. This fraction was nominally 54% (75% FQ) to 52% (95% FQ) for E3, and 42% (75% FQ) to 48% (95% FQ) for E15. Flow rate and screw speed were determined to be the most significant factors ($P<0.05$) affecting both particle size and granule fracture strength whereas binder MW had a minor influence on granule properties and foam quality showed no significant influence.

The negligible influence of foam quality on granule properties was reconciled with the earlier mentioned attributes of foam stability and visual observations of the process to present a proposed two-stage nucleation regime inside the extruder. The talk will present this nucleation mechanism and how foam properties can control the spreading length of the binder inside the extruder.

CONCLUSIONS: A robust continuous process for wet granulation using a twin screw extruder was presented using a foamed binder. Lower material temperatures resulted and a broader set of operating conditions were possible using the foamed binder in comparison to direct liquid addition to the extruder. The negligible influence of foam quality to granule properties points to a consistent wetting mechanism where the foam collapses at a similar spreading length regardless of foam stiffness as it is introduced to the extruder. This regulation of wetting rate is related to the environment of the extruder where foams of lower stiffness favor collapse over stagnant solids and foams of higher stiffness, near the high shear planes of the barrel. In comparison to foam properties, operating conditions (flow rate and screw speed) most strongly affected granule properties suggesting a mechanical dispersion dominated nucleation mechanism.

(402d) Powder Feeder Pairing with Downsteam Continuous Mixer

William E. Engisch Jr. and Fernando J. Muzzio, Chemical and Biochemical Engineering, Rutgers, The State University of New Jersey, Piscataway, NJ

In pharmaceutical powder handling, as well as other continuous powder processing industries, the challenge to consistently feed powder can be one of the most important parts of the overall processing. Inconsistent flow from a powder feeder can lead to inconsistencies in subsequent unit operations, such as mixing. There are methods to improve feeding consistency, but due to the intrinsic nature of powder, feedrate variability cannot be entirely removed. The dilemma then arises of what is an acceptable level of variability for the process.

The focus of this work is developing a method to pair a subsequent unit operations with a powder feeder or vice versa. By using pulse experiments to determine the characteristic residence time distributions (RTD) of a mixer, it is possible to develop a transfer function model that can be used to predict the effect of fluctuations and/or setpoint deviations in an inlet feed stream. By changing the configuration or design of either the mixer or feeders, it is possible to greatly improve overall performance of the feeders-mixer system.

(402e) Powder Lubrication in a Continuous Mixing Process

Daniel O. Blackwood[1], Bruce C. MacDonald[1], Joseph Kushner IV[1], Bruno C. Hancock[1] and Yang A. Liu[2], (1)Drug Product Design, Pfizer Worldwide Research and Development, Groton, CT, (2)Analytical Development, Pfizer Worldwide Research and Development, Groton, CT

In traditional batch processes, a lubricant, such as magnesium stearate, is incorporated into the blend during the final processing step before tablet compression. Recent models, from Kushner and Moore, have proposed a process scaling parameter for a number of different formulations to facilitate scale up of these batch-based processes. In the current work, a series of powder lubrication experimental trials have been completed using an in-line powder mixer. The influence of mass throughput rate, hold up mass of powder in the mixer, and the impeller rotational speed is studied. A new process scaling term, based on the tip speed of the impeller and the mean residence time is proposed for this continuous mixing process and compared to the model proposed by Kushner and Moore for batch processes. Additionally, two new sensors (NIR and torque) are evaluated as part of these experimental trials and introduced as potential on-line PAT techniques for monitoring powder lubrication in a continuous mixer.

(402f) Effect of Paddle Shape On Flow of Powder Through Feed Frames On Rotary Tablet Presses

G. Singh[1], J. Givand[2], P. Rajniak[2], R. Chern[2], S.D. Reynolds[2] and I.C. Sinka[1], (1)University of Leicester, Leicester, United Kingdom, (2)Merck, Sharp & Dohme, West Point, PA

Effect of paddle shape on flow of powder through feed frames on rotary tablet presses

G. Singh*, J. Givand**, P. Rajniak**, R. Chern**, S.D. Reynolds** and I.C. Sinka*

*Department of Engineering, University of Leicester, University Road, Leicester, LE1 7RH, UK
**Merck, Sharp & Dohme, Sumneytown Pike, West Point, PA19486, USA

A commercial scale high speed rotary tablet press can produce in excess of 0.5 million pharmaceutical tablets per hour. This involves the same number of die fill operations, which must be executed consistently to ensure weight and content uniformity from tablet to tablet. High productivity can only be achieved if the formulation of the powder and the selection of process parameters are optimised.

Die fill is a complex and dynamic process and consists of a number of distinct stages that present unique features: delivery of the powder to the press (in batches or continuously), powder flow from hopper to feed frame, powder flow through feed frame to die opening, and die fill. In this paper we examine the flow of powders through a lab scale rotary feeder which was developed at the University of Leicester to explore the mechanisms present in the feed frames of rotary tablet presses. The feeder pictured below consists of an inlet hopper, a disk shaped frame housing a paddle wheel and a powder outlet.

The flow regimes were examined for a poorly flowing placebo formulation and the following parameters were considered:

· Paddle wheel speed: 25, 50, …, 400 rpm

· Outlet diameter: 5, 10, 15, 20 mm

· Shape of paddle wheel spoke cross section: round, square and inclined

It is shown that the powder delivery is not always as continuous as the figure below implies and that poor flowing materials can perform surprisingly well under the conditions imparted by the rotary feeder. The results are quantified in terms of mass flow rate through the feeder, which increases and saturates as the freed frame speed is increased. It was interesting to note, however, that the mass flow rate per unit area of the outlet can increase or decrease as the outlet diameter is reduced, depending on powder. For all cases a significant effect of the shape of the paddle cross section was observed.

Experimental set-up of lab scale rotary feeder

(402g) Supply Chain Analysis of Continuous Pharmaceutical Manufacturing

Arul Sundaramoorthy, James M.B. Evans and P.I. Barton, Department of Chemical Engineering, Massachusetts Institute of Technology, Cambridge, MA

In traditional pharmaceutical supply chains, production takes place in two major steps: 1) primary production, and 2) secondary production. In primary production, the raw materials are converted into active pharmaceutical ingredients (APIs) through multiple intermediate production stages. It is common in practice to carry out the first few stages of production in one location, and the other stages in a different location. In secondary production, the APIs are formulated into different drug products (DPs) through various unit operations such as granulation, molding, coating, and packaging. The secondary production facilities are usually located closer to the major markets. Finally, the DPs are supplied to warehouses or retailers for distributions. The above traditional supply chain design has several disadvantages such as high costs of inventory and logistics, long supply-chain cycle times, high degree of supply-chain disruptions, and less resilience. Furthermore, the pharmaceutical industry has been dominated by the batch mode of manufacturing for decades, which greatly contributes to the above inefficiencies in the supply chain. Recently, the Novartis-MIT Center for Continuous Manufacturing has embarked on an innovative project to shift from the batch mode of manufacturing to the continuous mode, where raw materials through the APIs to the finished products are processed seamlessly in an integrated facility. Such an integrated end-to-end production scheme is more promising for several reasons: a) low costs of inventory, logistics, and production, b) short supply-chain cycle times, c) less exposure to supply-chain disruptions, and d) resilient supply chain. Thus, the option of continuous manufacturing brings in additional challenges and opportunities in the context of pharmaceutical supply chains.

In this work, we develop an optimization framework to analyze various supply chain configurations based on continuous production facilities. Since the option of storing APIs contributes to increased degrees of freedom, we also analyze supply chain configurations based on the decoupled production scheme, where primary and secondary production are separated by storage units. However, the mode of production in the decoupled scheme is still continuous. We consider three different supply chain configurations: 1) centralized configuration, where APIs and DPs of all products are produced in a single location, 2) semi-centralized configuration, where APIs of all products are produced in one location, and DPs are produced in different locations closer to the markets, 3) localized configuration, where both APIs and DPs of various products are produced in different locations closer to the markets. We present a mixed-integer linear programming (MILP) model for the analysis of the various supply chain configurations. The key decisions of the proposed framework are which production scheme is economical, which supply chain configuration is efficient, and what should be the sizes of facilities given the demands of products over the planning horizon. We illustrate the application of the proposed framework using three different drug molecules of different dosage levels and different demand volumes.

(408a) Stimulus-Responsive Controlled Release System by Covalent Immobilization of Proteins Into Mesoporous Silica Nanoparticles

Kai H. Griebenow, Chemistry, University of Puerto Rico, Rio Piedras, San Juan, PR

Modern treatment options in cancer and other therapies increasingly rely on passive and active targeting of the drug. In this context, mesoporous silica nanoparticles (MSN) have emerged as an attractive class of drug delivery carriers for therapeutic agents. We recently explored the covalent immobilization of model proteins into MSN to generate stimulus-responsive controlled release systems. The model proteins carbonic anhydrase (CA) and cytochrome c (Cyt-c) were coupled to sulfosuccinimidyl 6-[3'(2-pyridyldithio)-propionamido]hexanoate (Sulfo-LC-SPDP) and subsequently covalently immobilized into MSN via thiol-disulfide interchange. The model proteins could be released from MSN with 10 mM glutathione which represents intra-cellular redox conditions while it remained bound to the MSN at extra-cellular redox conditions represented by 1 μM glutathione. Bioactivity of the model proteins was maintained well during immobilization and release. Studies using human cervical cancer (HeLa) cells showed good biocompatibility. MSN containing immobilized Cyt-c induced apoptosis in HeLa cells. We will also highlight recent data on endocytosis of Cyt-c loaded MSN and endosomal escape obtained by confocal microscopy. Furthermore, we will present data on active targeting using MSN by binding of of suitable ligands.

(408b) Engineering and Characterizing Aggregation-Resistant Antibodies

Joseph M. Perchiacca, Shantanu V. Sule, Ali Reza A. Ladiwala, Jayapriya Jayaraman, Moumita Bhattacharya and Peter M. Tessier, Dept of Chemical & Biological Engineering, Rensselaer Polytechnic Institute, Troy, NY

Protein aggregation is one of the most significant challenges in successfully developing and delivering monoclonal antibodies for therapeutic applications. We are investigating how to engineer the sequences of antibodies to promote high solubility without altering biological activity, as well as how to identify optimal formulation conditions that maximize antibody solubility using high-throughput measurements of antibody self-association. In this presentation, we will first discuss our work on investigating the impact of the complementarity determining regions (CDRs) of antibodies on their aggregation behavior for a library of homologous antibody fragments (Perchiacca et al., Proteins, 2011). Importantly, we find that charged mutations near the edges of hydrophobic CDR loops significantly increase antibody solubility without reducing binding activity. Second, we will discuss our development of a high-throughput method (self-interaction nanoparticle spectroscopy, SINS) for measuring antibody self-association (Sule et al., Biophys J, 2011). We find that SINS can reliably and rapidly identify sequence variants and formulation conditions that minimize monoclonal antibody aggregation.

(408c) Characterization of Monoclonal Antibody Conformations and Self-Associations At High Concentrations Using Neutron Scattering Techniques

Yun Liu[1,2], Eric J. Yearley[1], Isidro Zarraga[3], Norman J. Wagner[1], Thomas M. Scherer[3], Steven J. Shire[4] and Yatin R. Gokarn[3], (1)Chemical and Biomolecular Engineering, University of Delaware, Newark, DE, (2)Center for Neutron Science, NIST, Gaithersburg, MD, (3)Late Stage Pharmaceutical and Processing Development, Genentech Inc., South San Francisco, CA, (4)Pharmaceutical R&D, Genentech Inc., South San Francisco, CA

Monoclonal antibodies (MAb) have become a crucial therapeutic agent. However, some applications such as a subcutaneous injection need highly concentrated protein solutions that can have undesired large viscosity. Therefore, it is of great importance to examine the relation between MAb self-associated nanostructure and its viscosity under various concentration and excipient conditions with techniques which can probe both structure and dynamics that span the length scale of an individual protein to a larger aggregate. We have investigated the protein-protein interactions and diffusive properties of highly concentrated MAbs using small angle neutron scattering (SANS) and neutron spin echo (NSE). Our results indicate that concentration, temperature, pH and surfactant do not have a strong effect on the individual MAb conformation for our MAb samples. The short-range attraction of some MAb proteins is found to be highly anisotropic in contrast to many other protein solutions at low ionic strength. Using neutron spin echo, we have accurately measured the short-time self-diffusion coefficients of MAbs at very large concentrations. One of MAb proteins is found to form small clusters at a wide range of concentrations. We speculate that the cluster formation is thus the underlying cause of the dramatic increase in viscosity as a function of concentration in this specific MAb protein.

(408d) Aggregation of Proteins Studied by Deep UV Resonance and Nonresonance Raman Spectroscopy

Sergey Arzhantsev[1], Connie Ruzicka[1], Vincent Vilker[2] and John Kauffman[1], (1)Division of Pharmaceutical Analysis, US FDA, Saint Louis, MO, (2)Office of Testing and Research, FDA, Silver Spring, MD

Aggregation of proteins is one of the degradation processes which occurs during the storage of pharmaceutical protein products, and can significantly decrease the potency of protein drugs. Several studies have shown that aggregation correlates with changes in the secondary structure of proteins. Raman spectroscopy is a nondestructive analytical method that is sensitive to the secondary structure of proteins. Deep UV resonance Raman (DUVRR) spectroscopy has the potential to be a good alternative technique to UV and vibrational circular dichroism for determination of secondary structure in formulated therapeutic proteins. The resonance excitation increases the Raman signal, specifically enhancing the intensity of peptide vibration bands, and completely removing fluorescence background. The advantage of DUVRR spectroscopy is that it does not usually require any sample preparation and is as sensitive as circular dichroism. We have developed a mathematical algorithm for mixture decomposition in order to remove excipient spectral contributions from the total Raman signal of formulated protein products. This presentation will discuss the use of DUVRR spectroscopy to study the aggregation of insulin and insulin variants, and will compare the capabilities of resonance Raman and nonresonance Raman spectroscopies.

(408e) Reactivity of Polymersome Encapsulated Hemoglobin with Physiologically Important Gaseous Ligands: Oxygen, Carbon Monoxide, and Nitric Oxide

Shahid Rameez[1], **Uddyalok Banerjee**[2], Jorge Fontes[3], Alexander Roth[2] and Andre Palmer[2], (1)Dept. of Biochemistry, The Ohio State University, Columbus, OH, (2)Department of Chemical and Biomolecular Engineering, The Ohio State University, Columbus, OH, (3)Dept. of Chemical and Biomolecular Engineering, The Ohio State University, Columbus, OH

Two distinct preparations of amphiphilic diblock copolymer vesicles (i.e., polymersomes), composed of (poly-(ethylene oxide)−poly(butadiene)) (PEO−PBD), with molecular weights of 1.8 and 10.4 kDa, offering different hydrophobic membrane thicknesses, were used to encapsulate the oxygen (O2) storage and transport protein hemoglobin (Hb) for possible application as a red blood cell (RBC) substitute. Key biophysical properties like p50 and cooperativity as well as the kinetics of polymersome encapsulated Hb (PEH) interaction with physiologically important gaseous ligands (O2, carbon monoxide, and nitric oxide) were measured as a function of the hydrophobic membrane thickness of the PEH particle. Taken together, the results of this work show that PEHs exhibit biophysical properties and retarded ligand binding/release kinetics (compared to cell-free Hb), which are similar to the behavior of RBCs. Therefore, PEHs have the potential to serve as safe and efficacious RBC substitutes for use in transfusion medicine. Detailed studies involving different amphilic diblock copolymers can be conducted in future to explore these possibilities.

(408f) Low-Viscosity Highly Concentrated Dispersions of Stable Protein Nanoclusters for Subcutaneous Injection

Aileen K. Dinin[1], Ameya U. Borwankar[2], Maria Andrea Miller[1], Tarik A. Khan[3], Brian Wilson[1], Kevin Kaczorowski[4], Jennifer A. Maynard[5], Thomas M. Truskett[6] and Keith P. Johnston[7], (1)Dept. Chemical Engineering, University of Texas at Austin, Austin, TX, (2)Chemical Engineering, University of Texas at Austin, Austin, TX, (3)Department of Chemical Engineering, The University of Texas at Austin, Austin, TX, (4)Chemical Engineering, Massachusetts Institute of Technology, Cambridge, MA, (5)Chemical Engineering and Materials Science, University of Minnesota, Minneapolis, MN, (6)Chemical Engineering and Institute for Theoretical Chemistry, The University of Texas at Austin, Austin, TX, (7)Chemical Engineering, The University of Texas, Austin, TX

Stabilizing proteins at high concentration is of broad interest in drug delivery, for treatment of cancer and many other diseases. Proteins have a tendency to undergo irreversible aggregation, gelation or precipitation at high concentrations due to unfolding caused by specific short ranged forces. Herein, we create highly concentrated antibody dispersions (up to 260 mg/ml) comprising of dense equilibrium nanoclusters of protein [monoclonal antibody (mAb) 1B7, polyclonal sheep Immunoglobulin G (IgG) and bovine serum albumin (BSA)] molecules, which upon dilution *in vitro* or administration *in vivo,* remain conformationally stable and biologically active. The nanoclusters are formed by adding trehalose as a cosolute which strengthens the short-ranged attraction between protein molecules. The protein cluster diameter was reversibly tuned from 50 to 300 nm by balancing short-ranged attraction against long–ranged electrostatic repulsion of weakly charged protein at a pH near the isoelectric point (pI). The high protein volume fraction within the cluster stabilizes the protein conformation through a self-crowding mechanism and the primarily repulsive electrostatic interactions between the clusters keep them colloidally stable, preventing gelation. Upon dilution of the dispersion *in vitro*, the clusters rapidly dissociated into fully active protein monomers as shown with biophysical analysis (SEC, dynamic light scattering (DLS), circular dichroism (CD) and SDS-PAGE) and sensitive biological assays. *In vivo* subcutaneous injection into mice results in indistinguishable pharmacokinetics *versus* a standard antibody solution. This tunablility of multi-scale interactions , specifically inter-monomer short range attraction and inter-cluster long range repulsion, to produce nanoclusters is not protein-specific. Therefore this nanocluster concept can be applied to a wide range of therapeutic proteins without the need to engineer modified proteins through changing the amino acid sequence. Stable protein dispersions with low viscosities may potentially enable patient self-administration by subcutaneous injection of antibody therapeutics being discovered and developed.

(423a) Enabling Transcriptional Control in Saccharomyces Cerevisiae Utilizing a Synthetic Hybrid Promoter Engineering Approach

John Blazeck, Department of Chemical Engineering, University of Texas at Austin, Austin, TX and Hal Alper, Chemical Engineering, The University of Texas at Austin, Austin, TX

Well-characterized promoters are essential for pathway engineering and synthetic biology efforts in the model eukaryotic yeast, *Saccharomyces cerevisiae*. Specifically, two main types of promoter elements are required: (1) a series of constitutive promoters exhibiting a dynamic range of expression capacities and (2) well-controlled inducible systems with defined expression outputs. In this talk, we describe recent advances using our synthetic hybrid promoter approach to create a wide range of well-controlled constitutive and inducible promoters in *S. cerevisiae*. Synthetic hybrid promoters are composed of two modular components— the enhancer element, consisting of tandem repeats or combinations of upstream activation sequences (UAS), and the core promoter element. We demonstrate the utility of the hybrid promoter engineering approach with three main case studies. First, we establish a dynamic range of constitutive promoters and in doing so expand transcriptional capacity of the strongest constitutive yeast promoter, P_{GPD}, by 2.5-fold in terms of mRNA levels. Second, we demonstrate the ability to augment regulation through hybrid promoters. Specifically, we regulate constitutive expression of endogenous promoters through the use of a galactose-inducible UAS element and decrease the repressive capacity of the native P_{GAL} promoter through incorporating additional constitutive UAS elements. Third, we create a library of inducible promoters that enable tunable galactose-inducible gene expression based on the strong P_{GAL} promoter with nearly a 50-fold dynamic range and likewise increase the transcriptional capacity of this promoter by more than 15%. Collectively, this work creates the strongest characterized promoters in *S. cerevisiae* and establishes hybrid promoter construction as a generic, synthetic approach to expand, enhance, and control promoter strength in yeast.

(423b) Coordinated Induction of Multi-Gene Pathways in Saccharomyces Cerevisiae

Jing Liang[1], Jonathan Ning[1] and Huimin Zhao[2], (1)Department of Chemical and Biomolecular Engineering, University of Illinois at Urbana-Champaign, Urbana, IL, (2)Department of Chemical and Biomolecular Engineering, Department of Chemistry and Biochemistry, Institute for Genomic Biology, University of Illinois at Urbana-Champaign, Urbana, IL

Bacterial operon has been an important tool in regulating and coordinating multi-gene expression in prokaryotes. It is also a gene architecture commonly employed in the biosynthesis of many pharmaceutically important compounds and industrially usefully chemicals. Despite being an important eukaryotic production host, *S. cerevisiae* never had such gene architecture. Here, we report the development of a system to assemble a multi-gene pathway into an operon-like structure using pre-made parts from a plasmid toolbox. Subsequently, through the use of a yeast strain containing a stably integrated gene switch, we can regulate the assembled pathway using a readily available compound – estradiol – with extremely high sensitivity. To demonstrate the utility of the system, we assembled the 5-gene zeaxanthin biosynthetic pathway in a single step and showed the ligand dependent, coordinated expression of all 5 genes as well as the tightly-regulated production of zeaxanthin. Using a previously reported constitutive zeaxanthin pathway as a benchmark, our inducible pathway is shown to have 50-fold higher production level.

(423c) Quantitative Analysis of Translational Coupling in *Escherichia Coli*

Shuyan Zhang, Chemical and Biomolecular Engineering, University of Illinois at Urbana Champaign, Urbana, IL, Lon Chubiz, Chemical and Biomolecular Engineering, University of Illinois, Urbana, IL and Christopher V. Rao, Chemical and Biomolecular Engineering, University of Illinois, Urbana Champaign, Urbana, IL

Multiple genes are often transcribed on the same mRNA transcript in bacteria. These multi-gene transcriptional units, called operons, are widespread in bacteria with nearly half of all annotated genes predicted to reside in them. As operons tend to contain functionally related genes, they are thought to have evolved for regulatory purposes so that the genes contained in them will have similar transcriptional expression patterns. However, not all co-regulated genes reside in the same operon. Furthermore, alternative hypotheses have been proposed for the existence and prevalence for operons in bacteria, namely that they are the consequence of horizontal gene transfer. One aspect of operon structure is clear: operons link multiple genes to the same promoter and transcriptional regulatory sequences.

How operons affect translational regulation is still not known. On one hand, adjacent genes are often translationally coupled, where translation of the downstream gene is conditional on the translation of the upstream gene. Presumably, this mechanism is used to ensure that these genes are expressed at similar levels. On the other hand, the relative expression of adjacent genes can vary by orders of magnitude with no clear correlation between expression and ordering within the operon. Given these alternatives and everything in between, any modeling approach applied to translational regulation clearly needs to account for operon structure.

As a step towards developing such a model, we have engineered a synthetic two-gene operon, where translation of the downstream gene is conditional on the translation of the upstream gene. This operon encodes two fluorescent proteins, where the upstream gene is translated by orthogonal ribosomes and the downstream gene by native ribosomes. This design allows us to precisely tune the translation of the upstream gene and then record its affect on the translation of the downstream gene. Using this general system, we show that the translation of these two genes can be tightly coupled. We have also determined the factors that determine the degree of coupling between them. A mathematical model for translational coupling is proposed based on our experimental data.

(423d) A Biophysical Model of Structured Ribosomal Standby Sites and Their Effect On the mRNA Translation Rate

Amin Espah Borujeni, Chemical Engineering, Pennsylvania State University, State College, PA and Howard Salis, Chemical Engineering / Biological Engineering, Pennsylvania State University, University Park, PA

Translation initiation is a key rate-limiting step during protein expression. It begins when the ribosome's 30S subunit initially binds to the mRNA at what is now called the ribosomal standby site. After initial contact, the 30S subunit shifts into position over the ribosome binding site, followed by initiation of polypeptide synthesis. We previously developed a thermodynamic model, called the Ribosome Binding Site Calculator, that predicts the Gibbs free energy change when the ribosome binds to the mRNA. Using statistical thermodynamics, this Gibbs free energy change is then related to the mRNA's translation initiation rate. The previous model made strong assumptions about the nature of the ribosomal standby site and how mRNA structures upstream of the ribosome binding site affect the rate of translation initiation. Accordingly, the model was less accurate when structured ribosomal standby sites were present. Structured standby sites are often present within bacterial operons, and can facilitate post-transcriptional regulation of protein expression.

We present an exhaustive quantitative characterization that establishes how a structured ribosomal standby site affects the rate of translation initiation. Using this experimental data, we develop a biophysical model that predicts the Gibbs free energy of ribosome binding to an arbitrarily structured ribosomal standby site. The thermodynamic model accounts for diverse mRNA structures with varying hairpin sizes and folding energies, and proximal and distal standby sites with varying lengths. The model's predictions were experimentally validated by employing flow cytometry to measure fluorescent protein expression from over 90 designed, synthetic mRNA sequences, in addition to mRNA level measurements with RT-qPCR.

We present two key findings: (i) the ribosome's 30S subunit does not unfold strong mRNA structures surrounding standby sites and will instead bind to constrained single-stranded regions via compression; and (ii) the ribosome can bind to distal standby sites and bypass mRNA structures via rolling or sliding to shift into position over a ribosome binding site. Our findings show that the ribosome is a flexible molecular machine that minimizes its binding free energy to mRNA.

(423e) Exploiting Thermodynamics of mRNA Secondary Structure for Antisense Design

Erik Johnson and **Ranjan Srivastava**, Chemical, Materials and Biomolecular Engineering, University of Connecticut, Storrs, CT

The ability to manipulate gene expression is one of the most important techniques in biomolecular engineering, with potential benefits spanning a wide spectrum of fields from gene therapy to metabolic engineering. One specific method of gene regulation, which has proven to be effective in natural and artificial systems alike, is the use of antisense RNA to inhibit protein synthesis. Despite biotechnological advances, the design of effective antisense sequences remains a formidable task. In an effort to remedy this issue, a novel method of predicting effective antisense is proposed. In recent years, the idea that an mRNA strand may not always take the form of a distinct fixed molecular structure has become much more prominent. It is believed that an mRNA molecule may actually be in a state of constant structural fluctuation, transitioning between different conformations near the minimum free energy (MFE) structure, particularly in an ever-changing cellular environment. Analyzing suboptimal mRNA structures with a thermodynamic stability comparable to that of the MFE structure may reveal that certain regions are more "volatile" than others. Since these regions have the ability to change conformation without significantly altering the Gibb's free energy of the entire molecule, they may have more freedom to alter their hydrogen bonding. Therefore, these regions would likely be the most accessible targets for antisense binding because of their incessant making and breaking of intramolecular hydrogen bonds. A computational framework, GenAVERT, was developed using the existing programs, UNAFold and RNAforester. UNAFold is used to predict the minimum free energy structure and a set of suboptimal structures, which are then structurally compared to each other using RNAforester, revealing regions of similarity and dissimilarity. Regions of dissimilarity are proposed to indicate sites in an mRNA structure that would be particularly accessible for antisense binding since they display a theoretically lower level of thermodynamic stability over the suboptimal range of free energy. Validation was carried out by predicting antisense sequences and comparing results to actual sequences found in natural toxin-antitoxin antisense systems of bacteria. GenAVERT predicted that the most volatile regions of those toxin-encoding mRNAs were essentially the same as those of the natural antisense target sites. The significance of these toxin-antitoxin systems is in the necessity for efficient inhibition of protein expression to avoid cell death. Translation levels must be brought to an extremely low level or blocked completely for a cell to continue to carry these suicide genes, therefore their corresponding antisense inhibitors must be exceedingly effective. A successful antisense prediction system should be able to predict sequences similar to these RNA antitoxins after analyzing their correlating toxin-encoding mRNAs. As a benchmark, other currently available programs designed for the development of inhibitory RNA were also used to predict potential antisense for the same mRNA sequences and compared. The results revealed that predicting antisense sequences using the concept of volatility in mRNA secondary structure provided a much more accurate prediction method without requiring that the user have detailed knowledge of rational antisense design.

(423f) Variants of the Quorum-Sensing Repressor Modulated Promoter, Pesar, Alter Amplitude, Signal Sensitivity and Switch-Like Behavior

Jasmine Shong and Cynthia H. Collins, Chemical and Biological Engineering, Rensselaer Polytechnic Institute, Troy, NY

Synthetic biology has advanced to but is also challenged by engineering microbial communities that exhibit coordinated, population-dependent behaviors. Studies of multicellular systems have been conducted incorporating acyl-homoserine lactone (AHL)-based quorum-sensing (QS) systems. To date, the most employed QS regulator has been a transcriptional activator, LuxR, but not limited. Expanding the use of QS components, we focus on a transcriptional repressor, EsaR from *Pantoea stewartii*, identified to repress its cognate DNA binding site, the *esa* box. In addition to protein engineering, promoter engineering also provides an alternative to fine-tune the regulatory systems. We have observed that placing the wild-type P*esa* promoter in a low copy-number plasmid, repression was detected in the absence of AHL. However, when P*esa* promoter is harbored by a high copy-number plasmid, it shows no detectable repression by EsaR. To address this issue, we developed a set of mutant promoters that include additional copies of the *esa* box. We hypothesized that providing two *esa* boxes in one promoter could increase EsaR affinity for the promoter and lead to switch-like behavior in high-copy plasmids. We will show that altering the position of a second *esa* box within P*esa* affects promoter behavior in a manner that depends on its position. In addition to identifying a novel promoter that shows AHL-inducible gene expression in our high copy plasmid, our new promoters show a range of AHL-dependencies and amplitudes. These promoters, in combination with our previously evolved EsaR variants with varying signal sensitivities, represent a new set of tools for enabling cell-cell communication.

(423g) Robustness and Energetic Analysis of an Extended Pluripotency MODEL of Embryonic STEM CELLS Transcriptional Network

Deepak Nagrath, Department of Chemical Engineering, Rice University, Houston, TX, Marco Avila, Massachusetts General Hospital/Shriners Burn Hospital/Harvard Medical School, Boston, MA and Martin L. Yarmush, Center for Engineering in Medicine, MGH/Harvard Medical School, Boston, MA

Embryonic stem cells (ESCs) are pluripotent because they can give rise to cells derived from all three germ layers. ESCs are considered a potential source of cells for human disease therapies due to their limitless capacity for self-renewal and proliferation, and their ability to differentiate into all major cell lineages. Octamer-4 (Oct4), Sox2 and Nanog are important markers of pluripotency, expressed by primitive embryonic cells both *in vivo* and *in vitro*. Oct4 and Nanog expression is downregulated during early differentiation. In embryonic stem cells (ESCs), the stem cell-ness is determined by the expression of three major transcription factors: Oct4, Sox2 and Nanog.

The transcriptional regulatory network model for pluripotency of embryonic stem cells consists of various autoregulatory, feedback and feedforward loops, single input modules and bifan motifs. The ESC regulatory network as seen in figure has multilevel network architecture. It can be divided into three levels: (1) Level-0, transcriptional regulatory circuit with three transcription factors Oct4, Sox2 and Nanog; (2) Level-1, inner core of transcriptional circuity with nine transcription factors Klf4, c-Myc, Zfp281, Sox2, Nanog, Oct4, Rex1, Dax1 and Nac1; and (3) Level-2, combined transcriptional regulatory circuit with target hubs of multiple transcription factors.

Not much has been understood about the occurrence of these motifs and levels in embryonic stem cells network biology. Thus, there has been lack of explanation on how a level architecture in ESC network influences other levels, functional role of various levels, dynamics and functional objectives of ESCs TRN. Moreover, there has been the lack of a regulatory network analysis criterion which can describe the underlying mechanisms behind the occurrence of these layers, motifs, as well as the way in which these motifs play a role in developmental biology and the way that both functional objectives and topological architecture may evolve.

Here, we postulate that the architecture of ESC-TRN can be explained based on a conceptual framework that integrates nonequilibrium thermodynamics with Pareto-optimality of the biological functions to be carried out by the embryonic stem cells. We present an energetic-cost theory that can explain cofactor occupancy as well as the topological arrangement of the ESC network. Through the developed framework, we have tried to answer the questions of why during development certain architecture is favored and what advantage different levels offer for steady state analyses. Our study also demonstrates that ESC network analyses using often-ignored energetics enables identification of new functionalities for various topological arrangements.

(423h) Dose-Dependent p21(Cip1/Waf1) Cell Cycle Regulation

K. Wesley Overton and Clifford L. Wang, Chemical Engineering, Stanford University, Stanford, CA

p21(Cip1/Waf1) is a member of the Cyclin-dependent kinase inhibitor (CKI) class of proteins involved in arresting the cell cycle in response to DNA damage. Previous studies have shown that the N-terminal domain of p21 is involved in binding to and inhibiting the activity of Cyclin-CDK complexes, and the C-terminal domain of p21 binds to and inhibits proliferating cell nuclear antigen (PCNA), a DNA polymerase auxiliary factor involved in DNA synthesis and repair.

Our lab has developed genetic tools and methods to evaluate cellular responses over varying levels of gene expression. Recently, we have developed a library of expression vectors that control the level of protein translation. We have used this vector library to express exogenous p21 levels in an Hct116 human colon cancer cell line in which the endogenous p21 has been inactivated by gene-targeting. We were able to generate a full range of p21 expression levels, including those greater than the p21 levels induced by DNA damaging agents and those below levels observed in undamaged wild-type Hct116 cells. This wide range of expression allows us to monitor the behavior of cells exposed to varying amounts of p21, including levels that are physiologically relevant.

Our results show that cells expressing exogenous p21 are not actively cycling through S phase. To explore the underlying mechanism of p21 expression strength affecting cell cycle arrest, we made two mutant versions of the p21 gene—one that does not bind to Cyclin/CDK complexes (p21 CDK-) and one that does not bind to PCNA (p21 PCNA-). While each of these mutant versions of p21 alters the cell cycle, neither has been able to fully eliminate S phase and arrest the cell cycle in the same manner as non-mutated p21.

(425a) High Throughput Screening the Effects of Antibiotic Delivery Rates On Biofilm Antibiotic Resistance

Jinzi Deng, Department of Chemical, Materials & Biomolecular Engineering, University of Connecticut, Storrs, CT and Leslie M. Shor, Department of Chemical, Materials and Biomolecular Engineering, University of Connecticut, Storrs, CT; Center for Environmental Sciences and Engineering, University of Connecticut, Storrs, CT

Bacterial biofilms are a heterogeneous community of bacteria distributed in a matrix comprised primarily of hydrated polysaccharides. Compared to bacteria dispersed in liquid cultures, biofilm-associated bacteria can tolerate antimicrobial concentrations orders of magnitude higher. Phenotypic differentiation and lateral gene transfer of antibiotic resistance genes in biofilm promotes antibiotic resistance in biofilm. High throughput technologies have been widely used to screen effective antimicrobial concentrations, where biofilms are usually exposed in constant antibiotic concentration or fixed mass of antibiotic. However, biofilms in nature are usually challenged with changing micro-chemical environment due to diffusion limitations and periodic dosing regimes. The effect of antimicrobial delivery rates on antimicrobial resistance in biofilms is infrequently measured. Here we introduce a novel microfluidic approach for high-throughput screening of respiration inhibition of bacteria in a biofilm array morphology (Figure 1). The device geometry and operating conditions allow antimicrobial concentrations and fluxes to vary systematically and predictably with space and time. In effect, one experiment can screen biofilm respiratory responses to many different antimicrobial delivery rates. Our results demonstrate the effect of antibiotic concentration on biofilm respiration inhibition. Biofilm experiencing slowly increasing antibiotic concentrations can withstand higher total concentrations compared with bacteria exposed to rapidly increasing concentrations. We anticipate our approach can be used as a starting point for investigating the effects of antimicrobial delivery rate on development of antimicrobial resistance in biofilms. Longer term, the work may help direct clinical work on effective antimicrobial therapies for the treatment of biofilm-associated diseases.

Figure 1:

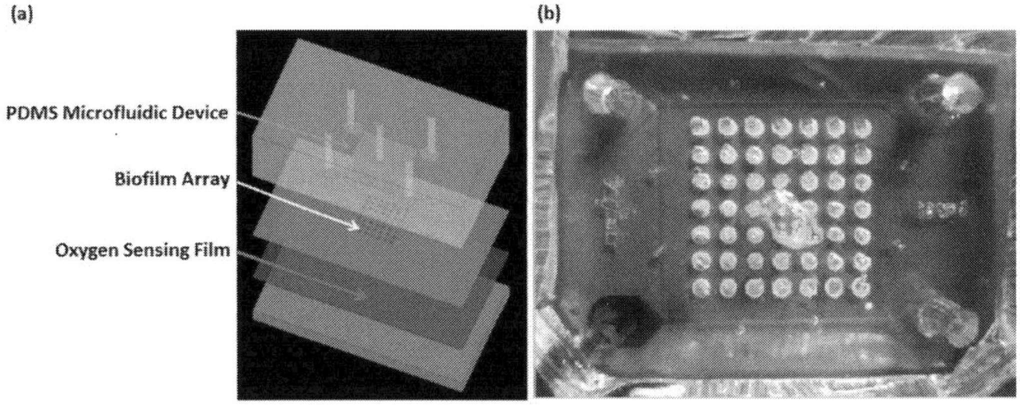

Figure 1. Schematic of the microfluidic diffusion device with biofilm array. (a) The microfluidic device consists of oxygen sensing film on a glass slide, a PDMS layer with a biofilm array patterned on top, and a PDMS microfluidic device bonded on top. (b) Photograph of an array enclosed in a microfluidic diffusion device, with red dye loaded in a perimeter source well.

(425b) High Throughput Genetic Screens of C. Elegans with Microfluidics and Computer Vision

Adriana San Miguel Delgadillo[1], Matthew Crane[2], Peri Kurshan[3], Kang Shen[3] and Hang Lu[4], (1)Chemical & Biomolecular Engineering, Georgia Institute of Technology, Atlanta, GA, (2)Interdisciplinary Program in Bioengineering, Georgia Institute of Technology, Atlanta, GA, (3)Stanford University, Stanford, CA, (4)Chemical and Biomolecular Engineering, Georgia Institute of Technology, Atlanta, GA

C. elegans, a soil dwelling nematode, is a highly studied multi-cellular organism that offers several experimental advantages, including a short life span, ease of culture and transparency, to name a few. Additionally, its full genome and neuronal wiring diagram are known. This, combined with current molecular biological techniques, make it an excellent model system for biological study, especially for neuroscience. Genetic screens in C. *elegans* have led to understanding the function of many relevant genes, but the identification of mutants with significantly altered phenotypes, easily identifiable by simple visual inspection, has reached a saturation point. Aside from the difficulty of typical screens that require manual handling and inspection of a very large number of animals, the current challenge lies on the identification of mutants with very subtle phenotypes difficult to identify by eye. Performing screens based on fluorescent reporters of very small features, such as synapses, present an exceptionally difficult scenario.

Here, we use computer vision algorithms to objectively quantify relevant information from synapses in *C. elegans* from a mutagenized animal population and thus, identify mutants with subtle phenotypes. Additionally, we incorporate microfluidic devices as a platform for automated worm imaging, handling and sorting. Computer algorithms allow the identification of worms which are in the correct orientation and are suitable for imaging, as well as those which have a slightly altered phenotype and are sorted as mutants. Integrating microfluidics and computer vision we have generated a platform for automated high-throughput imaging and sorting of thousands of worms based on quantitative synapse-related features. This method enables imaging, phenotyping and sorting about 100 times faster than manual handling. Not only does this method allow performing genetic screens in a simple, automated and fast manner, but it also provides a platform for discovery of mutants which would otherwise be overlooked in a typical manual screen.

(425c) Live-Cell Screens for Studying Regulatory Networks in Human Mesenchymal Stem Cell Differentiation

Roshan Padmashali[1], Mao-Shih Liang[1], **Panagiotis Mistriotis**[1] and Stelios T. Andreadis[1,2], (1)Chemical and Biological Engineering, University at Buffalo, State University of New York, Amherst, NY, (2)Center of Excellence in Bioinformatics and Life Sciences, Buffalo, NY

Mesenchymal stem cells (MSCs) are abundant and multipotent cells with great potential for regenerative medicine and tissue engineering. The conventional tools for investigating gene expression such as qPCR are laborious, cost intensive and they cannot provide functional information on the pathways involved in biological processes. On the other hand, LentiViral microArrays (LVA) provide live-cell, high-throughput monitoring of activation of biological pathways during biological processes such as MSC differentiation. In this paper, we employed LVA with a library of lentiviral vectors to monitor pathway activation during MSC differentiation into bone, fat or cartilage. In addition, we hypothesized that use of LVA in combination with a small molecular chemical library might identify pathways that control MSC fate commitment.

To address this hypothesis, first we engineered a dual-promoter lentiviral vector—LVDP containing a specific transcriptional regulatory element encoding for green fluorescence protein for monitoring gene regulation. Additionally this vector contains a constitutive promoter driving red fluorescence protein providing an internal normalization strategy, thereby allowing measurements of gene or pathway dynamics independent of the gene transfer efficiency. We developed a library of lineage specific reporters containing promoters such as adiponectin, SOX9 and Runx2 for adipogenesis, chondrogenesis and osteogenesis, respectively as well as transcription factor response motifs e.g. Vitamin D, SMAD, C/EBP etc. By employing automated fluorescence microscopy, we determined the kinetics of lineage-specific pathway activation in MSC derived from two different anatomic locations i.e. hair follicle and bone marrow, (Fig 1(a)). Furthermore, we screened a small array of chemical inhibitors to monitor the pathways that may control differentiation into one or more lineages.

Our results demonstrate differential reporter activation between the two MSC sources, suggesting that the regulatory pathways controlling differentiation may be dependent on the source of MSC. To identify key molecules affecting MSC differentiation, several inhibitors were selected to illustrate the effects on lineage specific reporters. We found that inhibition of TGF-β receptor promoted adipogenesis as evidenced by increased activity of the adiponectin promoter. Similarly, inhibition of p38 increased the activity of both chondrogenic (SMAD2/3) and osteogenic (OCBOX) reporters, while inhibition of TGF-β or FGF receptors down-regulated chondrogenic or osteogenic reporters, respectively. Inhibition of the PI3 kinase and glucocorticoids readily suppressed the activity of adipogenic pathways. Also, inhibition of JNK and ERK blocked the expression of osteo-specific reporters such as Vitamin-D and OCBOX. Finally, we identified novel pathways differentially regulating MSC differentiation along different lineages. The complete results were summarized in Fig 1(b).

We developed a LVA system that enabled efficient monitoring of gene and pathway activation in MSC differentiation. This approach also allowed us to identify differences in the kinetics of pathway activation between different MSC lines, despite their similar differentiation potential. Notably, the LVA enabled screening of small molecule libraries that identified pathways that were necessary for differentiation along particular lineages. Our results suggest that the LVA is a tool with great potential to enhance understanding of stem cell fate specification as well as pathway regulation and drug discovery.

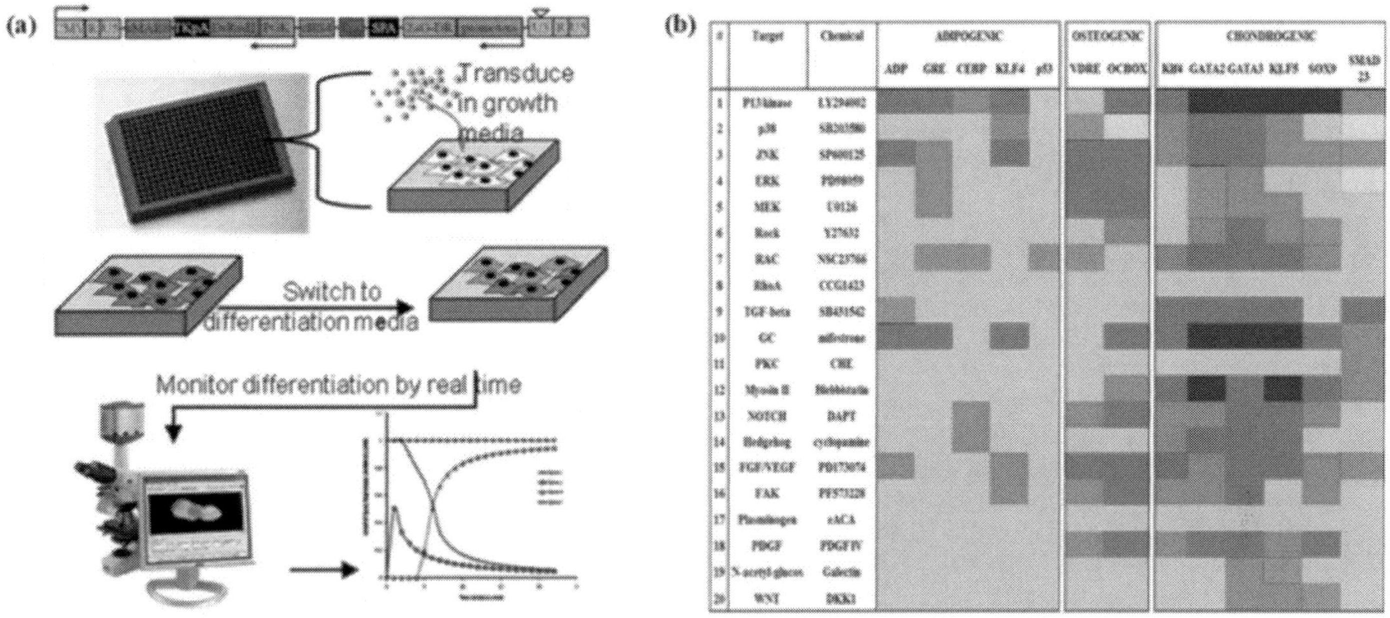

Figure 1 (a) In the experiment setup, MSCs were transduced with various LVDP viruses and the dynamics of lineage-specific reporters were monitored real-time with fluoresunce microscopy during differentiation. (b) Heat map summarizing the effects of chemical inhibitors on lineage-specific reporters during differentiation.

(425d) Scaling Down the Size and Increasing the Throughput of Glycosyltransferase Assays: Activity Changes On Stem Cell Differentiation

Shilpa A. Patil[1], E. V. Chandrasekaran[2], Khushi L. Matta[2], Abhirath Parikh[1], Emmanuel S. Tzanakakis[1] and Sriram Neelamegham[1], (1)Chemical and Biological Engineering, State University of New York at Buffalo, Buffalo, NY, (2)Cancer Biology, Roswell Park Cancer Institute, Buffalo, NY

Glycosyltransferases (GlycoTs) catalyze the transfer of monosaccharides from nucleotide-sugars to carbohydrate, lipid and protein based acceptors. We examined strategies to scale down and increase the throughput of glycoT enzymatic assays since traditional methods require large reaction volumes and complex chromatography. Approaches tested used: i) microarray pin-printing. an appropriate method when glycoT activity was high; ii) microwells and microcentrifuge tubes, a suitable method for studies with cell lysates when enzyme activity was moderate; or iii) C_{18} pipette tips and solvent-extraction, a method that enriched reaction product when the extent of reaction was low. In all cases, reverse phase-thin layer chromatography (RP-TLC) coupled with phosphorimaging quantified reaction rate. Studies with mouse embryonic stem cells (mESCs) demonstrate an increase in overall β(1,3)galactosyltransferase and α(2,3)sialyltransferase activity, and a decrease in α(1,3)fucosyltransferases when these cells differentiate towards cardiomyocytes. Enzymatic and lectin binding data suggest a transition from Le^X type structures in mESCs to sialylated Galβ1,3GalNAc type glycans upon differentiation, with more prominent changes in enzyme activity occurring at later stages when embryoid bodies differentiated toward cardiomyocytes. Overall, simple, rapid, quantitative and scalable glycoT activity analysis methods are presented. These use a range of natural and synthetic acceptors for the analysis of complex biological specimen that have limited availability.

(425e) The p53Pro72Arg Polymorphism Specifies a Balance Between Stem Cell Renewal and Cancer Progression

Marc R. Birtwistle[1,2], Emily Cloessner[2], Josephine Tidwell[2], Megan Clendenning[2] and Philip Buckhaults[2], (1)Pharmacology and Systems Therapeutics, Mount Sinai School of Medicine, New York, NY, (2)Cancer Research Center, Georgia Health Sciences University, Augusta, GA

We studied the common Pro72Arg polymorphism of the p53 transcription factor, which is mutated or lost in over half of human cancers, and found that the Pro/Pro genotype, which is enriched in African American populations, is significantly associated with a 6 year decrease in the mean age of colon cancer diagnosis, independent of race. To explore the hypothesis that Pro- and Arg-specific gene expression patterns may contribute this decrease, we used somatic cell knockout techniques to construct p53 hemizygous or knockout variants of the RKO colon carcinoma cell line, subjected these cells to the DNA damaging agent etoposide, and then performed next generation RNA sequencing experiments. Our analysis of the results revealed not only a comprehensive list of 632 p53-dependent transcripts, but also 83 Arg-specific and 65 Pro-specific transcripts. Bioinformatic analysis showed that Pro-specific gene expression patterns are significantly associated with poorly differentiated cancer and stem cell function, whereas those of Arg are associated with differentiation and downregulation of DNA repair. The transcriptional co-repressor PRDM1b, which recruits H3K9 and H3K27 trimethyl repressive chromatin marks and is necessary for proper differentiation of stem cells, was identified and validated as an Arg-specific target. Taken together, our results suggest that the p53 codon72 polymorphism defines a balance between increased protection from cancer at the expense of stem cell number and longevity.

Finding Murine t-Cell Receptor Repertoire Shifts Due to Ovalbumin Challenges Using High Throughput Sequencing

Zachary P. Frye, Dept. Chemical Engineering, University of Texas at Austin, Austin, TX, Benjamin M. Roy, Dept. Chemical Engineering, University of Texas at Austin, Austin, TX and Jennifer A. Maynard, Dept. Chemical Engineering, University of Texas at Austin, Austin, TX

Abstract

Thymocyte driven cellular immunity utilize membrane-bound receptors to recognize specific disease antigens and initiate the signaling cascade to provide the appropriate immune response. An individual may have billions of thymocyte cells (t-cells) each with a unique receptor comprise their repertoire and with newly developed high throughput sequencing (HTS) technologies these expressed protein receptors can be sequenced. Analyzing the shifts in t-cell receptor (TCR) repertoires due to disease challenges can elucidate how immune systems respond and adapt to cues from the environment. The TCR is formed from specific chromosomal regions surrounding a highly variable, antigen interrogating portion of the receptor. Challenging an individual with a particular antigen known to give a specific immune response is thought to shift or expand the repertoire to include new TCRs directly targeting the foreign disease antigen. These responses are observed as changes in the distribution and diversity of TCR sequences when compared with repertoires in healthy and unchallenged individuals.

HTS of the TCR mRNA harvested from sorted $CD4^+$ & $CD8^+$ t-cells shows the conserved nature of endogenous and exogenous antigen responses and how the immune system selects different t-cell lines to combat diseases and antigens. Examination of the TCR-gene recombinations, VJ-regions in the α-chain and VDJ-regions of the more complex β-chain, gives insight into the mechanisms of t-cell maturation and selection process. This can also explain evolutionarily how the modern immune system has developed whether through co-evolution with common diseases and viruses or through other mechanisms.

Preliminary experimental data show proper recognition of utilized genes and recombination sequence with distinct TCR repertoires existing in the control groups and cell lines. Presently, error rates in the 454 sequencing reactions and t-cell gene recognition recover approximately 40% of the all TCR specific sequences harvested from lymph node t-cells. Developing better mRNA isolation techniques and sequence analysis algorithms to increase recognition accuracy and anticipate erroneous miscalled nucleotides will increase the total number of functional sequences collected in the experiment and create a platform technology for analyzing and characterizing the immune system on an individual basis.

Background

The human body's immune system consists of a complex network of cells proactively monitoring and responding to disorders, infections, and diseases. The more prominent Cytotoxic T Lymphocytes (CTL) cells, Natural Killer, $CD4^+$, and $CD8^+$ cells, play an important role in the short- and long-term immune response to infections. New infections inducing a primary immune response cause the maturation, expansion, and proliferation of naïve $CD4^+$ and $CD8^+$ cells to create a repertoire of CTL and effector cells with antigen sensing αβ receptors and other pertinent immune receptors in an attempt to produce a subset of cells which recognize the peptide antigen produced by the disease and presented by the major histocompatibility complex (MHC). These cells have a unique αβ dimer receptor selected to recognize almost exclusively a particular disease-peptide presented by antigen presenting cells (APC). Successful CTL interaction initiates signaling for further downstream immune responses leading to fighting and clearing the disease.

Structurally, TCRs are highly analogous to antibody Fab fragments, consisting of an alpha-beta hetero-dimer, each of which includes a conserved constant domain and a less conserved variable domain. Like antibodies, the variable domain consists of a beta sheet structure with three hyper-variable loops, termed complementarity determining regions (CDRs), which are responsible for direct contact with ligand. The MHC molecule is also a hetero-dimer with discrete constant and variable domains, but here, the variable domain presents a grooved surface to accommodate a wide variety of short peptides (9-17 amino acids in length) produced by intracellular digestion of native or exogenous proteins. In general, CDR3 loops are the most variable, as considerable sequence and structural diversity is required to recognize the vast array of potential peptides.

The development of the immune system specifically the t-cell repertoire is an important process that protects an individual from environmental challenges. Measuring the diversity, distribution, and frequency of the t-cell receptor sequences with high throughput sequencing technology allows for a deeper analysis of the various TCRs selected to specifically target a given disease antigen. Analyzing repertoire formation mechanisms will allow for the better development of vaccines and cellular therapy techniques.

Methods

Mouse inoculation and t-cell collection

Four 25-27 week old BALB/C mice immunized with 100 μg of OVA. Two weeks after inoculation, their spleen and lymph nodes were harvested collecting 100 mg of organ material, which was then placed in a solution of RNAlater for one hour. The tissue was then homogenized the organs using a sonicator at the highest setting for 1 minute to suspend individual t-cells. The suspended lymph node cellular solution was then sorted using magnetic Dynabeads (Invitrogen #114.45D, 114.47D) for separating $CD4^+$ and $CD8^+$ t-cells by their respective markers.

Primer Design

Primers were designed to amplify the TCR mRNA using nearly consensus sequences in both the upstream V-region and constant region. This was accomplished by aligning the germline sequences of the various V-region alleles and making degenerate primers to incorporate the various sequences into the primer set for the amplification of the gene region. Table 1 below shows the 49 primers used in this work to recover and isolate the TCR sequences. These primers were tested and validated on mouse lymph nodes, spleens, CTL hybridomas and subsequently sorted $CD4^+$ and $CD8^+$ CTLs as shown in the gel pictures in Figure 1 (A – I). In optimizing the cDNA PCR amplification a variety of temperatures were tested in their ability to effectively amplify each individual V-region. The primers were further validated by TOPO cloning and sequencing the amplicons from the individual V-region PCRs. The quality scores of the bands in the gel are assigned based on observations in what the amplicon band(s) looked like. A strong single band received a **(0)**, likewise a weak single band **(1)**, a doublet **(2)**, a triplet **(3)**, four or more bands **(4)**, and a smear received the worst score of **(5)**.

Table 1: Detailed list of the V-region primer set used to amplify the cDNA. The columns are: primer name, V-region gene targets, melting temperature in °C, nucleotide sequence, validated by high throughput sequencing, validated with TOPO cloning, and amplicon quality at 50°C and 53°C annealing temperature respectively.

Primer	Gene Targets	Tm (°C)	Sequence	HTS	TOPO	50°C	53°C
TRAV01	1	52.7	GGT TTG TTC TCT ATC TCT TCC TG		X	2	5
TRAV02	2	53.7	ATG AAG CAG GTG GCA AAA G			0	5
TRAV03	3-1,3,4; 3D-3	54.3±0.4	GAC CTT TGT TSC TGT GCT TC	X	X	0	2
TRAV04	4-3,4; 4D-3,4	55	GCT GTG CTG GGG ATT CT	X		2	5
TRAV04-2	4-2	59.2	CTG CTG TTG GTG CCG CT	X		4	2
TRAV05-1	5-1	54.8	CTT CTG GCT ACA GAT GGA CT			0	5
TRAV05x-4	5-4;5D-4	51.5	ATT ATT CAT GTT TCT ATG GCT GC			0	5
TRAV06-1	6-1:3; 6D-3	53.8	GCT TTA GTG ACT GTG ATG CTG	X		2	2
TRAV06-4	6-4; 6D-4	56.5±2.1	GTT TTA GTR ACT GYG ATG CTG CTG	X		2.5	4
TRAV06x-5	6-5; 6D-5	55.4	CCT GAA CTG GGK ATT CTA CTC TTC	-	-	-	-
TRAV06-6	6-6; 6D-6	54.4±0.8	CTY TTC TCC AGG CTT CGT G		X	2	5
TRAV06-7	6-7; 6D-7	52.5±1	ATG AYT GTG ATG CTC CTC AT	X		0	5
TRAV07	7-1:5; 7D-2:5	54.3±0.3	CTA GTG GTC CTG TGG CTY	X		2	5
TRAV07x-6	7-6; 7D-6	52.2	CTA GTG TTC CTC TGG CTT C			0	4
TRAV08x-1	8-1; 8D-1	53.9	GGT TGT TGA TGG TGT CAC TG		X	2	3

TRAV08x-2	8-2; 8D-2	52.8	GGA ATA TCT TTG GTG ACT CTA TGG		X	0	2
TRAV09x-1	9-1; 9D-1	54.3	TCG TTC CTC GGG ATA CAT TTC			0	5
TRAV09x-2	9-2:4; 9D-2:4	53.7±0.5	CAG TSC TGG GGA TAC ACT	X	X	0	5
TRAV10	10; 10D	54.5	ACA TCC CTT CAC ACT GTA TTC C			0	5
TRAV11	11; 11D	57.9±1.5	GCC TKA GTG CCT GCT GG	X		2	3
TRAV12	12-1:3; 12D-1:3	56.5±1.2	SCT CAG TTC TYG TGC TCC TC	X		0	5
TRAV13	13-1:5; 13D-1:4	56.9±2.6	TCT CTG YTG GGG CTY CTG	X		5	5
TRAV14	14-1:3; 14D-1:3	54.5±1.5	GAC AMG ATC CTG ACA GCA WC	X		2.5	5
TRAV15	15-1:2; 15D-1:2	57.1±0.3	ATG CCT CCT CAS AGC CTG	X		3	4.5
TRAV16	16; 16D	54.2±1.1	CTG ATT CTA AGC CTG YTG GG			1.5	4
TRAV17	17	55	AGT GAC CAT TCT GCT GCT C			0	4
TRAV18	18	54.5	ATG CTC CTG AAA CTC TCT GTG		X	1.5	4
TRAV19	19	54.4	GCC TTG CTG TTG GTT CTG			3	5
TRAV21	21	54.1	GAT GTG TGA GTG GAA TTG CC		X	0	5
TRBV01	1	56.6	TTC TGT GCC TCT GTG TAC TCA T			0	0
TRBV02	2	56.6	GGC TCC ATT TTC CTC AGT TGC	X	X	0	0
TRBV03	3	53.9	CTG GCT TCT AGG TTG GAT AAT TTT TAG			0	3
TRBV04	4	56.6	GCT GTA GGC TCC TAA GCT GT			0	4
TRBV05	5	55.4	GCA GGC TTC TCC TCT ATG TTT C	X	X	0	4
TRBV12	12-1:2	56.4±1.5	CTG CTA TCT TGG GTT RCT STC TTT C			0	3
TRBV13	13-1:3	56.2±2	RCA AGG TGR CAG TAA CAG GAG	X	X	2	0
TRBV14	14	56.7	CTT GGC TGG GCA GTG TTC	X		1.5	0
TRBV15	15	54.2	AGA CCC TCT GTT GTG TGA TC			0	0
TRBV16	16	56.4	CTT TTC TGT CTG GTT CTT TGC TTC		X	0	0
TRBV17	17	55.4	CTA GAC TTC TTT GCT GTG TGA TCT TC			0	3
TRBV19	19	56.1	ATG AAC AAG TGG GTT TTC TGC TG			1.5	4
TRBV20	20	55.2	ATG TTM CTG CTT CTA TTA CTT CTG G		X	0	0
TRBV23	23	53.7	CAC GGC TCA TTT GCT ATG TAG			1	0
TRBV24	24	55.4	CAA GAC TGC TCT GCT GTG TAG			1	0
TRBV26	26	56.7	ATG GCT ACA AGG CTC CTC TG			2	5
TRBV29	29	56.6	TTA GGC TCA TCT CTG CTG TGG	X		2	2
TRBV30	30	53.5	GAC ATT CCT GCT ACT TCT TTG G		X	2	4
TRBV31	31	56.5	TGC TGT ACT CTC TCC TTG CC		X	1.5	4

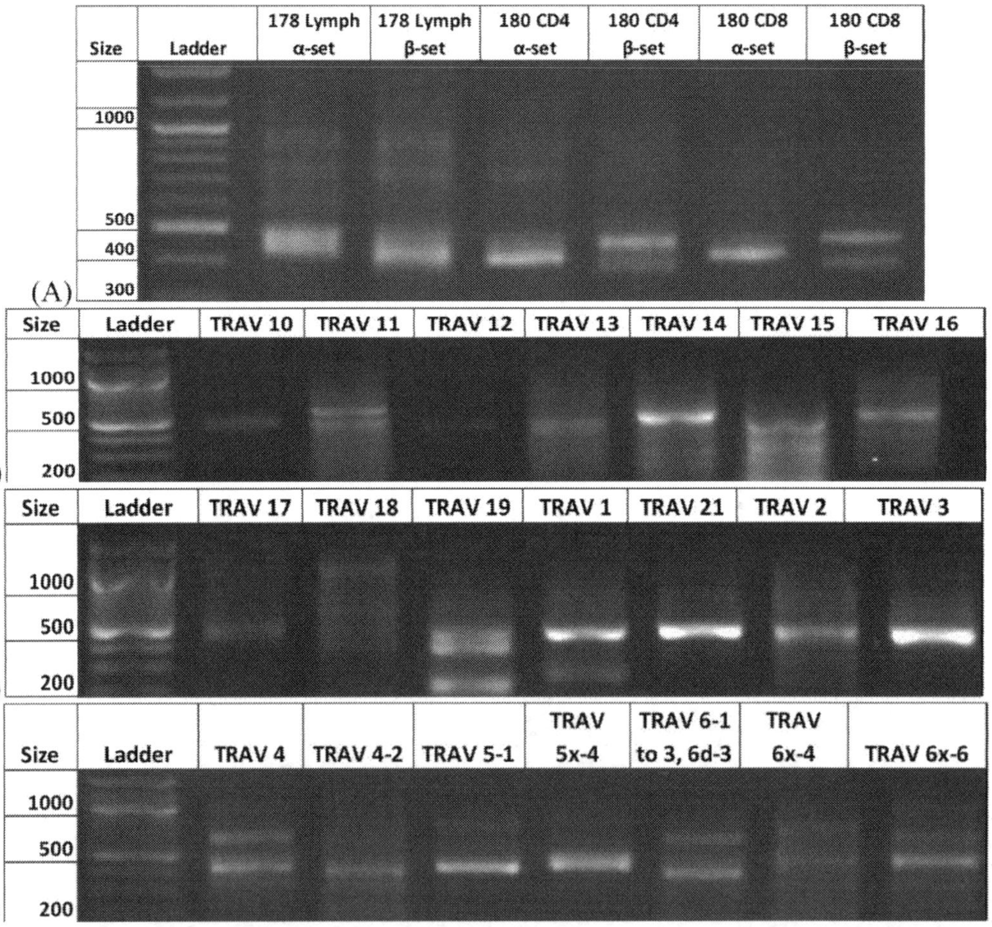

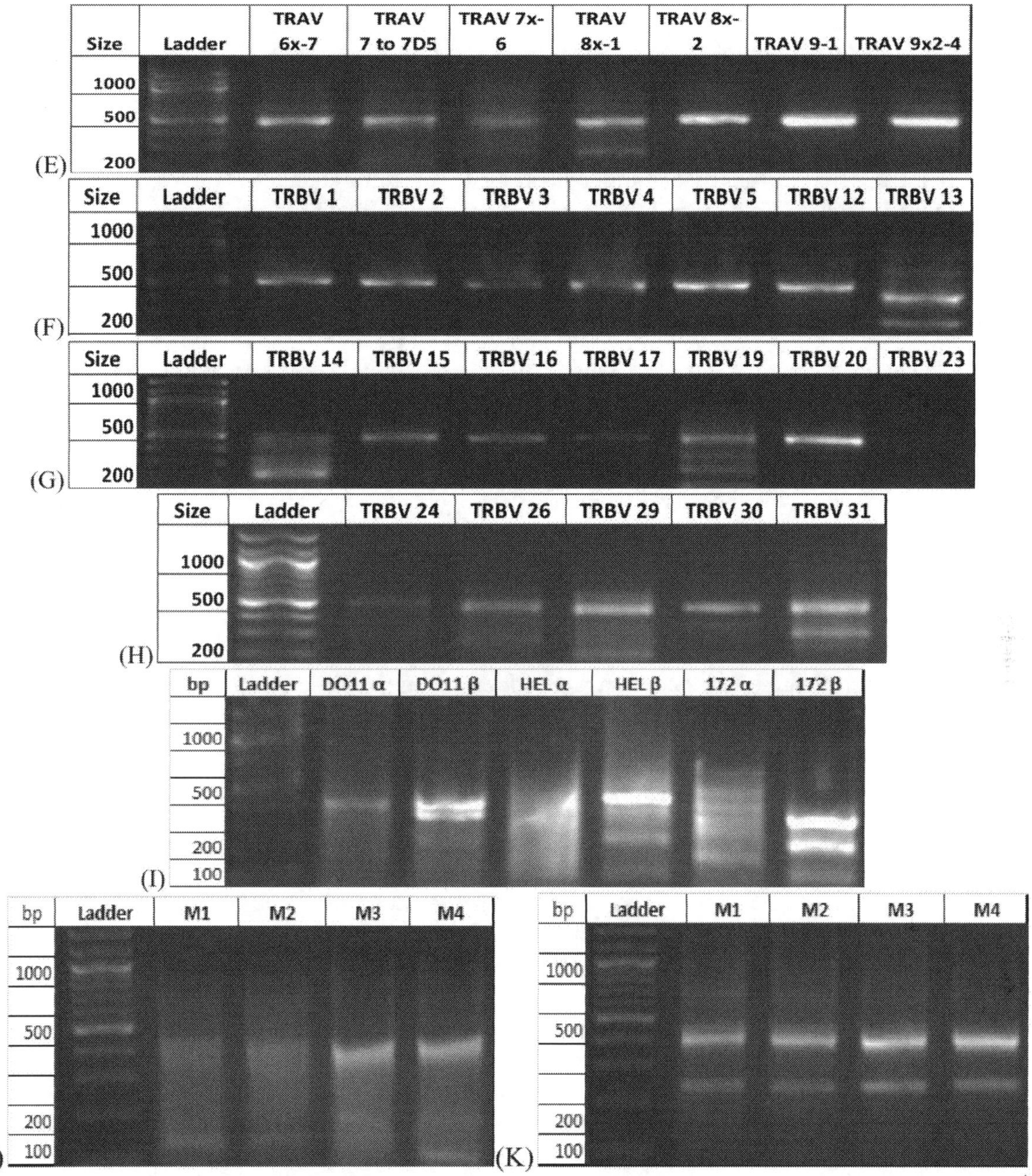

Figure 1: Gels showing: (A) amplification of α and β V-regions from lymph node and CD4$^+$ CD8$^+$ sorted cellular mRNA, (B – H) amplified V-regions of all the primers in the set, (I) V-region amplification from Tcell hybridomas, (J, K) amplified V-regions from the experimental mice using the α and β primers respectively.

cDNA Preparation, Amplification and Sequencing

Total mRMA was purified from the sacrificed mice using Qiagen RNeasy Mini Kit under the standard protocols. Samples were further purified from possible V-region chromosomal contamination with an Ambion (Applied Biosystems) DNase treatment as follows use 1 µl rDNaseI (2 units) for up to 10 µg of RNA in a 50 µl reaction. Combine and gently mix 17 µl of total RNA sample, 2 µl volume of 10X DNaseI Buffer, and 1 µl rDNaseI. Incubate solution at 37°C for 30 minutes, then add and mix 2.2 µl resuspended DNase inactivation reagent. Incubate solution for 2 minutes at room temperature mixing occasionally then centrifuge at max speed for 2 minutes and transfer the RNA solution to a fresh micro centrifuge tube.

The DNase treated RNA was then reverse transcribed with an Invitrogen Superscript II kit following the standard protocol given by Invitrogen for a 20-µl reaction volume. The reaction mixture consists of 0.5 µl PolyA Tail Primer (5'-TTTTTTTTTTTTTTTTTTTTTTTTVN-3') , 1 µg total RNA, 1 µl dNTP Mix (10 mM each), and sterile, deionized water to 12 µl. Heat sample to 65°C for 5 min and chill on ice. Spin to collect sample and add to the reaction 4 µl 5X First-Strand Buffer and 2 µl M DTT. Mix tube gently and incubate at 42°C for 2 min then add 1 µl SuperScript™ II RT and1 µl of sterile water. Incubate at 42°C for 50 min then inactivate the reaction by heating at 70°C for 15 min. Then immediately use the cDNA for PCR amplification.

PCR amplification of the cDNA from the total mRNA was performed using a 50 µl reaction volume of 5 µl of cDNA from fresh reverse transcription reaction, 5 µl pfx buffer, 0. 5 µl pfx enzyme, 0.5 µl the α/β V-region primer mixes and the flanking constant region primer, and 38.5 µl of sterilized water. Samples then ran at 94°C for 2 minutes then 50 °C for 20 seconds and then 68 °C for 30 seconds for 25 cycles to optimize the lowest amount of replicate amplicons while still having enough material to purify for sequencing reactions. PCR amplifications were purified using a Qiagen PCR purification kit with 3 supplemental PE wash steps to minimize the amount of residual salts left in the sample after the final elution.

Sequence Analysis

PCR amplicons were sent to the University of Texas Core Facilities where the samples underwent 454 pyrosequencing sequencing to collect 460,796 (276,024 effective) individual sequences. The sequence data were then processed using immunogene recognition software with IMGT's online high volume sequence analyzer which recognizes and isolates the V-, D-, and J-gene regions and details the recombination event site centered over CDR3. The results were processed in large batches using programming tools in Microsoft's Server Query Language with various queries to elucidate the subtle differences between the populations of TCR sequences between the control groups.

Results and Discussion

As of September 24th these results are incomplete with further analysis still underway.

From the high throughput sequencing experiment 13 additional primers were validated adding to the 15 found during preliminary experiments and TOPO cloning to isolate and recover TCR sequences. The sequencing reactions found 460,796 individual sequences with 276,024 effective sequences having an identifiable V-region, J-region, and properly formed CDR3 region. Many of the sequences labeled as unproductive were incomplete, prematurely stopping short of either full gene recognition or an indeterminate junction region. Furthermore IMGT analysis tended to return recognition errors when coupled with the 454sequencing error rate. These errors mainly included a false "unproductive" reading requiring no real modification of the data and a larger amount of frame shifts tolerated in the germline sections, but not tolerated in the recombination portion of the sequence in CDR3. A final anomaly from the IMGT data analysis involves the J-regions TRAJ 44 and TRAJ 47 which should both be open reading frames, yet the majority of our "unknown" labeled sequences had these two J-regions. Beyond the found errors within the IMGT sequence processing the remaining gene-region identification data should be accurate.

Sequences were considered identical if they had the same V-region, J-region, and CDR3 nucleotide sequences as determined by both IMGT and a Markov search of likely CDR3 candidates. Looking at how well the TCR populations intersect the control mouse had 57863 unique sequences and the challenged mouse had 70162 unique sequences with both groups sharing 14745 sequences between all four mice. There are 1894 identical sequences found in all 4 mice. Further investigation into possible identical sequence bias from the cDNA PCR amplification step showed that minimal duplication bias occurred even in sequences with higher frequencies. Bias introduced for certain V-regions such as Vα7 (152k) and even some VJ combinations Vα 11 – Jα18 (4.2k). If bias were present there would have been found throughout all the mice more frequent numbers of identical sequences caused by the various PCR reactions driven by primer association rates and other PCR parameters.

Next the diversity of amino acids at certain positions in the CDR3 sequence for prevalence and frequency between the mice. To do this the first and last half of the CDR3 sequences was separated from one another, then the number of times a particular amino acid appears at a certain position divided by the total number of sequences was calculated for each length CDR3. Overlaying the various lengths of CDR3 sequences with their respective counterparts with other lengths with the start and end amino acids in the same position revealed the frequency of trends forming in how common certain residues were. We were able to show that the formation of CDR3s between the three mice were unique and that the prevalence of certain amino acids and the N- and -C end of CDR3.

An unexpected result of our sequencing analysis showed no sequences that looked strongly like the well-studied DO11.10 TCR. The DO11.10 TCR uses gene regions Vα5D-4 and Vβ13-2 and a distinct CDR3 sequence. While this experiment doesn't guarantee the recovery of sequences identical to those used in the majority of the DO11.10 and OVA Tcell work there was the expectation of some base level of consensus sequences appearing in the group of challenged mice but missing in the control group. The lack of a clear distinction between the two groups may be evidence of accumulated bias and noise added to the immune repertoire signal throughout our methods.

Additional Remarks

This abstract is meant to summarize and discuss the early aspects of the data and preliminary results and is lacking a more thorough conclusive analysis on the date of submission (September 24, 2012). The data for this work are still being analyzed and reviewed for presentation and publication. Additional results and conclusions will be available at the day of presentation and can be provided upon request by the presenter of work if they are not updated within the proceedings of the conference (Zachary Frye).

References

1. David Wu and Anna Sherwood (2012), "High-Throughput Sequencing Detects Minimal Residual Diseases in Acute T Lymphoblastic Leukemia," *Sci. Transl. Med. 4,* 134ra63.
2. Raphael Genolet (2012), "Highly Diverse TCRα chain repertoire of pre-immune CD8$^+$ T cells reveals new insights in gene recombination," *EMBO*, Febrary, pg. 1666-1678.
3. Nicole Gruta (2008), "Epitope-specific TCRβ repertoire diversity imparts no functional advantage on the CD8$^+$ T cell response to cognate viral peptides," *PNAS*, February, pg. 2034-2039.
4. Chunlin Wang (2010), "High throughput sequencing reveals a complex pattern of dynamic interrelationships among human T cell subsets," *PNAS*, January, pg. 1518-1523
5. George Wang and Pradyot Dash (2012), "T Cell Receptor αβ Diversity Inversely Correlates with Pathogen-Specific Antibody Levels in Human Cytomegalovirus Infection," *Hum Immun*. April, 128ra42
6. Rowe AH, Xiao Y, Scales J, Linse KD, Rowe MP, et al. (2011) Isolation and Characterization of CvIV4: A Pain Inducing a- Scorpion Toxin. PLoS ONE 6(8): e23520. doi:10.1371/journal.pone.0023520ref

(425g) Quantitative Profiling of Active Transcription Factors in Parallel

Betul Bilgin[1], Li Liu[2], Christina Chan[1,3] and S. Patrick Walton[1], (1)Chemical Engineering and Materials Science, Michigan State University, East Lansing, MI, (2)Microbiology and Molecular Genetics, Michigan State University, East Lansing, MI, (3)Biochemistry and Molecular Biology, Michigan State University, East Lansing, MI

Transcription factors (TFs) are cell regulatory proteins facilitating proper cell function by controlling gene expression in response to intracellular and extracellular stimuli. TFs mediate cellular responses through binding to specific sites of chromosomal DNA. TF activation can occur from many stimuli, and, conversely, a single stimulus can activate many TFs. Thus, measurement of the activity of a single TF provides only a limited picture of the instantaneous state of cell function. However, given the ~1850 different human TFs, parallel quantification of the levels of active TFs even under basal conditions remains technically challenging.

Several methods currently exist to measure the levels of active TFs in parallel. These include electrophoretic mobility-shift assays (EMSA), reporter gene assays, or protein binding microarrays. However, these techniques are limited in scale, accuracy, and sensitivity. Using magnetic bead separation, we have developed a new, scalable, flexible, sensitive, and complementary method for screening of multiple TFs in parallel. In this proof-of-principle work, we successfully analyzed NF-kB and Ap1 levels quantitatively in parallel with ~5-10 fold improved sensitivity over existing approaches. We have also quantified differences in the levels of NF-kB and Ap1 from cells following TNF-alpha stimulation relative to control cells. In this presentation, current results will be described along with further directions, including expansion to larger sets of TFs and analysis of complex biological processes.

(425h) High-Throughput Production of HIV-Specific Antibodies Identified Via Microengraving in the Yeast Pichia Pastoris

Timothy J. Politano, Brittany A. Thomas, Sangram Bagh, Kerry Routenberg Love and J. Christopher Love, Department of Chemical Engineering, The David H. Koch Institute for Integrative Cancer Research, Massachusetts Institute of Technology, Cambridge, MA

Efforts to develop a vaccine that confers sterilizing immunity against the human immunodeficiency virus type 1 (HIV-1) have thus far been unsuccessful, due in part to the virus' ability to rapidly mutate and evolve to evade the host immune response. However, it is known that a subset of HIV-infected individuals can develop broadly-reactive antibodies over time that are capable of neutralizing the virus, despite its attempts to escape the immune response. Identifying and studying such antibodies in large numbers from a broad range of individuals would provide greater details regarding their characteristics and the evolution of these responses within individuals. Such information could ultimately inform more effective vaccine designs. Direct identification of B cells producing HIV-specific antibodies by flow cytometry has made it possible to profile the antibody responses of small numbers of patients, but the secondary validation of the antibodies is slow and tedious. Molecular cloning of each antibody by traditional methods of plasmid construction and propagation in bacteria limits the throughput and scale of evaluating recombinant antibodies from a subject. Here we present an integrated method for identifying, producing, and characterizing such broadly-neutralizing antibodies from HIV+ patient samples in high-throughput. Our approach combines a microengraving method established by our lab with overlap-extension PCR (OE-PCR) to rapidly assemble antibody-expression constructs for expression in the yeast *Pichia pastoris*. Validation of the approach and scalability will be presented as well as feasibility of automation for production of manifold antibodies from subjects for secondary characterization such as binding affinities and neutralizing activity.

(436a) Comparison of the Isoprenoid Pathways in Marine Diatoms Using Isotope Assisted Metabolic Flux Analysis and Genome Scale Modeling

Andrew Quinn[1], Yuting Zheng[1], Steven Hutcheson[2] and Ganesh Sriram[1], (1)Chemical and Biomolecular Engineering, University of Maryland, College Park, MD, (2)Cell Biology and Molecular Genetics, University of Maryland, College Park, MD

Isoprenoids are a diverse class of hydrocarbons formed from five carbon molecules that have high value as therapeutics and biofuel precursors. Large-scale production of isoprenoids through metabolic engineering techniques has received much attention. The production of isoprenoids occurs in two organelles (the plastid and the cytosol) of most photosynthetic organisms. Isoprenoid production in the cytosol is exclusively the result of the mevalonate pathway, while plastidic production occurs via the 2-*C*-methyl-D-erythritol 4-phosphate (MEP) pathway.

We will report the metabolic flux and network analysis toward determining carbon flux through each of these parallel pathways in the model diatom *Phaeodactylum tricornutum* (Pt). We chose Pt for this study due to its high photosynthetic efficiency and its ability to synthesize large quantities of lipids and other reduced compounds. Prior isotope labeling studies have shown that the mevalonate and MEP pathways utilize different carbon substrates (Cvejic, J., Rohmer, M., Phytochemistry, 53, 21-28, 2000). We will expand on this work, using isotope-assisted metabolic flux analysis (MFA). This technique is widely used to analyze the labeling patterns of intracellular metabolites from cultures grown on substrates strategically labeled with the ^{13}C isotope of carbon. Knowledge of the carbon rearrangements between metabolites along with the measured ^{13}C enrichment data for each metabolite allows for the calculation of the flux through each reaction. To compare the mevalonate and MEP pathways, which share a number of common metabolites, we will design a set of ^{13}C labeled tracer substrates that create distinct labeling patterns within each pathway. Optimal tracer design is a non-trivial problem, because substrate labels can produce ambiguous labeling patterns, which prevent the determination of certain fluxes. Additionally, we will report the maximum theoretical flux through each pathway by using flux balance analysis employing a model such as those previously reported for Arabidopsis and *Zea mays*. This computational effort will produce a theoretical flux map showing the fluxes for maximal isoprenoid production, subject to the organism's metabolic constraints.

We anticipate that this work will provide tools to investigate the relative activity of the two isoprenoid pathways in organisms such as Pt under different conditions. These results, in tandem, will provide the necessary information for any future work focused on rational engineering of Pt to maximize the production of target isoprenoids.

(436b) Recent Computational Tools for the Analysis of Complex Biochemical Reaction Networks

Daniel Knight, Chemical and Biomolecular Engineering, Ohio State University, Columbus, OH, Haixia Ji, Mathematics, Ohio State University, Columbus, OH and Martin Feinberg, Ohio State University, Columbus, OH

Recent Computational Tools for the Analysis of Complex Biochemical Reaction Networks

Daniel Knight, Haixia Ji, Martin Feinberg

Chemical and Biomolecular Engineering, Ohio State University, Columbus, Ohio

When studying biological systems, understanding of their dynamical properties is more readily inferred from knowledge of only the reaction network, as detailed kinetic information is generally not available with confidence. There are exemplary complex biochemical reaction networks with rich behavior (multiple steady states, oscillatory behavior, etc.), while others have relatively dull, stable dynamics. Recent theoretical developments[1, 2] allow for powerful statements to be made about the dynamics of even very complex networks, provided the kinetics falls within a very general and natural class. In particular, it is possible to distinguish in a precise way between networks that might provide the basis for a bistable switch, and networks – even very intricate ones – that cannot. These recent theoretical developments have been incorporated into a freely available computer program, Chemical Reaction Network Toolbox (CRNT)[3]. The theory will be discussed, and examples of its implementation via the Toolbox will be given.

1. G. Shinar and M. Feinberg, *Concordant Chemical Reaction Networks,* 2011, available at www.arXiv.org/pdf/1109.2923v1.pdf.

2. G. Shinar and M. Feinberg, *Concordant Chemical Reaction Networks and the Species-Reaction Graph,* 2012, available at www.arXiv.org/pdf/1203.6560v3.pdf.

3. H. Ji, P. Ellison, D. Knight, and M. Feinberg, *The chemical reaction network toolbox, version 2.1*, 2011, available at http://www.chbmeng.ohiostate.edu/ feinberg/crntwin/.

(436c) Runge-Kutta Tau-Leaping Methods for Accelerating Stochastic Simulations of Biochemical Reaction Networks

Leonard A. Harris and James R. Faeder, Computational and Systems Biology, University of Pittsburgh School of Medicine, Pittsburgh, PA

Tau leaping [1] is a promising approach for accelerating the stochastic simulation of biochemical reaction networks. In its simplest form, the tau-leaping algorithm is analogous to the simple forward-Euler method for numerically integrating ordinary differential equations (ODEs) [1]. Since many higher-order ODE integration methods exist with properties far superior to forward Euler, it is natural to ask whether analogous tau-leaping methods are possible.

Building on prior work in this area [1-4], we have developed a general framework for implementing higher-order Runge-Kutta variants of tau leaping. The approach is novel in that it considers aspects of the tau-leaping algorithm that have not been considered previously within a higher-order context, such as the selection of time steps (tau selection) and post-leap checking. We have applied the method to the "partitioned-leaping algorithm" [5,6], a tau-leaping variant, and implemented it within the open-source modeling and simulation platform BioNetGen [7].

We present results that illustrate the advantages of higher-order tau leaping in computational systems biology. These include applications to models of gene regulation, intracellular signaling, and cell cycle progression. Specifically, we demonstrate the large gains in efficiency that can be achieved relative to exact-stochastic simulation methods [8] as well as improved accuracy relative to lower-order tau-leaping approaches.

[1] Gillespie DT (2001) Approximate accelerated stochastic simulation of chemically reacting systems. *J. Chem. Phys.* **115**, 1716-1733.

[2] Burrage K, Tian T (2004) Poisson Runge-Kutta methods for chemical reaction systems. In *Advances in Scientific Computing and Applications* (Lu YY, Sun WW, Tang T, Ed) Science Press, Beijing/New York, pp. 82–96.

[3] Rathinam M, Petzold LR, Cao Y, Gillespie DT (2003) Stiffness in stochastic chemically reacting systems: The implicit tau-leaping method. *J. Chem. Phys.* **119**, 12784–12794.

[4] Cao Y, Petzold LR, Rathinam M, Gillespie DT (2004) The numerical stability of leaping methods for stochastic simulation of chemically reacting systems. *J. Chem. Phys.* **121**, 12169–12178.

[5] Harris LA, Clancy P (2006) A "partitioned leaping" approach for multiscale modeling of chemical reaction dynamics. *J. Chem. Phys.* **125**, 144107.

[6] Harris LA, Piccirilli AM, Majusiak ER, Clancy P (2009) Quantifying stochastic effects in biochemical reaction networks using partitioned leaping. *Phys. Rev. E* **79**, 051906.

[7] Faeder JR, Blinov ML, Hlavacek WS (2009) Rule-based modeling of biochemical systems with BioNetGen. *Methods Mol. Biol.* **500**, 113-167.

[8] Gillespie DT (1976) A general method for numerically simulating the stochastic time evolution of coupled chemical reactions. *J. Comput. Phys.* **22**, 403–434.

(436d) Mechanism of a Hotdog Fold Thioesterase-Catalyzed Reaction Proposed by QM/MM Metadynamics Simulation

David C. Cantu[1], Albert Ardevol[2], Carme Rovira[3] and Peter J. Reilly[1], (1)Chemical and Biological Engineering, Iowa State University, Ames, IA, (2)Faculty of Informatics, Università della Svizzera italiana, CH-6904 Lugano, Switzerland, (3)Parc Científic Barcelona, 08028 Barcelona, Spain

Members of a number of thioesterase and hydroxyacyl dehydratase families have HotDog structural folds, while members of other families have other folds. These enzymes are mainly involved in fatty acid synthesis, but they catalyze other reactions as well. We have tabulated the primary and tertiary structures of all these enzymes as part of the ThYme database of enzymes involved in fatty acid synthesis and related reactions. To date many putative mechanisms have been proposed, none conclusively, for HotDog fold thioesterases based on existing crystal structures. We employ QM/MM metadynamics molecular simulation techniques to explore putative mechanisms of this type of thioesterase. An acid-base two-step mechanism is found where an aspartate residue orients and activates a water nucleophile that attacks the thioester carbon atom. A serine residue donates a proton to break the thioester bond, and it receives a proton from a histidine residue. The transition states and energy barriers are elucidated.

(436e) Splitting Hill-Type Kinetics: Reduced, Modular, Stochastic Simulations

Patrick Smadbeck, Chemical Engineering and Materials Science, University of Minnesota - Twin Cities, Minneapolis, MN and Yiannis Kaznessis, Department of Chemical Engineering and Materials Science, University of Minnesota, Minneapolis, MN

In recent years it has become evident that while stochastic simulations are necessary for many biological models, these simulations are often difficult to implement. Stochastic simulations fully account for the "randomness" or "noise" associated with a system. Such probability-based methods are necessary when the system exists far from the thermodynamic limit, considered a common condition in biological simulations. Unfortunately, many biological models include a large number of components and reactions over wide-ranging timescales. As the system grows, many of the unknown kinetic constants begin to become difficult or impossible to determine.

A common solution to reduce the number of unknown parameters is to bundle many unknowns into a non-linear reaction rate using an approximating method like the Quasi-Steady State Approximation. Such approximations are common in deterministic systems, but implementation in stochastic systems may only be applicable across a narrow set of conditions. Outside of these conditions the approximations fail to accurately describe higher-order statistics (e.g. the variance). This problem makes approximating methods difficult to implement in stochastic systems.

Modification of the non-linear reaction rates allows for higher-order statistics to be accounted for without significantly increasing the complexity of the system. In our particular work, transcription-based repression in a small gene network can be described by Hill-type kinetics. The Hill-type reaction rate has two parameters that can fully account for the mean protein output in the presence of repressor protein molecules. However, while the mean output is accurately described, the 2nd-order statistics do not match well. The Hill reaction rate law is then modified by "splitting" the rate law, effectively adding an additional parameter to take into account higher-order statistics.

Two systems were produced using the Hill-reduction method and the Split-Hill method to show how accuracy is improved and modularity preserved by accounting for higher-order statistics. A negative-feedback loop illustrates the ability of the Split-Hill model to fully and accurately describe results. A bistable switch constructed from two repressor systems demonstrates a tendency for modularity in the Split-Hill systems.

Higher-order statistics are a vital part of stochastic simulations and any reduction method or approximation must take these statistics into account to ensure accuracy and modularity. The Split-Hill model as presented is a modification to the Hill-type reaction rate designed for transcription-based control in gene networks. The results show that the Split-Hill network is accurate and modular in simple gene networks, without significantly increasing the complexity of the model.

(436f) Use of 3-Dimensional Response Surface Plots for Heat Transport in the Human Eye to Predict Time of Death

Jimmy L. Smart, Chemical & Materials Engineering, University of Kentucky, Paducah, KY

Historically, estimation of time of death in humans has always been a valuable forensic tool. Over the years, there have been numerous models proposed to estimate time of death [1,2]. Results of these models can be useful to crime scene investigators, but are not strictly admissible into courts of law like DNA analyses, due to much error and variability associated with these current models.

Most "time of death" models that are developed rely upon torso modeling and rectal temperature profiles. This approach is untenable since there is too much variability in tissue size and distribution across human bodies in the general population. Every body is different, so a single reliable a priori model is not practical. A more reliable approach is believed to be based upon temperature decay in the human eyeball. The size of the eyeball and distribution of surrounding tissues (skull, brain, etc.) are fairly consistent throughout the human population.

COMSOL Multiphysics® finite element software program was used to model convective/radiation heat transfer from the human eyeball to the surrounding air. Postmortem temperature decay curves were collected in eyeballs of several human corpses over a period of 2 - 24 hours. Of course, each of these curves represent only a portion of the complete temperature decay curve, since the pathologist was able to start collecting temperature data only after some time had elapsed from when the individual had died. These portions of the actual cooling curves were compared to complete model cooling curves developed with COMSOL. Theoretically, if a model cooling curve can be superimposed exactly upon the actual cooling curve, time of death can be reliably predicted.

Based upon the preliminary results of earlier work [3], 3-dimensional plots have been prepared for constant ambient temperature conditions of 10 - 33 °C, for convection/radiation heat transfer coefficients of 7.0 - 13.0 W/m^2 K, over a period of 0 - 24 hrs. Given a single constant ambient temperature, a model response surface of eyeball temperature has been generated as a function of time and heat transfer coefficient. Therefore, there are 24 model response surfaces, each described by a mathematical relationship, to address the range of constant ambient temperature conditions of 10 - 33 °C.

For practical purposes, once a coroner or a medical examiner arrives at the scene, the ambient temperature is recorded and the fact verified that the victim has not been previously exposed to widely varying temperatures. A recording device is used to record the temperature within the eyeball and the ambient temperature every 5 min for a 1 hr period. At this point, the coroner or medical examiner has captured a small portion of the actual postmortem temperature decay curve (PTDC). This is called the *field PTDC*. The 3-dimensional response surfaces generated from the software model is called the *model PTDC*. The small response surface of the field PTDC is mathematically matched to the larger comprehensive response surface of the model PTDC. Once the match is complete, through reverse engineering, the model can be used to extrapolate back to time zero to estimate time of death.

Sensitivity studies were also performed to evaluate how sensitive is the estimated time of death to variations in ambient temperature conditions and values of the convective/radiation heat transfer coefficient.

1. Henssge, Claus, The Estimation of the Time Since Death in the Early Postmortem Period, 2nd Ed., London: Arnold Publishing, 2002.
2. Smart, Jimmy L, Estimation of Time of Death With a Fourier Series Unsteady State Heat Transfer Model, J. Forensic Sci, June 2010.
3. Jimmy L. Smart, Michal Kaliszan. Use of a Finite Element Model of Heat Transport in the Human Eye to Predict Time of Death, J. Forensic Sci, submitted Feb 2011, accepted Jan 2012, published Mar 2013.

(436g) Mapping the Cell Cycle in GS-NS0: Developing a Cyclin Blueprint As a Tool for Optimizing Productivity

D. G. Garcia Münzer, Biological Systems Engineering Laboratory - Centre for Process Systems Engineering, Department of Chemical Engineering, Imperial College, Imperial College London, London, United Kingdom, A. Mantalaris, Biological Systems Engineering Laboratory, Department of Chemical Engineering, Imperial College London, London, United Kingdom and E. N. Pistikopoulos, Centre for Process Systems Engineering, Department of Chemical Engineering, Imperial College London, London, United Kingdom

Mapping the Cell Cycle in GS-NS0: Developing a cyclin blueprint as a tool for optimizing productivity

D.G. García Münzer [*, **], A. Mantalaris[*], E.N. Pistikopoulos[**]

[*]Biological Systems Engineering Laboratory, Department of Chemical Engineering, Imperial College, London SW7 2AZ, UK

[**]Centre for Process Systems Engineering, Department of Chemical Engineering, Imperial College, London SW7 2BY, UK

The use of mammalian cells for the production of high value bio-pharmaceuticals (biologics), such as monoclonal antibodies, has growth rapidly. It is expected to reach a market value of $239 billion by 2015 [1]. Mammalian cell factories are complex physical and chemical structures whose productivity (and product quality) is under the control of a large number of coordinated chemical reactions (metabolism) influenced by culture parameters. The cell cycle is at the centre of growth, productivity and cell death, which vary during the different phases of the cell cycle. Specifically, cell productivity is cell cycle, cell-line and promoter dependant [2]. Consequently, knowledge of the cell cycle-associated production profile will assist in determining the optimization strategy towards improving productivity [3].

Cyclins are key regulators of the cell cycle. They activate their partner cyclin-dependent kinases (CDKs) and target specific proteins to drive the cell through particular processes, check points and phases. Although a number of studies have dealt with cell cycle regulation in various human cell lines [4-6], to our knowledge, there is no information on cyclin phase-dependant expression profiles and thresholds of industrial relevant mammalian cells. Therefore, there exists a need to identify and quantify major landmarks of the cell cycle that will allow for the systematic study of cell productivity.

We have studied the timing of expression of the three cyclins, under both perturbed and unperturbed growth using the GS-NS0 cell line by flow cytometry. The perturbed systems involved arresting the cell using two different DNA synthesis inhibitors, thymidine and dimethyl sulfoxide (DMSO). This approach allows establishment of characteristic cyclin profiles, thresholds and unscheduled production. In particular, we looked at two G1 class cyclins, namely cyclin D and E, and one G2 cyclin, namely cyclin B. The observed patterns (including thresholds and unscheduled production) provide a blueprint of the cell line's cell cycle, which can be used for cell cycle modelling. Briefly, expression of cyclin B showed a clear cell cycle phase specific pattern whereas Cyclin D expression was fairly invariable throughout the cycle (our data indicate an unrelated entrance to the S phase with respect to cyclin D). Similarly, cyclin E was expressed during all phases, in progressively decreasing manner from G1 towards G2, suggesting that cyclin E is being degraded.

Cyclin B

Control

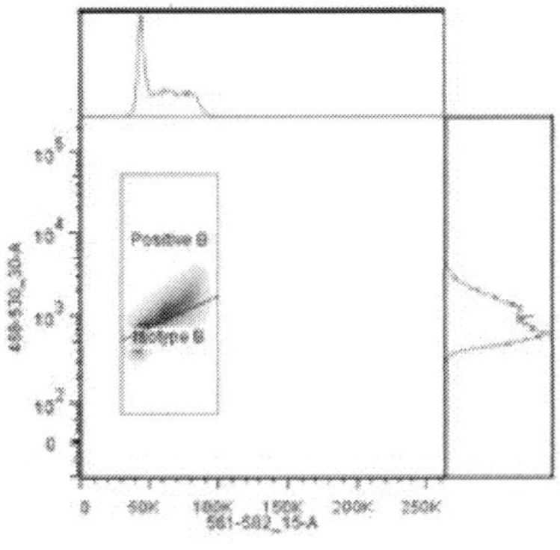

Arrest

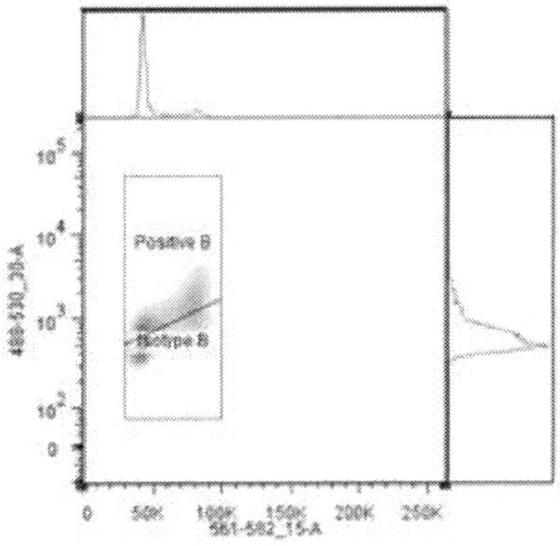

Cyclin D

Control

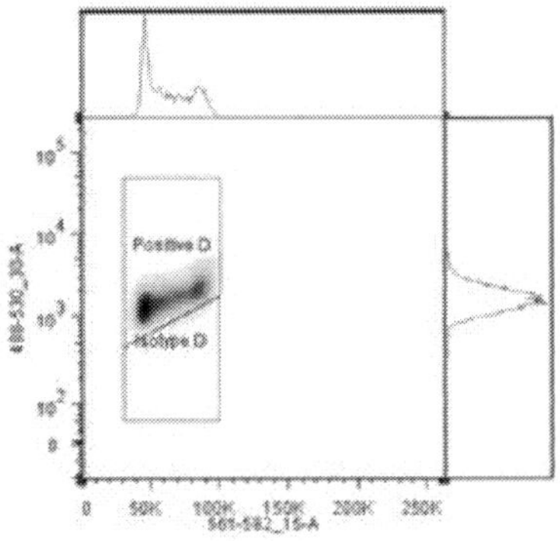

Arrest

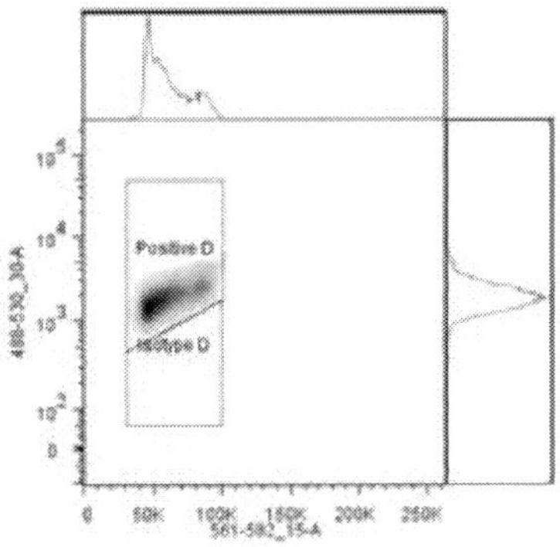

Cyclin E

Control

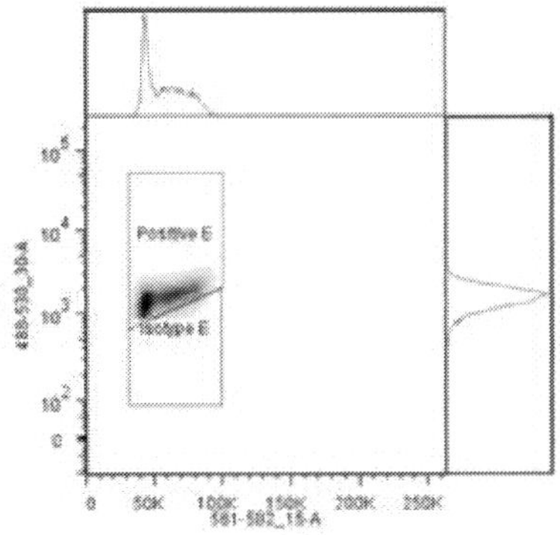

Arrest

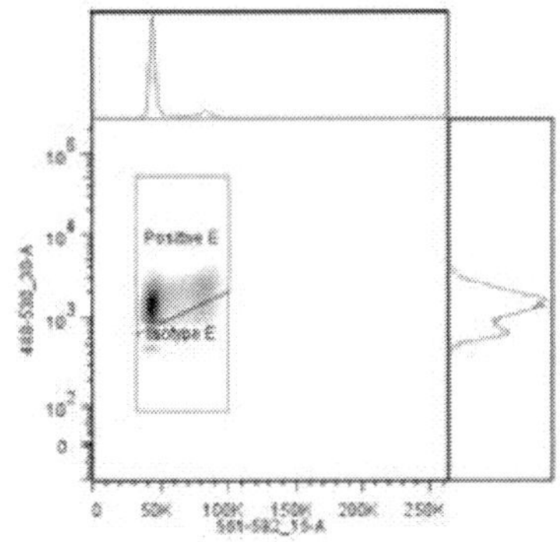

A key feature of cell cycle modelling in cell culture systems is the ability to account for important differences between the populations. There is a need for developing cell cycle models that can capture accurately the complexity of the system while being computationally tractable. Cyclins represent excellent cell cycle modelling variables as they not only provide information regarding the proliferating potential of the cell population, but also serve as landmarks throughout the cycle. Therefore, cyclin distributed models represent the next step in cell cycle modelling. The use of cyclins as distributed variables can be experimentally validated (quantitatively) and will avoid the use of weakly supported variables such as age, volume or mass. The development of a biological relevant cell cycle model, while keeping it tractable, is possible via the model building framework [7]. Ultimately, the development of these models will pave the way for the systematic study of the cell culture system, the improvement of productivity and product quality.

References

[1] Bbc Research. http://www.bccresearch.com/report/biologic-therapeutic-drugs-bio079a.html

[2] Alrubeai, M. and A. N. Emery (1990). "Mechanisms and Kinetics of Monoclonal-Antibody Synthesis and Secretion in Synchronous and Asynchronous Hybridoma Cell-Cultures." Journal of Biotechnology 16(1-2): 67-86.

[3] Dutton, R. L., J. M. Scharer, et al. (1998). "Descriptive parameter evaluation in mammalian cell culture." Cytotechnology 26(2): 139-152.

[4] Darzynkiewicz, Z., J. P. Gong, et al. (1996). "Cytometry of cyclin proteins." Cytometry 25(1): 1-13.

[5] Tomasoni, D., M. Lupi, et al. (2003). "Timing the changes of cyclin E cell content in G(1) in exponentially growing cells." Experimental Cell Research 288(1): 158-167.

[6] Frisa, P. S. and J. W. Jacobberger (2009). "Cell Cycle-Related Cyclin B1 Quantification." Plos One 4(9).

[7] Kiparissides, A., M. Koutinas, et al. (2011). "'Closing the loop' in biological systems modeling - From the in silico to the in vitro." Automatica 47(6): 1147-1155.

(436h) Modeling Impact of Lipids On Orally Delivered Drug Dissolution

Selena Di Maio[1], David E. Budil[2] and Rebecca L. Carrier[1], (1)Chemical Engineering, Northeastern University, Boston, MA, (2)Chemistry and Biological Chemistry, Northeastern University, Boston, MA

Despite the recognized capability of ingested lipids to significantly impact oral absorption of drugs, the fate of co-administered compounds remains unclear and not predictable *a priori*. The aim of this study is to quantitatively investigate and model the effects of ingested lipids and lipid digestion on dissolution and partitioning of orally delivered drugs by means of *in vitro* lipolysis models incorporating simulated intestinal fluids.

An updated *in vitro* digestion model was designed to closely mimic fed state intestinal conditions. The paramagnetic compound TEMPOL benzoate (TB) was selected as a model for poorly water-soluble moderately lipophilic drug (log P = 2.46). Digestion kinetics was investigated by titration and compared with measured dissolution kinetics in the presence and absence of lipids to assess processes likely to significantly impact overall drug absorption kinetics. Drug partitioning between colloidal phases (oil, micellar, aqueous) formed during *in vitro* digestion was studied by electron paramagnetic resonance (EPR), which provided real-time compound tracking measurements and quantification.

A film-equilibrium drug dissolution model was developed and utilized to predict drug dissolution kinetics in simulated intestinal fluids with and without lipids. In the model, the influence of colloids interacting with drug on drug transport rates was explicitly taken into account, and drug partitioning with micelles was considered as a pseudo-equilibrium process. The impact of lipids and their digestion on dissolution, experimentally evaluated based on EPR measurements pre- and during *in vitro* lipolysis, was compared with model predictions. Appropriate rate expressions describing lipid digestion, drug dissolution, drug partitioning, and drug transport across the intestinal membrane were assessed and incorporated into a model to predict overall drug absorption. The significance of ingested lipid-drug interactions during digestion to the overall drug absorption is not always obvious and still unpredictable. However, a systems-based model incorporating drug dissolution, drug partitioning, and lipid digestion kinetics might enable prediction of the fate of orally administered drugs during the lipid digestion process and provide insight into their anticipated overall effect on oral absorption. The ability to predict the impact of lipids on oral drug delivery would shed considerable light on the "food effect" on oral drug absorption, a phenomenon of tremendous significance to the pharmaceutical industry and drug development.

(447a) A Rapid Approach to the Development of Affinity Reagents Using Yeast Surface Display

James A. Van Deventer, Koch Institute for Integrative Cancer Research, Massachusetts Institute of Technology, Cambridge, MA, Sachdev Sidhu, Banting and Best Department of Medical Research, University of Toronto, Toronto, ON, Canada and K. Dane Wittrup, Department of Biological Engineering, Massachusetts Institute of Technology, Cambridge, MA

Affinity reagents form a key part of many biological assays, diagnostic tests, and therapeutics, but their generation can require considerable time, effort, and expense. In this work, we present an approach to the rapid isolation of affinity reagents from a naive, synthetic antibody library using yeast surface display. A billion-member single chain variable fragment (scFv) library has been constructed from a minimalist library design based on a common framework region and restricted diversity in the complementarity determining regions. This library enables rapid, yeast-based isolation of affinity reagents from the naive library in approximately two weeks using a combination of bead- and flow cytometry-based sorting. Initial characterizations suggest that isolated clones exhibit low double-digit nanomolar affinities toward target antigens, comparable to many commercially available antibodies. Furthermore, the isolated scFv clones can be converted to a soluble antibody-like structure using a single ligation-free cloning step, enabling rapid validation in standard biological assays. Current efforts are focused on cloning-free conversion between displayed and secreted proteins and flow cytometry-free approaches to isolating individual binding proteins. We expect this approach to be useful to the scientific community for the development of custom affinity reagents for use in immunofluorescence microscopy, western blots, immunoprecipitations, flow cytometry, and other standard biological assays.

(447b) Cloning Nuclear Hormone Receptors to Develop Biosensors Using the Gibson Assembly

Miriam Shakalli Tang[1], Richard A. Lease[2] and David Wood[1], (1)Department of Chemical and Biomolecular Engineering, Ohio State University, Columbus, OH, (2)Department of Chemical and Biomolecular Engineering, Ohio State University, Columbus, OH

Cloning Nuclear Hormone Receptors to develop Biosensors using the Gibson Assembly

Miriam Shakalli Tang, Richard A. Lease, David W. Wood

Nuclear hormone receptor (NHR) protein domains comprise one of the largest classes of protein targets for drug ligands because they can bind tightly to a wide variety of endocrine disruptors in addition to their native hormones. We are developing high-throughput methods for detection and characterization of hormone-like compounds, to be used in applications such as toxicology screening, insecticide development, and drug discovery. Current compound screening strategies for hormone mimics are complex, time consuming, and expensive. To address this, we have developed a bacterial biosensor containing a genetically engineered reporter protein. This reporter protein incorporates a hormone ligand-binding domain (LBD) of an animal nuclear hormone receptor within a bacterial intein-protein platform technology. These biosensors can detect compounds that target NHRs, where the readout is a simple growth assay. These biosensors can distinguish positive or negative effects of a test compound on the receptor's activity, as well as agonist versus antagonist activities.

The goal of this project is to optimize the rapid construction of nuclear hormone binding sensors using the Gibson Assembly method, and to apply the Gibson method to the development of protein biosensors based on NHRs from *Aedes aegypti* (mosquito vector for dengue fever) and *Tribolium confusum* (red flour beetle). Gibson Assembly is a robust in vitro recombinant technology, which allows efficient construction of recombinant fusion proteins. In this case, the Gibson strategy allowed the insertion of the LBD of the insect NHR gene into a larger fusion protein gene. The fusion protein gene includes solubilization and stabilization domains (maltose-binding protein plus an intein linker domain), as well as a thymidylate synthase (TS) enzyme domain to generate the bacterial growth phenotype. The resulting reporter protein links binding activity of the LBD to TS activity and subsequence bacterial growth phenotype. The constructed sensors were subsequently tested to discern the presence of hormones or hormone-like compounds through changes in growth phenotype of TS knockout *E. coli* strains in thymineless media. Following the success of the Gibson Assembly in cloning NHR biosensors, its capability of joining DNA fragments will be harnessed for the rapid (HTS) production of recombinant protein products and for screening of endocrine disruptor ligands as potential insecticides.

(447c) Engineering Cystine Knot Peptides As a New Class of Molecular Imaging Agents

Cheuk Lun Leung[1], Sarah J. Moore[2], Heidi Norton[3] and Jennifer R. Cochran[2], (1)Chemical Engineering, Stanford University, Stanford, CA, (2)Bioengineering, Stanford University, Stanford, CA, (3)Materials Science and Engineering, Stanford University, Stanford, CA

Cystine knots, also known as knottins, are a class of peptides that consist of a structural core of three disulfide bonds interwoven into a molecular "knot" confirmation. This disulfide-bonded structure confers knottins with high chemical, thermal, and proteolytic stability. We are using knottins as novel molecular scaffolds to engineer peptides that can target and illuminate tumors. Using yeast surface display, our lab demonstrated that two knottin scaffolds, based on the Ecballium elaterium trypsin inhibitor (EETI) and the Agouti-related protein (AgRP), can be engineered to bind with high affinity to $\alpha v\beta 3$ integrins expressed on tumor vasculature. Non-invasive in vivo optical imaging of the engineered knottin peptides based on both EETI and AgRP scaffolds elicited high tumor signals and rapid clearance from non-target tissues. However, AgRP-based peptides had high kidney uptake and retention, while EETI-based peptides showed minimal kidney uptake. To further investigate and elucidate these differences amongst knottin scaffolds, we engineered a third knottin from spider venom, Agatoxin (AgTx), to bind to $\alpha v\beta 3$ integrins using a loop grafting approach. We also found mutations in the AgTx scaffold that assisted in folding of the engineered protein. We then compared the tissue biodistribution observed between wild-type scaffolds and integrin-binding EETI, AgRP, and AgTx knottin peptides in mouse tumor models. We showed that desirable in vivo biodistribution properties are inherent to the three knottin scaffolds tested, but particular amino acid sequences within the engineered binding loop of AgRP and AgTx result in high imaging signals in the kidneys. Thus, we demonstrate that AgTx can be engineered for new molecular recognition properties, and that tissue biodistribution of fluorescent knottin peptides can be tuned by modifying their amino acid composition.
Funded by NIH NCI P50 CA114747: In Vivo Cellular and Molecular Imaging Center at Stanford

(447d) Engineering Hetero-Bivalent Peptidic Ligands to Inhibit Mast Cell Degranulation in a Heterotetravalent Allergy Model

Michael W. Handlogten[1], Tanyel Kiziltepe[2] and Basar Bilgicer[1], (1)Chemical and Biomolecular Engineering, University of Notre Dame, Notre Dame, IN, (2)Advanced Diagnostics and Therapeutics / Chemical and Biomolecular Engineering, University of Notre Dame, Notre Dame, IN

An allergic response results from the activation of the adaptive immune system directed against noninfectious environmental substances (allergens). Cross-linking of immunoglobulin E (IgE) bound to its high affinity receptor, FcεR1, on the surface of basophils and mast cells initiates a complex signaling cascade that results in degranulation which is the release of preformed chemical mediators associated with allergic symptoms. The majority of the current therapies used to treat allergies and related adverse immunologic responses, such as autoimmune diseases, involve non-specific suppression of significant parts of the immune system resulting in the increased risk of infection. The development and use of selectively targeted therapeutics would decrease these risks. Our overall goal in this project was to engineer hetero-bivalent peptidic ligands that would competitively and selectively inhibit allergen binding to IgEs on mast cells, thereby inhibiting IgE clustering and preventing an allergic response. This was accomplished via designing hetero-bivalent inhibitors (HBIs) that simultaneously targeted two nearby sites located on the Fv domain of an IgE, the antigen binding site, and the less renowned "unconventional nucleotide binding site". Simultaneous bivalent binding to both these sites provided the hetero-bivalent inhibitors with enhanced avidity and selectivity for the target IgE, and enabled competitive inhibition of allergen binding to IgE. We previously demonstrated the utility of the HBI design with the most commonly used allergy model – the rat basophilic leukemia cell (RBL)/IgE^{DNP} system. In this system, a saturating concentration of a single monoclonal IgE is used to prime the mast cells, then the allergen, typically BSA presenting multiple copies of DNP, is added to the system and the degranulation response is observed. This system has been very useful for studying allergy, but lacks much of the complexity present in the natural system. First, in humans, mast cells are primed with serum derived polyclonal IgE resulting in mast cells presenting IgE specific for a wide range of antigens, of which only 0.1 – 20% is specific for any given allergen. Second, natural allergens typically contain 2 – 9 distinct immunodominant epitopes that possess a range of affinities for the polyclonal IgEs. Third, the method of conjugating multiple copies of a single hapten to scaffolds such as BSA results in poorly defined allergens that contain a wide range in the number of haptens per allergen. A matter that is further complicated as not all the haptens in these synthetic allergens are capable of binding to mast cell bound IgE due to steric constraints.

The limitations in the current model led us to develop a new system to study allergy and determine the effectiveness of our HBIs. The system we developed was based on the design of a series of heterotetravalent allergens (HtTAs) that were used to stimulate the degranulation of mast cells that were primed with 3 monoclonal IgEs each with a different target antigen. The HtTAs contained two sets of two haptens that were specific for a different IgE antibody. The third IgE antibody was used to model the orthogonal IgE (IgE of other specificity) on mast cells. The use of the third IgE allowed us to test the HtTAs with physiological relevant amounts of allergen specific IgE on the mast cell surface. The HtTAs proved to be potent stimulators of degranulation only when the mast cells were primed with both of the IgEs specific to the allergen. This was expected as bivalent allergens are poor stimulators of degranulation.

With our new more advanced allergy model we were able to successfully and specifically inhibit allergen-mediated mast cell degranulation using our HBIs. Specifically, we designed an HtTA possessing two sets of two haptens each specific for a different monoclonal IgE. The first IgE-hapten interaction had a high affinity (K_d = 55 nM) while the second IgE-hapten pair had a weak affinity (K_d = 41 μM). The HtTA stimulated degranulation, following a bell-shaped dose response curve, thus illustrating the importance of weak binding epitopes in allergy. An HBI was designed to inhibit the binding of only 1 of the hapten-IgE pairs by conjugating an antigen binding moiety to a nucleotide binding moiety via a short ethylene glycol linker. The increased avidity and specificity of simultaneous bivalent binding to the antigen binding region and the nucleotide binding site on a single Fv of an IgE, inhibited allergen binding to the IgE. When the HBI was co-incubated with the allergen, degranulation was inhibited in a dose dependent manor. The binding of the HBI to one of the allergen specific IgEs essentially reduced the HtTA to a bivalent allergen which are not effective stimulators of degranulation. This result demonstrated that it is possible to completely inhibit allergen-mediated degranulation through the specific targeting of only a fraction of the allergen specific IgE on the mast cell. The long-term goal of this project is to develop a broadly applicable technology that can selectively inhibit the allergic responses induced by an allergen via competitively inhibiting its binding to the IgE antibodies with engineered hetero-bivalent ligands.

(447e) Targeted Drug Delivery to the Brain Using Transferrin-Binding Peptides

Divya Chandra and Pankaj Karande, Chemical and Biological Engineering, Rensselaer Polytechnic Institute, Troy, NY

Drug-delivery to the brain is a significant challenge due to the presence of the formidable blood-brain barrier (BBB). Current therapeutic interventions to treat brain disorders include surgical implants or catheters, both of which are highly invasive and carry the risk of long term neurological damage. A treatment strategy that is non-invasive provides uniform distribution of the drug in the brain at therapeutically relevant concentrations, and carries a low risk of neurological damage is highly desirable but not clinically available. We are focusing on a specific pathway called receptor-mediated transcytosis (RMT) that is facilitated by proteins expressed on the surface of endothelial cells that form the BBB and are dedicated for ferrying transport proteins across the BBB. This pathway has previously shown great promise for transport of antibodies and drug conjugates of transport proteins like transferrin (Tf), insulin etc. We are developing a novel drug delivery strategy based on short peptides as drug carriers to the brain, whereby these peptides will bind to human Tf (hTf) and deliver the attached drug via the RMT pathway of hTf. The transferrin receptor (hTfR) is of particular importance here as it is over-expressed on brain capillaries and is involved in the transfer of iron-carrying hTf to the brain. Given the high circulation half-life (~8 days) of hTf, designing peptides that enable drugs to stay conjugated to hTf increases the *in vivo* residence time and efficacy of the drugs as well as reduces their required dosage and side effects. This addresses another major challenge of obtaining desirable pharmacokinetics when designing therapeutics for brain diseases. High abundance (1-3 mg/ml) of hTf in blood also facilitates high drug dosing. Additionally, chaperone peptides eliminate the need for covalent conjugation of drugs to hTf outside the body and can be tailored to improve solubility and stability of drugs in circulation.

In order to design high-affinity hTf–binding peptides that can act as drug chaperones we adopted a rational design approach that relied on naturally occurring receptors for Tf. It is known that several bacterial species have evolved surface proteins that bind to hTf in serum. These proteins, called transferrin binding proteins (Tfbp) enable the pathogens to scavenge iron from circulating hTf and use it for their own metabolic requirements. The binding interfaces of Tfbps on different pathogens do not bear any significant structural or sequence similarity to hTfR and therefore are expected not to interfere with binding of hTf to hTfR. Thus, they offer a potential search space for designing short peptides that bind to hTf with a high-affinity and enable the delivery of a drug cargo across the BBB.

Using high-throughput peptide synthesis and microarray screening, several potential peptide candidates that mimic Tfbps have been identified from an initial library of ~1000 peptides. These peptides show very high affinity (1-10nM) for hTf in the presence of physiologically relevant concentrations of human serum albumin (HSA) – the most abundant protein in human serum. This ensures that the peptides not only have high affinity but also high selectivity for hTf. As a first step, while peptides have been identified that bind to hTf, the future validation of this drug delivery strategy *in vivo* in a mouse model would require peptides that bind to mouse Tf (msTf). Keeping this in mind, the entire library has also been screened with msTf in the presence of HSA in order to select peptides that bind to msTf with high affinity and selectivity. A set of candidates that bind to both hTf and msTf has been selected for *in vitro* and *in vivo*studies.

We are currently testing a subset of these peptides in an *in vitro* BBB cell culture model. This, followed by studies in an animal model would further validate the lead peptide candidates as drug carriers. Finally, the fact that these peptides come from a non-human protein and therefore less likely to compete with hTfR-hTf interaction will have a significant impact on designing better peptide chaperones for drug delivery to the brain.

(447f) Engineered Monobody Inhibitors of Erk-2 Dependent Signaling

Jasdeep Mann[1], Anne-Fleur Stephan[1], Jordan Wood[2], Denise Ferkey[2], E. Manolis Tzanakakis[1] and Sheldon Park[1], (1)Chemical and Biological Engineering, State University of New York at Buffalo, Buffalo, NY, (2)Biological Sciences, State University of New York at Buffalo, Buffalo, NY

The phosphorylation networks are central to intracellular signaling and orchestrate a broad range of cellular responses to environmental cues. The mitogen activated protein kinases (MAPK) form evolutionarily conserved pathways that are activated by stimuli at the cell surface and induce changes in gene expression. In human, there are at least four MAPK pathways, including the Erk1/2, p38, JNK, and Erk5 pathways, that mediate cellular responses to growth factors, inflammatory cytokines, and stress signals, and regulate embryogenesis, and cell fate determination. All MAPK pathways are organized in a three tier system, in which a MAPK is phosphorylated by a dual specificity MAPK kinase, or MEK, which is in turn phosphorylated by a MEK kinase (MEKK). Dysregulation of MAPK pathways is associated with various diseases, including cancer, diabetes, neurological and inflammatory diseases. Therefore, engineered MAPK inhibitors would be useful tools for studying the molecular basis of signaling in model organisms and for developing therapeutics against overactive signaling networks. In the current study, we describe the engineering of monobody inhibitors that selectively block the interaction between Erk-2 and its *in vivo* substrates, activators, and negative regulators. MAPKs use the docking of a D-motif peptide found on their interactors and conserved "CD" docking domain to confer specificity of interaction and increase the efficiency of signaling. Since the docking interaction is essential for the function of Erk-2, the designed monobody should efficiently reduce the efficiency of Erk-2 activation and its catalytic efficiency. We screened a combinatorial fibronectin type III (FN3) monobody library expressed on the yeast surface using magnetic and fluorescence activated cell sorting. To isolate the monobodies that specifically target the CD domain, affinity maturation was alternated with a negative sort using mutant Erk-2 containing mutations in the docking site. The clones identified from the sort bind Erk-2 at the docking site with low nM affinity and inhibit its phosphorylation by MEK2 as well as phosphorylation of the transcription factor Elk1. Some of the clones exhibit high target specificity and can discriminate against p38 and JNK with > 100 fold selectivity. The inhibitors strongly inhibit the Erk-2 pathway in transfected HEK293 cells and affect the subcellular location of Erk-2 by disrupting its interaction with upstream kinases. Finally, we show that the monobodies can modulate the physiological processes in yeast and worms by inhibiting the activity of Erk-2 orthologs in each organism, demonstrating that the engineered inhibitors are useful in a variety of cellular context.

(447g) Structure-Based Design of Antibodies Specific for Misfolded Proteins

Joseph M. Perchiacca, Moumita Bhattacharya, Ali Reza A. Ladiwala, Swarnim Ranjan and Peter M. Tessier, Dept of Chemical & Biological Engineering, Rensselaer Polytechnic Institute, Troy, NY

Conformation-specific antibodies that recognize aggregated proteins associated with several conformational disorders (e.g., Parkinson's disease) are invaluable for diagnostic and therapeutic applications. We have developed a novel approach for designing conformation- and sequence-specific antibodies against misfolded proteins that is inspired by the homotypic interactions governing protein aggregation. We demonstrate that grafting small hydrophobic peptides from several amyloidogenic polypeptides (e.g., the beta-amyloid peptide associated with Alzheimer's disease) into the complementarity determining regions of antibody fragments generates antibody variants that selectively recognize aggregated proteins containing the cognate peptide motifs (Perchiacca et al., *PNAS*, 2012). We refer to these antibodies as gammabodies for Grafted AMyloid-Motif AntiBODIES. Gammabodies display strict sequence specificity and scrambling their grafted sequences eliminates antibody reactivity. We also find that gammabodies inhibit the cytotoxicity of protein aggregates. We will discuss how gammabodies can be designed against diverse misfolded proteins, as well as how they can be used as site-specific probes to analyze the structures of toxic misfolded proteins.

(447h) Synthetic Cell-Based Devices for Diagnosis, Treatment and Discovery

Joshua N. Leonard, Chemical and Biological Engineering, Chemistry of Life Processes Institute, R.H. Lurie Comprehensive Cancer Center, Northwestern University, Evanston, IL

The ability to engineer customized mammalian cellular functions would enable the construction of sophisticated cell-based therapeutics and transformative tools for fundamental biological research. Such capabilities could overcome persistent barriers to treatment in applications ranging from cancer immunotherapy to regenerative medicine. The emerging field of synthetic biology provides such an approach for building novel cellular functions from the bottom up, and the "toolbox" of biological parts that operate in mammalian cells is rapidly expanding. To date, however, we lack the ability to construct synthetic cell-based biosensors that detect and respond to exclusively extracellular cues. Because many species of biological relevance, including cytokines, chemokines, cell-surface antigens, and many pathogens are exclusively extracellular, engineering cell-based devices that interface robustly with host physiology will require sensors for extracellular species. Here, I will discuss our work developing novel synthetic biology tools that address this need, with a focus on engineering protein components that enable the construction of cell-based biosensors. I will also discuss our application of these technologies to probe interactions between cancer and the immune system, and to develop effective therapies that harness the immune system to control or eradicate cancer.

(448a) Effect of Histone Tail Modifications On Chromosome Structure and Activity

Nate Nurse, Chemical Engineering, Purdue University, West Lafayette, IN and **Chongli Yuan**, School of Chemical Engineering, Purdue University, West Lafayette, IN

The N-terminal regions of histone H3 and H4 proteins play an important role in gene regulation. Abnormal modifications on H3 and H4 tails are associated with various types of cancer diseases, such as lung, prostate and breast cancer. Drug molecules targeting at specific modification sites on H3 and H4 tails provides a viable route for editing aberrant gene transcription pattern. It is therefore of critical importance to understand the role of each histone tail in regulating gene transcriptional activity. We constructed model nucleosomes and nucleosomal arrays with modified H3 and H4 N-terminal tails and examined their respective effects on nucleosome conformation and dynamics using various biophysical experiments. Under different ionic strengths, H3 tails participate primarily in intra-nucleosomal interactions by constraining the DNA breathing motion and limiting its conformational variations. Histone H4 tails, on the other hand, are less effective in regulating the stability of a nucleosome. H4 tails are more involved in inter-nucleosomal interactions. Our results revealed the detailed mechanism as to how H3 and H4 tails are involved in intra- and inter- nucleosomal interactions under different ionic conditions. The findings will facilitate the screening of drug targets on histone tails.

(448h) A Zipper-Like Mechanism Steers the Recognition of Cyanobacterial Split Inteins

Mirco Sorci[1], Bareket Dassa[2], Shmuel Pietrokovski[2], Marlene Belfort[3] and Georges Belfort[1], (1)Rensselaer Polytechnic Institute, Troy, NY, (2)Weizmann Institute of Science, Rehovot, Israel, (3)University at Albany, Albany, NY

The formation of a peptide bond catalyzed by two halves of a split intein is called protein trans-splicing. The potential to use this process synthetically offers many new applications, but little is known about the molecular recognition between the two polypeptide split-intein partners. In this study, we investigated the association between native (wild-type) and mixed pairs of C- and N-terminal partners of naturally occurring split inteins from three *cyanobacteria*: *Oscillatoria limnetica*, *Thermosynechococcus vulcanus* and *Nostoc species PCC7120*. Each split partner was attached to an affinity tag, making possible the direct immobilization of the protein using monoclonal antibodies. Atomic force microscopy (AFM) was then used in force mode to measure the binding forces between the different pairs. The results show that native and mixed pairs exhibit similar binding forces within the error of the measurement technique (~52 pN). Control experiments (e.g. using an engineered non interacting protein) were performed to support our finding. In addition, these non-equilibrium measurements of the work of binding are compared with equilibrium measurements of the free energy of binding confirming the Jarzynski's inequality. Using a bioinformatic sequence analysis and computational structural prediction, we have discovered the presence of highly conserved interspersed electrostatic and non-polar pairs in triplet interacting along the binding surface of the C- and N- terminal partners. This complementary zipper explains the indistinguishable intermolecular forces between native and mixed pairs measured with AFM.

(448c) Forcefield Ptm: Development and Testing of a First Generation AMBER Forcefield for Post-Translational Modifications

George A. Khoury, Jeff P. Thompson and Christodoulos A. Floudas, Chemical and Biological Engineering, Princeton University, Princeton, NJ

Currently the Protein Data Bank (PDB) [1] contains over 80,000 structures. Over 21,000 structures contain at least one modification, of which ~6,000 contain modifications other than disulfide bridges. There is no general forcefield that can accurately simulate how these ubiquitous chemical alterations affect the micro-environments of the proteins they reside in as well as their interactions with each other.

Therefore, in this work we develop and validate a consistent set of AMBER [2, 3] forcefield parameters for the 27 most-common post-translational modifications [4] (**FFPTM**). Calculations consistent with the AMBER forcefield were derived using the procedure of Duan et al. [5]. A two-stage RESP-fitting procedure [6] using α-helical and β-strand conformations of dipeptides (ACE-XXX-NME) was used to fit charges to reproduce the quantum mechanically calculated electrostatic potential (ESP) for each PTM. Utilizing Antechamber [7] for atom-type perception, the minimum set of missing Van der Waals, bond, angle, and dihedral parameters were added for each PTM from the General Amber Forcefield (GAFF) [8] to maintain consistency and to minimize the number of new parameters added to define each new modified amino acid type.

The parameters derived in **FFPTM** were validated using over 10 CPU years of stress-testing via all-atom Langevin molecular dynamics simulations in AMBER for 22 pairs of modified/unmodified systems. Unrestrained local minimizations starting from the initial PDB coordinates for each of the modified proteins yielded an average C_α RMSD of 0.410±0.173Å, comparable to the average C_α RMSD of 0.421±0.163Å for the unmodified structures. For each system, hundreds of local minimizations were performed starting from random initial points created along the MD trajectory. The forcefield parameters yielded an $R^2 = 0.880$ between the average C_α RMSD of the modified structure and its initial point compared to the unmodified structure and its corresponding initial point. Further validation was done via analysis of secondary structure preservation, stability in energies, as well as correlations between the between C_α RMSDs and energetics during the trajectories from the unmodified to the modified structures. Although the initial structures were dissimilar for about 1/5 of the pairs, the average energies during the modified/unmodified structure simulations correlated with $R^2 = 0.997$, whereas average C_α RMSDs to their corresponding minimized PDB structures were correlated with $R^2 = 0.725$, indicating the new parameters did not largely perturb the energetics while the structures dynamically evolved according to the equations of motion. We extend the analysis to validate multiple modifications on a single structure. All modifications parameterized were conformed to the PDB atom-naming convention and were tested using starting coordinates contained in the PDB for each modification for ease of user portability. Insights for applications of the forcefield for blind structure prediction of proteins containing PTMs as well as *de novo* protein design will also be discussed. Further insights regarding structural implications of post-translational modifications observed as well as PTM Density will be presented.

References

1. Berman, H.M., et al., *The Protein Data Bank.* Nucleic Acids Res, 2000. **28**(1): p. 235-42.

2. Weiner, P.K. and P.A. Kollman, *AMBER: Assisted model building with energy refinement. A general program for modeling molecules and their interactions.* Journal of Computational Chemistry, 1981. **2**(3): p. 287-303.

3. Cornell, W.D., et al., *A Second Generation Force Field for the Simulation of Proteins, Nucleic Acids, and Organic*

Molecules. Journal of the American Chemical Society, 1995. **117**(19): p. 5179-5197.

4. Khoury, G.A., R.C. Baliban, and C.A. Floudas, *Proteome-wide post-translational modification statistics: frequency analysis and curation of the swiss-prot database.* Sci. Rep., 2011. **1**.

5. Duan, Y., et al., *A point-charge force field for molecular mechanics simulations of proteins based on condensed-phase quantum mechanical calculations.* Journal of Computational Chemistry, 2003. **24**(16): p. 1999-2012.

6. Bayly, C.I., et al., *A well-behaved electrostatic potential based method using charge restraints for deriving atomic charges: the RESP model.* The Journal of Physical Chemistry, 1993. **97**(40): p. 10269-10280.

7. Wang, J., et al., *Automatic atom type and bond type perception in molecular mechanical calculations.* Journal of Molecular Graphics and Modelling, 2006. **25**(2): p. 247-260.

8. Wang, J., et al., *Development and testing of a general amber force field.* Journal of Computational Chemistry, 2004. **25**(9): p. 1157-1174.

(448d) Alterations in Endothelial Barrier Function Differentially Regulate the Transport of Adiponectin Oligomers

Joseph M. Rutkowski and Philipp E. Scherer, Touchstone Diabetes Center - Department of Internal Medicine, UT Southwestern Medical Center, Dallas, TX

Introduction: Inversely expressed with obesity, the adipokine adiponectin has positive overall insulin-sensitizing and metabolic effects via signaling in various tissues such as liver and skeletal muscle. Adiponectin is unique in its post-translational modifications that form oligomers from the 30kDa monomer up to 540kDa 18-mers found in circulation. These oligomers possess differential efficacy on insulin sensitivity with the high molecular weight (HMW) oligomers, or HMW:Trimer ratio, correlating best with metabolic health. Due to potential transport limitations of HMW adiponectin, we propose that the endothelium presents a transport barrier to adiponectin in a tissue-specific manner, that endothelial barriers are altered with metabolic health, and that endothelial cell function modulates adiponectin's positive effects.

Materials and Methods: Full-form murine adiponectin was produced from stably transfected HEK293T cells and purified by ion exchange; mammalian cells were necessary for HMW formation. HMW, low molecular weight (LMW), and trimeric adiponectin were purified by gel filtration FPLC. Stokes' radii and diffusion rates were estimated by dynamic light scattering for each fraction. Murine endothelial cell lines were used in transwell permeability assays to determine the relative transport rates of adiponectin across different monolayers *in vitro*. IR800 dye-labeled individual fractions were labeled and injected intravenously into several mouse models to assess the role of endothelial barrier function *in vivo*. To quantify adiponectin circulatory fates with changes in endothelium, mice were analyzed in different metabolic states or treated with isosorbide dinitrate and a PPARγ agonist. Transgenic mice lacking endothelial caveolar transport (Cav1 null) were also used. Serum collected at various time intervals allowed quantification of adiponectin circulatory half-life.

Results and Discussion: Adiponectin oligomers possess Stokes' radii of approximately 5, 10, and 25nm making physiologic transport phenomena non-trivial events. Ms1 and b.End3 murine cells form tight cell-cell junctions that limit adiponectin transport and permit little passage of the HMW form. The leaky contacts of EOMA cells, however, permit equivalent oligomer transport. Commercially available, bacterially-produced globular adiponectin – smaller than the mammalian trimer – was transported equivalently across all cells highlighting the physiologic relevance of our preparation. Adiponectin clearance from circulation was oligomer size dependent and altered with endothelial dysfunction. In obesity, adiponectin clearance from circulation is reduced, but can be increased with fasting. Acute isosorbide dinitrate treatment increased HMW adiponectin flux with no effect on LMW clearance. Chronic treatment with a PPARγ agonist or caveolin-1 null mice slowed HMW clearance, but hastened LMW uptake. Endothelial dysfunction and the resultant active and passive changes to transendothelial fluxes reduced tissue access to metabolically positive HMW adiponectin.

Conclusions: Access of adiponectin oligomers into tissues of metabolic relevance may be dependent on the endothelial barrier within each tissue. Adiponectin bioavailability for target cells under metabolically challenged conditions is reduced due to reduced transendothelial transport that, in turn, reduces tissue insulin sensitivity. Adiponectin transport into and across these cells modulates the link between adipose tissue health and cardiovascular disease.

Acknowledgement: JMR was supported in part by NIH F32-DK085935

(448e) Structural and Functional Characterization of Photosynthetic Proteins in Marine Algae

Jing Jiang[1], Hao Zhang[2], Robert Blankenship[2,3] and Cynthia S. Lo[1], (1)Energy, Environmental and Chemical Engineering, Washington University in St. Louis, St. Louis, MO, (2)Chemistry, Washington University in St. Louis, St. Louis, MO, (3)Biology, Washington University in St. Louis, St. Louis, MO

The water-soluble peridinin-chlorophyll *a*-protein (PCP) and the membrane bound intrinsic light-harvesting complex (acpPC) are major light harvesting complexes in photosynthetic dinoflagellates, which are a group of marine algae. PCP contains the carotenoid peridinin and chlorophyll *a*, while acpPC contains an additional pigment chlorophyll c_2. In this study, we identified and characterized the PCP and acpPC proteins in *Symbiodinium* sp. CS-156. The apoPCP has a molecular mass of 32.7 kDa. Despite the heterogeneity of PCP gene tandem repeats, we identified a single form of PCP, which was sequenced by LC-MS/MS analysis of tryptic digested PCP. Protein sequence comparison with the *Amphidinium* main form PCP (MFPCP) shows that the chl *a* and peridinin binding pockets are conserved in these two genera, which indicates the high degree of spectroscopic similarity and the same pigment stoichiometry; these hypotheses were confirmed by experiments. The spectroscopic characterization revealed that the peridinin-to-Chl a Q_y energy transfer efficiency is 95 % in this complex, which suggests that PCP may be promising for use in solar energy capture and eventual biofuel production in engineered devices.

(448f) NMR Characterization of an Engineered Allosteric Enzyme

Abigail H. Laurent[1], Yuchao Chen[2], Luke Arbogast[3], Jason Labonte[1], Ananya Majumdar[4], Jeffrey J. Gray[1], Joel Tolman[3] and Marc Ostermeier[1], (1)Chemical and Biomolecular Engineering, Johns Hopkins University, Baltimore, MD, (2)Biophysics, Johns Hopkins University, Baltimore, MD, (3)Chemistry, Johns Hopkins University, Baltimore, MD, (4)Biomolecular NMR Center, Johns Hopkins University, Baltimore, MD

Allosteric enzymes have activity that is modulated by the presence of an effector that binds to the protein at a site distal from the active site and is transmitted by a conformational change. Engineered allosteric enzymes, or protein-based molecular switches, have many potential applications including as biosensors and locally targeted therapeutics. One way of developing switches is by fusing an enzyme with a ligand binding proteins in such a manner that the ligand modulates the enzyme activity.

RG13 is a maltose activated beta-lactamase (BLA) enzyme identified from a combinatorial library in which circular permutants of TEM-1 BLA were inserted into maltose binding protein (MBP). RG13's catalytic activity is compromised in the absence of maltose and increases 25-fold in the presence of maltose, approaching that of wild-type TEM-1. The primary structure of RG13 consists of a circularly permuted TEM-1 (that has a GSGGG linker between the original N- and C-termini and new termini W227 and G226 respectively) inserted in place of residues 317 and 318 of MBP, with a serine inserted at the end of the TEM-1 sequence. This corresponds to the insertion of TEM-1 near the hinge region of the MBP domain, opposite the maltose binding pocket.

In this work, nuclear magnetic resonance (NMR) is used to study the allosteric mechanism of RG13. NMR spectra of ^{15}N-^{2}H-labeled RG13 confirmed the similarity of the MBP and TEM-1 domains to the spectra of their wild-type parental proteins. They also provided evidence for larger structural differences in the TEM-1 domain in RG13 in the absence of maltose, consistent with the depressed enzymatic activity of RG13. Further work on ^{13}C-^{15}N-^{2}H-labeled protein is allowing for assignments of NMR measurements to specific residues. Mutations at the linkage points between the TEM-1 and MBP that affect the switching provide evidence for the linker's role in the transmission of the conformational change from the MBP domain to the BLA domain. Results have implications for the natural evolution of allostery and for construction of engineered protein switches.

(448g) Biophysical Characterization of Mutated Fiber Adenoviruses

Grit Kupgan and Joshua Ramsey, Chemical Engineering, Oklahoma State University, Stillwater, OK

Adenovirus is one the most widely used vectors in clinical trials for gene therapy. The majority of adenovirus internalizes into cells *via* highly specific interactions between the fiber protein on the virus and the Coxsackie-adenovirus receptor (CAR) on the surface of target cells. This broad natural tropism results in difficulty targeting adenovirus to specific diseased cells due to the ubiquitous presence of CAR receptors on various cells and tissues. Additionally, the fiber and capsid proteins on the surface of adenovirus are recognized by neutralizing antibodies (NAbs), which severely hamper the efficacy of adenovirus as a gene delivery and vaccine vector. The fiber protein consists of three domains that include the tail, shaft, and knob. The shaft domain consists of 22 repeating motif of approximately 15 residues and the pattern of hydrophobic residues can be aligned, and the knob domain is primarily responsible for interaction with the CAR receptor.

Our approach to overcoming issues of promiscuous tropism and immunogenicity is to remove the fiber protein from adenovirus, and replace its function with synthetic materials. The fiberless virus, however, is unstable and unable to assemble into correctly matured viral particles. To optimize the stability of the vector, while also reducing the natural tropism and immunogenicity, the knob protein was removed entirely and the length of the fiber protein was varied from 7-repeats, 14-repeats, and 21-repeats. The relative stability of these mutants was compared with the native adenovirus and fiberless adenovirus using circular dichroism, static and dynamic light scattering, intrinsic tryptophan fluorescence, and extrinsic fluorescence. The results indicate the minimum number of pseudo-repeats required for assembly of stable fiber-mutants adenoviral particles and are being used to identify the best candidate for future gene delivery studies.

(448b) Acyl Carrier Protein Structural Classification and Normal Mode Analysis

David C. Cantu, Michael J. Forrester, Katherine Charov and Peter J. Reilly, Chemical and Biological Engineering, Iowa State University, Ames, IA

Acyl carrier proteins (ACP) bind fatty acids and polyketides during their synthesis. They are short proteins composed of antiparallel beta-sheets. All ACPs were gathered into the ThYme database, in which enzyme primary and tertiary structures involved in fatty acid and polyketide synthesis and related reactions are tabulated. ACPs are classified into 16 families by amino acid sequence similarity, with members of the different families having sequences with statistically highly significant differences. These classifications are supported by tertiary structure superposition analysis. Tertiary structures from a number of families are very similar, suggesting that these families may come from a single distant ancestor. Normal vibrational mode analysis was conducted on experimentally determined freestanding structures, showing greater fluctuations at chain termini and loops than in most helices. Their modes overlap more so within families than between different families. The tertiary structures of three ACP families that lacked any known structures were predicted as well.

(451a) Building Tools for Predicting Allosteric Regulation Pathways in Proteins

Galen Collier and Vanessa Ortiz, Chemical Engineering, Columbia University, New York, NY

Allosteric regulation of protein function is a mechanism by which a structural event occurring at one site within a protein's overall structure causes an effect at another site, and allostery plays a central role in a huge variety of mechanistic pathways in biological systems. The prediction and modulation of allosteric responses in proteins holds promise for applications in a diverse range of research areas, including drug design and materials engineering. The structural changes involved in protein allostery include deformations where local elastic moduli play a major role. Knowledge of the elastic modulus at each atom position or within specific regions in a protein can provide valuable information about the local mobility and stability under conformational deformations. However, most methods used for calculating local elastic moduli cannot be applied to proteins in a straightforward manner because they require second derivatives of the potentials that describe the interactions amongst a protein's atoms. To address this issue, we have begun the development of computational methods that enable the measurement of elastic moduli at different positions in a protein in order to build a 3D map of local mechanical properties that can be used to identify pathways of allostery in proteins and protein networks. Presented here are the results from a preliminary set of protein structural analyses completed using these methods to calculate elastic moduli in an atomistic fashion.

(451b) Computer Aided Design of Bio-Reactor for Bone Tissue Engineering

Marina Campolo, Center for Fluid Dynamics and Hydraulics, University of Udine, Udine, Italy, Francesco Curcio, Dept. Biological and Medical Sciences, University of Udine, Udine, Italy and Alfredo Soldati, Department of Mathematics and Informatics, University of Udine, Udine, Italy

In this work we propose a modelling approach to identify sets of culture conditions to promote homogeneous growth of cells in perfusion bioreactors equipped with regular shape scaffolds.
Homogeneous growth is the outcome of an initial homogeneous cell seeding and of an homogeneous cell proliferation which depend on a number of variables including pH, temperature, oxygen tension, nutrient supply, catabolites removal and also biochemical and mechanical stimuli. Consequently, the entire reactor and its operating protocols should be carefully designed to achieve a homogeneous distribution of cells during seeding, to maintain conditions which promote homogeneous proliferation and to prevent conditions which may lead to cell necrosis.
Providing the right mechanical stimulation and adequate nutrient supply are key issues in cell culturing: at optimal operating conditions the level of shear stress should be large enough to stimulate cell proliferation/differentiation (Porter et al., 2005, Ma et al., 2007, Olivares et al., 2009) and, yet, not too large to cause cell damage or detachment from the scaffold; in addition, nutrient feeds should be large enough to replenish continuously species consumed by cellular metabolic activity and removal of catabolites should be efficient enough to avoid local accumulation of species (e.g. gas bubbles) which may affect the culture microenvironment.
We use a zero dimensional mass transport and reaction model to identify cases in which dynamic culturing is necessary to avoid nutrient limitation. Then, based on the three dimensional rendering of the flow field inside the bioreactor, we (1) identify regions where cellular growth may become critical due to the low/high value of local shear stress and (2) we evaluate the risk of accumulating catabolites in the form of gas bubbles which may lead to clogging of scaffold microchannels; finally, using a one dimensional mass transport and reaction model we calculate the minimal perfusion flow necessary to maintain the cellular growth rate above a target threshold and we identify the conditions (bio-reactor region, flow rate, limiting bubble size) for which bubbles may adhere stably to the scaffold, potentially leading to bubble clogging.
The developed approach is used to analyse culturing conditions inside an indirect perfusion bioreactor equipped with a lattice scaffold. Regions where the perfusion flow is inadequate to foster cellular growth at the desired rate are identified and the probability of local clogging is evaluated. The perfusion flow required to maintain the target growth rate inside the bioreactor is calculated.

(451c) Modelling and Robust Model Predictive Control of Insulin Delivery in Type 1 DM

Stamatina Zavitsanou[1], Nicki Panoskaltsis[2], Athanasios Mantalaris[1], Michael C. Georgiadis[1] and Efstratios N. Pistikopoulos[3], (1)Department of Chemical Engineering, Centre for Process Systems Engineering, Imperial College, London, United Kingdom, (2)Department of Haematology, Northwick Park Hospital, London, United Kingdom, (3)Chemical Engineering, Imperial College London, Centre for Process Systems Engineering, London, United Kingdom

The determination of effective strategies to treat Type 1 Diabetes Mellitus is of paramount importance since it can be characterised as one of the most prevalent severe chronic diseases in childhood. Motivated by the challenge to ensure the patients' safety and to improve their quality of life the development of a closed loop system of optimal insulin delivery has been the most widely studied approach in the field (Kovatchev et al., 2010), (Grosman et al., 2010). In this framework, a detailed physiologically based compartmental model is presented, extended from previous studies (Zavitsanou, et al. 2011). It represents personalised glucose-insulin interactions in the fasting, prandial and postprandial state while insulin is administered exogenously with an insulin pump and mimics the mechanism of pancreatic response. Glucose coming from food consumption and endogenously produced is distributed in blood and tissue compartments, representing the involved organs and glucose regulation is achieved with the appropriate subcutaneous insulin infusion to compensate for high blood glucose concentrations. Patient specific characteristics such as age, gender, weight, height, hematocrit and insulin sensitivity are included and the model can represent a virtual patient. The individual insulin delivery (dosing and scheduling) is optimised in terms of maintaining blood glucose concentration within the safe range. Closed loop studies in the context of explicit/multi-parametric model predictive control (Dua et al., 2006) are presented to evaluate the performance of the controller that takes into account safety constraints of desired blood glucose concentration and operational constraints of available insulin delivery. Case studies of closed loop insulin delivery of patients subjected to external disturbances are presented to highlight the challenges of the system such as high intra patient variability and time delays.

Dua, P., Doyle, F. J., & Pistikopoulos, E. N. (2006). Model-based blood glucose control for Type 1 diabetes via parametric programming. *IEEE transactions on bio-medical engineering*, *53*(8), 1478-91.

Grosman, B., Dassau, E., Zisser, H. C., Jovanovič, L., & Doyle, F. J. (2010). Zone model predictive control: a strategy to minimize hyper- and hypoglycemic events. *Journal of diabetes science and technology*, *4*(4), 961-75.

Kovatchev, B., D, P., Cobelli, C., Renard, E., Anderson, S., Breton, M., et al. (2010). Multinational study of subcutaneous model-predictive closed-loop control in type 1 diabetes mellitus: summary of the results. *Journal of diabetes science and technology*, *4*(6), 1374-81.

Zavitsanou, S., Panoskaltsi, N., Mantalaris, A., Georgiadis, C, M., & Pistikopoulos, E. N. (2011). Modelling of the Insulin Delivery System for patients with Type 1 Diabetes Mellitus. *21st European Symposium on Computer Aides Process Engineering*.

(451d) Deriving 2D Velocity Profile Using Streamlines Image Velocimetry (SIV)

Eliezer Keinan[1], Elishai Ezra[1] and Yaakov Nahmias[1,2], (1)Bioengineering Program, Benin School of Computer Science and Engineering , The Hebrew University of Jerusalem, Jerusalem, Israel, (2)Department of Cell and Developmental Biology, Silberman Institute of Life Sciences, The Hebrew University of Jerusalem, Jerusalem, Israel

Introduction

Fluid dynamics plays a fundamental role in environmental, biological, and industrial processes ranging from mass transport to separations and heat exchange. Investigation of sand dunes, wind waves, heat transfer, fouling, and aneurysm failure are a few examples where an exact solution of the fluid velocity field is required. Current methods to investigate this velocity field are cumbersome and hence, a rapid, accurate and user friendly analysis tool is required.

Current direct velocimetry methods include Particle Imaging Velocimetry (PIV), Particle Tracking Velocimetry (PTV), and Laser Doppler Velocimetry (LDV). These methods require significant investment in equipment and computational power. For example, a commercial PIV system consists of a laser diode and an air-cooled superfast camera.

Here, we introduce a novel method termed Streamline Image Velocimetry (SIV) which allows a rapid quantification of velocity fields in arbitrary geometries. We use a single long-exposure image to capture multiple particles streaking in individual streamlines across the field. Streamlines confine streamtubes, in which the volumetric flow is constant for incompressible fluid. Using the explicit analytical solution as an initial condition, we quantify the velocity field throughout the entire field. We compiled our results into a computer program that enables the rapid derivation of velocity profiles from a single input image, which can be taken using a simple household camera. Our results show an excellent correlation to numerical simulations and are able to quantify fluid velocities across complex porous barrier, where numerical simulations are difficult at best.

Results

Analytical Solution for Rectangular Channel

The analytical solution for the velocity field in the middle of a low aspect ratio (h<<w) rectangular channel is approximated as a Fourier series in equation 1:

$$(1)\ V_x(y)=48Q/(\Pi^3*(w-0.63h))*\Sigma^{\infty}_{n,odd}*n^{-3}*(1-\cosh(n\Pi y/h)/\cosh(n\Pi w/2h))*\sin(n\Pi/2)$$

Where v_x is velocity vector, h is the channel height, w is the channel width, y is horizontal location inside the channel and Q is the volumetric flow rate.

Streamline Imaging

To experimentally quantify the fluid velocity field we introduced fluorescent microbeads into a microfabricated channel which contained distortions. Long exposure images were taken using a fluorescent microscope, taking care to trace complete paths of multiple particles across the entire field of view. The average velocity between two streamlines is defined by volumetric flow rate divided by stream tube cross section. Polynomials were fitted to the visualized streamlines. A mean polynomial was calculated between every two neighboring streamlines. The velocity direction was determined by the derivative of the mean polynomial. The distance between the streamlines was calculated as the length of a line perpendicular to the mean polynomial drawn from the upper polynomial to the lower one.

Computational Image Analysis

The image processing and velocity profile calculation were compiled into a single MATLAB program (code is available on request) with a graphical user interface (GUI). The program enables to load streamlines snapshot, preliminary image processing (alignment, cropping, binary conversion, PSF deconvolution). The program is loaded with macroscopic parameters, as fluid viscosity, density and "anchor" region geometry. Based on the entered parameters, the program can fit a set of polynomials to the streamlines, and calculate the velocity field. Data can be presented as either numerical values, or an arrow field.

SIV Analysis of Flow in Microfluidic Device

We fabricated a PDMS device with a negative arched distortion and took snapshots of the streamlines using fluorescent microscopy. We analyzed the streamlines snapshots in MATLAB and calculated the velocity profile in cross sections. The velocities were compared to COMSOL derived numerical simulations: in the "anchor" region where the program calculates the analytical solution, in the middle of the arch where the streamlines are parallel and there is no perpendicular velocity component and in the quarter of the arch. The results show high correlation (0.97). The correlation between SIV and the numerical calculation is high (0.96) even at the quarter of the distortion where the perpendicular component of the velocity is maximal presenting a rapid change in velocity as function of distance from the channel wall.

SIV Analysis of a Microfluidic Channel with Porous Elements

In order to verify our method with a different flow irregularity, characterized with unknown boundary conditions, we produced the velocity field of a visualized streamlines snapshot, taken from a recent publication. The visualized streamlines were taken from microfluidic device integrated with nanoporous forests of vertically aligned carbon nanotubes. In such devices the complexity of the velocity field calculation increases due to the porous boundary conditions. However, for SIV, the boundary conditions are the streamtubes enabling the derivation of the velocity field based solely on the visualized streamlines. The results indicate that SIV produces high-precision results in complex flow regions where the fluid velocity magnitude and direction rapidly changes. SIV analysis presenting velocimetry at low cost and reduces computational resources.

Materials and Methods

Chemicals

Poly(dimethyl)siloxane (PDMS) was purchased from Dow Corning (SYLGARD). SU8-2050 was purchased from Microchem (Newton, MA). Fluorescent green microbeads, Pluronics F68, and Phosphate Buffered Saline (PBS) were purchased from Sigma Aldrich (Rehovot, Israel).

Numerical Modeling

Numerical simulations were performed using COMSOL multiphysics simulation platform v3.5A with a direct linear system solver and extra fine mesh (PARADISO). Fluid density was defined as $1x10^3$ kg/m^3, with a dynamic viscosity of $1x10^{-3}$Pa·s.

Microfluidic Channels

Microfluidic devices were fabricated by soft lithography. Briefly, molds were fabricated by photolithography of SU8 on silicon wafers at the Hervey Krueger Center of Nanoscience and Nanotechnology at the Hebrew University of Jerusalem. Channels were replica molded in PDMS and bonded to glass using oxygen plasma bonding. The device is 100 μm high. Main channel width is 3000 μm and the disturbance depth is 100 μm.

Streamline Imaging

Prior to the experiment the device channel was coated with Pluronic F-68 for 1 hour at room temperature to prevent non-specific adhesion of fluorescent particles. The experiments were performed with microbeads diluted in 1% Pluronic F-68 at a concentration of $2{\cdot}10^6$ particles/ml. Microbeads were perfused using a Chemyx Fusion 200 syringe pump and imaged on a Zeiss Axiovert LSM700 Microscope.

Streamlines Image Analysis

Images processing was carried out in MATLAB. Image were imported and deconvolved using Gaussian filter generated from the theoretical point spread function (PSF). Streamlines were traced using MATLAB 2011b algorithm. Polynomial functions were fitted to traced streamlines and transformed into a binary 2D matrix.

Streamlines Image Velocimetry

A streamline is a curve parallel to the velocity vector which can be visualized using long exposure snapshots of fluorescent microparticles in fluid. In steady flow, the streamlines coincide with the paths of the fluid particles. A streamtube is a surface confined between streamlines. The volumetric flow rate in a streamtube is conserved; therefore, the product of the streamtube cross-section (A) and the average velocity (w) is constant as described in Equation 2:

$$(2)\ A_1 w_1 = A_2 w_2 = Aw$$

Our method relies on the fact, that in a streamtube that is formed in a simple geometry-shaped vessel, the volumetric flow rate is conserved when entering into a more complex geometry and hence, the volumetric flow rate in the complex region can be derived from flow in the simple geometry. Hence, our method requires that the analyzed vessel will include a simple geometry fluid entrance region, which will serve as an "anchor" for further calculations. Our analytical solution is suitable for rectangular cross section channels, however it can be easily modified for any other geometry.

(451e) Non-Dimensional Analysis of Retinal Microaneurysms: Critical Threshold for Treatment

Elishai Ezra[1], Eliezer Keinan[1], Yossi Mandel[1], Michael Boulton[2] and Yaakov Nahmias[1,3], (1)Bioengineering Program, Benin School of Computer Science and Engineering, The Hebrew University of Jerusalem, Jerusalem, Israel, (2)Department of Anatomy and Cell Biology, University of Florida, (3)Department of Cell and Developmental Biology, Silberman Institute of Life Sciences, The Hebrew University of Jerusalem

Introduction: Diabetic retinopathy is responsible for 5% of the cases of blindness worldwide and for up to 17% of the cases in the western world. Retinal microanuerysms reduce vision quality due to local loss of endothelial barrier function, causing leakage and retinal edema. Localized leakage from microaneurysms can be detected and blocked using laser ablation, slowing the progression of diabetic blindness. Our approach has been to gain a critical understanding of the fluid dynamics of the problem, which will enable us to predict risk associated with the diseased morphology. However, analytical solutions of Navier-Stokes equations governing fluid dynamics exist solely for restricted conditions, while numerical simulations can only be preformed on one set of initial and boundary conditions at a time. We derived a non-dimensional expansion parameter from an approximate analytical solution and generalized it using an experimentally validated numerical solution, uncovering a non-dimensional behavior, which allows for rapid calculation of critical properties. Such properties could be used to rapidly calculate changes in fluid velocity, shear force, or fluid pressure due to microvessel expansion, without the need for complex analytical or numerical solutions. In order to validate our findings, we used high-resolution confocal images of diabetic retinas and observed that microaneurysms above our critical threshold demonstrated a dramatic, size-dependent increase in von Willebrand Factor (vWF) expression, a marker of endothelial dysfunction.

Materials and Methods: Numerical simulations were performed using COMSOL multiphysics simulation platform v3.5A with a direct (PARADISO) linear system solver and extra fine mesh. The experimental validation of the numerical model was performed in microfluidic devices, which were fabricated by soft lithography at the Harvey Krueger Center of Nanoscience and Nanotechnology at the Hebrew University of Jerusalem. Postmortem donor eyes from diabetic patients with early retinopathy characterized by the presence of microaneurysms were obtained from the National Disease Research Interchange (Philadelphia, USA). Retinas were examined from the vitreous surface using Biorad MRC 600 confocal laser scanning attachment mounted on a Zeiss Axiovert microscope using a 25 mW argon laser (Ion Laser Technology). Maximum projection images were analyzed using the Image J software (NIH) to quantifying microaneurysm geometry and average vWF intensity.

Results and Discussion: Endothelial dysfunction occurs when shear force drops below 0.4 Pa, enabling us to calculate a critical non-dimensional expansion ε of 0.38. Our analysis sets a critical aneurysm diameter of 15 μm for capillaries, 25 μm for venules and 104 μm for arterioles, above which endothelial dysfunction is predicted to occur. To test this hypothesis we obtained high-resolution confocal images of human diabetic retinas and quantified microaneurysm's size distribution. The percentage of microaneurysms above our threshold ranged from 10% to 60%. Remarkably, while vWF expression in microaneurysms below our threshold was not significantly different than the surrounding microvasculature ($p=0.076$), it was 60±10% lower in microaneurysms above our threshold ($p=7\times10^{-15}$).

In contrast to the high correlation between vWF to the non-dimensional expansion, there was a weak correlation with the aneurysm size (R2=0.59). The rapid advance of optics over the past decade permits high-resolution imaging of the retina in vivo with functional markers, such as vWF. A combined imaging of structure and function can lead to optimization and customization of preventive treatment before the onset of retinal edema. Our method will allow experts to identify and treat microaneurysms posing a high-risk of leakage, prior to edema, minimizing damage and saving vision.

Conclusions: Our approach has been to gain a critical understanding of the fluid dynamics in retinal microaneurysms, which will enable us to predict the risk associated with the diseased morphology by uncovering a non-dimensional behavior, allowing rapid calculation of critical propermmmDa sdsdly LE in the quarter i while tified ties.

(451f) Engineered Macromolecules As Inhibitors to Oxidized Low Density Lipoprotein by Macrophage Scavenger Receptors: Simulation of Structure – Function Relationships

Michael Tomasini, Biomedical Engineering, Rutgers University, Piscataway, NJ and M. Silvina Tomassone, Chemical and Biochemical Engineering, Rutgers University, Piscataway, NJ

Atherosclerosis begins when high levels of circulating low-density lipoproteins (LDL) are transported into the extravascular space and become trapped in the extracellular matrix where they are subject to oxidative modification. Oxidized LDL causes damage to surrounding endothelial cells and initiates an inflammatory signaling cascade which results in the recruitment of monocytes to the site of injury. Monocytes differentiate into macrophages that express surface scavenger receptors active in the non-specific uptake of modified LDL. The excessive uptake of oxidized LDL results in the formation of foam cells which become engorged, die, and subsequently form atherosclerotic plaques. One possible method to combat plaque formation is through the use of inhibitors to prevent uptake of oxidized LDL by the scavenger receptors. Tian et al. 2004 [1] developed a class of micelle-forming molecules termed nanolipoblockers (NLB) comprised of a mucic acid head group with attached aliphatic chains and a long polyethylene glycol (PEG) tail used to block uptake of oxidized LDL by scavenger receptors. In this work, we use coarse-grained molecular dynamics simulations to explore the structural and electrostatic properties of NLBs with the scavenger receptor A (SR-A) to determine optimal interaction and thus inhibition of oxidized LDL uptake.

We use a coarse-grained model based on the MARTINI force field [2] to describe the NLB molecules as well as the collagen-like domain of SR-A containing the binding pocket for oxidized LDL [3]. We include four different NLB structures varying the charge, charge location, and the NLB ability to form micelles. We find that for NLB molecules containing aliphatic groups near the mucic acid head, micelles form within 300 ns with a micelle diameter of 7.6 – 12.8 nm. As an indirect assessment of binding affinity, we identified the number of contacts that each type of NLB made with SR-A both at 0.8 nm and 2.0 nm from the receptor. The number of close contacts (< 0.8 nm) were similar for each NLB, but aligned with experimental trends showing anionic micelle-forming NLBs as the best inhibitors of oxidized LDL uptake. In calculating the specific contacts between the charged portions of NLBs and the binding pocket of SR-A indicated that the addition of a negative charge on the PEG tails of NLBs did not act to enhance interaction with the important residues of the SR-A, however an anionic charge at the NLB head increased interaction with the positively charged residues of the binding pocket, confirming experimental findings

References:

[1] Tian L, Yam L, Zhou N, Tat H, and Uhrich KE. Amphiphilic Scorpion-like Macromolecules: Design, Synthesis, and Characterization. *Macromolecules*. 2004; **37**: 538-543.

[2] Monticelli L, Kandasamy SK, Periole X, Larson RG, Tieleman DP, and Marrink SJ. The MARTINI Coarse-Grained Force Field: Extension to Proteins. *J. Chem. Theory and Comput.* 2008; **4**(5): 819-834.

[3] Kodama T, Freeman M, Rohrer L, Zabrecky J, Matsudaira P, Krieger M. Type I macrophage scavenger receptor contains alpha-helical and collagen-like coiled coils. *Nature*. 1990; **343**(6258): p. 531-535.

(451g) Mechanistic Modeling of DNA Hybridization On Surfaces for Improved Microarray Design

Kyle E. Pratt, Ryan C. Welling, Terry J. Schmitt and Thomas A. Knotts IV, Department of Chemical Engineering, Brigham Young University, Provo, UT

DNA microarrays are chip-based, analysis tools which can perform hundreds of thousands of assays in parallel. They can be used to determine the identity of genes or gene expression levels present in a sample and have been identified as a key technology in genomic sciences and emergent medical techniques. Despite their abundant use in research laboratories, microarrays have not been used in the clinical setting to their fullest potential due to the difficulty of obtaining reproducible results. Microarrays work on the principle of DNA hybridization, and can only be as accurate as this process is robust, yet fundamental, molecular-level understanding of hybridization on surfaces is lacking. Specifically, a detailed knowledge of the various factors affecting hybridization of probes to targets is needed to further refine these devices.

Using a carefully-parameterized, coarse-grain model of DNA, we have spent the last several years studying the thermodynamics of hybridization on surfaces for simple probe/target pairings. This has led to several important insights, but a detailed mechanism of the process has never been reported. In this presentation, we outline efforts to fill in several gaps in understanding. A detailed description of the hybridization process is presented first. These results were obtained using extensive computer simulations to create a projection of the free energy of hybridization onto two reaction coordinates: the probe/target separation distance and the angle of approach between the two strands. Several cases, such as strands of equal and unequal lengths in the bulk and on the surface, are presented. The free-energy projections were obtained from two-dimensional umbrella sampling, and each required over 11,000 simulations. Though computationally demanding, this work has created an unprecedented picture of the hybridization process.

In addition to the mechanisms described above, we also present results on how factors found in real microarrays affect the hybridization process. Specifically, we show how the presence of multiple probes on the surface affects hybridization, and describe how the number and location of mismatched base pairs affect the stability of the duplex. Finally, we describe how the hybridization process is influenced by the locations of probe/surface connections. As a whole, the results give the most complete picture to date of the biophysics involved in microarray performance and how this knowledge can be used to improve next-generation devices.

(451h) Modeling Nonspecific Interactions in Biological Systems

Andrew D. White, Ann K. Nowinski, Wenjun Huang, Andrew J. Keefe, Fang Sun and Shaoyi Jiang, Chemical Engineering, University of Washington, Seattle, WA

Many of the difficulties in biomedical applications arise from the complexities of multiple interactions in biological systems. These interactions can be broadly broken into two categories: those which are important to function (high activity for a specific target) and nonspecific binding which is detrimental to function (low activity binding to many targets). Nonspecific interactions are central to many failures of drugs, devices and materials in the biomedical field. Despite this fact, there is little modeling work on studying nonspecific interactions. Modeling all possible interactions within a biological system is difficult, yet there are ways to both indirectly study nonspecific binding and directly study many possible interactions using bioinformatics. In this work, we describe recent work using bioinformatics, phenomenological modeling, molecular simulations, and experiments to study three types of nonspecific interactions. They are the weak nonspecific interactions between proteins in crowded environments, irreversible nonspecific protein adsorption, and nonspecific stabilization of proteins in molecular chaperones. These disparate systems have allowed us to generalize our conclusions into a small set of principles that other researchers should consider when designing specific activity in order to maintain resistance to nonspecific interactions.

(451i) Exploring the Relationship Between Helicobacter Pylori's Caga Sequence and Affinity with Host's Receptors: A Proposal for Molecular Diagnostic Tools

Paula Andrea Delgado Pinzón[1], Carlos Alberto Jaramillo Henao[2], María del Pilar Delgado Perafán[2], Natalia Melisa Peñaranda Fajardo[2], Nathalia Garces Ferrreira[3], Harold Enrique Castro Barrera[3] and Andrés Fernando González Barrios[1], (1)Chemical Engineering, Universidad de los Andes, Bogotá, Colombia, (2)Biological Sciences , Universidad de los Andes, Bogotá, Colombia, (3)Systems and Computing Engineering , Universidad de los Andes, Bogotá, Colombia

Helicobacter pylori is a human pathogen reported to be present in nearly 50% of world's population[1,2]. Moreover the virulence of the strain is one of the major determinants for disease developing among infected people. One of the major factors associated to *H. pylori* virulence is the cytotoxin associated gene A (*cag*A) which encodes the CagA protein[3,4]. This protein is injected directly from the microorganism through type 4 secretion system (T4SS) to gastrointestinal epithelial cells, where disrupts the cellular signaling mechanism[5]. It is known that CagA has a 5' region highly conserve while 3' region is variable due to the presence of different types and/or numbers of repeat sequences which contains an EPIYA motif location where the phosphorylation of the protein takes place[2,6]. These differences in sequence may have a relationship with the affinity of CagA with host's receptors and hence with the pathogenicity of *H. pylori*. Here in this work we obtained several sequences of different Colombian CagA strains and determine the tertiary structure using I-Tasser[7] server based on multiple threading complemented with Replica Exchange Dynamics technique[8] using opportunistic Cloud Infraestructure. Then, these structures were docked[9,10] with different receptors including SCHP-2 and Src. Free energy calculations were correlated with the pathogenicity of the strains which were previously categorized by its pathogenicity level (cancer, chronic ulcer, chronic gastritis etc) in order to determine a potential tool for molecular diagnostics.

References

1. Higashi H. SHP-2 Tyrosine Phosphatase as an Intracellular Target of *Helicobacter pylori* CagA Protein. Science. 2001;295(5555):683-686.

2. Acosta N. *Helicobacter pylori* CagA protein polymorphisms and their lack of association with pathogenesis. World Journal of Gastroenterology. 2010;16(31):3936.

3. Backert S, Tegtmeyer N, Selbach M. The versatility of *Helicobacter pylori* CagA effector protein functions: the master key hypothesis. Helicobacter. 2010;15(3):163-176.

4. Jones KR, Whitmire JM, Merrell DS. A Tale of Two Toxins: *Helicobacter pylori* CagA and VacA Modulate Host Pathways that Impact Disease. Frontiers in Microbiology. 2010;1.

5. Handa O, Naito Y, Yoshikawa T. CagA protein of *Helicobacter pylori*: A hijacker of gastric epithelial cell signaling. Biochemical Pharmacology. 2007;73(11):1697-1702.

6. Salehi Z, Jelodar MH, Rassa M, Ahaki M, Mollasalehi H, Mashayekhi F. *Helicobacter pylori* cagA Status and Peptic Ulcer Disease in Iran. Digestive Diseases and Sciences. 2008;54(3):608-613.

7. Roy A, Kucukural A, Zhang Y. I-TASSER: a unified platform for automated protein structure and function prediction. Nature protocols. 2010;5(4):725-738.

8. Sugita Y, Okamoto Y. Replica-exchange molecular dynamics method for protein folding. Chemical Physics Letters. 1999;314(1-2):141-151.

9. Qu K, Brooijmans N. Structure-Based Drug Design Computational Methods for Protein Structure Prediction and Modeling. In: Xu Y, Xu D, Liang J, eds: Springer New York; 2007:135-176.

10. Pierce B, Weng Z. Structure prediction of protein complexes. Computational methods for protein structure prediction and modeling. 2007:109-134.

(453a) Improving Stem Cell Transplantation Through Fluid Dynamics and Polymer Physics

Sarah C. Heilshorn, Materials Science and Engineering, Stanford University, Stanford, CA

Stem cell transplantation is a promising therapy for a myriad of debilitating diseases and injuries; however, current delivery protocols are inadequate. Transplantation by direct injection, which is clinically preferred for its minimal invasiveness, commonly results in less than 5% cell viability, greatly inhibiting clinical outcomes. We demonstrate that a significant cause of cell death occurs during transport through the syringe needle, where extensional stretching forces experienced by the cells results in mechanical membrane disruption. We further demonstrate that a variety of clinically relevant cell types including adipose-derived stem cells (ASC), marrow stromal cells (MSC), and neural progenitor cells (NPC) can be protected from these damaging forces by encapsulation within hydrogels of specific viscoelastic properties. We hypothesize that these shear-thinning hydrogels may undergo plug-flow through the syringe needle, whereby bands of shear-thinned polymer are located at the needle walls and lubricate the flow of individual gel "plugs" through the needle. These gel plugs may serve as transport vehicles for the encapsulated cells, thereby shielding them from disruptive mechanical forces.

Building on these fundamental studies, we have designed a reproducible, bio-resorbable, customizable hydrogel using protein-engineering technology. In our Mixing-Induced Two-Component Hydrogel (MITCH), network assembly is driven by specific and stoichiometric peptide-peptide binding interactions. By integrating protein science methodologies with polymer physics models, we manipulate the polypeptide chain interactions and demonstrate the direct ability to tune the network crosslinking density, sol-gel phase behavior, and gel mechanics. These MITCH materials enable stem cell and growth factor encapsulation upon simple mixing at constant physiological conditions, making them well suited for use in the surgical suite. *In vivo* studies in a murine model demonstrate that transplanted ASCs achieve significantly better retention at the subcutaneous injection site compared to cells delivered in collagen, alginate, or saline alone. These results provide mechanistic insight into the role of mechanical forces during cell delivery and support the use of protective hydrogels in future clinical stem cell injection studies.

(453b) 2D Culture and 3D Material Arrays to Define Optimal Conditions for Tenogenic MSC Differentiation

Steven R. Caliari[1], Emily A. Gonnerman[1] and Brendan A. Harley[1,2], (1)Chemical & Biomolecular Engineering, University of Illinois at Urbana-Champaign, Urbana, IL, (2)Institute for Genomic Biology, University of Illinois at Urbana-Champaign, Urbana, IL

The repair of tendon injuries is critically limited by the inability to produce large numbers of tendon fibroblasts, or tenocytes. Multipotent cells, such as mesenchymal stem cells (MSCs), are a potential cell source for tendon repair and can be driven to differentiate down osteogenic, chondrogenic, adipogenic, and other lineages using established protocols. While previous work has highlighted the possible roles of mechanical stimulation, cell confinement/alignment, and co-culture with primary tenocytes, there is no standard protocol for robust tenogenic MSC differentiation. One reason for this is the lack of molecular markers to distinguish tendon from other mesenchymal tissues. Recently, scleraxis (SCX) and thrombospondin-4 (THBS4) have been suggested as tendon-specific markers. Here we present 2D and 3D culture systems to evaluate combinations of biochemical and biophysical factors to drive tenogenic MSC differentiation.

We first used a 2D culture system to assess the effects of a wide range of biochemical factors on the expression of tenogenic markers SCX and THBS4. Human MSCs were seeded in 96-well plates at a density of 1 x 10^3 cells/well and cultured for 14 days in the presence of single or binary combinations of tenogenic soluble factors bFGF, IGF-1, GDF-5, and GDF-7. Additionally, the influence of standard differentiation agents dexamethasone and ascorbic acid was tested. MSCs were fixed, incubated overnight with primary antibodies against SCX and THBS4, tagged with fluorescently labeled secondary antibodies, and then assayed on a spectrophotometer. Fluorescence was normalized to cell number (DAPI). We found that the presence of dexamethasone and ascorbic acid resulted in increased levels of both SCX and THBS4 on a per cell basis for nearly every soluble factor combination assayed. Additionally, the IGF-1 and GDF-7 groups elicited the greatest increase in SCX and THBS4 production per cell.

In parallel, we have developed arrays of 3D collagen-GAG (CG) scaffolds. Tenocytes are known to de-differentiate in standard 2D culture so we used this system to interrogate the combined effect of scaffold microstructure and biochemistry on tenogenic MSC differentiation. Scaffold arrays with two distinct microstructural regions were fabricated by controlling local heat transfer during the freeze-drying process. A specialized mold was created for this purpose, consisting of a polysulfone chip (2 mm thick) with circular holes (6.5 mm diameter) mounted on a removable base. The geometry/spacing of the nodes was designed to be identical to the dimensions of a 96-well plate. The removable base contained an aluminum section and a polysulfone section ($k_{aluminum}/k_{polysulfone} \sim 900$); this disparity in thermal conductivity was intended to control local heat transfer, and subsequently ice crystal growth kinetics, during freezing. The mean pore sizes for scaffolds were 88 ± 28 μm within the aluminum section and 157 ± 47 μm within the polysulfone section of the array for a freezing temperature of -40°C. Cells seeded in the scaffold array and stained with calcein showed no significant differences in fluorescence compared with cells seeded in a standard 96-well plate. Within the array we could also pattern near-linear ($R^2 = 0.91$) step-wise gradients of a model protein (biotinylated concanavalin A) using carbodiimide chemistry. Ongoing work with these systems is continuing to define optimal biochemical and biophysical cues for guiding tenogenic MSC differentiation.

(453c) Effect of Early Endoderm Induction On Late Stage Pancreatic Maturation of Differentiating Human Embryonic Stem Cells

Maria Jaramillo[1], Shibin Mathew[2] and Ipsita Banerjee[2], (1)Depatment of Bioengineering, University of Pittsburgh, Pittsburgh, PA, (2)Chemical and Petroleum Engineering, University of Pittsburgh, Pittsburgh, PA

Pancreatic differentiation of mouse and human embryonic stem cells has received considerable attention over the last decade. While there has been some success in deriving insulin positive cells from embryonic stem cells, typically the differentiated cells are limited in yield and functionality. Most of the successful differentiation protocols involve a step-wise directed differentiation strategy starting with definitive endoderm (DE) induction; followed by pancreatic progenitor induction and finally islet maturation. While extensive studies have established multiple pathways of endoderm induction, the effect of these pathways on late stage pancreatic maturation remains elusive. The aim of this study is to develop a thorough understanding of the effect of pathway of early endoderm induction on late stage pancreatic maturation.

The human embryonic stem cells were differentiated to mature insulin producing cells following a directed differentiation approach. The first stage of definitive endoderm was induced following four alternative pathways, based on most promising reports by different groups. Namely, we used Activin A in combination with one of the following molecules and growth factors: Fgf2, Bmp4, Wnt3a or PI3K Inhibitor. DE derivatives obtained from these four pathways were subjected to same maturation protocol: pancreatic progenitor induction using sonic hedgehog inhibition and retinoid signaling; islet maturation through notch inhibition. The differentiation potential of each of the endoderm derivatives were compared at different stages of differentiation.

Our results clearly indicate that late stage pancreatic maturation is distinctly governed by the initial pathway of endoderm commitment. From the results obtained from 6 independent studies, we found that DE induction by treatment with PI3KI resulted in at least two times higher upregulation of DE markers than any other conditions; however, at later stages of differentiation these cells obtained a moderate 800-fold upregulation of insulin. Conversely, cells induced towards DE by treatment with Wnt3a and Fgf2 showed between 10,000 and 20,000 fold upregulation of insulin. Consistent with PCR analysis, Wnt3a led to highest yield of C-peptide expressing cells with around 25% positive cells. Interestingly, while the Bmp4 derived DE cells resulted in only about 2-fold increase in insulin expression, they showed the highest upregulation of glucagon: around 4,000 times higher than undifferentiated cells, leading to the conclusion that inducing DE with Bmp4 leads to more alpha cells than beta cells. Detailed analysis using computational methods showed Wnt3a and Fgf2 cultures to show high similarity in gene patterning, and Wnt3a to follow closer resemblance to pancreatic organogenesis in respect to pancreatic marker progression.

We conclude that using Activin A in combination with Wnt3a at the definitive endoderm stage leads to the highest insulin upregulation as well as higher yield of C-peptide expressing cells. Moreover, our analysis shows that mature cell type specification is affected by early differentiation events. Namely our results suggest that while Wnt3a derivatives differentiate into beta-like cells, the use of Bmp4 preferentially directs cells toward alpha cell commitment. To the best of our knowledge, no other study compares methods of definitive endoderm induction and their effect at later stages of development. This study is important since it provides a better understanding of pathways involved in early events of pancreatic differentiation and allows for standardization of these stages in order to achieve better yield and functionality of differentiated human embryonic stem cells

(453d) Directing Pluripotent Stem Cell Differentiation Using Hydrogel Microspheres

Samuel S. Chang, Auburn University, Auburn, AL, Alexander J. Hodge, Chemical Engineering, Auburn University, Auburn, AL and **Elizabeth A. Lipke**, Department of Chemical Engineering, Auburn University, Auburn, AL

Pluripotent stem cells are a promising cell source for treatment of diseased or damaged tissues, including repairing the human heart. Current pluripotent stem cell differentiation protocols generally focus on deriving cardiac tissue with high purity and function, yet few of these methods are highly scalable and also tightly control the cellular microenvironment. Using a water-in-oil emulsion technique, this research has created a scalable and easily modifiable hydrogel microsphere system for pluripotent stem cell differentiation. Microspheres supported the proliferation of the encapsulated pluripotent stem cells and subsequent cardiac differentiation. MESCs were successfully encapsulated into the PEG-fibrinogen microspheres and maintained for over 23 days. mESCs proliferated and initiated formation of embryoid bodies (EBs) within the microspheres during the first week of differentiation. It was found that proliferation and EB formation occurred more readily when cells were encapsulated at high densities (45 and 60 million cells/mL precursor solution) rather than low densities (below 15 million cells/mL precursor). Cell encapsulation concentration did not have a significant impact on initial mean sphere diameter. EBs typically degraded the microsphere and attached to the culture vessel within 10 days of differentiation; these attached tissues initiated spontaneous contraction as early as Day 11 of differentiation. qPCR analysis of gene expression confirmed the down-regulation of pluripotent marker Oct4 and the up-regulation of Nkx2.5. Frequency of spontaneous contraction ranged from 1.5 to 2 Hz.

(453e) Analysis of the Proliferation Potential of Mesenchymal Stem Cells As a Function of Potency with a High-Capacity Clonal Assay

Kim OConnor[1], Katie Russell[1], Michelle Lacey[2], Alan Tucker[3] and Donald Phinney[4], (1)Chemical and Biomolecular Engineering Dept., Tulane University, New Orleans, LA, (2)Department of Mathematics, Tulane University, New Orleans, LA, (3)School of Medicine, Tulane University, New Orleans, LA, (4)Department of Molecular Therapeutics, The Scripps Research Institute, Jupiter, FL

Human mesenchymal stem cells (MSCs) from bone marrow are a heterogeneous ensemble of progenitors and lineage-committed cells, with a broad range of regenerative properties. Ex vivo expansion to produce sufficient quantities of MSCs is essential for most therapeutic applications. The present study resolves the relationship between proliferation potential of MSCs and their potency. Clonal analysis with a unique high-capacity assay generated single-cell derived colonies of MSCs that were classified according to their trilineage potential to exhibit adipo- (A), chondro- (C), and osteogenesis (O) as a measure of potency. Multipotent OAC clones were highly proliferative with colony-forming efficiencies that ranged from 35% to 90%; whereas, O clones formed colonies with an efficiency of 5% or less ($P<0.01$). Similar trends were evident during ex vivo expansion: for example, the median specific growth rate was 0.85 per day (20 h doubling time) for cultures inoculated with OAC clones and was 5-fold less for inocula of O clones ($P<0.01$). OA and OC clones had similar proliferation potentials. More than 75% of cells in subconfluent cultures inoculated with O clones stained positive for senescence-associated beta-galactosidase activity vs. less than 10% for OAC clones ($P<0.001$). Apoptotic cells were in the minority for all potency groups. Preliminary data generated during clonal analysis suggest that osteogenic potential of MSCs to produce mineralized matrix is a function of potency, as well. These results are discussed in the context of the preparation of efficacious MSC therapies by ex vivo expansion.
This research was sponsored by grants to Prof. O'Connor from the National Institutes of Health (R03EB007281) and Tulane University.

(453f) Hydrogel Systems to Examine Diffusion-Mediated Paracrine Signaling On Hematopoietic Stem Cell Fate

Bhushan Mahadik, Sara Pedron, Paul J.A. Kenis and Brendan A.C. Harley, Chemical and Biomolecular Engineering, University of Illinois at Urbana-Champaign, Urbana, IL

Hydrogel systems to examine diffusion-mediated paracrine signaling on hematopoietic stem cell fate

Bhushan Mahadik[*], Dr. Sara Pedron[*], Dr. Paul Kenis[*,#], Brendan Harley[*,#]

[*]Dept. of Chemical and Biomolecular Engineering, University of Illinois at Urbana-Champaign, Urbana, IL 61801

[#]Institute for Genomic Biology, University of Illinois at Urbana-Champaign, Urbana, IL 61801

Introduction: The stem cell niche provides extrinsic signals that regulate stem cell fate. The hematopoietic stem cell (HSC) niche in the bone marrow (BM) contains cells, the extracellular matrix (ECM), and ECM-bound or soluble biomolecules that present chemotactic, durotactic, and paratactic cues as well paracrine and juxtacrine signals to drive HSC fate. This niche regulates HSC fate decisions and hematopoiesis, the physiological process responsible for generating the body's blood and immune cells. While multiple cell types have been proposed as key regulators of HSC fate, the direct influence of heterotypic cell interactions via paracrine and juxtacrine signaling remain relatively unknown. We have developed a microfluidic platform to create microgels containing counter gradients of cell and matrix environments to determine differences in primary mHSC fate decisions in response to niche cell co-cultures using monolithic and counter-gradient 3D hydrogel environments designed to alter diffusion-mediated paracrine signaling.

Materials and Methods: Murine hematopoietic stem/progenitor cells (HSPCs) were isolated via FACS as $Lin^-ckit^+Sca1^+$ (LSK) or $LSKCD34^-$ BM cells. HSPCs were co-cultured with Lin^+BM cells under liquid as well as collagen, gelatin, and HA hydrogel cultures; hydrogel density as well as both HSC density and the HSPC:niche cell ratio (1:10, 1:100, 1:500) were varied to alter cell-cell communication. Cell-laden microgel constructs were polymerized thermally or via UV exposure and cultured over a period of 7 days with liquid co-culture used as a non-limited diffusion control. Hydrogel mechanical and diffusive properties were determined via compression and FRAP assays, respectively. HSPC proliferation was tracked via a fluorescent cell division assay while post culture HSPC fate was analyzed via FACS and gene expression analysis. A previously described microfluidic mixing device was used to create counter-gradient hydrogels comprised of HSPCs as well as multiple niche cells. Systematically varying HSPC:niche cell ratio as well as hydrogel type and density provides a rich landscape for examining the functional role of paracrine signaling on HSC fate.

Results and Discussion: Cell viability was not acutely affected by hydrogel gelation; higher HSPC viability and expansion (~50%) was observed for smaller gel volumes (50μL vs. 100μL). After 2 days of culture, FACS analysis indicates significant difference in HSPCs quiescence/maintenance (LSK) vs. myeloid lineage (CMP) differentiation due to both gel environment and niche cell co-culture. HSPCs showed increased specification towards myeloid lineages with increasing Lin+ niche cell co-culture. Significant differences were also observed between liquid and hydrogel conditions, with reduced niche cell effects observed in hydrogel environments that limited biomolecule diffusion. These results suggest a strong influence of niche cell interactions on HSPC fate.

Conclusions: We have established a biomaterial platform to explore the influence of niche cell mediated paracrine signaling on HSPC fate decisions. Ongoing experiments are examining longer culture periods and incorporating increasingly primitive niche cell populations (vascular endothelial cells, mMSCs) to alter the balance between lineage specification and quiescence. Counter gradients of quiescent and active niche populations are being incorporated into a single heterotypic microgel environment as an ectopic BM culture mimic to examine the dynamics of HSC fate decisions.

Acknowledgements: American Cancer Society, Illinois Division, Inc. (#160673, #189782).

(453g) Characterization of Acellular Matrix Derived From ESC Aggregates As Bioactive Scaffolds

Sébastien Sart, Teng Ma and Yan Li, Chemical and Biomedical Engineering, Florida State University, Tallahassee, FL

In recent years, embryonic stem cells (ESCs) have emerged as a powerful tool for the potential treatment of various incurable diseases. For therapeutic perspectives, mimicking stem cell niche or microenvironment, including *ad equate* cell-cell interactions and combination of extracellular matrix (ECM) molecules and biochemical factors, is essential for efficient pluripotent stem cell (PSC) expansion and differentiation. Acellular matrices derived from ESC aggregates, undifferentiated or embryoid bodies (EBs), can serve as potential scaffolds providing a more controlled bioactive three-dimensional microenvironment.

In this study, different chemical acellularization methods (i.e. Triton X-100, ammonia solution, SDS, and their combinations) are compared in association with DNAse treatment. The optimized protocol efficiently preserves ECM expression and improves DNA removal of ESC aggregates. These results highlight that the various defined structures generated from ESC aggregates could be efficiently acellularized and used as intact ECM-derived scaffolds. Next, ECM (fibronectin, laminin, collagen type IV, and vitronectin) expression is kinetically assessed for ESC monolayer, undifferentiated aggregates, and differentiated EBs with or without acellularization. Confocal microscopy and in situ ELISA are used to characterize ECM expression qualitatively and quantitatively. It is observed that ECM expression pattern is dynamically remodeled during expansion and differentiation. These results indicate that various culture conditions enable to generate different ESC-derived structures composed of specific ECM molecules composition. Finally, to prove the effectiveness of such scaffolds for ESC expansion and differentiation, undifferentiated ESCs are seeded on these ESC-ECM derived scaffolds and the markers of pluripotency and lineage differentiation are kinetically monitored to measure the degree of stemness and commitment of the newly seeded cells.

In summary, these results open the way towards the development of characterized bioactive scaffolds for more rapid and homogeneous ESC expansion and specific lineage differentiation.

(453h) Implantation of Vascular Grafts Made From Small Intestinal Sub-Mucosa and Hair Follicle Stem Cells in an Ovine Animal Model

Sindhu Row[1], Evan M. Schlaich[1], Hao-Fan Peng[1], Daniel D Swartz[2,3] and Stelios T. Andreadis[1,3,4], (1)Chemical and Biological Engineering, University at Buffalo, The State University of New York, Buffalo, NY, (2)Pediatrics, Women and Childrens Hospital of Buffalo, University at Buffalo, The State University of New York, Buffalo, NY, (3)Center of Excellence in Bioinformatics and Life Sciences, University at Buffalo, The State University of New York, Buffalo, NY, (4)Department of Biomedical Engineering, University at Buffalo, State University of New York, Buffalo, NY

Introduction: Cardiovascular disease is the leading cause of mortality worldwide. Regarded as the therapeutic gold standard, treatment by autologous vein grafting has several technical and patient-related risks. Alternatively, engineered grafts have shown to be one promising solution. Synthetic grafts have complications with compatibility, whereas natural, cell-seeded biomaterial scaffolds have been proven to perform better at the blood-material-interface. In this study, we aim to engineer fully functional and implantable arterial grafts made from human hair follicle derived smooth muscle cells (HF-SMC), ovine pulmonary artery endothelial cells, and small intestine submucosa (SIS).

Materials and Methods: Tubular tissues constructs were prepared from SIS and HF-SMC were incorporated in the vessel wall using fibrin glue. In order to promote living tissue development and improve the function of our TEVs, we mechanically stimulated the HF-SMCs within the construct using biomechanical forces and the stimulatory effects of TGFβ-1, insulin, and ascorbic acid. Using a customized bioreactor, arterial shear stress was achieved within our constructs. Prior to implantation, we investigated *in vivo* effects on parameters sustaining long-term patency of our grafts with the use of an arterio-venous shunt bypass model that we developed in our laboratory. Long-term patency of SIS grafts was evaluated when implanted as an interpositional graft in an ovine model, and followed for up to 12 weeks.

Results and Discussion: The grafts exhibited high burst pressure (1123327 mmHg, n=5) and similar compliance as compared to the native artery within physiological range (SIS=8.42+/-4.51, n=7; artery=8.62+/-2.99%, n=4, P>0.05, 0-120mmHg). Under dynamic strain, HF-SMC aligned circumferentially and showed collagen and elastin synthesis, as evidenced by histological examination. As expected, application of shear stress induced alignment of endothelial cells in the lumen in the direction of flow. The implanted tissues were monitored biweekly using Doppler ultrasound to document patency and measure blood flow rates through the grafts. At 4 and 12 weeks we also performed angiography to examine patency and measure blood flow rates through the grafts. After angiography the tissues were explanted and processed for histology and immunostaining. We found that all implants remained patent, contained a uniform endothelial monolayer and a robust vascular wall with numerous smooth muscle cells. Explanted grafts showed infiltration of host cells, robust ECM remodeling and vascular contractility similar to that of ovine arteries, demonstrating that our strategy was successful in engineering implantable arterial substitutes.

Conclusions: Our results demonstrate that SIS based vascular grafts with SMC from human HF-derived mesenchymal stem cells can be implanted in the arterial system of a large animal model. The patency of the grafts and their capacity for remodeling and development of vascular function suggest that this approach is very promising and may be of clinical significance.

(454a) Leveraged Process Development for Biomass to Ethanol

Brandon Emme, Novozymes North America, Franklinton, NC and M. Torry-Smith, Novozymes North America Inc., Franklinton, NC

Overview: Visionary talks (invited) will outline commercial, governmental, or academic efforts to advance multi-product biorefineries. Topics likely to be addressed include: development or improvement -- including conceptual analysis and sustainability assessments -- of renewable feedstocks-based multiproduct biorefineries, i.e., based on starch or oil crops and/or lignocellulosic biomass; development of enabling catalytic and/or separative processes; discovery or development of enabling enzyme or fermentative/microbial technologies or bio-based materials.

Appropriate integration of process development steps in the biomass to ethanol process is a challenge. Building a fully flexible pilot plant to allow for multiple process configurations is cost prohibitive for most technology developers. For lack of fully integrated data, technoeconomic models can be employed to integrate smaller scale or individual unit operation data into the context of a full process to determine the best operational configuration and production cost. Furthermore, these models can be used to compare competing process concepts, both for state of technology as well as at n-th state conditions.

This talk will present the Novozymes 2G flexible cost model and illustrate through several Novozymes process development examples how it has been used to assimilate results from Pretreatment through Fermentation to establish winning process concepts. The latest small and large scale hydrolysis data using the new commercial cellulase enzyme Cellic CTec3 on several different feedstocks will be presented, as well as some insights into what the models can tell us about the likely short and long term impacts of enzyme development on the Minimum Ethanol Selling Price.

(454b) The Ethics of Biofuels in a World of Nine Billion People

John J. Sheehan, NREL, Golden, CO

Our global agriculture system is failing us:

- Food prices are up.
- A billion people face obesity.
- A billion people face starvation and malnutrition.
- Nitrogen and phosphate pollution from farms continue to threaten the health of our waterways and other natural ecosystems.
- Greenhouse gas emissions from agricultural expansion and land clearing continue to contribute a third to our total emissions.
- Agriculture remains the largest single consumer of land and water on the planet.

And in the midst of all this, we are struggling to realize a vision for a truly sustainable bio-economy that will put more demands on the global agriculture system.

This talks focuses on the ethical as well as technical challenges facing the nascent bio-economy, especially with respect to the challenge of feeding nine billion in the year 2050. Never has there been a time when it was so urgent to untangle these issues. We need a straightforward discussion of the ethics of sustainable development and sustainable biofuels.

Built off a recent paper by Foley et al in *Nature* on how we will feed more people demanding more food, the talk extends this question to include feeding and fueling the world. It focuses on four key strategies that are vital to maintaining a sustainable agricultural system: 1) Elimination of expansion of agricultural lands globally; 2) Reduction of the agricultural yield gap around the world; 3) Improvement of resource utilization (water and fertilizer); and 4) Reduction in meat consumption and food system waste. Successfully pulling off these strategies *may* provide room for agriculture to deliver food, fuel and chemicals to an increasingly wealthy population of nine billion people in the year 2050.

(454c) Biobased Chemical Production Through a Generalized Technology Platform

Brent H. Shanks, Chemical and Biological Engineering, Iowa State University, Ames, IA

Replacement of crude oil-derived chemicals with biorenewables chemicals will necessitate developing capabilities to remove oxygen selectively from carbohydrates. The serial approach of targeting one biorenewable chemical product is both time intensive and expensive. Since biorenewable chemicals must compete against a highly efficient petrochemical production system there is a need to create a generalized oxygen-removal framework in which a range of biorenewable chemicals can be produced from a common technological platform. One such generalized technological framework, being developed by *CBiRC*, depends on the exploitation of a common metabolic pathway leading to a diversity of intermediate chemicals that are subsequently converted to chemical products using chemical catalysts. An overview of the technical strategy being used by *CBiRC* to achieve a generalized chemical production platform will be discussed as well as several specific examples involving varying chain length carboxylic acids with decarboxylation to olefins and biological-produced pyrones with subsequent ring opening or aromatization.

(454d) Progress Toward Sustainable Biofuels — Pilot-Scale Demonstration of Integrated Cellulosic Ethanol Production

Daniel J. Schell, National Renewable Energy Laboratory, Golden, CO

In 2007, the U.S. Department of Energy's (DOE) Office of the Biomass Program, in collaboration with the National Renewable Energy Laboratory (NREL), established the goal of achieving cost competitive cellulosic ethanol production from corn stover by 2012. The baseline conversion process to be improved consisted of dilute sulfuric acid pretreatment followed by enzymatic cellulose hydrolysis and cofermentation of biomass sugars to ethanol using a metabolically-engineered fermentation strain. The technology improvement strategy was based on achieving yearly increases in overall process conversion yields, along with associated reductions in modeled ethanol production costs, from 2007 to 2012. Progress toward this goal was measured using n^{th}-plant technoeconomic models to estimate ethanol production cost based on conversion yields obtained in integrated pilot and bench-scale experiments. The final objective was to demonstrate at pilot scale integrated process conversion yields commensurate with the 2012 cost target.

This presentation will briefly review the history of this technology development effort and highlight results achieved by the DOE Biomass Program at NREL from 2000 to the present. Much of the progress was facilitated by DOE-supported, industry-led research to develop improved hydrolytic enzymes for cellulose conversion and more robust microorganisms for ethanol production from a broader range of biomass sugars. Specifically, improved cellulases increased cellulose-to-glucose yields during enzymatic hydrolysis of pretreated biomass in the presence of high levels of biomass-derived sugars, and improved cofermenting microorganisms were able to convert more of the available biomass-derived sugars to ethanol in the presence of inhibitory compounds. Work at Idaho National Laboratory to improve the efficiency of the corn stover collection and supply chain also lowered the cost of delivered feedstock.

The presentation will highlight work that led to improved process understanding, higher sugar and ethanol titers, and ultimately increased conversion yields and associated cost reductions culminating in pilot scale demonstration runs. Preliminary results from bench-scale experiments are achieving yields above 75 gal ethanol/ton feedstock, ethanol titers above 68 g/L and a modeled cost below $2.20/gal ethanol. Pilot scale results will be presented along with bench-scale testing results performed in parallel characterizing integrated conversion performance over a wider range of operating conditions than could be demonstrated at pilot scale.

(462a) Towards Development of an Optimal Modular Cell for Production of Biochemicals and Biofuels

Cong T. Trinh[1], Michael Wierzbicki[1], Donovan S. Layton[1], Adam Thompson[2] and Narajan Niraula[1], (1)Dept. of Chemical and Biomolecular Engineering, University of Tennessee, Knoxville, TN, (2)Energy Science and Engineering, University of Tennessee, Knoxville, TN

Relevant hosts such as *Escherichia coli* and *Saccharomyces cerevisiae* are commonly engineered and optimized to produce target products with different genetic modifications. Thus, an engineered host that is optimized to produce one target product may not be suitable to function as a host to efficiently produce other target compounds. To address this bottleneck, we have applied the metabolic pathway analysis tool based on elementary mode analysis to design an optimal modular cell that can metabolically couple with a diverse class of chemicals- and biofuels-producing pathways as exchangeable and tunable modules to efficiently produce a diverse set of biochemicals and biofuels at high yields, titers, and productivities. In this talk, we will present the design, construction, and characterization of an optimal modular cell that can couple with different types of exchangeable and tunable modules to produce short-chain length (<C6) alcohols and a diverse spectrum of their derived esters under anaerobic conditions.

(462b) Epathbrick Directed Modular Pathway Engineering for Improved Fatty Acids Production in *E. Coli*

Peng Xu and Mattheos A.G. Koffas, Chemical and Biological Engineering, Rensselaer Polytechnic Institute, Troy, NY

As an emerging discipline, synthetic biology is becoming increasingly important to design, construct and optimize metabolic pathways for producing biofuels and pharmaceuticals in genetically tractable organisms. To date, a number of synthetic biology approaches have been applied to metabolic pathway optimization including modification of plasmid copy number, promoter strength and gene codon usage. Nevertheless, most of these approaches are not modular and require time-consuming work tweaking the individual pathway components. From an engineering perspective, we need a more efficient approach that streamlines the process of pathway construction and optimization. Based on central pathway architectures, *E.coli* fatty acids pathway were decomposed into three modules: (I) the upper module includes the glycolytic pathway encoded by pgk, gapA, aceE, aceF and lpdA; (II) the intermediary module includes the acetyl-CoA carboxylation pathway encoded by accA, accB, accC, accD; (III) the lower module includes the fatty aicds formation pathway encoded by fabA, fabD, fabG, fabI and tesA' (or other plant fatty-acyl thioesterase). By balancing the expression level of the three module pathway, we have successfully constructed an *E. coli* recombinant strain exhibiting 25-fold increase in fatty acids production. It was found that fatty acids composition was dramatically changed as a result of expressing different fatty acyl-ACP thioesterase. Expressing the *E coli* native thioesterase (tesA') shifted the fatty acids to short-chain C14:0; whereas expressing the plant-derived fatty acyl-ACP thioesterase (BnFatA, CnFatB2, EGOTE) shifted the major fatty acids to palmitic acid compared with the wild type strain.

(462c) An Integrated Computational and Experimental Study for the Overproduction of Fatty Acids in Escherichia Coli

Sridhar Ranganathan[1], **Anupam Chowdhury**[2], Ali R. Zomorrodi[3], Ting Wei Tee[4], Jong M. Yoon[5], Yanfen Fu[5], Jacqueline V. Shanks[5] and Costas D. Maranas[6], (1)Huck Institutes of the Life Sciences, The Pennsylvania State University, University Park, PA, (2)Chemical Engineering, Pennsylvania State University, University Park, PA, (3)Chemical Engineering, The Pennsylvania State Univeristy, University Park, PA, (4)Chemical and Biological Engineering, Iowa State University, Ames, IA, (5)Department of Chemical and Biological Engineering, Iowa State University, Ames, IA, (6)Department of Chemical Engineering, The Pennsylvania State University, University Park, PA

An Integrated Computational and Experimental Study for the Overproduction of Fatty Acids in *Escherichia coli*

Sridhar Ranganathan [1], Anupam Chowdhury[2], Ali R. Zomorrodi[2], Ting Wei Tee [3], Jong Moon Yoon [3+], Yanfen Fu [3], Jacqueline V. Shanks [3], and Costas D. Maranas[2]

[1] Huck Institutes of Life Sciences, Pennsylvania State University, University Park, PA

[2] Department of Chemical Engineering, Pennsylvania State University, University Park, PA

[3] Department of Chemical and Biological Engineering, Iowa State University, Ames, IA

Fatty acids of varying chain lengths ($C_6 - C_{16}$) naturally synthesized in many organisms are promising starting points for the catalytic production of industrial chemicals and diesel-like biofuels. However, bio-production of fatty acids from plants and other microbial production hosts relies heavily on manipulating tightly regulated fatty acid biosynthetic pathways. In addition, precursors for fatty acids are used along other central metabolic pathways for the production of amino acids and biomass, which further complicates the engineering of microbial hosts for higher yields. In this study, we demonstrate an iterative metabolic engineering effort that integrates computationally driven predictions and metabolic flux analysis techniques to meet this challenge. The OptForce procedure, our recently developed computational strain design algorithm, was used for suggesting and prioritizing genetic manipulations that overproduce fatty acids of different chain lengths from C_6 to C_{16} starting with wild-type *E. coli*. We identified some chain-specific genetic interventions alluding to the possibility of fine-tuning overproduction for specific fatty acid chain lengths. In accordance with the OptForce prioritization of interventions, *fabZ* and acyl-ACP thioesterase were upregulated and *fadD* was deleted to arrive at a strain that produces 1.70 g/L and 0.14 g fatty acid/g glucose (~ 39% maximum theoretical yield) of C_{14-16} fatty acid in minimal M9 medium. OptForce suggested additional modifications distal from the fatty acid synthesis pathways that led to additional improvements in yield. Our study reinforces the advantage of integrating computational and experimental tools for the design and engineering of microbial strains to overproduce value-added chemicals or biofuels.

(462d) Metabolic Engineering of E. Coli Species for Fatty Acid Biosynthesis Using Acetic Acid

Yi Xiao and Yinjie Tang, Washington University in St. Louis, Saint Louis, MO

Acetic acid (HAc) can be cheaply generated from syngas by microbial and chemical catalysts or acidogenic fermentation using organic wastes. This study focuses on utilization of HAc for fatty acid (precursor for biodiesel) production. Firstly, an *E. coli* strain containing plasmids with acyl-ACP thioesterase (TE) and acetyl-CoA carboxylase (Acc) is created, producing 1.5 g/L free fatty acid from glucose. Then, we overexpress two native HAc metabolism pathways (*pta-ack* and *acs*) in *E. coli*. The results indicate that overexpression of the *acs* pathway significantly increases cell growth in the presence of acetate, while knock-out of *poxB* or *pta* gene impairs HAc-based cell growth. To synthesize free fatty acid using HAc, the *acs* pathway and TE are overexpressed and the *fadE* gene involved in fatty acid degradation is deleted. The engineered strain achieves ~0.5 g/L free fatty acid from HAc in the shaking flask culture. To significant improve the titer of fatty acids, we build a fed batch experiment and employ a kinetic model to describe and simulate the fatty acid production using HAc as the main carbon source.

(462e) Comparative Genomic Analysis of a High n-Butanol Tolerant and Producing Mutant Strain of Clostridium Acetobutylicum

Mengmeng Xu, Chemical and Biomolecular Engineering, The Ohio State University, Columbus, OH, Jingbo Zhao, William G. Lowrie Department of Chemical and Biomolecular Engineering, The Ohio State University, Columbus, OH and Shang-Tian Yang, The Ohio State University, Columbus, OH

Butanol is an important organic solvent and has been considered as potential fuel or fuel additive. Biological production of butanol through bacterial fermentation has many advantages over petrochemical process; however, so far, it cannot compete with the petrochemical process economically because of the relatively low titer of butanol. To improve the butanol tolerance and final titer, a mutant strain of *Clostridium acetobutylicum* ATCC 55025 has been obtained through efficient strain evolution in the fibrous bed bioreactor. Compared with the parental strain, the mutant showed better butanol tolerance, much higher butanol titer, improved autolysis resistance and more complete substrate utilization. Further research showed that there are remarkable changes of membrane fatty acid composition and proteomic expression profiling. To better understand the differences between the parental and the mutant strain at the molecular level, complete genomic analysis was conducted. The result revealed 7 mutations in the mutant genome, which might have contributed to its improved butanol tolerance and fermentation performance. Among the 7 mutations, the C-terminal truncated histidine kinase and the C-terminal extended membrane protein were identified as targets for further inverse metabolic engineering to investigate their functions in affecting butanol biosynthesis and tolerance. The results from this study will provide better understanding of butanol tolerance, which can lead to the development of high butanol producing strains for industrial applications, and will be presented in this paper.

(462f) An Examination of Lipid Production by Chlorella Protothecoides Using Compartmentalized Metabolic Flux Analysis

John O'Grady and John A. Morgan, School of Chemical Engineering, Purdue University, West Lafayette, IN

With increasing energy demands, many forms of alternative and sustainable energy production are receiving increased attention. Biodiesel from algae continues to be a promising source of renewable liquid fuel. *C. protothecoides* has demonstrated the highest lipid productivity of any strain of algae when it is grown heterotrophically. However, much still remains to be learned about its primary metabolism. A deeper understanding of the metabolism of this lipid-producing algal strain will set the stage for further improvements in its lipid production, as well as the lipid production of other algal strains with alternate desirable characteristics.

^{13}C metabolic flux analysis is a powerful tool to examine the distribution of metabolic rates within a cell. An examination of the labeling patterns in intermediary metabolites makes it possible to determine the flux rates of many key reactions in carbon metabolism. Using subcellular fractionation we are able to deconvolute the labeling patterns of metabolites located in the plastid and the cytosol. Thus we are able to separate the fluxes of identical reactions which occur in both of these locations in the cell. We have examined these flux rates using both glucose and glycerol to determine the differences in heterotrophic metabolism on each substrate with a focus on determining their impact on lipid metabolism.

(462g) Isotope-Assisted Metabolic Flux Analysis Reveals Efficient Photosynthetic Pathways in the Unicellular Diatom (alga) Phaeodactylum Tricornutum

Yuting Zheng and Ganesh Sriram, Chemical and Biomolecular Engineering, University of Maryland, College Park, MD

Photosynthesis is crucial in converting solar energy and atmospheric CO_2 into fuel and renewable carbon. A rapidly growing human population and an increasing need for renewable fuel and carbon warrant a substantial increase in photosynthetic output over the next few decades. This motivates research into metabolic engineering for enhanced photosynthetic productivity. However, the overall efficiency of native photosynthesis is low because of the low processing ability of its central enzyme ribulose bisphosphate carboxylase/oxygenase (RuBisCO). This is primarily because of the difficulty of concentrating CO_2 around RuBisCO at high enough concentrations to facilitate fast reaction. A small number of photosynthetic organisms overcome this problem by concentrating or "pumping" CO_2 around RuBisCO through biophysical and biochemical mechanisms. Although the orchestration of such a mechanism requires a multicellular architecture (as in maize or certain grasses), recent circumstantial evidence including genome sequences, transcriptome profiling and biophysical measurements has suggested that diatoms, a class of unicellular, marine algae may operate an efficient photosynthetic pathway. This may explain why diatoms are responsible for 20% to 40% of global photosynthetic output despite surviving in CO_2 -depleted environments. However, direct, metabolic evidence of such a pathway and the measurements of flux through it are lacking.

We have employed metabolic engineering and systems biology tools such as isotope-assisted metabolic flux analysis to address this gap in knowledge. We performed isotope-assisted metabolic flux analysis on the model diatom *Phaeodactylum tricornutum* and used mass spectrometry to measure incorporation of these carbon sources into soluble metabolites and biomass components. We then employed metabolic network modeling to convert the isotopomer abundances to metabolic flux maps. The results of these analyses reveal signatures of efficient photosynthetic pathways by which *P. tricornutum* apparently assimilates HCO_3^- (the abundant form of CO_2 in marine environments) into its biomass. In this presentation, we will discuss how our isotope labeling data and its analysis led to this conclusion. We anticipate that the knowledge obtained from this investigation will have important implications toward metabolic engineering of diatoms and other organisms for improved CO_2 fixation, and ultimately toward engineering synthetic CO_2-sequestering devices.

(462h) Isotopically Nonstationary 13C Flux Analysis of Isobutyraldehyde Production in S. Elongatus

Lara Jazmin[1], Yao Xu[2], Carl Johnson[2] and Jamey Young[1], (1)Chemical and Biomolecular Engineering, Vanderbilt University, Nashville, TN, (2)Biological Sciences, Vanderbilt University, Nashville, TN

The development of fuel substitutes from renewable resources has become significantly important as a result of current global energy and environmental problems. Recent studies have demonstrated the feasibility of converting energy from sunlight and carbon from CO_2 directly into biofuels using photosynthetic microorganisms. Despite the advances made in cyanobacterial biofuels production, the productivities achieved by cyanobacterial fermentations are currently too low for industrial feasibility and few tools are available that specifically address the challenges of redirecting and enhancing metabolic flux in photosynthetic microbes.

The ability to perform quantitative studies using isotope tracers and metabolic flux analysis (MFA) is critical for accurately assessing *in vivo* regulation of photoautotrophic metabolism, as well as crucial to identifying pathways that will maximize carbon flux from Calvin cycle intermediates into biofuel-producing pathways. Although ^{13}C is the preferred isotope tracer for mapping central carbon metabolism in heterotrophic organisms, photoautotrophs assimilate carbon solely from CO_2 and therefore produce a uniform steady-state ^{13}C-labeling pattern that is insensitive to fluxes. However, transient measurements of isotope incorporation following a step change from labeled to unlabeled CO_2 can be used to map carbon fluxes under autotrophic growth conditions. This involves quantification of intracellular metabolic fluxes based upon computational analysis of dynamic isotope labeling trajectories, which applies the technique of isotopically nonstationary MFA (INST-MFA).

We have recently introduced an INST-MFA approach to quantify intracellular fluxes in the cyanobacterium Synechocystis sp. PCC 6803, a model photosynthetic organism, under photoautotrophic conditions *(1)*. This involved using both GC/MS and LC/MS/MS to quantify the trajectories of metabolite labeling that result from introduction of 13C-labeled bicarbonate. Comparison of the INST-MFA flux map to theoretical values predicted by a linear programming (LP) method revealed inefficiencies in photosynthesis due to oxidative pentose phosphate pathway and malic enzyme activity, despite negligible photorespiration. Our ongoing work is focused on applying the INST-MFA approach to several engineered strains of *Synechococcus elongatus* PCC 7942 to enable the systematic improvement of cyanobacterial production of isobutyraldehyde (IBA, a direct precursor of isobutanol). Quantification of photosynthetic carbon fluxes in cyanobacteria will pinpoint pathway bottlenecks that can be subsequently removed in further rounds of metabolic engineering, thus leading to maximal IBA productivity by redirecting flux into IBA-producing pathways.

References

1. Young, J.D., Shastri, A. a, Stephanopoulos, G., et al. (2011) Mapping photoautotrophic metabolism with isotopically nonstationary (13)C flux analysis. *Metabolic engineering*. **13**, 656-665.

A Safer, Greener Grignard

AIChE Annual Meeting

Pittsburgh, PA

31-Oct-12

Answers That Matter.

Grignard Reaction

- Versatile synthetic transformation with wide application
- Best method for the preparation of chiral tertiary alcohols
- Atom economy
- Safer than analogous lithium chemistry
- More stable complex than analogous lithium chemistry
- Simplify a synthetic route
- Inexpensive

RDS

Answers That Matter.

Grignard Reaction

Process Engineer

- Magnesium activation
- Grignard initiation
- Potential thermal events
- Oxygen and water sensitive
- Magnesium quench
- Hydrogen liberation

RDS

Lilly
Answers That Matter.

How do we bridge this diversity of thought?

RDS

Lilly
Answers That Matter.

Option #1?

RDS

Lilly
Answers That Matter.

Option #2

The Flow Grignard enables a powerful and useful chemical transformation while minimizing the process safety risks.

How does it achieve that?

1. By reducing the Grignard scale
2. By using continuous stirred tank reactors (CSTR)
3. By sequestering the magnesium

RDS

Lilly
Answers That Matter.

Edivoxetine Flow Grignard

Edivoxetine·HCl, a highly selective norepinephrine reuptake inhibitor, under clinical evaluation for the treatment of Depression and ADHD.

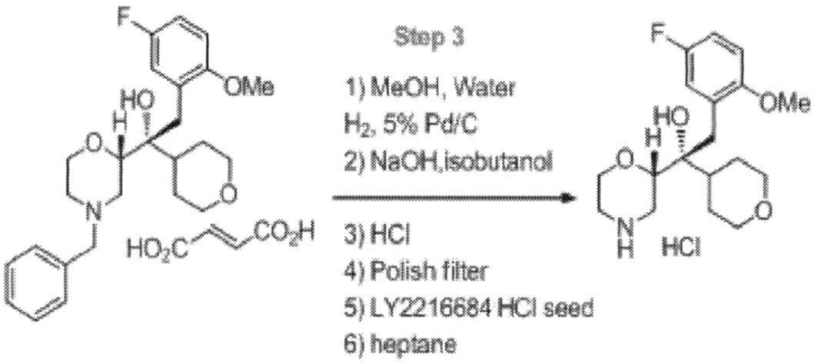

RDS

Lilly Answers That Matter.

Edivoxetine Flow Grignard

- Edivoxetine Step 2

    Preparation of the free base of the penultimate fumarate salt.

- Classic Grignard
    1. Grignard complex Rx
    2. Coupling Rx
    3. Quench Rx
    4. Wash Rx

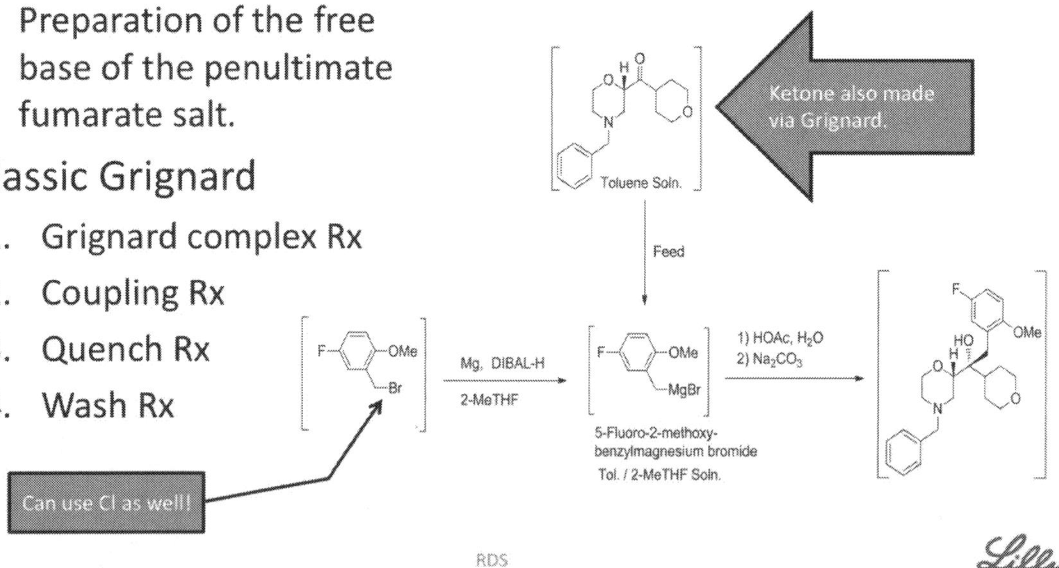

RDS

Lilly Answers That Matter.

Edivoxetine Flow Grignard

Process Flow Sheet

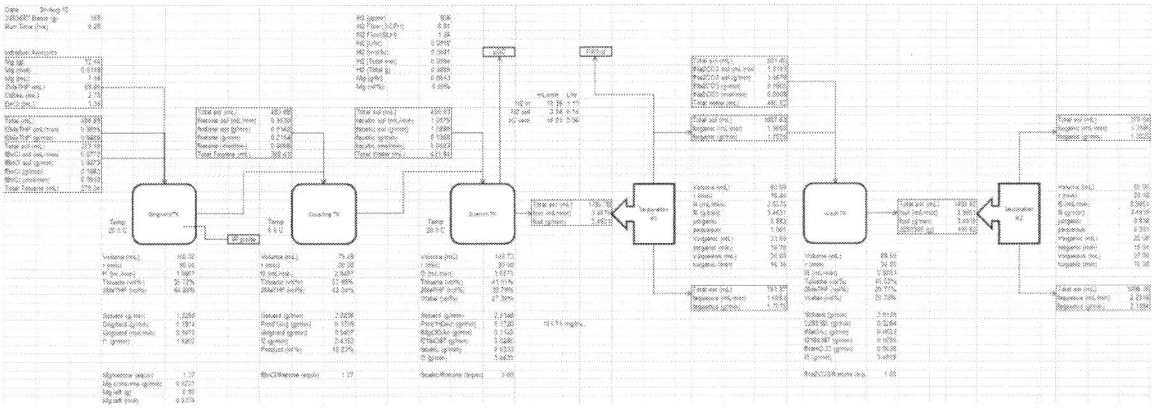

RDS

Lilly
Answers That Matter.

The Rig – Part I

Cart 162A by Heller, Lewis, etal.

RDS

Lilly
Answers That Matter.

The Rig – Part II

1. Benzyl Chloride
2. 2-MeTHF
3. Ketone
4. Aqueous acetic acid
5. Na_2CO_3 solution
6. Product collection

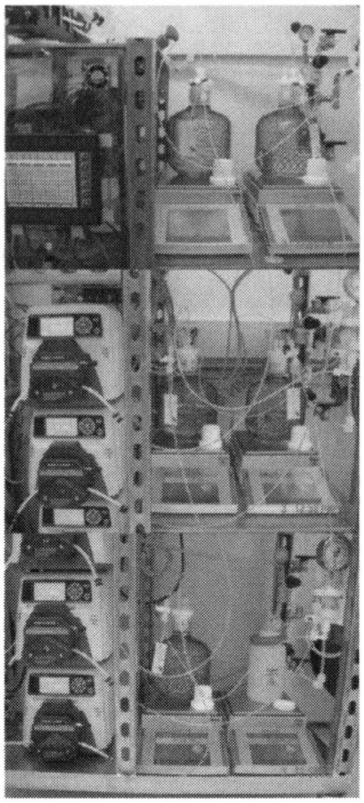

Cart 162B by Heller, Lewis, etal.

RDS

Lilly
Answers That Matter.

Edivoxetine Flow Grignard

Grignard > Coupling > Quench

Grignard
- τ = 90 min
- Mg, BnX, 2MeTHF
- 20 C

Coupling
- τ = 30 min
- Ketone
- 0 C

Quench
- τ = 30 min
- Aqueous acetic acid
- 20 C

RDS

Lilly
Answers That Matter.

Edivoxetine Flow Grignard

Separation → Wash → Separation

Separation	Wash	Separation
• τ = 16.5 min • Organic = product • Aqueous = waste	• τ = 30 min • Sodium carbonate • Ambient	• τ = 16.5 min • Organic = product • Aqueous = waste

RDS

Lilly
Answers That Matter.

Sequestered Mg

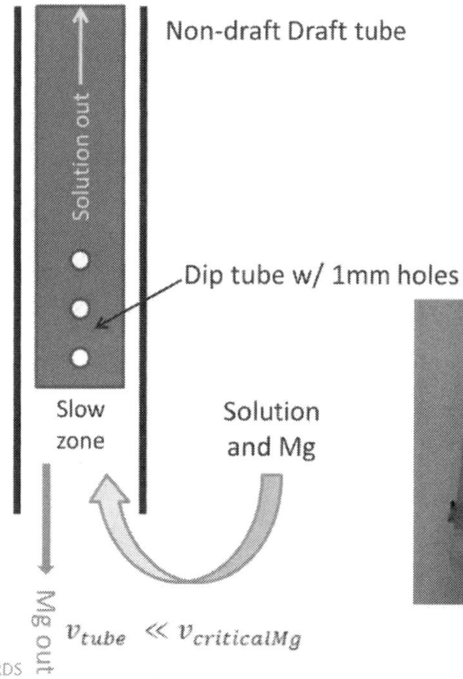

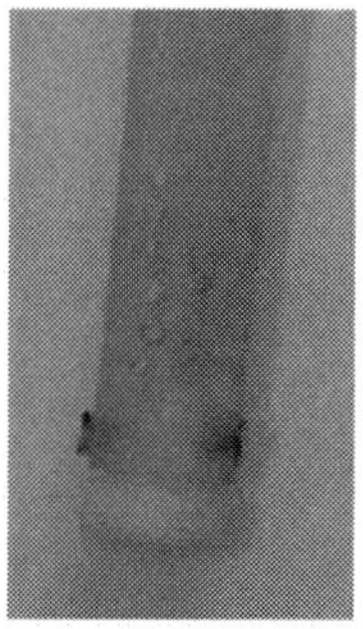

RDS

Lilly
Answers That Matter.

τ in Continuous Flow

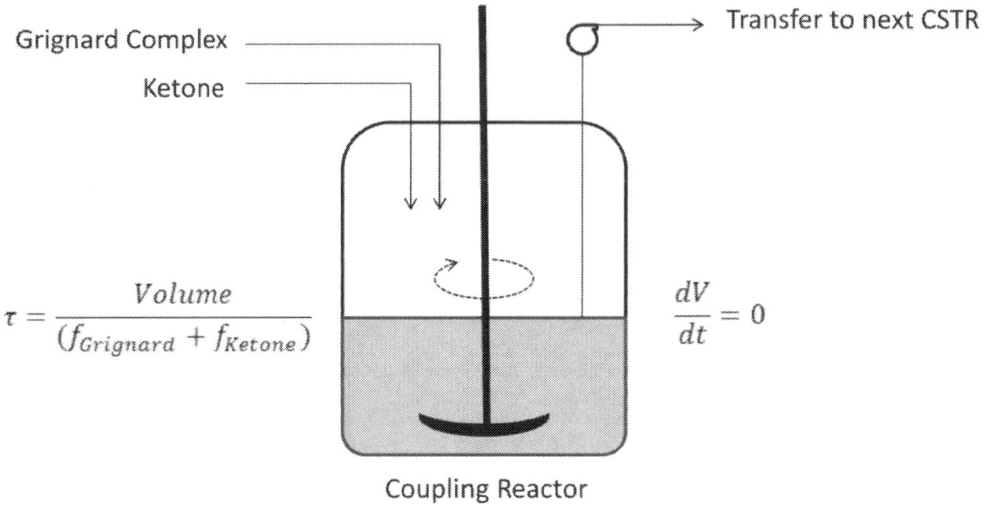

$$\tau = \frac{Volume}{(f_{Grignard} + f_{Ketone})}$$

$$\frac{dV}{dt} = 0$$

Reagent Feed Flow

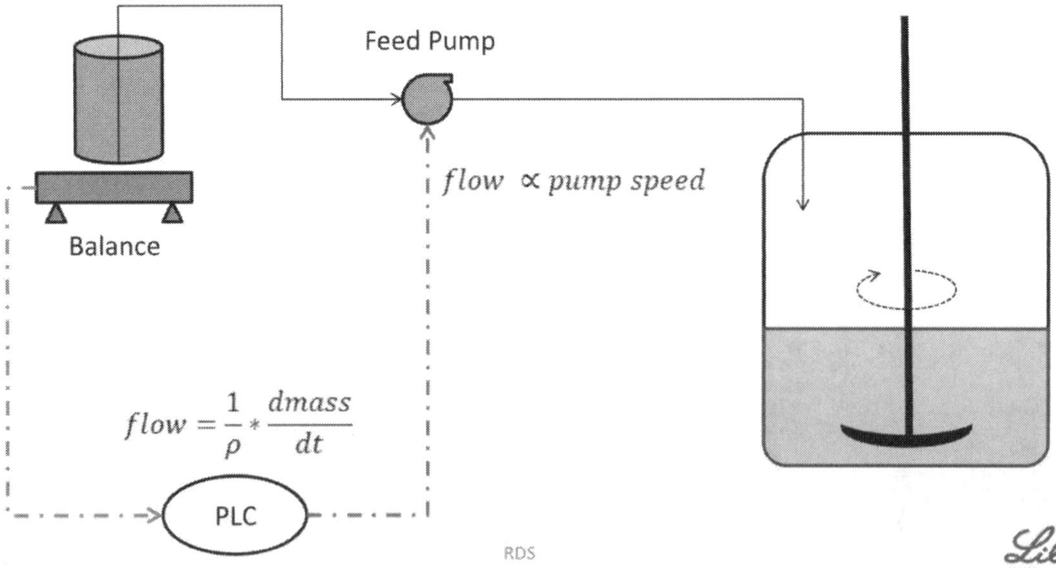

$$flow \propto pump\ speed$$

$$flow = \frac{1}{\rho} * \frac{dmass}{dt}$$

In-situ Raman IR

Grignard Complex Reaction

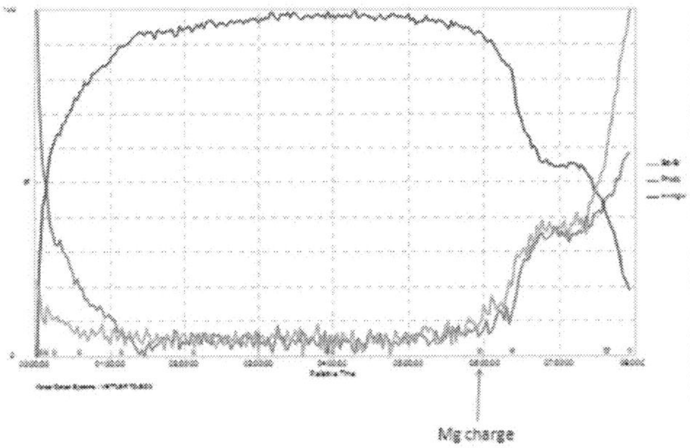

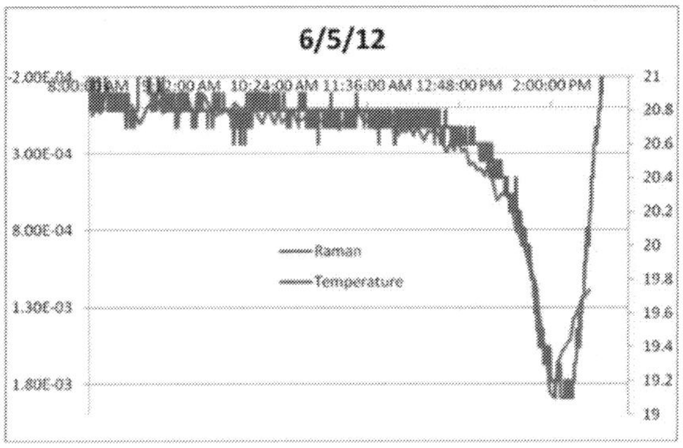

Data by Adam McFarland

RDS

Answers That Matter.

On-Line PATROL UPLC®

Quenched Coupling Reaction

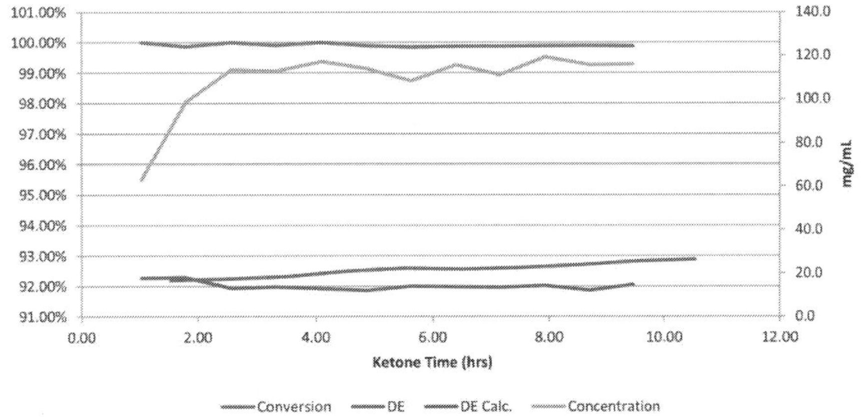

Data by Ed Sheldon

RDS

Answers That Matter.

On-Line PATROL UPLC®

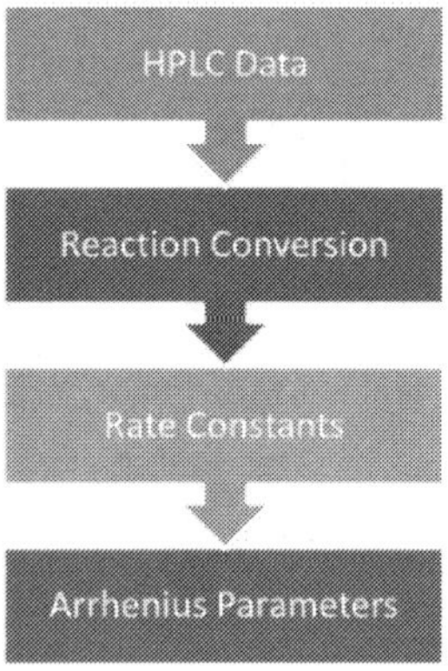

$$x = 1 - \left(\frac{1}{1+\tau k}\right)$$

$$k = Ae^{-\frac{E}{RT}}$$

$$\ln k = \ln A - \frac{E}{RT}$$

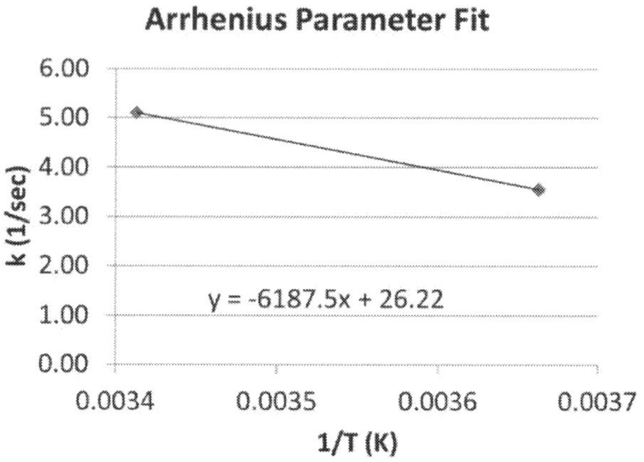

Conv	Temp (K)	k (min-1)	k (sec-1)	1/T (K)	lnk			
94.6%	273	0.58	35.00	0.0037	3.56	lnA	26.22	A = 2.44E+11sec-1
98.8%	293	2.74	164.40	0.0034	5.10	E/R	6187.50	E = 51443J/mol

RDS

Answers That Matter.

On-line Mass Spec.

Quench Reaction

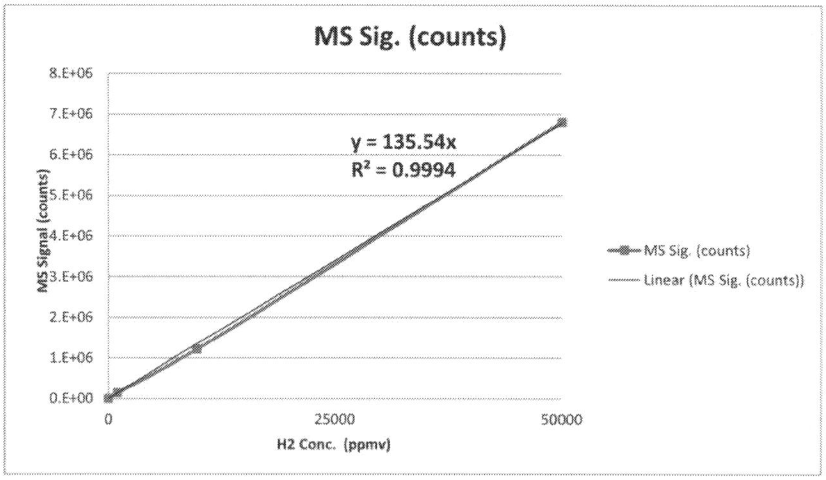

All experimental data was below the instrument LOD of ~1000 ppm (0.10%) or 1/40th of LEL.

Data by Steve Doherty

RDS

Answers That Matter.

On-Line GC

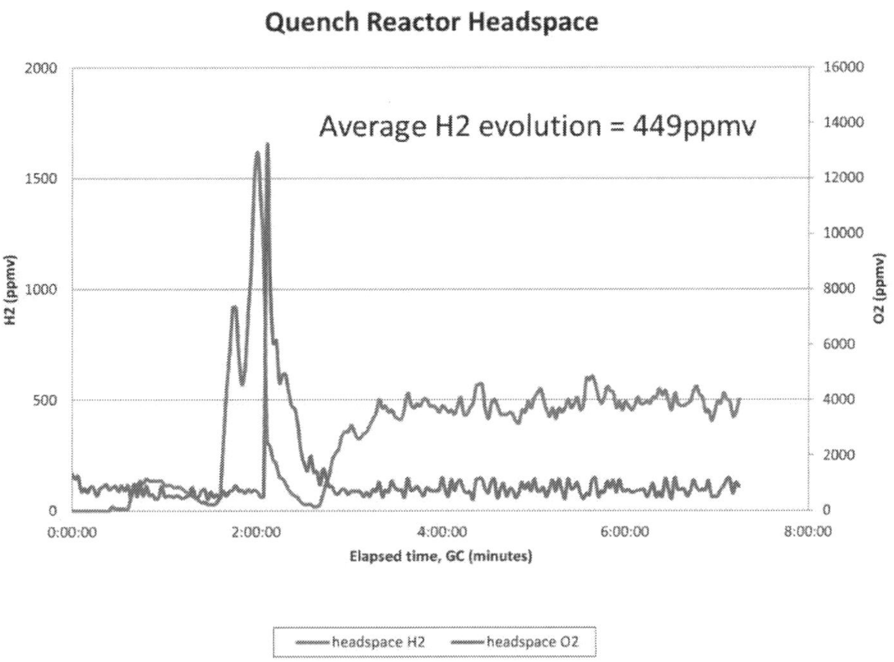

Data by Steve Doherty

RDS

Answers That Matter.

Safety Side by Side

Batch Grignard

- DIBAL Mg activation every batch
- All Mg is activated at once
- Grignard complex initiation every batch
- 20% charge of the BnX
- Use BnBr as halide
- Excess Mg to quench every batch
- 0.25equivalents of hydrogen liberated every batch

Flow Grignard

- Single DIBAL Mg activation
- Portion of Mg activated at once
- Single Grignard complex initiation
- 0.07% charge of the BnX
- Use BnCl as halide
- Complete Mg consumption possible
- 0.002 equivalents of hydrogen liberated

RDS

Answers That Matter.

Green Side by Side

Batch Grignard

- DIBAL Mg activation every batch
- 1.50 equivalents BnX
- 1.75 equivalents Mg
- Intermittent dry boundary
  - Inter batch vessel drying
- Higher energy consumption from multiple temperature changes

Flow Grignard

- Single DIBAL Mg activation
- 1.27 equivalents BnX
- 1.37 equivalents Mg
- Continuous dry boundary
  - No inter batch vessel drying
- Fixed temperature CSTR's

RDS

Lilly
Answers That Matter.

Complexity Side by Side

Batch Grignard

- Single reactor at multiple temperatures
- Intermittent heat transfer
- One reactor feed at a time
- Intermittent dry boundary
  - Inter batch vessel drying
- Once coupling starts, process must continue through quench

Flow Grignard

- Four CSTR's at fixed temperatures
- Constant heat transfer
- Five continuous feed streams
- Continuous inter-vessel transfers
- Continuous dry boundary
  - No inter batch vessel drying
- Mg recharge
- Stops and restarts possible

RDS

Lilly
Answers That Matter.

Manufacturing Interests

- Quality
  - Comparable to batch Grignard
  - Process robustness
- Safety
  - Inherently safer process
    - Less potential energy in the system
    - Reduced number of events
  - 148X reduction in hydrogen evolution
  - Use BnCl instead of BnBr
- Cost
  - Capital Cost
    - €1MM for 100L CSTR's
    - €3.6MM for scale-up to 2000 gallon rig
  - Operating Cost
    - 15% reduction in BnCl
    - 22% reduction in Mg

RDS

Lilly
Answers That Matter.

Manufacturing Capacity

Batch Grignard

- 250 gallon Batch†
  - 49 kg product/48 hours
  - 1.0 kg/hr
- 2000 gallon Batch††
  - 392 kg product/43 hours
  - 9.1 kg/hr

Flow Grignard

- 10 L CSTR
  - 28 kg product/24 hours
  - 1.2 kg/hr
- 100 L CSTR
  - 280 kg product/24 hours
  - 11.7 kg/hr

†Jul 2012 250 gal production data

†† Assumes 10% cycle time reduction

RDS

Lilly
Answers That Matter.

RDS

Lilly
Answers That Matter.

References

- Oroskar, A.R. and Turian, R.M., “The Critical Velocity in Pipeline Flow of Slurries”, AIChE Journal, Volume 26, pp550-558, 1980.
- Fogler, H. S., “Elements of Chemical Reaction Engineering”, Prentice-Hall, p60, p114, 1986.

RDS

Lilly
Answers That Matter.

(467b) Continuous High Pressure Vapor/Liquid PFR Sampling System with On-Line HPLC

Wei-Ming Sun, Chemical Product Research and Development, Eli Lilly &Co., Inc., Indianapolis, IN, Martin D. Johnson, Chemical Product Research and Development, Eli Lilly and Company, Indianapolis, IN and Edward Sheldon, Analytical Technologies Design & Development , Eli Lilly & Co., Indianapolis, IN

Continuous High Pressure Vapor/Liquid PFR Sampling System with On-line HPLC

Wei-Ming Sun, Martin Johnson, Edward M. Sheldon

Abstract:

An automated sampling approach utilizing on-line HPLC had been designed for continuous thermal tube and hydrogenation tube reactors. The system is an integrated part of the automation system used to control continuous plug flow tube reactors. A six port, two position column switching valve is used in conjunction with a fixed volume loop to introduce the sample. There are 2 major challenges for directly connecting HPLC with continuous flow reactor. First, the inlet flow of HPLC needs to be only liquid phase without any vapor phase so that an accurate and precise dilution can be accomplished. Second, the control reaction back pressure is critical; therefore, the interface with the HPLC must be accomplished without reaction pressure perturbation. This approach utilizes a system that can directly connect to wide range of pressure (0-1,000 psig) continuous flow reactor without affecting the back pressure of the reactor. In addition, this system is able separate the liquid and vapor phases for sampling to the HPLC loop. The small amount of product can be diluted to the desired concentration, and automatically mixed, then, the homogenous sample solution is directly transferred to the HPLC loop for injection. In summary, the system can automatically take a 1 ml vapor/liquid sample from a continuous reactor outlet stream at pressures up to 1000 psig, separate the liquid from the vapor, precisely and accurately dilute the liquid phase, mix the diluted sample, and inject into and HPLC that can be located up to 25 meters from the continuous reactor.

(467c) Characterization and Selection of Continuous Plug Flow Reactors in Pharmaceutical Development

Brian D. Haeberle, Eli Lilly and Company, Indianapolis, IN, Martin D. Johnson, Chemical Product Research and Development, Eli Lilly and Company, Indianapolis, IN, Scott A. May, Chemical Product Research and Development, Eli Lilly & Co., Indianapolis, IN, Timothy Braden, Chemical Product Research and Development, Eli Lilly &Co., Inc., Indianapolis, IN and Ed Plocharczyk, Flow technology, D&M Continuous Solutions LLC, Greenwood, IN

The use of plug flow tube reactors (PFR) in the pharmaceutical industry offers many advantages to drug development and scale up of processes. PFRs can enable chemistry that would be difficult to perform batch, increase safety vs. batch reactors, and allow for rapid condition screening to accelerate early phase development. One of the main parameters used to characterize PFRs is the residence time distribution, which can be described by axial dispersion. The extent of dispersion can affect the reaction time as well as influence the impurity profile and must be understood to effectively scale a process. Dispersion in over twenty PFRs was characterized at a variety of Reynolds numbers, length to diameter ratios (L/d), and coil diameters. Reactor volume spanned 0.00175 to 73 L and resulted in vessel dispersion numbers (D/uL) ranging from 5.4×10^{-6} to 3.85×10^{-2}. For small volume PFRs a lower Reynolds number resulted in a lower D/uL, however at larger volume reactors the trend reversed and a higher Re resulted in a lower D/uL. Reactors with a larger L/d had lower axial dispersion between reactors of constant volume and residence time. Both liquid only and liquid-gas regimes were investigated, and it was found that for a given reactor and constant residence time D/uL was significantly lower with liquid-gas flow.

(467d) Multi Scale Flowsheet Simulation for the Purification and Processing of Active Pharmaceutical Ingredients

Maitraye Sen, Anwesha Chaudhury, Joyce John, Ravendra Singh and Rohit Ramachandran, Department of Chemical and Biochemical Engineering, Rutgers, The State University of New Jersey, Piscataway, NJ

The quality of the final pharmaceutical products depends on the raw material (API) properties and purity. So an efficient production, separation and purification process through which the desired API properties and purity can be achieved is highly desired. Continuous mode of operation makes it possible to follow state of the art techniques to satisfy the final product requirements [1]. Continuous manufacturing leads to improved QbD (Quality by design) and also result into better implementation of the PAT (Process analytical technology) tools for real time monitoring and control of product quality attributes. Therefore transition from batch to continuous mode of production has been one of the prime objectives in the pharmaceutical industries. However, continuous manufacturing and separation of APIs is still a challenging task because of different level of complexities involved because the downstream separation and purification processes involves complex phenomena and are still least understood. It is important to understand each and every unit operation which is a part of the continuous process so that the desired operational level can be achieved. On one hand the complex particle interactions involved in the operations need to be understood and on another hand the flow behavior during operation is also important to analyze. Therefore, an efficient hybrid mathematical model, so called integrated continuous flow sheet model coupled with CFD model is highly desired.

In the work reported here, an integrated flow sheet model for continuous separation, purification and processing of APIs has been developed. The flow sheet is based on the first principle models of each unit operations involved in separation and purification steps of API (crystallization, filtration, drying, mixing). A flow sheet is designed by integrating all these unit operations. As a result an actual plant operation can be approximately represented with the help of a robust flow sheet model [2]. The developed flow sheet model has several applications such as to understand, design and optimize the process and its control system that will regulate the process and also to identify and fix the bottlenecks in design and control objectives. The flow sheet, coupled with CFD for crystallization (Computational fluid dynamics) and DEM (discrete element models) for mixing, provides a multi- scale framework which can incorporate information from particle level to the unit-operation level.

The objective of this work is to highlight the scope of integrated flow sheet modeling coupled with CFD and to demonstrate the developed multi-scale framework for downstream processing of active pharmaceutical ingredient (API). Downstream processing of API in pharmaceutical industries includes formation of API crystals during the crystallization operation, filtration of the API crystals from the mother liquor and drying of the crystals. Crystals of different size are formed during crystallization of API. There is further size change during drying when the particles shrink due to moisture removal. Hence a population balance model has been formulated in order to represent each the unit operations with size changes. A flow sheet has been developed by integrating each of these unit operations. The population balance model has been built using gPROMS (Process System Enterprise) which is a robust and fast equation oriented software allowing both steady state and dynamic simulation runs. A CFD model can be coupled with the population balance model in order to account for the mixing of the fluid phase in an agitated crystallizer and agitated dryer.

References

[1] A. Gorsek, P. Glavic, Design of batch versus continuous processes: part 1: single-purpose equipment, Chemical Engineering Research and Design 75 (1997) 709-717

[2] F. Boukouvala, V. Niotis, R. Ramachandran, F. Muzzio, M.G. Ierapetritou, An integrated approach for dynamic flowsheet modeling and sensitivity analysis of a continuous tablet manufacturing process: an integrated approach, Computers and Chemical Engineering, doi:10.1016/j.compchemeng.2012.02.015

(467e) Adaptive Continuous Template Based Novel Manufacturing Technique for Faster Manufacturing of New APIs for Clinical Trials

Ravendra Singh, Krist V. Gernaey, Rafiqul Gani and John M. Woodley, Department of Chemical and Biochemical Engineering, Technical University of Denmark, Kgs. Lyngby, Denmark

The discovery of an effective and safe pharmaceutical product is based on success in clinical trials. Often, several candidate compounds targeting the same disease area are tested in order to identify the most efficacious products. This involves the manufacture of small quantities of compounds (APIs) for early delivery campaigns. Of these candidates only a few will be successful such that further development is required to scale–up the process. These API candidates are often new and complex molecules, and typically little or no information/data and (kinetic) models are available. Conventionally, for each API candidate the required data/information, model and batch process including operating conditions is developed based on an extensive experimental effort. However, this approach is time and resource intensive, especially when the number of API candidates is large, which is often the case for discovery of new pharmaceutics. Therefore, a common manufacturing platform that can be adapted for production of a series of similar API candidates with less consumption of time and resources is highly desired.

In the work reported here, a common manufacturing platform, a so-called *process template*, has been developed and used together with a systematic substrate adoption methodology (SAM) to adapt the template for a series of APIs with similar molecular functionality. The adaptation procedure includes the hazard assessment to quickly check whether the substances involved are hazardous or not, an initial solubility screen to select the right solvent, reactivity & selectivity assessment to identify the right reducing agent and catalyst, flowsheet generation to generate a process flowsheet from the template, process analysis to generate the operating conditions. The significant saving in time and resources for manufacturing of new APIs with the proposed manufacturing concept is due to several novel characteristics, for example: (i) the process template is a kind of superstructure and also consists of generic process equipments with wide operating range making it adaptable for a series of similar products; (ii) Based on a literature survey and experimentation, few candidates for the required solvent, catalyst and reducing agent which can be feasible for a series of similar products are identified beforehand which makes the sreening procedure faster; (iii) a windows of operation method has been developed for quick identification of suitable operating conditions; (iv) predictive methods and tools are employed for faster property prediction (e.g. solubility); (v) the systematic susbtrate adaption methodology provides the stepwise guidelines to the user that makes the process template adaption easier and faster. With the template based new manufacturing technique, a series of API candidates can be produced quickly for clinical trials. Therefore, on the one hand a wider range of API candidates can be tested, thus increasing the chance of identifying the right product, and on the other hand also shortening the time to market for a given compound. In addition, the template is compatible with continuous manufacturing, which is typically more efficient than batch operation.

The objective of this presentation is two-fold: First to highlight the value of the process template based new manufacturing concept and the associated methods and tools, and second to demonstrate its application using a pharmaceutical case study involving a nitro reduction.

***Current address of first author:** Department of Chemical and Biochemical Engineering, Rutgers, The State University of New Jersey, 98 Brett road, Piscataway, NJ 08854, USA

(467f) Sustainable and Continuous Reduction of Amides

Kathryn Rix[1], Geoffrey Kelsall[2], **Klaus Hellgardt**[3] and Mimi Hii[1], (1)Chemistry, Imperial College London, London, United Kingdom, (2)Imperial College London, London, United Kingdom, (3)Chemical Engineering, Imperial College London, London, United Kingdom

Sustainable and Continuous Reduction of Amides

Kathryn Rix, Geoff Kelsall, Mimi Hii, Klaus Hellgardt

Amines are important organic intermediates for the chemical industry; used in the production of agrochemicals, dyes, pharmaceuticals, surfactants and plastics. Amines constitute important core structures of many biologically active molecules, and are particularly prevalent in central nervous system (CNS) drugs, which make up the largest sector of pharmaceuticals sold worldwide.

The reduction of amides to amines can be effected electrochemically by a route summarised below:

Cathode: $R\text{-}CONH_2 + 4H^+ + 4e^- \ ^{\text{reduction}}\ R\text{-}CH_2NH_2 + H_2O$ (1.0)

Nafion membrane: H^+ (anolyte) H^+ (catholyte) (1.1)

Anode: $2H_2O\ ^{\text{oxidation}}\ O_2 + 4H^+ + 4e^-$ (1.2)

Overall reaction: $R\text{-}CONH_2 + H_2O\ R\text{-}CH_2NH_2 + O_2$ (1.3)

An electrochemical flow-cell reactor has been constructed using boron doped diamond cathodes (BDD) to reduce a range of amides, demonstrating chemo-selectivity, stereo-selectivity and product selectivity by forming the corresponding amine.

For example, maleimide was selectively reduced to succinimide rather than to the corresponding amine, demonstrating chemoselectivity. This was developed further to show that the cyclic double bond can be preferentially reduced over an allyl group.

A conversion of >99% and a current efficiency of 96% were achieved. After neutralisation and extraction, ^1H NMR indicated that the cyclic double bond had been reduced while the allyl group remained intact (Figure 1). As the cyclic double bond is more electron deficient, the addition of two electrons and two protons is more favourable than addition to the allyl group.

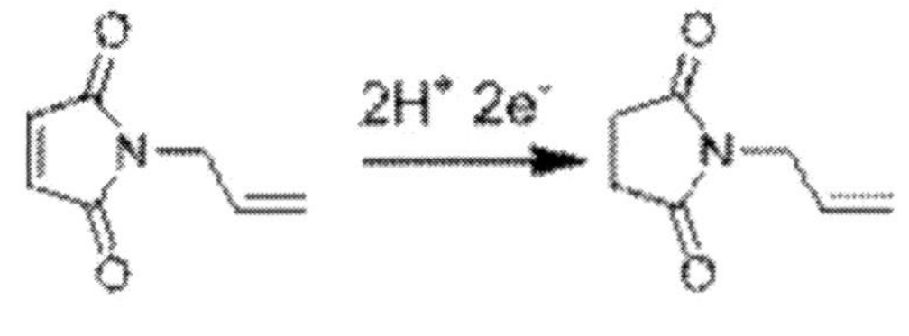

Figure 1: Electrochemical reduction of *N*-allylmaleimide

In this presentation we will discuss further examples of the facile continuous electrochemical reduction of amides, avoiding the use of hydride agents.

(467g) Benign Alkylation of Amines by Alcohols (AAA)

Luka Tallon[1], **Klaus Hellgardt**[2] and Mimi Hii[1], (1)Chemistry, Imperial College London, London, United Kingdom, (2)Chemical Engineering, Imperial College London, London, United Kingdom

Benign Alkylation of Amines by Alcohols (AAA)

Luka Tallon, Klaus Hellgardt, Mimi Hii

In recent years the increasing concern for the environment and concepts such as 'e-factor' and 'atom economy', have catalysed the integration of sustainable development and green chemistry. In particular, the activation of alcohols for nucleophilic substitution was identified by the American Chemical Society (ACS) Green Chemistry Institute (GCI) Pharmaceutical Roundtable as one of the key reactions where improvements are required. For example, it has been estimated that 64% of all nitrogen substitutions constitute amine alkylations in the pharmaceutical industry.

Two commonly used approaches for the activation of an OH group towards nucleophilic substitution are by protonation or conversion to a halide/sulfonate. However, protonation often causes the incoming nucleophile to become deactivated in acidic environment and halogenation generates a large quantity of waste products. In both cases, over-alkylation (formation of tertiary amines from primary amines) is also a recurring problem.

A 'hydrogen borrowing' methodology offers an alternative, benign approach to the activation of alcohols; in this presentation we will discuss the ***A***lkylation of ***A***mines by ***A***lcohols (*AAA*) using this process.

Hydrogen borrowing is a form of transfer hydrogenation; a catalyst, [M], is employed both as a hydrogen acceptor and donor (Figure 1). During the reaction, hydrogen is abstracted from the alcohol to form the more reactive carbonyl, which is then condensed with the amine to form an imine intermediate, to which the catalyst then returns hydrogen to form the newly alkylated amine.

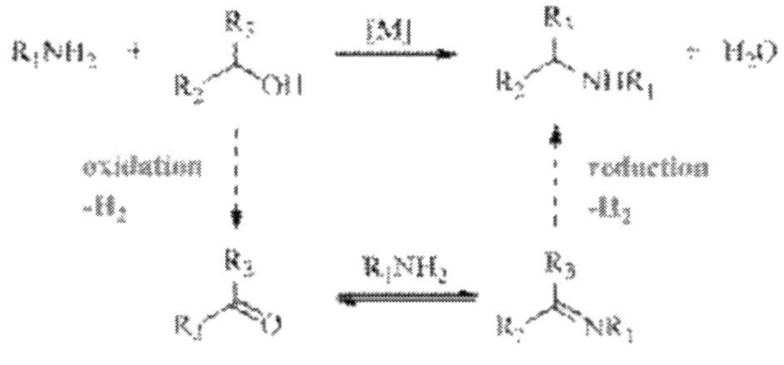

Figure 1: *N- Alkylation of amine using hydrogen borrowing methodology*

This system has clear advantages over previous routes for the activation of alcohols. Only water is generated as a by-product therefore achieving an enhanced environmental profile.

Very recently the work from our research group has shown that by using heterogeneous Au/TiO_2, highly selective direct alkylation of a range of amines under aerobic conditions can be achieved using a flow reactor (Figure 2).

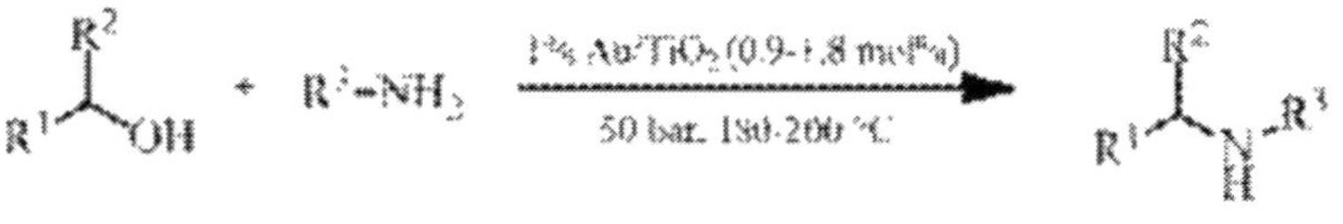

Figure 2: Continuous flow alkylation of amine over heterogeneous Au catalysts

The scope of this system will be demonstrated, including details on the synthesis of Piribedil. Although higher temperatures and 50 bar pressure are required, the method involves no use of co-catalysts such as $[RuCl_2(p\text{-cymene})]_2$.

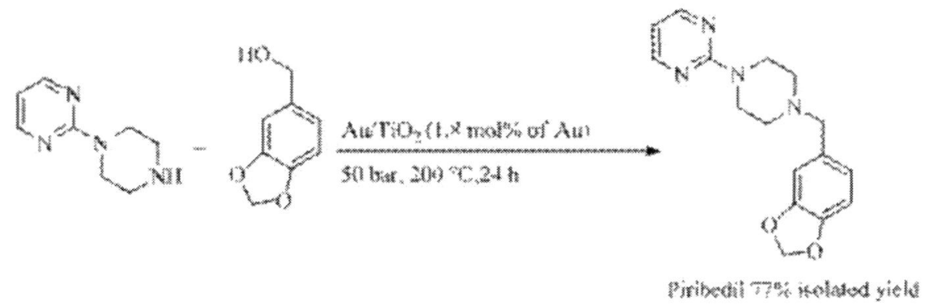

Figure 3: Facile Piribedil synthesis over Au based catalyst

Additionally, we wish to highlight that the reaction of optically pure α-methylbenzylamine and benzyl alcohol proceeds with complete retention of configuration to give the substituted product (Figure 4).

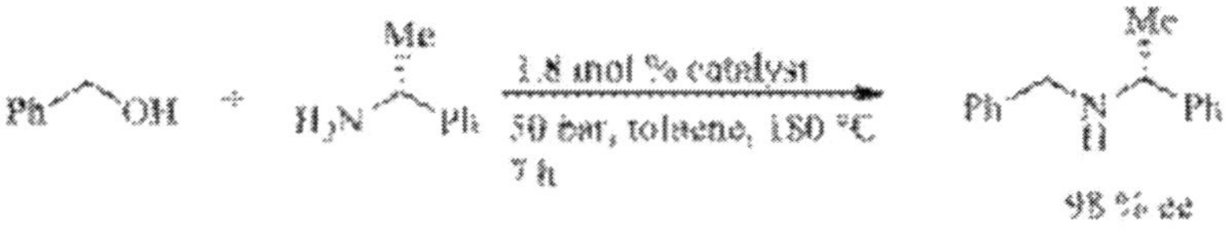

Figure 4: *Reaction of optically pure α-methylbenzylamine and benzyl alcohol.*

(483a) Active Stealth Signaling with a Synthetic 'self' Peptide

Dennis E. Discher, Chemical and Biomolecular Engineering, University of Pennsylvania, Philadelphia, PA

Foreign particles and cells are rapidly cleared from the body by professional phagocytes that incessantly encounter and recognize 'self' cells. The membrane protein CD47 is reportedly a 'Marker of Self' in mouse that impedes phagocytosis of self cells, with signaling through the highly polymorphic receptor CD172a on phagocytes. Here, minimal 'Self' peptides were computationally designed from $CD47^{human}$, synthesized with anchoring groups, and attached to virus-size nanoparticles for intravenous injection into NOD.SCID (NSG) mice that express a unique $CD172a^{NSG}$ (SIRPA) compatible with $CD47^{human}$. 'Self' peptides delay splenic macrophage clearance of particles with an exponential advantage in persistent circulation, which we use to enhance near-infrared imaging of human tumor xenografts by >10-fold. The affinity of $CD47^{human}$ for $CD172a^{NSG}$ proves weak but within the broad range (0.1 ~ 5 micro-Molar) measured for ten constructed variants of $CD172a^{human}$; several versions of 'Self' peptide are likewise shown to bind and potently inhibit nanoparticle uptake by an unexpected cytoskeletal mechanism. The reductionist approach reveals the importance and utility as well as some limits of a human 'Self' peptide.

(483b) Exploring Peptide-Based Nanostructures As Effective Drug Carriers

Ran Lin, Andrew G. Cheetham, Pengcheng Zhang and Honggang Cui, Chemical and Biomolecular Engineering, Johns Hopkins University, Baltimore, MD

Peptides can assemble into a variety of discrete nanostructures that could serve as biodegradable drug carriers for cancer chemotherapy. In this work, we demonstrate our rational design of short peptide conjugates to create various nano-objects such as spheres, cylinders, and cylinders with a multicompartment core. The critical micellization concentrations (CMC) of these conjugates can be tuned by molecular design, ranging in the nanomolar to micromolar scale. We evaluated the potential use of these nanostructures as drug carriers. In our drug encapsulation experiments, higher loading capacity is often achieved using peptides with relatively larger volume of the hydrophobic segments. In addition, by directly conjugating the anticancer drugs such as paclitaxel to a rationally designed short peptide, a quantitative and high drug loading can be achieved, and the resulting nanostructures can be used to encapsulate a different type of drugs. Our cytotoxicity experiments reveal that these supramolecular nanostructures can work effectively against a number of cancer cell lines including breast cancer, prostate cancer and ovarian cancer.

(483c) Dendron-Based Micelles: A Potential Nanocarrier Platform

Ryan Pearson, Jin Woo Bae, Hao-Jui Hsu, Sayam Uddin and Seungpyo Hong, Biopharmaceutical Sciences, University of Illinois at Chicago, Chicago, IL

Dendron-based amphiphilic copolymers possess unique features such as well-defined structure, self-assembling capability, and precise controllability over functionalities and high hydrophilicity. To assess the ability of dendron-based copolymers for use in drug delivery, we have synthesized novel, highly PEGylated dendron copolymers (PDC) with controlled hydrophilic-lipophilic ratios (HLs) ranging from 52:48 to 91:9. We then systematically compared their self-assembly properties with those of linear-block copolymer (LBC) counterparts at similar HLs. PDCs were synthesized via a multistep synthetic procedure. A hydroxyl-terminated generation 3 (G3) polyester dendron bearing a focal acetylene moiety was conjugated with poly(e-caprolactone) (PCL) by 'click' chemistry followed by PEGylation of the dendron surface with multiple methoxy poly(ethylene glycol) (mPEG) moieties. Four types of both PCL-G3-mPEG PDCs and PCL-mPEG LBCs with different block lengths were prepared and their structures were confirmed using ^{1}H NMR, FT-IR, and GPC. The critical micelle concentration (CMC) of PDCs was measured as low as 1.17 X 10^{-8}. The CMC values of the LBCs were up to 2 orders of magnitude higher than those of the PDCs at the same HLs, demonstrating superior thermodynamic stability of the PDC micelles. TEM images revealed that the PDCs formed spherical micelles (20-30 nm in diameter) with narrow size distributions, which was further supported by dynamic light scattering measurements. To understand in molecular detail the differences in self-assembly between PDCs and LBCs we utilized atomistic molecular dynamics (MD) simulations. Individual PDC copolymers were observed to adopt a stable conical shape in water, which is ideal for precise packing into spherical structures. Interestingly, the core was more completely covered by PEG in PDC micelles. The higher PEG surface coverage of PDC micelles may result in longer circulation half-lives due to decreased opsonization and enhanced accumulation at the therapeutic target site. Each micelle formulation was found to exhibit controlled release properties over 3 days using indomethacin as a model drug. PDCs exhibited no apparent toxicity up to concentrations as high as 100 mM evaluated by MTS assay. The low CMC at high HLs, controlled micelle morphologies, controlled release properties, and high PEG surface coverage all indicates the promising potential of the PDCs as a nanocarrier platform.

AUTHOR INDEX

AUTHOR INDEX

AUTHOR INDEX

AUTHOR INDEX

AUTHOR INDEX

AUTHOR INDEX

AUTHOR INDEX

AUTHOR INDEX

AUTHOR INDEX

AUTHOR INDEX

AUTHOR INDEX

AUTHOR INDEX

AUTHOR INDEX

AUTHOR INDEX

AUTHOR INDEX

AUTHOR INDEX

AUTHOR INDEX

AUTHOR INDEX

AUTHOR INDEX

AUTHOR INDEX

AUTHOR INDEX

Food, Pharmaceutical and Bioengineering Division

Core Programming Topic at the 2012 AIChE Annual Meeting

Pittsburgh, Pennsylvania, USA
28 October - 2 November 2012

Volume 2 of 2

ISBN: 978-1-62276-728-1

Printed from e-media with permission by:

Curran Associates, Inc.
57 Morehouse Lane
Red Hook, NY 12571

Some format issues inherent in the e-media version may also appear in this print version.

Printed by Curran Associates, Inc. (2013)

For permission requests, please contact AIChE
at the address below.

AIChE
3 Park Avenue
New York, NY 10016-5991

Phone: (203) 702-7660
Fax: (203) 775-5177

www.aiche.org

Additional copies of this publication are available from:

Curran Associates, Inc.
57 Morehouse Lane
Red Hook, NY 12571 USA
Phone: 845-758-0400
Fax: 845-758-2634
Email: curran@proceedings.com
Web: www.proceedings.com

TABLE OF CONTENTS

Volume 1

Volume 2

(483d) Microfluidic Cell Deformation As a Robust, Vector-Free Method for Cytosolic Delivery of Macromolecules

Armon Sharei[1], Janeta Zoldan[1], Andrea Adamo[2], Woo Young Sim[1], Nahyun Cho[1], Emily Jackson[1], Shirley Mao[1], Sabina Schneider[1], Abigail Lytton-Jean[1], Jungmin Lee[1], Daniel A. Heller[3], Robert S. Langer[4] and Klavs F. Jensen[5], (1)MIT, Cambridge, MA, (2)Chemical Engineering, Massachusetts Institute of Technology, Cambridge, MA, (3)Department of Chemical Engineering, Massachusetts Institute of Technology, Cambridge, MA, (4)Massachusetts Institute of Technology, Cambridge, MA, (5)Chemical Engineering, MIT, Cambridge, MA

Introduction

Intracellular delivery of macromolecules is a critical step in therapeutic and research applications. Nanoparticle mediated delivery of DNA and RNA, for example, is being explored for gene therapy[1], while protein delivery is a promising means of affecting cellular function in both clinical[2] and laboratory[3] settings. Other materials, such as small molecules, quantum dots, or gold nanoparticles, are of interest for cancer therapies[4], intracellular labeling[5], and single molecule tracking[6].

The cell membrane is largely impermeable to macromolecules. Many existing techniques utilize polymeric nanoparticles[7], or chemical modifications of the target molecule, such as cell penetrating peptides (CPPs) [8], to facilitate membrane poration or endocytotic delivery. In these methods, the delivery vehicle's efficacy is often dependent on the structure of the target molecule and the cell type. These methods are thus efficient in the delivery of structurally uniform materials, such as nucleic acids, but often ill-suited for the delivery of more structurally diverse materials, such as proteins[9] and some nanomaterials[10]. Moreover, the endosome escape mechanism that most of these methods rely on is often inefficient; hence, much material remains trapped in endosomal and lysosomal vesicles.

In this presentation, we detail a method for cytosolic delivery based on rapid mechanical deformation of the cell to produce transient membrane disruptions that facilitate the passive diffusion of material into the cell cytosol. This method was developed with the aim of delivering almost any macromolecule of interest to almost any cell type, at high throughput. Although shear based delivery methods have been demonstrated previously, they are unsuitable for some applications due to low viability and/or delivery efficiency[11]. The proposed technique, however, is distinct because it uses a physical constriction to deform and shear the cells in a controlled, reproducible manner. Unlike most current methods, this approach does not rely on electric fields, exogenous materials, endocytosis or chemical modification of the target molecule. Our data indicate that this method could be particularly advantageous for applications involving nanomaterials, proteins, or difficult-to-transfect cell types, such as immune cells and stem cells - all of which are often underserved by current methods.

Delivery method

We hypothesize that the rapid mechanical deformation of a cell, as it passes through a constriction with a minimum dimension smaller than the cell diameter, results in the formation of transient membrane disruptions or pores (**Fig. 1a**). The size and frequency of these pores would be a function of the shear and compressive forces experienced by the cell during its passage through the constriction. Material from the surrounding medium may then diffuse directly into the cell cytosol throughout the lifespan of these pores. Such an approach could theoretically enable the diffusive delivery of any macromolecule small enough to fit through the pores. To implement this approach, we generated a family of microfluidic devices with different constriction dimensions and numbers of constrictions in series.

Each device consists of 45 identical, parallel microfluidic channels, containing one or more constrictions, etched onto a silicon chip and sealed by a Pyrex layer. The width and length of each constriction range from 4-8um and 10-40um respectively. The current design is typically operated at a throughput rate of 20,000 cells/s, yielding close to one million treated cells per device prior to failure, due to clogging. The parallel channel design was chosen to increase throughput, while insuring uniform treatment of cells, because any clogging or defects in one channel cannot affect the flow speed in neighboring channels (the device is operated at constant pressure). Prior to use, the device is first connected to a steel interface that connects the inlet and outlet reservoirs to the silicon device. A mixture of cells and the desired delivery material is then placed into the inlet reservoir and Teflon tubing is attached at the inlet (not shown). A pressure regulator is then used to adjust the pressure at the inlet reservoir and drive the cells through the device. Treated cells are collected from the outlet reservoir.

An enabling platform

In order to illustrate our method's potential in addressing current delivery challenges, we conducted a number of proof-of-concept experiments using various nanomaterials and proteins of interest. Antibodies to tubulin, for example, were delivered (**Fig. 1b**) using this technique, yielding a diffuse distribution throughout the cell that would be consistent with cytosolic delivery. PEG1000 coated, 15nm gold nanoparticles were also delivered as measured by tunneling electron microscopy (TEM) of fixed cells (**Fig. 1c**). These nanoparticles appear to be mostly un-aggregated and are not visibly sequestered into endosomes. We were also able to verify the successful delivery of carbon nanotubes[12] (encapsulated by a DNA oligonucleotide) by flow cytometry and Raman spectroscopy. The aforementioned materials are currently difficult to deliver to the cell cytosol and each material often requires a specialized modification to facilitate delivery. In our work, all four materials were delivered to HeLa cells using the same set of conditions on a 10um-6umx5 device. In addition, this method has demonstrated the potential to accommodate a diverse range of cell types, including embryonic stem cells and primary lymphocytes (**Fig. 1d,e**), thereby signaling its potential for use across a range of research and perhaps clinical applications.

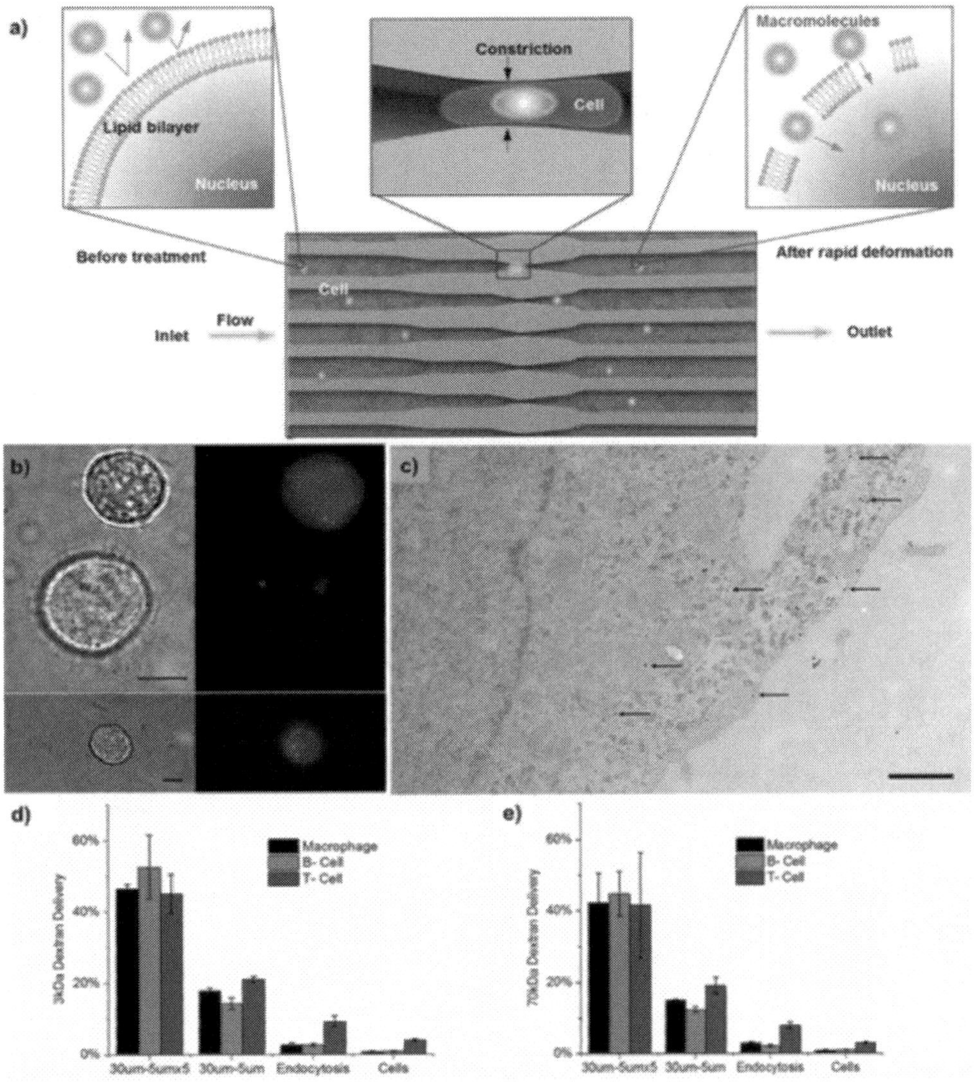

Figure 1. An overview of the delivery system and its capabilities. a) Illustration of delivery hypothesis whereby the rapid deformation of a cell, as it passes through a microfluidic constriction, generates transient membrane pores. Includes an electron micrograph of current parallel channel design with blue cells as an illustration. b) Fluorescent micrographs of HeLa cells immediately after delivery of antibodies to tubulin with an Alexa Fluor 488 tag. Scale bars at 5um. c) Tunneling electron microscopy (TEM) images of gold nanoparticles (some indicated by arrows) in a cell fixed ~1s after treatment. Scale bar at 500nm. d) Delivery of pacific blue labeled 3kDa dextran and e) fluorescein labeled 70kDa dextran to Macrophages (CD11b+), B lymphocytes (CD19+) and T lymphocytes (TCR+) derived from the blood of BL6 mice. The 30um-5umx5 and 30um-5um conditions denote two different device designs.

Summary

We have detailed a new method for cytosolic delivery that relies on the rapid mechanical deformation of a cell to induce transient membrane poration. This technique has demonstrated the potential to deliver a broad range of materials, some of which are challenging to use with current methods, to a variety of difficult-to-transfect cell types, including stem cells and immune cells. By providing such flexibility in application and obviating the need for exogenous materials or electrical fields, this method could potentially enable new avenues of disease research and treatment. Indeed our work has demonstrated this systems ability to deliver carbon nanotubes and antibodies to live cells – applications that could enable new sensing and imaging modalities.

References

1. Whitehead, K.A., Langer, R. & Anderson, D.G. Knocking down barriers: advances in siRNA delivery. *Nat Rev Drug Discov* **8**, 129-138 (2009).

2. Leader, B., Baca, Q.J. & Golan, D.E. Protein therapeutics: a summary and pharmacological classification. *Nat Rev Drug Discov* **7**, 21-39 (2008).

3. Kim, D. et al. Generation of Human Induced Pluripotent Stem Cells by Direct Delivery of Reprogramming Proteins. *Cell Stem Cell* **4**, 472-476 (2009).

4. Dhar, S., Daniel, W.L., Giljohann, D.A., Mirkin, C.A. & Lippard, S.J. Polyvalent Oligonucleotide Gold Nanoparticle Conjugates as Delivery Vehicles for Platinum(IV) Warheads. *Journal of the American Chemical Society* **131**, 14652-14653 (2009).

5. Michalet, X. et al. Quantum Dots for Live Cells, in Vivo Imaging, and Diagnostics. *Science* **307**, 538-544 (2005).

6. Dahan, M. et al. Diffusion Dynamics of Glycine Receptors Revealed by Single-Quantum Dot Tracking. *Science* **302**, 442-445 (2003).

7. Pack, D.W., Hoffman, A.S., Pun, S. & Stayton, P.S. Design and development of polymers for gene delivery. *Nat Rev Drug Discov* **4**, 581-593 (2005).

8. Heitz, F., Morris, M.C. & Divita, G. Twenty years of cell-penetrating peptides: from molecular mechanisms to therapeutics. *British Journal of Pharmacology* **157**, 195-206 (2009).

9. Yan, M. et al. A novel intracellular protein delivery platform based on single-protein nanocapsules. *Nat Nano* **5**, 48-53 (2010).

10. Derfus, A.M., Chan, W.C.W. & Bhatia, S.N. Intracellular Delivery of Quantum Dots for Live Cell Labeling and Organelle Tracking. *Advanced Materials* **16**, 961-966 (2004).

11. Hallow, D.M. et al. Shear-induced intracellular loading of cells with molecules by controlled microfluidics. *Biotechnology and Bioengineering* **99**, 846-854 (2008).

12. Heller, D.A., Baik, S., Eurell, T.E. & Strano, M.S. Single-Walled Carbon Nanotube Spectroscopy in Live Cells: Towards Long-Term Labels and Optical Sensors. *Advanced Materials* **17**, 2793-2799 (2005).

(483e) Self-Dispersing Drug Carriers for Pulmonary Delivery: Spreading of Aqueous Surfactant Solutions On Model Airway Surface Liquid Subphases

Amsul Khanal[1], Ramankur Sharma[2], Roomi Kalita[2], Fan Gao[3], Timothy Corcoran[4], Ellen Peterson[5], Todd M. Przybycien[6], Stephen Garoff[3] and Robert D. Tilton[6], (1)Department of Biomedical Engineering, Carnegie Mellon University, Pittsburgh, PA, (2)Department of Chemical Engineering, Carnegie Mellon University, Pittsburgh, PA, (3)Department of Physics, Carnegie Mellon University, Pittsburgh, PA, (4)Department of Medicine, University of Pittsburgh, Pittsburgh, PA, (5)Department of Mathematical Sciences, Carnegie Mellon University, Pittsburgh, PA, (6)Department of Chemical Engineering and Department of Biomedical Engineering, Carnegie Mellon University, Pittsburgh, PA

Persistent pulmonary infections are a leading cause of mortality in cystic fibrosis lung disease. This condition is associated with over-secretion of abnormally viscoelastic mucus that creates pulmonary airway obstructions. Current aerosolized antibiotic delivery techniques rely entirely on aerodynamic mechanisms to disperse and distribute these medications inside the lung after inhalation. The elements of obstruction common to cystic fibrosis and other lung diseases can decrease ventilation in portions of the lung, preventing aerosol drugs from reaching these zones. The unusual aerodynamics associated with obstructive disease can also cause highly non-uniform aerosol deposition patterns, further limiting distribution of drug throughout the lung. Poor distribution ultimately limits their efficacy. The inhaled antibiotics used to treat bacterial infections associated with cystic fibrosis lung disease often provide successful suppression of infection but rarely provide eradication. Drug resistance has also been associated with these therapies, likely due to the consistent delivery of sub-therapeutic doses at sites of infection. This group is investigating the potential to develop self-dispersing aerosol medications to improve the uniformity of drug delivery to obstructed lungs. An active agent is formulated with aqueous surfactant solutions that harness surface tension gradients to promote spreading on the airway surface liquid after aerosols deposit. Experiments on model systems show that surfactants significantly enhance spreading relative to saline controls as expected. Surfactant-laden aqueous drops when placed on entangled aqueous polymer solutions spread to form static lenses that persist on top of the subphase for tens of minutes, despite their complete miscibility. Our interests are focused on the relationship between formulation composition and the final spread area of deposited drops, a surrogate for the extent of drug dispersal after droplet deposition on the airway surface liquid. We use entangled aqueous poly(acrylamide) and porcine gastric mucin (PGM) solutions as subphases to mimic the physical properties of airway mucus plaques that are the main obstacles to uniform aerosol deposition in the lungs of patients with obstructive lung diseases. Solutions of the anionic surfactant sodium dodecyl sulfate (SDS), cationic surfactant cetyltrimethylammonium bromide (CTAB) or non-ionic surfactant tyloxapol contain dyes (absorptive or fluorescent) that serve as both model "drugs" and as tracers to measure the extent of drop spreading on the subphase. Regardless of the type of surfactant, when a drop of surfactant solution is placed on the polymer subphase, we observe that the dye spreads with the drop and that the surface tension of the subphase decreases in regions far outside the deposited drop, indicating that surfactant escapes across the drop contact line. This phenomenon is not observed with surfactant-free aqueous drops. The spread drop forms a finite, static lens that exists for on the order of 10 minutes while its contents slowly diffuse into the subphase. While Marangoni stresses initiate flow, a capillary force balance terminates the convective spreading process, resulting in the confinement of the spread solution to a well-defined static area. Spreading times are on the order of tens of seconds, considerably shorter than the timescales for diffusion into the subphase bulk or for subphase disentanglement. Spreading of aerosolized surfactant solutions is compared to the extent of spreading of macroscopic surfactant solution drops. The final spread areas for a macroscopic drop and the same deposited volume of aerosolized surfactant solution (aerosol size 2-10 um) are comparable, suggesting a similar spreading trend for aerosolized surfactant solutions and surfactant-laden drops on entangled polymer subphases that model the airway surface liquid.

(483f) Cathepsin B Degradable Peptidic Dendrimers for Drug Delivery

Rohit Kolhatkar, Ravi Shankar and Abhilash Samykutty, Biopharmaceutical Sciences, University of Illinois Chicago, Rockford, IL

INTRODUCTION: Polymers are widely used as drug delivery vehicles due to their ability to accumulate passively in tumor. Polymers can also be functionalized to achieve active targeting. Dendrimers offer several advantages over other conventional polymers. Here we report the use of peptidic dendrimers as drug delivery vehicles. These dendrimers are designed to degrade in the presence of an enzyme cathepsin B which is overexpressed at tumor site. A tetrapeptide *glycyl-phenylalanyl-leucyl-glycine* (GFLG) is used as a building block. The tetrapeptide GFLG is known to be recognized by cathepsin B. Two polymers with positive and negative surface charge were synthesized and characterized for their degradation profile in the presence of cathepsin B. By conjugation of a model drug 2-methoxyestradiol (2ME) to these polymers we show that peptidic dendrimers have the potential to be used as drug delivery vehicles.

EXPERIMENTAL METHODS: Positively and negatively charged peptidic dendrimers were synthesized using a convergent approach where tetrapeptide (GFLG) was added to the central core having four arms. Degradation profile of these polymers in the presence of enzyme cathepsin B and stability in the presence of PBS and acetate buffer was evaluated using HPLC. Glycine derivative of 2ME was attached to the tetrapeptide (GFLG) followed by conjugation of this unit (GFLGG2ME) to the central core to yield 2ME containing peptidic dendrimer. The activity of monomer and the polymer was evaluated by performing cytotoxicity studies using breast cancer cell line MDA-MB-231.

RESULTS AND DISCUSSION: Both positively and negatively charged peptidic dendrimers showed time dependant degradation in the presence of cathepsin B. Negatively charged dendrimer showed faster rate of degradation compared to positively charged amine terminated dendrimer. Polymers were stable in the presence of PBS and acetate buffer for 24 h. GI_{50} values (concentration to inhibit cellular growth by 50 %) for free drug 2ME, monomer GFLGG2ME were 0.89 μM and 7.6 μM respectively. Drug containing polymer (38 % wt/wt of drug) showed GI_{50} value of 8.5 μM while control polymer did not show any cytotoxicity. At equivalent drug concentrations, peptidic dendrimers were at least 8 times more active than *N*-(2-hydroxypropylmethacrylamide) (HPMA) copolymers containing 2ME (GI_{50} value of 114 μM).

CONCLUSION: Peptidic dendrimer has the potential to be used as carrier for 2-methoxyestradiol for breast cancer treatment.

ACKNOWLEDGEMENTS: This research was supported by Department of Biopharmaceutical Sciences, University of Illinois Chicago and Department of Defense (Breast Cancer Research) concept award (W81XWH-09-1-0687) to R. Kolhatkar.

(483g) Antitumor Efficacy Following the Intracellular and Interstitial Release of Liposomal Doxorubicin

Amey Bandekar, Biomedical Engineering and Chemical and Biochemical Engineering, Rutgers University, Piscataway, NJ and **Stavroula Sofou**, Biomedical Engineering, and Chemical and Biochemical Engineering, Rutgers University, Piscataway, NJ

pH-triggered lipid-membranes designed from biophysical principles are evaluated in the form of tar- geted liposomal doxorubicin with the aim to ultimately better control the growth of vascularized tumors. We compare the antitumor efficacy of anti-HER2/neu pH-triggered lipid vesicles encapsulating doxo- rubicin to the anti-HER2/neu form of an FDA approved liposomal doxorubicin of DSPC/cholesterol-based vesicles. The HER2/neu receptor is chosen due to its abundance in human breast cancers and its connection to low prognosis. On a subcutaneous murine BT474 xenograft model, superior control of tumor growth is demonstrated by targeted pH-triggered vesicles relative to targeted DSPC/cholesterol- based vesicles (35% vs. 19% decrease in tumor volume after 32 days upon initiation of treatment). Superior tumor control is also confirmed on SKBR3 subcutaneous xenografts of lower HER2/neu expression. The non-targeted form of pH-triggered vesicles encapsulating doxorubicin results also in better tumor control relative to the non-targeted DSPC/cholesterol-based vesicles (34% vs. 41% increase in tumor volume). Studies in BT474 multicellular spheroids suggest that the observed efficacy could be attributed to release of doxorubicin directly into the acidic tumor interstitium from pH-triggered vesicles extravasated into the tumor but not internalized by cancer cells. pH-triggered liposome carriers engi- neered from gel-phase bilayers that reversibly phase-separate with lowering pH, form transiently defective interfacial boundaries resulting in fast release of encapsulated doxorubicin. Our studies show that pH-triggered liposomes release encapsulated doxorubicin intracellularly and intratumorally, and may improve tumor control at the same or even lower administered doses relative to FDA approved liposomal chemotherapy.

(483h) Targeted Nitric Oxide Pretreatment Alters p53 and O6-Methylguanine-DNA Methyltransferase Activity Resulting in Enhanced Chemosensitivity in Glioma Cells

Shahana Safdar, Chemical and Biomolecular Engineering, Georgia Tech, Atlanta, GA, Courtney A. Payne, Spelman College, Atlanta, GA, Nam H. Tu, Georgia Institute of Technology, Atlanta, GA and Lakeshia Taite, Chemical and Biomolecular Engineering, Georgia Institute of Technology, Atlanta, GA

Introduction: Carmustine, 1,3-bis(2-chloroethyl)-1-nitrosourea (BCNU) and temozolomide (TMZ) are chemotherapeutics commonly used to treat glioblastoma multiforme. However a significant number of patients are resistant to these therapeutics resulting in poor patient survival. In the past, high doses of nitric oxide (NO) have been used to induce chemosensitivity in glioma cells. However the high doses coupled with the non-specific delivery of NO result in toxic effects to healthy tissue. Thus, to limit these side effects we have developed a targeted NO donor to enhance the efficacy of TMZ and BCNU on human malignant glioma cells.

Materials and Methods: To synthesize glioma-targeting NO donor, Chlorotoxin (CTX), a glioma targeting protein isolated from the venom of the Deathstalker scorpion, was reacted with NO gas for 24 hours in a hypoxic environment transforming CTX in to CTX-NO. Two malignant human glioblastoma cell lines T98 and U-87MG, normal human astrocytes (NHA) and human brain and microvascular endothelial cells (HBMEC) were used for *in vivo* studies. The cells were incubated for 2 hours with CTX-NO, after which they were incubated with BCNU (75 μM) or TMZ (50 μM) for 48 hours. At the end of the incubation period, cell viability, intercellular MGMT levels, or p53 expression was quantified.

Results and Discussion: TMZ and BCNU did not significantly change the viability of T98G cells. However, pretreatment with 2 μM of NO followed by TMZ or BCNU significantly decreased T98G cell viability to 76.8 ± 8.5% and 67.1 ± 1.8% of the control. Conversely, U-87MGs were more sensitive to both chemotherapeutics and NO pretreatment further enhanced this chemosensitivity (Fig. 1). CTX-NO resulted in no significant changes in the chemosensitivity of non-tumor control cells (Fig. 1). The resistance of tumor cells TMZ and BCNU is partially due to the presence of the DNA repair protein O^6-methylguanine-DNA methyltransferase (MGMT) therefore we investigated the effect of CTX-NO on MGMT levels in glioma cells. When T98G cells were treated with NO at concentrations of 1 μM or 2 μM, MGMT levels were decreased to 73.1 ± 5.9% and 60.1 ± 9.9% of the control. Another biomolecule that plays an important role in determining chemosensitivity is the protein p53 which a pivotal role in maintaining the balance between DNA repair mechanisms and induction of apoptosis. In T98G cells, which express mutant p53, CTX-NO treatment was able to significantly decrease in a dose dependant manner. Conversely, CTX-NO treatment did not affect p53 levels in U-87MG cells which express wild-type p53.

Conclusion: In this study we used small doses of a targeted NO donor, CTX-NO, to chemosensitize malignant glioma cells to chemotherapeutic agents, BCNU and TMZ. Furthermore the data showed that low levels of NO were able to decrease MGMT and p53 levels, both of which contribute to chemoresistance in glioma cells. Using targeted NO to sensitize cells towards chemotherapeutics holds great potential as part of a treatment plan for patients with various tumors that are normally unresponsive to chemotherapy.

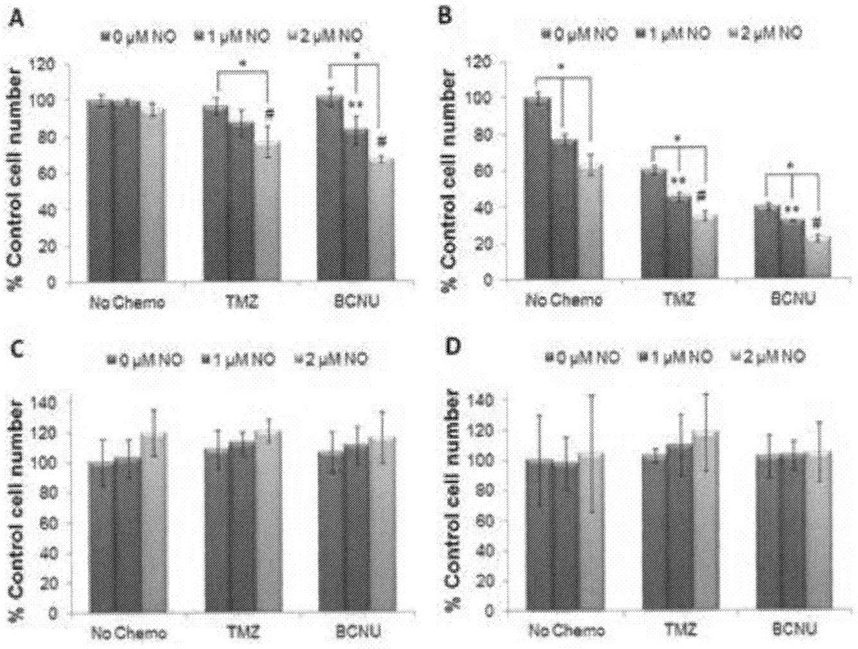

Figure 1: Cell viability of (A)T98G cells, (B)U-87MG, (C)NHA and (D)HBMEC. Cells counted are presented as a percentage of the number of cells that received no treatment. * compared to no pretreatment, ** compared to pretreatment with CTX-NO (1 µM), # compared to pretreatment with CTX-NO (2 µM) $p < 0.05$, n = 3

(483i) Strategies for Convection-Enhanced Drug Delivery

William L. Olbricht, School of Chemical and Biomolecular Engineering, Cornell University, Ithaca, NY

Convection-enhanced delivery (CED) is a novel method of delivering chemotherapeutics and other compounds to the brain for the treatment of primary brain tumors and certain neurological disorders. In CED, compounds are infused directly into the interstitium of brain tissue through a needle or cannula implanted in the brain. Because CED bypasses the blood-brain barrier, it can be used to deliver compounds that cannot be delivered to the brain by systemic means, including proteins, viral vectors, nucleotides, and molecules packaged in nanoparticles and liposomes.

The challenges in CED that we address are 1) achieving high infusion rates into the brain, and 2) controlling the spatial and temporal distribution of the infused material. High flow rates are important, because transport of the infused compounds near the needle is dominated by convection. For the drugs to penetrate far into the surrounding tissue, the convection rate must be sufficient to overcome the clearance of infused compounds due to a variety of elimination mechanisms.

To achieve high infusion rates into the brain, we have developed a series of microfabricated catheters containing microfluidic circuits that permit higher infusion rates and have several other advantages over needles currently used in CED. The microfabricated devices are much smaller than needles, they can be designed to deliver several therapeutic fluids in a programmed sequence, and they can be made flexible so that they can be implanted in the brain for extended times.

To control the spatial distribution of compounds once they have been infused into the brain, we expose regions of the brain tissue to safe levels of ultrasound. Ultrasound has been shown to increase mass transfer in tissue in several applications. By focusing the ultrasound on particular tissue volumes, the flow of infused fluid can be directed toward specific tissue targets identified from imaging studies. A series of experiments are described in brain phantoms and in anesthetized rodents to illustrate these CED strategies.

(489a) Multivalent Effectors to Control Stem Cell Differentiation

Anthony Conway and David V. Schaffer, Department of Chemical and Biomolecular Engineering, University of California, Berkeley, Berkeley, CA

Multivalent effectors to control stem cell differentiation

Numerous cellular signaling systems that regulate stem cell self-renewal and differentiation involve the assembly of multivalent ligands – oligomeric entities that present multiple binding sites and thereby bind multiple receptors on a target cell or stem cell – such as morphogens that naturally oligomerize, extracellular matrix protein engagement of integrins, and juxtacrine signaling between membrane-associated ligands and receptors. The resulting multivalent binding can be more potent than corresponding monovalent interactions, potentially through initiating a process of cellular receptor clustering that may promote enhanced signal transduction. However, the ability to control and monitor these naturally occurring multivalent interactions within a living cell is currently limited, since it is difficult to modulate the valency of naturally-occurring ligands, especially membrane-associated proteins. We have developed a means to synthetically conjugate the binding domains of normally cell membrane-associated ligands to the soluble biopolymer hyaluronic acid (HA). Specifically, using EDC/NHS chemistry, we functionalized the high molecular weight HA backbone and attached recombinantly produced ephrin-B ligand ectodomains via an added terminal cysteine. To precisely estimate the valency and molecular weight of the bioconjugates, we characterized the molecules using size exclusion chromatography and multi-angle light scattering.

To determine the effects of bioconjugate multivalency on the ability to cluster cell surface receptors, we incubated adult hippocampal neural progenitor cells (AHNPCs) with the multivalent molecule *in vitro.* Multivalent ephrin-B showed enhanced clustering of cognate Eph receptors as compared to antibody-clustered ligands. Next, to determine if increased receptor clustering could affect stem cell differentiation, cultures of AHNPCs were incubated solely with either antibody-clustered ligands or high conjugation ratio conjugates for six days. A maximum three-fold increase over the antibody-clustered ligand for the highest ratio conjugate was observed. To determine if the effect of multivalency in directing neuronal stem cell fate could be applied more generally to other cell types, cultures of human embryonic and induced pluripotent stem cells were incubated with the factors for 2-4 weeks. In both cultures a 2-fold increase in the fraction of neurons formed and a 4-fold increase in the fraction of dopaminergic neurons compared to antibody-clustered controls was observed. Cultures also showed increases in the midbrain-specific marker En1 and had significantly higher levels of dopamine. Finally, in an effort to validate the enhanced activity of this multivalent construct *in vivo,* we injected it directly into the brain of adult rats via intracranial stereotaxic injection. Brains injected with bioconjugates showed a three-fold increased ability to induce neuronal differentiation as compared to the antibody-clustered control, thus indicating that highly multivalent ligands potently enhance the fraction of new neurons formed compared to using ligands clustered by conventional means. These results have a variety of biomedical implications, in that they establish a general platform for creating highly bioactive, defined, and reproducible forms of protein-based ligands, which have the ability to more potently activate downstream effectors for use in basic research and therapeutic applications.

(489b) Engineered Microenvironment for Osteogenic Differentiation of Stem Cells

Ameya Phadke, University of California, San Diego, La Jolla, CA, Yu-Ru Shih, Bioengineering, University of California, San Diego, La Jolla, CA and **Shyni Varghese**, Bioengineering, University Of California, San Diego, La Jolla, CA

Bone is one of the most common transplanted tissues, second only to blood. Currently, autografts and allografts are the primary choice for treating compromised bone tissue, but both suffer from drawbacks such as donor site morbidity, scarcity, immunorejection, and disease transmission. In contrast, synthetic bone grafts with intrinsic osteoinductivity and osteoconductivity could provide an easy-to-manufacture and cost-effective therapeutic strategy for treating bone defects. However, employing biomaterial-based bone grafts to accelerate regeneration of critical bone defects requires a detailed understanding of bone extracellular matrix (ECM), its role in the differentiation of stem cells into osteoblasts, and tissue homeostasis. Recently, we have developed novel, synthetic bone-like ECM structures using principles of biomineralization. Thus-formed biomineralized matrices recapitulate various attributes of native bone ECM. Our ongoing studies involving human mesenchymal stem cells indicate that the biomimetic matrix directed their osteogenic differentiation both in vitro and in vivo. Furthermore, we have employed the biomineralized matrices as an "artificial niche" to understand the mechanism by which mineralized matrices induces osteogenic differentiation commitment of stem cells.

(489c) Variation of Oxygen in a Controlled Manner Markedly Enhances Multi-Stage Differentiation of Embryonic Stem Cells to Insulin Producing Cells

Amanda R. Dilenno, Jeffrey R. Millman, Anna Kokensparger and Clark K. Colton, Chemical Engineering, Massachusetts Institute of Technology, Cambridge, MA

Pluripotent stem cells (PSC) hold promise for cell replacement therapy, but efficient differentiation to desired cell types remains a major obstacle. Although most PSC research is performed in high, non-physiological O_2 environments, cells during embryonic development are actually exposed to much lower O_2. Here we report that variation of O_2 in a controlled manner markedly enhances multi-stage differentiation of embryonic stem cells to insulin producing cells. We differentiated CyT49 human embryonic stem cells (hESC), using a modification of the 5-stage protocol published by ViaCyte, Inc (D'Amour et al., Nat Biotech 2006, Kroon et al., Nat Biotech 2008) under different sequences of well-characterized pO_2 environments. We controlled cellular oxygen exposure through adhesion culture on highly O_2-permeable silicone rubber membranes so that the cells in contact with the membrane were exposed to the same pO_2 as in the gas phase.

Our strategy in selecting O_2 patterns to investigate was based on two hypotheses: (1) cells in the developing embryo are initially exposed to low O_2 levels that increase as the tissues become vascularized. (2) The effects of low O_2 on the early stages of differentiation can be manifested in later stages of differentiation. At each stage there were two steps. We started with stage 1at six different O_2 levels (1%, 3%, 5%, 8%, 10%, & 20%). Samples at the end of stage 1 were analyzed with qPCR and flow cytometry for markers of definitive endoderm (DE). In the second step, we carried the differentiation to completion. The one to three best oxygen levels that produced the highest fraction of DE cells were then used in stage 1, and the remaining stages were carried out at 20% O_2. Each set of complete differentiation experiments employed a control condition of constant 20% O_2 for comparison. The best stage 1 O_2 level was selected on the basis of the fraction c-peptide+ and/or insulin+ cells in the stage 5 preparation as well as the extent of gene expression of relevant pancreatic endoderm (PE) and beta cell markers, and this O_2 level was used for all subsequent experiments. This two-step process was repeated sequentially with stage 2, 3, and 4; the condition for stage 5 was fixed at 20% O_2. Beginning with stage 3 an additional O_2 condition of 40% was added to the conditions tested. All together, 12 different sequences of O_2 levels in successive stages were evaluated.

The fraction of cells containing insulin or c-peptide was similar. Results for 9 of the 12 sequences are summarized in Table 1 in terms of the pooled mean of the fractions of c-peptide+ and insulin+ cells.

Table 1: Fraction of cells that were c-peptide+ and/or insulin+ (Mean ± SD)

Conditions*	C-peptide+/ Insulin+ (%)
20-20-20-20	4 ± 1
5-20-20-20	9 ± 4
5-5-20-20	8 ± 2
5-5-5-(1 to 3)	8 ± 7
5-5-5-(5 to 40)	34 ±13

* percent O_2 in sequential stages

With normoxic conditions in all of the stages (designated 20-20-20-20), the fraction was about 4%. Reduction of O_2 to 5% in stage 1caused the fraction to double. A similar additional reduction of O_2 in stage 2 had no further effect. The effect of another sequential reduction to 5% in stage 3 depended on the oxygen level in stage 4. If O_2 was further reduced below 5% in stage 4, there was no change. However, in the range of 5 to 40% O_2 for stage 4, the fraction of c-peptide+/insulin+ cells jumped by a factor of four to a mean of 34%. At two conditions, the fraction was about 50%, but there was not a monotonic trend with O_2 level.

These results demonstrated that modulation of the O_2 level to which cells are exposed can have a pronounced effect on the outcome of directed differentiation protocols. In the present example, culture at reduced oxygen for the first three stages, followed by an increase in O_2 levels for the last two stages, produced an eight-fold increase in the fraction of c-peptide+/insulin+ cells.

(489d) Mesenchymal Stem Cell Therapeutics for Protection and Repair of Injured Tissues & Vital Organs

Martin L. Yarmush, Biomedical Engineering, Rutgers University and Massachusetts General Hospital, Piscataway, NJ

Recently there has been a paradigm shift in what is considered to be the therapeutic promise of mesenchymal stem cells (MSCs) in diseases of vital organs. Originally, research focused on MSCs as a source of regenerative cells by differentiation of transplanted cells into lost cell types. It is now clear that trophic modulation of inflammation, cell death, fibrosis, and tissue repair are the primary mechanisms of MSC therapy. This has been clarified in studies where delivery of growth factors, cytokines, and other signaling molecules secreted by MSCs is often sufficient to obtain the therapeutic effects. In this presentation, we provide a summary of the current knowledge on trophic mechanisms of MSC therapy in disease models of vital organs using models of acute liver failure, acute kidney failure, and spinal cord injury. Important issues regarding the optimal delivery methods of MSC therapy are discussed and critical gaps in our knowledge hampering experimental progress and clinical implementation are discussed.

(489e) Endothelial Cells Mediate Maturation of Human Embryonic Stem Cell Dervied Pancreatic Progenitors Into Insulin Expressing Cells

Maria Jaramillo[1], Saik Kia Goh[1] and Ipsita Banerjee[2], (1)Bioengineering, University of Pittsburgh, Pittsburgh, PA, (2)Chemical and Petroleum Engineering, University of Pittsburgh, Pittsburgh, PA

Diabetes and its complications affect 25.8 million people in the US and is a growing healthcare problem worldwide. While transplantation of pancreatic islets offer potential therapy, it is restricted by scarcity of donor islets. Effective use of hESC could overcome this limitation, however current approaches to differentiating hESCs to mature islets have been restricted in yield and functionality of the differentiated phenotype. In our work we propose to mimic *in-vivo* development during maturation of hESC derived pancreatic progenitor cells. Over past years developmental studies have established the critical involvement of endothelial cells in maturation and functionality of pancreatic islets, which is not surprising given the islet's dense vasculature.

We have developed a multi-stage protocol for directed differentiation of hESCs involving definitive endoderm induction, followed by pancreatic progenitor specification and final maturation into insulin producing cells. For the last step, we propose to co-culture the hESC-derived pancreatic progenitor cells with endothelial cells in accordance with developmental studies. We used two different cell types in our studies: Rat Heart Microvascular Endothelial Cells (RHMVEC) and Human Umbilical Vein Endothelial Cells (HUVEC). As a control experiment we co-cultured the pancreatic progenitor cells with NIH3T3 fibroblast to verify sensitivity to cell types. All direct contact co-cultures were FACS sorted and analyzed for gene and protein expression levels. Two different non-contact co-culture configurations were also investigated: transwell co-culture and endothelial cell conditioned media. Results were evaluated with respect to conventional maturation using notch inhibition by DAPT.

We evaluated each stage of differentiation using qRT-PCR, immunostaining and flow cytometry. Our protocol resulted in 89% Foxa2 positive endoderm and high, uniform nuclear staining for Pdx-1 at pancreatic progenitor stage with roughly 70% of the cells being Pdx1 positive as observed by flow cytometry. Maturation with DAPT resulted in 17,500 fold upregulation of insulin mRNA with 8% commitment, while parallel RHMVEC co-culture resulted in 20,000 fold upregulation of insulin mRNA with 34% commitment. HUVEC co-culture resulted in an even higher insulin expression, 7 times higher than RHMVEC. Control cultures with fibroblast did not show appreciable insulin upregulation, indicating the effect to be specific to endothelial cells. Alternative co-culture configurations also show promising results, with transwell dominating over conditioned media, but still lower than direct contact configuration.

Our results establish an alternate strategy for hESC maturation into islet cell types. Current investigation is underway to elucidate the mechanism of endothelial cell mediated maturation.

(489f) Human Induced Pluripotent Stem Cells Differentiate Into Contractile Vascular Smooth Muscle Fate Via Mesenchymal Stem Cell Intermediates: Implication for Cardiovascular Regeneration

Vivek K. Bajpai, Chemical & Biological Engineering, State University of New York at Buffalo, Amherst, NY and Stelios T. Andreadis, Department of Chemical and Biological Engineering, University at Buffalo, the State University of New York, Amherst, NY

Smooth muscle cells (SMC) maintain vascular homeostasis and play prominent role in vascular diseases. Induced pluripotent stem cells (iPSC) can provide an untapped resource of functional SMC for autologous cell based therapies and for creating models for studying vascular diseases. This would require efficient differentiation of iPSC towards SMC fate. However existing protocols of pluripotent stem cell differentiation are inefficient and primarily give rise to terminally differentiated SMC that critically limits their in-vitro expansion that is unequivocally desirable for regenerative medicine applications. To address this challenge, we developed a robust, stage-wise monolayer protocol of human (h)iPSC differentiation towards functional SMC fate. Employing this protocol, we demonstrate that hiPSC differentiate into functional smooth muscle cell fate via intermediate mesenchymal stem cell (MSC) stage where they are highly proliferative and amenable to differentiate along all mesenchymal cell types including osteogenic, chondrogenic and adipogenic lineages. Thus using this protocol, MSC can be kept in a proliferative state for 45 population doublings and tuned into contractile SMC with short-term induction treatment when required. In the first stage of differentiation hiPSC lines (generated from two different set of transcription factors) were induced to undergo epithelial to mesenchymal transition (EMT) by culturing them on matrigel-coated dishes in presence of EMT medium (MEF conditioned medium human embryonic stem cell medium supplemented plus FGF2). EMT was followed in real time using smooth muscle alpha actin (ACTA2) promoter driven ZsGreen lentiviral reporter. To ensure that ACTA2-ZsGreen reporter reliably captures hiPSC undergoing EMT, we derived stable ACTA2-ZsGreen lentiviral reporter hiPSC clones and by immunocytochemistry clearly demonstrated that undifferentiated OCT4+ colonies also lacked ACTA2-ZsGreen expression. However, as early as day 3 after induction of EMT cells in the periphery of hiPSC colonies assumed fibroblastic morphology and expressed ZsGreen, suggesting upregulation of ACTA2 promoter and EMT. Indeed, EMT was confirmed by RT-PCR for EMT markers like SNAIL, TWIST, SLUG, FOXC2, FSP1, CDH1, CDH2, fibronectin and vimentin. EMT cells (ZsGreen+) were enriched (stage 2) into MSC fate by culture on gelatin-coated surfaces in the presence of enrichment medium (M231 supplemented with 5%FBS, bFGF, EGF, insulin and heparin). During the enrichment stage, the fraction of ZsGreen+ cells was significantly increased for the both hiPSC lines tested as quantified by flow cytometry. Enriched cells displayed remarkable similarity with bone marrow derived MSC as they displayed MSC surface immunophenotype (CD105+/CD73+/CD44+/CD49b+/CD90+/CD45-/CD34-) and could be coaxed to differentiate into adipogenic, osteogenic and chondrogenic lineages as shown by oil red O, Von Kossa and alcian blue stains, respectively. Enriched cells were highly proliferative cells as they were kept in culture for at least 45 population doublings without reaching cellular senescence. While enriched cells were highly proliferative, they were weakly contractile, perhaps due to immature contractile machinery as evidenced by immunostaining for ACTA2, CNN1 and MYH11. To induce contractile phenotype (stage 3), enriched MSC were further treated with maturation medium (M231 plus TGF-β1 and heparin). Immunostaining of mature cells showed markedly increased filamentous organization of ACTA2, CNN1 and MYH11 proteins. Contractile phenotype was further confirmed by excellent contractility in response to receptor and non-receptor mediated vasoagonists exhibited by cylindrical tissue engineered vessels (TEV) made from mature cells.

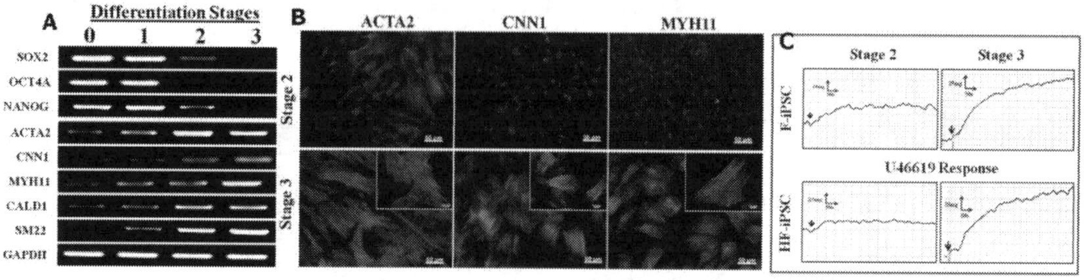

Figure 1. RT-PCR analysis of SMC specific and pluripotency markers during differentiation (A) Immunocytochemistry for SMC specific markers show filamentous organization of these contractile proteins after maturation treatment (B) Vasoconstriction of TEV made of enriched and mature cells for two hiPSC lines (C).

In summary, this study demonstrate that hiPSC undergo functional SMC fate through an intermediate stage of multipotent MSC. The iPSC derived SMC may be useful for studying the pathology of vascular diseases especially in context of SMC phenotypic conversion between synthetic and contractile fates and engineering SMC-containing tissues such as vascular grafts.

(500a) A Transplantable Liver Graft with Improved Blood Compatibility

Yeonhee Kim[1], Sinan Özer[1], Tim Berendsen[2], Korkut Uygun[2], Martin L. Yarmush[2] and Basak Uygun[2], (1)Harvard Medical School/Massachusetts General Hospital, Boston, MA, (2)Center for Engineering in Medicine, MGH/Harvard Medical School, Boston, MA

The only definitive treatment for end-stage liver disease is orthotopic liver transplantation and it is limited by the shortage of available donor organs. To address the issue, we recently developed a novel method to prepare transplantable liver grafts using perfusion decellularized liver matrix (DLM). While we successfully showed that the engineered liver grafts were functional in vitro and could be transplanted in rats, the post-transplantation survival time of the recipient animals was significantly low (~8 h) due to complications related to the high thrombogenicity of the grafts. An ideal engineered liver graft should be anti-thrombogenic and should enable undisrupted blood circulation through the graft in order to allow for its therapeutic testing in a liver failure model. Therefore, we hypothesized that the immobilization of active heparin, an anti-thrombogenic molecule, on the DLM scaffold surface will render the DLM blood compatible, hence will prevent blood clotting when exposed to blood upon transplantation and prolong the post transplantation survival time of the recipient animal. In this study, we deposited heparin on the DLM scaffold surface using layer-by-layer polyelectrolyte multilayer deposition technique. Heparin deposition was confirmed by histological and chemical assays. The ability of the heparinized DLM (hDLM) to prevent blood coagulation was tested via 2-h ex vivo diluted blood perfusion and there was significant reduction in the blood clot formation in hDLM scaffolds. The macroscopic observation indicated that the extent of reduction was proportional to the amount of heparin deposited. Heparin deposition on the DLM did not appear to affect the function of engineered grafts as evident by sustained secretion of albumin and urea during 5 d of perfusion culture. The heterotopic transplantation using hDLM and recellularized hDLM is under investigation in order to assess the feasibility of the tissue-engineered liver graft as a potential replacement for liver transplantation. In the preliminary study the survival over 2-wk period was observed post-transplantation in our animal model; that is much more promising than 8-h survival limited in the previous study. The generation of transplantable and functional engineered liver grafts that can serve as a reliable alternative for liver transplantation will offer a new avenue for treatment of liver diseases.

(500b) Whole Organ 3D Microenvironment As a Regulatory Cue for Pancreatic Differentiation of Embryonic Stem Cells

Saik Kia Goh[1], Suzanne Bertera[2], Phillip Olsen[1], Lei Yang[3] and Ipsita Banerjee[4], (1)Bioengineering, University of Pittsburgh, Pittsburgh, PA, (2)Department of Pediatrics, Children's Hospital of Pittsburgh, Pittsburgh, PA, (3)Developmental Biology, University of Pittsburgh, Pittsburgh, PA, (4)Chemical and Petroleum Engg, University of Pittsburgh, Pittsburgh, PA

Introduction: Type I Diabetes affects over 1 million people in the United States. While islet transplantation has proved to be a promising therapeutic strategy, it is still hindered by lack of donor tissue. Embryonic stem cells (ESCs) have emerged as an alternative cell source owing to its virtually unlimited replicative capacity and the potential to differentiate into a variety of cell types, including islet-cells. However, ESC derived islet-cells are currently limited in their yield and functionality. Current differentiation strategies primarily involve various growth factor/ inducer/ repressor concoctions with less emphasis on the substrate. Governed by the understanding that the extracellular matrix (ECM) proteins and architecture significantly influences specific cellular behaviors, we hypothesize that organ specific ECM microenvironment plays an important role in organogenesis. In this project we are testing our hypothesis by inducing pancreatic differentiation of hESCs in 3 different whole organ derived ECM scaffold: native organ – pancreas; alternate endoderm derived organ – liver; mesoderm derived organ – heart.

Materials and Method: Cadaveric pancreata, livers and hearts were isolated from adult mice (n=8) and mounted on a perfusion apparatus. Ionic detergent, 0.1% SDS was perfused until tissues were translucent and white in color (2 hours). Cell nuclei post-decellularization was quantified via DAPI staining and PicoGreen DNA assay. Immunofluorescence and histology were done to characterize the retention of the extracellular matrix (ECM) protein before and after decellularization. Ultrastructure and the macroscopic three-dimensional architecture of organ-derived ECM scaffolds were characterized using electron microscopy. To examine cell-matrix interaction, whole organ decellularized pancreas, liver or heart were seeded with hESC-derived endodermal cells ($3x10^6$ cells) and cultured in pancreatic differentiation medium with growth factors induction. Recell constructs (n=3) were then mounted on a perfusion apparatus to allow dynamic culture for 4 days and 12 days. Whole organ recell constructs were analyzed using immunohistochemistry, TUNEL, and qRT-PCR.

Results and Discussion: Perfusion decellularization with 0.1% SDS resulted in complete decellularization of pancreas, liver and heart. DAPI staining and DNA quantification confirmed the removal of cells and residual DNA. Microstructure and major ECM components such as collagen I, collagen IV, fibronectin and laminin were preserved. In all the repopulated whole organ constructs, hESC derived DE cells were found engrafted - evidenced by positive hNuclei staining from IHC. Low apoptotic cells were detected (<10% TUNEL positive) in all three repopulated constructs. qRT-PCR of hESC differentiated on pancreatic matrix showed 52.7±10.8 fold increase in PDX1, pancreatic gene expression, and 92.7±12.7 fold increase in insulin gene expression compared to Matrigel (Fig. 1). Cells found engrafted in the pancreatic matrix were positive for PDX1 and C-peptide from immunostaining - confirming the differentiation of engrafted hESC into insulin-expressing cells. Compared to pancreatic matrix, qPCR of recellularized liver and heart construct gave lower expression of pancreatic gene expression. Immunostaining results were consistent with qPCR result – with low PDX-1 and C-peptide expression found in liver and heart recell constructs.

Conclusions: Perfusion-decellularization of cadaveric pancreas, liver and heart efficiently removes cells while retaining ECM protein and microstructure. Tissue-specific three-dimensional spatial organization and ECM composition drives cell fate differently with pancreatic matrix being the preferred ECM substrate for pancreatic differentiation. Further proteomic studies are underway to establish comprehensive and mechanistic differences between the ECM composition of endoderm organs (pancreas and liver) and mesoderm organs (heart) to better understand the cell-matrix interactions that contributed to the specific differentiation of hESC.

(500c) Development of a Pancreatic Substitute Based On Genetically Engineered Intestinal Endocrine Cells

Aubrey Tiernan[1], Kiranmai Durvasula[1] and Athanassios Sambanis[1,2,3], (1)School of Chemical & Biomolecular Engineering, Georgia Institute of Technology, Atlanta, GA, (2)Wallace H. Coulter Department of Biomedical Engineering, Georgia Institute of Technology/Emory University, Atlanta, GA, (3)Georgia Tech/Emory Center for the Engineering of Living Tissues, Georgia Institute of Technology, Atlanta, GA

Introduction: A tissue engineered pancreatic substitute constitutes a promising cell-based insulin therapy for achieving tight glycemic regulation in diabetes patients. In particular, non-beta cells genetically engineered to secrete insulin can address the major issues of immune rejection and donor cell availability through their potential autologous nature. The objective of this study was to generate a recombinant system of luminescent and insulin-secreting intestinal L-cells that allows treatment of diabetic mice and direct monitoring of the graft over time. Recombinant insulin-secreting GLUTag-INS cells were previously generated by stably transfecting murine intestinal endocrine GLUTag cells with human B10 insulin [1]. Although insulin secretion was achieved, *in vivo* therapeutically relevant levels were not obtained [2]. Further genetic engineering via transduction with a lentivirus containing the wild-type human insulin gene followed by a GFP reporter gene, generated cell line GLUTag-eINS with enhanced insulin secretion and promise for therapeutic application. As these cells constitute allogeneic therapy in diabetic mice, we are pursuing cell microencapsulation in alginate material as a protective barrier against *in vivo* immune rejection. In addition, we are investigating luciferase incorporation for non-invasive monitoring via bioluminescence of *in vivo* graft survival.

Materials and Methods: Sequential lentiviral transductions were performed on GLUTag-INS cells: the first contained the wild-type insulin gene followed by a GFP reporter gene, and the second contained the luciferase reporter gene. Fluorescence-activated sorting was employed to enrich the transduced population to 98% fluorescence; bioluminescence images and signals were captured with the IVIS Lumina (Xenogen). Insulin concentrations in secretion tests were measured by radioimmunoassay (Millipore, MA). Cell microencapsulation was performed in 3.3 wt% alginate (FMC Biopolymer, Norway) cross-linked with barium using an electrostatic droplet generator [3].

Results and Discussion: The first lentivirus transduction generated the enhanced cell line GLUTag-eINS and significantly improved insulin secretion from GLUTag-INS, making them promising for therapeutic application. The second lentivirus transduction generated the cell line GLUTag-eINS-fluc with no negative effects on insulin secretion (Figure 1). Strong bioluminescence signal of 1.5-1.8x10^8 photons/(s·cm^2·sr) was obtained from GLUTag-eINS-fluc monolayers in 12-well plates and was stable upon expansion and passaging.

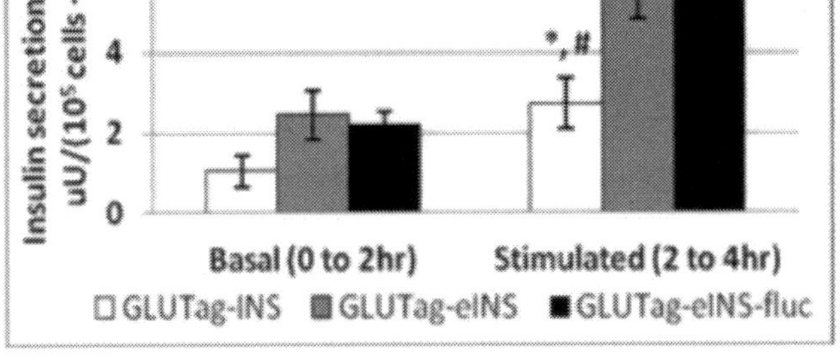

Figure 1: Effect on insulin secretion after sequential transduction of GLUTag-INS with lentiviruses. Stimulation medium was composed of basal medium (5mM Glucose) + 2% Meat Hydrolysate. Wild-type insulin gene incorporation resulted in significant insulin secretion enhancement (GLUTag-eINS). Luciferase gene incorporation had no negative effects on insulin secretion (GLUTag-eINS-fluc). *,#p<0.05

Insulin secretion from GLUTag-eINS cells in barium alginate microcapsules was comparable to that from cell monolayers; a protective barrier can therefore be provided without impeding insulin secretion. Bioluminescence signal from microencapsulated GLUTag-eINS-fluc cells was also similar to monolayers, thus offering promise for *in vivo* monitoring capabilities. *In vivo* experiments with microencapsulated GLUTag-eINS-fluc cells are currently in progress in normal mice at subtherapeutic levels as a first step in characterizing this new system toward *in vivo* efficacy studies.

Conclusion: A system of recombinant luminescent and insulin-secreting intestinal L-cells can potentially serve as a pancreatic substitute for the treatment of diabetic animal models, as well as offer the ability for direct monitoring of graft survival over time.

References:

1. Bara, H. and A. Sambanis, *Insulin-secreting L-cells for the treatment of insulin-dependent diabetes.* Biochem Biophys Res Commun, 2008. **371**(1): p. 39-43.

2. Bara, H., P.M. Thule, and A. Sambanis, *A cell-based approach for diabetes treatment using engineered non-beta cells.* J Diabetes Sci Technol, 2009. **3**(3): p. 555-61.

3. Goh, F. and A. Sambanis, *In vivo noninvasive monitoring of dissolved oxygen concentration within an implanted tissue-engineered pancreatic construct.* Tissue Eng Part C Methods, 2011. **17**(9): p. 887-94.

(500d) In Vitro Recapitulation of Organ Growth for a Model Genetic Tissue, the Drosophila Wing Imaginal Disc

Jeremiah J. Zartman[1], Simon Restrepo[2] and Konrad Basler[2], (1)Chemical and Biomolecular Engineering, University of Notre Dame, Notre Dame, IN, (2)Institute of Molecular Life Sciences, University of Zurich, Zurich, Switzerland

Inspiration in developing novel methods in tissue and stem cell engineering has often come from a deeper understanding of developmental biology. To form an important positive feedback loop, advances in bioengineering can also be applied toward solving fundamental biological questions in pattern formation and growth control. The wing imaginal disc of *Drosophila melanogaster* is an important model system that has greatly contributed to our understanding of genetics and organ development. However, current studies have begun to be limited by relying on end-point analysis provided by fixed tissues. In vitro methods that recapitulate the development and growth of an intact wing imaginal disc will help to uncover new biological questions and solve long-standing unsolved biological questions. Here we describe a high-throughput pipeline based on design-of-experiments methodology to develop an optimized culture medium that doubles the proliferation rate of wing discs in short-term cultures over multiple hours. The optimized protocol has identified critical medium components and synergies and, coupled with an improved live-imaging system, is providing further insight into cell-cell dynamics within the context of tissue growth. Future efforts will focus on using cultured wing discs as a screening platform that harnesses the flexibility of in vitro methods with the power of Drosophila genetics to answer questions regarding the regulation of final tissue size.

(500e) The Role of the Cytoskeleton in Focal Adhesion Development and Migration of Cells Attached On 3D Aligned Fibrous Scaffolds

Kevin Sheets, School of Biomedical Engineering and Sciences, Virginia Tech, Blacksburg, VA and Amrinder S. Nain, Mechanical Engineering, Virginia Tech, Blacksburg, VA

The Role of the Cytoskeleton in Focal Adhesion Development and Migration of Cells Attached on 3D Aligned Fibrous Scaffolds

Kevin Sheets[1], Amrinder Nain[1,2]

1. STEP Lab, School of Biomedical Engineering & Sciences, Virginia Tech, Blacksburg, VA

2. Department of Mechanical Engineering, Virginia Tech, Blacksburg, VA

Introduction

In the human body, cells attach to and receive biophysical and biochemical cues from their immediate fibrous microenvironment known as the extracellular matrix (ECM), which helps drive overall cell behavior. Cells form attachments to the ECM through multi-protein complexes known as focal adhesions (FACs), whose size and orientation dynamics (spatial and temporal) adapt with changes in surrounding geometry[1–3]. Likewise, other aspects of the cytoskeleton including f-actin stress fibers, myosin IIa motors, and microtubules adjust according to the makeup of their extracellular environment and cooperatively interact to produce locomotive force against the ECM via FACs[4,5]. While this phenomenon is well-understood on flat substrate conditions, recent work has shown that cell attachment profiles contain considerable differences when cultured on three dimensional fibrous substrates mimicking the native ECM[6–8]. Using the STEP (Spinneret-based Tunable Engineered Parameters) platform, which generates fibrous scaffolds of high aspect ratio nano/micro fibers with tightly controlled diameters and spacing, we examine FAC behavior on ECM-mimicking scaffolds in both normal cells and cells with cytoskeletal knockdowns[9–12]. By administering the drugs blebbistatin, nocodazole, and cytochalasin-D, cytoskeletal elements myosin IIa, microtubules, and actin are depolymerized respectively, causing notable changes to migration speed and FAC behavior on these scaffolds. The efficacy of these drugs in changing the FAC dynamics can be used for developing drug testing platforms, arresting cellular migration of metastatic cancerous cells and developing more effective tissue engineering scaffolds.

Materials and Methods

In addition to using flat glass substrates, suspended polystyrene (PS) fibers of approximately 500 nm diameter were spun onto hollow STEP substrates. Scaffolds were coated with fibronectin and incubated overnight prior to cell seeding using standard cell culture techniques. Seeded mouse C2C12 myoblasts attached to and spread on the fibers within 4-6 hours. Cells on fiber networks were then time-lapse imaged using an incubating microscope (Zeiss) with a digitally-controlled X-Y-Z stage. Cell migration speeds were calculated via time-lapse by tracking the position of the cell's nucleus every hour and noting the maximum distance traveled per hour in a 12-hour window. Blebbistatin (50 μM), nocodazole (10 μM), and cytochalasin-D (5 μM) were added directly to media at typically-noted concentrations 6 hours after initial seeding and were time-lapse imaged immediately[13].

Fluorescence microscopy was used to locate focal adhesion proteins as well as the f-actin stress fibers and cell nuclei. Cells were fixed in 4% paraformaldehyde for 15 minutes at room temperature, permeabilized in a 0.1% solution of

Triton X100, and then blocked in 10% goat serum for 30 minutes. Primary goat antibodies against paxillin (Invitrogen) were diluted 1:250 followed by GFP secondary antibodies also at 1:250 to visualize FACs. Rhodamine phalloidin was used 1:100 to stain stress fibers, and DAPI was used to stain the nuclei.

Results and Discussion

PS fibers of 500 nm diameter were deposited in single, parallel and orthogonal configurations with cells responding to change in orientation and topography accordingly (Fig.1A-C). After seeding cells on STEP scaffolds, cells were found to attach within six hours. Immediately following this time, cells were treated with the drugs in aforementioned concentrations and then observed for changes to migration behavior and cytoskeleton arrangement. As can be seen in Figure 1, migration speeds were found to vary according to both cell shape and administered drug.

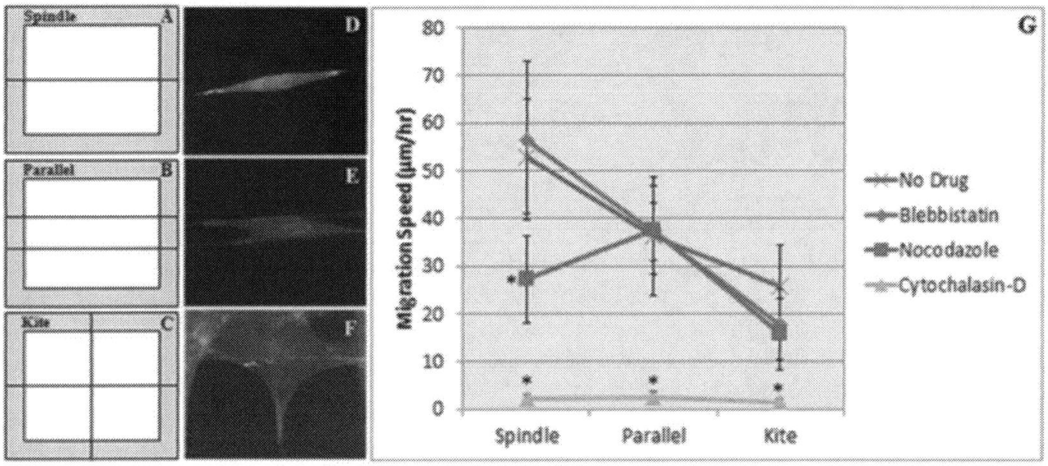

Fig. 1: A-C) STEP fiber layouts used to get cells to attach in D-F) spindle, parallel, and kite configurations, respectively. (G) C2C12 migration speeds as a function of cell shape and administered drug. With the exception of cytochalasin-D, the only significant change to migration speed was to spindle-shaped cells treated with nocodazole. *indicates statistical significance compared to no drug case ($p<0.001$, $n = 5$ cells each).

Spindle cells, which attach and elongate along a single fiber axis, form stable adhesion patches (~ 11 µm in length) at the cell periphery (Fig.1D). Due to their high polarity, these cells are capable of migrating at high speeds (50+ µm/hr). However, when actin is depolymerized with cytochalasin-D, the cell is no longer able to extend its membrane and form stable adhesion patches, which causes cells to be non-migratory. On the other hand, nocodazole, which depolymerizes microtubules, reduced spindle cell speed to approximately half the original value. Since the cell no longer contains microtubules, this drug prevents the cell from splitting into daughter cells despite progressing through the cell cycle. The end result is a more massive cell containing multiple nuclei which remains confined to the same attachment profile. In theory, because the mass of the cell increases without any visible change to the FAC, migration speed decreases. Lastly, blebbistatin-treated cells migrated at the same speed as the wild type cells. In the absence of myosin IIa, cells are unable to sense the stiffness of the fiber matrix and thus do not develop mature adhesions. These temporally unstable FACs ultimately allow the cell to maintain normal migration speeds.

Parallel cells were found on STEP scaffolds which contained fibers that were spaced relatively close together (<20 µm). At this separation distance, the cells were often found spanning the gap between the two fibers and forming prominent adhesion clusters at the cell periphery – two per fiber for a total of four main clusters (Fig 1E). Parallel cells migrated about 40 µm/hr. Cytochalasin-D again stopped migration altogether. Blebbistatin did not alter migration speed and the effect of nocodazole was dampened for this shape, most likely since parallel cells travel slower to begin with and have additional adhesion sites compared to spindle cells, which only have one fiber to elongate on.

Kite-shaped cells, which attach to intersecting fibers and form two adhesion clusters along each fiber axis, migrate the slowest of the three shapes at about 30 µm/hr (Fig.1F). Again, cytochalasin-D stopped migration whereas blebbistatin and nocodazole did not significantly change migration speed. Cells travel slower in this shape presumably due to the formation of four nearly equal FACs which each pull towards the cell center.

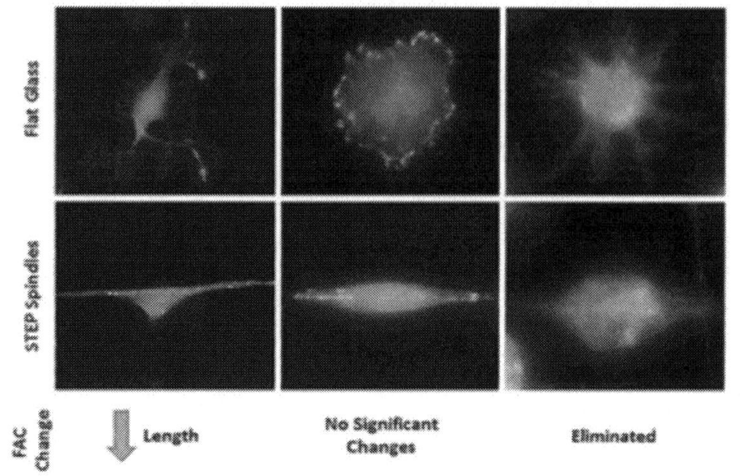

Fig. 2: Fluorescence images showing paxillin (FACs, green), phalloidin (actin, red), and DAPI (nucleus, blue) on drug-treated cells for both the flat glass and spindle shape conditions. Blebbistatin causes shorter FACs to form. Nocodazole FACs are very similar to untreated cells. Using cytochalasin-D, FACs are no longer able to form.

To further probe the effect of drugs on the cytoskeleton, cells were fixed and stained to measure changes in focal adhesion lengths, stress fiber formation, and nucleus positioning (Fig 2). This was performed on both flat glass as a control substrate and on STEP fibers. There were several notable changes to the structure of the cells that were treated with drugs which confirm changes seen in migration speeds. First, blebbistatin-treated cells contained FACs that were significantly shorter than untreated cells. This is likely due to the inability of the latent adhesions to mature by responding to the fiber beam stiffness (N/m). Cells exposed to nocodazole were frequently observed to contain multiple nuclei within the same cell body. Since this drug depolymerizes microtubules, a structural component key to mitosis, the cell continued to progress throughout the cell cycle without forming daughter cells. Despite this extra cell matter, FACs were essentially unchanged. Lastly, cytochalasin-D treated cells contained no traces of actin or FACs whatsoever, confirming its significance in the cell migration process.

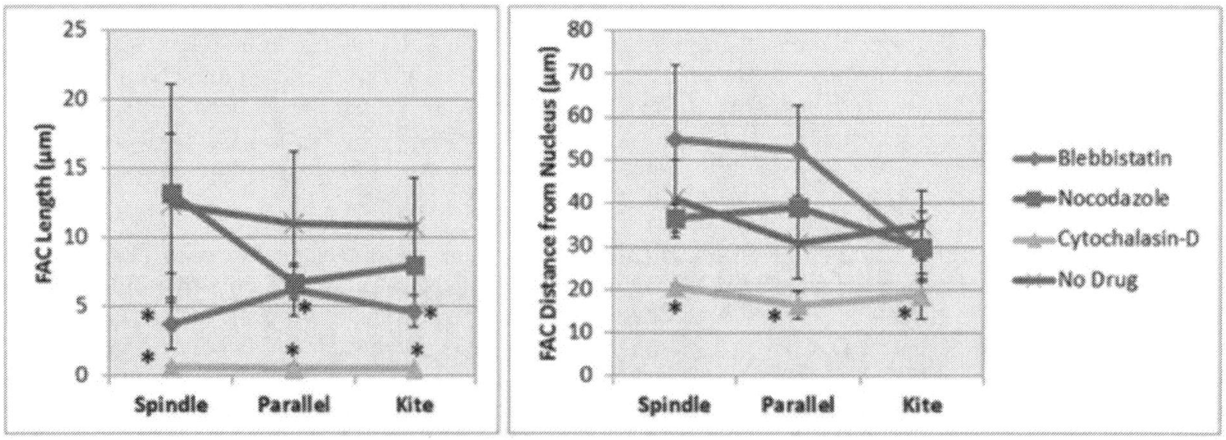

Fig. 3: FAC lengths and distances from nucleus for each cell shape and drug combination. * indicates statistical significance compared to no drug case ($p<0.05$, $n = 3$ for each cell/drug combination)

Conclusions

Using the STEP platform for fiber manufacturing, cellular migratory behavior and cytoskeletal arrangements were tracked for both normal and drug conditions. While fluorescence images demonstrated definitive changes to cytoskeletal appearance and arrangements compared to the 2D case, overall migratory speeds were observed to remain constant with treatment of blebbistatin, and nocodazole only significantly affected spindle shaped cells. Cytochalasin-D caused migration to stop altogether. In these cases, changes to the cytoskeletal proteins caused subsequent changes to focal adhesion formation and function, especially compared to the traditionally-studied 2D cases. This information will be used in the future to further understand and predict cell motion based on attachment profiles with the hopes of designing structures to either speed cell migration for applications such as wound healing, or slow migration in cases such as metastatic cancer cells.

References

(1) Gallant, N. D.; Michael, K. E.; Garcı, J. *Molecular Biology of the Cell* **2005**, *16*, 4329-4340.

(2) Deguchi, S.; Tsubasa, M. S.; Lio, K. *Cytoskeleton* **2011**, *651*, 639-651.

(3) Rape, A. D.; Guo, W.-H.; Wang, Y.-L. *Biomaterials* **2011**, *32*, 2043-51.

(4) Ryan, G. L.; Petroccia, H. M.; Watanabe, N.; Vavylonis, D. *Biophysical Journal* **2012**, *102*, 1493-1502.

(5) Lam Hui, K.; Wang, C.; Grooman, B.; Wayt, J.; Upadhyaya, A. *Biophysical Journal* **2012**, *102*, 1524-1533.

(6) Engler, A. J.; Sen, S.; Sweeney, H. L.; Discher, D. E. *Cell* **2006**, *126*, 677-89.

(7) Lazopoulos, K. a; Stamenović, D. *Journal of biomechanics* **2008**, *41*, 1289-94.

(8) Lock, J. G.; Wehrle-Haller, B.; Strömblad, S. *Seminars in cancer biology* **2008**, *18*, 65-76.

(9) Nain, A. S.; Sitti, M.; Jacobson, A.; Kowalewski, T.; Amon, C. *Macromolecular rapid communications* **2009**, *30*, 1406-12.

(10) Nain, A. S.; Phillippi, J. a; Sitti, M.; Mackrell, J.; Campbell, P. G.; Amon, C. *Small (Weinheim an der Bergstrasse, Germany)* **2008**, *4*, 1153-9.

(11) Ker, E. D. F.; Nain, A. S.; Weiss, L. E.; Wang, J.; Suhan, J.; Amon, C. H.; Campbell, P. G. *Biomaterials* **2011**, *32*, 8097-107.

(12) Bakhru, S.; Nain, A. S.; Highley, C.; Wang, J.; Campbell, P.; Amon, C.; Zappe, S. *Integrative biology : quantitative biosciences from nano to macro* **2011**, *3*, 1207-14.

(13) Kraning-Rush, C. M.; Carey, S. P.; Califano, J. P.; Smith, B. N.; Reinhart-King, C. a *Physical biology* **2011**, *8*, 015009.

(500f) Migration Dynamics of Mouse C2C12 Myoblasts On Single Suspended Fiber Mimicking ECM Fibril Beam Stiffness (N/m)

Sean Meehan[1], Kevin Sheets[2] and Amrinder S. Nain[1], (1)Mechanical Engineering, Virginia Tech, Blacksburg, VA, (2)School of Biomedical Engineering and Sciences, Virginia Tech, Blacksburg, VA

Migration Dynamics of Mouse C_2C_{12} Myoblasts on Single Suspended Fiber Mimicking ECM Fibril Beam Stiffness (N/m)

Sean Meehan[1], Kevin Sheets[2], Amrinder S. Nain[1,2], Department of Mechanical Engineering[1], School of Biomedical Engineering and Sciences[2], Virginia Tech, Blacksburg VA

Introduction

In vitro testing platforms are one of the particularly promising approaches to characterize the dynamic roles of biophysical and biochemical cues on cellular growth, development, and behavior. Numerous studies have shown that mechanical stimuli in a cell's external environment can have a profound effect on migration, proliferation, differentiation, and apoptosis. Nanoscale topographical features have been shown to not only increase migration and directional persistence time, but also decrease proliferation in human endothelial cells [1]. Additionally, it has also been shown that mesenchymal stem cell differentiation can be influenced using custom topographical and geometric cues [2,3]. In addition to topographical cues, the stiffness (measured in N/m^2) of the commonly used *flat* substrate used for cell studies has been shown to have a significant effect on cellular migration and differentiation. As little as a four kilopascal change in stiffness can elicit significant changes in cellular proliferation, migration speed, and differentiation in both single cells [4] and confluent sheets [5]. However, flat substrates do not exhibit many of the fiber specific qualities that cells encounter when interacting with the native extra cellular matrix (ECM), which is a three dimensional, dense, fibrous mesh immediately surrounding the cell. Cells on the ECM fibers attach, spread and eventually migrate with the fiber beam stiffness (measured in N/m) playing an integral role in cellular behavior. Furthermore, it has been shown that the mechanical properties of the extracellular matrix are tissue-specific, and these different properties can dictate migration speed, cell's ability to remodel the ECM, and many other important cellular behaviors [6,7]. Hence, it is obvious that the mechanical properties of the ECM, especially stiffness (both modulus and beam stiffness), have an important role in the life and behavior of cells within the body. However, creating a mechanistically tunable system to easily explore these properties is a challenging task due to lack of understanding regarding cell-ECM relationships and difficulties to fabricate precise nano-micro fibrous scaffolds of native proteins mimicking the native environment.

In order to explore this relationship between cellular activity and fiber substrate, a novel testing platform is presented in this study. Using a combination of the previously reported STEP (Spinneret Based Tunable Engineering Parameters) fiber manufacturing platform [8-11] with soft-lithography based micron scale PDMS pillars, we explore the link between cellular behavior and fiber beam substrate stiffness. The importance of this study lies in the unique ability of the STEP platform to provide precise control over fiber diameter and spatial deposition, thus providing a tunable mechanistic environment to explore cellular behavior. Current studies using electrospinning platforms have been used to study cell behavior on a collection-mat of fibers. However, single cell-fiber studies using electrospinning platforms are hard to accomplish due to lack of control on fiber alignment, spacing and diameter. [12] In this study, the role of substrate fiber beam stiffness (N/m) on cellular behavior, particularly migration, is envisioned to provide new insights into the dynamics of cellular polarizability, focal adhesion complex and cytoskeleton arrangement, and cellular mechanics as a whole.

Methods and Materials

Standard photolithography practices were used to create SU-8 2025 (Microchem) molds for the fabrication of micropillars. In short, the SU-8 photoresist was spun onto a 4 inch diameter silicon wafer (University Wafers), prebaked, exposed using a mask depicting pillar size and placement, and then developed using SU-8 Developer (Microchem). Molds were silanized and then coated with Sylgard 184 PDMS and cured at 80°C for 1 hour. The thin PDMS films with attached pillars were then "glued" to glass substrates using uncured PDMS as an adhesive, and then were cured again at 80°C. Deposition of single nano-scale polystyrene (PS 8% w/w) fibers was performed using the previously reported STEP technique, ensuring even spacing and homogenous fiber parameters. A small amount of epoxy adhesive was used to ensure fixed boundaries at each end of the fibers. Scanning Electron Microscope (SEM) images were taken to verify fiber homogeneity and fixed boundaries (Fig 1). C_2C_{12} Mouse Myoblast cells were seeded onto the fibers and migration was characterized using time lapse microscopy (Fig 2(A)). After time lapse studies were completed, samples were immunostained for F-actin stress fibers, nuclei, and focal adhesions. Scanning electron microscope images were also taken of some samples for complete characterization of micropillar and nanofiber geometry.

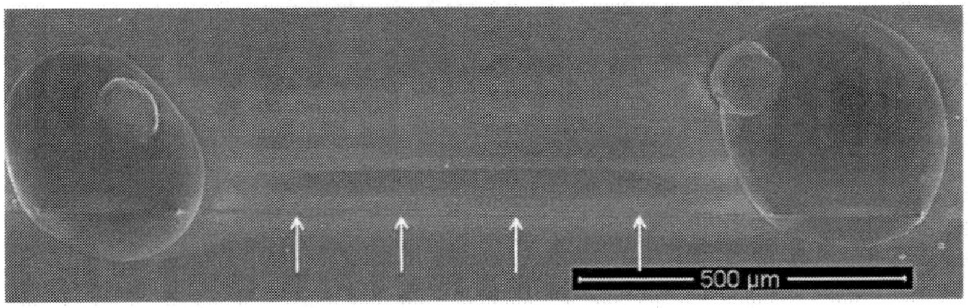

Figure 1: SEM image showing fiber and micro-scale pillars with fixed boundaries. Arrows indicate suspended fiber

Results and Discussion

Overall, cells were largely observed to migrate away from the center of the fiber, where fiber beam stiffness is lowest, with average migration speed decreasing near the edges of the fiber, where fiber stiffness is the highest (Fig 2(B)). Suspended fibers were characterized for stiffness using an Atomic Force Microscope. Preliminary data has shown that fiber stiffness is highly variable along the length of the fiber (Figure (2(C)). Cells were observed to attach to single fibers and elongate along the fiber axis with immunostaining studies showing that focal adhesion sites concentrated at the poles of the cells with a higher level of focal adhesion at the leading edge of the cell. F-actin stress fibers were found to be localized along the periphery of the cell with a higher concentration at the leading edge of the cell. The nucleus was found to be stretched along the fiber axis and enclosed within stress fibers (Fig 2(D)). This phenotype suggests a high level of polarizability of cells on the single fibers, with migration occurring in a single direction until an external stimuli (decreasing fiber beam stiffness) causes a change in direction. Quantification of the leading and trailing edge spatio-temporal dynamics is of immense interest and part of our future efforts.

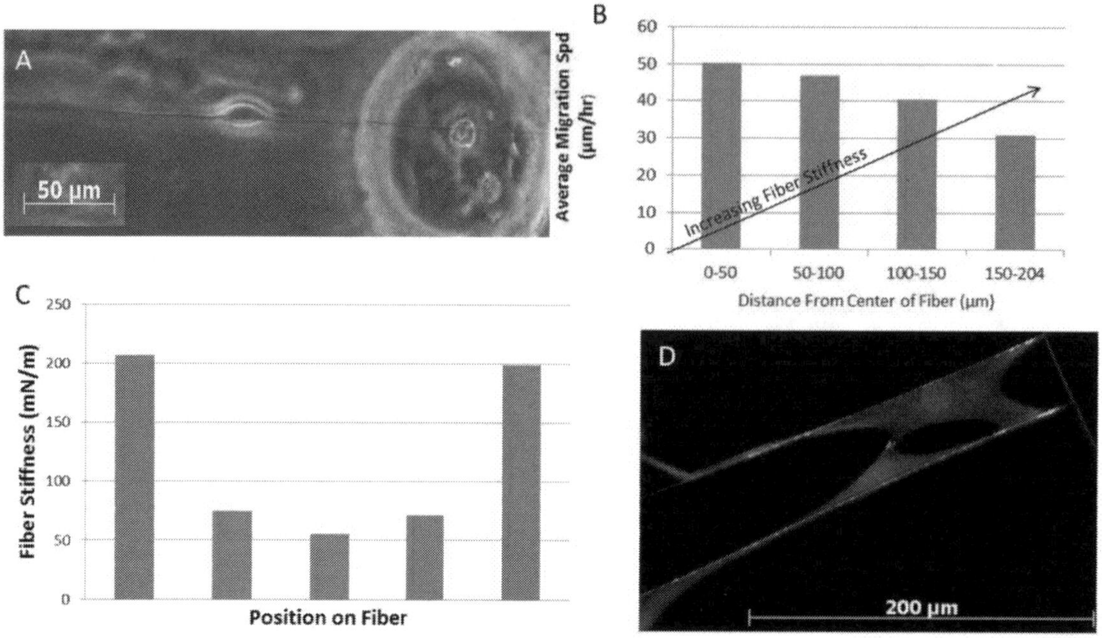

Figure 2: A) Phase image of a cell traveling along a single fiber. B) Average cellular migration speed as a function of distance from the center of the fiber (where stiffness is lowest) C) Average fiber stiffness along the fiber. Stiffness dramatically increases near the pillars D) Immunostained image showing f-actin (red), nuclei (blue), and focal adhesions (green)

Conclusions

The complete characterization of the influence of fiber substrate stiffness on cellular migratory behavior will be an important step in understanding many complex processes occurring within cells. The use of the STEP technique with simple photolithography techniques provides suspended single fibers that mimic single ECM fibrils in order to provide conditions that are as similar to *in vivo* conditions as possible. This technique allows for accurate measurement of cellular migration while maintaining simple, low cost fabrication as well as the possibility of modification to mimic the natural microenvironment of cells in various conditions. Cells attached on single suspended fibers were observed to preferentially migrate towards the stiffest region of the fiber with migration speed decreasing with increase in substrate stiffness. Additionally, immunostaining showed a high level of polarization and organization of F-actin stress fibers along the fiber axis. The complete characterization of migratory behavior on this combined system will allow for a more complete view of the *in vivo* interaction of cells with the extra cellular matrix and it's relation to cellular mechanics as a whole, which will be invaluable in the design of accurate wound healing scaffolds, drug testing platforms and design of accurate tissue engineering scaffolds.

References

1. McKee, Clayton T., et al., *The Influence of a Biologically Relevant Substratum Topography on Human Aortic and Umbilical Vein Endothelial Cells.* Biophys J, 2012. **102**(5): p. 1224-1233.

2. Dalby, M.J., et al., *The control of human mesenchymal cell differentiation using nanoscale symmetry and disorder.* Nat Mater, 2007. **6**(12): p. 997-1003.

3. Kilian, K.A., et al., *Geometric cues for directing the differentiation of mesenchymal stem cells.* Proc Natl Acad Sci U S A, 2010. **107**(11): p. 4872-7.

4. Wong, H.C. and W.C. Tang, *Correlating Simulation and Experimental Data of Traction and Cell Speed as Functions of Substrate Stiffness.* 2011: p. 240-244.

5. Wang, Y., et al., *Substrate stiffness regulates the proliferation, migration, and differentiation of epidermal cells.*

Burns, 2012. **38**(3): p. 414-20.

6. Ehrbar, M., et al., *Elucidating the Role of Matrix Stiffness in 3D Cell Migration and Remodeling.* Biophys J, 2011. **100**(2): p. 284-293.

7. Reilly, G.C. and A.J. Engler, *Intrinsic extracellular matrix properties regulate stem cell differentiation.* J Biomech, 2010. **43**(1): p. 55-62.

8. Nain, A. S.; Sitti, M.; Jacobson, A.; Kowalewski, T.; Amon, C. *Macromolecular rapid communications* **2009**, *30*, 1406-12.

9. Nain, A. S.; Phillippi, J. a; Sitti, M.; Mackrell, J.; Campbell, P. G.; Amon, C. *Small (Weinheim an der Bergstrasse, Germany)* **2008**, *4*, 1153-9.

10. Ker, E. D. F.; Nain, A. S.; Weiss, L. E.; Wang, J.; Suhan, J.; Amon, C. H.; Campbell, P. G. *Biomaterials* **2011**, *32*, 8097-107.

11. Bakhru, S.; Nain, A. S.; Highley, C.; Wang, J.; Campbell, P.; Amon, C.; Zappe, S. *Integrative biology : quantitative biosciences from nano to macro* **2011**, *3*, 1207-14.

12. Wang, H.B., et al., *Varying the diameter of aligned electrospun fibers alters neurite outgrowth and Schwann cell migration.* Acta Biomater, 2010. **6**(8): p. 2970-8.

(500g) Contact Guidance Differs On Micropatterned Collagen Substrates and Organized Collagen Fibers

Ian Schneider, Nick Romsey and Carin Lightner, Chemical and Biological Engineering, Iowa State University, Ames, IA

Before metastasis occurs, collagen fibers are reorganized around the tumor by surrounding stromal cells. The fibers transition from a circumferential pattern with collagen fiber alignment parallel to the tumor margin to a radial pattern with collagen fiber alignment perpendicular to the tumor. This change in organization directs cells to migrate along the radially organized collagen fiber bundles to surrounding tissue in a process called contact guidance. While qualitative knowledge of contact guidance is well-established, quantitative details associated with how local collagen organization promotes directed migration during contact guidance is lacking. We have been using two approaches to induce directed migration on 2D surfaces. In the first approach collagen is patterned on glass using microcontact printing. This approach results in potent directional cues that can be moderately tuned based on line spacing and background adhesion strength. In the second approach collagen fibers are grown on atomically smooth mica surfaces. By tuning buffer conditions, fibers of different sizes and fields of fibers with different degrees of alignment can be generated. This approach results in two interesting behaviors. First, large diameter collagen fibers induce random migration even when aligned, whereas small diameter collagen fibers induce potent directed cell migration. Second, the random migration on large fibers appears to be distinct from random migration on glass. These large organized collagen fibers appear to stimulate migrational behavior similar to that seen in 3D environments, where cells make many directional changes by extending protrusions that split and retract. Consequently, while microcontact printed surfaces generate a fairly potent approach for directing migration, assembled collagen fibers that organize into aligned patterns on mica surfaces constitute a better mimic of 3D environments and stimulate rich behavior that is highly sensitive to the underlying collagen organization.

(500h) Unique Hepatic Responses to Burn, Sepsis and Trauma: The Adaptability of Innate Immunity in the Face of Different Stimuli

John Mattick[1], Mehmet A. Orman[1], Qian Yang[1], Marianthi G. Ierapetritou[1], Francois Berthiaume[2] and Ioannis P. Androulakis[1], (1)Department of Chemical and Biochemical Engineering, Rutgers, The State University of New Jersey, Piscataway, NJ, (2)Department of Biomedical Engineering, Rutgers, The State University of New Jersey, Piscataway, NJ

The systemic immune response is considered to be a core response to external insults upon mammalian systems, and consists of a wide variety of effects, including dilation of the vasculature, hypermetabolism and lean body mass loss, acute phase protein synthesis, and macrophage activation., which, if untreated, can eventually lead to multiple organ failure and death [1]. Because of its central role in the metabolic aspects of this systemic response, as well as protein synthesis and innate immune activation, the liver represents a central player in this response whose transcriptional activity following injury is an indicator of critical pathways which are affected by the response, as well as a biomarker for the long term resolution of the host response. Traditionally, the acute phase of the inflammatory response is considered to be critically dangerous, however recent evidence suggests that the long term response to injuries can also result in serious patient complications, including mortality [2]. In order to understand the long term effects of injuries on pathways involved in the inflammatory response, we utilized sepsis and/or burn injury in rats, to assess their whole liver transcriptional response to multiple different injury conditions over the course of 8 days. We used statistical and clustering algorithms to investigate the genetic signatures of hepatic response to this double-injury model and evaluated the progression of the response in a temporal manner, in order to resolve critical pathways involved in the long term, as well as patterns of resolution in each injury.

The rat model receiving a 20 % total body surface area burn (B) followed 2 days later by cecal ligation and puncture (CLP) to induce sepsis was used. Control groups were given a sham-burn (S) treatment, which is identical to burn treatment but the animals were immersed in 37°C water instead of boiling water, or sham cecal ligation and puncture treatment (SCLP), which utilized an identical procedure to the CLP treatment, but did not ligate or puncture the cecum (SCLP condition). Animals were sacrificed at different time points during the first 10 days to collect liver samples for microarray analysis.

Co-expressed and injury-responsive gene groups in the microarray analysis were identified between CLP and SCLP using a "consensus clustering" approach [3] , which selected for temporal variations in the gene expression profiles of both control and treatment groups. Principle component analysis was then used to identify key differences in temporal progression between CLP and SCLP over the entire time course. In order to assess the combined effects of the BCLP injury, a technique known as SAM (Significant Analysis of Microarrays) was utilized [4] to determine a set of significantly varying genes following injury. This method uses multiple gene specific t-tests to verify the significance of the fold change of the double hit injury relative to sepsis alone, followed by a false discovery rate test, which aims to exclude genes whose variations are the result of random chance.

The consensus clustering method identified a total of 3 major responses in the CLP condition when compared directly to the SCLP control. These clusters are all strongly related to the inflammatory response, with two of the major clusters representing co-expressed sets of pro inflammatory genes, while the third cluster represents co-expressed anti inflammatory genes. Principle component analysis of the concatenated data set (Figure 1) reveals that CLP and SCLP have polar opposite responses along the first, major principle component, while they have almost identical responses along the second. Analysis of genes contributing to each of the principle components reveals that principle component 1 is mapping out a completely unique inflammatory response profile for each injury: the septic injury (CLP) reveals an early pro inflammatory response followed by severe immunosuppression, while the trauma injury (SCLP) shows the opposite dynamics, with early immunosuppression followed by severe pro inflammatory activity.

The SAM analysis identified a total of 957 genes that were either reduced in expression level or increased by a factor of 2 between the CLP condition, and the CLP condition influenced by prior burn injuries. Critical genes can be seen in Figure 2, which divides the functional differences into metabolic and immune changes: one can observe an up regulation in amino acid degradation over the first day of the response, as well as an early immunosuppression that quickly resolves into increased innate immune function. The long term response appears to be affected in its method of recuperation through a shift towards amino acid biosynthesis and wound repair.

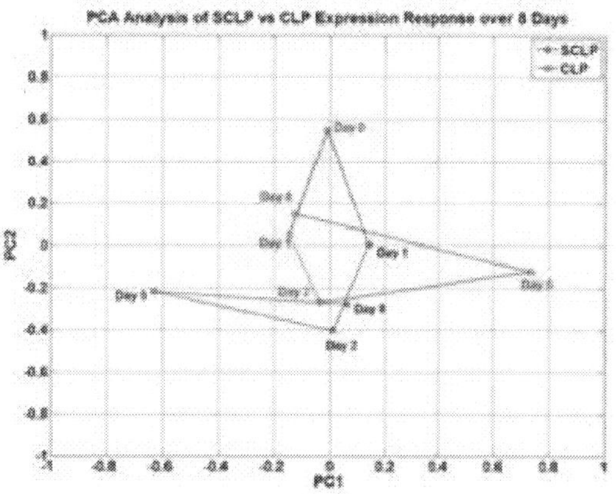

Figure 1: Principle component analysis of the progression of the long term CLP response compared to SCLP control

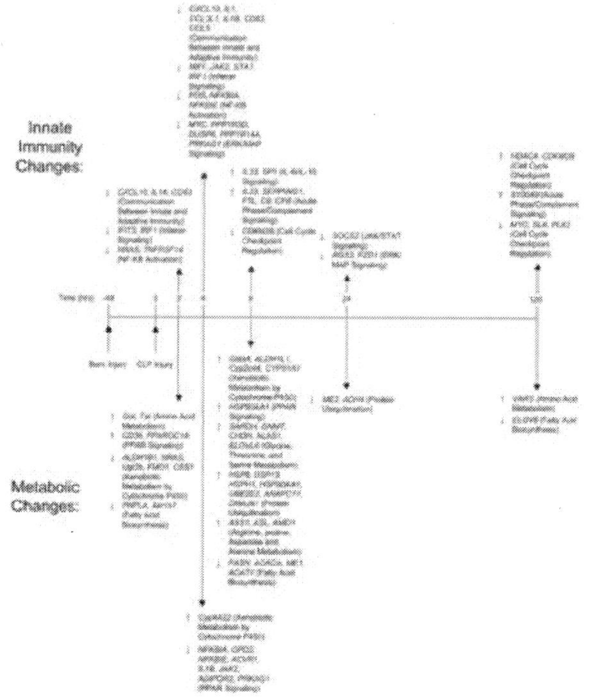

Figure 2: Critical genes altered by the burn injury over the progression of the subsequent CLP response

In conclusion, the differences in the responses of the innate immune response to a variety of stimuli necessitate a new paradigm of studying these conditions: inflammation is not a single condition, but rather a collection of responses unique to the injury. By utilizing transcriptional analysis, we were able to observe that the trauma injury has a completely different hepatic response to the septic injury, while burn priming does not appear to change the fundamental underlying response. Clinical strategies aiming to mitigate the disruptive effects of SIRS should therefore be tailored to account for the adaptability of the innate immune system in the face of various stimuli.

References:

1 . Beal, A. L. and F. B. Cerra, *Multiple organ failure syndrome in the 1990s. Systemic inflammatory response and organ dysfunction.* Jama, 1994. 271(3): p. 226-33.

2 . Sasse, Kent C., et al., *Long-term survival after intensive care unit admission with sepsis.* Critical Care Medicine, 1995. 23(6): p. 1040-1047.

3 . Nguyen, T. T., R. S. Nowakowski, and I. P. Androulakis, *Unsupervised selection of highly coexpressed and noncoexpressed genes using a consensus clustering approach.* Omics, 2009. 13(3): p. 219-37.

4 . Tusher, Virginia Goss, Robert Tibshirani, and Gilbert Chu, *Significance analysis of microarrays applied to the ionizing radiation response.* Proceedings of the National Academy of Sciences, 2001. 98(9): p. 5116-5121.

(507a) Combinatorial Design of Hydrolysate-Tolerant E. Coli Mutants

Tirzah Y. Glebes, Nicholas R. Sandoval, Nanette R. Boyle and Ryan T. Gill, Chemical and Biological Engineering, University of Colorado at Boulder, Boulder, CO

Lignocellulosic biomass is a cheap, renewable, and sustainable feedstock for biofuel production. For routes utilizing biological conversion of the lignocellulosic hydrolysate, microbial strains need to be engineered for the efficient production of the desired product, as well as designed for tolerance to the variety of inhibitors in hydrolysate (e.g., acetate and furfural). Here, we provide data from genome-wide searches based on the TRackable Multiplex Recombineering (Warner et al. 2010) approach. TRMR utilizes two barcoded-libraries, with ~ 4,000 members each, coding for either an increased or decreased expression of virtually every gene within *E. coli*. Selections were performed using TRMR libraries in the presence of acetate, furfural, and hydrolysate. Mutations that were enriched during the course of any selection were tracked via microarray. Targets that were conserved for enrichment among these selections were identified and individually confirmed for improved tolerance to these inhibitors. In addition, select targets were combinatorially designed into mutant strains to further improve tolerance to hydrolysate. Based on these conserved enrichments, hypotheses about toxicity mechanisms of hydrolysate were formulated and tested. We will provide a summary of these individual mutations and the combinatorial construction in this work, highlighting the aspects of designing targeted pools to enhance epistatic effects of combinatorial mutants.

(507b) Complete-MFA: Complementary Parallel Labeling Experiments Technique for Metabolic Flux Analysis

Maciek R. Antoniewicz, Department of Chemical and Biomolecular Engineering, University of Delaware, Newark, DE and Robert Leighty, Department of Chemical Engineering, Univerisity of Delaware, Newark, DE

Metabolic flux analysis (MFA) has emerged as a practical tool for measuring fluxes in complex biological systems. In ^{13}C-MFA, fluxes are obtained by introducing a labeled tracer into a cell culture at metabolic steady state, measuring the relative labeling of metabolic intermediates and by-products at isotopic steady-state, and computationally processing these measurements to quantify fluxes. In this work, we describe a novel, more advanced method for ^{13}C-MFA that we called COMPLETE-MFA, which stands for complementary parallel labeling experiments technique for high-resolution metabolic flux analysis.

The COMPLETE-MFA approach was developed specifically for high-resolution flux measurements. Until now, metabolic fluxes have been estimated from single tracer experiments. However, it is well know that in realistic network models there is often not one optimal isotopic tracer for determining all fluxes at high resolution. The COMPLETE-MFA method that we present here relies on using complementary isotopic tracers in parallel cell cultures, where each tracer is optimal for a specific part of metabolism. While each tracer may be suboptimal for elucidating the complete network model, we demonstrate that combined analysis of the parallel labeling experiments by simultaneously fitting all data sets to a single flux model provides flux results that are superior compared to any single tracer experiment. As proof of concept, we have quantified high-resolution fluxes for wild-type *E. coli* in batch culture. Using our COMPLETE-MFA approach we have resolved metabolic fluxes with 10-fold better precision (i.e. smaller confidence intervals) compared to current best method.

(507c) Metabolic Engineering of Rhizopus Oryzae: Effects of Overexpressing Fumr Genes On Fumaric Acid Biosynthesis From Glucose

Baohua Zhang, WILLIAM G. LOWRIE Department of Chemical and Biomolecular Engineering, The Ohio State University, Columbus, OH, **Kun Zhang**, Chemical and Biomolecular Engineering, The Ohio State University, Columbus, OH and Shang-Tian Yang, Department of Chemical and Biomolecular Engineering, Ohio State University, Columbus, OH

Fumaric acid, a dicarboxylic acid extensively used as a food acidulant and in the manufacturing of synthetic resins, can be produced from glucose in aerobic fermentation by the filamentous fungus *Rhizopus oryzae*. However, the fumaric acid yield is limited by the co-production of ethanol and other fermentation byproducts. Also, the fumaric acid accumulation mechanism in *R. oryzae* has not been fully resolved. Fumarase, which converts malate to fumarate in cytosol and the reverse reaction in the TCA cycle, is the key enzyme regulating fumaric acid biosynthesis. In order to understand the mechanism for fumaric acid biosynthesis and optimize the fermentation, effects of overexpressing the endogenous fumarase (*fumR*) gene were investigated. *R. oryzae* M16, a uracil auxotrophic mutant of *R. oryzae* 99880, was used as the parental strain. The uracil auxotroph has a mutation in *pyrF* gene encoding OMP pyrophosphorylase (*pyrF*). Only transformants carrying plasmids containing the *pyrF* gene could grow on the selective medium. The *fumR* gene was amplified from *R. oryzae* genomic DNA and cloned into the expression plasmids containing *pyrF* gene. Then, the expression plasmids were transformed into *R. oryzae* M16 spores using microprojectile particle bombardment. After culturing for 5-7 days, several *fumR* tranformants were obtained from the resulting colonies on the agar plates. These transformants were stable in maintaining the recombinant plasmids even after culturing for over 10 days on PDA plates without the selective pressure. Enzyme activity assay showed that the specific fumarase activity in the *fumR* mutants increased 5- to 10-fold compared to the wild type. However, in shake-flask fermentations, fumaric acid production by *fumR* mutants produced about the same amount of fumaric acid, but twice more malic acid (0.28-0.38 g/g glucose vs. 0.14 g/g for the wild type). The result suggested that the cloned fumarase catalyzed the reversible hydration of fumaric acid to malic acid. This is the first metabolic engineering study of fumaric acid biosynthesis in *R. oryzae*. Although overexpressing *fumR* did not increase fumaric acid production, the mutants produced much more malic acid and are promising for industrial production of malic acid.

(507d) Flux Regulation At a Primary Metabolic Node: Lessons for Acetyl-CoA Derived Products

Karthik Sekar and **Keith E.J. Tyo**, Chemical and Biological Engineering, Northwestern University, Evanston, IL

Managing carbon flux in cellular metabolism is a critical problem in metabolic engineering. Of particular interest is the acetyl-CoA node, as it is the precursor for many important bio-products. Past methods have focused on improving product yield by reducing acetate overflow through gene deletion, electron carrier manipulation, pathway enzyme overexpression and C-limited fed-batch cultivations. However, little work has directly investigated how flux distribution changes in the context of pathway engineering. Understanding how flux changes in response to heterologous pathways is an important step toward improving product yields through metabolic engineering.

We have estimated intracellular flux distributions over a wide range of heterologous pathway expression levels, growth rates, and media compositions to better understand how acetyl-CoA consuming pathways are regulated. We have engineered strains of *E. coli* with various copy numbers of the polyhydroxybutyrate (PHB) pathway using chemically-induced chromosomal engineering (CIChE). The PHB pathway uses acetyl-CoA as a precursor and therefore competes with native acetyl-CoA consuming reactions. We tested both N-limited and C-limited conditions in chemostats and measured exchange fluxes. Using these measured fluxes, we estimated internal fluxes using flux balance analysis (FBA). These results suggest a highly integrated tradeoff between PHB, acetate, and lactate that is constrained by ATP and NADH balances in the cell.

Our study highlights the interdependence of heterologous pathways on global ATP and NADH levels and constrains the strategies that will be effective in improving product flux. FBA was essential to interpret observed metabolic regulation and formulate strategies for improving production. The results should be useful to others engineering acetyl-CoA derived products, and the strategy should be of general use in metabolic engineering.

(507e) Estimation of Genome-Scale Metabolic Fluxes From Tn-Seq Single Mutant Growth Measurements

Hong Yang[1], Elias W. Krumholz[1], Evan D. Brutinel[2], Jeffrey A. Gralnick[2] and Igor G.L. Libourel[1], (1)Department of Plant Biology, University of Minnesota, St. Paul, MN, (2)Department of Microbiology, University of Minnesota, St. Paul, MN

Metabolic flux is the key to understand both wild-type and mutant metabolic phenotypes. Minimization of metabolic adjustment (MOMA) has been extensively used to estimate mutant flux maps by choosing the nearest feasible flux solution to a wild-type flux map. However, a genome-wide wild-type flux map is never available, and an FBA solution space is usually used instead of a unique solution. MOMA predictions are therefore a solution space as well, where each prediction is a function of the location in the FBA solution space from which the MOMA projection is made.

To overcome this limitation, we developed a novel algorithm to calculate the wild-type flux map by minimizing the difference between predicted mutant growth rates (using MOMA) and measured mutant growth rates from Tn-seq data. Using the estimated wild-type flux solution, we compared mutant growth predictions for a control set to predictions made by sampling the FBA solution space. The Tn-seq data was acquired for the Shewanella oneidensis strain. More generally, this work provides a new approach to calculate the unique wild-type flux map which provides better predictions for mutant phenotypes.

(507f) Reconstruction of Integrated Metabolic and Transcriptional Regulatory Network Models As a Platform for Metabolic Engineering

Joonhoon Kim and Jennifer L. Reed, Department of Chemical and Biological Engineering; DOE Great Lakes Bioenergy Research Center, University of Wisconsin-Madison, Madison, WI

Computational models of metabolic and transcriptional regulatory networks have been useful for metabolic engineering of microbial systems. While genome-scale metabolic network models are being rapidly developed for many organisms, the reconstruction of genome-scale transcriptional regulatory networks has been challenging due to the limited availability of experimental data and computational tools. A number of computational approaches based on statistical inference have been developed to reconstruct transcriptional regulatory network models, but these models often cannot be directly integrated with metabolic network models. Among different types of models, a Boolean model of transcriptional regulatory networks provides mechanistic details and can be easily combined with a metabolic model to yield a functional integrated model [1]. We previously developed an automated approach to integrate and refine metabolic and transcriptional regulatory network models using large-scale growth phenotype data [2], and also developed a computational method to identify metabolic engineering strategies using these integrated models [3]. Here, we describe an automated approach for reconstruction of Boolean transcriptional regulatory network models by integrating genome-scale metabolic models and high-throughput datasets including protein-DNA binding, gene expression, and growth phenotypes. The resulting integrated metabolic and transcriptional regulatory network models are therefore consistent with multiple datasets, and can be easily updated as new experimental data becomes available. The developed approach will facilitate the reconstruction of such models for different organisms and their applications in metabolic engineering.

1. Covert MW, Knight EM, Reed JL, Herrgard MJ, Palsson BO (2004) Integrating high-throughput and computational data elucidates bacterial networks. Nature 429: 92-96.

2. Barua D, Kim J, Reed JL (2010) An automated phenotype-driven approach (GeneForce) for refining metabolic and regulatory models. PLoS Computational Biology 6: e1000970.

3. Kim J, Reed JL (2010) OptORF: Optimal metabolic and regulatory perturbations for metabolic engineering of microbial strains. BMC Systems Biology 4: 53.

(507g) Flux and Reflux: Metabolite Back-Mixing in Plant Cell Suspensions and Its Implications On Isotope-Assisted Metabolic Flux Analysis

Xiaofeng Zhang[1], Ashish Misra[2], Shilpa Nargund[2], Gary D. Coleman[3] and Ganesh Sriram[2], (1)Department of Chemical and Biomolecular Engineering, University of Maryland, College Park, College Park, MD, (2)Chemical and Biomolecular Engineering, University of Maryland, College Park, MD, (3)Plant Science and Landscape Architecture, University of Maryland, College Park, College Park, MD

Isotope-assisted metabolic flux analysis (isotopic MFA) is a powerful tool for studying plant metabolism as it quantifies carbon flow (fluxes) through metabolic networks. It involves feeding an isotopically labeled substrate to cell or tissue culture and measuring the labeling patterns of the products using nuclear magnetic resonance (NMR) or mass spectrometry (MS). Metabolic fluxes are estimated by fitting these measurements to a mathematical flux-isotopomer model.

An intrinsic problem in steady state isotopic MFA, particularly in plants and eukaryotic systems, is that the metabolites in the initially present naturally labeled (1.1% ^{13}C) biomass (e.g. the initial cell culture or embryo) back-mix with those in the *de novo*-synthesized biomass (usually >> 1.1% ^{13}C) in a nontrivial manner. This back-mixing affects the final labeling patterns of the biomass components and confuses their interpretation. Therefore, it has to be accounted for to obtain accurate flux estimates. Previous MFA studies on plant systems accounted for this ^{13}C dilution by introducing an additional parameter amongst the parameters estimated in the flux evaluation procedure. Here, we endeavor to accurately model the back-mixing by incorporating it into the metabolic network model.

We performed this study on cell suspensions of poplar, a potential biofuel crop that efficiently recycles nitrogen from its leaves during winter senescence. Our isotope labeling experiment results clearly point to significant back-mixing of *de novo*-synthesized biomass by unlabeled initial biomass. Therefore, we established a metabolic model that included back-mixing fluxes of initially present proteinogenic amino acids and sugars, as well as the pool sizes (concentrations) of these metabolites. Simulation results using metabolic network models with the back-mixing fluxes show a better fit of MS-derived isotopomer data than models without back-mixing. Besides, the results surprisingly indicate that different amino acids may have very different extent of cycling, even when they originate from the same metabolic precursor(s). We anticipate that our methodology that incorporates back-mixing of initial biomass will benefit plant metabolic flux analysis by helping us obtain more accurate flux estimation and a more holistic view of plant metabolism.

(507h) A Second-Generation Genome Scale Model for Clostridium Acetobutylicum

Patrick F. Suthers and Costas D. Maranas, Department of Chemical Engineering, The Pennsylvania State University, University Park, PA

The development of genome scale models (GSM) for a variety of microbial, eykaryotic and multi-cellelar organisms [1] is proceeding with an accelerating pace. Automated reconstruction tools such as ModelSEED [2], Pathway Tools [3], and others are increasingly being utilized to quickly arrive at draft compilations of metabolic models followed by detailed curation to improve upon modeling components crucial for answering the queries relevant to the specific organism and study. Typically, simulations are carried out for an idealized growth substrate typically containing a single carbon substrate (e.g., glucose). Regulation is customarily handled by shutting off specific reactions in the presence of a specific nutrient environment. For example, in the iAF1260 *E. coli* model, there are specific reaction restrictions for anaerobic, aerobic, and aerobic with glucose growth conditions [13].

In this talk, we present a second-generation genome scale metabolic model for *Clostridium acetobutylicum*. The model contains over 750 genes and 1200 reactions. The reactions are all charge and elementally balanced. We made extensive use of MetRxn (http://metrxn.che.psu.edu), a web-based knowledgebase that includes standardized metabolite and reaction descriptions and integrates information from 8 databases and 44 metabolic models into a single unified data set. In addition to metabolic biotransformations, the model also includes regulatory information. Establishment of detailed gene-protein-reaction associations enables us to properly localize regulatory interactions at the transcriptional, protein or metabolite level and rely on the gene-protein-reaction (GPR) connections to propagate their action through metabolism. We categorize regulatory interactions into either up or down and of moderate or high strength. We make use of transcriptional data obtained under various stress conditions. In analogy to GPR associations, we define and populate stressor-regulator-operon-gene (SROG) associations that pinpoint which genes are affected by a specific stressor. We discuss the use of gene expression data throughout model development such as during the GapFill process. The model allows for phenotype predictions for growth under various media and conditions and enables the generation of testable hypotheses on the response to substrate changes and genetic alterations on biofuel production.

(518a) Flow Cytometry-Based High-Throughput Screening Systems for Directed Evolution of Proteases

Ran Tu, Huiling Yuan, Lin Sun, Qinhong Wang and Yanhe Ma, Tianjin Institute of Industrial Biotechnology, Chinese Academy of Sciences, Tianjin, China

Proteases are industrially important enzymes and advances in screening technologies such as flow cytometry (FC) based *in vitro* compartmentalization (IVC) offer novel opportunities to tailor protease properties by directed evolution. Here we present two flow cytometry-based screening systems for proteases, cell-free based ProCF-FCIVC and whole-cell based Pro-FCIVC, with a high-throughput (>7500 per second). The systems are based on the linkage between genotype and phenotypic activity by encapsulation of single cell together corresponding fluorogenic substrate in *in vitro* emulsion compartments, and subsequently analyzing and sorting those emulsions by flow cytometry machine. In the study, system related technologies such as generation of water-in-oil-in-water double emulsions, selection of suitable substrate and calibration of the sorting performance of flow cytometry were investigated and optimized. For the cell-free based ProCF-FCIVC, the system was developed using a model protease M57 isolated from a metagenomic library. Different parameters like *in vitro* expression of protease, enzymatic reaction in emulsions, and recovery of sorted DNA variants were investigated. Sorting and subsequent activity analysis of the M57 protease reference library demonstrated a proof of concept of the developed ProCF-FCIVC system. For the whole-cell based Pro-FCIVC system, an extracellular proteases-deficient *Bacillus* strain together with a subtilisin Carlsberg (SC) secretion system was used. Several main challenges such as avoiding the high background of natural occurring proteases, generating mutagenesis library in high diversity using *Bacillus* expression system, and optimizing the sorting strategies were studied. The developed whole-cell Pro-FCIVC system was validated by screening an epPCR library of SC for increased inhibitor resistance. In these two cases mutants with improved activities and interesting changes were found. These approaches exemplify the application of flow cytometry based HTS for reengineering of proteases for additional useful properties in future.

(518b) Pfunkel: Efficient, Expansive, User-Defined Mutagenesis

Elad Firnberg and Marc Ostermeier, Chemical and Biomolecular Engineering, Johns Hopkins University, Baltimore, MD

An efficient and high-throughput mutagenesis strategy is an integral part of protein studies, directed evolution and biological studies. The mutagenesis toolbox currently lacks a method for creating extensive DNA libraries with a researcher-defined mutational composition spanning across an entire gene. For example, until now there has been no efficient method to make a library consisting of all 18,900 possible single codon substitutions of a 300 amino acid long protein, nor is there a method to make a researcher-prescribed subset of only 2000 of these 18900 mutations. We have developed a method termed PFunkel mutagenesis by which such libraries are readily created. Additionally, the protocol has been adapted for the introduction of multiple defined mutations per gene with very high efficiency. The method is rapid, taking a single day and performed in a single tube. In short, PFunkel is a versatile and efficient mutagenesis technique that enables creation of gene libraries or individual constructs where each member receives one or multiple user-defined mutations at any position.

Using PFunkel we constructed a comprehensive codon substitution library of nearly all of the ~18,000 possible codon substitutions in the *TEM-1* gene, which encodes the TEM-1 b-lactamase protein. TEM-1 has high activity for degrading the beta-lactam antibiotic ampicillin, but negligible activity on the antibiotic cefotaxime. By subjecting this library to a selection for increased cefotaxime resistance we have identified nearly all single-codon mutations that provide increased activity on cefotaxime. This extensive set of improved variants of a single gene revealed interesting patterns in the nature of gain-of-function mutations. In a second experiment, we constructed a *TEM-1* library in which four distal positions that are known to be important for activity on cefotaxime were simultaneously randomized to all possible codons. From this library we discovered new variants that confer greater resistance to cefotaxime than any of those identified in seven previous directed evolution studies.

(518c) Probing Human Transferrin-Receptor Interactions Using Linear Epitope Mapping

Divya Chandra and Pankaj Karande, Chemical and Biological Engineering, Rensselaer Polytechnic Institute, Troy, NY

Transferrin (Tf) is a carrier protein involved in the transport of iron (Fe3+). Tf binds to transmembrane transferrin receptors (TfRs) expressed on cell surfaces and transports Fe3+ by receptor mediated transcytosis (RMT). The Tf-TfR complex is an interesting system as it provides a natural biological pathway for delivery of Tf-bound cargo across cells and tissues. It has also been demonstrated that cancer cells over-expressing TfR can be selectively targeted with Tf-conjugated cytotoxic drugs. Understanding the interactions between Tf and TfR is therefore valuable for designing better therapeutic modalities for a broad range of physiological diseases. A few studies have highlighted putative amino acid residues involved in the formation of the Tf-TfR complex. The crystal structures of the N-lobe of Tf and the ectodomain of TfR have also been resolved independently. We propose to use peptide microarrays for epitope mapping of the Tf-TfR complex to probe the interactions between Tf and TfR. Linear epitope mapping has been widely used as a technique for identifying antigenic regions in proteins in a fast, reliable high-throughput fashion. We have employed peptide microarrays of short peptides that are linear, continuous fragments of TfR to capture the interactions between TfR and its ligand. Microarrays printed with peptide fragments that span the entire sequence of TfR were tested against Tf to screen for regions on TfR that are involved in binding to Tf. Our results have confirmed the prior findings on the Tf-TfR interaction and complex formation. Additionally, in order to probe what regions of Tf interact with TfR we have prepared microarrays of peptides that are short fragments of Tf. We are currently conducting cell-based screening assays where these microarrays are incubated with varying densities of live cells over-expressing TfRs on their surface. Not only will this cell-based screening technique provide insight into the regions of Tf involved in interaction with TfR but also serve as a platform for probing ligand-receptor interactions of other cell-surface receptors. We anticipate that our studies along with the existing knowledge about the Tf-TfR system will further advance the understanding about this important transport and regulatory pathway.

(518d) Dihydropyridines Inhibit Amyloid-β Aggregation and Alter the Morphology of Amyloid-β Fibrils Associated with Alzheimer's Disease

Jui-Heng Tseng[1], James Chapman[2] and Melissa A. Moss[1], (1)Chemical Engineering, University of South Carolina, Columbia, SC, (2)College of Pharmacy, University of South Carolina, Columbia, SC

Alzheimer's disease (AD), one of the most widespread neurodegenerative diseases, is characterized by the accumulation of intracellular neurofibrillary tangles and extracellular amyloid plaques. Symptoms include confusion, language skill breakdown, long-term memory loss and other cognitive impairments. Unfortunately, current therapies cannot halt or slow AD pathogenesis. It has been proposed that amyloid plaque formation initiates pathological events that culminate in neuronal loss. These amyloid plaques are composed primarily of fibrillar amyloid-β protein (Aβ), formed as a result of the aggregation of Aβ monomer. Thus, understanding of the process of Aβ aggregation as well as the structures of Aβ fibrils formed is necessary to comprehend AD pathology and provide a rational design of compounds to prevent formation of pathogenic Aβ species.

Dihydropyridines (DHPs) are a group of small molecules containing at least one pyridine ring and one phenyl ring. These compounds, which can act as calcium channel blockers, are FDA approved for treatment of hypertension. DHPs are promising drugs for AD as a result of their aromatic nature. Numerous in vitro studies have reported that small molecules containing aromatic centers are potential candidates to alter Aβ fibril formation. In this research, four DHPs, as well as their metabolic products, are investigated for their ability to alter different stages of Aβ aggregation.

The DHPs studied all showed some ability to inhibit aggregation of 20 μM Aβ monomer. For most compounds, this activity was observed at a compound concentration in excess of Aβ. However, nicardipine exhibited an IC_{50} value of 5.7 μM, demonstrating its strong inhibitory capability. The ability of DHPs to inhibit growth of pre-formed Aβ aggregates was also examined. Only three DHPs showed significant inhibition at later stages of Aβ aggregation. Results were confirmed by transmission electron microscopy (TEM), which demonstrates that the densities of fibrils correlate closely with monomer aggregation results. Furthermore, variations of fibril morphology were observed and could be correlated with the effect of DHPs in later stages of aggregation. Recent studies showed that by changing Aβ aggregation mechanisms, Aβ fibrils exhibit morphological differences, and these differences are predicted to alter their physiological activity. In summary, DHPs are promising compounds that reduce aggregate formation and alter fibril structures, thus potentially attenuating physiological activity and disease pathogenesis.

(518e) The Exploitation of S. Cerevisiae - Improved Understanding and Optimal Yields of Single-Chain Antibody Fragment (scFv) 4-4-20

Carissa L. Young[1,2], Ronald W. Maurer III[2], Jeffrey Caplan[1], Kirk J. Czymmek[1] and Anne S. Robinson[3], (1)UD Bio-Imaging Center, Delaware Biotechnology Institute, Newark, DE, (2)Department of Chemical and Biomolecular Engineering, University of Delaware, Newark, DE, (3)Department of Chemical and Biomolecular Engineering, Tulane University, New Orleans, LA

Recombinant antibody fragments – for example, the classic monovalent single chain antibody (scFv) – have emerged as credible alternatives to monoclonal antibody (mAb) products. scFv fragments maintain a diverse range of potential applications in biotechnology, specifically as therapeutic and diagnostic agents. As such, a variety of hosts have produced antibody fragments with varying degrees of success. Yeast, *Saccharomyces cerevisiae*, is an attractive microbe due to similarities in the secretory pathway of eukaryotic organisms including analogous mechanisms for protein synthesis, translocation, maturation, and secretory trafficking. However, the expression of recombinant proteins in yeast is not trivial; neither are cellular quality control pathways simplistic. The endoplasmic reticulum (ER) is a dynamic organelle, capable of sensing and adjusting its folding capacity in response to increased demand. When protein abundance or terminally misfolded proteins overwhelm the ER's capacity, the unfolded protein response (UPR) is activated. Elucidating the role of ER stress, both physiological and pathological, will enable the design of new therapeutic modalities aimed at stress reduction.

We have established methodologies for investigating the role of cellular quality control and its modulation during heterologous protein expression of scFv 4-4-20, focusing specifically on the UPR, autophagy, and ER-associated degradation (ERAD). By implementing DNA recombination strategies combined with high-resolution imaging techniques, we have identified the intracellular localization of scFv and determined the extent of protein interactions with ER folding factors. In pursuit of a thorough analysis of protein distribution at the subcellular level, yeast expression cassettes [1] were designed to test the effects of codon-optimized fluorescent variants, small epitope tags (reviewed in [2]), polylinker length for N- and C-terminal tags, and the inclusion of essential retrieval sequences for ER luminal chaperones and foldases [3]. In fact, we have assessed many ERQC proteins in yeast that significantly effect cell physiology and disease pathology [4]. *S. cerevisiae* strains were further engineered to express fluorescent proteins targeted to various organelles [5]. To investigate discrete subpopulations of tagged proteins using live-cell imaging methods and super-resolution techniques (e.g. Fluorescence-Photoactivation Localization Microscopy, F-PALM), a photoconvertible GFP variant (i.e. mEos2) and six-residue tetracysteine motif required for FlAsH (fluorescein arsenical helix binder)-based technology were implemented. For yeast recombinant protein expression, versatile constructs were designed to regulate trafficking effects. Consequently, the implementation of motifs and elimination of retrograde transport have resulted in improved secretion of the model antibody fragment, scFv 4-4-20 [6]. Furthermore, we have assessed the UPR at the cDNA and protein levels, as well as quantified total scFv production versus secretion utilizing novel purified standards.

Time course analysis, quantitative PCR, co-immunoprecipitation of select proteins, and yeast deletion strains in combination with high-resolution imaging techniques have enabled us to evaluate different expression conditions, minimize UPR, and determine co-localization with organelles and sub-compartments. Combined with microarray studies, our systems have facilitated an improved understanding of the pivotal role of cellular quality control, specifically the UPR, autophagy, and ERAD [6].

Publications

1. C. L. Young, D. Raden, J. Caplan, K. Czymmek, A. S. Robinson *Optimized Cassettes for Live-Cell Imaging of Proteins and High Resolution Techniques in Yeast*, Yeast, 2012 doi:10.1002/yea.2895. [Epub 2012 Apr 4]

2. C. L. Young, Z. T. Britton, A. S. Robinson *Recombinant Protein Expression and Purification: A Comprehensive Review of Affinity Tags and Microbial Applications*, Biotechnology Journal, 7(4), Jan 10 2012 doi:10.1002/biot.201100155. [Epub ahead of print]

3. C. L. Young, D. L. Raden, A. S. Robinson, *Analysis of Endoplasmic Reticulum Resident Proteins in S. cerevisiae: Implementation of H/KDEL Retrieval Sequences,* 2012 (submitted).

4. C. L. Young, D. L. Raden, J. Caplan, B. Chung, K. Czymmek, A. S. Robinson *Dynamics of Endoplasmic Reticulum Resident Proteins and Organelle Morphology in S. cerevisiae,* 2012 (in preparation).

5. C. Young, Z. Britton, J. Caplan, K. Czymmek, A. Robinson *Exploiting S. cerevisiae: Cellular Systems & Techniques Aimed at Identifying the Localization of Targeted Proteins,* 2012 (in preparation).

6. C. Young, R. Maurer, P. Xu, A. Robinson *scFv: the model single-chain antibody fragment,* 2012 (in preparation)

(518f) Directed Evolution of Influenza Hemagglutinin (HA) Reveals Novel pH-Sensing Mutants

J. Vincent Price, Chemical and Biomolecular Engineering, University of Tennessee, Knoxville, TN, Jeong H. Lee, Chemical and Biomolecular Engineering, University of Pennsylvania, Philadelphia, PA, Morgan R. Baltz, Chemical and Biomolecular Engineering, University of Tennessee and Eric T. Boder, Department of Chemical and Biomolecular Engineering, University of Tennessee, Knoxville, TN

Hemagglutinin (HA) is a viral fusion protein that undergoes an irreversible conformational change upon acidification to catalyze the fusion of endosomal and viral membranes. Given the ubiquity of membrane fusion processes in nature, engineering a fusion protein for use in a drug or gene delivery vesicle, possibly promoting liposome fusion, would be a worthwhile endeavor. We have employed directed evolution to identify novel mutants with activation pH across a range of 4.8 – 6.0; wild-type HA activates around pH 5.2. On the surface of transduced cells, a small number of engineered HA with altered pH sensing phenotypes can trigger activation of an excess of wild-type HA (or vice versa) by an autocatalytic mechanism, suggesting the ability to develop modular protein-based materials with separately tuned sensing and effector modules. Examination of library mutants has enabled us to identify individual amino acids responsible for phenotypic changes and thus provides additional structure-function insight.

(518g) Engineering and Characterization of a C-Terminal Cleaving Intein for Use in Protein Purification

Miguel Ramirez, Najla Valdes, Dongli Guan and Zhilei Chen, Chemical Engineering, Texas A&M University, College Station, TX

A highly efficient C-terminal cleaving split DnaE intein from *Nostoc punctiform* was engineered through rational design. Npu is a thio-induced intein. A single mutation in the C-fragment of Npu, D118G, was introduced based on a sequence alignment with the mini-Mtu *recA* intein from *Mycobacterium tuberculosis* to curb the trans-splicing reaction, giving rise to the mutant split intein Npu*. At neutral pH and in the presence of reducing agent (e.g. DTT), Npu* exhibits no detectable trans-splicing between the N- and C-exteins and instead undergoes a single cleavage reaction at the C-extein. The C-extein cleavage reaction is rapid in the presence of the N-intein and reducing agent, with ~80% cleavage occurring in just 1 hr at 37°C. To achieve the same 80% cleavage, 2 and 4 hours were needed for samples incubated at room temperature (22°C) and 16°C, respectively. We demonstrate the application of Npu* for the rapid purification of 6 proteins.

(518h) Protein Purification Using Lytag and Q Sepharose

Gina Pietro, Chemical and Biomolecular Engineering, The Ohio State University, Columbus , OH, Michael J. Coolbaugh, Chemical and Biomolecular Engineering, The Ohio State University, Columbus, OH, Miriam Tang, Chemical and Biomolecular Engineering, The Ohio State University , Columbus, OH and David Wood, Department of Chemical and Biomolecular Enginnering, Ohio State University, Columbus, OH

Protein Purification using LYTAG and Q Sepharose

Gina L. Pietro, Michael J. Coolbaugh, Miriam S. Tang, David W. Wood

A method of purifying recombinant proteins using affinity chromatography with Q Sepharose and a LYTAG-intein affinity tag was developed. The LYTAG protein affinity tag is currently used with a specifically designed resin for recombinant protein purification. Inteins are proteins that have been engineered to cleave on a shift in pH or temperature, facilitating their use as a component of self-cleaving tags. The gene that creates LYTAG is joined with the gene for a self-cleaving intein, and this combined gene is joined to the gene for an interchangeable product protein. The cell then reads the genes and creates proteins, called fusion proteins, made of LYTAG connected to intein and an interchangeable product protein to be purified. The LYTAG-intein fusion proteins can be attached to the Q Sepharose resin. The intein can then be forced to cleave leaving pure product to be eluted from the column.

Q Sepharose resin is usually used as an anion exchanging resin that separates proteins based on charge. This method utilizes the special affinity that LYTAG has to quaternary amines, with the controlled cleaving of intein using pH shift and temperature change. Initially, the LYTAG-intein fusion protein is bound to the Q Sepharose column utilizing its specific affinity for quaternary amines rather than ionic interaction. The fusion protein is bound to the column using high salt concentrations to force off other proteins that might have bound to the column with ionic interactions. Equilibrating the column to a lower pH and raising the temperature induces intein self-cleavage. The pure product can then be eluted from the column. From qualitative analysis of SDS-PAGE gel, the purity of the product protein appears to be high.

Because of the ability of Q Sepharose to be scaled up, this technology could mean lab scale research and drug development would more easily translate to industry. It would also be a step toward introducing intein technology to industry. This process could be implemented in the bio-pharmaceutical industry using existing Q Sepharose columns with minimal regulatory approval or change to current setup.

(531a) Genome-Scale Model and Pathway Analysis for Optimization of Butyric Acid Fermentation by Clostridium Tyrobutyricum

Ying Jin, Chemical engineering, OSU, columbus, OH and Shang-Tian Yang, Department of Chemical and Biomolecular Engineering, Ohio State University, Columbus, OH

Understanding the metabolic behavior of an organism or a living cell is essential for fermentation process optimization. Predictions of metabolic behavior and flux distribution in batch cultures enable better regulation of fermentation in terms of biomass yield and productivity. Flux balance analysis (FBA) is a constraint-based modeling approach using linear programming to study cellular metabolism. It can not only develop models based on annotated genomic information, but also predict or validate current knowledge of specific microorganism's metabolic network. In this study, FBA was performed using Pathway Tool for systematic analysis of butyric acid fermentation by *Clostridium tyrobutyricum*. A comprehensive genome-scale model was developed for this organism, and flux analysis was then performed to evaluate the constraint factors of fermentation, optimize the process, and predict mutant strain metabolic behavior. The results from this analysis will be presented in this paper.

(531b) Synergistic Action of Chitosan Nanoparticles and Low Voltage Electric Field in Bacterial Inhibition

Glareh Azadi[1], Shiwen Sun[2], Adam Cham[2], Germaine Tsui[3], Nina Shapley[2], Karl Matthews[3] and Anubhav Tripathi[1], (1)School of Engineering, Brown University, Providence, RI, (2)Department of Chemical and Biochemical Engineering , Rutgers University, Piscataway, NJ, (3)Department of Food Science, Rutgers University, New Brunswick, NJ

Techniques to inhibit gram-negative bacteria such as shiga-toxin producing *Escherichia coli* are valuable as the prevalence of large-scale industrial food preparation increases the likelihood of contamination. Chitosan, a deacetylated derivative of chitin, is widely used for drug delivery as a biocompatible, biodegradable polymer with inherent antimicrobial characteristics. Chitosan inhibits bacteria growth in acidic environments, but is significantly less effective at pH > 7. Pulsed electric fields, constituting another method of bacteria inhibition, are difficult to generate at sufficient strength due to the high electric potentials required. Here, we combine the adsorption of Chitosan nanoparticles in a very low electric field for an increased inhibition of gram-negative bacteria at mild treatment conditions. Our results show that the synergistic action of these particles with a low voltage electric field applied in two minute time intervals results in significant inhibition of bacteria. Combination of sustained pulses at 100 V/cm and 0.3mg/ml Chitosan reduces the bacteria viability by 2-2.5 orders of magnitude. In addition, we study the inhibition effect of mixtures containing antimicrobial substances (e.g. nisin, polylysine, or lauric arginate) and Chitosan nanoparticles in the presence of applied electric fields.

Industrial Platform Design for Large Scale Production of Probiotic Yeast

Hesham A. El Enshasy
Institute of Bioproduct Development (IBD), Universiti Teknologi Malaysia (UTM), Johor Bahru, Malaysia

Abstract

Nowadays, probiotics are widely used biotechnology products and considered as one of the main components of wellness industries. These microorganisms have strong influence on human health by improving food digestion, increasing immune system and acting against many pathogentic microorganisms. In the present work, we represent complete industrial platform for high cell mass production of probiotic/biotherapeutic yeast *Saccharomyces boulardii.* A special type of yeast adapted to dryness and temperature stress to increase the cells shelf life time and also to decrease the production cost by replacing the freeze drying step by spray drying (the most fast efficient and economic drying process in downstream in biotechnology industries).

Introduction

Probiotics are live microorganisms or microbial mixture administrated to improve human microbial balance, particularly the environment of the gastrointestinal tract and the vagina. For many years, different bacterial strains belong to *Lactobacillus* sp., *Lactococcus* sp. *Streptococcus* sp. and *Bifidobacterium* sp. are widely used as efficient probiotic strains. In addition, yeast strain such as *Saccharomyces boulardii*, a non-pathogenic yeast that grows optimally at body temperature, has been applied as an efficient probiotic strain in the prevention of microbial associated diarrhea. *S. boulardii* is a non-pathogenic biotherapeutic and probiotic yeast, widely prescribed in a lyophilized form woldwide. *S. boulardii* acts as a shuttle liberating effective enzymes, proteins and trophic factors during tis interstinal transit that improve host immune defenses, digestion and absroption of nutrienst (1-3). Moreover, *S. boulardii* secretes during its intestinal transit polyamines, mainly spermine and spermidine that regulate gene expression and protein synthesis (4). *in vitro* studies have demonstrated that *S. boulardii* exerts antagonistic activity against various bacterial pathogens. This type of yeast has high ability to adhere to *Salmonela enterica* serovars Typhimurium and Enteritis, enteropathogenic *Escherichia coli* (EPEC) and enterohemorrhagic *E. coli*. Therefore, it protect the intestinal tract from diarrheal pathogens and prevent intestinal lesions (3,5). Beside the different pathogen prevention activity, *S. boulardii* shows also high potency for the prevention of Antibiotic Associated Diarrhea (AAD) which is resulted from long term antibiotic treatment (6,7). In spite of many literatures published concerning the importance of *S. boulardii* as probiotic strain, little information is available concerning the cell mass production of this type of yeast. The production of this strain in industrial level faces two main challenges. First, the significant low biomass production when cells cultivated in large scale due to oxygen limitation and alcohol production. Second, high sensitivity of cells to elevated temperature and dryness, and thus, the down-stream process to obtain cells in dry form requires freeze-drying step instead of the cheap spray drying. To overcome the first bottle-neck of this process, cells were cultivated in an optimized semi-defined industrial medium using exponential fed-batch cultivation strategy in 15-L stirred tank bioreactor to achieve high cell density more than 100 g/L with minimal alcohol production. This process was scaled up to 70-L and 300-L bioreactors to investigate the scalability and to complete the industrialization cycle of this process. To overcome the second bottle neck of this process, cells were initially adapted to dryness and to short time exposure to high temperature. Thus, spray drying was applied in downstream process instead of the conventional drying method of this yeast using the expensive freeze drying technique.

Materials and Methods

Microorganism

The yeast strain used in this study was *Saccharomyces boulardii* HA-8. This strain was derived from ATCC-MYA-796 (American Type Culture Collection, Manassas, VA, USA) after successive adaptation to dryness and elevated temperature up to 45 °C (El Enshasy and Shereef, 2008). This dryness/thermo adapted strain was kept at -80 °C in cell bank until used.

Growth medium

The frozen cells were initially propagated on yeast-peptone-dextorse (YPD) agar medium of the following composition (g/L): glucose, 10; yeast extract, 3; peptone, 3 and agar, 20. The pH of medium was adjusted to 5.5 before sterilization.

Production medium

Unless otherwise mentioned, the yeast production medium applied was semi-synthetic medium (SSM). This medium was composed of (g/L): Glucose, 15.0; $(NH_4)_2SO_4$, 5.0; KH_2PO_4, 3.0.; $MgSO_4.7H_2O$, 1.0; yeast extract, 3.0. The pH was adjusted to pH 5.5 before autoclaving. Glucose was sterilized separately by autoclaving at 110 °C and added aseptically to the fermentation medium before inoculation.

Shake flask cultivation and inoculum preparation

Shake flask cultivations were carried out in 250 ml Erlenmeyer flasks containing 50 ml. of growth medium. After inoculation, the flasks were incubated aerobically on a rotary shaker (Annova 4330, New Brunswick, NJ, USA) at 200 rpm and 30°C for 24 h. Cells were used thereafter to inoculate the bioreactor with inoculum concentration of 5% (v/v)

Bioreactor cultivation

Cultivations were conducted in three types of stirred tank bioreactors of difference capacities: First, cultivations were carried to in 15-L stirred tank bioreactor (Biostat C-15, Sartorius, Germany), 70-L stirred tank bioreactor (Biostat-D-70) and 300-L stirred tank bioreactor (Biostat D-300). Agitation was conducted in all bioreactor using standard three 4-bladed Rushton turbine impellers. Unless otherwise mentioned, aeration was adjusted to constant rate of [1 v/v/min]. During cultivation, both of pH and DO were measured and controlled using on-line electrodes (Mettler Toledo, Urdorf, Switzerland)

Sample preparation and cell dry weight determination

In case of bioreactor cultivations, aliquots in form of 20 ml of the culture were taken from the bioreactor vessel through a sampling system. At first, cells were separated from broth by centrifugation and the supernatant was frozen at -20°C and used for glucose and alcohol determination. The cell pellets were resuspended and washed/centrifuged twice using distilled water. The optical density of washed cells was determined immediately by spectrophotometer (Pharmacia Biotech, Cambridge, Englnad) at 600 nm after dilution to optical density range of 0 -0.5. The OD of culture was converted to dry cell mass by using previously developed linear correlation equation. Biomass concentration is defined as dry cell mass per unit volume of culture (g/L). For this strain, 1 OD was equal to 0.28 $\pm$0.03 g/L.

Determination of glucose

The glucose concentration in culture was determined enzymatically by glucose odixase-peroxidase determination kit (Biocon GmbH, Germany). The intensity of developed color was determined at 500 nm using spectrophotometer (Pharmacia Biotech, Cambridge, Englnad).

Results and Discussion

Batch cultivation in 15-L stirred tank bioreactor under controlled DO in culture

In this experiment, cultivation was carried out in 15-L stirred tank bioreactor with working volume of 9 L in batch mode under agitation speed of 400 rpm, aeration rate of 1 v/v/min and constant pH of 5.5. As shown in figure 1, cells grew exponentially with specific growth rate of about 0.1 h^{-1}. During cell growth phase, dissolved oxygen decreased gradually in culture until 10 hours. After that time, the DO sharply increased by time and reached about 90% saturation at the end of cultivation. This indicates directly the active growth and cellular aerobic metabolic pathways during the first 10 hours of cultivation. The date obtained from this batch culture was necessary to set up the bioprocess kinetic data required for fed-batch cultivation when operation carried out using exponential feeding strategy.

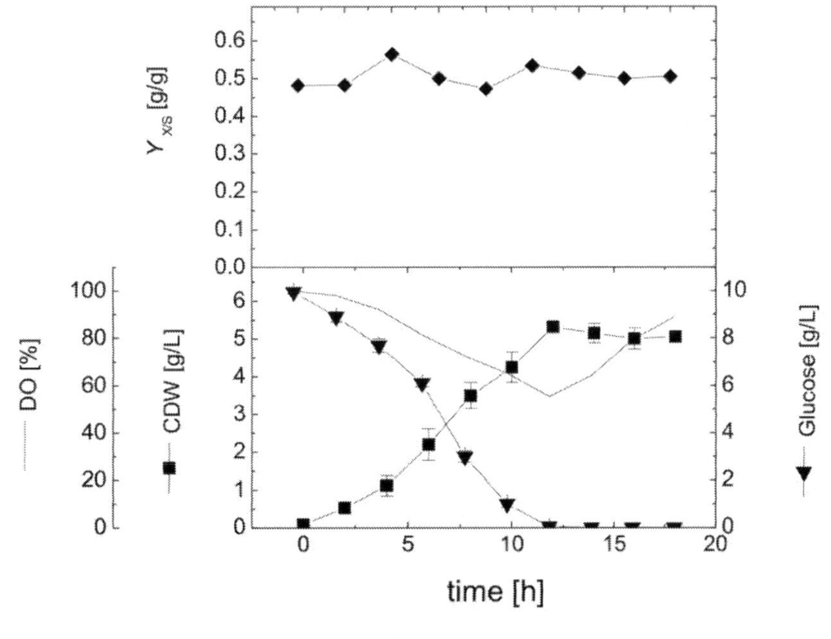

Figure 1. Cell growth profile of *S. boulardii* in 15-L stirred tank bioreactor under the following cultivation conditions: temperature: 30°C, agitation: 200 rpm and aeration: 1 v/v/min.

Based on this batch cultivation, the following data were obtained: maximal cell dry weight, X_{max}= 5.3 g/L; glucose consumption rate, ds/dt= 0.95 g/L/h; specific yield coefficient $Y_{X/S}$ $_{(average)}$= 0.52. These kinetic parameters were necessary to set proper exponential feeding strategy.

Fed-batch cultivations in 15-L stirred tank bioreactor using exponential feeding strategy

Based on the date obtained from the previous batch culture, fed-batch cultivation was designed using exponential feeding strategy. This strategy of feeding was successfully applied for high cell density cultivation of different types of microorganisms (8,9). Cultivation was started as in the previous experiments in batch mode. After 8 hours, feeding started using the following exponential feeding equation:

$$m_s(t) = \left[\frac{1}{Y_{X/S}} . \mu_{set} + m_E \right] . V_L(t) . X_F . e^{\mu_{set}(t - t_F)}$$

Where:

m_s: Mass flow of substrate [g/h]

t: Cultivation time [h]

t_F: Start time of feeding phase [h] = *8 h**

μ_{set}: Adjusted specific growth rate [h^{-1}] = *0.1 h^{-1}**

m_E: Maintenance coefficient [g/g/h]

$Y_{X/S}$: The biomass/substrate yield coefficient [g/g] = *0.52 g/g**

X_F: The biomass concentration at the start time of feeding phase [g] = *5.3 g/L**

V_L: The culture volume [L] *(start with 8 at the time of feeding)**

(*) data obtained from the previous batch culture.

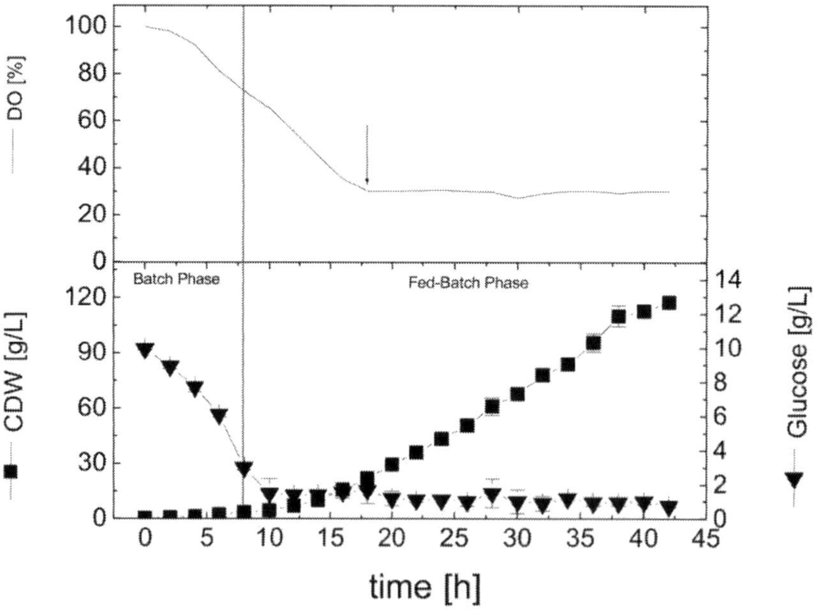

Figure 2. High Cell Density Cultivation (HCDC) of *S. boulardii* in 15-L STR. Arrow shows the time of MFCS operation by increasing of agitation/aeration to keep DO value at 30% saturation.

The feeding solution was composed on concentrated glucose/yeast extract solution in balanced ratio based on the initial medium composition. To prevent oxygen limitation in culture, DO was set at 30% and cascaded to aeration/agitation of the bioreactor using supervisory control system (MFCS, Sartorius BBI system GmbH, Melsungen, Germany). Figure 2 shows typical profile for high cell density cultivation of *S. boulardii* in 15-L stirred tank bioreactor. This bioprocess was successfully scaled up to 300-L bioreactor. 15-L and 70-L bioreactors were used for two stages inoculums preparation and the final 300-L bioreactor was inoculated with inoculums size of 15% (v/v) to reduce the time of the initial batch phase to only 6 hours. When conducting fed-batch cultivation in large bioreactor according to the proposed exponential feeding strategy, the biomass obtained was between 100-110 g/L with insignificant production of alcohol. The cell yield obtained in this process was almost the same or even higher than those obtained by many authors for HCDC of other yeast probiotics such as *S. cerevisiae* (10) and *Kluyveromyces* strains (11).

Industrial Platform for large scale production of dried active cells of *S. boulardii*

Based on the obtained results of high cell density cultivation of this important probiotic/biotherapeutic yeast strain, complete industrial platform for this process was developed. Figure 3 shows simplified industrial platform for large scale production of probiotic yeast. Process started by activation of cells, stored in working cell bank in glycerin culture at -80 °C, by one step cell cultivation in shake flask for 24 hours. The obtained cells were used to inoculate 15-L bioreactor. After 15 hours cultivation, the obtained cells were transferred to 70-L bioreactor as intermediate inoculums propagation step for only 10 hours before final transfer to 300-L STR. Based on this inoculums propagation strategy, the inoculums was in highly active form and thus fed-batch phase started after only 5-6 hours. In this culture maximal cell mass of about 110 g/L was obtained. After fermentation, cells were harvested by CIP/SIP self clean separator with and the obtained cells were re-suspended and centrifuged again for complete cell washing. Two times washing of cells were enough for complete removal of medium residues before spray drying. After spray drying, cells viability was tested using plate count method. For this type of cells, the average loss in cell viability after spray drying was between 10 to 20% in cell viability.

Conclusions

This study demonstrated complete, efficient and economic industrial platform for large scale production of the biotherapeutic/probiotic yeast. Combination between the improvement of cell tolerance to heat and dryness and improved cell mass production in large scale make this process attractive to reduce the production cost in both upstream and downstream.

Figure 3. Simple industrial platform for biomass production of the biotherapeutic yeast *Saccharomyces boulardii.* [1] Working cell bank; [2] Shake flask; [3] 15-L Bioreactor ; [4] 70-L Bioreactor; [5] 300-L Bioreactor; [6] Self clean separator; [7] Washing tanks; [8] Spray dryer ; [9] Finished product in powder form.

References

1. Buts, JP, Bernasconi P, Vaerman JP, Dive C. (1990). Stimulation of secretory IgA and secretory components of immunoglobulin in small intestine of rats treated with *Saccharomyces boulardii*. *Dig. Dis. Sci.* **35,** 251-256.

2. Elmer GW, Surawicz CM, McFarland LV. (1996). Biotherapeutic agents: a neglected modality for the treatment and prevention of selected intestinal and vaginal infections. *JAMA* **275,** 870-876.

3. Czerucka D, Dahan S, Mograbi B, Rossi B, Rampal P. (2000). *Saccharomyces boulardii* preserves the barrier function and modulates the signal transduction pathway induced in enteropathogenic *Escherichia coli*-infected T84 cells. *Infection and Immunity,* **68,** 5998-6004.

4. Buts, J.P. and De Keyser, N. (2006). Effects of *Saccharomyces boulardii* on intestinal mucosa. *Dig. Dis. Sci.* **51,** 1485-1492.

5. Rodrigues ACP, Nardi RM, Bambirra EA, Viera EC, Nicoli JR. (1996). Effect of *Saccharomyces boulardii* against experimental oral infection with *Salmonella typhimurium* and *Shigella flexneri* in conventional and gnotobiotic mice. *J. Appl. Bacteriol.* **91,** 251-256.

6. McFarland, L.V. (1998). Epidemiology, risk factors and treatments for antibiotic-associated diarrhea. *Dig. Dis.* **16,** 292-307.

7. Szajewska H, Mrukowicz J. (2005). Meta-analysis: non-pathogenic yeast *Saccharomyces boulardii* in the prevention of antibiotic-associated diarrhea. Aliment *Pharmacol. Ther.* **22,** 365-372.

8. Korz DJ, Rinas U, Hellmuth K, Sanders EA, Deckwer W-D. (1995). Simple fed-batch technique for high cell density cultivation of *Escherichia coli*. *J. Biotechnol.* **39,** 59-65.

9. El Enshasy H. (1998). Optimization of glucose oxidase production by recombinant *Aspergillus niger*. Ph.D. thesis. Technische Universität Carolo-Wilhelminia, Braunschweig.

10. Knoll A, Bartsch S, Husemann B, Engel P, Schroer K, Ribeiro B, Stöckmann C, Seletzky J, Büchs J. (2007). High cell density cultivation of recombinant yeasts and bacteria under non-pressurized and pressurized conditions in stirred tank bioreactors. *Journal of Biotechnology*. **132,** 167-179.

11. Pinheiro R, Belo I, Mota M (2000). Air pressure effects on biomass yield of two different *Kluyveromyces* strains. *Enz. Microb. Technol.* **26,** 756-762.

(531d) Effect of Ionic Liquid Pretreatment On Acid Hydrolysis of Cordyceps Polysaccharide

LI Na, WANG Zhao-mei and **Xiao Kaijun**, South China University of Technology, Guangzhou, China

Effect of pretreatment with Ionic liquid 1-butyl-3-methylimidazolium chloride ([BMIM]Cl) on acid hydrolysis of cordyceps polysaccharide was examined by polarized optical microscope, X-ray diffraction(XRD), FTIR and 3,5-dinitrosalicylic acid method. The polarizing area of cordyceps polysaccharide gradually disappeared with the increase of pretreatment time in Ionic liquid. Results showed that the hydrogen bonding of polysaccharide molecular was weakened and the crystallinity of cordyceps polysaccharide declined dramatically with [BMIM]Cl pretreatment. Meanwhile, the reducing sugar content increased with the increase of acid hydrolysis time, and then decreased after 4 hours. The yielding rate of the reducing sugar for cordyceps polysaccharide with [BMIM]Cl pretreatment increased 44 % more than that of the untreated sample in 4 mol/L trifluoroacetic acid at 100°C for 4h, indicating that [BMIM]Cl pretreatment could significantly improve the hydrolysis rates of cordyceps polysaccharide. In our study, [BMIM]Cl pretreatment could break the orderly structures and hydrogen bonding of cordyceps polysaccharide, and make it easy to enhance the efficiency of hydrolysis.

(531e) Pilot Scale Chromatographic Separation of Poly Unsaturated Ethyl Esters Using Liquefied Gases and Supercritical Fluids

Steve J. Tallon, Fernando Montanes, Kevin Mitchell and Paul Rose, Industrial Research Ltd, Lower Hutt, New Zealand

This paper describes the scale-up, development and commissioning of a pilot scale preparative chromatography plant suitable for use with liquefied gases and supercritical fluids as the mobile phase, including the use of flammable gases. The pilot plant has 2 by 10L columns capable of operation up to 350 bar pressure and 85°C temperature. The columns can be operated in series or independently, and in single or semi-continuous automated operation. A UV-Vis detection system is used, and up to 4 fractions can be separately recovered. A co-solvent gradient can also be applied.

A laboratory scale process for separation of the polyunsaturated fatty acids EPA and DHA from mixtures of ethyl esters derived from fish and algal sources has been developed, and a comparison of results between laboratory and pilot scale is given. Very good separation, exceeding 90% concentration of EPA and DHA, and very good recovery was able to be achieved in a single pass operation using CO_2 as the mobile phase and a silica column. Conditions were optimised over the range 150 to 250 bar pressure, temperatures from 40 to 60°C, and using a range of ethyl ester and CO_2 flow rates. Results were able to be reproduced on the pilot scale using loadings between 10g and 100g of ethyl esters.

(531f) Scheduling Considerations in Process and Package Operations

Charles Siletti, INTELLIGEN, INC., Mount Laurel, NJ and Demetri P. Petrides, INTELLIGEN, INC., Scotch Plains, NJ

A number of food preparation processes follow a simple sequence of blend, process and package. While simple from a processing perspective, such processes can pose scheduling challenges. The products of these processes are generally defined by material type and packaging size. The blending is often done in batches, while the packaging is semi-continuous. Equipment cleaning times depend on the product type sequence, e.g. chocolate to vanilla requires more cleaning than vanilla to chocolate, and equipment setup-times and changeover times generally depend on the packaging size, e.g. 16 oz. bottles to 32 oz. bottles. This presentation shows how organizing the products into type families and packaging sizes can keep the sequence-dependent durations manageable. Further, it is sometimes convenient to partially decouple blending and processing from the packaging. These concepts are illustrated through the following three examples: ice-cream manufacture, yogurt production, and bottled beverage manufacture. Finally some aspects of managing real-time schedule upkeep are addressed.

(550a) Drug Encapsulated Polymeric Microspheres for Temporally-Staged, Localized Brain Tumor Therapy

J. Alaina Floyd[1], Anna Galperin[2], Rohan Ramakrishna[3], Robert Rostomily[3] and Buddy Ratner[4], (1)Chemical Engineering, University of Washington, Seattle, WA, (2)Bioengineering, University of Washington, Seattle, WA, (3)Neurological Surgery, University of Washington, Seattle, WA, (4)Chemical and Bioengineering, University of Washington, Seattle, WA

The American Cancer Society 2012 estimate is that 22,910 malignant tumors of the brain or spinal cord will be diagnosed in the United Sates. Approximately 59.8% will be fatal. Despite current treatments, survival remains low with a median survival of approximately one year despite maximal therapy. Complete tumor removal cannot always be achieved because of the potential for brain damage, leading to cancerous regrowth and ultimately, patient death. In addition, these tumors can develop drug resistance, rendering chemotherapy ineffectual. We believe that the development of a topical, slow release, multi-drug delivery system (M-DDS) that is applied post-surgically directly in the tumor cavity could significantly increase life expectancy. Our M-DDS consists of multi-drug loaded poly(lactic acid) (PLA), poly(lactic-co-glycolic acid) (PLGA), and poly(ε-caprolactone) (PCL) microspheres suspended in an aqueous solution of a degradable, thermoresponsive poly(*N*-isopropylacrylamide) (PNIPAM) synthesized in our laboratory. At room temperature, PNIPAM suspends drug encapsulated microspheres that can be directly "sprayed on" the post-surgical site. The 37°C temperature of the brain would pass PNIPAM through its lower critical solution temperature, causing the polymer matrix to solidify and form an adherent gel layer with entrapped drug loaded microspheres on the surface of the brain. This would provide intimate contact with the remaining tumor cells, conforming to the convoluted morphology of the brain surface. Over time PNIPAM would degrade along with PLGA, PLA, and then PCL microspheres, releasing multiple chemotherapeutics at different rates directly to the tumor remnants to inhibit cancerous re-growth. This work focuses on microsphere characterization, model drug (rhodamine B (RB)) and chemotherapeutic encapsulation, and RB release from microspheres and microspheres entrapped in PNIPAM.

Both double and single emulsions have been applied to form PLA, PCL, and PLGA microspheres. Smooth, nonporous microspheres were produced with an adjustable size dependent on impeller speed during the emulsification process. RB, a hydrophilic dye and the model drug, was successfully encapsulated within PLA, PCL, and PLGA based microspheres and microsphere morphology remained unaffected. The encapsulation efficiency was polymer dependent; PLGA was most efficient at RB incorporation ($84.6 \pm 4.8\%$) due to its hydrophilic nature. An *in vitro* release from PLGA microspheres alone and PLGA microspheres entrapped in degradable PNIPAM was characterized in PBS at 37 °C. A biphasic release profile was seen for both conditions, with an initial, moderate, continuous release (passive diffusion) followed by a significant increase in the release curve slope (polymer degradation) before plateauing. Microspheres entrapped in PNIPAM had a two day delay of release compared to microspheres alone, with the release plateau occurring at day 22 versus day 20. In both instances, smaller microspheres (22.5 ± 7.8 μm) released faster than the larger microspheres (42.2 ± 15.4 μm) during the diffusion period due to a higher surface area to volume ratio, showing release profile tunability.

Our M-DDS has also been tested to visualize local and topical delivery on brain tissue. RB encapsulated PLGA microspheres suspended in degradable PNIPAM were "sprayed on" a rat brain that was preserved at 37 °C. The PNIPAM solution underwent a phase transition on the rat brain, showing stable adhesion to the tissue and efficient entrapment of drug loaded microspheres to a localized site (Figure 1).

The chemotherapeutic gefitinib has also been successfully encapsulated in PLGA microspheres (Figure 2). These gefitinib loaded PLGA microspheres will be tested against a C6 glioma cell line to explore drug efficacy along with PCL microspheres encapsulating the chemotherapeutic temozolomide. A pilot animal study will be conducted to test the effect of the delivery system on a rat model and the diffusion of the chemotherapeutics through the brain. Our M-DDS has the potential to provide a localized, tunable time release of multiple chemotherapeutics that will directly impact tumor remnants to inhibit cancerous re-growth, overcome tumor drug resistance, and increase patient life expectancy.

a) b)

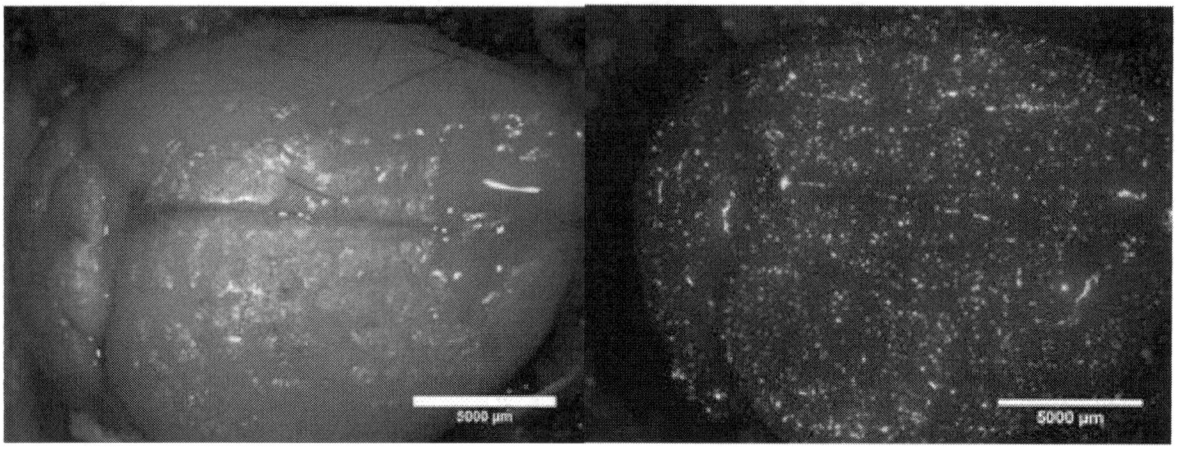

Figure 1: Sprayed RB encapsulated PLGA microspheres entrapped in degradable PNIPAM on a rat brain at 37 °C viewed with a) bright field and b) fluorescent microscopy

a) b)

Figure 2: Gefitinib encapsulated PLGA microspheres viewed by a) fluorescent and b) scanning electron microscopy

(550b) Redox Responsive Polymeric Nanocapsules for Protein Delivery

Muxun Zhao, Chemical and Biomolecular Engineering, UCLA, Los Angeles, CA and Yi Tang, Chemical and Biomolecular Engineering, University of California, Los Angeles, Los Angeles, CA

Virtually all human cancer cells have elaborate anti-apoptotic strategies to overcome apoptosis, which is a vital anti-neoplastic mechanism to obstruct tumor progression. Therefore, the ability to rapidly resurrect the apoptosis circuitry in tumor cells can be a highly effective option for cancer treatment. Although different strategies have been developed to counter the individual inactivation mechanisms, a more potent chemotherapy option is to directly arm the cancer cells with executioner proteins or apoptosis-inducing proteins that are not targeted by these aforementioned anti-apoptotic maneuvers. From a therapeutic perspective, protein-based approaches are safer than gene therapy because no random or permanent genetic changes are involved, and only transient actions of proteins are needed for the desired results. However, compared to small molecule therapeutics, proteins suffer from serum instability and are inefficient or unable to penetrate the cell membrane. As a result, protein-based antitumor options still remain elusive, despite its numerous therapeutic targets and vast potential. We have developed a simple and effective method of delivering proteins intracellularly. Using this method, the avian virus-derived protein Apoptin, which is a selective killer of various tumor cells, have been delivered to cancer cells in cell culture and in xenograft mouse tumors.

Apoptin is a small protein (121 amino acids) isolated from chicken anemia virus (CAV) that induces p53-independent apoptosis in a tumor-specific way. In a variety of tumor cell lines, Apoptin becomes phosphorylated, enters the nucleus, and induces apoptosis. In sharp contrast, Apoptin is unphosphorylated in normal cells and stays in the cytoplasm. An important feature of Apoptin is that it can recognize early stages of oncogenesis and it can induce apoptosis. Currently known Apoptin targets include DEDAF, Nur77, Nmi, and Hippi, some of which are p53-independent signaling proteins in the apoptotic pathway. Due to its high selectivity and potency, Apoptin has become an attractive antitumor target for gene therapy approaches.

To deliver apoptin protein directly, we first encapsulated recombinant maltose-binding-prortien (MBP) apoptin fusion protein in a degradable polymeric nanocapsule. The nanocapsule is formed by first electrostatically depositing positively charged monomers on to the surface of the MBP-apoptin complex (MBP-apoptin forms a 30-40 mer multimeric complex with an estimated size of 35 nm). Followed by addition of a redox-sensitive crosslinker, polymerization is initiated to form a thin, positively charged polymer layer that encapsulates the MBP-apoptin. The particles are found to be uniform in size and charge. When added to cancer cell cultures, the nanocapsule readily penetrated the cell membranes, disintegrated to release the MBP-apoptin and induced rapid apoptosis of the cells. In contrast, the nanocapsule had no apoptotic effect on primary cell lines such as HFF, highlighting the differential targeting on tumor cells only. Confocal microscopy of rhodamine labeled MBP-Apoptin showed that the protein entered the nucleus of cancer cells only, while remained in cytosol of HFF. Additional controls using nondegradable nanocapsules and apoptin mutants were performed to confirm the proposed mechanism of action of the nanocapsulated MBP-apoptin. When injected intratumorally into mice carrying xenografted MCF7 tumor. The nanocapsule significantly inhibited the progression of tumor growth in comparison to the controls, demonstrating the potential utility of this polymer-protein complex as a cancer therapeutic.

(550c) Stabilization of Pneumoccoccal Surface Protein A in Polyanhydride Nanoparticles: Consequences for the Design of a Pneumonia Vaccine

Shannon Haughney[1], Latrisha Petersen[1], Janice King[2], Amanda Ramer-Tait[3], Amy Schoofs[3], David Briles[2], Michael J. Wannemuehler[3] and Balaji Narasimhan[4], (1)Chemical and Biological Engineering, Iowa State University, Ames, IA, (2)Department of Microbiology, University of Alabama Birmingham, Birmingham, AL, (3)Department of Veterinary Microbiology and Preventive Medicine, Iowa State University, Ames, IA, (4)Department of Chemical and Biological Engineering, Iowa State University, Ames, IA

Pneumoccoccal surface protein A (PspA) is a choline-binding protein found on the surface of all pneumococcal strains and has been shown to be critical to the virulence of infections with *Streptococcus pneumoniae*. PspA plays two different roles in invasive infection and nasopharyngeal carriage. In invasive, systemic infections with *S. pneumoniae*, PspA prevents the deposition of complement, an immune protein that assists antibodies and phagocytic cells. PspA also plays a role in carriage by binding the bactericidal glycoprotein, apolactoferrin. Vaccination with PspA has been shown to be protective against a lethal challenge with *S. pneumoniae* making it a promising candidate for use in vaccines. Nanoparticles and microparticles made from polyanhydrides have been shown to successfully preserve antigenicity of proteins throughout the manufacture, storage, and release steps. These polymers have excellent biocompatibility and have been shown to have non-toxic, non-mutagenic degradation products. Another important benefit that leads to increased stabilization of protein is the use of a non-aqueous particle fabrication process, which not only prevents protein degradation during exposure to water, but also the loss of encapsulation efficiency observed in aqueous fabrication processes. Amphiphilic polyanhydrides are ideal candidates for vaccine delivery because of their ability to: stabilize and protect fragile antigens; provide a sustained release of antigen; and modulate the immune response. The encapsulation and release of pneumococcal surface protein A from four polyanhydride chemistries based on sebacic acid (SA), 1,6-bis-(*p*-carboxyphenoxy)hexane (CPH) and 1,8-bis-(*p*-carboxyphenoxy)-3,6-dioxaoctane (CPTEG) demonstrated sustained drug delivery over a month. The protein released from amphiphilic chemistries was shown to retain its structure through a combination of SDS-PAGE and circular dicroism. The antigenicity of protein released from amphiphilic polyanhydride nanoparticles was preserved as confirmed with ELISA. We also assessed the preservation of the biological functionality of PspA through a functional assay which takes advantage of the ability of secretory PspA to bind to apolactoferrin and prevent its bactericidal effects. In these studies, we used a PspA deficient strain of *streptococcus pneumoniae* and demonstrated that PspA released from amphiphilic polyanhydride nanoparticles retained the ability to bind to apolactoferrin and prevent the killing of PspA deficient *S. pneumoniae*. Together, these studies provide a framework for the rational design of a pneumonia vaccine based on amphiphilic polyanhydride nanoparticles.

(550d) Oral Mucosal Vaccination Using Coated Microneedles

Yunzhe Ma, Wenqian Tao and Harvinder Gill, Chemical Engineering, Texas Tech University, Lubbock, TX

The oral cavity has significant potential as a mucosal vaccination site because it is easily accessible and is immunologically attractive due to the presence of several lymphoid tissues in the oral cavity that could potentiate a robust immune response. Despite these advantages, the oral cavity remains under-utilized for vaccination due to the difficulty of delivering vaccine antigens across the tough transport barrier formed by the stratified epithelium covering the oral cavity tissues and the poor residence time due to salivary flow. We present the first study demonstrating the use of coated microneedles for enhancing vaccine delivery to the oral cavity using a rabbit animal model. In this study, we tested the ability of coated microneedles to penetrate the oral cavity tissues, specifically the inner lower lip and the dorsal surface of the tongue. Penetration of microneedles into tissues was confirmed by histological analyses. Delivery efficiency of coatings was found to be 63.9% and 91.2% for the lips and tongue respectively despite the moist environment in the oral cavity. Ovalbumin-coated microneedles were inserted into the lip and the tongue of rabbit to evaluate the immune response by direct delivery of antigens to the oral cavity tissues. Significantly higher ($p<0.05$) ovalbumin-specific IgG antibodies were found in serum. However, the increase in mucosal IgA antibody secretions was marginal ($p>0.05$). In addition, no significant difference ($p>0.05$) was observed between the lip or the tongue as the site for vaccination. In conclusion, this study provides the first validation of the use of microneedles for delivery to the oral cavity with specific focus on vaccination to induce both systemic and mucosal immunity using antigen-coated microneedles as minimally invasive devices for oral cavity vaccination. This methodology could form the basis of delivery of safe adjuvants to the oral cavity to further potentiate mucosal immunity.

(550e) Transdermal Delivery of Biopharmaceuticals Using Microsecond Thermal Ablation

Jeong Woo Lee, Chemical and Biomolecular Engineering, Georgia Institute of Technology, Atlanta, GA, Priya Gadiraju, Georgia Institute of Technology, Jung-Hwan Park, Bionano Technology, Gachon University, Seongnam, South Korea, Mark G. Allen, Georgia Institute of Technology, Atlanta, GA and Mark R. Prausnitz, School of Chemical and Biomolecular Engineering, Georgia Institute of Technology, Atlanta, GA

SEQ CHAPTER \h \r 1**TRANSDERMAL DELIVERY OF BIOPHARMACEUTICALS USING MICROSECOND THERMAL ABLATION**

Jeong Woo Lee[1], Priya Gadiraju[1], Jung-Hwan Park[3], Mark G. Allen[1,2], Mark R. Prausnitz[1]

[1]School of Chemical and Biomolecular Engineering, Georgia Institute of Technology, Atlanta, GA

[2]School of Electrical and Computer Engineering, Georgia Institute of Technology, Atlanta, GA

[3]Department of BioNano Technology and Gachon BioNano Research Institute, Gachon University,

Sungnam, Republic of Korea

Introduction

The conventional transdermal patch has provided advantages of non-invasive and simple administration, but it is generally limited to only small molecular weight and lipophilic therapeutics due to the protective function of the skin, which mostly resides in the top superficial layer of epidermis, stratum corneum. Macromolecular biopharmaceuticals cannot be delivered through the skin at therapeutic rates due to the stratum corneum barrier, so that various technologies have been investigated to overcome the skin barrier property such as mechanical disruption, radio frequency-based heating, ohmic heating, and laser ablation. The key to success with skin ablation technology is to develop methods to remove the stratum corneum barrier selectively without causing deeper skin damage [1]. Here, we present transdermal delivery of biopharmaceuticals with microsecond thermal skin ablation technology assessed by modeling of heat penetration into the skin, fabrication and characterization of a novel thermal ablation device, and skin permeability measurement.

Methods and Materials

Heat penetration into the skin was simulated using ANSYS software employing a one-dimensional transient semi-infinite model with the boundary condition [2] that the temperature of the skin surface was held constant at various temperatures up to 1100 ° during the thermal exposure of 1 μs to 100 ms. Simulations were carried out with and without a conductive mask used for guiding thermal energy only and with and without a "windows" mask used for localized ablation effect as shown in Fig. 1.

The thermal ablation device was designed by modifying a laminated gas generator actuator developed previously for aeronautical purposes [3]. A microchamber consisting of five patterned and integrated layers (two brass electrodes, two PMMA backing layers, and one microchamber layer) was fabricated using laser micromachining and lamination of each layer as shown in Fig. 2A. Two different mask designs were fabricated to facilitate mechanistic study and device optimization. A conductive mask was designed to efficiently conduct heat without allowing physical contact between the steam and the skin. A windows mask was designed to allow heat transfer to the skin through 100 μm-diameter holes in the mask and to insulate the skin everywhere else.

For the time characterization of thermal ablation, two voltage driver circuits were placed on either side of the microchamber to measure the voltage drop across the microchamber and the MOSFET switch. A ring-type piezoelectric force sensor was attached behind the microchamber to measure the force exerted by the actuation of the device. During device actuation, in some cases a high-speed video was taken at a rate of 100,000 frames per second (EKTAPRO 4540, Kodak).

To visualize the selective and localized removal of stratum corneum, a piece of full-thickness pig cadaver skin prepared with IACUC approval was ablated and then a hydrophilic sulforhodamine solution was applied to the skin surface. To measure the permeability of thermally abated skin, human cadaver skin was obtained with approval from the Georgia Tech Institutional Review Board and either left intact or exposed to thermal ablation. The skin permeability was measured for up to 48 h using a Franz diffusion cell, the donor compartment of which was filled with either 10^{-3} M sulforhodamine B or 10^{-4} M Texas Red-labeled bovine serum albumin (BSA) in PBS.

Results and Discussion

Our goal was to identify the longest thermal exposure that does not allow significant heat to reach the viable epidermis using finite element modeling. In Case I shown in Fig. 1A, when the skin is directly exposed to superheated steam at 1100 °, the viable epidermis is predicted to have no significant temperature rise after exposures of 1 ms, 10 ms and 100 ms as shown in Fig. 1B. Because heat transfer is most delayed in the stratum corneum layer with the lowest thermal conductivity of all the skin layers, heat barely penetrates into the viable epidermis during these short exposures. After a 1 ms exposure, the viable epidermis is predicted to be heated to a temperature of 180 °, which is expected to injure cells and cause pain as shown in Fig. 1E.

The mask design used in this study as in Case II and III shown in Fig. 1A permits sufficient heat transfer that the skin surface temperature is almost the same as the ejectate temperature after millisecond exposures, but is significantly lower after microsecond exposures as shown in Fig 1C and 1D. These data suggest that heating the skin for on the order of 100 ms may be optimal because that exposure time is long enough to allow the skin surface to be substantially heated, but still short enough to avoid heating below the stratum corneum.

SHAPE * MERGEFORMAT

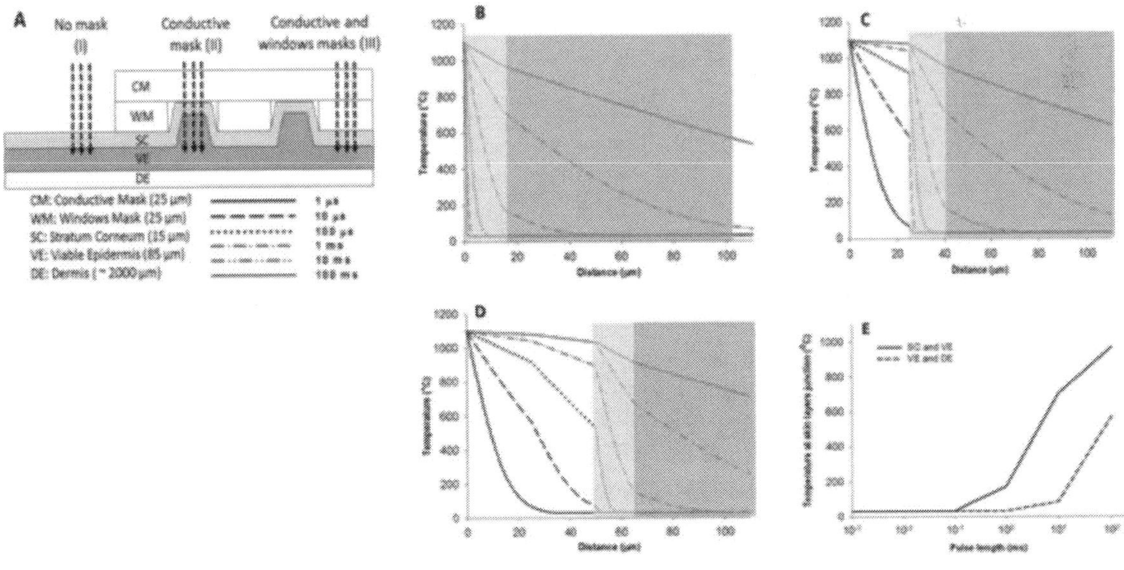

Figure 1. Simulation of heat penetration across skin heated on its surface to 1100°C for 1 ms to 100 ms. (A) Schematic showing skin heating with different schemes. Simulation predictions for heating (B) as direct exposure, (C) through a conductive mask and (D) through a conductive and a windows mask. (E) The time required to increase skin temperature by 10°C at the interface.

Guided by the heat transfer simulation, we designed a thermal ablation device to achieve thermal exposures on the order of 100 μs as shown in Fig. 2. After the microchamber is connected to a power supply circuit (Fig. 2B), the electrical energy stored in the capacitor (Fig. 2C) is discharged rapidly with the MOSFET switch on, generating an electrical discharge between the two electrodes in the microchamber. This nearly instantaneous event heats, vaporizes and ejects the saline from the microchamber through the nozzle, indicating the rapid transformation of electrical energy into thermal and mechanical energy that can be used for skin ablation.

We found that the thermal ablation device operates on a timescale on the order of 100 μs. The video image shown in Fig. 3A is the image of ejectate captured in a single frame, indicating that the steam jet was ejected within 10 μs at the most. As shown in Fig. 3B, the voltage across the microchamber rapidly rises and peaks at 7 μs, which is followed by a sudden voltage drop until 50 μs, which is the primary electrical discharge occurring in the microchamber. As shown in Fig. 3C, the force exerted during electrical discharge peaks at 100 μs, which corresponds to the timescale over which the primary steam ejection occurs. The force curve then oscillates due to resonance of the apparatus and dissipation of the reaction force energy, but is not believed to reflect forces generated directly by the steam jet ejection.

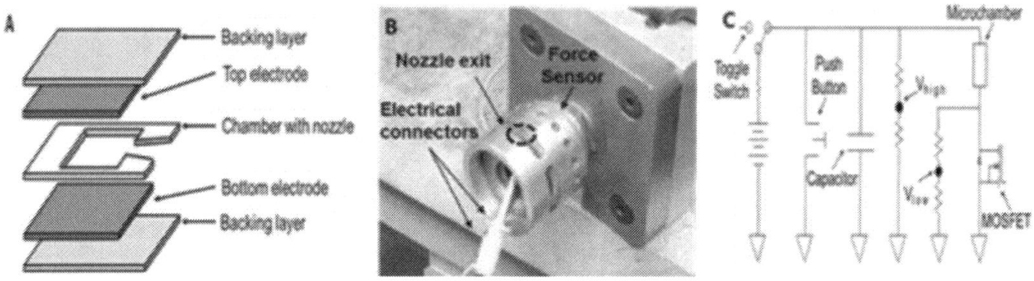

Figure 2. Microsecond thermal ablation system. (A) Schematic diagram of the disintegrated layers of the microchamber, (B) Image of a microchamber connected to a power supply circuit, (C) Diagram of the electrical circuit used to transform electrical energy into thermal and mechanical energy in the microchamber

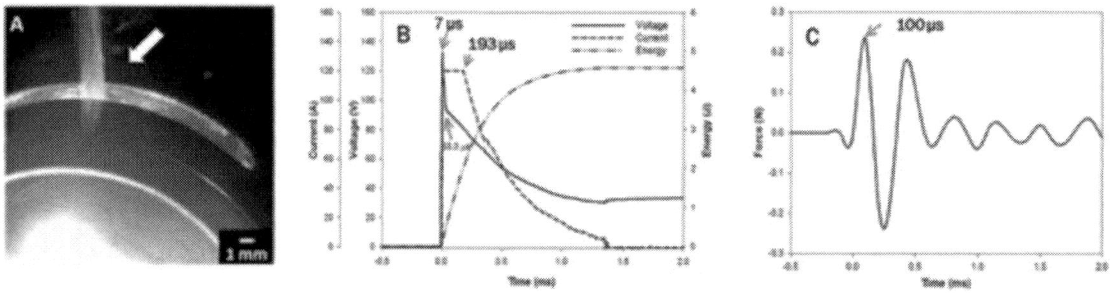

Figure 3. (A) View of the ejectate from the device, (B) Voltage change across the microchamber, and (C) Force measured during electrical discharge.

Guided by the design, fabrication, and characterization of microsecond thermal ablation system, we ablated the skin to cause selective removal of the stratum corneum as shown in Fig 4A and 4B. The selective removal of stratum corneum led to a remarkable increase in the transdermal flux of both molecules studied, as shown in Fig 4C and 4D. After a lag time, skin permeability to sulforhodamine and BSA was 3.1 (± 0.9) • 10^{-2} cm/h and 3.6 (± 0.8) • 10^{-3} cm/h, respectively. This corresponds to a 10^4-fold increase in permeability to sulforhodamine and a 10^3-fold increase in permeability to BSA after thermal ablation in comparison with intact and untreated skin.

SHAPE * MERGEFORMAT

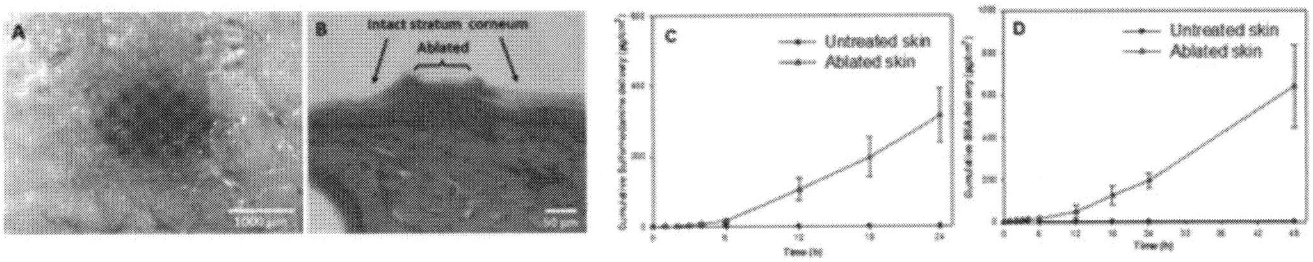

Figure 4. (A) The surface of skin ablated with microsecond thermal ablation device using a windows mask, (B) Histological image of the skin sample in (B), Cumulative delivery of (C) sulforhodamine and (D) Texas Red-labeled BSA.

Conclusion

Through modeling, we determined that thermal ablation should occur on the 100 µs time scale in order to ablate the stratum corneum selectively. To generate the high temperatures needed for skin ablation on this time scale, we designed a microdevice that rapidly heats water by an electrical discharge, thereby ejecting a superheated steam jet at the skin surface. We found that this device was able to selectively ablate stratum corneum and, when combined with a windows mask, enabled three-dimensional control over tissue removal, resulting in the increase of skin permeability to sulforhodamine and BSA. This study was funded by NIH and by Tyco Healthcare (now Covidien).

Reference

[1] J. W. Lee, P. Gadiraju, J. H. Park, M. G. Allen, M. R. Prausnitz, *J Control Release* **2011**, *154*, 58-68.

[2] M. L. Cohen, *Journal of Investigative Dermatology* **1977**, *69*, 333-338.

[3] B. English, in *School of Mechanical Engineering*, Vol. Doctor of Philosophy, Georgia Institute of Technology, Atlanta **2006**.

(550f) 28-Day Ocular Delivery of Brimonidine Tartrate From Rationally Designed Degradable Microparticles In a Rabbit Model

Morgan Fedorchak[1,2,3,4], Jeremy Wingard[2], Carlos A. Medina[2], Eiyass Albeiruti[2], Joel S. Schuman[2,3,4,5] and Steven R. Little[1,4,5,6], (1)Chemical Engineering, University of Pittsburgh, Pittsburgh, PA, (2)Ophthalmology, University of Pittsburgh, Pittsburgh, PA, (3)Louis J. Fox Center for Vision Restoration, Pittsburgh, PA, (4)McGowan Institute for Regenerative Medicine, Pittsburgh, PA, (5)Bioengineering, University of Pittsburgh, Pittsburgh, PA, (6)Immunology, University of Pittsburgh, Pittsburgh, PA

Glaucoma is the second leading cause of blindness worldwide, expected to affect up to 3 million Americans by 2020 (Quigley HA. Lancet. 2011;377[9774]:1367-77). One of the main risk factors in glaucoma is an unsafe increase in intraocular pressure (IOP). IOP reduction in patients with glaucoma is typically accomplished through the administration of medicated eye drops several times daily, the difficult and frequent nature of which contributes to patient compliance rates estimated to be as low as 50% (Hermann MM. Int Ophthalmol. 2010;30[4]:385-90). Medications such as brimonidine tartrate (BT), which requires dosing every 8-12 hours, have yet to be adapted into controlled-release formulations that could drastically improve compliance. The purpose of this study was to develop and test a controlled release BT formulation that provides one month of therapeutic levels of drug, thereby eliminating compliance issues inherent to eye drop administration.

BT was loaded into poly(lactic-co-glycolic) acid (PLGA) microparticles, which were fabricated using a standard double emulsion procedure. Both blank and drug-loaded microparticles were characterized for surface morphology and volume average diameter. In vitro release of BT from the drug-loaded microparticles was quantified by incubating a known mass of microparticles in buffer and measuring the absorption using a spectrophotometer. For our in vivo studies, groups of 3 healthy Dutch belted rabbits were randomized to receive blank microparticles (no drug), BT-loaded microparticles, or 0.2% BT drops (Alphagan¨, Allergan, Irvine, CA). Rabbits in both microparticle groups received a subconjunctival injection of microparticles suspended in sterile saline on Day 0 in one eye only. Rabbits in the positive control group received a single drop of Alphagan¨ solution in one eye twice a day each day of the study. Intraocular pressure was monitored over 28 days in both eyes and slit lamp examination was performed periodically to assess the effect, if any, of the microparticles on the surrounding tissue. Following sacrifice on Day 28, both eyes were enucleated for histological analysis. All slides were masked prior to performing histopathological analysis for evidence of irritation or foreign body response.

Microparticles were confirmed to have a diameter of 7.5±2.9 μm with a primarily poreless morphology. They released an average of 0.62±0.34 μg BT/mg particles/day in our in vitro setup, within the calculated therapeutic range of Alphagan¨ drops. Our in vivo study of the BT-loaded microparticles demonstrated that the decrease in IOP was significantly lower ($p<0.05$) in the treated eye for BT drops versus BT microparticles for all time points. There was no significant difference in ΔIOP for the untreated eye between BT drops and BT microparticles. In contrast, IOP steadily increased in rabbits injected with the blank microparticles. The microparticle bleb was visible using light microscopy and no evidence of microparticle migration was seen. Histological analysis showed no signs of foreign body response due to the microparticles or migration of the microparticles from the periocular area.

The BT-loaded microparticles presented in this study are capable of releasing a therapeutically relevant amount of a common glaucoma medication for over 30 days. Additionally, treatments based on BTMPs demonstrate a significantly greater drop in IOP over the entire 28-day study compared to the baseline IOP in a healthy rabbit model. Subconjunctival injections of the microparticles resulted in neither foreign body response nor infection after the full 28-day experimental period, and microparticles were confined to the injection area. We are currently adapting this controlled-release BT delivery system into a simple, patient-administered formulation that provides the same IOP reduction potential and biocompatibility without the need for intraocular injections.

(550g) Preclinical Evaluation of Treg Recruiting Microparticles for the Treatment of Periodontitis

Andrew J. Glowacki[1,2,3], Sayuri Yoshizawa[4,5,6], Siddharth Jhunjhunwala[7], Gustavo P. Garlet[8], Charles Sfeir[2,4,6] and Steven R. Little[1,7,9,10,11], (1)Chemical Engineering, University of Pittsburgh, Pittsburgh, PA, (2)The Center for Craniofacial Regeneration, University of Pittsburgh, Pittsburgh, PA, (3)McGowan Institute for Regenative Medicine, University of Pittsburgh, Pittsburgh, PA, (4)Oral Biology, University of Pittsburgh, (5)The Center for Craniofacial Regeneration , University of Pittsburgh, (6)McGowan Institute for Regenative Medicine, University of Pittsburgh, (7)Bioengineering, University of Pittsburgh, Pittsburgh, PA, (8)Department of Biological Sciences, School of Dentistry of Bauru, University of Sao Paulo (FOB/USP), Bauru, Brazil, (9)McGowan Institute for Regenerative Medicine, Pittsburgh, PA, (10)Immunology, University of Pittsburgh, Pittsburgh, PA, (11)The Center for Craniofacial Regeneration, University of Pittsburgh

Periodontal disease (gum disease) affects an estimated 78 million Americans, and is the leading cause of tooth loss according to the American Dental Association[1]. Current treatments for periodontal diseases focus on the removal of invasive bacterial species through non-surgical scaling and root planing procedures (sometimes accompanied by antibiotics)[2]. Although bacteria initiate the disease, tissue destruction in the periodontium (ligament and tooth supporting bone) is perpetuated by the inflammatory immune response. Recently, a subset of lymphocytes known as regulatory T cells (Tregs) has been shown to play a critical role in the regulation of harmful inflammatory immune responses[3]. Furthermore, the presence or absence of Tregs has been correlated with the presence or absence of aggressive periodontitis[4]. Therefore, we hypothesized that resolution of inflammatory conditions through the recruitment of Tregs may alleviate periodontal disease symptoms and prevent alveolar bone loss.

In support of this hypothesis, here we show Treg recruitment to poly(lactic-co-glycolic) acid microparticles encapsulating CCL22 placed in the subgingival pocket, lead to the resolution of periodontal disease symptoms in both a murine and canine model. Specifically, upon injection of CCL22 loaded microparticles in the mouse model, we observed an increase in the expression of FOXP3 (Treg transcription factor), and anti-inflammatory cytokines IL-10 and TGFβ. Indeed, we also observed a corresponding decrease in the expression of inflammatory cytokines IL-1, TNF, and IFNγ, without affecting the protective host response (we observed comparable bacterial loads, Fig. 1A). Furthermore, in the most widely accepted preclinical model for periodontitis (dog model), we observed decreased clinical scores for inflammation, probing depth (Fig. 1B), and bleeding on probing. In addition, we observed a marked decrease in alveolar bone loss in the CCL22 treated animals (observed and quantified using microCT, Fig. 1C).

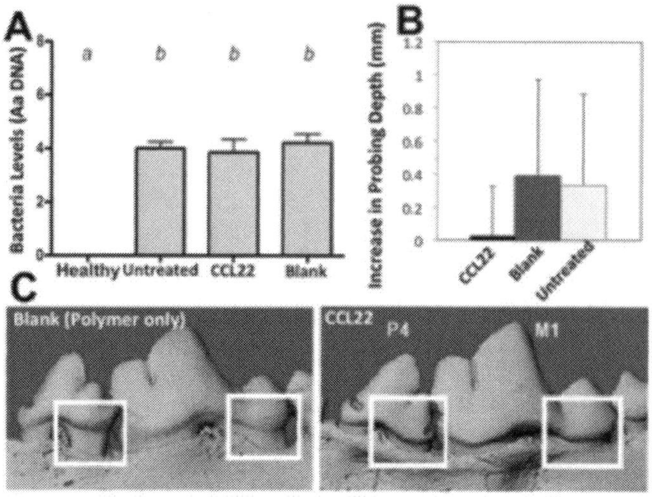

Figure 1: CCL22 Microparticles reduce periodontal disease symptoms, and maintain a protective host immune response. (A) Periodontal pathogen *Actinobacillus actinomycetemcomitans* DNA extracted from periodontium of mice. **(B)** Periodontal probing depths 4 weeks after disease induction in beagle dogs. **(C)** MicroCT 3-D reconstruction of the 4th premolar and carnassial of the left mandible of beagle dogs.

[1] P. I. Eke, R. J. Genco, *Journal of Periodontology* **2007**, 78, 1366.

[2] R. C. Williams, D. W. Paquette, S. Offenbacher, D. F. Adams, G. C. Armitage, K. Bray, J. Caton, D. L. Cochran, C. H. Drisko, J. P. Fiorellini, W. V. Giannobile, S. Grossi, D. M. Guerrero, G. K. Johnson, I. B. Lamster, I. Magnusson, R. J. Oringer, G. R. Persson, T. E. Van Dyke, L. F. Wolff, E. A. Santucci, B. E. Rodda, J. Lessem, *Journal of Periodontology* **2001**, 72, 1535.

[3] Q. Tang, J. A. Bluestone, *Nature Immunology* **2008**, 9, 239.

[4] G. P. Garlet, C. R. Cardoso, F. S. Mariano, M. Claudino, G. F. de Assis, A. P. Campanelli, M. J. Avila-Campos, J. S. Silva, *Journal of Clinical Periodontology* **2010**, 37, 591.

(550h) Development of Two Steps Tissue Allograft with Sequential Growth Factor Delivery to Promote Revascularization in Denudated Pneumonectomy Bronchial Stump

Eva M. Martín del Valle[1], Cristina Rodriguez Rivero[1], Gonzalo Varela[2], Dolores Ludeña[3], Marta Regueiro-Purriños[4] and Miguel A. Galán[1], (1)Department of Chemical Engineering, University of Salamanca, Salamanca, Spain, (2)Thoracic Surgery, Universitary Hospital of Salmanca, Salamanca, Spain, (3)Anatomic Pathology, Universitary Hospital of Salamanca, Salamanca, Spain, (4)Animal Surgery , University of Leon, Leon, Spain

Devascularization due to large lymph node dissection is one of the causative factors of bronchial stump fistula (BF). In this work we have developed and analyzed the utility of a tissue allograft composed of alginate polymer (AP), autologous fibrin (AF), tissue-engineered fibroblasts (TEF), vascular endotelial growth factor (VEGF) and encapsulated fibroblast growth factor (FGF) to prevent BF and promote revascularization of the bronchial stump in a pig pneumonectomy model. To validate the system a left pneumonectomy was performed in 15 pigs, leaving a long and exhaustively devascularized bronchial stump. The animals were allocated in groups of five to receive no bronchial stump coverage (group A), coverage with AP + AF only (group B) or coverage with AP + AF + TEF + VEGF + FGF (group C). After being euthanized on postoperative days 3,7, 14 , 21 and 30, bronchial stumps for each group were tested for BF occurrence and subsequently sectioned for histology and stained with hematoxylin-eosin, CD-31 and α-smooth muscle actin antibodies to evaluate blood vessels formation and maturation, capillary density and TEF proliferation by means of a semiquantitative method.

From the analysis done it can be concluded that: No BF could be detected for any group, probably due to a high spontaneous healing capacity of pig tissues. In group A, fibroblastic proliferation and revascularization were scarce; group B showed some immature neovessels besides a moderate TEF proliferation while in group C, in addition to an intense TEF proliferation, a large number of vascular yolks could be detected on postoperative day 7, continuing to complete vascular structures on postoperative day 14 within a highly dense capillary network. Therefore, the avascular scaffold highly promotes revascularization of the denudated bronchial stump. Further studies are being done to clarify the clinical application of these findings.

(557a) Cancer Cell Migration in 3D

Stephanie I. Fraley[1], Yunfeng Feng[2], Pei-Hsun Wu[3], Gregory Longmore[4] and Denis Wirtz[1], (1)Chemical and Biomolecular Engineering, Johns Hopkins University, Baltimore, MD, (2)Cell Biology, BRIGHT Institute, Washington University, St. Louis, MO, (3)Chemical and biomolecular Engineering, Johns Hopkins University, Baltimore, MD, (4)Medicine, Cell Biology, BRIGHT Institute, Washington University, St. Louis, MO

Although much is known about mechanical and spatial signal transduction in cells migrating on conventional two-dimensional (2D) substrates, drastic functional differences are apparent in the molecular machinery driving cell migration through more physiologically-relevant, fibrilar, three-dimensional (3D) environments. Taking an inside-out approach to understanding the function of cell adhesion proteins in 3D cell migration, we use shRNA constructs to silence expression of a variety of focal adhesion proteins and demonstrate that external dimensional cues are critical to protein localization and signaling. We show that 3D cell motility parameters are not predicted by the same parameters in 2D. Likewise, the focal adhesion protein zyxin plays a distinctive role in the temporal and dimensional aspects of cell trajectories in 3D and 1D microenvironments, but not on 2D substrates. Lastly, taking an outside-in approach, we demonstrate that changes in 3D matrix microstructure alone modulates modes of cell motility. This occurs via microstructure induced spatial positioning of protrusions and matrix metalloproteinase (MMP) activity. Thus, cell motility may only be abrogated by small molecule inhibition of MMPs in distinct matrix microstructures.

References

Fraley *et al*. Nature Cell Biology 12: 598-604 (2010)

Fraley *et al*. Nature Cell Biology 13: 5-7 (2011)

Fraley *et al*. Nature Communications 3:719-731 (2012)

(557b) Engineering Chimeric Antigen Receptors Targeting an Endogenous Murine Tumor Associated Antigen

Cary F. Opel[1], Matthias T. Stephan[2] and K. Dane Wittrup[1,3], (1)Chemical Engineering, Massachusetts Institute of Technology, Cambridge, MA, (2)Clinical Research Division, Fred Hutchinson Cancer Research Center, Seattle, WA, (3)Biological Engineering, Massachusetts Institute of Technology, Cambridge, MA

Adoptive cell transfer (ACT) therapy for the treatment of cancer is the act of administering lymphocytes to a patient with the goal of achieving an active immune response against tumor cells. A variety of methods have emerged to modify the lymphocytes in order to enhance their anti-tumor potency. One increasingly effective method is the use of chimeric antigen receptors (CAR) to target the lymphocytes specifically to tumor associated antigens. CARs are surface bound proteins consisting of an intracellular activation domain fused to an extracellular targeting domain. These receptors enable the lymphocytes to bypass many of the restrictions on the endogenous activation pathway, potentially allowing for a more effective immune response to the target tumor cells.

Despite many recent successes using lymphocytes expressing CARs in a clinical setting, some unforeseen side effects have been reported, and the design criteria for the most effective CAR domain structure have yet to be clearly established. Of critical importance to the design of a CAR is the selection of internal signaling domains. Weakly activating domains may not elicit a strong enough response to eradicate the cancer, while potent activating domains may cause an autoimmune response with severe side effects. Additionally, some activating domains may enhance other beneficial properties such as the survival, persistence, or proliferation of the transferred lymphocytes. First generations CARs used only the activating domain of CD3, while more recent second and third generation CARs combine two or three internal activating domains to enhance the effectiveness of the transferred lymphocytes. Unfortunately, the exact benefits and side effects of the various domain combinations remain unclear. This work aims to elucidate some of the design parameters that enable the most effective targeting of tumor cells by CAR expressing lymphocytes.

To study the effects of different activating domains, a model system was developed using the poorly immunogenic murine melanoma cell line, B16F10, and an antibody, TA99, targeting gp75, an endogenous tumor associated antigen expressed on the surface of B16F10 cells. The targeting domain of a CAR was created using an scFv constructed from the variable regions of the TA99 antibody. Preliminary cytotoxic assays using a first generation CAR showed promising results, so several CARs were constructed using various combinations of internal signaling domains and their effectiveness was tested *in vitro*.

Unlike many other model systems for adoptive cell transfer (ACT), this model will be expanded to use immune competent C57BL/6 mice. Typically, the effectiveness of CARs is tested on immunodeficient mice using human cell lines and targeting human antigens. In these models, off-target effects as well as interactions with the immune system are not evaluated, creating potentially misleading results. By using a fully endogenous system, a more accurate measurement of the effectiveness of various activating domains may be achieved, providing a more accurate prediction of how treatments will perform in a clinical setting.

(557c) Genetically Modified Bacteria Actively Secrete Recombinant Protein Biomarker for Early Detection of Solid Tumors

Jan Panteli, Chemical Engineering, University of Massachussetts Amherst, Amherst, MA, Britanny Forkus, Chemical Engineering, University of Massachussetts Amherst, Amhest, MA and Neil S. Forbes, Chemical Engineering, University of Massachusetts, Amherst, MA; Chemical Engineering, University of Massachusetts, Amherst, Amherst, MA

An attenuated strain of *Salmonella typhimurium* has been engineered to actively secrete *Zoanthus sp.* green fluorescent protein (ZsGreen) in a controlled manner for use in *in vitro* and *in vivo* tumor models for early detection of cancer. Current methods of cancer detection involve a primary screen for abnormal levels of native biomarkers, which might indicate cancer and other diseases. These preliminary screens often yield false results, requiring an invasive biopsy that is often painful and discomforting, to confirm diagnosis. Secretion of an exogenous recombinant protein from within tumor tissue allows cancer detection through noninvasive blood sampling without uncertainty in the results. It has been shown previously that systemically administered *Salmonella typhimurium* localize in solid tumor tissue 2000 fold more than in healthy tissue (Low *et al.* 1999). Once colonization of the bacteria in the tumor is achieved, triggered production and secretion of a protein biomarker is transported back to the blood stream for detection. Fluid transport in tumor interstitial space has been shown to have outward flowing pressure gradients hindering systemic administration of small molecule chemotherapeutics. These transport limitations act in favor of export of protein from within tumor tissue to the blood stream. A secreted biomarker, produced from the bacteria, within tumor tissue, would indicate the presence of cancer without invasive procedures and require a simple blood draw for detection. In this study a gram negative type III secretion system(T3SS) signal was fused to the n-terminus of ZsGreen protein (MvZsGreen) yielding a 3-fold increase in secretion in wild type strain (SL1344) and 5-Fold increase in an attenuated strain (VNP20009) of *S. typhimurium*. Over half of the protein produced is secreted from the attenuated strain (pBAD_MvZsG_VNP) under control of the L-arabinose inducible promoter (pBAD). Time-lapse fluorescent microscopy with viability staining revealed that a majority of cells (85%) are viable after 36hrs of culture and release protein at twice the rate than dead cells. Immunoblotting measurements and fluorescence measurements of secreted recombinant MvZsGreen diluted in blood was shown to have detection limits of 1:10 and 1:1000 respectively. A previously developed micro-fluidic 4T1 tumor model (Toley *et al.* 2011) was used to determine transport rates of protein through tissue and detection limit on the secreted protein from actual tumor tissue under triggered induction using the L-arabinose inducible system. Increased secretion from VNP20009 over the wild type *S.* typhimurium is extremely beneficial in reducing virulence while improving protein production and transport for detection. Viability studies revealed two mechanisms of protein release in culture with an increased rate in active secretion through T3SS compared to cell death related diffusion. This allows production and secretion of biomarker, MvZsGreen, as the bacteria proliferate within the tumor tissue, which is key for monitoring cancer growth for prolonged periods of time. MvZsGreen has higher sensitivity for detection via fluorescence than immunoblotting allowing for more accurate determination of the presence of a tumor mass. Micro-fluidic studies revealed favorable MvZsGreen transport through solid tumor tissue and subsequent detection. This system demonstrates the ability of bacteria to produce recombinant protein in a controlled manner deep within tissue, secrete that protein, and have the protein transport through tumor tissue to the blood stream for detection. These studies demonstrate the feasibility of a engineered bacterial tool for early detection and diagnosis of cancer.

(557d) A Decision Tool for the Design of Optimal Personalised Chemotherapy Protocols for the Treatment of Acute Myeloid Leukaemia (AML)

Eleni Pefani, Chemical Engineering, Imperial College of London, London, United Kingdom, Nicki Panoskaltsis, Department of Haematology, Northwick Park Hospital, London, United Kingdom, Athanasios Mantalaris, Centre for Process Systems Engineering, Department of Chemical Engineering, Imperial College, London, United Kingdom, Michael C. Georgiadis, Centre for Process Systems Engineering, Chemical Engineering, Imperial College London, London, United Kingdom and Efstratios N. Pistikopoulos, Chemical Engineering, Imperial College London, Centre for Process Systems Engineering, London, United Kingdom

AML is an aggressive cancer of the blood and bone marrow resulting in failure of the blood and immune systems. Immature white blood cells which are not able to develop into normal functioning blood cells are overproduced and build up in the bone marrow and blood. This inhibits the development of healthy blood and immune cells due to space restrictions and inhibitory and clonal factors specific to the disease (Panoskaltsis, 2003; Panoskaltsis, 2005).

The most common treatment for AML is intensive chemotherapy requiring high doses of anticancer drugs to eradicate cancer cells. Although effective for debulking of the leukaemia tumor burden, the use of these chemotherapy regimens is also highly destructive to normal tissues and cells, often to a life-threatening extent. Currently, the choice of chemotherapy drugs, dose and schedule only take few patient-specific factors into consideration and the choice of treatment often depends on the treating physician's experience. However, a more systematic approach for the design of treatment protocols that will depend on specific inter-patient and intra-leukaemia factors is required. This systematic treatment design will lead to drug individualization and treatment dose optimisation so as to balance the benefits of higher dose therapy against the increase in toxicity in normal tissue. As presented in previous work (Dua et. al., 2005; Dua et. al., 2008; Pefani et. al. 2011), this can be achieved by modelling the key biological mechanisms as a means to gain insight into the chemotherapy procedure, which can then be used as a predictive tool for patient response during treatment.

In this work, a mathematical model is derived by combining the actions on the cell cycle, (drug target), with pharmacokinetic and pharmacodynamic aspects. This combination should provide a comprehensive description of drug diffusion and action after administration. The derived model depends on patient characteristics such as age, sex, body mass index and on disease characteristics such as tumor burden at the point of diagnosis and individual AML cell cycle characteristics.

Simulation and optimisation results are presented using the derived model for two patient case studies. The studied patients are of same sex, age and body mass index characteristics. However, their initial tumour burden and cell cycle times differ. They are both treated with the same treatment protocol in the model as they would be in clinical practise (a combination of cytarabine and daunorubicin). At completion of the first chemotherapy cycle one patient displays a bone marrow which contains a higher number of normal cells, whilst the other exhibits a higher number of AML cells. An optimisation problem is formulated thereafter for the latter patient resulting in the normal cell population incrementing enough to allow for normal bone marrow tissue recovery prior to the next chemotherapy cycle to further decrease AML cells.

Model analysis results of these two patient case studies reveal the potential future applicability of the proposed tool and indicate the need for an interdisciplinary approach between clinicians, experimental- and model-based experts to collaborate in enabling personalised schedules for chemotherapy delivery in cancers such as AML.

References

Panoskaltsis N, Reid CDL, Knight SC. Quantification and cytokine production of circulating lymphoid and myeloid cells in acute myelogenous leukemia (AML). *Leukemia* 2003; 17:716-730.

Panoskaltsis N. Dendritic cells in MDS and AML – cause, effect or solution to the immune pathogenesis of disease? *Leukemia* 2005; 19:354-357.

Dua P., Dua V., Pistikopoulos E. N.: *Optimal delivery of chemotherapeutic agents in cancer*. Computers & Chemical Engineering 32(1-2): 99-107 (2008).

Dua P., Pistikopoulos E. N.: *Modelling and control of drug delivery systems*. Computers & Chemical Engineering 29(11-12): 2290-2296 (2005).

Pefani E., Panoskaltsis N., Mantalaris A., Georgiadis M. C., Pistikopoulos E. N.. Modelling and Simulation of Drug Delivery Systems for the treatment of Acute Myeloid Leukemia. Proceedings of the 5th European Conference of International Federation for Medical and Biological Engineering (IFMBE), Budapest, Hungary, 2011.

Acknowledgment

This work is supported by European Research Council (MOBILE, ERC Advanced Grant, No: 226462), The Richard Thomas Leukaemia Fund and the CPSE Industrial Consortium.

(557e) Pharmacokinetic/Pharmacodynamic Model Predicts the Response to Cancer Therapeutics Targeting VEGF

Stacey D. Finley and Aleksander Popel, Biomedical Engineering, Johns Hopkins University, School of Medicine, Baltimore, MD

Angiogenesis, the formation of new blood vessels from pre-existing vasculature, is a complex biological process involved in physiological conditions, as well as in pathological cases such as cancer. Vascular endothelial growth factor (VEGF) is a key regulator of angiogenesis and has been targeted in order to inhibit angiogenesis and prevent tumor vascularization. This study applies a molecular-detailed compartment model of VEGF kinetics and transport in the human body [1, 2] to investigate the effects of anti-angiogenic therapies targeting the VEGF pathway.

The model includes normal tissue (represented by skeletal muscle), blood, and tumor. We include molecular interactions between two major VEGF isoforms $VEGF_{121}$ and $VEGF_{165}$, receptors VEGFR1 and VEGFR2, and co-receptors NRP1 and NRP2. Macromolecular transport between compartments occurs via transendothelial microvascular permeability and lymphatic flow. Additionally, free VEGF is subjected to protein degradation in the tissue compartments, and species are cleared from the blood. VEGF receptors and co-receptors are localized on the abluminal and luminal endothelial surfaces, as well as on parenchymal cells. Receptor density is based on *in vitro* and *in vivo* quantitative flow cytometry [1, 3]. The model is represented by 125 ordinary differential equations and has been validated against experimental data.

We apply the model to predict the effects of therapeutics that inhibit VEGF or its receptors. We also investigate combination therapies and targeted delivery of the anti-angiogenic agents. The response to anti-angiogenic treatment is characterized by the level of free VEGF in the body and the number of VEGF/VEGFR2 complexes (considered to be pro-angiogenic). Following neutralization of VEGF with an antibody, plasma free VEGF and the number of VEGF/VEGFR2 complexes in the blood increase tenfold. The increase in plasma VEGF has been observed clinically. Interestingly, the model predicts that VEGF in the tumor interstitium can increase or decrease following administration of the VEGF antibody, depending on properties of the tumor microenvironment, including the relative rate at which $VEGF_{121}$ and $VEGF_{165}$ are secreted by the tumor and the expression level of the VEGF receptors on tumor cells. Additionally, we find that targeting VEGFR2 is effective in inhibiting the formation of pro-angiogenic complexes, while mediating the concomitant increase in free VEGF that can occur with inhibition of VEGF or VEGFR1.

We have developed a framework for investigating the systemic effects of anti-angiogenic drugs that inhibit VEGF-mediated signaling, which depend on tumor-specific properties and are difficult to quantify clinically. The model should be instrumental in optimizing personalized cancer treatment strategies that target the VEGF pathway.

This work was supported by NIH grant R01 CA138264 (ASP), NIH fellowship F32 CA154213 (SDF), and UNCF/Merck Postdoctoral Fellowship (SDF).

References

1. Finley, S.D., et al., *Pharmacokinetics and pharmacodynamics of VEGF-neutralizing antibodies.* BMC Systems Biology, 2011. **5**:193.

2. Finley, S.D. and A.S. Popel, *Predicting the effects of anti-angiogenic agents targeting specific VEGF isoforms.* AAPS Journal, 2012. In press.

3. Imoukhuede, P.I. and A.S. Popel, *Quantification and cell-to-cell variation of vascular endothelial growth factor receptors.* Experimental Cell Research, 2011. **317**(7):955-965.

An integrated systems-based modelling framework for investigating the effect of anticancer drugs on solid tumours

Cong Liu, J. Krishnan, Xiao Yun Xu

Department of Chemical Engineering, Imperial College London, UK

Abstract

Background. The effectiveness of clinical chemotherapy is dependent on the penetration of anticancer agents in tumour tissues and the responsiveness of tumour cells towards exposure to administered drugs. Most blood-borne anticancer drugs gain their access to tumour cells by going through a series of physical and biological processes, including convection in the tumour vasculature, extravasation across the capillary wall, penetration through the tumour interstitium and finally translocation inside tumour cells. However, these transport processes are often impaired by the abnormal tumour microenvironment (i.e. irregular tumour vasculature, elevated interstitial fluid pressure, and extracellular matrix composition), resulting in limited drug concentration available at the target site. Upon interaction with their intracellular targets, anticancer drugs induce multiple cellular signalling pathways, through which they exert therapeutic effects with apoptosis (programmed cell death) being of primary interest. The way by which cellular signalling functions is extremely complex as cellular pathways are carried out in a complex and interconnected network and usually display highly non-linear input-output relationships. Given the complexities associated with drug transport in tumour tissues and intracellular signal transduction, understanding the effect of anticancer drugs on solid tumours presents a highly challenging problem.

Methods. In this study, an integrated systems-based modelling framework is developed which is capable of (i) capturing relevant biological processes and their interconnections, and (ii) providing predictive and mechanistic insight into the effects of chemotherapeutic agents. As a first attempt, the model starts with basic descriptions of the essential elements with a focus on their integration in the system. The essential modules include drug transport, intracellular apoptosis signalling and tumour cell density dynamics.

Drug transport in tumour tissue is described by a diffusion-convection-reaction equation, with doxorubicin as the anti-cancer drug for modelling purpose. Velocity field is obtained by solving tumour blood flow, which is governed by the Navier-Stokes equations in the tumour vasculature, Starling's law for flow across the capillary wall and Darcy's law for flow in the tumour interstitium. Intracellular drug concentration serves as the stress signal triggering apoptosis pathway. Here, two types of coarse-grained intracellular apoptosis models, monostable and bistable apoptosis models, extracted from the systems biology literature are examined. Both simplified models capture key features of apoptosis- threshold effect and irreversibility. The cellular response is reflected on the tissue level in terms of tumour cell density which is described by a sharp decrease in the tumour growth rate in the logistic formulism of tumour growth. Re-distribution of tumour cell density is examined upon different pulse drug injections. The integrated systems-based model is applied to a simplified tumour cord geometry.

Results. Simulation results show that insufficient drug transport in tumour interstitium is a major limiting factor in inducing drug effect. Despite qualitatively different dynamics of signal transduction, the monostable and bistable apoptosis models exhibit similar trends of tumour cell density distribution for most drug stimuli encountered except for a case where drug-induced apoptosis is involved in enhancement in interstitial drug penetration due to lower tumour cell density with the monostable apoptosis model. The sensitivity analysis of drug-specific properties indicates that (a) there exists an optimal drug diffusivity to balance interstitial drug transport and the specific acquirement of apoptosis switch, and (b) higher vascular permeability does not lead to significant improvement in drug effect, demonstrating the interstitial drug transport being a limiting factor.

Conclusion. Despite a number of simplified assumptions employed, the present modelling framework is capable of providing solid conclusions of reasoning-consequence for some aspects of the complex problem and offers clear-cut insights into the effect of various drug injections and other contributing factors on drug efficacy. It serves as a credible platform, which is modular and transparent to allow for systematic incorporation of additional layers of complexities and is expected to bridge the gap between the progress in systems biology of cellular signalling processes and its application in improving the efficacy of anticancer drugs.

Key words: drug transport, solid tumour, intracellular signalling, systems-based modelling.

(557g) Development and Characterization of Novel, Micelle-Based Parthenolide Delivery Systems

Michael Baranello[1], Hannah Watkins[2], Craig Jordan[3] and Danielle Benoit[2,4,5], (1)Chemical Engineering, University of Rochester, Rochester, NY, (2)Department of Biomedical Engineering, University of Rochester, Rochester, NY, (3)Department of Medicine, Hematology/Oncology, University of Rochester Medical Center, Rochester, NY, (4)Department of Chemical Engineering, University of Rochester, Rochester, NY, (5)Center for Musculoskeletal Researcch, University of Rochester Medical Center, Rochester, NY

INTRODUCTION:

Parthenolide (PTL) is effective at eradicating a number of cancer cells, including leukemia stem cells (LSCs) but its low solubility in blood limits its use therapeutically[1,2]. One way to circumvent this problem is through the use of a polymeric delivery system[3]. We are utilizing core-shell micelles to enable robust PTL loading within the micelles' hydrophobic interior, and we aim to increase PTL's solubility and circulation time *in vivo* with a protective hydrophilic corona. Using reversible addition-fragmentation chain transfer (RAFT) polymerization, amphiphilic diblock copolymers of poly(styrene-*alt*-maleic anhydride)-b-poly(styrene) (PSMA-b-PS) and poly(styrene-*alt*-maleic anhydride)-b-poly(butyl acrylate) (PSMA-b-PBA) were synthesized, and subsequently assembled into core-shell micelles for PTL delivery. A variety of micelles made from PSMA-b-PS and PSMA-b-PBA diblocks of different overall molecular weights and relative hydrophobic:hydrophilic chain lengths have been characterized with respect to size, morphology, PTL loading capabilities, stability in aqueous solution, and *in vitro* cytotoxicity.

MATERIALS AND METHODS:

Diblock copolymers of PSMA-b-PS were synthesized in a one-step polymerization by providing an excess of styrene monomer. Various molar ratios (18:1, 9:1, 4:1) of styrene:maleic anhydride monomers were combined in dioxane with 4-cyano-4-(dodecylsulfanyltrithiocarbonyl) sulfanylpentanoic acid (DCT) chain transfer agent and azobisisobutyronitrile (AIBN) thermal initiator. The same RAFT chain transfer agent and thermal initiators were used to synthesize diblocks of PSMA-b-PBA in two-step polymerizations. ^{1}H NMR was used to confirm conversion of monomer and formation of polymer, and gel permeation chromatography (GPC) was used to measure polymer molecular weight and polydispersity index (PDI).

Diblock copolymers were self-assembled into micelle nanoparticles by adding water drop-wise to well-mixed polymer solutions in dimethylformamide. After self-assembly, micelle size and morphology was analyzed by dynamic light scattering (DLS) and transmission electron microscopy (TEM). Micelles were then loaded with PTL using solvent replacement techniques, and the PTL loading efficiency and capacity was determined using high performance liquid chromatography.

MV4-11 leukemia cells were cultured in untreated 24-well styrene plates at $5*10^5$ cells/mL with Iscove's Modified Dulbecco's Media (IMDM) containing 10% heat-inactivated fetal bovine serum, and 1% penicillin/streptomycin. Cells were dosed with free PTL at 2.5, 5, and 10 μM, and cell counts were taken 24 hours after the dose. The cytotoxicity profile of free PTL was compared untreated control groups, as well as PTL-loaded micelles at the same PTL concentrations.

RESULTS AND DISCUSSION:

PSMA-b-PS and PSMA-b-PBA polymers with molecular weights ranging from 18 to 48 kDa and estimated hydrophobic:hydrophilic molecular weight ratios between 0.5 – 1.2 were synthesized. All polymers self-assembled into spherical micelles with diameters between 20 - 200 nm. Interestingly, distinct morphological differences can be observed between micelles formed from diblocks of different hydrophobic:hydrophilic ratios (shown in Figure 1).

PSMA-b-PS micelles successfully loaded PTL with efficiencies ranging between 20% and 40%. These PTL-loaded micelles produced a 50% reduction in MV4-11 leukemia cell viability at PTL doses of 5 μM. Importantly, micelles in the absence of PTL exhibited no significant cytotoxicity towards these cells, when delivered at the same concentration as PTL-loaded micelles.

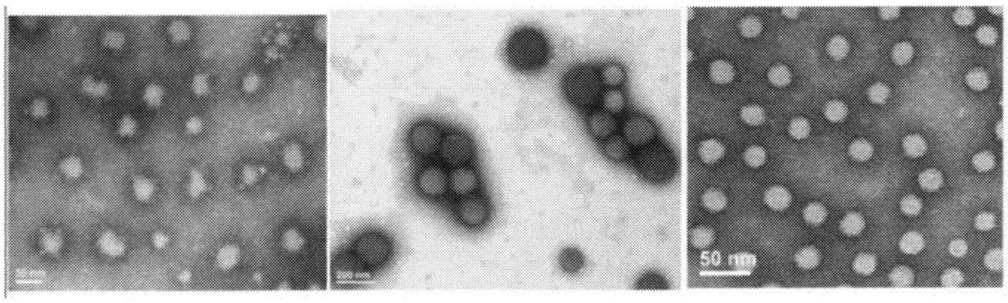

Figure 1: Representative TEM Images of Polymer Micelles. From left to right: $PSMA_{186}$-b-PBA_{107} (brush-like structures, 50 nm in diameter), $PSMA_{59}$-b-PS_{63} (spherical micelles, 200 nm in diameter), $PSMA_{100}$-b-PS_{250} (dense spherical micelles, 20 nm in diameter).

CONCLUSIONS AND FUTURE DIRECTIONS:

PSMA-b-PS and PSMA-b-PBA diblock copolymers were made using RAFT polymerization, and subsequently self-assembled into core-shell polymer micelles for drug delivery applications. A potent leukemia therapeutic, PTL, was successfully loaded into these micelles and exhibited robust LC cytotoxicity *in vitro* when delivered via polymer micelles, where unloaded micelles were non-toxic. Ongoing experiments are examining PTL release rate *in vitro*, as well as *in vivo* circulation time and biodistribution of different micelle architectures.

REFERENCES:

1. Guzman, M.L. et al. The sesquiterpene lactone parthenolide induces apoptosis of human acute myelogenous leukemia stem and progenitor cells. *Blood* 105, 4163-4169, 2005.

2. Jordan, C.T. The leukemic stem cell. *Best Pract Res Clin Hematology* 20, 13-18, 2007.

3. Torchilin, V.P. Structure and design of polymeric surfactant-based drug delivery systems. *J Control Release* 73, 137-172, 2001.

(558a) Towards Ex Vivo Platelet Manufacturing: The Importance of Matrix Elasticity and Shear Force On Megakaryocytic Differentiation

Jinlin Jiang, Stephan C. Lindsey and Eleftherios T. Papoutsakis, Department of Chemical and Biomolecular Engineering & Delaware Biotechnology Institute, University of Delaware, Newark, DE

Matrix elasticity (Engler et al., Cell **126**: 677 (2006)) and biomechanical (shear) force (Baruch et al., Blood **114**:1975 (2009)) have been identified as important differentiation cues for stem cells. In this study, we investigated how these biomechanical/biophysical signals affect megakaryocytic (Mk) differentiation from hematopoietic stem and progenitor cells (HSPCs), aiming to uncover signaling mechanisms behind these effects, and establish better culture environment for Mk differentiation and platelet biogenesis. Application of shear stress (2.5 dyne/cm^2; physiological range: 1.3~4.1 dyne/cm^2) to CHRF cells, a well-established Mk cell line, adherent to fibronectin-coated surface for two hours resulted in retraction, shedding and formation of cytoplasmic extensions (proplatelet-like structures), whereas there were no morphological changes in static culture. Primary human Mks derived from HSPCs *in vitro* also deformed their cell shape and formed proplatelet structures in response to flow application (1 dyne/cm^2). In addition, shear force accelerated endomitosis as assessed by flow-cytometry based analysis of BrdU incorporation: CHRF cells exposed for two hours to shear flow had 16% more BrdU positive cells compared to cells under static conditions. The stress responsive p53 is a regulator in both cell-cycle arrest by laminar flow (Zeng et al., JBC **278**: 24594 (2003)) and Mk differentiation (Fuhrken et al., JBC **283**: 15589 (2008)). To determine if p53 is involved in the response of CHRF cells to shear stress, Western analysis was used to show that two hours of exposure to shear (2.5 dyne/cm^2) increased the ratio of K382-acetylated p53 to total p53 by more than 188% ($n=3$, $p<0.05$) while total p53 protein expression did not change significantly. These data suggest that p53 acetylation is mediating, partially at least, the response of Mk cells to biomechanical forces. In order to examine the effect of matrix elasticity on Mk differentiation, polyacrylamide gels were used to control elasticity. We coated the gel surface with fibronectin to provide holding points for CHRF cells to sense gel elasticity. As matrix elasticity increased, the % of CHRF cells with cytoplasmic extensions increased first and decreased later (Fig. 1, n=3, Error bar: SEM). The highest percentage (17%) of CHRF cells displaying cytoplasmic extensions was observed on a gel with elasticity (ca. 8.2 kPa) mimicking the *in vivo* vascular niche. Non-muscle myosin IIA (NMMIIA) has been shown to serve as elasticity sensor in mesenchymal stem cells. Thus, we aimed to investigate the role of NMMIIA in the response of CHRF cells to matrix elasticity. Blebbistatin, an inhibitor of NMMIIA, minimized the differences of the percentage of cells with cytoplasmic extensions between gels with different elasticity (Fig. 2), thus suggesting that NMMIIA is involved in the response of CHRF cells to matrix elasticity and may also serve as a biological elasticity sensor. As shown by fluorescent microscopy, the cellular distribution of NMMIIA was directed by matrix elasticity (Fig. 3). On soft gels (75 Pa; n=3), NMMIIA was concentrated in the center of the cytoplasm. On a moderately stiff gel (5.6 kPa; n=3), NMMIIA was distributed across the whole cell, while on glass (the stiffest condition tested; n=3), NMMIIA moved into the nucleus and the tips of cytoplasmic extensions. Others have shown NMMIIA is involved in regulation of gene expression when it moves into the nucleus, so our observation is that NMMIIA translocates into the nucleus as gel elasticity increases, thus strongly suggesting that NMMIIA may regulate gene transcription during proplatelet formation and platelet biogenesis. We are working to further investigate and prove this hypothesis. Our data shed light on the largely unexplored role of biomechanical and biophysical parameters (of physiological significance) on Mk differentiation. The impact of these parameters is both of physiological and practical significance in the context of platelet biogenesis and ex vivo platelet generation.

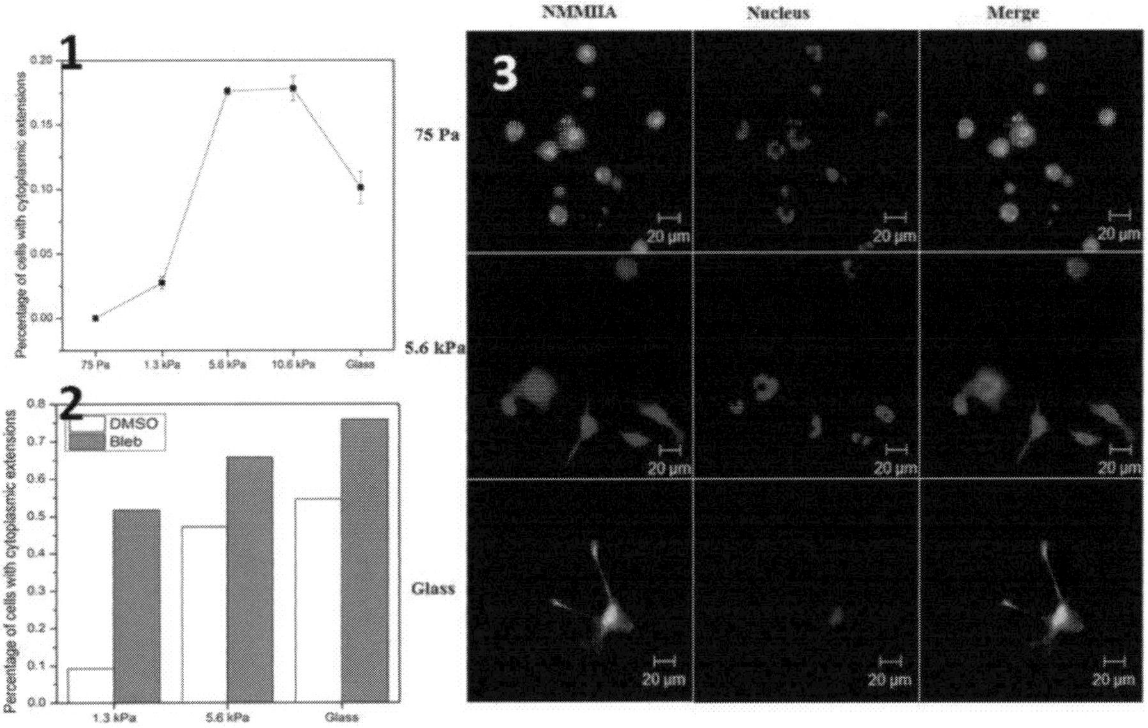
1
Percentage of cells with cytoplasmic extensions
0.20
0.15
0.10
0.05
0.00
75 Pa
1.3 kPa
5.6 kPa
10.6 kPa
Glass
2
Percentage of cells with cytoplasmic extensions
0.8
0.7
0.6
0.5
0.4
0.3
0.2
0.1
0.0
DMSO
Bleb
1.3 kPa
5.6 kPa
Glass
3
NMMIIA
Nucleus
Merge
75 Pa
5.6 kPa
Glass
20 µm

(558b) Cardiac Tissue Engineering Using Human Pluripotent Stem Cells

Tung Ying Lu, Bo Lin and **Lei Yang**, Developmental Biology, University of Pittsburgh, Pittsburgh, PA

Approximately one person dies in every 34 seconds as a consequence of heart diseases in the United States, and more than 5 million patients in the US suffer from heart failure, a disabling and frequently fatal disease condition. Without organ transplantation, end stage heart failure is irreversible. In addition, approximately 50% of heart disease patients do not respond to regular therapies including medication, device implantation and eventually require heart transplantation. Thus heart disease therapy urgently requires the development of novel disease-specific drugs, of new cell resource for replacement therapy, as well as the donor hearts for transplantation. However, those treatments largely remain underdeveloped or inadequate. Heart tissue engineering holds great promise for heart disease therapy by building patient and disease-specific engineered heart tissues for drug screening, drug testing and myocardium implantation, as well as the ultimate engineered hearts for transplantation. Recent progress in human induced pluripotent stem (iPS) cells provides the required cell resource to engineer patient-specific heart constructs for the future heart disease therapies. In this study, we engineered 3D heart tissues with human pluripotent stem cells, including human iPS cell and human embryonic stem (ES) cell derived cardiovascular cells. The engineered heart constructs exhibited spontaneous contractions and were functionally synchronized with electrical stimulation. To explore the potential pharmaceutical applications, we challenged the engineered heart constructs with various chemicals and found the constructs showed similar responses as those from the *in vivo* hearts.

(558c) Stem Cell Senescence: Nanog Reverses the Effects of Organismal Aging On Proliferation and Myogenic Differentiation Potential of Mesenchymal Stem Cells

Juhee Han[1], Panagiotis Mistriotis[2] and Stelios T. Andreadis[1], (1)Department of Chemical and Biological Engineering, University at Buffalo, the State University of New York, Amherst, NY, (2)Chemical and Biological Engineering, University at Buffalo, State University of New York, Amherst, NY

Although the therapeutic potential of mesenchymal stem cells (MSC) is widely accepted, loss of cell function due to donor aging or culture senescence are major limiting factors hampering their clinical application. Our laboratory recently showed that MSC originating from older donors suffer from limited proliferative capacity and significantly reduced myogenic differentiation potential. This is a major concern, as the patients most likely to suffer from cardiovascular disease are elderly. Here we tested the hypothesis that a single pluripotency associated transcription factor, namely Nanog, may reverse the proliferation and differentiation potential of BM-MSC from adult donors. Our microarray analysis showed that adult (a)BM-MSC expressing Nanog clustered close to Nanog-expressing neonatal cells. Nanog markedly upregulated genes involved in cell cycle, DNA replication and DNA damage repair and enhanced the proliferation rate and clonogenic capacity of aBM-MSC. Notably, Nanog enhanced the myogenic differentiation potential and reversed the effects of organismal aging on the contractile function of aBM-MSC to a similar level as that of neonatal (n)BM-MSC. The effect of Nanog on contractility was, in part, mediated through activation of the TGF-b pathway by diffusible factors secreted in the conditioned medium of Nanog-expressing BM-MSC. In addition, Nanog reversed senescence-dependent decline of alpha-smooth muscle actin (α-SMA) by synergizing with the TGF-b1 pathway and upregulating serum response factor (SRF). Overall, our results suggest that Nanog may be used to overcome the effects of organismal aging on aBM-MSC, thereby increasing the potential of MSC from aged donors for cellular therapy and tissue regeneration.

(558d) Multiplexed Single Cell Analysis of Embryonic Stem Cells and Induced Pluripotent Stem Cells

Jun Wang[1], Ritchie Hor[2], Kathrin Plath[2] and James Heath[3], (1)Chemistry, California Insitute of Technology, Pasadena, CA, (2)UCLA, Los Angeles, CA, (3)Chemistry and Chemical Engineering, California Institute of Technology, Pasadena, CA

Embryonic stem cells (ESCs) are derived from the inner cell mass of a developing blastocyst, propagated in *in vitro* culture systems, and defined by the ability to self-renew their pluripotent state. The ability to grow ESCs in culture relies heavily on the media components in which they grow, where specific cytokines and growth factors are necessary to maintain the self-renewal of ESCs and prevent spontaneous, terminal differentiation into heterogeneous somatic cells. Typically, human ESCs need to be grown on a layer of fibroblasts, known as feeder cells, where exogenously supplied FGF2 induces the feeders to secrete the signaling proteins TGFB1 and IGF2 to ESCs, instructing them to stably maintain their pluripotent state. ESCs can also be grown in feeder-free cultures, on the condition that the growth media are still supplemented with FGF2, TGFB1, and IGF2. It has been reported that in the absence of exogenously supplied feeders, a minor population of an ESC colony will differentiate into fibroblast-like cells in order to fulfill the role of feeders in establishing an *in vitro* niche. While the contents of basal media and supplemental growth factors are well defined, the profile of molecules secreted by either ESCs or feeder cells are not well understood. Thus the microenvironment, defined as the interface where ESCs within a colony exchange signals with each other as well as with feeder cells, remains to be further characterized.

While immunofluorescent labeling of ESC surface receptors can suggest a signal dependence on the cognate ligand for those receptors, it provides limited insight as to whether this pathway is autocrine or paracrine in nature. Bulk analysis of media conditioned by ESCs and feeders have been profiled with proteomic studies, but the distinct molecules within a heterogenous mixture cannot be traced back to the secretory cell of origin. Our platform uniquely enables the high thoroughput, quantitative profiling of candidate signaling proteins secreted by a single ESC or ESC-derivative. Furthermore, the ability to simultaneously assay a multiplex of secreted proteins can define and classify subpopulations of cells with distinct secretomes, several of which may be characteristic of pluripotent cells. Lastly, this novel diagnostic tool will also be useful in the field of regenerative medicine, particularly in studies related to the generation and application of induced pluripotent stem cell (iPSC) technology. The reprogramming of a somatic cell to a pluripotent cell is highly inefficient, and the mechanism underlying this process remains largely unknown. Probing the signaling aspects of a single cell undergoing the reprogramming process will yield novel insights into molecular mechanisms and thereby enhance the efficiency of the process. In terms of the clinical application of patient-specific iPSCs for cell replacement therapies, the safety of reprogrammed cells must be assessed prior to transplantation. Assaying for single cell secretomes anti-correlative to tumorigenic secretomes provides an additional parameter to ensure the safety of iPSCs used for regenerative medicine.

(558e) Neural Stem Cell 3D Neuronal Differentiation in Fluorinated Methacrylamide Chitosan Hydrogels

Hang Li[1], Asanka Wijekoon[2] and Nic D. Leipzig[1], (1)Department of Chemical and Biomolecular Engineering, The University of Akron, Akron, OH, (2)ARL BioPharma Inc., Oklahoma City, OK

Oxygen plays a vital role in the metabolism of living organisms and in tissue regrowth. Tissue engineering offers a potential solution to regenerate many tissues in the body including the central nervous system (CNS). Within the CNS, neural stem cells (NSCs) have been isolated that offer potential cellular treatments for neurodegenerative diseases, stroke, traumatic brain injury, and spinal cord injury. We recently invented a biocompatible photocrosslinkable chitosan hydrogel modified with perfluorocarbons (PFCs), which we call perfluorocarbon methacrylamide chitosan (MACF) that is able to uptake and release beneficial levels of oxygen to cells and tissues. The overall objective of this work is to demonstrate how MACFs can be utilized in 3D tissue engineered scaffolds to enhance cellularity and NSC differentiation.

In this study a series of novel, biocompatible, oxygen rich hydrogel scaffolds have been developed and synthesized by incorporating several PFCs to photopolymerizable methacrylamide chitosan hydrogels via Schiff base, nucleophilic substitution followed by radical polymerization to form hydrogels. The synthesized fluorinated methacrylamide chitosan (MACF) hydrogels were characterized by high resolution ^{19}F NMR and demonstrated the desired degree of fluorine substitution (35-45%). MACF oxygen adsorption and desorption studies demonstrated that MACF can be easily regenerated by reloading with oxygen once release is complete for at least three cycles with minimal change in properties. To study how oxygen rich hydrogels affect NSC survival and differentiation, NSCs were seeded and cultured in oxygenated and unoxygenated MACF hydrogels (MAC(Ali5)F with 5 aliphatic fluorine substitutions, MAC(Ar5F) with 5 aromatic fluorine substitutions and MAC(Ali15) with 15 aliphatic fluorine substitutions) and compared to nonfluorinated MAC hydrogels. After 8 days of differentiation and culture (in neuron specific medium), total cell number and neuronal differentiation was quantified. Results have revealed that MACFs support more cells than unfluorinated controls, especially in the interior of the scaffolds. Significant neuronal differentiation has been observed in all scaffolds, and we are currently studying neuronal extension and process formation.

(558f) Cytosolic Delivery of Reprogramming Factors with a Microfluidic Device

Nahyun Cho, MIT, Cambridge, MA

An efficient, robust cell delivery method for proteins has always been a daunting task. Current methods include electroporation, which creates pores in cell membranes to allow material to enter the cell, cell penetrating peptides, which modify the proteins to signal endocytosis, and polymers and liposomes, which encapsulate proteins and induce endocytosis. Each has its own flaws and are not considered universal delivery methods. For example, all three methods have been shown to be potentially toxic and the last two methods have been shown to be protein-specific.

By use of our microfluidic device, cells are forced through a constriction which creates pores on the membrane of the cell. This approach has many benefits, one of which is that our method is not material specific. Cells are suspended in solution with the protein of interest to deliver the protein. The second benefit of this method is the robustness to cell type. This microfluidic device works with established cell lines, immune cells, stem cells, as well as skin cells.

While this device is applicable to more than just protein delivery, this specific project focuses on the delivery of reprogramming factors, which are proteins, to skin cells to induce reprogramming into a pluripotent state. These induced pluripotent stem cells (iPS cells) have many benefits such as patient-specific tissues and the creation of immortalized model cell lines for diseases such as Parkinson's. The benefits of a fast and efficient method of creating iPS cells do not just extend to research, but also to potential clinical and tissue engineering applications as well.

(563a) Hydrothermal Carbonization of Lignocellulosic Biomass Using Salts

Joan G. Lynam[1], M. Toufiq Reza[1], Md. Helal Uddin[2], Charles J Coronella[2] and Victor R. Vásquez[3], (1)Chemical and Materials Engineering, University of Nevada, Reno, Reno, NV, (2)Chemical & Materials Engineering, University of Nevada, Reno, Reno, NV, (3)Chemical Engineering, University of Nevada, Reno, Reno, NV

As we progress toward a paper-free society, vast quantities of valuable renewable lignocellulosic biomass become available for other uses. However, there are challenges to transporting and using biomass as an energy source. As harvested, biomass has a high water content, making it perishable during shipping. Drying biomass can be energy intensive and expensive. In addition, its fuel value (HHV) is low compared to that of coal. With hydrothermal carbonization, high moisture content lignocellulosic biomass is heated in water to temperatures of about 260C, at a pressure high enough to keep the water liquid. This process renders the biomass hydrophobic, and thus less subject to decomposition during transport, as well as increasing its fuel value.

The reactions which occur in hydrothermal carbonization increase the fuel value of the solid product. The addition of salts to the biomass and water reactants can enhance these reactions so that HHV is further increased. Salt addition also decreases the minimum pressure required to maintain liquid water.

Ca propionate, Ca acetate, Mg acetate, Ca lactate, Li chloride, Ca chloride, and Ca formate were tested in hydrothermal carbonizations at 260°C for 5 min using loblolly pine as a biomass. Li chloride, Ca chloride, and Ca lactate were found to increase the HHV of the solid product, as well as reduce reaction pressure. The addition of Ca in coal co-firing has been found to reduce SO_x and NO_x emissions in the literature. Therefore, adding extra calcium to a solid fuel product may reduce the need for emission scrubbing in a power plant. Adding Ca chloride or Ca lactate to a biomass co-firing fuel may enhance the power generation process.

(563b) Water Soluble Pigments From Biomass for Energy Applications

Maria Teresa Gutierrez-Wing[1], Asmita Phadke[1], Alfredo Prudente[2] and Kelly A. Rusch[1], (1)Civil and Environmental Engineering, Louisiana State University, Baton Rouge, LA, (2)Food Science, Louisiana State University, Baton Rouge, LA

The increasing interest in microalgal-based fuel feedstock such as lipids for biodiesel and starches for alcohols provides an economic opportunity based on the large amounts of biomass needed for these biofuels. Valuable co-products can help make these fuels economically viable. Among these products are some polymers and pigments that can provide added value to the biomass.

Most current microalgal operations use single species cultures of eukaryotic organisms generally from collections, that are difficult to maintain and can be contaminated by competing or predatory species. The use of mixed cultures of microalgae and cyanobacteria can provide some added benefits, including production of vitamins and probiotics that microalgae need and cannot synthesize. Additionally, the use of native species reduces the risk associated with the introduction of non-native species.

In this work, mixed cultures of cyanobacteria and microalgae isolated from natural water bodies in Baton Rouge, Louisiana were used to produce lipids and pigments. The water soluble pigments are of particular interest to this work, as they can be extracted from the microalgal/cyanobacteria biomass with aqueous based solvents. These solvents do not affect the lipids in the biomass. The target lipids are non-polar fatty acids, as these can be readily converted to biofuels.

The objectives of this work are to identify pigments that can provide an added value to native aquatic microbial biomass used as feedstock for biofuels. Water soluble pigments were the main targets of this research, due to their compatibility with the lipid extraction process and the high value of these pigments in diverse markets that go from feed and cosmetics to analytical and biotechnological products.

To reach the objective, single and mixed species of aquatic photosynthetic organisms were isolated from a Louisiana natural water body. The cultures were isolated based on their resistance to environmental changes and their growth rate. One mixed culture of a microalgae (*Chlorella vulgaris*) and a cyanobacteria (*Leptolyngbya sp.*), called in this document LA coculture, was selected for this work.

The LA coculture was maintained in Erlenmeyer flasks with a culture volume of 500 ml at temperatures of 25°C±2, under PAR irradiance of 400 μmol cm^{-2} s^{-1} with aeration of 14 lph (0.5 scfh). The light was supplied with HPS (high pressure sodium) 400 watts lamps. The biomass was collected at the onset of the stationary phase of the cultures, where lipid accumulation occurs. An aliquot of 150 mL of the culture was extracted with water or with phosphate buffer 0.1, 0.5 and 1 M. The extracted pigments were analyzed by spectrophotometry and the absorbance spectra of the pigments solution obtained with the each one of the aqueous solvents was compared. The maximum absorbance of the pigments was at 620 nm, characteristic of the C-phycocyanin. The ratio A620/A68 (an indicator of the purity of phycocyanin) from each solution was calculated. The pigment was further purified by rinsing it with different solvents. The highest purity was obtained on the pigment extracted by freezing and thawing with phosphate buffer 0.1 M, after rinse with hexane (purity 3.11-3.67). The lowest purity was btained with pigment extracted with phosphate buffer 1.0 M. Filtration and centrifugation of the pigment after extraction increased its purity in all cases. A second phycobilin in minor quatities was also identified in the extract.

The lipids were extracted with Folch's method at room temperature, with chloroform:methanol. The lipid content was 33 % with a SD of 5%. No significant differences were observed with the extraction of pigments, compared with the non-extracted biomass.

(563c) Carbon Negative Production of Hydrogen From Biomass Using an Alkaline Thermal Treatment: Kinetic and Mechanistic Studies for the Investigation of the Reaction Pathways

Camille Petit, Department of Earth and Environmental Engineering and Department of Chemical Engineering, Columbia University, New York, NY, Thomas Ferguson, Earth and Environmental Engineering, Columbia University, New York, NY and Ah-Hyung Alissa Park, Earth and Environmental Engineering & Chemical Engineering, Columbia University, New York, NY

Using biomass as a feedstock in place of fossil fuels for the production of energy and fuels can represent a sustainable alternative since this energy source is renewable, carbon neutral, and globally distributed. In this study, the one-step H_2 production from biomass using an alkaline hydroxide (i.e., NaOH) under mild reaction conditions (i.e., 373 – 573 K, 1 bar) was investigated using a model biomass system. This method has great potential since it enables the production of a high-purity H_2 stream with increased yield. Moreover, it is carbon negative since the carbon in the biomass is converted to a solid, environmentally benign carbonate. Owing to the suppressed CO formation in this process, the produced gas can be directly used in subsequent energy conversion systems such as PEM fuel cells without additional gas cleanup steps. The formation of gaseous products (i.e., H_2, CH_4, CO and CO_2) during the reaction was monitored using gas chromatography, and the characterization of the derived solid products was carried out using ^{13}C NMR and Raman spectroscopies, while investigating the effect of the reaction conditions (i.e., final reaction temperature, reactants ratio). The findings of this study provided important information regarding the mechanistic pathways of the alkaline thermal treatment of biomass for sustainable energy conversion.

(563d) Recycle of Water and Nutrients Using Anaerobic Digestion of Algal Hydrolysate Obtained by Flash Hydrolysis

Jose Garcia, Old Dominion University, Norfolk, VA, Sandeep Kumar, Civil & Environmental Engineering, Old Dominion University, Norfolk, VA and James W. Lee, Department of Chemistry and Biochemistry, Old Dominion University, Norfolk, VA

Anaerobic digestion is considered as an effective alternative to convert biomass into energy. The process aims to transform the complex compounds present in organic matter in much simpler compounds; through the whole process gas is generated comprised approximately by 60% methane and 40% carbon dioxide called biogas, and also a solid byproduct is obtained with fertilizer characteristics.

In previous studies it was demonstrated that by means of *in house* developed flash hydrolysis about 60-80 wt% of proteins in microalgae could be extracted within few seconds in the temperature range of 200-325¢ªC under subcritical water conditions. After the flash hydrolysis of algae slurry a liquid algae hydrolysate is obtained that is rich in nitrogen (peptides and amino acids) and organic carbon and is near-neutral (pH 6.2-7.0) liquid, making it an attractive feedstock for the anaerobic digestion process. In order to maximize the biogas production, as well as the water and nutrients recycle to algal farm, factors such as the feedstock C/N ratio, pH, solids load, solids retention time, temperature, COD and BOD will be monitored and modified. After the anaerobic digestion process most of the organic compounds originally present in the algal hydrolysate will be transformed into much simpler compounds and inorganic salts mostly dissolved in water. The recirculation of the anaerobic digestion final product will be evaluated as a useful source of dissolved nutrients for algae cultivation.

(563e) Evaluation of Food Waste and Paper-Cardboard Waste Blend for Biohydrogen and Methane Production Using Mixed Microbial Consortia

Brahmaiah Pendyala[1], Subba Rao Chaganti[2], Dr. Jerald. A. Lalman[3] and Daniel D. Heath[2], (1)Civil and Environmnetal, University of Windsor, Windsor, ON, Canada, (2)Great Lakes Institute for Environmental Research, University of Windsor, Windsor, ON, Canada, (3)Department of Civil and Environmental Engineering, University of Windsor, Windsor, ON, Canada

Evaluation of food waste and paper-cardboard waste blend for biohydrogen and methane production using mixed microbial consortia

Brahmaiah Pendyala[1], Subba Rao Chaganti[2], Jerald A. Lalman[3*] Daniel D Heath[4]

[1]Department of Civil and Environmental Engineering, Essex Hall, University of Windsor, 401 Sunset Ave., Windsor, ON, Canada, N9B 3P4. Email: pendyala@uwindsor.ca

[2]Great Lakes Institute for Environmental Research, University of Windsor, 401 Sunset Ave., Windsor, ON, N9B 3P4 Email: chaganti@uwindsor.ca

[3]*Department of Civil and Environmental Engineering, Essex Hall, University of Windsor, 401 Sunset Ave., Windsor, ON, Canada, N9B 3P4. (Corresponding Author): +1 519 253 3000 (ext.2519), +1 519 971 3686 (fax), Email: lalman@uwindsor.ca

[4]Great Lakes Institute for Environmental Research, University of Windsor, 401 Sunset Ave., Windsor, ON, N9B 3P4 Email: dheath@uwindsor.ca

Abstract

A food waste and paper-cardboard waste blend was used to evaluate the H_2 production using a mixed microbial community. At a feed concentration of 10 g COD/L, the maximum H_2 production (34±1 mL/g COD) was observed for linoleic acid (LA) treated cultures. The microbial diversity data showed that *Methanosarcina* sp. was dominant in the 5 g/L COD untreated culture. In comparison, methanogens were inhibited and the dominant populations were *Lactobacillus* sp. and *Clostridium* sp in the LA treated cultures fed with 10 g/L COD of the blended filtrate (food plus paper cardboard waste). For the LA treated cultures fed with 5 g/L COD and 10 g/L COD, the H_2 yield were statistically the same.

Introduction

Urban waste management is a key global concern because of environmental concerns such as local pollution issues and green house gas emissions. According to 2010 data from the United States Environmental Protection Agency (USEPA), the estimated annual production of municipal solid waste (MSW) in the United States (U.S.) was 250 million tonnes with approximately 136 million tonnes (54.2 %) sent to landfills. Conventional landfilling practices are unsustainable because it leads to the production of leachate and uncontrolled greenhouse gas emissions. The organic fraction of MSW (OFMSW) in landfills is a large renewable energy asset. OMSW can be converted into renewable fuels such as hydrogen and ethanol using bioprocessing technologies. Developing bioprocesses to produce fuels and chemicals from OMSW will reduce the environmental pollution impact and add-value to a waste product. Optimizing conditions to extract free sugars and minimize the formation of microbial inhibitory byproducts from treating OMSW using different technologies is challenging. In the proposed work, steam explosion is used to extract free sugars from a synthetic OMSW and paper cardboard waste.

Materials and Experimental Methods

Degrading a food waste plus a paper-cardboard blend by mixed anaerobic communities into H_2 and methane was assessed at 37°C, at a pH set at 5.5 and different organic concentrations. The characteristics of municipal food waste from collection facilities are variable; hence, a synthetic food waste was prepared to minimize compositional variation. The waste was prepared based on unused portions of food pieces which consisted of protein, carbohydrates, vegetables and fats. The diet composition was in accordance with the Canada Food Guide (CFG, 2007) and thus, it was representative of Canadian kitchen waste. The paper and cardboard slurry samples were prepared using 50 g paper and 50 g corrugated cardboard in 1.5 L water. The solid waste samples were liquefied using a domestic food blender and water was added to prepare a slurry. The slurries were divided into three portions of which two were heated in a steam explosion reactor in batch mode with a working volume of 1 L. The two portions were heated separately for 1 h at 160°C, 180°C and the third portion not heat treated.

Results and Discussion

The filtrate from heat treated and untreated samples were analysed for COD, free reducing sugars and furfurals (Table 1). When compared to untreated samples, the COD concentration, the free reducing sugars and furfurals levels increased in the steam exploded samples. With increasing temperature conditions, the COD decreased and the free reducing sugar concentration reduced; however, the furfural (furfural and hydroxyl methyl furfural) quantities increased in the food waste steam exploded filtrate samples. In contrast, the paper-cardboard waste filtrate samples showed an increase in the COD and furfural concentration and a slight decrease in the free reducing sugar concentration.

Table 1. Analysis of treated and untreated food and paper-cardboard waste samples

Sample	COD (mg/L)	Free reducing sugar (mg/L)	Furfural (mg/L)	Hydroxy methyl furfural (mg/L)
FW180 °C	98,800	17,357	469	6,100
FW160 °C	120,000	29,696	75.4	1,043
PW180 °C	5,207	2,584	44.8	50
PW160 °C	4,460	2,617	4.5	42.7
UTFW	63,867	21,495	N.A.	N.A.
UTPW	1,787	958	N.A.	N.A.

Notes: FW180°C = food waste treated at 180°C; FW160°C = food waste treated at 160°C; PW180°C = paper-cardboard waste treated at180°C; PW160°C = paper-cardboard waste treated at160°C; UTFW = untreated food waste; UTPW = untreated paper-cardboard waste

Because of inhibited microbial growth, decreasing H_2 production was detected in cultures not treated with LA but fed with the steam explosion samples treated at 180 °C and containing higher furfural concentrations. According to Modig et al. (2002), furan derivatives are known to inhibit many metabolic pathways. The higher concentration of furan derivatives in the samples treated at 180°C can behave as inhibitors and decrease the H_2 yield. The lower COD content of 4.46 g COD/L for the paper-cardboard waste treated at 160°C was unable to produce a maximum H_2 yield (Ginkel et al., 2001). To eliminate the inhibitory problem, a compromise option was developed by blending the food and paper-cardboard waste treated at 160°C to enhance the H_2 production potential. This blend was used as a feedstock for the anaerobic cultures.

Mixed microbial anaerobic cultures at an initial pH value of 5.5 and at 37 °C were treated with linoleic acid (LA) and without LA and fed with the filtrate blend to assess their H_2 producing capacity. For the untreated cultures fed with 5 and 10 g COD/L of the blend, the H_2 yield reached 8.9±1 and 17±2 mL/g COD, respectively. In comparison, when the cultures were treated with LA and 5 and 10 g COD/L of the steam exploded samples, the H_2 production was

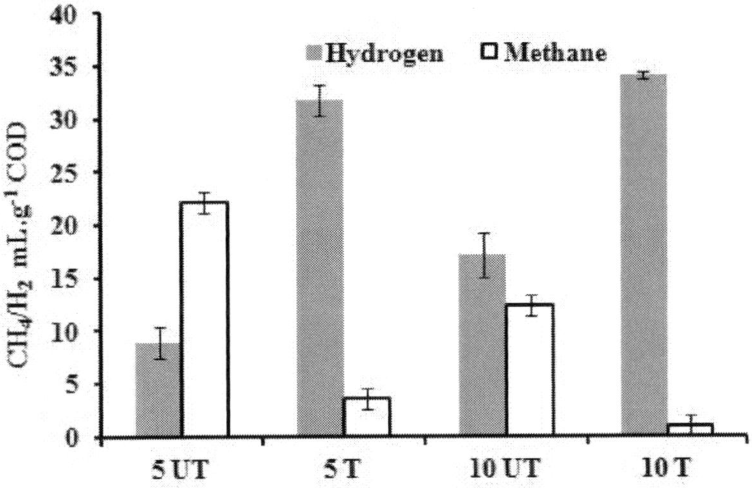

Fig 1. Biohydogen and methane production potential of both treated and untreated cultures

Notes: 5UT = 5 g COD untreated culture; 5T = 5 g COD treated culture (LA); 10UT = 10 g COD untreated culture; 10T = 10 g COD treated culture (LA).

32±2 and 34±1 mL/g COD, respectively. The data clearly revealed that compared to higher methane producing untreated cultures, the LA treated cultures significantly inhibited methane production by suppressing methanogenic growth and subsequently, increasing H_2 production (Ray et al., 2008; Chaganti et al., 2011; Saady et al., 2012).

Conclusions

The optimum COD concentration for higher H_2 production (34±1 mL/g COD) was observed at a feed concentration of 10 g COD/L. Microbial diversity data showed that *Methanosarcina* sp., a methanogen, was the dominant microbial population in the 5 g/L COD untreated culture. In comparison, in cultures fed with 10 g/L COD of the blended filtrate (food plus paper cardboard waste) and treated with LA, methanogens were inhibited and the dominant populations were *Lactobacillus* sp. and *Clostridium* sp. The H_2 yield were statistically the same for the LA treated cultures fed with the two different COD concentrations (Figure 1).

REFERENCES

USEPA, 2010. Municipal Solid Waste Generation, Recycling and Disposal in the United States: Facts and Figs. for

2010. <http://www.epa.gov/msw/facts.htm>.

CFG, 2007. <www.hc-sc.gc.ca/fn-an/alt_formats/hpfb-dgpsa/pdf/food-guidealiment/ view_eatwell_vue_bienmang_e.pdf>.

Modig T., Liden G., Taherzadeh M.J. (2002). Inhibition effects of furfural on alcohol dehydrogenase, aldehyde dehydrogenase and pyruvate dehydrogenase. Biochem J, 363, 769–776.

Ginkel S.V., Sung S., Lay J.J. (2001). Biohydrogen Production as a Function of pH and Substrate Concentration. Environ. Sci. Technol., 35 (24), 4726–4730.

Ray S., Chowdhury N., Lalman J.A., Seth R., Biswas N. (2008). Impact of initial pH and linoleic acid (C18:2) on hydrogen production by a mesophilic anaerobic mixed culture. J Environ Eng134, 110–117.

Chaganti S.R., Kim D.H., Lalman J.A. (2011). Flux balance analysis of mixed anaerobic microbial communities: Effects of linoleic acid (LA) and pH on biohydrogen production. Inter J Hydrogen Energy, 36(21), 14141-14152.

Saady N.M.C., Chaganti S.R., Lalman J.A., Heath D. (2012). Impact of culture source and linoleic acid (C18:2) on biohydrogen production from glucose under mesophilic conditions. Int J Hydrogen Energy. 37, 4036-4045.

ISOLATION OF A LIGNOLYTIC FACULTATIVE ANAEROBE FROM FORMER GOLDMINE IN HOMESTAKE, LEAD, SOUTH DAKOTA

Anu Shende, Rajneesh Jaswal and Rajesh V Shende
South Dakota School of Mines and Technology, Rapid City, SD

Abstract

The former goldmine at Homestake, Lead, South Dakota had wooden support infrastructure. Some of the wooden logs have been decomposing in the water logged soil since closing of the mine in 2001. The hot (between 40°C-43°C) and humid mine conditions devoid of sunlight mimic the environment in a rain forest and has led to rapid decomposition of the wooden logs. Soil underneath decomposing logs was collected to screen for lignin degrading anaerobes. Microflora from the soil was enriched on minimal media containing Kraft lignin as sole carbon source. The soil cultures were incubated under anaerobic conditions without shaking at 43°C. After 5 successive transfers in 60 days, a gram negative lignolytic bacterial species was isolated. The bacteria were found to completely degrade Kraft lignin in 76 days. The fermentation products formed were identified by GC-MS. Succinic acid was identified as one of the products formed within 48 hrs of inoculation. The biochemical characteristics and 16S rRNA sequence analysis identified the lignolytic microbe as a strain from the Enterobacter cloacae .

Introduction

The lignin macromolecule due to its large size, complex aromatic structure and stable linkages has low biodegradability [1,2] and therefore has been an underutilized biomass resource. Kraft lignin (kl), a byproduct of the paper and pulp industry constitutes 85% of the total world production of technical lignins [3]. During kraft processing, the polymeric lignin undergoes extensive cleavage of its aryl-alkyl bonds, side chain modifications and condensation reactions [3,4], forming a heterogeneous, yet potential feedstock of biofuels and platform chemicals. Although, several million tons of kl is produced annually, an average of only about 100,000 tons is utilized to generate commercially applicable products [5]. Moreover, in the near future, advances in biofuel technology will result in bio-refineries regularly processing large amounts of cellulose and hemicellulose fractions of lignocellulosic biomass while accumulating lignin as a major waste product. Most of the technical lignin produced today is treated as residual and incinerated to produce process heat. Alternatively, lignin can be reformed by several processes [6], however, microbial bioconversion of lignin is more appealing because it is less energy intensive and eco-friendly. Search for novel lignolytic microbes and enzymes has thus become imperative to up-grade lignin from a low level fuel to a feedstock for biofuels and value added chemicals

In our search for novel lignolytic bacteria, we targeted the subsurface soil surrounding decomposing wood lying dormant at 4950 feet below ground in the Homestake gold mine, Lead, South Dakota, USA.

The aim of this study was to screen soil samples for lignolytic bacteria and demonstrate their ability to reform technical lignin (kl) into value added chemicals. Here, we describe isolation, characterization and lignin degrading abilities of a novel *Enterobacter cloacae* strain RC202 isolated from soil underneath decomposing wooden logs in the subsurfaces of the Homestake mine.

Materials and methods

Sample Collection and Strain Isolation

Soil samples underneath decomposing wooden logs at 4950 feet underground in the mine in Lead, South Dakota (44°35′2074″N, 103°75′082′W) were collected into sterile 50 ml Falcon tubes and processed within 2 hours of collection. A suspension of 1 gm soil was made in 20 ml KLM (Kraft Lignin Mineral) media described by Shende et al [7]. The pH of the media was adjusted to 7.6 using 0.1 N HCl. The media was filled in 50 ml serum bottles (Wheaton), sealed with butyl rubber stoppers fitted with aluminum crimps and purged with mixture of 80% N_2 and 20% CO_2 for 15 min. The cultures were incubated without shaking at 42°C for 5 days. Enriched cultures of lignolytic bacteria were obtained by 5 successive transfers to fresh KLM media. The enriched culture was streaked on nutrient agar and incubated at 42°C to obtain pure cultures.

Morphology and Biochemical characteristics of the Isolate

Morphology of cells grown in nutrient broth and KLM media was studied by gram staining and scanning electron microscopy (SEM). To prepare samples for SEM, cells grown in nutrient broth were centrifuged at 5000 x g for 5 min. The cell pellet was fixed for 60 min with 10 vol% of ice cold glutaraldehyde made in 100 mM cacodylate buffer. The cell suspension was centrifuged and the pellet was dehydrated with increasing concentrations of ethanol. The fixed cells were analyzed by a JEOL JSM-840A scanning electron microscope. Motility was tested by observing growth pattern in SIM media. Sporulation was determined using the pasteurization test [8]. Standard protocols for carbohydrate utilization, IMViC, amino acid decarboxylation and enzyme tests to identify members of the Enterobacteriacae family were followed [9].

Phylogenetic analysis

The primers 27F and 1492R were used to amplify the 16S rRNA gene from the total DNA extracted from overnight cultures. Gel purified PCR products were sequenced by MCLAB, California, USA. The 16S rRNA sequence was deposited in GenBank under the accession number JX220175. The Basic Local Alignment Search Tool (BLAST) available with the NCBI, USA, was used to obtain homologous 16S rRNA sequences from the bacteria and archae database. Sequences of closely related representative genera were imported into the MEGA 4.0 and multiple alignment was done using CLUSTAL_X version 1.83 [10]. The aligned data was used to construct a phylogenetic tree using the software package MEGA 4.0 [11]. The neighbor-joining method was used and robustness was tested by a bootstrap analysis of 1000 replicates.

GC-MS Analysis of kl Degradation Products

Based on data obtained from UV/visible spectroscopic monitoring of kl degradation, the sample withdrawn after 5 days of incubation with RC202 was selected for further analysis by GC-MS. Equal volumes of the abiotic and inoculated media samples were centrifuged at 10,000 x g for 10 min. Trimethyl silyl derivatives of the ethyl acetate extracts were prepared following protocol described by Lundquist and Kirk [12]. An aliquot of 1.0 μl was injected into GC-MS (GC Model 7890/ MS Model 5975, Agilent Technologies Inc, CA, USA) equipped with HP-5MS capillary column. Helium (UHP grade, 99.99%) was used as carrier gas at rate of 2 ml min^{-1}. The column temperature was held at 50°C for 1 min and then increased to 250°C at the rate of 10°C min^{-1}. A solvent delay time of 5 min was set to eliminate solvent peaks. Compounds were identified by comparing their mass spectra with National Institute of Standard and Technology (NIST) library Version 2.0f supplied with the GC-MS.

Results and Discussion

Isolation and Characterization of RC202

A pure culture of a facultative anaerobe was obtained by enrichment on kraft lignin supplemented media. The phenotypic and biochemical characteristics of the isolate are described in Table 1. The strain was gram negative, motile, non-sporulating and capsulated. The biochemical properties of isolate were in line with those observed for the *Enterobacter cloacae* species. Hoffmann et al. [12] reported that a single biochemical test of esculin metabolism differentiates the *Enterobacter cloacae* subspecies *cloacae* from *Enterobacter cloacae* subspecies *dissolvens*. The RC202 was unable to metabolize esculin thus confirming its relatedness with the *Enterobacter cloacae* subspecies *cloacae*. The SEM image (Fig.1) shows rods varying from 0.75-1.5 μm in length and 0.5 μm in width.

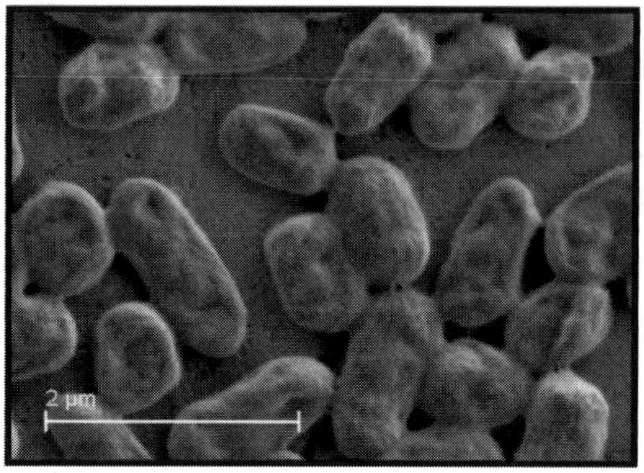

Figure 1 SEM image shows rod shaped Enterobacter cloacae RC202

Phylogenetics and FAME Analysis

In the BLAST analysis, 16S rRNA gene sequence of RC202 showed 99% homology with *E. cloacae* strain 279-56 and *E. cloacae* subspecies *dissolvens* strain LMG 2683. Phylogenetic tree constructed using the 16s rRNA gene sequences is shown in Figure 2. The RC202 isolate, *E. cloacae* and *E. cloacae* subspecies *disslovens* were found to segregate in a single clade, demonstrating relatedness between the three species. However, since the esculin test for RC202 was negative, it was concluded that RC202 strain is not a member of the *E. cloacae* subspecies *dissolvens* [12]

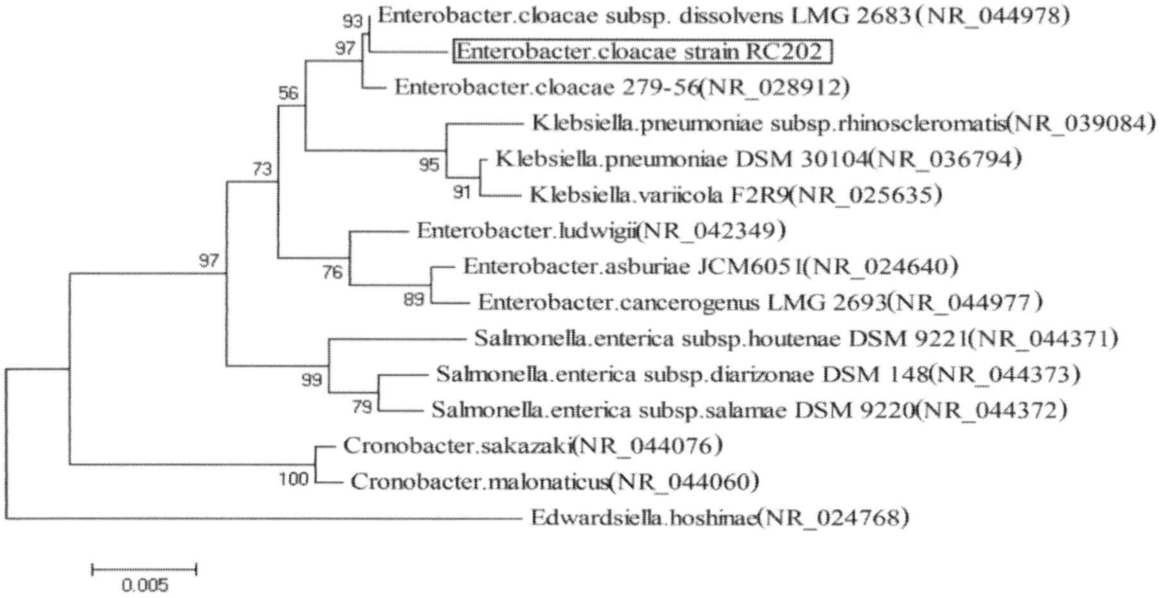

Figure 2 Neighbor joining method based phylogenetic tree demonstrating similarity between species from closely related taxa of *Enterobacter*, *Klebsiella,Salmonella* and *Cronobacter* was constructed using Mega 4.0 software. Bootstrap values obtained from 1000 iterations are shown. *Edwardsiella hoshinae* was used as an outgroup in the analysis. The rDNA sequences were procured from the Genbank. The corresponding accession numbers are in parentheses

The members of the Enterobacteriaceae family have long known to play a role in the fermentative conversion of aromatic compounds [14]. Presently, *Enterobacter* species are emerging as important lignolytic microbes inhabiting soil ecosystem [15] which is consistent with the observation made in this study. However, the *Enterobacter* isolates are metabolically and genetically diverse and the extant methods for categorization of the genus *Enterobacter* are inadequate [12]. Based on16S rRNA gene sequence data, the *Enterobacter lignolyticus* was incorrectly classified as an *Enterobacter cloacae* species. After extensive genomic analysis, the bacterium was reclassified as a novel Enterobacter species [8]. In our study, we have used data obtained from 16S rRNA sequence analysis and biochemical properties, to classify RC202 as a distinct strain of the *E. cloacae.*

GC-MS Analysis of kl Degradation Products

GC-MS chromatograms of TMS derivatized compounds in samples withdrawn from the abiotic control and inoculated cultures after 120 hours of incubation are shown in Figure 3 GC-MS spectrum of abiotic control sample, three main peaks corresponding to vanillic acid and

Table 1 Morphology and biochemical characteristics of *Enterobactor cloacae* RC202.

Morphology/Biochemical Characteristics	**RC202**
Shape	Rods
Gram reaction	G-ve
Dimensions	0.75-1.5 μm x 0.5 μm
Motility	+
Spore formation	-
Capsulation	+
Oxidase	-
Urease	-
Gelatinase	-
Amylase	-
Indole	-
Lysine decarboxylase	-
Methyl red	-
Voges-Proskauer	+
H_2S (Triple Sugar Iron)	-
Gas (Triple Sugar Iron)	+
Citrate	+
Arginine	+
Ornithine	+
Glucose, Lactose, Sucrose, Mannose, Galactose, Xylose, L-Arabinose	**+(A/G)**

+ = Positive, - = Negative, A/G = acid and gas produced

succinic acid were observed at retention times of 14.56 min and 9.09 min, respectively. It has been previously reported [16] that vanillic acid is formed during kraft processing. However, their levels were much higher in the inoculated media. GC-MS chromatogram of sample withdrawn after 5 days shows the formation of significantly higher quantities of vanillic acid and succinic acid. In addition other organic acids such as propanedioic acid, methyl maleic acid and pthalic acid were also detected. No acidic products were detected in samples withdrawn after prolonged incubation of 76 dats (data not shown. Kraft lignin was the sole source of carbon in the growth media, it can be concluded that the organic acids detected were derived by degradation of the side chain and the benzene rings of the lignin monomers.

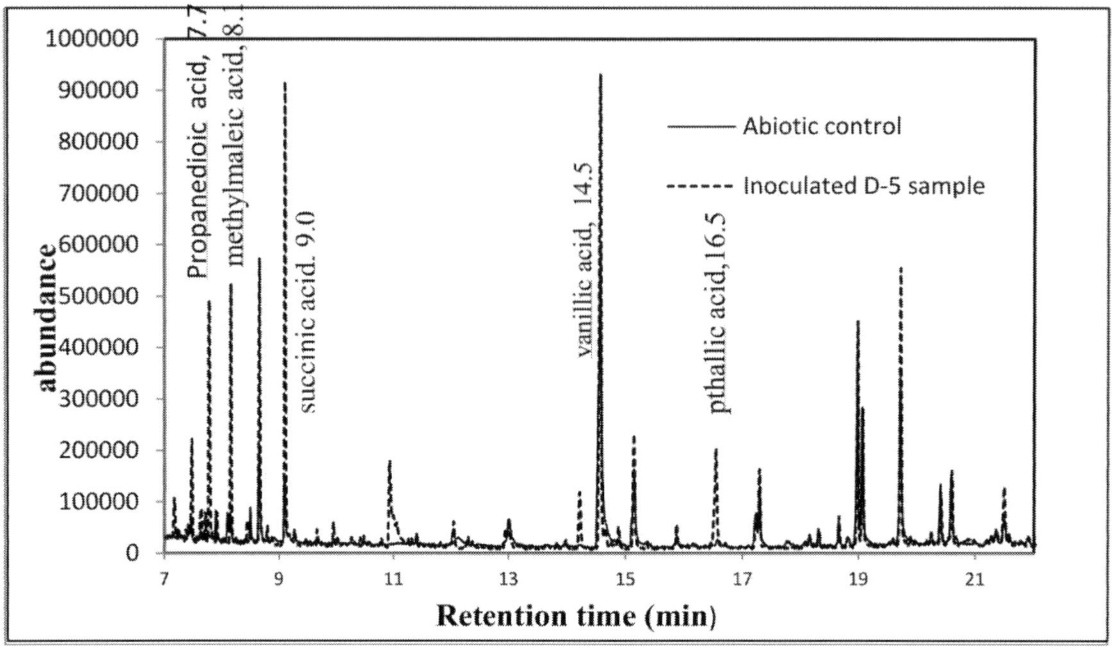

Figure 3 GC-MS chromatogram of abiotic control and kl after 5 days of incubation with RC202.The retention times follow the name of the identified compound.

The data clearly demonstrated the lignolytic acapabilities of the Enterobacter cloacae RC 202. Further studies have been undertaken to identify the lignolytic enzymes in this bacteria.

Acknowledgements

The authors gratefully acknowledge Jaret Heise and Tom Tracynger from the Sanford Underground Laboratory at Homestake, SD for providing training and access to the 4950 feet subsurface level of the Homestake mine. Ph.D. student fellowship support by CBE, SDSM&T is gratefully acknowledged.

References

1. Ruiz-Duenas FJ, Martinez AT (2009) Microbial degradation of lignin: how a bulky recalcitrant polymer is efficiently recycled in nature and how we can take advantage of this. Microbial Biotechnol 2: :164-177
2. Mohan D, Pittman CU, Steele PH (2006) Single, binary and multicomponent adsorption of copper and cadmium from aqueous solutions on kraft lignin- a biosorbent. J Colloid Interf Sci 297(2):489-504
3. Tejado A, Pena C, Labidi J, Echeverria JM, Mondragon I (2007) Physicochemical characterization of lignin from different sources for use in phenol formaldehyde resin synthesis. Bioresource Technol 98(8):1655-1663

4. Chakar SF, Ragauskas JA (2004) Review of current and future softwood kraft lignin process chemistry. Ind Crops Prod 20:131–141

5. Bandounas L (2011) Isolation and characterization of novel bacterial strains exhibiting lignolytic potential. BMC Biotechnology 11:94-105
6. Holladay JE, Bozell JJ, White JF, Johnson D (2007) Top value-added chemicals from biomass. Volume II- Results of screening for potential candidates from biorefinery lignin. A report available via:http//www1.eere.energy.gov/biomass/pdfs/pnnl-16983.pdf
7. Shende A, Jaswal R, Harder-Heinz D, Sellers E and Shende R (2012) ZnO/ZrO_2 doped TiO_2 mediated photocatalytic pretreatment of lignin for microbial conversion to value added products Manuscriot submitted to Naomaterials and Energy, September 2012
8. DeAngelis KM, Haeseleer PD, Chivian D et al (2011) Complete genome sequence of "*Enterobcater lignolyticus*"SCF1.Standards in Genomic Sciences 5:69-85
9. Williams ST, Sharpe ME, Holt HG (1993) Bergy's manual of systemic bacteriology Vol 4 SA. Williams and Wilkins
10. Thompson J, Gibson T, Plewniak F, Jeanmougin F, Higgins D (1997) The CLUSTAL_X windows interface: flexible strategies for multiple sequence alignment aided by quality analysis tools. Nucleic Acids Res 25:4876-4882
11. Tamura K, Dudley J, Nei M, Kumar S (2007) MEGA4: Moelcular evolutionary Genetics Analysis (MEGA) software version 4.0.Mol Biol Evol 24:1596-1599
12. Lundquist K, Kirk TK (1971) Acid degradation of lignin. Acta Chem Scand 25:889–89416. Manter DK, Hunter WJ, Vivanco JM (2011) *Enterobacter soli* sp. nov.: a lignin -degrading gamma proteobacteria isolated from soil. Curr Microbiol 62(3):1044-1049
13. Hoffmann H, Stidl S, Ludwig W, Stumf A, Mehlen A, Hessemann J, Monget D, Schleifer KH, Roggenkamp A (2005) Reassignment of *Enterobacter dissolvens* to *Enterobacter cloacae* as *E.cloacae* subspecies dissolvens comb. nov. and emended description of *Enterobacter asburiae* and *Enterobacter kobei*. Syst Appl Microbiol 28(3):196-205
14. Grbic-Galic D (1986) O-demethylation, dehydroxylation, ring reduction and cleavage of aromatic substrates by Enterobacteriaceae under anaerobic conditions. J Appl Bacteriol 61(6):491-497
15. Manter DK, Hunter WJ, Vivanco JM (2011) *Enterobacter soli* sp. nov.: a lignin -degrading gamma proteobacteria isolated from soil. Curr Microbiol 62(3):1044-1049
16. Chen YH, Chai LY, Zhu YH, Yang ZH, Zheng Y, Zhang H (2012) Biodegradation of kraft lignin by a bacterial strain *Comamonas* Sp.B-9 isolated from eroded bamboo slips. J Appl Microbiol 112:900-906

(563g) High-Temperature Molecular Dynamics Simulations of Carbohydrates

Jessica D. Murillo, Chemistry, Environmental Sciences, Tennessee Technological University, Cookeville, TN, Joseph J. Biernacki, Department of Chemical Engineering, TennesseeTechnological University, Cookeville, TN and Scott Northrup, Department of Chemistry, Tennessee Technological University, Cookeville, TN

Biofuels are still the only source of short-term renewable liquid fuels. For this reason, advancements in biomass conversion technologies continue to be of utmost priority. In addition to providing renewable fuels and chemical feedstock, developments in the biofuels industry will aid in the mitigation of global climate change and may provide opportunities to revitalize global economies.

Pyrolysis is a thermochemical conversion route where biomass is heated at high temperatures in an inert environment to produce permanent gases, bio-oil, and char. Pyrolysis products can further be upgraded to renewable fuels and chemicals. To this end it is important to understand the high-temperature behavior of the starting material. Cellulose, hemicellulose, and lignin comprise the bulk of plant matter. Extensive experimental and computational research is needed on the individual biomass components and raw feedstock to determine reaction mechanisms, product distributions, and ultimately kinetic models for the development of reliable conversion equipment. More recently, molecular dynamics (MD) modeling has been used to probe the decomposition of α-cyclodextrin during pyrolysis[1]. The results indicate that the product distribution of α-cyclodextrin closely resembles that of cellulose, and thus might be a suitable small, surrogate compound for cellulose decomposition. This prior work further suggests that α-cyclodextrin (and cellulose) decomposes through various simultaneous reactions involving homolytic cleavage of glycosidic and pyran ring bonds.

In the present study, semi-empirical quantum mechanical methods were used in an attempt to validate the aforementioned hypotheses. High-temperature MD simulations of α-cyclodextrin show susceptibility to cleave bonds near the α-glycosidic linkage joining the individual glucose monomers. To establish a point of reference for using the semi-empirical method, MD simulations of α- and β-linked disaccharides were performed. When compared, the β-coupled disaccharide showed bond lengthening at the C-O bonds of the glucose monomer, whereas bond cleavage is more likely to occur at the glycosidic link of the α-coupled disaccharide, as observed in MD simulations of α-cyclodextrin. Model compounds, such as α-cyclodextrin, may offer a fast and accurate alternative to modeling large molecules when using such advanced computational methods, which will at last allow for a deeper understanding of the fundamental chemistry governing pyrolytic decomposition of biomass.

[1] Mettler, M. S.; Mushrif, S. H.; Paulsen, A. D.; Javadekar, A. D.; Vlachos, D. G.; Dauenhauer, P. J. Revealing pyrolysis chemistry for biofuels production: Conversion of cellulose to furans and small oxygenates. (2011) *Energy Environ. Sci.* DOI:10.1039/c1ee02743.

(570a) Incremental Parameter Estimation and Ensemble Kinetic Modeling of Metabolic Networks

Gengjie Jia, Chemical and Bimolecular Engineering, Singapore MIT Alliance, Singapore, Singapore, Gregory N. Stephanopoulos, Chemical Engineering, Massachusetts Institute of Technology, Cambridge, MA and **Rudiyanto Gunawan**, Institute for Chemical and Bioengineering, ETH Zurich, Zurich, Switzerland

The estimation of unknown kinetic parameters from time-series measurements of biological molecules is a major bottleneck in the ODE model building process in systems biology and metabolic engineering (Chou and Voit, 2009). The majority of current estimation methods involve simultaneous (single-step) parameter identification, where model prediction errors are minimized over the entire parameter space. These methods often rely on global optimization methods, such as simulated annealing, genetic algorithms and other evolutionary approaches (Daisuke and Horton, 2006; Gonzalez, et al., 2007; Kikuchi, et al., 2003; Kimura, et al., 2005; Noman and Iba, 2007). The problem of obtaining the best-fit parameter estimates however is typically ill-posed due to issues related with data informativeness, problem formulation and parameter correlation, all of which contribute to the lack of complete parameter identifiability (Srinath and Gunawan, 2010). Not to mention, finding the global minimum of model residuals over highly multidimensional parameter space is challenging and can become prohibitively expensive to perform on a computer workstation, even for tens of parameters.

Here, we consider the modeling of cellular metabolism using the canonical power-law formalism, specifically the generalized mass action (GMA) systems (Savageau, 1969a; Savageau, 1969b). The power-law formalism has many advantages, which have been detailed elsewhere (Chou and Voit, 2009; Voit, 2000). Notably, power laws have a relatively simple structure that permits algebraic manipulation in the logarithmic scale, but nonetheless is capable of describing essentially any nonlinearity. Regulatory interactions among metabolites can also be described straightforwardly through the kinetic order parameters, establishing an equivalence between structural identification and parametric estimation. However, the number of parameters increases proportionally with the number of metabolites and fluxes, leading to a large-scale parameter identification problem, one where single-step estimation methods often struggle to converge.

As ODE integrations constitute a major part of the computational cost in the typical ODE parameter estimation (Voit and Almeida, 2004), alternative formulations have been proposed that avoid these integrations either completely (Tsai and Wang, 2005; Voit and Almeida, 2004) or partially (Jia, et al., 2011; Kimura, et al., 2005; Maki, et al., 2002). Particularly, computational cost could be significantly reduced by decomposing the estimation problem into two phases, starting with the calculation of dynamic reaction rates or fluxes from the slopes of concentration data, followed by the least square regressions of kinetic parameters (Bardow and Marquardt, 2004; Goel, et al., 2008; Marquardt, et al., 2006). In this case, the final parameter estimation is done one flux at a time, each involving only a handful of parameters and thus, the global minimum solution can be either computed analytically (for example, when using log-linear power-law flux functions) or determined efficiently. Moreover, as the first estimation phase (flux estimation) depends only on the assumption of the topology of the metabolic network, the flux estimates can subsequently be used to guide the selection of the most appropriate flux functions for the second phase or to detect inconsistencies in the assumed topology of the network separately from the flux equations (Goel, et al., 2008). However, the application of this method requires the number of metabolites to be equal to or larger than that of fluxes, so that the flux estimation can result in a unique solution. Since the reverse situation is more commonly encountered in the typical metabolic networks, a generalization of this incremental estimation approach is the main focus in this study.

As noted above, the new parameter estimation method in this work is built on the concept of incremental identification (Bardow and Marquardt, 2004; Marquardt, et al., 2006) or dynamical flux estimation (DFE) method (Goel, et al., 2008; Voit, et al., 2009), and designed to handle the degrees of freedom associated with the flux estimation from time-course concentration data. Specifically, two parameter estimation formulations are proposed with two different purposes. The first follows the most common formulation in which model prediction errors are minimized and a single "optimal" parameter set is produced. On the other hand, the second formulation is created toward the generation of an ensemble of parameter estimates and models, where each parameter combination will be able to fit the noisy concentration data equally well. The application of the first formulation to GMA models of a generic branched metabolic pathway and L. lactis glycolytic pathway (Voit, et al., 2006) demonstrates the significant improvement in the numerical efficiency offered by the proposed incremental approach, where the computational times were reduced by over two orders of magnitude in comparison with simultaneous (single-step) estimation methods. Meanwhile, in the second formulation, the ensemble of parameter estimates for the two GMA models above are constructed efficiently using an adaptive Monte Carlo and multiple ellipsoid sampling (Zamora-Sillero, et al., 2011). During this construction, the time-varying concentration simulations corresponding to the ensemble of models (or parameter estimates) are also produced. The ability to create such ensemble will necessitate the development of new framework for its use for metabolic engineering, which is a topic of our future work.

Bardow, A. and Marquardt, W. (2004) Incremental and simultaneous identification of reaction kinetics: methods and comparison, Chemical Engineering Science, 59, 2673-2684.

Chou, I.C. and Voit, E.O. (2009) Recent developments in parameter estimation and structure identification of biochemical and genomic systems, Math Biosci, 219, 57-83.

Daisuke, T. and Horton, P. (2006) Inference of scale-free networks from gene expression time series, Journal of bioinformatics and computational biology, 4, 503-514.

Goel, G., Chou, I.C. and Voit, E.O. (2008) System estimation from metabolic time-series data, Bioinformatics, 24, 2505-2511.

Gonzalez, O.R., et al. (2007) Parameter estimation using Simulated Annealing for S-system models of biochemical networks, Bioinformatics, 23, 480-48.

Jia, G., Stephanopoulos, G.N. and Gunawan, R. (2011) Parameter estimation of kinetic models from metabolic profiles: two-phase dynamic decoupling method, Bioinformatics, 27, 1964-1970.

Kikuchi, S., et al. (2003) Dynamic modeling of genetic networks using genetic algorithm and S-system, Bioinformatics, 19, 643-650.

Kimura, S., et al. (2005) Inference of S-system models of genetic networks using a cooperative coevolutionary algorithm, Bioinformatics, 21, 1154-1163.

Maki, Y., et al. (2002) Inference of genetic network using the expression profile time course data of mouse P19 cells, Genome Informatics, 13, 382-383.

Marquardt, W., Brendel, M. and Bonvin, D. (2006) Incremental identification of kinetic models for homogeneous reaction systems, Chemical Engineering Science, 61, 5404-5420.

Noman, N. and Iba, H. (2007) Inferring gene regulatory networks using differential evolution with local search heuristics, IEEE/ACM transactions on computational biology and bioinformatics / IEEE, ACM, 4, 634-647.

Savageau, M.A. (1969a) Biochemical systems analysis. I. Some mathematical properties of the rate law for the component enzymatic reactions, J Theor Biol, 25, 365-369.

Savageau, M.A. (1969b) Biochemical systems analysis. II. The steady-state solutions for an n-pool system using a power-law approximation, J Theor Biol, 25, 370-379.

Srinath, S. and Gunawan, R. (2010) Parameter identifiability of power-law biochemical system models, J Biotechnol, 149, 132-140.

Tsai, K.Y. and Wang, F.S. (2005) Evolutionary optimization with data collocation for reverse engineering of biological networks, Bioinformatics, 21, 1180-1188.

Voit, E.O. (2000) Computational analysis of biochemical systems : a practical guide for biochemists and molecular biologists. Cambridge University Press, New York.

Voit, E.O. and Almeida, J. (2004) Decoupling dynamical systems for pathway identification from metabolic profiles, Bioinformatics, 20, 1670-1681.

Voit, E.O., et al. (2006) Regulation of glycolysis in Lactococcus lactis: an unfinished systems biological case study, Syst Biol (Stevenage), 153, 286-298.

Voit, E.O., et al. (2009) Estimation of metabolic pathway systems from different data sources, IET Syst Biol, 3, 513-522.

Zamora-Sillero, E., et al. (2011) Efficient characterization of high-dimensional parameter spaces for systems biology, BMC Syst Biol, 5, 142.

(570b) Rational Design of 13C-Labeling Experiments for Metabolic Flux Analysis Using Elementary Metabolite Unit-Basis Vectors (EMU-BV)

Maciek R. Antoniewicz, Department of Chemical and Biomolecular Engineering, University of Delaware, Newark, DE and Scott B. Crown, Chemical Engineering, University of Delaware, Newark

^{13}C-Tracer selection is an important step in conducting ^{13}C-metabolic flux analysis (^{13}C-MFA) studies; however, current methods are restricted to trial-and-error approaches, which commonly focus on an arbitrary subset of the tracer design space. To effectively select optimal tracers, there is a pressing need for new rational approaches. Towards this end, we have introduced the concept of elementary metabolite unit basis vectors (EMU-BV). We demonstrate that any metabolite in a network model can be described as a linear combination of EMU-BVs, where the corresponding coefficients indicate the fractional contribution of the EMU-BV to the product metabolite. The strength of our approach is the decoupling of the substrate labeling, i.e. the EMU-BVs, from the dependence on free fluxes, i.e. the coefficients. We demonstrate that flux observability depends inherently on the number of independent EMU-BVs and the sensitivities of coefficients with respect to free fluxes.

The EMU-BV methodology was applied to design tracer experiments in microbial and mammalian systems of interest. First, we applied it to a model of HEK-293 metabolism to select optimal tracers for elucidating two key fluxes, the oxidative pentose phosphate pathway (oxPPP) and pyruvate carboxylase (PC) fluxes. Through efficient grouping of coefficient sensitivities, simple tracer selection rules were derived for high-resolution of the fluxes. The approach resulted in a significant reduction of the number of possible tracers and the feasible tracers were evaluated using numerical simulations. Two optimal, novel tracers were identified, specifically [2,3,4,5,6-13C]glucose for elucidating oxPPP flux and [3,4-13C]glucose for elucidating PC flux. The novel tracers improved confidence intervals of the estimated fluxes by 3- to 10-fold compared to currently used tracers. Secondly, we investigated flux branch points in microbial metabolism. In particular, we developed a high-precision method to estimate the oxidative PP pathway, Entner Doudoroff, and glycolysis fluxes with a single tracer experiment. Using glycolysis intermediates as measurements, candidate tracers were identified and numerical simulations narrowed the candidate list to a handful of optimal tracers. Experimental determination of the split ratios was conducted in two model organisms, E. coli and S. cerevisiae, to validate the modeling predictions. In summary, our work highlights the power of rational selection of isotopic tracers and provides a roadmap for high resolution flux measurements.

(570c) Thermodynamics-Based Flux-Balance Analysis: Incorporation of Thermodynamic and Metabolomic Data Into Genome-Scale Constraint-Based Models

Joshua J. Hamilton and Jennifer L. Reed, Department of Chemical and Biological Engineering, University of Wisconsin-Madison, Madison, WI

Genome-scale metabolic reconstructions provide a concise, mathematical representation of cellular metabolism, and can be used as tools to probe a variety of biological questions via constraint-based methods. These methods have been successfully used to predict cellular phenotypes, guide metabolic engineering, and provide context to high-throughput ('-omics') data. The predictive accuracy of constraint-based methods relies on an accurate representation of the underlying biochemical network, including both the network stoichiometry and the quantitative assignment of reaction directions. Existing approaches to quantitatively assign reaction directionality rely on group contribution theory [1] to calculate thermodynamically feasible directions of individual reactions based on a standard metabolite concentration [2], or to enforce thermodynamic feasibility via the imposition of additional reversibility constraints for reactions identified as qualitatively reversible [3].

We have used a thermodynamic-based flux balance analysis approach, TFBA, to ensure thermodynamically feasible directions for all reactions in a genome-scale metabolic network. We used group contribution theory estimates and errors of Gibbs free energies of formation for all metabolites, and calculate Gibbs free energies of reaction in response to metabolite concentrations. Metabolite concentrations were allowed to vary freely, or were constrained based on experimental measurements. We compared TFBA predictions for growth phenotypes of *Escherichia coli* single-gene knockout mutants to both experimental data and to existing flux balance analysis predictions. We also examined the sensitivity of predicted growth rate to uncertainties in intracellular metabolite concentrations and Gibbs free energies of formation. We subsequently identified a set of metabolites for which precise concentration and thermodynamic data are necessary for accurate *in silico*predictions. These results emphasize the need for precise metabolomic data and experimental measurements of thermodynamic properties of key metabolites.

References

[1] Jankowski, et al. "Group contribution method for thermodynamic analysis of complex metabolic networks." *Biophysical Journal*2008. 95:1487-1499.

[2] Fleming, et al. "Quantitative assignment of reaction directionality in constraint-based models of metabolism: Application to Escherichia coli." *Biophysical Chemistry*2009. 145:47-56.

[3] Henry, et al. "Thermodynamics-Based Metabolic Flux Analysis." *Biophysical Journal* 2007. 92:1792-1805.

(570d) Discriminating Significant From Insignificant Model Parameters: The Case of a Dynamic CHO Cell Model

Hana Sheikh, Department of Chemical and Biological Engineering, Tufts University, Medford, MA, Kyongbum Lee, Chemical and Biological Engineering, Tufts University, Medford, MA and Christos Georgakis, Systems Research Institute and Chemical and Biological Engineering, Tufts University, Medford, MA

Kinetic modeling for metabolic networks, formulated as a set of ordinary differential equations for intracellular species concentrations, provides the ability to simulate the dynamic behavior of cellular metabolism. Such models aim to predict cellular response to various external stimuli, allowing an investigator to develop a detailed fundamental understanding of the phenomena studied. Investigators often choose to include more than the necessary model details rather than risk the error of including less than necessary details. However, this increases the number of parameters that need to be identified from experimental data and introduces a substantial mathematical limitation in the identification of the important parameters. As more details are added to the model, the increased number of parameters implies an increased amount of experimental data. Most importantly, it is not clear whether all the model details and the corresponding parameters are necessary for a desired set of model predictions. The present paper presents a computationally efficient methodology to identify the model parameters that are significant for the model predictions and thus distinguish them from the insignificant ones.

The proposed approach is inspired by the classic Design of Experiments (DOE) techniques, performed *in silico* using a preliminary model. We start by defining the possible range of each of the unknown model parameters and designing a set of experiments that are simulated through the preliminary model. Utilizing analysis of variance (ANOVA) and response surface model (RSM) linear regression tools we develop a simplified nonlinear meta-model in which only the significant parameters are retained. We applied this methodology to a dynamic model of Chinese hamster ovary (CHO) cell metabolism [Nolan et al. 2010]. This model, comprising 51 parameters and 34 reaction fluxes, was able to provide a reliable preliminary prediction of the effects of fed-batch process variables such as temperature shift, specific productivity, and nutrient concentrations. A D-optimal design of experiments was used to sample the parameters across their ranges, and a RSM was obtained with antibody flux as the output. Investigating linear, linearly interactive, and quadratic RSMs, we efficiently eliminated approximately 90% of the terms as being insignificant, shedding light on the importance of each of the 51 original model parameters in the predictions of the metabolic model.

(570e) Metabolic Flux-Based Modularity Using Shortest Retroactive Distances

Gautham V. Sridharan[1], Michael Yi[1], Soha Hassoun[2] and Kyongbum Lee[1], (1)Chemical and Biological Engineering, Tufts University, Medford, MA, (2)Computer Science, Tufts University, Medford, MA

Modularity analysis offers a route to better understand the organization of cellular biochemical networks as well as to derive practically useful, simplified models of these complex systems. While there is general agreement regarding the qualitative properties of a biochemical module, there is no clear consensus on the quantitative criteria that may be used to systematically derive these modules. Recently, we proposed a round trip distance metric, termed Shortest Retroactive Distance (ShReD), to characterize the cyclical interaction between any two reactions in a biochemical network and to group together network components that mutually influence each other [1]. Applied to a liver metabolic network, the ShReD partitions found hierarchically arranged modules, many of which contained reactions that were cyclically related by the shared consumption and production of cofactors. One limitation of this analysis was that it relied on uniform edge weights, implicitly assuming that all the reactions in the network are equally engaged. In this work, we update the ShReD-base partitioning algorithm with a novel edge-weighting scheme that enables the use of metabolic flux data for metabolic state-dependent modularity analysis.

In a reaction-centric graph model of a metabolic network, a quantitative measure of the interaction between a pair of connected reaction nodes can be obtained from the flux of the intermediary metabolite. We define the weight of an edge between a connected pair of reaction nodes as the *inverse of the fraction of the intermediate metabolite production flux that is directed towards the destination node*. This definition is consistent with the intuitive notion that a large metabolite flux from one reaction to another corresponds to a strong interaction, and thus is represented by a relatively short edge, whereas a weak interaction is represented by a long edge. To examine the impact of the flux weights, we applied the modified partitioning algorithm to flux data describing adipocyte differentiation and enzyme inhibition [2]. Our results indicate that the metabolic state of the adipocyte significantly impacts the modular assignment of two key upstream reactions in fatty acid synthesis and glycerogenesis, pyruvate carboxylase (PCX), and phosphoenolpyruvate carboxykinase (PEPCK), which connect carbohydrate metabolism to lipogenesis. Our analysis also identifies several robust reaction pairs that consistently partition together into the same module regardless of metabolic state. Examples include reaction pairs that couple the pentose phosphate shunt and palmitate synthesis through the production and consumption of NADPH. Lastly, we show that the modular organization of adipocyte metabolism is relatively stable with respect to the inhibition of an enzyme compared to a major physiological change such as cellular differentiation.

[1] Sridharan GV, Hassoun S, Lee K (2011) Identification of Biochemical Network Modules Based on Shortest Retroactive Distances. PLoS Comput Biol 7(11): e1002262. doi:10.1371/journal.pcbi.1002262

[2] Si, Yaguang, Hai Shi, and Kyongbum Lee. "Impact of perturbed pyruvate metabolism on adipocyte triglyceride accumulation." *Metabolic engineering* 11, no. 6 (November 2009): 382-90.

(570f) Analysis of Critical Transitions in a Model of Human Endotoxemia

Jeremy D. Scheff, Biomedical Engineering, Rutgers University, Piscataway, NJ, Steve E. Calvano, Department of Surgery , UMDNJ-Robert Wood Johnson Medical School, New Brunswick, NJ and Ioannis P. Androulakis, Department of Chemical and Biochemical Engineering, Rutgers, The State University of New Jersey, Piscataway, NJ

Outcomes from inflammation-linked diseases are determined not only by the severity of injury or the virulence of a pathogen, but also by the body's endogenous inflammatory response. An appropriate inflammatory response leads to healing and recovery, but either too strong or too weak of a response can lead to complications. This is particularly relevant in critically ill patients, whose inflammatory networks are at heightened states of activity as their bodies attempts to recover from their precarious states. A critically ill patient can undergo transitions to different states, for instance representing a return to healthy homeostasis, the development of a concomitant disease, or death, depending on the interactions between their immune response and their illness. Identification and analysis of "tipping points" between states could not only allow for more informed therapy of inflammation-linked diseases, but could also have significant diagnostic and prognostic implications (An, Nieman et al. 2012). Discrete tipping points as a stimulus is gradually increased have been observed in a variety of computational, experimental and clinical environments (An, Nieman et al. 2012). In the context of inflammation, feedforward loops between damage and inflammatory mediators can become activated in response to severe injury, leading to bistability and the potential for abrupt switching between states when a threshold is crossed. Therefore, the identification and analysis of inflammation-driven tipping points is extremely important, particularly if a forthcoming tipping point can be predicted ahead of time to allow for more focused intervention. We approached this issue based on a model of human endotoxemia (Scheff, Mavroudis et al. 2011) to investigate critical transitions within the context of inflammation and study how tipping points emerge; under what conditions the resolution of inflammation and the restoration of homeostasis are possible; how critical transitions between states can be predicted ahead of time; and how circadian homeostatic control systems factor into the transition between states. We evaluated the hypothesis that a gradually changing disturbance provokes a critical transition between steady states of the inflammatory response, and furthermore that this transition can be predicted by stability analysis and that there are underlying circadian patterns in these responses.

Human endotoxemia, an experimental model of systemic inflammation, consists of the administration of low doses of endotoxin (lipopolysaccharides, LPS) to healthy human volunteers. This provokes physiological changes that, in part, mimic those occurring in inflammation-linked diseases such as sepsis and trauma. We previously designed a mathematical model of human endotoxemia (Scheff, Mavroudis et al. 2011) which incorporates a feedforward loop in the inflammatory response that leads to bistability, with one steady state representing healthy homeostasis with low levels of inflammatory mediators, and the other steady state exhibiting heightened levels of inflammatory mediators that persist even when LPS is removed. In this work, we studied the response to chronic LPS exposure, mimicking a slowly growing bacterial infection. Under these conditions, the system initially remains in an intermediate transition state, where state variables are very gradually evolving, until there is an abrupt change and the heightened inflammatory state is reached. To assess the stability of this model as a function of time, the Jacobian matrix was numerically estimated. Given the eigenvalues of a linearizable system, it is possible to determine its stability, has been used an indicator of an impending transition in ecological models (Lade and Gross 2012). All simulations were performed multiple times throughout the day to evaluate the impact of circadian rhythms on the system.

We identified that, after the system becomes unstable, there is a delay before the trajectory begins to significantly move towards the heightened inflammatory state, as shown in Figure 1. This represents a critical time window where identification of an impeding state transition is possible, but intervention still allows the system to return to homeostasis. Additionally, there are circadian dependences on both the time of onset of this time window as well as its size, illustrating how homeostatic control systems can play a role in the response to stress. These results indicate how critical information about a system, information which may be of significant translational value, can be buried within the dynamic behavior of that system.

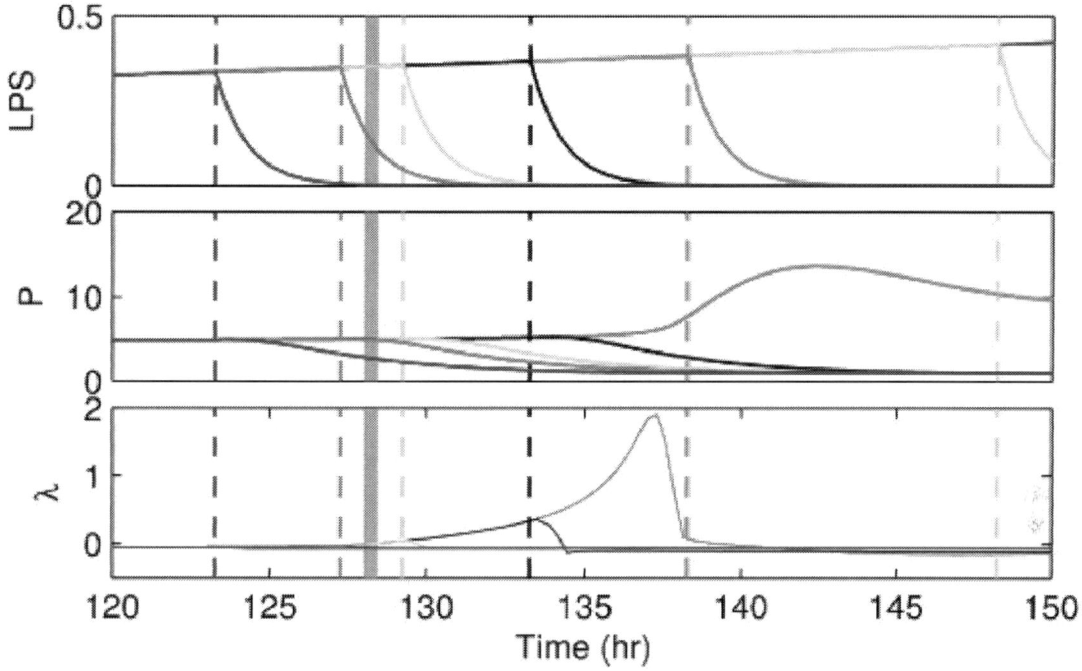

***Figure 1**: LPS exposure is initiated at t=0hr. The three panels respectively show two different model variables (LPS and P, pro-inflammatory signaling) and the largest eigenvalue of the Jacobian (λ) as functions of time. Each thin solid curve represents a separate simulation, each with identical beginnings, but with LPS removed at different time points, denoted by the vertical dashed lines. The thick vertical grey bars mark the time at which the maximum eigenvalue of the Jacobian matrix became positive, signaling the impending transition between steady states.*

This type of analysis is a first step towards identifying how the dynamical components of the inflammatory network can be leveraged to predict forthcoming tipping points as well as to more broadly understand system stability properties. As progress is made towards addressing the issues raised here, we will grow ever-closer to the ideal of building personalizable models based on clinically available biomarkers. A model calibrated on an individual patient could provide a personalized trajectory that could be analyzed for its stability, allowing for more focused care when a critical transition is imminent.

References

- An, G., G. Nieman, et al. (2012). "Computational and systems biology in trauma and sepsis: current state and future perspectives." Int J Burn Trauma 2(1): 1-10.
- An, G., G. Nieman, et al. (2012). "Toward Computational Identification of Multiscale "Tipping Points" in Acute Inflammation and Multiple Organ Failure." Ann Biomed Eng.
- Lade, S. J. and T. Gross (2012). "Early warning signals for critical transitions: a generalized modeling approach." PLoS Comput Biol 8(2): e1002360.
- Scheff, J. D., P. D. Mavroudis, et al. (2011). "Modeling autonomic regulation of cardiac function and heart rate variability in human endotoxemia." Physiol Genomics 43(16): 951-64.

(570g) A Global Sensitivity Approach for the Analysis of Intracellular PI3K/AKT Signaling Pathway During Definitive Endoderm Induction of Human Embryonic Stem Cells

Shibin Mathew, Department of Chemical and Petroleum Engineering, University of Pittsburgh, Pittsburgh, PA and Ipsita Banerjee, Chemical and Petroleum Engg, University of Pittsburgh, Pittsburgh, PA

Introduction: Differentiation of human embryonic stem cells (hESCs) to the definitive endoderm lineage is an important step towards important functional cell types of the pancreas and liver. *In vitro*differentiation efforts are directed towards the activation of cellular signaling pathways which are active during embryonic development in vertebrates. As an important step in the chemical induction of definitive endoderm, modulation of the growth factor mediated PI3K/AKT pathway has been implicated in the control of the outcome of other pathways by making early fate decisions between the maintenance of pluripotency and initiation of differentiation [1]. Specifically, PI3K/AKT pathway suppresses the differentiation signals and sustains the undifferentiated state, while inhibition of the pathway results in up-regulation of Smad and Wnt signaling pathways [2]. Mathematical models can help in the formulation of informative experiments to identify the precise role of these interactions and the conditions under which the cell commits to a particular fate. However, by nature, stem cell systems exhibit intricate signaling pathways; and with the added uncertainty in physical parameters, mechanistic models are either incomplete or difficult to build. One possible solution to construct predictable models is to perform sensitivity analysis on models validated in other cellular contexts and identify regions of interest relating to an experimentally observed output. Given the multivariate nature of the problem, such a sensitivity analysis has to be global in approach to capture the nonlinear interactions between the system parameters. The goal of our study was the use of global sensitivity analysis to design experiments to analyze the role of PI3K/AKT pathway following the induction of hESCs towards definitive endoderm.

Methods: IGF mediated activation of the PI3K/AKT pathway was analyzed using the model developed by Sedaghat *et al* [3]. The model encompasses receptor binding dynamics followed by activation of intracellular signaling cascade resulting in the phosphorylation of a key node in the pathway, i.e. AKT. The activation of Akt has been correlated to various phenotypic outcomes of the cell related to survival, differentiation, stemness, migration etc. We based our global sensitivity analysis on random sampling high dimensional model representation (RS-HDMR) developed by Rabitz *et al* [4]. The technique constructs a meta-model of the original signaling pathway and by using orthogonal function decomposition techniques, evaluates the well-known Sobol' sensitivity indices of the system input parameters to the model output. The advantage of using RS-HDMR lies in the reduction in number of random samples required ($\sim 10^3$ - 10^4) to estimate the sensitivity indices as compared to computationally expensive brute-force Monte Carlo methods ($> 10^6$). Through our analysis, we identified key rate parameters to which the phospho-Akt concentrations are sensitive to, in the presence and absence of feedback mechanisms. Our analysis revealed interesting behaviors of phopho-Akt and its functional dependence on the various system parameters. This knowledge was used to design an *in vitro* strategy to inhibit the PI3K/AKT pathway and study the phospho-Akt protein dynamics during the initial stages of differentiating hESCs.

Results: According to our sensitivity analysis, PTEN (phosphatase and tensin homolog), which is a negative regulator of the PI3K/AKT pathway and parameters associated with receptor activation were the most sensitive parameters in the absence of feedback. However, with feedback, the *in silico* phopho-Akt population segregated into high phospho-Akt and low phospho-Akt regions. RS-HDMR in each of these regions revealed differences in sensitive parameters, particularly to negative feedback loops which inhibit the ability of tyrosine phosphorylated IRS (Insulin receptor substrate) to complex with PI3K and activate downstream Akt. Interestingly, even after PI3K inhibition, our experiments showed increase in phospho-Akt in the initial stages of differentiation and Smad activation and it hints towards a possible cross-talk with other pathways like Smad and Wnt involved in differentiation which maintain higher phopho-Akt levels by modulation of the likely sensitive parameters.

Conclusion: RS-HDMR with polynomial approximation was successful in identifying the most sensitive parameters in the insulin signaling pathway and compared well to the brute-force Monte Carlo estimates. Although different variants of RS-HDMR exist, our analysis proves that even simple orthonormal polynomial approximations can give accurate ranking of the sensitive parameters and often, such a ranking is what we seek to prioritize experimental resources to reducing the uncertainty of the sensitive inputs. Our integrated mathematical and experimental approach was successful in explaining the important parameters affecting the PI3K/AKT pathway in differentiating hESCs.

References

1. McLean AB, D'Amour KA, Jones KL, Krishnamoorthy M, Kulik MJ, Reynolds DM, Sheppard AM, Liu H, Xu Y, Baetge EE: **Activin A efficiently specifies definitive endoderm from human embryonic stem cells only when phosphatidylinositol 3-kinase signaling is suppressed.** *Stem Cells* 2007, **25:**29-38.

2. Singh AM, Reynolds D, Cliff T, Ohtsuka S, Mattheyses AL, Sun Y, Menendez L, Kulik M, Dalton S: **Signaling Network Crosstalk in Human Pluripotent Cells: A Smad2/3-Regulated Switch that Controls the Balance between Self-Renewal and Differentiation.** *Cell Stem Cell* 2012, **10:**312-326.

3. Sedaghat AR, Sherman A, Quon MJ: **A mathematical model of metabolic insulin signaling pathways.** *American Journal of Physiology-Endocrinology and Metabolism* 2002, **283:**E1084-E1101.

4. Li G, Rosenthal C, Rabitz H: **High dimensional model representations.** *The Journal of Physical Chemistry A* 2001, **105:**7765-7777.

(570h) A Mathematical Model of Tumor-Induced Bone Disease Based On the Vicious Cycle Concept

Junhwan Jeon, Deptartment of Chemical and Biomolecular Engineering, Vanderbilt University, Nashville, TN, Scott A. Guelcher, Department of Chemical and Biomolecular Engineering, Vanderbilt University, Nashville, TN, Julie A. Sterling, Cancer Biology, Vanderbilt University, Nashville, TN and Peter T. Cummings, Center of Nanophase Materials Sciences, Oak Ridge National Laboratory, Oak Ridge, TN

Tumor cells grow in specific metastatic sites because of molecular interactions that occur between tumor cells and their microenvironment. Several tumor cells such as breast and prostate cancers frequently metastasize to bone where they release factors that cause bone destruction thwarting normal bone remodeling. During the bone destruction, growth factors are released from bone matrix that stimulate further growth of the tumor cells and subsequently more bone resorbing factors are released from the tumor cells. In order to understand underlying mechanism of this "vicious cycle", a mathematical model has been developed based on a recent mathematical model of bone remodeling by Komarova et al. [1]. Our model describes the dynamics of cell population and cell-cell interactions among osteoclasts, osteoblasts, and tumor cells within a single bone modeling unit where the net effects of local factors on the rates of cell populations have been approximated using power law relations. Tumor burden and bone mass change calculated from the model are compared to values obtained from experiments in which MDA-MB-231 cells, a human cell line that preferentially metastasizes to bone, were injected directly into the tibiae of mice. It is found that the model captures the autocatalytic mechanism of bone destruction observed experimentally, demonstrating at least qualitative validation of the model.

[1] Komarova SV, Smith RJ, Dixon SJ, Sims SM, Wahl LM. Mathematical model predicts a critical role for osteoclast autocrine regulation in the control of bone remodeling. BONE 2003;33: 206-215.

(571a) Engineering Yeast for Advanced Biofuels

Jose L. Avalos[1,2], Gerald Fink[1] and Gregory Stephanopoulos[2], (1)Whitehead Institute, Cambridge, MA, (2)Department of Chemical Engineering, Massachusetts Institute of Technology, Cambridge, MA

The development of renewable liquid transportation fuels is a key component of the national strategy to reduce CO_2 emissions, slow the progression of global climate change, and achieve energy self-sufficiency. Although bioethanol is the most commercially successful biofuel to date, its physicochemical properties and current methods of production make it an inadequate long-term solution to these problems. We are developing yeast strains that produce advanced biofuels that overcome ethanol's limitations. In order to accelerate strain development, we have developed a set of molecular tools and strategies that facilitate the cloning, assembly, integration, and optimization of engineered pathways. We demonstrate the use of this molecular toolkit to overcome roadblocks in the molecular and cellular biology of yeast that present bottlenecks in metabolic engineering, such as the screening of isoenzymes, vectors, promoters and strains. These molecular tools also facilitate new strategies for the assembly of multigenic pathways, which we used in the combinatorial assembly of multiple "isopathways" for the production of advanced biofuels. Using these tools we have engineered yeast strains that produce three different advanced biofuels with better fuel properties than ethanol, and better compatibility with prevalent infrastructures.

(571b) Phosphate Optimization for Economical Lipid Production From Lipomyces Starkeyi Grown On Starch

Sharif M. Rahman, **Ramalingam Subramaniam**, Stephen Dufreche, Mark Zappi and Rakesh Bajpai, Chemical Engineering, University of Louisiana at Lafayette, Lafayette, LA

Phosphate optimization for economical lipid production from *Lipomyces starkeyi* grown on starch

Sharif M. Rahman ,Ramalingam Subramaniam, Stephen Dufreche, Mark Zappi and Rakesh Bajpai, Chemical Engineering, University of Louisiana at Lafayette, Lafayette, LA

Microbial lipids can be obtained by the oleaginous yeast *Lipomyces starkeyi* using sweet potato starch as carbon source which is available largely from processors during cutting and canning process. But high cost of production is the major barrier for industrial production of microbial lipids. This cost can be reduced by optimizing nutrient components. Phosphorus is one of the highest cost contributing elements for microbial culture; and it is mostly incorporated with phospholipids and nucleic acids into the cells. It is under investigation that reduction of phosphate concentration affects cell growth as well as lipid accumulation. However, it is predicted that there is an optimum point of phosphate level where lipid accumulation does not change significantly. This work mainly focuses on optimizing phosphate concentration of the medium since it is a vital factor for economical microbial lipid production using waste materials.

(571c) Platform Pathway for the Synthesis of 3-Hydroxyacids As Value Added Products Derived From Biomass

Himanshu H. Dhamankar and Kristala L.J. Prather, Chemical Engineering, Massachusetts Institute of Technology, Cambridge, MA

The dwindling nature of petroleum and other fossil reserves has provided an impetus to microbial synthesis of fuels and value added chemicals from biomass as a renewable resource. The use of biomass as a feedstock requires the design and engineering of novel pathways connecting biomass derived simple sugars to such value added chemicals and fuels. 3-Hydroxyacids (3-HAs) are a versatile class of molecules that can serve as value added products derived from biomass. 3HAs such as 3-hydroxybutyrate and 3-hydroxyvalerate serve as monomers for biodegradable polyhydroxyalkanoates (PHAs) and chiral building blocks for pharmaceuticals and chemicals. We have established a novel platform pathway in *E. coli* that allows stereospecific synthesis of straight and branched chain 3HAs using glucose and small acid molecules as starting materials. Using this pathway we have been successful in establishing the first biosynthetic route towards the valuable pharmaceutical building block 3-hydroxy-γ-butyrolactone (3-HBL) and demonstrated synthesis of two novel PHA monomers: 3-hydroxyhexanoic acid and 4-methyl-3-hydroxyvaleric acid. Exploiting the full potential of this pathway requires exploring alternative pathway enzymes and engineering them to enhance pathway productivity and improve versatility for the synthesis of other valuable 3HAs. In particular, we are interested in enhancing pathway productivity for synthesis of 3-HBL which features as one of the top ten value added chemicals from biomass in a 2004 Department of Energy report and serves as a versatile chiral building block for the synthesis of drugs like Zetia®, cholesterol reducing synthetic statins such as atorvastatin (Lipitor ®) and rosuvastatin (Crestor ®), antibiotics such as Zyvox® and the nutritional supplement L-Carnitine. Ongoing efforts to identify alternative pathway enzymes for extending the 3HA pathway towards additional value added products as also enhancing enzyme activity and specificity through protein engineering towards specific products will be discussed.

(571d) Engineering a Yeast Conversion System for the Novel Production of Alkanes

John Blazeck, Department of Chemical Engineering, University of Texas at Austin, Austin, TX and Hal Alper, Chemical Engineering, The University of Texas at Austin, Austin, TX

Bio-derived alkanes are attractive second generation biofuels that circumvent some of the disadvantages associated with traditional biofuels. Engineering alkane production in yeasts relies upon the metabolic engineering of fatty acid metabolism. Fatty acids consist of a long aliphatic carbon chain decorated with only one carboxylic acid group, and thus serve as excellent substrates for alkane pathway engineering. An attractive host for alkane production is *Yarrowia lipolytica*, an oleaginous yeast adept at producing high levels of fatty acids. Here, we report the novel heterologous production of pentane in a *Y. lipolytica* fungal system by assembling a plant-derived pentane-production pathway. Pentane production was optimized using environmental and genetic techniques including increasing precursor availability through genetic knockouts and media formulation and by employing promoter engineering and codon optimization for high enzyme expression levels. Here, we will present the results of this optimization and discuss strategies for improved pentane production and for long chain alkane production in this yeast.

(571e) Engineering of Fatty Acid Pathway and Hydrocarbon Production in E. Coli

Fengming Lin, chemical engineering, University of michigan, ann arbor, MI, Yu Chen, Department of Chemical Engineering, University of Michigan, Ann Arbor, MI, Robert Levine, Chemical Engineering, University of Michigan, Ann Arbor, MI, Neil Marsh, chemistry, University of michigan, ann arbor, MI and Xiaoxia Nina Lin, Chemical Engineering, University of Michigan - Ann Arbor, Ann Arbor, MI

To improve the efficiency of producing fatty acid derived hydrocarbons, such as alkane, alkene, and alkyl ester, in model microorganism *Escherichia coli,* we are engineering pathways related to fatty acid biosynthesis using several strategies. First, a metabolic network bi-level optimization model was developed and applied to strain design for maximum fatty acid production. The gene manipulations suggested by this model were performed and led to an increase of total fatty acid from 58.5 mg/g dry cell weight in the wild type to 84.9 mg/g dry cell weight in a multiple knockout strain. Second, the triacylglycerol biosynthesis pathway was successfully introduced into *E. coli* by heterogeneous expression of wax ester synthase/acyl-coenzyme A: diacylglycerol acyltransferase (WS/DGAT) enzymes together with the knockout of dgkA, which not only increases the total amount of fatty acids in *E. coli*, but also improves the fuel quality of fatty acids. Third, an alkane biosynthesis pathway of cyanobacterial origin is being explored by utilization of synthetic protein scaffolds which spatially organize the enzymes involved. All combined, these strategies should lead to the improvement of titer and yield of producing fatty acid derived hydrocarbons in microorganisms.

(571f) Metabolic Engineering of Clostridium Tyrobutyricum for Isopropanol Production

Wenyan Jiang, Chemical Engineering, The Ohio State University, Columbus, OH, Mingrui Yu, Department of Chemical and Biomolecular Engineering, The Ohio State University, Columbus, OH and Shang-Tian Yang, The Ohio State University, Columbus, OH

Isopropanol is one of the most widely used solvents in the world. It is also one of the secondary alcohols that can be used as a direct or partial replacement for gasoline. Isopropanol can be produced by anaerobic microorganisms such as *Clostridium beijerinckii*. However, the native metabolic pathways are limited by tight regulation and the isopropanol production is low, less than 30 mM, in these strains. *Clostridium tyrobutyricum* is an anaerobic acidogen which can convert glucose and xylose to acetate and butyrate. Previously, our group had engineered this bacterium to produce n-butanol at a high rate, yield, and titer that are even higher than most of those reported native solventogenic *Clostridia*. In this work, we studied the feasibility of enabling C. *tyrobutyricum* to produce isopropanol. A plasmid was constructed with *C. acetobutylicum ctf* (acetoacetyl-CoA transferase), *adc* (acetoacetate decarboxylase) and *C. beijerinckii adh* (secondary alcohol dehydrogenase) genes under *C. tyrobutyricum* ATCC 25755 *thl* (acetoacetyl-CoA transferase) promoter. The plasmid was transformed into both *C. tyrobutyricum* wild type and *ack* (acetate kinase) knockout mutant strains. Isopropanol production by these mutants were confirmed in batch fermentation with glucose as the carbon source. The effects of isopropanol production on cell growth and fermentation kinetics were studied and will be presented in this paper.

(571g) Renewable Production of 5-Carbon Polyamide Building Blocks Using Engineered E. Coli

Jake Adkins, Chemical Engineering, Arizona State University and **David R. Nielsen**, Chemical Engineering, Arizona State University, Tempe, AZ

Polyamides are a class of homopolymers and copolymers with properties that balance excellent mechanical strength with durability and chemical resistance. Common examples include nylons which are produced from amino acid or diamine and diacid monomer building blocks. Today, such monomers are principally derived from petroleum resources. In recent years, however, the microbial production of 4- through 6-carbon monomers such as succinic acid, adipic acid, putrescine, and cadaverine have been reported. Here we report how, through *de novo* pathway engineering, our lab has been exploring the 'bottom up' development of microbial biocatalysts to produce other remaining polyamide monomer building blocks from renewable feedstocks. Systematic over-expression of the natural lysine degradation pathway enzymes of *Pseudomonas* sp. in a lysine over-producing *Escherichia coli* strain has allowed us to produce, as individual products, both 5-aminovaleric acid and glutaric acid (precursors to nylon-5 and copolymers such as nylon-5,5, for example). To date, genes from the *davBADT* operon of *P. putida* KT2440 has proven to be support the highest productivities in shake flask cultures. Alternative expression strategies been designed to compare the utility of the natural operon versus the individualized expression of pathway enzymes with the latter proving most effective. In shake flask cultures containing glucose in minimal media, our initial strains have already achieved 5-aminovaleric acid and glutaric acid titers as high as at least 350 mg/L. Here we report an overview of our most recent findings in this project. In all cases of strain development, systematic pathway development will be described through the recruitment and screening of candidate isoenzymes from different genetic sources for each step. In addition, to support greater overall productivity, strategies aimed at increasing precursor availability, improving pathway flux, and overcoming product toxicity are also described.

(580a) Performance of a Pilot Scale Membrane Aerated Biofilm Reactor for the Treatment of Landfill Leachate

Eoin Syron, School of Chemical and Bioprocess Eng,, University College Dublin, Belfield, Dublin, Ireland and Eoin Casey, School of Chemical and BioProcess Engineering, University College Dublin, Dublin, Ireland

Performance of a pilot scale membrane aerated biofilm reactor for the treatment of landfill leachate

Eoin SYRON, Eoin CASEY,

School of Chemical and Bioprocess Engineering, University College Dublin, Belfield, Dublin 4, Ireland

Keywords: Oxygen Permeable Membranes, Bubbleless Aeration, Biofilm, Wastewater treatment

Aeration accounts for over 60% of the total energy spend in wastewater treatment plants, and even with recent technical improvements, bubble aeration can only achieve 50% oxygen transfer efficiency. Due to increasing energy costs and increasing regulatory pressure, there is a substantial growing interest in the application of diffusive gas transfer membranes for bubbleless aeration. These membranes avoid the mass transfer limitations associated with bubble-liquid interactions and increase the gas residence time within the reactor allowing more of the supplied oxygen to be transferred. Unlike conventional membranes in wastewater treatment, the concept of biofouling is not discouraged and the resultant biofilm on the membrane plays an active role in the degradation of COD and Nitrogenous constituents in wastewater.

Landfill leachate is water which has percolated through the waste bed in a landfill and leached organics and nutrients from the solid waste. This wastewater must be treated before being discharged to a water course to prevent pollution of the receiving water. Typically leachate has a high ammonia concentration due to the anaerobic degradation which takes place in the landfill. The Membrane Aerated Biofilm Reactor (MABR) is ideally suited for treatment of a high strength ammonia wastewater firstly as to the slow growing nitrifiers are attached to the membranes there is no need of an extra biomass retention operation, secondly the location of the nitrifying organisms on the membrane surface and their high oxygen demand exploit the oxygen transfer advantages of the MABR. While previous studies have reported on the MABR for ammonia nitrification, use of the MABR for the treatment of leachate has not previously been undertaken.

A 60 litre pilot plant (figure 1) was constructed to treat leachate from Arthurstown landfill, Co Kildare, Ireland. The pilot uses hollow fibre PDMS membranes supplied by SilPro (Broadstairs, UK) which have an internal diameter of 300mm and an external diameter of 500 mm giving a wall thickness of 100 mm and a total surface area of 4.2 m^2 . PDMS membranes were chosen for this application due to their durability, ease of supply, and high oxygen permeability. In contrast to porous membranes, PDMS does not suffer from problems such as intra membrane wetting and pore clogging. Although the surface roughness of PDMS is quite low, there have been few reports of poor adhesion or biofilm formation on these membranes. The hollow fibres were potted in bunches of 60 fibres 0.4m long into 10mm PVC piping using polyurethane as the potting agent. Bunches were then arranged in cassettes of 11 bunches and 10 cassettes were placed in the reactor. The pH is controlled via a feedback loop and maintained between 7.5 and 8 with the addition of Sodium Hydroxide or Hydrochloric acid. Initial Oxygen Transfer Rates OTR experiments determined by re-oxygenation test gave results of $50gO_2\ m^{-2}\ bar^{-2}\ day^{-1}$.

Currently the reactor is removing 2.5g $N\text{-}NH_4\ m^{-2}\ day^{-1}$which is comparable with previously reported systems Syron & Casey 2008, but it is foreseen that this rate of reaction will continue to increase with time and thicker biofilm growth, a theoretical maximum of 15 g $N\text{-}NH_4\ m^{-2}\ day^{-1}$ if all of the oxygen supplied is used for oxidation to nitrite. The current pilot is one of the few pilot scale studies of a MABR. Previous attempts have encountered problems with the control of excess biofilm resulting in clogging in the membrane module. In the present study the biofilm control strategy is allowing the pilot to continue with stable performance over a prolonged period.

References

Casey, E.; Glennon, B.; Hamer, G. Oxygen mass transfer characteristics in a membrane-aerated biofilm reactor. *Biotechnol.*

Bioeng. **1999**, *62* (2), 183–192.

Syron, E.; Casey, E. Membrane-aerated biofilms for high ratebiotreatment: Performance appraisal, engineering principles,scale-up, and development requirements. *Environ. Sci. Technol.***2008**, *42*, 1833–1844.

(580b) Symbiotic Hollow Fiber Membrane Photobioreactor for Microalgae Growth and Bacterial Wastewater Treatment

Linh Vu T. K., Department of Chemical and Biomolecular Engineering, National University of Singapore, Singapore 117576, Singapore and Kai-Chee Loh, Department of Chemical and Biomolecular Engineering, National University of Singapore, Singapore, Singapore

For decades, microalgae have gained much interest for production of high value compounds for food, feed, cosmetic and pharmaceutical products. With rapid depletion of fossil fuels and severe effect of greenhouse gas emissions on global climate change, many microalgae strains are potential sources for biodiesel production. Much attention is now focused on the development of a suitable photobioreactor that can assure mass production of high quality microalgal biomass. One of the crucial factors in designing a photobioreactor is the efficient gaseous transfer of CO_2 supply and O_2 removal.

Symbiotic microalgal-bacterial processes which incorporate microalgae and bacteria in one closed system could provide a potential solution for CO_2 supply and O_2 removal with low energy consumption. In this system, microalgae can produce sufficient O_2 for bacteria to biodegrade organic matter, while bacteria concomitantly release CO_2 for microalgae during photosynthesis. This process, so called photosynthetic oxygenation, is a promising approach to reduce the energy cost for intensive mechanical aeration and potentially eliminate CO_2 emission. For years, photosynthetic oxygenation has been applied not only for economical treatment of wastewater but also for degradation of hazardous contaminants. However, when growing microalgae and bacteria together, slow growth rate of microalgae and unexpected interactions between microalgae and bacteria may result in process failure. More importantly, microalgal biomass from the microalgal-bacteria consortia cannot be used flexibly for the production of food, feed or other high value compounds.

In this research program, we proposed to develop a hollow fiber membrane photobioreactor (HFMP) for symbiotic microalgal growth and bacterial wastewater treatment. In the proposed bioreactor, a gas exchange membrane will be used to physically separate microalgae and bacteria cultures as well as to solely facilitate transfer of CO_2 and O_2. The overall benefits of the proposed membrane bioreactor are to reduce the energy cost for mechanical aeration in an activated sludge process, reduce CO_2 emission to the atmosphere and most importantly, harvest microalgal biomass for extracting other value-added products.

Results obtained from the developed hollow fiber membrane photobioreactor (HFMP) demonstrate a proof-of-concept of the symbiotic relationship that existed between microalgae and activated sludge in wastewater treatment. In the absence of aeration in the synthetic waste water, BOD of synthetic wastewater was completely degraded while microalgae grew continuously in the absence of the external CO_2 supply. This concept hence can be applied for existing activated sludge treatment tanks in order to reduce the demand of aeration energy, and reduce CO_2 emission to the atmosphere. Most importantly, additional produced microalgal biomass can be harvested for other purposes such as for food, feed or biodiesel production.

(580c) Increased Operational Flexibility and Robustness of Anaerobic Wastewater Processes Via Bioreactor Coupling

Satish J. Parulekar, Chemical and Biological Engineering, Illinois Institute of Technology, Chicago, IL

Anaerobic wastewater purification processes have been increasingly used in the recent years. Anaerobic digestion is a biological process in which organic matter is transformed by a community of microorganisms into biogas (primarily methane and carbon dioxide) in the absence of oxygen. This process appears to be a promising method to solve some energy and ecological problems in agriculture and industry. Advantages this process has over the other common waste treatment process, the aerobic activated sludge process, include (i) much lower energy requirement; (ii) generation of less sludge, curtailing thereby the cost of energy-intensive post-treatment and disposal; and (iii) generation of biogas, which can be used as an energy source. In the past few years, a number of experimental studies aimed at improving the performance of anerobic digestion and kinetic models for the process have been reported. The process involves a consortium of acid generating bacteria (host population) and methane generating bacteria (commensal population). The initiation of this process requires hydolysis of complex insoluble organics by extracellular enzymes synthesized by the host population to produce simple soluble organics, which can be utilized by the host population. A comprehensive analysis of static and dynamic behavior of a mixed culture of involving acidogens (X) and methanogens (Y) in two coupled bioreactors is presented here. The two-bioreactor system is considered here to make the activated sludge process more robust with respect to wide fluctuations in feed conditions. A single continuous culture may operate at up to seven steady states, including up to four coexistence steady states, with only one coexistence steady state being locally stable. The one-way interaction between X and Y allows for compartmentalization of the system for a stand-alone bioreactor and two coupled bioreactors into two subsystems, which facilitates the analysis of steady state types and stability characteristics of these and classification of dynamic behavior. The bioreactors in the two-reactor system are identical only in terms of feed composition and bioreactor space time. The reactors need not have identical volumes and volumetric feed, effluent, and exchange rates. A two-reactor system may admit up to forty nine steady states, which are comprised of up to forty coexistence steady states, at least at very low interaction rates. The interaction rates between the two bioreactors, R_1 and R_2, are considered to be unequal. The steady states are comprised of symmetric steady states and asymmetric steady states. The symmetric steady states correspond to identical composition in the two bioreactors. The remaining steady states, the asymmetric steady states, correspond to different composition in the two bioreactors and are admissible over a range of interaction rates. The static and dynamic analysis of the two-reactor system is facilitated by appropriate grouping of large number of steady states arising for very low R into nine clusters. Numerical illustrations reveal the rich steady state structure of the bioprocess in coupled bioreactors. While a single bioreactor can operate at only one locally stable coexistence steady state, the coupled bioreactors can operate at up to five locally stable coexistence steady states over certain ranges of R_1 and R_2. The two-reactor system is operationally more flexible and more robust vis-a-vis single reactor as concerns maintenance of mixed culture. Emergence of four additional steady state clusters and additional coexistence and partial washout steady states at intermediate values of interaction rates reveals that the coupled bioreactors are an example of a complex system.

(580d) Two Phase Biodegradation of Phenol in a Hollow Fiber Supported Liquid Membrane Bioreactor

Prashant Praveen, Chemical and Biomolecular Engineering, National University of Singapore, Singapore, Singapore and Kai-Chee Loh, Department of Chemical and Biomolecular Engineering, National University of Singapore, Singapore, Singapore

During biodegradation of toxic aromatic compounds, microorganisms often experience substrate inhibition wherein the cell growth and metabolism are inhibited at higher concentration of the toxic substrate. Several bioreactors have been proposed for the alleviation of substrate inhibition. Most of these are based on the use of immobilized cells but with the advent of two phase partitioning bioreactor (TPPB), this is set to change. TPPBs are based on the equilibrium distribution of pollutants between cell culture medium and an immiscible, biocompatible and non-biodegradable organic solvent with high affinity for the substrate. Although TBBPs attain very high growth and biodegradation rates and would have made an ideal system; but various problems such as foaming and emulsification, arising from phase dispersion limit their applicability.

The operational challenges encountered in conventional TPPBs can be obviated by non-dispersive contacting between the two phases using semi-permeable membranes. In this research, a hollow fiber supported liquid membrane bioreactor (HFSLMB) was developed for simultaneous extraction and biodegradation of phenol from wastewater using *Pseudomonas putida*. In a semi-dispersive approach, extraction was carried out by dispersing the organic solvent into feed solution whereas the transport of substrate from solvent to the cells through the liquid membrane was non-dispersive. The dispersed solvent droplets facilitated liquid membrane renewal and resulted in a stable mass transfer flux while in the absence of any solvent, cell growth environment resembled that in a monophasic system. *P. putida* in suspension was able to biodegrade inhibitory phenol concentrations at 1000-4000 mg/L without experiencing severe substrate inhibition. For example, biodegradation of 4000 mg/L was achieved within 76 hours while the specific growth rate and biomass yield were 0.31 hr^{-1} and 0.26 g/g, respectively. Phenol biodegradation in the HFSLMB occurred in two stages: an exponential removal rate was observed in the beginning which subsequently assumed a linear profile under diffusion limitation. The specific growth rates during the exponential growth phase could be controlled by increasing the volume of the solvent while the removal rate under mass transfer limitation improved by changing the operating conditions and the interfacial area. In evaluating the long-term stability of the HFSLMB, the bioreactor was repeatedly operated in batch mode with 1000 mg/L phenol. The biodegradation performance improved during the first few runs due to the presence of attached cells on the membrane. But the performance deteriorated subsequently at higher biofilm thickness. The biofilms could be removed by alkaline washing lasting less than 5% of the operation time and resulting in the restoration of the biodegradation performance. With two washing cycles, HFSLMB operation could be sustained for 416 hours in which 20 batch runs were conducted.

The HFSLMB offered many advantages that included a better growth environment for the microorganisms, lower energy demand, high separation factor, high removal rates and elimination of secondary waste. The modular design is easier to scale-up and can also be used for continuous operation. Modeling the HFSLMB will provide further insights into substrate mass transfer and metabolism.

The novel anaerobic-circulating fluidized bed bioreactor (A-CFBBR) For High rate biological nutrient removal from high strength wastewater

MEHRAN ANDALIB , GEORGE NAKHLA *, JESSE ZHU

Introduction

High strength wastes produced by fertilizer production, explosive manufacturing, and recovery of nuclear fuels as well as landfill leachate contain in addition to organic matter, nitrogen in the forms of ammonia, nitric and nitrous acids (Chen and Lin, 1993). Similarly, many food processing wastewaters such as rendering wastes also contain high ammonia concentrations as a result of protein digestion. Nitrogen removal from such wastes, which require anaerobic processes to recover energy, can be challenging.

The Circulating Fluidized Bed Bioreactor (CFBBR), introduced and developed by Nakhla and his coworkers (Cui et al., 2004), was tested for biological nutrient removal (BNR) from municipal wastewaters in both lab and pilot scales at OLR of 5 kg COD/m^3·d and NLR of 0.5 kg N/m^3·d (Chowdhury et al., 2010). The CFBBR consists of an anoxic riser and an aerobic downer with fast and conventional fluidization regimes respectively. More than 90% organic, 70%-80% total nitrogen and 50%-70% phosphorous removal were reported, at hydraulic retention times (HRTs) of 2-3 h and an observed biomass yield of 0.12-0.16 g VSS/g COD. The CFBBR was reported to treat high strength wastewater such as landfill leachate and septage as well at the aforementioned OLR and NLR (Eldyasti et al., 2010). However, in order to accomplish high rate biological nutrient removal from high strength waste containing nitrogen, a newly developed bioreactor referred to henceforth as anaerobic-circulating fluidized bed bioreactor (A-CFBBR) was designed and tested to treat synthetic high strength wastewater with TCOD of 10600±270 mg/L, NH_3-N of 250±10 mg/L. The advantages of the A-CFBBR over conventional anaerobic treatment is simultaneous high rate nitrogen and carbon removal as well as energy recovery as opposed to only methanation.

Materials and Methods

System description and operating conditions

The A-CFBBR (**Figure 1)** is comprised of a conventional anaerobic fluidized bed (AF) with an overall height of 3.6 m followed by a CFBBR comprising anoxic and aerobic fluidized bed bioreactors (FBRs) with heights of 2 m and 3.6 m respectively. Zeolite particles (3 kg) with an average diameter (d_m) of 425-610 μm were used as carrier media in the anaerobic column and 3 kg zeolite particles with average diameter of 610-825 μm, were circulated between the aerobic and anoxic fluidized bed columns of CFBBR. The particle weight hold up in the aerobic column of the CFBBR was maintained at 2.4 kg continuously. In the riser, heterotrophic bacteria grow on the media and the biofilm becomes thicker. At a certain biofilm thickness, depending on the superficial liquid velocity, the biofilm-coated particles reach the height where they can be transferred to the downer through the inclined pipe. An intermediate graduated container was placed between the two columns, as shown in **Figure 1,** to monitor the particle transfer rate. After exposure to the high shear force in the gas-liquid-solid phase in the downer, the biofilm detaches and leaves the system along with the effluent. Particles from the bottom dense phase of the downer with a thin biofilm (< 100 μm) were transferred back to the riser manually through a gate valve that was occasionally opened at the time of particle transfer and to make up the particles in the riser. The hydrodynamic pressure difference between the riser and downer facilitated particle transfer.

The system was operated for about six months on synthetic wastewater (SWW). **Table 1** shows the operating conditions of the A-CFBBR system in Mode 3 in three different phases of VI-I to VI-III. Complementary information regarding the operating conditions in other Modes and Phases can be found Elsewhere (Andalib et al., 2012). At the maximum volumetric loading rate (Mode 3), the system was operating at an overall hydraulic retention time (HRT) of 12.8 h, empty bed contact time (EBCT) of 3.2 h and AF and CFBBR estimated sludge retention times (SRTs) of 32 d, and 28 d, respectively. The AF and CFBBR were operated at 37°C and 20°C, respectively.

Acclimatization and start-up

Initially, the clean media was fluidized in the anaerobic, anoxic, and aerobic columns at superficial liquid velocities (u_l) of 1.4 cm/s, 1.2 cm/s and 1.6 cm/s respectively. The anaerobic column was seeded with 16 L of secondary anaerobic digester sludge (ADS) from the St. Mary wastewater treatment plant (Ontario, Canada) with TSS and VSS concentrations of 15900 mg/L and 12180 mg/L respectively. The CFBBR was seeded with enriched nitrifiers, acclimatized in the lab using 10 L of returned activated sludge (RAS) from the Adelaide Pollution Control Plant, London, Canada, with TSS and VSS concentrations of approximately 7000 and 4800 mg/L respectively. Thereafter, the SWW, with the composition shown in **Table 2a**, was fed at a flow rate of 1.8 L/d and 0.1 L/d into the anaerobic and anoxic columns on day eight corresponding to OLR 0.28 kg $COD/m^3{}_{AF}\cdot d$ and NLR of 0.01 kg $N/m^3{}_{AF}\cdot d$.

Analytical methods

Samples from the feed tank, the anaerobic column, the anoxic column, and the final effluent were collected for analysis. Total suspended solids (TSS), volatile suspended solids, (VSS) were analyzed in accordance with Standard Methods 2540D, 2540E respectively (APHA 1998). HACH methods and testing kits (HACH Odyssey DR/2800) were used to analyze total and soluble chemical oxygen demand (TCOD and SCOD), total nitrogen (TN), NH_4-N, NO_2-N, NO_3-N, and PO_4. Alkalinity was measured by titration with 0.01 N H_2SO_4 in accordance with the Standard Method no 2320 (APHA 1998). DO and ORP were measured onsite using an Oakton DO 6 meter, and an Oakton ORPTestr 10 (Oakton, Singapore). Based on Standard Method no 2540G (APHA, 1998), the attached biomass on the carrier media was measured and expressed as mg VSS/g clean particles. The rate of biogas produced in the methanogenic column was measured by a wet tip gas meter connected to the top of anaerobic column. Methane, nitrogen, and hydrogen content of the biogas were determined using a GC.

Results and discussion

System biomass inventory

Figure 2a illustrates the buildup of attached biomass in the anaerobic, anoxic, and aerobic columns. The process of biomass buildup comprises two steps: the attachment of biomass during acclimatization and the growth of biomass during the process. The biomass attachment during acclimatization under the same hydrodynamic conditions should result in relatively the same biomass attachment. As shown in **Figure 2a**, the attachment of biomass in anaerobic, anoxic, and aerobic column, labeled as A, B and C, are 12.5 mg VSS/g media, 14 mg VSS/g 10.8 mg VSS/g media respectively. The aforementioned attached biomass densities correlated inversely with u_l of 1.4, 1.2, and 1.6 cm/s respectively. As shown in **Figure 2a**, the concentration of immobilized biomass in the anaerobic reactor increased slowly between days 1 and 120 as a result of the gradual start-up strategy adopted which limited the availability of substrate in the reactor, and the low growth rate associated with the methanogenic bacteria.

Nutrient removal

The anaerobic fluidized bed (AF) was operated as a mesophilic methanogenic reactor at a temperature of 35.7±0.8 °C and alkalinity of 3034±228 as mg $CaCO_3$/L. **Table 2b** illustrates the steady-state characteristics of the anaerobic, anoxic and aerobic effluents in Mode 3 when A-CFBBR was operational. The complementary information regarding the steady-state characteristics in Mode 1 when only the AF was operational and Mode 2 when the AF and only the aerobic fluidized bed were operational were depicted in **Table B** of the SD. As shown in **Table 2b,** although the feed pH was 4.6, pH inside the AF remained constant at 7 throughout the experiment due to the high rate of acetic acid degradation in the column. ORP in the AF was observed in the range of -158 mV to -193 mV.

Figure 2b illustrates the temporal biogas production in AF during phases I to VI. As shown in **Figure 2b**, the percent methane of the produced biogas fluctuated between 45% and 52% resulting in methane production rates of 103±6 L/d, 144±5 L/d, and 188±6 L/d in Phases IV, V and VI respectively. Based on the CH_4 equivalent of COD converted by aceticlastic methanogens at standard conditions (0°C and 1 atm) of 0.35 L CH_4/g COD, the theoretical methane production at 37°C was also calculated and depicted in **Figure 2b**. Discrepancies between the theoretical and measured methane of 3.8%-6.4% were observed. It should be noted that the ratio of biogas produced in the system to the volume of the reactor in Phase VI was 24 L_{gas}/L_{rea}.d which was handled without the need for a dedicated solid/liquid/gas separator at the top of the reactor.

Figure 2c depicts the COD removal rate in AF in different phases. As shown in **Table B** of the **SD**, the influent TCOD concentration was constant at 10600±270 mg/L. The effluent SCOD in phases I to IV, when the anaerobic attached biomass was not fully developed, was observed to be in the range of 218 mg/L to 260 mg/L. In phase IV, SCOD decreased to 124 mg/L at OLR of 26 kg COD/m^3·d. As seen in **Table 2b**, in Phase VI at the maximum OLR of 35 kg COD/m^3·d the system reached 98.3% TCOD removal, with effluent SCOD increasing marginally to 180 mg/L. TSS concentrations in the bulk liquid after seeding in phase I decreased gradually from 927 mg/L to the minimum level of 124 mg/L in Phase V at OLR of 26 kg COD/m^3_{AF}·d. However, increasing the OLR to 35 kg COD/ m^3_{AF}·d in Phase VI resulted in an increase in TSS to 188 mg/L. TCOD decreased from 10600 mg/L to an average of 50±20 mg/L in the A-CFBBR effluent throughout all phases, corresponding to the overall TCOD removal of 99.3%. The influent TCOD to the CFBBR comprised the AF effluent, which were 181 mg/L and 390 mg/L in phases V and VI respectively and 5% feed with a TCOD of 10600 mg/L. The TCOD influent to the CFBBR was 740±20 mg/L in phase VI corresponding to OLR of 3.3-4.8 kg COD/m^3_{CFBBR}·d based on the overall volume of the CFBBR in Phase VI (**Table 1**). The final effluent SCOD of the A-CFBBR was 21±10 mg/L and 35±10 mg/L in phase V and VI respectively. In Phase IV, the suspended VSS in the anaerobic column averaged 45 mg/L, which increased to 65 mg/L and 146 mg/L in Phases V and VI. The increase in the anaerobic effluent VSS as a result of higher VSS detachment rates in Phases V and VI (**Table 1** and **Table A** of the SD) coincided with higher OLRs in those Phases.

In Phase V, the anoxic column was not operational and nitrogen removal was solely based on biomass assimilation in the AF where NH_3-N decreased from 250 mg/L in the feed to 213±10 mg/L in the AF effluent, and nitrification in the aerobic column reduced ammonia from 213±10 mg N/L to < 1 mg NH_3-N/L in the system effluent. Assuming the overall anaerobic yield of 0.016 g VSS/g COD and 9±1.1% nitrogen content of biomass (**Figure 2d**), the nitrogen assimilated into biomass was calculated as 20±3 mg /L. The nitrate and nitrite concentrations in AF remained below 3.9 m Ng/L and 0.2 mg N/L in Phase V respectively. The NH_3-N remaining in the AF effluent was sequentially completely nitrified in the aerobic column to an average concentration of 0.9±0.5 mg N/L at NLR of 0.7 kg N/m^3_{CFBBR}·d. SNR batch tests in Phase V showed nitrification rate of 0.25-0.27 g N/g VSS·d, which confirmed the aerobic online nitrification rate within 10% discrepancy. The influent STN to the aerobic column in Phase V was 217±11 mg N/L in the form of NH_3-N and the effluent STN from the aerobic column was 205±9 mg/L in the of NO_x-N. Nitrogen assimilation in the aerobic biomass was estimated at 1.4 mg N/L in Phase V with OLR of 0.5±0.1 kg COD/m^3_{CFBBR}·d and biomass nitrogen content of 9%, thus, indicating that a partial denitrification, of 10 mg N/L may have occurred simultaneously in the aerobic column. The anoxic column was operated with three aerobic to anoxic recirculation flows to feed ratios of 2.3, 4, and 6.2 in Phases VI-I, VI-II and VI-III respectively (**Table 1**). The influent NH_3-N and TN to CFBBR in Phase VI were 215 mg N/L and 232 mg N/L corresponding to NLR of 1.1 kg N/m^3_{CFBBR}·d. As shown in **Table 2b**, the effluent NH_3-N throughout Phase VI remained 1-3 mg N/L, within 13% of the batch SNR range of 0.25-0.31 g N/g VSS·d. The recirculated nitrate to the anoxic column was denitrified to effluent concentrations of 76±22 mg NO_3-N/L, 39±2 mg NO_3-N/L and 88±5 mg NO_3-N/L in Phases VI-I, VI-II and VI-III respectively. In Phase VI-II when the effluent NO_3-N and NO_2-N were 39 mg N/L and 2 mg N/L, total nitrogen removal of 82.3% was achieved with effluent NH_3-N of 1.3 mg N/L. The batch SDNR test showed a range of 3.4-3.8 g NO_3-N/g VSS·d. In Phase VI-III, when the aerobic to anoxic recirculation flows to feed ratio increased to 6.2, aerobic DO dropped to 5.6 mg O_2/L from the initial value of 6.3 mg O_2/L and anoxic DO increased to 0.7 mg O_2/L in Phase VI-II, resulting in incomplete nitrification in the aerobic column as reflected by the average nitrite concentration of 20 mgN/L (**Table 2b**) as well as insufficient denitrification in the anoxic column. Therefore, effluent nitrite and nitrate increased to 20 mg N/L and 88 mg N/L, respectively (**Table 2b**). Phosphorus removal occurred as a result of biomass assimilation in the three columns. As shown in **Table 2b**, in Phase VI, 2.8±0.2 mg P/L and 3.0±0.3 mg P/L, were removed in AF and CFBBR respectively as compared with 3.3 mg P/L and 0.8 mg P/L estimated based on 1.8% biomass phosphorus content by weight of TSS.

At the optimum performance, Phase VI-II of Mode 3, the CFBBR in this work demonstrated 85% COD removal and 84% nitrogen removal at OLR of 4.8 kg COD/m^3_{CFBBR}·d, NLR of 1.1 kg N/m^3_{CFBBR}·d, EBCT of 1.6 h and HRT of 5 h. In comparison, Chowdhury et al. (2010) reported 90% COD and 80% nitrogen removal from municipal wastewater (MWW) using CFBBR at OLR of 4.2 kg COD/m^3_{CFBBR}·d, NLR of 0.26 kg N/m^3_{CFBBR}·d, EBCT of 1.5 h and HRT of 2.4 h. Elsewhere Eldyasti et al. (2010) observed 87% COD and nitrogen removal from landfill leachate using the same system at OLR of 2.6 kg COD/m^3_{CFBBR}·d, NLR of 0.8 kg N/m^3_{CFBBR}·d, EBCT of 11 h and HRT of 18 h. Other versions of the CFBBR developed by the same research group such as CFBBR with a bigger anoxic column called TCFBBR (Andalib et al., 2011b) and enhanced bioparticle circulation CFBBR (Li et al., 2012) also showed 88% and 95% COD as well as 84 and 88% nitrogen removal from MWW at OLR of 4.3 and 3.4 kg COD/m^3_{CFBBR}·d, NLR of 0.5 and 0.44 kg N/m^3_{CFBBR}·d, EBCT of 1.0 h and HRT of 2.2 h respectively. Thus, the A-CFBBR achieved comparable organics and nitrogen efficiencies, albeit at 2-4 times the NLR of MWW at roughly the same EBCT. Chen et al. (2011) also reported 87% COD and nitrogen removal from MWW using an A2/O system incorporating a biological aerated filter at OLR of 1.23 kg

COD/m^3·d, NLR of 0.23 kg N/m^3·d and HRT of 8 h. A longer HRT in the CFBBR in this study (of 5 h versus 2.2 h) and the study conducted by Eldyasti et al. (2010) seemed to be conducive to higher nitrogen loading rates of 1.1 and 0.8 kg N/m^3_{CFBBR}·d respectively. On the other hand, using a finer aeration system in the aerobic column of this work resulted in a lower biomass detachment rate of 0.035 1/d and a higher attached biomass of 20 mg VSS/ gmedia (**Table 1**) compared to the detachment rate of 0.093 1/d and attached biomass of 7 mg VSS/ gmedia in the work of Chowdhury et al. (2010), and the detachment rate of 0.18 1/d and attached biomass of 4.7 mg VSS/ gmedia in work conducted by Andalib et al. (2011b), which resulted in a higher nitrification rate.

Biomass yield

Figure 2d illustrates the linear regression of cumulative VSS produced, based on the sum of the effluent biomass, the net change in attached biomass and biomass wasted, versus cumulative COD removed in anaerobic fluidized bed, CFBBR and A-CFBBR. The highest observed yield for the AF utilizing acetic acid was 0.026 g VSS/g COD in phase II at OLR of 5.3±0.2 kg COD/m^3_{AF}·d. Along with increasing the OLR to 34.5 kg COD/m^3_{AF}·d in AF, the observed yield decreased to 0.016 g VSS/g COD. The very low observed yield of the anaerobic column was expected, due to the estimated long sludge retention time (SRT) in the column of 27-32 days (**Table 1**) and the true yield for methanogens utilizing acetate of 0.05 g VSS/g COD (Metcalf and Eddy 2003). Very low observed yields of 0.116 g VSS/g COD to 0.138 g VSS/g COD were measured in Phase VI-I to VI-III in the aerobic/anoxic CFBBR. Due to the fact that 95% of the overall COD was consumed in the anaerobic fluidized bed, the overall observed yield of A-CFBBR was calculated as 0.017 g VSS/g COD, close to the yield of AF.

Conclusions

- The newly developed A-CFBBR affected 99.7% COD, 84% nitrogen removal with a very low sludge yield of 0.017 g VSS/g COD while treating a high strength SWW containing 10700 mg COD/L and 250 mg NH_3-N/L. The system performance at a high OLR of 35 kg COD/m^3_{AF}·d and 1.1 kg N/m^3_{CFBBR}·d with an overall HRT of less than 12 h is remarkable.
- Microbial communities analysis using DGGE confirmed the presence of AOB and NOB in both the aerobic and anoxic columns and *Pseudomonas putida* responsible for SND in the aerobic column.

List of Tables and Figures

Table 1- Operating condition of the system in Mode 3 (A-CFBBR under operation)

Table 2- **(a)** Composition of the synthetic wastewater **(b)** Steady-state characteristics of anaerobic, anoxic and aerobic effluents in Mode 3 (A-CFBBR under operation)

Figure 1- Schematic of anaerobic-circulating fluidized bed bioreactor (A-CFBBR)

Figure 2 - **(a)** Biomass inventory **(b)** methane production in AF **(c)** COD removal in AF **(d)** biomass yields in AF, CFBBR and A-CFBBR

References

1. Andalib, M., Nakhla, G., Zhu, J., 2010 b. Dynamic testing of the twin circulating fluidized bed bioreactor (TCFBBR) for nutrient removal from municipal wastewater. Chem. Eng. J. 162, 616-625.
2. Andalib M, Nakhla G, , Zhu J. 2012. High-rate biological nutrient removal from high strength wastewater using novel Anaerobic Circulating Fluidized Bed Bioreactor (A-CFBBR). Bioresource Technology 118, 526-535.
3. APHA; AWWA; WEF. 1998. Standard methods for the examination of water and wastewater. 20th Edition, American Public Health Association, Washington D.C.
4. Chen, K.C., Lin, Y.F., 1993. The relationship between denitrifying bacteria and methanogenic bacteria in a mixed culture system of acclimated sludges. Wat. Res. 27(12), 1749-1759.
5. Chowdhury, N., Nakhla, G., Zhu, J., Islam, M., 2010. Pilot-scale experience with biological nutrient removal and biomass yield reduction in a liquid-solid circulating fluidized bed bioreactor. Wat. Environ. Res. Doi: WER 08-11-1541RR.
6. Cui, Y., Nakhla, G., Zhu, J., Patel, A., 2004. Simultaneous carbon and nitrogen removal in anoxic-aerobic circulating fluidized bed biological reactor (CFBBR). Environ. Technol. 25, 699-712.
7. Eldyasti, A., Chowdhury, N., Nakhla, G, Zhu, J., 2010. Biological nutrient removal from leachate using a pilot liquid–solid circulating fluidized bed bioreactor (LSCFB). J of Haz. Mat. 181, 289–297.
8. Metcalf and Eddy, 2003. Wastewater Engineering: Treatment and reuse, Fourth ed., McGraw-Hill.
9. Chen, Y., Peng, C., Wang, J., Ye, L., Zhang, L., Peng, Y., 2011, Effect of nitrate recycling ratio on simultaneous biological nutrient removal in a novel anaerobic/anoxic/oxic (A2/O)-biological aerated filter (BAF) system. Bioresource Technol. 102(10), 5722-5727.

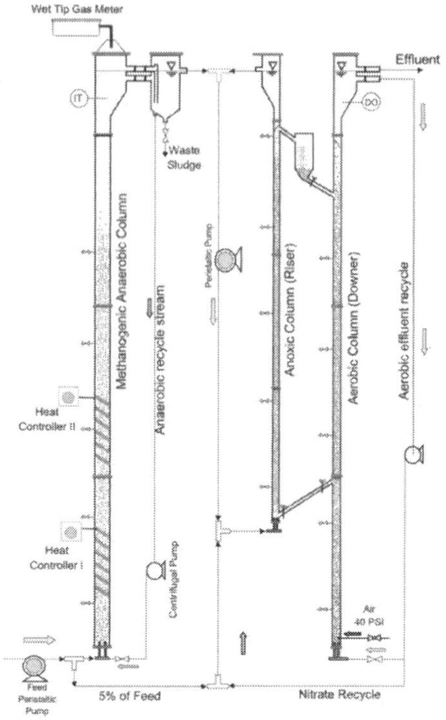

Figure 1

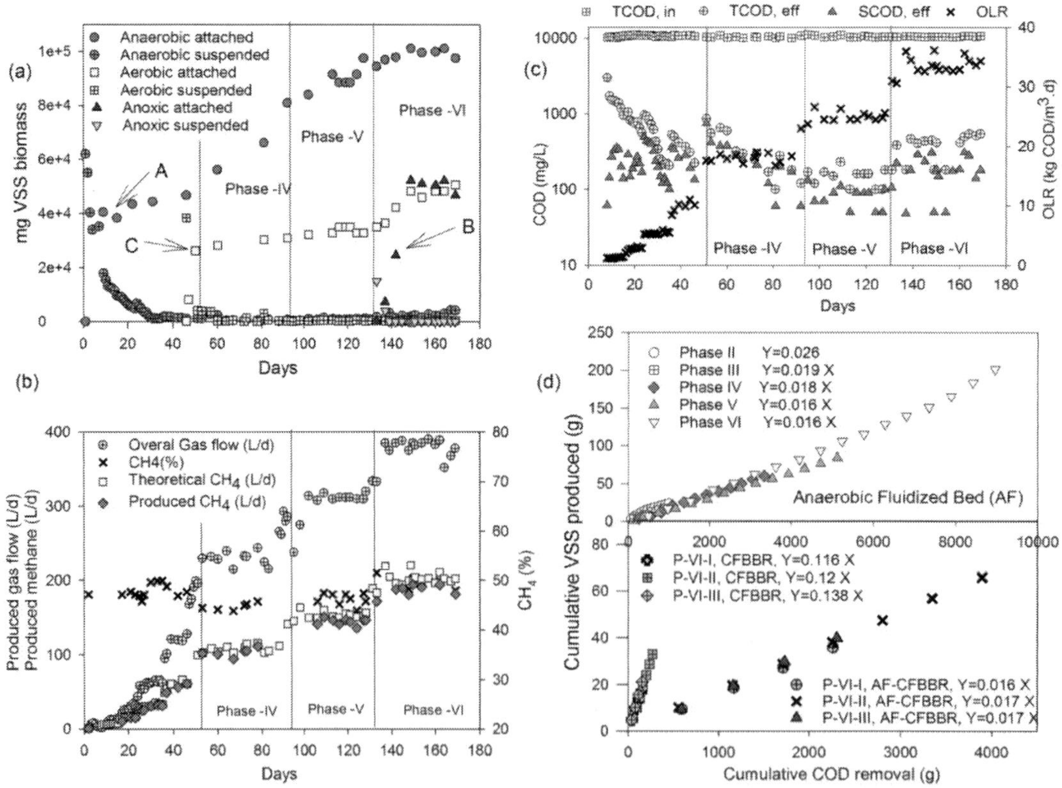

Figure 2

Table 1- System operating conditions in Mode 3 (A-CFBBR under operation)

	A-CFBBR under operation (**Mode 3**)		
	Phase-VI-I	**Phase-VI-II**	**Phase-VI-III**
Day	129-145	145-160	160-172
Feed flow rate (L/d)	47-55[a]	55[a]	55[a]
Anaerobic-Anaerobic circulation flow (Q/Q_{in})	109	107	105
Anoxic-Anoxic Circulation flow (Q/Q_{in})	2	2	2
Aerobic-Aerobic circulation flow (Q/Q_{in})	64	59	64
Aerobic-Anoxic Circulation flow (Q/Q_{in})	2.3	4	6.2
OLR based on anaerobic reactor (kg $COD/m^3.d$)	33.3±1.5	33.5±1.2	34.5±0.8
NLR based on anaerobic reactor (kg $N/m^3.d$)	0.8±0.1	0.8±0.1	0.85±0.0
OLR based on CFBBR reactor (kg $COD/m^3.d$)	3.8±0.7	3.3±0.1	4.8±0.2
NLR based on CFBBR reactor (kg $N/m^3.d$)	1.11±0.0	1.12±0.0	1.15±0.0
Anaerobic EBCT (h)=$V_{compact}/Q_{in}$	1.6	1.6	1.6
Aerobic EBCT (h)=$V_{compact}/Q_{in}$	1.3	1.3	1.3
Anoxic EBCT (h)=$V_{compact}/Q_{in}$	0.27	0.26	0.26
Anaerobic HRT(d)	0.32	0.32	0.32
Aerobic HRT(d)	0.17	0.17	0.17
Anoxic HRT (d)	0.04	0.04	0.04
Anaerobic attached biomass (mg VSS/g media)	32±0.5	33±0.2	33±0.4
Aerobic attached biomass (mg VSS/g media)	16.5±1.7	20.6±0.5	20.5±0.5
Anoxic attached biomass (mg VSS/g media)	1-41	85±1.5	82.2±2
Total anaerobic attached biomass (g VSS)	96.4	100.2	99.3
Total aerobic attached biomass (g VSS)	37.9	47.5	47.8
Total anoxic attached biomass (g VSS)	0.3-24.6	51.2	49.5
F/M in AF (g COD/g VSS.d)	5.5	5.4	5.5
F/M in CFBBR (g COD/g VSS.d)	0.66	0.35	0.55
Anaerobic col. detachment rate (1/d)		0.035	
Aerobic col. detachment rate (1/d)		0.035	
Anoxic col. detachment rate (1/d)		0.02	
Aerobic DO (mg/L)	6.2±0.1	6.3±0.1	5.6±0.0
Anoxic DO (mg/L)	0.35±0.1	0.4±0.1	0.7±0.2
AF estimated SRT (d)		32	
CFBBR estimated SRT (d)		28	
Superficial liquid velocity in anaerobic col. (cm/s)		1.42±0.04[b]	
Superficial liquid velocity in aerobic col. (cm/s)		1.7±0.1	
Superficial liquid velocity in anoxic col. (cm/s)		1.2±0.1	
Specific nitrification rate (**SNR**) (g NH_3-N/g VSS·d)-Downer		0.25-0.31	
Specific nitrification rate (**SNR**) (g NH_3-N/g VSS·d)-Riser		0.05±0.01	
Specific denitrification rate (**SDNR**) (g NO_3-N/g VSS·d)-Riser		3.4-3.8	
Specific denitrification rate (**SDNR**) (g NO_3-N/g VSS·d)-Downer		0.3±0.1	

[a] 5% of overall feed was directly fed to the bottom of the anoxic column.
[b] The terminal settling velocity of Zeolite with d_m=600 μm and 1000 μm biofilm thickness is 2.8 cm/s, still significantly lower than the operating u_L (Andalib et al., 2010 a).

Table 2a. Composition of the synthetic wastewater

Feed composition [a]	CH_3COOH (mL/L$_{Feed}$)	NH_4Cl (g/L$_{Feed}$)	K_2HPO_4 (g/L$_{Feed}$)	$MgSO_4.7H_2O$ (g/L$_{Feed}$)	$CaCl_2.2H_2O$ (g/L$_{Feed}$)	Yeast extract (g/L$_{Feed}$)	$NaHCO_3$ (g/L$_{Feed}$)	Trace element (mL/L$_{Feed}$)
Concentrations	9.5	1.7 and 0.93	0.1	0.03	0.03	0.03	5.7-6.7	1
Trace element Composition	$FeCl_2.4H_2O$ (mg/L$_{Trace\ E}$)	$MnCl_2.4H_2O$ (mg/L$_{Trace\ E}$)	H_3BO_3 (mg/L$_{Trace\ E}$)	$ZnCl_2$ (mg/L$_{Trace\ E}$)	$CuCl_2$ (mg/L$_{Trace\ E}$)	$AlCl_3$ (mg/L$_{Trace\ E}$)	$CoCl_2.6H_2O$ (mg/L$_{Trace\ E}$)	$NiCl_2$ (mg/L$_{Trace\ E}$)
Concentrations	2000	500	50	50	30	50	50	50

[a] All the above reagent were ACS Grade, 99.5% min.

Table 2b. Steady-state characteristics of anaerobic, anoxic, and aerobic effluents in Mode 3 (A-CFBBR operational)

		A-CFBBR operated (Mode 3)						
Parameter (mg/L)		**Phase-VI-I** to **VI-III**	**Phase-VI-I**		**Phase-VI-II**		**Phase-VI-III**	
Days (d)		*129-172*	*129-145*		*145-160*		*160-172*	
	Influent	Anaerobic Effluent	Anoxic	Effluent	Anoxic	Effluent	Anoxic	Effluent
Produced biogas(L/d)	-	333 to 405	-	-	-	-	-	-
TCOD	10600 ± 270	390± 140	61±14	49±8	80±5	58±7	79±24	67±1
SCOD		181± 86	30±11	31±5	50±16	40±18	46±16	34±12
TSS	-	188±85	28±8	22±10	42±15	32±10	29±8	30±10
VSS	-	146±56	21±5	16±8	27±12	19±7	23±6	23±8
NH_3-N	250	215±8	70±1	3±1	45±12	1.3±0.3	57±19	3.1±0.6
NO_3-N	2	1.8±1	4.7±4	76±22	2±0.5	39±2	16±11	88±5
NO_2-N	1	1.3±1	1.3±0.6	0.9±0.1	2±0.5	2±0.5	38±2	20±2
TN	255(430)[b]	231±12	-	78±21	-	45±10	-	113±10
PO_4-P	26.3	23.5±2.6	19±0.5	18.8±0.8	20.5±2	20±2	21±1	19.3±0.5
Alkalinity[a]	-	3065±115	2115±21	1875±35	2630±180	2017±280	2095±100	1885±120
ORP(mV)	-	-193±8	-	38±13	-	. 12±2	-	2±1
pH	4.6±0.1	7.09±0.0	8.2±0.1	8.4±0.1	8.2±0.2	8.4±0.1	8.1±0.1	8.3±0.0
Temp (°C)	20	35.7±0.8	22.4±0.1					

[a] as mg $CaCO_3$/L   Nitrogen and phosphorus content of biomass were measured 9±1.2% and 2.1±0.4% respectively

Reduction of VOC Emissions in High Purity Oxygen Activated Sludge Wastewater Treatment Process: Toxchem™ Based Fate & Modeling Case Study

Malcolm Fabiyi, Richard A. Novak, Praxair, Inc., Burr Ridge, IL
Rajeev Goel, Spencer Snowling, Hydromantis, Inc., Hamilton, ON

ABSTRACT

In the last decade, stringent regulations have been imposed on the permissible level of Volatile Organic Compound (VOC) emissions from industrial facilities. The clean air act amendments of 1990 and related regulatory extensions (e.g., National Emissions Standards for Hazardous Air Pollutants, NESHAPs) have significantly reduced the acceptable level of Volatile Organic Compound (VOC) emissions from industrial facilities (Woodward & Curran, 2006). A total of 188 organic compounds have been designated Hazardous Air Pollutants (HAPs); facilities which generate or handle these compounds are subject to permitting, monitoring and reporting requirements. Although extensive emissions controls have been implemented in the production process of many industrial facilities, significant amounts of VOCs may appear in the wastewater. These VOCs can be stripped into the air during conveyance through collection systems or during biological treatment.

Fate and Transport models are used for a variety of tasks including design, emissions analysis and regulatory reporting. The development of fate & transport models in wastewater collection and treatment systems were spurred by regulatory drivers like the Clean Air Act (Melcer, 1994). A variety of computer-based fate & transport models such as BASTE, EPA Water Models (versions 7, 8 and 9), CORAL, PAVE, SIMS, TORONTO, INTERCEPTOR and Toxchem™ have been applied for addressing the fate of volatile contaminants in collection systems, drop structures, weirs, quiescent surfaces, and wastewater treatment processes (Quigley et al, 2006; Melcer, 1994).

EPA Water Models (7, 8 & 9) and Toxchem™ are the most commonly utilized fate and emissions transport modeling platforms in the wastewater industry. In both models, the default aeration options exclusively model the mass transfer behavior as diffused air processes or surface aeration style mechanical mixers. Attempts at modeling the performance of High Purity Oxygen (HPO) based devices have typically required finding the most appropriate approximation to the mass transfer and VOC stripping behavior of HPO aeration systems by modifying default parameters in the diffused air or surface aeration modules (NYSERDA, 2000; Levine et al, 2010; Rodieck et al, 2001). In contrast, an improved modeling approach is employed in the recently released Toxchem™ 4.1 (Hydromantis, 2012), which extends the simulation capabilities of the platform to include the modeling HPO and Sequencing Batch Reactor (SBR) process systems.

This paper discusses results for VOC emissions modeling obtained by applying Toxchem™ 4.1 to VOC reduction projects that utilize SBR and HPO systems. We provide a methodology for modifying the default parameters in diffuser and mechanical surface aeration modules in fate and transport models in order to enable VOC emissions characteristics associated with HPO systems to be modeled, and provide comparative analyses on the effect of the approach adopted on simulation results.

KEY WORDS

VOC emissions, mass transfer, mechanically agitated contactors, Toxchem, Fate & Transport models, High Purity Oxygen

BACKGROUND

The mass transfer models in fate and transport emissions models are based on ASCE standard model parameters (Chern & Yu, 1995). They model the three major mechanisms through which VOC removal from wastewater occurs: (i) biosorption of VOCs onto activated sludge floc and subsequent removal via waste sludge, (ii) volatilization and stripping of VOCs into the atmosphere during aeration, and (iii) biological degradation.

Mechanisms for VOC Stripping

There are two distinct mechanisms for VOC stripping in aerated activated sludge biological wastewater treatment processes: (i) *turbulence induced stripping*, in which turbulence generated by the aeration system induces surface renewal leading to the transfer of VOCs from the bulk liquid across the air-water interface; and (ii) *bubble aided stripping,* in which VOCs in the wastewater diffuse into un-dissolved gas bubbles and are stripped as the bubbles leave the water surface.

Aeration Systems and Their effect on VOC Stripping

Three general classes of aeration devices exist (Melcer, 1994; Mueller et al, 2002): (i) mechanical surface aerators (ii) diffused aeration systems, and (iii) subsurface mechanical aerators with gas sparging.

Mechanical Surface Aerators: Mechanical surface aerators spray water droplets into the headspace above the aerator; gas-liquid exchange occurs as gases diffuse from the surrounding atmosphere into the droplets, as well as from the droplets into the atmosphere. In these devices, VOC emissions occur in three ways: (i) from the droplets to the air, (ii) from the bulk liquid into air bubbles that have been entrained under the surface of the water, as the air bubbles rise and reach the liquid surface, and (iii) across the turbulent free surface of the basin.

Diffused Aeration Systems: Fine Bubble Diffuser systems are commonly utilized for aeration in wastewater treatment processes because they provide relatively high oxygen transfer efficiencies (OTE) (in the range of 20%-30% for most common applications) relative to alternatives like coarse bubble systems and surface aerators (Mueller et al, 2002). Considering the fact that air typically has about 21% oxygen by volume, OTE values for aeration systems in the range of 20%-30% implies that only about 4%-6% of the volumetric air flow supplied to a basin using a diffused aeration system is actually dissolved. The undissolved gases, which make up about 94%-96% of the gas supplied, provide a large volumetric flow of gas which strips VOCs from the basin.

Subsurface mechanical aerators: In subsurface mechanical aeration systems, process gases such as oxygen are injected below the water surface in combination with a mechanical mixer which serves to maintain efficient gas-liquid contacting. Oxygen diffuses into the wastewater, and VOCs transfer from the liquid phase to the gas bubbles. The bubbles eventually rise to the surface and VOC release occurs as the bubbles break the surface. Subsurface mechanical aerators generate some turbulence at the water surface, and also provide subsurface sparging of gases. The mass transfer and VOC stripping behavior of such systems therefore combines elements of both mechanical surface aeration (surface renewal aided VOC diffusion across the air-water interface due to turbulence) as well as diffuser aeration (off-gas VOC stripping.)

High Purity Oxygenation Systems for VOC Emissions Reduction

The VOC stripping potential of the large un-dissolved volume of gases (approximately 94%-96% of gas supplied) associated with the conventional aeration process has long been recognized. Process options that have high oxygen transfer efficiencies, such as high purity oxygenation systems, have been selected for several wastewater treatment applications where reduction in VOC emissions from the wastewater treatment process is a key project goal (e.g., Rodieck et al, 2001; Lubkowitz-Bailey et al, 2002; Levine et al, 2010; NYSERDA, 2000; Melcer, 1994). High purity oxygen (HPO) systems have been used in wastewater treatment applications for over four decades (Mueller et al, 2002). High efficiency oxygenation systems such as Praxair's In-Situ Oxygenation (I-SO™) system (Mueller, 2002) have been demonstrated to provide up to 97% reduction in VOC emissions relative to a comparative diffused air based aeration system (Rodieck et al, 2001).

Praxair's In-Situ Oxygenation (I-SO™) system is a subsurface, down-pumping mixing and oxygenation system used for providing oxygen in open basin HPO operations (Figure 1). The down-pumping action of the helical impeller generates a vortex which entrains pure oxygen gas supplied to a floating hood. The high shear in the draft tube region breaks up the entrained gas into fine bubbles. The intimate gas-liquid contact in the draft tube significantly enhances mass transfer. Dissolved oxygen is dispersed by the mixing action induced by the high velocity liquid jet ejected from the draft tube. Undissolved gas bubbles detach from the liquid flow and rise; some bubbles are recycled through the draft tube, allowing high oxygen transfer efficiencies to be attained. Standard aeration efficiencies of up to 10 Ibs O_2/HP-hr and oxygen transfer efficiencies of up to 90%-95% can be obtained using the I-SO™ mixing system (Praxair, 2003; NYSERDA, 2000). HPO systems have high aeration efficiencies (8-10 Ibs O_2/HP-hr), low mixing intensities (as low as <10 w/m^3) and high standard oxygen transfer efficiencies (+90%), and have been associated with significant reductions in VOC emissions when used in the wastewater treatment process (Mueller et al, 2002; Levine et al, 2010; Lubkowitz-Bailey et al, 2002; Rodieck et al, 2001; and NYSERDA, 2000).

METHODOLOGY

Given that commercially available modeling platforms offer only diffused aeration and mechanical surface aeration options as the default modules, the approach taken in this paper is to adapt aeration models (based on air) for use in modeling the behavior of high purity oxygen systems. Using Toxchem 4.1 as the fate & transport modeling platform, critical model parameters such as comparative gas flow rates, oxygen transfer efficiencies, O_2 mass fraction in gas source, ratio of the gas and liquid film mass transfer coefficients (K_g/K_l), aeration power, standard oxygen transfer rate (lbs O_2/HP-hr), and mass transfer correction factors (alpha) are varied for both diffused air and mechanical surface aeration systems to establish which variables to modify to adequately model the VOC emissions characteristics of pure oxygen systems.

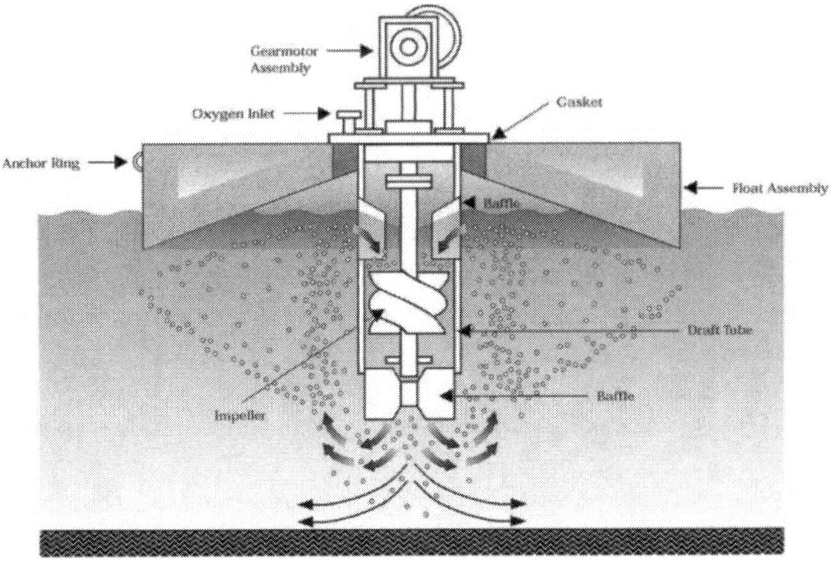

Figure 1: Praxair's In-Situ Oxygenation (I-SO™) system

Published data on comparative VOC emissions reductions from pilot and full scale applications using HPO vs. conventional air-based aeration systems are used as the data source for the analysis (Lubkowitz-Bailey et al, 2002; Rodieck et al, 2001).

In Lubkowitz-Bailey et al, 2002 and Rodieck et al, 2001, pilot and full scale data are presented for a SBR activated sludge process, used for the treatment of Active Pharmaceutical Ingredient (API) wastewater from a pharmaceutical plant. The data evaluated in this study is from the pilot system, which comprised a single SBR system operated in either diffused aeration or HPO mode (see Tables 1 & 2). The pilot scale SBR was operated in 6 hour cycles with 5 phases (Mix/Fill, React/Fill, React, Settle, and Decant). Wastewater flow of 250 gpd was supplied to the process. The reactor diameter was 10 ft, with a height of 15 ft, giving a total SBR volume of 6,000 gals. Fill volume during the tests was 4,650 gals, giving an effective liquid depth of about 11.6 ft. Gas flowrates supplied during the tests ranged from 0.32-0.67 cfm during HPO operations, and 28.5 to 32.5 cfm for diffused aeration operations. Under all aeration modes, enough oxygen was provided to ensure that COD removal targets were met. The VOCs contained in the API wastewater included Acetone, Toluene, MTBE, MEK, Methylene Chloride, 1,1-Dichloroethane, Ethanol, Isopropanol, Methanol, 1,1,1-Trichloroethane, Butyl Acetate and Heptane. During pilot testing, comparative VOC treatment performance was carried out in three baselines tests, and three spike tests (Rodieck et al, 2001). The spike tests were used as the basis for the modeling exercise in this study. Average MLSS was approximately 3,500 mg/l. Detailed monitoring of VOC emissions was undertaken by monitoring VOC in influent, effluent and off-gas samples.

Acetone and toluene were the major HAPs in the influent, with concentrations ranging from 7to 5,2200 and 49 – 320 mg/l respectively at pilot scale. Influent COD ranged from 4,069 mg/l (Rodieck et al, 2001) to 4,800 mg/l (Lubkowitz-Bailey et al, 2002).

The full scale system was designed for 240,000 gpd peak daily flow. Treatment was carried out in three (3) HPO SBR systems, with a design F/M basis of 0. 1 Ib BOD/lb MLSS-day. High Purity Oxygen supply to each SBR was provided using a single 40 HP Praxair I-SO™ system in each reactor, capable of transferring about 300 Ibs O_2/hr.

Model Configuration

VOC fate modeling was undertaken using the new Toxchem™ 4.1 modeling platform. This latest version of Toxchem™ (Hydromantis, 2012) includes High Purity Oxygen and Sequencing Batch Reactor Modules. Given that SBRs were utilized with pure oxygen and conventional aeration in the studies that were used as a basis for this paper (Rodieck et al, 2001; Lubkowitz-Bailey et al, 2002) the updated Toxchem™ platform enables the effective simulation of VOC and odor emission in treatment processes that involve the application of pure oxygen in subsurface mixing and dissolution systems. An earlier version of Toxchem™ which did not have custom SBR and HPO modules was used to support the modeling work in Rodieck et al (2001).

Since an integrated SBR and HPO module does not exist in Toxchem™ 4.1, the HPO SBR system was modeled as a SBR with modified diffused aeration parameters. To capture the dynamics of HPO operations, the effective oxygen transfer efficiencies and gas flowrates that apply to HPO operations were utilized in the diffuser models. Previous work has demonstrated that this approach to modeling HPO systems is effective (Fabiyi, 2012; Rodieck et al, 2001; NYSERDA, 2000). The three Spike tests reported in Rodieck et al (2001) were used as the basis for the simulations. The simulations were set up to evaluate HPO vs. diffused aeration system performance in parallel trains. Toxchem™ also allows visual displays that match emissions values around process units (emissions hotspots) to preset threshold values that are characterized by representative color codes (see Figure 2).

Model Calibration

Test data for the spike test runs were input into the representative model layouts and run using test values for process flowrates, oxygen mass transfer efficiencies and reactor conditions. A systematic analysis of the impact of various process parameters on the predictive accuracy of the model was undertaken using the Sensitivity Analysis tool in Toxchem™. This tool allows the evaluation of the impact of four broad classes of factors on model output, including: contaminant properties (e.g., Van't Hoff coefficients for estimation of Henry's Law Constant, adsorption characteristics, bio-kinetic properties of the compounds); site properties (elevation, wind speed, pH); and biological treatment system properties (e.g. liquid depth, oxygen transfer efficiencies, aeration power input, DO levels, Kg/Kl, aeration, etc.)

The sensitivity analysis indicated that the Van't Hoff estimates were the most important factor for ensuring that the model output was reliable. The default values for the Van't Hoff variables were reviewed and updated where necessary. Given that the reactor was operated at 80-100°F (Rodieck et al, 2001), default Henry's law constant data that were reported outside of this temperature range were dropped and replaced with the Van't Hoff estimates, which allow adjustment of Henry's Law constants for temperature.

A simplified form of the Van't Hoff relationship is given as

$$LnH_{cc,T} = A - \frac{B}{T}$$

Where $H_{cc,T}$ is the Henry's Law Constant, while A and B are the Van't Hoff coefficients. The value for A varies with the exact unit specifications used for Henry's Law Constant, but the numerical value estimated for B does not usually change for the commonly utilized unit expressions for $H_{cc,T}$.

Henry's Law Constant (HLC) values can be affected by a variety of factors such as temperature, pH, compound hydration, compound concentration, co-solutes, co-solvents, dissolved salts, suspended solids, dissolved natural organic materials, surfactants and water composition (Staudinger & Roberts, 1996). Starting Van't Hoff estimates were obtained using average values from a comprehensive compilation of published studies reporting experimental determinations of Henry's Law Constant temperature dependence relationships (Staudinger & Roberts, 2001). Table 3 summarizes the HLCs for the VOCs utilized in the model.

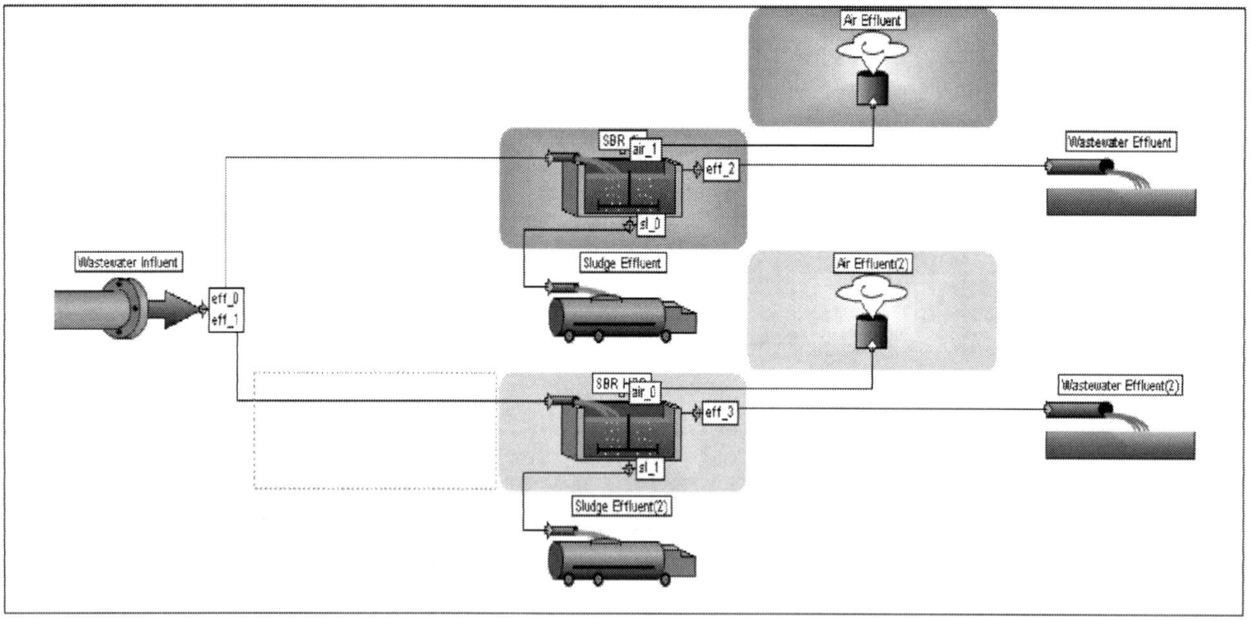

Figure 2: Layout depicting parallel trains used for evaluating VOC emissions behavior of HPO vs conventional diffused aeration SBR processes.

Table 1: Summary of Comparative VOC Emissions from pilot tests used in this study (Source: Rodieck et al, 2001).

FIELD MEASUREMENT RESULTS FOR TOLUENE

Testing Event	Percent of Influent Toluene Load Emitted to the Atmosphere			
	Measured		Model	
	HPO	DA	HPO	DA
HPO Baseline Test #1	0.1	---	0.3	---
HPO Baseline Test #2	0.5	---	0.4	---
DA Baseline Test	---	10.5	---	7.6
Spike Test #1	0.5	18.7	0.2	4.9
Spike Test #2	0.3	9.5	0.2	3.7
Spike Test #3	0.2	11.5	0.2	4.2
Average	0.3	12.6	0.2	5.1

Table 2: Model input variables (Source: Rodieck et al, 2001).

FATE AND TRANSPORT MODEL INPUT PARAMETERS

Category	Parameter	Units	HPO	DA
Ambient Conditions	Elevation	ft	329	
	Temperature	°F	80	
	Wind speed	mph	6.2	
Influent	VOC concentration	mg/L	variable	
	Suspended solids (SS)	mg/L	10	
	Volatile SS Ratio	%	10	
	Temperature	°F	80-100	
Reactor	Reactor Depth	ft	8.33	
	Reactor Surface Area	ft^2	30.6	
	Mixed liquor suspended solids (MLSS)	mg/L	3,000 – 4,400	
	MLSS volatile SS ratio	%	50	
	MLSS dissolved oxygen content	mg/L	2.0	
	Oxygen supply gas flow rate	cfm	0.33-0.80	15-35
	Oxygen transfer efficiency	%	95	10
	Mole fraction of oxygen source	---	1.0	0.2

Table 3: HLC = Henry's Law Constant. S&R (2001) refers to Staudinger, J & Roberts, P.V. A Critical Compilation of Henry's Law Constant Temperature Dependence Relations for Organic Compounds in Dilute Aqueous Solutions. Chemosphere 44 (2001).

		Default			S&R (2001)		Modified	
		HLC	Van't Hoff A	Van't Hoff B	Van't Hoff A	Van't Hoff B	Van't Hoff A	Van't Hoff B
Toluene			5.133	3024	5.271	1745	5.271	2800
	°C/St.Dev				(±1.11)	(±436.3)		
Methyl-Tertiary-Butyl Ether		0.022688			9.07	3178	6	3178
	°C/St.Dev	(@25°C)						
Methylene Chloride		0.122618			4.561	1644	4.561	2900
	°C/St.Dev	(@25°C)			(±0.50)	(±148)		
Butyl Acetate,n-		0.00671			6.4	2486	6.4	3846
	°C/St.Dev	(@25°C)						

RESULTS & OBSERVATIONS

a. Effect of HPO Operations on VOC Emissions

Rodieck et al (2001) carried out three spike tests using numerous VOCs that were intended to stress test the HPO system and evaluate its capacity for VOC reduction versus a comparative air based system. Those spike tests have been modeled in this paper, and good agreement was found between model predictions and test results. The VOCs of particular interest across all the spike tests that were carried out are Toluene, MTBE, $MeCl_2$ and n-Butyl Acetate. The model provided reliable estimations of the effect of aeration mode (conventional vs. HPO) on VOC stripping behavior (see Figures 3 – 6).

Spike Test 1 (Figure 3). 97% and 93% reductions in Toluene and MTBE emissions respectively were observed in Spike Test 1. The model also provides reliable estimates of VOC emissions reduction, with 98% and 97% reductions in Toluene and MTBE emissions respectively. . Comparative gas supply ratio V_{Air}/V_{HPO} (Air volume vs. HPO volume) is 91.2 (i.e., 77 ft3 HPO, 7024 ft3 air).

Spike Test 2 (Figure 4). 97% and 96% reductions in Toluene and Methyl Chloride emissions respectively were observed. Model estimates of VOC emissions reduction tracked test results at 97% for Toluene and 96% for Methyl Chloride. Comparative gas supply ratio V_{Air}/V_{HPO} (Air volume vs. HPO volume) is 54.6 (i.e., 125 ft3 HPO, 6830 ft3 air)

Spike Test 3 (Figure 5). 96% and 98% reductions in Toluene and Butyl Acetate emissions respectively were observed during the test. Model estimates of VOC emissions reduction was 97% for Toluene and 93% for Butyl Acetate. Comparative gas supply ratio V_{Air}/V_{HPO} (Air volume vs. HPO volume) is 48.4 (i.e., 161 ft3 HPO, 7800 ft3 air)

The data and simulation results across all the spike tests consistently demonstrate the significant reduction in VOC emissions obtained by using HPO instead of air as the source of oxygen in the process gas supply. Toluene was common to all the spike tests, and the average reduction in

Toluene related VOC emissions across all tests was 97% for HPO vs. conventional diffused aeration based systems (see Figure 6). Differences in the comparative volumes of gas introduced into the system during air versus HPO operations ranged from 48-92.

b. Comparison of HPO and SBR Modules

The predictive capacity of the new HPO module in Toxchem™ 4.1 was tested by comparing the results from simulations of the Batch 3 spike tests of Rodieck et al (2001) using the SBR and HPO modules. As previously indicated, the modeling of HPO operations in the SBR is done using diffused aeration principles. Although the I-SO™ system is a mechanically agitated mixing and gas dissolution system, it differs fundamentally from surface aerators because it is a subsurface mixer. The stripping behavior of subsurface mixing systems in the wastewater treatment process is modeled most accurately by the off-gas volumes. The mass transfer characteristics of the new HPO module in Toxchem™ is modeled on the hydrodynamics and mass transfer behavior of the I-SO™ system (Praxair, 2003).

Effect of Depth and Surface Area: The default HPO and Activated Sludge Diffusion Aeration modules in Toxchem™ are complete mix systems. In order to model an SBR system using these modules it could be necessary to modify the sequencing batch steps in a SBR as separate process units. We investigated the effect of depth and surface area on the stripping behavior in the HPO and Diffuse Aeration process layouts to determine whether the modification of either of these two parameters could be tolerated when trying to model the SBR using complete mix systems (see Figure 7). For the sensitivity analysis, depth ranged from 2 to 20 ft, and surface area from 30 to 200 sqft. Actual SBR depth and surface area were 11.6ft and 78.5 sqft respectively. The analysis indicated that while depth and surface area had no major effect on the stripping behavior in the HPO system, the diffused aeration system was significantly affected by these parameters. The sensitivity of the VOC stripping behavior in diffused aeration systems to surface area and depth, suggest that the modification of either of these parameters when trying to model the SBR using complete mix systems could lead to issues with the interpretation and accuracy of the model results.

Due to this observation, a simplified layout was selected, which comprised only of the HPO and Diffused Aeration module with no additional process units (Figure 8). The process units had the same depth and surface area values as the actual SBR. A key limitation to the method could stem from high VOC losses during the fill cycle of the SBR. Rodieck et al (2001) minimized VOC losses and volatilization during the fill cycle by feeding the liquid below the liquid surface (Geiling, 2012).

The results obtained for VOC reduction for HPO vs. conventional aeration processes using this approach compared well with results from the SBR modeling approach (see Figure 9).

c. *Effect of Gas Flow rate on VOC Emissions*

There are significant differences in gas flows between HPO and conventional aeration systems. As previously observed, the use of HPO systems can reduce gas flow as much as 50-100 times compared with air based process. Reduced gas supply and off gas rates in HPO systems is a key factor in the enhanced VOC reduction observed in HPO systems (Fabiyi, 2012, Rodieck et al, 2001; Levine et al, 2010).

A sensitivity analysis was run in Toxchem™ to determine the effect of gas flowrates on VOC stripping. A linear correlation was observed between VOC stripping and the volumes of airflow supplied to the wastewater treatment process within the range of flowrates analyzed (see Figure 10). The linear relationship observed between gas flowrates and VOC stripping provides strong support for the identification of bubble-aided stripping as a key mechanism for VOC emissions in diffused aeration and subsurface mechanically aerated systems (Levine et al, 2010).

CONCLUSIONS

VOC emissions are affected by a variety of factors including mixing intensity, gas flowrates, surface renewal rates and surface turbulence (Chern & Yu, 1995). This implies that VOC emissions can be reduced by using aeration devices that (i) have a relatively low power input, hence minimizing excessive turbulence and surface renewal, and (ii) provide high oxygen transfer efficiencies (OTE), thereby minimizing the amount of un-dissolved gas bubbles that will strip VOCs from the liquid.

We find that the VOC emission behavior of the HPO system is predicted accurately by modeling the mass transfer characteristics as that of a fine bubble diffuser system, with adjustments for gas flows (actual O_2 feed to the HPO unit is used), O_2 purity and oxygen transfer efficiency.

Attempts to utilize the default mechanical aeration module indicated that the mass transfer characteristics of surface aeration systems are not well suited to model HPO systems (Fabiyi, 2012). The ready adaptability of diffuse aeration models for capturing the dynamics of VOC emission from mechanical HPO oxygenation systems suggests that the dominant VOC emission mechanism in both diffused air and subsurface aeration systems is similar – VOC transfer to un-dissolved gas bubbles. In contrast, mechanical surface aeration systems introduce a third mechanism into VOC stripping dynamics. In addition to volatilization due to surface renewal from turbulence, and VOC emissions due to volatile transfer to un-dissolved gas bubbles, VOC losses also occur due to exchange between the air and volatiles in the airborne water droplets. Gross air and water conditions such as temperature and wind velocities therefore become more important in accurately estimating VOC emissions in mechanical surface aeration systems (Chern & Chou, 1999; Roberts & Dandliker, 1983).

Despite the potential for VOC emission reduction afforded by high purity oxygen systems for wastewater treatment, commonly available fate & transport models do not have default modules that enable their modeling (Levine et al, 2010; Lubkowitz-Bailey et al, 2002; Rodieck et al, 2001). Toxchem™ 4.1 is the first commercially available fate modeling tool that has a default HPO module. Our work has demonstrated that this new tool is robust and reliable, accurately modeling both pilot and full scale application of HPO aeration for VOC emission reduction.

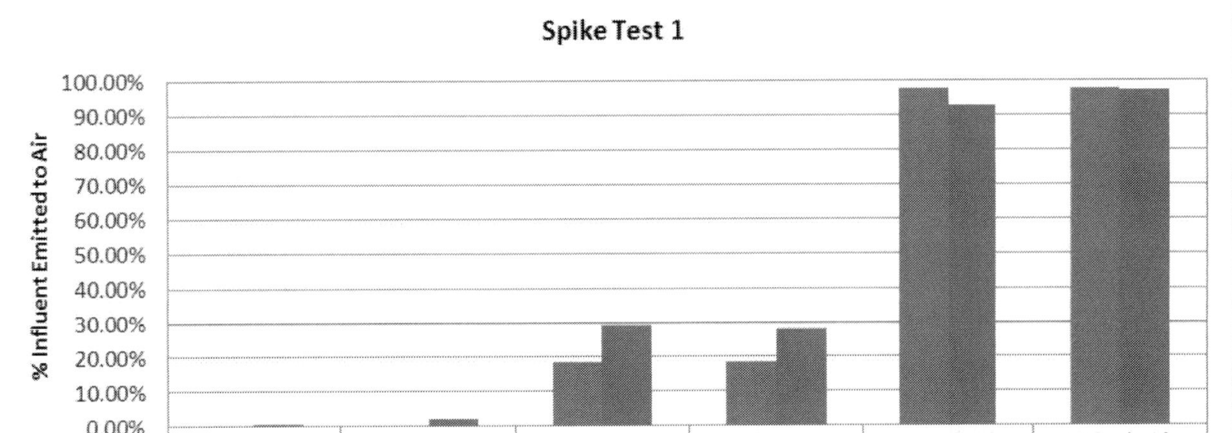

	HPO (Model)	HPO (Data)	Air (Model)	Air (Data)	VOC Reduction (Data)	VOC Reduction (Model)
Toluene	0.45%	0.50%	18.67%	18.70%	97%	98%
MTBE	0.81%	2.10%	29.24%	28.30%	93%	97%

Figure 3: Effect of aeration mode on VOC Stripping in Spike Test 1

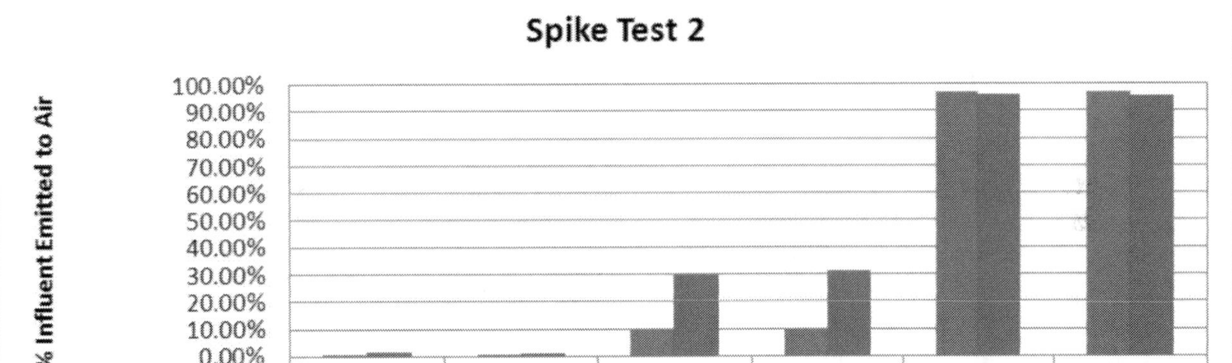

	HPO (Model)	HPO (Data)	Air (Model)	Air (Data)	VOC Reduction (Data)	VOC Reduction (Model)
Toluene	0.31%	0.30%	9.99%	9.50%	97%	97%
Methylene Chloride	1.34%	1.30%	29.83%	31.20%	96%	96%

Figure 4: Effect of aeration mode on VOC Stripping in Spike Test 2

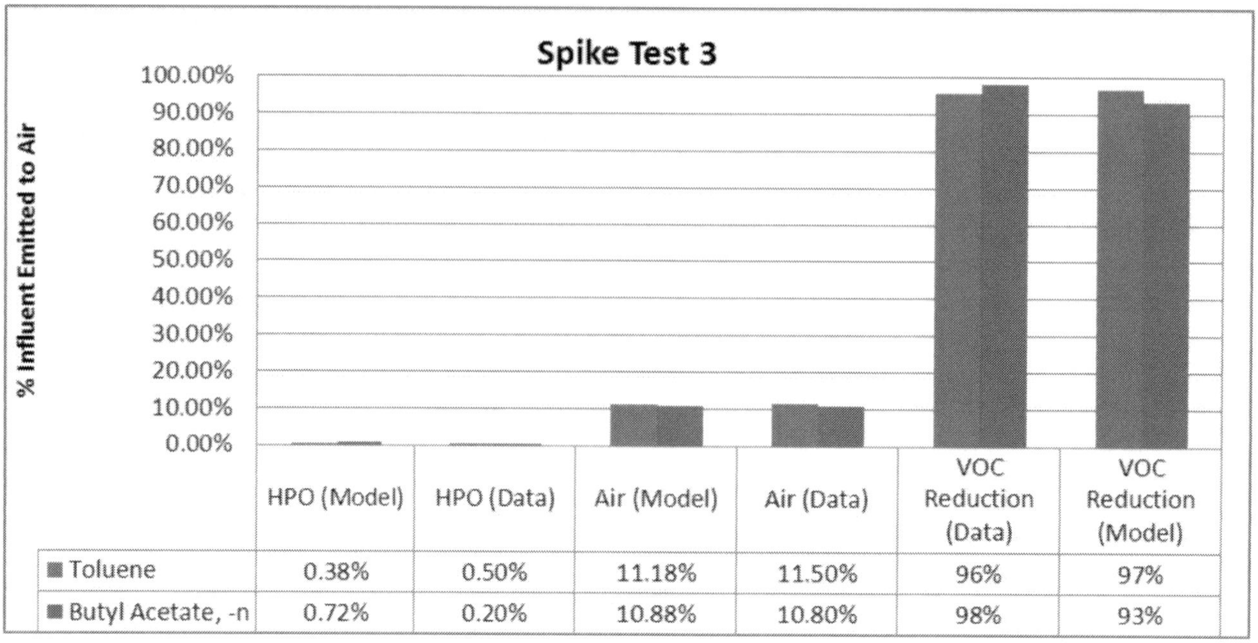

Figure 5: Effect of aeration mode on VOC Stripping in Spike Test 3

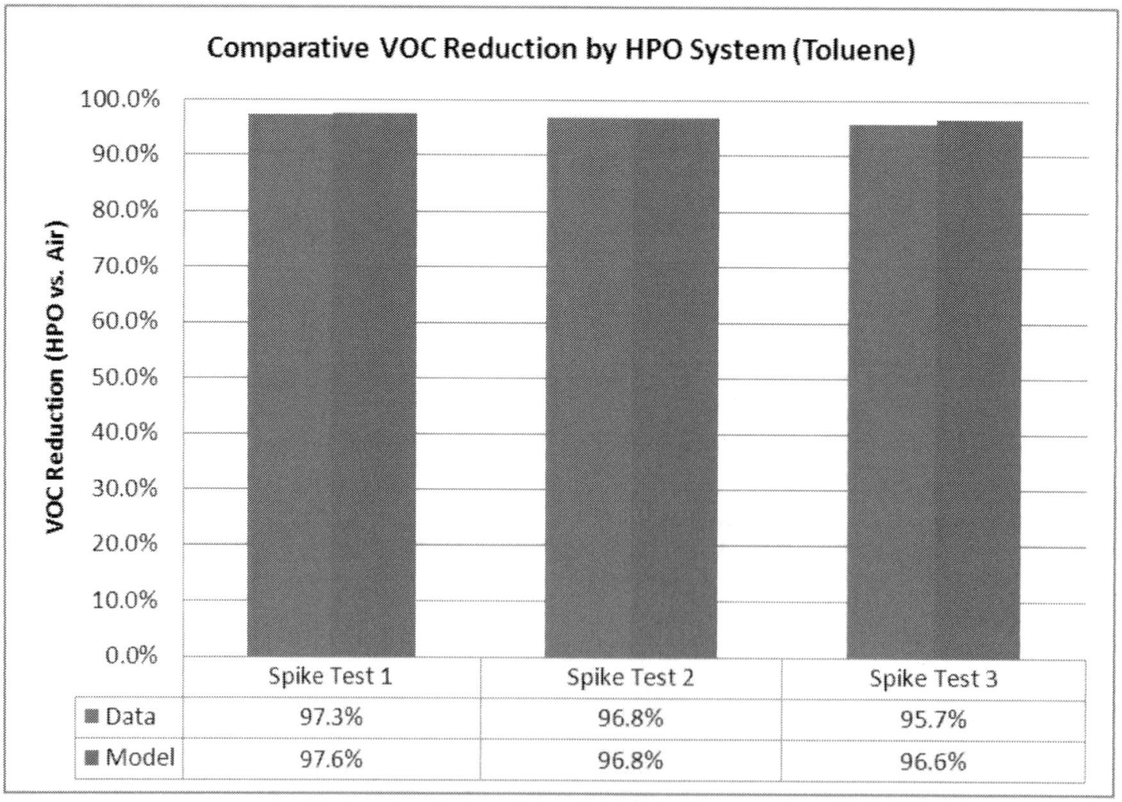

Figure 6: Summary of Modeling and Empirical Values for VOC Emissions reduction obtained using I-SO™ HPO Systems

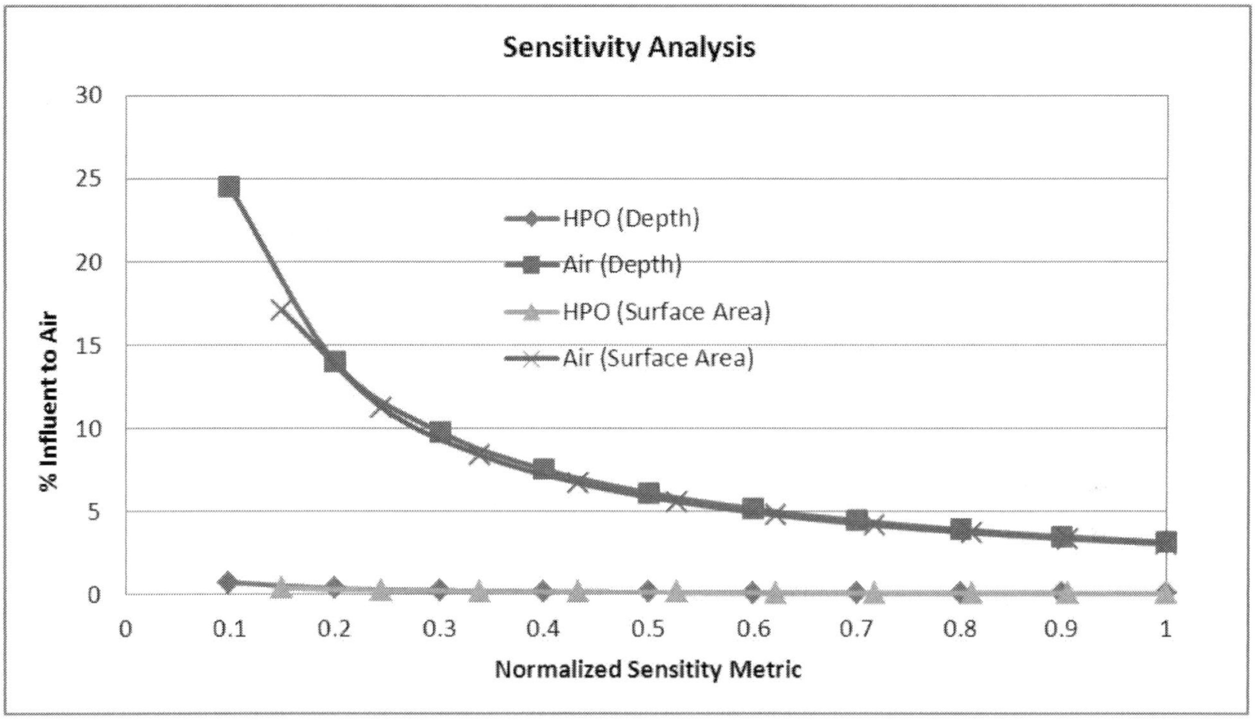

Figure 7: Effect of depth and surface area on VOC emissions for Toluene.

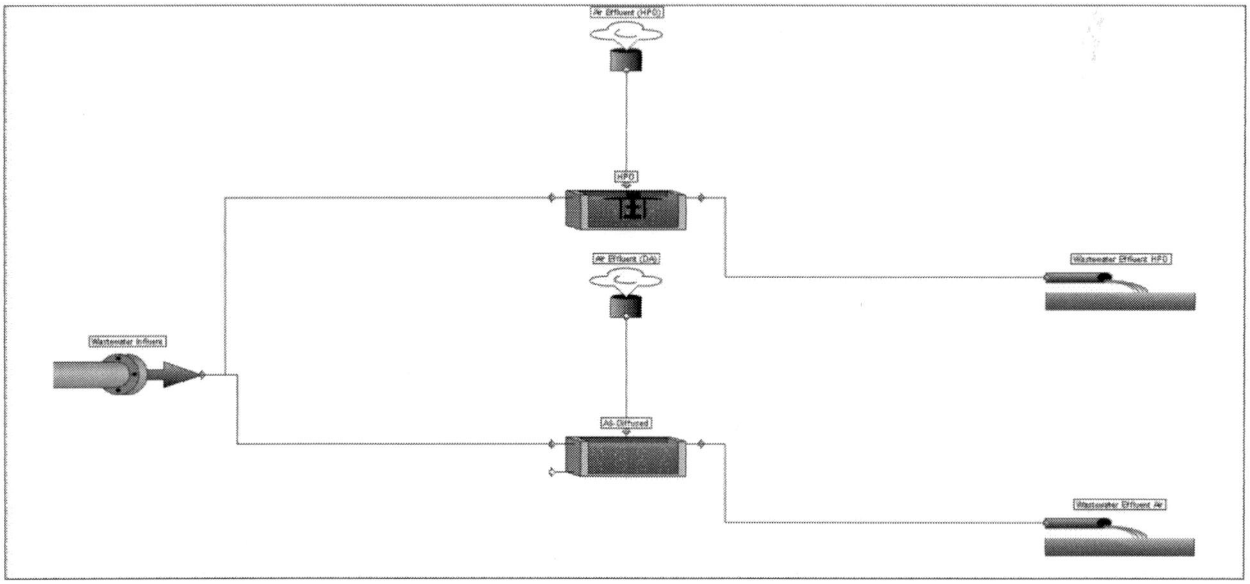

Figure 8: Simplified process layout with HPO and Diffused Aeration modules.

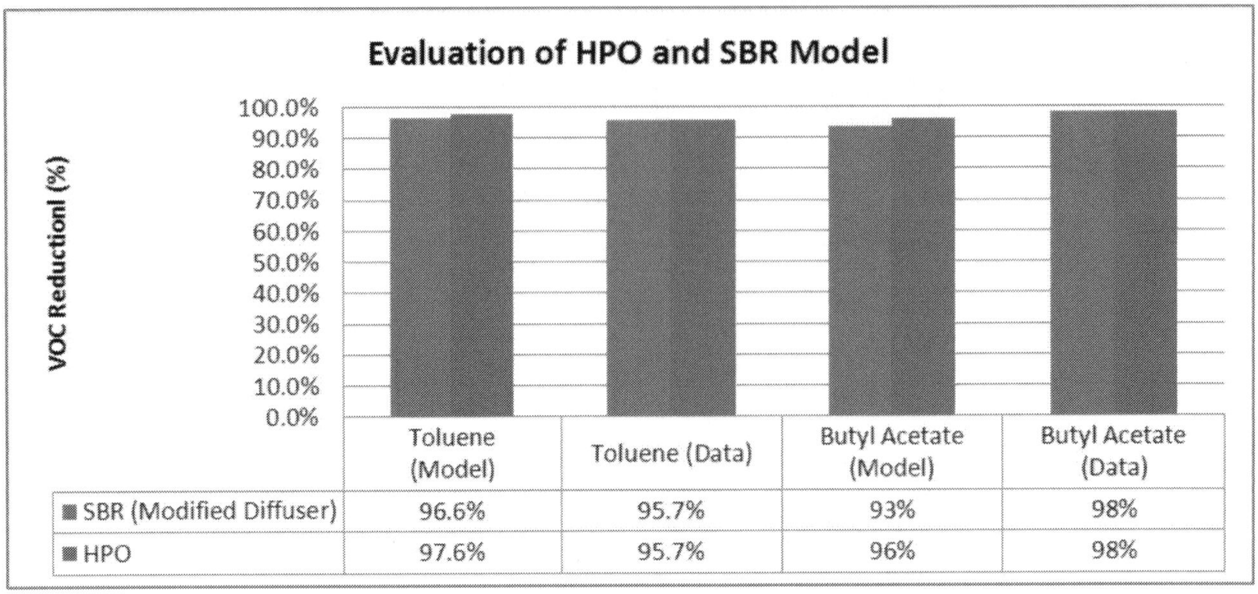

Figure 9: VOC reduction for HPO vs. Air based processes using the SBR model with modified diffused aeration parameters and the new HPO process layout.

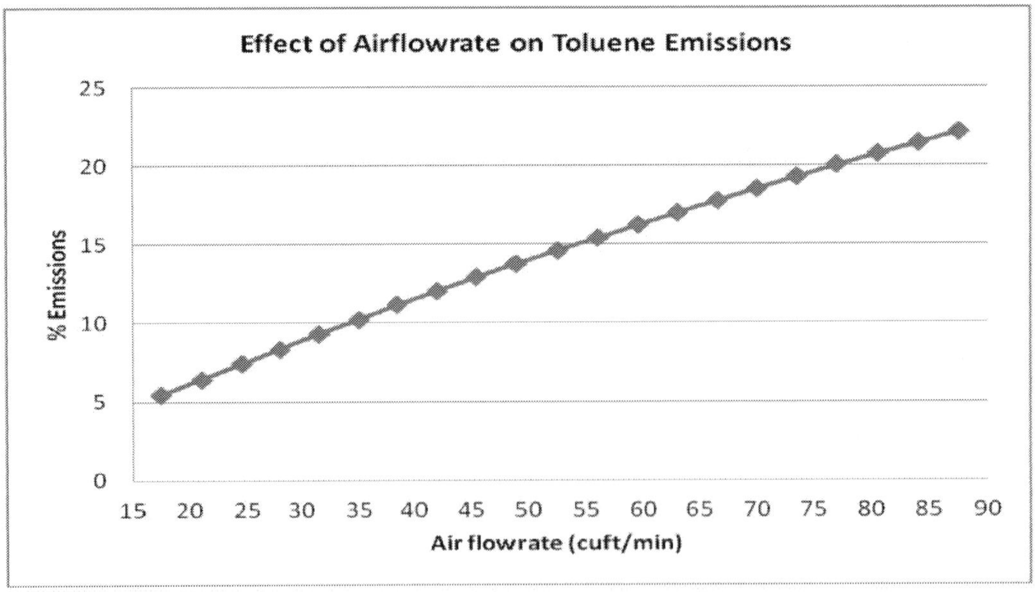

Figure 10: Effect of gas flowrates on Toluene emissions in API pharmaceutical wastewater

REFERENCES

1. Geiling, C. Personal Communications. May 2012.
2. Hydromantis, 2012. Toxchem Version 4.1 Release Notice. http://www.hydromantis.com/news_new.html
3. Jia-Ming Chern and Cheng-Fu Yu. Volatile Organic Compound Emission Rates from Diffused Aeration Systems. 1. Mass Transfer Modeling. Ind. Eng. Chem. Res. 1995, 34, 2634-2643.
4. Jia-Ming Chern and Shun-Ren Chou. Volatile Organic Compound Emission Rates from Mechanical Surface Aerators: Mass Transfer Modeling. Ind. Eng. Chem. Res. 1999, 38, 3176-3185.
5. Levine et al, 2010. Pure Oxygen Activated Sludge Unit Effectively Controls Volatile Organic Compound Emissions from a Mixed Petrochemical Wastewater, WEFTEC 2010
6. Lubkowitz-Bailey et al, 2002. Start-up of a High Purity Oxygen Sequencing Batch Reactor System for Treatment of an Active Pharmaceutical Ingredient Wastewater. WEFTEC 2002
7. Melcer, H. Monitoring and Modeling VOCs in Wastewater Facilities. Environmental Sci. Tech. Vol 28, No. 7, 1994.
8. Mueller et al, 2002. Aeration: Principles & Practice.
9. NYSERDA Report, 2000. In-Situ Oxygenation (I-SO™) for Volatile Compound Emission Control.
10. Praxair Report of Clean Water Tests at Philadelphia Mixing Solutions, 2003.
11. Quigley, C., Card, T., Monteith, H., Witherspoonm, J., Adams, G., Pettit, W., & Torres, E. A Comparison of Three Wastewater Collection System VOC Emissions Models. WEFTEC, 2006.
12. Roberts, P. V., & Dandliker, P. G. Mass Transfer of Volatile Organic Contaminants from Aqueous Solution to the Atmosphere during Surface Aeration. Environmental Sci. Tech. Vol. 17, No. 8, 1983.
13. Rodieck et al, 2001. Alternative Treatment Strategy for High Strength Pharmaceutical Wastewater. WEFTEC 2001.
14. Staudinger, J & Roberts, P.V. 1996. A Critical review of Henry's Constant for Environmental Applications. Critical Rev. Environ. Sci. Technol. 26 (3), 205-297
15. Staudinger, J & Roberts, P.V. A Critical Compilation of Henry's Law Constant Temperature Dependence Relations for Organic Compounds in Dilute Aqueous Solutions. Chemosphere 44 (2001) pp 561 – 576.
16. Woodward & Curran, 2006. Industrial Waste Treatment Handbook

(583a) Design of Better Particle Coatings with Respect to Attrition Resistance

Gabrie Meesters, Chem Eng, Delft Univ Technology, Delft, Netherlands; DSM Biotechnology Center, DSM, Delft, Netherlands, Giacomo Perfetti, DSM Composite Resins, Zwolle, Netherlands and P. Van Hee, Food Specialties, DSM, Delft, Netherlands

Many coated particles show wear due to transport after production. In this research we have focused on assessing this wear, in particular attrition and have related this attrition resistance to the particle coating properties. By measuring the coating properties on the particle, as casted films but also as spray coated films a clear correlation is found between resistance to attrition and the visco-elastic properties of the coating. This finding will make the optimization easier when attrition wear needs to be reduced, since polymer will not have to be tested randomly, which is often the case, but it can be done much more direct by looking into the visco-elastic properties of the coating. Several HPMC and PVA coatings were tested and the most of the results will be presented. The research opens a new field of assessing coating properties without applying them on to coated particles, which can reduce the time of product development a lot.

(583b) Improved Recovery and Dissolution of Poorly Water-Soluble Drug Nanoparticles From Dried Nanocomposite Microparticles

Anagha Bhakay, Chemical engineering, New Jersey Institute of Technology, Newark, NJ, Mohammad Azad, Chemical, Biological, and Pharmaceutical Engineering, New Jersey Institute of Technology, Newark, NJ, Rajesh N. Dave, Chemical Engineering, New Jersey Institute of Technology, Newark, NJ and Ecevit Bilgili, Chemical, Biological and Pharmaceutical Engineering, New Jersey Institute of Technology, Newark, NJ

Incorporation of poorly water-soluble drug nanoparticles into solid dosage forms entails drying of nanosuspensions as an intermediate step, which can lead to loss of their large surface area and may lead to poor bioavailability. In this study, nanoparticle recovery of poorly water soluble drugs from nanocomposite microparticles (NCMPs) redispersed in water was investigated as a function of various excipients. Nanosuspensions of griseofulvin (GF), considered as a model poorly water soluble drug, were prepared by wet media milling and subsequently dried through coating on Pharmatose® carrier particles in a fluidized bed processor. A polymer (hydroxypropyl cellulose, HPC), a surfactant (sodium dodecyl sulfate, SDS), their combinations at various concentrations and a superdisintegrant (croscarmellose sodium, CCS) were used as stabilizers during milling. The NCMPs obtained from the fluidized bed coating were redispersed in water using various methods of agitation to recover the GF nanoparticles. The redispersion methods studied had a slight impact on the recovery of the nanoparticles when SDS was present in the formulation. The HPC–SDS combination resulted in a synergistic effect in the formation of finer nanoparticles as well as in fast redispersion of the nanoparticles of five different poorly water soluble drugs from the NCMPs. Poor redispersion of the nanoparticles and slower dissolution were observed when SDS was removed from the formulations and addition of dispersants like mannitol also could not recover the GF nanoparticles. However, incorporation of a novel additive such as milled CCS and a mixture of milled/unmilled CCS led to enhanced GF nanoparticle recovery and dissolution rate from surfactant-free NCMPs.

(583c) One-Step Fabrication of Agent-Loaded Biodegradable Microspheroids for Drug Delivery and Imaging Applications

Michael J. Heslinga, Gabriella Willis and **Omolola Eniola-Adefeso**, Chemical Engineering, University of Michigan, Ann Arbor, MI

Non-spherical particles may offer advantages over conventional spherical systems for vascular wall-targeted imaging and drug delivery applications for the treatment of several diseases. This work describes the novel fabrication of agent-loaded poly(lactic-co-glycolic acid) (PLGA) spheroid carriers via oil-in-water (O/W) and water-in-oil-in-water (W/O/W) emulsion solvent evaporation (ESE) methods, respectively. The versatility of this technique is demonstrated via the loading of a variety of therapeutics including paclitaxel, bovine serum albumin, and cadmium sulfide fluorescent nanospheres into PLGA prolate spheroids. The encapsulation efficiency for spheroids made via O/W emulsions is found to be highest at low aqueous phase surfactant concentrations and high oil phase volume fractions while encapsulation efficiency for spheroids made from W/O/W is highest at high aqueous phase surfactant concentrations and basic aqueous phase pH values. We show that rods fabricated via the W/O/W method display less polydispersity than ones via the O/W. Overall, particle volume polydispersity in the ESE method can be minimized with high stir rate and high PLGA concentration while aspect ratio polydispersity can be minimized via use of high stir rate, as well as high aqueous phase Tris concentration and pH. The ESE technique is an attractive alternative over recently described methods for fabrication of non-spherical particles due to its simplicity in setup, high particle yield and adaptability to a variety of biodegradable polymers and therapeutics.

(583d) Particle Engineering Via Dry Coating of Micronized API Powders for Improved Dissolution of Directly Compacted Tablets with High Drug Loading

Xi Han, Otto H. York Department of Chemical, Biological and Pharmaceutical Engineering, New Jersey Institute of Technology, Newark, NJ, Chinmay Ghoroi, Department of Chemical, Biological and Pharmaceutical Engineering, New Jersey Center for Engineered Particulates, Newark, NJ and **Rajesh N. Dave**, Chemical, Biological and Pharmaceutical Engineering, New Jersey Institute of Technology, Newark, NJ

Our recent study showed that simultaneous micronization and surface modification through fluid energy milling (FEM), a continuous process, can improve both the flow and dissolution rate of the active pharmaceutical ingredient (API) powders. In this paper, the improved performance of direct compacted tablets with high drug loading of micronized dry coated API powders (20 µm) is examined. As a benchmark comparison, larger, 50 µm API powders dry coated without micronization using a continuous comil process are also considered. Blends with 30, 60 and 70% drug loading, containing API powders that are micronized or non-micronized and dry coated or uncoated are prepared and evaluated. The results show that the blends containing dry coated API powders, even micronized ones, have excellent flowability and high bulk density, which are required for direct compaction. In contrast, blends containing uncoated APIs have poor flow and lower bulk densities. As the drug loading increases, the difference between dry coated and uncoated blends is more pronounced, which is clearly illustrated through a bulk density-FFC phase map. Powder compressibility tests showed improved performance of dry coated powder blends, indicating a good potential for direct compaction. Consequently, tablets are formed through direct compaction for micronized API blends; tablets prepared from dry coated APIs exhibit superior compactibility and dissolution profiles, particularly for higher drug loadings. Dissolution time, t_{80}, for 30 or 60 % drug loaded tablets from 20 µm dry coated ibuprofen, is well under 5 min, illustrating the advantages of the dry coating during API micronization.

(583e) Controlled Release From Self-Assembled Polymer Microparticles That Are Responsive to pH and Temperature

James M. Myrick, Department of Chemical & Biomolecular Engineering, Clarkson University, Potsdam, NY, Sitaraman Krishnan, Department of Chemical and Biomolecular Engineering, Clarkson University, Potsdam, NY and Frederick A. Sexton, New World Pharmaceuticals, Rumson, NJ

Controlled release of biologically active molecules is of significant interest in the pharmaceutical and biomaterials fields. Benefits of controlled release mechanisms include adjustable release rate, improved pharmacokinetics, and reduction of undesirable side-effects. Among controlled delivery systems, oral delivery using biocompatible polymer particle excipients has received significant attention [1]. In this study, particles were formed from a selectively water-soluble cellulose ether with a lower critical solution temperature (LCST) close to body temperature [2]. Above the LCST, hydrophobic polymer interactions dominate resulting in self-aggregation of polymer chains. The polymer chains were cross-linked above the LCST, and formed hydrated microgel particles when cooled below the phase transition temperature. Particles were encapsulated with a carboxylated polymer, resulting in a pH-responsive core-shell configuration. Controlled release from the particles was investigated using hydrophilic small-molecules as entrapped solutes. The efficacy of particle formulations was measured through release kinetics performance in comparison to aqueous solutions of the hydrophilic molecules. Sustained release from the particles and control was analyzed *in vitro* at temperatures above and below the LCST using a horizontal static diffusion cell. The empirical data were found be in good agreement with a mathematical model developed for *in vitro* release kinetics experiments.

REFERENCES

1. Kost, J.; Langer, R. Responsive polymeric delivery systems. *Adv. Drug Deliv. Rev.* **2001**, *46*, 125-148.
2. Sexton, F. A.; Krishnan, S.; Vendra, V. K. Sustained release of nutrients *in vivo*. *U. S. Pat. Appl. Publ.*, US 20090155409 A1 20090618 (2009).

(583f) Wet Coating of Geldart-C Type Particles in a Rotating Fluidized Bed in a Static Geometry

Philippe Eliaers and Juray De Wilde, Materials and Process Engineering, UCL, Louvain-la-Neuve, Belgium

Fine particle coating has important applications in the pharmaceutical and food/feed industry. Two main routes can be used: wet- and dry coating. Fluidized bed technology is widely used for wet particle coating. The performance of conventional (i.e. gravitational) fluidized beds is, however, limited when coating cohesive Geldart-C type particles. In this work, a rotating fluidized bed in a static geometry (RFB-SG) [1] was used for coating fine particles with an aqueous polymer solution. Proof of concept was given and it was shown that high-G operation allows removing important limitations of conventional fluidized beds.

A 24 cm diameter, 5 cm length chamber, equipped with 72, 0.2 mm gas inlet slots was used. The particles to be coated were fed via the front side of the chamber, whereas the liquid solution was injected using a 15° spray mounted centrally in the chamber and directing toward the outer wall of the reactor (Figure 1). The particles with a mean diameter of 70 micron and a density of 260 kg/m^3 were fed at 2 g/s. The liquid solution was heated up to 90°C prior to injection and the liquid droplets leaving the spray nozzle were on average 65 micron. The liquid flow rate and liquid-solid contact time were varied. The liquid solution was injected after establishing a stable rotating particle bed. Experiments were carried out at different air flow rates, 250 and 400 Nm3/h. The air fed was heated by means of an electric resistance to average feed temperatures between 55°C and 70°C. Batch-wise and continuous particle coating was studied.

The operating conditions were found to be critical for both the particle bed stability and the quality of the particle coating process. Within a range of operating conditions, coated particles could be produced in a reproducible way. Figure 2 shows an example of coated powder that could be produced and its particle size distribution. A comparison with the original powder is made. Some agglomeration was observed, but the agglomeration could be controlled by means of the gas, liquid and solids flow rates, as illustrated in Figure 3. The role of the operating conditions and the chamber geometry are discussed in detail.

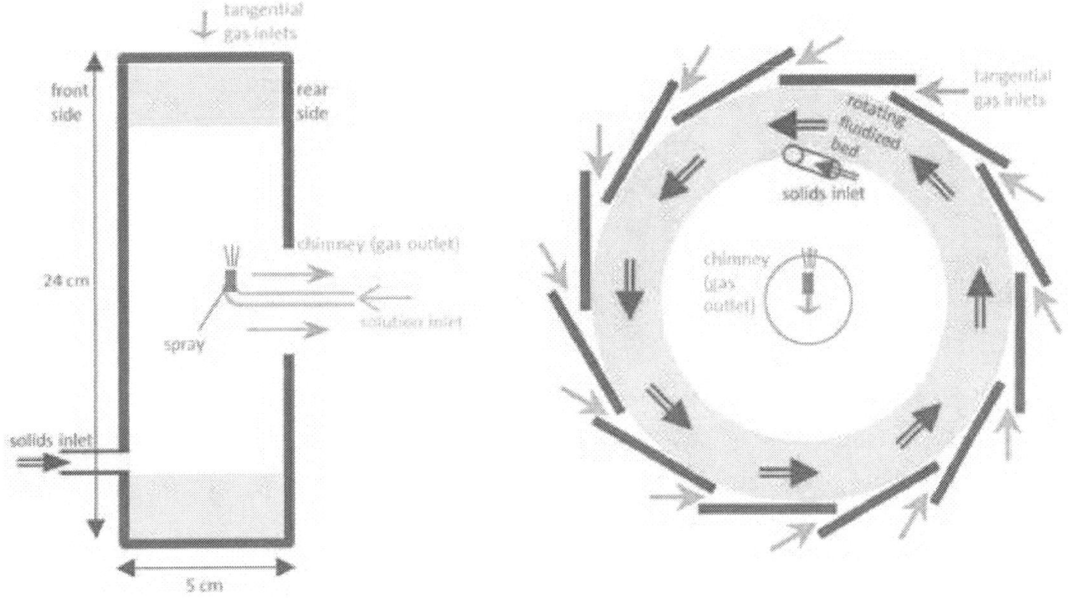

Figure 1. Schematic representation of the RFB-SG particle coater, top and front view.

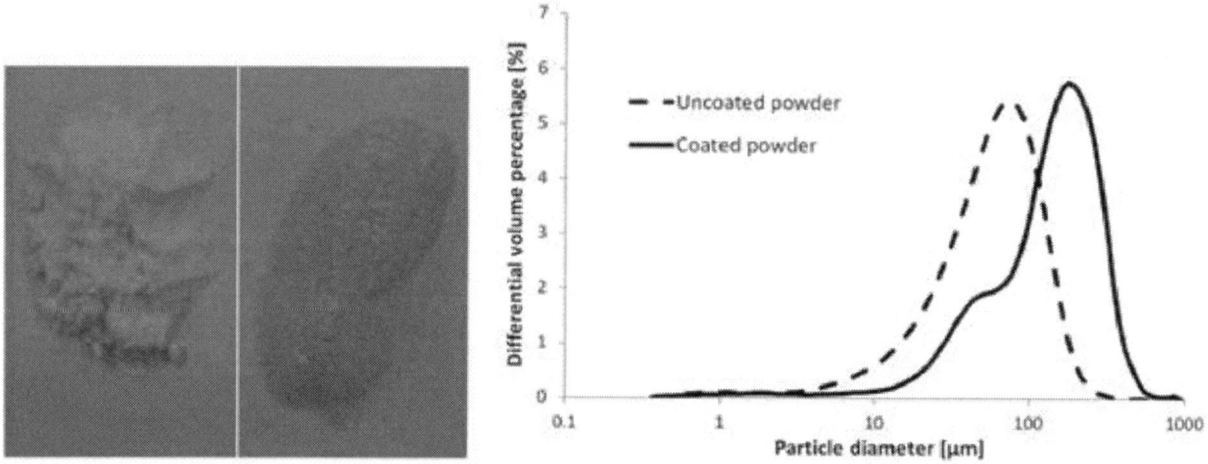

(a) (b)

Figure 2. (a) Visual comparison of the uncoated (left) and coated (right) powder; (b) Particle size distribution of the uncoated and coated powder. Solids feeding rate: 2 g/s, liquid solution feeding rate: 1 g/s, air flow rate: 250 Nm3/h at 55°C.

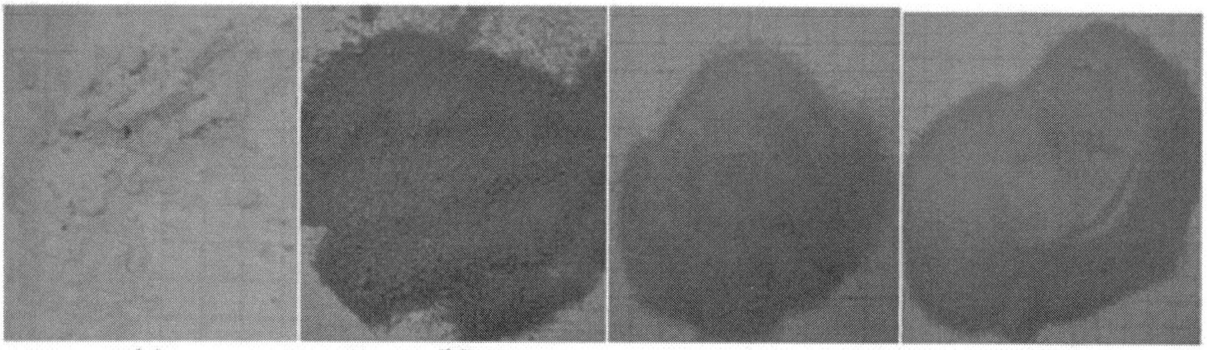

(a) (b) (c) (d)

Figure 3. (a) Uncoated powder; (b)-(d) Coated particles of different granulometry produced with decreasing liquid-solid contact time.

[1] J. De Wilde and A.de Broqueville: Rotating fluidized beds in a static geometry: Experimental proof of concept. AIChE Journal, 53, p. 793–810, 2007.

(583g) Solvent-Free Beta-Carotene Nanoparticle Manufature

Phong Huynh, Rutgers University, New Brunswick, NJ and Paul Takhistov, Food Science, Cook College, Rutgers, The State University of New Jersey, New Brunswick, NJ

Most nutraceutical compounds are poorly-water soluble. Their low solubility decreases the adsorption rate in living organisms leading to their low bioavailability. Utilization of nanoparticles is a promising way to improve the solubility of hydrophobic compounds. Some traditional methods for decreasing particle size include pearl or jet milling, where particles are broken down through grinding or collisions under high pressure. These mechanical processes not only require high energy input but also raise a concern of milling media residues. The high pressure homogenizer approach applies implosion forces and collision of particles to generate nanosuspensions. This method requires microsuspensions as starting material and consumes high energy. Among several emulsion-based techniques for preparing nanoparticles, solvent diffusion practice is a novel approach in which a poorly-water soluble compound is transferred into nanoemulsion droplets of a partially water-soluble organic solvent. The compound then crystallizes because the solvent diffuses out of the emulsion droplets. The key point of proposed emulsion-diffusion technology is that the phase transition occurs within an isolated nanoemulsion droplet. We propose a "green" and scalable method for preparing nanosuspensions of highly hydrophobic compounds using FDA GRAS ingredients to create nanoparticles of poorly-water soluble material. b-carotene is selected as a model hydrophobic nutraceutical. Triacetin, a partially-water soluble triacetate compound, is used as the dispersed phase of nanoemulsions. The influence of surfactant, water concentration, and homogenization time on particle size and stability is investigated. The impact of surfactant on diffusion flux of triacetin is studied. Kolmogorov theory is applied to reveal the breakup mechanism of emulsion droplets under shear and predict their size. A mathematical model is built to discover the size of emulsion droplet during the formation of nanosuspensions.

EFFECT OF PH AND AERATION ON PLASMID STABILITY AND PHYTASE EXPRESSION IN *Escherichia coli* BL21(DE3) DURING BATCH CULTIVATIONS IN SEMI-SCALE BIOREACTOR.

Othman, N.Z[1], Ramli.S[1]., Masri, J.H[1]., Sarmidi, M.R[1], Aziz, R[1], Tran, T.T[2], Hatikaul.R[2]., El Enshasy, H.A[1,3].

1.Institute of Bioproduct Development, Universiti Teknologi Malaysia, Skudai Johor, Malaysia
2. Department of Biotechnology, Center for Chemistry and Chemical Engineering, Lund University, Sweden.
3.Bioprocess Development Department, City for Scientific Research and Technology Applications (CSAT), New Burg Al Arab, Alexandria, Egypt;

Introduction

Phytase (myo-inositol hexakisphosphate) has become important feed additive for monogastric livestock. The enzyme action activity is by hydrolyzing phytic acid to myo-inositol derivates and inorganic phosphorus. Since the digestive tract of monogastric animal including pigs and poultry lack phytase, organic phytic acid that naturally exist in the feed cannot be utilized by these animals and thus inorganic phosphate has to be added to meet sufficient phosphate supply (11). Furthermore *E. coli* was widely used bacterial platform nowadays for expressing simple heterologous proteins such as phytase (8). *E. coli* is facultative anaerobic bacterium which capable to live either in the presence of oxygen or absence of oxygen (24). Unfortunately, under normal growth condition *E. coli* can only express very low phytase activity (6). Growing the *E. coli* to high cell density and improving the volumetric productivity is the major objective of any *E. coli* based process. Since expression of heterologus protein by recombinant *E. coli* is relative to the final cell density and the specific productivity of the host cells (4). One of the key factors in obtaining the aim is aeration supply. Scale-up processes for HCDC of recombinant *E.coli* for protein expression from shake flasks to fermentors generally have low protein expression great difference between shake flasks and fermentors is the supply and limitation of oxygen (15, 18, 23, 25). However, oxygen supply has been reported to have variable effects on heterogeneous gene expression in recombinant *E.coli* cells would increase or decrease the expression of heterogeneous proteins that depends on the strain and types of plasmid used.To obtain high cell density cultivation (HCDC) and recombinant phytase production, a high capacity of oxygen through aeration supply is required. The formation of acetate during aerobic growth is the main problem. Overproduction of acetate, also known as metabolism overflow eventhough cultivation is done under aerobic condition with excess glucose. This occurs when conversion of glucose to acetate is high compared to the conversion of glucose to biomass and carbon dioxide (18). Assimilation of acetate as low as 0.5 g L^{-1} in the fermentation broth is the main problem in the HCDC cultivation of recombinant *E.coli* or wildtype *E.coli* strain due to low cell growth and volumetric yield of the expression protein (10, 19,20,26). Previous researchers found that the suitable aeration supply to the bioreactor culture can improve the cultivation performance of *E. coli* and also the production of the expression. Oxygen supply can be controlled by manipulating the aeration rate and agitation speed (12).The main purpose of the investigations reported here was to study the growth kinetics of *E. coli* BL21(DE3) in different pH condition and aeration rate in the semi scale 16L bioreactor by analyzing cell growth, acetate accumulation, plasmid stability, glucose consumption and distribution of phytase.

Materials and Methods

Microorganism

E.coli BL21 (DE3) harboring pE10C2 that carries thermophile phytase gene from *Bacillus* sp. MD2 under the control of *lac*-O promoter to express thermostable phytase enzyme was used for this research. The strain was originally from of Biotechnology, Center for Chemistry and Chemical Engineering, Lund University and was delivered and stored in Wellness Industries Culture Collection (WICC), Institute of Bioproduct Development, Universiti Teknologi Malaysia. Cells were first propagated on LB agar plate containing 100 μg/mL ampicillin, was incubated at 30 °C for 24 hours. The arisen colonies were harvested by glycerol solution (50%) and put in series of 2 mL cryogen vials. The vials were frozen at -20°C for 24 hours followed by further storage as working cell bank at -80°C for further use.

Inoculum preparation of E. coli BL21 (DE3)

The *E. coli* BL21 (DE3) cells harbouring the recombinant vector pE10C2 were grown on a LB-agar plate containing 100 μg mL^{-1} of amphicillin at 30°C. A single colony was picked from the plate and cultured for 24h in 5 ml LB broth containing 100 μg mL^{-1} of amphicillin at 30°C and 200 rpm. The shake flasks cultivations were carried out in 250 ml Erlenmyer flasks containing 50 ml broth and incubated in a rotary shaker (4230 Innova, New Brunswick, NJ, USA).

Production medium and Bioreactor cultivations

For this study, optimize medium were used the medium medium composed of [g L^{-1}]; 15 g glucose, 4 g $(NH_4)_2$ SO_4, 1g $(NH_4)_2HC_6H_5O_7$, 6.5 g K_2HPO_4 and 1.541 g $NaH_2PO_4.H_2O$, 1.44g $MgSO_4$, 2 mL trace elements solution and a mixture of 30 mM lactose as inducer with 10mM $CaCl_2$. The pH was adjusted to 7 before sterilization. Glucose, trace elements and $MgSO_4$ were sterilized separately and added to medium before inoculation.

Cultivations with growing cells were carried out in a 16-L stirred-tank bioreactor with working volume of 6-L (BioEngineering, Wald, Switzerland). The bioreactor is equipped with pH probe, oxygen probe, foam sensor, and stirrer of two–four bladed Rushton turbines. For controlled pH cultivations, the pH was maintained at 7.0 by addition of 4 M NaOH and 2 M HCl solution. For uncontrolled pH experiments, the pH was also adjusted to 7 before inoculation and monitored only during the cultivation time. During the experiments, temperature, aeration rate and the agitation speed were controlled at 30°C under controlled and uncontrolled pH condition at 400 rpm with aeration rate of 1.0 vv^{-1} min^{-1}. Then followed by studied in different aeration rate 1.0, 2.0, 3.0 vv^{-1} min^{-1} with agitation rate at 400 rpm respectively. The dissolved oxygen was adjusted to 100% saturation before inoculation and kept uncontrolled during cultivation process.

The total time of fermentation was approximately 28 hours. Samples were taken at approximately 2 hour intervals and analyzed immediately. Antifoam reagent (Silicone antifoam, Sigma, USA) was added to suppress the foaming when necessary. A Polarographic probe (Metler Toledo-Ingold) was used to detect DO concentration that was calibrated before and after calibration with air (100%) at atmospheric pressure.

Analysis

Optical Density and Dry Cell Weight Determination

Immediately after sampling the optical density was determined at OD_{600} after proper dilution. The obtained OD_{600} value of the cells was correlated with the cell dry weight via the calibration curve. The sample was then centrifuged at 5,000 rpm for 20 min. and a small fraction was frozen at -20°C for phytase activity determination.

Plasmid stability

To analyze plasmid stability, a replicate plating method was used to analyze plasmid instability by comparing the percentage of viable colonies on LB agar plate, with and without 100 μg mL^{-1} of amphicillin.

Determination of enzyme activity intracellular and extracellular fractions

The extracellular activity was determined in the cell-free supernatant obtained after centrifugation of the sample at 8,000 g for 10 min at 4°C. The cell pellet was washed with physiological saline the remaining cell pellet was re-suspended in 0.1 M Tris–HCl buffer, pH 7, containing 5 mM $CaCl_2$, and the cytoplasmic content was released by four rounds of sonication. All the assays were performed in triplicates.

Phytase activity assay

The extracellular activity was determined using supernatant obtained after centrifugation of the sample. Base on the method studied by Tran, *et al.*,(27), a mixture of appropriate diluted enzyme sample and sodium phytate (1.5 mM) was dissolved in 100 mM Tris-HCl buffer, pH 7 containing 5 mM $CaCl_2$. Then, the solution was incubated at 70 °C for 20 min. The reaction is stopped by adding an equal volume of 15% trichloroacetic acid (TCA) solution. The precipitate obtained is removed by centrifugation, and 500 μl of the clear supernatant transferred to an Eppendorf tube followed by addition of 500 μl ammonium molybdate reagents (containing 1:4 mixtures of 2.7% $FeSO_4$ and 1.5% ammonium molybdate in 5.5% H_2SO_4). After 5 min incubation at room temperature, absorbance at 700 nm will be read and used to calculate the concentration of P_i from a standard curve made using sodium dihydrogen phosphate solution in the concentration range of 0-600 μmol/l P_i. All the samples were assayed in triplicates. One unit of phytase activity was defined as the amount of enzyme that releases 1 μmole of P_i per minute under the assay conditions.

Glucose and Acetic acid analysis

For analysis of glucose, the sample was centrifuged at 5000 rpm for about 7 min. The supernatant was diluted with distilled water with the dilution factor of 1:10. 1 mL sample then was pipetted into a triplicate set of test tubes and 1.0 mL 3, 5-dinitrosalicylic acid (DNS) reagent was added into each test tube. The samples were incubated at 99°C for 15 min, were cooled with ice and 8.0 mL deionized water was added. The intensity of brown color was read at 540 nm. Acetic acid was determined using High Performance Liquid Chromatography.

Results and discussion

Effects of uncontrolled pH on the kinetics of cell growth of E. coli BL21(DE3)and total phytase production in 16- L bioreactor.

The growth kinetics of *E. coli* BL21(DE3) was studied in a 16-L pilot scale stirred tank bioreactor using the optimized medium without additional of amphicilin with an inoculum size of 10% (v/v). Bioreactor cultivations with a working volume of 6 L were carried out under uncontrolled and controlled pH conditions. **Figure 1** shows cell growth, total phytase, acetate production and glucose consumption in the optimized medium with uncontrolled pH during a batch cultivation of *E. coli* BL21(DE3). As shown, cells grew exponentially during the first 6 hours in a slower rate at a growth rate of $0.292h^{-1}$ reaching a maximal cell mass of 1.17 g L^{-1}. Stationary growth phase was observed after 8 hours of cultivation. At the beginning of the experiments were performed completely exposed to 100% oxygen under 1 v v^{-1} min^{-1} aeration rate. After 4 hours of inoculation the dissolved oxygen fell rapidly and the cell growth under zero percent of dissolve oxygen for almost 6 hours before induced with lactose. As acetic acid produced, the pH of the medium decreased and causes the cell growth in the stationary phase throughout the post-induction phase. The culture reached pH 5.0 at the beginning of the stationary growth phase and kept more or less constant for the rest of cultivation time. The drop of pH in uncontrolled cultivations was mainly due to the formation of acetic acid concomitantly with the limitation of dissolved oxygen in the fermentation broth causes lost of plasmid from the host cells after 4 hours of post-induction phase. Therefore the maximum production of phytase activity only reached 1.198 U mL^{-1} (**Table 1**) after 2 hours induced with 30 mM lactose. The low expression of total phytase activity impaired by acetate accumulation and limited dissolved oxygen in the fermentation broth. However studied by Wang *et al.*,(28) shown that the high acetate stress had no significant desructive impacts on the protein expression but the expression was greatly influenced on the cell oxygen utilization capacity and finally influenced the cell metabolism intensity.

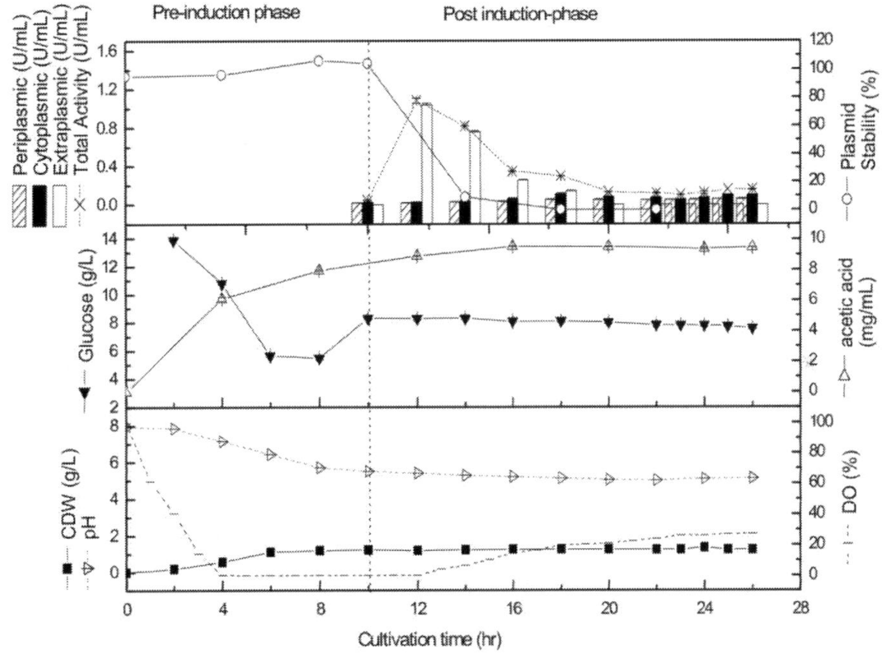

Figure 1: The cell growth kinetics, plasmid stability, acetate and total phytase production in uncontrolled pH batch cultivation of *E. coli* BL21(DE3) with 30 mM lactose as an inducer when aeration at 1 vv^{-1} min^{-1}.

Effects of controlled pH on the kinetics cell growth of E. coli BL21(DE3)and total phytase production in 16- L bioreactor.

In this study, the growth kinetics of *E. coli BL21(DE3)* for high cell mass production in large scale 16-L bioreactor with controlled pH strategy was conducted. Under controlled pH, the growth profile as shown in **Figure 2**. From this profile, the culture growth rapidly in the exponential growth phase for almost 16 hours. The stationary phase was observed after 18 hours of growth profile. Optimized medium under controlled pH conditions of fermentation medium resulted specific growth rate of 0.126 h^{-1} .The results showed that the growth and expression of the phytase by *E. coli BL21(DE3)* was greatly influenced by the pH (28). The growth performance was dramatically improved when the culture pH was held constant at 7.0. A maximum biomass of 2.14 g L^{-1} was achieved when the pH was maintained at 7.0. It's observed that after 24 hours of fermentation the biomass production was constant. The glucose consumption rate in the controlled pH cultivation slightly higher which was 1.25 g $L^{-1}h^{-1}$ when compared to uncontrolled pH cultivation only 1.0 g $L^{-1}h^{-1}$. Meanwhile, 2.507 U mL^{-1} of total phytase activity was achieved after 4 hour of the post-induction phase when the culture pH was held constant at 7.0. The pH condition applied did not affect the dissolved oxygen profiles (3). As shown in **Figure 2** the trend of dissolved oxygen dropped same as the cultivation under uncontrolled pH. Therefore due to limitation of dissolved oxygen concomitantly causes accumulation of acetic acid in the fermentation broth almost 8.0 g L^{-1} causes the lost of plasmid that carried phytase gene after 4 hours of post-induction phase. High concentration of acetate reduces the growth rate of the culture and lead to low protein expression (7, 10, 17). Since *E. coli* B (BL21) a lower acetic acid producing strain, the accumulation of acetic acid inversely correlated to the dissolve oxygen levels (22). Consequently, the lost of plasmid that carried phytase gene may influenced the expression of phytase by *E. coli* BL21(DE3) eventhough the culture growth two times more than cultivation in the uncontrolled bioreactor. On the other hand, the lost of plasmid may related to the metabolic load of recombinant protein synthesis to cellular capacities (21).

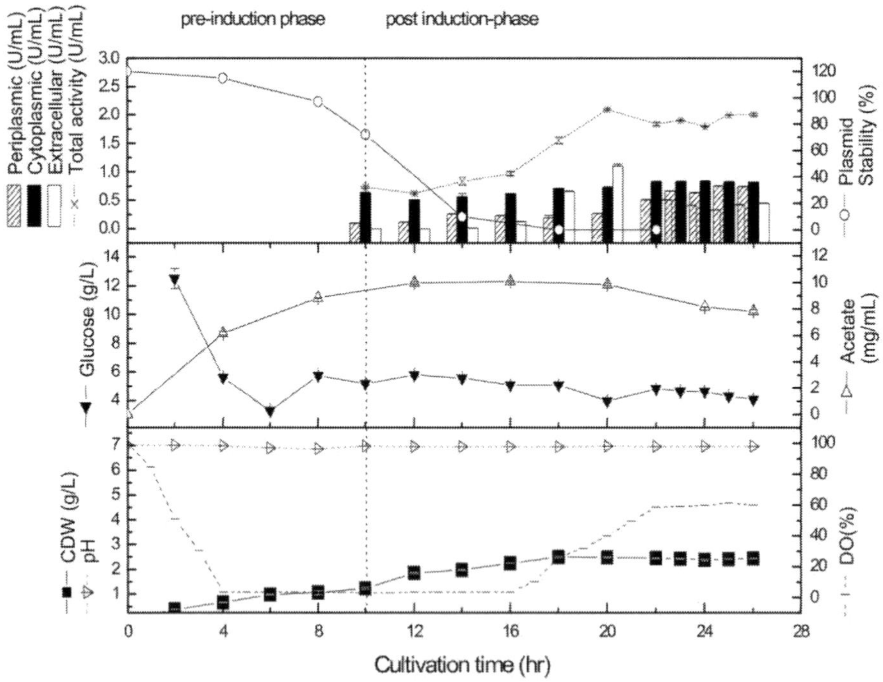

Figure 2: The cell growth kinetics, plasmid stability, acetate and total phytase production in controlled pH batch cultivation of *E. coli* BL21(DE3) with 30 mM lactose as an inducer when aeration at 1 vv^{-1} min^{-1}.

Growth kinetics under different aeration rate during cultivation of E. coli BL21(DE3) in a 16-L pilot scale stirred tank bioreactor

Glucose, acetic acid production, cell concentration and plasmid stability profiles

As shown **Figure 2** for aeration rate operation of 1 vv^{-1} min^{-1} for the glucose consumption rates was low only 1.25 g L^{-1}h^{-1}. Glucose concentration, at an initial concentration of 15 g L^{-1} decreased to 3 g L^{-1} until 8 hours of cultivations. During the post-induction phase, the glucose residuals in the fermentation broth increased approximately 4-5 g L^{-1} after induced with lactose. Then the concentration of glucose residual showing modest decreased of the residual glucose. This is contributed by the catabolised of lactose to glucose and galactose when induced with 30 mM lactose. Comparing the aeration rate at 2 and 3 v v^{-1} min^{-1} growth profiles, it can be observed that they have similar glucose consumption rate that was 1.67 g L^{-1}h^{-1}. The same observation was observed during the post-induction phase for the glucose residual which was increase to 10 g L^{-1} and gradually decreased too approximately 5 g L^{-1} until the end of post-induction phase. On the other hand, cultivation in the 16-L stirred-tank bioreactor under controlled pH and aeration of 1 v v^{-1} min^{-}1, 2.0 v v^{-1} min^{-1}and 3 v v^{-1} min^{-1} supported cell growth of 3.43 g L^{-1} and 3.82 g L^{-1}, respectively. Unfortunately, both cultivation in aeration rate of 2 v v^{-1} min^{-1}and 3 v v^{-1} min^{-1} showing the accumulation of acetic acid in the fermentation broth only 2 g L^{-1} after 16 hours of cultivation.

Comparing between the different aeration rates to the cell growth is low in aeration of 1 v v^{-1} min^{-1} with only 2.14 g L^{-1} of cell biomass when compared to the growth in high aeration rate. However, increasing the aeration rate from 2 to 3 v v^{-1} min^{-1} may not increased the cell biomass and the cell growths enter the stationary growth phase after 8 hours of cultivation. At the end of the cultivation process, the highest cell biomass value was attained at 3v v^{-1} min^{-1} as X_{max} =3.81 g L^{-1} with the specific growth rate, μ = 0.333 h^{-1}. While the X_{max} =3.43 g L^{-1} was achieved when the aeration rate operation was 2 v v^{-1} min^{-1}. It's clearly indicated that the cell biomass was not dependent on the aeration rate during the cultivation. Consequently, **Figure 2**, **3** and **4** has demonstrated that the concentration of the dissolved oxygen in the cultivation broth may influence the plasmid stability. As shown from **Figure 2**, the plasmid that carried phytase gene was decreased slowly and finally lost of the plasmid after 12 hours of cultivation. The plasmid with the phytase gene can sustained below than 30% after 2 hours induction with lactose and totally lost the plasmid after 4 hours in the post-induction phase. The observation of the plasmid stability for aeration rate of 2 v v^{-1} min^{-1} can sustained the gene almost 20% after 14 hours induced with lactose. The plasmid stability was gradually decreased and totally lost at the end of the post-induction phase. The different observation was found for aeration of 3 v v^{-1} min^{-1} where at the beginning of the post-induction phase that was after 6 hours of induction the plasmid stability was decreased to only 20% after that the percentage of the plasmid stability gradually decreased and can sustained almost 60% of the plasmid that carried the phytase gene after 11 hours of induction. This may influenced expression of phytase by *E. coli* BL21(DE3) with lactose as inducer without additional of amphicillin in the fermentation broth. Plasmid stability is one of the most important issues affecting the yield of recombinant protein production in recombinat host cells cultivation (2).

Dissolve oxygen profiles

The variation of aeration rate that is 1, 2 and 3 v v^{-1} min^{-1} with controlled pH is shown in **Figure 2**, **3** and **4** respectively. At the beginning of each experiments cultivation was completely exposed to 100% oxygen with agitation rate remain constant at 400 rpm throughout the cultivation time. At the beginning of the experiments were performed completely exposed to 100% oxygen. The dissolved oxygen was rapidly decreased from 100% to almost zero percent during 4 hours cultivation for 1 v v^{-1} min^{-1} due to high oxygen demand. As shown in **Figure 2**, at this aeration rate the demand of the dissolved oxygen is high and the cell growth under zero percent of dissolved oxygen for almost 12 hours. Then the profiles gradually increase until the end of the cultivation. The dissolved oxygen profile for aeration rates of 2 and 3 v v^{-1} min^{-1} showing the similar trend. However aeration rate of 3 vv^{-1} min^{-1} after 6 hour of cultivation, the dissolved oxygen concentration in the medium was slightly higher than 2 v v^{-1} min^{-1} which can maintain almost 30

% for 2 hours before decreased to zero percent after 8 hour of cultivation. This can help sustain optimal cellular health, which is required to overcome the metabolic stress associated with recombinant protein expression (14).

Phytase activity

From **Figure 2**, **3** and **4** clearly demonstrated that effect of aeration rate to the phytase production. As increasing the aeration rate from 1 to 3 v v^{-1} min^{-1} may increase the total phytase activity form 1.198-14.25 U mL^{-1}. As mention by Phue and Shiloach (22), oxygen limitation seems to repress the constitutive expression of the glyoxylate shunt and gluconeognesis of glucose metabolism. This may contribute to accumulation of acetate in the fermentation broth and supressed expression of heterologus protein. At aeration rate of 3 v v^{-1} min^{-1} showing the activity of cytoplasmic, periplasmic and extracellular activity (Table 1) that contributed to the highest total phytase activity that is 5.753, 0.94, 8.941 and 14.25 U m L^{-1} respectively. This is coupled with the specific activity ($Y_{p/x}$) of the phytase is higehst among the aeration rate with $Y_{p/x}$ = 5088.21 U g^{-1}. Volumetric phytase production rate was also higher among the different aeration rate almost Q_p = 949.8 U L^{-1} h^{-1}. The higest productivity of total phytase activity mainly because of the plasmid stability during cultivation that was correlated with the dissolved oxygen and accumulation of acetic acid in the fermentation broth. Therefore, *E.coli* BL21 (DE3) that harboring pE10C2 for the expression of phytase in this studied were influenced by the concentration of the dissolved oxygen that may causes cumulative oxidative damage (9) to the host cells or acetate stress due to acetate accumulation in the fermentation broth(13) and plasmid stability (5).

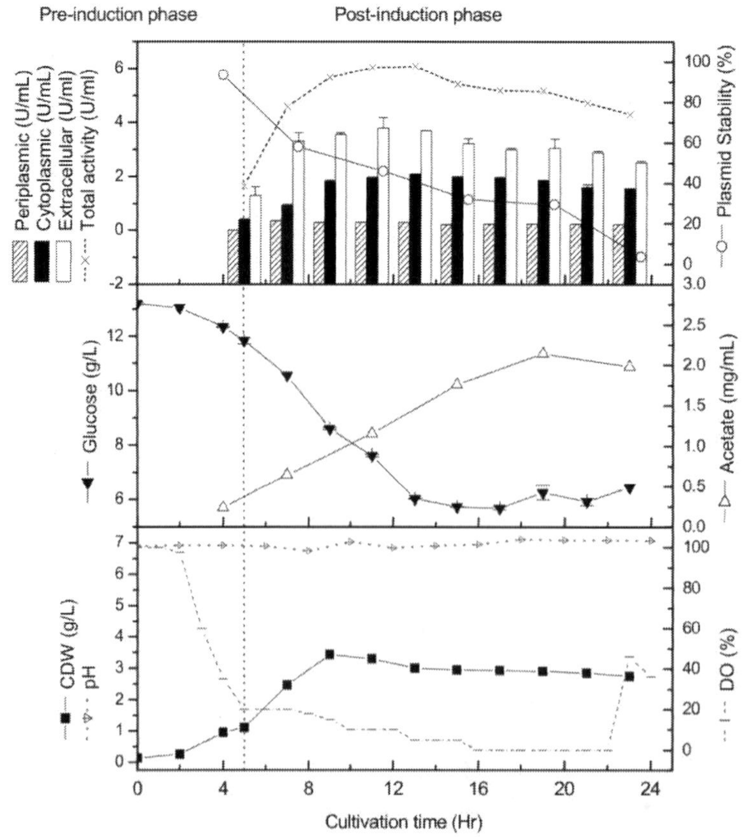

Figure 3: The cell growth kinetics, plasmid stability, acetate and total phytase production in controlled pH (left) and uncontrolled pH (right) batch cultivation of *E. coli* BL21(DE3) with 30 mM lactose as an inducer when aeration at 2 vv^{-1} min^{-1}.

Table 1 : Growth inetic parameters for cell growth and phytase production by *E. coli* BL21 (DE3) under uncontrolled, controlled pH and different aeration rate.

Parameters	Aeration rate			
	1 vv^{-1} min^{-1}		2 vv^{-1} min^{-1}	3 vv^{-1} min^{-1}
	Uncontrolled pH	Controlled pH		
Xmax (g/L)	1.17	2.41	3.43	3.81
μ (h^{-1})	0.292	0.126	0.5	0.333
$P_{max(extra)}$ [U mL^{-1}]	1.045	1.478	3.79	8.941
$P_{max(cyto)}$ [U mL^{-1}]	0.132	0.656	1.95	5.753
$P_{max(peri)}$ [U mL^{-1}]	0.021	0.376	0.294	0.94
$P_{max(total)}$ [U mL^{-1}]	1.198	2.507	6.084	14.25
$Y_{p/x}$ [U g^{-1}]	1937.703	1286.384	2028.0	5088.21
Q_p [U L^{-1} h^{-1}]	99.83	105.57	468.0	949.8

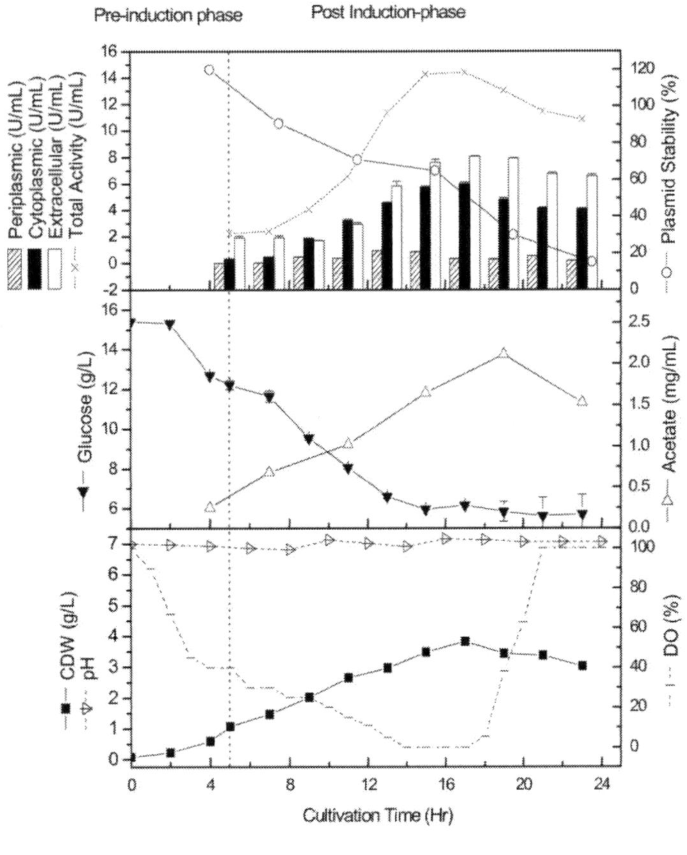

Figure 4: The cell growth kinetics, plasmid stability, acetate and total phytase production in controlled pH (left) and uncontrolled pH (right) batch cultivation of *E. coli* BL21(DE3) with 30 mM lactose as an inducer when aeration at 3 vv^{-1} min^{-1}.

Acknowledgments
The authors would like to thanks the Ministry of Science, Technology and Innovation (MOSTI), Malaysia for her studied scholarship.

References:

1. Anderson K.B., and Meyenburg K. (1980), Are Growth Rates of *Escherichia coli* in Batch Cultures Limited by Respiration, Journal Bacteriology, 144:114-123.
2. Babaeipour V., Shojaosadati S.A., Robatjazi S.M., Khalilzadeh R., Maghsoudi N., (2007), Over-production of human interferon-γ by HCDC of recombinant *Escherichia coli.* Process Biochemistry, 42:112-117.
3. Calik P., Yilgor, P., Demir S.A., (2006), Influence of controlled-pH and uncontrolled pH operations on recombinant benzaldehyde lyase production by E.coli. Enzyme and Micribial Technology, 38:617-627.
4. Choi J.H., Keum, K.C,, Lee, S.Y.,(2006), Production of recombinant protein by high cell density culture of *Escherichia coli,* Chem Eng Sci, 61:876-885.
5. Grabherr R., Nilsson E., Striedner G., Bayer K., (2002), Stabilizing plasmid copy number to improve recombinant protein production, Biotechnol Bioeng ,77:142–147.
6. Greiner R., Konietzny U., Jany K.D., (1993), Purification and characterization of two phytases from *Escherichia coli,* Arch Biochem Biophys, 303:107-103.
7. Han K., Lim, H.C., Hong, J., (1992), Acetic acid formation in *Escherichia coli* fermentation, Biotechnol Bioeng, 39:663-671.
8. Huang C.J., Chen X.Y. (2005), High-level expression and characterization of two chitinases, ChiCH and ChiCW, of Bacillus cereus 28-9 in *Escherichia coli*, Biochem. Biophys. Res. Commun, 327:8-17.
9. Imlay J.A., (2003), Pathways of oxidative damage. Annu Rev Microbiol,57:395–418
10. Jensen E.B., Carlsen, S., (1990), Production of recombinant human growth hormone in *Escherichia coli.* Expression of different precursors and physiological effects of glucose, Biotechnol Bioeng, 36:1-11.
11. Kim Y.O., Lee, J.K., Kim, H.K., Yu, J.H., Oh, T.K., (1998), Cloning of the thermostable phytase gene (phy) from *Bacillus* DS11 and its over expression in *Escherichia coli*, FEMS Microbiol Lett, 162:185-191.
12. Kleist S., Miksh G., Hitzman B., Arndt M., Freish K., Flaschel E., (2003), Optimization of the extracellular production of a bacterial phytase with *Escherichia coli* by using different fed-batch fermentation strategies, Appl Microbiol Biotechnol, 61:456-462.
13. Kleman G.L., Strohl, W.R., (1994), Acetate metabolism by *Escherichia coli* in high-cell-density fermentation, Appl Environ Microbiol, 60:3952–3958.
14. Kushoo A., Pal Y., Mukherjee K.J., (2005), Optimization of extracellular production of recombinant asparaginase in *Escherichia coli* in shake-flask and bioreactor, Appl Microbiol Biotechnol, 68:189-197.
15. Lee S.Y., (1996), High-cell density culture of *Escherichia coli*, Trends Biotechnol 14:98–105.
16. Lin H.Y., Neubauer P., (2000), Influence of Controlled Glucose Oscillations on A Fed-Batch Process of Recombinant *Escherichia coli*, Journal of Biotechnology, 79:27-37.
17. Pan J.G., Rhee J.S., Lebeault J.M., (1987), Physiological constraints in increasing biomass concentration of *Escherichia coli* B in fed batch culture, Biotechnol Lett 9:89-94.
18. Losen M, Frölich B, Pohl M, Büchs J (2004), Effect of oxygen limitation and medium composition on Escherichia coli fermentation in shake-flask cultures, Biotechnol Prog 20:1062–1068.
19. MacDonald H.L., Neway J.O. (1990), Effects of medium quality on the expression of human interlukins-2 at high cell density in fermentor cultures of *E.coli* K-12, Applied Environmental Microbiology, 56:1004-1011.

20. Nakano, K., Rischke, M., Sato, S., Märkl . H., (1997), Influence of Acetic Acid on The growth of *E.coli* K12 during high cell density cultivation in a dialysis reactor. Applied Microbiology Biotechnology, 48:597-601.
21. Neubauer P., Hofmann, K., (1994), Efficient use of lactose for the *lac* promoter controlled overexpression of the main antigenic protein of the foot and mouth disease virus in *Escherichia coli* under fed-batch fermentation conditions, FEMS Microbiol Rev 14: 99-102.
22. Phue J.N., Shiloach, J.(2005), Impact of dissolved oxygen concentration on acetate accumulation and physiology of *E.coli* BL21, evaluating transcription levels of key genes at different dissolved oxygen conditions, Metabolic Engineering, 7:353-363.
23. Qoronfleh M.W., (1999), Dissolved oxygen concentration affects the accumulation of HIV-1 recombinant proteins in *Escherichia coli*, Appl Biochem Biotechnol, 80:107–120.
24. Rusli F.M., Mohamed M.S., Mohamad R., (2009), Kinetics of Xylanase Fermentation by Recombinant *Escherichia coli* DH5α in Shake Flask Culture, American Journal of Biochemistry and Biotechnology, 5(3): 110-118.
25. Ryan W., Parulekar S.J., Stark B.C., (1989), Expression of β-lactamase by recombinant *Escherichia coli* strains containing plasmids of different sizes - effects of pH, phosphate, and dissolved oxygen, Biotechnol Bioeng, 34:309–319.
26. Shimizu N., Fukuzono S., Fujimori K., Nishimura N., Odawara Y., (1988), Fed-batch Cultures of Recombinant *E.coli* with inhibitory substance concentration monitoring, Journal Fermentation Technology, 64:197-210.
27. Tran T.T., Mamo G., Mattiasson B., Hatti-Kaul R., (2010), A Thermostable Phytase From *Bacillus sp*. MD2: Cloning, Expression and High-Level Production in *Escherichia coli*, Journal Industrial Microbiology Biotechnology, 37(3):279-87.
28. Wang, H., Wang, F., Wei, D., (2009), Impact of oxygen supply on rtPA expression in *Escherichia coli BL21(DE3)*:ammonia effects, Applied Microbiol Biotechnol, 82:249-259.
29. Yang C., Hua Q., Baba, T., Mori H., Shimizu K., (2003), Analysis of *Escherichia coli* anaplerotic metabolism and its regulation mechanisms from the metabolic responses to altered dilution rates and phosphoenolpyruvate carboxykinase knockout, *Biotechnol Bioeng,* 84: 129–144

(593aa) Engineering Global Regulator cAMP Receptor Protein (CRP) of E. Coli to Improve Strain Performance Under Stress

Hongfang Zhang[1], Huiqing Chong[1], Souvik Basak[1], Lei Huang[2] and **Rongrong Jiang**[1], (1)School of Chemical and Biomedical Engineering, Nanyang Technological University, Singapore, Singapore, (2)School of Chemical & Biomedical Engineering, Nanyang Technological University, Singapore, Singapore

Strain engineering has long been an intensively studied topic in both industry and academic. In general, there are two major categories in strain engineering: 1) "random" approach by using mutagens (UV, N-methyl-N'-nitro-N-nitrosoguanidine, and etc.), which is often time and resource intensive; 2) "rational"approach that uses metabolic engineering tools. Since most microorganisms lack of detailed metabolism knowledge and genotype-phenotype information, the use of metabolic engineering tools is often limited. Even engineering the simplest microbes like *E. coli* can be very complicated.

Here, we want to demonstrate that engineering global regulator cAMP receptor protein (CRP) of *E. coli* can improve cell performance under stressful conditions. CRP is one of the global regulatory proteins that can regulate the transcription of over 400 genes in *E. coli*. Our first target phenotype was strain osmotolerance. Amino acid mutations were introduced to CRP by error-prone PCR and DNA shuffling, and the random mutagenesis libraries were subjected to enrichment selection under NaCl stress. Five CRP mutants (MT1– MT5) were selected from the error-prone PCR libraries with enhanced osmotolerance. DNA shuffling technique was employed to generate mutant MT6 with MT1 – MT5 as templates. All of these variants showed much better growth in the presence of NaCl compared to wild type, and MT6 presented the best tolerance towards NaCl. In the presence of 0.9M NaCl, the growth rate of MT6 was 0.113 h^{-1}, while that of WT is 0.077h^{-1}. MT6 also exhibited resistance to other osmotic stressors, such as KCl, glucose and sucrose. Microarray analysis demonstrated that mutations in CRP could influence the expression level of genes not regulated by CRP. Scanning electron microscopy images confirmed the elongation of both WT and MT6 when exposed to NaCl but the cell surface of MT6 was relatively smooth.[1]

We have also applied the same technique to improve strain biofuel (1-butanol) tolerance[2], tolerance towards oxidative stress, and organic solvent tolerance. All selected mutant *E. coli* strains have shown much better performance under stress. In conclusion, transcriptional engineering via CRP can provide an efficient alternative for *E. coli* strain engineering.

1. Zhang, H.; Chong, H.; Ching, C.; Jiang, R. ***Biotech. Bioeng***. 2012, 109(5), 1165-1172
2. Zhang, H.; Chong, H.; Ching, C.; Song, H.; Jiang, R. *Appl. Microb. Biotech. 2012, 94 (4), 1107-1117*

(593ab) Homeoviscous Response of Clostridium Pasteurianum to Butanol

Yogi Kurniawan[1], Keerthi P. Venkataramanan[2], Judy J. Boatman[3], Casandra H. Haynes[3], Lenore M. Martin[4], Geoffrey D. Bothun[1], Katherine A. Taconi[5] and Carmen Scholz[6], (1)Department of Chemical Engineering, University of Rhode Island, Kingston, RI, (2)Biotechnology Science and Engineering Program, University of Alabama in Huntsville, Huntsville, AL, (3)Department of Chemical and Materials Engineering, University of Alabama in Huntsville, Huntsville, AL, (4)Department of Cell and Molecular Biology, University of Rhode Island, Kingston, RI, (5)University of Alabama in Huntsville, Huntsville, AL, (6)Department of Chemistry, University of Alabama, Huntsville, Huntsville, AL

Clostridium pasteurianum has been shown to ferment pure and partially purified biodiesel-derived crude glycerol into butanol at higher yields than obtained when sugars are used as the carbon source. n-Butanol is an attractive biofuel with favorable physicochemical properties. However, n-butanol is lipophilic and, when produced by fermentation, fluidizes biological membranes and leads to cellular inhibition. This study was focused on investigating the homeoviscous response of *C. pasteurianum* during n-butanol challenge experiments. Membrane lipid composition was analyzed by GC-MS and NMR, and membrane structure was analyzed by fluorescence anisotropy. *C. pasteurianum* was found to display two different homeoviscous responses that altered membrane lipid composition in an attempt to counteract the n-butanol toxicity. Addition of exogenous n-butanol to a fermentation, when *C. pasteurianum* produced endogenous n-butanol, led to an increase in the ratio of saturated to unsaturated fatty acids. On the other hand, addition of exogenous n-butanol to fermentation, when *C. pasteurianum* did not produce endogenous n-butanol, led to a decrease in the ratio of saturated to unsaturated fatty acids. This differential response for exogenous n-butanol during the presence and absence of n-butanol production indicates that *C. pasteurianum* is a versatile micro-organism that has the potential to be engineered as an industrial n-butanol producer using crude glycerol, a promising low cost feedstock for n-butanol production.

(593ac) Immobilization of Lipase B From Candida Antarctica B On Poly (methyl methacrylate) Epoxylated Support Aiming Biodiesel Production

Leonardo J.B.L. de Matos[1], José C.S. dos Santos[1], Bruna B. Pinheiro[1], Paulo W. Tardiolli[2], Raquel L.C. Giordano[2] and Luciana R.B. Gonçalves[1], (1)Chemical Engineering, Universidade Federal do Ceará, Fortaleza, Brazil, (2)Chemical Engineering, Universidade Federal de São Carlos, São Carlos, Brazil

Immobilization of enzymes and proteins on activated supports may be used to improve some enzyme properties in addition to allowing the simplification of reactor design. In this sense, supports containing epoxy groups may be very useful tools to stabilize proteins via uni or multipoint covalent attachment if the immobilization is properly designed. In this work, the support used for immobilization was Immobead 350-Chiralvision, which is a poly (methyl methacrylate) epoxylated with nonpolar characteristics support, diameter ranging between 300- 700μm. For the unipunctual immobilization, assays were performed varying the pH between 6-8 and ionic strength, an important factor, between 100-700mM. Best immobilization results were obtained at pH 7.5 and 700 mM, where recovery activity was around 25%. For multipoint immobilization, conducted at pH 10.05, the epoxylated support was modified by introducing amino groups, since the presence of amino groups allows a very rapid ionic adsorption of the proteins onto the support at low ionic strength. The new support has a layer of epoxy groups over a layer of ethylenediamine that is covalently bound to the support. Thus, this support has a great anionic exchanger power and a high number of epoxy groups. Immobilization was conducted at low ionic strength using amino-Immobead 350, while high ionic strength using conventional monofunctional epoxy supports was required. Furthermore, immobilization is expected to be much more rapid using amino-epoxy supports than employing conventional epoxy supports. Last but not least, the performance of both biocatalysts will be investigated in the synthesis of the ethyl oleate, aiming biodiesel production by esterification of acidic substrates.

(593ad) Riboswitch-sRNA for Dual Transcript Control by a Ligand

Richard A. Lease, Department of Chemical and Biomolecular Engineering, Ohio State University, Columbus, OH

Butanol is an alternative to gasoline and an excellent biofuel, which can be used to run cars without modifying their engines. Butanol derived from fermentation of polysaccharides by Clostridium bacteria is an established technology (ABE fermentation), similar to ethanol fermentation by yeast. However, "side reactions" from bacterial enzymes produce contaminant metabolites such as acetate and butyrate (butyric acid), increasing the economic costs of butanol recovery from scaled fermentations. One successful metabolic engineering approach has been to delete the clostridial genes encoding enzymes that produce acetate. However, deletions in genes encoding butyrate-synthesis enzymes also starve the bacteria of energy required for growth, and are lethal to the bacteria.

To solve this gene expression problem and to make industrially fermented butanol more economically viable, we are re-engineering native bacterial RNA sequences. Specifically, existing RNA genetic control elements within bacteria, called sRNAs, are repurposed here to specifically target and decrease the messenger RNA (mRNA) produced from the butyrate synthesis genes. This hybrid sRNA is essentially a metabolite-controlled molecular device, a "dimmer switch" riboswitch that which will be used to diminish, not turn off, specific butyrate gene expression, ultimately as a dynamic feedback of specific metabolite levels. This system presents an advantage over gene deletions because, when not artificially diminished, butyrate gene expression will default to its natural state, improving the health of clostridial cultures during fermentation. Initially the system is being designed to work in E. coli with well-characterized sRNAs, riboswitch aptamer sequences and reporter genes. The long-term goal is to evolve a metabolite-binding riboswitch by directed evolution, for incorporation into a retargeted sRNA as a metabolic feedback in the gram-positive Clostridia. By reducing butyrate contaminants this project will make butanol and other locally produced biofuels a more economically feasible alternative to gasoline.

Here we report our progress in repurposing a model sRNA regulator, the *E. coli* riboregulator DsrA, to build (1) a hybrid riboswitch-DsrA and (2) a retargeted DsrA that binds to the desired target mRNAs by antisense base pairing. Our initial approach has been to set up an mRNA reporter gene system that verifies and quantitatively reports the sRNA-mRNA interactions with a known mRNA target, prior to sRNA-riboswitch integration studies. We designed an arabinose-inducible hns-lacZ reporter fusion to be expressed from a plasmid, such that DsrA expressed from its native promoter on a second plasmid binds the hns RNA sequences naturally regulated by DsrA and diminishes LacZ production, as indicated by a colorimetric assay. A mutant DsrA control that is produced stably (Lease et al 2004, J Bact 186:6179) but which cannot base pair and thus cannot regulate hns RNA, does not interfere in the production of LacZ from our reporter gene, which confirms the specificity of the sRNA-mRNA interaction seen previously (Lease et al. 1998, PNAS USA 95:12456) in this reporter gene context. We will further demonstrate two simultaneous reporter gene activities separately controlled by DsrA.

For improved control of DsrA levels and to facilitate both riboswitch integration into DsrA and retargeting of DsrA, we are using an inducible DsrA which is produced from a lac promoter on a plasmid (Mandin & Gottesman, 2010 EMBO J). We have designed DsrA variants using natural restriction sites within the dsrA gene as well as engineered sites both upstream and downstream. These plasmid constructs permit ready insertion or replacement of short patches of DsrA sequence, such as insertion of riboswitch sequences, and retargeting of the sRNA by altering the nucleotide sequences that base-pair with specific mRNA targets. We will present the results of integrating a riboswitch into different positions within DsrA on its activity. Also, we will present the initial results in the reprogramming of DsrA to bind directly to reporter gene sequences that lack the natively regulated (e.g. hns RNA) sequences. These results are the initial steps towards achieving specifiable, dual RNA transcript control with small metabolite ligand-based metabolic feedback, in order to optimize fermentation culture health for butanol biofuel production at scale.

(593ae) Simulation-Based Optimization for Learning Parameters of Viral Self-Assembly Systems

Lu Xie[1,2], Gregory Smith[3], Xian Feng[3] and Russell Schwartz[2,3], (1)Joint Ph.D. Program in Computational Biology, Carnegie Mellon/University of Pittsburgh, Pittsburgh, PA, (2)Lane Center for Computational Biology, Carnegie Mellon University, Pittsburgh, PA, (3)Biological Sciences, Carnegie Mellon University, Pittsburgh, PA

Simulation-based Optimization for Learning Parameters of Viral Self-Assembly Systems

Lu Xie, Carnegie Mellon/University of Pittsburgh Joint Ph.D. Program in Computational Biology

Greg Smith, Department of Biological Sciences, Carnegie Mellon University

Xian Feng, Department of Biological Sciences, Carnegie Mellon University

Russell Schwartz, Department of Biological Sciences and Lane Center for Computational Biology, Carnegie Mellon University

The assembly of icosahedral viral capsids, or protein shells, has become a key model system for understanding complicated self-assembly processes, attracting considerable attention from various theoretical modeling communities. Simulation methods have proven a valuable tool for these studies due to the difficulty of directly experimentally observing the assembly process. For example, simulation models have proven useful for characterizing possible mechanisms and kinetics of capsid assembly, understanding potential pitfalls to efficient assembly, and learning strategies by which viruses overcome them. While such models have been useful in exploring the space of possible assembly models for simple icosahedral assembly systems, though, they have been of limited value to date in understanding the assembly of any specific real viruses. This limitation arises in large part because of the difficulty of experimentally determining the physical parameters needed to instantiate a simulation, specifically sets of rate constants for possible association and dissociation reactions between coat proteins by which the capsids can assemble. We address this problem by using simulation-based optimization to determine rate parameters consistent with indirect experimental measures of bulk assembly progress.

Our methods depend on stochastic simulations of virus capsid assembly that sample possible assembly trajectories from by coarse-grained "local rule" models that describe possible binding interactions among coat proteins [1]. The structure of any given virus is described by a set of such rules, each of which specifies positions and specificities of binding sites by which coat proteins may attach to one another. Adding association and dissociation rates to these rules allows them to implicitly specify a distribution of possible reaction pathways by which the coat proteins might assembly into viruses. Sampled reaction trajectories can then be converted into simulations of experimental measures assembly, which can then be fit to actual measurable data on bulk assembly progress. We specifically seek to fit viral systems to light scattering measurements, which track average assembly sizes over time in an assembly system.

We previously described a strategy for minimizing the deviation between true and simulated light scattering measurements designed to deal with particular challenges of the viral assembly system [2]. These challenges include a typically high cost of individual simulations, making large numbers of function evaluations infeasible, and a computational necessity for stochastic simulations, making it difficult to accurately estimate gradients or response surfaces. We previously addressed these problems with a local optimizer that interpolated between gradient-based and response surface strategies to help minimize function evaluations on smooth regions of the search space while adjusting as needed to more challenging regions. This local optimizer was then built into a heuristic global search strategy. In the present work, we have improved on that prior method by introducing extensions to simultaneously fit multiple real curves in order to reduce redundancy in solutions and developing new strategies to better utilize parallel computation. We have applied the new methods to three viral assembly systems ¨C human papillomavirus (HPV), hepatitis B virus (HBV), and cowpea chlorotic mottle virus (CCMV). The results indicate distinct kinds of assembly pathways for the three viruses, showing a diversity of assembly mechanisms available in nature for seemingly similar viral structures. This work demonstrates the value of simulation-based optimization for understanding virus assembly and suggests the potential of such methods for broader use in learning experimentally inaccessible features of complicated reaction systems.

<>[1] T. Zhang, R. Rohls and R. Schwartz (2005), Implementation of a discrete event simulator for biological self-assembly systems, *Proceedings of the 2005 Winter Simulation Conference*

[2] M. S. Kumar and R. Schwartz (2010), A parameter estimation technique for stochastic self-assembly systems and its application to human papillomavirus self-assembly, *Physical Biology*, Volume 7:045005

(593ag) Engineering Chimeric Antigen Receptors for Logical Computation

Yvonne Y. Chen, Harvard Society of Fellows, Cambridge, MA, Michael C. Jensen, Center for Childhood Cancer Research, Seattle Children's Research Institute, Seattle, WA and Pamela Silver, Systems Biology, Harvard Medical School, Boston, MA

Adoptive T-cell therapy is a promising cancer treatment strategy that seeks to strengthen and redirect the patient's immune system against tumor cells. In particular, autologous T cells genetically modified to express tumor-targeting receptors known as chimeric antigen receptors (CARs) have demonstrated safety and efficacy in treating recalcitrant malignancies including leukemia and metastatic melanoma. However, current CAR designs remain single-input, single-output signal-processing units that are vulnerable to significant obstacles, including mutational escape by tumor cells and off-target killing of normal cells. Here, we present the development of next-generation CARs that incorporate Boolean-logic computation to address these challenges. We have applied synthetic biology techniques to the modular design and rapid construction of OR-, AND-, and NOT-gate CARs. Results of functional studies in primary human T-cell culture will be presented. By conferring engineered T cells with the ability to compute multiple input signals before triggering cell-mediated killing of target cells, these novel CAR constructs will enable more precise tumor eradication, leading to a safer and more effective cell-based cancer therapy.

Wireless capnograph for respiratory function diagnosis and management

Di Zhao, Center for Bioelectronics and Biosensors, Biodesign Institute, Arizona State University, Tempe, AZ.
Ranganath Krishnan, Center for Bioelectronics and Biosensors, Biodesign Institute, Arizona State University, Tempe, AZ.
Dylan Miller, Center for Bioelectronics and Biosensors, Biodesign Institute, Arizona State University, Tempe, AZ.
Francis Tsow, Center for Bioelectronics and Biosensors, Biodesign Institute, Arizona State University, Tempe, AZ.
Erica Forzani, Center for Bioelectronics and Biosensors, Biodesign Institute, Arizona State University, Tempe, AZ.
Nongjian Tao, Center for Bioelectronics and Biosensors, Biodesign Institute, Arizona State University, Tempe, AZ.

Chronic respiratory diseases, such as chronic obstructive pulmonary disease (COPD) and asthma, affect more than 30 million Americans, cost over 30 billion dollars, and rank as the fourth death cause in the U.S. Many of these patients rely on long-term oxygen therapy, the only clinically proven and non-surgical therapy to treat their pulmonary dysfunctions and to extend their lives. Effective oxygen therapy requires real-time capnography, a widely used method that can detect deficiency of pulmonary functions, such as poor gas exchange, abnormal ventilation conditions, and insufficient lung blood flow even before detrimental associated symptoms. The existing capnography equipment not only is expensive and bulky, but also requires substantial training for correct operation and maintenance. The goal of the present project is to develop a low cost and wireless capnography device for personal, home-based oxygen therapy or for resource deprived emergency care conditions. The device takes advantage of a highly selective and humidity immune nanocomposite sensing material for carbon dioxide, which enables real-time breath-by-breath carbon dioxide detection. It uses a novel integrated sample collection and detection principle synergically combined with an intelligent signal-processing algorithm to maximize analysis accuracy. The technology features compactness, low cost, user friendliness, and robustness. In addition, the device can be wirelessly paired to the user cell-phone, computer, notebook, netbook or tabloid, providing a seamless information gateway between patients and medical care professionals and maximizing its benefit to the large and growing chronic respiratory disease patients.

(593ai) Effect of Carbohydrates On the Interaction Between Lysozyme and Procyanidin

Miao Liang, Rongxin Su, Rui Liu, Mengfan Wang, Wei Qi and Zhimin He, School of Chemical Engineering and Technology, Tianjin University, Tianjin, China

The interaction of procyanidins with proteins has aroused extensive attention due to its important relationship with many biologically relevant processes and the astringent property of polyphenols in red wine [Mcrae and Kennedy, 2011]. Moreover, this interaction often leads to the formation of complex aggregates, disruption of multilevel structures of proteins and the loss of biological activity [Goncalves et al., 2010]. Therefore, a deep understanding of the interaction mechanism is critical to control the aggregation behavior and improve the quality of red wine.

This present work investigated the interaction between procyanidin B3 and lysozyme using various biophysical methods. Firstly, we characterized the multilevel structural changes (namely, quaternary, tertiary and secondary) of lysozyme during its interaction with procyanidin B3 by light scattering techniques, fluorescence measurement, and circular dichroism. Then, the effect of two carbohydrates, gum arabic and sucrose, on the aggregation between lysozyme and B3 was also studied inspired by the phenomenon during the fruit ripening stage. The aggregation mechanism and carbohydrate inhibition behavior were analyzed. Moreover, the change of lysozyme activity was correlated with its structural alterations during aggregation. The main ideas and conclusions are summarized as follows.

(1) Light scattering techniques were employed to investigate the formation of lysozyme/procyanidin B3 aggregates. Dynamic light scattering results showed that a dramatic increase in the size of aggregates was initially observed after the addition of procyanidin B3. The average size increased slowly with continuously increasing procyanidin B3 concentration. These results clearly indicated the complex formation produced by the strong binding affinity of procyanidin B3 to lysozyme. Nephelometry technique also confirmed the increased size of the dispersed insoluble aggregates with increasing B3 concentrations

(2) Fluorescence measurements were conducted to analyze the tertiary structure changes of lysozyme. A progressive decrease in the fluorescence intensity was caused with increasing B3 concentrations, which indicated an interaction between lysozyme and B3. The quenching process analysis according to the modified Stern–Volmer equation indicated a static quenching which suggested the ground state complex formation between B3 and lysozyme. Moreover, synchronous fluorescence spectroscopic studies show the maximum emission wavelength of Tyr remained unchanged, whereas the emission maxima of Trp had a faint red shift (1.0 ± 0.1 nm). This observed red shift indicating a change in the microenvironment of Trp from a relatively hydrophobic core to a polar aqueous environment resulted from unfolding of the compact tertiary structure of lysozyme. In addition, the quenching of lysozyme fluorescence may be responsible for the loss of enzyme activity, because the mainly intrinsic fluorescence residues (Trp 62 and Trp 108) are arranged close to the substrate-binding site and play important roles in binding with substrate [Rothwarf and Scheraga, 1996].

(3) Circular dichroism spectroscopy was performed to evaluate the effect of B3 on the secondary structural changes of lysozyme. The gradually increased intensity at 208 nm demonstrated the binding of B3 to lysozyme disrupts the protein secondary structure upon complexation. The α-helical content of lysozyme decreased from 24.66% to 10.9% at a molar ratio of lysozyme to B3 of 1:10. Meanwhile, lysozyme activity against *M. lysodeikticus* gradually decreased from 4.465×10^4 U/mg to 3.186×10^4 U/mg. This result combined with the structural changes of lysozyme during aggregation confirmed that the conformation of a protein determines its biological function.

(4) Sucrose and gum arabic have the ability to inhibit the formation of insoluble protein/procyanidin aggregates. A rapid decrease in hydrodynamic radius of the aggregates was observed with the addition of gum arabic, while sucrose had a lesser reductive effect on complex formation. Moreover, the interaction of lysozyme/B3 with gum arabic does not restore the fluorescence and activity, combined with the reduction in aggregate size, support the idea that gum

arabic partially disassociates lysozyme/B3 complexes and induces the formation of a more soluble ternary complex between lysozyme/B3 and gum arabic [Mateus et al., 2004]. For sucrose, the aggregate size decreases concomitantly with the recovery of fluorescence and enzymatic activity may be explained by the association of the low molecular weight carbohydrates with the polyphenols during incubation. This molecular association in solution prevents polyphenols from further binding to the protein, thus inhibiting aggregate formation. These findings provide deep insights into the mechanism of aggregation and carbohydrate inhibition between lysozyme and procyanidin B3.

This work was supported by the Natural Science Foundation of China (No. 20806057 and 31071509), the Ministry of Science and Technology of China (Nos. 2012BAD29B05 and 2012AA06A303), and the Ministry of Education (No. NCET-11-0372).

References

1. Mcrae J. M. and Kennedy J. A. Wine and Grape Tannin Interactions with Salivary Proteins and Their Impact on Astringency: A Review of Current Research, *Molecules*, 16, 2348-2364, 2011
2. Goncalves R., Mateus N. and de Freitas V. Biological Relevance of the Interaction between Procyanidins and Trypsin: A Multitechnique Approach, *Journal of Agricultural and Food Chemistry*, 58, 11924–11931,2010
3. Rothwarf D. M. and Scheraga H. A. Role of Non-Native Aromatic and Hydrophobic Interactions in the Folding of Hen Egg White Lysozyme, *Biochemistry*, 35, 13797–13807, 1996
4. Mateus N., Carvalho E., Luis C. and de Freitas V. Influence of the tannin structure on the disruption effect of carbohydrates on protein–tannin aggregates, *Analytica Chimica Acta*, 513, 135–140, 2004

(593aj) Modular Protein Switches Employing Fibronectin-Derived Monobodies As Input Domains

Amol Date, Manu Kanwar and Marc Ostermeier, Chemical and Biomolecular Engineering, Johns Hopkins University, Baltimore, MD

Biological sensor platforms often combine a recognition element (e.g. an antibody) that binds to a target molecule and an output scheme to produce a readable signal. There is a need to develop biosensors with a simple, direct connection between target binding and output signal in a platform that is readily adaptable for the detection of any biological targets. We have developed target-activated enzyme switches that employ a fibronectin type III-based synthetic binding domain (called a monobody) as the target recognition domain and TEM-1 β-lactamase (BLA) as the output domain. Monobodies have potential as antibody mimetics and have been engineered by others to specifically bind a variety of protein targets including maltose binding protein (MBP).

We implemented a novel, PCR-based combinatorial approach to create a library in which a circularly permuted β-lactamase was inserted at all possible locations in the monobody MBP-74 that specifically binds MBP. This library was subjected to a bacterial band-pass genetic selection system to identify gene-fusions that confer increased resistance to β-lactam antibiotics when MBP is coexpressed. A number of switch genes conferring this phenotype were identified, with the best genes conferring a 16-fold increase in β-lactam antibiotic resistance in the presence of MBP. The switching activity resulted from a combination of allosteric effects (i.e. MBP allosterically activates the BLA domain's enzyme activity) and cellular accumulation affects (i.e. the switch protein accumulates at higher levels in the presence of MBP). We tested the modularity of these switches by introducing mutations to the monobody domain that are known to convert their binding specificity to new target proteins such as yeast small ubiquitin-like modifier protein (ySUMO) and green fluorescent protein (GFP). The altered switches demonstrated both allosteric switching and the ability to confer a switching phenotype to cells. These experiments demonstrate the versatility and modularity or monobody-based protein switches.

BACTERIAL BIOSURFACTANT PRODUCTION FROM BIOMASS-DERIVED SUGARS AIMING BIOREMEDIATION OF MARINE ECOSYSTEMS CONTAMINATED BY HYDROCARBONS

I. W. L de FRANÇA, D. W. F. de OLIVEIRA, V. M. M. MELO, H. B. DE SANT'ANA, L. R. B. GONÇALVES
Universidade Federal do Ceará, Departamento de Engenharia Química, Fortaleza, CE, Brazil.

Introduction

Surfactants are amphipathic molecules constituted by one polar and one nonpolar portion. The portion polar, or hydrophilic, might be ionic, nonionic or amphoteric. A nonpolar, or hydrophobic portion, is usually a hydrocarbon, insoluble in water but soluble in lipids (5). These molecules are an important class of chemical compounds widely used in various industrial sectors, due to its characteristics, such as, easy biodegradability, low toxicity, environmental acceptability and stability in drastic conditions of temperature or pH (5, 4, 15, 23). The biosurfactants can be produced by bacteria, yeasts and filamentous fungi, and it presents a great difference (advantage) compared to synthetic surfactants due to the variability in its hydrophilic and hydrophobic nature. It can be produced by biotransformation of renewable raw materials, showing the advantage of lower environmental impact compared to conventional chemical detergents (1).

The potential of application of the biosurfactant produced by microorganisms is based on their functional properties, including emulsifying, separating, wetting, solubilizing, corrosion inhibition, reduction of viscosity and liquid surface tension reduction. These properties are applied in several fields of agriculture, construction, food industry, beverage, oil, paper, metal, textile, pharmaceutical and cosmetics (2, 10, 16). The interest of using surfactants obtained as metabolic products by bacteria, fungi and yeasts (biosurfactants) is growing due to the advantages that these compounds have similar relative to their chemically synthesized, and also because of its wide range of applications (6, 3). The most effective lipopeptides biosurfactants are produced by bacteria of the genus *Bacillus*, especially the surfactin produced by *Bacillus subtilis* (5).

Surfactants have the ability to reduce surface and interfacial tensions and promote the formation of microemulsions, where water can solubilize hydrocarbons (9). It has critical role in the stability of emulsions, in other words, reduction of interfacial tension associated with the formation of an emulsion (21). The surfactants act to facilitate the emulsification through the formation of an adsorbed film around the dispersed droplets, which hinders the flocculation and coalescence (3). The emulsification process is accompanied by an increase of the interfacial area, which produces a sudden increase of surface free energy.

One of the uses for the biosurfactants which has received attention is known constant MEOR (Microbial-Enhanced Oil Recovery), whose process involves introducing microorganisms in crude oil reservoirs to enhance the extraction of oily sludge and recover hydrocarbon fractions. This method is also being proposed and tested in some parts of the world to combat environmental pollution generated by accidental oil spills. This process prevents the persistence layer of oil on the environment, which are responsible for the depletion of oxygen and poisoning of marine life (7). Accidents with oil have been very frequent, so this area has a vast field for the application of biosurfactants.

The main commercial use of biosurfactants is in bioremediation of environments contaminated with hydrocarbons, because of its ability to stabilize emulsions. This causes of an increase in solubility and availability of hydrophobic contaminants, increasing the potential for biodegradation. In the formulated emulsion, it is easier to occur the degradation of hydrophobic compounds by microorganisms, thus decreasing the concentration of these substances, and consequently reducing the social and environmental impact of potential oil spills (18).

The oil industry is the biggest market for biosurfactants, which are used directly in the production of petroleum or are incorporated in formulations of lubricants (7). Another use relates to the potential for recovery of oil in tank cleaning, preparation of oil-alcohol mixtures for fuels and dispersing spilled oil in aquatic ecosystems (17).

Bacillus subtilis LAMI005 isolated from the tank of Chlorination at the Wastewater Treatment Plant on Campus do Pici in Federal University of Ceará, Brazil, is a promising biosurfactant producer, which uses glucose and fructose as carbon source, main sugars present in cashew apple juice.

A previous evaluation of the microorganism's capability of degrading the organic pollutants is the first step to set up a field scale remediation project based on addition of microorganisms (17). The largest application of biosurfactants is in the oil industry, in view of this factor, the objectives of this work are related to the study of biosurfactant production and direct applications in marine ecosystems contaminated with hydrocarbons.

Experimental Part

Microorganism and Description of the Fermentation Process

The strain of *Bacillus subtilis* LAMI005 used in this study was isolated from the tank of chlorination at the Wastewater Treatment Plant on 'Campus do Pici' in the Federal University of Ceará, Brazil. Its rRNA 16S sequence is deposited in the Genbank with the following access number: FJ143046. The strain was maintained on APGE medium (consisting of peptone, 5.0 g L^{-1}, glucose 5.0 g L^{-1}, yeast extract 2.5 g L^{-1}, and agar 15.0 g L^{-1}).

Surfactin production was conducted utilizing mineral medium (MM), which contained (in grams per liter) (NH4)2SO4 1.0, Na2HPO4 6.0, KH2PO4 3.0, NaCl 2.7, and Mg2SO4.7H2O 0.6 (14). Glucose and fructose in equimolar proportions were used as carbon sources. It was added 0,1% of a trace element solution (consisting of ZnSO4.7H2O 10.95 g L^{-1}, FeSO4 .7H2O 5.0 g L^{-1}, MnSO4.H2O 0.39 g L^{-1}, CuSO4.5H2O 54 g L^{-1}, Co(NO3)2.6H2O 0.25 g L−1, and Na2B4O7.10H2O 0.17 g L−1).

Bacillus subtilis LAMI005 was inoculated on APGE plates and incubated at 30 °C for 24 h. After this period, three colonies were transferred to 250-mL Erlenmeyer flask containing 100 mL of MM supplemented with glucose, as carbon source (Inoculum). The flask was incubated in rotary shaker (Tecnal-TE240) at 180 rpm, 30 °C for 40 h. Afterward, optical density of this culture was adjusted to 0.1 at 600 nm, and an aliquot of 10 mL (10% of inoculum) was transferred to a 250-mL Erlenmeyer flask containing mineral medium and 0.1% of trace element solution, with mineral medium (MM) and glucose at a concentration of 10 g L^{-1}. The culture media and inoculum were adjusted to pH 7.0 and autoclaved at 110 ° C for 10 minutes.

Extraction of surfactin

The extraction of surfactin was conducted in accordance with the methodology described by (11). Cells were separated by centrifugation at 10,000×g for 15 min at 4 °C. The supernatant was filtrated using a 0.45 μm membrane (cellulose esters, Millipore), and the pH of the filtrate was adjusted to 2.0 by adding 3 M HCl. The resultant solution was maintained in repose to allow precipitation of surfactin by 12 h. The precipitate was collected by centrifugation at10,000×g for 15 min at 45 °C to obtain the crude surfactin. For further purification, the crude surfactin was dissolved in deionized water, and it was extracted three times by using an equal volume of dichloromethane (Vetec, São Paulo, Brazil) (22). The solvent layer was harvested and evaporated to dryness. The resulting brown-colored paste was dissolved in deionized water, and it was centrifuged to remove insoluble substances. After further filtration and drying, the purified surfactin was obtained.

Analytical methods

Determination of Biomass: The cell growth was obtained by indirect method of turbidimetry by measuring the optical density using a UV visible spectrophotometer (*Genesys* series 20) to 600 nm, as described in (11).

Concentration of total reducing sugars: The concentration of glucose and fructose (reducing sugars - TRS) was determined by High Performance Liquid Chromatography (HPLC) on Waters chromatograph (Waters, Milford, MA, USA) coupled to a refractive index detector (Model 2414) and Supelcogel C610H column (30 cm x 7.8 mm) from Sigma Aldrich using 0.1% H3PO4 in water (H2O MiliQ) as the mobile phase , at a flow rate of 0.5 mL min-1. The sample injection volume was 10 mL and the analysis time was 20 minutes at room temperature (19).

Surfactin concentration: Surfactin concentration was measured by HPLC using a Waters high-performance liquid chromatographer equipped with a UV detector (Model 2487, Waters, Milford, MA, USA) at 205 nm and a Symmetry C18 column (150×4,6 mm, 5 μm, Waters, Ireland). The mobile phase consisted of 20% trifluoroacetic acid (3.8 mM) and 80% acetonitrile. The elution rate was 1 mL min−1 at 30 °C and the sample size was 20 μL. The identity of the purified surfactin was done using commercially available 95% pure surfactin (Sigma-Aldrich) as the authentic compound (22).

Surface tension: Surface tension was determined with a Tensiometer (Torsion Balance of White Electrical Instrument) at 30 °C, according to the De Nöuy ring method. The surface tension determination was replicated and it was performed using cell free supernatants (20).

Critical Micelle Concentration (CMC): It was performed by quantifying the tension superficial of successive dilutions of an initial solution rich in surfactin (biosurfactant produced) of known concentration, according to the methodology described by (20).

Salt Stability: The stability of surface-active properties of the biosurfactant was studied by varying the concentration of NaCl (0.5% to 20% w/v) (13) in a solution containing surfactin. It was used the semi-purified biosurfactant in this analysis. Surface tension was determined with a Tensiometer (Torsion Balance of White Electrical Instrument) at 30 °C, according to the De Nöuy ring method.

Test of the redox indicator 2,6-dichlorophenol indophenol - degradation of hydrocarbons: The biodegradability experiments were carried out using a technique based on the redox indicator 2.6 - dichlorophenol indophenol (DCPIP) (12). It has been changed the original methodology, presenting a resemblance to the methodology used by MARIANO (2008). The inóculom of *Bacillus subtlis* LAMI005 were added (125 μL, O.D = 0.55 at l = 610 nm, separately, to test tubes (duplicates) that contained sterile Bushnell-Hass (BH) medium (7.5 mL) and 50 μL of hydrocarbons. The concentration of DCPIP was 27 mg/mL. Test tubes and Erlenmeyer flasks were kept under agitation (240 rpm) at room temperature (27±2°C). The BH medium consists of, $g.L^{-1}$: $MgSO_4$: 0.2; $CaCl_2$: 0.02; KH_2PO_4: 1.0; K_2HPO_4: 1.0; NH_4NO_3: 1.0; $FeCl_3$: 0.05 (9).

The principle of this technique is that during the microbial oxidation of hydrocarbons, electrons are transferred to electron acceptors such as O_2, nitrates and sulphate. By incorporating an electron acceptor such as DCPIP to the culture medium, it is possible to ascertain the ability of the microorganism to utilize hydrocarbon substrate by observing the colour change of DCPIP from blue (oxidized) to colourless (reduced). This technique has been employed in several works (12, 17).

Results and Discussion

Previous studies (8) have evaluated the influence of the initial concentration of substrate (glucose and fructose) (S_0) in the formation of product (surfactin) by *Bacillus subtilis* LAMI005. Aerobic cultivations were performed with concentrations of total reducing sugars of 16.55, 30.07 and 43.10 g L-1. It was noted that an increase in the initial concentration of sugars caused a decrease in the production of surfactin, which may indicate a substrate inhibition. It was observed that the increased production of surfactin occurred in the test with $S_0 = 16.55$ g.L^{-1}, resulting in a concentration of biosurfactant around 250mgL^{-1} and critical micelle concentration (CMC) of 14mg.L^{-1}. This production of surfactin promoted the reduction of surface tension of water from 72 to 28 dyna.cm^{-1}, which shows the efficiency of the surfactant produced by *Bacillus subtilis*.

It was conducted a study of the salt stability of the surface-active proprieties of the biosurfactant produced in this work. A solution rich in surfactin was submitted to addition of varied concentrations of NaCl, and in each addition of salt, it was measured the surface tension of the solution. It was observed that as the NaCl is added (variation 0-20% w/v), not occurred considerable variation in surface tension, showing that the surfactin produced presents saline stability. The biggest concentration of NaCl (20% m/v) used is similar to the one observed in synthetic sea water (24% m /v), showing the correspondence of experiments with a real application. This result is interesting when it's considered the use of biosurfactant in bioremediation in saline environments, as for example, in the recovery of oil spills in marine environments (19).

The redox indicator 2,6-dichlorophenol indophenol in presence of hydrocarbons presents blue coloration. When hydrocarbons are degraded, so oxidized or consumed, there is a reduction of this characteristic color. So, this indicator was used in order to identify the degradation capacity of diesel, gasoline, hexane, naphthalene, n-hexadecane, motor oil, soybean oil, pyrene and petroleum by *Bacillus subtilis* LAMI005.

The figure 13 presents some results obtained on the assay.

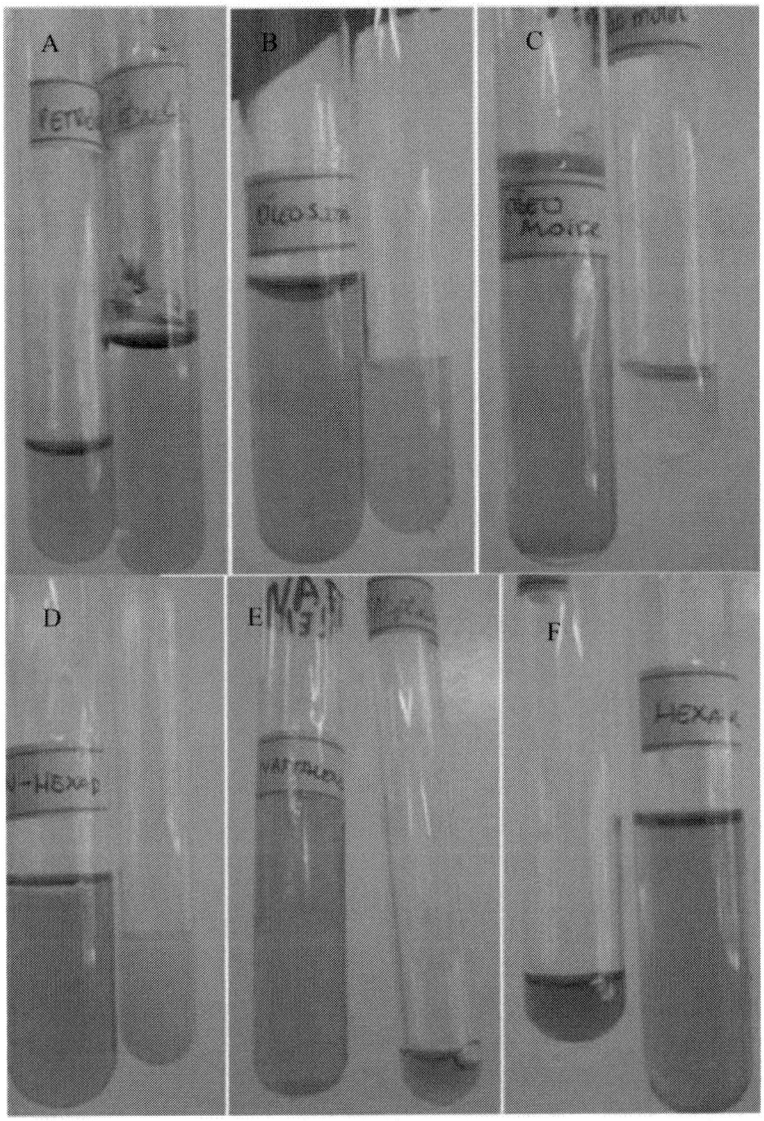

Figure 13 - Results for redox indicator 2,6-dichlorophenol indophenol to identify the degradation of hydrocarbons: Petroleum (A), Soy oil (B) Motor oil (C), N-hexadecane (D), naphthalene (E) and hexane (F).

The blue colored tubes show the HB medium and the hydrocarbon, and the colorless shows the result of the addiction of *Bacillus subtilis*.

It was observed some change in the coloration of some hydrocarbons, such as gasoline, n-hexadecane, motor oil, soybean oil and oil, so it was verified that *Bacillus subtilis* was able to degrade these hydrocarbons, possibly with the contribution of the biosurfactant produced during the degradation process of these hydrocarbons. This fact presents some evidences for its application in bioremediation *in situ*. The process of degradation occurred about 140 hours. It wasn't observed a loss of coloration of hydrocarbons which have aromatic ring, such as naphthalene and pyrene, possibly justified by their long and stable carbonic chains.

Other studies (17) obtained similar results when studying the degradation of hydrocarbons by *Pseudomonas aeruginosa* LBI, which there was a loss of coloration of the medium, for diesel and petroleum after 23 hours of assay, 68 hours for kerosene and a not complete loss of coloration of oily sludge after 72 hours. *P. aeruginosa* is a well know producer of the rhamnolipid biosurfactant. The strain *P. aeruginosa* LBI was capable of producing biosurfactant using soapstock, mannitol and glycerol and kerosene, diesel oil, crude oil and oily sludge. *O. anthropi* enhanced the biodegradation of oily sludge and was capable of biodegrading hexane, heptane, hexadecane and commercial diesel oil. It observed that the contaminated soils or groundwaters already had a microbiot adapted to degrade recalcitrant hydrocarbons.

Conclusion

The results obtained has presented that the biosurfactant produced by *Bacillus subtilis* LAMI005 is promising product, due to the salt stability and the tension reduction. The biosurfactant produced has showed salt stability and contributed to the degradation of oil and other hydrocarbons. Results have indicated that surfactin can be used in bioremediation of marine ecosystems contaminated by hydrocarbons.

Acknowledges

The authors acknowledge the Agência Nacional do Petróleo, Gás Natural e Biocombustíveis (ANP) (PRH-31); CAPES, CNPQ and UFC.

References

1. BOGNOLO, G. Biosurfactants as emulsifying agents for hydrocarbons. Collo. and Surf. v. 152, p. 41-52, 1999.
2. BANAT, I. M., MAKKAR, R. S., CAMEOTRA, S. S. Potential Commercial Applications of Microbial Surfactants. App. Microb. Biotech., 53, p.495-508, 2000.
3. Cameotra, R.S., Makkar, R.S., and Banat, I.M. (2011): Advances in utilization of renewable substrates for biosurfactant production. AMB Exp. 1-5.
4. COOPER, D. G.; MAEDONALD, C. R.; DUFF, S. J. B.; KOSARIC N. Enhanced production of surfactin from *Bacillus subtilis* by continuous product removal and metal cation additions. Appl. and Environ. Microb., v. 42, p. 408-412, 1981.
5. DESAI, J. D.; BANAT, I. M. Microbial production of surfactants and their commercial potencial. Microbiol. Mol. Biol. Rev., v. 61, p. 47-64, 1997.
6. DUBEY, K.; JUWARKAR, A. Distilley and curd whey wastes as viable alternative sources for biosurfactants production. Worl. Journ. Microb. Biot., 17, p.61-69, 2001.
7. DYKE, M. I.; LEE, H.; TREVORS, J. T. Applications of Microbial Surfactants. Biotech. Advan. v. 9, p. 241-252. 1991.
8. FRANÇA, I. W. L; OLIVEIRA, D. W. F.; MONTENEGRO, D.L.; MARTINS, J.J.L.; GIRO, M.E.A., Melo, V.M.M.; GONÇALVES, L.R.B. (2010), Influência da concentração de substrato na produção de biossurfactante por Bacilus subtilis LAMI005. XVIII Congresso Brasileiro de Engenharia Química.
9. FERNANDES, A. C. L. B.; PEREIRA, M. D. ; KRONEMBERGER, F. A.; FREIRE, D. M. G. Produção de biossurfactantes para aplicação em remediação de solos contaminados por

Petróleo: estudo dos fatores de virulência. In: RIO OIL & GAS Expo and Conference, 2008, Rio de Janeiro - Rio de Janeiro. RIO OIL & GAS Expo and Conference, 2008.

10. FIECHTER, A. Biosurfactants: moving towards industrial application, TIBTECH., v. 10, p.208 – 217, 1992.

11. GIRO, M. E. A.; MARTINS, J. J. L.; ROCHA, M. V. P.; MELO, V. M. M.; GONÇALVES, L. R. B. Clarified cashew apple juice as alternative raw material for biosurfactant production by *Bacillus subtilis* in a batch bioreactor. Biotechnol. J., v. 4, p. 738–747, 2009.

12. Hanson, K.G.; Desai, J.D.; Desai, A.J. (1993). A rapid and simple screening technique for potential crude oil degrading microorganisms. *Biotechnol. Tech.*, 7, 745-748.

13. ILORI, M. O., AMOBI, A. C., ODOCHA, A. C., Factors affecting biosurfactant production by oil degrading *Aeromonas spp.* isolated from a tropical environment. Chemosph. 2005, 61, 985-992.

14. MORÁN, A. C.; MARTÍNEZ, A. M.; SIÑERIZ, F. Quantification of surfactin in culture supernatants by hemolytic activity. Biotechnol. Lett., v. 24, p. 177–180, 2000.

15. MULLIGAN, C. N.; YONG, R. N.; GIBBS, B. F. Surfactant-enchanced remediation of contaminated soil: a review. Eng. Geol., v. 60, p. 371, 2001.

16. MULLIGAN, C. N. Enviromental aplications for biosurfactants. Envirom. Pollution, v .133 p.183-198, 2005.

17. MARIANO, Adriano Pinto et al. Biodegradability of commercial and weathered diesel oils. *Braz. J. Microbiol.* [online]. 2008, vol.39, n.1 [cited 2012-09-20], pp. 133-142 .

18. NITSCHKE, M.; PASTORE, G. M. Production and properties of a surfactant obtained from *Bacillus subtilis* grown on cassava wastewater. Bior. Tech., v. 97, p. 336-341, 2006.

19. ROCHA, M. V. P., GOMES-BARRETO, R. V., MELO, V. M. M., GONÇALVES, L. R. B., Evaluation of Cashew Apple Juice for Surfactin Production by *Bacillus subtilis* LAMI008. Appl. Biochem. Biotechnol. 2008.

20. ROCHA, M. V. R.; OLIVEIRA, A. H. S.; SOUZA, M. C. M.; GONÇALVES, L. R. B. Natural Cashew apple juice as fermentation medium for biosurfactant production by *Acinetobacter calcoaceticus*. Wor. J. Microb. and Biotech., v. 22, p. 1295-1299, 2006.

21. SHAHIDZADEH, N.; BONN, D.; MEUNIER, J.; NABAVI, M.; AIRIAU, M.; MORYAN, M. Dinamics of Spontaneous Emulsification for Fabrication of Oil in Water Emulsions, Langmuir. 16, p. 9703-9708, 2000.

22. YEH, M. S.; WEI, Y. H. CHANG, J.S. Bioreactor design for enhanced carrier-assisted surfactin production with *Bacillus subtilis*. Process Biochem., v. 41, p. 1799-1805, 2006.

23. YEH, M. S.; WEI, Y. H.; CHANG, J. S. Enhanced Production of Surfactin from *Bacillus subtilis* by Addition of Solid Carriers. Biotechnol. Prog., v. 21, p. 1329-1334, 2005.

(593al) Carbon Nanofibers: Rosette Nanotubes Injectable Scaffolds for Myocardial Application

Xiangling Meng[1], David A. Stout[2], Linlin Sun[3], Rachel Beigessner[4], Hicham Fenniri[5] and Thomas J. Webster[3], (1)Chemistry, Brown University, Providence, RI, (2)Biomedical Engineering, Brown University, Providence, RI, (3)School of Engineering, Brown University, Providence, RI, (4)University of Alberta, Edmonton, AB, Canada, (5)National Institute for Nanotechnology and Department of Chemistry, University of Alberta, Edmonton, AB, Canada

Carbon Nanofibers: Rosette Nanotubes Injectable Scaffolds for Myocardial Application

Xiangling Meng[1], David A. Stout[2], Linlin Sun[2], Rachel L. Beigessner[3], Hicham Fenniri[3]and Thomas J. Webster[2,4]

[1]Department of Chemistry, Brown University, Providence, RI 02912, USA

[2]School of Engineering, Brown University, Providence, RI 02912, USA

[3]National Institute for Nanotechnology and Departments of Chemistry and Biomedical Engineering, University of Alberta, 11421 Saskatchewan Drive, Edmonton, AB T6G 2M9, Canada

[4]Department of Orthopaedics, Brown University, Providence, RI 02912, USA

Introduction:

Heart failure is a leading cause of death in the world. The lack of blood supply caused by a sudden obstruction in the major blood vessel leads to damage of the cardiac muscle. The rapid wound healing process forms fibrosis instead of cardiac muscle, which may lead to another myocardial infarction in the future.Therefore, the purpose of this in vitro experiment was to design, fabricate and test the ability of a novel injectable biomimetic and conductive material for the regeneration heart tissue.

Materials and Methods:

Combinations of three different materials were tested here: rosette nanotubes (RNTs), carbon nanofibers (CNFs) and pHEMA-poly (2-hydroxyethyl methacrylate). RNTs are soft nanotubes which self-assemble from a synthetic DNA base analog when placed in aqueous solutions. The great cytocompatibilityproperties of RNTs have been proved in osteoblast and endothelial engineering [1,2]. CNFs are conductive and were used here to increase the conductivity of the composites to more closely emulate the conductivity of heart tissue. CNFs used in this study were modified with –OH and –COOH in order to enhance hydrophilicity properties and their dispersion in solution.pHEMA was chosen here since it has been widely used in drug delivery and orthopedicapplications.

CNFs were well dispersed in RNT and HEMA solutions by sonication. The samples were polymerized in an oven under 75 for 3h. The samples were washed with water (15min×3) in order to remove the unreacted monomers. Samples were sterilized in 70% and 100% ethanol for 5min and dried over night. Samples were rinsed with PBS three times before cell seeding.

Human cardiomyocytes (Celprogen Inc.)were seeded on the composites at a density of 3000 cells $cm^{-2}$for 1day and 3 day proliferation assaysand 3500 cells $cm^{-2}$for 4h adhesion assaysin Human Cardiomyocyte Complete Media supplemented with 10% fetal bovine serum (FBS) under standard cell culture conditions (humidified, 5% CO_2/95% air environment at 37). Live cells were counted using the MTT assay.A scanning electron microscope (SEM, LEO 1530-VP) operating at a 10 kV voltage was used to image the composite. Water contact angles on the surface were analyzed by a drop shape analyzer to determine the hydrophilicity of the composites. 3 μl of deionized water was dropped onto the surfaces. The electrical resistance of the samples was determined using a multimeter by touching the surface with two alligator clips. The resistance was tested dry at room temperature and the probes were 10mm apart. An atomic force microscope in contact mode was used to determine the surface roughness of the samples. A sharp tipped cantilever that was 10-50N/m in stiffness was used here and it was calibrated at 32 N/m every time.An Instron Model 5882 Frame with an extension rate of 5mm/min was used to determine the tensile mechanical properties of composites with different concentrations of CNFs and hydrogels controls. Type IVdog shaped scaffolds were used for tensile tests.

All experiments were performedin triplicate, data were plotted as the mean ± standard error of the mean, and statistical analyses were performed. When data were compared, ANOVA software and a student T-test were used. A p-value of < 0.05 was considered to be significant in this study.

Results and Discussion:

Representative SEM pictures showed that RNTs and CNFs were well dispersed in pHEMA via sonication. An increase of surface roughness was observed in samples with greater amounts of CNFs.CNFs increased the conductivity of composites(from 0 to $1.8\times10^{-3}\Omega^{-1}$), while RNTs didn't play an obvious role in increasing conductivity. The contact angle of the samples also decreased from 51.6° to 38.1°, which means the hydrophilicity increased, when adding more CNFs and RNTs. Tensile stress of composites decreased when the concentration of CNFs increased, which is convenient for injecting the composites into the body. Results of the cardiomyocyteadhesion assays and proliferation assays demonstrated greater cardiomyocyte density on the composites with greater amounts of CNFs and RNTs.

Conclusions: HRNs and CNFs were successfully used to design a novel biomimetic composite to improve cytocompatibility properties for cardiomyocytes and thus, should be further studied for cardiovascular tissue engineering applications.

Acknowledgements: This study was supported by Hermann Foundation, the National Research Council of Canada, the Natural Science and Engineering Research Council of Canada, and the University of Alberta, the National Science FoundationGraduate Research Fellowship Program for David Stout (NSF #1058262).

References:

[1] Zhang L, Rakotondradany F, Myles AJ, Fenniri H, Webster TJ. Arginine-glycine-aspartic acid modified rosette nanotube–hydrogel composites for bone tissue engineering. Biomaterials 2009;30:1309-20.

[2] Fine E, Zhang L, Fenniri H, Webster TJ. Enhanced endothelial cell functions on rosette nanotube-coated titanium vascular stents. Int J Nanomedicine 2009;4:91-7.

(593am) Comparison of Estrogenic Properties of Small Molecules Across Several Species Using Biosensors

Derek Reichel[1], **Angela Chen**[1], Lily Glick[1], Jeevan Baretto[2] and David W. Wood[2], (1)The Ohio State University, Columbus, OH, (2)Chemical and Biomolecular Engineering, The Ohio State University, Columbus, OH

Estrogenic compounds have important pharmaceutical and toxicological properties. As pharmaceuticals, these compounds can be used to treat several types of cancer, including breast, pancreatic and testicular cancer. As trace contaminants in the environment, estrogens have been linked to health risks, such as decreased ages of puberty onset in girls. Using engineered *Escherchilia coli* biosensors, we have developed an assay that eliminates the initial need for animal testing, in evaluating potentially estrogenic compounds. The bacterial biosensor relies on an engineered protein consisting of an intein with an inserted estrogen receptor beta (ERβ) ligand-binding domain, and a thymidylate synthase (TS) reporter enzyme. Culturing the *E.coli* biosensor strain in thymine-less media allows for identification of estrogenic compounds based on growth phenotype. The assay is able to distinguish between ERβ agonists and antagonists. By comparing the effect of a compound on different biosensors with different animal ERβ receptors, the difference in estrogenicity between several test compounds across several species can be determined. The current set-up is a 96-well plate system, but this setup can be adapted for high-throughput screening of unknown toxins. We verified this system by constructing dose-response curves using thirty compounds, determining the EC50 for each compound, and comparing it to literature regarding the compound. The EC50 value was then used to make cross-strain comparisons. In all we tested the estrogenic responses for our test library of 30 compounds against eight animal species. Results indicate that the potential risk of endocrine disruption for some compounds may vary from species to species, while other compounds have a more consistent risk level.

(593an) Comprehensive, Longitudinal T Cell Functional Proteomic Analyses Correlated with the Clinical Outcome of a Melanoma ACT Immunotherapy Trial

Chao Ma[1], Antoni Ribas[2] and James Heath[1], (1)California Institute of Technology, Pasadena, CA, (2)University of California Los Angeles, Los Angeles, CA

Background: Adoptive cell transfer (ACT) of transgenic T lymphocytes expressing tumor antigen specific T cell receptor (TCR) can mediate objective responses in 30%-50% of metastatic melanoma patients and result in long term tumor regression. However, little is known about the fate of engineered T cells once transferred and the dynamics of host immunity, which is important for understanding the biological mechanism of the therapy and differing long term efficacy between patients.

Methods: Utilizing newly developed, highly multiplexed clinical immune diagnostics methods (Ma, C. *et al*. Nature Medicine 17, 738-743 (2011)), we designed a comprehensive study that monitored the engineered and acquired immune response in 8 metastatic melanoma patients over 8-10 time points in the context of an ACT therapy with transgenic TCR T cells. (1) By integrating a clinical microchip platform with florescent activated cell sorting, we quantified 19 functional proteins secreted from single cells to evaluate the functional proteomics of 6 phenotypically defined T cell subsets; (2) We enumerated the abundance of 35 melanoma antigen specific T cell subsets using a peptide/MHC tetramer microarray; (3) We measured a panel of 37 blood proteins that incorporates melanoma associated markers, cytokines and chemokines by an ELISA-like assay. These assays yielded millions of high dimensional and highly informative readouts and we developed high dimensional machine learning algorithm to analyze the data.

Results: In all patients, there was an initial clinical antitumor response followed by disease progression. The change of blood marker abundance reflected this general process, starting from lymphodepletion pre-conditioning, immune system recovery, immune response, to tumor regression. This clinical course correlated with changes observed in T cell functional proteomics. The transgenic MART-1 T cells expanded rapidly after infusion, producing multiple cytotoxic and antitumor cytokines. However, their abundance and polyfunctionality diminished within 2-3 weeks. At around 30 days post infusion, a multi-clonal expansion of melanoma specific CTLs emerged in patients with regressing metastatic lesions. This wave of acquired endogenous cellular immunity exhibited antitumor functions initially, but gained inhibitory and non-specific functions and diminished when tumor started to progress. The patient with the most endogenous melanoma specific T cell populations emerged and with the strongest functions in the *in vivo* acquired immune response showed the most effective long term tumor control; weaker second wave response was associated with weaker therapeutic efficacy.

Conclusions: The melanoma-specific T cells within the ACT went through a process of *in vivo* functional changes that correlated with clinical antitumor activity. The appearance and the persistence of the *in vivo* acquired immune response featured by engineered and newly appeared endogenous antitumor cellular immunity provided indications of long term therapeutic efficacy. Therefore, the study points to the need to develop combinatorial therapeutic approaches that could maintain T cell functionality to improve the efficacy of ACT therapy.

EFFECT OF STRETCHING ON TRANSPORT ACROSS THE HUMAN STRATUM CORNEUM

Monica Hwang, Kristina Runas, Annie Jensen, Bruce Yan, Olivia Warren, and Nancy K. Lape
Harvey Mudd College, Claremont, CA, USA

Abstract

The effect of mechanical stretching of skin on drug transport rates across the stratum corneum was investigated *in vivo*. A stretching device was designed, built and tested to apply constant strain to the ventral forearm of human subjects. An *in vivo* tape stripping study and trans epidermal water loss (TEWL) measurements were performed using this device. The transport of Methylparaben (MPB) and TEWL were compared between a test site under 40% strain to an un-stretched control site. Intersubject results show that in some cases stretching seems to have increased MPB transport and TEWL, however in other cases transport has been decreased or left unaffected. Intrasubject results show that stretching increased transport for the first trial but decreased for the second trial for some participants, or vice versa. Other participants showed similar transport behavior between trials. The current data does not account for the reduction in surface area from un-stretched to stretched skin, which could explain the decrease in relative transport. PRIMOS imaging equipment was used to examine differences in surface area and normalize the transport data.

Introduction

The outermost layer of the human epidermis, known as the stratum corneum (SC), is responsible for the majority of the barrier function of human skin, keeping chemicals out of the body while preventing dehydration [1]. The SC consists of corneocytes, often considered impermeable, imbedded in a permeable lipid bilayer. Research for improving transdermal drug delivery through increased permeation of the SC often focuses on altering the barrier behavior through processes such as hydration and chemical enhancers. To our knowledge, no previous work has been done on the effects of skin extension on the permeability of the SC.

The SC has a reported Young's Modulus (E; a measure of stiffness) of 420kPa [2], whereas for an isolated corneocyte, E was found to be 450MPa [3]. Young's Modulus for the lipid bilayer has not yet been determined. Given that $E_{corneocyte} >> E_{SC}$, it is likely that the lipids will stretch more than the corneocytes under applied strain. An extension of the SC would thereby increase the lateral surface area of the lipid bilayer, which is generally considered to be the main route of drug penetration across the SC. We therefore hypothesized that the application of uniaxial strain would increase permeability and thus drug transport through the SC.

To investigate the effects of uniaxial strain on transdermal transport, a stretching device was designed, constructed and implemented. The device applied a constant strain to a test site on the forearm. An *in vivo* tape stripping study was performed on human participants to analyze the effect of drug transport on stretched skin compared to non-stretched skin. An *in vivo* trans epidermal water loss (TEWL) study was also performed to determine the difference in water flux out of the skin between stretched and un-stretched skin.

Experimental Methods

The stretching device applied a constant strain to a test site on the ventral side of the forearm. The stretching mechanism consisted of two tabs that pulled outwards from the test site on the ventral forearm of the subject's arm. The position of the tabs was controlled by an Arduino Decimilia, which operated two GWS S03N STD Servo motors. Strain was applied the skin on each side of the test site by attaching 3M Healthcare tape between the tabs and test site.

An *in vivo* experiment was performed on human participants (n=20). The stretching device was used to stretch a 2cm x 3cm test site on the subjects' right forearm to 40% strain. The left forearm was used as an un-stretched control site. 2mL of a saturated aqueous Methylparaben (MPB) solution was placed on the test site for 1 hour. Ten tape strips were successively applied to the test sites to determine the amount of MPB that had diffused through the SC. The amount of SC per tape strip was determined using a SquameScan 850. MPB was extracted from the tape strips using a 75:25 acetonitrile:H_2O solution. Capillary electrophoresis (CE) with a 70mM borate running buffer was used to quantify the amount of MPB on each tape strip. A 0.10mg/ml benzoic acid solution was used as an internal standard. Another *in vivo* experiment study was performed on human subjects (n=20) to explore intrasubject variability, where each participant was tested twice with a period of at least one week in between the first and second trial.

A second *in vivo* experiment was performed on human subjects using the stretching device to investigate the effect of stretching on trans epidermal water loss. Five repeated TEWL measurements were taken on 7 participants with a Delfin™ Vapometer and three repeated TEWL measurements were taken on 3 particpants with a BioX™ Aquaflux instrument. With both instruments the measurements were taken before the skin was stretched as a control, and again after 40% strain was applied to the test site.

Results and Discussion

The cumulative amount of MPB was plotted against the cumulative amount of SC removed for the control and stretched sites of each participant obtained from the tape stripping study as shown in Figure 1. Slopes from this plot can be correlated to the MPB permeation rate. The ratio between the slope of the stretched site to the slope of the control site was calculated for each participant as a measure of the change in permeability as shown in Figure 2. A permeability ratio that is greater than one (marked by the red line in Figure 2) indicates a higher permeation rate for the stretched skin.

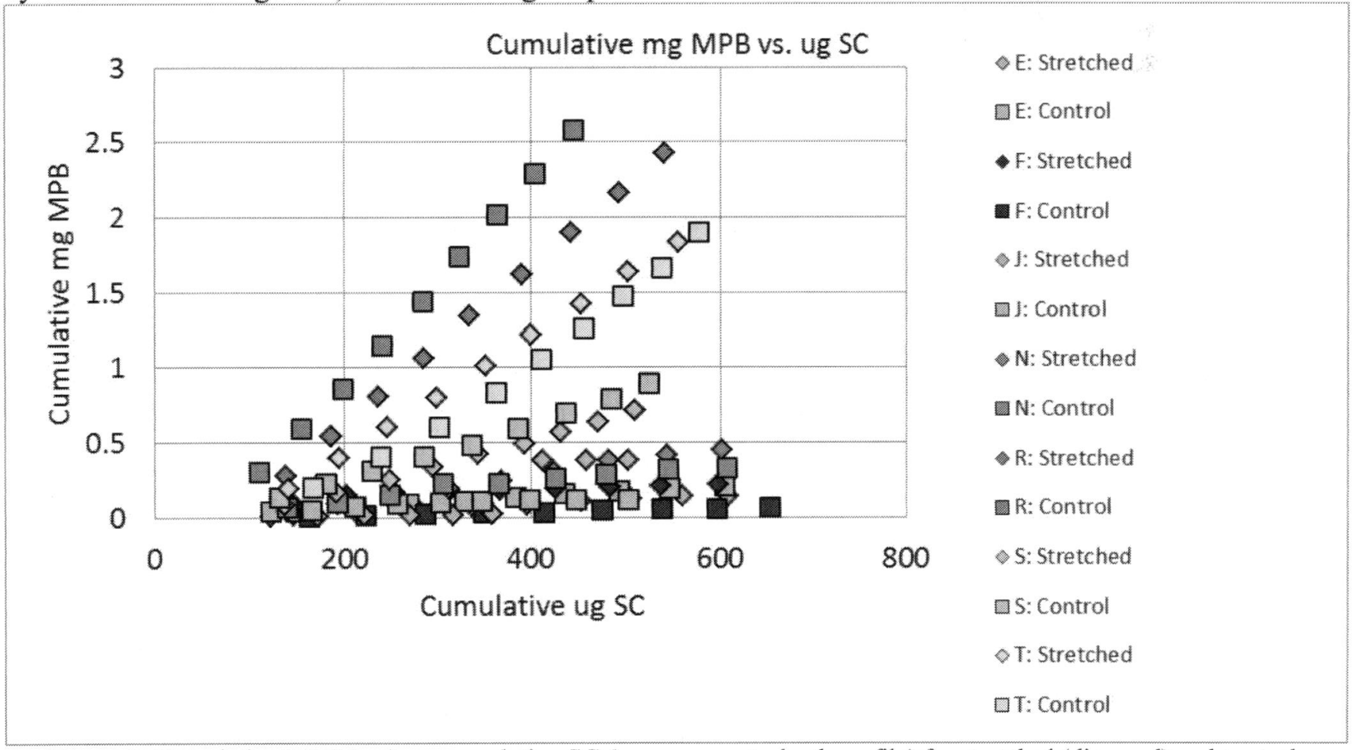

Figure 1. Cumulative MPB (mg) vs. cumulative SC (μg; represents depth profile) for stretched (diamond) and control (square) sites. Each letter corresponds to a different participant.

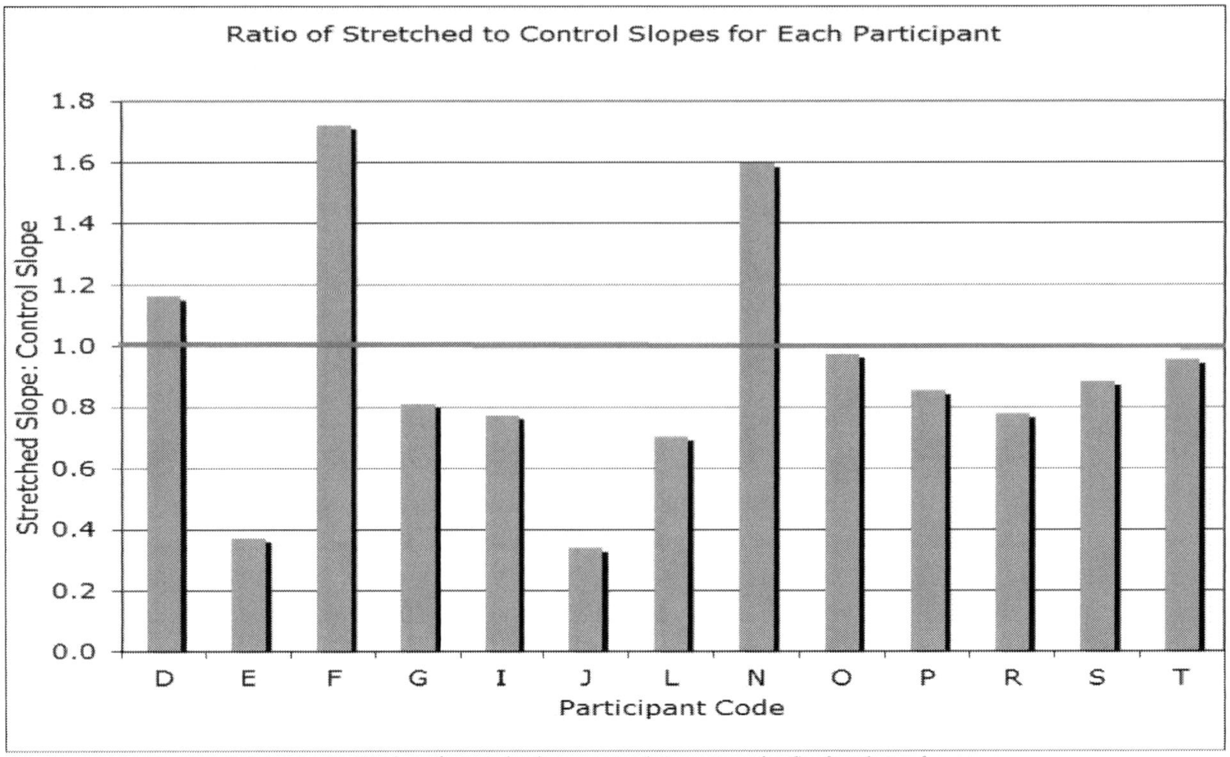

Figure 2. Ratio of stretched to control (un-stretched) site data slopes.

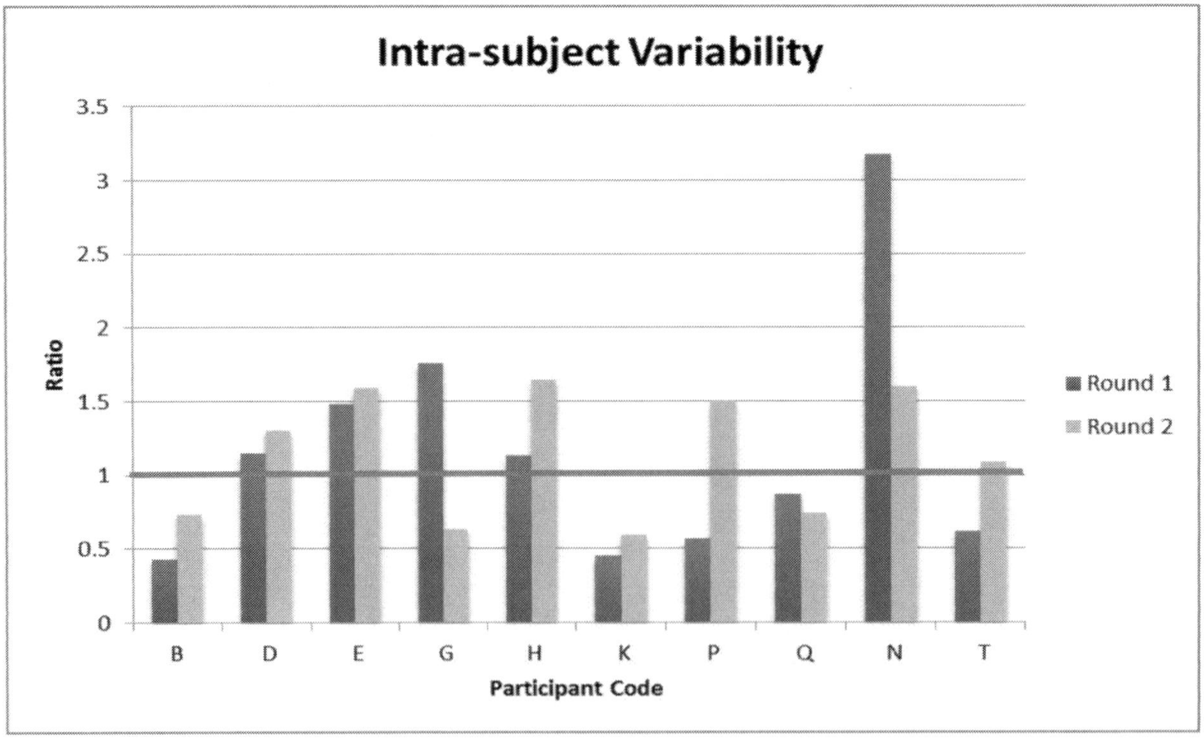

Figure 3. Ratio of stretched to control (un-stretched) site data slopes. Each participant was tested twice with at least one week in between trials.

The results show that in some cases stretching seems to have increased MPB transport, however in other cases transport has decreased or been left unaffected. The intrasubject variability study results show that some participants exhibit similar transport behavior between trials, but other participants show extreme differences between trials as shown in Figure 3. The difference in transport behavior for each participant could be attributed to the length of time between each trial. Some participants were tested the second time after more than one week from the first trial, whereas other participants were tested exactly after one week from the first trial. The length of time between tape stripping sessions could be related to the stratum corneum recovery rate after tape stripping. More tests are currently being completed to determine if the time between trials is a large effect on transport behavior.

Trans epidermal water loss (TEWL) measurements taken using the *in vivo* stretching device for both stretched and un-stretched skin indicate a similar trend in results as shown in Figure 4, where there is an increase in transport for most of the participants, but there is one participant that showed a decrease in transport. The TEWL measurements made by the Aquaflux showed a decrease in flux for the stretched site (Figure 5). More extensive TEWL studies are currently in progress in exploring intersubject and intrasubject variability.

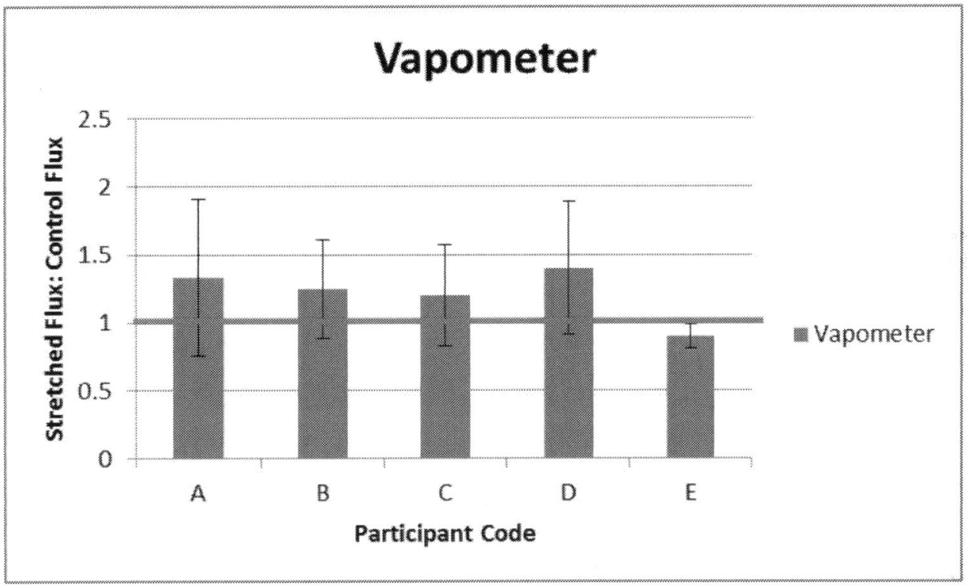

Figure 4. Ratio of stretched to control (un-stretched) site TEWL measurements from the Vapometer.

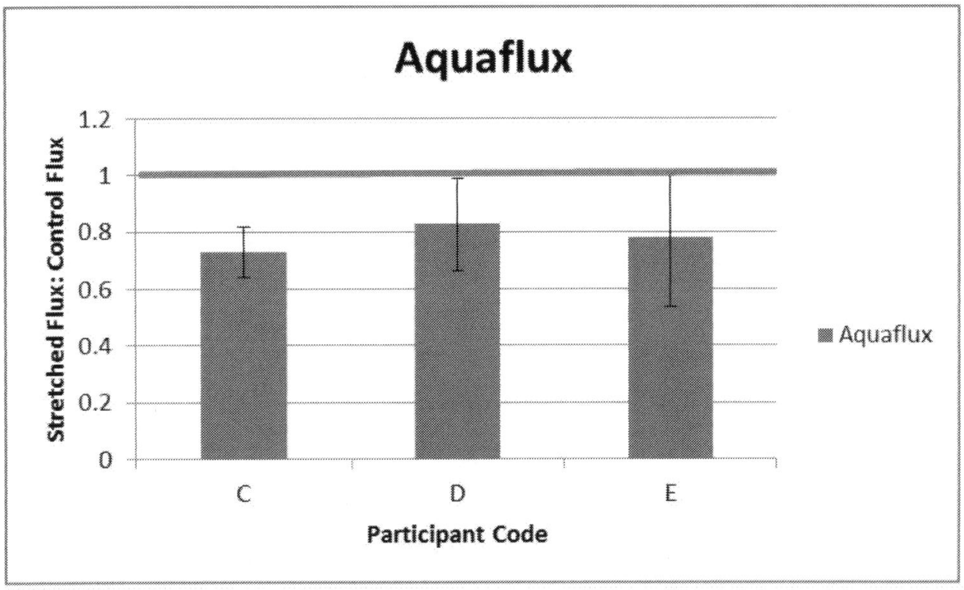

Figure 5. Ratio of stretched to control (un-stretched) site TEWL measurements from the Aquaflux.

The current data in both the tape stripping and TEWL measurements does not account for the change in surface area between stretched and un-stretched skin. When skin is stretched the area of application decreases as shown in Figure 6. This reduction of area in contact with MPB during the experiment could reduce the amount of relative transport. When this change is corrected for it is expected that there will be a statically significant increase in drug transport. Using PRIMOS imaging equipment the surface profile was obtained *in vivo* using the constant strain skin stretching device for both stretched and un-stretched skin. A polynomial fit will be used to determine the surface area for both the stretched and un-stretched skin in order to obtain the area offset. Analysis of this data is currently being performed.

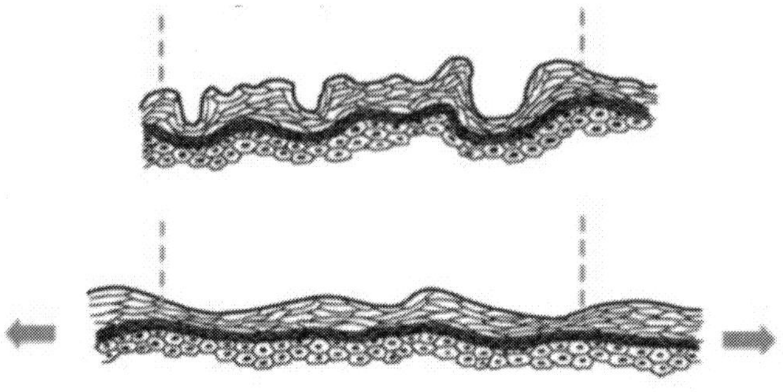

Figure 6. Sketch of skin cross-section, with natural folding before extension (top) and a more level profile after extension (bottom). The dotted lines represent the same projected area on each surface.

Additionally, a new stretching device that can apply constant stress, or force, to the test site is being currently developed. The new device uses LC201 load cells to measure force feedback, along with PQ12 Actuators, operated by an Arduino Decimilia, to supply linear movement. This device will be used for a future *in vivo* TEWL measurement and MPB tape stripping study, where constant stress will be applied to the same test site on the ventral forearm.

Conclusion

The results indicate that effect of mechanical stretching of skin generally causes a decrease in both MPB transport and TEWL. Due to the increased surface area of the lipid bilayer during stretching it is expected that the effect of stretching the skin will be an increase in the rate of transport. It is expected that correcting for the change in surface area over the contours of the skin might lead to support for the original hypothesis. In addition, the intrasubject variability results show that transport behavior for a subject can greatly differ between trials which can contribute to research in conducting clinical drug trials. Further investigation in the length of time between trials involving tape stripping experiments could help increase understanding of stratum corneum recovery rate after tape stripping as well.

References

[1] Nitsche JM, Wang TF, Kasting GB. 2006. A Two-Phase Analysis of Solute Partitioning into the Stratum Corneum. *J Pharm Sci*. 95: 649-666.

[2] Agache PG, Monneur C, Leveque JL, De Rigal J. 1980. Mechanical Properties and Young's Modulus of Human Skin in Vivo. *Archives of Dermatological Research*. 269: 221-232.

[3] Leveque JL, Poelman MC, De Rigal J, Kilgman AM. 1998. Are Corneocytes Elastic? *Dermatology*. 176: 65-69.

Acknowledgements

The authors are grateful for financial support from the Patton and Claire Lewis Fellowship in Professional Practice, The Howard Hughs Medical Institute (HHMI), and Johnson & Johnson, as well as assistance from Russell Walters (Johnson & Johnson), Catherine Mack Correa (Johnson & Johnson), Prof. Hal Van Ryswyck (Harvey Mudd Collge), and Paul Stovall (Harvey Mudd College).

(593ap) Elastin Based Nano Particles for Treatment of Chronic Wounds

Yuan Yuan, Chemical and Biomedical Engineering, University of South Florida, Tampa, FL and Piyush Koria, Department of Chemical and Biomedical Engineering, University of South Florida, Tampa, FL

Introduction: Chronic wounds affect 5.7 million patients and cost an estimated 20 billion dollars annually. Recently, we developed a growth factor delivery platform based on elastin like peptides (ELPs) for growth factor delivery to chronic wounds. Elastin like peptides are repeats of the pentapeptide (VPGXG, X=guest residue can be anything but Proline) derived from natural elastin. They are attractive delivery vehicles due to their unique phase transition properties and are genetically encodable and biocompatible. The unique phase transition property of ELPs is a direct function of the chain length and guest residue, which allows us to design biomaterial delivery systems with desirable physical properties. Interestingly, we also observed that ELPs were not only useful for delivery of growth factors but also enhanced granulation in the wound through modulation of fibroblast growth and function. In this study, we aim to elucidate the effect of chain length and guest residues on the ELP induced modulation of fibroblast function.

Materials and Methods: We created several ELPs with different chain lengths (number of VPGVG repeats) and having different cysteines in the guest residue position. These ELPs were recombinantly produced in bacteria using standard molecular biology and protein expression techniques. The protein was purified to high homogeneity using inverse temperature cycling which exploits the phase transitioning property of ELPs. The purity was tested using SDS-PAGE and the physical properties of ELPs were characterized using dynamic light scattering and transmission electron microscopy. Normal human dermal fibroblasts were treated with different ELPs and proliferation was evaluated using Hoechst assay. We used RT-PCR to assess the effect of ELPs on several genes associated with normal fibroblast functions such as collagen production, matrix metalloproteinase activity and fibrillin expression.

Results: We observed that ELPs self-assembled into nanoparticles with different hydrodynamic diameters depending on their chain length and proportion of cysteines. We further found that ELPs increased fibroblast proliferation significantly. Interestingly, this increase was dependent on the chain length and proportion of cysteines. This increase of proliferation was successfully blocked using lactose suggesting that ELPs bind to the elastin receptor to induce proliferation.

Conclusions: Our study suggests that ELPs induce fibroblast proliferation. This biological activity of ELPs is dependent on chain length as well as guest residues which also determine the physical phase transition properties of ELPs. Furthermore, our data suggest that ELPs induce their biological effect on fibroblasts through interaction with the elastin receptor complex present on the fibroblast surface. An understanding of the exact interaction of ELPs and identifying the determinants of this interaction (i.e. chain length, guest residue) will help us in the design of unique ELP based biomaterials having the desired physical and biological properties that are conducive to the healing of chronic wounds. Thus they will not only be useful as drug delivery vehicles but also may induce healing synergistically through their intrinsic biological activities.

Enzyme Immobilization onto Various Nanosupports: A Critical Study

Alan Campbell[1], Chenbo Dong[1], Chengcheng Xiang[2], Nianquiang Wu[2], Jonathan Dordick[3], and Cerasela Zoica Dinu[1,*]

[1]Department of Chemical Engineering, West Virginia University
[2]Department of Mechanical and Aerospace Engineering, West Virginia University
[3]Department of Chemical and Biological Engineering, Rensselaer Polytechnic Institute
*Corresponding author: Cerasela-Zoica.Dinu@mail.wvu.edu

Due to their high catalytic properties and selectivity, enzymes are widely used in multiple industries including food and paper processing, brewery, and biofuel generation. However, enzymes possess low storage and operational stability, and may leach into solution. Thus, there is a critical need to improve enzyme stability and retention when such applications are envisioned. Immobilization of enzymes onto nanosupports, such as nanotubes or nanoparticles, is a possible solution. These nanosupports posses high surface-area-to-volume ratios and high aspect ratios, which allow for high enzyme loadings upon immobilization and enzyme retention when the support is exposed in solution. The attachment of enzymes to the nanosupport surface has also been shown to enhance enzyme stability under storage and operational conditions in some cases. However, the process of immobilization will decrease the activity of the enzymes due to interactions at the protein-nanomaterial interface. Herein we are presenting a critical study focused on outlining what properties influence activity loss. Briefly, we have immobilized various enzymes (i.e. soybean peroxidase (SBP), chloroperoxidase (CPO), and glucose oxidase (GO_X)) onto multiple nanosupports (i.e. multi-walled carbon nanotubes (MWNTs), single-walled carbon nanotubes (SWNTs), titanium dioxide nanobelts (TiO_2), tungsten oxide nanoparticles (WO_3), platinized tungsten oxide nanoparticles ($PtWO_3$), and graphene oxide nanosheets (GON)) using either physical or covalent binding. In the cases of MWNTs and TiO_2, covalent binding and covalent binding with the utilization of an amino-dPEG12-COOH (PEG) linker were also tested. In order to test the interactions at the protein-nanomaterial interface, the derived conjugates were characterized in terms of enzyme loading, retained activity, and thermostability. Protein concentration after attachment, or loading, was determined using a standard colorimetric assay. The retained activity refers to the activity of the immobilized enzyme, determined via colorimetric reactions, compared to the activity of free enzyme in solution at the corresponding concentration. The Michaelis-Menten parameters of all conjugates were also studied and compared to those of free enzymes to determine any influence of the nanosupport on the enzyme kinetics. Thermal inactivation was tested at 75°C. Our data show that enzyme activity loss upon immobilization can be attributed to enzyme denaturation at the interface, reduced enzyme surface area available for substrate interaction, and unwanted protein-protein interactions due to unspecific binding. It was demonstrated that as the diameter of the nanosupport increases (i.e. rate of curvature decreases) protein inactivation is more dramatic because more enzyme surface area comes into contact with the nanosupport; such observations are also dependent on the surface properties of the enzyme and the interface reactions with the nanosupport. Our studies can be implemented to improve upon current uses of enzymes as well as new applications such as active surface decontamination and biosensors.

(593ar) Extracellular Chemical Stimuli and Mechanotransduction

Kris Noel Dahl[1], Stephen T. Spagnol[2] and **James S. Weltz**[2], (1)Chemical Engineering, Carnegie Mellon University, Pittsburgh, PA, (2)Department of Chemical Engineering, Carnegie Mellon University, Pittsburgh, PA

Abstract

Mechanical force transduction, or mechanotransduction, has been shown to play an important role in cellular response. In endothelial cells, mechanical force, in the form of shear stress, has been shown to cause cytoskeletal reorganization with an increase in the size and number of actin fibers. More recently, the mechanical connection between the cytoskeleton and the nucleoskeleton has been elucidated.

We have shown that removal of the actin cytoskeleton, via latrunculin A, results in altered nuclear morphology. Increasing latrunculin A concentration results in a decrease in actin cytoskeleton. With decreasing actin cytoskeleton, we observed decreasing nuclear area and decreasing nuclear circularity. Between the control and the highest treatment concentration (100nM Latrunculin A, 3 hours), the difference in both nuclear area and circularity was found to be statistically different with an alpha of 0.001. Nuclear morphology was measured with automated image processing, and allowed data sets in excess of 600 nuclei.

Collaboration with the Bioimage Informatics Department will allow precise quantitation of actin stress fibers in treated and untreated cells, using fiber quantity and length as metrics. This will allow combination of the large nuclear morphology data with a quantitative measurement of actin.

(593as) High-Throughput Screening System for the Determination of Estrogenic Properties of Small Molecules

Derek Reichel[1], Lily Glick[1], Angela Chen[1], Brian Saunders[1], **Elif Miskioglu**[1] and David W. Wood[2], (1)The Ohio State University, Columbus, OH, (2)Chemical and Biomolecular Engineering, The Ohio State University, Columbus, OH

Estrogenic compounds are an important group of chemicals with significant properties influencing human health. They are used to treat several types of cancer, including breast, pancreatic and testicular cancer. Estrogenic compounds are also linked to negative health effects, such as decreased age of onset in puberty in women. Therefore, determining the estrogenic potential of a compound is of pharmaceutical and toxicological relevance. Currently, estrogenic properties of a compound are determined using animal assays. These assays involve the growth, treatment, sacrifice and autopsy of laboratory animals to determine the effect of the compound in question on animal health. This is a costly procedure, making it infeasible as a high-throughput screening system.

We have developed an assay using thymidylate synthase (TS) knockout *Escherchilia coli* cells, in which we have inserted an intein containing the human estrogen receptor beta (ERβ) and thymidylate synthase, to create a bacterial biosensor. The assay can discriminate between ERβ agonists and antagonists. The current setup uses 96-well plates, but can be scaled to 384- or 1596-well plates for high-throughput screening. To validate the assay as an effective method for non-animal testing, dose-response curves for thirty compounds were determined and compared to expected behaviors from literature. These compounds have been recommended by the (ICCVAM) for the validation of non-animal endocrine disruptor assays. Our results indicate that our bacterial strains are capable of reliably indicating the potential estrogenicity of most of the test compounds. This method, which is much cheaper and more animal friendly then many conventional assays, can therefore provide an early screen for new therapeutic compounds, as well as the identification of potentially dangerous environmental estrogen disruptors.

(593av) Substrate Specificity and Stability of a Novel Alginate Lyase (AlgL) From Stenotrophomonas Maltophilia

Liam T. Smith, Chemical Engineering, Lehigh University, Bethlehem, PA, Sean O'Keefe, Bioengineering, Lehigh University, Bethlehem, PA, Emily L. Wong, Division of Infectious Diseases, Lehigh Valley Hospital, Allentown, PA and Bryan W. Berger, Chemical Engineering and Program in Bioengineering, Lehigh University, Bethlehem, PA

Stenotrophomonas maltophilia is a gram-negative bacteria that is an increasing important nocosomial pathogen in immunocompromised patients. Approximately 1% of all nosocomal bacteraemias are due to *S. maltophilia* infection, with the attributed mortality rate at nearly 28%, placing it among the highest attributed mortality rates observed for nosocomal bacteraemias. A major reason for the high mortality associated with *S. maltophilia* is its ability to produce exopolysaccharides (EPS) that form a biofilm resistant to antimicrobials, as well as enable binding to stents and ventilators. Our goal is to characterize an alginate lyase (AlgL) discovered in our group for selectivity to degrade the EPS present in *S. maltophilia* biofilms. I developed a high-yield *E. coli* expression system for AlgL, and measured the kinetic parameters for EPS degradation as a function of temperature. My results indicate the discovered AlgL is robust to high temperatures, and may be a useful starting point for removing biofilm-resistant infections.

(593aw) Thermodynamic Properties of Nuclear Lamin Proteins

Kelli Coffey[1], Matthew Biegler[2], Agnieszka Kalinowski[3] and Kris Noel Dahl[1], (1)Chemical Engineering, Carnegie Mellon University, Pittsburgh, PA, (2)Materials Science and Engineering, Carnegie Mellon University, (3)Biomedical Engineering, Carnegie Mellon University, Pittsburgh, PA

The cell contains many filamentous network systems: the cytoskeleton is found throughout the cellular interior and the nucleoskeleton is found inside the inner nuclear membrane. The nucleoskeleton provides mechanical stability to the nucleus and aids in expression of genes. The nucleoskeleton is formed from independent networks of intermediate filaments primarily made of lamin A and lamin B proteins. There are numerous disease-causing mutations in lamin A, including the premature aging disorder Hutchison Gilford progeria syndrome (HGPS). A duplication of the lamin B gene LMNB1 causes adult-onset autosomal dominant leukodystrophy (ADLD), which is progressive neurological disorder. We have studied thermodynamic changes of the tail domain conformation of these proteins, since the tail domain is responsible for most of the protein-protein assocation. We examine mature lamin B, mature lamin A, prelamin A and the mutant form of lamin A associated with HGPS, Δ50, by stepwise increases in temperature or changes in salt, and we analyzing the changes of the fluorescence of the tryptophan residues inside the the Ig-fold found in the tail domains. Temperature- and salt-induced changes the conformation of both prelamin A and Δ50 from a more to less compact form, and this conformational change is necessary for nuclear membrane association. The effective changes in thermodynamic stability may provide molecular insight into progression of nucleoskeletal disorders.

(593ax) Utilization of Drop-On-Demand Technology in the Manufacture of a Novel Nicotine Transdermal Patches

Marlena Brown, Chemical and Biochemical Engineering/Food Science, Rutgers University, Ocean, NJ and Paul Takhistov, Food Science, Cook College, Rutgers, The State University of New Jersey, New Brunswick, NJ

The skin represents an ideal avenue for transport of APIs to obtain enhanced functional and therapeutic effects; however, the successful application of transdermal drug delivery systems remains limited and faces many challenges. To address this issue, we utilize Drop on Demand technology to construct API formulations and polymer films creating personalized transdermal patches with controllable architecture. As result, one can increase the dissolution rates of API(s) by creating a novel drug-in-adhesive layer, negating the need for a separate adhesive layer for use in transdermal delivery. We propose to create distinct transdermal dosage forms where (1) the adhesive is printed adjacent to the API droplet (2) the interdroplet spacing and volume of the API droplet is varied, and (3) two different drugs & adhesive will be printed onto the same layer. Additionally a mathematical model using partial differential equations was developed to examine the flux of a drug through the layers of a traditional transdermal patch compared to the modified patch developed using Drop on Demand technology. We found that the flux of the drug in our modified transdermal patch was much greater as compared to a traditional patch where the drug was immersed in adhesive (drug in adhesive patch) or where the adhesive layer was completely separate from the drug layer.

(593ay) A Model-Based and a Multi-Objective Optimisation Framework for Incremental Scale-up of Bioreactors

Miguel Mauricio-Iglesias, CAPEC, Department of Chemical and Biochemical Engineering, Technical University of Denmark, 2800 Kgs. Lyngby, Denmark and Gürkan Sin, Department of Chemical and Biochemical Engineering, Technical University of Denmark, Kgs. Lyngby, Denmark

A model-based and a multi-objective optimisation framework for incremental scale-up of bioreactors

Mauricio-Iglesias, Miguel[1]; Sin, Gürkan[2]

CAPEC - Department of Chemical and Biochemical Engineering, Technical University of Denmark, Søltofts Plads, Building 229, DK-2800 Kgs. Lyngby, Denmark

[1] mim@kt.dtu.dk; [2] gsi@kt.dtu.dk

Scale-up is an important stage in development and commercialization of new process technologies in many industries, for production of chemicals such enzymes, antibiotics or materials such as bioplastics. Moving a fermentation process from a lab-scale to a commercial/production scale remains still challenging due to a number of factors that affects biological response of cells to changing conditions from cultivation. Scale-up problems may arise then due to inadequate interphase mass transfer, heat removal and non-uniform temperature and concentration gradients in the reactor (Villadsen et al. 2011). As a result, many large-scale fermentation processes give a lower yield than achieved in the laboratory.

Traditionally, scale-up problem is tackled using a step-wise process development approach: from lab-scale to pilot-scale and ultimately to production-scale by following certain empirical criteria based on dimensionless numbers or ratios critical to performance of the process. The difficulty with this approach lies in several aspects as there is just a limited number of empirical criteria than one can keep constant during the scale-up. Additionally, the controlling conditions (bottleneck) may change at different scales.

This contribution aims at developing a framework for scale-up of bioreactors based on data and information translated into quantitative knowledge using models. The framework allows using multiobjective scale-up principles and various degree of models, from first-principles to empirical (e.g. response surface type models based on experimental data) to hybrid models.

The proposed methodology is illustrated as a flowchart in Figure 1 and assumes that a successful bioreactor design and operation protocol has been developed at a lab/bench scale. The methodology consists of the following steps:

1) **Definition of objectives and constraints**: where the objective of the scaled-up operation is defined, including the scaling-up factor. It is important to note that the goal of scaling-up is not to reproduce the bench scale reactor at a larger scale but to accomplish the objective defined. To this aim, the bioreactor configuration can be modified in the large scale

2) **Model development.** Depending on the available information and the goal of the process, a model with the corresponding assumptions is formulated at the small scale and is assumed to be translatable to the large scale. The model complexity can range from complete first-principle model including the mass, heat and momentum balances and a structured metabolic kinetic description to simple meta-models, e.g. obtained from an experimental campaign response surface methodology.

3) Design scale-up. Using the model developed, the problem of design at a large scale is formulated as an optimization to achieve the defined objective. If a full model is available, the large scale design can be done without taking into account the bench scale protocol. Otherwise, sensitivity analysis of the bench scale model is used to determine the operating parameters that affect most the performance of the reactor. In the vast majority of cases the scale-up will represent a trade-off since not all the operation parameters/ratios can be kept constant. A multiobjective optimization is proposed as a systematic method to carry out the scale up, hence keeping constant the most significant operating conditions and parameters (according to the sensitivity analysis) and whereas the least important ones can be varied. Hence, the use of multiobjective optimization together with an assessment of the model in order to choose the most relevant phenomena for the process provides a rational method for scale-up.

4) **Scale-up validation**. The goal of this step is to check whether the assumptions formulated in step 2 are applicable at the new scale and thereby, the validity of the model and the scale-up. The most spread and straightforward tool is the regime analysis of all the phenomena taking place in the reactor. The limiting steps can be therefore inferred and, the hypotheses used to build the model can be validated (Sweere et al. 1986). Otherwise, a more complex model is needed in order to properly describe the operation in the reactor, and the hypotheses are reformulated.

5) **Evaluation**. This step consists on checking whether the objectives initially defined are actually fulfilled at the large scale. If not, the model and/or the objectives may have to be redefined if the model cannot capture the bioreactor description properly or if the defined objectives are not feasible at large scale.

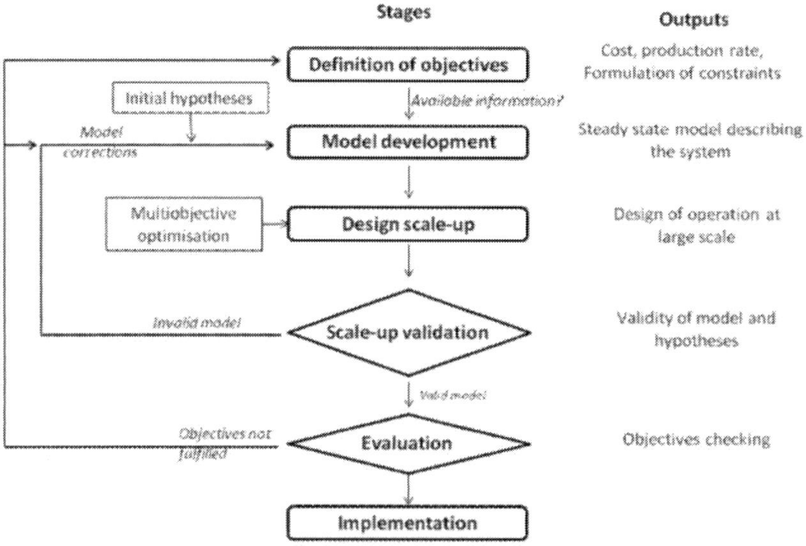

Figure 1. Methodology proposed for incremental scale-up of bioreactors based on incremental model development

The methodology is illustrated step by step through a case-study consisting on a bacterial fermentation scale-up (*E. coli* in a rich glucose environment, Villadsen et al. 2011) assuming different levels of available information: i) a full model, ii) a partial model consisting on mass balances and kinetic rates but failing to include hydrodynamics and iii) a meta-model obtained from simulated experimental data. The results using the methodology were benchmarked with other widely used methods based on rules of thumbs (e.g. keeping the power/volume ratio constant at all scales) and dimensional analysis (e.g. keeping the Sherwood number constant at all the scales). The proposed methodology was proven to provide a rational framework for scaling up, considering both simple and complex models.

References

Villadsen J. et al. (2011) *Bioreaction Engineering Principles.* Springer Science

Sweere, A.P.J. Luyben, K.C.A. M., Kossen, N.W.F.. (1986), Regime analysis and scale-down: tools to investigate the performance of bioreactors, *Enwyme Microb. Technol. vol.9, 386*

(593b) Coupling of Gene Expression and Growth Rate Determines Selection of Transcriptional Regulation Mechanisms

Priya Rao, Jatin Narula and Oleg Igoshin, Rice University, Houston, TX

Coupling of Gene Expression and Growth Rate Determines Selection of Transcriptional Regulation Mechanisms

Priya Rao, Jatin Narula, Oleg Igoshin

Department of Bioengineering, Rice University, Houston, TX

Gene expression in bacterial cells can be regulated by either activators or repressors but it is not clear how evolution selects one mode of regulation over the other. From a molecular biology perspective, selection of a particular mode of regulation could be based upon optimization of functional performance if these modes differ functionally [1]. Alternatively, even functionally equivalent modes of regulation may be selected for or against based upon robustness to evolutionary forces of mutation and drift [2]. Previously, both of these approaches were applied separately to this problem to show that the mode of regulation is selected such that the regulator is active for greater part of the cell's lifetime (i.e. activators are selected to control often-needed genes and repressors to control seldom-needed genes). In reality, both types of criteria probably interact and together affect the selection of the appropriate modes of gene regulation. Here we combine mathematical modeling of gene regulation with a theoretical population biology framework to explain how evolution and functional performance together influence the choice between activator and repressor modes of control.

Methods

We modeled inducible gene regulation by activators and repressors using the typical ODE formulation. Coupling between gene expression and cell growth was modeled following the results of [3]. In this formulation, we used a system of two coupled ODEs for the concentration of the transcription factor and cell volume respectively. Cells were assumed to divide when they doubled their initial volume. Stochasticity of gene expression was included in this model by adding a Gaussian white-noise type term to set up a Langevin equation for the concentration of the transcription factor. For the population biology simulations we used a continuous-time version of the Wright-Fisher genetic drift simulation analogous to the Moran process [4]. The two coupled ODE system described above was propagated for a population of cells in this framework. We ignored the effects of mutations for the small time-scales considered in our simulations.

Results

Activators and Repressor are functionally equivalent modes of inducible gene regulation

Inducible gene expression control can be achieved by both activators and repressors if the activity of these transcription factors is modulated by binding of a small molecule inducer. We used a mathematically controlled comparison to determine if these modes of regulation differ functionally. We show that with appropriate choice of parameters, both modes of control can result in identical gene expression over the whole range of inducer concentrations. Thus these modes of control are functionally equivalent in terms of their control over gene expression.

Feedback between cell growth and gene expression depends on mode of regulation

Several recent reports have shown that gene expression is associated with costs and benefits that couple it to cell growth [3, 5]. In an environment when the gene is needed, gene expression increases growth rate. In an environment when the gene is not needed, gene expression decreases growth rate. We show that this coupling between gene expression and growth results in a feedback loop between the concentration of transcription factors and cellular growth rate. While the increase in growth rates leads to higher effective dilution rate of stable proteins, the resulting decrease in protein concentration may either increase or decrease growth rates resulting in either negative or positive feedback loops. Specifically, when the gene of interest is needed for growth, increases in gene expression increase growth rate. In this scenario, activators increase gene expression and thus increase growth rate resulting in a negative feedback loop whereas repressors have the opposite effect on growth and are coupled to it in a positive feedback loop. The feedback signs for activators and repressor modes are reversed if the expression of the gene of interest is not essential and decreases growth. Therefore, we show that the choice between activator and repressor is a choice between negative and positive feedback coupling of the regulator and growth rate.

We found that a positive feedback loop between growth and gene expression increases the time required to dilute stochastic fluctuations in protein concentrations. This creates a greater variance in protein levels within a cell population, which then leads to a greater variance in growth rates. However, the mean growth rate is unaffected by the choice of feedback loop. As a result, if the fluctuations are not heritable, the two modes of control are functionally equivalent in their control over both gene expression and cell growth.

Heritable fluctuations in gene expression can influence the selection of activators and repressors

If the fluctuations in gene expression are heritable (dilution of the fluctuation takes longer than a single cell generation), then the resulting heritable variability in growth rates affects both the mean and variance of gene expression in a cell population. Specifically, gene expression fluctuations that increase growth rate are selected for and the population becomes biased over time. This effect is more prominent in the case of positive feedback since it takes longer to dilute out fluctuations and effectively has greater fluctuation heritability. As a result, when we set up a competition between different modes of gene regulation in a Wright-Fisher type genetic drift simulation, we found that the otherwise equivalent modes of control differed in their probability of fixation. In particular, modes of regulation that were coupled with cell growth in a positive feedback loop always had a higher probability of fixation in the population.

This result was quite surprising since it predicts that activators and repressors should control seldom-needed and often-needed genes respectively and contradicts a rule for selection of mode of regulation based on mutational robustness that has been proposed previously [2]. As a follow-up, we compared positive and negative feedback modes when fluctuations of gene expression only adversely affect growth. In this case, we found that modes of regulation coupled with cell growth in a negative feedback loop have greater probability of fixation. This result agrees with the previously proposed rules of Savageau [2] for selection of mode of regulation and predicts that activators and repressors should control often-needed and seldom-needed genes respectively.

The two scenarios regarding the effect of gene expression fluctuations on cell growth described above refer to situations where gene expression is optimized for growth and where gene expression is not optimally controlled respectively. The first scenario is relevant for the regulation of many essential metabolic enzymes and in this case activators and repressors indeed control often-needed and seldom-needed genes respectively. However we also found several stress-response related, seldom-needed, genes that are sub-optimally regulated (expressed even in the absence of stress) and controlled by activators. These examples are significant because they contradict the mode of regulation rule based on mutational robustness but can be explained by evolutionary selection based on these growth feedback effects as described here.

Conclusions

We have shown that evolutionary selection between functionally equivalent modes of regulation can be explained when the coupling between gene expression and cell growth is taken into account. Strikingly, activators and repressors differ in their feedback coupling with cell growth. We have found that this feedback can affect the heritability of fluctuations in cell growth and thereby influence the selection of the mode of regulation. These results illustrate the importance of evaluating design principles for gene regulatory networks in a population dynamics framework rather than basing them entirely upon principles of functional optimization.

[1] Shinar G, Dekel E, Tlusty T, Alon U. Rules for biological regulation based on error minimization. *Proc Natl Acad Sci USA*. 2006;103(11):3999-4004.

[2] Savageau M A. Design of molecular control mechanisms and the demand for gene expression. *Proc Natl Acad Sci USA*. 1977;74(12):5647-5651.

[3] Dekel E, Alon U. Optimality and evolutionary tuning of the expression level of a protein. *Nature*. 2005;436:588-592.

[4] Hamilton, M. *Population Genetics*. 2009.

[5] Klumpp S, Zhang Z, Hwa T. Growth Rate-Dependent Global Effects on Gene Expression in Bacteria. *Cell*. 2009;139(7):1366-1375.

(593c) Serum Protein Adsorption Onto Functionalized Polyanhydride Nanoparticles Influences Particle Interactions with Dendritic Cells

Jonathan Goodman[1], Julia Vela-Ramirez[1], Brenda R. Carrillo-Conde[1], Amanda Ramer-Tait[2], Michael J. Wannemuehler[2] and Balaji Narasimhan[1], (1)Chemical and Biological Engineering, Iowa State University, Ames, IA, (2)Department of Veterinary Microbiology and Preventive Medicine, Iowa State University, Ames, IA

Polyanhydride-based nanoparticles are of increasing interest for use as vaccine adjuvants and delivery vehicles. Recent work has demonstrated that nanoparticles, whose surface is chemically functionalized with carbohydrates, target C-type lectin receptors on antigen presenting cells (APCs) efficiently. This results in enhanced nanoparticle internalization and activation of APCs. The objective of this study was to analyze how mouse serum proteins adsorb onto the surface of carbohydrate-functionalized nanoparticles *in vitro*. Additionally, it was desired to determine the effects of protein adsorption on the uptake and activation of bone marrow-derived dendritic cells (DCs). Quantum dot (QD)-loaded nanoparticles based on sebacic anhydride (SA), 1,6-bis-(*p*-carboxyphenoxy)hexane (CPH) and 1,8-bis-(*p*-carboxyphenoxy)-3,6-dioxaoctane (CPTEG) were fabricated by an anti-solvent nanoprecipitation method. Di-mannose, lactose, and galactose, were the three carbohydrates that were studied and glycolic acid was used as a linker. For each trial nanoparticle suspensions were incubated with mouse serum, and 2D gel electrophoresis was used to analyze the adsorbed proteins. Complement component 3 (C3)-specific ELISA measurements on adsorbed proteins resulted in the identification of C3 levels on the particle surface. Sham- and serum-coated nanoparticles were then incubated with DCs for 48 hours and flow cytometry was used to measure MHC I, MHC II, CD40, and CD86 expression. Multiplex cytokine assays were performed to analyze the total amount of IL-1β, IL-10, TNF-α, IL-6, and IL-12p40 secreted by the cells after stimulation with nanoparticles. Results indicated that less serum proteins were adsorbed on more hydrophilic polymers. In general, linker-functionalized particles were internalized more effectively and all the functionalized particles enhanced dendritic cell activation. These results provide new insights into the rational design of nanoparticles for use as adjuvants and as vaccine delivery vehicles.

(593d) Smart Multi-Objective Global Optimization for Metabolic Engineering

Carlos Pozo[1], **Gonzalo Guillén-Gosálbez**[1], Laureano Jiménez[1] and Albert Sorribas[2], (1)Departament d'Enginyeria Química, Universitat Rovira i Virgili, Tarragona, Spain, (2)Basic Medical Sciences Department, Unversity of Lleida, Lleida, Spain

Optimization models in metabolic engineering and systems biology focus typically on optimizing a unique criterion, usually the synthesis rate of a metabolite of interest or the rate of growth. Connectivity and non-linear regulatory effects, however, make it necessary to consider multiple objectives in order to identify useful strategies that balance out different metabolic issues. This is a fundamental aspect, as optimization of maximum yield in a given condition may involve unrealistic values in other key processes. Due to the difficulties associated with detailed non-linear models, analysis using stoichiometric descriptions and linear optimization methods have become rather popular in systems biology. However, despite being useful, these approaches fail in capturing the intrinsic nonlinear nature of the underlying metabolic systems and the regulatory signals involved. Targeting more complex biological systems requires the application of global optimization methods to non-linear representations. In this work we address the multi-objective global optimization of metabolic networks that are described by a special class of models based on the power-law formalism: the generalized mass action (GMA) representation. Our goal is to develop global optimization methods capable of efficiently dealing with several biological criteria simultaneously. In order to overcome the numerical difficulties of dealing with multiple criteria in the optimization, we propose a heuristic approach based on the epsilon constraint method that reduces the computational burden of generating a set of Pareto optimal alternatives, each achieving a unique combination of objectives values. To facilitate the post-optimal analysis of these solutions and narrow down their number prior to being tested in the laboratory, we explore the use of Pareto filters that identify the preferred subset of enzymatic profiles. We demonstrate the usefulness of our approach by means of a case study that optimizes the ethanol production in the fermentation of *Saccharomyces cerevisiae*.

(593e) Modular Production and Cloning of Error-Prone Genomic Libraries

Benjamin G. Freedman, Biological Systems Engineering, Virginia Tech, Blacksburg, VA and Ryan S. Senger, Biological Systems Engineering Department, Virginia Tech, Blacksburg, VA

A genomic library produced from cloning and over-expressing sheared DNA is an important tool in combinatorial metabolic engineering. This technique has led to several improved phenotypes in response to environmental stress, and efforts have been implemented to enrich genomic libraries to optimize traits such as product secretion that are not tied to growth. However, traditional methods of DNA shearing, polishing, and blunt-end cloning are highly inefficient and time-consuming. Newer TA-cloning methods provide improvements but are also far from ideal. Here, we present a method for generating genomic libraries through random priming PCR and modular Gibson-inspired cloning method that allows extremely high (i) control over product insert size and (ii) cloning efficiency. This method is also capable of generating non-native fragments through error-prone PCR methods. We demonstrate these methods in detail and provide examples of library enrichment leading to new strains of *Clostridium cellulolyticum* optimized for cellulose consumption in consolidated bioprocessing conditions.

(593f) Optimization of High Throughput Process Development Methodologies for Chromatographic Process Development

Jessica Lewis, Downstream Process Development, Biopharmaceutical Process Development, GlaxoSmithKline, King of Prussia, PA

The desire to be 'first to market' combined with the need to reduce production costs drives the need for novel and improved high throughput methods that yield rapid and predictive process development. To this end, much effort in recent years and across the industry has been placed into adapting robotic liquid handling and high throughput methods to streamline chromatographic process development. Here, we demonstrate an optimized protocol using miniaturized "resin plates" (96-well plates containing defined volumes of chromatography media) that reduces error and helps to better characterize adsorption isotherms and binding trends as a function of solution conditions (binding maps). We focus primarily on static binding experiments over dynamic experiments because the latter can be measured directly on ÄKTA FPLC systems using mini-columns if automation approaches are fully exploited. We propose standard methodologies with respect to high throughput chromatographic process development on robot liquid handlers for both monoclonal antibodies and domain antibody fragment process development. By using high throughput techniques in this focused manner, we have been able to increase overall process understanding, reduce the duration of downstream development time, and minimize the amount of manual labor required for understanding the binding behavior of proteins of interest.

(593g) Evaluation of Transmembrane Proteins for Stabilizing Emulsions

Luz Stella Tautiva[1], Oscar Alberto Alvarez[2], Harold Enrique Castro Barrera[3], Nathalia Garces Ferreira[3], Watson L. Vargas[4] and Andrés Fernando González Barrios[5], (1)Chemical Engineering, Universidad de los Andes, Bogotá D. C., Colombia, (2)Chemical Engineering, Universidad De Los Andes, Bogotá, Colombia, (3)Systems and Computing Engineering , Universidad de los Andes, Bogotá, Colombia, (4)Department of Chemical Engineering, Universidad de los Andes, Bogotá, Colombia, (5)Chemical Engineering, Universidad de los Andes, Bogotá, Colombia

The majority of surfactants in the industry are compounds synthesized from hydrocarbons and many of them are highly toxic to the environment. They can be difficult to degrade and their sub-products and synthesis process may generate a high environmental impact. Biosurfactants are produced by living cells and may substitute some of the surfactants synthesized by the petrochemical industry, contributing to the reduction of the release of pollutants and environmental risk. They are biodegradable, less toxic,non-hazardous, have better foaming properties, higher selectivity, and can be active at extreme temperatures, pH and salinity. Moreover, they can be produced from industrial wastes and by-products. These advantages demonstrate the importance of studying this type of surfactants and to produce it from biological origin. There are different industrial applications for the Biosurfactants, such as stabilizers of emulsions and surface active

The bacteria *Escherichia coli* contain large amounts of outer membrane protein in their membranes that can be used as biosurfactants. The Outer membrane protein X (OmpX) is an *E. coli* outer membrane protein, which is conformationally stable. The crystal structure of OmpX consists of eight all- next neighbor antiparallel -strands that fold into a cylindrical barrel with polar residues on the inside and hydrophobic residues on the outside facing the membrane environment. The height of the cylindrical barrel is approximately 32 A° and its diameter 20 A°, and four of the -strands extend out of the predicted membrane boundaries to a total height of 50 A°. The strands are connected by three loops on the perisplamic side. The structure and the stability of the OmpX in water-dodecane solution and in pure water are investigated by molecular dynamics simulations on the 3000 ns timescale. The crystal structure (Protein Data Bank (PDB) entry 1QJ9, resolution 2.1 Amstrong has been taken as the starting structure. The simulations allow the determination of protein-lipid mixed micelle and the calculation of the emulsion´s viscosity. Two different types of simulations of OmpX in water-dodecane are carried out: one type starts from the protein solvated in a box of TIP4 and the other starts from the OmpX solvated in water-dodecane mixture in a box of 6 x 6 x 6 nm^3, using a Gromos 54 force field, the number of dodecane´s molecules added randomly is 75,92 which corresponds to the optimal percentage of lipids used in the industry . The system is equilibrated under conditions of NVT (300 K) and NPT; they are neutralized with NaCl molecules. Each simulation starts with an energy minimization. All MD simulations are carried out using the GROMACS 4 5 4.

In order to validate and compare the MD results experimentally, the OmpX is expressed with IPTG (2 mM), purified and then concentrated from the *E. coli* K-12 ASKA library using heterologous expression and affinity chromatography. The OmpX expression is confirmed by SDS PAGE to finally determine the stability of this protein in water - dodecane emulsions and evaluate the potential of the OmpX as biosurfactant

(593h) Medium Design for Ethanol Production Through Syngas Fermentation

Jie Gao, John R. Phillips, Hasan K. Atiyeh, Mark R. Wilkins and Raymond L. Huhnke, Biosystems and Agricultural Engineering, Oklahoma State University, Stillwater, OK

Biomass derived syngas (CO, H_2 and CO_2) can be fermented into ethanol using *Clostridium ragsdalei*. The use of low cost medium is critical to make syngas fermentation commercially viable. Elimination and reduction in concentrations of expensive nutrients such as morpholinoethane sulfonic acid (MES) buffer and yeast extract (YE) from the medium were examined. Syngas fermentations were done in 250 ml bottles at 37°C using 20 CO: 15 CO_2: 5 H_2 mol% syngas mixture. The syngas was supplied to cells every 24 h for 15 days. Samples were collected every 24 h and analyzed for pH, cell mass, product concentrations and gas utilization. Results showed that MES buffer can be completely removed from the medium and pH can be maintained at desired values using sodium bicarbonate. YE was necessary for *C. ragsdalei* cell growth; however, lower YE concentrations preferentially generated more ethanol. A defined medium was designed in which YE and MES were eliminated and concentrations of minerals and trace metals were adjusted. Cell growth was slow in the defined medium without YE; however, ethanol production was comparable in both media. The cost of the defined medium was less than one third that of the YE medium. The feasibility of alternative nutrients to replace YE will be presented. The cost of the developed defined medium was reduced by 97% compared to the standard YE medium with MES, offering significant cost benefits for potential industrial ethanol production.

(593i) Multivariate Analysis of TOF-SIMS Data for Applications in Tissue Engineering and Quantifying Biomolecules

Mary L. Kraft, Department of Chemical & Biomolecular Engineering, University of Illinois at Urbana-Champaign, Urbana, IL, Robert L. Wilson, Chemistry, University of Illinois, Urbana, IL, Jessica F. Frisz, Chemistry, University of Illinois at Urbana-Champaign, Urbana, IL, Ji Sun Choi, Chemical and Biomolecular Engineering, University of Illinois at Urbana-Champaign, Urbana, IL and Brendan A.C. Harley, Chemical & Biomolecular Engineering, University of Illinois at Urbana-Champaign, Urbana, IL

The ability to map the chemical composition at the surface of the sample with as good as sub-μm lateral resolution and without the need for labels renders TOF-SIMS a powerful approach for studying the compositions of cells and other biological materials. Molecular ions and nearly intact molecular fragment ions that are unique to a single molecule are often employed to unambiguously identify components of interest in the sample. Though more difficult to interpret, compositional information is also encoded in the low-mass ($m/z<300$) peaks that are abundant in the spectra of biological materials. Multivariate analysis of TOF-SIMS data enables utilizing the abundant, low-mass for sample classification or the identification of sample components. Here we present our use supervised multivariate analysis of TOF-SIMS data for two applications: classifying cells according to differentiation stage, and measuring cholesterol concentration in model lipid membranes. The factors that affect the outcome of the analysis, and the improvements that would expand its utility will be discussed.

(593j) Effect of Substrate and Temperature On Biogas Production From Anaerobic Digestion

Matthew J. Metzger[1], Tonderayi S Matambo[2], Michelle Low[3], Ralph Muvhiiwa[3], Phillip Chafa[3], David Glasser[4] and Diane Hildebrandt[4], (1)Chemical and Biochemical Engineering, Rutgers University, Piscataway, NJ, (2)Chemical and Metallurgical Engineering, University of the Witwatersrand, Johannesburg, Johannesburg, South Africa, (3)School of Chemical and Metallurgical Engineering, University of the Witswatersrand, Johannesburg, South Africa, (4)Centre of Material and Process Synthesis (COMPS), University of the Witwatersrand, Johannesburg, South Africa

Anaerobic digestion of waste has been shown to be a potential route to produce sustainable energy. Organic waste is decomposed by anaerobes in the absence of oxygen to produce biogas, a mixture of mainly carbon dioxide and methane, which can be used to generate electricity or run gas powered devices such as refrigerators and lights. This technology has emerged as an attractive energy source for rural areas without access to electricity due to its simplicity and relatively low cost. In this work, we have attempted to determine the effect of temperature and substrate type on the bench scale in the hopes of maximizing methane production. Temperatures within the supposed optimal operating regime (20°C-40°C) are tested, along with a variety of substrates ranging from discarded kitchen waste, to cow mature and dog feces over a period of 2 months. Concentration measurements of methane, hydrogen, carbon dioxide, nitrogen and oxygen show that the mixture of cow manure and dog feces yields the highest amount of combustible biogas over the time period tested. Results and suggestions will be used to develop a larger scale household demonstration unit to provide light to a rural township in Johannesburg, South Africa.

OVERPRODUCTION OF PLEURAN: AN IMMUNEMODULATOR EXOPOLYSACCHARIDE PRODUCED IN SUBMERGED CULTURE OF PLEUROTUS OSTREATUS IN STIRRED TANK BIOREACTOR

Maftoun, P[1], Masri, M.H.J[1], Malek, R.A[1*], Ismail, N.S[1], Othman, N.Z[1], Sarmidi, M.R[1], Aziz, R.A[1], El Enshasy, H.A[1,2]

[1]Institute of Bioproduct Development (IBD), Universiti Teknologi Malaysia (UTM), 81310 Skudai, Johor, Malaysia
[2]City for Scientific Research and Technology Applications, New Burg Al Arab, Alexandria, Egypt

Introduction

Pleurotus ostreatus is the third largest commercially produced mushroom in the world due to its food, nutraceutical and medicinal applications (1). Recently, extensive research on this fungus is conducted with the aim of extracting biological active substances. The medicinal properties of *P. ostreatus* exopolysaccharides are studied by several research groups and showed antitumor, immunomodulating, antiviral and antibacterial activities (2, 3, 4). Pleuran can significantly stimulate the human immune system. The efficacy and immunomodulating ability of β-glucan depends on the source, the mode of production (isolation) and most importantly, its purity. It stimulates the body's defense system against infections (bacterial, viral, yeast, and parasitic (5). Pleuran is a carbohydrate which contains a large number of β-glucose. In fact, it is β-(1→3), (1→6)-D-glucans with a typical structure of β-glucans which enclosed in cell wall (6,7). It is an alkali insoluble exopolysaccharide with molecular formula of $(C_6H_{12}O_6)_n$ and molecular weight of 762 kDa (6). The molecular weight of 2.4×10^4 Da was stated by (8) for this water soluble polysaccharide. The conventional method for cultivation of mushrooms is solid substrate under controlled temperature and humidity. This method of cultivation is time consuming and labor intensive without significant concentrate of active metabolite along with high possibility of contamination (9). Thereby submerged culture of mushrooms for efficient production of medicinal compounds is studied by several authors (10, 11 12). This study focused on maximizing exopolysaccharide (EPS) production in submerged culture by *P. ostreatus* in shake flask and bioreactor (controlled and uncontrolled) via altering process parameters such as aeration rate.

Materials and Method

Microorganism

Pleurotus ostreatus (WICC-F18) kindly obtained from wellness industry culture collection (WICC) at institute of bioproduct development (IBD), UTM, Malaysia..

Working of cell bank preparation

P. ostreatus was cultivated on Yeast-Malt agar plates for 7 days at 26°C. The surface mycelia with spores were harvested from culture in sterile 50% glycerol solution. The harvested mycelia and spores were stored in sterile vials as working cell bank at -80 °C.

Shake flask culture cultivation

The shake flask culture experiments were performed in 250 ml Erlenmeyer flasks with working volume of 50 ml after inoculating with 5% (v/v) of the inoculum. The culture was incubated at 26°C on a rotary shaker incubator (Innova 4080, New Brunswick Scientific, NJ, USA) at 200 rpm. Samples were collected for analyzing morphology, cell dry weight, EPS and glucose content

Growth Curve study

The growth curve study in shake flask scale was carried out with the aim of analyzing pellet forming and cell morphology of *P. ostreatus* during its growth and investigating the EPS production. Growth curve study was conducted for 21 days. Samples were taken daily with 2 replications.

Bioreactor Cultivation condition

The submerged cultivation was carried out in16-L stirred tank bioreactor (BioEngineering, Wald, Switzerland) with working volume of 8 L for 20 days. Medium which was used was as same as shake flask cultivation. Temperature was kept at 26° C and agitation was adjusted at 200 rpm using two 6 bladed Rushton turbine impellers (d_i,impeller diameter = 85 mm; d_t, tank diameter = 214 mm, $^{d_i}/_{d_t}$=0.397). For suppress foaming, silicone antifoam grade A (Sigma-Aldrich Inc., USA) was added when it was required. During the cultivation process, pH value and dissolved oxygen concentration were determined using pH and DO polarographic electrodes, respectively (Ingold, Mittler-Toledo, Switzerland).

Two different experiments have been carried in bioreactor scale. In the first experiment the importance of keeping pH constant at 5.5 was studied by running ph controlled and uncontrolled bioreactor. In case of pH controlled culture, the medium was adjusted to 5.5 by cascading the pH controller with acid/base feeding peristaltic pumps connected with 2.0M of HCL and NaOH, respectively. Aeration rates for these bioreactors were 1 v/v/min. In the next experiment different aeration rates of 0.5, 1, 1.5 and 2 v/v/min were applied. Samples of 40 ml were taken daily through sterile sampling system

Cell dry weight determination

Pre-weight filter paper No. 1 (Whatman International Ltd, Kent, England) was used to separate cells from supernatant. Later, supernatant was used for sugar and EPS determination. The filter papers with cells were transferred to oven (LDO 250F, Daihan LabTech Co., Ltd., Korea) at 60°C for drying purpose and weighted daily till the constant weight was achieved. Cell dry weight (CDW) was calculated by subtracting the dried samples weight from the initial weight of filter paper.

EPS determination

Before analysis, the cells were separated from the fermentation broth using filter paper. Then, pre-weighted centrifuge tubes were filled with 10 ml of supernatant and 30 ml of cold ethanol. Latter, they were kept for 24 h at 4°C. EPS precipitated during this time as it is alcohol insoluble. Thereby, the precipitated EPS can be separated by centrifugation at 4000 rpm for 30 min (D78532, Hettich,Tuttlingen,Germany). The supernatant was removed from tubes and precipitated EPS was dried at 60°C in oven and weighted daily till the constant weight was achieved. EPS concentration was calculated by subtracting the dried samples weight from the initial weight of centrifuge tube.

Glucose and ethanol determination

Glucose content from the culture was determined by using Dinitrosalicylic acid or DNS method(13,14).

Result and Discussion

Growth kinetics in shake flask

Cells were cultivated in shake flask culture to evaluate the potency of EPS production as well as changes in pellet forming and morphology analysis (data not shown) in submerged culture. As shown in Figure 1 EPS was produced for the first time after 24 hour and increased gradually in parallel to cell growth until 16 day, reaching maximal volumetric production of 1.25 g/l with specific EPS production of about 0.19g/g. Moreover, specific growth and production rate for the exponential phase were 7.4 h^{-1} and 4.9 mg/L/h, respectively. During this time glucose was consumed with a rate of 0.06 g/L/h and limited after 16 days cultivation. The growth curve obtained in this experiment demonstrates that EPS production is growth associated product by *P. ostreatus*

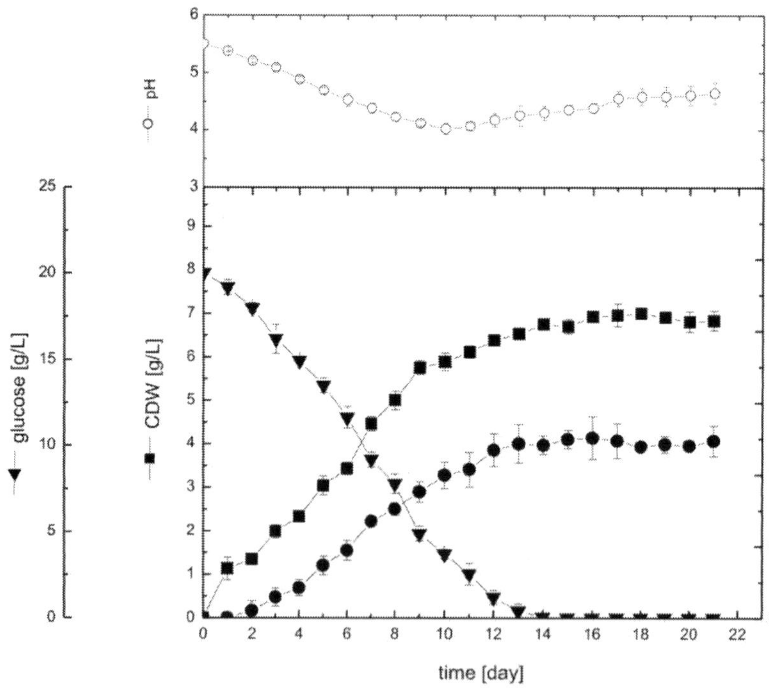

Figure 1: Growth kinetic and EPS production by *P. ostreatus* in shake flask cultivation

Cultivation in uncontrolled and controlled pH bioreactors in submerged culture

Based in this experiment, *P. ostreatus* was cultivated in batch culture mode in 16-L pilot scale stirred bioreactor, under controlled pH (5.5) and uncontrolled pH condition. Figures 2 and 3 illustrate the trend of cell growth, EPS production, glucose consumption and pH values change during 20 days of cultivation in both bioreactors. According to figures 2 and 3, typical exponential cell growth reached to 4.48g/L after 288 hour in controlled pH bioreactor and 3.195g/L after 276 hour in uncontrolled one. Specific cell growth rates during were 0.0115 h^{-1} and 0.0063 g/L/h in controlled and uncontrolled pH bioreactor, respectively. Moreover, glucose was almost limited in controlled pH bioreactor, while such limitation was not occurred in uncontrolled pH bioreactor. Higher biomass as well as EPS production could be the causes of almost complete glucose consumption in controlled pH condition. Complete consumption of glucose by *P. ostreatus* in 15L stirred bioreactor under controlled pH was also observed before by (15). In addition, kinetic parameters related to cell growth and EPS production and glucose consumption are summarize in table 1.

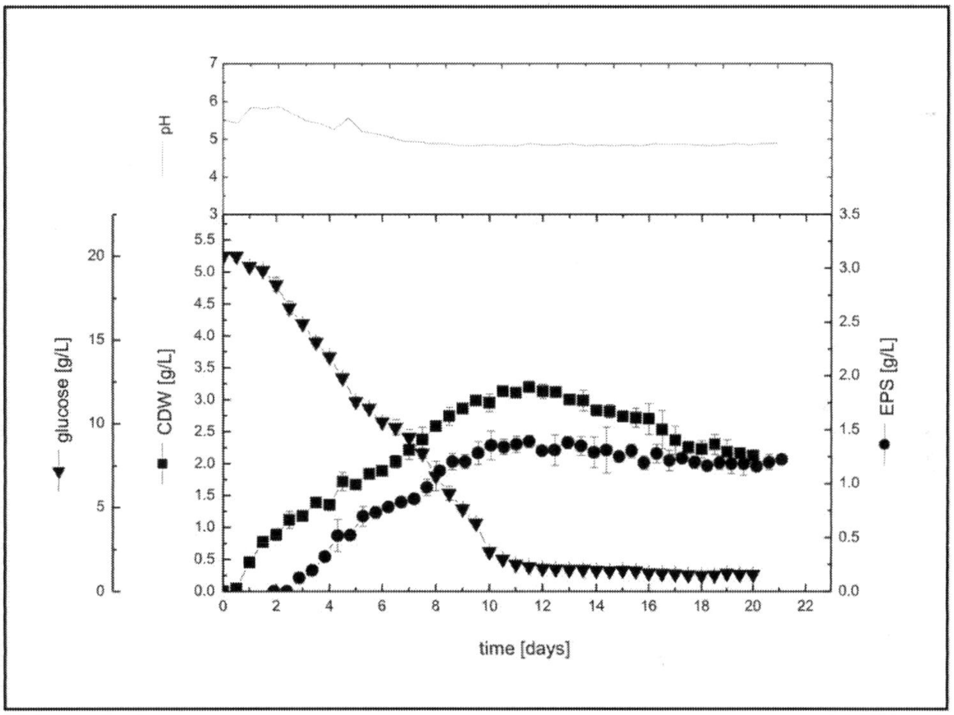

Figure 2: Growth kinetic and EPS production by *P. ostreatus* in uncontrolled pH bioreactor.

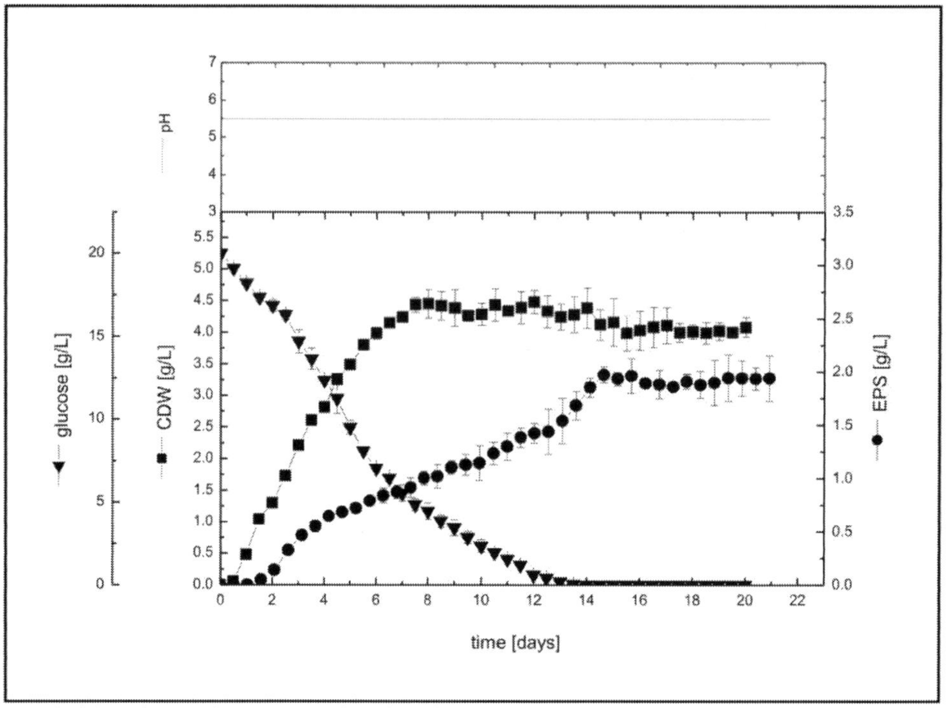

Figure 3: Growth kinetic and EPS production by *P. ostreatus* in controlled pH bioreactor (pH:5.5).

Table 1: Kinetic parameters related to cell growth and EPS production by *P. ostreatus* in bioreactor with and without pH control.

Parameters / Bioreactors	X_{max} g/L	P_{max} g/L	μ h^{-1}	Q_p g/L/h	Q_s g/L/h	$Y_{x/s}$ g/g	$Y_{p/s}$ g/g	$Y_{p/x}$ g/g
Controlled pH	4.48	1.975	0.0115	0.0063	0.067	0.23	0.098	0.45
Uncontrolled pH	3.195	1.39	0.0063	0.0052	0.0753	0.172	0.0788	0.43

Comparison of different aeration rates in bioreactor

To compare performance of different aerated culture regarding EPS and cell biomass production, parameters related to their growth kinetics are summarized in table 2. In this experiment, oxygen limitation was observed in low aerated rate cultivation, leading to less biomass and EPS production. At higher aeration rate cultures, oxygen was not limited during the cultivation process. Further increase in aeration rate resulted in significant increase in cell growth up to 7.1 g/L in 2 v/v/min aerated culture. This result is in contrary with study by(16) who found reduction of cell production in 1.5 v/v/min aerated culture, in comparison to 1v/v/min aerated culture by *P. ostreatus*. However, studies on other filamentous mushroom growth and EPS production revealed increase in both EPS and cell concentration through raising up the aeration rate(17,18). On the other hand, the maximal volumetric EPS production of about 3.1 g/L was obtained in 1.5 v/v/min, aerated cultures. Moreover, the highest coefficient yield of EPS substrate and EPS to cell biomass of 0.16 g/g and 0.673 g/g were achieved in this cultivation, respectively.

Table 2: Kinetics parameters related to cell growth and EPS production by *P. ostreatus* during cultivation in 16 –L stirred tank bioreactor at different aerations

Parameters / Aeration rate	X_{max} g/L	P_{max} g/L	μ h^{-1}	Q_p g/L/h	Q_s g/L/h	$Y_{x/s}$ g/g	$Y_{p/s}$ g/g	$Y_{p/x}$ g/g
0.5 v/v/min	3.25	0.94	0.013	0.005	0.057	0.254	0.07	0.305
1.00 v/v/min	4.48	1.975	0.013	0.006	0.068	0.23	0.099	0.45
1.5 v/v/min	5.275	3.285	0.123	0.047	0.073	0.35	0.16	0.673
2.00 v/v/min	7.225	1.275	0.011	0.005	0.012	0.36	0.063	0.17

Conclusion

In the present work, exopolysaccharide (EPS) production was conducted in submerged cultivation of *P. ostreatus* in shake flask and in bioreactor level. In shake flask culture, when mushroom was cultivated using semi-defined optimized medium, the maximal polysaccharide produced was 1.25 g/L after 13 days of cultivation. It is found that *P. ostratus* pellets forms from aggregation of mycelium .After this stage, cultivation process was scaled up to 16-L *in situ* pilot scale stirred tank bioreactor under controlled and uncontrolled pH conditions. The maximal cell growth of 4.48 g/L was obtained in controlled pH culture (pH: 5.5). This value was almost 40% higher than those obtained in uncontrolled pH culture. On the other hand, the maximal volumetric and specific polysaccharide production of about 1.98 g/L and 0.45 g/g, respectively, was obtained in uncontrolled pH culture. Further studies have been carried out to investigate the effect of aeration rate on cell growth and EPS production. Low aeration rate of 0.5 v/v/min lead to oxygen limitation after 5 days of cultivation. At higher aeration rate cultures, oxygen was not limited during the cultivation process. The maximal volumetric and specific EPS production were 3.28 g/L and 0.67 g/g, respectively, were obtained in 1.5 v/v/min aerated culture. On the other hand, increase in aeration rate resulted in significant increase in cell growth up to 7.1 g/L in 2 v/v/min aerated culture with significant decrease in EPS production by 61% compared to 1.5v/v/min aerated culture.

Reference

1. Yildiz, S., E. D. Gezer, Ü. C. Yildiz, A. Temiz and E. Dizman (2005). Effects of Different Substrate Combinations on the Mycelial Growth Period of *Pleurotus ostreatus*. *The 5th International Conference on Mushroom Biology and Mushroom Products*. 8-12 April. Shanghai.
2. Bobek, P. and S. Galbavy (2001). Effect of pleuran (beta-glucan from *Pleurotus ostreatus*) on the antioxidant status of the organism and on dimethylhydrazine-induced precancerous lesions in rat colon. *British Journal of Biomedical Science*. 58 (3), 164-168.
3. Sarangi, I., D. Ghosh, S. K. Bhutia, S. K. Mallick and T. K. Maiti (2006). Anti-tumor and immunomodulating effects of *Pleurotus ostreatus* mycelia-derived proteoglycans. *International Immunopharmacology*. 6 (8), 1287-1297.
4. Yoshioka, Y., R. Tabeta, H. Saitô, N. Uehara and F. Fukuoka (1985). Antitumor polysaccharides from *P. ostreatus* (Fr.) quél.: Isolation and structure of a [beta]-glucan. *Carbohydrate Research*. 140 (1), 93-100.
5. Synytsya, A., K. Mícková, A. Synytsya, I. Jablonský, J. Spevácek, V. Erban, E. Kováríková and J. Copíková (2009). Glucans from fruit bodies of cultivated mushrooms *Pleurotus ostreatus* and *Pleurotus eryngii*: Structure and potential prebiotic activity. *Carbohydrate Polymers*. 76 (4), 548-556.
6. Karácsonyi, S. and L. Kuniak (1994). Polysaccharides of *Pleurotus ostreatus*: Isolation and structure of pleuran, an alkali-insoluble [beta]-d-glucan. *Carbohydrate Polymers*. 24 (2), 107-111.
7. Novak, M. and V. Vetvicka (2008). β-Glucans, History, and the Present: Immunomodulatory Aspects and Mechanisms of Action. *Journal of Immunotoxicology*. 5 (1), 47-57.
8. Sun, Y. and J. Liu (2009). Purification, structure and immunobiological activity of a water-soluble polysaccharide from the fruiting body of *Pleurotus ostreatus*. *Bioresource Technology*. 100 (2), 983-986.
9. Zhong, J.-J. and Y.-J. Tang (2004). *Submerged Cultivation of Medicinal Mushrooms for Production of Valuable Bioactive Metabolites Biomanufacturing*. J.-J. Zhong. *Advances in Biochemical Engineering/Biotechnology*. (25-59). Heidelberg: Springer Berlin.
10. El-Enshasy, H. (2010). *Immunomodulators Industrial Applications*. M. Hofrichter. (165-194). Springer Berlin Heidelberg.
11. Gern, R. M. M., E. Wisbeck, J. R. Rampinelli, J. L. Ninow and S. A. Furlan (2008). Alternative medium for production of *Pleurotus ostreatus* biomass and potential antitumor polysaccharides. *Bioresource Technology*. 99 (1), 76-82.
12. Téllez-Téllez, M., F. Fernández, A. Montiel-González, C. Sánchez and G. Díaz-Godínez (2008). Growth and laccase production by *Pleurotus ostreatus* in submerged and solid-state fermentation. *Applied Microbiology and Biotechnology*. 81 (4), 675-679.
13. Enshasy, H., K. Hellmuth and U. Rinas (1999). Fungal morphology in submerged cultures and its relation to glucose oxidase excretion by recombinant *Aspergillus niger*. *Applied Biochemistry and Biotechnology*. 81 (1), 1-11.
14. Miller, G. L. (1959). Use of dinitrosalisylic acid reagent for determination of reducing sugar. *Analytical Biochemistry*. 31 426-442.
15. El-Enshasy, H., A. Daba, M. El-Demellawy, A. Ibrahim, S. El Sayed and I. El-Badry (2010). Bioprocess Development for Large Scale Production of Anticancer Exo-Polysaccharide by *Pleurotus ostreatus* in Submerged Culture. *Journal of Applied Sciences*. 10 (21), 2523-2529.

16. Márquez-rocha, F. J., G. K. Guillén, J. E. Sánchez and R. Vázquez-duhalt (1999). Growth characteristics of *Pleurotus ostreatus* in bioreactors. *Biotechnology Techniques*. 13 (1), 29-32.
17. Lee, B. C., J. T. Bae, H. B. Pyo, T. B. Choe, S. W. Kim, H. J. Hwang and J. W. Yun (2004). Submerged culture conditions for the production of mycelial biomass and exopolysaccharides by the edible Basidiomycete *Grifola frondosa*. *Enzyme and Microbial Technology*. 35 (5), 369-376.
18. Xu, C.-P. and J. W. Yun (2004). Influence of aeration on the production and the quality of the exopolysaccharides from *Paecilomyces tenuipes* C240 in a stirred-tank fermenter. *Enzyme and Microbial Technology*. 35 (1), 33-39.

SEMI-INDUSTRIAL SCALE PRODUCTION OF BIO- CONTROL FUNGUS: *HENDERSONIA* SP., A POTENTIAL AGENT AGAINST *GANODERMA* OIL PALM DISEASE

Azman,M.D[1], Malek,R.A[1*], Othman,N.Z[1], Peng,T[2], Idris,A[3], Then,C[2], Chang, H.W[2], Sukumaran,S[2], Wan Mustapha,W.A[2], Ramli,S[1], Sarmidi, M.R[1], Aziz, R[1], El Enshasy, H.A[1,4]

[1]Institute of Bioproduct Development (IBD), Universiti Teknologi Malaysia, (UTM), 81310 UTM Skudai, Johor, Malaysia.
[2]All Cosmos Industries Sdn Bhd, Pasir Gudang, Johor Bahru, Malaysia
[3]Malaysian Palm Oil Board (MPOB), Ministry of Plantation Industries and Commodities, Malaysia
[4]Bioprocess Development Departments, City for Scientific Research (CSAT), New Burg Al Arab, Alexandria, Egypt.

Introduction

Oil palm (*Elaeis guinensis*) is one of the important commercial commodities and considered as a Malaysian national treasure grown in Malaysia. Being one of the world's largest producers and exporters of palm oil products, Malaysia exported 24.27 million tonnes of palm oil products in 2011. These oil palm products, which include palm oil, palm kernel oil, palm kernel cake, oleochemicals, biodiesel and finished products, account for RM 80.4 billion in total export revenue (1). However, the industry experiences losses between RM 2 billion and RM 3 billion (approximately USD 0.65 billion to USD 0.97 billion) annually from oil palm disease especially Basal Stem Rot caused by *Ganoderma*. However, like most crops, oil palm is susceptible to diseases. One of the major diseases of oil palm is known as basal stem rot (BSR), which is caused by pathogenic fungal species especially *Ganoderma* (2,3).This disease poses a serious threat to the cultivation and the production of oil palm, which leads to severe economic losses. Over recent years, a lot of attempts have been done to control basal stem rot in oil palm. The usage of chemical-based fungicide (4) as control showed only mediocre results due to the characteristics of *Ganoderma* which is soil borne (5). Thus, researchers are currently focusing in studying the biological control of the disease using endophytic microorganisms including *Pseudomonas, Trichoderma,* and *Hendersonia(6,7,8)*.Biological control by antagonistic microorganisms is widely recognized as a promising and potential method for controlling plant diseases and to replace the use of chemical fungicide. *Hendersonia* sp., a novel strain successfully isolated from the root of oil palm tree, is effective in controlling *Ganoderma* infection. In this study, we aimed to optimize the *Hendersonia* cultivation process platform for the industrialization of *Hendersonia*.

Materials and Method

Microorganisms

In this study, *Hendersonia* sp. was used in all experiments. The strain is an industrial strain owned by All Cosmos Industries Sdn Bhd, Johor Malaysia. The strain is currently undergoing characterization for its specific species name.

Bioreactor 16L Cultivations

In this study, the initial bioreactor batch cultivation was carried out in a 16-L and 150L stirred tank bioreactor (Bioengineering, Wald, Switzerland). The bioreactor was equipped with dissolved oxygen probe, pH probe, and stirrer with two Rushton turbines. Apart from that, several parameters and indicators including pH, agitation speed, aeration rate, dissolved oxygen, and anti-foam can be accessed via its respected controllers.The composition of the medium used for this bioreactor composed of (g L)$^{-1}$:Malt extract, 10 ;Glucose, 20; Peptone,1 and Yeast, 5. The carbon source was autoclaved separately from the rest of the components. The other components were sterilized *in situ* in the bioreactor. The bioreactor as well as the medium were sterilized at 121 °C for 15 minutes prior to inoculation.

Studies on the effects of agitation speed in 150-L bioreactor

The effect of agitation speeds on the cell growth has been studied. Three different setups utilizing different agitation speeds (i.e. 150 rpm, 200 rpm, and 250 rpm) were experimented in 150-L bioreactor cultivation. Similar to the previous 150-L bioreactor study, the inoculum was prepared beforehand in the 16-L bioreactor with a working volume of 10 L. It was then transferred into the 150-L bioreactor containing the sterilized media with same composition as the previous 150-L bioreactor study listed in 16L and with a working volume of 100 L. In each cultivations, two rushton turbines were used to agitate the cells. Furthermore, the antifoam agent used was Struktol J673A at 30 percent concentration. 2 L of 4M HCl and 2 L of 4M NaOH were also used as the buffer. In all three setups, the pH was maintained at 5.5 but was left uncontrolled throughout the cultivations. The temperature was maintained at 28 °C, while the aeration rate was fixed at 10 L/min. In each batches, the agitation speed was maintained at 150 rpm, 200 rpm, 250 rpm.The cells were cultivated in the bioreactor for 70 hours and samplings were done in every 5 hours. Samples were stored in the refrigerator for further analysis.

Analytical methods

Generally, three types of analysis were done in both shake flask and bioreactor studies which were; pH determination, cell dry weight determination, and total carbohydrate determination (for shake flask study) or glucose determination (in bioreactor study). Additionally, the morphology of the strain had also been studied by using light microscope.

pH determination

For the shake flask study, the pH of the media was determined once the flasks had been taken out from the incubator shaker. The pH was determined by using the table pH meter (Mettler-Toledo Delta 320, Greifensee, Switzerland).For the bioreactor study, the pH reading was taken directly from the probe in the bioreactor itself.

Cell dry weight determination

Celloluse filter papers (Whatman, 20-25 μm, United Kingdoms) were used for the determination of the cell dry weight. Once the samples were filtered, the filter papers were dried at 70 °C for at least 48 hours until the weight stabilized. The supernatant from the samples were kept chill for analysis.

Glucose determination

Glucose content from the culture was determined by using Dinitrosalicylic acid or DNS method.350 πL of citric acid buffer and 600 πL of DNS reagent were added to 50 πL of diluted supernatant (at 10 times dilution) (9). The mixture was mixed well and heated in the water bath at 95 °C for 3 minutes. After heating, the solution was rapidly cooled in ice bath for 3 minutes before the absorbance was measured at 540 nm.

Results and Discussions

Shake flask media optimization using response surface methodology

In this study, medium M4 which composed of (g/L): Sucrose, 40; K_2HPO_4, 1; $MgSO_4.7H_2O$, 1; KCl, 0.5; $FeSO_4$, 0.01; Yeast extract, 2, was selected for further optimization. For this purpose, Minitab 16 statistical software was used to design and analyze the experiment. From the medium components, only three factors were experiented and optimized. The components were Sucrose, K_2HPO_4, and Yeast extract, which represented the carbon source, phosphate source, and nitrogen source, respectively. The rest of the components were unchanged in all flasks.Response Surface Methodology (RSM) (10,11,12) has been used to optimize the medium which was selected previously in our study. From the results, the Estimated Regression Coefficients for cell dry weight were generated by the Minitab software. Apart from that, the following statistical parameters were also generated by the software:

S = 0.302420, PRESS = 4.42752

R-Sq = 95.95%, R-Sq(pred) = 90.20%, R-Sq(adj) = 94.13%

The analysis for regression coefficients was generated to analyze the relationship between the variables. This are used to estimate the expected cell dry weight should the variables changes. In order to observe the relationship between the factors, several contour plots have been generated by the Minitab software. Figure 1, showed the contour plot of the cell dry weight against the relationship between K_2HPO_4 and Sucrose. From the contour plot, as the Sucrose concentration increases, so did the cell dry weight. However, apparently, higher concentration of K_2HPO_4 resulted with lower cell dry weight. From the figure, maximal cell dry weight was achieved when the Sucrose concentration was at 60 g/L and K_2HPO_4 at 0.5 g/L.

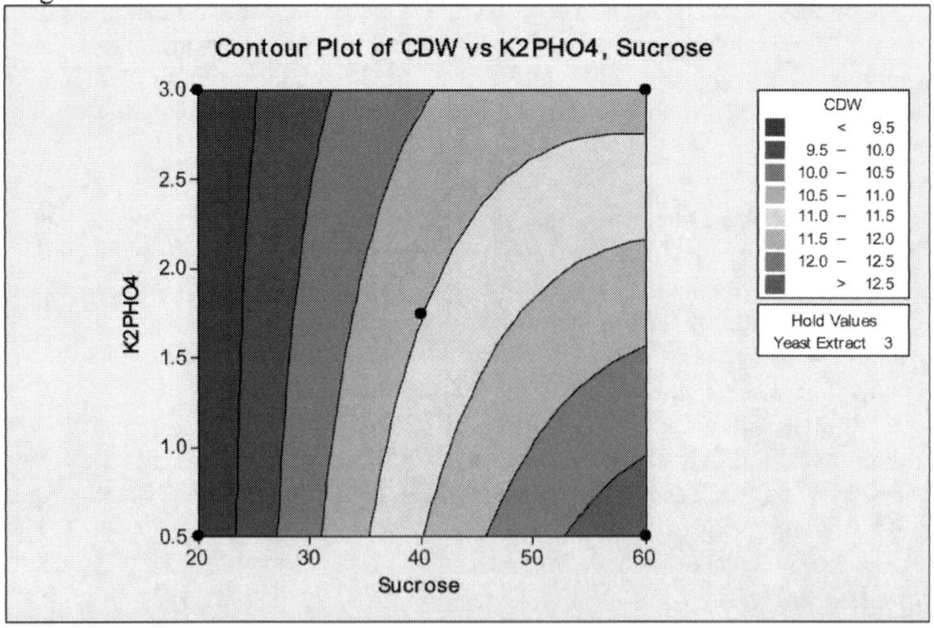

Figure 1 : Contour plot of cell dry weight versus K_2HPO_4 and Sucrose

From the contour plot, as the Sucrose concentration increases, so did the cell dry weight. However, apparently, higher concentration of K_2HPO_4 resulted with lower cell dry weight. From the figure, maximal cell dry weight was achieved when the Sucrose concentration was at 60 g/L and K_2HPO_4 at 0.5 g/L.Apart from that, the contour plot of the cell dry weight against the relationship between Yeast extract and Sucrose was observed. The contour plot is shown in Figure 2. Based from the contour plot, similar to the previous contour plot shown in Figure 2, higher amount of Sucrose resulted with higher cell dry weight. On the other hand, 3 g/L of Yeast extract seemed to be providing the optimum cell dry weight of the culture. It was observed that if the concentration of the Yeast extract was too high or too low, it does not result with higher cell dry weight.

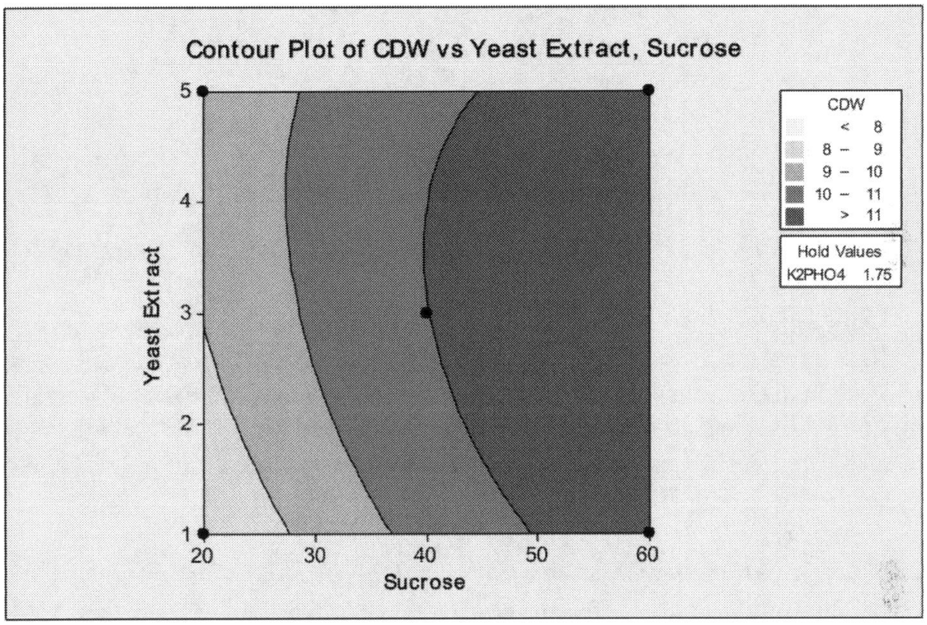

Figure 2: Contour plot of cell dry weight versus Yeast extract and Sucrose

Last but not least, the contour plot of the cell cry weight against the relationship between Yeast extract and K_2HPO_4 was analzyed and showed in Figure 3. From the contour plot, it was observed that lower amount of K_2HPO_4 resulted with higher cell dry weight. However, optimum cell dry weight was obtained when the concentration of Yeast extract was between 3 to 4 g/L. If the concentration of the Yeast extract was too high (exceeding 5g/L) or too low (below 2 g/L), it did not gave the optimum cell dry weight. Based from the analysis, the Minitab software had generated the global solution for the respose optimization of the culture as followed, (g/L): Sucrose, 60; K_2HPO_4, 0.5; Yeast Extract, 3; $MgSO_4.7H_2O$, 1; KCl, 0.5; $FeSO_4$, 0.01.

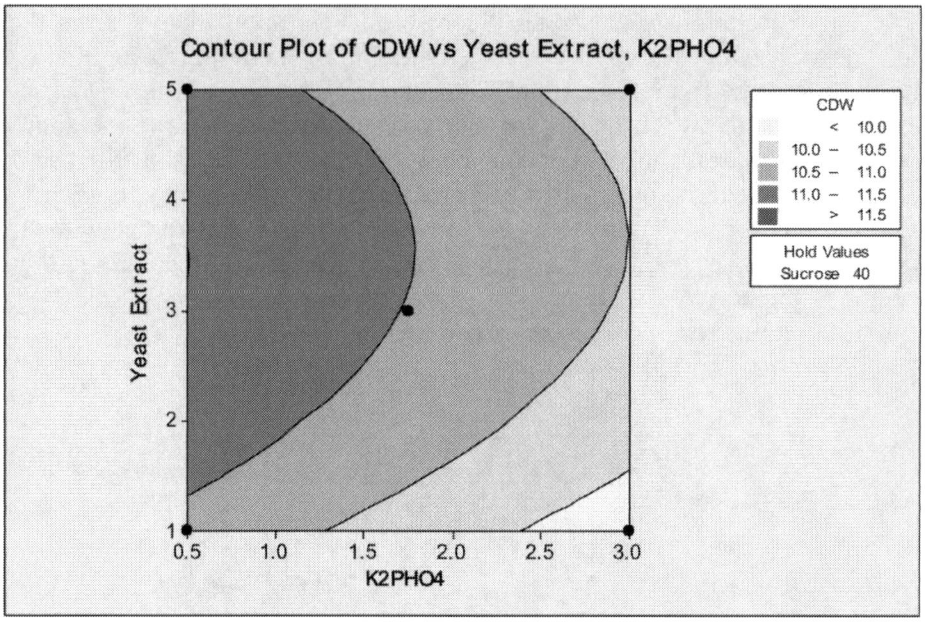

Figure 3 : Contour plot of cell dry weight versus Yeast extract and K_2HPO_4

Comparison of growth kinetics for optimized and un-optimized medium

Based from the results (data not shown), at 96 hours, maximal cell dry weight of 8.35 g/L was obtained from the optimized medium in comparison to un-optimized at 7.44 g/L. The optimized medium has shown an increase of 11 percent in cell dry weight compared to the un-optimized medium. Furthermore, the new formulated medium gave a maximum growth rate (μ max) of 0.015 g/L/hour, while the un-optimized gave 0.014 g/L/hour.

Cell cultivation under controlled and uncontrolled pH conditions

In this study, two different configurations were used for the 150-L bioreactor cultivation. Under controlled pH condition, the pH was maintained at 5.5 throughout the cultivation. On the other hand, under uncontrolled condition, the pH was left uncontrolled until the end. Based from the results in Figure 4, cultivation using uncontrolled pH showed higher cell dry weight. At 70 hours, maximal cell dry weight of 13.65 g/L was obtained from the uncontrolled pH cultivation as opposed to 12.65 g/L under controlled pH conditions.

Cell cultivation under different agitation speed

In the study, three different agitation speeds of 150 rpm, 200 rpm, and 250 rpm have been used in the 150-L bioreactor cultivation. The pH was set at 5.5 but left uncontrolled throughout the cultivation. Based from the results in Figure 5, by maintaining the agitation speed at 150 rpm, maximal cell dry weight of 12.8 g/L was obtained at 70 hours. Apart from that, at 65 hours, maximal cell dry weight of 13.3 g/L was obtained by utilizing the agitation speed at 200 rpm. At 250 rpm agitation speed, maximal cell dry weight of 15.6 g/L was obtained at 70 hours. From the results, it can be concluded that by increasing the agitation speed, higher cell dry weight could be obtained.

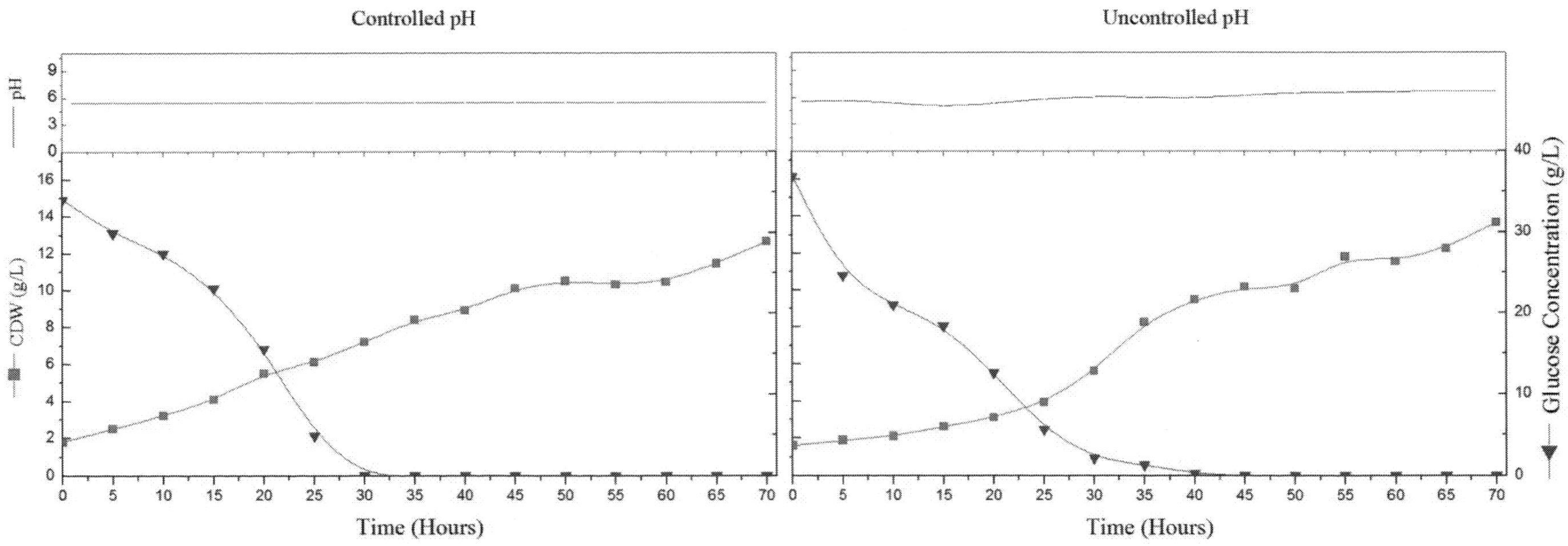

Figure 4: Comparison of bioreactor cultivations under controlled pH and uncontrolled pH

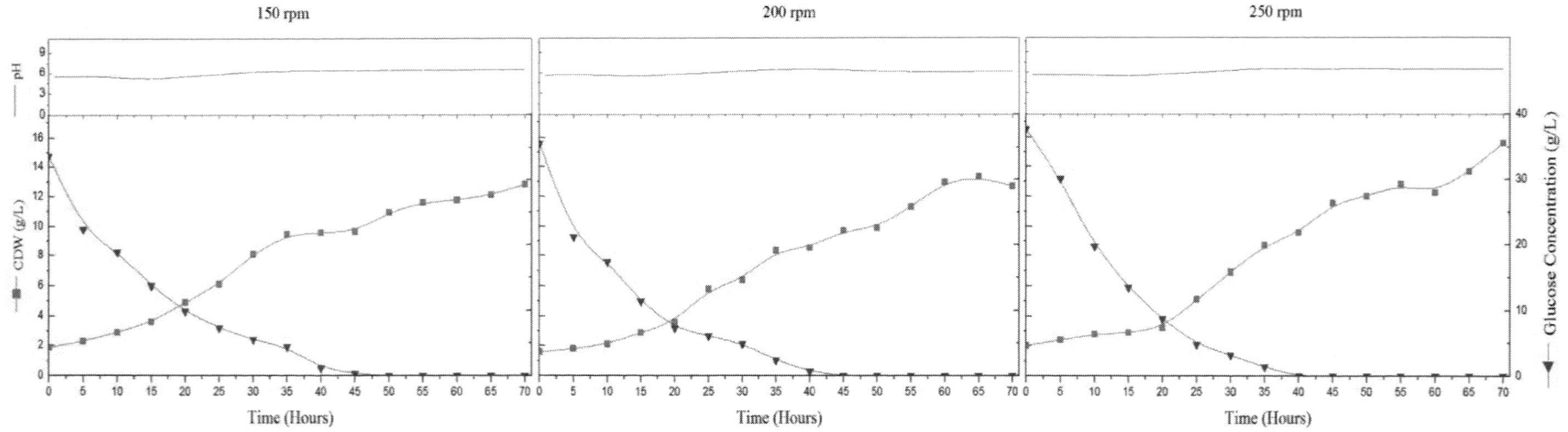

Figure 5: Comparison of bioreactor cultivations using different agitation speeds

Conclusion

The new optimized medium has shown impressive results in terms of higher cell dry weight and growth rate, while at the same time, remaining cost effective for industrial purposes. For the bioreactor cultivation, under uncontrolled pH condition, the cell dry weight obtained was higher compared to controlled pH condition. This bioreactor setup could also be beneficial since it does not require acid and base to be pumped continuously. In addition, higher cell dry weight was obtained from higher agitation speed. This could be resulted from greater oxygen transfer and better mixing.

The present study does not only propose a reliable and cost effective industrial formulation of the medium, it also offers a promising strategy for high cell mass production of *Hendersonia* sp. for semi-industrial scale.

Reference

1. Malaysia Palm Oil Board. (2011). *Overview of the Malaysian Oil Palm Industry 2010.* Malaysian Palm Oil Board.
2. Pilotti, C. A. (2005). Stem rots of oil palm caused by *Ganoderma boninense*: Pathogen biology and epidemiology. *Mycopathologia.* 159, 129-137.
3. Thompson, A. (1931). Stem rot of oil palm in Malaya. *Bulletin of the Department of Agriculture.*
4. Soepena, H. Purba, R.Y., and Pawirosukarto, S. A control strategy for basal stem rot (Ganoderma) on oil palm. In Flood. *et al.* (Ed.) *Ganoderma Disease of Perennial Crops.* United Kingdom: CAB International. 83-88, 2000
5. Susanto, A., Sudharto, P. S., and Purba, R.Y. (2005). Enhancing biological control of basal stem rot disease (*Ganoderma boninense*) in oil palm plantations. *Mycopathologia.* 159, 157-157.
6. Chincholker, S.B. and Mukerji, K.G. (2007). *Biological Control of Plant Diseases.* New York; Haworth Press.
7. Ellis, M.B. (1997). *Microfungi on Land Plants: An Identification Handbook.* United Kingdom: The Richmond Publishing Co. Ltd.
8. Idris, A.S., Nurrashyeda, R., Maizatul, S.M., Madihah, A.Z., Tarmizi, A.M., Kushairi, A., Wan Azha, W.M., and Tony Peng, S.H. (2012). Biofertiliser *Hendersonia* GanoEF as biological control of *Ganoderma* in oil palm. *Malaysia Palm Oil Berhad Information Series.* ISSN 1511-7871.
9. Enshasy, H., K. Hellmuth and U. Rinas (1999). Fungal morphology in submerged cultures and its relation to glucose oxidase excretion by recombinant *Aspergillus niger. Applied Biochemistry and Biotechnology.* 81 (1), 1-11.
10. Montgomery D.C. (2000), *Design and analysis of experiments*, 5th edition, NewYork: John Wiley & Sons Inc.
11. Ferreira S.L.C, Bruns R.E., Ferreira H.S., Matos G.D., David J.M., Brand G.C., da Silva E.G.P., Portugal L.A., dos Reis P.S., Souza A.S. and dos Santos W.N.L.(2007). Box-Behnken design: An alternative for the optimization of analytical methods, *Analytica Chimica Acta,* 597, 179 – 186.
12. Navaneeth S., Bhuvanesh S., Bhaskar V., Vijay Kumar P., Kandaswamy S.K.J. and Achary A. (2009).Optimization of medium for the production of subtilisin from *Bacillus subtilis* MTCC 441.*African Journal of Biotechnology,* 8, 6327– 6331.

HIGH CELL DENSITY CULTIVATION OF *Azotobacter vinelandii* IN PILOT SCALE STIRRED TANK BIOREACTORS FOR BIOFERTILIZER APPLICATIONS

Then,C[1], , Othman, N.Z[2], Malek, R.A[2], , Peng, T[1], Chang, H.W[1], Sukumaran,S[1], Wan Mustapha, W.A[1],Ramli, S[2], Saat, M[2], Sarmidi, M.R[2], Aziz, R[2], El Enshasy, H.A[2,3]

1. All Cosmos Industries Sdn Bhd, Pasir Gudang, Johor Bahru, Malaysia
2. Institute of Bioproduct Development (IBD, Universiti Teknologi Malaysia, (UTM), UTM Skudai, Johor, Malaysia;
3. Bioprocess Development Departments, City for Scientific Research (CSAT),New Burg Al Arab, Alexandria, Egypt.

Introduction

Nitrogen is one of the most important elements supplied by chemical fertilizers to increase the productivity of crops. The FAO estimates the world nitrogen fertilizer demand to increase at an annual rate of about 1.4% until 2011/2012, which is an overall increase of 7.3 million tonnes. Excessive usage of inorganic N fertilizers are linked with hazards such as nitrate leaching, ammonia volatilization, greenhouse gas emissions, water pollution and soil nutrient content degradation (Fisher and Newton, 2002; Hernandez, 2002; Mosier *et al.*, 2004). Therefore, biological nitrogen fixation, BNF is seen as a sustainable alternative to inorganic N fertilizers whereby the N in the atmosphere is made available to plants via N-fixing microorganisms or diazotrophs. *Azotobactervinelandii* is a model free-living N-fixing bacterium capable of converting stable nitrogen gas into ammonia,a nitrogen source accessible to plants (Dalton and Kramer; 2007; Newton, 2007). Although much research has been done on the genetics, metabolism and alginate production of *A. vinelandii*, little information onhigh cell mass production for biofertilizer applications is available. Therefore, the objective of this research is to develop a cultivation strategy for the mass production of *A. vinelandii* in a semi-industrial scale.

Materials and Methods

Batch cultivation in 16-L pilot scale stirred tank bioreactor

*Azotobactervinelandii*NRRL B-14641 was used throughout the study. Batch cultures were carried out in 16-L stirred tank bioreactors (BioEngineering, Wald, Switzerland) with a working volume of 8 L. Inoculum cultures were prepared using a semi-defined culture medium in 250 ml Erlenmeyer shake flasks. The agitation speed and aeration during batch cultures were maintained at 400 rpm and 1 vvm respectively throughout the cultivation. The pH of the culture medium in the bioreactor was adjusted to 7.2 prior to inoculation.

Fed-batch cultivation in 16-L pilot scale stirred tank bioreactor

Fed-batch cultures of *A. vinelandii* NRRL B-14641 in 16-L stirred tank bioreactors were carried out under conditions similar to batch cultures with the exception of continuous nutrient feeding during the cultivation process and modifications to the dissolved oxygen control protocol. The feed medium was fed continuously at a constant pre-determined rate using a peristaltic pump.

Fed-batch cultivation in 150-L pilot scale stirred tank bioreactor

Fed-batch cultures of *A. vinelandii* NRRL B-14641 in 150-L stirred tank bioreactors were carried out under conditions similar to fed-batch cultures in 16-L stirred tank bioreactors with the exception of a working volume of 70 L and a DO-stat configuration. Inoculum cultures were prepared using a 16-L stirred tank bioreactor.

Results and Discussion

Cultivationin 16-L stirred tank bioreactor

Batch cultures of *A. vinelandii* under controlled and uncontrolled pH yielded cell mass of 7.52 g L^{-1} and 15.86 g L^{-1} respectively indicating that pH control may not be crucial during cultivation. Saribay, (2003) noted that *Azotobacter* was able to grow in culture media with pH ranging from 4.5 to 9.0. Maximal alginate concentration obtained from batch cultures with and without pH controlled were 0.657 g L^{-1} and 0.987 g L^{-1} respectively. However, when the experiment was repeated under fed-batch conditions without dissolved oxygen control the maximal cell mass was almost similar to that of batch cultures. Alginate production on the other hand increased in both fed-batch cultures thus indicating that dissolved oxygen control may be a crucial factor in improving the growth of *A. vinelandii* during bioreactor cultivations. Maximal cell mass obtained from fed-batch cultures with and without pH control was 7.97 g L^{-1} and 15.51 g L^{-1} respectively while the maximal alginate concentration obtained was 4.51 g L^{-1} and 1.42 g L^{-1} respectively. It has been reported that the optimum pH for crucial enzymes involved in the biosynthesis of alginate such as phosphomannoseisomerase, GDP mannose dehydrogenase and GDP mannose pyrophosphorylaserange from 7.0 – 8.0 (Horan *et al.*, 1981).A maximal cell mass of 40.65 g L^{-1} and alginate concentration of 8.40 g L^{-1} was obtained in fed-batch cultivation of *A. vinelandii* aerated with a mixture of filtered air and oxygen under controlled pH and dissolved oxygen conditions. As indicated by previous authors effective oxygenation strategies are capable of increasing microbial growth (Keevil*et al.*, 1979; Alexander *et al.*, 1988; Jones *et al.*, 2007). Table 1 summarizes the cell mass and alginate concentration obtained in batch cultures of *A.vinelandii* while Table 2 summarizes the cell mass and alginate concentration obtained in fed-batch cultures of *A. vinelandii*.

Cultivation in 150-L stirred tank bioreactor

Fed-batch culture of *A. vinelandii* in a 150-L stirred tank bioreactor under controlled pH at 7.2 and controlled dissolved oxygen concentration at 30% saturation yielded a maximal cell mass of 28.35 g L^{-1} and a maximal alginate concentration of 18.60 g L^{-1}. The increased size of the vessel resulted in a lower oxygen transfer rate (Schmidt, 2005). This in turn resulted in oxygen limiting conditions which increased the production of alginate. Hence, the use of dissolved oxygen carriers may be necessary to alleviate dissolved oxygen limitations in pilot scale cultivations of *A.vinelandii* (Ntwampe*at al.*, 2010).

Conclusion

A fed-batch cultivation process combined with pH and dissolved oxygen control coupled with oxygen supplementation is required to maximize cell growth of *A. vinelandii* while minimizing alginate production.

Table 1: Summary of cell mass and alginate concentration obtained from batch cultures of *Azotobactervinelandii*

Parameters	Batch culture	
	Uncontrolled pH	**Controlled pH (at 7.2)**
Biomass, X (g L^{-1})	15.86	7.52
Alginate, P (g L^{-1})	0.985	0.657

Table 2: Summary of cell mass and alginate concentration obtained from fed-batch cultures of *Azotobactervinelandii*

Parameters	Uncontrolled pH		Controlled pH at 7.2		
	Without DO-control	**With DO-control**	**Without DO-control**	**With DO-control**	**With DO-control & O_2 aeration**
Max Biomass, X (g L^{-1})	15.51	16.20	7.97	24.45	40.65
Max Alginate, P (g L^{-1})	1.42	2.11	4.51	8.10	8.40

References:

[1] Alexander M.A., Chapman T.W., Jeffries T.W., (1988), Xylose metabolism by *Candida shehatae* in continuous culture, *Applied, Microbiology and Biotechnology***28**, 478 – 486.

[2] Dalton D.A., Kramer S., (2007), Nitrogen-fixing bacteria in non-legumes, *Plant-Associated Bacteria*, The Netherlands, Springer, 105 – 130.

[3] Fisher K., Newton W.E., (2002), Nitrogen fixation – A general overview, *Nitrogen Fixation at the Millenium*, B.V., Amsterdam, The Netherlands, Elsevier, 1 – 34.

[4] Hernandez G., (2002), *Chair's Comments: Biological nitrogen fixation and sustainable agriculture, nitrogen fixation: global perspectives*, CAB International, Wallingford, UK.

[5] Horan N.J., Jarman T.R., Dawes E.A, (1981), Effects of carbon source and inorganic phosphate concentration on the production of alginic acid by a mutant of *Azotobacter*

[6] Jones H.L., Margaritis A., Stewart R.J., (2007), The combined effects of oxygen supply strategy, inoculum size and temperature profile on very-high-gravity beer fermentation by *Saccharomyces cerevisiae*, *Journal of the Institute of Brewing***113**, 168 – 184.

[7] Keevil C.W., Hough J.S., Cole J.A., (1979), Regulation of respiratory and fermentative modes of growth of *Citrobacterfreundii* by oxygen, nitrate and glucose, *Journal of General Microbiology***113**, 83 – 95.

[8] Mosier A.R., Syers J.K., Freney J.R., (2004), *Agriculture and theNitrogen fertilizer: An essential component of increased food, feed and fiber production, agriculture and the nitrogen cycle*, Island Press, Washington, DC 20009, USA

[9] Newton W.E., (2007), Physiology, biochemistry, and molecular biology of nitrogen fixation, *Biology of the Nitrogen Cycle*, Elsevier, Amsterdam, The Netherlands, 109 – 130.

[10] Ntwampe S.K.O., Williams C.C., Sheldon M.S., (2010), Water-immiscible dissolved oxygen carriers in combination with Pluronic F 68 in bioreactors, *African Journal of Biotechnology***9**, 1106 – 1114.

[11] Saribay G.F., (2003), Growth and nitrogen fixation dynamics of *Azotobacterchroococcum* in nitrogen-free medium and OMW containing medium, master thesis, The Middle East Technical University.

[12] Schmidt F.R., (2005), Optimization and scale up of industrial fermentation processes, *Applied, Microbiology and Biotechnology***68**, 425 – 435.

(593n) Co-Immobilization of Multiple Cascade Enzymes On the Surface of Magnetic Nanoparticles: Easy Separation and Substrate Channeling

Suwan Myung, Biological Systems Engineering, Virginia Polytechnic Institute and State University, Blacksburg, VA; Institute for Critical Technology and Applied Science, Virginia Polytechnic Institute and State University, Blacksburg, VA, Chun You, Biological Systems Engineering, Virginia Polytechnic Institute and State University and Y.-H. Percival Zhang, Biological Systems Engineering Department, Virginia Polytechnic Institute and State University; Institute for Critical Technology and Applied Science, Virginia Polytechnic Institute and State University

Cell-free synthetic pathway biotransformation (SyPaB) is *in vitro* assembly of numerous enzymes for implementing complicated biological reactions that microbes and chemical catalysts cannot do. To decrease labor for the purification of numerous enzymes and facilitate easy separation of biocatalysts and products, we are developing to a new approach to co-immobilizing three cascade enzymes on the surface of magnetic nanoparticles. Three enzymes are *Thermus thermophilus* triose phosphate Isomerase (*Ttc*TIM), *Thermotoga maritima* fructose bisphosphate aldolase (*Tm*ALD), and *Thermotoga maritima* fructose 1,6-bisphosphatase (*Tm*FBP), which are key enzymes in non-reductive pentose phosphate pathway. First, we add polysaccharides on the surface of magnetic nanoparticles. Second, the carbohydrate-binding module-containing mini-scaffoldin is bound on the surface of polysaccharides. Third, three dockerin-containing enzymes can bind with the mini-scaffoldin through the high-affinity interaction among the dockerin and cohesin modules. Such self-assembled enzyme complex can be separated easily by using magnetic force and their proximate organization resulted in drastic increases in reaction rates compared to the free enzyme mixture at the same enzyme concentration.

(593o) Evaluation of the Potential of Biosurfactant Production by Bacillus Strains in Different Culture Media

Darlane Wellen Freitas de Oliveira[1], Ítalo Waldimiro Lima França Sr.[1], Diana Pereira Bezerra[2], Vania Maria M. Melo[3] and Luciana R. B. Goncalves[1], (1)Chemical Engineering, Universidade Federal do Ceará, Fortaleza, Brazil, (2)Food Engineering, Universidade Federal do Ceará, Fortaleza, Brazil, (3)Department Biology, Universidade Federal do Ceará, Fortaleza, Brazil

Evaluation of the potential of biosurfactant production by *Bacillus* strains in different culture media

Darlane Wellen Freitas de Oliveira, Ítalo Waldimiro Lima de França, Diana Pereira Bezerra, Vânia Maria Maciel Melo, Luciana Rocha Barros Gonçalves*

Universidade Federal do Ceará, Department of Chemical Engineering, Av. Mister Hull, 2977, Block 709 – Campus do Pici – Zip-code 60.455-760, Fortaleza, Brazil

* *lrg@ufc.br*

Biosurfactants are surface-active molecules synthesized by microorganisms and have advantages as biodegradability, more ecological acceptance, greater structural diversity and low toxicity to the environment when compared to synthetic surfactants. The greatest challenge in producing biosurfactants is to develop a bioproduct economically compatible with the synthetic surfactan. In this study, the potential in producing biosurfactant by *Bacillus subtilis* Jag 248, a strain that was selected from the collection of bacteria belonging to the Laboratory of Microbiology and Immunology, Federal University of Ceará, was evaluated. Three different culture mediums were studied: clarified cashew apple juice, mineral medium and mineral medium supplemented with a solution of micronutrients. When *Bacillus subtilis* Jag 248 was grown in clarified cashew apple juice, surfactin and biomass concentrations around 27.0 $mg.L^{-1}$ and 1.0 $g.L^{-1}$ were achieved, respectively. On the other hand, when mineral medium was used, surfactin and biomass concentration was 161.0 $mg.L^{-1}$ and 1.5 $g.L^{-1}$, respectively. Furthermore, the supplementation of mineral medium with a solution of micronutrients did not enhance biomass concentration, that was around 1.3 $g.L^{-1}$. Regarding substrate consumption, a significant amount of residual carbon was observed when *B. subtilis* Jag 248 was cultivated in clarified cashew apple juice. On the contrary, the strain consumed the carbon source present in in mineral medium after 20 hours of culture. The surface tension of all mediums was reduced up to 29.0 $dyna.cm^{-1}$, which indicates biosurfactant production. The pH of all media studied was between 6.0 and 7.0. These results show that the strain evaluated in this work is a potential biosurfactant producer and further experiments will be performed in order to optimize production.

(593p) Cinnamyl Acetate Synthesis by Lipase-Catalyzed Transesterification in a Solvent-Free System

Mengfan Wang, Bo Geng, Wei Qi, Rongxin Su and Zhimin He, School of Chemical Engineering and Technology, Tianjin University, Tianjin, China

Flavor esters have various applications in food, perfume, flavor and pharmaceuticals since their aroma qualities. Recently, application of enzyme in the synthesis of esters has aroused widely interests for its high efficiency, moderate reaction condition and environmental friendly. The enzymatic synthesis of esters was mostly carried out through the esterification between alcohols and acids. However, it usually obtained low yield due to the severe deactivation of lipase activity under the high concentration of acid [G¨¹venç, 2002].The transesterification route using alcohols and esters as substrates could eliminate this problem. Vinyl esters are usually used as acyl donors in transesterification to obtain the higher conversion since the produced vinyl alcohol can tautomerize to acetaldehyde, which could promote the synthesis process. However, the co-product acetaldehyde has an unfavorable deactivation effect on lipase since it would bond to lysine residues by the formation of Schiff-base [Villeneuve, 2000]. Also, most of these lipase catalyzed reactions are carried out in the organic solvent such as n-hexane, heptane or toluene which are toxic and inflammable.

In this work, a commercial flavor ester, cinnamyl acetate, was synthesized by lipase (Novozym435) through transesterification route in a solvent-free system (Fig A). Ethyl acetate was chosen as the solvent as well as the acyl donor instead of vinyl acetate in order to avoid the deactivation of acetaldehyde on enzyme. 90.06% of conversion was achieved after 3 h when the transesterification was carried out at ethyl acetate/cinnamyl alcohol=15:1, 2.67 g L^{-1} of lipase loading and 40°C (Fig B). The excellent stability and reusability of enzyme resulted from the moderate reaction system. The immobilized lipase can be reused at least 6 cycles with few decline of activity. Kinetic studies indicated that the reaction abode by the Ping-Pong Bi-Bi mechanism with the inhibition of cinnamyl alcohol on lipase (Fig C) .The Michaelis constant for ethyl acetate and cinnamyl alcohol were calculated as 2.241 mmol L^{-1} and 206.81 mmol L^{-1}, respectively (Fig D). The inhibition constant of cinnamyl alcohol was 0.461 mmol L^{-1}, and the maximum reaction rate was 41.807 mmol L^{-1} min^{-1}.

(1) G¨¹venç, A., Kapucu, N. and Mehmetoğlu, Ü. (2002) The production of isoamyl acetate using immobilized lipases in a solvent-free system. Process. Biochem. 38, 379-386.

(2) Villeneuve, P., Muderhwa, J. M., Graille, J. and Haas, M. J. (2000) Customizing lipases for biocatalysis a survey of chemical, physical and molecular biological approaches. J. Mol.Catal.B-Enzym.9113-148.

This work was supported by the Program for New Century Excellent Talents in Chinese University (NCET-08-0386; NCET-11-0372), the 863 Program of China (2008AA10Z318), the Natural Science Foundation of China (20976125; 31071509; 51173128) and Tianjin (10JCYBJC05100), and the Program of Introducing Talents of Discipline to Universities of China (No. B06006).

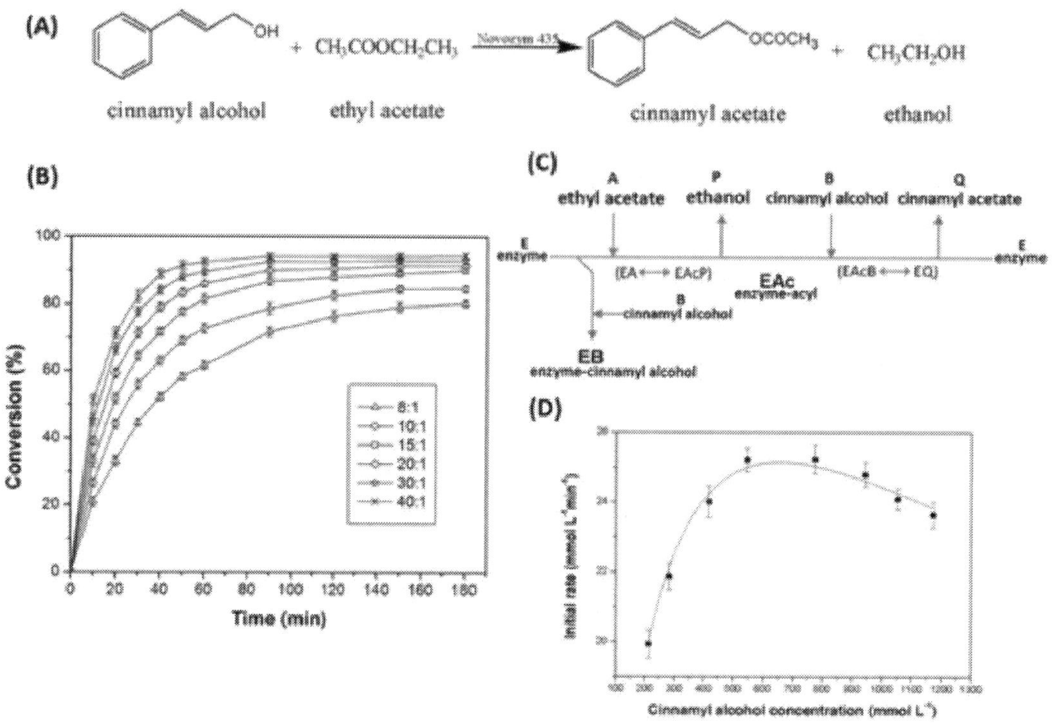

Figure(A)Reaction equation, (B) Reaction curve, (C) Transesterification mechanism, (D) Kinetic curve for synthesis of cinnamyl acetate

(593q) Systems Metabolic Engineering of Polyhydroxyalkanoates

Anu Raghunathan, Chemical Engineering, National Chemical Laboratory, Pune, India and Abhishek Gupta, National Chemical Laboratory, Pune, India

Polyhydroxyalkanoates (PHAs) form a class of natural polyesters that many organisms accumulate as intracellular granules to store carbon and reducing equivalents in response to specific environmental conditions. Thus, using microbial route to producing them may be an economically viable and sustainable alternative to current processes. Metabolic pathway analysis and flux balance analysis have been used to analyze the *in silico* biosynthesis of such polyesters. Using *bilevel optimization* we identify strategies for knocking in and knocking out genes for PHB production in common recombinant hosts- *E. coli* and Yeast that can be metabolically engineered in the laboratory. In E. coli, bilevel optimization returns only combination of 5 knockouts for increased production of PHA including several central metabolism genes. Concurrence with litearture related to microaerophilic conditions being better than highly aerobic conditions. We use in parallel, elementary mode analysis, to understand the dominant modes that will allow simultaneous production of biomass and biopolyester.

Using the previously published E.coli genome-scale model iAF1260, knocking out a speciifc set of genes results in a dramatic reduction of possible modes and also removes alternate by-product formation like succinate. These modes could help rationally design a pathway that could potentially be synthetically constructed in the laboratory. Results from the experiments and simulations will be discussed.

(593r) Investigation of Some Characters of Proteins Using Three Different Fractal Approaches

Xin Peng, Wei Qi, Rongxin Su and Zhimin He, School of Chemical Engineering and Technology, Tianjin University, Tianjin, China

Like other biological macromolecules such as polysaccharides and nucleic acids, proteins are essential parts of organisms and participate in virtually every process within cells. So that the researches of protein structure and function are extremely important. However, the spatial structures of proteins are so complicated and changeable that it is unrealistic to use the basic units (spheres, cubes and other regular shapes) to describe the complexity of an protein, i.e, protein molecules cannot be simply described in terms of the Euclidean geometry. Actually, protein has been described as a 'complex mesoscopic system' and characterized by self-similarity [Banerji, 2011]. On the other hand, fractal theory, as a very active mathematic branch of modern nonlinear science, has been used widely to describe irregular and non-differentiable geometric shapes existing in both natural world and man-made objects. So we can use the fractal method to characterize the complicated spatial and dynamical strctures of proteins.

In this paper we calculated the fractal dimensions of four proteins, chymotrypsin, elastase, trypsin and subtilisin, which are made up of about 220-275 amino acids and belong to the family of serine proteinase [Kraut, 1977; Hedstrom, 2002] by using three definitions of fractal dimension i.e. the chain fractal dimension (D_L), the mass fractal dimension (D_m) and the correlation fractal dimension (D_c). We analyzed the relationship between fractal dimension and space structure or secondary structure contents of proteins. The results showed that the more similar structures, the more equal fractal dimensions, and if the fractal dimensions of proteins are different from each other, the three dimensional structures should not be similar. Furthermore, we analyzed the differences between the fractal dimensions of complete proteins and that of their active sites. It is found that the values of fractal dimensions are almost same for the global mammalian enzymes (chymotrypsin, elastase and trypsin), but different for the global subtilisin. On the other hand, the detailed structures and fractal dimensions of the active sites of four enzymes are extraordinarily similar. So the multifractal theory is a very useful tool to solve this problem. To sum up, the fractal method can be applied to the elucidation of protein evolution and the fractal analysis can be used to depict some intrinsic characteristics of proteins. It is also demonstrated that proteins are a kind of fractal object with self-affinity and self-similarity.

(1) Banerji, A., Ghosh, I.. (2011) Fractal symmetry of protein interior: what have we learned? Cel. Mol. Life Sci. 68, 2711-2737.

(2) Kraut, J.. (1977) Serine Proteases: Structure and Mechanism of Catalysis. Annu. Rev. Biochem. 46, 331-358.

(3) Hedstrom, L.. (2002) Serine Protease Mechanism and Specificity. Chem. Rev. 102, 4501-4523.

This work was supported by the Program for New Century Excellent Talents in Chinese University (NCET-08-0386), the 863 Program of China (2008AA10Z318), the Natural Science Foundation of China (20976125; 31071509; 51173128) and Tianjin (10JCYBJC05100), and the Program of Introducing Talents of Discipline to Universities of China (No. B06006).

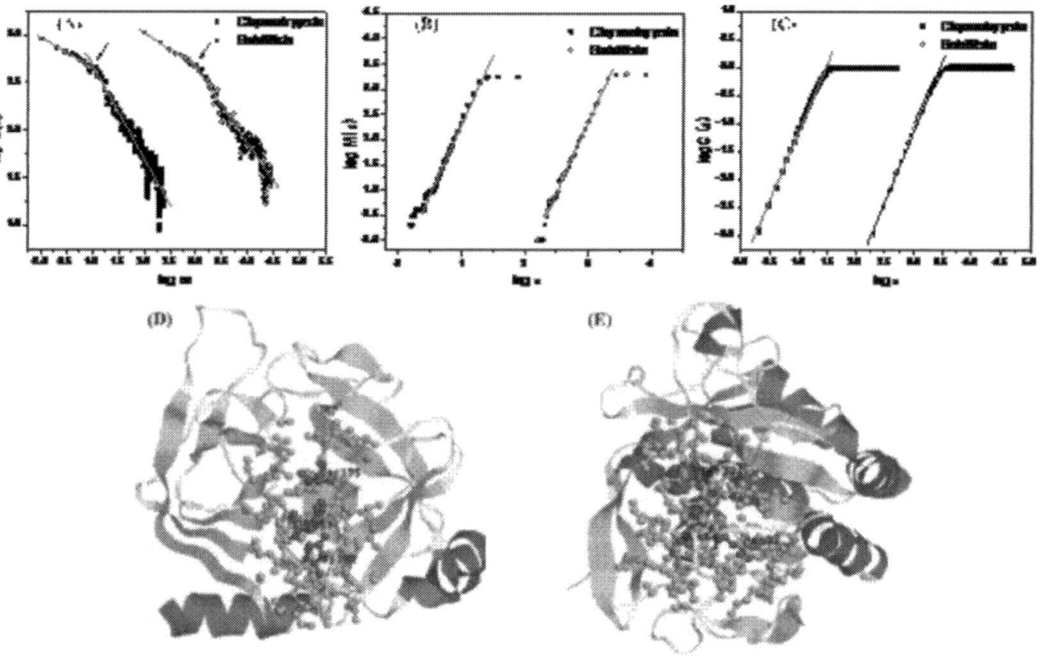

Figure (A) The chain fractal dimension, (B) The mass fractal dimension, (C) The correlation fractal dimension. The three dimensional structure of chymotrypsin (D) and subtilisin (E)

(593s) Deconstruction of Insulin Amyloid Fibrils Upon Laser Irradiation and Small Molecules and Probed by Time-Resolved Fluorescence Spectroscopy

Rui Liu[1], Renliang Huang[2], Rongxin Su[1], Wei Qi[1] and Zhimin He[1], (1)School of Chemical Engineering and Technology, Tianjin University, Tianjin, China, (2)School of Environmental Science and Engineering, Tianjin University, Tianjin, China

The accumulation of misfolded proteins as amyloid fibrils is the hallmark of a large number of human degenerative diseases [Gotz, 2009]. Recent evidence indicates that protein can induce photophysical changes to the amyloid indicator dye thioflavin T, indicating that the active sites of different amyloid proteins may be the same [Mishra, 2008; Arora et al, 2004]. The photoinhibition mechanism of different ThT analogues on amyloid proteins is therefore of great importance. Here, we targeted insulin fibril formation by small aromatic molecules, with different numbers of aromatic rings and various substituents, and studied the requisite probes for the identification of amyloid fibrils and the natural plant polyphenols for inhibition of protein aggregation in view of their aromatic structures.

The interactions between small aromatic molecules and insulin amyloid fibrils were monitered by Florescence Resonance Energy Transition (FRET) method. Steady-state fluorescence results indicate that insulin amyloid fibrils are acceptors and tyrosine is the active site, transferring energy to the around donors. The inhibition effects of small aromatic molecules on insulin fibrils were observed by laser irradiation, which were further confirmed by the time-resolved fluorescence studies. The docking results further indicated that congo red, ThT, and resveratrol showed high inhibitory effects on protein aggregation when conjugated with laser irradiation, whereas procyanidin and trehalose could not disrupt insulin amyloid fibrils. Our FRET results may provide an effective way to design potential inhibitors of amyloid fibrils.

This work was supported by the Natural Science Foundation of China (No. 20806057 and 31071509), the Ministry of Science and Technology of China (Nos. 2012BAD29B05 and 2012AA06A303), and the Ministry of Education (No. NCET-11-0372).

References

1. Gotz J.; Ittner L. M.; Lim Y. A. Common features between diabetes mellitus and Alzheimer's disease. *Cellular* and Molecular Life Sciences, 66: 1321-1325, 2009
2. Mishra R.; Bulic B.; Sellin D. Small-molecule inhibitors of islet amyloid polypeptide fibril formation. *Angewandte Chemie-International Edition*, 47: 4679-4682, 2008
3. Arora A.; Ha C.; Park C. B. Inhibition of insulin amyloid formation by small stress molecules. *FEBS Letters*, 564: 121-125, 2004

(593t) Characterization of Photosynthetic Proteins in Marine Algae

Jing Jiang[1], Hao Zhang[2], Robert Blankenship[2,3] and Cynthia S. Lo[1], (1)Energy, Environmental and Chemical Engineering, Washington University in St. Louis, St. Louis, MO, (2)Chemistry, Washington University in St. Louis, St. Louis, MO, (3)Biology, Washington University in St. Louis, St. Louis, MO

The water-soluble peridinin-chlorophyll *a*-protein (PCP) and the membrane bound intrinsic light-harvesting complex (acpPC) are major light harvesting complexes in photosynthetic dinoflagellates, which are a group of marine algae. PCP contains the carotenoid peridinin and chlorophyll *a*, while acpPC contains an additional pigment chlorophyll c_2. In this study, we identified and characterized the PCP and acpPC proteins in *Symbiodinium* sp. CS-156. The apoPCP has a molecular mass of 32.7 kDa. Despite the heterogeneity of PCP gene tandem repeats, we identified a single form of PCP, which was sequenced by LC-MS/MS analysis of tryptic digested PCP. Protein sequence comparison with the *Amphidinium* main form PCP (MFPCP) shows that the chl *a* and peridinin binding pockets are conserved in these two genera, which indicates the high degree of spectroscopic similarity and the same pigment stoichiometry; these hypotheses were confirmed by experiments. The spectroscopic characterization revealed that the peridinin-to-Chl *a* Q_y energy transfer efficiency is 95 % in this complex, which suggests that PCP may be promising for use in solar energy capture and eventual biofuel production in engineered devices.

(593u) A Purification Method for Thermostable Recombinant Carbonic Anhydrase Proteins Produced in E. Coli

Geoffrey Kleimeyer[1], Michael J. Coolbaugh[2], Iraj Ghazi[1] and David W. Wood[2], (1)The Ohio State University, Columbus, OH, (2)Chemical and Biomolecular Engineering, The Ohio State University, Columbus, OH

As technologies using thermostable carbonic anhydrases, an enzyme that catalyzes the reversible reaction of carbon dioxide and water to form carbonic acid, are being developed, there is a growing need for an economical, efficient and scalable method for purifying recombinant thermostable carbonic anhydrases from E. coli. We have been developing such a method that purifies thermostable carbonic anhydrases in high yield using only heat and ammonium sulfate. We have used this purification method to produce variants of the carbonic anhydrase from *Methanosarcina thermophilia* with tags to be used in technologies that require the immobilization of the enzyme. Currently we are investigating the effects of various types of covalent immobilization on the activity and stability of these enzymes. The applications of immobilized carbonic anhydrases include their use in carbon capture and storage technologies. The development of a large-scale purification method for these enzymes is critical for the viability of these technologies.

(593v) Adiponectin Levels Modulate Recovery of Renal Function in a Novel Model of Podocyte Ablation

Joseph M. Rutkowski and Philipp E. Scherer, Touchstone Diabetes Center - Department of Internal Medicine, UT Southwestern Medical Center, Dallas, TX

Objective: Adiponectin is altered in numerous diseases with renal pathologies: low levels are associated with the metabolic syndrome and Type II diabetes while heightened levels are often concurrent with chronic kidney disease and Type I diabetes. To explore the effects of adiponectin on kidney disease progression, we have created a novel murine model of podocyte apoptosis using a unique *POD*ocin promoter driven cassette for *A*poptosis *T*hrough *T*argeted *A*ctivation of *C*aspase 8 ("*POD-ATTAC*").

Methods and Results: The podocyte-expressed ATTAC fusion protein contains a mutant FKBP domain that binds specifically to the low-molecular weight AP20187 "dimerizer". Dimerization of the ATTAC fusion protein induces caspase-8 activation resulting in rapid podocyte-specific apoptosis. Podocyte loss is the first step in many clinically-important nephropathies. The number of podocytes ablated, overall renal functional insult, and kidney disease progression are highly dimerizer dose-dependent providing a titratable model: dosing permits POD-ATTAC mice to mimic many levels of human nephropathies and potentially restore podocyte number and function. POD-ATTAC mice lacking adiponectin develop irreversible albuminuria and renal failure over time. Adiponectin over-expressing POD-ATTAC mice, conversely, demonstrate rapid recovery and exhibit limited interstitial fibrosis despite similar numbers of podocytes ablated. Unlike overexpression, moderately raising serum adiponectin levels with an agonist for the nuclear receptor PPARγ fails to confer sufficient renal protection or recovery potential.

Conclusions: The novel POD-ATTAC mouse model provides an excellent platform for further studies of precisely timed podocyte injury, kidney disease progression, and functional regeneration. Our data also highlights the powerful protective role of the adipokine adiponectin in cardio-renal physiology.

Acknowledgement: JMR was supported in part by NIH F32-DK085935

(593w) Improve Acetyl-CoA Level in Baker's Yeast

Yarong Gao, Liang Wang, Jianping Wang, Wenshan Liu and **Rongrong Jiang**, School of Chemical and Biomedical Engineering, Nanyang Technological University, Singapore, Singapore

Acetyl CoA is an important precursor for many valuable compounds, and participates in the biosynthesis of fatty acids and many amino acids. Acetyl-CoA can be generated from pyruvate or acetate. In this project, we aim to examine the over-expression effect of several upstream rate-controlling enzymes on acetyl-CoA synthesis in *Saccharomyces cerevisiae.*

In yeast, cytosolic pyruvate is converted into cytosolic acetyl-CoA with the help of pyruvate decarboxylase, cytosolic acetaldehyde dehydrogenase, and acetyl-CoA synthase. In order to increase the acetyl-CoA level in yeast, we first tried to enhance the CoA level by over-expressing pantothenate kinase (Pank), a key regulatory enzyme in CoA biosynthesis in both prokaryotes and eukaryotes. Acetyl-coenzyme A synthetase (ACS) can catalyze the formation of acetyl-CoA from acetate and acetaldehyde dehydrogenase (ALD) is able to enhance the carbon flux into pyruvate dehydrogenase. Hence, we have over-expressed PanK, ACS, and ALD to increase the acetyl-CoA level in yeast. We have identified the gene sequences from the yeast strain, amplified them with PCR, and cloned them to 2-micron yeast expression plasmids with the fusion of GAL1 promoter and terminator. The recombinant plasmids were transferred into *Saccharomyces cerevisiae* BY4742. The cell lysate samples from the engineered yeast strains were analyzed by HPLC. We found that the amount of acetyl–CoA amount had been increased by two fold with the overexpression of PanK, compared to the control, but not with the overexpression of ACS alone.

(593x) Surfactant Improved Fatty Acids Production From Babassu Oil Using Rizomuchor Miehei Lipase

José Cleiton Sousa dos Santos, Leonardo José Brandão Lima de Matos, Maria Cristiane Martins de Souza, Hosiberto B. Sant'ana Sr. and Luciana R. B. Gonçalves, Departament of Chemical Engineering, Universidade Federal do Ceará, Fortaleza, Brazil

Enzymes are attractive for applications in biochemical and industrial fields because of its specificity and its ability to catalyze reactions. Lipases (glycerol ester hydrolases) are present in detergent formulations, the oleochemical and pharmaceutical processing, as well as catalyze the synthesis of biosurfactants, dairy and agrochemical industries, and the hydrolysis of triglycerides. The hydrolysis of oils is the natural function of lipases. Hydrolysis of triglycerides is an essential reaction for the production of fatty acids or special glycerides. The main advantage of lipase-catalyzed hydrolysis over conventional processes is the ability that enzymes have to perform these reactions at room temperature and pressure. In order to enhance conversion in lipase catalyzed hydrolysis, surfactants can be utilized. Therefore, in this study, fatty acids were produced from babassu oil (*Attalea speciosa*) using *Rizomuchor miehei* lipase (RML) in the presence of different surfactants: SDS (anionic surfactant), CTAB (cationic surfactant), Span 20, Triton X-100, Igepal, Brij 30 and Tween 60. The concentration of surfactant used was up the critical micelle concentration (CMC). The hydrolysis reaction was carried out in thermostated batch reactors, under stirring, using 10 mL of reactional media with 1%v/v lipase concentration, 55% v/v oil concentration and distilled water, at 37°C, 200 rpm. The acidity was determined by titration of free fatty acids (FFAs) with 0.25N NaOH and the result was expressed in grams of FFAs per 100 g of the sample. The surfactant that enhanced the formation of FFAs was chosen as the most suitable surfactant. Preliminary results in the presence of surfactant and absence free lipase showed a maximum value for free fatty acids of 1.5 ± 0.1 mgKOH/g. When RML catalyzed the hydrolysis, FFA was 50.89 ± 2.5 mgKOH/g and 28.87 ± 1.7 mgKOH/g, respectively, in the presence nonionic detergent TWIN and SPAN. When anionic/cationic surfactants (SDS) were used together with RML, the amount of free fatty acids in the samples low, 0.61 ± 0.0 mgKOH/g and 0.45 ± 0.1 mgKOH/g, respectively, in the presence of CTAB (cationic) or SDS (anionic).

(593y) Production, Purification and Characterization of b-Mannanase From *Bacillus Subtilis* TJ-101 and Its Application in the Preparation of Gluco-Mannooligosaccharides

Shuaishuai Zhang, Mengfan Wang, Wei Qi, Zhaohui Liu, Weina Wu, Rongxin Su and Zhimin He, School of Chemical Engineering and Technology, Tianjin University, Tianjin, China

b-mannanase (b-1,4-D-mannanmannohydrolase, EC 3.2.1.78) is an important hemicellulase, which can catalyze the hydrolysis of internal b-1,4-D-mannopyranosyl linkages in various b-mannan such as galactomannan, glucomannan, galactoglucomannan. b-mannanase has been widely used in the food, feed, paper and oil industries. In the food industry, b-mannanase is an excellent biocatalyst for the production of gluco-mannooligosaccharides (GMOS). Several studies have verified that GMOS with the degree of polymerization (DP) of 2~6 have the outstanding bioactivities on human being. For example, GMOS could increase the growth of intestinal microorganisms, decrease pathogenic bacteria and improve the integrity of intestinal mucosa [Zhang et al., 2009]. Among various microorganisms being reported as b-mannanase producers, *Bacillus subtilis* is recommended because of its safety, fast growth and the high secretion ability of b-mannanase [Jiang et al., 2006]. b-mannanase of *Bacillus subtilis* is an inducible and extracellular enzyme whose activity and production is greatly influenced by media components, such as carbon sources, nitrogen sources and metal ions.

In this work, strain TJ-101 was firstly identified by 16S rDNA as *Bacillus subtilis* (Fig. A). The response surface method was applied to improving and enhancing the enzyme production. The optimized media components were obtained (g/L): konjac 34.68, $(NH_4)_2SO_4$ 4.69, Na_2HPO_4 $12H_2O$ 4.00, KH_2PO_4 0.30, $CaCO_3$ 2.69, $CaCl_2$ 1.00, $NaCO_3$ 1.00, $MgCl_2$ $6H_2O$ 0.60 (Fig. B). Under these conditions, the maximum enzyme activity increased 27.68% from 120.42U/mL to 153.75U/mL. Then, b-mannanase was produced in a 7-L fermentor and 7.39-fold purified through the salting out, ultrafiltration, anion-exchange and size-exclusion preparative chromatography with a recovery of 21.41% and a specificity of 125.36U/mg proteins. b-mannanase was optimally active at pH 5.0-8.0 and 60°C (Fig. C) and its molecular weight was 38kDa determined by SDS-PAGE (Fig. D). The enzyme activity was slightly stimulated by some metal ions, such as Fe^{3+}, Cu^{2+} and Cr^{2+}. The purified b-mannanase showed excellent catalysis efficiency in hydrolysis of konjac flour to prepare GMOS. The GMOS yield of 57.76% has been achieved with 8.71% of mannose and 14.49% of glucose (Fig. E), demonstrating the potential use of b-mannanase in food industry.

(1) Jiang, Z., Wei, Y., Li, D., et al. High-level production, purification and characterization of a thermostable beta-mannanase from the newly isolated *Bacillus subtilis* WY34, Carbohydrate Polymers, 66, 88-96, 2006.

(2) Zhang, Y.Z., Zhang, M., Chen, X.L., et al. Purification and functional characterization of endo-beta-mannanase MAN5 and its application in oligosaccharide production from konjac flour, Applied Microbiology and Biotechnology, 83, 865-873, 2009.

This work was supported by the Program for New Century Excellent Talents in Chinese University (NCET-08-0386), the 863 Program of China (2008AA10Z318), the Natural Science Foundation of China (20976125; 31071509; 51173128) and Tianjin (10JCYBJC05100), and the Program of Introducing Talents of Discipline to Universities of China (No. B06006).

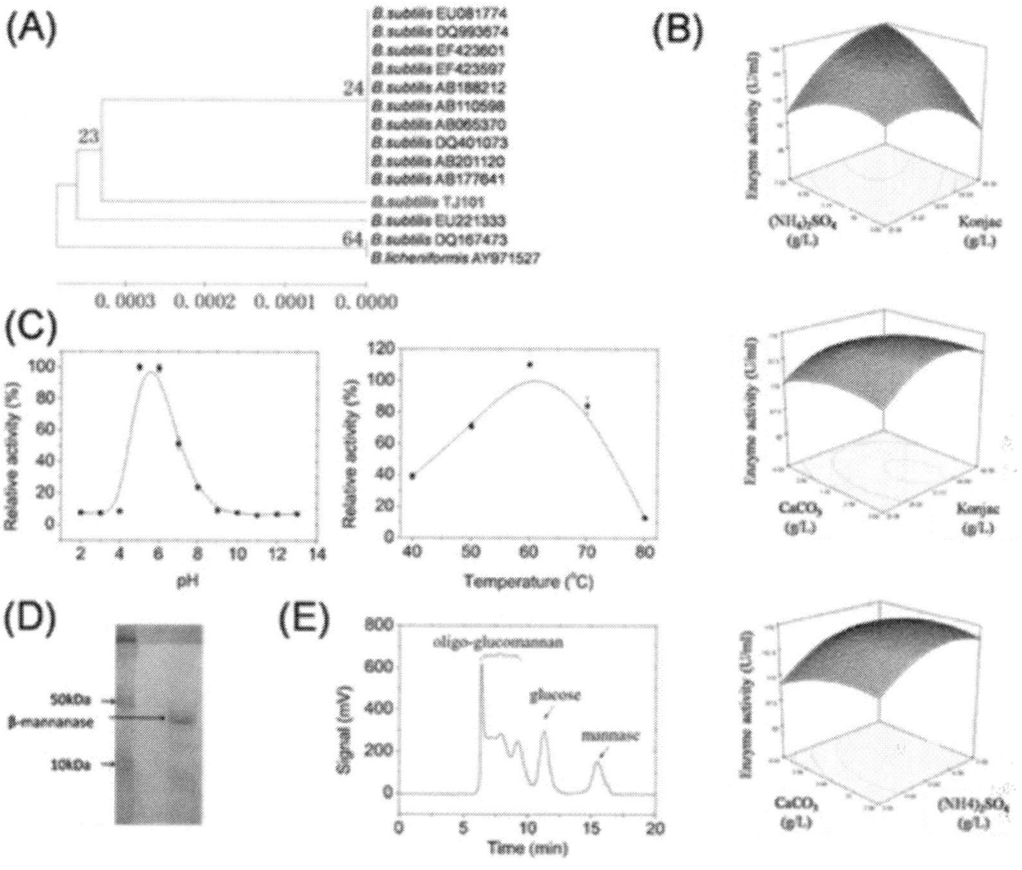

Figure (A) Phylogenetic tree of *Bacillus subtilis* TJ-101, (B) 3D surface plots of RSM, (C) Optimum pH and temperature of b-mannanase, (D) SDS¨CPAGE of b-mannanase, (E) HPLC of gluco-mannooligosaccharides.

(593z) Novel Endogenous Molecular Biosensors From Engineered Regulatory Proteins AraC and TetR

Christopher S. Frei, Joseph A. Gredell and Patrick C. Cirino, Chemical & Biomolecular Engineering, University of Houston, Houston, TX

NOVEL ENDOGENOUS MOLECULAR BIOSENSORS FROM ENGINEERED REGULATORY PROTEINS ARAC AND TETR

Christopher Frei, Joseph Gredell, Shuang-Yan Tang, Patrick C. Cirino

University of Houston, S294 Engineering Bldg 1, Houston, TX, 77204-4004, USA
T: 832-842-6306, cfrei@uh.edu

Endogenous molecular biosensors for specific, sensitive, and high-throughput detection/monitoring of small molecules are emerging as invaluable tools in metabolic engineering and synthetic biology. Regulatory proteins controlled by "effector" molecules naturally couple molecular recognition to changes in gene expression, providing a platform for endogenous molecular reporters by linking *in vivo* molecular synthesis to a readily detectable phenotype (e.g. GFP expression). Numerous advances have been made in the detection of endogenous small molecules through whole-cell regulatory protein-based biosensors [1], but limitations arise due to the narrow range of natural effectors recognized by regulatory proteins. We are developing customized molecular biosensors by engineering regulatory protein effector recognition through directed evolution techniques. In general, these customized molecular biosensors are applicable to high-throughput screening of combinatorial enzyme libraries to improve enzymatic or microbial production of metabolites.

For these purposes, our initial focus was the well-studied L-arabinose-responsive AraC dual regulatory protein from *E. coli*. Saturation mutagenesis libraries, targeting up to 5 residues (~10^7 mutants), were generated based on analysis of the AraC ligand binding pocket. Using a pBAD-GFP construct expressed in *E. coli* and FACS, AraC variants responsive to D-arabinose, mevalonate, and triacetic acid lactone (TAL) were isolated [2-4]. The mevalonate and TAL reporter systems were subsequently used to screen for improved production of these compounds from *E. coli* expressed pathway-specific mutant libraries. Following the success of the previously engineered AraC-based biosensors, further AraC libraries were screened for a response to p-coumaric acid, trans-cinnamic acid, and vanillin. Continued engineering of AraC effector specificity is now aimed at exploring the range of ligands accessible with this motif as well as improving the understanding of the molecular determinants of ligand binding and regulatory switching.

We next sought to engineer reporter systems for considerably larger and more complex natural products such as polyketides. The regulatory protein TetR from the Tn10 transposon was selected due to its well-characterized functions and ability to naturally recognize large polycyclic molecules (i.e. tetracycline). Combinatorial mutant libraries of TetR were created through restricted saturation mutagenesis. Due to the sophisticated hydrogen bonding network found within the ligand binding domain of TetR and the limitations of FACS, our selection of specific target residues was critical in our library design. Restricted saturation mutagenesis allowed us to scan more residues per library while maintaining potential cooperative interactions and a workable library size (~10^7). The TetR mutants were expressed in *E. coli* cells containing a TetR-regulated GFP reporter construct. The libraries are being screened for specificity toward various polyketide compounds. Improved understanding of effector recognition will help optimize the mutational composition of various libraries generated, resulting in more rapid screening and isolation of mutants responsive to compounds of interest. These customized reporters will prove valuable in engineering heterologous polyketide biosynthesis pathways.

1. Gredell, J.A., C.S. Frei, and P.C. Cirino, *Protein and RNA engineering to customize microbial molecular reporting.* Biotechnology Journal, 2012. **7**(4): p. 477-499.

2. Tang, S.Y., H. Fazelinia, and P.C. Cirino, *AraC regulatory protein mutants with altered effector specificity.* Journal of the American Chemical Society, 2008. **130**(15): p. 5267-5271.

3. Tang, S.-Y., O. Akinterinwa, and P.C. Cirino, *A novel endogenous reporter of triacetic acid lactone (TAL) enables screening for improved TAL production by E. coli.* submitted.

4. Tang, S.-Y. and P.C. Cirino, *Design and Application of a Mevalonate-Responsive Regulatory Protein.* Angewandte Chemie International Edition, 2011. **50**(5): p. 1084-1086.

ADDIN EN.REFLIST

(594a) Ultrasonic Depolymerization of Schizophyllan

Juan Zeng, College of Light Industry and Food Sciences, South China University of Technology, Guangzhou, China, Huina Li, College of Light Industry and Food Sciences, Sourh China University of Technology, Guangzhou, China and Bisheng Zheng, Collegea of Light Industry and Food Sciences, South Chia University of Technology, Guangzhou, China

Ultrasonic depolymerization of Schizophyllan

Juan Zeng Huina Li Bisheng Zheng*

College of Light Industry and Food Sciences, South China University of Technology, Guangzhou, China 510640

Abstract.

Schizophyllan (SPG), a polysaccharide extracted from *Schizophyllum commune Fr.*, has antitumor and immune activities. However, its physical properties, such as large molecular weight, high viscosity and little solubility in water, make *Schizophyllan*not be absorbed easily by human body, and not easy to process. In this research, ultrasonic method was applied to depolymerize the Schizophyllan.

Schizophyllan was depolymerized by ultrasonication, and changes in molecular weight and its distribution were observed by gel permeation chromatograghy (GPC). The influences of sonication time, power, temperature, initial SPG concentration were investigated systematically. As the ultrasonication time prolonged, the average molecular weight of ultradonified Schizophyllan (us-SPG) decreased. (Fig. 1). Moreover, when the ultrasonication time was less than 200 minutes, the molecular weight of Schizophyllan decreased fastly, but when the ultrasonication time was more than 200 minutes, the molecular weight of Schizophyllan decreased much more slowly (Fig 3). After SPG was ultrasonic treated for 200 minutes, the molecular was decreased from about 6×10^6 to 1.5×10^5Da. The polydispersity of ultrasonified Schizophyllan were between 1.5 and 2.5. The polydispersity randomly changes when the ultrasonication time was less than 200 minutes, but when the ultrasonication time was more than 200 minutes, the polydispersity almost remains 1.7(Fig.3).

* corresponding author

(594b) An Effective and Green Method for the Extraction and Purification of Aglycone Isoflavones From Soybean

Jian Guo, Mengfan Wang, Wei Qi, Rongxin Su and Zhimin He, School of Chemical Engineering and Technology, Tianjin University, Tianjin, China

Isoflavones, belonging to phytoestrogen and with significant amounts in soybean [Elizabeth JZ 2007], have obvious effects in the prevention and therapy of chronic diseases such as heart diseases, osteoporosis and cancers (prostrate, breast, thyroid and colon) [Cohen LA 2000]. Especially the aglycone isoflavones (genistein and daidzein) exhibit higher biological activity, so the production of aglycone isoflavones has been attracting more and more attention in recent years. However, the high energy consumption, poor efficiency, the use of toxic solvents are the problems in the extraction and purification of aglycone isoflavones.

In this paper, an effective and green strategy was developed to obtain aglycone isoflavones from soybean through the extraction, enzymatic hydrolysis and purification steps (Fig A). In the extraction step, a novel ethanol-alkaline extraction method was designed to promote the conversion of conjugated forms (acetyl glycoside isoflavones and malonyl glycoside isoflavones) into their non-conjugate forms (glycoside isoflavones) [Ismail B 2005]. A high extraction yield (94.34%) of isoflavones was achieved at the optimum conditions (pH 9.0, 65% ethanol, 70°C, solid to liquid ratio (m:v) 1:15, 1h and 400rpm stirring). Then, in the enzymatic hydrolysis step, cellulase (GC-220) instead of traditional b-glucosidase was used to convert the glycoside isoflavones into aglycone isoflavones with an excellent conversion of 94.80%. Finally, the crude aglycone isoflavones were purified by an environmental-friendly procedure comprised of ethanol precipitation, ethyl acetate extraction and water rinse. The final amount of aglycone isoflavones was 1.279mg from every gram of defatted soybean, and the final recovery and purity of aglycone isoflavones calculated based on HPLC (Fig B), could achieve 80.38% and 87.21%, respectively. Additionally, the organic solvent of ethanol and acetic ether used in the whole process could be recycled with the recoveries of 87.60% and 90.00%. The whole extraction, enzymatic hydrolysis and purification process is an environment-friendly, economical and simple process, which will increase the yield of aglycone isoflavones and make the production safe and green.

(1) Elizabeth JZ, Ka MN and Kathy QL. Extraction and purification of isoflavones from soybeans and characterization of their estrogenic activities. J. Agr. Food Chem 55:6940-6950 (2007).

(2) Cohen LA, Zhao Z, Pittman B and Scimeca JA. Effect of intact and isoflavone-depleted soyprotein on NMU-induced rat mammary tumorigenesis. Carcinogenesis 21:929-935 (2000).

(3) Ismail B and Hayes K. b-Glycosidase activity toward different glycosidic forms of isoflavones. J. Agr. Food Chem 53:4918-4924 (2005).

This work was supported by the Program for New Century Excellent Talents in Chinese University (NCET-08-0386; NCET-11-0372), the 863 Program of China (2008AA10Z318), the Natural Science Foundation of China (20976125; 31071509; 51173128) and Tianjin (10JCYBJC05100), and the Program of Introducing Talents of Discipline to Universities of China (No. B06006).

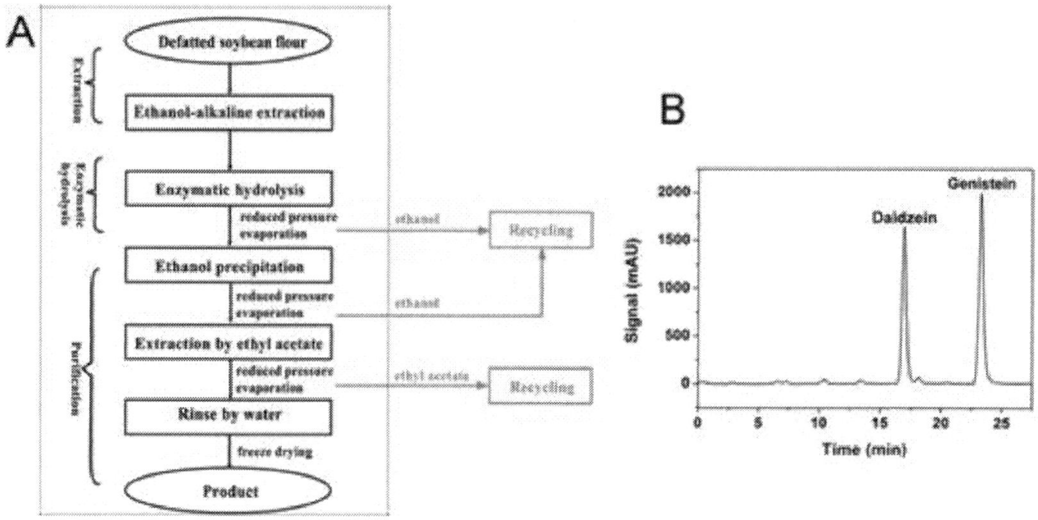

Figure (A) Extraction, hydorlysis and purification of aglycone isoflavones. (B) The HPLC chromatogram of final products

(594c) Effect of Cooling-Induced Blooming On Hardness of Milk Chocolate

Karen M. Tschinkel and Eric C. Huang, Chemical Engineering, Manhattan College, Riverdale, NY

Milk chocolate is a common household sweet produced on a large production scale. Precise control of temperature during production can affect material properties tied to product quality. One such material property is hardness, which is dependent on crystal structure and, therefore, affected by varying cooling rates.

Hershey's Milk Chocolate bars were melted in a double boiler and allowed to transition back to a solid phase either at room temperature (~20°C) or in a refrigerator at max setting (<5°C). Both sample sets were then stored at room temperature and tested for hardness over the course of thirty hours. Results indicated a consistent difference in Brinnell hardness numbers between refrigerated and ambient-cooled samples ($p<0.02$). The samples demonstrated a consistent increase in hardness over time that correlated with visual observation of fat bloom. Hardness of the refrigerated samples also showed a negative correlation with temperature, which is consistent with previous results found in literature.

SEMI-INDUSTRIAL PRODUCTION OF PROBIOTIC/BIOTHERAPEUTIC YEAST *SACCHAROMYCES BOULARDII* IN HIGH CELL DENSITY CULTURE USING DIFFERENT CULTIVATION STRATEGIES

Malek, R.A[1*], Ishak, A.F[1], El Sayed, S[2], El Badry, I[2], Othman, N.Z[1], Sarmidi, M.R[1], Aziz, R.A[1], El Enshasy, H.A[1,2]

[1]Institute of Bioproduct Development (IBD), Universiti Teknologi Malaysia (UTM), 81310 Skudai, Johor, Malaysia
[2] Bioprocess Development Department, City for Scientific Research and Technology Applications (CSAT), New Burg Al Arab, Alexandria, Egypt

Introduction

The term of "Probiotics" or "biotherapeutic" agent has been defined as a live non pathogenic microbial feed of food supplement which beneficially affects the host by improving its intestinal microbial balance (1). The meaning of "biotherapeutic agents" has been extended recently to human health as "living microorganism used to prevent human diseases by interacting with the natural microecology of the host (2,3). Probiotic was proven and effectives against a range of gastrointestinal diseases and disorders (4,5,6,7). *Saccharomyces boulardii*, is yeast which is isolated from lychee and mangosteen fruit and grows optimally at body temperature of 37°C (8). *S. boulardii*, is non-pathogenic yeast which has been used as a preventive and therapeutic agent for the treatment of a variety of diarrhea disease (9, 10, 11,12,13). Since *S. boulardii*, has many applications in medical areas, therefore *S. boulardii* is especially useful for the large scale production of probiotic. There are numerous literatures reviewed on the importance of *S. boulardii* and its benefits in medical areas especially in treatment and prevention of gastrointestinal diseases, however there are not many literatures published on scale up production of *S. boulardii* to semi industrial .Thus high cell mass production of this type of yeast as high value biotherapeutic/probiotic product was attractive topic of many researchers nowadays. For industrial propagation, probiotic cell should be produced in high cell density culture to improve process economics. In the present work, cultivation of *S. boulardii* was carried out using different cultivation strategies to achieve high cell density cultivation with minimal alcohol production.

Materials and method

Microorganism

The yeast strain *Saccharomyces boulardii* ATCC-MYA-796 obtained from (American Type Culture Collection, Manassas, VA, USA) was used in this study. This strain was adapted to dryness by successive adaptation method. Cells were grew on solid medium and allowed to dry on agar plate. This strain was cultivated in Yeast-peptone-dextrose (YPD) agar After complete drying of culture, broth medium was added and the cells grew thereafter were isolated for further growth on solid medium. This process was repeated for more than 8 times.

Master cell bank.

For cell propagation on solid medium, yeast-peptone-dextrose (YPD) medium was used. This medium composed of (g l^{-1}): glucose, 10; yeast extract, 3; peptone, 3 and agar, 20. The pH was adjusted to 5.5 before sterilization.

Production medium

The first medium used in this study was complex (YMG) medium. This medium was composed of (g L^{-1}): Malt extract, 10.0; yeast extract, 4.0; glucose, 4.0. The pH was adjusted to 5.5 before sterilization. Unless otherwise mentioned, the yeast production medium applied was complete synthetic medium (CSM). This medium was composed of (g L^{-1}): Glucose, 15.0; $(NH_4)_2SO_4$, 5.0; KH_2PO_4, 3.0.; $MgSO_4.7H_2O$, 1.0. The pH was adjusted to pH 5.5 before autoclaving. Glucose was sterilized separately by autoclaving at 110 °C and added to the medium aseptically before inoculation

Cultivation conditions

Shake flask cultivation and inoculums preparation

For shake flask cultures, 250 ml Erleymyer flasks with 50 ml working volume were used. After inoculation, flasks were incubated on rotary shaker (Innova 4080, New Brunswick, NJ, USA) at 200 rpm and 37°C for 24 h for biomass production. For bioreactor cultures, cultivations were run in pilot scale bioreactors of 16 L with working volume of 8 liters (BioEngineering, Wald, Switzerland). Cultivation was conducted at 400 rpm, 37°C and pH was controlled at 5.5 by addition of 2.5 M $(NH_4)OH$.

Bioreactor cultivation

For bioreactor experiments, cultivations were carried out in 3 L stirred tank bioreactor, Bioflow 3000 (New Brunswick Scientific Co., New Brunswick, NJ, USA), with a working volume of 2.0 L. Agitation was performed using a three 4-bladed rushton turbine impellers ($d_{i(impeller\ diameter)}$ = 65 mm; $d_{t(tank\ diameter)}$ = 135 mm, $d_i d_t^{-1}$ = 0.48). Different agitation speeds of 200, 400 and 600 rpm were used. Aeration was performed by filtered sterile air [1 v v^{-1} min^{-1}]. Dissolved oxygen concentrations were analyzed by polarographic electrode (Ingold, Switzerland)

. Foam was suppressed, when necessary, by the addition of silicon antifoam reagent (Fluka, Switzerland). During fed-batch culture, pH was controlled at 5.0 by addition of 2.5 mol l^{-1} NaOH. Addition of 200 ml of concentrated sterile solution composed of sugar cane molasses 120 ml and yeast extract 10 g was carried out after 20 and 40 hours.

Analytical Method

Sample Preparation

For both shake flasks and bioreactor, samples were taken at different time intervals. Every sample was prepared in a proper method in order to get accurate answers.

Cell dry weight determination

For both shake flasks and bioreactors, samples (in form of 20 ml) were taken at different time intervals and the optical density was measured spectrophotometrically against un-inoculated medium at 600 nm using single beam spectrophotometer (HACH, Loveland, CO.). The OD of culture was converted to dry cell mass through pre-determined standard curve between the OD value and cell dry weight.

Glucose and ethanol determination

Both of glucose and ethanol were determined by HPLC method. The falcon tube filled with 10.0 ml sample was being centrifuged using centrifuge machine (Hettich, ZENTRIFUGEN) at 3000rpm for 10 minute. After the centrifugation, the clear supernatant was kept in a new universal bottle for ethanol determination. Percentage of ethanol produced was determined using Gas Chromatography (GC).

Results and Discussions

Kinetics of cell growth and alcohol production by S. boulardii

For better understanding on the kinetics of cell growth and alcohol production by *S. boulardii*, cells were cultivated in batch culture in shake flask levels. Cell growth, alcohol production and change in culture pH were followed hourly. As shown in figure 1, cells grew exponentially during the first 10 hours cultivation reaching cell mass of 2 g L^{-1}. Cells grew thereafter with very low rate until the end of cultivations.

On the other hand, alcohol was accumulated in culture throughout cultivation time and the maximal alcohol production of about 0.28% was obtained after 24 h. Thus we can conclude that, the growth was limited after 10 h cultivations and the rest of cultivation time was used for alcohol production.

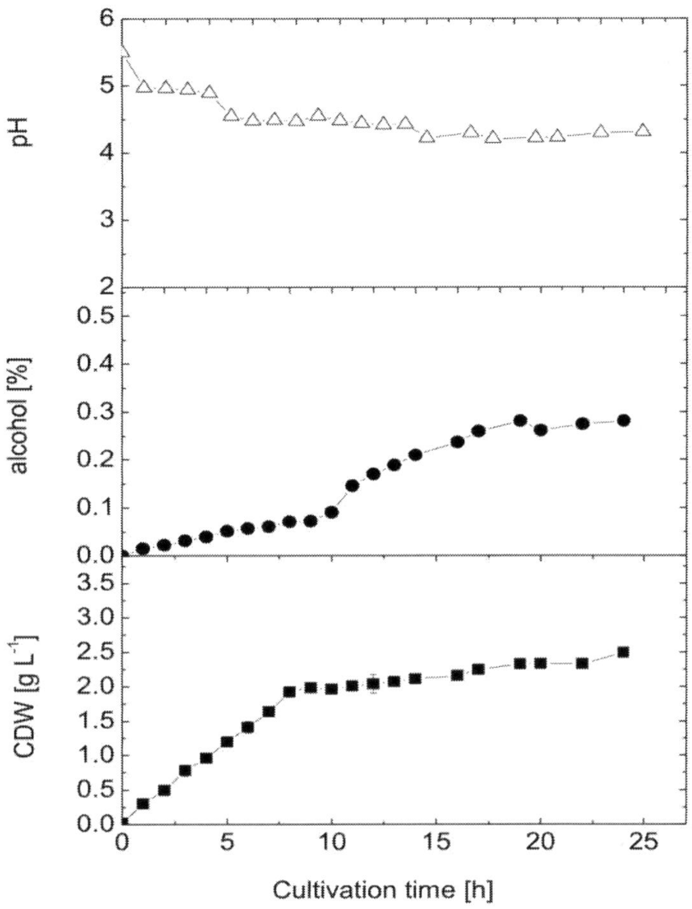

Figure 1: Kinetics of cell growth and alcohol production by *S. boulardii*

Fed-batch cultivations of cells in high cell density culture in 16-L Stirred Tank pilot scale Bioreactor (STR)

Based on the new medium formula, cultivations were conducted in pilot scale *in situ* sterilizable pilot scale bioreactor for study the scalability and industrialization of the production process. At first, batch cultivation was conducted to study the kinetics of growth and the effect of well mixed environment creased by high aeration/agitation in STR. The glucose consumption rate in batch culture was about 2 g L^{-1} h^{-1} and glucose was almost consumed at the end of exponential phase (data not shown).

Based on this, fed-batch cultivation strategy was designed with increased feeding strategy of glucose to support cell growth and to minimize catabolite repression and alcohol formation during cell cultivation as well. Thus, glucose was fed with the same consumption rate to prevent substrate limitation. However, the glucose feeding rate was increased by 2 g L^{-1} h^{-1} every 5 hours to compensate the increased volumetric consumption rate of glucose as a result of further cell growth.

Figure 2 shows clearly the significant improvement in cell mass production in fed-batch cultivation in STR. The maximal cell dry mass of about 70 g L^{-1} was obtained in fed-batch culture after 35 h. This amount of cell produced was almost 23 times of those obtained in shake flasks using the same medium. It has been also observed that, oxygen was not limited during the fed-batch cultivation.

The DO was kept in range between 15-20% saturation by stepwise increase of both aeration and agitation. This was done by automatic cascading system of the bioreactor control unit. On the other hand, nitrogen was also not limited during the cultivation since ammonium hydroxide was used to keep the pH at 5.5 during cultivation.

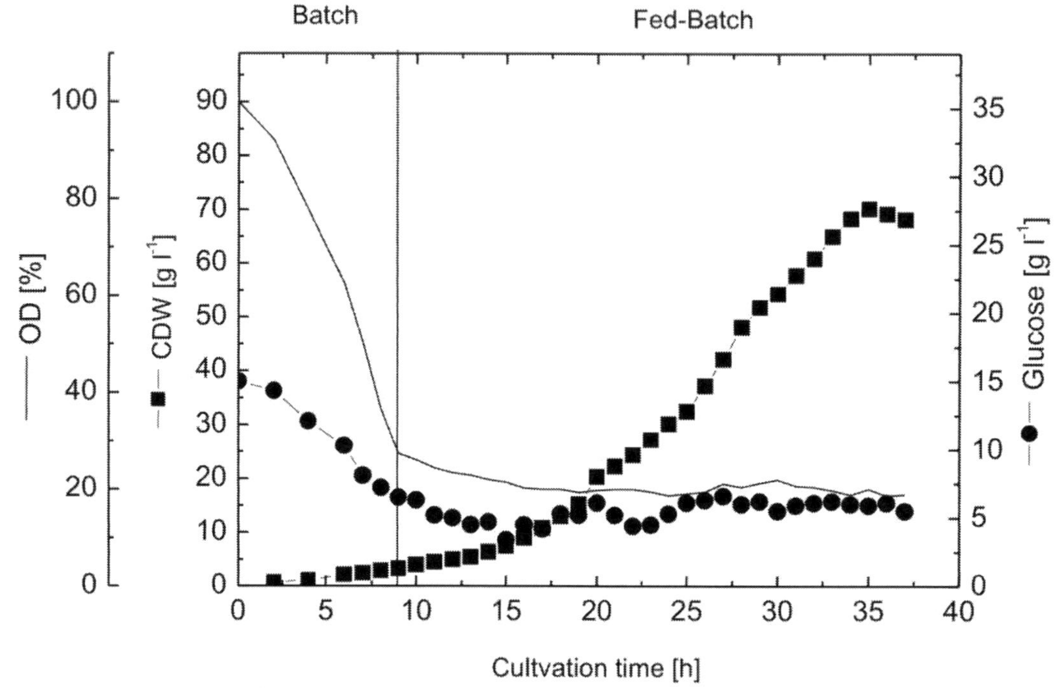

Figure 2: Kinetics of cell growth, glucose consumption and other cultivation parameters during *S. boulardii* cultivation in 16-L pilot scale STR in fed-batch

Conclusion

In the present work new industrial medium formula was developed for high cell density cultivation of biotherapeutic/probiotic yeast *S. boulardii*. Furthermore, cultivations were successfully carried out in pilot scale STR in both batch and fed-batch for high cell density production. Glucose feeding with increased rate combined with the use of bioreactor control system to cascade between stirrer speed and airflow was find to be good cultivation strategy for yeast high cell mass production.

Acknowledgement

The authors would like to thanks the MOHE and Universiti Teknologi Malaysia (UTM) for GUP grant no Q.J.130000.7125.01H26 that have been supported this research

Reference

1. Fuller, R (1991). Probiotics in human medicine. Gut.32:439-442
2. Elmer, G.W and McFarland, L.V (2001). Biotherapeutic agents in the treatment of infectious diarrhea. Gastroenterology Clinics of North America. 30:837-854
3. Elmer, G.W., Surawicz, C.M and McFarland, L.V (1996). Biotherapeutic agents. A neglected modality for the treatment and prevention of the selected intestinal and vaginal infections. Journal of the American Medical Association. 275:870-876.
4. Del Piano, M., L. Morelli, G.P. Strozzi, S. Allesina, M. Barba, F. Deidda, P. Lorenzini, M. Ballaré, F. Montino, M. Orsello, M. Sartori, E. Garello, S. Carmagnola, M. Pagliarulo and L. Capurso (2006). Probiotics: from research to consumer. Digestive and Liver Disease. **38,** 248-255.
5. Park, J. and M. H. Floch (2007). Prebiotics, probiotics, and dietary fiber in gastrointestinal disease. Gastroenterology. **36,** 47-63
6. Zubillaga, M.R, Weill,E,Postaire, C, Goldman,R, Caro and J.Boccio (2001).Effects of probiotics and functional foods and their usew in different diseases. Nutrition Research 21, 569-579.
7. Fioramonti,J.V., Theodorou and L.Bueno(2003). Probiotics:what are they?What are their effects on gut physiology? Best Practice and Research Clinical Gastroenterology. 17, 711-724
8. McFarland, L.V and Bernasconi, P (1993). *Saccharomyces boulardii:* A review of an innovative biotherapeutic agent. Microbial Ecology in Health and Disease. 6: 157-171
9. McFarland, L.V (1998). Epidemiology, risk factors and treatments for antibiotics associated diarrhea. Diagnostic Microbiology and Infectious Disease. 16: 292-307
10. McFarland, L.V (2007). Meta analysis of probiotics for prevention of travellers diarrhea. Travel Medicine and Infectious Disease. 5:97-105
11. *Elmer, G.W., C.M. Surawitz, L.V. McFarland*: Biotherapeutic agents. A neglected modality for the treatment and prevention of selected intestinal and vaginal infections. JAMA **275,** 870-876 (1996).
12. *Elmer, G.W., L.V. McFarland, C.M. Surawicz (Eds.)*: Biotherapeutic agents and infectious diseases. Totowa, N.J. Humana Press (1999).
13. *Mansour-Ghanaei, F., N. Dehbashi, K. Yazdanparast* and *A. Shafaghi*.: Efficacy of *Saccharomyces boulardii* with antibiotics in acute amoebiasis. World J. Gastroenterol. **9,** 1832-1833 (2003).

BIOPROCESS OPTIMIZATION FOR CELL MASS PRODUCTION AND FUNCTIONAL CHARACTERISATION OF *Lactobacillus plantarum*

Othman, N.Z[1,] Chan,Y.S[1]., Malek, R.A[1], Ramli, S[1], Ismail, N.S[1], Sarmidi, M.R[1], Aziz, R[1], El Enshasy, H.A[1,2].

[1]Institute of Bioproduct Development, Universiti Teknologi Malaysia, Skudai Johor, Malaysia.
[2]Bioprocess Development Department, Mubarak City for Scientific Research and Technology Applications (MuCSAT), New Burg Al Arab, Alexandria, Egypt

Introduction

Probiotics are defined as "living microorganisms, which upon ingestion in certain numbers, exert health benefits beyond inherent basic nutrition"(21). *Lactobacillus plantarum* as Lactic Acid Bacteria (LAB) a member of the genus *Lactobacillus and* Gram-positive bacterium that can produce lactic acid. It lives in many different environments, including some food such as fermented food products, plant materials and human gastrointestinal tract (GIT). This strain which one of the most widespread strains used in food industries has excellent adherence characteristics (14) and potential to break down proteins and fats which indicate by its activities of protease, peptidase, esterase and lipase (7). The most common and effective supplement in fermentation broth for cultivation of *L. plantarum* has been found to be yeast extract which is rich source of vitamins, essential amino acids and nucleic acids (1). *L. plantarum* need amino acids for growth and proteolytic systems. Moreover, *L. plantarum* also show it ability to withstand in concentrated bile stress and has the antibiotic-resistance genes in nature such as vancomycin, erythromycin and chloramphenicol (3,9). Unfortunately, application of probiotics in the food industries should be susceptible to antibiotic to prevent or create the chances of transmitting the antibiotic resistance genes to another pathogenic bacteria in intestine (4). Thus, increasing cell mass and activity especially in the economical point of view are the major objectives of fermentation in research and industry (19). However, cultivation of *L. plantarum* produces high lactic acid as a product that influences growth rate and productivity which is initially proportional on cell growth and final biomass (15). This study was conducted to maximizing the *L. plantarum's* biomass production through the reduction of products inhibitory effect using different media and processing parameters in both shake flask and bioreactor. Besides, the optimization of *L. plantarum's* growth kinetics in bioreactor by controlling processing parameter and its functionality characteristic also have been studied.

Materials and Method

Media and Cell Bank Preparation

Lactobacillus plantarum NRRL B-227 was first cultivated in MRS and the pH was adjusted to 6.2 $\pm$ 0.2. Glucose was autoclaved separately and mixed together with other medium ingredients after cooling to avoid Maillard browning reaction. Cells formed were harvested by using 50 % sterile glycerol solution. Cells obtained from the master cell bank were used to make working cell bank through preservation of cells in glycerol culture and stored at -80°C until further use. The cell bank then was used for inoculum preparation.

Growth and production media

Based on literature review for the media applied for *L. salivarius* cultivations, seven different types of broth media were used in this study.

Inoculum preparation and cultivations conditions in shake flask

For the initial bacterial cell propagation, 1 cryogenic vial from stock culture that stored at -80°C, was taken and used to inoculate 250 ml Erlenmeyer flask of 50 ml working volume. For inoculum preparation, MRS broth was used for vegetative growth. After inoculation, flasks were incubated on rotary shaker (Innova 4080, New Brunswick, NJ, USA) at 200 rpm and 37°C for 24 h. The obtained cells were used to inoculate either shake flask or bioreactor to obtain cell concentration of 0.1 OD_{600} in culture.

Bioreactor cultivations

Cultivations with growing cells were carried out in a 16-L stirred-tank bioreactor with working volume of 8-L (BioEngineering, Wald, Switzerland). The bioreactor is equipped with pH probe, oxygen probe, foam sensor, and stirrer of two–four bladed Rushton turbines. For controlled pH cultivations, the pH was maintained at 7.0 by addition of 4 M NaOH and 2 M HCl solution. For uncontrolled pH experiments, the pH was also adjusted to 6.5 before inoculation and monitored only during the cultivation time. During the experiments, temperature, aeration rate and the agitation speed were controlled at 37°C, 1 vv^{-1} min^{-1} and 400 rpm, respectively. The dissolved oxygen was adjusted to 100% saturation before inoculation and kept uncontrolled during cultivation process. The total time of fermentation was approximately 24 hours. Samples were taken at approximately 1 hour intervals and analyzed immediately for optical density and lactate concentration. Antifoam reagent (Silicone antifoam, Sigma, USA) was added to suppress the foaming when necessary. Glucose was prepared and autoclaved separately in 1000 ml and added aseptically to the bioreactor before inoculation.

Survival of L. salivarius in simulated gastric juices and bile solution

The Simulated Gastric Juice (SGJ) was prepared by suspending of 3.5 g D-glucose, 2.05 g NaCl, 0.6 g KH_2PO_4, 0.11 g $CaCl_2$, 0.37 g KCl, 0.05 g oxgall bile (Difco, Lab., Detroit, MI, USA) and 13.3 g pepsin in 1000 ml distilled water according to the method of Kim *et al.* (13). The artificial gastric juice was adjusted to different pH values (1, 2, 3, and 4) using 1M HCl. One percent cell suspension was used to inoculate MRS medium without bile salts as control. The prepared salt solution was sterilized by autoclaving at 121°C for 10 min. Both SGJ and bile solution were prepared fresh daily Samples were withdrawn and enumeration of viable bacteria was counted on MRS agar (pH 7.2) after 48 hours incubation at 37°C. 0.1 ml aliquot of bacterial suspensions was inoculated into 10 ml sterile artificial gastric juice of different pH ranged between 1 and 4. Samples were withdrawn periodically at 0, 30, 60, 90 and 120 min. The enumeration of viable bacteria was conducted on MRS agar (pH 7.2) after 48 hours incubation at 37°C. Following incubation, colony forming units were counted and recorded. Control samples without acidification were also prepared. All experiment for cell viability was conducted in triplicates.

Optical density determination

The optical density was measured by using spectrophotometer (Model DR/2500, Hach Company, Loveland, CO, USA) at 600 nm after proper dilution. For all samples the cultivated broth were diluted to give values less than (1 OD_{600}) for better accuracy. The OD of culture was converted to dry cell mass through a linear correlation standard curve. 1 OD_{600} was almost equivalent to 0.3 g L^{-1}.

Glucose and lactate determination

Both glucose and *lactate* were determined by High Performance High Chromatography (HPLC) obtained from Waters. For lactic acid, a 250 mm × 4.6 mm ID Spherisob Octyl Column (Waters, Milford, MA, USA) and a UV detector (210 nm) were used. The adsorbed substances were eluted with 0.2 M H_3PO_4 at flow rate of 0.8 ml/min at room temperature. For glucose, a 300 × 4 mm ID μ Bondapak/Carbohydrate column (Waters, Milford, MA, USA) with IR detector were used. The mobile phase use was acetonitrile: water (80:20) at a flow rate 1.0 ml min^{-1} at room temperature.

Results and Discussion

Effect of different medium composition on cell mass and lactate production

Figure 1, illustrates that production of biomass and lactate in the 7 different cultivation media. From this figure, it can be concluded that Medium 3 produced the best results. The highest biomass produced by medium 3 was 1.86 gL^{-1} which contains main components (g L^{-1}) of peptone from casein (10), meat extract (10), yeast extract (5), glucose (20). This particular medium produced the highest density of cell dry weight. However it also produced lactate in the highest concentration (9.18 g L^{-1}). Therefore, optimization study was essential to increase the production of biomass while reducing formation of lactate.

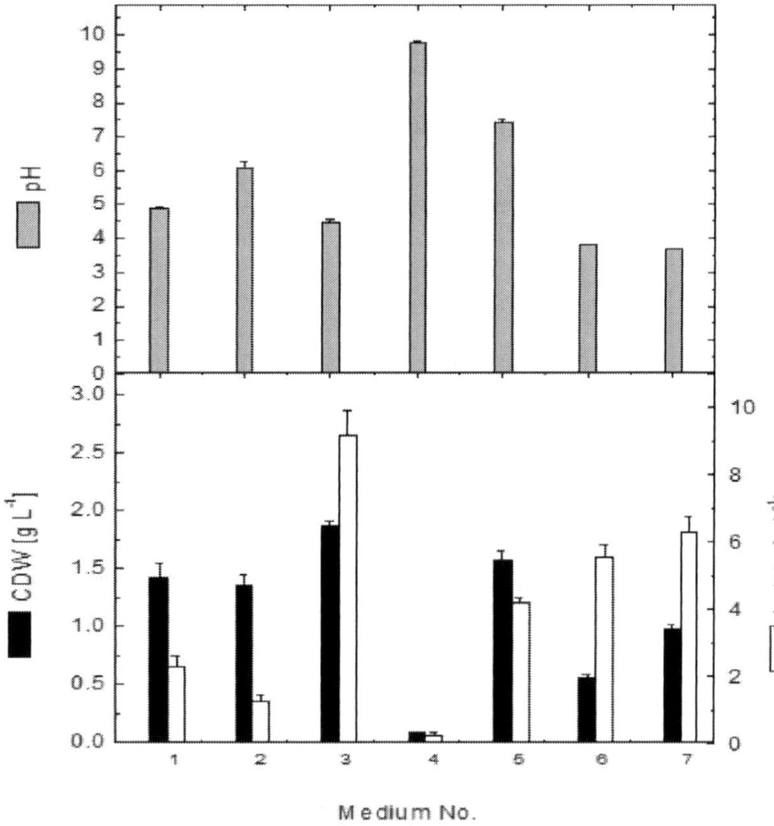

Figure 1: Cell Growth, pH and Lactate Production in 7 Different Media

Kinetic Cell Growth and lactate production in un-optimized medium and optimized medium

In this section, *L. plantarum* was first cultivated in a shake flask to study the cell growth in the medium 3, lactate production and the change in pH, as represented in **Figure 2**. The growth profile of *L. plantarum* and lactate production followed a typical Monod equation. The rapid exponential growth phase started after 3 hours of cultivation. The cell density increased rapidly from 0.11 -2.12 g L^{-1} after 18 hours of cultivation, which was the optimum stage of biomass production with the specific growth rate, μ_{max} of *L. plantarum* was 0.319 h^{-1}. It was observed that lactate was produced simultaneously with microbial growth and can be classified as growth-associated products. Production of lactate increased from 0.74-8.94 g L^{-1} after 2 hours of cultivation until it reached a steady state in the next 12 hours. As lactate concentration increased, the pH of the cultivation broth decreased. Both lactate concentration and pH were maintained after 16 hours of cultivation, namely at 8.94 g L^{-1} and pH 4.41 respectively. However, during the rapid propagation of biomass, it was noticed that, glucose consumption had also reduced rapidly from 17.05 g L^{-1} to 0.1 g L^{-1}. According to Shular and Kargi (18), glucose is one of the main carbon sources for cell cultivation.

The optimized medium of medium 3 that supports high cell biomass was used to study the growth kinetics of *Lactobacillus plantarumin* in the shake flask. with the following composition (g L^{-1}): glucose, 30; meat extract, 4; yeast extract, 7; sodium acetate, 5; K_2HPO_4, 2; $(NH_4)_2SO_4$, 2; $MgSO_4.7H_2O$, 0.2; $MnSO_4.4H_2O$, 0.05 after optimization studied. The research was then followed by studying the kinetic growth behavior of *L. plantarum* in optimized medium. It was found that, biomass produced by cultivating *L. plantarum* in optimized medium increased by 63.30% from 2.1205 g L^{-1} to

3.4627 g L^{-1}. Furthermore, an increased in specific growth rate from 0.3198 h^{-1} for un-optimized medium to 0.3707 h^{-1} for optimized medium was observed. The increase in the production of lactate gradually slowed down after 12 hours of cultivation and the concentration maintained in the range of 10.76 g L^{-1} to13.13 g L^{-1}. The initial pH of medium decreased gradually and reached the minimal value of 4.02 after 24 hours of cultivation. Glucose determination had shown that the amount of glucose decreased rapidly from 30 g L^{-1} to 2.85 g L^{-1} after 16 hours of cultivation.

Effects of uncontrolled and controlled- pH fermentation on kinetics of cell growth and lactate production in 16- L bioreactor

Figure 3 represents biomass production of *L. plantarum* in a 16L bioreactor under un-controlled and controlled pH condition. During cultivation of *L. plantarum*, lactate produced would indirectly reduce the pH of cultivation broth. According to Fu and Mathews (5) and Giraud *et al.* (6), in batch cultures, low pH (pH 4) inhibits the cell growth as compared to higher pH. The growth of *L. plantarum* in a 16-L stirred tank bioreactor under un-controlled condition showed a short lag phase (2 hours) followed by an exponential growth phase in the next 3 hours. A maximal cell growth of 2.78 g L^{-1} was observed during the stationary phase after 14 hours of cultivation consistent with drastically dropped of dissolved oxygen as the cells grew during the exponential phase. However, the percentage of dissolved oxygen increased during the stationary phase because cells started to die and the oxygen consumption was reduced. Lactate was produced at maximal of 12.27 g L^{-1} during the stationary phase. The increased in the concentration of lactate corresponded to the decrease of pH from 6.5 to approximately 4.54 pH. Meanwhile, a sharp decrease in the concentration of glucose occurred in the first 12 hours of cultivation as it was used up during the process. The specific growth rate of *L. plantarum* in a un-controlled pH bioreactor was 0.61 h^{-1} with doubling time of 1.13 h. The growth kinetics of *L. plantarum* in an optimized medium in a 16-L pilot scale stirred tank bioreactor shows that lactate production was suppressed at 4.815 g L^{-1} when the pH was maintained at 6.5.

Maintaining pH may increased the biomass production with a maximal production of 16.05 g L^{-1} after 14 hours of cultivation. Therefore, cultivation in controlled pH bioreactor will give higher biomass production (6). Glucose concentration also decreased from 30 g L^{-1} to 3.55 g L^{-1} after 16 hours of cultivation; however, this was slower compared to the cultivation in the un-controlled pH. The specific growth rate of 0.6369 h^{-1} with doubling time of 1.088 h was achieved. It was also observed that after 14 hours of cultivation, the biomass production was approximately constant in the range of 15.75 – 16.06 g L^{-1} before the cell started to enter the death phase. During the rapid growth of *L. plantarum* in the first 16 hours, dissolved oxygen decreased drastically from 100% to 29.4% and increased to 100% during the stationary phase.

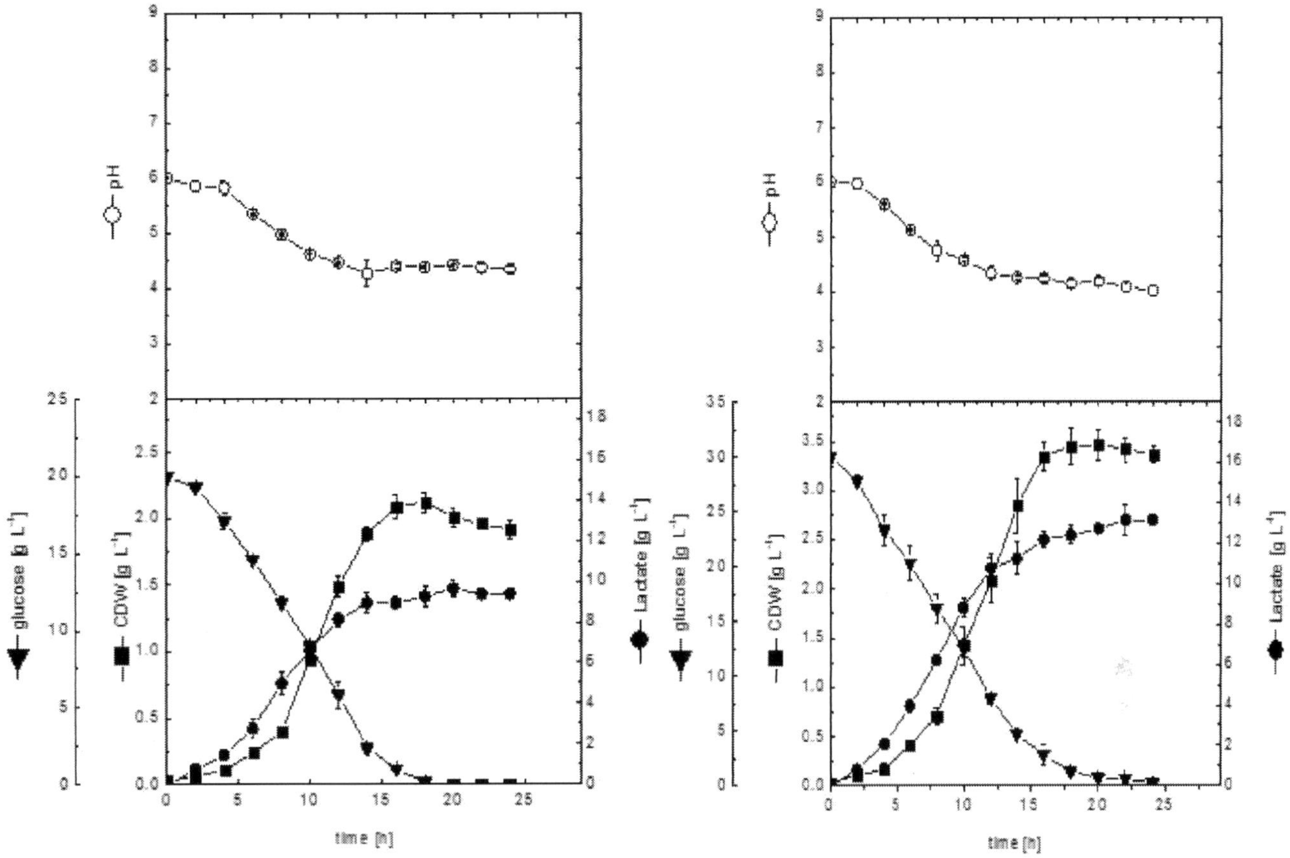

Figure 2: Growth Kinetics and pH of *L. plantarum* in un-optimized (left) and optimized medium (right) in the shake flask study.

Viability of L. plantarum in Simulated Gastrointestinal tract conditions

In this section, viability and stability of *L. plantarum* was determined by incubating 1% cell paste in artificial simulated gastric juice solution of pH 1, pH2, pH3, and pH4 (37°C). pH 1 to pH 4 of simulated gastric juice were prepared because the pH of stomach varies in this range depending on the food intake. The survival rate of *L. plantarum* was recorded after 30 minutes intervals for 2 hours. According to Food and Drug Administration, to be classified as probiotic is its abilities to multiply and survive in human intestines. Sung-Me and Soon Im (20) mentioned in their research that a probiotic strain will give health-promoting effects if they had successfully arrived to human colon by passing through gastrointestinal tracts. The strain viability in probiotic commercial preparation is usually varied from 41% (*Lactobacillus rhamnosus* GG and *Lactobacillus johnsonii* La1) to more than 70% (*Lactobacillus caseimunitass)* (16). In this experiment (**Table 2**), *L. plantarum* was found to survive in the first 90 minutes in SGJ+ pH 1 solution (100% survival rate) before it died rapidly (31.28% survival rate). When the pH of SGJ was increased, the survival rate, however, increased from 31.28% in pH 1 to 54.33% in pH 4 after 2 hours of cultivation.

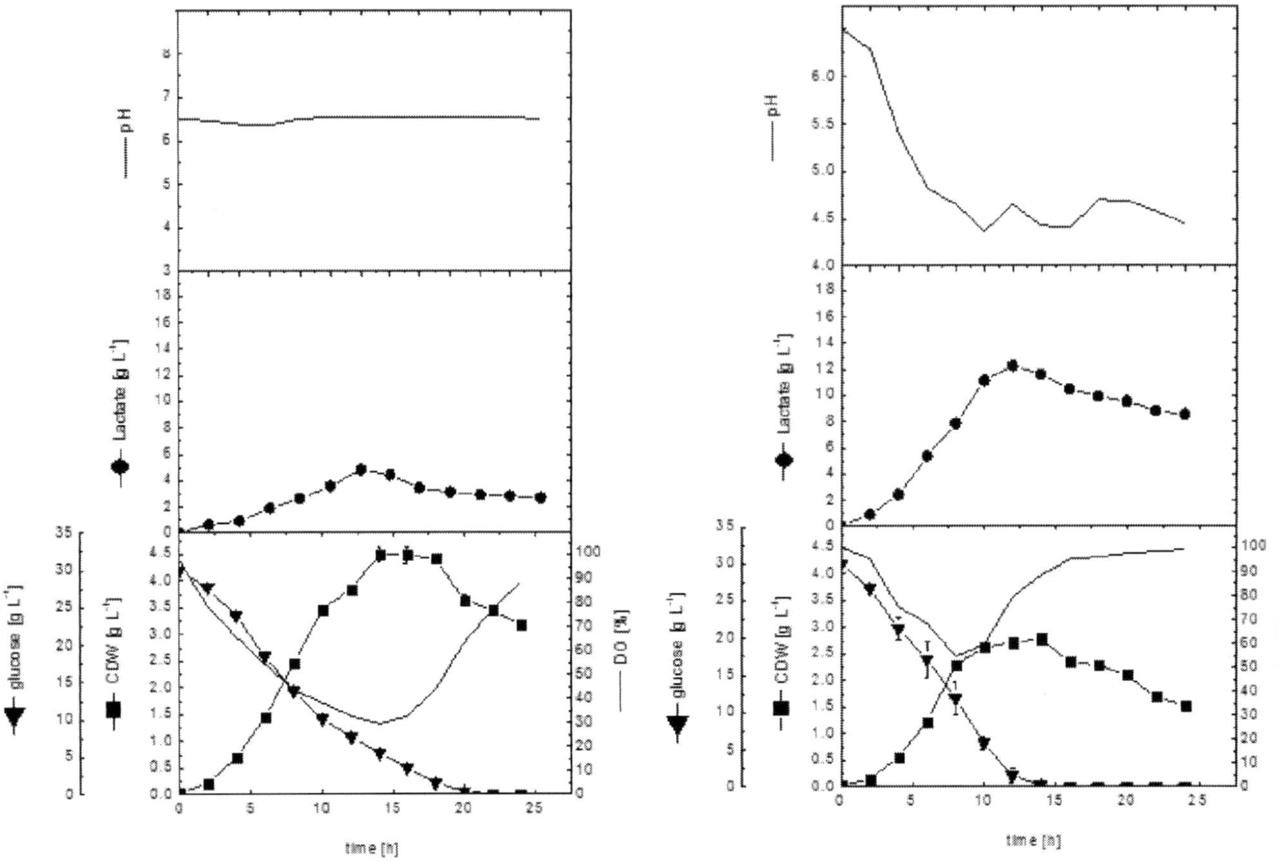

Figure 3: Kinetics of cell growth, lactate production and change in pH for optimization medium in controlled (left) and un-controlled pH (right) in pilot scale bioreactor for 24 hours incubation

Table 2: Viability of *L. plantarum* in Synthetic Gastric Juice of Different Condition

SGJ+ varied pH	Log cfu ml^{-1} (% survival)					% survival after 2 h
	0 min	30 min	60 min	90min	120min	
SGJ+pH1	15 (100%)	15 (100%)	15 (100%)	15 (100%)	4.69 (31.28%)	31.28%
SGJ+pH2	15 (100%)	15 (100%)	15 (100%)	5.90 (39.35%)	5.63 (37.56%)	37.56%
SGJ+pH3	15 (100%)	15 (100%)	7.74 (51.63%)	6.74 (44.96%)	6.43 (42.86%)	42.86%
SGJ+pH4	15 (100%)	10.27 (68.49%)	9.05 (60.32%)	8.40 (55.98%)	8.15 (54.33%)	54.33%

% Survival = log (no of viable cells survived (CFU ml^{-1}) / log (no of initial viable cells inoculated (cfu ml^{-1}) x 100%

***Viability of L. plantarum salivarius* in bile salts conditions**

Bile tolerance is one of the essential properties required for lactic acid bacteria to survive in the small intestine and to be functionally effective in intestine (10, 17). A probiotic stain has synergetic effects when it can withstand ox gall concentration of between 0.15% and 0.3%. Golin and Gorbach (8). In this study, *L. plantarum* was tested in its tolerance of 2 hours of incubation in MRS broth containing 0.5% to 4% of oxgall. *L. plantarum* showed 100% of survival rate when it was incubated at low concentration of bile salt (0.5%, 1% and 2%) as shown in **Table 3**. Further increase in bile salt concentration decreased the survival rate from 100% to 87.55% at 3% and 53.92% at 4% concentration of bile salt. These results had proven that *L. plantarum* was more tolerant to bile salt.

Table 3: Cell Viability of *L. plantarum* in bile salt

MRS+ varied bile salt conc.	Log cfu ml^{-1}					% Inhibition after 2 h	% Survival
	0 min	30 min	60 min	90min	120min		
MRS + 0.5%	TMTC	TMTC	TMTC	TMTC	TMTC	0	100
MRS + 1%	TMTC	TMTC	TMTC	TMTC	TMTC	0	100
MRS + 2%	TMTC	TMTC	TMTC	TMTC	TMTC	0	100
MRS + 3%	8.40	8.30	8.10	7.91	7.49	12.15	87.55
MRS + 4%	9.70	8.27	7.65	7.03	6.64	46.08	53.92

Antibiotic resistance

Table 4 shows the summary of different types of antibiotic inhibition on the growth of *L. plantarum*. It can be concluded that, ampicillin gave no inhibition effect on *L. plantarum* growth. This may be due to the function of ampicillin as an anti-inflammation agent, according to data obtained from Lyme Disease Association of Southeastern Pennsylvania, Inc. Another antibiotic that had permitted the growth of *L. plantarum* in low concentrations was gentamicin. Gentamicin allowed *L. plantarum* to grow at low dosages (5 and 10μg ml^{-1}). The inhibition zone became larger when increasing gentamicin dosages (μg ml^{-1}) were applied on top of the cells. Among the antibiotic listed, only gentamicin and ampicillin had allowed the growth of *L. plantarum*. The strains of *L. plantarum* were susceptible to streptomycin, tetracycline, erythromycin , and rifampicin. This is also in agreement with those results obtained by Boyle *et al.* (2) and Ishibashi and Yamakazi (11). According to D' Aimmo *et al.* (4), a probiotic strain must be susceptible to antibiotic to prevent antibiotic resistance strain from transforming into pathogenic bacteria.

Table 4: Determination of Antibiotic Susceptibility

Antibiotics	Concentrate (μg ml^{-1})	Average Inhibition Zone (mm)
Streptomycin	5	15.5
	10	20.0
	15	21.0
	20	16.0
Ampicillin	5	6.0
	10	6.0
	15	6.0
	20	7.0
Gentamicin	5	6.0
	10	7.0
	15	9.5
	20	12.5
Tetracycline	5	12.0
	10	14.5
	15	16.0

	20	23.0
Erythromycin	5 10 15 20	23.0 24.5 25.0 29.5
Rifampicin	5 10 15 20	14.5 15.0 15.5 20.0

***References*:**

1. Aeschlimann A., and Von Stockar, U. (1990), The effect of yeast extract supplementation on the production of lactic acid from whey permeate by *Lactobacillus helviticus*, Applied Microbiology and Biotechnology, 32: 398-401.
2. Boyle R., J, Robins-Browne R., M., and Tang, M. L. K ., (2006), Probiotics use in clinical practice: What are the risks? The American Journal Clinical Nutrition, 83:1256-64
3. Bron P. A., Marco, M., Hoffer, S. M., Van Mullekon, E., de Vos, W.M., and Kleerebezem, M. (2004), Genetic Characterization of the Bile Salt Response in Lactobacillus plantarum and Analysis of Responsive Promoters in vitro and in situ in the Gastrointestinal Tract. Journal of Bacteriology. **18:** 7829-7835.
4. D'Aimmo M.R., Modesto M., Biavati B,.(2007), Antibiotic resistance of lactic acid bacteria and *bifidobacterium* spp. isolated from dairy and pharmaceutical products, International Journal of Food Microbiology, 2007;115:35-42
5. Fu W., and Mathews A. P., (1999), Lactic Acid Production from Lactose by Lactobacillus plantarum: Kinetic Model and Effects of pH, Substrate, and Oxygen. Biochemical Engineering Journal, 3: 163-170
6. Giraud E., Lelong, B., and Raimbault, M., (1991), Influence of pH and Initial Lactate Concentration on the Growth of Lactobacillus plantarum, Applied Microbiology and Biotechnology, 36: 96-99.
7. Gobbetti M., Fox P. F., Smacchi E., Stepamoal L., and Damiani P., (1996), Purification and Characterization of a Lipase from *Lactobacillus plantarum* 2739, Journal of food Biochemistry, 20: 227-246.
8. Goldin B. R., and Gorbach S. L., (1992), Probiotics for humans. In Probiotics ± the Scienti®c Basis, ed. Fuller, R., pp. 355-376.London: Chapman & Hall, ISBN 0-412-40850-3; 1
9. Hamilton-Miller J.M.T., Shah S., (2002), Deficiencies in microbiological quality and labelling of probiotic supplements, Int J Food Microbiol, 72:175–176.
10. Ibrahim S. A., Benzkorovainy A,. (1993), Survival of bifidobacteria in the presence of bile salts, Journal of the science of Food and Agriculturec, 62:351-354.
11. Ishibashi N, and Yamakazaki S.(2001), Probiotics and safety,The American Journal of Clinical Nutrition, 73: 465S-470S.
12. Kemp T. L., Nazmul K. M., and Linden J. C., (1989), Response Surface Optimization of *Lactobacillus plantarum* Batch Growth, Biotechnology Letters, 11: 817-820.
13. Kim P., Young J. M., Hyo C. Y., Kim S., Jae-Kim S, Ha P. Y., (2007), Probiotic properties of *Lactobacillus* and *Bifidobacterium* strains isolated from porcine gastrointestinal tract, Applied Microbiology and Biotechnology, 74:1103-1111.
14. Klarin B., Johansson M.L., Molin G., Larsson A., Jeppsson B., (2005), Adhesion of the probiotic bacterium *Lactobacillus plantarum* 299v onto the gut mucosa in critically ill patients: a randomised open trial, Crit Care 9: Jun;9(3):R285-93.

15. Luedeking R., and Piret E. L. ,(2000), A Kinetic Study of the Lactic Acid Fermentation Batch Process at Controlled pH, Biotechnology Bioengineering, 61: 393-400.
16. Martin R., Jimenez E., Olivares M., Marin M.L., Fernanderz L., Xaus J., Rodriguez J.M., (2006), *Lactobacillus salivarius* CECT 5713, a potential probiotic strain isolated from iInfant feces and breast milk of a mother-child pair, International journal of Food Microbiology. 122:35-43.
17. Park H. S., Lee S. H., Uhm T. B., (1998), Selection of microorganism for probiotics and their characterization, Kor.J.Food Nutr, **27**:433-440.
18. Riesenberg D., Guthke R. ,(1999), High cell density cultivation of microorganisms, Apply Micrbiol Biotechol, (51):422-430
19. Shuler M. L., and Kargi F., (2002), Bioprocess Engineering Basic Concepts (2nd ed.). Pearson Prentice Hall.
20. Sung-Mee L., and Soon I. M. D., (2009), Screening and characterization of probiotics lactic acid bacteria isolated from korean fermented foods. Journal of Microbiology and Biotechnology.19:178-186
21. Tannock G. W., (2002), Probiotics and Prebiotics*: Where are We Going?* Curr. Issues Intest. Microbiol, (2003) 4: 33-42.

(594f) Evaluation of Ethanol Production From Renewable Cellulosic Resources Using Process Simulation Tools

Demetri P. Petrides, INTELLIGEN, INC., Scotch Plains, NJ and **Charles Siletti**, INTELLIGEN, INC., Mount Laurel, NJ

Over the past three decades there has been intense investigation on the development of fuel producing processes that are based on the use of renewable agricultural materials as feedstock. This activity is driven primarily by the quest for fuel self-reliance and carbon oxides emission reductions. The main effort has been concentrated on bio-ethanol and bio-diesel which have been shown to give motor engine performance similar to that of conventional petroleum based fuels. In addition to product characteristics, however, process economics play an equally important role in any successful product commercialization. In this work, realistic process simulation models have been developed in order to analyze the economics of corn-stover to ethanol conversion. This poster illustrates how such models can guide R&D work and facilitate process optimization. Specifically it illustrates the entire process from corn stover to dehydrated ethanol. The information includes highlights of the material and energy balances and overall process economics.

(594g) The Role of Simulation and Scheduling Tools in Bioprocess Development and Manufacturing

Demetri Petrides, INTELLIGEN, INC., Scotch Plains, NJ and **Charles Siletti**, INTELLIGEN, INC., Mount Laurel, NJ

The successful scale up and commercialization of biopharmaceuticals, bio-fuels and biomaterials is a challenging task that requires collaboration of professionals from many disciplines. Process simulators and other computer aids can facilitate this task by assisting scientists and engineers to answer the following and other related questions: What is the impact of product titer increase on the capacity load of the downstream section, the overall throughput of a plant, and the cost of goods? What changes are required in an existing multi-product facility to accommodate the process of a new product? What is the range of variability that a process can accommodate if it operates under a tight cycle time? What is the impact of single-use systems on the demand for utilities, the environment and the cost of goods? Our experience in addressing the above questions will be presented using industrial examples in which we evaluated alternative technologies for producing therapeutic monoclonal antibodies, vaccines, second generation bio-fuels and commodity bio-chemicals.

(594h) Production and Application of Bioactive Food-Derived Peptides

Li Yan and **Xiao Kaijun**, South China University of Technology, Guangzhou, China

There's an increased interest in bioactive food-derived peptides (BFP) nowadays because of their function against many human diseases and maintaining human health, such as antihypertention, anticancer, antioxidant, anti-inflammatory properties, as well as regulating immune function and liver disease treatment. These bioactive peptides can be released from plant or animal proteins by enzymatic /chemical hydrolysis, fermentation, and other food processing methods, of which enzyme proteolysis is the most commonly used method producing BFP. This paper summarized the mostly used enzyme and protein sources reported in relevant research papers. Bioactivity of the peptides depends on many factors including size and amino acid composition of the resulting peptides, which in turn affected by processing conditions and posthydrolysis process to isolate the BFP. BFP can be fractionated and purified based on peptide size, net charge and hydrophobicity using membrane ultrafiltration, chromatography, and electrodialysis ultrafiltration technologies. There are wide applications of the BFP depending on product purify and market, such as food, supplement, and pharmaceutical areas. This paper reviewed the functional properties of different BFP when applied in food products including solubility, emulsion stability, water and fat binding capacities and the challenges were revealed including bitter taste and decreased *in vivo* bioavailability. The information presented in this paper provided background for scale-up production of BFP by utilizing a variety source of food protein and food protein processing byproducts.

A NOVEL ORGANIC SOLVENT STABLE SERINE PROTEASE FROM A NEWLY ISOLATED *Serratia* sp.

*Lipsy Chopra, Ramit Rikhi, Debendra K. Sahoo**
Biochemical Engineering Research and Process Development Centre
CSIR - Institute of Microbial Technology, Chandigarh, INDIA

Abstract

During our screening for novel proteases, a *Serratia* sp. was isolated from alkaline soils the Himalayas, a biodiversity rich hotspot and identified according to polyphasic approach which includes morphological, biochemical and physiological characteristics and 16S rDNA sequence analysis. The protease was purified to homogeneity by acetone precipitation followed by gel filtration using Sephacryl S-100. The intact molecular mass of purified protease was determined by MALDI TOF and SDS-PAGE. The enzyme was highly active in the pH range of 6.0 to 11.0 and the optimum pH and temperature were found to be 7.5 and 50°C, respectively. The protease was characterized as serine protease based on the studies on class specific inhibitors and retained 92 % and 99 % of its activity in the presence of sodium perborate (0.5 %, w/v) and dodecyl benzene sulphonate (0.1 %, w/v), respectively. High stability in presence of detergents such as sodium dodecyl sulphate (0.1 %, w/v), Tween 80 (1.0 %, w/v), Triton X-100 (1.0 %, w/v) and organic solvents such as dimethyl sulfoxide, ethyl acetate and hexane (25 % and 50 % v/v) and its performance in stain removal makes this *Serratia* protease as an ideal choice for industrial applications such as in detergent, leather, food, pharmaceutical and chemical synthesis industries.
Keywords: *Serratia* sp., Serine protease, Purification, Organic solvent stable

Introduction

Microorganisms secrete a wide variety of proteolytic enzymes, a group of hydrolytic enzymes constituting about 50% of the total worldwide enzyme market and having applications in diverse areas including detergent, leather, food, pharmaceutical and textile processing industries and in bioremediation processes. It constitutes a complex group of enzymes whose common feature is that they catalyze the cleavage of peptide bonds but may differ in some of their properties, such as substrate specificity, active site or catalytic mechanism. Though, microbial proteases have been extensively studied since the advent of enzymology because of their importance in physiological field and industrial applications, screening of alkalophiles from naturally occurring habitats in different parts of the world is expected to result in isolation of new protease producing organisms potentially useful for many novel applications.[1] Enzymatic conversions in non-aqueous media are drawing attention due to its various potential applications in food and pharmaceuticals to specialty chemicals and hence, search for solvent stable proteases has been an important area of research. Alkaline proteases are currently receiving increased attention in view of their inherent stability at high pH, temperature, and in the presence of surfactants, organic solvents and denaturing agents, that enable their use in processes that restrict conventional enzymes.[2]

* Corresponding author. Tel: +91 172 6665324; Fax: +91 172 2690585
Email: debsahoo@imtech.res.in

Experimental

Materials

Azocasein, Trizma Base, tricholoroacetic acid, SDS and protease inhibitors were procured from Sigma (St.Louis, MO, USA). A commercial grade partially hydrolyzed soyaprotein, PH soyatose, was obtained from Workem Laboratories, Mumbai and sodium perborate was obtained from Fluka Chemie (Buchs, Switzerland). All other reagents and media components used were of highest purity grade available in India.

Methods

The microorganism was isolated from alkaline soils of Leh and Ladakh (India) by plating on skim milk agar plate. The isolate was characterized by polyphasic approach that includes morphological, physiological, biochemical characteristics and 16S rDNA sequence analysis. Identification of phylogenetic neighbours and the calculation of pairwise 16S rRNA gene sequence similarities were achieved using the EzTaxon server http://www.eztaxon.org/.[3] Sequences of closely related taxa were retrieved; aligned using Clustal X programme[4] and the alignment was manually corrected. For the neighbour-joining analysis[5], the distances between the sequences were calculated using Kimura's two-parameter model.[6] Bootstrap analysis was performed to assess the confidence limits of the branching.[7]

The organism was maintained on skim milk agar plates (at pH 7.0) by sub-culturing it every four weeks and glycerol stocks (20 %) of the same were also stored at - 80°C. The medium for pre-inoculum, inoculum as well as enzyme production contained (g/l): glucose, 10.0, biopeptone, 5.0, yeast extract, 5.0, KH_2PO_4, 1.0, $MgSO_4.7H_2O$, 0.2, and Na_2CO_3, 10.0. The solution of Na_2CO_3 was sterilized separately and then added to the medium. Preinoculum was obtained by transferring a loopful of 24 h old culture from skim milk agar plate into 250 ml Erlenmeyer flask containing 50 ml of sterile media and incubating the flask at 30°C and 200 rpm in an orbital incubator shaker. For alkaline protease production studies, 10 ml of inoculum (10% v/v) was transferred to 500 ml Erlenmeyer flasks containing 90 ml of sterile media and these were incubated in an orbital shaker at 30°C, 200 rpm for 24 h. The samples from the flasks were collected at different time intervals and centrifuged at 12,000xg for 15 min at 4°C. The supernatant was used as the crude enzyme sample.

The crude enzyme was also concentrated by adding chilled acetone to the centrifuged supernatant to 70 % saturation and was kept at 4°C overnight. The precipitate thus obtained was centrifuged at 10000xg for 20 min at 4°C. The pellet was then dissolved in 50 mM Tris HCl buffer, pH 7.2 and filtered through 0.22 μ syringe filter. The filtered crude enzyme was then loaded onto Hiprep (16/60) Sephacryl S-100 gel filtration column (GE Healthcare Biosciences AB, Sweden). The column was equilibrated with 50 mM Tris HCl buffer, pH 7.2 and eluted with the same buffer. The chromatography was operated on AKTA chromatography system (GE Healthcare, USA) at 0.3 ml/min and 3 ml fractions were collected for elute. The purity of the active fractions was verified by SDS PAGE analysis and the protein bands were visualized by silver staining.[8]

Analytical

Proteolytic activity was estimated at pH 7 using azocasein as a substrate and the activity in azocaseinase units (AU)/ml was defined as the increase in absorbance caused by 1.0 ml of enzyme per hour at 420 nm under the given assay conditions.[1] Total protein content of the samples was estimated by Bicinchoninic acid method using bovine serum albumin as standard.[9]

The effect of temperature and pH on the protease activity was determined by assaying the enzyme activity at different temperatures ranging from 30°C - 65°C and pH in the range of 4.0 -12.0 according to protocol of Singh et al.(1999).[10] The relative activities were based on the ratio of the activity obtained at certain pH to the maximum activity obtained and expressed as percentage.

The effects of metal ions (K^+, Ca^{2+}, Fe^{2+}, Mg^{2+}, Cu^{2+},and Mn^{2+}), surfactants (SDS, Tween 80, Triton X 100, DBS and SPB), organic solvents (isopropanol, DMSO, acetone, ethyl acetate, benzene and hexane) and inhibitors such as (phenylmethylsulfonyl fluoride (PMSF), tosyl L-lysine chloromethyl ketone (TLCK), tosyl L-phenylalanine chloromethyl ketone (TPCK), ethylenediaminetetraacetic acid (EDTA), N-CBZ-L-phenylalanine chloromethyl ketone (ZPCK), iodoacetic acid (IAA), 1,10-phenanthrolein and benzamidine) on the enzyme activities were evaluated according to Singh et al.(1999).[10] The protease sample was pre-incubated in assay buffer containing an inhibitor at the specified concentrations for 2 h at 37°C and the residual activity was estimated under standard assay conditions after addition of azocasein and in case of organic solvents, the protease sample was pre-incubated with organic solvents for 1 h at 37°C and 150 rpm and the residual activity was determined under standard assay conditions.

Results and Discussion

Isolation and identification of the organism

The strain was isolated by screening the soil samples of Leh and Ladakh region of The Himalayas on skim milk agar plates. The isolates showing larger zone of clearance were screened in liquid culture. On the basis of higher proteolytic activity in liquid culture, this strain was selected and identified as *Serratia* sp. according to polyphasic approach. The colony morphology was round, raised, moist, off-white colored with entire margin and the bacterium was found to be gram negative rods, positive for catalase, citrate utilization and arginine dihydrolase.

Purification

Serratia sp. extracellular protease was purified from fermentation broth by acetone precipitation and gel filtration chromatography. After precipitating with acetone and subsequent purification with Sephacryl S-100 gel filtration column chromatography, the active fractions collected were run on SDS PAGE. The enzyme was purified 5.70 fold with a 30 % yield and a single band observed on silver staining established the purity of the enzyme (Fig. 1 inset). The intact mass determination of the purified protease by MALDI TOF revealed its size to be 66.71 KDa (Fig. 1) which was in good agreement with SDS PAGE data of this protease (Fig. 1 inset).

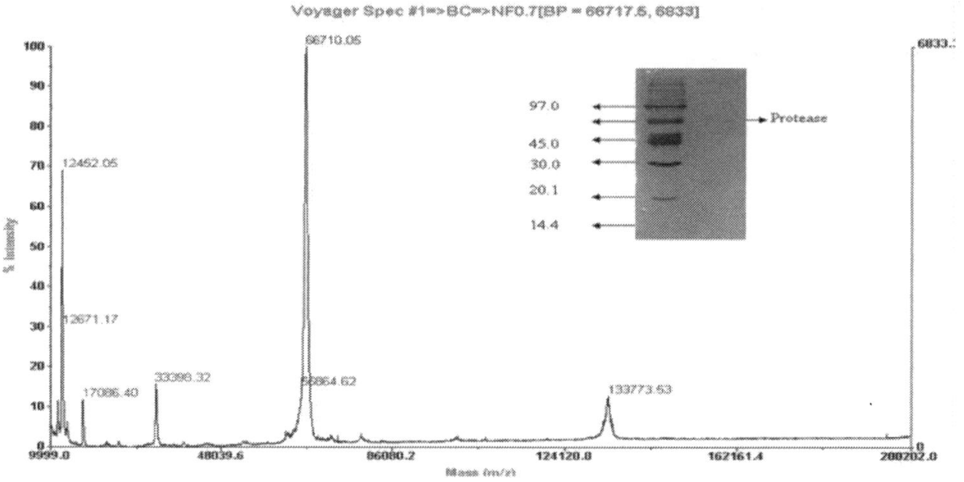

Figure 1: Mass spectrum of pure *Serratia* protease fraction. SDS-PAGE of purified protease from *Serratia* sp. is shown as inset. Lane 1: Low molecular weight marker, Lane 2: Purified protease fraction collected from Sephacryl S-100 run.

Effect of temperature and pH on activity and stability of protease

The effect of temperature and pH on the activity of purified protease is shown in Fig. 2. The protease was active at temperature range of 30°C to 65°C and the maximum activity was observed at 50°C. The thermal stability experiment showed the present enzyme to be stable at 37°C up to 2 h under experimental conditions. The effect of pH on protease activity was investigated at pH range of 6.0-11.0 and the maximum activity was observed at pH 7.5. Previously, a *Serratia marcescens* protease with similar molecular mass of 66.5 KDa had been reported to have a pH and temperature optima at 9.5 and 45-50°C, respectively[11] and another *Serratia* sp. protease had pH and temperature optima of pH 8.0 and 45°C, respectively.[12]

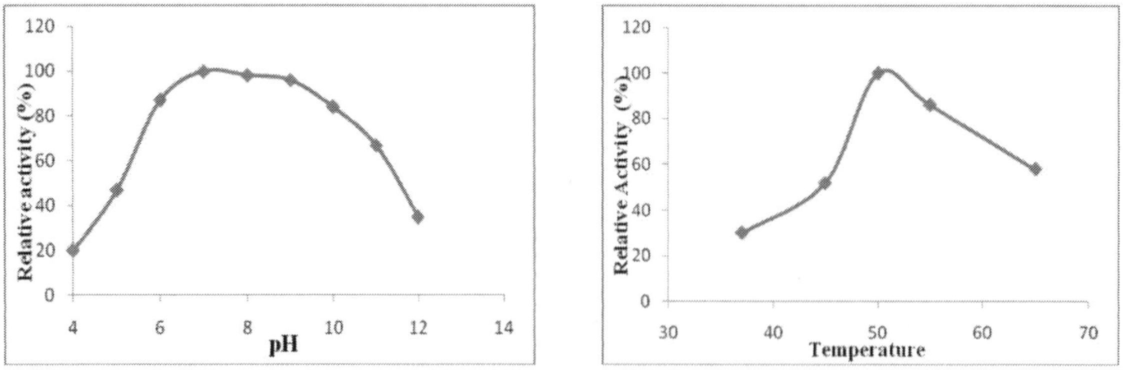

(A) (B)

Figure 2: Effects of pH (A) and temperature (B) on *Serratia* protease. The protease activity was assayed in the pH range of 4.0 – 12.0 and temperature range of 30°C – 65°C. The maximal enzyme activities observed at pH 7.0 and 50°C were set as 100% relative activity.

Effect of metal ions and surfactants on protease activity

The effect of various metal ions at 1mM concentration on the protease activity was studied at 50°C. It was observed that $Ca^{2+}$and Mn^{2+} increased the protease activity to 183.33 % and 165.63 % of the control value respectively. However, K^{+}, Fe^{2+}, Mg^{2+}, and Cu^{2+} at a concentration of 1 mM decreased the activity to 90.77, 52.59, 65.56, and 28.38 % of the control value, respectively. These results indicated that Ca^{2+} is the best activator for this *Serratia* protease. Though, our finding was in contrast to the results where Ca^{2+} not only reduced serine protease activity of a *Serratia* sp. SYBC H, but also significantly decreased its thermostability[13], similar results of the effect of Ca^{2+}(1mM) on alkaline proteases from another *Serratia* sp. and *Bacillus*[1, 10, 14] have been reported. The effect of Ca^{2+} on the thermal stability of enzyme could be explained on the basis of the role of Ca^{2+} in strengthening the interactions inside protease molecules and by the binding of Ca^{2} to the autolysis site. The strong inhibitory effects of Fe^{2+}, Mg^{2+}, and Cu^{2+} on the protease activity showed that these metal ions might correlate with the active center of the protease and/or with the toxic effects on the protease directly.[15] Stability studies in the presence of surfactants, both ionic and non-ionic, and oxidizing agents showed this protease to be highly stable in presence of 0.1 % SDS, 1 % Tween 80, and 1 % Triton X-100, 0.1 % DBS, 0.1 % SPB whereas at 0.5 % SDS and 0.5 % SPB the activity of the enzyme was found to decrease to 51.59 % and 92.75 %. In comparison, the activity of *S. marcescens* MH6 protease was almost completely inhibited in presence of 1 % SDS.[12]

Effect of organic solvents on protease activity

When organic solvents are used as media for enzyme reactions, the reaction equilibrium of hydrolytic enzymes could be shifted towards completion of reverse reaction, that is, synthetic reaction. However, enzymes are usually inactivated or denatured in the presence of organic solvents, especially in water miscible solvents. As the effect of organic solvents on protease activity differs from protease to protease, the screening of the stability of proteases in the presence of organic solvents has potential to find novel solvent stable enzymes. A number of recent reports show the enhancement of peptide synthesis in presence of organic solvents in the reaction mixture, especially in water miscible solvents.[16] However, most of the reported solvent stable proteases are produced by *Pseudomonas* sp.[17-19] and *Bacillus* sp..[20-21] In the present study, the effect of organic solvents such as isopropanol, DMSO, acetone, ethyl acetate, benzene and hexane (50 % (v/v) level) on *Serratia* protease was investigated and the results demonstrated strong hydrophilic solvent stability of this novel enzyme as shown in Table 1. Recently, Wan et al. (2010)[12] have reported a *Serratia marcescens* MH6 metalloprotease retaining 50 % and 10 % activity in DMSO and isopropanol respectively, each at 50 % (v/v) concentration. In comparison, the protease from *Serratia* sp in the current study retained over 90% of its activity in the presence of isopropanol and showed excellent stability (~100%) in the presence of DMSO. This enzyme also showed stability of its activity in the presence of acetone, ethyl acetate and hexane at 50 % (v/v) concentrations of these solvents when incubated for 1 h at 37°C and retained more than 95 % of its activity in the presence of benzene. The toxicity of the organic solvent with log P value below 4 is considered extremely toxic as in such cases, essential water molecules present on enzyme surface can be easily replaced with solvent molecules.[12, 22] The strong hydrophilic solvent-stable properties of this *Serratia* protease make it an ideal catalyst for kinetic and equilibrium- controlled synthesis.

Table 1 Effect of organic solvents on the stability of the alkaline protease from *Serratia* sp.

Solvent	Log P	% Relative Residual Activity (at 50% V/V)
Control		100
Isopropanol	0.074	95
Dimethylsulfoxide (DMSO)	-1.378	103
Acetone	-0.208	93
Ethyl acetate	0.711	113
Benzene	2.0	101
Hexane	3.5	107

Effect of class- specific protease inhibitors on protease activity

The effect of chelating agents and class- specific inhibitors on purified protease was determined by measuring residual activity following 2 h pre-incubation of the enzyme with various inhibitors as summarized in Table 2. Inhibition studies primarily give an insight into the nature of an enzyme, its cofactor requirements and the nature of its active site.[15] Protease was strongly inhibited (91.6 %) by chelating agent EDTA which is a potent inhibitor of metalloprotease, however, the other metalloprotease inhibitor 1,10 phenanthrolein had no inhibitory effect. It should be noted that the basis for classification of metalloproteases is the presence of metal ion (usually zinc) which participates in catalysis and 1,10 Phenanthroline is usually preferred as a diagnostic indicator of metalloproteases because of its higher stability constant for zinc than calcium.[23] Several reports on serine proteases susceptible to EDTA inhibition are also available in literature.[23-24] The activity of the protease was significantly inhibited by ZPCK (87.4 %), and TPCK (71.6 %), known to inactivate subtilisins and chymotrypsin, respectively. No inhibition of protease activity by thiol reagent iodoacetic acid and TLCK, a specific inhibitor for trypsin and strong inhibition of this enzyme activity by PMSF, a serine protease inhibitor, at 1 mM concentration, indicated *Serratia* sp. protease to be a serine protease.

Table 2 Effect of inhibitors and surfactants on the activity of *Serratia* sp. alkaline protease

Inhibitor	**Concentration**	**% Inhibition**
PMSF	1 mM	88.6
EDTA	1 mM	91.6
1,10	10 mM	NIL
Phenanthroline	1 mM	NIL
TLCK	1 mM	77.2
TPCK	1 mM	87.4
ZPCK	1 mM	NIL
Iodoacetic acid	1 mM	NIL
Benzamidine		

Conclusion

We report a novel organic solvent stable serine protease from a newly isolated *Serratia* sp.. The enzyme, purified 5.57 fold with a specific activity of 348.5 AU/mg and 30 % recovery, was characterized. The protease has molecular mass of 66.71 KDa, determined by MALDI TOF and SDS PAGE and possesses temperature and pH optima at 50°C and 7.5, respectively, with stability in the pH range of 6 to 11. The properties of this *Serratia* protease such as high proteolytic activity and stability at high alkaline pH and in the presence of wide range of solvents, detergents and oxidizing agents strongly suggest its potential for industrial applications.

Acknowledgement

Financial support from Council of Scientific and Industrial Research, New Delhi under CSIR Network Project (NWP 0044) is greatly acknowledged.

References

1. Singh J, Gill N, Devasahayam G, Sahoo DK (1999a) Studies on alkaline protease produced by *Bacillus* sp. NG 312. Appl Biochem Biotechnol 76:57-63
2. Doddapaneni KK, Tatineni R, Vellanki RN, Gandu B, Panyala NR, Chakali B, Mangamoori LN (2007) Purification and characterization of two novel extracellular proteases from *Serratia rubidaea*. Proc Biochem 42:1229-1236
3. Chun J, Lee JH, Jung Y, Kim M, Kim S, Kim BK, Lim YW (2007) EzTaxon: a web-based tool for the identification of prokaryotes based on 16S ribosomal RNA gene sequences. Int J Syst Evol Microbiol 57:2259-2261
4. Thompson JD, Gibson TJ, Plewniak F, Jeanmougin F, Higgins DG (1997) The CLUSTAL_X windows interface: flexible strategies for multiple sequence alignment aided by quality analysis tools. Nucleic Acids Res 25(24):4876-82
5. Saitou N and Nei M (1987) The neighbor-joining method: a new method for reconstructing phylogenetic trees. Mol Biol Evol 4:406-25
6. Kimura M (1980) A simple method for estimating evolutionary rate of base substitutions through comparative studies of nucleotide sequences. J Mol Evol 16:111-120
7. Felsenstein J (1985) Confidence limits on phylogenies: An approach using the bootstrap, Evolution 39:783-791
8. Laemmli UK (1970) Cleavage of structural proteins during assembly of head of bacteriophage T4. Nature 227:680-685
9. Smith PK, Krohn RI, Hermanson GT, Mallia AK, Gartner FH, Provenzano MD, Fujimoto EK, Goeke NM, Olson BJ, Klenk DC (1985) Measurement of protein using bicinchoninic acid Anal Biochem 150:76-85
10. Singh J, Vohra RM, Sahoo DK (1999b) Alkaline protease from a new obligate alkalophilic isolate of *Bacillus sphaericus*. Biotechnol Lett 21:921-924
11. Romero FJ, Garcia LA, Salas JA, Diaz M, Quiros LM (2001) Production, purification and partial characterization of two extracellular proteases from *Serratia marcescens* grown in whey. Proc Biochem 36:507–515

12. Wan MH, Wu B, Ren W, He BF (2010) Screening, characterization, and cloning of a solvent- tolerant protease from *Serratia marcescens* MH6. J Microbiol Biotechnol 20: 881-888
13. Li GY, Cai YJ, Liao XR, Yin J (2011) A novel nonionic surfactant and solvent -stable alkaline serine protease from *Serratia* sp. SYBC H with duckweed as nitrogen source: production, purification, characteristics and application. *J. Indus. Microbiol. Biotechnol.* 38:845-853
14. Khardenavis AA, Kapley A, Purohit HJ (2009) Processing of poultry feathers by alkaline keratin hydrolyzing enzyme from *Serratia* sp. HPC1383. Waste Management 29: 1409-1415
15. Moradian F, Khajeh K, manesh HN, Sadeghizadeh M (2009) Isolation, purification and characterization of a surfactants, laundry detergents-and organic solvents resistant alkaline protease from *Bacillus* sp. HR-08. Appl Biochem Biotechnol 159:33-45
16. Geok LP, Razak A, Rehman RNZRA, Basri M, Salleh AB (2003) Isolation and screening of an extracellular organic solvent tolerant protease producer. Biochem Eng J 13:73-77
17. Gupta A, Roy I, Khare SK, Gupta MN (2005) Purification and characterization of a solvent stable protease from *Pseudomonas aeruginosa* PseA. J Chromatogr *A* 1069:155-161
18. Ogino H, Yasuik, Shiotani T, Ishihara T, Ishikawa H (1995) Organic solvent-tolerant bacterium which secretes an organic solvent stable proteolytic enzyme. Appl Environ Microbiol 177:4258-4262
19. Ranos JL, Duque E, Huertas MJ, Haidover A (1995) Isolation and expansion of the catabolic potential of a *Pseudomonas putida* strain able to grow in the presence of high concentrations of aromatic hydrocarbons. J Bacteriol 77:3911-3916
20. Mehrota S, Pandey Pk, Gaur R, Darmwal NS (1999) The Production of alkali protease by a *Bacillus* sp. isolate. Biores Technol 67:201-203
21. Ghorbel B, Kamoun AS, Nasri M (2003) Stability Studies of protease from *Bacillus cereus* BGI. Enzyme Microb Technol 32:513-518
22. Ogino H, Ishikawa H (2001) Enzymes which are stable in the presence of organic solvents. J Biosci Bioeng 91:109-116
23. Abraham LD and Breuil C (1996) Isolation and characterization of a subtilisin-like serine protease secreted by the sap-staining fungus *Ophiostoma pieceae*. Enzyme Microb Technol 18:133-140
24. Singh J, Vohra RM, Sahoo DK (2001) Purification and characterization of two extracellular alkaline proteases from a newly isolated obligate alkalophilic *Bacillus sphaericus*. J Indus Microbiol Biotechnol 26:387-393

(594k) Microbial Load Control by Intermittently Delivered Pulsed Electric Fields

Alexander Golberg, UC Berkeley, Berkeley, CA

Microbial load control by intermittently delivered pulsed electric fields.

Alexander Golberg

Department of Mechanical Engineering, Etcheverry Hall 6124, University of California at Berkeley, Berkeley, CA 94720, USA.

Abstract: Microbial contamination prevention by intermittent delivery of pulsed electric field (IDPEF) is a new concept and method for turbid products storage. The treatment may be delivered at time intervals that is prescribed by PEF inactivation efficiency and microorganisms' growth kinetics, when known. This approach facilitates sterile storage without the need for chemical preservatives, additives, radiation or the complex infrastructure demanded by constanct connection to the electrical grid, which is not available in many developing countries. Unlike ultraviolet radiation, IDPEF can be used in turbid media. Furthermore, our previous work showed that IDPEF microbial control is comparable with refrigiration. Here, we first discuss the efficient IDPEF protocol desing principals. Second, we present the preliminary, laboratory scale studies of IDPEF application for milk storage. In this work we investigated the impact of IDPEF on *Lysteria monocytogines* in milk. U*sing the L. monocytogines kinetics growth data, we designed an IDPEF protocol that consists of* 2 sequences of 10 square wave pulses, 50 μs duration, 12.5 kV/cm electric field strength, delivered at 0.5 Hz and 1 min pause between the sequences *applied every 1.5 h. In a demonstration, 12 hours experiment, at* 32^0C storage temperature, *L. monocytogines* concentration of untreated samples *reached* $(9.1\pm0.6)\cdot10^7$ CFU/ml *and* $(7.1\pm0.3)\cdot10^8$ CFU/ml for *initial contamination levels of* $(1.4\pm0.2)\cdot10^3$ CFU/ml and $(3.1\pm0.3)\cdot10^6$ CFU/ml, respectively; while the final concentration in the IDPEF treated samples was 120±44 CFU/ml and $(1.1\pm0.3)\cdot10^5$ CFU/ml. The IDPEF storage method is of particular importance for parts of the world lacking continuous electricity supply, where today's pasteurization technologies use biomass burning that contributes to 4% of global CO_2 emissions. Prelimenary life-cycle analyses reveals that an energy needed for IDPEF storage of milk in India can be generated by a small scale 2kW solar energy system operating 5.5 hours per day in combination with small scale energy storage systems.

(594l) Studies In Fermentative Production and Downstream Processing of Polylysine

Sandip Bankar and Rekha Singhal, Food Engineering and Technology Department, Institute of Chemical Technology, Mumbai, India

Poly-ε-lysine (ε-PL), an unusual naturally occurring homopolyamide of L-lysine having linkage between ε-amino and α-carboxyl groups, is biodegradable, edible and non-toxic towards human and environment. It is preferentially produced by streptomyces species. Potential applications of ε-PL and its derivative have been of multifarious interest in foods, pharmaceuticals and medicine. In spite of commercial production and availability of ε-PL in a few countries, there are knowledge gaps still existing on many aspects of its production and downstream processing. This work was undertaken to look into a few of these aspects.

Initially, various marine strains were screened for the ability to produce ε-PL among which a strain labelled as G 349 was selected for ε-PL production. Biochemical tests and 16S rRNA results showed the ε-PL producer to belong to streptomyces species. Optimization of medium components by classical optimization and with metabolic simulators was carried out. In parallel set of experiments, the production of ε-PL by *Streptomyces noursei* NRRL 5126 in shake-flask culture was optimized by identifying the most significant medium components which affect ε-PL production by Placket–Burman design and subsequently by a statistical design, *viz.* evolutionary operation (EVOP) to determine the optimal concentrations of these components. Various metabolic precursors such as amino acids, tricarboxylic acid cycle intermediates and cofactors were investigated for improved production of ε-PL. The marine strain did not produce appreciable amount of ε-PL as compared to standard strain. Hence all further work was done with the standard strain itself.

The logistic and Luedeking-Piret equations have been proposed to describe the time course of ε-PL formation, substrate consumption and cell growth in shake flask level as well as in a fermenter. The growth kinetics of *S. noursei* NRRL 5126 was investigated under different aeration and agitation combinations in a 5.0 L stirred tank fermenter. Optimized oxygen supply (300 rpm and 2 vvm) in to the stirred tank fermenter shifted mixed growth associated biosynthesis of ε-PL to growth associated biosynthesis. A constant DO at 40% in the growth phase and 20% in the production phase increased the ε-PL yield as well as cell mass to their maximum values of 2 g/l and 20 g/l, respectively. The oxygen transfer rate (OTR), oxygen utilization rate (OUR), and specific oxygen uptake rates (qO_2) in the fermentation broth increased in the growth phase and remained unchanged in the stationary phase.

Use of resting cell cultures to increase the stability and shift in metabolism toward ε-PL biosynthesis was also implied. Optimization of non growth medium to achieve uniform culture density and higher ε-PL titre with minimal degradation was aimed with the help of Box-Behnken design and artificial intelligence. A comparative study of Box-Behnken design and artificial intelligence lead to higher production of ε-PL (1 g/l) during shake flask level.

Purification of ε-PL was done by using various chromatographic techniques such as ion exchange and gel permeation chromatography. Partial characterization of the ε-PL has been done. Further characterization of the same is in progress.

Potential applications of ε-PL to increase the stability of a model enzyme (α-amylase) and to increase the solubility of pharmaceutical drug have been evaluated.

(594m) Strain Engineering to Control a Product Quality Attribute in Escherichia Coli Recombinant Protein Production Process

Karthik Veeravalli, Late Stage Cell Culture, Genentech, Inc., South San Francisco, CA

Strain engineering to control a product quality attribute in *Escherichia coli* recombinant protein production processes

Continuous or bolus feeding of nutrients is commonly used in *E. coli* recombinant protein production processes to control different product quality attributes (PQAs). Although these strategies are effective in controlling PQAs, there are several operational disadvantages associated with continuous or bolus feeding. Additions to the culture increase the operational complexity and the overall cost of the fermentation process. In addition, these nutrient feeds lead to undesirable dilution of the fermentation medium resulting in lower cell densities and possibly lower product yields. An attractive alternative to using these feeding strategies, where applicable, is to engineer hosts that can control particular PQAs. In this work, we engineered the genome of an *E. coli* host expressing a recombinant protein product to control a PQA in the absence of the nutrient feeding. Furthermore, we show that the fermentation process performance utilizing the engineered host was comparable to the standard host for the recombinant protein production process.

(594n) Contribution of Maltose Binding Protein to Ligand-Allosteric Control of Chimeric Intein Cleavage

Jamie Mills, Chemical & Biomolecular Engineering, The Ohio State University, Columbus, OH, Jingjing Li, Department of Chemical and Biomolecular Engineering, the Ohio State University, Columbus, OH, David W. Wood, Chemical and Biomolecular Engineering, The Ohio State University, Columbus, OH and Richard A. Lease, Department of Chemical and Biomolecular Engineering, Ohio State University, Columbus, OH

We are interested in downstream protein purification using affinity tag technology coupled with a self-cleaving protein platform for bioseparations. Our model system is derived from a natural genetic element called an intein which, similarly to intron RNA sequences, may be inserted into a specific "host" protein-coding gene sequence (at the level of DNA) so as to interrupt the host gene amino acid sequence. Upon protein expression, intein sequences are capable of catalyzing their own excision and removal from the host peptide, leaving a re-joined, contiguous host protein sequence and the free intein peptide. In its amino acid sequence, the intein encodes several catalytic functions, including peptide-cleavage reactions at its N-terminal and C-terminal ends. We have repurposed the intein C-terminal cleaving activity into a bioseparations platform technology. In combination with existing affinity tag technology, mutant variants of the Mtu intein protein of *M. tuberculosis* now serve as a controllable, self-cleaving protein platform, which can be purified with any convenient affinity tag. The desired target product protein can be controllably cleaved off after the affinity purification step to produce purified protein. Presently, mutant variants of the engineered Mtu intein convey significant pH and temperature control over the cleavage reaction used to purify proteins. In principle the sequence for any protein desired for purification may be fused with the affinity-tagged intein at the DNA level and expressed as a protein for purification. Advantages of the intein-based system include the omission of protease cleavage to separate the desired protein from the affinity tag platform after expression and purification. Further, a variety of affinity tags have been used in conjunction with the engineered intein for chromatographic purification (e.g. chitin-binding domain, CBD, and maltose-binding protein or MBP) and for non-chromatographic purification via reversible aggregation tags (e.g. the elastin-like peptide or ELP tag; Wu et al. 2006, Nat Protocols **1**: 2257).

One shortcoming of this intein technology is that premature target protein cleavage may occur during cell growth and protein expression, decreasing yields and thus limiting its application at an industrial scale. Accordingly we sought to limit premature cleavage by incorporating small-ligand control of intein cleavage. We incorporated nuclear hormone receptor (NHR) ligand-binding domains (LBDs), e.g. thyroid receptor TR-beta and estrogen receptors, inserted within the inteins. Constructs containing maltose-binding protein (MBP) fused to the intein containing the NHRs apparently lost their intein-cleavage activities. Nevertheless, these intein fusion protein constructs have served as a basis for another project, namely our hormone-binding biosensor constructs (Skretas and Wood, 2005, JMB **349**:464; Li et al. 2011, Biosens Bioelectron **29**:132; Gierach et al., 2011, Environ Toxicol, *in press*) which transduce hormone binding into a growth phenotype, apparently by steric conformational mechanisms rather than via intein cleavage.

Recently we observed a variant of one of these biosensor constructs in which substitution of MBP with another protein (here, Cyan Fluorescent Protein or CyFP) seems to partially restore intein cleavage in an intein containing the thyroid receptor TR-beta. In contrast to its role as an affinity and solubility tag, we considered whether MBP is also a functional repressor of the intein cleavage reaction. Therefore, we proposed to replace MBP with other affinity tag domains (e.g. CBD or ELP tags) to determine whether CyFP promotes intein cleavage, or whether MBP blocks intein cleavage in the TR-beta-intein constructs. Ligand-binding control might be thus conveyed upon the intein cleavage reaction for the protein purification platform by careful choice of N-terminal protein domain in the TR-beta-intein context. We also discuss deletion analyses and mutagenesis studies to determine which specific regions or amino acid residues activate or inactivate NHR-intein cleavage.

(594o) Characterization of Flavin-Binding Fluorescent Proteins As a New Class of Oxygen-Independent Biological Imaging Probes

Arnab Mukherjee, Kevin B. Weyant, Joshua Walker and Charles M. Schroeder, Chemical and Biomolecular Engineering, University of Illinois at Urbana-Champaign, Urbana, IL

In this work, we report comprehensive biochemical and biophysical characterization of a new class of oxygen-independent flavin-binding fluorescent proteins (FbFPs) to enable their robust application as versatile biological imaging probes. Genetically encodable fluorescent reporters have catalyzed significant advances in our understanding of biological processes. By enabling live cell imaging with exquisite spatiotemporal resolution, fluorescent reporter proteins can monitor in a minimally invasive fashion, intracellular protein expression, localization, trafficking, and protein-protein, and protein-nucleic acid interactions. However, existing fluorescent proteins are almost exclusively derived from the green fluorescent protein (GFP) family. GFPs require molecular oxygen for fluorescence, which excludes their application to investigating anaerobic bioprocesses. Recently, an alternative class of fluorescent imaging probes was developed that bind flavin mononucleotide (FMN) as their chromophore and are capable of fluorescence in an oxygen-independent manner [1,2]. Therefore, flavin-binding fluorescent proteins (FbFPs) are attractive candidates for live cell imaging in microaerobic and anaerobic conditions.

In previous work, we have engineered FbFPs by directed evolution to improve brightness of fluorescence emission [3]. Here, we cloned, expressed, and purified existing FbFPs identified in *B.subtilis, P.putida*, and *A.thaliana* and characterized their performance as fluorescent reporters with respect to several key parameters. Specifically, we determined quantum yield, oligomeric state, fraction of fluorescent holoprotein in solution, intracellular stability, and denaturation/maturation kinetics. In addition, we evaluated sensitivity of fluorescence in FbFPs to environmental conditions such as temperature, pH, and redox. Furthermore, we investigated their potential as transcriptional reporters by employing FbFP-tagged promoters to monitor gene expression in *E.coli* cells under different conditions of growth. Based on our results, we identified iLOV, (from *A.thaliana*), as the most suitable FbFP for biological imaging applications. Overall, our characterization data brings to light several properties of FbFPs that are highly desirable in fluorescent reporter proteins, *e.g.*, thermal tolerance, broad pH range, redox insensitivity, small size, and fast maturation. Combined with oxygen-independent fluorescence, FbFPs represent a new and powerful class of fluorescent reporter proteins with potential applications in cancer therapy, bioremediation, and metabolic engineering for the production of high value biomolecules like bio-derived fuels and polyhydroxyalkanoates (PHA). We expect our work to enable the widespread use of FbFPs as reporters as well as provide a biochemical and biophysical basis for further protein engineering to optimize their spectral properties.

References:

1. T. Drepper, **et al**. *Nature Biotechnology*, 25: 443:445 (2007).
2. S. Chapman, **et al**. *PNAS*, 105(50):20038:20043 (2008)
3. A. Mukherjee, K.B. Weyant, and C.M. Schroeder, **submitted.**

(594p) Effect of Thermoplastic Extrusion and Protease Supplementation in Red Sorghum, Decorticated Sorghum and Maize On the Enzymatic Hydrolysis Efficiency As Pretreatment for Time Process Reduction

Mayeli Peralta-Contreras[1], Sergio Serna-Saldivar[2], Esther Perez-Carrillo[2], Erandi Escamilla-García[3] and Andrea Alcázar-Pizaña[4], (1)Biotechnology Center, ITESM, MOnterrey, Mexico, (2)Biotechnology Center, ITESM, Monterrey, Mexico, (3)Centro de Investigación y Desarrollo en Ciencias de la Salud, UANL, Universidad Autonoma de Nuevo León, Monterrey, Mexico, (4)Unidad de Odontología Integral y Especialidades, Centro de Investigación y Desarrollo en Ciencias de la Salud (CIDCS), Monterrey, Mexico

The growing demand for efficient processes has been the driving force for the implementation of thermoplastic extrusion in food and bio-fuel industries. This technology allows a quick starch gelatinization and promotes structural changes. The aim of this research is to compare the yield of extruded cereal with dried ground meals in terms of reducing sugars and free amino acid profile obtained during the starch hydrolysis. Split plot was the proposed experimental design to evaluate the cereal type, the conditions processing, and the protease supplementation effect in the reducing sugars and the prolamins. The extruded treatments increased sugars production during the first 45 minutes of hydrolysis (140 g/L) in comparison with the grounded treatments (45 g/L). The prolamins proteins were studied with a solvent extraction preparation changing the polarity of the solvent, before and after the enzymatic hydrolysis, preliminary results show that extruded materials modified the extraction yield and it was found a decrease in the protein content was also found. Protease use modifies the free amino acid and prolamins content during extruded treatments which can be correlated with starch breakdown. The research has demonstrated that the use of extrusion decreases time hydrolysis process and modifies the protein matrix.

(594q) Sub Micron Grinding of a Preservative for Food Surfaces

Gabrie Meesters, Chem Eng, Delft Univ Technology, Delft, Netherlands; DSM Biotechnology Center, DSM, Delft, Netherlands, Stephen Hennart, DSM Food Specialties / TUDelft, Delft, Netherlands and P. Van Hee, Food Specialties, DSM, Delft, Netherlands

This paper will describe how the activity of a poorly water soluble preservative product used in food coatings can be optimized with respect to microbial protection of the food. The paper focuses on food coatings. Preservation of food coatings is needed to prevent the growth of micro organisms.

The starting material is a particulate product with a $d_{(4,3)}$ of around 15 um. This product is ground down in a stirred media mill to a $d_{(4,3)}$ of about 180 nm. This process will be described, as well as the analysis done on the particle size measurements. Particle size analysis proved to be quite difficult. Next to this the product is tested in application on prevention of microbial growth. The sub micron preservative showed to be reaching its maximum concentration much faster than the non grounded material. The material was tested on a cheese surface. The results show a better protection against microbial growth of these smaller particles.

(594r) Cell Culture Process Development for pH Sensitive CHO Culture

Silvana D. Arevalo, Genentech Inc, San Francisco, CA

Initial 2L bioreactor studies for a high demand CHO product resulted in low titers and high lactate accumulation compared to previous shake flask screening studies. Trouble-shooting studies were performed to test differences between the bioreactor studies and the shake flask studies. Mimicking the shake flask uncontrolled pH profile in both the inoculum train and production culture was found to increase titer and reduce residual lactate concentration. Optimization of pH conditions in all cell culture stages improved titers and metabolic behavior in bioreactors and resulted in more comparable performance to initial cell line screens.

(594s) Bioconversion of Coffee Pulp by the Lignoculous Fungus Alternaria Alternata

Patricia Ortiz-Bermúdez, Department of Chemical Engineering, University of Puerto Rico, Mayaguez, PR

María del Pilar Sierra-Gómez[1], Mónica Medina-Quintana[2], Erick Zorrilla-Ramos[1] and Patricia Ortiz-Bermúdez[1]

1) Department of Chemical Engineering, University of Puerto Rico, Mayagüez Campus

2) Department of Civil and Environmental Engineering, University of Puerto Rico, Mayagüez Campus

E-mails: maria.sierra@upr.edu, monica.medina@upr.edu, erick.zorrilla@upr.edu, patricia.ortiz3@upr.edu

Efforts devoted to the efficient use of agro-industrial residues have increased, especially if their re-utilization contributes to the reduction of pollution problems associated with industrial processing. Coffee pulp is the main byproduct generated during the wet processing of coffee. To reduce the pollution from this process it is necessary to establish new treatments to bio-transform these wastes, which contain various toxic substances (e.g. caffeine and lignin). The presence of these substances limits their recycling potential and makes their management an environmental challenge. This study focuses on the identification and characterization of fungal strains that might be able to detoxify coffee processing wastes. Among the activities we screened for are decaffeination and delignification. We have identified *Alternaria alternata* as a potential candidate to degrade caffeine. This isolate was obtained from a ligniculous fungi collection provided by the Forest Products Laboratory in Madison, Wisconsin. *A. alternata* was evaluated in terms of caffeine degradation as carbon and nitrogen source. Under these conditions we observed a decrease in the caffeine content of 45.33% in a purified culture system. After addition of sucrose the caffeine concentration was reduced to 60.92%, in the same period of time (69 days). Analysis of solid-state fermentations of coffee pulp using *Alternaria alternata* demonstrated a significant removal of caffeine in 20 days, obtaining a 44.5% caffeine reduction. Analyses of dry weights of fermented coffee pulp coupled to ergosterol determinations confirmed fungal growth on the coffee pulp. In addition these analyses suggested that the fungus might be degrading the coffee pulp as well. This is the first report of the detoxification potential of *A. alternata* with respect to coffee-processing waste.

(596a) Engineering Cell-Material Interfaces for Long-Term Expansion of Human Pluripotent Stem Cells

Shyni Varghese, Bioengineering, University Of California, San Diego, La Jolla, CA and **Yongsung Hwang**, Bioengineering, University of California, San Diego, La Jolla, CA

Developing cost-effective, and scalable synthetic matrices for long-term expansion of human pluripotent stem cells (hPSCs) is important to realize their applications, ranging from drug screening platforms to regenerative medicine. Here, we report the development of a synthetic matrix containing heparin mimetic moieties to support long-term expansion of hPSCs (>20 passages) in a chemically defined Stempro medium. HPSCs expanded on these hydrogels maintained their characteristic morphology, colony forming ability, karyotypic stability, and differentiation potential. The synthetic matrix-expanded hPSCs exhibited pluripotent markers comparable to those cultured on Matrigel. Employing the hydrogel-based synthetic platform, we also investigated the effect of various physicochemical properties of the matrix (*e.g.,* functional group, hydrophobicity, and matrix rigidity) on adhesion, growth, and self-renewal of hPSCs. Our findings suggest that the synthetic hydrogels having an optimal interfacial hydrophobicity (or hydrophilicity) and matrix rigidity support long-term self-renewal of hPSCs. The observed cellular responses are explained in terms of matrix mediated binding of extracellular matrix proteins and growth factors, which provide a conducive microenvironment for the initial adhesion and growth of hPSCs. Such a matrix-assisted initial instructive environment would then be harnessed by the cells to activate the necessary integrin profile as well as ECM synthesis to create a niche that support self-renewal in contrast to differentiation. The synthetic matrices comprising of "off-the shelf" components are easy to synthesize and do not require any sophisticated processing thus making them cost-effective and translational. Furthermore, synthetic matrices with defined bulk and interfacial properties will be an ideal tool to understand the molecular mechanisms that control fate and commitment of hPSCs.

(596aa) Confocal Fluorescent Microscopy Vs. Direct Enumeration for Quantification of Biofilm Infection Deactivation

Ann O'Toole and Eric Nuxoll, Chemical and Biochemical Engineering, University of Iowa, Iowa City, IA

Development of strategies to mitigate bacterial biofilm infections is hampered by the difficulty in observing and quantifying biofilm viability. We are investigating thermal deactivation of *Pseudomonas aeruginosa* biofilms at a variety of time and temperature combinations. Eradication has been evaluated both through confocal fluorescent microscopy using membrane permeability dyes and through direct enumeration of colony forming units cultured from sonically homogenized post-treatment biofilms. Microscopy, while providing much more rapid, direct feedback, is limited by the ambiguity of intermediate light intensities, where it is unclear whether the light is due to fluorescing biomass in the confocal plane, or whether the light is simply background from extracellular or out-of-plane material. This problem is particularly acute at low signal levels, i.e. when only tiny but important fractions of viable bacteria remain. This presentation will discuss the applicability of image processor-based quantification of bacterial biofilms using both subjective and objective light thresholding, as compared to direct enumeration.

(596ab) Investigating the Effects of Dynamic External Stimuli On Single Cell Fitness and Gene Expression in E. Coli

Utsav Agrawal[1], Eric Johnson Chavarria[2], Arnab Mukherjee[1], Melikhan Tanyeri[1] and Charles M. Schroeder[1], (1)Chemical and Biomolecular Engineering, University of Illinois at Urbana-Champaign, Urbana, IL, (2)Center for Biophysics and Computational Biology, University of Illinois Urbana-Champaign, Urbana, IL

In this work, we study the bacterial cellular response to time-dependent external stimuli in single living cells. We developed a microfluidic platform for single cell analysis that allows for dynamic control of well-defined environmental growth conditions and culture media. Using this platform, we studied the effect of small molecule inducers on gene expression in the *lac* operon using fluorescent reporter proteins and cell growth rates as a proxy of cellular fitness. We applied temporally varying inducer concentrations by translating single cells between two adjacent fluid streams containing either growth medium or growth medium and inducer. We observed that single cell gene expression depends on growth rate and frequency of exposure to inducer concentrations. Single cell induction experiments are compared to control experiments with and without continuous fluid flow in microfluidic channels.

A microfluidic-based hydrodynamic trap recently developed in our lab facilitates single cell analysis for these experiments. The hydrodynamic trap enables confinement and manipulation of single cells in free solution using the sole action of fluid flow. Automated feedback control is integrated into the device using an "on-chip" valve, which allows for precise confinement of cells in free solution. Using this device, cells are confined at a fluid stagnation point generated in a cross-slot microfluidic geometry, thereby enabling non-perturbative trapping of cells for long time scales.

Using optical microscopy, we observe single *Escherichia coli* cells growing and dividing both at room temperature and at 37^0C, and cell division rates in the microfluidic trap compare favorably to the growth rates of *E. coli* measured in bulk studies. We demonstrated proof-of-principle induction of T5 promoter system with IPTG using the hydrodynamic trap. We are currently using this system to research the effect of alternating inducer concentrations at different frequencies on gene expression in order to determine a cutoff frequency for transcription of the lac operon. We anticipate that the microfluidic-based trap is an ideal platform to study cellular regulation and gene network dynamics of single living cells in free solution.

(596ac) Tunable Synthetic Divergent Promoters

Kang Wu, Energy Biosciences Institute, University of Illinois, Urbana, IL

In bacteria, many gene pairs are transcribed in opposite directions from divergent promoters. The promoters are characterized as having closely spaced and/or overlapping binding sites for RNA polymerase. In *E. coli*, divergent promoters are extremely common and actually outnumber standard, unidirectional promoters. Of these divergent promoters, roughly half have overlapping RNA polymerase binding sites.

In previous work we found that divergent promoters enable novel control in gene regulation by coupling the transcription of the regulator and structural genes. In this work, we investigated the ingredients involved in engineering divergent promoters. To do this, we constructed three libraries of overlapping divergent promoters based on a previously developed synthetic promoter library for unidirectional promoters. Clones with varying promoter strengths in both directions were obtained and analyzed to uncover the factors determining the behavior of overlapping divergent promoters.

• Three libraries of divergent promoters were constructed using degenerate oligonucleotide sequences with consensus sequences for *E. coli* promoters. All show a wide range of promoter strength in both directions.

• Ten promoters have mutations in the -10 or -35 consequence sequence. However, these mutations are not necessarily linked to reduced promoter strength.

• Phylogenetic analysis can qualitatively indicate the importance of certain positions. However, we did not observe a strong correlation based on the phylogenetic tree.

• Further analysis is needed to uncover the factors involved in varying promoter strengths.

(596ad) Supercooling: An Alternative Biopreservation Scheme

O. Berk Usta[1], Basak Uygun[2], Yeonhee Kim[1], Bote Bruinsma[3], Tim Berendsen[4], Jung Woo Lee[5], Korkut Uygun[2] and Martin L. Yarmush[6], (1)Surgery, Massachusetts General Hospital / Harvard Medical School, Boston, MA, (2)Center for Engineering in Medicine, Massachusetts General Hospital, Harvard Medical School, Shriners Hospital for Children, Boston, MA, (3)Surgery, Massachusetts General Hospital, Boston, (4)Center for Engineering in Medicine, MGH/Harvard Medical School, Boston, MA, (5)Surgery, Massachusetts General Hospital/ Harvard Medical School, Boston, MA, (6)Massachusetts General Hospital, Harvard Medical School, Boston, MA

Shortage in donor organs is increasing every year in part due to discarded organs for lack of extended preservation protocols; tissue engineered products face a similar problem where successful technologies cannot be translated from the lab to the clinic due to a bottleneck in suitable dissemination technologies. This presentation will focus on the development of a supercooling preservation method for liver cells and tissues with a long-term goal of whole liver preservation. Supercooling involves the sunzero-nonfreezing storage of cells, tissues and organs and strives to alleviate problems faced by current solutions such as cryopreservatoin and hypothermic storage. The long term goal with these studies is to extend preservation times of cells, tissues and eventually organs to more than a week which would alleviate challenges and constraints faced by both organ transplantation studies and tissue engineered technologies. The project also aims to furnish a fundamental understanding of cell damages via molecular simulations with the long term goal of aiding and improving experimental studies.

(596ae) Size-Dependent Cryopreservation of Pluripotent Stem Cell Aggregates

Sebastien Sart, Teng Ma and Yan Li, Chemical and Biomedical Engineering, Florida State University, Tallahassee, FL

Size-dependent Cryopreservation of Pluripotent Stem Cell Aggregates

Sebastien Sart, Teng Ma, Yan Li*

In recent years, pluripotent stem cells (PSC) are expanded as aggregates in suspension culture, which enables the large scale production in stirred bioreactors. During the production, PSCs are usually thawed from one vial of cells from Master Cell Bank (MCB) or Working Cell Bank (WCB). Integrating the culture system with cryopreservation process improves the feasibility of large-scale PSC banking and expansion. The optimized cryopreservation method also improves cell expansion and the differentiation. In this study, cryopreservation of PSC aggregates as dissociated single cells and intact aggregates was compared. It was found that intact aggregates quickly recovered after thawing in suspension. The size of aggregates affected the recovery and small aggregates (80-90 μm in diameter) recovered better than medium and large aggregates (average diameter 180 μm and 300 μm). Labeling with CellTracker Red showed the aggregate size-dependent diffusion. The production of reactive oxygen species (ROS) for the aggregates post-thaw was also investigated. It was observed that ROS production was initiated from aggregate center and small aggregates generated significant less ROS compared to medium and large aggregates. Different cryoprotectants were compared and PSC aggregate can be frozen in defined xeno-free formulation. The post-thawed aggregates maintained extracellular matrix (ECM) secretion, pluripotent marker expression, and the ability to differentiate into three-germ layers. The developed method can be used for establishing MCB and WCB for large scale PSC-derived cell production. This study also provides the insight for microtissue cryopreservation toward tissue engineering applications.

(596af) Role of Transferrin Receptor in Breast Cancers

Suhas Rao, Patrick T. Underhill and **Pankaj Karande**, The Isermann Department of Chemical and Biological Engineering, Rensselaer Polytechnic Institute, Troy, NY

To design effective treatments for breast cancer, it is important to elucidate the key molecular events and/or pathways that determine the progression of breast cancer and its response or resistance to molecular therapies. ~70% of breast cancers express estrogen receptor (ER), and ER-directed therapy has become mainstay treatment for breast cancers. Unfortunately, a substantial proportion of patients despite being ER and/or PR (progesterone receptor) positive, are either primarily resistant to hormone therapies or develop endocrine resistant breast cancers (ERBC). Additionally, ~15% of breast cancers do not express ER, PR or HER2 (human epidermal growth factor receptor 2) and are termed triple negative breast cancers (TNBC). It is well-established that transferrin receptor (TfR) is overexpressed in breast cancer. We will discuss our efforts to correlate the expression and distribution of TfR in breast cancers to their progression and resistance to therapies. This information will guide the future design of combination therapies to overcome the limitations of current therapies.

(596ag) Kinetic Models of Tissue Tropism in Breast Cancer Metastasis

Lauren E. Barney[1], Erinn C. Dandley[1], T. J. Mountziaris[2] and Shelly R. Peyton[1], (1)Chemical Engineering, University of Massachusetts, Amherst, Amherst, MA, (2)Department of Chemical Engineering, University of Massachusetts, Amherst, MA

Metastasis is the leading cause of fatality for women diagnosed with breast cancer. The most common anatomical sites of distant tumor growth include the brain, lung, liver, and bone, and it is well known that this metastatic spread in breast cancer is not random. Rather, different clinical subtypes of breast cancer exhibit unique patterns of metastatic site preference, called tissue tropism. Given the physical and chemical diversity of these secondary tissue sites, we hypothesize that there is a relationship between the biophysical and biochemical properties of the tissue, and the ability of cells within a particular subtype of breast cancer to adhere, migrate, and grow. Here, we present Engineered Metastatic Microenvironments (EMMs), which capture the integrin-binding properties of the tissues of several common secondary sites *in vivo* (brain, lung, and bone). Using these EMMs, we have observed differences in cell behavior based upon the extracellular matrix (ECM) of these tissues, which can be used to determine specific mechanisms in metastatic site preference. We are building a kinetic model of the ability of cells to adhere, spread, and polarize in different subtype-EMM combinations, in order to eventually build a predictive model of tissue tropism in breast cancer patients.

The EMMs we have designed consist of ECM proteins coupled to glass coverslips using silane chemistry. The composition of ECM varies with each EMM: **brain:** 1ug/cm^2 of 50% fibronectin, 25% vitronectin, 20% tenascin and 5% laminin (wt%), **lung:** 2ug/cm^2 of 3% laminin, 33% collagen IV, 15% collagen I, 15% fibronectin, and 4% tenascin (wt%), and **bone**: 5ug/cm^2 of 99% collagen I and 1% osteopontin (wt%). We have quantified the real-time adhesion kinetics of four human breast cancer cell lines on each of these EMMs: MDA-MB-231, BT-549, MCF7 and SKBr3, which are classified into three distinct clinical subtypes: claudin low, luminal, and estrogen (ER) and progesterone receptor (PR) negative/human epidermal growth factor receptor-2 (HER2)-enriched. These clinical subtypes each have a known human, clinical metastatic profile [1]. We have imaged individual cell spreading on EMMs at high magnification, and observed qualitative differences in initial spreading. To quantify this, we developed an equation based upon Michaelis-Menten enzyme kinetics to describe cell spreading. This allows for the determination of two biologically relevant adhesion parameters: K, which is the time to half of the maximum cell spreading, and A_{max}, which is the maximum cell area.

Each of the cell lines shows distinct differences in their initial spreading on the three EMMs. The highest K values occur for all four cell lines on the bone, indicating that cells spread at a slower rate in the bone environment than in the lung or brain overall. We also observe faster rates of spreading and higher variability in maximum area in cell lines that are known to generally be more metastatic *in vivo*. This is a significant result, because an individual cell's immediate reaction to the secondary tissue site may be critical to the eventual formation of a successful metastatic lesion. We are currently investigating whether integrin blocking will shift these rate curves, to understand mechanisms of how initial integrin-binding kinetics control the metastatic process. In addition to these well-established cell lines, we are now comparing these results with the adhesion kinetics of breast cancer samples from a local tissue bank. Our preliminary observations suggest that even very aggressive, metastatic patient cell samples have much lower spreading kinetics than these well-established cell lines, challenging the translational use of cell lines in lab settings. We envision that this simple kinetic model of cell adhesion and spreading, combined with our work on polarization, migration, and proliferation, can be used to predict probable metastatic sites in patients, leading to patient-specific therapy and relapse site probability.

1. Kennecke, H., et al., J Clin Oncol. **28(20):** p. 3271-7.

(596ah) A Distinct Phenotypic Signature for Pancreatic Cancer Metastasis

Pei-Hsun Wu[1], Jude Phillip[1], Wei-Chiang Chen[1], Sonal Gupta[2], Jeffery T Leek[3], Anirban Maitra[2,4] and Denis Wirtz[1], (1)Chemical and Biomolecular Engineering, Johns Hopkins University, Baltimore, MD, (2)Department of Oncology, Johns Hopkins University School of Medicine, Baltimore, MD, (3)Department of Biostatistics, Johns Hopkins University, Baltimore, MD, (4)Departments of Pathology and the Sol Goldman Pancreatic Cancer Center, Johns Hopkins University School of Medicine, Baltimore, MD

Ninety percent of all cancer-related deaths are caused by metastatic disease, i.e. the spreading of a subset of cells from a primary tumor in an organ to distal sites in other organs. Understanding this progression from localized to metastatic disease is essential for further developing effective therapeutic and treatment strategies. However, despite research efforts, no distinct genetic, epigenetic, or proteomic signature of cancer metastasis has been identified so far. Metastasis is a physical event: through invasion and migration through the dense, tortuous stromal matrix, intravasation, shear forces of blood flow, successful re-attachment to blood vessel walls, migration, the colonization of a distal site, and, finally, reactivation following dormancy, metastatic cells may share precise physical properties. Cell morphology is the most direct physical property that can be measured. In this work, we develop a high throughput cell phenotyping process and investigate the morphological signature of primary tumor cells and liver metastatic pancreatic cancer cells.

(596ai) Mechanism-Based Pharmacokinetic Model to Predict Endogenous Autoantibody Clearance Levels

Venkat R. Pannala[1], Dilip Challa[2], Sally Ward[2] and Leonidas Bleris[1], (1)Department of Bioengineering, University of Texas at Dallas, Richardson, TX, (2)Department of Immunology, University of Texas Southwestern Medical Center, Dallas, TX

Immunoglobulin G (IgG) is the major antibody in the serum which is involved in secondary immune response against foreign microorganisms. The neonatal Fc receptor, FcRn, is responsible for controlling the half-life of antibodies [1] by protecting them from degradation via recycling. A range of autoimmune diseases are characterized by the involvement of self-reactive antibodies in pathogenesis. As a result, engineered antibodies have been developed to inhibit the FcRn function to modulate IgG half-lives [2, 3]. In particular, engineered antibodies with increased affinities for FcRn (Abdegs) can enhance the clearance rates of endogenous IgGs by inhibiting the FcRn function. As such, Abdegs have potential for the treatment of autoimmunity [4].

In this study, we developed a mechanism-based pharmacokinetic model that describes the FcRn-mediated IgG recycling pathway and reveals the effect of Abdegs on radiolabeled mIgG1pharmacokinetics. The developed FcRn trafficking pathway is based on experimental observations and includes the following novel aspects: (i) a weak interaction of FcRn and IgG at neutral pH, (ii) 1:2 stoichiometric interaction of IgG with FcRn both at neutral and acidic pH, (iii) synthesis of endogenous FcRn, (iv) degradation of free and bound FcRn complexes only in the lysosomes. The parameters governing the model were obtained from experimentally determined pharmacokinetics of radiolabeled mIgG1 in mice.

Our model accurately captures the serum radiolabeled mIgG1 data (concentration/time) in the absence and presence of Abdegs. The simulations can predict the effect of Abdegs on the clearance of endogenous IgG in a mouse. We show that endogenous IgG levels drop to 70%, 58% and 46% levels with the injection of 0.5, 1 and 2 mg/mouse concentrations of Abdegs, respectively. These simulation results are in agreement with the experimentally observed endogenous clearance levels. Finally, we show that our model can reliably predict pharmacokinetics of mutated IgGs that have varied affinities towards FcRn and therefore aid in the engineering of novel antibodies towards customized control of endogenous IgG half-lives.

References

1. Ghetie V, Ward ES: **Multiple roles for the major histocompatibility complex class I-related receptor FcRn**. In., vol. 18; 2000: 739-766.

2. Low SC, Mezo AR: **Inhibitors of the FcRn:IgG protein-protein interaction**. *AAPS Journal* 2009, **11**(3):432-434.

3. Ward ES, Ober RJ: **Chapter 4: Multitasking by Exploitation of Intracellular Transport Functions the Many Faces of FcRn**. *Adv Immunol* 2009, **103**:77-115.

4. Patel DA, Puig-Canto A, Challa DK, Montoyo HP, Ober RJ, Ward ES: **Neonatal Fc receptor blockade by Fc engineering ameliorates arthritis in a murine model**. *Journal of Immunology* 2011, **187**(2):1015-1022.

(596aj) Three Dimensional Tumor Spheroids Through Cell-Instructive Polyamine Based Hydrogels

Thrimoorthy Potta, Chemical Engineering, Arizona State University, Tempe, AZ, Taraka Sai Pavan Grandhi, Biomedical Engineering, Arizona State University, Tempe, AZ and Kaushal Rege, Chemical Engineering, Biomedical Engineering, Arizona State University, Tempe, AZ

Recently significant efforts have been made to develop in vitro 3D tumor models with structural resemblance to *in vivo* cancers of interest to study the tumor biology and drug treatment. In tumors, cancer cells co-exist with several other cell types and form complex organ like structures with significant heterogeneity. One popular 3D tumor model is comprised of tumor spheroids, which are generally obtained by culturing cells suspended in spinner flasks. Though this method of developing spheroids is popular, but it suffers from many limitations such as controlling the spheroids size and its collection from culture media. Herein we developed a novel hydrogels from a chemical reaction between polyamines and diglycidyl ethers, which supported spheroid formation and easy separation. We developed spheroids with controllable size from 100 microns to 900 microns. We developed three dimensional breast cancer tumor model containing fibroblast and breast cancer cell lines. Confocal imaging microscopy was used to study the tumor heterogenicity and spheroid cell viability. These preliminary studies strongly advocate the potential of these hydrogels to support three dimensional disease models.

(596ak) Doing Cancer Surgery One Cell At a Time

Joseph A. Zasadzinski, Department of Chemical Engineering and Materials Science, University of Minnesota, Minneapolis, MN and Dmitri Lapotko, Biochemistry and Cell Biology, Rice University, Houston, TX

Cancer chemotherapy is indiscriminate, and is often limited by damage to the heart, liver, kidneys, etc. The goal of our research is to be able to discriminate cell by cell, that is, to deliver overwhelming doses of a drug to cancer cells while sparing healthy cells from even minor exposure. To do this, we use antibody targeting to selectively deliver gold nanoparticles to cancer cells surrounded by normal cells in culture. The selectivity of the antibody insures that the targeted cells receive a higher dose of nanoparticles, which are sequestered in endosomal compartments inside the cells. Fewer nanoparticles at a lower density are found in normal cells. The cell culture is then irradiated with with pixosecond-long pulses of near-infrared light. The light energy is stongly absorbed by the gold nanoparticles and the cells are essentially transparent to the NIR light, making the irradiation completly non-toxic to the cells. The light energy absorbed by the gold nanoparticles converts instantaneously to heat, raising the nanoparticle temperature sufficiently to melt the gold. As heat begins to dissipate to the surrounding liquid environment, a small amount of water is vaporized to generate transient nanobubbles. Like the cavitation events in sonication, these nanobubbles are unstable, and rapidly grow, then collapse on themselves, generating substantial liquid and vapor jets. These tiny jets tear apart *individual* cancer cells mechanically; with no need for diffusion of heat or a chemical chemotherapy agent. By tuning the incident light energy, only cancer cells that have received multiple nanoparticles initiate the nucleation of nanobubbles and are destroyed, greatly enhancing the selectivity of the antibody. Nanobubbles can be combined with a 10-fold reduced dose of conventional chemotherapy agents such as doxorubicin or cisplatin to enhance the delivery of the drug to the mechanically damaged cells. The combination of chemotherapy and mechanical damage destroyed even drug-resistant cancer cells without significant damage to normal cells.

(596al) Transdifferentiation of Mesenchymal Stem Cells on Micropatterned Polymeric Substrates

Anup Sharma, Department of Genetics, Development and Cell Biology / Chemical and Biological Engineering, Iowa State University, Ames, IN

Peripheral nerve injuries are cause of serious neurological disorders and can result in paralysis and sensory/motor dysfunctions. Schwann cells (SCs) can enhance myelin production capabilities for insulating neuronal axons which may also help to improve the motor function of the damaged nervous tissue. Researchers in our group and elsewhere have extracted Schwann cells and seeded them in conduits and have shown that they facilitate and enhance nerve regeneration in various animal models. However, there is no good source of Schwann cells, and they have to be obtained by sacrificing a healthy nerve which may result in donor site morbidity. Additionally, since SCs are not easy to maintain in a cell culture environment, we are focusing on producing cells having SC like phenotype using adult stem cells. Mesenchymal stem cells (MSCs) are multipotent cells capable of transdifferentiating into various cell types including a Schwann cell like phenotypes. The focus of this work is to transdifferentiate MSCs into SC-like phenotypes using micropatterned polymer substrates. Micropatterning has been used for aligning various cell types in the past, and our hypothesis is that alignment would promote transdifferentiation into SC-like phenotypes. We have shown that MSCs align themselves in the direction of microgrooves when grown on micropatterned substrates. As a proof of principle, we have used polystyrene (PS) films for testing the transdifferentiation of MSCs and their alignment, as PS is biocompatible and a good substrate for cell culture. Immunocytochemistry experiments have been done on the MSCs grown on PS films to determine the alignment and transdifferentiation into SC-like phenotypes. The transdifferentiated cells align along the microgrooves, which can help the regenerating axons grow in particular directions. In the future, we are going to use FDA approved biodegradable polymers, which could be engineered into a conduit and can be inserted at the site of damaged nerve to promote peripheral nerve regeneration.

(596am) Physiologically Based Pharmacokinetic and Pharmacodynamic Approach to Nanoparticle Biodistribution and Tumor Uptake: A Comparison Between Gold Nanoparticles and Spions Nanoparticles

Angel A. Galvis, Ruben D. Vargas and Watson L. Vargas, Department of Chemical Engineering, Universidad de los Andes, Bogotá, Colombia

Nanomedicine and nanostructured materials research have been in exponential growth during the last decade, several efforts have been focused on the nanoparticle design. Nevertheless, the knowledge about nanoparticles absorption, distribution, metabolism and excretion (ADME) has been growing more slowly, in fact these ADME parameters are required in order to obtain the pharmacokinetic and pharmacodynamic of these nano-sized materials.

Physiologically based pharmacokinetic models (PBPK) are mathematical strategies developed in order to analyze ADME properties for pharmaceutical compounds, treating the organs such as compartmental structures with several interconnections modeled by differential equations. Therefore, the appeal of PBPKs arises from accurate prediction models for concentration profiles by varying: doses, exposure duration and routes of administration. In addition these models have been useful for interspecies scalability.

The aim of the current study is to provide *in silico* physiological based pharmacokinetic models for the dynamic biodistribution of gold nanoparticles and superparamagnetic iron oxide nanoparticles (SPION´s) both coated with different surface modifiers typical for each kind of particles (dextran, polyetileneglicol, and citrate), analyzing surface charge and size effects into the model. To date, there are few studies on how the surface charges of nanocarriers impact their biodistribution, pharmacokinetics and tumor uptake.

The proposed models are focused on dynamic cancer treatment strategies, a comparison between gold nanoparticles and SPION´s is developed, and issues such as liver or lymphatic system bioaccumulation and metabolism are specially treated, using experimental data for the most relevant organs (compartments). In addition, exploratory models for tumor evolution are included into the model to provide a PBPK model for chemotherapy agents and tumor-targeted substances coupled to gold nanoparticles.

The PBPK models includes separation between venous and arterial blood pools, and the main organs for nanoparticles biodistribution have been modeled in detail (i.e. Liver, lungs, kidneys, spleen, hearth, brain), Additionally, the parameters for this model have been adjusted and validated using different kinds of preclinical data in mice.

(596an) Selectin-Mediated Adhesion in Shear Flow Using Micropatterned Substrates

Luthur Siu-Lun Cheung[1], ZiQiu Tong[1], Kathleen J. Stebe[2] and Konstantinos Konstantopoulos[1], (1)Chemical and Biomolecular Engineering, Johns Hopkins University, Baltimore, MD, (2)Chemical & Biomolecular Engineering, University of Pennsylvania, Philadelphia, PA

Receptor-ligand adhesive interactions play a pivotal role in diverse biological processes including inflammation and cancer metastasis. Cell adhesion is mediated by the molecular recognition of membrane-bound receptors by their cognate ligands on apposing cells. Cell-cell binding is regulated by distinct parameters such as the receptor-ligand binding kinetics, the tensile strength of individual bonds, the involvement of multiple bonds and their modulation by hydrodynamic shear. This work aims to investigate the interplay of these parameters on selectin-mediated cell adhesion in shear flow. We designed a microfluidic device that delivers cells in a single file over a receptor-functionalized substrate, thereby permitting accurate determination of the cell flux. The selectin(s) was presented on striped patches of fixed width and varying length. We identified the critical patch lengths of P- and L-selectins for the initiation of HL-60 cell binding in shear flow. This characteristic length is governed by the time required to form multiple-bond interactions, as revealed by a multiple-bond mathematical model. The number of bonds required to support cell binding increases with the applied shear stress (0.5-2 dyn/cm^2) for L- but not P-selectin. This finding is explained by differences in the tensile strength of P- and L-selectin for PSGL-1. Our integrated experimental and mathematical approach advances our understanding of receptor-mediated cell adhesion in the vasculature. Detailed knowledge of how molecular interactions modulate macroscopic cell binding behavior pertinent to inflammation and metastasis would facilitate the development of promising diagnostic tools to combat these diseases.

(596ao) Neutrophil Chemotaxis in Multiple Chemoattractant Gradients Using Microfluidic Platforms

Matthew Byrne[1], Yuki Kimura[1], Ashish Kapoor[1], Fei Wang[2], Paul J.A. Kenis[3] and Christopher V. Rao[1], (1)Chemical and Biomolecular Engineering, University of Illinois, Urbana Champaign, Urbana, IL, (2)Cell and Developmental Biology, University of Illinois, Urbana Champaign, Urbana, IL, (3)Department of Chemical & Biomolecular Engineering, University of Illinois at Urbana Champaign, Urbana, IL

Neutrophils, the most abundant type of white blood cells, play an essential role in the primary stages of the cellular immune response. These cells efficiently sense and migrate up concentration gradients of chemical attractants toward sites of infection and inflammation in a process termed chemotaxis. Such chemoattractants include end-target chemoattractants produced at or proximal to the source of infection, as well as intermediary (endogenous) chemoattractants produced by inflamed host tissues. While much is known about the molecular interactions and signaling pathways that regulate cellular responses to individual cues, little is known about how cells process multiple cues to effect migration in the appropriate direction *in vivo*.

The goal of this work is to understand how neutrophils integrate and prioritize multiple chemotactic signals to navigate toward pathogens. To this end, we investigated neutrophil migration in response to multiple chemoattractant gradients through the generation of precisely controlled microenvironments using microfluidics. By analyzing the migration behavior of the cells in response to single and dual gradients of chemoattractants (IL8, LTB4 and fMLP), we have begun to elucidate how neutrophils respond to complex overlapping chemotactic cues. The chemotactic behavior of primary neutrophils suggests the role of directional persistence and cellular memory in guiding cells to their targets. These results provide insight into a key component of the inflammatory response and should aid in the design of novel therapeutic strategies to treat certain immune-based health disorders.

(596ap) A Three Compartment Model for Pharmacodynamics and Anomalous Diffusion of Engineered Nanoparticles

Lina Ubaque, Ruben D. Vargas and Watson L. Vargas, Department of Chemical Engineering, Universidad de los Andes, Bogotá, Colombia

Application of nanocarriers as a solution for targeted drug delivery is a topic of increasing interest [1-4]. Transport of these nano-particles through biological tissues has been a problematic area since describing its dynamics has proven to be inefficient using classical approaches due to the anomalous diffusion (Podlubny, 1999) that often times is observed in such complex media.

Anomalous diffusion is characterized by a nonlinear evolution of the solute Mean Square Displacement (MSD) in time as is supposed by classical transport laws (e.g. Fick law). As a solution to these kind of anomalies in MSD evolution, fractional calculus has emerged as a new branch of applied mathematics capable of involving memory in transport processes, an important number of works has been recently published incorporating fractional temporal derivatives [6] over the classical Flick's Law, therefore, arriving to the so-called Time Fractional Diffusion Equation (TFDE).

In this work a three compartment Physiologically Based Pharmacokinetic Model (PBPK) is presented with the incorporation of the TFDE that takes into account memory effects in transport through biological tissues. The compartments are connected in series with only mass transfer by diffusion as the only mechanism available for transport. The first of the compartments is a liquid cavity assumed to be perfectly well mixed and initialized with a high particle concentration of nanoparticles. The second is a biological tissue where one-dimensional anomalous diffusion takes place and with an initially null particle concentration. Finally the third compartment is another perfectly mixed liquid cavity with no particle concentration at the beginning of the process.

Parameters of the model (including the fractional derivative order) are adjusted to experimental data from a Franz diffusion cell via lest squares optimization over the evolution of the third compartment particle concentration. Future directions for Improving the PK of nanoparticles in complex tissues are discussed.

Keywords: Time delay, pharmacodynamics, anomalous diffusion, fractional derivatives.

References:

1. Kohji Ohno, Tatsuki Akashi, Yoshinobu Tsujii, Masaya Yamamoto, and Yasuhiko Tabata. Blood Clearance and Biodistribution of Polymer Brush-Afforded Silica Particles Prepared by Surface-Initiated Living Radical Polymerization. *Biomacromolecules* 2012, 13, 927−936
2. Rochelle R. Arvizo, Oscar R. Miranda, Daniel F. Moyano, Chad A. Walden, Karuna Giri, Resham Bhattacharya, J. David Robertson, Vincent M. Rotello, Joel M. Reid, Priyabrata Mukherjee. Modulating Pharmacokinetics, Tumor Uptake and Biodistribution by Engineered Nanoparticles *PLoS ONE* | www.plosone.org September 2011 | Volume 6 | Issue 9 | e24374.
3. Mingguang Li, Zoi Panagi, Konstantinos Avgoustakis, Joshua Reineke. Physiologically based pharmacokinetic modeling of PLGA nanoparticles with varied mPEG content. *International Journal of Nanomedicine* 2012:7 1345–1356
4. Shyh-Dar Li and Leaf Huang. Pharmacokinetics and Biodistribution of Nanoparticles. *Molecular pharmaceutics* 2008, VOL. 5, NO. 4, 496–504
5. Podlubny, Fractional Differential Equations, Academic Press, San Diego (1999).
6. Ivo Petraš, Richard L. Magin. Simulation of drug uptake in a two compartmental fractional model for a biological system. *Commun Nonlinear Sci Numer Simulat* 16 (2011) 4588–4595

(596aq) Physiologically Based Three-Compartment Model for Anomalous Diffusion of Gold Nanoparticles with a Pulsed Inlet

Laura M. Hernandez, Angel A. Galvis, Ruben D. Vargas and Watson L. Vargas, Department of Chemical Engineering, Universidad de los Andes, Bogotá, Colombia

The enhanced permeability and retention (EPR) effect has been explored as an appealing route for nanoparticle transport into tummoral tissues, due to the high rate of angiogenesis in cancerous lymphomas among other neoplasms. The EPR effect allows nanoparticles to penetrate a high number of blood vessels in comparison to non-cancerous tissues. Nevertheless, classical models for diffusive and convective transport of nanoparticles through biological tissues do not achieve the level of detail needed, because there are several restrictions related to the behavior of the mean square displacement (MSD) in such complex media (*i.e.* anomalous diffusion). In addition, today little is known on how nanoparticles properties affect individual transportation kinetic parameters [1].

This study provides a model and its numerical solution for nanoparticle diffusion in an emulated biological blood vessel with some simplifications, modeled by a three-compartment model, where the nanoparticles mass transport within the tissue is assumed to take place by an anomalous diffusion process, the first compartment is a reservoir with high concentration of gold nanoparticles and with a pulsed inlet, the second compartment is an emulated biological tissue that allows only a one-dimensional diffusion initially with a null concentration of nanoparticles, through this compartment nanoparticles reach a well mixed liquid cavity with no particle at the onset of the experiment.

The effect of the pulsed behavior in the first compartment is an interesting factor that is studied in this model due to the fact that it is not straightforward to develop a prediction of how a pulsed inlet condition could influence the behavior of the particles MSD within the tissue. Fractional calculus strategies arise as the natural model for these kinds of transport phenomena because the time fractional diffusion equation (TFDE) allows the inclusion of memory effects into the model.

The experimental set-up consists of permeation experiments conducted in a Franz diffusion cell with rigorous controlled conditions, the nanoparticles are used without any marker substance in order to provide an accurate prediction of the chemical interaction between the particles and the emulated biological tissue (two types of gold nanoparticle with surface functionalization are used: a biocompatible polymer modified surface and citrate modified surface), the concentration inside the third compartment is measured via HR-UV-VIS spectroscopy. Finally, a finite differences type method is used for obtaining theoretically predicted concentrations in the third compartment, this concentrations are contrasted with the experimental data available through a lest squares optimization which allows the evaluation of the transport parameters related to the anomalous MSD behavior (fractional derivative order, relaxation time and diffusivity).

Key Words: anomalous diffusion, nanoparticle diffusion, transport phenomena, gold nanoparticles.

References:

1. Mingguang Li, Zoi Panagi, Konstantinos Avgoustakis andJoshua Reineke. Physiologically based pharmacokinetic modeling of PLGA nanoparticles with varied mPEG content. *International Journal of Nanomedicine* 2012:7 1345–1356

(596ar) Interstitial Permeation of Human Blood Clots Formed Under Flow Using Controlled Pressure Gradients in a Microfluidic Model of Bleeding

Ryan W. Muthard and Scott L. Diamond, Chemical and Biomolecular Engineering, University of Pennsylvania, Philadelphia, PA

Traumatic injuries and genetic bleeding disorders provide a sustained clinical need to understand the complex transport properties of intraluminal thrombi. We developed a microfluidic device to measure intraluminal thrombi permeabilities under controlled pressure gradients. Unlike many previous *in vitro* permeability methods, this device incorporates the critical mechanical and biological thrombi properties that can only be obtained under hemodynamic conditions. The PDMS microfluidic device maintains blood pressure and flow rate by utilizing two constant volume syringe pumps and a LabVIEW interface. The main channel is perfused with anti-coagulated human whole blood at an initial shear rate of 1100 s^{-1}. After flowing blood contacts the localized collagen or collagen/lipidated tissue factor scaffold; the controlled pressure gradient forces blood to permeate through the structure. The pressure gradient is maintained by a downstream infusion of Ca^{2+} buffer controlled by LabVIEW and measured using three pressure transducers. Permeate velocity was measured using a pulse of texas red dye and a numerical model was employed to determine the permeabilities of the collagen scaffold, platelet layer, and platelet/fibrin layer. The device was validated by calculating a permeability of collagen ($1.98 \times 10^{-11} \pm 6.01 \times 10^{-12}$ cm^2) over varying pressures (12-23.4 mm Hg) and comparing to previous literature values. The resulting permeabilities for a platelet ($5.45 \times 10^{-14} \pm 9.66 \times 10^{-15}$ cm^2) and platelet/fibrin layer ($2.71 \times 10^{-14} \pm 3.56 \times 10^{-15}$ cm^2) formed over 10 minutes of flow provide insight into the relative resistance that each supply to the cessation of bleeding. Noticeably, the formation of fibrin and presence of thrombin provided a 50% reduction in permeability under hemodynamic conditions. In addition to permeability measurements, we observed decreased platelet accumulation with increased pressure drop (13.8-23.4 mm Hg) across the intraluminal thrombi in PPACK whole blood. This response can be explained by the increased permeation of ADP and TXA_2 through the forming thrombi. The design and validation of our microfluidic device allowed for permeability measurements of intraluminal thrombi with and without thrombin. In the absence of thrombin, PPACK whole blood formed a platelet mass which was >350 fold less permeable than collagen alone. An additional 50% reduction of permeability was observed when thrombin was present and platelet/fibrin clots were formed.

(596as) Calcium-Alginate Mediated Nucleic Acid Delivery in a Microfluidic Device

Jordan F. Betz[1,2], Yi Cheng[2,3], Chen-Yu Tsao[1,4], Hsuan-Chen Wu[1,4], Gregory F. Payne[1,4], William E. Bentley[1,4] and Gary W. Rubloff[2,3], (1)Fischell Department of Bioengineering, University of Maryland, College Park, MD, (2)Institute for Systems Research, University of Maryland, College Park, MD, (3)Department of Materials Science and Engineering, University of Maryland, College Park, MD, (4)Institute for Bioscience and Biotechnology Research, University of Maryland, College Park, MD

Calcium alginate entrapment of cells has been used extensively in cell culture for cellular and tissue engineering applications. We recently described a method to co-deposit and culture cells inside a calcium alginate hydrogel by means of a DC electric signal [1] that also allows for the study of intercellular signaling phenomena in a microfluidic device [2]. We report an advance that enables the site-specific genetic reprogramming of cells- introducing a plasmid into the cells by electrically-assisted chemical transformation as they are entrapped within the calcium alginate hydrogel. In addition, this system allows for direct optical access to the cells throughout the entire hydrogel, giving researchers a new level of control over cellular engineering experiments and the delivery of nucleic acids to cell populations as well as the ability to study spatial distribution, migration, and responses of reprogrammed cells through the use of confocal microscopy.

A mixture of 1% (w/v) sodium alginate solution and 0.5% (w/v) $CaCO_3$ nanoparticles was electrodeposited at a constant current density of 10 A/m^2 for 30 seconds at each of four electrodes inside a microfluidic device whose fabrication is described previously [2]. A mixture of 0.5% (w/v) sodium alginate solution and 20 ng/μL plasmid was mixed gently with an equal volume of electrocompetent *Escherichia coli* cells and electrodeposited at a constant current density of 6 A/m^2 for 3 minutes at the same electrodes. The transformed cells were then allowed to recover at 37°C for an hour in LB media before flowing 1 μL/min LB selection media supplemented with appropriate antibiotics and 40 mM $CaCl_2$ to maintain hydrogel integrity. Induction media, which is identical to the selection media except for the addition of 1 mM IPTG, was added at 1 μL/min after bacteria reached mid-log phase for experiments requiring induction.

The deposition of a small calcium alginate hydrogel containing an excess of $CaCO_3$ enables the electrodeposition of a calcium alginate hydrogel that is free of $CaCO_3$ particles, ensuring optical access to the cells throughout the entire hydrogel that was observed by transmitted light (Fig 1a) or fluorescence (Fig 1b). An optical density of the cells in the hydrogel was calculated by transmitted light (Fig 1c) and fluorescent protein production by the cells was monitored by fluorescence microscopy (Fig 1d). Confocal microscopy demonstrates this method is superior to previous methods [1, 2] by allowing optical access through the entirety of the hydrogel rather than the ~100 μm closest to the objective lens. Control experiments indicate cells are transformed by a synergistic combination of electric field, Ca^{2+}, and Joule heating. IPTG can be used to induce protein expression at a desired time point, as demonstrated in Figure 2. Transformed cells were cultured and used to produce protein constitutively or selectively induced by environmental factors, effectively acting as sensors. The most exciting aspect of this simple, rapid, and reagentless technique is the site-selective reprogramming of cells as they are entrapped within a calcium alginate hydrogel, offering researchers both spatial and temporal control over the genetic makeup of cell populations within the cell culture environment.

E. coli cells have been genetically reprogrammed by an electric field-assisted chemical transformation mechanism while they are simultaneously entrapped and cultured within a calcium alginate hydrogel at an electrode surface. This process enables selective transformation of desired cell populations at a specific spatial location. Cell growth and fluorescent protein expression were easily monitored optically within the microfluidic device. Transformed *E. coli* can act as cell-based sensors of chemicals in the environment. This powerful and versatile method is currently being extended into the study of intercellular communication phenomena and is being adapted for eukaryotic cell lines such as Sf9 and HEK293, which are important to the biomedical engineering field for basic research, vaccine development, and therapeutic protein production.

Acknowledgements: This work was supported by a grant from the Robert W. Deutsch Foundation.

References:

1) Shi, X.-W., et al., *Adv. Funct. Mater.*, **2009**, 19, 2074-2080.

2) Cheng, Y., et al., *Adv. Funct. Mater.*, **2012**, 22, 519-528.

Figure 1. (a) Transmitted light image of *E. coli* entrapped in a calcium alginate hydrogel. (b) False color fluorescent image of DsRed expressed by *E. coli* in (a). (c) Optical density measured from transmitted light images of hydrogels over 24 hour period. (d) DsRed expression measured by fluorescence images of hydrogels over 24 hour period.

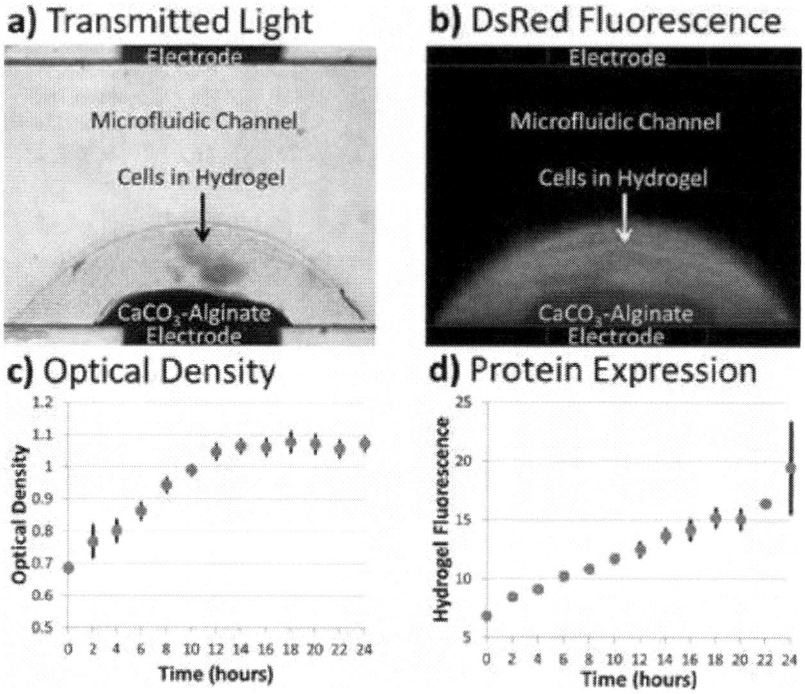

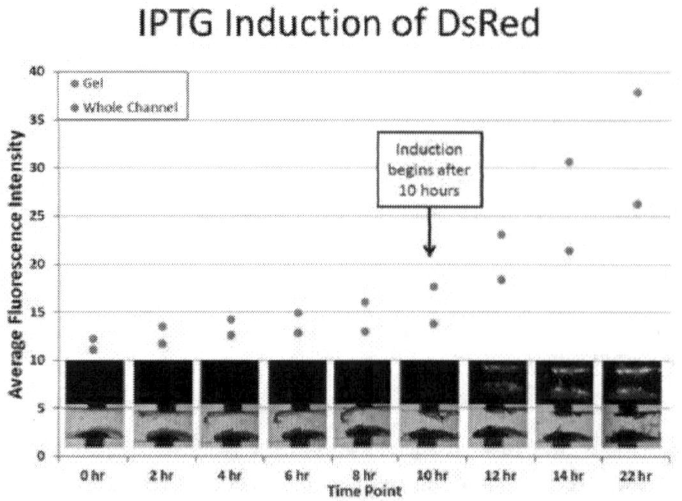

Figure 2. IPTG induction of DsRed expression in *E. coli* transformed during electrodeposition. Protein expression is minimal until IPTG induction begins at 10 hours post transformation, at which point it begins to increase at a rate much higher than previously measured.

(596at) Modeling Hematopoiesis: Clinical Applications of Population Balance Models

Jayachandran Devaraj and Doraiswami Ramkrishna, School of Chemical Engineering, Purdue University, West Lafayette, IN

In humans, blood cells (erythrocytes, leukocytes, platelets) are the most dynamic population of cells with high growth rates. Blood cells are derived from the long-term pluripotent stem cells residing in the bone marrow through hematopoiesis. Hematopoiesis is a complex process during which the stem cells are committed to a particular cell lineage, proliferate and mature to a specialized cell type according to the biochemical signals received from the periphery. Aberrations in these signaling pathways lead to several acute and chronic cyclical blood disorders which can be treated by growth factor cytokines. As the stem cells mature in an age-dependent manner, they progressively lose their proliferating capability and enter the peripheral blood as a fully functional cell. Thus, bone marrow is one of the most active and tightly regulated systems in the human body.

In addition to the blood disorders mentioned above, hematopoiesis is severely affected by chemotherapy, given the nature of most of the cytotoxic drugs in targeting fast growing cells. In both instance, i.e. therapy with growth factors and chemotherapy, peripheral blood count monitoring has become a routine to assess the bone marrow response. However, this is a reactive rather than a proactive approach because the bulk of the effect is sustained by the early bone marrow precursors which usually take between 5 to 12 days to present themselves in the periphery. This marked delay in the realization of the actual treatment effects in the periphery hinders the decision making process during therapy administration. Thus, accurate prediction of normal and diseased bone marrow cellular dynamics would help to forecast the oncoming spike/decline in the cell population and adopt corrective measures before the patient succumbs to the disease/side-effect.

Population balance modeling (PBM) provides an ideal framework for representing this age-dependent maturation and death process together with dose-dependent proliferation (with growth factor) and death process (with chemotherapy) during various stages of hematopoiesis. In this work, we employ PBM framework to model the hematopoiesis process under normal as well as diseased conditions. Influence of growth factors and chemotherapy on hematopoiesis is also considered. The model is validated using data from actual patients undergoing treatment with growth factor cytokines and chemotherapy. Based on the average patient response, dosing conditions are defined such that the cell population is maintained above a critical level. Utilization of such predictive models in clinical settings will augment the decision making capabilities of the treating physicians.

(596au) Model-Based Individualized Treatment for Acute Lymphoblastic Leukemia

Jayachandran Devaraj[1], Ann E. Rundell[2], Robert E. Hannemann[3] and Doraiswami Ramkrishna[1], (1)School of Chemical Engineering, Purdue University, West Lafayette, IN, (2)Biomedical Engineering, Purdue University, West Lafayette, IN, (3)Chemical Engineering, Purdue University, West Lafayette, IN

Acute Lymphoblastic Leukemia (ALL) constitutes a major type of cancer among children and childhood ALL accounts for more than 60% of all ALL cases. Although 5–year survival rates are impressive in ALL, long-term childhood survivors endure several medical and socio-economic miseries related to their earlier treatment. It has been known for many decades that individuals differ genetically and phenotypically which significantly influence the way they respond to treatment. Thus, inter-patient variation in drug response presents a great challenge in determining the optimum dose for an individual patient.

In the present form of chemotherapy administration, patients are started with an average dose of drug and 'titrated' to a maximum tolerated dose through trial and error approach. During this process, patients undergo several dose modifications and encounter severe episodes of acute side-effects. Many retrospective studies show that such endeavors are responsible for the long-term medical complication. A possible alternative would be a system theoretic framework adequately empowered by statistical and optimal control theory. Such a quantitative approach, if implemented carefully, will greatly enhance the decision making capabilities of the treating physicians.

In this work, we propose a model-based approach for individualizing the treatment for ALL patients using a chemotherapeutic agent 6-Mercaptopurine (6-MP). Pharmacogenomic and other cell count measurements obtained from individual patients are used to individualize the model. Models account for the drug disposition in the body, metabolic conversions and its effect on various cell populations. A Bayesian approach is adopted to identify the parameters for individual patients. The models can be utilized to predict the expected treatment response and optimal treatment condition for a given patient with few measurements. This approach holds a great potential to tailor the treatment based on patients' genetic and phenotypic make up as opposed to the current practice. Ultimately, this will help to reduce the therapy burden and improve the quality-of-life among ALL patients.

(596av) Direct Observation of Von Willebrand Factor Unfolding and Elongation On Collagen During Acute Whole Blood Exposure to Pathological Shear Rates

Thomas Colace and Scott L. Diamond, Chemical and Biomolecular Engineering, University of Pennsylvania, Philadelphia, PA

When blood moves through a severe arterial stenosis, von Willebrand Factor (vWF) experiences acute millisecond-scale exposures to pathological wall shear rates ($g_w > 5000\ s^{-1}$). Distinct from viscometers or recirculation systems that shear blood or its components for many seconds, we deployed microfluidic devices for single-pass perfusion of whole blood or platelet free plasma (PFP) over fibrillar collagen type 1 surfaces (< 10 msec transit time) at pathological g_w or spatial wall shear rate gradients (*grad* g_w). Long vWF fibers (>20 mm) bound to collagen were observed with fluorescent anti-vWF at $g_w > 30{,}000\ s^{-1}$ during perfusion of PFP (citrate, EDTA, or recalcified with PPACK to inhibit thrombin), a process occurring at constant g_w and enhanced at zero Ca^{2+} possibly through vWF A2 domain destabilization. Rapid deceleration of EDTA-PFP perfusion from 125,000 s^{-1} at *grad* $g_w = -3.7 \times 10^7\ s^{-1}/cm$ quenched vWF fiber formation. At 19,400 s^{-1}, EDTA-whole blood perfusion resulted in unstable and rolling vWF-platelet nets, while whole blood perfusion (normal Ca^{2+}) generated large vWF/platelet aggregates that repeatedly grew and embolized, a process blocked by GR144053 ($a_{IIb}b_3$ inhibitor) or anti-GP1b. PPACK-inhibited whole blood perfusion at venous shear rate of 200 s^{-1} produced a stable, non-growing platelet deposit that was a substrate for massive but unstable vWF-platelet aggregates when flow was suddenly increased to 7800 s^{-1}. Triggered by surface collagen and enhanced by platelets via GP1b and $a_{IIb}b_3$, vWF unfolding and fiber formation occurred during acute exposures to pathological shear rates and did not require spatial shear rate gradients.

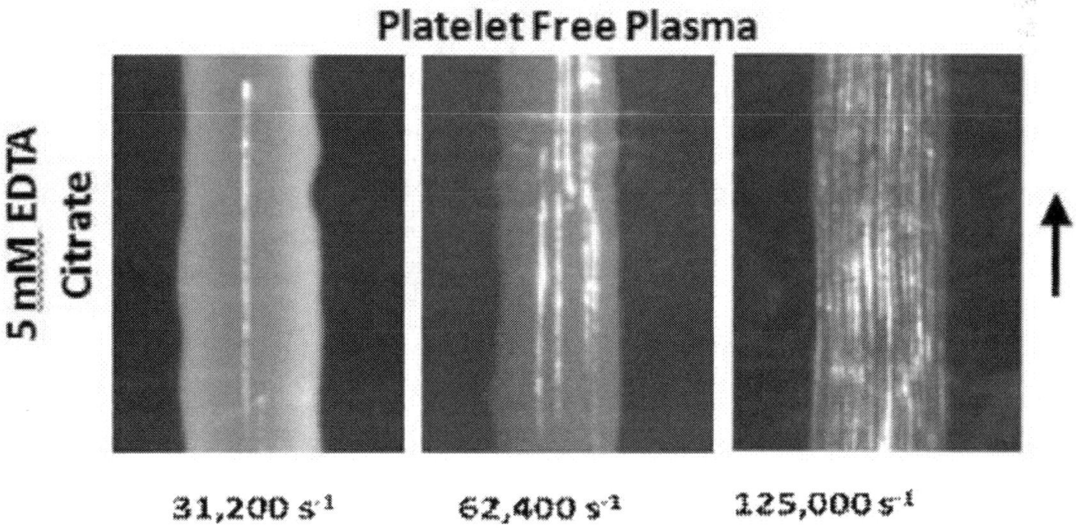

Figure 1, Platelet free citrated-plasma was treated with 1 μg/mL fluorescently labeled anti-vWF and 5 mM EDTA. The plasma samples were perfused over a collagen type 1 surface at the indicated wall shear rates. Long fibers of vWF (>20 μm) appeared at shear rates above ~30,000 s^{-1}, with more fibers appearing at higher shear rates. The bar indicates 15 μm.

(596aw) Interaction Between Coenzyme Q10 and Human Serum Albumin: Spectroscopic Approach

Yinhe Sun, Xin Peng, Wei Qi, Mengfan Wang, Rongxin Su and Zhimin He, School of Chemical Engineering and Technology, Tianjin University, Tianjin, China

Coenzyme Q_{10} (CoQ_{10}) exists in all body cells, and it is vital to a series of activities related to energy metabolism. Nowadays, CoQ_{10} is widely consumed by humans since it is well known as an important nutrient and a therapeutic agent in supporting health and as an inhibitor of ageing. Moreover, CoQ_{10} also has good effect in inhibiting the growth of human cancer cells. Human serum albumin (HSA) is the most prominent protein within blood plasma, and it plays an important role in the binding of many kinds of ligands, such as drugs. Due to its significantly physiological functions, HSA has been widely used as a model protein in evaluating protein¨Cdrug interactions.

So far, different spectroscopic approaches have been broadly used to detect the binding of protein with various of drugs. However, the interaction of CoQ_{10} with HSA has not been explored. Therefore, in this paper, UV-vis absorption, fluorescence and circular dichroism (CD) spectroscopies were employed to reveal the mechanism of the binding between coenzyme Q_{10} (CoQ_{10}) and human serum albumin (HSA) under physiological conditions (pH=7.4). The binding parameters have been calculated by fluorescence quenching method. The results demonstrated that the fluorescence quenching of HSA by CoQ_{10} was mainly static quenching due to the formation of HAS-CoQ_{10} complex, and the number of binding sites (*n*) equal to 1. The thermodynamic parameters (DH = - 43.18 kJ mol^{-1}, DS = - 47.05 J mol^{-1} K^{-1}, DG = - 29.15 kJ mol^{-1}) indicated that the enthalpy-driven binding process was spontaneous, and the main binding forces between CoQ_{10} and HSA were hydrogen bonds and Van der Waals forces. The competitive experiments using different site markers indicated that subdomains IIA (site I) of HSA was the primary binding site for CoQ_{10}. The average binding distance (*r*) between HSA and CoQ_{10} was 4.29 nm, which was estimated according to the Forster's theory of non-radiation energy transfer. The CD spectra, UV-vis absorption and synchronous fluorescence spectroscopy data showed the slight conformational changes of HSA in the presence of CoQ_{10}.

This work was supported by the Program for New Century Excellent Talents in Chinese University (NCET-08-0386), the 863 Program of China (2008AA10Z318), the Natural Science Foundation of China (20976125; 31071509; 51173128) and Tianjin (10JCYBJC05100), and the Program of Introducing Talents of Discipline to Universities of China (No. B06006).

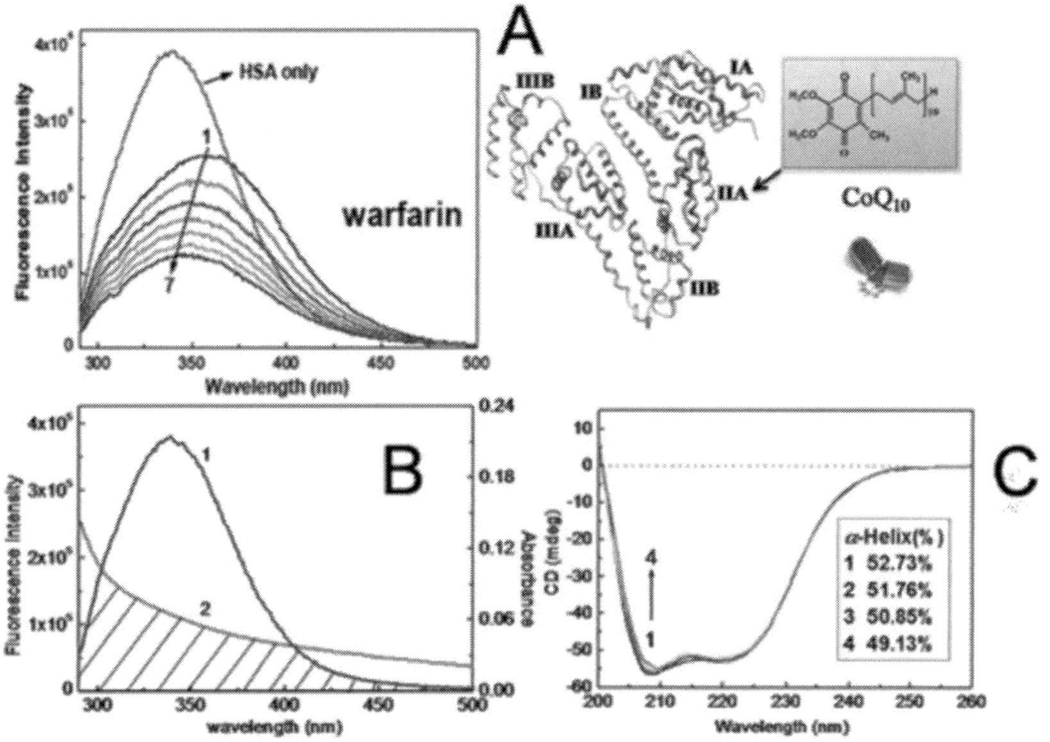

Figure (A) Binding site of CoQ_{10} in the HSA, (B) Energy transfer between HSA and CoQ_{10}, (C) CD spectra of HSA in the presence of CoQ_{10}.

(596ax) PEI/Nucleic Acid Polyplex Preparation for the Optimized Size, Gene Transfection, and Cytotoxicity: Implications in Nonviral Gene Delivery

Soo Kyung Cho, Chemical Engineering and Materials Science, University of California, Irvine, Irvine, CA, Chris Dang, Biological Sciences, University of California, Irvine, Irvine, CA, Jones Tsai, Pharmaceutical Sciences, University of California, Irvine, Irvine, CA and Young Jik Kwon, Pharmaceutical Sciences, Chemical Engineering and Material Science, Biomedical Engineering, University of California, Irvine, Irvine, CA

The "bottom-up" approach of biological nanostructure fabrication has been commonly utilized in the development of non-viral gene carrier systems. The design and use of self-assembling molecules has created a library of different polymeric gene delivery systems that has seen varied levels of successes both *in vitro* and *in vivo*. Among these, polyethyleneimine (PEI) is a highly cationic polymer that holds a prominent position in the area of nonviral gene delivery, lending to its exceptional ability to complex nucleic acids via electrostatic interactions and promoting endosomal escape inside the cell. Unfortunately, the use of PEI polyplexes has produced large variations in transfection efficiencies and toxicities depending on numerous variables such as cell type, presence of serum, incubation duration and polymer to nucleic acid ratio (represented by amines in the polymer to phosphates in nucleic acids, N/P ratios), which have been well documented. Contrarily, formulations of these PEI/nucleic acid polyplexes may also play important roles in both the stability and transfection efficiencies; and to our best knowledge, remain unclear. Polyplex formulations are generally controlled kinetically by numerous variables ranging from extensive pipetting, vortexing, and sonication, followed by a period of incubation. The complexation behavior of polyplexes is dependent on several major characteristics such as molecular weight, charge density, structure of the polymer, ionic strength of buffers, and N/P ratios. In addition, we found that the sequence of addition of components during polyplex formation significantly influences the resulting polyplex size as well as the transfection efficiency.

In this study, reversing the adding sequence of DNA and PEI was observed to result in significant difference in polyplex formation. It was found that adding PEI to DNA solution (PtoD) exhibited much larger polyplexes than those formed via adding DNA to PEI solution (DtoP) polyplexes. It was hypothesized that a single copy of DNA is complexed into a polyplex when DNA is added to PEI solutions, whereas multiple copies of DNA is incorporated with PEI in the opposite mixing order. PEI/DNA polyplexes prepared by two different mixing orders were incubated with NIH 3T3 cells to assess the transfection efficiency and cell viability. The transfection rate of the cells incubated with the polyplexes prepared by PtoD formulation was slightly higher than that those prepared by DtoP formulation. However, a significant increase in gene expression level was observed with the cells incubated with the polyplexes prepared by PtoD formulation, which implicates the hypothesis of varying copy numbers of complexed DNA in the polyplexes via different formulations. Further polyplex characterizations using HPLC, TEM, and AFM will also be presented.

(596ay) A Multiplexed Microfluidic Platform for Antibiotic Susceptibility Screening

Ritika Mohan, Arnab Mukherjee, Jaebum Lee, Emre Sevgen, Charles M. Schroeder and Paul J. A. Kenis, Chemical and Biomolecular Engineering, University of Illinois at Urbana-Champaign, Urbana, IL

Microfluidic Platform for Antibiotic Susceptibility Screening

In this work, we report a multiplexed microfluidic platform for rapid characterization of antibiotic susceptibilities in microbial pathogens for healthcare applications. Antibiotic abuse over the last few decades has resulted in an escalating emergence of multidrug resistant pathogens that respond poorly to conventional antibiotic therapy [1]. As resistance to single antibiotics evolves very rapidly, drug resistant pathogens are usually treated with combinations of antibiotics. However, existing techniques to decide the right therapeutic combinations are time-consuming, laborious, and require substantial volumes of patient body fluids like sputum and blood. Consequently, there is a pressing need for expedited diagnostic platforms. To address this, we designed and fabricated a 4x6-array microfluidic chip capable of confining cells and reagents in nanoliter volume compartments. Leak-free isolation of contiguous compartments is achieved with elastomeric valves that are closed at rest [2]. Actuation of the normally closed valves with negative pressure enables controlled exposure of cells to antibiotics that are loaded in adjacent compartments. The miniaturized device design is highly convenient for biological assays as it facilitates portability, uses microliter volumes of samples and reagents for analysis, and is readily amenable to automation. Most importantly, the microfluidic diagnostic platform considerably accelerates assay time due to increased detection sensitivity since it enables analysis at single cell resolution.

We employed the microfluidic platform to investigate susceptibilities of *E. coli* and *P. aeruginosa* cells to several commonly used antibiotics. Specifically, we assessed the effects of ampicillin, tetracycline, chloramphenicol, cephalexin, ciprofloxacin, and combinations thereof on *E. coli* MG1655 and *P. aeruginosa* PAO1 cells. Furthermore, we quantified synergistic effects of several antibiotic combinations. As an example, although ampicillin at 10 µg/mL and tetracycline at 1 µg/mL are ineffective individually, they exhibit significant bactericidal activity against *E. coli* when administered in combination. Finally, we extended the platform to characterize antibiotic resistance in a mixed microbial community of *E. coli* and *P. aeruginosa* cells. Overall, these results emphasize the utility of the microfluidic platform for rapidly characterizing antibiotic susceptibilities over a wide range of antibiotic concentrations. The platform improves significantly on existing diagnostic methods (disc diffusion and broth dilution assays) in terms of several key advantages including turn-around time, analysis cost, ease-of-use, detection sensitivity, and sample volumes required for diagnosis [3].

References:

[1] K. Churski *et al.*, *Lab Chip*, **2012**,12.

[2] R Mohan *et al.*, *Sensors and Actuators B*, **2011**, 1, 1216-1223.

[3] JP Torella *et al.*, *PLoS Comput Biol*, **2010**, 6.

(596az) Characterizing Fluid Dynamics of Impinging Jet Mixing with Ultrasonic Excitation Using Particle Image Velocimetry

David M. Rophael[1], **John C. Consiglio**[1], Christian Beck[2], David Wootton[1] and Rajesh N. Dave[3], (1)The Cooper Union for the Advancement of Science and Art, New York, NY, (2)Department of Chemical, Biological and Pharmaceutical engineering, New Jersey Institute of Technology, Newark, NJ, (3)Chemical, Biological and Pharmaceutical Engineering, New Jersey Institute of Technology, Newark, NJ

The development of reproducible pharmaceutical products is inherently attributed to the complex interactions between transport and mixing processes. Quite often, failure to duplicate mixing behavior causes batch-to-batch variability. Moreover, the improper scaling of these processes to varying batch size contributes to unpredictable outcomes. There is a significant need in the pharmaceutical industry to study and model the fluid flow behavior and mixing characteristics of chemical reactions, and thereby establish reliable production methods. Additionally, it is necessary to establish tools to characterize the impact of mixing variability during scaling.

A novel reaction chamber with impinging jet mixing and ultrasonic excitation was designed to aid in the development of a new pharmaceutical product. A Particle Image Velocimetry (PIV) system was used to measure velocity profiles of the fluid flow while the effects of the ultrasonic excitation on the flow characteristics were studied. Endoscopes were used to position the laser sheet and camera lens close to the centerline of the flow. A method to focus the camera on a small field of interest was developed, which allowed for accurate measurement of the velocity profiles. As the reactants in the pharmaceutical process studied are corrosive, water was used as a substitute medium in order to avoid damaging the imaging equipment. The flow was characterized at various volumetric flow rates and ultrasonic excitation intensity, as well as at various lengths away from the excitation source. From the flow fields, local shear rates were determined using the DaVis LaVision® PIV imaging software. From shear rates, energy dissipation was characterized. Local energy dissipation was also studied at various radial and axial positions from the excitation source.

Early results indicate that with increasing excitation at the sound tip, higher energy dissipation rates are achieved near the surface of the ultrasonic probe. The flow behavior transitions from a turbulent to a laminar regime along the reactor. These dissipation rates help in determining reaction kinetics, which allow for future characterization and improved pharmaceutical development.

(596b) Controlled Rupture of Drug-Encapsulated Ultrasound Contrast Agents

Yoonjee Park[1], Tuan Pham[1], Carl Beigie[1], Robin Cleveland[2], Jon O. Nagy[3] and Joyce Y. Wong[1], (1)Biomedical Engineering, Boston University, Boston, MA, (2)University of Oxford, Oxford, United Kingdom, (3)NanoValent Pharmaceuticals, Bozeman, MT

Ultrasound contrast agents using micron-sized bubbles (less than 8 μm in diameter) have the potential to be drug delivery vehicles in addition to providing targeted molecular imaging. However, several challenges remain in terms of producing monodisperse microbubbles stable against gas dissolution for improving blood circulation times. In the case of drug delivery, it is crucial to control destruction to release encapsulated contents *only* to the targeted area with minimal prior passive leakage of drug. By using a photopolymerizable diacetylene lipid as a component in a microbubble shell matrix, the surface properties such as elasticity or resistivity against gas dissolution can be tuned,[2] which therefore enables imaging at one frequency and drug release in the area of interest at a different frequency. Drug-encapsulated monodisperse bubbles are created using a double emulsion microfluidic technique to increase drug loading capacity.

Polymerizable lipid mixtures, consisting of 10 mol% of ethylene glycol diacetylene lipids (h-PEG_1PCDA), 15 mol% of PEG-diacetylene lipids (m-PEG_{5000}-PCDA) and L-α-phosphatidylcholine, hydro soy PC were used. The microfluidic flow focusing device designs were adapted from Hettiarachchi et al.[3,4] Paclitaxel in triacetin (0.1mg/ml) was used as a test drug of which 1 mol% was fluorescently labeled (Oregon Green 488-paclitaxel).

The size distribution of the polymerized shell microbubbles (PSM) produced by microfluidic flow focusing technique and commercially available microbubbles (Vevo MicroMarker, VMM) was observed to be 2.2 μm with 8% polydispersity, and 2.3 μm with 36% polydispersity, respectively. The PSM showed a significantly slower decrease in intensity of gray-scale ultrasound image brightness than VMM or nonpolymerizable soy PC shell microbubbles (NSM). The half-life (t_H the time for the ultrasound signal to drop 6 dB in brightness) for the 25%DA was $t_H = 5$ min, and for both VMM and NSM, $t_H < 1$ min. The brightness of the 25%DA decreased by 14 dB after 15 min, whereas the VMM brightness decreased by 36 dB, suggesting rapid microbubble destruction. Additionally, bubbles polymerized to different extents (t_{polym}) showed variable destruction rates at different ultrasound power levels. This suggests that polymerization can not only provide passive bubble longevity but also tunable rupture capabilities. The double emulsion microfluidic device produced the paclitaxel encapsulated microbubbles. The FITC-filtered image shows the location of paclitaxel within the shell. These results indicate that the dissolution of microbubbles in the bloodstream or under ultrasound stimulation is tunable by varying the fraction of polymerizable lipid or polymerization time. Therefore, these tunable microbubbles have the potential to be customized ultrasound contrast agents for targeted molecular imaging and therapeutic treatment applications.

References: [1] Park et al. *Langmuir* **2012**, *28*, 3766-3772. [2] Hettiarachchi, K.et al. *Lab Chip* **2007**, *7*, 463-8. [3] Hettiarachchi, K.et al. *Biotechnol. Prog.* **2009**, *25*, 938-945.

(596ba) Lead Identification, Optimization, and Characterization of Novel Cancer Treatment Strategies Using Repositioned Drugs

David J. Taylor, Chemical Engineering, Arizona State University, Tempe, AZ, Jeongyun Kim, Artie McFerrin Department of Chemical Engineering, Texas A&M University, College Station, TX, Arul Jayaraman, Department of Chemical Engineering, Texas A&M University, College Station, TX and Kaushal Rege, Chemical Engineering, Biomedical Engineering, Arizona State University, Tempe, AZ

Cancer is the second leading cause of death in the United States and novel methods of treating advanced malignancies are of high importance. Of these deaths prostate cancer and breast cancer are the second most fatal carcinomas in men and women respectively, while pancreatic cancer is the fourth most fatal in both men and women. Developing new drugs for the treatment of cancer is both a slow and expensive process. It is estimated that it takes an average of 15 years and an expense of $800 million to bring a single new drug to the market. However, it is also estimated that nearly 40% of that cost could be avoided by finding alternative uses for drugs that have already been approved by the Food and Drug Administration (FDA). The research presented describes the testing and identification of novel methods for treating many human carcinomas using drugs previously approved by the FDA. Approved drugs that can be used in combination with the protein TRAIL to induce apoptosis selectively in cancer cells will be identified in well plate based experiments and leads that are identified will be optimized using high-throughput microfluidic devices. Identification of FDA approved drugs that demonstrate the ability to depolarize the mitochondria of cancer cells will also be carried out coupled with the testing of these drugs in potential delivery vehicles. Mechanistic evaulations of lead candidates are also presented.

(596bb) Using Vibrational Spectroscopy to Evaluate Solid Disperision Stability

Ryanne N. Palermo, Carl A. Anderson and James K. Drennen III, Graduate School of Pharmaceutical Sciences, Duquesne University, Pittsburgh, PA

Formulation of amorphous (non-crystalline) drug products has been suggested in order to counter the ever-present need to make drug products sufficiently soluble. However, a major concern with amorphous pharmaceuticals is physical instability, where the higher free-energy form will spontaneously revert to the more stable (and less soluble) crystalline form. Stabilization of these products via solid dispersions have been proposed to prolong the kinetic stability of these drugs. Selection of appropriate analytical methods is essential to both monitor the physical stability, as well as gain knowledge of the effect of polymer dispersions and material characteristics (i.e., polymer type, drug loading, and compaction force) on the inhibition of crystallization in an intact dosage form under various environmental conditions. Raman and near-infrared spectroscopic methods have been evaluated and compared to powder x-ray diffraction. Net analyte signal theory was used to generate multivariate figures of merit and provide a platform to compare methods directly. Spectroscopic methods were found to be more accurate and have significantly lower detection limits than the standard x-ray analysis while providing much faster analysis times. Moreover, the estimated kinetics and spectral changes occuring during storage for different formulations could be related to mechanisms of instability. Indomethacin is used as a model drug.

(596c) Expansion of Functional Human Amniotic Fluid Stem Cells in Serum Free Medium

Meimei Liu, Chemical and Biomolecuar Engineering, The Ohio State University, Columbus, OH, Ning Liu, Stem Cell Culture, Irvine Scientific, Santa Ana, CA and Shang-Tian Yang, Department of Chemical and Biomolecular Engineering, Ohio State University, Columbus, OH

Due to their easy and safe accessibility, abundant cell numbers, and lack of ethical concerns, amniotic fluid stem cells (AFSCs) have emerging as an important cell source for tissue engineering and regenerative medicine.However, little is done about serum free medium for the expansion and differentiation of hAFSCs for clinical applications. In this study, the proliferation of hAFSCs cultured in different serum free media and serum contained complete medium for 3 passageswas investigated. For each passage, cell number, cell viability and cell morphology were studied and compared. In the end of the third passage, cells were harvested and characteristic surface markers were measured by flow cytometry analysis. Additionally, osteogenic and adipogenic differentiation of expanded cells were carried out. Our results indicated that one serum free medium can extensively expand hAFSCs, and the produced hAFSCs maintained their immunophenotypic profile and multilineage differentiation capacity. Therefore, this serum free medium could be a very promising medium to culture hAFSCs for clinical applications.

(596d) Oxygen Sensing and Control of Engineered Tissue

Seema M. Ehsan, Chemical Engineering and Materials Science, University of California, Irvine, Irvine, CA and Steven C. George, Biomedical Engineering, Chemical Engineering, University of California, Irvine, Irvine, CA

Introduction

To properly model physiological processes with heterogeneous oxygenation conditions, an experimental platform is needed that can dynamically control and measure the oxygenation of 3D cell culture. Many studies have demonstrated that ambient oxygen concentrations can differ significantly from the concentrations that cells actually experience in 2D culture. The inherent diffusion limitation of 3D culture, particularly at clinically relevant dimensions (> 1 mm), adds an additional hurdle for oxygen transport. Here we present an experimental platform that offers simultaneous dynamic oxygen control and measurement of oxygen diffusion through 3D engineered tissue. The platform, in conjunction with a non-steady state mathematical model for oxygen diffusion, enables quantification of the oxygen diffusion coefficient and cellular oxygen consumption rate, while also predicting the temporal and spatial distributions of oxygen in a thick tissue subject to time-varying oxygenation conditions.

Materials and Methods

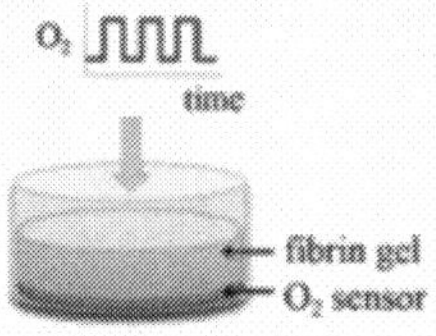

Figure 1

An oxygen control chamber system was fabricated to function as a miniature cell culture incubator with programmable oxygen tension in the gas phase. An oxygen sensor patch (Presens, Germany) was placed at the base of a cylindrical well, and a 2.5mg/mL fibrin tissue (3 million normal human lung fibroblasts per mL) was constructed on top of the patch (Figure 1). The well was then placed inside the oxygen control chamber, where an external oxygen probe non-invasively provides real-time measurements of the oxygen concentration at the base of the tissue. Michaelis-Menten kinetics was used to model oxygen consumption, and numerical techniques were performed to determine values for the oxygen consumption rate. Finite element analysis (FEA) software was then used to predict detailed spatial and temporal oxygen concentration.

Results

The oxygen control system achieves gas equilibration (90% response time) within 60 seconds, and can be programmed to deliver 0-100% O_2. The Michaelis Menten parameters were found to be: V_{max}= 1.1e-4 mol/m^3/s and K_m= 0.036 mol/m^3. FEA modeling provides temporal (Figure 2A) and spatial (Figure 2B) oxygen diffusion information.

Discussion and Conclusions

Through the simultaneous control and measurement of local tissue oxygen levels at the boundaries (top and base), a system has been developed that can predict the temporal and spatial oxygenation of 3D culture. While most mathematical models of 3D culture have relied on steady state assumptions and constant boundary conditions, the present model employs non-steady state analysis and allows for time varying boundary conditions consistent with *in vivo* observations of intermittent hypoxia.

A) B)

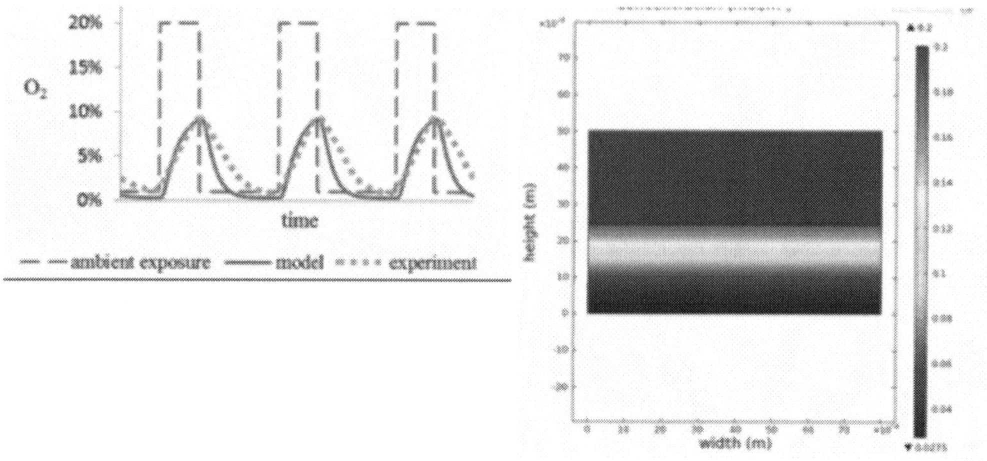

Figure 2: A) A square wave oxygen exposure profile at the top of the tissue (dashed line) has a significantly attenuated effect on the oxygenation at the base of a 2mm thick cellular tissue (solid, dotted lines). **B)** Sample spatial distribution of oxygen in model tissue (top layer: ambient gas, bottom layer: fibrin gel).

(596e) Cell Communication in Three Dimensional Microenvironments

Matthew Byrne[1], Lisa Trump[2], Amit V. Desai[3], Lauretta A. Rund[4], Lawrence B. Schook[5] and Paul J. A. Kenis[1], (1)Chemical and Biomolecular Engineering, University of Illinois, Urbana Champaign, Urbana, IL, (2)Animal Sciences, University of Illinois at Urbana-Champaign, Urbana, IL, (3)Department of Chemical & Biomolecular Engineering, University of Illinois at Urbana-Champaign, Urbana, IL, (4)Department of Animal Sciences, University of Illinois Urbana-Champaign, Urbana, IL, (5)Laboratory of Comparative Genomics, University of Illinois Urbana-Champaign, Urbana, IL

The ability of cells to signal one another is an important aspect in tissue function, activation of immune functions, and other bodily functions. Long range cell communication occurs via paracrine and endocrine signaling factors. Understanding how different cell types communicate in response to pathogens or neighboring cells is essential for developing new therapies and drugs. Additionally, cellular communication networks hold information for understanding cancer and disease states, and can help lead to effective culture and modulation of cell and stem cell maintenance and function.

The goal of this work is to co culture an engineered model system in a microfluidic device. This model system allows for real time monitoring of cellular communication via fluorescent reporter systems. Lipopolysaccharide released by *E. coli* stimulates mouse macrophage (RAW) cells, which then secrete TNFα. The TNFα activates cellular signaling cascades in nearby human embryonic kidney (HEK) cells to fluoresce. To control the cell signaling, cells were encapsulated within a Matrigel scaffold and seeded in a microfluidic device. The system was then monitored over 16 hours to evaluate the cellular communication.

(596f) Single-Particle Assay for Characterizing the Binding Kinetics of X31 Influenza Virus and the Efficacy of Antiviral Compounds

Donald Lee, Chemical Engineering, Cornell, Ithaca, NY and Susan Daniel, Chemical and Biomolecular Engineering, Cornell University, Ithaca, NY

Influenza viruses must bind to their target cell long enough to be internalized by the cell through endocytosis. The duration of binding is determined by the interaction between hemagglutinin proteins (HA) on the virus and the sialic acid receptor (SA) in the target bilayer. Determining the relationship between the receptor molecule structures and their binding duration time with each virus particle can provide new insight towards understanding factors that affect the binding avidity and the tropism of a virus. In addition, this research could aid the development of antiviral compounds that inhibit virus binding and the development of sensors that can identify the target receptor of a new virus strain. Using TIRF microscopy, the binding residence time of individual virus particles with receptors embedded in a supported lipid bilayer has been determined, and the binding time distribution of X31 (H3N2) influenza viruses with alpha2-3 linked sialic acid shows multiple peaks. These peaks suggest that some residence times occur more frequently than others, which may reflect the discrete number of contact points formed between the HA proteins and receptor molecules. This level of detail was not achievable before using standard surface plasmon resonance techniques, which tend to average out these fine details. Further testing shows that GM1 (1 SA per lipid molecule) has shorter binding times than Gd1a (2 SA per lipid molecule), despite the fact that both molecules have the same alpha2-3 linked SA groups. Gd1a may be increasing binding times by increasing the probability of forming more contacts per virus particle. The binding avidity of a virus to receptors has mainly been attributed to the sialic acid type, such as an alpha2-6 linked SA versus an alpha2-3 linked SA, but the results here show that other factors can also significantly affect avidity.

(596h) Effects of Histone Tail Modifications On Chromosome Structure and Activity

Nate Nurse, Chemical Engineering, Purdue University, West Lafayette, IN and **Chongli Yuan**, School of Chemical Engineering, Purdue University, West Lafayette, IN

The N-terminal regions of histone H3 and H4 proteins play an important role in gene regulation. Abnormal modifications on H3 and H4 tails are associated with various types of cancer diseases, such as lung, prostate and breast cancer. Drug molecules targeting at specific modification sites on H3 and H4 tails provides a viable route for editing aberrant gene transcription pattern. It is therefore of critical importance to understand the role of each histone tail in regulating gene transcriptional activity. We constructed model nucleosomes and nucleosomal arrays with modified H3 and H4 N-terminal tails and examined their respective effects on nucleosome conformation and dynamics using various biophysical experiments. Under different ionic strengths, H3 tails participate primarily in intra-nucleosomal interactions by constraining the DNA breathing motion and limiting its conformational variations. Histone H4 tails, on the other hand, are less effective in regulating the stability of a nucleosome. H4 tails are more involved in inter-nucleosomal interactions. Our results revealed the detailed mechanism as to how H3 and H4 tails are involved in intra- and inter- nucleosomal interactions under different ionic conditions. The findings will facilitate the screening of drug targets on histone tails.

(596i) Design, Synthesis and Characterization of Protein-Based Surfactants As Functional Drug Stabilizers

Rachael Barton[1], Kevin Sledziewski[2], Philo Morse[2], Robert Lee[2] and **Bryan W. Berger**[3], (1)Chemical Engineering, Lehigh University, Bethlehem, PA, (2)Particle Sciences, Inc., Bethlehem, PA, (3)Chemical Engineering and Program in Bioengineering, Lehigh University, Bethlehem, PA

Over the past 2 decades, surfactants have grown into a $20 billion annual business, with approximately 60% dedicated to industrial processes, 25% to household products and 15% in specialty applications. Most noteworthy is the emergence of specialty applications in the past 10 years, driven in large part by new areas such as biofuels, food preservatives and design of biocompatible interfaces for diagnostics and delivery systems, which require higher-cost, biodegradable materials whose phase behavior can be controlled through a molecular 'switch' or tunable property. Our group has been investigating the hydrophobins, which are a class of small (< 10 kDa), cysteine-rich peptides secreted by filamentous fungi in response to pesticides, organic solvents or elevated temperatures, creating a water- and pesticide-repellant surface. Despite their unique combination of biological (low immune response, biodegradable) and chemical (low CAC, water-repellant) properties, relatively little is known in terms of the structural features that give rise to their unique biological and chemical properties. I will describe our efforts to develop high-yield methods for overproduction of hydrophobins, as well as a detailed biophysical characterization of hydrophobin structure in solution to understand the molecular basis for their self-assembly. Our more recent efforts to engineer novel, 'switchable' functions into designed variants of hydrophobins will also be discussed

(596j) Effect of Fibril Alignment On Macromolecule Diffusion in Fibrin Gels

Armen H. Mekhdjian, Arjun S. Adhikari and Alexander R. Dunn, Chemical Engineering, Stanford University, Stanford, CA

Fibrin-containing blood clots experience substantial strain due to platelet contraction, fluid shear, and mechanical stress at the wound site. Recently, we and others [Adhikari et al., *Biomacromolecules* (2012); Varju et al., *J. Thromb. Haemostasis* (2011)] reported that strain protects fibrin clots from proteolysis, and moreover leads to marked decreases in macromolecular diffusivity perpendicular to the strain axis. In this work, we examine the physical origins of strain-induced diffusional anisotropy in fibrin networks. We use confocal fluorescence recovery after photobleaching (FRAP) to measure macromolecular diffusivities in fibrin clots of varying density and fibril alignment. We then present analytical theory and modeling that account for the pronounced diffusional anisotropy that is observed in response to fibril alignment. The resulting model is potentially medically significant, given that the diffusion of proteolytic factors into clots is essential during the treatment of a myocardial infarct (heart attack) and deep vein thrombosis. In addition, anisotropic diffusion in strained tissues may play presently unrecognized roles in guiding embryonic development and tissue regeneration.

(596k) Biomedical Potentials of Bacterial Fibrinolytic Enzymes

Sudhir Rai, LGRD, NICHD, NIH, BETHESDA, MD

Many common disorders are related to thrombosis , of which myocardial infarction and stroke are the most common. Thrombosis is also a clinical problem in various cancers and after surgery, especially orthopedic. Blood clots, mostly comprising fibrin, are formed from fibrinogen by thrombin (EC 3.4.21.5); its cross-linking is catalyzed by factor XIIIa. Fibrin is hydrolyzed by plasmin (EC 3.4.21.7), which is activated from plasminogen by plasminogen activators. Fibrin formation and fibrinolysis are kept in balance by homeostasis. However, when fibrin is not hydrolyzed because of some disorder, thrombosis occurs.

Fibrinolytic agents that are available for clinical use are mostly plasminogen activators, such as a tissue-type plasminogen activator (t-PA), urokinase, and the bacterial plasminogen activator streptokinase. Despite their widespread use, these fibrinolytic agents present important shortcomings, including short half-lives, high cost, and the risk of allergic reactions and bleeding complications. Therefore, research for fibrinolytic agents from other sources continues. To date, fibrinolytic enzymes have been discovered in snake venom, insects, marine creatures, and fermented foods such as *Bacillus subtilis* natto, Chungkook-jang, and Tempeh. In particular, oral administration of natto or its enzyme (nattokinase) can enhance the release of an endogenous plasminogen activator in both animal models and human subjects.

Recently, serine proteases have attracted interest for their hydrolytic activity toward recalcitrant animal proteins such as collagen, keratin, blood clot and amyloid prion proteins from the perspective of beneficial use of industrial waste and medical applications. We recently found a serine protease from *Bacillus* sp., through screening of 130 environmental samples. Among them, few isolates showed high hydrolytic activity toward recalcitrant proteins.

In the present study, we demonstrated the remarkable fibrinolytic ability of purified serine protease *in vitro* and *in vivo*. First, purified serine protease was analyzed for its hydrolytic activities toward mimics of plasminogen and fibrin. Fibrinolytic activity of purified serine protease was compared with those of known fibrinolytic enzymes such as thrombin, prothrombin or coagulation factor Xa. The fibrin/fibrinogen degradation pattern demonstrated that purified serine protease preferentially degraded the Aa- and Bb-chains of fibrin and fibrinogen, suggesting it as alpha,beta-fibrinogenase. Further, purified serine protease is devoid of acute toxicity in a mice model and is non-toxic to mammalian cells, suggesting suitable candidate for development of a thrombolytic drug to prevent and combat the various thrombosis-associated diseases and cardiovascular disorders

(596l) Encapsulation of Adjuvant in Acetalated Dextran Microparticles for Production of Subunit Vaccines Using an Electrospray Method

Anthony Duong[1], Sadhana Sharma[2], Eric M. Bachelder[2], Barbara E. Wyslouzil[1] and Kristy M. Ainslie[1,2],
(1)Chemical and Biomolecular Engineering, The Ohio State University, Columbus, OH, (2)Division of Pharmaceutics, College of Pharmacy, The Ohio State University, Columbus, OH

Vaccines are arguably the most effective tool for prevention of disability and disease. Subunit or protein based vaccines are among the safest types of vaccines, due to the fact that they contain elements of the pathogen but not the entire pathogen. However, they also tend to have low immunostimulatory capabilities. One solution has been to add an immunostimulatory molecule as an adjuvant, which can enhance the immune response to the subunit protein antigen. For example, agonists of toll like receptors (TLRs) have the capacity to promote an immune response. In the case of TLR-7, the agonist must be delivered intracellularly, which can be achieved by encapsulating the drug in a delivery vehicle.

Imiquimod, a TLR-7 agonist, has previously been encapsulated in acid-sensitive acetalated dextran (Ac-DEX) microparticles, using a single emulsion method. The encapsulated drug has demonstrated improved immunostimulatory capabilities over that of free drug. Resiquimod is similar to imiquimod, but has greater potency for promoting cytokine production. However, the emulsion method yields low encapsulation efficiency for resiquimod, since it is partially soluble in water. Furthermore, it is a batch process, making high scale production difficult. This work investigates the use of electrohydrodynamic spraying (electrospray) to encapsulate resiquimod in Ac-DEX microparticles.

A single capillary electrospray apparatus was used to generate an aerosol from a solution of resiquimod and Ac-DEX in a solvent. Six straight chain alcohols were tested as the solvent, and the resulting particle size and morphology were analyzed by scanning electron microscopy (SEM). The Peclet number (Pe) was calculated and used to explain the observed collapsed particle morphology. Ethanol was chosen as an ideal solvent, and solutions of Ac-DEX and resiquimod in ethanol were electrosprayed to generate approximately 2 μm Ac-DEX particles containing resiquimod with an encapsulation efficiency of 85%. For comparison, resiquimod loaded Ac-DEX particles were prepared using a single emulsion method, with a resulting encapsulation efficiency of 6%. To improve the particles' ability to disperse in an aqueous medium, Tween 80 was blended with the Ac-DEX at ~10% w/w of total polymer and particles containing resiquimod were formed via electrospray with encapsulation efficiencies between 40% and 60%. In vitro release profiles of resiquimod from Ac-DEX/Tween 80 particles exhibited the acid-sensitive nature of Ac-DEX, with 100% drug release after 8 h at pH 5 (phagosomal pH) and after 48 h at pH 7.4 (physiological pH). RAW macrophages that were treated with resiquimod loaded Ac-DEX/Tween 80 particles displayed a significantly greater immune response than macrophages treated with the same dose of free drug. Overall, we present a scalable method that incorporates the blending of Tween 80 to produce Ac-DEX microparticles for vaccine applications. This is an elegant method for the encapsulation of adjuvant into an acid sensitive delivery vehicle. This work opens future opportunities for application of resiquimod for the production of effective subunit vaccines.

TEMPERATURE DEPENDENT RELEASE FROM LIPOSOMES IMMOBILIZED IN CALCIUM ALGINATE GEL

Martin Ullrich, Jiří Dohnal, František Štěpánek

Institute of Chemical Technology, Chemical Engineering Department, Technicka 5, 166 28 Prague 6, Czech Republic, tel. +420 220 443 048, e-mail: ullrichm@vscht.cz

Introduction

Liposomes are small vesicles based on phospholipids that forms a bilayer (Fig. 1a). Their size ranges from 50 nm to tens of micrometers depending on the preparation method[1]. Their hollow structure enables to encapsulate a wide spectrum of substances. Hydrophobic parts of phospholipids molecules ("tails") make a diffusion barrier for hydrophilic substances. Hydrophobic molecules are dissolved within this hydrophobic part of the bilayer. The non-polar (hydrophobic) part is formed by fatty acids and the polar part by glycerol, phosphoric acid and usually another molecule such as choline, ethanolamine etc. (Fig. 1b). Liposomes are also biocompatible so they are used in medicine as carriers of drugs[2].

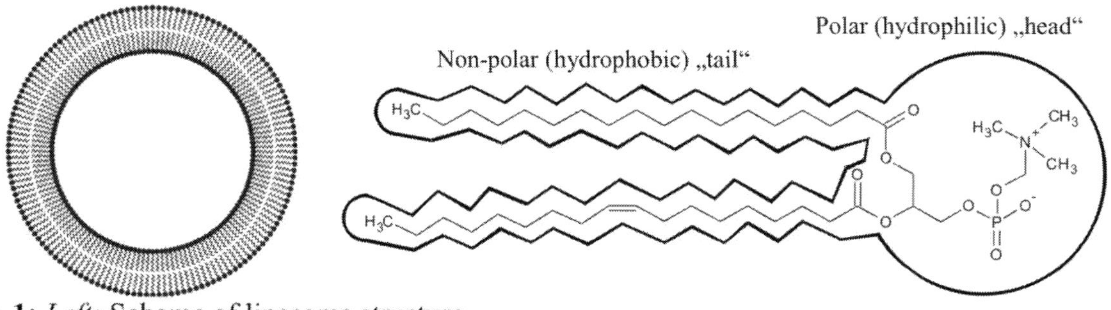

Fig. 1: *Left:* Scheme of liposome structure.
Right: A phospholipid molecule (*1*-palmitoyl-*2*-oleoyl-*sn*-glycero-*3*-phosphocholine) with hydrophobic and hydrophilic part highlighted.

The early liposome-based medicaments utilized the idea of slow release and a prolongation of the medicinal effect[3]. However, the liposomes have more potential. Depending on its composition, the phospholipid bilayer undergoes a phase transition that makes the membrane much more permeable[4]. The membrane is solid-like (gel) and less permeable at lower temperatures. At a phase transition temperature, the bilayer melts into a liquid-like structure characterized by more mobile phospholipid molecules, which causes the bilayer to be much more permeable and more flexible[5]. The phase change temperature is characteristic to every composition of the bilayer; it depends on a length and a saturation of fatty acids chains and it is also different for a particular polar head type[6, 7]. This phase change of the bilayer makes the release from liposomes controllable by temperature.

Despite all these features, it is not so easy to use liposomes in practice. Liposomes are relatively fragile, they could be destroyed by a human immune system[8], they are leaking while in contact with an organic phase, etc. Their immobilization in an alginate gel is one of many ways of their protection[9]. The resulting liposome-in-alginate particle has properties of both materials, liposomes react to the temperature and mechanical and chemical resistance is provided by an alginate gel.

Materials and methods

Chemicals

Liposomes were prepared from *1,2*-dipalmitoyl-*sn*-glycero-*3*-phosphocholine = DPPC (Corden Pharma LLC, Switzerland) and cholesterol = Chol (Sigma). Chloroform (Lach-Ner, Czech Republic), isotonic Tris-HCl buffer with pH of 7.4 and Sephadex® G-50 (Sigma) were also used for the preparation of liposomes. Tris-HCl buffer was prepared from Tris (*2*-Amino-*2*-hydroxymethyl-propane-*1,3*-diol) that was purchased from Aldrich and HCl (hydrochloric acid) from Lach-Ner, Czech Republic. 5(6)-carboxyfluorescein = CF (Sigma) and surfactant Triton® X-100 (Sigma-Aldrich) were used for a measurement the release rate. Sodium alginate (SAFC®) and calcium chloride (Lach-Ner, Czech Republic) were used for alginate particles preparation. Aqueous solutions were prepared using deionized water filtered by ion exchanger Aqual 25 (Aqual, Czech Republic).

Preparation of liposomes

10 mg of the phospholipid (DPPC or its mixture with cholesterol) was dissolved in approx. 1 mL of chloroform in a round-bottom flask. The organic solvent was then removed by a rotatory evaporator (IKA® RV10) at 60°C whilst pressure was slowly decreased to 200 mbar. A thin lipid layer was created at the bottom of the flask after complete evaporation. The film was then left for at least 2 hours under vacuum (~5 mbar) to remove any residues of chloroform.

The dried layer was hydrated by 1 mL of a solution of CF in Tris (50 mM CF, final osmolarity = isotonic, pH = 7.4) in the next step. The mixture was heated to 60°C while mixing and vortexing until all the phospholipids were transferred into the liquid.

Liposomes created by this procedure were then homogenized by 30 passes of extrusion through a polycarbonate membrane with a defined size of pores (first 400 nm, then 100 nm). The pore size defines directly the diameter of the liposomes. Avanti Mini-Extruder (Avanti Polar Lipids, USA) was used for this procedure. The extrusion was done at 60°C. The final size of the liposomes was measured by dynamic light scattering (ZetaSizer Nano-ZS, Malvern Instruments, UK) to check a correctness of the extrusion process.

The final step is the removal of non-encapsulated CF. Gel permeation chromatography was utilized using a 1.5 cm wide, 20 cm high glass column (Lenz Laborglas, Germany). Sephadex® G-50 was used as a static phase and isotonic Tris-HCl as a mobile phase. An orange liposome fraction was collected while non-encapsulated dye was delayed in the gel.

Alginate gel particles fabrication

1 mL of 1% (w/w) sodium alginate mixture with liposomes (1:1, v/v) was dripped into 10 mL of 2% (w/w) aqueous solution of calcium chloride solution. A syringe with needle was used for this. After 1 hour of gelation the small beads were washed by deionized water.

Release kinetics measurement

Carboxyfluorescein (CF) was used as a model substance for release rate determination. It is a fluorescent dye that is fluorescent while diluted, but it forms dimer in concentrated solutions and the dimer is not fluorescence active. The dimer is inside the liposomes and the dye starts to fluoresce after dilution (releasing outside). The released amount is then correlated to the fluorescence intensity hence fluorescence spectrophotometer (Cary Eclipse Fluorescence

Spectrometer, Agilent Technologies, USA) was used for release rate measurement. 10-15 μl of liposome solution was dispersed in 3 ml of Tris-HCl buffer in a glass cuvette. 20 μl of 10% Triton® X-100 was added at the end of the measurement, because it causes collapsing of liposomes into micelles hence release of all the dye[10]. Alginate particles were disintegrated by homogenizer (IKA T10 basic, IKA, Germany) before adding Triton® X-100.

Results and discussion

Leakage from liposomes entrapped within an alginate gel

Let us first study a stability of liposomes during cross-linking process of the alginate gel. Liposomes with various content of cholesterol were prepared according to the protocol in previous chapter. It is known, that cholesterol can improve mechanical and chemical stability of liposomes[11]. The liposomes were dispersed in sodium alginate. 1 mL of 1% aqueous sodium alginate solution containing liposomes (filled with CF) was dripped into 10 mL of 2% $CaCl_2$ by a syringe with a needle. The experiment was done for 4 different concentrations of cholesterol within the bilayer; pure DPPC liposomes and molar ratios of DPPC:Chol of 8:1, 4:1 and 2:1. 100 μL of $CaCl_2$ solution above the alginate beads was sampled and diluted by 3 mL of an isotonic Tris-HCl buffer and a fluorescence intensity of this mixture was determined.

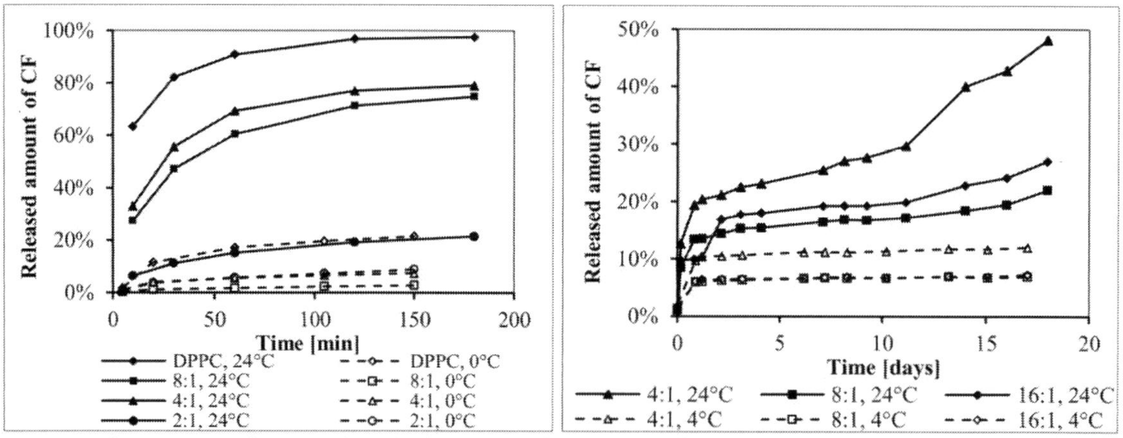

Fig. 2: *Left*: Preparation - CF leakage during cross-linking proces. *Right*: Storage - CF leakage from particles after cross-linking. Ratios in the legend are molar ratios of DPPC:Chol.

A liposomal stability in alginate beads during the cross-linking process was rather lowered (Fig. 2, left). In case of DPPC:Chol molar ratios of 4:1 and 8:1, more than 50% of CF was released in the first hour at laboratory temperature. For the pure DPPC liposomes it was even more than 90% of encapsulated dye which is inappropriate, because majority of the dye leaked before gel beads are created. The only acceptable leakage had "DPPC:Chol = 2:1" liposomes. The following experiment was done at lowered temperature. All the chemicals were cooled in a refrigerator at the beginning and the process was done in an ice-cold bath. The result was much more auspicious. The pure DPPC liposomes were the most unstable again, but no more than 20% of the dye leaked out. The liposomes with cholesterol were even more promising. Molar ratio of DPPC:Chol of 8:1 released slightly less of the dye than 4:1 and 2:1 liposomes. Apparently the release from DPPC liposomes during the preparation of liposome-in-alginate beads could be lowered by an addition of cholesterol and it is necessary to carry out the cross-linking process at lower temperature to avoid a loss of majority of an inner content of liposomes.

Based on the findings presented above, DPPC liposomes with cholesterol (molar ratios of 16:1, 8:1 and 4:1) were encapsulated into alginate beads at a temperature of an ice-cold bath. After

1 hour of cross-linking the beads were washed by deionized water and the leakage of the dye from entrapped liposomes was studied. At the beginning the volume of each sample was 20 mL. Every point in the chart represents a fluorescence intensity of 200 μL of the solution diluted in 3 mL isotonic Tris-HCl buffer. The measurement was simultaneously done at 2 different temperatures; at laboratory temperature (~24°C) and in a refrigerator (4°C).

The results showed that the storage temperature and also cholesterol concentration have influence on a stability of liposomes. The liposomes with DPPC:Chol ratio of 4:1 were the least stable in the alginate beads at both temperatures. There was found an acceleration of the release for all 3 concentrations after about 11 days at a laboratory temperature. This phenomenon was the most significant for the DPPC liposomes with 20 mol.% of cholesterol (ratio 4:1). The acceleration was also observed by Dhoot and Wheatley[12]. The release was slower in a refrigerator. After few days, the fluorescence intensity stayed almost constant in all samples and no secondary acceleration of release was found in 14 days at this temperature. Low temperature does help during the preparation process and also improves the stability of liposomes hence storage in a refrigerator is proper way to keep the most of the inner content inside the liposomes.

Choleterol effect on temperature-induced release kinetics from free liposomes

At this level it is known, that addition of cholesterol to the bilayer improves a stability of liposomes in calcium alginate beads. The bilayer phase transition is a key factor for a release kinetics study as mentioned in Introduction. So it is a question, how much cholesterol affects the temperature-induced release. Phase behavior (and its dependence on the release rate) of free liposomes with different content of cholesterol was studied. A method of "temperature ramps" was utilized for this study. Fluorescence intensity was determined with the respect to temperature, while the temperature was increased linearly. The rate was set to be 0.3°C.min^{-1}. Five samples of liposomes with various concentration of cholesterol was prepared, molar ratios of DPPC:Chol of 16:1, 8:1, 4:1, 2:1 and 1:1 (that is 5.9, 11.1, 20.0, 33.3 and 50.0 mol.% of cholesterol). A temperature-induced release was studied in an isotonic and also diluted buffer; inside the liposomes there was isotonic solution of CF and outside 170 mM mM Tris-HCl buffer (isotonic = 300 mOsm/L), resp. 10 mM Tris-HCl buffer (~17.5 mOsm/L). A presence of the buffer is necessary for a direct fluorescence measurement, because the fluorescence intensity of CF is not stable in pure water.

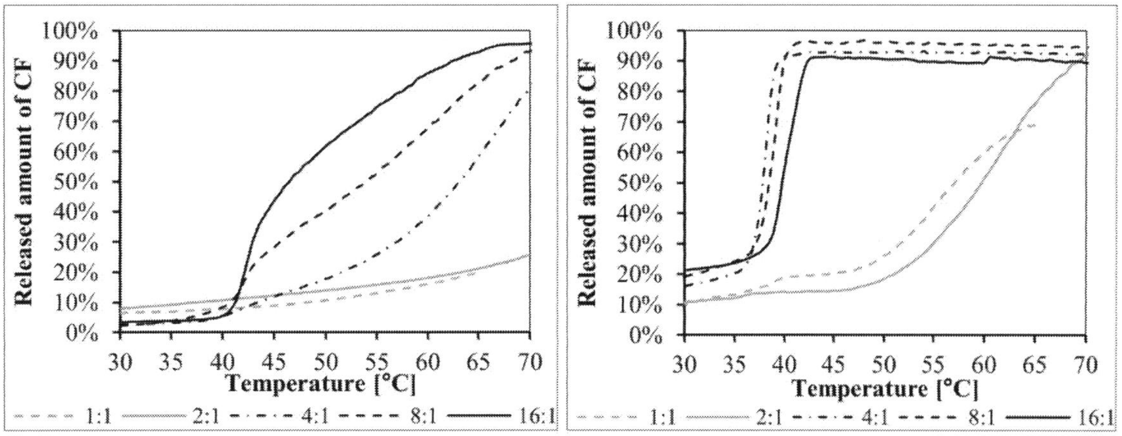

Fig. 3: Temperature induced release from DPPC liposomes with cholesterol in: isotonic 170mM (*left*); 10mM (*right*) Tris-HCl buffer (pH=7.4). Ratios in the legend represent molar ratios of DPPC:Chol.

The results in Fig. 3 (left) revealed two important things among others. The higher cholesterol content was, the more temperature-resistant the liposomes were in the isotonic buffer.

The sensitivity of the liposomes to temperature was lower and lower when increasing the cholesterol concentration. Higher concentrations of cholesterol (33 and 50 mol.%) even prevent liposomes to release high amount of CF above 50°C. The data were recalculated to the initial temperature, because a temperature dependence of the fluorescence intensity was significant in this experiment.

The same experiment was done in the diluted buffer to see the differences when the osmotic gradient across the bilayer is present (Fig. 3, right). The release acceleration caused by the phase change was much higher than in the isotonic buffer buffer. The liposomes with less than 20 mol.% released almost all the dye in few minutes after reaching the phase transition temperature. A liquid-like bilayer apparently cannot hold out the osmotic difference. The liposomes with molar ratio DPPC:Chol = 2:1 and 1:1 survived heating to 40°C in diluted buffer and the release was accelerated at higher temperatures.

This behavior could be explained by changes in a phase behavior of DPPC bilayer when cholesterol is added[13]. A cholesterol-rich *β*-phase starts to form, when reaching approx. 6 mol.% of cholesterol. Between 6 mol.% and 21 mol.%, the *β*-phase is at equilibrium with a liquid-like phase and above 21 mol.%, it exist only the *β*-phase. That can explain the decrease of sensitivity with addition of cholesterol and the significant change of the behavior, especially in the diluted buffer.

Thermal behavior of liposomes in alginate beads

Previous findings show importance of low temperature during formation of gel particles and helpfulness of cholesterol for stability of liposomes in the gel. On the other side, high amount of cholesterol make liposomes less sensitive to temperature. Therefore the molar ratio of DPPC:Chol of 8:1 was chosen for the next experiment It shows a potential of the liposome-containing gel particles for the controlled release. The alginate particles with liposomes were prepared according to protocol mentioned in the previous section. The beads were then left in a refrigerator for 24 hours to stabilize the fluorescence intensity after the initial burst.

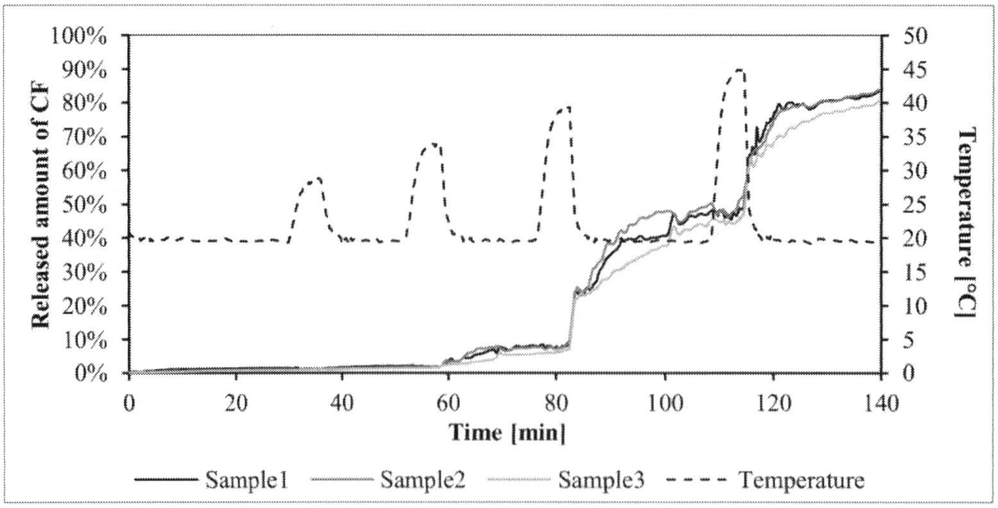

Fig. 4: On-demand release from composite liposomes-alginate particles induced by short bursts of temperature.

Each of 3 samples contains 3 alginate beads in 3 mL of 10 mM Tris-HCl. The beads were approximately 1 mm in a diameter. The default temperature for the measurement was set to be 20°C. The temperature was repeatedly raised up in 5-minute pulses and the temperature of these pulses was increasing (30°C, 35°C, 40°C and 45°C). A stepwise release from the liposome-in-alginate particles was achieved by this thermal program (Fig. 4). Fluorescence intensities were recalculated to 20°C, due to significant changes in fluorescence intensity as mentioned above. In first 30 minutes the temperature was kept at 20°C and no significant release was observed. The first temperature pulse of 30°C did not cause any visible acceleration of release rate. Approximately 5% was released after second pulse (35°C), but it was delayed. This delay is caused by diffusion barrier made by alginate gel[14]. Approximately 45% of CF was released into the gel beads after 5 minutes at 40°C (near the main phase transition temperature of DPPC) and the diffusion of CF out of the gel beads took almost 30 minutes. The next 45°C pulse caused a release of the residual dye.

Conclusions

Factors affecting the stability and the leakage during a preparation of an alginate gel containing liposomes were studied. The next step was investigation of liposomal stability during preparation of alginate beads. The release kinetics was measured for DPPC liposomes (with various content of cholesterol) entrapped within the alginate beads in $CaCl_2$ (cross-linking agent). Also the release of CF from prepared beads was investigated. Results shows, that cholesterol in the bilayer is desirable and the preparation at lower temperature (ice-cold bath, refrigerator) is necessary to prevent major leakage. The lowest leakage of CF from all tested samples had the liposomes with molar ratio DPPC:Chol of 8:1, respectively 2:1.

Bilayer phase change is a key factor affecting the liposome temperature-induced release kinetics. An exposure of liposomes to a linearly increasing temperature and a measurement of the dependence of the fluorescence intensity on temperature ("temperature ramps") turned out as sufficient and quick thermal characteristics of liposomes. The measured curve reveals 3 basic facts: the point when the bilayer starts to melt, the "width" of the phase change and the acceleration rate caused by the transition. The temperature ramps were used for a characterization of DPPC liposomes containing cholesterol. Cholesterol decreased the transition acceleration of release under isotonic conditions. The more cholesterol was contained in the membrane, the smaller was the acceleration. Actually, no acceleration was found for DPPC liposomes with 33 mol.% and 50 mol.% in the isotonic buffer. The same experiment was done in a 17-times diluted buffer (10 mM Tris-HCl buffer, pH = 7.4). The osmotic difference between the inner and the outer space of the liposomes caused a great acceleration and they released over 80% of the encapsulated dye in few minutes after reaching the phase transition temperature when the concentration of cholesterol was up to 20 mol.%. The liposomes with 33% and 50% of cholesterol released also almost all the dye, but higher temperature was needed.

The knowledge of the phase behavior of DPPC+cholesterol liposomes and ideal preparation conditions were combined for preparation of liposome-in-alginate beads suitable for temperature-induced release. "8:1 liposomes" inside calcium alginate gel were repeatedly heated to 30°C, 35°C, 40°C and 45°C at 5-minute pulses. It was shown, that the release is mostly accelerated around 40°C, like in case of non-encapsulated liposomes. It was released approx. 45% of CF after a 5-minute pulse of 40°C. The release was delayed, because of a higher diffusion coefficient of CF inside the gel particle. Following 5 minutes at 45°C was sufficient to release the rest of the dye.

We believe that these particles have a great potential. The bilayer can be destabilized by many factors. For example the release is faster when they are in contact with a hydrophobic surface (plastic cuvette, magnetic stirrer...). In our case, the liposomes were protected by an alginate gel. Liposomes in the alginate gel survived even in hydrophobic substances (e.g. heptane), because they are enclosed inside the hydrophilic gel. Immobilized liposomes in the alginate gel are little bit less stable than free liposomes, but they remain their phase behavior; temperature and "sharpness" of the transition. The delay caused by the diffusion barrier of the gel could be undesirable for more precision of the release control. This problem should be removed by using smaller particles, e.g. prepared by Drop-on-Demand Ink-Jet Technology[15]. The size of these microparticles approaches that of single-cell organisms and therefore opens also the possibility to study hybrid artificial-living multicellular systems.

References

1. Lasic, D. D., The Mechanism of Vesicle Formation. *Biochemical Journal* **1988,** *256* (1), 1-11.
2. Lasic, D. D.; Papahadjopoulos, D., *Medical Applications of Liposomes*. Elsevier Science: 1998.
3. Gabizon, A.; Catane, R.; Uziely, B.; Kaufman, B.; Safra, T.; Cohen, R.; Martin, F.; Huang, A.; Barenholz, Y., Prolonged Circulation Time And Enhanced Accumulation In Malignant Exudates Of Doxorubicin Encapsulated In Polyethylene-Glycol Coated Liposomes. *Cancer Research* **1994,** *54* (4), 987-992.
4. Nagle, J. F.; Tristram-Nagle, S., Structure of lipid bilayers. *Biochimica et Biophysica Acta-Reviews on Biomembranes* **2000,** *1469* (3), 159-195.
5. Eeman, M.; Deleu, M., From biological membranes to biomimetic model membranes. *Biotechnologie Agronomie Societe et Environnement* **2010,** *14* (4), 719-736.
6. Lewis, R. N. A. H.; Mak, N.; McElhaney, R. N., A Differential Scanning Calorimetric Study of the Thermotropic Phase-Behavior of Model Membranes Composed of Phosphatidylcholines Containing Linear Saturated Fatty Acyl Chains. *Biochemistry* **1987,** *26* (19), 6118-6126.
7. Silvius, J. R.; Read, B. D.; McElhaney, R. N., Thermotropic Phase-Transitions of Phosphatidylcholines with Odd-Numbered N-Acyl Chains. *Biochimica et Biophysica Acta* **1979,** *555* (1), 175-178.
8. Torchilin, V. P., Recent advances with liposomes as pharmaceutical carriers. *Nature Reviews Drug Discovery* **2005,** *4* (2), 145-160.
9. Dai, C. Y.; Wang, B. C.; Zhao, H. W.; Li, B.; Wang, H., Preparation and characterization of liposomes-in-alginate (LIA) for protein delivery system. *Colloids and Surfaces B-Biointerfaces* **2006,** *47* (2), 205-210.
10. delaMaza, A.; Parra, J. L., Changes in phosphatidylcholine liposomes caused by a mixture of Triton X-100 and sodium dodecyl sulfate. *Biochimica Et Biophysica Acta-Lipids and Lipid Metabolism* **1996,** *1300* (2), 125-134.
11. Liang, X. M.; Mao, G. Z.; Ng, K. Y. S., Mechanical properties and stability measurement of cholesterol-containing liposome on mica by atomic force microscopy. *Journal of Colloid and Interface Science* **2004,** *278* (1), 53-62.
12. Dhoot, N. O.; Wheatley, M. A., Microencapsulated Liposomes in controlled drug delivery: Strategies to modulate drug release and eliminate the burst effect. *Journal of Pharmaceutical Sciences* **2003,** *92* (3), 679-689.
13. Vist, M. R.; Davis, J. H., Phase equilibria of cholesterol/dipalmitoylphosphatidylcholine mixtures: deuterium nuclear magnetic resonance and differential scanning calorimetry. *Biochemistry* **1990,** *29* (2), 451-464.
14. Dohnal, J.; Štěpánek, F., Inkjet fabrication and characterization of calcium alginate microcapsules. *Powder Technology* **2010,** *200* (3), 254-259.
15. Haufová, P.; Dohnal, J.; Hanuš, J.; Štěpánek, F., Towards the inkjet fabrication of artificial cells. *Colloids and Surfaces A: Physicochemical and Engineering Aspects* **2012,** *410* (0), 52-58.

Scaffold Fabrication from Cadaver Goat Lung Tissue and Its Biocompatibility Study for Tissue Engineering Applications

Sweta Gupta, Department of Polymer & Process Engg, Indian Institute of Technology Roorkee, India, **Amit K. Dinda**, Pathology, All India Institute Of Medical Sciences, New Delhi, India and **Narayan C. Mishra**, Department of Polymer & Process Engg, Indian Institute of Technology Roorkee, Roorkee, India

Abstract

Fabricating scaffold from cadaver tissue and its application to regenerate various tissues is a current challenge in the area of tissue engineering. Our objective is to fabricate scaffold from cadaver goat lung, and to test its biocompatibility for application in tissue engineering. Scaffold was fabricated by decellularizing cadaver goat-lung by physical and chemical methods, and characterized by scanning electron microscopy (SEM), X-ray diffraction (XRD) and energy dispersion spectroscopy (EDS). Cell viability was performed by MTT assay after 24 hours in culture, by seeding Hep G2 cell lines onto the scaffold. Gross morphology of decellularized and recellularized lung tissue scaffold was demonstrated by Hemotoxylin and Eosin staining. SEM images of decellularized lung revealed porous interconnected structure as expected for a lung scaffold. XRD study indicates the amorphous nature of the scaffold matrix and EDS study proves the composition of the scaffold to be organic. MTT assay resulted in 99.2% viability of the cells in the scaffold. The histological examination revealed no remaining cells in the decellularized lung-tissue scaffold with negative cell nuclei staining, but nuclear counterstaining was detectable in the recellularized lung scaffold. The results demonstrated a novel but simple and economically feasible method of decellularization of cadaver lung. The biomimetic nature of the scaffold and high cell viability in the scaffold shows it's potential to be a good candidate for tissue engineering application and for studying cellular responses for drug discovery.

1. Introduction

The loss or failure of an organ or tissue is a costly problem in health care and is quite a frequent issue occurring in millions of patients every year [1, 2]. Tissue engineering is an interdisciplinary field, applying a set of tools at the interface of biomedical and engineering sciences that uses living cells or attract endogenous cells to aid tissue formation or regeneration, to restore or maintain tissue function [3,4]. Earlier scaffolds were used to be fabricated from natural or synthetic polymers, but as the xenogenic/cadaver tissue mimics the extracellular matrix of the human tissue (structural homology), it is commonly being used these days. . Irrespective of application, the foremost essential criteria for using these tissues are that - it must be made acellular and sterile, because the xenogenic component may have adverse immunogenic reaction in the body after in vivo implantation of the tissue engineered construct. It has already been proved that the components of the ECM are generally conserved among species and are tolerated well even by xenogeneic recipients [5]. Biologic scaffolds derived from decellularized tissues and organs have been successfully used in both pre-clinical animal studies and in human clinical applications [6]. ECM from a variety of tissues, including heart valves [7], blood vessels [8], skin [9], nerves [10], skeletal muscle [11], tendons [12], ligaments [13], small intestinal submucosa (SIS) [14], urinary bladder [15], and liver [16] have been studied for tissue engineering and regenerative medicine applications.

The aim of this study was to develop a scaffold from the cadaver tissue in a simple and economical way and to characterize and evaluate the developed scaffold for cell culture studies.

2. Material and methods

2.1. Scaffold development

Scaffold was developed using chemical and enzymatic method of decellularization. Goat lung tissue was at first rinsed in water and then cut into very thin rectangular sections of equal size (around 1 mm x 1 mm). The tissue sections were incubated with 0.25% trypsin containing 0.02% ethylenediaminetetraacetic acid (Trypsin-EDTA, Himedia, India) in 1XPBS having 1% Pen-Strep solution (Himedia, India) for 24 hours at 37°C. They were then further treated with 0.1% SDS in 1X $PBS^{-/-}$ at 37°C for 6-8 hr and then in solution containing RNase (20μg/ml) and DNase (0.2mg/ml) for 24 hr at 37°C, followed by an intensive wash cycles with $1XPBS^{-/-}$(1X Dulbecco's Phosphate Buffered Saline solution without Ca^{2+} and Mg^{2+} , Himedia, India). The tissue sections were then disinfected by agitating in an aqueous solution of 0.1% (v/v) hydrogen peroxide and 4% (v/v) ethanol for 2 hours. The developed decellularized matrix was stored in 1X PBS at 4^0C. Before using these scaffolds, samples were exposed to unfiltered UV light from a mercury arc lamp source under the laminar flow hood (Micro-filt, India).

2.2. Scaffold characterization

Before doing cell culture studies, the developed scaffold was characterized by using environmental scanning electron microscope (ESEM) (Quanta 200F Model, FEI, Netherland), Energy Dispersion Spectroscopy (EDS), X-ray diffraction and Hemotoxylin and eosin (H&E stain, Sigma) staining. To study the surface morphology and integrity of the developed scaffold, the surface of the scaffold was examined through ESEM, while hydrated at the saturation pressure of water vapor (1torr) and an accelerating voltage of 15 KV. Wet scaffolds were observed by ESEM technique, as it does not require dehydration processing or any coating of samples. Chemical analysis of scaffold samples was performed with a LEO 1525 scanning electron microscope equipped with an EDS detector (Carl Zeiss SMT Ltd., Hertfordshire, UK). To quantify the chemical content of small areas (~60 μm^2) on the sample surface, distinctly different topographical areas were chosen from the sample and examined at 10 kV using INCA Energy 3000 software. X-ray diffraction was used to detect the presence of phase composition analysis in the scaffold sample using a wide angle X-ray diffraction apparatus. A Siemens D-5000 diffractometer coupled to a copper anode tube was used to obtain the X-ray diffraction pattern and identify the crystalline compound. The K_a wavelength was selected with a diffracted beam monochromator. X-ray scans were collected over a 2θ range of 10° to 80° with a dwell time of 1s and increments of 0.2°. Histological analysis was done by hemotoxylin and eosin staining by fixing the decellularized tissues in 4% paraformaldehyde, paraffin embedding and staining it using hemotoxylin and eosin (H&E stain, Sigma). The sections were mounted on slides to identify the collagen fibrous structure in the scaffold matrix.

2.3. Cytotoxicity studies of the decellularized scaffold (MTT assay)

HepG2 cell line was maintained in DMEM with 10% FBS at 37°C in a CO_2 incubator (BINDER, Germany). Cells were plated in 96 well plate at a density of 1 x 10^3 cells/well (200 µL volume) and allowed to grow to 60% confluency. Scaffold were placed in the wells in triplicate and incubated for 24 h at 37°C in CO_2 incubator. After the incubation point, the scaffold was removed from wells and the media was discarded. 90 µL of fresh complete media was added to the wells and 10 µL of MTT solution (5 mg/ml stock in PBS) was added to the media for final volume of 100 µL. The plate was incubated at 37°C for 4 h until purple formazan crystals were formed due to reduction of MTT by viable cells. The media was discarded and 200 µL of dimethyl sulphoxide (DMSO) was added to the wells to dissolve the formazan crystals. Absorbance was taken in a Biorad plate reader at 490 nm with the substraction for plate absorbance at 650 nm. Cell viability was expressed by the following equation:

$$\% \text{ Cell Viability } = \frac{I_{sample}}{I_{control}} \times 100$$

where, I_{sample} = absorbance of the wells incubated with sample taken in triplicate, and
$I_{control}$ = Mean absorbance of control wells (wells incubated with cells without scaffold).

The results were compared to the control wells incubated with cells in which no scaffold was present.

3. Result and Discussion

3.1. Scaffold characterization

The ESEM image (Figure 1a) of the decellularized scaffold shows that the architecture of scaffold is highly porous and has good interconnectivity. The high porosity will provide adequate space for cell growth as well as will promote ECM production. It (high porosity) will also help in transporting of nutrients from nutrient media and in excretion of waste products. The EDS spectrum of the developed scaffold (Figure 1b) shows that it contains Carbon and Oxygen as the major elements in the scaffold. Other elements were present in very minute amount. This proves that the scaffold was organic in nature and can be degraded in the body after in vivo implantation of the tissue engineered construct. The XRD spectrum (Figure 1c) of the scaffold shows a broad, ill-defined peak, which suggests semi-crystalline phase composition of the scaffold. The histological analysis of the developed scaffold (Figure 1d) shows that

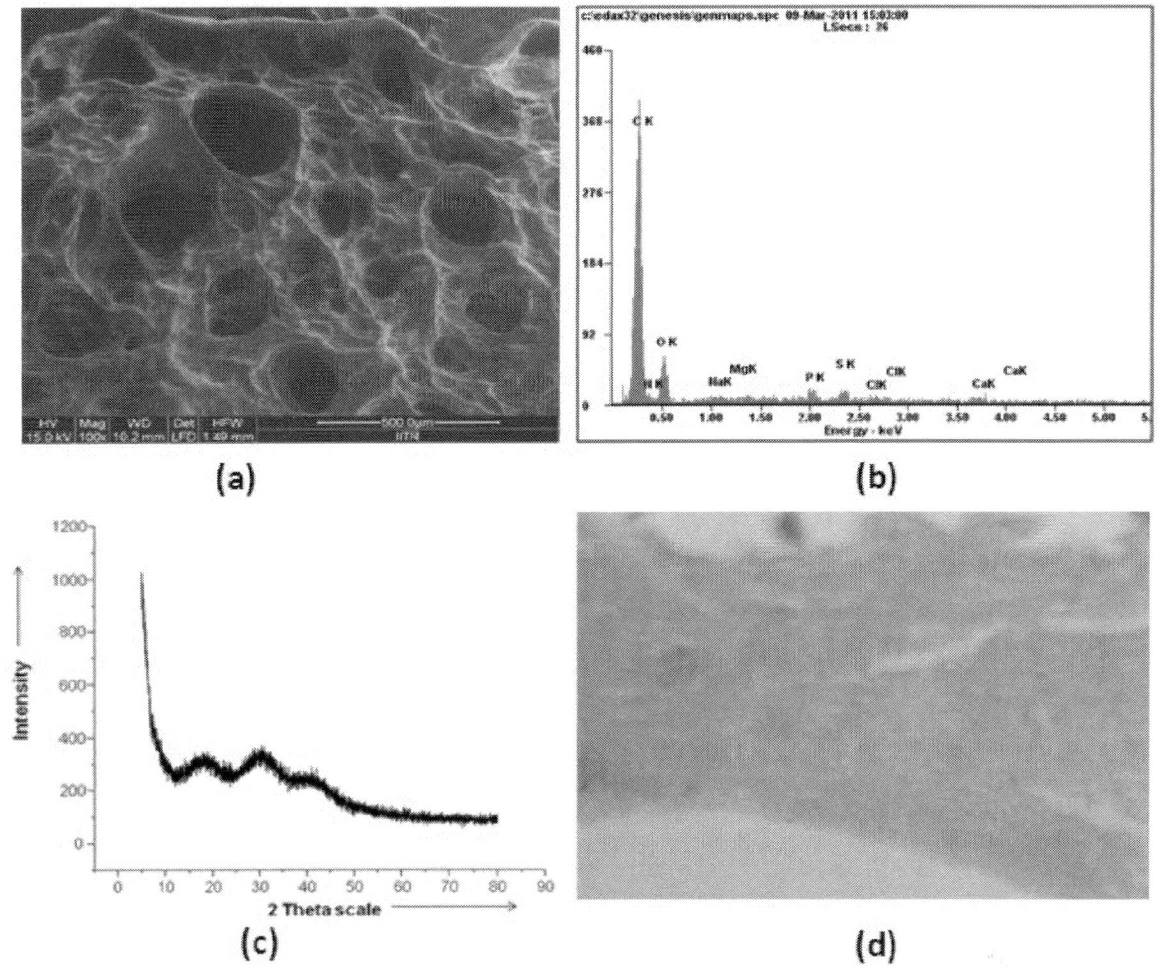

Figure 1. (a) SEM image of the developed scaffold showing porous nature. (b) EDS spectra of the developed scaffold showing Carbon and Oxygen as the major constituent of the scaffold. (c) XRD spectra of the developed scaffold having ill-defined peaks revealing its semi-crytalline nature. (d) Hemotoxylin and Eosin staining of the decellularized scaffold, showing the collagenous structure wit negative cell-nuclei staining.

3.2. Cytotoxicity studies: MTT assay

MTT assays showed that HepG2 cells cultured in DMEM in presence of decellularized scaffold matrix showed a similar proliferation rate with that of positive control, from where the scaffold was omitted. The production of resultant purple colored formazan is directly proportional to the level of energy metabolism in the cells, which was same in positive control and in experimental column. Cells showed 99.2% viability in presence of scaffold (Figure 2). HepG2 cells grew up to the decellularized matrix with no evidence of contact inhibition, providing evidence that the decellularized matrix was not cytotoxic. But in negative control, where the triton X was present, all the cells died and there was no formation of purple colored formazan crystals.

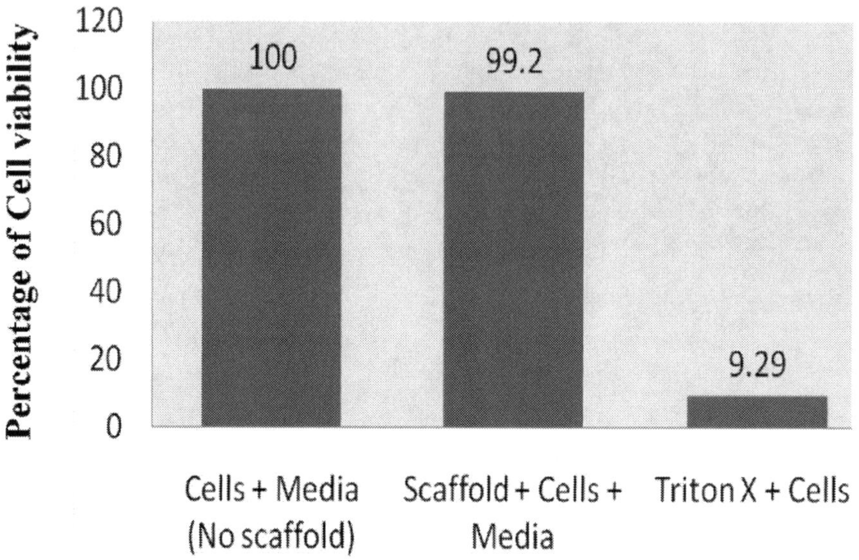

Figure 2. Viability percentage of HepG2 cells on the developed scaffold for 24 horus. HepG2 cells showed 99.2% cell viability in the presence of the developed scaffold.

4. Conclusion

Goat lung tissue was successfully decellularized by chemical and enzymatic method of decellularization. The develop scaffold showed porosity and good interconnectivity which is found to be suitable for the adhesion and proliferation of HepG2 cells. The scaffold also exhibit good biocompatibility. More than 99% of the HepG2 cell viability in the presence of the developed scaffold proves that the scaffold does not contain any leachable toxins which might be harmful for the growth of cells. Hence, the scaffold developed from the cadaver tissues can be a promising candidate for tissue engineering applications.

5. Acknowledgements

The authors are grateful to the Council of Scientific and Educational Research (Grant No. 27(0222)/1 O/EM R-II dated 31.05.10), India, for funding this research work.

6. References

1. Langer R, Vacanti JP. Tissue engineering. *Science* 1993; 260:920-926.

2. Uematsu K, Hattori K, Ishimoto Y, Yamauchi J, Habata T, Takakura Y, Ohgushi H, Fukuch, T, Sato M. Cartilage regeneration using mesenchymal stem cells and a three-dimensional poly-lactic-glycolic acid (PLGA) scaffold. Biomaterials 2005; 26: 4273-4279.

3. Vacanti JP, Vacanti CA. 'Principles of tissue engineering' in The history and scope of tissue engineering, eds. Lanza RP, Langer R, Vacanti JP, J Acad Press (3-4), San Diego, 2000.

4. Rabkin E, Hoerstrup SP, Aikawa M, Mayer JE, Schoen FJ. Evolution of cell phenotype and extra cellular matrix in tissue engineered heart valves during in-vitro maturation and in-vivo remodeling. J Heart Valve Dis 2002; 11: 308-314.

5. Bernard MP, Chu ML, Myers JC, Ramirez F, Eikenberry EF, Prockop DJ. Nucleotide sequences of complementary deoxyribonucleic acids for the pro alpha 1 chain of human type I procollagen. Statistical evaluation of structures that are conserved during evolution. Biochemistry 1983;22:5213–23.

6. Harper C. Permacol: clinical experience with a new biomaterial. Hosp Med 2001;62:90–5.

7. Schenke-Layland K, Vasilevski O, Opitz F, Konig K, Riemann I, Halbhuber KJ, et al. Impact of decellularization of xenogeneic tissue on extracellular matrix integrity for tissue engineering of heart valves. J Struct Biol 2003;143:201–8.

8. Schmidt CE, Baier JM. Acellular vascular tissues: natural biomaterials for tissue repair and tissue engineering. Biomaterials 2000;21:2215–31.

9. Chen RN, Ho HO, Tsai YT, Sheu MT. Process development of an acellular dermal matrix (ADM) for biomedical applications. Biomaterials 2004;25:2679–86.

10. Kim BS, Yoo JJ, Atala A. Peripheral nerve regeneration using acellular nerve grafts. J Biomed Mater Res A 2004;68:201–9.

11. Borschel GH, Dennis RG, Kuzon JrWM. Contractile skeletal muscle tissue-engineered on an acellular scaffold. Plast Reconstr Surg 2004;113:595–602 [discussion 603-4].

12. Cartmell JS, Dunn MG. Effect of chemical treatments on tendon cellularity and mechanical properties. J Biomed Mater Res 2000;49:134–40.

13. Woods T, Gratzer PF. Effectiveness of three extraction techniques in the development of a decellularized bone–anterior cruciate ligament–bone graft. Biomaterials 2005;26:7339–49.

14. Kropp BP, Eppley BL, Prevel CD, Rippy MK, Harruff RC, Badylak SF, et al. Experimental assessment of small intestinal submucosa as a bladder wall substitute. Urology 1995;46:396–400.

15. Gilbert TW, Stolz DB, Biancaniello F, Simmons-Byrd A, Badylak SF. Production and characterization of ECM powder: implications for tissue engineering applications. Biomaterials 2005;26:1431–5.

16. Lin P, Chan WC, Badylak SF, Bhatia SN. Assessing porcine liver-derived biomatrix for hepatic tissue engineering. Tissue Eng 2004;10:1046–53.

(596o) Effect of Poly(ethylene glycol) Spacers On the Vascular Targeting of Particles to Endothelium in Blood

Peter Onyskiw and Omolola Eniola-Adefeso, Chemical Engineering, University of Michigan, Ann Arbor, MI

Vascular targeting is an intriguing means for therapeutic treatment and diagnosis of cardiovascular diseases such as atherosclerosis. To target diseased tissue particulate carriers are modified to contain antibodies or epitopes, such as anti-intercellular adhesion molecule-1 (aICAM-1) and Sialyl Lewis^{a}. Along with targeting moieties, poly(ethylene glycol) spacers are used to improve circulation time by reducing protein adsorption on the surface of particles. Though the use of poly(ethylene glycol) to reduce protein adsorption is well studied, little work has been done on examining the effect of poly(ethylene glycol) on adhesion dynamics; specifically in human blood flow. Here we exam the influence of different molecular weight poly(ethylene glycol) spacers on particle adhesion dynamics to inflamed human endothelial cells in blood. Polystyrene nano/microspheres were grafted with poly(ethylene glycol) spacers of different lengths and densities to serve as a linker between particle surface and targeting ligand – anti-ICAM-1 or Sialyl Lewis^{a}. Targeted particles in human blood were then perfused over a monolayer of IL-1β stimulated human umbilical chord endothelial cells using a parallel plate flow chamber at shear rates of 200 s ^{1}, 500 s ^{1}, and 1000 s ^{1}. Poly(ethylene glycol) spacers were shown to improve binding flux of antibody-coated microspheres to activated cells but had no effect on the binding flux of Sialyl Lewis^{a} –coated microspheres and.the dynamics appear to be a function of PEG surface conformation and blood shear rate

(596p) Food-Associated Stimuli Impact On Particle Penetration Through Gastrointestinal Mucus Barrier

Hasan M. Yildiz and Rebecca L. Carrier, Chemical Engineering, Northeastern University, Boston, MA

Food-associated Stimuli Impact on Particle Penetration through Gastrointestinal Mucus Barrier

Hasan Yildiz, yildiz.h@husky.neu.edu

Rebecca L. Carrier, rebecca@coe.neu.edu

Introduction:

Mucus is a complex gel network comprised of mucins, lipids, salts, proteins, macromolecules and cellular debris. A heterogeneous mucus layer protects the gastrointestinal (GI) tract and other mucosal surfaces. Orally delivered drug and gene vectors must traverse through mucus in order to enter the circulatory system. However, the transport of these vectors and general barrier properties of mucus are not well characterized. The main objective of this project is to analyze the impact of food-derived lipids and bile salts as well as physiologically relevant stimuli, pH and [Ca^{+2}], on particle transport across GI tract mucus. These post-prandial stimuli could potentially be used to control drug carrier transport or prevent pathogen invasion.

Materials and Methods:

Particle transport across porcine intestinal mucus from pig jejunum was investigated. 200 nm amine-, carboxylate-, sulfate- and 100 nm carboxylate-modified particles were diluted in maleate buffer, bile salts (NaTDC) and simulated fed intestinal contents ("FED state") including maleate buffer, bile salts/phospholipids (lecithin) and a lipid mixture containing soybean oil and monoglycerol for a particle concentration of 0.0025 wt.-%. Calcium ion composition was adjusted by changing $CaCl_2$ levels in maleate buffer, and medium pH was controlled by adjusting NaOH concentration. Particle diffusion was measured by tracking the positions of diluted microspheres using real- time multiple particle tracking technique (MPT). Olympus IX51 was used to detect particles at 40 X magnification and videomicroscopy. Positions of particle centroids were used to calculate time-averaged mean squared displacements (MSD) and effective diffusivities (D_{eff}): $MSD = [x(t+\tau)-x(t)]^2 + [y(t+\tau)-y(t)]^2$ and $D_{eff} = MSD/(4\tau)$ where x and y are positional data and τ is the time scale. Particle effective diffusion coefficients were then used to estimate the fraction of nanoparticles expected to penetrate an intestinal mucus layer with a given thickness using a numerical integration of Fick's second law : $dC/dt = D_{eff}\, d^2C/dx^2$ where C is the concentration of particles, t is time and x is position.

Results:

Lipids associated with fed state intestinal contents reduced particle transport rate through gastrointestinal mucus. Estimated fraction of dosed particle penetration was reduced by half for microspheres dosed with lipids. Surprisingly, particles dosed in Bile Salts have higher transport rates (~ 1.5 times faster diffusion rate) than particles dosed in Maleate Buffer. This could be related to BS adsorption changing the electrostatic interactions with the mucus network. Lower diffusion rates were also observed at higher calcium ion concentration. Acidity also plays important role on particle diffusion rates. pH elevation induced a dramatic decrease in effective diffusivities of microspheres.

Conclusions:

The observed impact of physicochemical stimuli: lipids, bile salts, pH, and [Ca^{+2}] on particle transport support the significance of food contents on mucus barrier properties and the use of naturally changing factors in the GI tract to improve the effective transport of drug carriers for oral drug delivery. In addition, these factors could contribute to food effect on oral drug absorption, an important factor to oral drug delivery.

(596q) Hyperbranched Temperature Responsive Copolymers for Drug Delivery Applications

Kai Chang, Chemical and Biomolecular Engineering, Georgia Institute of Technology, Atlanta, GA and Lakeshia J. Taite, Georgia Institute of Technology, Atlanta, GA

Introduction: Poly(*N*-isopropylacrylamide) (pNIPAAm) has been incorporated into a variety of biomaterials due to its biologically inert nature and thermally responsive properties. Changing the pNIPAAm topology from a linear to a hyperbranched scheme provides additional functionalities such as the abilities to encapsulate small molecule drugs, target cancer cells, and release the drugs upon exposure to external stimuli. To utilize this material as a controlled-release drug delivery system, several functions must be optimized. In this study, we engineer the properties of hyperbranched pNIPAAm for optimal controlled drug release through the modification of its lower critical solution temperature (LCST) from a native 28 °C to ~45 °C via copolymerization. While the behavior of linear analogues of these copolymers is well-known, the use of a hyperbranching scheme can significantly shift the LCST. To our knowledge, this is the first extensive study of hyperbranching on these copolymers. The three most common copolymers for this application were chosen for this study: -co-acrylamide (AAm), -co-dimethylacrylamide (DMA), and –co-acrylic acid (AAc).

Materials and Methods: The hyperbranching agent and copolymers were synthesized similar to the procedure set forth by Carter et al.[1] Molecular weights and PDIs were determined using a GPC system with a refractive index detector. ^{1}H NMR was conducted on a Varian Mercury Vx 400 spectrometer using $CDCl_3$. A Cary 50 UV-Vis Spectrophotometer with a single cell Peltier accessory was used to measure the LCST of the polymers.

Results and Discussion: The three different copolymers exhibit widely differing LCST temperatures and behaviours. As seen in Figure 1, even with consistent 10% copolymer content, the effects varied dramatically. DMA and AAm copolymers show sharp transitions but require extensive amounts of copolymer content. In order to achieve a 45 °C transition for drug delivery, a copolymer content of 30% was required for an AAm copolymer system while a copolymer content of 40% was required for a DMA copolymer system. Since large amounts of hydrophilic co-monomer can reduce the efficacy of the polymer collapse at the transition, these high percentages may have adverse effects for drug delivery system. Acrylic acid, the most common pNIPAAm copolymer for biological systems, only requires 7% copolymer content to cause an LCST shift to 45 °C; however, it also increases the range of temperature transition from < 1 °C to > 20 °C depending upon the amount of AAc added. A trade-off is clearly seen between a sharp transition at the desired temperature and efficacy.

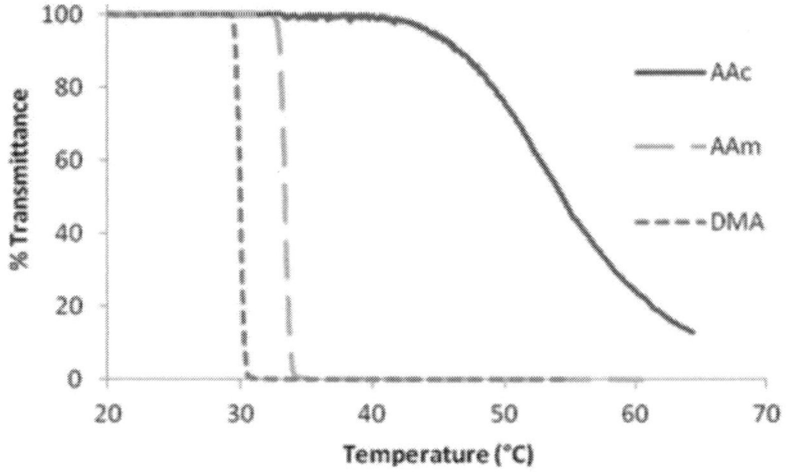

Figure 1. Different temperature transition profiles for different copolymers. Copolymers had 10% co-monomer content. AAc exhibited a broad transition at 55 °C. AAm and DMA exhibited sharp transitions at 33 °C and 30 °C respectively.

Conclusions: This study indicates that while synthesizing similar copolymers may be sufficient for hydrogel or linear pNIPAAm systems, this method exhibits significant challenges in highly branched systems. A significant amount of -co-DMA or -co-AAm needs to be included for a sharp transition at 45 °C. AAc is only a viable substitute if a large range for transition is acceptable. Understanding these effects allow engineers to make informed decisions in the engineering of controlled release drug delivery vehicles that use pNIPAAm.

References:

1. Carter, S.; Rimmer, S.; Sturdy, A.; Webb, M. *Macromolecular Bioscience* 2005, 5, 373.

(596r) Delivery Systems for Controlled and Sustained Release of Malarial Inhibitors

Christina Yacoob[1], Kayode K. Ojo[2], Steven M. Johnson[3], Dustin J. Maly[3], Wesley C. Van Voorhis[2,4] and Hong Shen[1], (1)Department of Chemical Engineering, University of Washington, Seattle, WA, (2)Department of Medicine, University of Washington, Seattle, WA, (3)Department of Chemistry, University of Washington, Seattle, WA, (4)Department of Global Health, University of Washington, Seattle, WA

Drug delivery systems are engineered-materials that dispense a constant supply of drugs to a patient for a desired dosage period at a level below the minimum toxic concentration (MTC) and above the minimum effective concentration (MEC). Since each molecule possesses unique hydrophobicity and size, which can alter overall drug retention or *in vivo* release kinetics, a known challenge is the adaptability of therapeutic molecules to the desired system. In this study, we develop a process to effectively incorporate slightly hydrophilic drugs into poly(lactic-co-glycolic acid) (PLGA) particles.

The model drug is a malaria transmission blocking inhibitor, BKI-1. Malaria is a mosquito-borne disease caused by a Plasmodium parasite, with 216 million reported cases world-wide and close to one million deaths per year according to the Center for Disease Control. Current treatments for malaria are becoming ineffective and allow gametocytes to infect mosquitoes for up to 3 weeks after successful therapy. A possible strategy through collaborative efforts is to develop compounds that break the malaria transmission circle. These compounds are predicted to inhibit malaria sexual stage in mosquitoes without noticeable toxic effects to mammals, potentially resulting in a decrease in the transmission and spread of malaria. In mouse PK studies, BKI-1 has a high blood clearance rate and is slightly hydrophilic. However, the extended presence of viable gametocytes in the mammalian host will require a prolonged drug bioavailability at a level between the MTC and MEC for effective transmission blocking to occur. To overcome this challenge, we have commenced the development of a low-cost, easy to fabricate, and easy to administer subcutaneous particulate injection formulated with PLGA to ensure slow release of BKIs into the blood stream over an extended period.

The main challenge of incorporating slightly hydrophilic drugs in PLGA particles by either single emulsion or double emulsion is that the loading efficiency is low. We hypothesized that the partition of drug in aqueous and organic phases during the fabrication process would be critical for the effective incorporation of drug into PLGA particles. To test our hypothesis, we varied the overall drug level, pH in aqueous phase and the composition of organic phase. We measured the loading of BKI-1 in PLGA particles. It was found that there existed an optimal overall drug level, pH and the composition of organic phase. We will then correlate the efficiency of drug loading with the partition coefficient.

Detection of interaction constants between biological clock proteins by Surface Plasmon Resonance

Burcu Kepsutlu[a], Hande Asimgil[a], Bilal Çakır[b], Doğan Gidon[a], Gozde Sultan Demirer[a], Rıza Kızılel[a], Halil Kavaklı[a], Seda Kızılel[a*]
[a] Koç Universitesi, İstanbul, 34450
[b] Washington State University, Pullman
* Seda Kızılel :*Koç Universitesi,İstanbul, 34450, skizilel@ku.edu.tr*

Abstract
Organisms adopt their behaviors and physiology to the appropriate time of the day to anticipate daily environmental changes and the circadian clock regulates their daily rhythms. In mammals, the clock is present in essentially every cell. A heterodimer of CLOCK and BMAL1 proteins binds to the E-box in *Per* and *Cry* promoters and activates their transcription. In this work, we have purified core clock proteins and characterized the affinity of previously identified clock-relevant transcription factors. We have investigated the mechanism of the clock complex and the interactions of clock proteins with and without DNA using Surface Plasmon Resonance (SPR). Kinetic parameters determined from real time data bring a solid insight into the interactions of the clock proteins with their cognate promoter.
Keywords— Circadian, Clock, Heterodimer, Surface Plasmon Resonance

Introduction

Circadian clock is an internal time keeping system recurring 24 hour daily rhythm in physiology and behavior of organisms. The biological clock is coordinated by suprachiasmatic nucleus as the main oscillator in order to correlate internal metabolism with external environmental changes. The master clock regulates the multiple circadian oscillator cells through neural and hormonal pathways. Those circadian oscillators are scattered throughout the body of mammals controlling the local rhythms in physiology and behavior. One of the input of the master clock oscillator is photic input and the output of the biological clock regulates the physiological and behavioral daily rhythm [1]. Photic input is an important signal for master clock. For instance, blind people are found to suffer from recurrent insomnia, daytime sleepiness because the internal rhythm periodically stays out of phase lacking the ability to synchronize itself to light-dark cycle [2].

Circadian clock contains transcription and translation feedback loop composed of four main gene and protein groups such as CLOCK/NPAS2, BMAL1, *Cry1/2* and *Per1/2*. Many different approaches defined for interactions between BMAL1-CLOCK heterodimer and PER and CRY repressors [3]. Heterodimer of CLOCK/NPAS2 and BMAL1 binds to E-box element in the promoter of *Per* and *Cry* genes in order to activate their transcription. According to Sancar and colleagues, CRY and PER making heterodimeric complexes enter the nucleus in order to inhibit CLOCK-BMAL1 activation of transcription of their own genes [4]. CRY1 proteins act as potent inhibitor of CLOCK-BMAL1 induced transcription and CRY proteins modify the interaction between PER2 and BMAL1 [3] [5]. Phosphorylation takes role in the transcription-translation loop and it is at its maximum value at time of negative feedback on BMAL1-CLOCK induced transcription *in vivo.* The phosphorylation is important for localization of CLOCK and BMAL1 proteins into the nucleus. Also mCRY proteins may be required for stability of phosphorylated mPER2 [1]. Because the biological clock is integrated with all major signalling and metabolic pathways, disruption of the biological clock may lead to unexpected results within the cellular mechanisms of the organism [5], such as cardiovascular function, thermoregulation and glucose metabolism. For instance, Gale suggests that disruption of circadian clock may increase the risk of type 2 diabetes mellitus due to decreased sensitivity of cells to insulin [6].

To date, interactions between core biological clock proteins have been studied biochemically. The only available quantitative data is equilibrium constant of BMAL1-CLOCK heterodimer and *mPer2* E2-box interaction which was reported as as approximately 59nM [7]. The aim of this study is to define binding affinities of clock proteins among each other on and off DNA modes through Surface Plasmon Resonance (SPR) technique.

Surface Plasmon resonance biosensors are utilized for real-time measuring molecular interactions without labeling requirements in a short time. The biosensor experiments involve immobilization of one reactant on the surface while the other reactant is injected through the immobilized surface. The kinetic data can be obtained from the real time data of repeated cycles of association, dissociation process of different concentrations of analyte on immobilized reagent. Reactions rates and equilibrium constants of interactions can be easily detected via time dependent observation of concentration changes in the reaction compartment [8-9].

The SPR biosensors exploit the surface plasmon resonance phenomenon in a micron-sized reaction chamber. SPR is an optical phenomenon occuring due to total internal reflection of light at metal film-liquid interphase. If the light encounters a dense medium while a light travels through an optically dense medium, the light is reflected through the dense medium at the interphase. The criteria is a greater angle of the incidence light than the critical angle for the optical media pair. When incident light is plane polarized, monochromatic light and the interphase between media is coated with a thin metal film, energy from incidence light is lost to metal film as a plasmon wave reducing the reflected light at the interface. The angle required for surface plasmon resonance is precise and defined by the refractive index of the less dense medium. The less dense medium is the reaction mixture where immobilized reagent reacts with the analyte. As reaction proceeds, refractive index of the medium is altered changing the resonance angle [10]. As surface plasmon angle shifts, a graph is obtained (Figure.1). Approximately, a change of 1000RU corresponds to 0.1° shift in the surface plasmon angle. Each line in the graph corresponds to different concentration of analyte. For globular proteins, 1000 RU is equivalent to surface concentration of 1 ng/mm^2 [10-11].

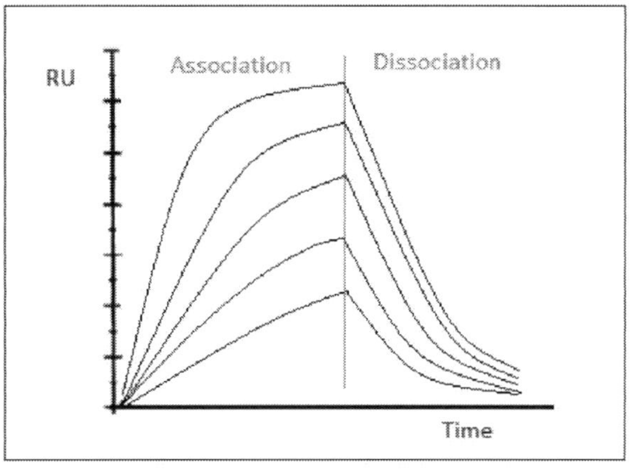

Figure.1 Example of a typical data obtained using SPR

If the ligand and analyte interacts with pseudo first order reaction kinetics, the binging process may be describes as

$$\frac{dR}{dt} = k_aC_a\,(R_{max}\text{-}R)\text{-}\ k_dR \qquad \text{Eq(1)}$$

where Ca, R, ka, kd and Rmax is analyte concentration, response unit signal, association rate constant, dissociation rate constant and response unit signal at maximum saturation of surface sites [12].

The integrated form of the equation is as in the following [12];

$$R_{association}(t) = \frac{R_{max}}{1+\frac{k_d}{k_aC_a}}\left[1-e^{(k_dC_a+k_d)t}\right] \quad Eq(2)$$

$$R_{dissociation} = R_{association}(t_d)e^{(-k_d(t\text{-}t_d))} \qquad Eq(3)$$

where t_d denotes the end of association phase. Eq(2) denotes the association phase while Eq(3) defines dissociation phase of SPR data [12].

Later, equilibrium constant of the reaction can be calculated by dividing on and off rate constants to each other as in the following equation [12].

$$K_D = \frac{k_d}{k_a} \qquad Eq(4)$$

Materials and Methods

Protein expression and purification:

mBmal1-His, Flag-mClock-His, Flag- mCryptochrome2-His and Flag-mPeriod2-His genes were previously cloned into pfast-bac donor plasmid. Then they are transposed to DH10Bac.GET buffer containing RNAase solution in order to isolate bacmid DNA. The recombinant plasmids from each construct were selected and transfected into 2*106 Sf21 insect cells. The cells were then incubated and purified. Large scale expressions in multiple T75 flasks were done in order to start purification process with high amount of protein. The proteins were purified by affinity chromatography using Ni-NTA resin. All proteins were eluted from the resin with 250mM imidazole.

SPR experiments:

During SPR experiments, the carboxy surface over polysaccharide covered gold metal is immobilized with the ligand through amine coupling. After surface immobilization, repeated injections of proteins at different concentrations performed to obtain time dependent response unit data similar to the graph depicted at Figure.1.

Kinetic data analysis of SPR:

The time dependent response unit data is divided into its association and dissociation parts. Eq(2) and Eq(3) is simplified by defining parameters a,b,c and d as in the following;

$$a=\frac{R_{max}}{1+\frac{k_d}{k_aC_a}} \qquad b= k_aC_a+k_d \qquad c=R(t_d) \qquad d=k_d$$

Later, best fit lines for association and dissociation data is defined using curve fitting tool of MATLAB Software. Then, using the parameters a,b,c and d values, rate constants and equilibrium contant is calculated. Each best fit line has its own a,b,c and d values.

Results and Discussion

Protein expression and Purification Results:

According to SDS-PAGE silver staining results for BMAL1 and CLOCK proteins, purification step of CLOCK and BMAL1 proteins is successfully achieved. (Figure.2.A) Also Western Blot Analysis exhibited successful purification of PER2 and CRY2 proteins (Figure.2.B).

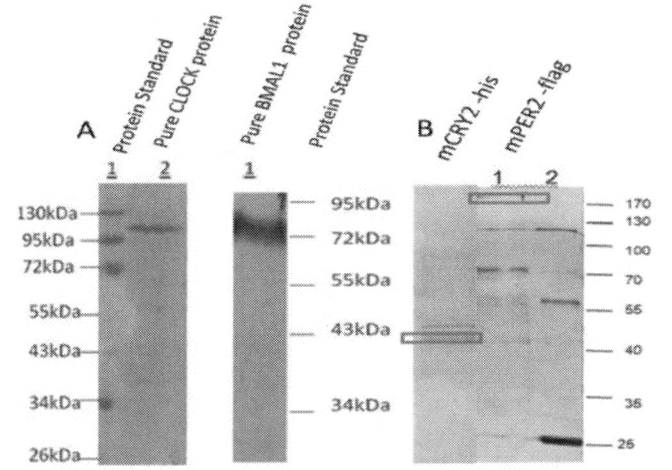

Figure.2 (A) SDS-PAGE silver staining results for CLOCK and BMAL1 proteins (B) Western Blot Analysis of expression of PER2 and CRY2 proteins

SPR Results:

At the end of repeated injection of analytes at different concentrations over immobilized protein, time dependent response unit graphs are obtained for protein pairs (BMAL1-CRY2, BMAL1-PER2, CRY2-PER2 and BMAL1-CLOCK). In the graphs, dotted curves demonstrate the real data, while smooth lines demonstrate the best fit data by curve fitting tool of MATLAB software. For each protein pairs, maximum response unit attained increases with increasing analyte concentration as expected. In CLOCK-BMAL1 interaction, the reproducibility of the experiments were shown by presenting duplicate data.

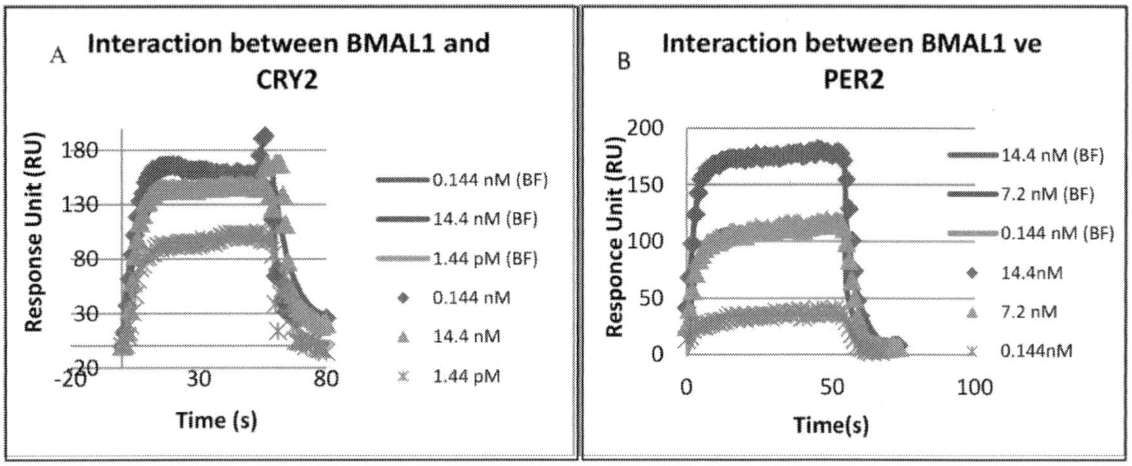

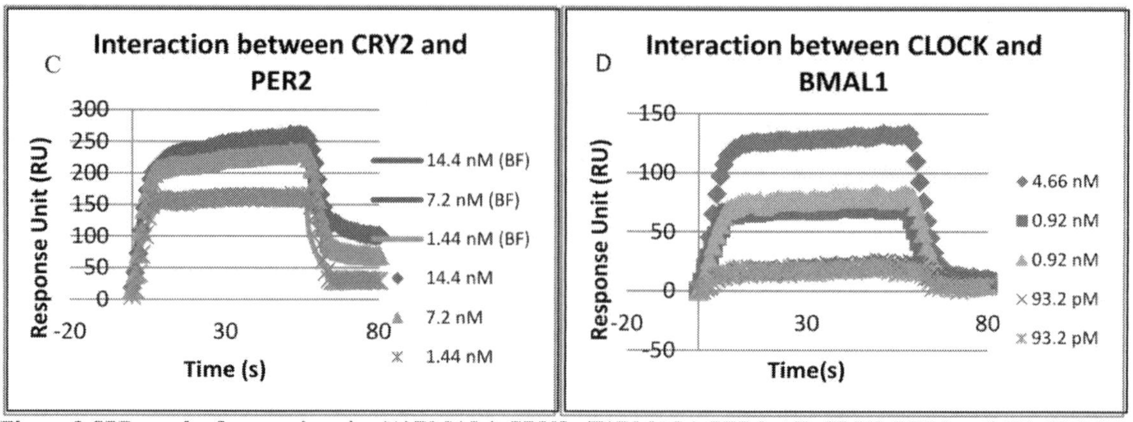

Figure.3 SPR results for protein pairs (A)BMAL1-CRY2, (B)BMAL1-PER2, (C) CRY2-PER2 and (D) CLOCK and CRY2

After analysis of the time dependent response unit curves for each protein pair, rate constant and equilibrium constant is calculated and tabulated in Table.1.

Protein Pairs	k_a (M^{-1} s^{-1})	k_d (s^{-1})	K_D (k_d/k_a) (M)
BMAL1 ve CRY2	6.80×10^{7}-9.02×10^{8}	0.2380-0.3500	3.93×10^{-9}-2.64×10^{-11}
BMAL1 ve PER2	1.24×10^{7}-7.08×10^{7}	0.2074-0.2342	1.88×10^{-8}-2.93×10^{-9}
CRY2 ve PER	2.10×10^{6}-1.38×10^{7}	0.2160-0.2720	1.03×10^{-7}-1.73×10^{-8}
CLOCK ve BMAL1	3.65×10^{6}-7.78×10^{7}	0.0851- 0.1630	4.11×10^{-8}-1.09×10^{-9}

Table.1 The rate constants and equilibrium constant for protein pairs

The lower the equilibrium constant is, the higher the binding affinity of the reacting agents. Therefore, according to the data, CRY2 protein has higher affinity to BMAL1 than PER2 which is parallel to biochemical data in literature [3]. The reason for any discrepancies between the experimental observations presented here previous literature findings could be due to the production of the proteins in insect cell instead of mammalian cells. Also phosphorylation which is a very important element for regulation of transcriptional and translation loop of circadian clock *in vivo* can not be carried in *in vitro* studies.

Acknowledgement

We acknowledge financial support from The Scientific and Technological Research Council of Turkey (TUBITAK-110T423).

References

1. Reppert, S.M. and D.R. Weaver, *Coordination of circadian timing in mammals.* Nature, 2002. **418**(6901): p. 935-41.
2. Sack, R.L., et al., *Circadian rhythm abnormalities in totally blind people: incidence and clinical significance.* J Clin Endocrinol Metab, 1992. **75**(1): p. 127-34.
3. Langmesser, S., et al., *Interaction of circadian clock proteins PER2 and CRY with BMAL1 and CLOCK.* BMC Mol Biol, 2008. **9**: p. 41.

4. Ye, R., et al., *Biochemical analysis of the canonical model for the mammalian circadian clock.* J Biol Chem, 2011. **286**(29): p. 25891-902.
5. Lee, J.H. and A. Sancar, *Circadian clock disruption improves the efficacy of chemotherapy through p73-mediated apoptosis.* Proc Natl Acad Sci U S A, 2011. **108**(26): p. 10668-72.
6. Gale, J.E., et al., *Disruption of circadian rhythms accelerates development of diabetes through pancreatic beta-cell loss and dysfunction.* J Biol Rhythms, 2011. **26**(5): p. 423-33.
7. Huang, N., et al., *Crystal structure of the heterodimeric CLOCK:BMAL1 transcriptional activator complex.* Science, 2012. **337**(6091): p. 189-94.
8. Rich, R.L. and D.G. Myszka, *Advances in surface plasmon resonance biosensor analysis.* Curr Opin Biotechnol, 2000. **11**(1): p. 54-61.
9. Kizilel, R., et al., *Investigation of Real-Time Photorepair Activity on DNA via Surface Plasmon Resonance.* PLoS One, 2012. **7**(8): p. e44392.
10. O'Shannessy, D.J., et al., *Determination of rate and equilibrium binding constants for macromolecular interactions using surface plasmon resonance: use of nonlinear least squares analysis methods.* Anal Biochem, 1993. **212**(2): p. 457-68.
11. Myszka, D.G., *Kinetic analysis of macromolecular interactions using surface plasmon resonance biosensors.* Curr Opin Biotechnol, 1997. **8**(1): p. 50-7.
12. Peter Schuck, L.B., Peter Andersen, *Measuring Protein Interactions by Optical Biosensors*, in *Current Protocols in Cell Biology*. 2004, John Wiley & Sons, Inc. p. Unit 17.6.1-17.6.22.

(596t) Exploring Beta-Amyloid's Alteration of Signaling Cascades Associated with Learning and Memory of Neuron-Like Cells and the Subsequent Implication in the Mechanism of Alzheimer's Disease

Arundhathi K. Venkatasubramaniam, Chemical and Biochemical Engineering, University of Maryland Baltimore County, Baltimore, MD and Theresa Good, Chemical and Bichemical Engineering, University of Maryland Baltimore County, Baltimore, MD

Alzheimer's disease (AD) is a form of dementia that affects memory, thinking and behavior, and gradually worsens with time. Memory impairment, as well as problems with language, decision-making ability, judgment and personality are symptoms of this disease. AD primarily affects people over 65 years of age. In view of the United States' ageing population, it is especially important to discover new therapeutic approaches to combat this disease.

Neurofibrillary tangles and extracellular plaques are observed in the brains of AD patients at the histopathological level. Beta-amyloid (Aβ) peptide is a major component of these senile plaques and is believed to play a role in the pathology of the disease. Aβ is generated from the Amyloid precursor protein (APP) by enzymatic digestion involving β and γ-secretases. The formation of amyloid fibrils, proto-fibrils and oligomers from the Aβ peptide represent a hallmark of Alzheimer's disease. Aggregated Aβ is known to be toxic to neurons and is implicated in Alzheimer's disease. However, the mechanism by which Aβ causes the apoptosis of cells is unclear and methods to prevent its toxicity are yet to be confirmed. Moreover, the role that Aβ plays in changes in neuron function in this chronic neurodegenerative disease that leads to changes in learning and memory is yet to be elucidated.

The first symptoms of Alzheimer's disease are loss of learning and memory and not cell death; hence we look at how Aβ specifically influences this early phenomenon in cells.Learning and memory in neurons is linked directly to LTP, which is a candidate for synaptic mechanism, i.e., how neurons communicate with one another. There are a number of factors that play a major role in influencing LTP. Its induction and expression is mediated by *N*-methyl-D-aspartate (NMDA) and α-amino-3-hydroxy-5-methyl-4 isoxazolepropionic acid (AMPA) receptors respectively.

Glutamate is one of the two major excitatory neurotransmitters in the cerebral cortex (along with acetylcholine). LTP is induced by glutamate receptor release and simultaneous depolarization, resulting in a postsynaptic calcium influx through the NMDA receptor. This Calcium influx through the NMDA receptor (NMDAR) is crucial for the induction of LTP and activates an intra-cellular signaling cascade that includes several kinases, and that, among other things, leads to an increased number of AMPA receptor. Hence phenomena such as LTP occur through multiple mechanisms many of which may involve protein phosphorylation.

Tyrosine protein phosphorylation is also believed to have an important role in the regulation of neuronal function. This is substantiated by previous work which has shown that protein tyrosine kinase (PTK) inhibitors can inhibit the induction of LTP. The tyrosine kinase Src was shown to regulate the activity of NMDA channels. Low levels of phosphorylation of the NMDAR appear to increase channel activity, while phosphorylation also targets the channel for ubiquitination and degradation. Protein kinase C (PKC) is one of the protein kinase groups believed to be important for maintaining LTP. In particular PKM-zeta, a constitutively active form of an atypical protein kinase C (PKC) isozyme, has been implicated in long term memory maintenance in recent research.

Calcium is a key signaling ion involved in many different intracellular and extracellular processes ranging from synaptic activity to cell-cell communication and adhesion. In the majority of synapses that support LTP, the postsynaptic increase in calcium is mediated by NMDAR. Excitotoxicity appears to play an important role in the pathogenesis of AD. The intricate relationship between Aβ, LTP and AD has been researched only to a certain extent and a number of questions remain unanswered. Observations from studies that Aβ reduces LTP and facilitates LTD is suggestive of a role for it in regulating glutamate receptors and postsynaptic phosphorylation , and playing an important role in modulating learning and memory during AD. Metabotropic glutamate reports are also reported to be involved in the Aβ induced toxicity pathway. Aβ increases the vulnerability of neurons to excitotoxicity mediated by NMDAR. Aβ oligomers have been found to cause an increase in NMDAR activity, which might require direct association between the oligomers and the NR1 subunit of NMDAR. Alternately, Aβ may lead to activation of a tyrosine kinase in the Src family which phosphorylates the NR2B subunit of the NMDAR, leading to its activation. Following that deranged calcium signaling in AD is probable. The exact mechanism of alteration of LTP, the concentration and time scale effects of Aβ are as of yet unknown.

Cholesterol or lipid rich regions known as lipid rafts in the cell membrane are favorable for Aβ binding. Cholesterol has been found to control certain interactions between Aβ and neuronal membranes that are regarded as decisive in the initiation of a neurotoxic cascade.

We hypothesize that Aβ , upon interaction with cells, influences a series of cell signaling events that eventually affect learning and memory capacity in the neurons. We will explore this alternate theory, wherein Aβ binds to cell receptors and facilitates pathways in the cell. To test this hypothesis, we investigate if Aβ interacts with cells leading to an increased phosphorylation of tyrosine proteins. We investigate if Aβ alters the receptor expression on the cell surface in this process. We determine if specific tyrosine proteins, the Src family of tyrosine kinases, PKM-zeta and the Protein Kinase C are targeted. We further investigate if activation of the NMDA receptor (via tyrosine phosphorylation) and activation of the AMPA receptor (via serine phosphorylation) are events that are induced by Aβ. We study if Aβ perturbs intracellular calcium signaling by affecting calcium homeostasis, and if this phenomenon occurs in tandem with the other events induced by Aβ. We investigate if lipid rafts have a role in facilitating Aβ's influence on these signaling pathways.

We will develop an Aβ time and concentration dependent model of the kinetics and mechanism of this signaling cascade, and with this model explore the complex regulatory/control mechanisms that Aβ alters, and how this may impact stability of signaling within the cell.

(596u) Developing a Quantitative Method to Investigate Receptor Oligomerization: Applications to Receptor for Advanced Glycation Endproducts (RAGE)

Pin-Chuan Su, Chemical Engineering, Lehigh University, Bethlehem, PA and Bryan W. Berger, Chemical Engineering and Program in Bioengineering, Lehigh University, Bethlehem, PA

Transmembrane (TM) receptors play a key role in cell-cell communication and signal transduction that regulates diverse biological processes including cell migration, cell proliferation, and cellular differentiation. In most cases, regulation of receptor signaling occurs through receptor oligomerization, in which the interplay between interactions involving multiple receptor domains are necessary for stabilizing the activated form of the receptor for transmembrane signaling to occur. Thus, there is strong correlation observed between specific interactions involving transmembrane and cytoplasmic domains and receptor activation. While there are several tools available to investigate receptor oligomerization in mammalian (BRET) and bacterial (TOXCAT) membranes, technical challenges associated with these assays can limit their utility in rapidly analyzing regions of receptors responsible for oligomerization. Therefore, we developed a robust, transcriptional activator-based method to measure receptor domain oligomerization in bacterial membranes, and benchmarked this method against integrin α_{IIb} TM-CYTO with known mutations that promotes homooligomerization. We then extended this method to identify regions of the receptor for advanced glycation endproducts (RAGE) that promote ligand-independent oligomerization. RAGE receptor has been associated with several chronic disorders such as Alzheimer's disease, atherosclerosis, diabetes, and cancers. Furthermore, several studies have shown that RAGE homodimerization is essential for ligand recognition and signal transduction. Our results point to a key role for the RAGE cytosolic domain in promoting ligand-independent oligomerization, in contrast to previous studies that focused primarily on the soluble, secreted ectodomain.

(596v) Nuclear Mechanics in Crawling Cells

Jun Wu[1], TJ Chancellor[1], Nandini Shekhar[1], Agnes Mendonca[1], Kyle Roux[2], Richard Dickinson[1] and **Tanmay Lele**[1], (1)Chemical Engineering, University of Florida, Gainesville, FL, (2)Sanford Children's Health Research Center, Sioux Falls, SD

On cellular length scales, the nucleus is massive (~10-15 microns in diameter) and stiff relative to the cytoplasm. Motion of such a large object in the crowded intracellular space requires a significant expenditure of energy and hence represents a significant task for the motile cell. Currently, one of the basic questions is whether the nucleus is primarily "pushed" into position by compressive cytoskeletal forces, or "pulled" by tensile cytoskeletal forces. In this study, we show that the nucleus in a single, polarized crawling fibroblast is pulled forward toward the leading edge. The pulling forces originate from actomyosin contraction between the leading edge and the nuclear surface; these forces are opposed by actomyosin pulling in the trailing edge. Microtubules serve to damp fluctuations in nuclear position, but are not required for directional nuclear motion. Our results indicate that the nucleus is under net tension in a crawling cell due to a competition between actomyosin pulling from the front and back of the crawling cell. This work offers new insight into force generation on the nucleus in migrating cells.

(596w) Single-Cell Characterization of CD19-Specific CAR^+ T Cells for Immunotherapy

Ivan Liadi[1], Jason Roszik[2], Amin Merouane[3], Gabrielle Romain[1], Badri Roysam[3], Laurence Cooper[2] and Navin Varadarajan[1], (1)Chemical and Biomolecular Engineering, The University of Houston, Houston, TX, (2)Division of Pediatrics, The University of Texas MD Anderson Cancer Center, Houston, TX, (3)Electrical and Computer Engineering, The University of Houston, Houston, TX

In the last decade cancer immunotherapy using adoptive cell therapy (ACT) has emerged as a highly effective treatment option for diverse human cancers. In particular, the use of chimeric antigen receptor (CAR) T cells rendered specific for CD19 have demonstrated significant anti-tumor effects in patients with $CD19^+$ chronic lymphocytic leukemia (CLL) that were considered untreatable, including the use of stem-cell transplantation. In spite of the clinical promise of ACT in achieving complete responses, their efficacy remains unpredictable and new approaches are needed to address *a priori* define the therapeutic potential of T-cell based therapies. In our current work, we have characterized the functionality of CD19-specific CAR^+ T cells using a novel methodology that determines the cytotoxic ability and cytokine secretion capability of these cells at the single-cell level. Using our assay, we have studied the interaction of CD19-specific CAR^+ T cells (CART19) with CD19-expressing EL4 target cells (EL4-$CD19^+$) and NALM-6 tumor cells by encapsulating them in arrays of nanowells (~100k wells/array). For each single CAR^+ T-cell the ability to engage and kill multiple target cells as well as the ability to secrete cytokines upon target conjugation has been scored. These composite functional profiles at the single-cell resolution are being used to determine the *in vitro* potency of CAR^+ T cells and in combination with clinical data will help design the next generation of clinical trials.

(596x) 13C Isotopic Tracers to Elucidate Metabolic Dependences of Renal Cell Carcinoma Cells

Paulo A. Gameiro[1,2], Juanjuan Yang[1], Othon Iliopoulos[3] and Gregory Stephanopoulos[4], (1)Cancer Center, Massachusetts General Hospital, Charlestown, MA, (2)Chemical Engineering, Massachusetts Institute of Technology, Cambridge, MA, (3)Cancer Center, Massachusetts General Hospital, Cambridge, MA, (4)Department of Chemical Engineering, Massachusetts Institute of Technology, Cambridge, MA

Currently evolving paradigms in cancer biology support the view that transcription factors and gene deregulation impinge on the metabolism of cancer cells to support biosynthesis and accelerated growth. This recognition calls for the use of stable isotopic tracers to identify metabolic switches that are associated with malignancy. Hypoxia is a hallmark of the tumor microenvironment and we previously showed using ^{13}C isotopic tracers that hypoxic cells, while diverting most of the glucose to lactate production, use glutamine to generate lipids through reductive carboxylation of alpha-ketoglutarate, or reductive TCA cycle. We will show that the reductive TCA cycle is a metabolic signature of hypoxia inducible factors (HIF) expression in renal cell carcinoma (RCC) cells and that targeting this pathway may provide a novel therapeutic strategy for the treatment of tumors deficient in the VHL tumor suppressor protein. This work highlights how ^{13}C isotopic tracers can elucidate the metabolic transformation of cancer cells.

(596y) Vessel Wall Transport: Can Increasing the Concentration of the Membrane Protein Aquaporin-1 Slow Down Pre-Lesion Atherosclerosis?

Shripad D. Joshi, Chemical Engineering, City College and Graduate Center of City University of New York, New York, NY, Kung-Ming Jan, College of Physician and Surgeons, Columbia University, Bronx, NY and David S. Rumschitzki, Department of Chemical Engineering, The City College of The City University of New York, New York, NY

Macromolecular (read: low-density lipoprotein cholesterol or LDL) transport across the vessel wall appears to be the earliest pre-atherosclerotic event. It is known to occur due to advection: Transmural (i.e., across the vessel wall) pressure differences (i.e., blood pressure) drive plasma (water and small solutes) across the monolayer of endothelial cells (EC) that line the inside of the vessel wall. One EC every few thousand has junctions that are temporarily widened, some because they are either dying or dividing, and the transmural flow can advect large tracers through them into the wall. A wall parameter, the hydraulic conductivity (L_p), defined as the ratio of the transmural water flux to hydrostatic pressure difference, is central to the understanding of this transmural water transport. On one hand, it advects macromolecules like LDL into the arterial subendothelial intima (SI) through these rare, widened EC leaks and spreads it there, thereby giving it the chance to bind to extracellular matrix (ECM) in the SI layer and possibly to trigger the start of atherosclerotic lesion formation. On the other hand, the overall water flow across the normal (non-leaky) endothelium dilutes the local SI LDL concentration, thereby likely slowing binding kinetics, and washing not-yet-bound lipid further into the wall. Our group's discovery of the presence of the ubiquitous water channel membrane protein Aquaporin-1 (AQP) in rat aortic endothelial cells suggests a new possibility of water transport across the normal (non-leaky) endothelium, alongside the generally accepted paracellular route. Interestingly, we found that chemically blocking AQPs changes rat aortic wall Lp in a strongly pressure dependent manner. We then proposed a new theory that agrees well with this otherwise perplexing experimentally-observed pressure-dependence. One upshot of this theory is that it suggests that AQPs contributes about 30% to the phenomenological endothelial Lp at low transmural pressures.

In the present work we investigate the effect of AQPs on the overall transport of various species, like Horse Reddish Peroxidase (HRP) and LDL, across the vessel endothelium and its further spread in the vessel wall using the advection-diffusion model developed by Huang et al. (*Journal of Biomechanical Engineering*, vol 116, 1994, pp. 430-445) and later improved by Zeng et al. (*American Journal of Physiology*, Heart Circ Physiol, 302, 2012, pp. 1683-1699). We extend this model by incorporating the trancellular water flow through AQPs and use some important parameters like the hydraulic conductivity of the endothelium plus intima (Lp_e) and the corresponding intima thicknesses from our local filtration model mentioned above that allows us to calculate these important parameters at different pressures, something which was not considered in earlier models. The goal of this model is to investigate if increasing the endothelial AQP expression – say, pharmacologically or via diet, leading to higher endothelial Lp (which reduces SI compression causing higher overall vessel wall Lp), can decrease the overall concentration of macromolecules in the sub-endothelial space and thereby slow down kinetics of its binding to SI ECM, the triggering event in lesion formation and subsequent pre-atherosclerotic events. Our preliminary findings suggest that increasing transcellular transport by increasing AQP expression lowers the concentration of HRP in sub-endothelial space and by October we should be able to make definitive statements as to the analogous effects that could be observed for more biologically relevant macromolecules like LDL.

(596z) A Long Wavelength Fluorescent Sphingosine As a Tool to Visualize Sphingolipid Transport and Location in Living Cells

Raehyun Kim[1], Kaiyan Lou[2] and Mary L. Kraft[2], (1)Chemical and Biomolecular Engineering, University of Illinois, Urbana-Champaign, Urbana, IL, (2)Chemical & Biomolecular Engineering, University of Illinois at Urbana-Champaign, Urbana, IL

Sphingolipids are cell membrane components that also play critical roles in multiple biological events, including cell growth and apoptosis signaling. Information on dynamic sphingolipid processes, such as trafficking, sorting, and organization in membranes, has been acquired by visualizing fluorescent sphingolipid analogs in living cells. Investigations of the transport and subcellular distributions of sphingomyelin, ceramide and gangliosides have been facilitated by the commercial availability of fluorescent sphingolipid analogs in which the fluorophore is attached to the N-acyl fatty acid side chain. To study the subcellular locations and functions of biologically important, unacylated sphingolipid metabolites, such as sphingosine and sphingosine-1-phosphate, the fluorophore must instead be incorporated into the sphingoid backbone. Despite the utility of such derivatives, only a small number of fluorescent sphingolipid analogs have been reported, and the fluorescence emissions of most of these analogs overlap with that of the most common genetically encoded fluorescent protein label, green fluorescent protein (GFP). Development of a fluorescent sphingosine analog that emits at a longer wavelength is desirable to permit visualizing the fluorescent sphingolipid analog in parallel with GFP-labeled proteins. Here, we report the use of a new fluorescent sphingolipid precursor in which a long wavelength borondipyrromethene fluorophore, BODIPY 540, is incorporated into the sphingosine backbone. The BODIPY 540 sphingosine analog and its fluorescent metabolites were detected in the organelles where sphingolipid metabolism occurs, and could be imaged in parallel with GFP-labeled proteins within living cells. Thin layer chromatography and mass spectrometry analysis confirmed that mammalian cells readily metabolized the BODIPY 540 sphingosine into multiple fluorescent sphingolipids. We expect that this BODIPY 540 sphingosine analog will be a useful tool to study dynamic sphingolipid events and may also facilitate the detection of abnormal sphingolipid regulation.

EFFECTS OF OSCILLATORY CONVECTIVE FLOW ON CHEMICAL SIGNAL PROPAGATION IN EPITHELIA

Marek Nebyla, Institute of Chemical Technology, Prague, Czech Republic,nebylam@vscht.cz
Michal Přibyl, Institute of Chemical Technology, Prague, Czech Republic,pribylm@vscht.cz

Abstract

We study effects of convective transport on chemical signal propagation in a simple (single layer) epithelium by means of mathematical modeling. Oscillatory flow regimes were considered. In our model, we assume that the movement of an extracellular medium is affected by harmonically oscillating convective flow. The oscillatory flow can be induced by heart action or walking. We have carried out numerical simulations using Comsol Multiphysics 3.5. Our results suggest that the signal propagation velocity increases with increasing amplitude, if the convective velocity harmonically oscillates around the zero value. The dimensionless propagation velocity also depends on the period. At lower period regimes, the velocity is almost the same as in the case with stagnant extracellular medium (with zero convective velocity). With longer periods, the velocity becomes significantly higher. These results were confirmed by independent simulations based on the finite difference method.

Introduction

Intracellular signaling induced by growth factors can result in secretion of these factors into an extracellular medium. This process produces a positive feedback loop between ligand binding and ligand release. This feedback can generate signal waves traveling at a constant speed when coupled with intracellular processes [1].

Cell communication can be mediated by the epidermal growth factor receptor (EGFR) and its ligands. EGFR is a transmembrane protein. Together with other three proteins of a similar structure and function, EGFR is a member of the c-erb family of the tyrosine kinase receptors. EGFR is important due to its role in the signaling that controls the behavior and fate of epithelial cells [2].

Many growth factors and cytokines that bind to EGFR are released by intracellular or surface proteases [3]. The ligand-releasing enzymes can be activated by the signaling pathways that are stimulated by the extracellular ligands. Hence, a positive feedback loop can be established when a cell effectively responds to and recaptures the secreted ligands [4-6]. Membrane receptors can reversibly bind the secreted ligand. The ligand-receptor complexes on the cell surface stimulate the intracellular processes leading to further ligand release. The ligand-receptor complexes can also be internalized into the cell by endocytosis. Here, we focus on such positive feedback loop. We consider one-dimensional epithelium formed by a single cellular layer of cells with surface EGFRs.

In healthy epithelial cells, the expression of EGFR ranges from 40 000 to 100 000 receptors per cell [7]. However, EGFR is over-expressed in the majority of solid tumors e.g., breast cancer, head-and-neck cancer, colon cancer etc [8]. Such over-expression can lead to permanent activation of downstream signaling pathways, which results in uncontrolled cell division with invasiveness characteristics [2, 9].

In this work, we focus on the effects of oscillatory flow of an extracellular medium on the velocity of ligand propagation. The convective velocity of the extracellular medium harmonically oscillates with certain amplitude. The time averaged convective velocity is equal to zero. Our work is motivated, for example, by pulsating blood flow in mammalian vessels. Oscillatory flow is also produced by the motion of the whole body, e.g., walking.

Mathematical model

Model domain

Here, we consider a simplified 2D domain. The main processes considered in our model are depicted in Fig 1. Ligand transport processes (convection, diffusion) are considered in the extracellular medium. The extracellular domain can be characterized by a finite thickness. The ligand-receptor complex formation is localized on the plasmatic membrane. If the complex concentration exceeds certain threshold, intracellular protease is activated and new ligand molecules are released into the extracellular space. The ligand molecules are transported to free receptors in the surrounding tissue. Then, we can observe propagation of ligand front above the epithelium. All intracellular processes are simplified into equation for the protease activation. For other details, please see Ref. [1].

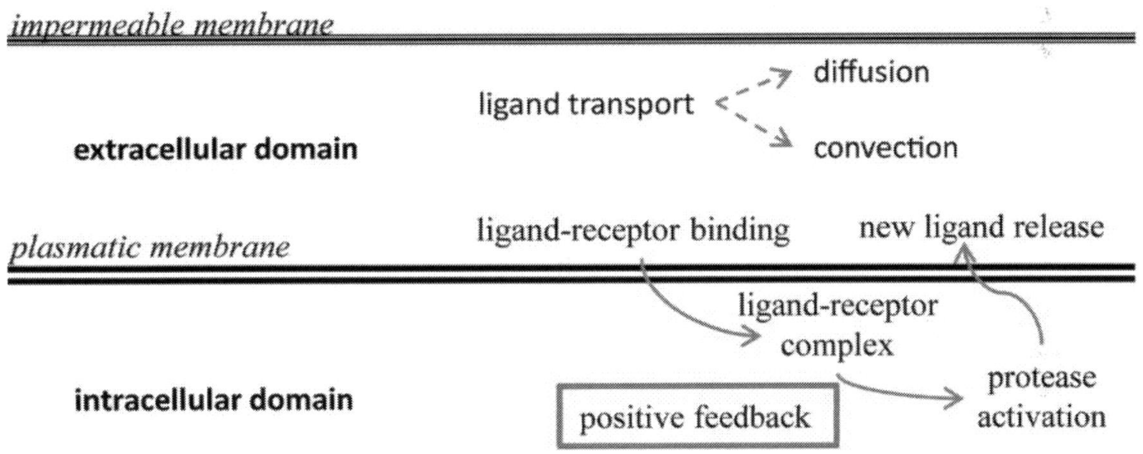

Fig. 1. Structure of the modeling domain and the main considered processes.

Dimensionless model equations

Model equations describing all processes taking place in modeling domain were adopted from [1, 6]. The mathematical model was transformed into a dimensionless form. See the Ref. [1] for the scaling factors. The mathematical model consists of the molar balances of the ligand S in the extracellular space together with boundary conditions, molar balances of receptors R and ligand receptor complexes C on the plasmatic membrane, and the equation for protease activation P. Tilde stands for dimensionless variables.

$$\tau_S \frac{\partial \tilde{S}}{\partial \tilde{t}} = -\tilde{v}_x \frac{\partial \tilde{S}}{\partial \tilde{x}} + \frac{\partial^2 \tilde{S}}{\partial \tilde{x}^2} + \frac{1}{\alpha^2} \frac{\partial^2 \tilde{S}}{\partial \tilde{y}^2} \tag{1}$$

$$-\left.\frac{\partial \tilde{S}}{\partial \tilde{y}}\right|_{\tilde{y}=0} = \alpha\left[(1-\beta)\tilde{C} - \tilde{S}\tilde{R} + \beta \tilde{P}\right] \tag{2}$$

$$-\left.\frac{\partial \tilde{S}}{\partial \tilde{y}}\right|_{\tilde{y}=1} = 0 \tag{3}$$

$$\tau_R \frac{\partial \tilde{R}}{\partial \tilde{t}} = 1 - \tilde{R} + \gamma\left[(1-\beta)\tilde{C} - \tilde{S}\tilde{R}\right] \tag{4}$$

$$\tau_C \frac{\partial \tilde{C}}{\partial \tilde{t}} = \tilde{S}\tilde{R} - \tilde{C} \tag{5}$$

$$\frac{\partial \tilde{P}}{\partial \tilde{t}} = -\tilde{P} + \sigma\left(\tilde{C} - C_T\right) \tag{6}$$

Physical meaning of all dimensionless parameters is described in Ref. [1]. Here, we focus on influence of the oscillatory flow on the velocity of signal propagation U. Hence, the dimensionless velocity of the convective flow is given by a harmonic function

$$\tilde{v}_x = A \sin\left(\frac{2\pi\tilde{t}}{T_p k_p}\right) \tag{7}$$

where A is the dimensionless amplitude of the convective velocity, k_p is the protease degradation constant used as the time scaling factor, and T_p is the dimensional period of the velocity oscillations.

Results and discussion

2D simulations were carried out in the Comsol Multiphysics 3.5 finite element package. The maximum element size at the cellular surface was set to 0.05. The dynamical femtime procedure used a standard odepack linear solver. The absolute and relative tolerances of the dynamical solver were set to 1×10^{-6}. Simulations for the velocity amplitude A = 0.5 and period varied from 2 s to 4×10^{4} s were carried out, Fig 2.

For shorter periods, results showed that the dimensionless velocity of signal propagation U is comparable to the case with a stagnant extracellular medium (dashed line in Fig 2). It means that there is no difference between the systems with zero convective velocity and the systems with convective velocity oscillating harmonically around zero. However, the velocity U increases with increasing period with maximum at period 1×10^{4} s. This finding is important from the view that the ligand signal can faster propagate in a tissue if periodic oscillations of the convective velocity (pressure) are imposed. Similar dependence of the dimensionless propagation velocity was obtained for other set of the model parameters, Fig 3.

One can see that the obtained velocity dependencies, Figs 2 and 3, contain several inflection points. We were aware of possible numerical errors. To confirm our results, simulations, which employed the finite difference method with three-point differential formulae, were carried out. Due to higher hardware and time demands of this method, the maximum discretization step was set greater to 0.25 and the absolute and relative tolerances of the solver were set to 1×10^{-4} (Fig 2, circles).

Simulations for the maximum element size were carried out also using the Comsol Multiphysics 3.5 (Fig 2, triangles). The results are similar to those carried out with higher precision (Fig 2, crosses, maximum step size 0.05).

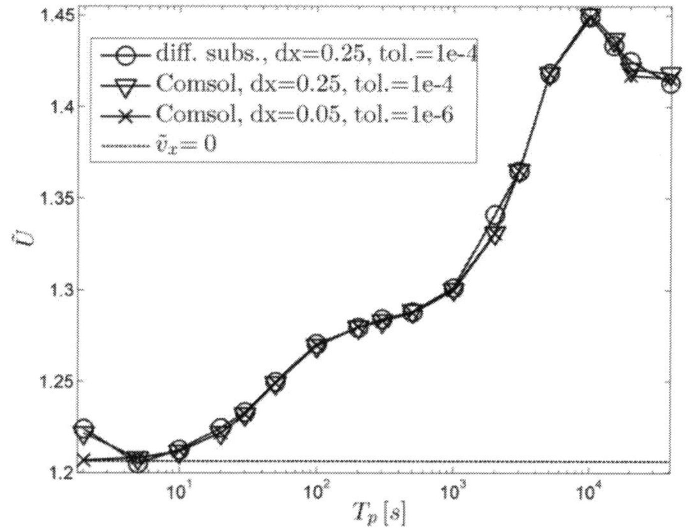

Fig. 2. Dependence of the dimensionless velocity of signal propagation U on the period of the oscillatory convective flow in the extracellular medium. Used parameters: $\alpha = 1$, $\beta = 0.5$, $\gamma = 1$, $\tau_C = 0.1$, $\tau_R = 0.5$, $\tau_S = 0.002$, $C_T = 0.3$, $k_p = 3.4\times10^{-4}$ s^{-1}.

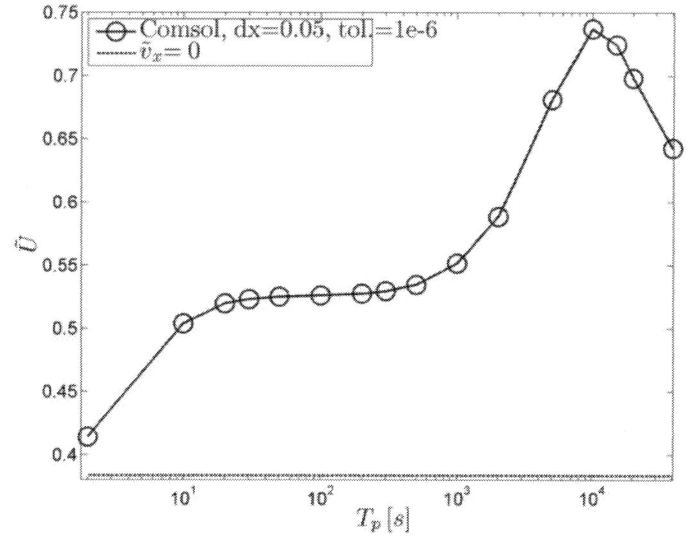

Fig. 3. Dependence of the dimensionless velocity of signal propagation U on the period of the oscillatory convective flow in the extracellular medium. Used parameters: $\alpha = 0.1$, $\beta = 0.5$, $\gamma = 1$, $\tau_C = 0.1$, $\tau_R = 0.5$, $\tau_S = 0.002$, $C_T = 0.3$, $k_p = 3.4\times10^{-4}$ s^{-1}.

Conclusions

We have found that the oscillatory convective transport above epithelial layers can significantly change the velocity of chemical signal transmission even if the velocity oscillates with the zero mean value. The oscillatory convection accompanies heart action, body movement or some circadian rhythms. Our simulations revealed the fact that the highest propagation velocity can be reached for long periods $\sim 1\times10^4$ s. This fact can lead to better understanding of effects of the nonstationary convection on developmental processes (vascular system development, tumor growth).

Nomenclature

EGFR epidermal growth factor receptor
C surface concentration of the complex
S ligand concentration
P protease activity
R surface concentration of the EGFR
U propagation velocity
t time
v_x convective velocity
x spatial variable
y spatial variable

Acknowledgements

The authors thank the Ministry of Education, Youth and Sports of the Czech Republic (projects KONTAKT ME10036 and Specific University research No. 21/2012) for financial support.

References

[1] Nebyla M, Přibyl M, Schreiber I. Effects of Convective Transport on Chemical Signal Propagation in Epithelia. *Biophysical Journal* 2012;102:990-1000.

[2] Herbst RS. Review of epidermal growth factor receptor biology. *International Journal of Radiation Oncology Biology Physics* 2004;59:21-26 Suppl.

[3] Dello S, Rovida E. Transduction of cell surface regulatory molecules via ectodomain shedding. *Biological Chemistry* 2002;383:69-83.

[4] Carpenter G. Employment of the epidermal growth factor receptor in growth factor-independent signaling pathways, *Journal of Cell Biology* 1999;146:697-702.

[5] Gschwind S, Zwick E, Prenzel N, Leserer M, Ullrich A. Cell communication networks: epidermal growth factors receptor transactivation as a paradigm for interreceptor signal transmission. *Oncogene* 2001;20:1594-1600.

[6] Přibyl M, Muratov CB, Shvartsman SY. Long-range signal transmission in autocrine relays. *Biophysical Journal* 2003;84:3624 – 3635.

[7] Carpenter G, Cohen S. Epidermal Growth-Factor. *Annual Review of Biochemistry* 1979;48:193-216.

[8] Herbst RS, Langer CJ. Epidermal growth factor receptors as a target for cancer treatment: The emerging role of IMC-C225 in the treatment of lung and head and neck cancers, *Seminars in Oncology* 2002;29:27-36.

[9] Ethier SP. Signal transduction pathways: The molecular basis for targeted therapies, *Seminars in Radiation Oncology* 2002;12:3-10 Suppl. 2.

(601a) Assessing the Impact of Experimental Designs On Pharmaceutical Calibration Models Based On near Infrared Spectroscopy

Robert Bondi Jr., Benoit Igne, Carl A. Anderson and James K. Drennen III, Mylan School of Pharmacy, Duquesne University, Pittsburgh, PA

The objective of this work was to determine how the predictive performance of partial least squares (PLS) models is affected by the calibration samples selected for model development. Experimental designs are widely used to produce a set of calibration samples for generating empirical models based on vibrational spectroscopy (e.g., near-infrared). There are numerous designs that can be utilized, but the choice often depends on the available resources and the intended use of the model. However, the predictive performance of the model may be directly influenced as concentration covariance is directly related to regression vector estimation on a theoretical level. It was hypothesized that the theoretical dependence of regression vector estimation on experimental designs is extended to practical applications. The following designs were evaluated for a model pharmaceutical composite system: 5-level full factorial, 3-level full factorial, central composite, D-Optimal, and I-Optimal. The factors for all designs were acetaminophen content and ratio of microcrystalline cellulose to lactose monohydrate. Other constituents included croscarmelose-sodium and magnesium stearate (content remained constant). Five 13-mm compacts were produced at each level (5 levels for a total of 150 compacts). Models were generated and independently optimized using data from individual design of experiments. The predictive performance of each model was evaluated based on an independent validation set. The effect of experimental design was tested by determining the statistical significance of the difference in bias and standard deviation for its model prediction performance.

EFFICIENT MICROWAVE-ASSISTED SYNTHESIS OF A SERIES OF POLYHYDROXYLFLAVONES

Baoping Dong, Yang Liu, Xiang Liu, Zhiqiang Wang, Jimiao Li*
School of Chemical and Material Engineering, Jiangnan University, Wuxi, China

Abstract

A series of polyhydroxylflavones were synthesized starting from substituted benzoic acid and 2-hydroxyl-4-methoxyacetophenone by acylation, esterification, Baker-Venkataraman rearrangement, cyclization and demethylation. Microwave irradiation was applied in cyclization which was the key step in synthesis of polyhydroxylflavones. The effects of the substrate concentration, microwave irradiation time and catalyst concentration on the yields were investigated in microwave-assisted cyclization. The use of microwaves was found to significantly improve product yields and shorten the reaction time of cyclization. This route is eco-friendly and high selectivity. Under the optimum reaction conditions, the target products were obtained in overall yields of 40% ~ 48% and purity of more than 96%. The structures of the compounds were confirmed by IR, ^{1}H NMR and MS.

Keywords: Polyhydroxylflavone; Cyclization; Microwave-assisted synthesis

Flavone is a kind of important natural organic compounds, have good scavenging ability to the super oxygen anion free radical and hydroxyl free radical that related to most diseases of human[1,2]. Therefore, flavones are used to cure hypertension, inflammatory, aggregatory diseases and cancer[3,4]. Recently investigated show that the position and number of hydroxyl in flavone have a significant effect to the radical scavenging ability of flavone, the more hydroxyl , the stronger radical scavenging ability, and the hydroxyl in *ortho* can enhance the capacity of scavenging radicals[5,6]. The synthesis of polyhydroxylflavones has attracted considerable attentions due to their significant pharmaceutical, biocidal, and antioxidant activities. At present, there have been various methods and routes for the synthesis of the flavones, but roughly divided into Fries rearrangement and Baker-Venkataraman rearrangement. These methods involve cyclization which is the key step in the synthesis of flavone compounds. Many cyclization methods have been reported, such as iodine catalytic cyclization, flavanol ketone oxidated cyclization, acid catalytic cyclization, solid phase load catalytic cyclization. However, we found that these methods all suffered either from involving a number of steps giving too low overall yields or from encountering considerable challenges due to long reaction times, the multifarious solvents and catalysts, the complex post-processing. Because microwave-assisted synthesis has the advantages of short reaction time, high yield, and eco-friendly [7], it is an effective way to use microwave irradiation in the synthesis of flavones.

An efficient and novel synthesis of functionalized polyhydroxylflavones was preformed by the reactions outlined in Scheme1. The substituted benzoyl chloride **2** was obtained by acylation of corresponding substituted benzoic acid according to the literatures [8]. The compound **3** was synthesized starting from **2** and 2-hydroxyl-4-methoxyacetophenone. The substituted 1-(2-hydroxy-4-methoxyphenyl)-3-aryl-1,3-propanedione **4** was acquired by Baker-Venkataraman rearrangement. The following intramolecular cyclization of substituted 1-(2-hydroxy-4-methoxyphenyl)-3-aryl-1,3-propanedione **4** was achieved under microwave irradiation[9] to gaving **5**. Finally, **5** was treated with 40% HBr/HOAc(1:1) for demethylation and the desired functionalized polyhydroxylflavones **1** was obtained in the good yields.

E-mail address: liuxiang@jiangnan.edu.cn (X. Liu)

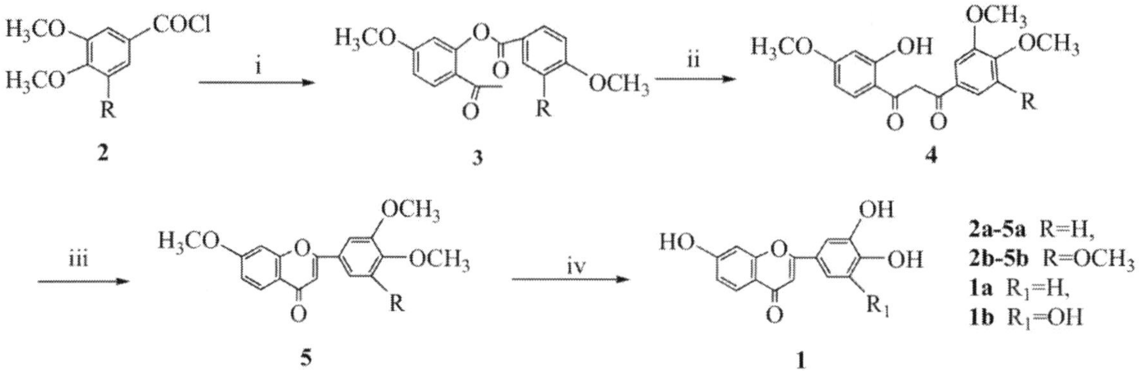

Scheme 1. Reagents and conditions: (i) 2'-hydroxy-4'-methoxyacetophenone, pyridine, heat; (ii) pyridine, KOH , heat; (iii) MW ,$CuCl_2$, heat; (iv) 40% HBr/HOAc, N_2, heat

Especially, microwave irradiation was applied in the cyclization of substituted 1-(2-hydroxy-4-methoxyphenyl)-3-aryl-1,3-propanedione **4** which was the key step. The effects of some factors, including substrate concentration, microwave irradiation time and catalyst concentration, on the yields of polymethoxylflavones **5** was examined and the results were showed in Fig1-3. The results of Fig 1 indicated that the yields of flavones gradually increased with the increase of the microwave irradiation time and reached the maximum value at a definite moment (15 min or 20 min), but after that the yields began to go down. The results of Fig 2 revealed that the yields reached the peak value at 1.0 g/L substrate concentration. When the substrate concentration exceeded 1.0g/L, the lower yields were obtained. The results of Fig 3 demonstrated that the yields reached the peak value at 0.08mmol/L catalyst concentration. When the catalyst concentration exceeded 0.08mmol/L, the remain yields were obtained. The appropriate reaction conditions were microwave irradiation for 15min, 1.0g/L substrate concentration, 0.08mmol/L catalyst concentration.

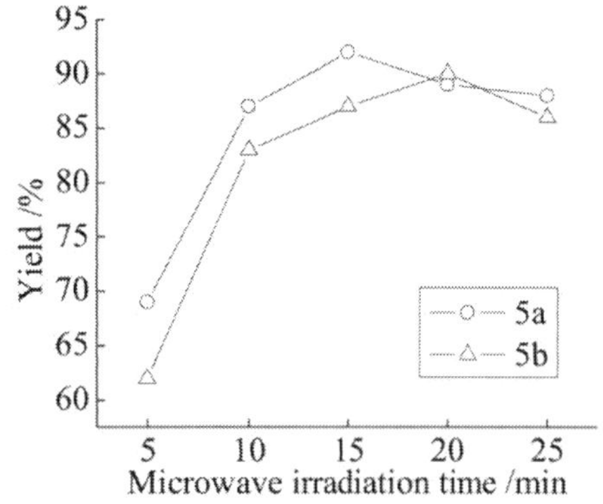

Fig 1 The effects of irradiation time on yields.

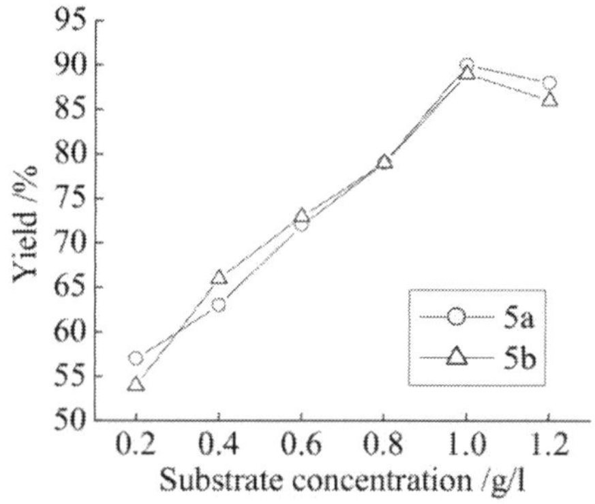

Fig 2 The effects of substrate concentration on yields.

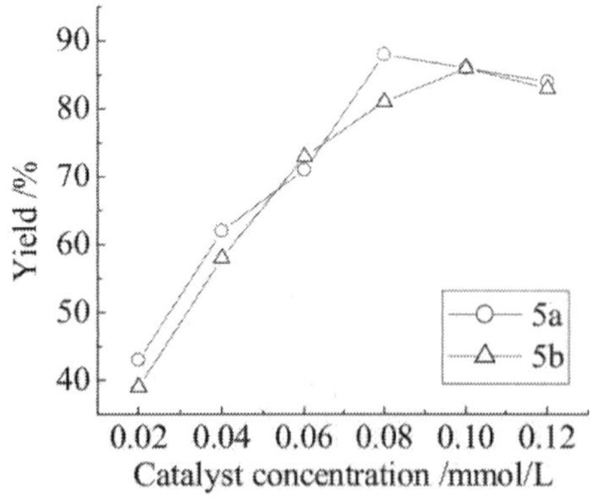

Fig 3 The effects of catalyst concentration on yields.

Experimental

All reagents were commercial materials and were used directly unless otherwise noted. All melting points were measured on a SPSIC WRS-1B melting apparatus and were uncorrected. IR spectra were recorded on a Boman FTLA2000 IR spectrometer. ^{1}H NMR spectra were recorded on a 400 MHz Bruker Avance III 400 NMR spectrometer. Mass spectra were recorded on a Waters Acquity UPLCSQD Liquid Mass spectrometer. Concentration were detected by Water 1525 HPLC （Ultraviolet detector）.

Synthesis of 3

50mmol of the substituted benzoyl chloride **2**, 50 mL anhydrous pyridine and 40mmol of 2'-hydroxy-4'-methoxyacetophenone were added into a 150 mL three-necked flask. The reaction mixture was stirred and heated for 4 h under reflux. After cooling, the mixture was poured into the water and neutralized with dilute HCl solution. The compound **3** was obtained by filtering, washing and recrystallization from EtOH.

3a: white crystal (11.9 g, 90%). m.p. 149～151℃; IR(KBr) ν/cm^{-1}: 3430, 3011, 2984, 2847, 1728, 1669, 1607, 1524, 1282, 1215, 1116, 867, 749; ^{1}H-NMR(DMSO-d_6) δ/ppm: 7.95 (1H ,d, J = 8.8 Hz, Ar-H), 7.77 (1H ,dd, J = 8.4, 2.0 Hz, Ar-H), 7.56 (1H ,d, J = 2.0 Hz, Ar-H), 7.16 (1H ,d, J = 8.5 Hz, Ar-H), 7.00 (1H , dd, J = 8.8, 2.5Hz,, Ar-H), 6.94 (1H , d, J = 2.5 Hz, Ar-H), 3.88 (3H ,s, OCH_3), 3.84 (6H , d, J = 5.7 Hz, OCH_3×2), 2.44 (3H ,s, CH_3); MS: $[M+1]^+$331.

3b: white crystals (12.5 g, 86.3%). m.p. 158～160℃; IR(KBr) ν/cm^{-1}: 3442, 3015, 2949, 2839,1728, 1673, 1607, 1509, 1266, 1211, 1129, 870, 753; ^{1}H-NMR(DMSO-d_6) δ/ppm: 7.98 (1H ,d, J = 8.8 Hz, Ar-H), 7.39 (2H ,s, Ar-H), 7.01 (1H, dd, J = 8.8, 2.5 Hz, Ar-H), 6.95 (1H ,d, J = 2.5 Hz, Ar-H), 3.86 (9H, d, J=3.2Hz, OCH_3×3), 3.78 (3H, s, OCH_3) ,2.46 (3H, s, CH_3); MS: $[M+1]^+$361.

Synthesis of 4

25 mmol **3**, 4 g KOH and 60 mL pyridine were added into a 150 mL three-necked flask and stirred for 5 h at 80℃. The reaction mixture was cooled. Water was added into the mixture. The mixture pH was adjusted to 5 with dilute HCl solution. The yellow precipitate was filtered and washed with water. The precipitate was recrystallized from EtOH to give **4**.

4a: yellow crystal (7.3 g,88.3%).m.p. 128～131℃; IR(KBr) ν/cm^{-1}: 3015, 2937, 2841, 1638, 1610, 1446, 1239, 960, 788; ^{1}H-NMR(DMSO-d_6) δ/ppm: 16.29 (1H, s, HO-C=C), 12.01(1H ,s, OH), 8.12(1H ,d, J = 9.0Hz, Ar-H), 7.72(1H ,d, J =8.2Hz, Ar-H), 7.55 (1H ,s, Ar-H), 7.23 (1H ,s , Ar-H), 7.11(2H ,m, Ar-H) , 6.53(2H ,m, Ar-H), 4.75 (1H ,s , -C=CH-),3.86 (6H , d, J =3.0Hz, OCH_3×2), 3.82 (3H ,m,OCH_3); MS: $[M+1]^+$ 331.

4b: yellow crystal (7.7 g,85.4%). m.p. 130～131℃; IR(KBr) ν/cm^{-1}: 3082, 2945, 2839, 1603, 1579, 1505, 1442, 1446, 1231, 1125, 965, 797; ^{1}H-NMR(DMSO-d_6) δ/ppm: 16.17 (1H , s , HO-CH=C),12.01(1H ,s,OH), 8.15 (1H ,d, J = 8.0Hz, Ar-H), 7.32 (2H ,d, J = 4.0Hz, Ar-H),7.24 (1H ,s, Ar-H), 6.59 (2H ,m, Ar-H), 4.18 (1H ,s, Ar-H), 3.89 (3H ,s ,OCH_3),3.83 (6H ,s, OCH_3) ,3.76 (3H ,s,OCH_3); MS: $[M+1]^+$361.

Synthesis of 5

A mixture of 8mmol **4**, 2mmol $CuCl_2$, 40mL ethanol was added in a two- necked flask and was irradiated with microwave for 25 min under reflux in an HWL08-A experimental microwave oven. After the irradiation was completed, the mixture was extracted with ethyl acetate (3×30 mL). The ester solution was washed with a saturated aq NaCl (2×45 mL). The ester solution was dried over Na_2SO_4 and concentrated to 10mL in a vacuum. The yield of **5** was detected by HPLC. The resulting oil was chromatographed on silica gel column to give **5**.

5a: white crystal (2.31 g, 92.1%). m.p. 174～176℃; IR(KBr) ν/cm^{-1}: 3008, 2950, 2839, 1634,

1603, 1520, 1442, 1270, 1169, 843, 770; ^{1}H-NMR(DMSO-d_6) δ/ppm: 7.92 (1H ,d, J = 8.8 Hz, Ar-H), 7.71 (1H ,d, J = 8.5 Hz, Ar-H), 7.59 (1H ,s, Ar-H), 7.33 (1H ,s, Ar-H), 7.13 (1H ,d, J = 8.6 Hz, Ar-H), 7.05 (1H ,dd, J = 8.8,2.1 Hz ,Ar-H), 6.98 (1H ,s, -C=CH-), 3.92 (3H,s, OCH_3), 3.89 (3H ,s, OCH_3), 3.85 (3H ,s, OCH_3); MS: $[M+1]^+$ 314.

5b: white crystal (2.5 g, 90.6%). m.p. 174～176℃; IR(KBr) ν/cm^{-1}: 3004, 2941, 2851, 1650, 1607, 1509, 1423, 1254, 1125, 840; ^{1}H-NMR(DMSO-d_6) δ/ppm: 7.93 (1H ,d, J = 8.8 Hz, Ar-H), 7.37 (3H ,s, Ar-H), 7.09 (1H,s, Ar-H), 7.06 (1H,dd, Ar-H), 3.92 (9H ,t, J = 7.6 Hz, OCH_3×3), 3.75 (3H ,s,OCH_3); MS: $[M]^+$344.

Synthesis of 1

A three-necked flask was charged with 3.2 mmol **5**, 25 ml 40% HBr and 25 mL acetic acid under an N_2 atmosphere. The reaction mixture was refluxed for 40h and then carefully poured into 500 g crushed ice. The resulting yellow precipitate was filtered and collected. The precipitate was recrystallized from ethanol to afford the desired polyhydroxylflavones **1**.

1a: yellow solid (0.69 g, 79%). IR(KBr) ν/cm^{-1}: 3328, 1626, 1606, 1446, 1277, 1165, 869, 816; ^{1}H-NMR(DMSO-d_6) δ/ppm: 10.75 (1H, s, OH), 9.87 (1H, s, OH),9.37 (1H, s, OH), 7.93 (1H, d, J = 8.8Hz, Ar-H), 7.86(1H, d, J= 8.6Hz, Ar-H), 7.39(1H, dd, J= 5.8,2.1 Hz, Ar-H), 7.24 (1H, d,J = 2.4 Hz, Ar-H), 7.05(1H, dd, J = 8.8,2.4Hz, Ar-H), 6.67 (1H, s,Ar-H),6.61 (1H, s,-C=CH-); MS: $[M+1]^+$ 271.

1b: yellow solid (0.68 g, 73.6%). IR(KBr) ν/cm^{-1}: 3445, 3284, 2980, 1646, 1606, 1555, 1508, 1446, 1199, 1041, 857; ^{1}H-NMR(DMSO-d_6) δ/ppm: 10.76 (1H, s, OH), 9.35 (2H, s, OH),9.05 (1H, s, OH), 7.94 (1H, d, J = 6.1Hz, Ar-H), 7.19(1H, d, J= 4.0Hz, Ar-H), 7.05(1H, dd, J= 2.0,8.6 Hz, Ar-H), 7.00 (2H, s, J = 2.0 Hz, Ar-H), 6.53 (1H, s, -C=CH-); MS:$[M+1]^+$287.

References

1. Rajendran M., Mahalakshmi M., Ramya R., etal(2011), "A semi-empirical study of flavone compounds with antioxidant efficiency," *Afr. J. Phar. Pharmaco.*, pp. 2140-2144.
2. Ayhan-Kilcigil G., Coban T., Tuncbilek M., etal(2004),"Antioxidant properties of flavone-6(4')-carboxaldehyde oxime ether derivatives," *Arch. Pharmacal. Res.*, 27(6), pp.610-614.
3. Balasubramanian R., Narayanan M., Kedalgovindaram L., etal(2007), "Cytotoxic activity of flavone glycoside from the stem of indigofera aspalathoides Vahl," *J. Nat. Med.*, 61(1), pp.80-83.
4. Ono M., Yoshida N., Ishibashi K., etal(2005), "Radioiodinated flavones for in vivo imaging of beta-amyloid plaques in the brain," *J. Med. Chem.*, pp.253-7260.
5. Woodman OL., Meeker WF., Boujaoude M(2005), "Vasorelaxant and antioxidant activity of flavonols and flavones: structure-activity relationships," *J. Cardiovasc. Phalrmacol.*, 46(3), pp.302-309.
6. Sadik CD., Sies H., Schewe T(2003), "Inhibition of 15-lipoxygenases by flavonoids: structure-activity relations and mode of action," *Biochem. Pharmacol.*, 65(5), pp.773-781.
7. Pelle L., Jason T., Bemard W., etal(2001), "Microwave assisted organic-a review," *Tetrahedron Lett.*, 9225-9283
8. Li X., Wang JL., Li JP., etal(2009), "Novel aminopeptidase N inhibitors derived from antineoplaston AS2-5 (Part I)," *Bioorg. Med., Chem.*, pp.3053－3060.
9. George W., Kabalka., Arjun R., Mereddy(2005), "Microwave-assisted synthesis of functionalized flavones and chromones," *Tetrahedron Lett.*, pp.6315-6317.

(601c) Salt and Form Selection of a Chiral API Prone to Epimerization: The Importance of Determining the Nature of the Crystalline Racemate

Benjamin Cohen, John Traverse and William Leong, Department of Process Chemistry, Celgene Corporation, Summit, NJ

Selection of the final salt and form of an Active Pharmaceutical Ingredient (API) early in development avoids expensive and time consuming bridging studies if it is changed later in development. The criteria for the selection are predominantly focused on having desirable physical attributes for the drug product (e.g. stability, flowability, solubility, low hygroscopicity). In the case of chiral APIs, it is advisable to also understand the ability of the proposed salt and form to upgrade enantiopurity through crystallization by determining the nature of the compound's crystalline racemate. This effort is particularly important for molecules that contain an epimerizable stereogenic center. A case study of the salt and form selection process of an epimerizable API for which an understanding of the compound's crystalline racemate was a significant contributing factor is presented.

To develop a thermodynamically controlled crystallization for enantioenrichment, it is important to determine the nature of the system's thermodynamically most stable racemic crystalline form. The most desired form for boosting enantiopurity is a conglomerate, however these comprise only ~10% of racemic crystals. Racemic compounds make up the majority. A third and relatively rare class, pseudoracemates, also exist.

The API of interest contains an epimerizable stereogenic center. The route was designed to delay the incorporation of the functionality leading to epimerization to the final bond forming step. Nonetheless, the route and process generates the API with moderate achiral and chiral purity, therefore a recrystallization is required to boost the assay, achiral, and chiral purity to an acceptable level. A preliminary salt and form screen identified a sulfonate salt as a good lead for the final API salt. However, the XRD, DSC, ternary phase diagram and crystallization data indicate that the RSO3H salt is a pseudoracemate (Type II). As a consequence, the enantiopurity can not be increased in this system. It was decided that efforts should be directed towards finding an alternate salt, due to the risk that the manufacturing process would not deliver the API with acceptable chiral purity.

The free base (ee ~94%) was subjected to an extensive salt screen with achiral and chiral acids. The ee of the solids and filtrate were monitored in addition to the XRD and DSC of the solids. One hit, the HX salt was found to have low hygroscopicity, high solubility, good stability and acceptable flowability. The XRD of the single enantiomer and racemic crystals of this salt are different, indicating that it is a racemic compound. TGA, DSC and DVS analyses show that while the single enantiomer form is anhydrous, the racemic crystal preferentially forms a hydrate and methanol solvate.

The eutectic ee of a racemic compound is typically independent of the solvent system except when one or both of the components is a solvate. Because the racemic HX crystals form solvates in the presence of water or methanol, the eutectic ee has a strong dependence on solvent composition and temperature. With the understanding of the eutectic's dependence on solvent and temperature, a proof of concept experiment was successfully run to demonstrate the ability of the methanol system to elevate chiral purity.

(601d) Effects of Particle Size and Particle Size Distribution On Fluidized Bed Drying of Pharmaceutical Materials

Xue Liu[1], Fernando J. Muzzio[2], Johannes G. Khinast[3] and Benjamin J. Glasser[2], (1)Chemical and Biochemical Engineering, Rutgers University, Piscataway, NJ, (2)Chemical and Biochemical Engineering, Rutgers, the State University of New Jersey, Piscataway, NJ, (3)Institute for Process and Particle Engineering, Graz University of Technology, Graz, Austria

Fluidized beds are extensively used in the pharmaceutical and other chemical industries either as a batch or continuous process for drying moist powders and granular solids because of good mixing of solids and intensive heat and mass transfer between the solid and hot gas phases in the system. Generally, high energy input is required during this process to provide the latent heat of water evaporation, and thus on-line measurements and computational simulations of fluidized bed drying become important to determine the optimal operation conditions in order to minimize energy usage. In the pharmaceutical industry, the drying step in many pharmaceutical production lines is often a bottleneck because it is difficult to determine the time required to dry the materials as well as analytically verify that a predetermined end point has been reached. If the active pharmaceutical ingredient has a low melting point, controlling the drying process becomes more critical due to the sensitivity of the formulation to the temperature and the risk of compromising product quality and performance during drying.

Particle size and particle size distribution are important parameters for pharmaceutical materials. These may greatly affect the heat and mass transfer, drying rate and drying time of the fluidized bed drying process. In this work, moist pharmaceutical powders and granules, such as lactose, avicel and dibasic calcium phosphate anhydrous (DCPA) with different particle size distribution, have been dried using a Glatt GPCG-5 spray fluidized bed dryer/granulator and a Mini- Glatt spray fluidized bed dryer/granulator. The pressure drop across the bed, the inlet and outlet air temperatures and the product temperature are measured as a function of time. The moisture content in the product during drying is measured using two approaches: 1) Taking samples from the bed every 3 minutes, drying the samples in an oven and calculate the sample moisture based on the mass loss before and after drying. 2) Using NIR to on-line measure the sample moisture. We have compared the NIR on-line measurement and the reference samples (dried in the oven), and a good match can be obtained. The NIR method can be used to determine the end point of the drying process. We varied the initial moisture content inside the pharmaceutical materials from 5% to 20% because in pharmaceutical manufacturing fluidized bed drying is generally carried out after the wet granulation step and the moisture content in the wet granules is normally in this range. We have classified the powders/granules into fines, medium, coarse, and examined the flow hydrodynamics, the drying rate and the agglomeration behaviors for each size cut as well as mixtures in the bed. This provides us useful information for the drying mechanisms given some specific powder/granule materials.

To better understand the drying process, we have carried out simulations for drying of a single particle with different sizes and geometries. This serves to provide physical insight into the fundamentals of drying of a porous particle. The long term goal is to characterize performance of fluidized bed dryer systems and to develop a predictive model for drying of granulations with different size distributions and different moisture contents.

(601e) Effect of Chitosan Molar Mass and Graft Chemistry On the Transport of Therapeutics Across a Model Airway Epithelium

Radovan Dimovski[1], Balaji S. Bharatwaj[1,2] and Sandro R. P. da Rocha[1], (1)Chemical Engineering and Materials Science, Wayne State University, Detroit, MI, (2)Respiratory Product Development, Merck & Co., Summit, NJ

Oral inhalation (OI) is not only the most sensible rout for the regional delivery of therapeutics to the lungs, but it is also a viable non-invasive route for the delivery of therapeutics to the systemic circulation. The lungs are an attractive pathway to the bloodstream as they have a large surface area, consist of a thin cellular barrier, and have lower proteolytic activity compared to those observed in traditional delivery routes. The cellular tight junctions populating the pulmonary epithelia highly modulate the transport into systemic circulation. Permeation enhancers such as chitosan (CS), a biodegradable polysaccharide that typically exhibits low cytotoxicity, may be potentially employed for opening the tight junctions, and thus for modulating the transport of therapeutics across the pulmonary epithelium.

In this work we discuss the effect of CS molar mass and graft chemistries on the transport of various model solutes, including a small hydrophilic molecule, a model protein, and a hydrophobic therapeutic across polarized Calu-3 monolayers, an established *in vitro* model of the human airway epithelium. The relevance of this work stems from the fact that the ability to modulate the transport of therapeutics across the pulmonary epithelium may improve the market share of OI formulations in the overall drug delivery market.

(601f) Understanding the Role of Ethanol As Co-Solvent in Hydrofluoroalkane Propellants

Jordan Grashik, Denise S. Conti, Lin Yang and Sandro R. P. da Rocha, Chemical Engineering and Materials Science, Wayne State University, Detroit, MI

Oral inhalation (OI) is not only the preferred route for the regional delivery of therapeutics to the lungs, but it may also offer opportunities for targeting the systemic circulation due to the lungs ample surface area, good epithelial permeability, and the extensive vascularization. Pressurized metered-dose inhalers (pMDIs) are the most common portable OI devices for the delivery of therapeutics to the lungs. Hydrofluoroalkanes (HFAs) are used as both solvent and propellants in pMDIs. There are many challenges in the development of excipients required for particle stabilization and valve lubrication in HFA-based pMDIs that arise due to the somewhat hydrophobic and lipophobic nature of HFAs. Ethanol has been widely used to enhance the solubility of HFA excipients, but little is understood about the ability of ethanol (co-solvent) to enhance the solvation of such chemistries in the propellants.

In this work we use chemical force microscopy (CFM) to evaluate the effect of ethanol as co-solvent for HFAs. We determine the interaction forces between various chemically modified tip/surface combinations in 2H,3H-perfluoropentane (HPFP), a model HFA, which is liquid at ambient conditions. The effect of varying volume fraction of ethanol on the resulting forces was also evaluated. Johnson–Kendall–Roberts (JKR) theory was used to model the results. Our results demonstrate that while the presence of ethanol indeed helps reduce the cohesive forces between the various chemistries, including those representatives of excipients commonly used in particle stabilization in HFAs, the reduction in cohesive forces within ethanol concentrations relevant for commercial pMDIs is relatively small. Ethanol, therefore, while enhancing the solubility of excipients, has limited effect in terms of solvation forces in HFAs, fact which is corroborated by the poor stability of colloidal particles in HFAs even in the presence of ethanol.

(601g) Characterization of Powder Mixing From Bench Scale to Commercial Scale with a Kinetic Model and NIR

Weixian Shi, Commercial Oral Technology & Operations, Bristol-Myers Squibb, New Brunswick, NJ, Rhye Hamey, PD, Bristol-Myers Squibb, New Brunswick, NJ and Dimuthu A. Jayawickrama, Bristol-Myers Squibb, New Brunswick, NJ

The current study presents a method to characterize powder mixing and provides quantitative prediction of mixing performance upon scaling up. The method unified a kinetic model (presented in 2010 Annual meeting) and a NIR on-line measurement model of a key ingredient (an active pharmaceutical ingredient or API, Compound X). Using this method, we quantified the driving force for mixing, or the shear energy, at various manufacturing scales. The quantification confirmed that that the shear energy increased upon scaling up from the bench scale to the pilot scale and to the commercial scale. Additionally, we monitored changes in the resistance for mixing, or cohesive energy, of Compound X at commercial scales with the same method. The variation in the cohesive energy of Compound X reflected the batch-to-batch fluctuation in the powder properties when the API was manufactured. With the current method and a small sample from a new batch of API, we can predict the impact of such variation on the mixing performance of the new batch of API at various scales. The cohesive energy of the new batch of API is measured at the bench scale and then plugged into the kinetic model with the shear energy at the scale of interest for prediction. The assumptions are 1) cohesive energy is an intrinsic material property and is scale independent; 2) shear energy is scale dependent but constant if mixing is operated with the same process parameters and formulation.

(601h) The Effect of Interdroplet Spacing and Droplet Volume On the Dissolution Profiles of Ink-Jet Printed Dopamine Hydrochloride

Marlena Brown, Chemical and Biochemical Engineering/Food Science, Rutgers University, Ocean, NJ and Paul Takhistov, Food Science, Cook College, Rutgers, The State University of New Jersey, New Brunswick, NJ

We propose to utilize Drop-on-Demand technology to manufacture transdermal dosage forms in an innovative and systematic approach. The primary goal is to fabricate a novel dosage form where two or more drugs can be printed onto the same layer to achieve desired efficacy effects. Additionally, the release profile can be manipulated and predicted with greater accuracy. Previous research has detailed that manipulation of substrate properties can be used to control the morphology of the API deposits and obtain API forms ranging from amorphous to crystalline structures. Data analysis also indicates that pattern inter-droplet spacing and droplet volume affect the dissolution rate. This study establishes the relationships between the API and polymer film properties, kinetics of the evaporation process, final stain morphology, solubility properties of the fabricated dosage, and API dissolution. Ultimately this research will enable the development of scientific fundamentals which allow us to accurately create novel API dosage forms where the final droplet structure exists as defined two and three dimensional architecture with a controllable release profile.

(601i) An Innovative Computer-Aided Molecular Design Approach to the Rational Design of Novel Small Molecule Inhibitors of Amyloid-β Aggregation

Donald P. Visco Jr.[1], Jui-Heng Tseng[2], Deborah Soto-Ortega[2], Chen Suo[2], Jie Gao[3], Shelby Chastain[2], Brandon P. Murphy[2], Mihyun Lim[4], Fang Xie[5], James Chapman[3], Qian Wang[5] and Melissa Moss[2], (1)University of Akron, Akron, OH, (2)Chemical Engineering, University of South Carolina, Columbia, SC, (3)College of Pharmacy, University of South Carolina, Columbia, SC, (4)Department of Biological Sciences, University of South Carolina, Columbia, SC, (5)Chemistry and Biochemistry, University of South Carolina, Columbia, SC

The 'amyloid cascade hypothesis', linking aggregation of the amyloid-β protein (Aβ) to the pathogenesis of Alzheimer's disease (AD), has led to the emergence of inhibition of Aβ aggregation as a therapeutic strategy for this currently unpreventable and devastating disease. The identification of Aβ aggregation inhibitors has proceeded primarily through the screening of drug-like molecules, leading to the discovery of hundreds of aggregation inhibitors belonging to multiple classes of compounds. However, only two small-molecule amyloid aggregation inhibitors have proven effective enough to advance to late clinical trials.

A different strategy to identify drug-like molecules than database screening is to design molecules (from sub-fragments) that are predicted to have an ideal property profile. This approach, called computer-aided molecular design (CAMD), has the potential to facilitate these drug discovery efforts. As it relates to the development of Aβ aggregation inhibitors, a bottle-neck in the development of these inhibitors is the lack of a solved molecular structure for Aβ aggregates. To that end, we propose the use of a novel, ligand-based approach to CAMD whereby the Signature molecular descriptor is used with appropriate quantitative structure-activity relationships (QSAR) in order to generate non-intuitive and novel structures without the need for detailed information about the drug target.

Data from a diverse library of small molecules, including dihydropyridines, naphthalimides, coumarin analogs, and polyphenols, was employed as an input to this model. Here, each compound was evaluated for its effect upon aggregation of Aβ monomer. Aggregation was initiated by incubating monomeric protein in the presence of NaCl and under continuous agitation. Progression from monomer to aggregates was evaluated using thioflavin T, a fluorescent dye that yields a shifted and enhanced fluorescence when bound to the β-sheet structure of amyloid aggregates. Uninhibited aggregation exhibited a lag time, characteristic of nucleation, followed by a period of rapid growth, and culminating with a plateau as monomer and aggregate species reached equilibrium. Compounds were characterized for both their ability to extend nucleation, indicated by a fold-increase in the lag time, and their ability to reduce the quantity of aggregate formed, indicated by a decrease in the fluorescence of the equilibrium plateau. The evaluated compounds exhibited inhibitory capabilities ranging from no effect to greater than 4-fold extension of the lag time and more than 90% reduction of the equilibrium plateau. These quantitative measures facilitated the development of informative structure-activity relationships (SARs) that were used to screen molecules generated from our CAMD approach. To ensure drug-likeness of model outputs, this SAR input was further combined with screens for synthetic feasibility, molecular stability, and BBB penetration via Lipinski's Rule. Output from CAMD demonstrates the generation of novel compounds that optimize the best features of known inhibitors and also possess desirable characteristics for potential therapeutics.

(601j) PAT for Multiphase Reaction Mixtures

Jacob Albrecht, Gregory Lane, Mark Lindrud and Srinivas Tummala, Chemical Development, Bristol-Myers Squibb Co., New Brunswick, NJ

Spectroscopic PAT techniques offer real-time monitoring of pharmaceutical processes, providing valuable process information and an opportunity for feedback control. However, reactions commonly encountered in API synthetic processes are frequently heterogeneous, containing two or more phases that can interfere with and reduce the sensitivity of techniques such as Raman and FTIR spectroscopy.

This paper reports the use of an intermediate filter layer to remove unwanted phases from real-time analysis. Removing this interference has been shown to dramatically reduce signal noise for binary and ternary phase systems by selectively passing only one of the phases to the sensing region of the PAT probe. Because of the local continuous phase separation [1], we evaluated the potential impact of sensor delay. Fortunately, we found the observed sensor delay to be insignificant relative to typical reaction times. For a desired analyte, the theory guiding the selection of filter material based on surface energy and porosity will be discussed, as well as opportunities for improvement and implementation on a manufacturing scale.

References:

[1] Kralj, J.G.; Sahoo,H. R.; Jensen, K. F., Integrated continuous microfluidic liquid-liquid extraction *Lab Chip* **2007**, 7, 256–263

(601k) Repeated Interactions Between ErbB1 Receptors and the Impact On Signal Initiation

Meghan M. McCabe[1], Shalini T. Low-Nam[2], Adam Halasz[3], Diane S. Lidke[4,5], Bridget S. Wilson[4,5] and Jeremy S. Edwards[1,5,6], (1)Chemical & Nuclear Engineering, University of New Mexico, Albuquerque, NM, (2)Chemistry and Biochemistry, South Dakota State University, Brookings, SD, (3)Mathematics, West Virginia University, Morgantown, WV, (4)Pathology, University of New Mexico Health Sciences Center, Albuquerque, NM, (5)Cancer Research and Treatment Center, University of New Mexico Health Sciences Center, Albuquerque, NM, (6)Molecular Genetics and Microbiology, University of New Mexico Health Sciences Center, Albuquerque, NM

Repeated Interactions between ErbB1 Receptors and the Impact on Signal Initiation

Meghan M. McCabe[1], Shalini T. Low-Nam[2], Adam Halasz[3], Diane S. Lidke[4,5], Bridget S. Wilson[4,5], and Jeremy S. Edwards[1,5,6]

[1]Department of Chemical and Nuclear Engineering, University of New Mexico, Albuquerque, New Mexico, USA.

[2]Department of Chemistry and Biochemistry, South Dakota State University, Brookings, South Dakota, USA

[3]Department of Mathematics, West Virginia University, Morgantown, West Virginia, USA.

[4]Department of Pathology, University of New Mexico Health Sciences Center, Albuquerque, New Mexico, USA.

[5]Cancer Research and Treatment Center, University of New Mexico Health Sciences Center, Albuquerque, New Mexico, USA.

[6]Molecular Genetics and Microbiology, University of New Mexico Health Sciences Center, Albuquerque, New Mexico, USA.

The ErbB family of receptors is a group of four tyrosine kinases (ErbB1/2/3/4) that are activated via ligand binding and subsequent formation of homo and heterodimers. The main function of the ErbB family is to mediate important cellular processes such as homeostasis, pathology, and development as well as the interactions between cells . Specifically, ErbB receptors play an important part in the regulation of cell growth, proliferation, differentiation, and apoptosis. Given these important roles in normal cell development, it is not surprising that in the early 1980's the ErbB receptor family was implicated in the progression of cancers such as breast, lung, and bladder. More specifically, the overexpression of ErbB1 is known to cause uncontrolled cell proliferation. Due to this strong link between ErbB1 and cancer, it is important to understand signal initiation and transduction mechanisms associated with this receptor. Traditionally, experimental methods are used to understand ErbB1 signaling. Single particle tracking (SPT) experiments are used to visualize ErbB1 behavior within the live cell context. The experimental team at University of New Mexico reported seeing repeated interactions between ErbB1 receptors during SPT experiments. We believe the repeated interactions are critical to receptor phosphorylation and signal initiation in ErbB1. An important caveat of these SPT experiments is that receptors must be labeled at a sufficiently low density to permit discrimination between individual proteins. Although SPT experiments have yielded useful kinetic and diffusion parameters, the contributions of the unlabeled (and therefore unseen) species remain unexplored. A model must be built in order to resolve these unlabeled species. The key to this model is ability to see when the labeled receptors interact with unlabeled receptors as well as other labeled receptors allowing us to investigate repeated interactions. While there are many models to simulate ErbB1 receptor behaviors, most lack a fundamental connection to experimental data. In a step towards making the connection between experiments and modeling stronger, the overall goal of this work is to develop a model using parameters from live cell imaging to investigate the impact of repeated interactions seen in SPT on ErbB1 phosphorylation and signal initiation.

(601l) Discrete Element Modeling (DEM) Validation of Tablet Motion in a Vibratory Coating Pan

Rahul Kumar and Carl Wassgren, Mechanical Engineering, Purdue University, West Lafayette, IN

Discrete Element Method (DEM) modeling holds great promise for simulating tablet coating processes. This study presents a DEM model for simulating the movement of true biconvex tablet shapes in a vibrating coater pan, along with the experimental methods used to validate the model. High speed video imaging is used to measure the circulation speed of tablets in the experiments and comparisons with DEM simulations are made for a range of vibration amplitudes and frequencies. Inputs to the DEM model, like tablet-tablet and tablet-wall friction coefficients are measured directly using a variation on the pin-on-disk apparatus and the coefficient of restitution values were obtained from literature. In addition, the sensitivity of the model to tablet-level parameters, including friction coefficients, coefficients of restitution, and contact stiffnesses, are presented. The tablet-wall friction coefficient was found to have a very strong influence on the circulation speed with the circulation speed increasing as the friction coefficient is increased. The coefficient of restitution was also found to have a significant effect on the circulation speed.

(608a) Sustainability of Biofuel Production Systems

Douglas J. Reinemann, Biological Systems Engineering, University of Wisconsin-Madison, Madison, WI

The Great Lakes Bioenergy Research Center is one of three DOE funded Bioenergy Research Centers in the US and the only one that has a major effort devoted to the study of the sustainability of biofuels production systems. We have used a Spatially-Explicit Integrative Modeling Framework (SEIMF) to evaluate the productivity and environmental impacts of biofuel production systems. The modeling framework was implemented to simulate feedstock supply to a cellulosic ethanol bio-refinery located at the center of each of two Regionally-Intensive Modeling Areas (RIMA). Feedstock sources modeled included crop residues and perennial species. Each production system was evaluated in terms of its effect on soil erosion, water-use efficiency, soil carbon stocks, nutrient losses, nitrate leaching, and nitrous oxide emissions. Perennial biomass crops grown on marginal lands can potentially contribute to meet cellulosic feedstock targets while avoiding food-vs.-fuel conflicts and maintaining or even enhancing ecosystem services. The SEIMF approach together with long-term environmental data and analysis was used to model potential cellulosic ethanol production from perennial mixed prairie systems grown on marginal lands across the 10-state U.S. North Central Region. These RIMA modeling results used to develop a spatially-explicit bio-economic model for the study of potential cellulosic biomass supply at regional scale.

We have applied ISO 14000 Life Cycle Assessment (LCA) methods to quantify the environmental impacts of a number of complex biofuels production scenarios including: a) four major ligno-cellulosic agricultural feedstocks (corn stover, switchgrass, poplar and mixed-prairie species) for RIMAs in southern Michigan and Wisconsin, b) biomass feedstocks from native forests in Northern Wisconsin, c) use of the primary AFEX pretreatment process for agricultural residues and SPORL and Dilute Acid pretreatment for woody biomass, d) final fermentation for ethanol production, e) logistics for transportation to regional preprocessing centers and final transport to a centralized ethanol production facility, and f) heat and electric power energy generation scenarios from co-products. Further, we have analyzed three potential breakthrough technologies a) creation of oil-rich grasses as dedicated energy crops, combining both the oil platform and cellulosic platform in one crop, b) feedstock pre-processing to potentially allow the use of high-diversity feedstocks (such as mixed prairie grasses) and other species, and c) use of AFEX-treated lignocellulosic biomass for cellulase production.

A paradigm shift to cellulosic biofuels seems unlikely unless several challenges can be met, including reconciling biofuels with food production, improving system environmental performance, avoiding harmful land use changes, and assembling the biomass supply chains. In particular, local economic and environmental sustainability of the supply system is vital to advance the farmers' interests. Appropriately priced, low supply risk feedstocks are needed to insure the growth of the nascent biofuel industry, as is the ability to inexpensively transport and store the biomass. Local economic and environmental concerns must be resolved in the feedstock supply chain, and diversifying the market for cellulosic feedstocks via multiple co-products would be advantageous as well. Unfortunately, few of these properties are currently demonstrated by cellulosic feedstocks. One promising means to address these and other issues is to process the biomass near where it is produced in so-called local biomass processing depots (LBPDs). RBPDs would increase the stability, uniformity, density, and economic value of a potentially wide variety of cellulosic biomass.

(608b) Neutron Reflectometry, QCM-D , and TIRF Study of the Interaction of Endoglucanases with Films of Amorphous Cellulose

Michael Kent, Joint BioEnergy Institute, Sandia National Laboratory, Emeryville, CA

Cellulase enzyme cocktails include exoglucanases that digest cellulose chain ends and endoglucanases that cleave randomly at interior points along the chains. While it is known that these enzymes work synergistically, the details are not fully understood. In addition, cellulose binding domains (CBDs) are known to play an important role in the digestion of crystalline cellulose but much less is known about the benefit of CBDs in the digestion of amorphous cellulose. Amorphous cellulose is of interest as pretreatment of biomass with ionic liquids, a promising next generation technology, results in a combination of amorphous cellulose and cellulose II. Determining the full effects of endoglucanase activity is challenging because these enzymes can alter the structure of insoluble cellulose in addition to releasing soluble oligomers. To unravel the actions of endoglucanases and the role of cellulose binding domains in enhancing activity on amorphous cellulose, we have combined studies of the profile of water through cellulose films during digestion by neutron reflectivity, measurements of changes in mass and film stiffness using a quartz crystal microbalance (QCM), and visualization of the motion of individual enzymes by total internal reflection fluorescence (TIRF) microscopy. A trend has emerged in that the enzymes possessing a CBM penetrate into and digest within the bulk of the films to a much greater extent than those lacking a CBM. In addition, much different results were observed for a classic endoglucanase and a processive endoglucanase that both possess CBMs, suggesting that penetration is driven by different mechanisms in the two cases.

(608c) Understanding and Redesigning Secondary Cell Wall Deposition in Plants

Dominique Loque, Joint BioEnergy Institute, Berkeley, CA; Physical Biosciences Division, Lawrence Berkeley National Laboratory, Berkeley, CA

The plant cell wall represents a large source of polysaccharides that could be used to substitute for sugar derived from starchy grains, which are currently used to feed and produce biofuels. This lignocellulosic biomass, largely under-utilized, is mainly composed of sugar polymers (cellulose and hemicellulose) embedded very strong aromatic polymer called lignin. Recalcitrant to degradation, lignin inhibits efficient extraction and hydrolysis of the cell wall polysaccharide and prevents cost-effective lignocellulosic-biofuel production. Unfortunately, lignin cannot simply be genetically removed without incurring deleterious consequences on plant productivity. The cost effectiveness of the conversion of the lignocellulosic biomass into sugars is still one of the major components to produce cheap biofuels. Therefore, strategies that can be used to reduce the lignin recalcitrance and that can increase polysaccharide deposition into the cell wall without altering plant growth should be developed.

We used synthetic biology to redesign cell wall biosynthesis and deposition without affecting the plant growth. We generated strategies to reduce lignin recalcitrance either by manipulating its spacio-temporal deposition or by manipulating its composition. We also reengineered the control of cell wall biosynthesis and used it to enhancing cell wall polysaccharides deposition and biomass density. Finally, several of these engineering strategies were combined and translated into an improvement in saccharification efficiency without affecting plant development. Developed approaches and preliminary data will be presented.

¶*This work conducted by the Joint BioEnergy Institute was supported by the Office of Science, Office of Biological and Environmental Research, of the U. S. Department of Energy under Contract No. DE-AC02-05CH11231*

(608e) Development of the Extractive-AFEX^{TM} (E-AFEX^{TM}) Pretreatment Process

Shishir Chundawat[1], Leonardo da costa Sousa[1], Venkatesh Balan[1], Vijay Bokade[2], Mingjie Jin[1], Nirmal Uppugundla[1], James Humpula[1] and Bruce E. Dale[1], (1)Chemical Engineering and Material Science, Great Lakes Bioenergy Center, Michigan State University, Lansing, MI, (2)Catalysis Division, National Chemical Laboratory (NCL), Pune, India

The ammonia fiber expansion or AFEX™ process is a leading thermochemical pretreatment process that facilitates the rapid deconstruction of lignocellulosic biomass to fermentable sugars in an economical fashion. Two major limitations of the conventional AFEX™ process are; (i) it is unable to reduce the inherent recalcitrance of crystalline cellulose I to enzyme biocatalysis, and (ii) it is unable to extract any lignin during the process that are potentially inhibitory to biological catalysts. In order to address these limitations, we have recently developed a novel ammonia based pretreatment called Extractive-AFEX™ (or E-AFEX™). One of the driving forces that led to the development of the E-AFEX™ process was based on our recent discoveries related to unique physicochemical properties of cellulose III (a less recalcitrant allomorph of cellulose; *Chundawat et al, 2011, JACS, 133:11163*). Fortuitously, the conditions that favor cellulose III formation also facilitate selective extraction of lignin from plant cell walls during the pretreatment. In this presentation, we will explore the salient features of the E-AFEX™ technology and its impact on the enzymatic and microbial processability of lignocellulosic biomass. The E-AFEX™ process is able to overcome several of the common bottlenecks to cellulosic biofuel production and can be readily integrated within a non-centralized biomass processing framework. Additionally, the lignin extracted during the process can be converted to valuable chemicals or materials. AFEX™ and E-AFEX™ processing technologies open up exciting new research avenues for cellulosic biorefineries, that allow selective conversion of the aromatic and carbohydrate fractions to a whole suite of fuels, materials and chemicals using desired catalytic or biocatalytic routes analogous to the petrochemical refineries.

(608f) ORNL: Initial Binding of a Cellulose Chain Into the Cellulase Catalytic Tunnel

Pavan GhattyVenkataKrishna, Computational Biology Group, ORNL, Oak Ridge, TN, Edward Uberbacher, Computational Biology Group, ORNL and Xiaolin Cheng, Center for Molecular Biophysics, ORNL

Cel7A proceeds along a single cellulose chain in a processive manner, cleaving one cellobiose unit per catalytic event. We have conducted extensive molecular dynamics (MD) simulations to understand the initial binding of a cellulose chain into the catalytic tunnel of the Trichoderma reesei Family 7 cellobiohydrolase (Cel7A). In unrestrained MD simulations, the cellulose chain diffuses into the tunnel by the length of a cellobiose unit. Among those residues lining the tunnel, the four tryptophan residues exhibit small structural fluctuations, and may thus play a role in guiding the substrate sliding, while the more dynamic polar residues are observed to promote the initial diffusion of the substrate into the tunnel. Further MD simulations show no evident preference for the reducing end or the non-reducing end binding of the cellulose chain, suggesting that the spontaneous initial binding is not driven by the dipole-dipole interaction between Cel7A and the cellulose chain. Rather, the shape complementarity of the cellulose chain end to the tunnel vestibule (involving local hydrophobic stacking and hydrogen bond interactions) plays a more dominant role. The potential of mean force (PMF) calculations, however, show a clear difference at -5 subsite between the reducing end and the non-reducing end binding, implicating that the selective hydrolysis of cellulose from the reducing end can be partially achieved through the preferential initial binding of cellulose to the Cel7A tunnel. The comparison of the PMFs for the wild type and the W40A mutant enzymes highlights the importance of the Trp40 residue in the initial threading of the cellulose chain to the catalytic tunnel.

(613a) Bio-Pulping Assisted Steam Pretreatment of Bagasse by Solid State Cultivation of White Rot Fungi

Jian Liu, Chuannan Long, Hailong Li, Xiaobing Wu and Minnan Long, School of Energy Research, Xiamen University, Xiamen, China

Pretreatment is an important processing step to increase the accessibility of cellulosic material in plant biomass, impacting efficiency of subsequent hydrolysis and fermentation. White rot fungi degrade lignin and have biotechnological applications in conversion of lignocellulose to valuable products. Steam pretreatment is considered to be one of the most practical and versatile pretreatment methods, but to be cost-effective it must produce more accessible fiber for enzymes and prevent the formation of by-products in lower temperature. This study investigated steam pretreatment assisted by bio-pulping of bagasse by solid state fermentation (SSF) using Ceripriopsis sp. to facilitate the conversion into fermentable sugar. The effects of substrate moisture content (68%, 75% and 82%), nutritional addition with metal ions and culture time (0–40 days) on ligninolytic activity, lignin degradation, total weight loss, pulp yield, sugar yield and sugars in soluble fraction were examined. Moisture content significantly affected lignin degradation, with 75% moisture content degrading approximately 2.9-3.7% more lignin than 68% and 82% moisture content after 40 days. Within the same moisture content, fermentations supplemented with metal ions nutrition were statistically better than those without nutrition for lignin degradation and sugar yield from both pulp and soluble fraction, but all 3 ligninolytic activities significantly decreased. Beyond the 21 days pretreatment, additional time resulted in greater lignin degradation, but indicated a decrease in pulp yield and sugar yield. Considering cost, solid state fermentation at 75% moisture content with nutrition was the most preferable pretreatment resulting in 19.1% lignin degradation, 68.2% pulp yield and 25.1% sugar yield over a period of 21 days. It was proved that microbial pretreatment by solid state fermentation has the potential to be a low cost, environmentally friendly alternative to chemical approaches. Moisture relationships will be significant to the design of an effective microbial pretreatment process using SSF technology. Microbial pretreatment employs microorganisms and their enzyme systems to breakdown lignin present in lignocellulosic biomass. This environmentally friendly approach has recently received increased attention and has potential advantages over the prevailing physicochemical pretreatment technologies due to reduced energy and material costs, simplified processes and equipment, and use of biologically based catalysts. Several basidiomycetes such as Phellinus sp., Ceriporiopsis subvermispora, Phlebia subserialis and Pleurotus ostreatus are capable of efficiently metabolizing lignin in a variety of lignocellulosic materials. Ceriporiopsis subvermispora, is a white rot fungus implicated in many diseases of broad-leaved and coniferous trees, causing heart rot, canker and root diseases, and also esca disease of grapevines. It produces oxidative enzymes that degrade lignin. Reducing the lignin content of biomass helps to expose the highly-ordered crystalline structure of cellulose and facilitates substrate access by hydrolytic enzymes. White rot fungi can decompose lignin effectively by excreting unique extra-cellular peroxidases (MnP) as secondary metabolites under nutrient starvation. It has been widely investigated because of its non-specific nature and exceptional oxidation potential for lignin degradation. The degradation efficiency and minimal use of cellulosic polymers relative to other white rot fungi make it a suitable candidate for pretreatment and lignin degradation. However, the efficiency of biodegradation varies with fermentation methods and environmental conditions. Solid state fermentation (SSF) is defined as the growth of microorganisms on solid materials in the absence of free liquid and the technology has been investigated for the development of bioprocesses. Compared to liquid fermentation, where microorganisms are submerged in free liquid, SSF possesses advantages such as simpler operation, lower raw material and input costs and reduced environmental impact from waste water. Moreover, solid state fermentation provides environmental conditions under which white rot fungi grow in nature, thus making it ideal for growth. Considering these merits, applying SSF technology to pretreat lignocellulosic biomass and biopulping may lead to better efficiency and reduce pretreatment costs. However, very limited resources report the application of SSF of Phellinus. sp. for pretreatment of lignocellulosic materials, especially to meet the need for generation of a sugar platform for ethanol production. In the present study, bio-pulping assisted steam pretreatment were applied to determine the efficiency in bagasse. This study examines the microbial pretreatment of wood powder by solid state fermentation. The natural compositions in the wood are not sufficient to meet critical nutrient requirements of fungi. The supplementation of extra nutrients(wheat bran and inorganic salt) is necessary in this study. Moisture content was assumed as a key factor that affected the lignin degrading performance. Moisture content, substrate characteristics, nutrient supplements, aeration and culture duration are important variables for performance of fungi in SSF, impacting growth and metabolite formation, but these processing parameters have not been fully explored. It is reported that moisture content is significant to microbial performance in SSF systems and affects growth as well as product formation. Water in SSF systems functions as a solvent for transport of nutrients, metabolites and other solutes as well as helps maintain stable cellular and molecular structures. Some research have found the quantity and kinetic properties of free water to be more influential to the diffusion of nutrients and metabolites and overall microbial activity. It is reported that the effectiveness of fermentation without nutrition addition on fungal growth and lignin degradation supports the recognized advantages of SSF systems in using wood substrates that provide necessary nutrients in medium preparation. Therefore, the effects of initial substrate moisture content (68%, 75% and 82% wet-basis), nutrition and culture time (0–40 days) on lignin degradation, pulp yield and sugar yield were investigated to evaluate process effectiveness.

(613b) Pretreatment of Corn Stover Using Propyl-Sulfonic Acid Functionalized Nanoparticles

Leidy Peña[1], Feng Xu[1], Keith L. Hohn[2] and Donghai Wang[1], (1)Biological and Agricultural Engineering, Kansas State University, Manhattan, KS, (2)Chemical Engineering, Kansa State University, Manhattan, KS

Propyl-sulfonic acid functionalized (PS) nanoparticles with various levels of functionalization were synthesized, characterized and evaluated as catalysts for pretreatment of corn stover. The degree of functionalization was measured as a function of the hydronium ion concentration, the total organic loading, and the elemental sulfur content. The acid concentration, organic loading, and sulfur concentration of the nanoparticles varied between 0.05 and 0.40 mmol H^+/g, between 5.0 and 30.0%, and between 1.0-18.0%, respectively. FTIR spectra revealed variation on the intensity of the peaks associated with the acid groups on the nanoparticles. The performance of PS nanoparticles on the pretreatment of corn stover was evaluated at 160 °C for 1 h. The biomass and catalyst loads used during the pretreatment were 2.5% and 0.25%, respectively. After pretreatment, the liquid and solid fractions were physically separated. The nanoparticles were magnetically separated from the liquid fraction. The concentration of monomer sugars and oligosaccharides as well as the concentration of degradation products in solution were measured by HPLC. The composition of the solid fraction was analyzed for determination of cellulose and hemicellulose recoveries. The solid fraction was digested with a commercial cocktail of cellulases at 50 °C for 120 h. Biomass digestibility was analyzed against the surface characteristics of the nanoparticles that were used during pretreatment. PS nanoparticles increased the biomass susceptibility to enzyme hydrolysis; therefore, they can be used as an effective catalyst for the pretreatment of corn stover.

Stimulating Cellulase Activity of *Microbulbifer Hydrolyticus* Using Hot-Water Wood Extract Hydrolisate as Substrate.

Karin Arens, Department of Paper & Bioprocess Engineering, State University of New York-College of Environmental Science & Forestry, Syracuse, NY and Shijie Liu, Department of Paper & Bioprocess Engineering, State University of New York- College of Environmental Science & Forestry, Syracuse, NY

ABSTRACT

Among the cellulase producing microorganisms, bacteria often present the advantage of high growth rate, which leads to high enzyme production, and the production of cellulosomes, providing increased function and synergy. In order to improve enzymatic production it is necessary to optimize the fermentation conditions and substrate utilization for a given microorganism. This study uses hydrolisate of sugar maple produced during hot-water extraction (hydrolisate) as substrate to stimulate the production of cellulase from *Microbulbifer Hydrolitycus* IRE-31 (ATCC 700072). Selective strain medium adaptation of the original *M. hydrolyticus* IRE31 was performed by gradual increment of hydrolisate concentration in the marine medium from 2% to 20% (v/v). To improve growth, the pH level was controlled using 5M sodium hydroxide (NaOH) and 0.02M calcium hydroxide ($Ca(OH)_2$) as bases. The effect of the bases on the bacterial growth and activity of the cellulases present in the broth resulted on higher cellulase activity in flask fermentation with 5M NaOH, while in agar the best growth appeared with 0.02M $Ca(OH)_2$ as base. A comparative study to determine the optimum 0.05M citrate buffer pH for enzymatic activity of the *M. hydrolitycus* cellulases present in marine broth supplemented with various hydrolisate concentrations showed best cellulase activity at pH of 4.8. Tests comparing flask fermentations using different substrates such as xylose, glucose, cellobiose, hydrolisate, sodium chloride (NaCl) and yeast extract, exhibited higher cellulase activity in sample broths containing hydrolisate and no discernible activity in broths containing glucose, yeast extract or xylose. The results indicate a positive correlation with the concentration of hydrolisate in the marine broth and the cellulase activity. The addition of 0.25M and 0.20M NaCl in the medium reduced the activity of cellulases in the broth. Scaling up the study to a1L stirred tank fermentation with 10% v/v hydrolysate, set at pH 7.3, 35°C, and 250rpm displayed a maximum enzymatic activity in the broth sample of 0.650 μmol/min-ml after 10 hours. The evidence from this study supports the inference that the use of hydrolysate as a substrate is promising in the stimulation of the production of cellulase by the marine bacteria. However, further study is required regarding optimization of the process.

INTRODUCTION

The current global energy demand and supply of crude oil, combined with projections of fossil fuel depletion, has highlighted the need to find an alternative energy source (EIA Outlook 2011).

The desire to find a renewable and sustainable source of energy has created an increased focus on production of biofuels and bioproducts from lignocellulosic biomass. In an effort to commercialize cellulosic biofuel, new technologies are being explored to increase product yield and reduce the cost in discrete process steps. This new commercial climate and direction for energy supply presents a large market potential for cellulase production (US Dept. of Energy 2011). Currently, the cost of cellulase enzymes accounts for 30 to 50% of the total cost in the production of lignocellulosic ethanol. This cost needs to be reduced in order to make biofuel production more competitive.

Research shows that the enzymatic production can be enhanced by isolating novel strains, improving existing strains, optimizing fermentation conditions and substrate utilization for a given microorganism (Wyman et. al. 2005; Zhang et. al. 2006; Moiser et.al. 2005; Hsu 1980; Eggeman and Elander 2005; Philippidis et.al.1993). Cellulases are mainly produced by fungi, bacteria and protozoans. Production of bacterial cellulases presents some advantages over cellulases produced by fungi. The bacteria growth rate is higher than fungi, leading to higher enzyme production. Moreover, bacterial cellulases are often complex enzyme structures organized in supramolecular complexes, called cellulosomes, which provide increased enzymatic function and synergy (Bayer et al., 1998, Bayer et. al. 2004; Maki et. al. 2009). In this study, hydrolisate of sugar maple produced during hot-water extraction (hydrolisate) is utilized as substrate to stimulate the production of cellulases from the marine bacteria *M. hydrolyticus*. The effect of hydrolisate, xylose, glucose, cellobiose, sodium chloride and yeast extract, on cellulase activity are measured and compared. Initially, a selective strain medium adaptation of the original *M. hydrolyticus* IRE3 was carried out by incremental addition of hydrolisate concentration in the marine medium, (agar and broth). The pH was controlled by using 5M sodium hydroxide (NaOH) and 0.02M calcium hydroxide ($Ca(OH)_2$) as bases, and the effect of the bases on the bacterial growth and the activity of the cellulases present in the broth was measured and compared. A comparative study was performed in order to determine the optimum 0.05M Citrate buffer pH to achieve the best activity for *M. hydrolitycus* cellulases present in broths supplemented with hydrolisate concentrations from 0 to 10% v/v.

METHODOLOGY AND MATERIALS

Microorganism growth, strain adaptation and flask fermentation

Original *Microbulbifer hydrolyticus* IRE31(ATCC700072) was grown in a 250ml Erlenmeyer flask containing 100ml of sterile marine broth 2216 (Difco Laboratories, Detroit, MI). The liquid culture was incubated in the dark at 30°C for 48 hours on a rotary shaker set at a rotation speed of 200rpm. After growth, a loop of liquid culture was transferred to marine agar 2216 (Difco) plates. The inoculated Petri dishes were incubated in the dark at 27°C for 24h. Agar plates were stored in the dark at 8°C for a maximum of 4 weeks.

The strain medium adaptation was performed by exposing the original *M. hydrolyticus* IRE31 to small increments of hydrolisate in the marine medium. The concentrated hydrolysate was filter-sterilized (0.22μm sterile filter, NALGENE, Rochester, NY). Both broth and agar were autoclaved at 121°C for 15 min, cooled (room temperature for broth and approximately to 45°C for agar) and mixed with hydrolysate. The initial selective mediums contained 2% hydrolisate (v/v). The largest colonies were selected, plated on selective mediums with the same concentration as the parent cells, and incubated in the dark at 27°C for 24h. Daughter colonies larger than parent colonies were transfer to a broth containing 3% hydrolisate (v/v) for fermentation tests. After fermentation, the cellulase activity was measured and the broth with the highest cellulase activity was selected for plating on 3% and 4% hydrolisate (v/v) selective medium. The selected strains underwent a gradual increase of hydrolysate concentration at the following increments: 2% to 3%, 4%, 5%, 6%, 10%, 15% and 20%.

The pH level was controlled using 5M NaOH, and 0.02M $Ca(OH)_2$. The effect of each of the bases on bacterial growth and cellulase activity was measured and compared at different concentrations of hydrolisate in flask inoculum and agar plates.

The flask fermentations were carried on 250ml Erlenmeyer glass flasks capped with cotton gauze containing 100ml of sterile fermentation medium. The fermentation was performed at 37°C for 48-72h on a rotary shaker, set at a speed of 300rpm. Samples were periodically drawn to measure the pH and cellulase activity. A comparative fermentation study was performed, maintaining all variables constant, except for the speed of rotation, to determine its impact on the cellulase activity. The fermentation was performed at 35°C for 40-64h on a rotary shaker set at a speed of 300 and 160rpm.

The *M. Hydrolitycus* cellulase activity was measured for microorganisms grown in various concentrations of hydrolisate by volume and compared to those grown in marine medium supplemented with NaCl (0.20M and 0.25M), yeast extract, cellobiose, glucose, and xylose. The marine broth 2216 was used as the blank. During the comparative study between hydrolisate, xylose, glucose and cellobiose, the average concentration of sugars in the hydrolisate was estimated to be $150 g \cdot L^{-1}$. The marine broths were supplemented with sugars at concentrations of 0%, 5% and 10% by volume.

A comparative study to determine the optimum 0.05M Citrate buffer pH for *M. hydrolitycus* cellulase activity, measured the activity of the cellulase present in marine broth with hydrolysate concentrations of 0%, 2%, 4%, 6% and 10% v/v. The cellulase activity was studied at various pH levels of 0.05M Citrate buffer, ranging from 4.8 to 6.0, in order to determine at which pH the enzyme displays a higher activity.

The study was scaled up to a1L stirred tank fermentation with 10% v/v hydrolysate, set at pH 7.3, 35°C, and 250rpm. Samples were periodically drawn to measure the growth and enzymatic activity of the cellulases present in the broth.

The cell growth was determined with a standard curve, in which the dry cell weight at different dilutions was measured and UV absorbance was read at 600nm with results ranging between 0.1 to 0.8A.

Filter paper assay for saccharifying cellulase

The cellulase activity was determined following the International Union of Pure and Applied Chemistry (IUPAC) guidelines, by measuring the enzyme's activity in terms of filter paper units (FPU) per milliliter of original (broth) enzyme solution (Ghose 1987). Dinitrosalicylic acid solution (DNS reagent) was used for color development (Miller 1959). The absorbance was measured against the reagent blank at 540nm with results ranging between 0.1 to 1.0A.

REAGENTS

Citric Buffer

The activity of the *Microbulbifer Hydrolitycus* cellulase was studied at various pH of 0.05M Citrate buffer, ranging from 4.8 to 6.0. The best results were observed at a pH of 4.8. A stock solution of 1M citrate buffer was prepared following the procedures provided by the National Renewable Energy Laboratory Technical Report (NREL/TP-510-42658). The 1M Citrate Buffer stock solution was diluted to 50mM, the pH was checked and adjusted as necessary to 4.8, 5.0, 5.2, 5.6, and 6.0. A glucose standard curve was created for each pH in order to determine the best pH for cellulase activity.

RESULTS AND DISCUSSION

The hydrolisate in the marine medium appears to stimulate cellulase production by *M. hydroliticus*. Initial trials indicate a direct and positive correspondence between the concentration of hydrolisate in the broth and the cellulase activity (see FIG. 1). The production of cellulase is present at specific points of the growth curve. These points of enzymatic activity coincide with a decrease of glucose concentration in the broth. In the various hydrolisate concentrations studied, the highest cellulase activity was shown within the first 20 hours of fermentation (see FIG. 2). Some spikes in cellulase activity are developed after at lower values (see FIG. 3). These results are in agreement to prior studies, which describe cellulase as a growth related enzyme.

FIG. 1. Initial pH 7.5. Results from flask fermentation incubated 36 hours at 30°C and 300rpm.

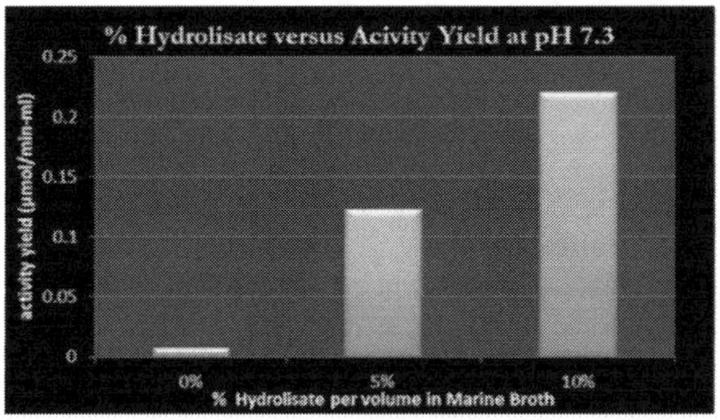

FIG. 2. Results of cell density and cellulase activity in the broth versus time in a 1L stirred tank fermentation with 10% hydrolysate (v/v), set at pH 7.3, 35°C, and 250rpm. Enzymatic activity was measured for broth sample.

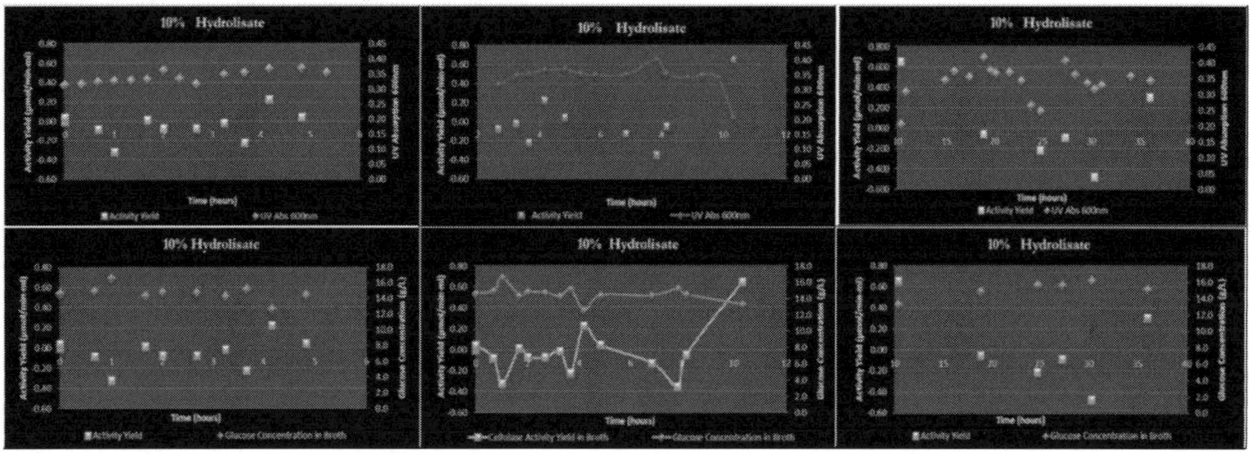

FIG. 3. Cellulase activity over time in a 1L stirred tank fermentation with 10% v/v hydrolysate set at pH 7.3, 35°C, and 250rpm.

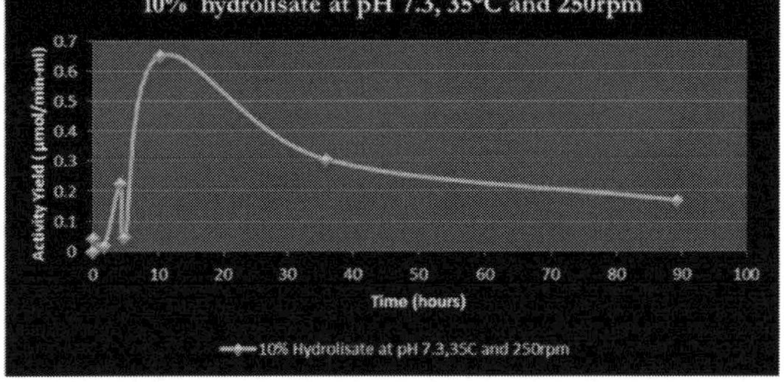

As the concentration of hydrolisate increased in the broth and agar, the importance of controlling the pH was critical. The cells were able to adapt up to 4% v/v of hydrolisate in agar without any addition of base. A further increase of hydrolisate concentration required the introduction of a base. In broth cultures, the cells were able to adapt up to a 20% v/v of hydrolisate. The pH level was controlled using 5M NaOH, and 0.02M $Ca(OH)_2$. The effect of each of the bases on the cellulose activity was measured and compared at different concentrations hydrolisate in flask inoculum. In agar the best results appeared with 0.02M $Ca(OH)_2$. One significant drawback to using 0.02M $Ca(OH)_2$ was identified: it requires greater base volume. For example, to increase the pH of a 100ml medium with 10% v/v hydrolisate to 7.5, a volume of 8ml 0.02M $Ca(OH)_2$ is needed, compared to 50μl of NaOH 5M. In the agar, this volume difference is managed more easily by adjusting the amount of water used during the agar preparation. This volume difference requirement can be problematic during stirred tank fermentation because it dilutes the hydrolisate concentration in the broth as the fermentation progresses with continuous need to add base to the medium, in order to maintain the pH. Another concern is the accumulation of calcium deposits in the feeding lines of the fermentor. Therefore, during the fermentation 5M NaOH is used to control the pH.

M. hydrolyticus optimal growth is at 37°C, pH of 7.5, 300rpm, and NaCl concentrations between 0.1 and 0.5 M (Gonzalez et. al. 1997). The effect of NaCl added to pure marine broth and agar and to marine broth with hydrolizate was also investigated. The addition of 0.25M and 0.20M sodium chloride (NaCl) in the attempt to further the growth of the bacteria in the agar medium resulted in smaller and fewer colonies after 24 hour incubation when compared to agar plates without NaCl. The NaCl addition reduced the pH and the activity of the enzyme in the broth (see FIG. 4). Overall, the pH seemed to be the key variable for an improved enzymatic activity.

FIG. 4. Comparison of *M. hydroliticus* cellulase activity yield and pH in 2% hydrolisate marine broth supplemented with NaCl.

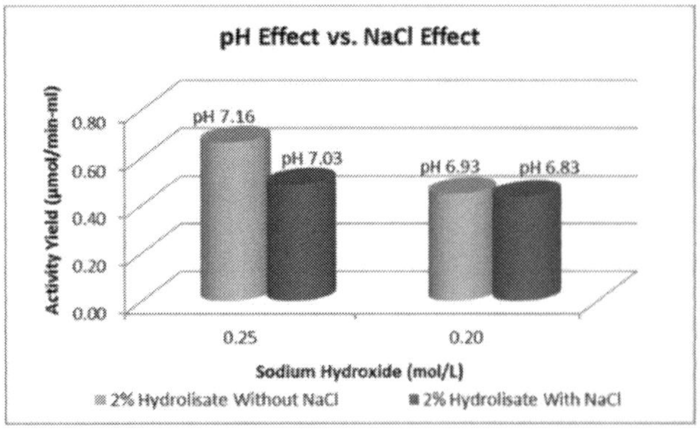

The cellulase activity was measured in citric acid buffer solutions pH of 4.8, 5.0, 5.2, 5.6, and 6.0. The best cellulase activity yield were observed at pH of 4.8 (See FIG. 5a and 5b).

FIG. 5a. Effect of citrate buffer in cellulase activity. Incubation of 72 hours at 30°C and 300rpm.

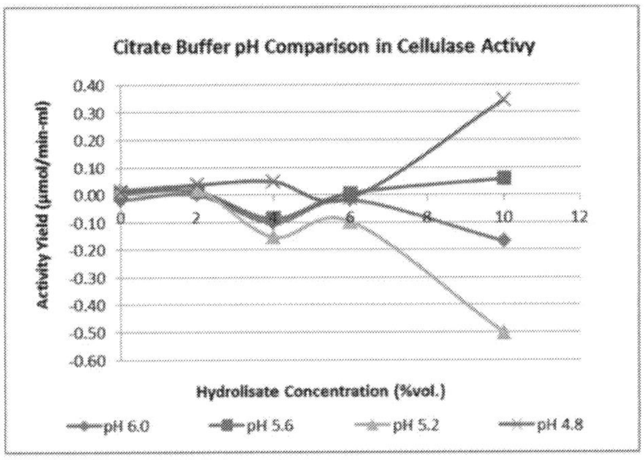

FIG. 5a. Effect of citrate buffer in cellulase activity.

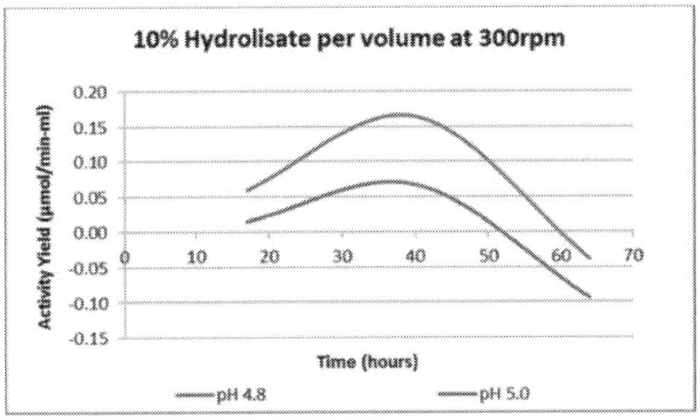

The supplement of the marine broth with glucose and xylose had an inhibitory effect on the production and activity of the enzyme. The addition of yeast extract to the marine broth did not seem to affect the enzyme activity. The presence of cellobiose in the broth stimulated the cellulase activity, but the activity results were less than when supplemented with hydrolisate (see Fig. 6). It was also noted that as the WE concentration increases, so does the activity yield. Better results for cellulase activity are obtained within the first 18hours of fermentation with pH control close to 7.5. At low agitation rate of 160rpm, the cell growth was diminished and did not show cellulase activity for broth supplemented with 10%v/v of hydrolisate. Higher agitation rate of 300rpm during incubation increased the oxygenation of the broth and thus increased the cell production and cellulase activity yield.

FIG.6 Effect of different substrates in the *M. hydroliticus* cellulase activity yield.

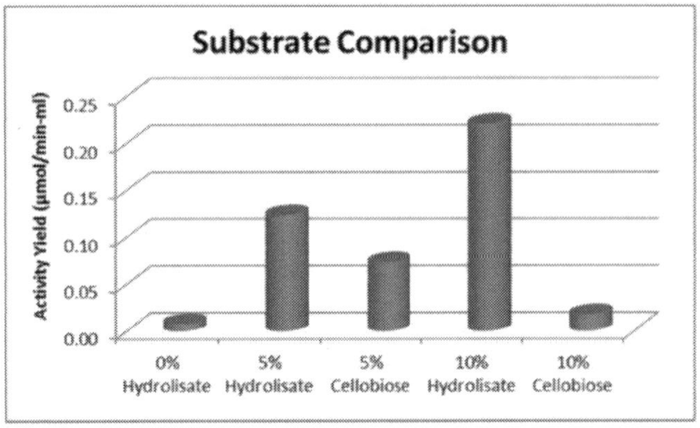

The cellulase activity yield seems to be improved when hydrolisate is used as substrate, and it displays a positive correlation between hydrolisate concentration and activity yield of the enzyme.

CONCLUSION

The marine bacteria *Microbulbifer Hydrolitycus* IRE-31 was adapted to 20% v/v of hydrolisate of sugar maple produced during hot-water extraction using selective strain medium adaptation and pH control. The hydrolisate in the marine medium appears to stimulate cellulase production by *M. hydroliticus*. Initial trials indicate a direct and positive correspondence between the concentration of hydrolisate in the broth and the cellulase activity. Spikes in the production of cellulases are present at specific points of the growth curve. These points of enzymatic activity coincide with a decrease of glucose concentration in the broth. In the various hydrolisate concentrations studied, the highest spikes of cellulase activity were uniformly shown within the first 20 hours of fermentation. The evidence in this study supports the inference that the use of hydrolysate as a substrate is promising in the stimulation of the production of cellulase by the marine bacteria. Such a result may be of significant import relative to decreasing the cost of cellulase enzymes during production of biological fuels. However, further study is required regarding optimization of the process.

REFERENCES

1. Bayer, E.A., H. Chanzy, R. Lamed and Y. Shoham. 1998. Cellulose, Cellulases and Cellulosomes. *Current Opinion Structural Biology* 8:548-557
2. Bayer, E.A., J.P. Belaich, Y. Shoham, and R. Lamed. 2004. The cellulosomes: multienzyme machines for degradation of plant cell wall polysaccharides. *Annu Rev Microbiol.* 58:521-554

3. Wyman, C.E., B.E. Dale, R.T. Elander, M. Hotzapple, M.R. Ladisch, and Y.Y. Lee. 2005. Coordinated development of leading biomass pretreatment technologies. *Bioresour Technol.* 96:1959-1966
4. Philippidis, G.P., T.K. Smith, and C.E. Wyman. 1993. Study of the enzymatic hydrolysis of cellulose for production of fuel ethanol by the simultaneous saccharification and fermentation process. *Biotechnol Bioeng.* 41(9):846-853
5. Lynd, L.R., W.H. van Zyl, J.E. McBride, and M. Laser. 2005. Consolidated bioprocessing of cellulosic biomass: an update. *Curr Opin Biotechnol.* 16(5):577-583
6. Eggeman, T., and R.T. Elander. 2005. Process and economic analysis of pretreatment technologies. *Bioresource Technology* 96: 2019–2025
7. Ghose, T.K. 1987. Measurement of Cellulase Activities. *Pure and Appl. Chem.* 59(2):257-268.
8. Gonzalez, J.M., F. Mayer, M.A. Moran, R.E. Hodson, and W.B. Whitman. 1997. Microbulbifer hydrolyticus gen. nov., sp. nov., and Marinobacterium georgiense gen. nov., sp. nov., Two Marine Bacteria from a Lignin-Rich Pulp Mill Waste Enrichment Community. *International Journal of Systemetic Bacteriology*. 47(2), 369-376
9. Hsu, T.A., 1996. Pretreatment of Biomass. In: Wyman, C.E. (Ed.), Handbook on Bioethanol, Production and Utilization. Taylor & Francis, Washington, DC.
10. Hsu, T.A., M.R. Ladisch, and G.T. Tsao. 1980. Alcohol from cellulose. *Chem. Technol.*, 10 (5), 315–319.
11. Kumar, P., D.M. Barrett, M.J Delwiche, and P. Stroeve. 2009. Methods for Pretreatment of Lignocellulosic Biomass for Efficient Hydrolysis and Biofuel Production. *Ind. Eng. Chem. Res.* 48: 3713–3729
12. Maki, M., K.T. Leung, and W. Qin. 2009. The prospects of cellulase-producing bacteria for the bioconversion of lignocellulosic biomass. *Int J Biol Sci.* 5(5):500-516.
13. Miller, G.C. 1959. Use of the Dinitrosalicylic Acid Reagent for the determination of reducing sugar. *Anal. Chem.* 31: 420–428.
14. Mosier, N., C. Wyman, B. Dale, R. Elander, Y.Y. Lee, M. Holtzapple, and M. Ladisch. 2005. Features of promising technologies for pre-treatment of lignocellulosic biomass. *Bioresource Technology.* 96: 673-686.
15. Zhang, Y.H.P., M.E. Himmel, and J.R. Mielenz. 2006. Outlook for cellulase improvement: screening and selection strategies. *Biotechnology Advances*. 24: 452-481
16. Corn Production: USDA National States Data - Crops query; corn grain, 1986-2006, United States www.nass.usda.gov/QuickStats/Create_FedeBiomass Program: US Department of Energy http://www1.eere.energy.gov/biomass/about.html
17. U.S. Billion Ton Update: Biomass Supply For A Bioenergy and Bioproducts Industry. http://www1.eere.energy.gov/biomass/pdfs/billion_ton_update.pdf
18. International Energy Outlook 2011. Figure 15. http://www.eia.gov/forecasts/ieo/pdf/0484(2011).pdf

(613d) Catalytic Conversion of Glucose Into Fuels and Chemicals Via 5-Hydroxymethylfurfural

Lu Lin, School of Energy Research, Xiamen University, Xiamen, China

With the depletion of fossil resources and growing concerns about global warming and environmental pollution, the efficient conversion of renewable biomass into fuels and chemicals has recently attracted considerable attention. Among the many possible biomass-derived chemicals, 5-hydroxymethylfurfural (HMF) has been recognized as a versatile intermediate that can be used for the production of a wide variety of fine chemicals, polymeric materials and biofuels. In the past few years, many researchers have selected fructose as feedstock and excellent HMF yields have been readily achieved using many methods. However, it should be pointed out that fructose is not abundant in nature and its cost is very high, which limit large-scale and sustainable production of HMF. In contrast, glucose is more abundant, available and cheaper than fructose, which is considered as the preferred feedstock for the production of HMF. Disappointingly, many catalytic systems in which fructose is readily converted into HMF are normally ineffective for glucose, which is attributed to stable pyranoside ring structure of glucose. Therefore, there is a strong incentive to develop an efficient process for the production of HMF using glucose as starting material. In this review, the state-of-the-art of catalytic conversion approaches is systematically summarized and discussed. Moreover, some emphasis is also put on the synthesis of HMF-based derivatives including levulinic acid (LA), 2,5-dimethylfuran (DMF) and 2,5-furandicarboxylic acid (FDCA). Furthermore, we also point out some promising research trends in further studies.

(613e) Comparison of Oleaginous Fungal Lipid Accumulation On Different Lignocellulosic Feedstocks

Zhenhua Ruan[1], Mike Zanotti[1], Yuan Zhong[2], Chad Ducey[3], Wei Liao[1] and Yan Liu[1], (1)Biosystems and Agricultural Engineering, Michigan State University, East Lansing, MI, (2)Biosystems and Agricultural Engineering, Michigan State Univeristy, East Lansing, MI, (3)Werks Management, LLC, Fishers, IN

The perennial grasses miscanthus, giant reed, and switchgrass, along with the agricultural residue corn stover, were evaluated for their bioconversion potential for fungal lipid accumulation. A co-hydrolysis process was adopted to convert lignocellulose into fermentable sugars. The process utilized dilute sulfuric acid pretreatment without detoxification and liquid-solid separation, directly followed by enzymatic hydrolysis. A factorial experiment in a randomized block design was employed to optimize the co-hydrolysis process. The influence of combined severity on the susceptibility of pretreated substrates by enzymatic hydrolysis was clearly discernible, showing that co-hydrolysis can be applied to lignocellulosic biomass to enhance the efficiency of sugar conversion, and reduce the amount of processing water.. The experimental results demonstrated that under the optimal conditions, corn stover had the highest sugar yield followed by switchgrass, giant reed and miscanthus. Then, an oleaginous fungus *Mortierella isabellina* ATCC42613 was selected and applied on hydrolysates to accumulate fungal lipids due to its capability of utilizing both C5 and C6 sugars. The cultivation using hydrolysates as the carbon source exhibited greater cell mass, and equivalent lipid production compared to synthetic medium with pure glucose and xylose. These results elucidated that combining fungal fermentation and co-hydrolysis to accumulate lipids could have potential to enhance utilization efficiency of lignocellulosic feedstocks for advanced fuels production.

(613f) Investigation of Accessory Hemicellulases and Pectinases for Polysaccharide Hydrolysis of Ionic Liquid Pretreated Biomass

Christopher J. Barr[1], Jeffery Mertens[2] and Constance Schall[1], (1)Chemical and Environmental Engineering, University of Toledo, Toledo, OH, (2)NCAUR, USDA-ARS, Peoria, IL

The polysaccharides, cellulose, hemicellulose, and other additional carbohydrate polymers of terrestrial biomass, comprise renewable feedstocks for carbon-based chemicals and fuels. Biomass pretreatment is required to overcome its recalcitrance to biochemical deconstruction to monomeric sugars for sugar platform fuels and chemicals. Ionic liquid (IL) pretreatment followed by enzymatic hydrolysis can be used to saccharify the biomass into its monomeric sugars at a rate much higher than many pretreatment methods. IL pretreatment creates unique biomass substrates consisting of primarily amorphous cellulose with largely intact hemicellulose and lignin. These substrates differ from many other pretreatments which remove hemicellulose and retain cellulose crystallinity. The unique substrate resulting from IL-pretreatment poses an interesting challenge for enzymatic deconstruction using traditional commercial mixtures.

Enzyme mixtures using *A. nidulans* mono-component enzymes with accessory hemicellulases and pectinases were formulated and used in hydrolysis experiments. Exploration of the accessory enzyme requirements yields insight into differences in hemicellulose structure and composition between biomass substrates. Accessory hemicellulases examined include acetyl xylan esterase (AXE), ferulic acid esterase, and arabinofuranosidase. In addition to mono-component enzyme mixtures, commercially-available enzyme mixtures were supplemented with accessory enzymes to optimize hydrolysis for lower enzyme loadings. As part of these studies, analysis of variance (ANOVA) was used to statistically assess and model the main and interaction effects of key enzyme activities.

(613g) Kinetics of Adsorption, Desorption, and Re-Adsorption of a Commercial Endoglucanase in Lignocellulosic Suspensions

Qianqian Wang, South China University of Technology, Guangzhou, China, J.Y. Zhu, USDA Forest Service, Forest Products Laboratory, Madison, WI and Chris Hunt, US Forest Service, Madison, WI

This study conducted quantitative kinetic modeling and in-situ and temporally resolved measurements of adsorption, desorption, and re-adsorption of a commercial endoglucanase in lignocellulosic suspensions. The study defined a cellulase adsorption and desorption competition parameter, a pseudo rate of binding and desorption, binding and desorption capacity, as well as cellulase binding reversibility (a thermodynamic property) and recyclability (a engineering parameter). The results indicate that both substrate chemical and physical structures play important roles in cellulase binding and desorption. Binding of a commercial cellulase onto a cellulosic substrate was reversible. Bindings to two different lignocellulosic substrates were almost irreversible. While lignin and its structure positively affect binding capacity to substrate, they negatively affect cellulase recyclability. Collapsing of substrate pores reduced cellulose accessibility and cellulase binding capacity and increased reversibility and recyclability. Increasing temperature and pH increase cellulase desorption and increased binding reversibility and capacity. This study lays the foundation for developing effective cellulase recycling strategies.

(615a) The Kinetics of Fibrin Clot Formation On Surfaces Immobilized with Cell and Fibronectin Binding Domains

Anand Ramanathan and Nancy Wangechi Karuri, Chemical and Biological Engineering, Illinois Institute of Technology, Chicago, IL

Fibronectin (FN) is essential for the formation of a fibrin clot, an extracellular matrix that arrests bleeding. In the clot, FN binds fibrin and provides a scaffold for wound healing to occur. Our laboratory had previously shown that the ratio between the amount of a FN and cell binding domains, III_{1-2} and III_{9-10} respectively, immobilized on a surface influences the amount of FN in the ECM of cells. We investigated the effect of surfaces with different amounts of III_{1-2} and III_{9-10} on the kinetics of fibrin clot formation in a cell free system. Affinity purified glutathione tagged III_{1-2} and III_{9-10} were coated on polystyrene dishes. ELISA assays were used to identify the limits of protein loading on the surface and to quantify the levels of the individual proteins in mixtures. Fibrinogen was added to surfaces immobilized with III_{1-2} and III_{9-10}, fibrin clot formation was proteolytically activated by the addition of thrombin and followed through dynamic light scattering. The kinetics of clot formation were studied in the presence and absence of full-length FN and with or without the covalent cross-linker, Factor XIIIA. We found that the rate of fibrin clot assembly was dependent on the ratio of III_{1-2} to III_{9-10} and higher rates were obtained when the ratio of III_{1-2} to III_{9-10} was high. Moreover, the effect of the III_{1-2} and III_{9-10} coated surfaces on the rate of clot formation was significantly enhanced by FN and factor XIIIA. These results in a cell free system coupled together with our previous findings provide new avenues for the design of materials that can be used to accelerate wound healing.

(615b) Injectable Thermo-Sensitive Hydrogel As an Adjuvant: In Vivo Modulation of Dendritic Cells for Cancer Vaccine

Kye-Il Joo, Liang Xiao, Yarong Liu and Pin Wang, Chemical Engineering and Materials Science, University of Southern California, Los Angeles, CA

Development of an effective vaccine for cancer involves eliciting a potent antitumor immune response by introduction of tumor antigens to trigger systemic immunity against the tumor. Dendritic Cells (DCs) are the most powerful antigen-presenting cells capable of initiating adaptive immune response by stimulating both T cells and B cells and therefore become one of the major target cells for the cancer vaccine development. Immunization by adoptive transfer of autologous DCs that are loaded with antigens *in vitro* is, however, labor-intensive and requires significant regulatory concerns. In addition, most transplanted DCs remain non-functional, and few home to the lymph nodes where the subsequent activation of T-cell immunity is induced. More cost-effective and efficient protocols involve delivering antigen to host DCs *in vivo* by direct injection of vaccine vectors to program DCs in situ. However, *in vivo* DC targeting approach is limited by low specificity and poor efficacy of antigen uptake to DCs. Thus, the ability to efficiently deliver antigens to host DCs *in vivo* is a crucial step towards generating a potent antitumor immunity.

In this study, we hypothesized that creating *in vivo* microenvironment to recruit and proliferate DCs, which mimics natural surroundings of infection in the body, may potentially provide a practical mean to enhance the uptake of antigens to host DCs and precisely control the timing of DC-trafficking and activation *in situ*. Therefore, we investigated the development of the functionalized biomaterials incorporating inflammatory cytokines as infection-mimicking microenvironment for recruiting and housing host DCs *in vivo* prior to introduce tumor antigens at the DC-enriched site. This approach involves the design and use of thermo-sensitive mPEG-PLGA hydrogel capable of solution-gel transition in the response to temperature, which facilitates drug loading and exhibits sustained release of the drug. The implantation of the hydrogel loaded with granulocyte-macrophage colony-stimulating factor (GM-CSF), which has been identified as a potent stimulator of DC recruitment and proliferation, could provide the subcutaneous microenvironment where host DCs could be recruited and proliferated by the sustained and localized release of GM-CSF at the hydrogel-injected site. Analysis of in vivo DC recruitment showed that the number of recruited DCs was peaked at around 7 to 10 days post-injection of GM-CSF hydrogel. The migration assay indicated that the recruited DCs were able to migrate to lymph node where T-cell response was triggered. Furthermore, our lymphocyte analysis suggested that the significantly improved antigen-specific immune responses were observed when immunogens (i.e. antigens) were administrated at the site of GM-CSF-hydrogel implanted at 7 days post-injection, compared to empty hydrogel or without hydrogel. In this presentation, this hydrogel adjuvant system with many different immunogens including lentiviral vector and liposome/protein will be discussed in more detail.

(615c) Control of Adult Stem Cells with Polyurethane Gel Matrix

Rajesh Krishnan[1], Sandeep Nalluri[2] and Debanjan Sarkar[1,2], (1)Biomedical Engineering, University at Buffalo, The State University of New York, Buffalo, NY, (2)Chemical & Biological Engineering, University at Buffalo, The State University of New York, Buffalo, NY

Control of stem cell fate in tissue engineering application is important for effective and functional tissue regeneration. Extracellular matrix provides structural and functional cues in a highly synchronized fashion to stem cells for controlling the cellular processes and cell fate. Synthetic matrices are designed to induce similar interactions between stem cell and microenvironment which can effectively guide stem cells with spatial and temporal control. Biocompatible and biodegradable polyurethanes with segmental composition can be designed as gel with nanostructured phases due to the presence of 'soft' and 'hard' segments. By tuning the structure and composition of polyurethanes, the inter- and intra-segmental interactions can be modulated to control the polyurethanes into gel like structure. We have designed polyurethanes which forms gel through physicochemical interactions. Specifically, degradable peptide-based polyurethanes were synthesized and their interactions with adult stem cells were examined in 2-D substrate and 3-D environment. We examined the effect of polyurethane gel structure by varying the soft and hard segment. Soft segment of the polyurethanes were altered by using polyethylene glycol of different molecular weight and hard segment was designed from aliphatic diisocyanate with dipeptide chain extender. Our results shows that the composition of polyurethane chains with segments induce specific gel structure. Thus, by controlling the segmental architecture of polyurethanes, morphology of the gels were controlled to provide signals to stem cells. Interaction of bone marrow derived mesenchymal stem cells (MSCs) with the polyurethane gels shows specific dependence of the polyurethane structure and gel morphology. MSCs interacted with the different segments of polyurethane gels to present distinct cellular structure and behavior. Specifically, cell adhesion, aggregation, cellular structure and morphology were influenced by the polyurethane gel matrix. Interaction of MSCs with polyurethane reveals that segmental composition and gel morphology of polyurethanes influences the cell function and behavior which can control the fate of MSCs, particularly the differentiation of the stem cells. Biosynthetic activities of MSCs within the polyurethane gels were modulated to produce extracellular material leading to formation of tissue like structure. This study indicates that molecular engineering of polyurethane gel provide exquisite control of stem cell behavior. Thus, controlling cell-material interactions with these matrices can modulate tissue regeneration.

(615d) Correlating Effects of Gel Microstructural Features with Specific Differentiation Patterning of Mouse Embryonic Stem Cells

Keith Task[1], Antonio D'Amore[2], Satish Singh[1], Maria Jaramillo[3], Prashant Kumta[2] and Ipsita Banerjee[4], (1)Chemical Engineering, University of Pittsburgh, Pittsburgh, PA, (2)Bioengineering, University of Pittsburgh, Pittsburgh, PA, (3)Depatment of Bioengineering, University of Pittsburgh, Pittsburgh, PA, (4)Chemical and Petroleum Engg, University of Pittsburgh, Pittsburgh, PA

Embryonic stem cells (ESC) have the potential to be used in many applications due to their ability to differentiate into cells of any of the three germ layers (ectoderm, endoderm, and mesoderm). There are several different routes by which ESC can be guided to differentiate into specific germ layers, including genetic manipulation, chemical cues, and mechanical stimuli. With respect to the latter, it has been observed that substrate stiffness plays a significant role in governing specific phenotype differentiation.

In our previous study we investigated the effect of mechanical stiffness of fibrin on ESC differentiation. Our results indicated that while ectoderm and mesoderm germ layers displayed weak response to the change in fibrin substrate stiffness in the chosen range (2 – 250KPa), endoderm markers were strongly responsive, with softer substrates up-regulating endoderm specific markers. Although cells respond differently to substrates of varying stiffness, it was also observed that gels with the same stiffness but fabricated under different conditions lead to different differentiation patterning. Therefore, this macroscopic property is perhaps not what the cell truly experiences; modification of fabrication conditions changes the gel microstructure, which in turn affects the stiffness, and these micro-characteristics directly interact with the cells. However, it is not clear what specific microstructural features are the most influential in inducing cellular differentiation, and correspondingly, how they affect ESC behavior.

In the current work, we investigate the effect of microstructural features of fibrin gels on the differentiation of mouse ESC. Twelve different fibrin gels were fabricated by varying the fibrinogen concentration and fibrinogen to thrombin crosslinking ratio. The fibrin gels were used to induce ESC differentiation employing both 2D and 3D cultures. After the differentiation protocol the ESC were analyzed for phenotypic commitment by performing qRT-PCR for the specific germ layer markers. Each of the 12 different fibrin gels was imaged with scanning electron microscopy. Microstructural features of each of these gels were quantified using an image analysis tool for the characterization of fibrous scaffolds [1]. Specific features which were characterized include fiber diameter, node density, fiber length, and pore size. A comparison of these attributes along with principal component analysis led to a subset of features which vary most across gel conditions. Gel microstructural features were correlated with the ESC differentiation patterning using regression analysis. The gels are heterogeneous in nature, and therefore a bootstrapping technique was used to account for the variability of the system. Interestingly, only a subset of genes shows a significant relationship with microstructural features, and that only a few of these features significantly affect differentiation, including fibrin pore size distribution and node density. These results are also compared to the correlation of gel stiffness, determined by both by rheological measurements and atomic force microscopy, to phenotype commitment. This analysis reveals the sensitivity of cellular phenotype commitment and differentiation patterning to each of the microstructural features. Moreover, such information can be used to help guide the design of scaffolds with specific properties for tissue engineering applications. Results of these studies will be presented and discussed.

[1] A. D'Amore *et al.* Biomaterials, 2010, Vol. 31, 5345-5354.

(615e) A Novel Application of Nanofiber Scaffolds for Gene Therapy

Mandula Borjigin[1], Bryan Strouse[1], Pawel Bialk[1], Chris Eskridge[1], Rohina Niamat[1], Jingwei Xie[2] and Eric Kmiec[1], (1)Chemistry, Delaware State University, Dover, DE, (2)Marshall Insititute for Interdisciplinary Research, Marshall University, Huntington, WV

Electrospun nanofiber scaffolds have been used in cell growth and tissue engineering. Here we expand the application of nanofiber scaffolds to the genetic technique known as gene editing. Gene editing is a process by which single base mutations can be corrected, in the context of the chromosome, using a short piece of single stranded DNA in the form of an oligonucleotides (ssODN). The ssODNs are introduced into the cell to correct the mutant nucleotide using the inherent pathways of DNA replication and repair. This technique is being developed rapidly for use in the treatment of inherited diseases such as Sickle Cell Disease (SCD) and Phenylketonuria (PKU). The survival and proliferation of the cells bearing modified genes however appear to be impeded by a phenomenon known as reduced proliferation phenotype (RPP); this is the key barrier to practical implementation of gene editing in the clinical setting. To overcome RPP, we have utilized nanofiber scaffolds as templates on which modified cells were allowed to recover, grow and expand after gene editing. Here, we present evidence that HCT116-19, bearing an integrated, mutated eGFP gene and corrected by gene editing, proliferate on polylysine or fibronectin coated polycaprolactone (PCL) nanofiber scaffolds. In contrast, cells from the same reaction protocol plated on both regular plain surfaces and polylysine (or fibronectin) coated plate surfaces do not proliferate. Therefore, growing genetically modified (edited) cells on electrospun nanofiber scaffolds promotes the reversal of the Reduced Proliferation Phenotype and increases the potential of gene editing as an *ex vivo* gene therapy application.

(615f) Topography Mediated Regulation of HER-2 Expression in Breast Cancer Cells

Amita Daverey, Austin Mytty and **Srivatsan Kidambi**, Dept of Chemical & Biomolecular Engineering, University of Nebraska-Lincoln, LINCOLN, NE

The breast tumor microenvironment is made up of heterogenous mixture of pores, ridges and collagen fibers with well defined topographical features. Although, significant progress has been achieved toward elucidating the biochemical and molecular mechanisms that underlie breast cancer progression, quantitative characterization of the associated mechanical/topographical properties and their role in breast tumor progression remains largely incomplete. Therefore, the aim of the study is to investigate the effect of topography on the adhesion and biology of breast cancer cells in *in* vitro cultures. Breast cancer cells were cultured on different topographies designed on polyelectrolyte multilayer (PEM) coated polydimethylsiloxane (PDMS) surfaces. We demonstrate that microtopography affects the cell adhesion and distribution depending on the topography on the PDMS surfaces. Further, we report that surface topography down-regulates the HER-2 gene and protein levels in breast cancer cells when compared to the tissue culture polystyrene surfaces (TCPS) control. Results from this study indicate that topography of different patterns modulates morphology of cells, their distribution and expression of HER-2 in breast cancer cells. The observations made by this study provide rational basis for understanding the native topography of breast in order to design bio-mimetic tissue culture substrate for better understanding of the disease.

(615g) The Effect of Length and Terminal Group of Poly(ethylene glycol)-Terminated Self-Assembled Monolayers (SAMs) On Dendritic Cell (DC) Maturation and Function

Christina Yacoob[1], Jung Park[2], Bingbing Sun[1] and Hong Shen[1], (1)Department of Chemical Engineering, University of Washington, Seattle, WA, (2)Department of Neuroscience and Behavioral Biology, Emory University, Atlanta, GA

Dendritic cells (DCs) are the most potent antigen presenting cells (APCs). They are the primary target for inducing effective immune responses or tolerance. *Ex-vivo* induced DCs have been used for immunotherapy of cancers. DCs are present around implanted devices though their functions remain to be elucidated. Oligomers or polymers of ethylene glycol (EG) have been grafted onto surfaces of various biomaterials for controlling protein adsorptions and increasing the biocompatibility of devices. In this study, immature DCs were exposed to poly(ethylene glycol) (PEG)-terminated self-assembled monolayers (SAMs). OH-terminated PEG SAMs activated immature DCs and upregulated surface molecules characteristic of mature DCs at a higher degree than COOH-terminated ones. Longer OH-terminated PEG SAM encouraged the maturation of DCs at a higher magnitude than shorter ones. Interestingly, none of SAMs induced cytokines characteristic of either inflammatory or tolerogenic DCs. Additionally, we demonstrated that OH-terminated PEG SAM activated DCs mainly through DAP-12 or Fc-gamma – associated surface receptors-dependent pathway. Our results give insights into biomaterial design for inducing activation or tolerance of DC-based immunity. Since DAP-12 or Fc-gamma – associated surface receptors-dependent signaling pathway is involved in many cell types, including macrophages, our results also suggest that PEG-terminated SAMs can potentially activate other cell types in a length and functional end-dependent manner.

(615h) An Interactive Engineered Protein Hydrogel: Controlling and Responding to Neurite Growth

Kyle J. Lampe and Sarah C. Heilshorn, Materials Science and Engineering, Stanford University, Stanford, CA

Neural regeneration within the central nervous system (CNS) is a critical unmet challenge as brain and CNS disorders continue to be the leading cause of disability nationwide. Common tissue engineering goals seek to customize cell-biomaterial interactions and guide cell behavior. Here we have developed a material that is both cell instructive and cell responsive, creating a dynamic interplay between neural cells and their engineered extracellular matrix (ECM). Using recombinant protein technology, we engineered a family of elastin-like protein hydrogels with multiple independently tunable properties. With these materials we can investigate individual and synergistic effects of elastic modulus, degradation rate, and cell adhesivity on cell behavior in a tunable microenvironment. Unlike natural matrices, the concentration of adhesive ligands such as the fibronectin-derived sequence RGD can be precisely controlled without altering the mechanical properties of the hydrogel. When dorsal root ganglion (DRG) neurons were cultured in 3D in these hydrogels, a cell-adhesive RGD density of 1.9×10^7 ligands/μm^3 promoted neuron-specific growth and more than doubled the rate of neurite extension compared to hydrogels without the adhesive sequence. Crosslinking density was tuned to create scaffolds with elastic moduli ranging from 0.5-2.1 kPa with constant adhesive ligand densities. The most compliant gels led to the greatest outgrowth from encapsulated DRGs with neurites extending over 1800 μm by day 7. In contrast, the stiffest gels permitted far fewer extensions and limited outgrowth to a maximum of 600 μm over the same time frame.

To render the materials cell-responsive, we designed the engineered proteins to contain a second set of bioactive sequences that specifically respond to changes in cell phenotype. Neural stem cells (NSCs) undergoing differentiation may change their production of the protease urokinase plasminogen activator (uPA), which has previously been found at the growth cones of extending neurites. By incorporating cell-mediated degradable subunits into the elastin-like proteins, we are able to mimic the natural remodeling of the ECM. We engineered multiple uPA target sites with different degradation kinetics into the elastin-like protein to allow neural cell-mediated control of the scaffold degradation dynamics. DRGs encapsulated in uPA degradable scaffolds extended neurites at a faster rate than in hydrogels of similar RGD density and initial stiffness without the uPA degradable sequence. This cell-responsive strategy was also used to enhance the functionality of the material by controlling delivery of multiple molecules with distinct release kinetics.

These crosslinked scaffolds are useful for directing the growth and differentiation of multiple cell types including clinically relevant NSCs. The tunable scaffolds are responsive to neuronal cells, which may be able to specifically self-modulate the release of multiple bioactive factors while undergoing differentiation. This work demonstrates the versatility and responsiveness of our modularly-designed protein hydrogels for neural cell culture and encourages continued development as a biomaterial tissue construct for treating spinal cord injury.

(621a) Development of Three-Dimensional Lung Multicellular Spheroids in Air and Liquid Interface Culture for the Evaluation of Anti-Cancer Therapeutics

Samantha A. Meenach[1,2], Alexandra N. Tsoras[1], Ronald C. McGarry[3], Heidi M. Mansour[2], J. Zach Hilt[1] and Kimberly W. Anderson[1], (1)Chemical and Materials Engineering, University of Kentucky, Lexington, KY, (2)Pharmaceutical Sciences - Drug Development Division, University of Kentucky, Lexington, KY, (3)Radiation Medicine, University of Kentucky, Lexington, KY

Applying *in vitro* outcomes to *in vivo* applications has limitations because conventional two-dimensional (2D) cell culture does not recreate a physiologically representative model for cells. This work investigated a three-dimensional (3D) cell culture technique to model lung tumors *in vitro*. A 3D lung cancer model was created by applying collagen to a cell culture Transwell, which allows for nutrient transfer through the collagen. Two lung cancer cells lines (H358, a bronchioalveolar carcinoma and A549, a lung adenocarcinoma) were seeded on the collagen. The non-adhesive collagen caused the cells to be more inclined to attach to one another rather than its surface allowing for the formation of multicellular spheroids (MCS). To better mimic the environment for lung cancer specifically, an air-interface culture (AIC) condition was created. For AIC conditions, the cell media on the apical side of the Transwell was removed, and the basolateral side with media provided effective nutrient transport to the MCS while still exposing the cells to air. 2D versus 3D cell behavior in response to paclitaxel and aerosol particles containing paclitaxel was evaluated using the viability of the cells.

Through the use of brightfield and fluorescence microscopy imaging, the AIC model yielded viable MCS at sizes similar to MCS formed in LCC conditions (100 to 150 μm in diameter). With the optimized 3D model, LCC cells were exposed to paclitaxel in media, and paclitaxel-loaded dry powder aerosol particles were delivered to AIC cells through direct application with an insufflator. This alternative delivery method using dry powder particles was used since paclitaxel could not be delivered through media as in LCC conditions. The particles containing paclitaxel were comprised of a PEGylated phospholipid excipient mixture which encapsulated the drug. Using viability analysis, it was shown that both applications of paclitaxel showed variance in efficacy when comparing 2D and 3D culture conditions (where the IC_{50} values for paclitaxel were higher for 3D compared to 2D). Transepithelial electrical resistance of A549 cell monolayers was evaluated before and after particle delivery to illustrate that the particle application does not affect the permeability of the cells, which infers that this form of drug therapy will not affect the permeability of lung tissue. Overall, a much more representative *in vitro* model has been developed that is expected to be an improved predictor of efficacy of alternative drug delivery methods such as direct pulmonary delivery for lung cancer, which could lead to more efficient drug therapies for lung cancer patients.

(621b) Treatment Tests of Breast Cancer in Mice Using a Vascular-Targeted Enzyme Prodrug Therapy

Brent D. Van Rite, Chemical, Biological and Materials Engineering, University of Oklahoma , Norman , OK, Vassilios I. Sikavitsas, Chemical, Biological, and Materials Engineering, University of Oklahoma, Norman, OK, Mohamad Cherry, Hematology-Oncology , University of Oklahoma Health Sciences Center , Oklahoma City, OK, Carla Kurkjian, Hematology-Oncology, University of Oklahoma Health Sciences Center, Oklahoma City, OK and **Roger G. Harrison**, Chemical, Biological and Materials Engineering, University of Oklahoma, Norman, OK

Metastatic breast cancer is usually not curable. The median survival of women with metastatic breast cancer is approximately two years after documentation of metastasis. Treatment for most patients with metastatic breast cancer is considered palliative in nature, although durable complete remissions have been achieved in a small number of patients (< 5%). Some disease response has been demonstrated with various chemotherapeutic agents; however, these agents are almost always associated with undesirable side effects such as nausea, vomiting, hair loss, cognitive dysfunction, and fatigue. Recent research has centered on therapies that are directed only towards malignant cells.

This work is on a novel enzyme prodrug therapy using a fusion protein containing annexin V to target only tumor vascular endothelial cells and cancer cells, reducing unwanted toxicity to healthy tissue. The L-methioninase enzyme will convert the inert prodrug selenomethionine to toxic methylselenol and also deplete the cancer cells of methionine necessary for protein synthesis and continued growth. The L-methioninase enzyme is linked to human annexin V protein in a fusion protein. This therapy takes advantage of the fact that phosphatidylserine (PS), an anionic phospholipid, is exposed almost exclusively on the outer membrane of cancer and endothelial cells of the tumor vasculature and that annexin V binds specifically to PS.

Recombinant technology was used to express and purify the methioninase-annexin V fusion protein. During purification, a $(His)_6$ tag was cleaved from the fusion protein, and endotoxin was removed. Pharmacokinetic testing in mice revealed complete clearance of the fusion protein from the bloodstream to occur within 12 hours following intraperitoneal (i.p.) injection. The enzyme prodrug system was tested in SCID (severe combined immune deficient) female mice with implanted with MDA-MB-231 human breast tumors and was found to significantly inhibit tumor growth compared to three control groups of mice (untreated, fusion protein only, and selenomethionine only). The fusion protein and the prodrug were delivered by i.p. injection. The MDA-MB-231 cancer cells had been transfected with green fluorescent protein (GFP) to follow the growth of the tumors using a small animal fluorescence imaging system. Results will also be reported of a separate study of this enzyme prodrug therapy using mice fed a diet deficient in methionine but supplemented with homocysteine.

(621c) Quorum-Sensing Salmonella Selectively Trigger Protein Expression Upon Colonization within Tumors

Charles A. Swofford, Adam T. St. Jean and Neil S. Forbes, Chemical Engineering, University of Massachusetts, Amherst, MA

Engineered *salmonella* possess unique capabilities that make them ideal drug delivery vectors for tumors. Current cancer chemotherapeutic regimes have limited efficacy due to therapeutic resistance, which decreases the success rate for treatment of late-stage cancer and metastases. This therapeutic resistance can be explained by a number of factors including 1) intra-tumoral transport limitations, 2) reduced susceptibility of quiescent cancer cells, and 3) protein pumps that expel therapeutics from cells. Bacteria are capable of overcoming these therapeutic limitations because they preferentially accumulate in tumors, actively penetrate tumor tissue, and can be engineered to produce anti-cancer drugs *in situ*. Due to systemic toxicity, however, bacteria that constitutively express an anti-cancer drug would not be an ideal alternative to standard chemotherapy. Therefore, we have engineered *salmonella* to secrete anti-cancer drugs after tumoral colonization has occurred through the use of the *lux* quorum sensing system. Quorum sensing is the ability for bacteria to change their gene expression based on changes in population density. Because *salmonella* preferentially accumulate in tumors, a quorum sensing genetic circuit, coupled with an anti-cancer agent, creates a robust expression switch that only turns on in bacterial tumor colonies. This unique bacterial drug delivery vector will allow for targeted delivery of a powerful anti-cancer drug to therapeutically resistant tumors and metastases while minimizing the risk of systemic toxicity. *Salmonella* transformed with the quorum-sensing system induce expression upon reaching a threshold bacterial density of 10^8 cfu/mL (OD = 0.5). Furthermore, in a tumor-on-a-chip device, quorum-sensing bacteria induce expression once colonies have formed within the tissue. These colonies have formed deep within the therapeutically resistant regions of tumors where most standard chemotherapeutic drugs cannot reach due to diffusion limitations. Additionally, 4T1 murine tumor models injected with the quorum-sensing system show co-localization of GFP and *salmonella* within tumor tissue while healthy tissue shows low-counts of *salmonella* without any GFP expression. Finally, quorum-sensing *salmonella* secrete therapeutic proteins that reduce tumor burden and enhance mouse survival. The culmination of this data demonstrates that *salmonella* integrated with a quorum-sensing expression vector can sense colonization within tumors and induce protein expression once a critical density has been reached. Bacteria transformed with a quorum-sensing system will also form colonies deeper into the therapeutically resistant regions of tumors allowing for treatment of tumor tissue previously unreachable by standard chemotherapeutic regiments. Overall, this system creates a robust on/off switch that will allow for targeted bacterial drug delivery to tumors and metastases while minimizing the risk of potentially harmful drugs reaching healthy tissue.

(621d) Inferring Tumor-Immune Interaction Networks Via Unbiased Secretome Profiling

Yogesh Kulkarni[1], **Kisheon Alexander**[1], Yueting Wu[1] and David J. Klinke[1,2,3], (1)Chemical Engineering, West Virginia University, Morgantown, WV, (2)Microbiology, Immunology and Cell Biology, West Virginia University, Morgantown, WV, (3)Mary Babb Randolph Cancer Center, School of Medicine, West Virginia University, Morgantown, WV

Despite the variability of the disease, cancer cells exhibit a set of common traits. Tumor-escape from the cytotoxic action of the immune system is an emerging hallmark of cancer but the interaction between the tumor cell and the host is complex. We employ experimental procedures in limited length and time scales to elucidate the mechanisms by which tumors escape immunosurveillance, and predict the effect on physiology. On understanding the cross-talk between cancer and immune cells, more comprehensive targeted therapies may be developed. This study aims to compare the biochemical signals secreted in vitro by malignant cells to normal human cells. Two human mammary epithelial cancer cell lines, BT474 (HER2+/ER+) and SKBR3 (HER2+/ER-) and one normal human mammary epithelial cell line, 184A1 were investigated. A proteomics workflow that includes two-dimensional gel electrophoreses, MALDI-TOF mass spectrometry and peptide mass finger-printing were used to analyze the secretome. Protein interaction networks and canonical pathways were extracted from the Ingenuity Knowledge Base to elucidate the role of tumor-derived secreted protein on immune cell function. The presence of tumor-derived exosomes was confirmed by SEM, and initial results suggest them as extra-cellular protein conveyers. Future studies will concentrate on inhibiting tumor cells from evading elimination. Proteomics provides a less biased approach to characterize the biochemical signals that malignant cells use to skew immunity and to better understand cell-to-cell communication in breast cancer.

(621e) Design of A Biomimetic Nanostructured Tissue Engineered Bone Model for in Vitro Breast Cancer Metastasis Study

Mian Wang, Department of Mechanical & Aerospace Engineering, George Washington University, Washington, DC, Sidney Fu, George Washington University, Washington, DC and Lijie Grace Zhang, Department of Mechanical and Aerospace Engineering and Institute for Biomedical Engineering, The George Washington University, Washington, DC

Breast cancer is the leading cause of cancer-related death in American women, accounting for more than 200,000 new cases and 40,000 deaths annually. Majority of advanced breast cancer patients will eventually develop bone metastasis. However, the cure and treatment of metastasis breast cancer is still very challenging because of insufficient knowledge about underlying metastasis mechanisms. Traditional tumor metastasis models (*in vivo* and *in vitro*) contain inherent limitations with regards to controllability, reproducibility, and flexibility. To solve the above problems, this research designs an easy-to-use and biomimetic 3D nanostructured bone model via natural chitosan and various hydroxyapatites (HA) for breast cancer metastasis study. 3D porous chitosan-HA scaffolds with different crystalline, size and concentration of HA (specifically, 10% and 20% nanocrystalline HA, microcrystalline HA and amorphous HA) were fabricated via a lyophilization procedure. Human MDA-MB231 metastasis breast cancer cells were seeded at 50000 cells/scaffold. After 24h cell culture, the cell adhesion and growth result reveals that concentration, size and crystallinity of HA have a significant impact on metastasis breast cancer cell behavior. 10% nanocrystalline HA can significantly increase breast cancer cell attachment when compared to all of other HA. Since natural bone matrix is made of rich nanocrystalline HA, our synthesized nanocrystalline HA in chitosan can create a more bone-like environment for breast cancer cell metastasis when compared to microcrystalline and amorphous HA. Furthermore, 20% nanocrystalline HA in chitosan shows the similar increasing trend for cancer cell attachment when compared to respective 20% microcrystalline HA in chitosan. In addition, the result reveals that lower concentration of HA (10%) can promote more breast cancer cell growth than higher concentration of HA (20%) in chitosan. Based on the result, 10 % nanocrystalline HA-chitosan scaffold can significantly improve more breast cancer cell growth than other size, concentration and crystallinity of HA chitosan scaffold, thus make it promising for further breast cancer cell metastasis study.

(621f) Mechanistic Investigation of the TRAIL-Sensitization Activity of Mitoxantrone

David J. Taylor, Chemical Engineering, Arizona State University, Tempe, AZ, Taraka Sai Pavan Grandhi, Biomedical Engineering, Arizona State University, Tempe, AZ, Arul Jayaraman, Department of Chemical Engineering, Texas A&M University, College Station, TX and Kaushal Rege, Chemical Engineering, Biomedical Engineering, Arizona State University, Tempe, AZ

In the United States, prostate cancer is the most commonly diagnosed cancer in men, and results in the second largest number of cancer-related fatalities (in men). Similarly, breast cancer is the most commonly diagnosed cancer in women, and results in the second largest number of fatalities (in women). Pancreatic cancer ranks as the fourth most fatal malignancy in both, men and women; the five-year survival rate is less than 5%. Recent studies from our laboratory resulted in the identification of an FDA-approved drug, mitoxantrone, for sensitizing prostate, pancreatic, and breast cancer cells to Tumor necrosis factor Related Apoptosis Inducing Ligand (TRAIL)-induced death. Here we report an investigation into the sub-cellular mechanisms by which mitoxantrone sensitizes resistant cancer cells to TRAIL. In the investigation, kinetic studies were performed to determine the period of time over which sensitization occurred. The concentrations of apoptosis specific proteins were also measured to identify changes following mitoxantrone treatments. Additionally, the kinetics of mitochondrial depolarization, generation of reactive oxygen species (ROS), and apoptosis induction were measured to determine the factors that contribute to the activity of the drug. Finally, nanoparticles were investigated as effective methods of delivering these potent combination treatments to cancer cells, with an eye toward enhancing efficacy and specificity.

(621g) Differential Effects of Chemokines and Growth Factors On Tumor Cell Migration Using a 3D Microfluidic in Vitro Model

Beum Jun Kim[1], Pimkhuan Hannanta-anan[1], Yoon Soo Kim[2], Michelle Chau[1], Melody A. Swartz[3] and Mingming Wu[1], (1)Biological and Environmental Engineering, Cornell University, Ithaca, NY, (2)Biomedical Engineering, Cornell University, Ithaca, NY, (3)Institute for Bioengineering, École Polytechnique Fédérale de Lausanne (EPFL), Lausanne, Switzerland

Tumor cell chemotaxis within the interstitial space is an important transport mechanism in cancer metastasis. Despite its clinical importance, the molecular mechanism which tumor cells use to migrate in chemokine or growth factor gradients is largely unknown. Using a recently developed 3D microfluidic gradient generating device, we studied tumor cell migration in well-defined and steady linear chemical gradients. We used chemokines, SDF-1α and CCL19, and a growth factor, EGF, which are three well-known cytokines present in tumor microenvironments, to generate chemical gradients; we also used a malignant breast tumor cell line, MDA-MB-231, embedded in type I collagen for 3D cell cultures. Our experimental results demonstrate (1) The major motility type in our 3D microfluidic culture is amoeboid motility: tumor cells also exhibit plasticity and heterogeneity in motility; (2) Both EGF and CCL19 promote tumor cell motility only: the speed increases synergistically when 0.25 nM EGF and 111 nM/mm SDF-1α are applied; (3) Breast tumor cells are chemotactic towards high concentrations of SDF-1α with enhanced motility: this chemotactic behavior follows a ligand-receptor binding kinetics (11~222 nM/mm); (4) The chemotactic sensitivity to a SDF-1α gradient of 111 nM/mm diminishes when either 0.25 or 8.33 nM EGF is supplemented. These results will eventually help us better understand the physical and chemical factors in tumor cell migration, in the context of microenvironmental interactions.

(621h) Analysis of Myc-Driven Metabolic Reprogramming in B-Cells by Isotopically Nonstationary 13C Flux Analysis

Taylor A. Murphy[1], Chi Dang[2] and **Jamey D. Young**[1], (1)Chemical and Biomolecular Engineering, Vanderbilt University, Nashville, TN, (2)Abramson Cancer Center

The transcription factor c-Myc is an oncoprotein whose expression is deregulated in over 30% of all cancers. A significant role of c-Myc (i.e., Myc) is its control of cell cycling and apoptosis, but it also exerts strong influence on central metabolism. Targeting metabolic phenotypes induced by Myc activation may offer a potential therapeutic strategy to selectively inhibit the proliferation of Myc-driven cancers. We have conducted ^{13}C metabolic flux analysis (^{13}C-MFA) studies on P493-6 B-cells to identify metabolic pathway alterations associated with Myc overexpression. These cells have a tetracycline-repressible genetic construct that enables three distinct c-Myc expression levels: High (oncogenic), Low (endogenous), and None. Using GC-MS profiling techniques on isotopically labeled protein and RNA samples, along with measurements of extracellular metabolite uptake and excretion fluxes, we have created comprehensive flux maps of the central carbon metabolism of P493-6 cells under both High and Low Myc conditions. These flux maps have revealed several key features of Myc metabolic regulation.

We observed that glycolysis is highly upregulated in both conditions, with lactate-to-glucose (L/G) ratios of 1.54 and 1.79 for High and Low Myc, respectively. However, 25% of the pyruvate synthesized via glycolysis in the High Myc condition entered the TCA cycle as compared to only 15% in the Low Myc cells. Most TCA cycle and amphibolic mitochondrial pathways exhibited 4-fold flux increases in High Myc cells, in contrast to modest increases in glucose uptake and lactate excretion. This was largely driven by increased catabolism of glutamine and other amino acids. The oxidative pentose phosphate pathway exhibited minimal activity under both High and Low Myc conditions. Oxygen uptake was significantly increased in High Myc cells as compared to Low Myc, confirming the increase in mitochondrial activity detected by MFA. We are currently conducting studies to quantify how hypoxia modulates these metabolic phenotypes as well as examining siRNA knockdowns of specific enzymes that are activated in High Myc cells.

(631a) Relationship Between Solution Metastable Limit and Solubility

Michael Tomasini, Biomedical Engineering, Rutgers University, Piscataway, NJ and M. Silvina Tomassone, Chemical and Biochemical Engineering, Rutgers University, Piscataway, NJ

In pharmaceutical crystallization, and especially in the creation of nanoparticles, the degree of supersaturation and the mechanism of producing supersaturation can have a significant impact on the resultant solid particle properties. For the creation of small particles it is beneficial to suppress the growth phase of crystallization in favor of the nucleation phase. One way to do this is to operate very close to the metastable limit. A difficulty that arises is that the metastable limit of a drug in a particular solvent is not an easy quantity to obtain. Therefore, we seek a method to determine the metastable limit such that we can operate in the metastable zone at a high level of supersaturation but not exceed the metastable limit. It has been empirically observed that the metastable limit curve (the metastable limit as a function of temperature) often mirrors the functionality of the solubility curve. If this were generally true, evaluation of the metastable limit curve would be a straightforward shift of the solubility curve. With the aim of exploring the functional relationship between the metastable limit curve and the solubility curve, we used gauge-cell Monte Carlo (MC) to simulate nucleation and determine the metastable limit and solubility of model drug compounds.

We performed simulations on four different model drug compounds with increasing solubility to determine the solubility curve and metastable limit curve for each compound. We determined that the mestastable limit curve and solubility curves show similar traits, both exhibiting a logarithmic functionality. We also computed the metastable zone width (MSZW) for each compound, the difference between the metastable limit curve and the solubility curve. When normalized to the drug solubility, the MSZW shows a decrease with increasing solubility indicating that as a more soluble drug will have more molecules in solution, it will therefore undergo more molecular collisions and nucleate at lower supersaturations. We also relate the computed collision frequency to the MSZW to allow for prediction of the metastable limit curve from the solubility curve. The results show similar trends to experimental results of drug nucleation.

(631b) Measuring the Energetic Heterogeneity of Pharmaceutical Powders and the Effects On Powder Processing and Formulation

Daniel J. Burnett, Surface Measurement Systems, Allentown, PA and Jiyi Khoo, Surface Measurement Systems, Alperton, United Kingdom

INTRODUCTION

Most pharmaceutical solids are energetically heterogeneous. This heterogeneity could be unintentional due to different crystal facets with different functional groups, co-crystal formation, impurities, defect sites, amorphous regions, or co-processed mixtures of materials. Additionally, surface energy heterogeneity or homogeneity could be engineered to produce desired powder properties. Therefore, it is often difficult or even misleading to characterize the surface energy of these materials with a single value. Recent advances in Inverse Gas Chromatography (IGC) theory and applications have developed new methods for determining surface energy distributions for powders. This relatively new approach now allows the direct measurement of dispersive surface energy at a specific surface coverage or determination of surface energy profiles over a range of surface coverages.

EXPERIMENTAL

The current study investigates the surface energetics of various pharmaceutically relevant materials using IGC technology. Finite concentration IGC experiments allow for the determination of surface energy distributions which more accurately describe the anisotropic surface energy for real materials. Various materials, including Indomethacin (with different processing routes), mannitol (with different surface treatments), lactose (with different amounts of fines), and aspirin (different milling conditions) were studied. The surface energy distributions were then correlated to powder performance properties including wet granulation, powder flow and dissolution behavior.

RESULTS

The affect of processing route on the surface properties of partially amorphous Indomethacin has been investigated. Quench-cooled, milled, and crystalline Indomethacin exhibited different surface energy profiles. In particular, amorphous Indomethacin was energetically most heterogeneous. Additionally, mannitol was studied to relate surface energy distributions to powder flow properties due to changes in surface chemistry. After surface treatment with a silanizing agent, mannitol showed a significantly more homogeneous and hydrophobic surface. These differences were reflected in their powder flow characteristics and wettability during granulation. Finally, the affect of milling was studied on aspirin powders. Milling significantly increased both the average surface energy and surface energy distribution, thus resulting in faster dissolution behavior.

CONCLUSIONS

Surface energy heterogeneity profiles were measured on a wide range of pharmaceutical materials including excipients and APIs. Processing route, surface treatment, and milling conditions showed a dramatic affect on the surface energy values for these materials. These changes in surface energy in turn showed a correlation with bulk power properties like powder flow, dissolution and wet granulation behavior. Therefore, measuring and controlling surface energy values can be an important characteristic for engineering pharmaceutical powders with desired performance properties.

(631c) Formation of Itraconazole-Succinic Acid Cocrystals by CO_2 Antisolvent Cocrystallization

Courtney A. Ober and Ram B. Gupta, Chemical Engineering, Auburn University, Auburn, AL

Cocrystals of itraconazole, an antifungal drug with poor bioavailability, and succinic acid, a water-soluble dicarboxylic acid, were formed by gas antisolvent (GAS) cocrystallization using pressurized CO_2 to improve itraconazole dissolution. In this study, itraconazole and succinic acid were simultaneously dissolved in a liquid solvent, tetrahydrofuran, at ambient conditions. The solution was then pressurized with CO_2, which decreased the solvating power of tetrahydrofuran and caused precipitation of itraconazole-succinic acid cocrystals. A fundamental understanding of cocrystallization in a CO_2-expanded liquid was obtained by modeling the equilibrium concentrations of drug and former in solution as a function of pressure using the Peng-Robinson equation of state and AspenTech software. The cocrystals prepared by GAS cocrystallization were compared to those produced using a traditional liquid antisolvent, n-heptane, for size and surface morphology, crystallinity, thermal behavior, chemical structure, clinical relevance, and stability. Scanning electron microscopy, powder X-ray diffraction, differential scanning calorimetry, and Fourier transform infrared spectroscopy analyses showed that itraconazole-succinic acid cocrystals with physical and chemical properties similar to cocrystals produced using a traditional liquid antisolvent technique can be prepared by CO_2 antisolvent cocrystallization. The dissolution profile of itraconazole was significantly enhanced through GAS cocrystallization with succinic acid, achieving over 90% dissolution in less than 2 hours. The cocrystals appeared stable against thermal stress for up to 4 weeks under accelerated stability conditions, showing only moderate decreases in their degree of crystallinity but no change in their crystalline structure. This study shows the utility of an itraconazole-succinic acid cocrystal for improving itraconazole bioavailability while also demonstrating the potential for CO_2 to replace traditional liquid antisolvents in cocrystal preparation, thus making cocrystal production more environmentally benign and scale-up more feasible.

(631d) Preparation of Co-Ground Drug-Superdisintegrant Nanosuspensions As a Precursor to the Production of Fast Dissolving Surfactant-Free Nanocomposite Microparticles

Mohammad Azad[1], Afolawemi Afolabi[1], Rajesh Davé[2] and Ecevit Bilgili[2], (1)Chemical, Biological, and Pharmaceutical Engineering, New Jersey Institute of Technology, Newark, NJ, (2)Chemical, Biological and Pharmaceutical Engineering, New Jersey Institute of Technology, Newark, NJ

About 40% of the newly discovered drug compounds in the pharmaceutical industry exhibit poor water solubility, which usually leads to poor drug bioavailability. The dissolution rate and bioavailability can be improved by producing nano-sized drug crystals as nanoparticle suspensions (nanosuspensions). In the manufacture of various solid dosage forms, nanosuspensions are dried to prepare nanocomposite microparticles, which are then incorporated into the dosage forms via standard pharmaceutical unit operations. During drying of the nanosuspensions, formation of drug nanoparticle aggregates, solid bridges and agglomerates, and fused (hard) aggregates leads to incomplete recovery of the nanoparticles and ensuing slower dissolution. As a class of dispersants, surfactants play a critical role in fast recovery and dissolution of drug particles from the nanocomposite microparticles (Bhakay et al., 2012). However, surfactants can cause several issues such as physical instability of the nanosuspensions via Ostwald ripening and toxicity especially in inhalation applications, etc. Therefore, surfactants should be eliminated or used sparingly to minimize all potential negative impact. Hence, production of surfactant-free nanosuspensions as a precursor material for nanocomposite microparticle production is of utmost interest to formulators.

Nano-sized superdisintegrants (SDIs) can be used in nanocomposite microparticles as a novel class of dispersants for fast recovery and dissolution of drug nanoparticles (Bilgili et al., 2012). SDIs are water-insoluble, swellable, cross-linked polymers that are traditionally used for fast disintegration of tablets. Unfortunately, the smallest size of SDIs available is in the several microns range. To the best knowledge of the authors, there is no information available in the literature about the nano-sized SDIs. In this study, two commonly used SDIs, sodium starch glycolate (SSG) and crospovidone (CP), were milled in a wet stirred media mill. Aqueous colloidal suspensions of these SDIs were produced in the presence or absence of stabilizers. To incorporate SDIs in the nanosuspensions of drug, SDIs were also co-ground with the drug. Fenofibrate (FNB) was used as a model poorly water-soluble drug. FNB was milled in the presence of an anionic surfactant (sodium dodecyl sulfate, SDS), and was also co-ground with SSG or CP in the absence of SDS. In both cases, a soluble biopolymer (hydroxypropyl cellulose, HPC) was used as a steric stabilizer. We have successfully demonstrated the feasibility of preparing stable SDI and FNB-SDI suspensions using HPC as the stabilizer. The particle size distribution of the FNB-SSG suspension was similar to that of the FNB-SDS suspension. This finding suggests that colloidal SDIs may replace the surfactants or minimize their use in nanocomposite microparticles. The potential implications of the findings are discussed in terms of the redispersion and dissolution of the drug nanoparticles.

References

1. Bhakay, A., Dave, R., Bilgili, E. 2012. Recovery of BCS Class II Drugs during Aqueous Redispersion of Core-Shell Type Nanocomposite Particles Produced via Fluidized Bed Coating. Powder Technol., doi: 10.1016/j.powtec.2011.12.066, in press.

2. Bilgili, E. A., Bhakay, A., Azad, M., Dave, R. 2012. System and Method for Superdisintegrant-Based Composite Particles for Dispersion and Dissolution of Active Pharmaceutical Agents. Provisional Patent, Application No. 61625082, Filed on April 16, 2012.

(631e) Wet-Media Milling of Colloidal Drug Suspensions Stabilized by Means of Charged Nanoparticles

Michael Juhnke, Technical R&D, Novartis Pharma AG, Basel, Switzerland and Edgar John, Technical R&D, Novartis Pharma AG, Basle, Switzerland

INTRODUCTION

Tailoring poorly water-soluble drugs into products with colloidal drug particle size is a proven formulation approach to improve general issues in drug delivery [1]. Colloidal drugs are often produced by wet-media milling technology, due to the availability from pre-clinical to production scale, reliable up-scaling and regulatory acceptance [2]. Colloidal drug suspensions obtained by wet-media milling can be converted by suitable techniques into regular drug products, e.g., tablets and injectables. Stabilization of colloidal drug suspensions is provided by steric and/or electrostatic stabilization using polymers and/or surfactants [2, 3]. Beside these conventional stabilization mechanisms, colloidal particles can be also stabilized by charged nanoparticles providing high electrical charge to the otherwise negligibly charged colloidal particles [4, 5].

EXPERIMENTAL

We investigated the stabilization of colloidal drug particles by charged nanoparticles in order to explore the mechanism of nanoparticle-stabilized colloids as a complementary method compared to conventional stabilization mechanisms. The stabilization mechanism was investigated with the simultaneous size reduction of drug particles by wet-media milling. Aqueous colloidal drug suspensions were produced for a poorly water-soluble drug compound with constant process parameters, pH and content of a non-ionic polymer to facilitate wetting of the hydrophobic drug compound. Stabilization of colloidal drug particles was investigated for positively and negatively surface functionalized charged nanoparticles from silica and polystyrene. Particle size and shape was characterized by dynamic light scattering (DLS) and scanning electron microscopy (SEM). Stabilization of colloidal drug suspensions was characterized without further preparation by rotational and oscillational rheometry. The structural arrangement of charged nanoparticles in relation to colloidal drug particles was determined by cryo transmission electron microscopy (cryo-TEM).

RESULTS

Colloidal drug particle size and shape were identified for all compositions investigated with comparable characteristics. Hence, stabilization of colloidal drug suspensions is obviously determined by the charged nanoparticles for the compositions investigated. Through variations in nanoparticulate matter, size, sign of charge and concentration, colloidal drug suspensions with pronounced stability were produced, in contrast to a strongly agglomerated colloidal drug suspension in gel state without addition of charged nanoparticles. Upon addition of negatively charged polystyrene nanoparticles up to a lower critical concentration, stabilization of negligible charged drug colloids is facilitated into liquid state and almost Newtonian shear behavior. A further increase in nanoparticle concentration resulted in pronounced stability of the colloidal drug suspension without any detectable thixotropy. Interestingly, the lower critical nanoparticle concentration to facilitate drug colloid stabilization is in good agreement with the results obtained by Chan and Lewis [5] for binary mixtures of negligibly charged colloidal silica stabilized by charged polystyrene nanoparticles. In contrast to Chan and Lewis [5] and Zhang et al. [6], we have identified no preferred enrichment of charged nanoparticles near colloidal drug particles for a suspension that undergoes pronounced stabilization.

CONCLUSIONS

We have demonstrated the successful application of charged nanoparticles to facilitate stabilization of a colloidal drug compound produced by an industrial relevant production technology. The application of charged nanoparticles can potentially replace conventional molecularly dissolved ionic stabilizers and opens up new opportunities in drug colloid stabilization without need for drug specific anchor segments.

REFERENCES

[1] Singh A, Worku ZA, van den Mooter G, *Expert Opin Drug Delivery* 8 (2011) 1361-1378.

[2] Merisko-Liversidge E, Liversidge GG, *Adv Drug Delivery Rev* 63 (2011) 427-440.

[3] Bhakay A, Merwade M, Bilgili E, Dave RN, *Drug Dev Ind Pharm* 37 (2011) 963-976.

[4] Tohver V, Smay JE, Braem A, Braun PV, Lewis JA, *Proc Nat Acad Sci* 98 (2001) 8950-8954.

[5] Chan AT, Lewis JA, *Langmuir* 24 (2008) 11399-11405.

[6] Zhang F, Long GG, Jemian PR, Ilavsky J, Milam VT, Lewis JA, *Langmuir* 24 (2008) 6504-6508.

(631f) Interactive Mixtures for Dissolution Rate Enhancement of Cefuroxime Axetil

Namita Dalal, Graduate School of Pharmaceutical Sciences, Duquesne University, Pittsburgh, PA and Ira Shea Buckner, Pharmaceutics, Duquesne University, Pittsburgh, PA

INTRODUCTION

The poor aqueous solubility of newly developed drug molecules is often associated with slower dissolution rate in biological fluids, insufficient and inconsistent biological exposure and sub-optimal efficacy in patients when delivered via the peroral route of administration. One of the methods commonly employed to enhance the dissolution rate is to increase the surface area available for dissolution. In principle, for a given dose of drug, increasing the surface area by decreasing the particle size will lead to a proportional increase in dissolution rate. However, formulations containing micronized drug particles do not always exhibit an improvement in dissolution rate; even reduced rates have been reported. This has been attributed to the agglomeration of fine, primary particles (due to electrostatic and van der Waals forces), which decreases the surface area available for dissolution.

One approach used to take better advantage of this particle size reduction procedure is the preparation of interactive mixtures. Interactive mixtures result from the adherence of fine particles of one constituent to the considerable larger particles of the other (carrier particle). On contact with the dissolution medium, the carrier particles (water-soluble or hydrophilic) dissolve rapidly and also facilitate wetting of the drug particles, releasing the dispersed drug particles into the dissolution medium. This method is believed to maximize the fraction of the particles surface that is in contact with the dissolution medium, thus improving the dissolution rate. This approach should increase the dissolution rate of poorly-soluble drugs such as cefuroxime axetil (CFA).

EXPERIMENTAL

The current study investigates the possibility of dissolution rate enhancement of CFA (commercially available, amorphous API, < 53 µm size range) by forming an interactive mixture with coarse carrier particles of sucrose (500-1000 µm size range). Interactive mixtures with 10% w/w drug loading were prepared using a high shear centrifugal planetary mixer (Thinky USA, Inc.) which was able to deagglomerate the drug particles and cause them to adhere to the sucrose carrier particles. Physical mixtures containing the same amount of drug were also made using a laboratory-scale end-over rotation V-blender for comparison. This mixing method did not effectively break up the CFA agglomerates and the absence of interaction was assessed using optical microscopy. Mechanical stability of the mixtures was determined by sieving methods. Dissolution studies were performed as per the USP 32/NF 27 method for cefuroxime axetil tablets using 0.07N hydrochloric acid (900 ml) as the dissolution medium with the basket apparatus. Samples were analyzed using a validated UV spectroscopic assay. Dissolution studies were performed for the CFA agglomerates as well. The intrinsic dissolution rate of the API itself was measured for reference using the stationary disk method.

RESULTS

The intrinsic dissolution rate (IDR) of amorphous CFA was found to be 0.0765±0.009 mg/cm^2/min. Interactive mixtures of the model drug, CFA and sucrose carrier particles showed a significant improvement in dissolution rate as compared to physical mixtures of the same and CFA agglomerates. The amount dissolved with respect to time was modeled using the Wagner model (assuming exponential surface area changes). The rate constant determined by this method for the CFA agglomerates, physical mixture and interactive mixture with sucrose (*n=3* each) were 0.0010±0.0000065 min^{-1}, 0.0014±0.000013 min^{-1} and 0.0254±0.0012 min^{-1}, respectively. In addition the effective specific surface area (SSA) was calculated using the rate constant and IDR values. The SSA was found to be 331.61±15.39 cm^2/g, 17.64±1.26cm^2/g and 12.96±0.84 cm^2/g for the interactive mixture, physical mixture and CFA agglomerates respectively. These results indicate that interactive mixing resulted in a significant increase in the fraction of the particles surface area that was available for dissolution leading to an increase in the dissolution rate. The considerably slower dissolution of CFA agglomerates and the physical mixture can be attributed to the fact that a large fraction of the particles surface was inaccessible to the dissolution medium. This problem was overcome by distributing primary particles of the API over the surface of the water-soluble carrier particles in the interactive mixture.

CONCLUSION

Agglomeration of fine particles is a particular problem for hydrophobic drugs which can limit the amount of drug that can dissolve and as a result contribute to the poor oral bioavailability of such drugs. In addition agglomeration can also to negate the particle size reduction procedure that is commonly employed to improve dissolution rate. This study demonstrated that deagglomeration of the CFA agglomerates and distribution of the primary particles of the drug over the surface of the carrier particles significantly enhanced the dissolution rate by increasing the effective surface area of the fine drug particles exposed to the dissolution medium and thus available for dissolution. It is expected that interactive mixtures can help to overcome limitations associated with the use of poorly-soluble, finely powdered drugs. This approach can be further extended to other poorly-soluble drugs in cases where it is critical to maximize the amount dissolved in the gastrointestinal tract.

(631g) Formation of Stable Nanocarriers by Tuning Pharmaceutically Active Ingredient Properties Via an *in Situ* Salt Precipitation

Nathalie M. Pinkerton[1], Arnaud Grandeury[2], Andreas Fisch[2], Jörg Brozio[3], Bernd Riebesehl[2] and Robert K. Prud'homme[1], (1)Chemical and Biological Engineering, Princeton University, Princeton, NJ, (2)Novartis Pharma AG, Basel, Switzerland, (3)Pharmaceutical & Analytical Development, Novartis Pharma AG, Basel, Switzerland

NanoCarriers (NCs) are a promising drug delivery route, which offer increased active pharmaceutical ingredient (API) circulation times and enhanced bioavailability. When formed via rapid precipitation techniques such as flashnanoprecipitation, NCs have higher API loading than their post-loaded delivery counterparts. The aqueous solubility and crystallinity of the API affects NC formation and stability. Generally, hydrophobic materials (logP > 3.5) can be successfully formulated as NCs because sufficient supersaturation can be achieved for nucleation-dominated particle growth. With weakly hydrophobic API, however, the required supersaturation cannot be achieved resulting in low encapsulation efficiency or loss of particle size control. Additionally, with weakly hydrophobic and highly crystalline API, instabilities can occur due to redissolution and recrystallization of the drug compound. To modify API properties for successful encapsulation into NCs, two common routes exist. The first is to modify the drug molecular structure by creating a prodrug. The prodrug is considered a new compound and must undergo FDA approval. The other is to create a salt of the drug, which does not create a new compound and thus circumvents the need for new FDA approval. The salt technique has been used for the formation of bulk salts to improve dissolution behavior, but never to aid in the formation and stability of NCs. We demonstrate the successful ion pairing of APIs to hydrophobic counterions to enable NC formation by rapid precipitation. No additional preparation steps prior to the precipitation process are needed, as the hydrophobic salt is formed *in situ* during the NC self-assembly. The rules for successful nanocarrier formation are presented using three model APIs, cinnarizine, clozapine and α-lipoic acid. This technique expands the types of APIs that can be successfully encapsulated in NC form for drug delivery.

(661a) Developing Improved Understanding of Spray Drying Through Process Modelling

Thoralf Hartwig[1], Francois Ricard[1], Ian C. Kemp[1], Mark A. Pinto[2] and Sean K. Bermingham[2], (1)Particle Generation and Control Engineering, GlaxoSmithKline Ltd, Stevenage, United Kingdom, (2)Process Systems Enterprise, London, United Kingdom

Spray drying is often used in the pharmaceutical industry to rapidly produce particles with narrow size distributions and with low moisture contents. In the spray drying process, a suspension or solution of a desirable product in a volatile solvent is converted to a largely dry solid product by contact with a drying medium. The process starts with an atomizer which creates droplets from the feed suspension or solution. These droplets are then mixed with a hot gas. Evaporation of the volatile solvent in the droplets takes place and a dry solid product is thus obtained.

Significant work has been done in the past to look at how individual droplets dry within a spray dryer. These models include detailed mass and energy balances and predict the droplet temperature and radius profile as drying occurs. In practice however, the atomisation process produces droplets of a range of sizes. As one of the factors influencing the rate of drying of a droplet is its initial droplet size, a model of a spray dryer should consider the distribution of droplet sizes generated in the atomiser in order to obtain a better prediction of product temperature and moisture content.

A model has been developed that considers this distribution of droplet sizes within the spray dryer. The model outputs include temperature and droplet size histories of droplets of various sizes as well as the evolution of the droplet size distribution within the spray dryer. The model has been used in a pharmaceutical application to look at how the mathematical modelling can aid in process development. The results obtained from this case study are promising and indicate that process modelling has the potential to significantly reduce process development times and costs.

(661b) Assessing the Robustness of a High-Shear Granulation Process to Meaningful Variability in Raw Material Physical Attributes

Sharvari Borkar, Pharmaceutical Commercialization Technology, Merck & Co., Inc., West Point, PA and Stephen L. Conway, Center for Material Science and Engineering, Merck & Co., Inc., West Point, PA

While much of a drug product development team's focus is applied to understanding the impact of composition and process parameters on the performance of manufacturing processes, utilizing tools such as risk analysis and multivariate experimental designs, difficulties have arisen when the same tools are inappropriately applied to assessment of process robustness to raw material variability. The drive for more efficient, flexible supply chains, and the use of both novel and more cost-effective ingredients exacerbates the need for clear assessment of process sensitivity to attributes known to impact performance, as well as variations in attributes previously unrecognized.

By analysis of granulation intermediate properties as surrogates for drug product performance, we illustrate how the sensitivity of a high-shear wet granulation process to meaningful variability in raw material attributes can be assessed. Use of experimental designs intended to rigorously probe process robustness to specified and unspecified raw material attributes, in the presence of inherent process noise, is described.

The results of granulation experiments and associated characterization measurements are presented. We identify performance risks due to shifts in API and excipient attributes that might reasonably be expected in commercial manufacturing. These statistical sensitivities are are related to granulation mechanistic understanding, and potential control strategies are explored. The case study highlights how one aspect of risk can be rigorously explored during development in anticipation of changes in raw material source and manufacturing process that will almost inevitably occur in a drug product lifecycle.

(661c) Development of a Laboratory and Mathematical Model to Predict Scale up Performance of an API Slurry Milling and Isolation Step

Edward Conder[1], Kevin D. Seibert[2], Carla Luciani[3], Eric Moher[4], Daniel Jarmer[5] and Michael Phillips[1], (1)Eli Lilly & Co, Indianapolis, IN, (2)Chermical Product Research and Development, Eli Lilly & Co., Indianapolis, IN, (3)CPR&D, Eli Lilly & Co, Indianapolis, IN, (4)CPR&D, Eli Lilly & Co., Indianapolis, IN, (5)CPRD, Eli Lilly & Co., Indianapolis, IN

Development of a Laboratory and Mathematical Model to Predict Scale up Performance of an Active Pharmaceutical Ingredient Slurry Milling and Isolation Step

Control of Active Pharmaceutical Ingredient particle size can be a significant challenge when designing a crystal that yields good flow properties for drug product formulation. This must also be balanced with a need for ensuring adequate bioavailability and content uniformity of the formulated product. When slurry milling can be employed for particle size control, a dry milling step and the associated costs and potential exposure can be significantly reduced. In this study, we examine an Active Pharmaceutical Ingredient crystallization requiring careful particle size control and the subsequent scale up for manufacturing. This analysis includes development of a laboratory model and a mathematical model to predict milling performance, including a prediction of system shear rate at full manufacturing scale to determine milling and attrition of particles for optimization of particle size and isolation cycle time.

(661d) Manufacture and Control of a Hygroscopic Anhydrous API

Nathan Domagalski[1], Brendan C. Mack[2], Amanda Rogers[1], Jose E. Tabora[3] and Lindsay Hobson[2], (1)Chemical Development, Bristol-Myers Squibb Co., New Brunswick, NJ, (2)Bristol-Myers Squibb Company, New Brunswick, NJ, (3)Chemical Development, Bristol Myers Squibb, New Brunswick, NJ

Control of the desired polymorphic form of an API is necessary to ensure consistent drug product manufacture and performance of the final drug product. In this paper an API crystallized as a di-Ethanolate solvate is dried and isolated as crystalline anhydrous API. We show how the form conversion may be monitored from the thermal profile of the pilot plant agitated drier which correlates strongly to the polymorphic state of the drying material. Once isolated, the anhydrous API may undergo conversion to a hydrate at relative humidity in excess of 30%. Hydration has a potential impact on the formulation of the drug product and requires control in the packaged material. Moisture uptake rates of the API were measured and the effective diffusivity of water vapor was used to estimate the risk for form conversion as a function of processing scale and define a control strategy to minimize the impact on the drug product manufacture.

(661e) Elucidation of Drying Mechanisms of De-Solvation and the Establishment of a Drying Model

Daniel Hsieh, Chenchi Wang, Shih-Ying Chang and San Kiang, Bristol-Myers Squibb Company, New Brunswick, NJ

To predict the drying performance of an API de-solvation process, a fundamental understanding of the drying mechanisms is required to establish a drying model that can predict the API drying time. A case study is presented in which the selected form of the API is a desolvated, crystalline form. This form was generated upon drying the wet cake of the solvated form in a commercial-scale dryer. The dynamic behavior of this drying process was monitored in the lab using in-situ Raman, mass spectrometry, and an automatic weighing device. SEM images of the dried product were used to define the geometry of diffusion and the diffusion length. The diffusion model established was scale independent and was used to predict the drying time in the pilot scale dryer. The drying time predicted from the model was in excellent agreement with the data obtained experimentally, and the desired API crystal form was generated.

(661f) Multi-Scale Modeling of Pharmaceutical Spray Drying

Pavol Rajniak, Pharmaceutical Commercialization Technology, Merck & Co., Inc., West Point, PA, Genong Li, Ansys Inc., Lebanon, NH, David Johnson, Merck & Co., Inc., West Point, PA, Rick Falk, Bend Research Inc, Bend, OR and Justin D. Moser, Pharmaceutical Development, Merck & Co., Inc., West Point, PA

Developing pharmaceutical product formulations in a timely manner and ensuring quality is a complex process which requires a systematic, science based approach and combination of information at multiple spatial and temporal scales. Advances in quantitative analysis techniques (e.g. mechanistic models) in combination with modern measurement tools enable engineers to design and scale-up pharmaceutical unit operations and processes with more confidence and reliability.

Spray drying is becoming an important unit operation in pharmaceutical industry to produce solid amorphous dispersions which are frequently used to improve the solubility and thus the bioavailability of poorly soluble active pharmaceutical ingredients.

Goal of the presentation is to share our experience how to develop, calibrate and combine mathematical models of spray drying at different levels/scales. This is accomplished through combination and solution of governing balance equations and using empirical data defining boundary conditions and input/output properties at following levels/scales of drying:

Scale 1: Single droplet. A detailed distributed parameter model was developed for simulation of a single droplet drying. The model describes heat and mass transfer inside the droplet/particle, shrinking of the particle, the skin formation followed by skin thickening, crust formation and the final drying of non-shrinking particle. The model also calculates exchange of heat and mass between the particle and external drying gas and the decreasing particle density needed for the momentum balance at scale 2. This model was calibrated against single droplet drying experimental results.

Scale 2: Spray dryer. A detailed Computational Fluid Dynamics (CFD) model was developed for simulation of drying of multiple droplets (sprays) using a mixed Eulerian/Lagrangian approach. Iterative coupling is employed to account for exchange of momentum, heat and mass between the particles and continuous phase. The single droplet model (Scale 1) is combined with the CFD code (Scale 2) in the form of a Fluent User Defined Function. The model (after calibration) also predicts the outlet gas composition needed for the dryer mass balance specification at scale 3.

Scale 3: Spray drying process. The Aspen Plus software was employed for development of a process model containing the different unit operations of the spray drying process. The key unit operations of the process (in addition to the stream dividers, mixers, etc.) are the simplified spray dryer, heater and condenser used for the recycle stream purification. Detailed simulations of the CFD model (Scale 2) are used for improvement of the Aspen Dryer model (Scale 3) predictability and the Aspen model is used for calculation of the solvent content in the recycle stream and mixing with the fresh drying gas before entering the spray dryer.

(662a) Real-Time Characterization of Exosomes Secreted From Single Cells by Tethered Lipoplex Nanoparticles and TIRF Microscopy

Yun Wu[1], Xi Zhao[2], Kwang Joo Kwak[3], Carlos E. Castro[4] and Ly James Lee[2], (1)Nanoscale Science and Engineering Center, The Ohio State University, Columbus, OH, (2)William G. Lowrie Department of Chemical and Biomolecular Engineering, The Ohio State University, Columbus, OH, (3)Chemical and Biomolecular Engineering, The Ohio State University, Columbus, OH, (4)Mechanical and Aerospace Engineering, The Ohio State University, Columbus, OH

Exosomes are small membrane vesicles with diameter of 50–100 nm secreted by a wide range of cells. Exosomes arise from the endocytic cellular pathway and are released when multivesicular bodies fuse with the plasma membrane. Exosomes can mediate communication between cells, facilitating processes such as antigen presentation. In 2007, exosomes were found to be natural carriers of over 1300 messenger RNAs (mRNA) and 121 non-coding microRNAs (miRNA), a major breakthrough in understanding exosome biology. Recently circulating exosomes have been demonstrated to be a potiental biosignature for cancer early diagnosis.

Although significant progress has been made, the biological function of exosomes is only just beginning to be understood. Currently, exosomes are isolated by ultracentrifugation and the encapsulated mRNA and miRNA expressions are characterized by qRT-PCR. These methods are time consuming and labor intensive. More importantly, information collected from these conventional methods is an average from a large number of exosomes. We have developed a tethered lipoplex nanoparticles based real-time imaging platform to visualize and characterize individual exosomes right after they are released by the cells. This platform was demonstrated with A549 non-small cell lung cancer cells and a microRNA, miR-21, was selected as the biomarker. Briefly, A549 cells were seeded on a glass substrate. Molecular beacons that detect miR-21 were encapsulated in cationic lipoplex nanoparticles tethered around the A549 cells. When exosomes were released from A549 cells, the electrostatic force between negatively charged exosomes and positively charged lipoplex nanoparticles facilitated the fusion between the naturally occurred nanoparticles and man-made nanoparticles. The molecular beacon encapsulated in the lipoplex nanoparticles then detected miR-21 in the exosomes. Total internal reflection fluorescence (TIRF) microscopy was used to capture the fluorescence signal emitted from the molecular beacon and detect the copy numbers of miR-21 in each exosome.

This is the first platform that is capable of visualizing and characterizing exosomes in real time. This technique may also be used to study cell-cell communication and cell response to the stimuli, such as anti-cancer drugs or the change of microenvironment.

(662b) Direct Chemical Imaging of the Sphingolipid and Cholesterol Distribution in Cell Membranes

Mary L. Kraft, Department of Chemical & Biomolecular Engineering, University of Illinois at Urbana-Champaign, Urbana, IL, Jessica F. Frisz, Chemistry, University of Illinois at Urbana-Champaign, Urbana, IL, Haley A. Klitzing, Chemistry, University of Illinois, Urbana, IL, Kaiyan Lou, Chemical & Biomolecular Engineering, University of Illinois at Urbana-Champaign, Urbana, IL, Joshua Zimmerberg, National Institute of Child Health and Human Development, National Institutes of Health, Bethesda, MD and Peter K. Weber, Lawrence Livermore National Laboratory, Livermore, CA

Domains of different protein composition are known to be present in the plasma membranes of mammalian cells. Different lipid species and cholesterol also seem to be organized into compositionally and functionally distinct plasma membrane domains. For example, membrane domains that are enriched with cholesterol and sphingolipids, which are often referred to as lipid rafts, are hypothesized to be present in the plasma membrane. However, the existence of lipid rafts and their significance to cell biology is controversial. In fact, even though the abundances of cholesterol and sphingolipids within the cell undoubtedly influence many biological processes, the distributions of cholesterol and sphingolipids within the plasma membrane have not been definitively established. Here we report the direct chemical imaging of the sphingolipid and cholesterol distribution in the plasma membranes of intact cells. We have used high resolution secondary ion mass spectrometry to map the isotope enrichment from metabolically incorporated ^{15}N-sphingolipids and ^{18}O-cholesterol on the entire dorsal surfaces of fibroblast cells. Statistical methods are used to definitively determine whether the sphingolipids and cholesterol were heterogeneously distributed and co-localized within the plasma membrane. Our results demonstrate that lipids are spatially organized within the plasma membrane.

(662c) Analysis and Visualization of Multiparameter, Single-Cell Data Using Self-Organizing Maps

Nicholas A. Graham[1], Ken-ichiro Kamei[2], Jing Sun[1], Michael Masterman-Smith[1], Jing Jiao[1], Minori Ohashi[1], Hsian-Rong Tseng[1] and Thomas G. Graeber[1], (1)Medical and Molecular Pharmacology, University of California, Los Angeles, Los Angeles, CA, (2)Kyoto University, Kyoto, Japan

Introduction: Technologies capable of multiple, quantitative, inexpensive, single-cell biomarker measurements (eg, microfluidics) are transforming biology by enabling systems analysis of microscopic samples at the single-cell level. Because the large data sets generated by these single-cell technologies are highly complex, meaningful interpretation of these multiparameter data sets requires novel bioinformatic tools. Here, we adapt self-organizing maps (SOMs), an unsupervised learning method which has found wide application in analysis of complex biological data sets (Tamayo et al, 1999), to analyze multiparameter, single-cell data sets including 1) proteomic measurements of signal transduction in clinical brain tumor specimens and 2) cytological features of human pluripotent stem cells.

Materials and Methods: To generate multiparameter, single-cell data sets, we utilized Microfluidic Image Cytometry (MIC), which combines the advantages of microfluidics and microscope-based cytometry to quantify multiple biomarkers at the single-cell level using immunocytochemistry (Sun et al, 2010). For clinical brain tumor specimens, we simultaneously quantified four signaling proteins (EGFR, PTEN, phospho-Akt and phospho-S6). For human embryonic (ES) and induced pluripotent stem cells (iPS), we quantified either protein biomarkers of pluripotency (OCT4, SSEA1) or 39 cytological features including nuclear morphology (DAPI) and cell cycle progression (incorporation of 5-ethynyl-2′-deoxyuridine, a marker of S-phase). All data sets included ~1,000 cells per sample.

SOMs were created using the Kohonen R package (Wehrens & Buydens, 2007). A SOM grid consists of a number of units each characterized by a multiparameter vector (eg, EGFR, PTEN, pAkt and pS6 levels). Vectors characterizing the SOM grid are trained so as to represent the global measurement space. Single-cell data are then mapped to the SOM grid based on similarity to the SOM units, and each sample is plotted as the frequency of cells in that sample that map to each SOM unit. SOM grid frequencies of each sample SOM grid are finally subjected to unsupervised hierarchical clustering.

Results and Discussion: To demonstrate the clinical application of SOMs in oncology, we used MIC to analyze the oncogenic PI3K/Akt/mTOR signaling pathway in a panel of 19 human brain tumor biopsies. SOM analysis of the MIC data revealed a diversity of signaling patterns that would have been masked by population-average measurements, including evidence of inter- and intra-tumoral heterogeneity. SOMs also stratified patients into molecularly-defined subsets that were predictive of tumor progression and patient survival.

To test the application of SOMs in stem cell biology, we measured either protein biomarkers of pluripotency (eg, OCT4 expression) or 39 cytological features (eg, nuclear morphology, cell cycle progression) in human ES and iPS cells. SOMs identified phenotypic changes dynamically occuring during differentiation. Additionally, in a panel of ES and iPS cells, SOMs revealed biomarker patterns that were predictive of the differentiated and pluripotent phenotypes.

Conclusions: The successful utilization of single-cell, quantitative diagnostic technologies including microfluidics will require bioinformatic tools to facilitate meaningful analysis of multiparameter, single-cell data. Here, we have demonstrated that SOMs are a robust, enabling tool for systems analysis of multiparameter, single-cell data.

References: Sun et al, *Cancer Res* (2010); 70:6128-38.

Tamayo et al, *Proc Natl Acad Sci U S A* (1999); 96: 2907-12.

Wehrens and Buydens, *J Stat Soft* (2007); 21: 1-19.

(662d) Double Layer CELL Wall MODEL for Yeast CELLS

Ruben Mercade-Prieto[1], Colin R. Thomas[1] and Zhibing Zhang[2], (1)Chemical Engineering, University of Birmingham, Birmingham, United Kingdom, (2)Department of Chemical Engineering, University of Birmingham, Birmingham B15 2TT, United Kingdom

In the last decade there has been extensive research on the mechanical properties of biological walls, using in particular atomic force microscope (AFM) and compression testing by micromanipulation. Small indentation experiments using AFM and sharp indenters consistently give an elastic modulus (E) of the yeast cell (Saccharomyces cerevisiae) wall of 0.2-1.6 MPa [1], whereas micromanipulation and other large deformation techniques generate values of 100-200 MPa [2]. No explanation has yet been reported to account for this difference of two orders of magnitude. It will be shown here using finite element modeling (FEM) that the Hertzian equation used to calculate E from AFM data is inappropriate for core-shell spheres like yeast cells. Secondly, a double-layer cell wall model is presented to explain the difference in E values obtained with the AFM and micromanipulation techniques.

Finite element modelling (FEM) was performed using ABAQUS/Standard 6.5. Half core-shell spheres were simulated in 2D under axis-symmetrical conditions, as performed previously for the compression of microcapsules with a core-shell structure [3]. The AFM indenter and the bottom substrate were modeled as rigid materials, while the cell wall was modeled using 7100 deformable CAX4RH elements; the mesh size was more refined close to the indentation area (see Fig. 1). For a double -layer wall, a second shell was introduced under the previous external layer using 1500 CAX4RH elements. The liquid core was modelled using FAX2 elements with a density of 1 kg L-1. The wall material was modeled as isotropic and linearly elastic; a Hookean approach has been considered as it has been shown to be a good mechanical model for the yeast wall up to very large deformations [1]. Different inner pressure values common for yeast cells were also modelled.

Single-layer cell wall model

AFM compression studies commonly calculate the mechanical properties of the cell wall using the Hertzian equation (1), assuming a spherical shape of the indenter:

$$F_{sphere} = \frac{4Er_{ind}^{0.5}}{3(1-\nu^2)} d_{ind}^{1.5} \qquad (1)$$

where r_{ind} is the indenter radius and d_{ind} is the indentation depth. However, this equation cannot be used for core-shell spheres, a possible model for yeast cells, because of the significant inward bending of the wall, as shown in Fig. 1, regardless of the inner pressure. The displacement of the indenter tip (d) is much larger than d_{ind}. In practice, what has been calculated is a pseudo Hertzian E assuming d_{ind} = d. Considering this, the E values reported with AFM for yeast cell are underestimated by 5-10 times.

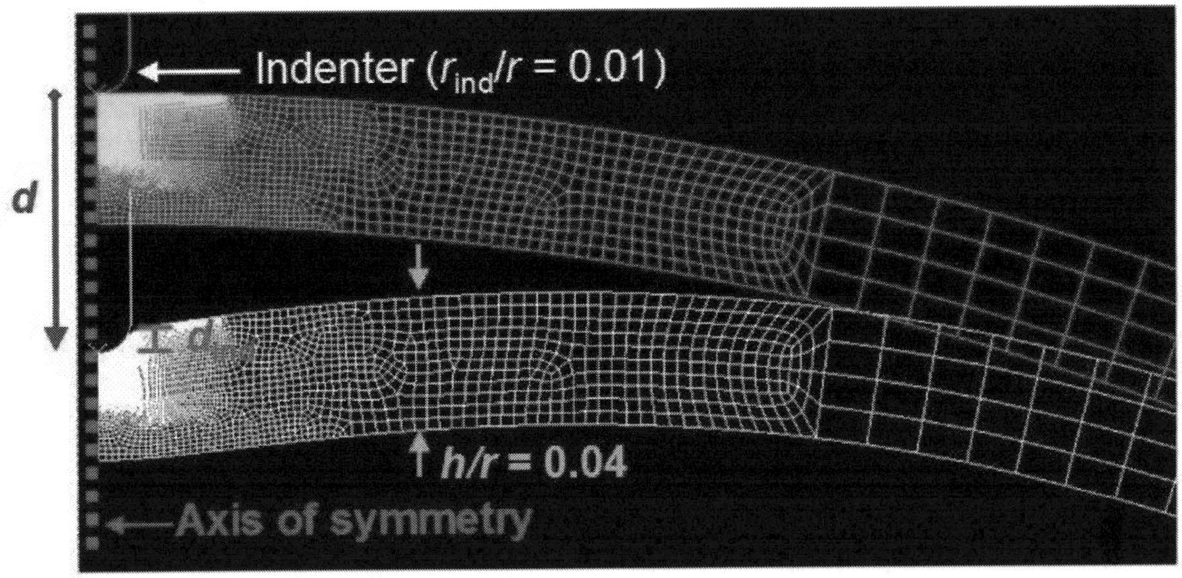

Figure 1: FEM results showing compression with a sharp indenter, for a displacement twice the wall thickness (d = 2h), yet the penetration indentation is d_{ind} ~0.18h.

Double-layer cell wall model

The common assumption of all previous biomechanical studies is that the yeast cell wall is homogeneous, although possibly having chitin-rich bud scars. However, there are two very different layers in the yeast cell wall encapsulating the plasma membrane. The outer layer is comprised of mannoproteins, reported to control the cell porosity, while the inner layer is composed of b1,3-glucan and chitin, which is believed to impart mechanical strength to the cell wall [4].

A FEM model was constructed for a double-layer wall with different thickness and elastic modulus values, assuming a soft external layer and a stiff inner layer. Micromanipulation experiments were conducted to compress single yeast cells until rupture, which usually occurred at fractional deformations e of 0.6-0.7, defined by the ratio of displacement to initial cell diameter (2r) [1]. Figure 2 shows that using $(Eh)_{total}$, defined as the sum of $(Eh)_{out}$ and $(Eh)_{in}$, a unique relationship for the normalized compression force versus deformation is obtained regardless of the actual elastic modulus values of the two layers within the chosen ranges. Therefore, assuming a double-layer model of the yeast cell wall, micromanipulation results provide $(Eh)_{total}$, found to be in the range of 11-15 N m^{-1}.

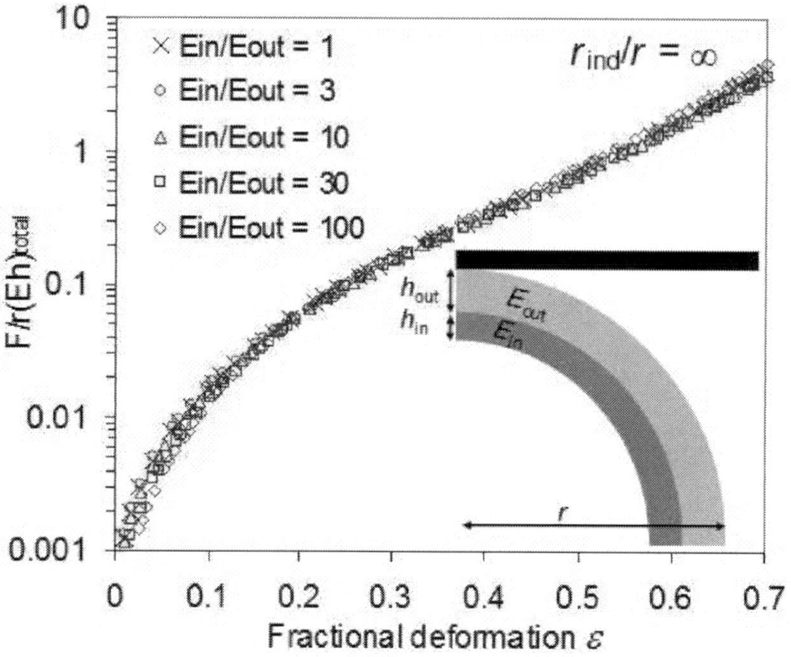

Figure 2: Normalized compression force with the total wall stiffness using parallel compression for a double-layer model with an inner wall of $h_{in}/r = 1\%$ and an outer wall of $h_{out}/r = 4\%$.

Under a double-layer model, the associated errors in using a Hertzian analysis, such as eq. (1), for the determination of E are smaller than those in the single-layer wall model due to the existence of the stiff inner layer that limits the inward bending of the cell wall. Figure 3 shows that Hertz-like curve fitting can still be performed at small deformations, as also observed for single layers, despite eq. (1) not being fully valid. In addition, the estimated E is close to that of the outer layer. Hence, AFM results using a sharp tip are reasonably correct only if a stiff inner layer is assumed.

AFM and micromanipulation results can be combined to determine the biomechanical properties of the two layers in the yeast cell wall. The external later is characterized with sharp tip indentation, E_{out} ~1 MPa. For a total wall thickness h of ~130 nm, the thickness of the b1,3-glucan fiber layer (h_{in}) is estimated to be around 10-40 nm. Finally, the elastic modulus of the stiff inner layer was calculated to be E_{in} ~ 0.3-1.3 GPa depending on chosen value of h_{in}.

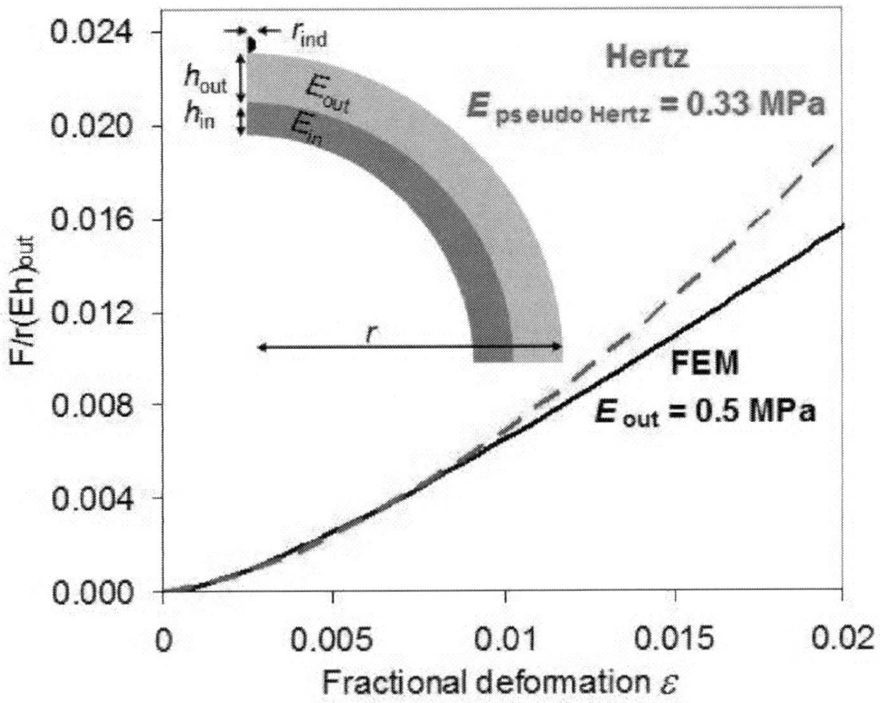

Figure 3: Normalized compression force with the outer wall stiffness using a sharp indenter for a double-layer model with $h_{in}/r = 0.8\%$, $E_{in} = 100$ MPa and $h_{out}/r = 4.4\%$. The FEM data has been fitted by Hertzian analysis.

Conclusions

A biomechanical double-layer cell wall model has been developed following the layered structure of the S. cerevisiae wall. Assuming an external soft layer plus an internal stiff layer, it is shown that the reported AFM elastic modulus corresponds to that of the external layer, while micromanipulation results provide the total wall stiffness. The results can be combined, using the thickness of the different layers, to estimate the elastic modulus of the stiff b1,3-glucan layer for the first time.

REFERENCES

1. Dague E, et al. Yeast 27, 673-684, 2010.

2. Smith AE, et al. PNAS USA 97, 9871-9874, 2000.

3. Mercadé-Prieto R, et al. Chem Eng Sci 66, 1835-1843, 2011.

4. Klis FM, et al. FEMS Microbiol Rev 26, 239-256, 2002.

(662e) CEBPA Mutant Regulates miR181a Expression in AML Cells: A Single Cell Study by Nanochannel Electroporation

Xi Zhao[1], Yun Wu[2], Xinmei Wang[2], Daniel Gallego-Perez[2], Pouyan E. Boukany[2], Xiaomeng Huang[2], Sebastian Schwind[3], Guido I. Marcucci[3] and L. James Lee[1], (1)William G. Lowrie Department of Chemical and Biomolecular Engineering, The Ohio State University, Columbus, OH, (2)Nanoscale Science and Engineering Center, The Ohio State University, Columbus, OH, (3)Division of Hematology, Department of Internal Medicine, Comprehensive Cancer Center, The Ohio State University, Columbus, OH

One of the major disadvantages of conventional electroporation is the compromise between viability and efficiency, especially with large plasmids. With certain combination of gene and cell type, the transfection could be so poor that it no longer remains a meaningful tool for the study of gene expression. We have previously reported a nanochannel electroporation (NEP) technology which provides high transfection efficiency and uniform dosage with minimal cell damage, thus high viability, for individual cells[1].

Clinically, mutation of CCAAT/enhancer-binding protein alpha (CEBPα) gene is observed in some population of acute myeloid leukemia (AML) patients[2], of which certain sub-groups show significant advantage of disease progression. Although the full and detailed effects of such mutation are yet to be established, we have previously found correlations between favorable clinical outcome, certain sub-type of mutation and increase of miR181a expression level in patient samples[3]. However, verification of such correlation *in vitro* in AML cell lines that do not express endogenous CEBPα was unsuccessful due to the low expression and viability after conventional electroporation.

Here we applied our NEP platform to transfect wild type and mutated CEBPα gene into individual AML KG1a cells as well as observing miR181a expression level afterwards. The cell viability and gene expression were verified by EthD-1/Calcein staining kit and the use of CEBPα-GFP encoded plasmid. At different time points after transfection, we analyzed the expression of miR181a by injecting a locked nucleic acid (LNA) molecular beacon (MB) targeting mature miR181a into the cells, and retrieving the transfected cells for single cell qRT-PCR. Our results shows that while the wild type CEBPα gene, of which the mRNA can be translated to two polypeptides with different length (p42 and p30, respectively), can increase the expression of miR181, the N-terminal mutant of this gene, which exclusively express p30, results in a statistically larger upregulation of miR181. In contrast, the C-terminal mutant, which may inhibits the function of CEBPα, has no significant effect on the expression of miR181a.

Our NEP technology demonstrated here for the study of gene expression/regulation successfully combines precise transfection with target detection, and shows great compatibility with small cell population. In the case of rare cell analysis, this platform will be the perfect tool.

[1] Boukany PE, Morss A, Liao WC, et al., Nanochannel electroporation delivers precise amounts of biomolecules into living cells, *Nature Nanotech*, 2011;6, 747-754

[2] Leroy H, Roumier C, Huyghe P, et al., CEBPA point mutations in hematological malignancies, *Leukemia*, 2005;19, 329-334

[3] Marcucci G, Maharry K, Radmacher MD, et al., Prognostic significance of, and gene and microRNA expression signatures associated with, CEBPA mutations in cytogenetically normal acute myeloid leukemia with high-risk molecular features: a cancer and leukemia group B study, J Clin Oncol. 2008;26, 5078-5087

(662f) Identifying Transcriptional Phenotypes Associated with Single Cell Variability in Hypertension

James Park[1,2], Anthony Brureau[2], Sonali Gulati[2], Carmen Nichols[3], Rajanikanth Vadigepalli[4], Babatunde A. Ogunnaike[5] and James S. Schwaber[2], (1)Chemical Engineering, University of Delaware, Newark, DE, (2)Daniel Baugh Institute for Functional Genomics and Computational Biology, Thomas Jefferson University, Philadelphia, PA, (3)Daniel Baugh Institute or Functional Genomics and Computational Biology, Thomas Jefferson University, Philadelphia, PA, (4)Daniel Baugh Institute for Functional Genomics and Computational Biology, Department of Pathology, Anatomy and Cell Biology, Thomas Jefferson University, Philadelphia, PA, (5)Department of Chemical and Biomolecular Engineering, University of Delaware, Newark, DE

Hypertension is characterized by a sustained elevation of arterial blood pressure; its pathology is multifactorial with many contributing genetic and environmental factors and is not completely understood. Consequently, various aspects of blood pressure regulation have been studied extensively in order to suggest mechanisms subtending the development of hypertension. Physiological evidence indicates a strong neurogenic component of hypertension pathology and implicates abnormal neural blood pressure control mechanisms, which involve A2 neurons located within the nucleus tractus solitarius (NTS). Additional evidence from gene expression studies also show that neuronal adaptations are strongly associated with hypertension development. These neuronal adaptations are believed to cause heterogeneous gene expression within the NTS and A2 neurons ("extrinsic biological noise") creating distinct phenotypic subtypes. Characterizing these phenotypic subtypes may inform strategies for preventative treatment of hypertension. Our objective is to distinguish these various subtypes though single cell analysis, with a specific focus on A2 neurons and cells responsive to an acute hypertensive challenge in the NTS (as indicated by cFos protein expression, a general marker of neural activation). Our strategy is to utilize a combination of experimental approaches to acquire gene expression measures, i.e. transcriptional signatures, and computational analysis to identify distinct transcriptional phenotypes. These phenotypes can be used to classify subgroups within these cell types associated with a hypertensive state.

In our approach, rats were infused with phenylephrine to induce an acute hypertensive challenge, pharmacologically increasing blood pressure ~40mmHg; corresponding control animals were infused with saline solution. NTS tissue samples were collected from the rats one hour post infusion. Single cell samples were categorized within the NTS by several factors including 3-dimensional rostro-caudal localization and immunohistochemistry (IHC) phenotype. IHC was used to identify the two distinct cell populations of interest: neurons expressing (1) tyrosine hydroxylase (TH), an enzyme involved in synthesis of catecholaminergic neurotransmitters found exclusively in A2 neurons in the NTS and (2) cFos, a sensitive marker of neural activation identifying cells responsive to the acute hypertensive perturbation. Laser Capture Microdissection (LCM) was used to lift single cells from the NTS tissue samples for analysis of their respective transcriptional signature.

We use a high-throughput quantitative PCR platform (BioMark^{TM}) to analyze gene expression of our single cell samples. This method enables us to assay 96 genes within 96 samples simultaneously, with low technical variability. Our 96 gene assays were refined from previous global gene expression studies that provide evidence for their co-regulated and coordinated response to blood pressure changes. This gene cohort is also relevant to the A2 phenotype and to angiotensin II type 1 receptor (AT1R) pathways, which are mediators of cellular responses associated with hypertension.

Our analysis involved the use of several multivariate analytical techniques. Through the use of Principal Component Analysis of single cell gene expression, we were able to identify distinct transcriptional signatures distinguishing the A2 neuronal (Th+) and responsive cell (cFos+) phenotypes within the NTS. Specifically Th, Dbh, Gal, Rgs4, Slc32a, Rgs1, Npy, Tac1, and Sst transcripts make significant contributions to the variability between these subphenotypes. Hierarchical clustering of the NTS cell phenotypes within specific regions of the NTS reveals anatomical subgroups within the cFos+ phenotype. Additionally, the transcriptional signatures of these single cells have been used to identify a subset of single cells that express both Th and cFos, forming a Th+/cFos+ subphenotype. This group may be used to identify specific A2 neurons that respond to an acute hypertensive challenge and may be a specific focus for future time-series and chronic studies of neurogenic hypertension.

Taken together, we have demonstrated a powerful methodology for studying mRNA expression variability within *in vivo* single cell samples in a hypertensive animal model. Our integrated experimental and computational approach has enabled us to characterize expression heterogeneity and identify specific transcriptional signatures that distinguish subphenotypes within the NTS. These subpopulations may play distinct roles in the neural pathology of hypertension development.

(662g) Investigating the Effects of Dynamic External Stimuli On Single Cell Fitness and Gene Expression in Escherichia Coli

Utsav Agrawal[1], Eric Johnson Chavarria[2], Arnab Mukherjee[1], Melikhan Tanyeri[1] and Charles M. Schroeder[1,2], (1)Chemical and Biomolecular Engineering, University of Illinois at Urbana-Champaign, Urbana, IL, (2)Center for Biophysics and Computational Biology, University of Illinois Urbana-Champaign, Urbana, IL

In this work, we study the bacterial cellular response to time-dependent external stimuli in single living cells. We developed a microfluidic platform for single cell analysis that allows for dynamic control of well-defined environmental growth conditions and culture media. Using this platform, we studied the effect of small molecule inducers on gene expression in the *lac*operon using fluorescent reporter proteins and cell growth rates as a proxy of cellular fitness. We applied temporally varying inducer concentrations by translating single cells between two adjacent fluid streams containing either growth medium or growth medium and inducer. We observed that single cell gene expression depends on growth rate and frequency of exposure to inducer concentrations. Single cell induction experiments are compared to control experiments with and without continuous fluid flow in microfluidic channels.

A microfluidic-based hydrodynamic trap recently developed in our lab facilitates single cell analysis for these experiments. The hydrodynamic trap enables confinement and manipulation of single cells in free solution using the sole action of fluid flow. Automated feedback control is integrated into the device using an "on-chip" valve, which allows for precise confinement of cells in free solution. Using this device, cells are confined at a fluid stagnation point generated in a cross-slot microfluidic geometry, thereby enabling non-perturbative trapping of cells for long time scales.

Using optical microscopy, we observe single *Escherichia coli* cells growing and dividing both at room temperature and at 37^0C, and cell division rates in the microfluidic trap compare favorably to the growth rates of *E. coli* measured in bulk studies. We demonstrated proof-of-principle induction of T5 promoter system with IPTG using the hydrodynamic trap. We are currently using this system to research the effect of alternating inducer concentrations at different frequencies on gene expression in order to determine a cutoff frequency for transcription of the lac operon. We anticipate that the microfluidic-based trap is an ideal platform to study cellular regulation and gene network dynamics of single living cells in free solution.

References:

1. Tanyeri M., Johnson-Chavarria E.M. and Schroeder C.M., "Hydrodynamic Trap for Single Particles and Cells", *Applied Physics Letters*, 96, 224101 (2010)
2. Johnson-Chavarria E.M., Tanyeri M. and Schroeder C.M., "A Microfluidic-based Hydrodynamic Trap for Single Particles", *Journal of Visualized Experiments*, 47, (2011).

(662h) Microrheology of Chromosomal Loci in Escherichia Coli

Zhicheng Long[1], Avelino Javer[2], Eileen Nugent[2], Pietro Cicuta[2], Bianca Sclavi[3], Marco Cosentino Lagomarsino[4] and Kevin D. Dorfman[1], (1)Chemical Engineering and Materials Science, University of Minnesota, Minneapolis, MN, (2)Cavendish Laboratory and Nanoscience Centre, Cambridge University, Cambridge, United Kingdom, (3)LBPA, UMR 8113 du CNRS, ENS-Cachan, Cachan, France, (4)Génophysique / Genomic Physics Group, UMR 7238 CNRS Génomique des Microorganismes, Universite Pierre et Marie Curie, Paris, France

Nucleoid shaping transcription factors (NSTFs) are DNA binding proteins that alter the shape of the bacterial nucleoid at both global and local levels. The activity of NSTFs depends on their environment, most notably the growth rate of the cell. Understanding how these proteins affect the shape of the bacterial nucleoid, in particular their role on the transcription network, is a challenging problem. One possible approach to study the binding is microrheology, where we track the mobility of fluorescently tagged chromosomal loci to obtain information about the local environment surrounding a DNA-bound NSTF.

We have taken a two-pronged approach towards these experiments. In the classical method, we grow colonies of E. coli cells from single parents on agar. While this approach is simple, the throughput is limited. More importantly, the nutrients become depleted as the population grows, leading to changes in the growth rate with time. To circumvent the limitations of the classic method, we designed a microfluidic chemostat that traps the *E. coli* cells, forcing them to grow in lines for many generations while maintaining a constant chemical environment. By carefully controlling the channel geometry, we can immobilize the *E. coli* cells without impacting their growth. We can thus monitor the growth rate at the single cell level and track the movements of fluorescently tagged loci in more than one thousand cells at a time using time-lapse microscopy.

Using both methods, we have obtained high resolution data for the mobility of 14 loci across the chromosome under four different growth conditions, two on our microchemostat and two on agar. At short-time scales (0.1-10s) we have observed a dependence of loci mobility on chromosomal position. Ter proximate loci have a decreased mobility with respect to Ori proximate loci, and the pattern of loci mobility is roughly symmetric with distance along the replichore arm. By examining loci dynamics as a function of sub-cellular position, we observed a tendency of Ter proximate loci to be located at the cell poles and at mid-cell and a corresponding decrease in the mobility of loci at these sub-cellular positions.

On the other hand, mean-square displacements (MSD) of all loci scale as $\gamma\times(\Delta t)^{\alpha}$, where the mean exponent α (despite its wide distribution) seems to be close to 0.4. The exponent is independent of the growth rate, cell cycle time, positions of loci, or the type of experiment (microdevice versus agar). We formulate the hypothesis that this universal dynamical property could be related to a viscoelasticity caused by the genome through a dense fractal organization, and not as a consequence of the action of extrinsic cytoskeletal elements as previously speculated.

(667a) A Two-Scale 13C-Based Method for Metabolic Flux Measurement and Prediction in Genome-Scale Models

Hector Garcia-Martin, Fuel Synthesis Division, Joint Bioenergy Institute, Emeryville, CA

Systems biology aims to provide a predictive and quantitative understanding of cell behaviour as the outcome of the interaction of its comprising parts. Metabolic flux profiles (i.e. the number of molecules traversing each biochemical reaction encoded in its genome per unit time) are not only a key phenotypic characteristic but also embody the essence of this complexity since they represent the final functional output of the interactions of all the molecular machinery studied by all the other "omics" fields. Two of the most popular methods for studying metabolic fluxes are Flux Balance Analysis (FBA) and 13C Metabolic Flux Analysis (13C MFA), each of them displaying its own advantages and disadvantages. In this talk I will present a new method: Two scale 13C Metabolic Flux Analysis (2S-13CMFA), which combines the advantages of FBA and 13C MFA. I will showcase its applications and possibilities with data from the KEIO knockout collection.

(667b) Phenotypic and Multi-Omic Approaches to Address Molecular Bottlenecks in the Fermentation of Lignocellulose Into Ethanol by *Saccharomyces Cerevisiae*

Trey Sato[1], Dana Wohlbach[2], Jeffrey Lewis[2], Yaoping Zhang[1], Mingjie Jin[3], Yury Bukhman[1], Wendy Schackwitz[4], Christa Pennacchio[4], David Hodge[3], Venkatesh Balan[3], Bruce E. Dale[3] and Audrey Gasch[2], (1)University of Wisconsin, Great Lakes Bioenergy Research Center, Madison, WI, (2)University of Wisconsin, Madison, WI, (3)Chemical Engineering and Material Science, Michigan State University, East Lansing, MI, (4)The Joint Genome Institute, Walnut Creek, CA

While cellulosic ethanol is being looked to for relief of the global energy demand, a number of molecular bottlenecks currently exist that prevent the efficient bioconversion of lignocellulose into ethanol. For example, it is well known that: 1) native *Saccharomyces cerevisiae* yeast strains cannot sufficiently ferment xylose, and 2) side products generated from pretreatment, including Ammonia Fiber Expansion (AFEX™) and alkaline hydrogen peroxide (AHP), of plant biomass illicit a cellular stress response, which further limits fermentation productivity. At the Great Lakes Bioenergy Research Center, we have taken a multi-comparative approach to facilitate the discovery and understanding of molecular bottlenecks in the fermentation of lignocellulosic hydrolysates by yeast. Through multi-phenotypic and bioinformatic analysis of 111 natural and domesticated isolates, we have identified a wild *S. cerevisiae* strain that maintains rapid growth and cell viability in a variety of distinctly prepared hydrolysates. Following directed engineering of a xylose metabolism pathway, we performed directed evolution that yielded mutants able to ferment 2 to 3-fold more xylose from AFEX™ corn stover hydrolysate (ACSH) than unevolved parents. Furthermore, we employed temporal profiling of gene expression levels during ACSH fermentations, which identified differences in cell physiology between evolved and unevolved strains. Analysis of extracellular metabolite, amino acid and metal concentrations additionally identified limiting nutrients during fermentation. Coupled with comparative genome resequencing of parental and evolved strains, this suite of Omic data is being integrated in metabolic network models to identify and understand genetic differences that impact xylose fermentation in stress-inducing lignocellulosic hydrolysates.

(667c) Towards Industrial Robustness: Understanding of Wild-Type and Populus Hydrolysate- Tolerant Mutant Strains of Clostridium Thermocellum

Jessica Linville[1], Miguel Rodriguez[2], Jonathan Mielenz[3] and Chris D. Cox[1,4], (1)Department of Civil and Environmental Engineering , University of Tennessee, Knoxville, TN, (2)Bioscieces Division, Oak Ridge National Laboratory, Oak Ridge, TN, (3)Biosciences Division, Oak Ridge National Laboratory, Oak Ridge, TN, (4)Center for Environmental Biotechnology, University of Tennessee, Knoxville, TN

An industrially robust microorganism capable of degrading lignocellulosic biomass in the presence of inhibitory compounds produced from the pretreatment process in a CBP scheme is needed to reduce the cost of cellulosic biofuels. *Clostridium thermocellum* has the ability to convert biomasses complex carbohydrates into ethanol by fermentation. However, *C. thermocellum* is inhibited by by-products from the biomass pretreatment process. To increase *C. thermocellum*'s industrial robustness, a 17.5% v/v *Populus* hydrolysate-tolerant mutant (PM) was produced by 117 successive transfers of *C. thermocellum* 27405 strain into increasing concentrations of the hydrolysate. A single colony culture of the PM was then produced and the genome has been sequenced. A study comparing the wild-type (WT) and PM strains of *C. thermocellum* was completed with growth on cellobiose examining sugar utilization and end product formation, and growth on Avicel examining protein concentration. The growth of the PM in 17.5% v/v *Populus* hydrolysate had a 20% higher optical density, a 33% faster sugar utilization rate (g glucan/L/hr), and produced 34% more ethanol (g ethanol/ g glucan) than the WT grown without hydrolysate. The PM retained these characteristics when grown without hydrolysate with a 37% higher optical density, a 40% faster sugar utilization rate and produced 20% more ethanol than the WT grown under the same conditions. The WT was severely inhibited when grown in the presence of 17.5% v/v Populus hydrolysate with a barely detectable optical density, 7% of sugar utilization rate, and produced 7.5% of the expected ethanol versus when it was grown without hydrolysate. A combination of the genomic mutations of the PM and gene expression profiles from triplicate fermentations in various concentrations of the *Populus* hydrolysate will be used to determine which mutations in the genome might be the cause of the increased tolerance to the hydrolysate.

(667d) Development of Advanced Technologies for Feedstock Chracterization and HTP Pretreatment Process Monitoring

Seema Singh, Joint BioEnergy Institute, Sandia National Laboratory, Emeryville, CA

Today, carbon-rich fossil fuels, primarily oil, coal and natural gas, provide 85 percent of the energy consumed in the United States. Fossil fuel use increases CO2 emissions, increasing the concentration of greenhouse gases and raising the risk of global warming. The high energy content of liquid hydrocarbon fuels makes them the preferred energy source for all modes of transportation. In the US alone, transportation consumes around 13.8 million barrels of oil per day and generates over 0.5 gigatons of carbon per year. This has spurred research into alternative, non-fossil energy sources. Among the options (nuclear, concentrated solar thermal, geothermal, hydroelectric, wind, solar and biomass), only biomass has the potential to provide a high-energy-content transportation fuel. Biomass is renewable resource that is carbon-neutral.

Currently, biofuels such as ethanol are produced largely from grains, but there is a large, untapped resource (estimated at more than a billion tons per year) of plant biomass that could be utilized as a renewable, domestic source of liquid fuels. Well-established processes convert the starch content of the grain into sugars that can be fermented to ethanol. Plant-derived biomass contains cellulose, which is more difficult to convert to sugars. The development of cost-effective and energy-efficient processes to transform cellulose in biomass into fuels is hampered by significant roadblocks, including the lack of specifically developed energy crops, the difficulty in separating biomass components, low activity of enzymes used to deconstruct biomass, and the inhibitory effect of fuels and processing byproducts on organisms responsible for producing fuels from biomass monomers.

The Joint BioEnergy Institute (JBEI) is one of three US Department of Energy Bioenergy Research Centers that will address these roadblocks in biofuels production. JBEI draws on the expertise and capabilities of three national laboratories (Lawrence Berkeley National Laboratory (LBNL), Sandia National Laboratories (SNL), and Lawrence Livermore National Laboratory (LLNL)), two leading US universities (University of California campuses at Berkeley (UCB) and Davis (UCD)), and a foundation (Carnegie Institute of Washington at Stanford University) to develop the scientific and technological base needed to convert the energy stored in cellulose into transportation fuels and commodity chemicals. Established scientists from the participating organizations are leading teams of researchers to solve the key scientific problems and develop the tools and infrastructure that will enable other researchers and companies to rapidly develop new biofuels and scale production to meet US transportation needs, and to develop and rapidly transition new technologies to the commercial sector. This talk will present a brief summary of the organization and efforts at JBEI, highlight the efforts on the development and deployment of novel technologies for biomass analysis (compositional and mechanical) and HTP methods for monitoring pretreatment process efficacies.

(667e) Discovery and Functional Determinations of Biomass-Degrading Enzymes by Robotic Cell-Free Translation

Brian G. Fox[1], Taichi E. Takasuka[1], Lai Bergeman[2], Walker Johnnie A.[3], Kai Deng[4] and Trent Northen[4], (1)Great Lakes Bioenergy Center, University of Wisconsin, Madison, WI, (2)Department of Biochemistry, University of Wisconsin-Madison, Madison, WI, (3)Department of Biochemistry, University of Wisconsin-Madison, mad, (4)Joint BioEnergy Institute, Emeryville, CA

Our use of a platform for functional determination of enzymes will be presented. Robotic wheat germ cell-free protein translation system was used for protein synthesis and enzymatic assays were performed using the translation reactions obtained directly from the robot. Fluorescent sugar analogs and purified polysaccharides were used for initial characterizations. Nanostructure initiator mass spectrometry (NIMS)-based assay provided additional functional assignments and also revealed diagnostic differences in time-dependent cascades of products given by different enzymes. The approach was used to evaluate a majority of secreted cellulosomal proteins from *C. thermocellum* in different enzyme formats, their comparative single reactivities, and their synergistic behavior for hydrolysis of various polysaccharide-containing materials. Combinatorial enzyme activity assays revealed synergistic activities between enzymes with different assigned functions

(667f) Characterization of the Interactions of Cellulose, Hemicellulose, and Lignin During Pretreatment Through the Use of Flowthrough Pretreatment

Heather L. McKenzie[1,2], Jaclyn D. DeMartini[1,2], Nancy L. Engle[1,3], Marcus Foston[1,4], Sivakumar Pattathil[1,5], Bruce A. Tomkins[1,6], Michael G. Hahn[1,5], Arthur J. Ragauskas[1,4], Timothy J. Tschaplinski[1,3], Gary A. Van Berkel[1,6] and Charles Wyman[2,7], (1)BioEnergy Science Center, Oak Ridge, TN, (2)Chemical and Environmental Engineering, University of California Riverside, CE-CERT, Riverside, CA, (3)Biosciences Division, Oak Ridge National Laboratory, Oak Ridge, TN, (4)Department of Chemistry and Biochemistry, Georgia Institute of Technology, Atlanta, GA, (5)Complex Carbohydrate Research Center, University of Georgia, Athens, GA, (6)Chemical Sciences Division, Oak Ridge National Laboratory, Oak Ridge, TN, (7)BioEnergy Science Center, Oak Ridge National Laboratory, Oak Ridge, TN

The conversion of lignocellulosic biomass to ethanol requires recovery of the sugars contained in hemicellulose and cellulose at high yields. Hydrothermal pretreatment can be used to hydrolyze and solubilize hemicellulose prior to enzymatic hydrolysis of the cellulose left in the residual solids to glucose. An understanding of the deconstruction of cell wall during pretreatment would support the development of improved energy crops and pretreatment strategies. The results of flowthrough pretreatment of native poplar and model substrates were examined for evidence of the interaction of cellulose, hemicellulose, and lignin during pretreatment. The liquid effluent was analyzed for oligomers and phenols, while the composition of the residual solids was tested using HPLC and NMR. Differences in the release of xylooligomers from native poplar, holocellulose, and isolated xylan demonstrated the importance of the interactions of cellulose-hemicellulose and lignin-hemicellulose on hemicellulose hydrolysis. Pretreatment of the isolated lignin indicates that depolymerization is the primary mechanism for lignin extraction while differences in the release of lignin from native poplar and the isolated lignin indicate that lignin-hemicellulose interactions influence relative reactivity. Recent work has identified the removal of specific cell wall structures, such as methyl glucuronoxylan, as key to improvements in enzymatic digestibility. The removal of these components during pretreatment will be examined closely.

INTEGRATION OF ACID-ENZYME HYDROLYSIS OF LIGNOCELLULOSIC BIOMASS TO FERMENTABLE SUGARS

Suping Zhang , Yan Li , Baozhen Zhang, Yongjie Yan

(East China University of Science and Technology, Shanghai, People's Republic of China)

Abstract

Although many previous studies determined dilute acid pretreatment conditions that maximize xylose yields from pretreatment or glucose yields from subsequent enzymatic hydrolysis of the pretreated cellulose, this paper emphasizes the use of different conditions to realize the highest total fermentable sugar yields from both stages. The acid pretreatment tests were performed under the conditions of acid concentration at 0.05%, 0.25%, 0.50%, 1.0% and the temperature in the range of 160□ to 180□. The maximum yield for just xylose in acid pretreatment occurred at 170□ with the acid concentration of 0.25%. However, the total fermentable sugar yields from the two stages occurred at the pretreatment conditions of temperature 180□ and acid concentration 0.25%. The results show that the conditions for the highest xylose yield in pretreatment step did not give the highest glucose yield in enzymatic hydrolysis step nor total fermentable sugars yield from both stages. Therefore, the effects of dilute acid hydrolysis conditions on the following enzymatic hydrolysis yields were investigated.
Keywords: Lignocellulosic Biomass; Total Fermentable Sugars; Dilute Acid Pretreatment

Introduction

With depletion of oil reserves and the green house effect caused by the consumption of fossil fuels, bio-ethanol production through conversion of biomass has attracted much attention in the whole world [1, 2]. The commercialized production of biofuel up-to-date is mostly based on sugar or grain as raw materials [3, 4]. However, the limited production capacity can not satisfy the large demands from the energy market. The abundant lignocellulosic biomass is a good substitute raw material, including agricultural residue, forest residue, bagasse from sugar industry and even municipal garbage.

Corn stalk is an abundant agricultural residue in China. It is estimated that annual harvest of crop stalk amounts to more than 700 million tons (dried base) [5]. On one hand, energy security can be realized by one more way; on the other hand, the problem of environment pollution caused by direct burning the crop stalks can find a solution.

Production of fermentable sugars from corn stalk is an important step for ethanol production [6, 7]. Cellulose is the main component in corn stalk which can be turned into glucose through hydrolysis. Glucose is the important monosaccharide in production of ethanol through fermentation. With the development of bacterial strains specific for fermentation of xylose and other pentose [8, 9] yielded from hydrolysis of hemicellulose, Xylose can thus be used to produce ethanol through fermentation. Utilization of xylose can greatly improve the economics of hydrolysis or saccharification process [10]. Therefore, research on improvement of the total fermentable sugar yield becomes especially important.

Sacharification can be realized through hydrolysis either acid or enzymatic; either one has advantages and disadvantages [11]. Dilute acid pretreatment is regard as one of the most promising and cost-effective pretreatment process [12].A number of previous studies determined dilute acid pretreatment conditions that maximize xylose yields from pretreatment or glucose yields from subsequent enzymatic hydrolysis [13]. The effect of acid pretreatment on subsequent enzymatic hydrolysis performance has not been studied in great detail, especially in regards to total fermentable sugars yield as the target [14]. This study was focused on the coupling acid-enzyme hydrolysis to maximize the total fermentable sugar yield in a percolation reactor and corn stalk as material.

Experimental

Materials

Corn stalk used in experiments was first washed clean, cut short and got naturally dried in air. After further crushing, the stalk dust through the sieve of 20-40 mesh was selected as materials and stored in dried place for standby. In Table 1, the composition of the material used in experiment is listed.

Table 1 The Composition of Corn Stover on Dry Basis.

Soluble Carbohydrates, Proteins and Tannins (%)	Hemicellulose (%)	Cellulose (%)	Lignin (%)	Ash (%)
10.34	25.78	43.65	17.21	3.02

Hydrolysis Process

Acid pretreatment was carried in a percolation reactor made of stainless steel (316L). The bulk of the reactor is of cylinder-shape with the inner diameter 10cm, total capacity of 1.5 L. The reactor is connected with a steam generator as shown in Fig 1

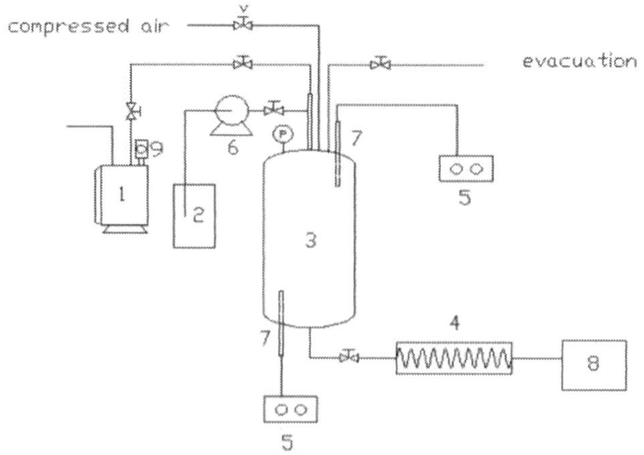

1.steam generator; 2.dilute acid tank; 3. hydrolyzer; 4.condenser; 5. temperature indicator; 6. metering pump; 7、thermal couple ; 8. container for hydrolysis products; 9. pressure controller; V. valve; P.pressure gauge

Fig. 1. Schematic diagram of acid pretreatment experimental apparatus

The residue from the acid pretreatment was subject to the sequent enzymatic hydrolysis by adding enzyme-containing solution and PH-adjusted hydrolysate from pretreatment acid hydrolysate to the predetermined solid-to-liquid ratio.

Analysis method

Monosaccharides were analyzed through external standard method on HPLC (SHIMADZU, LC-20AT) and the detector was Polymer Laboratories PL-ELS 2100 (move phase: acetonitrile: water=23:77(φ), total flow 0.8 mL/min, nebulization temperature 40□, gasification temperature 70□, column: intersil-NH_2, 4.6 mm×250 mm, 5-Micron).

Results and Discussions

Effects of Acid Concentration

The effects of acid concentration on xylose yield were investigated with the acid concentrations varied from 0.05% through 0.1%, 0.25% to 0.50% under the determined temperature 160℃ and the fixed liquid to solid ratio 10:1.

Data in Fig.2 indicates that while the acid concentration reduces to 0.05%, the xylose concentration is very lower due to a low level of hemicellulose hydrolysis rate under the lower acid concentration. It should be noted that the acid concentration in the reactor practically not equal to the acid concentration that is fed because of the buffer action from some natural cellulosic substances [15]. Maloney [16] and coworkers have concluded from their research that the buffer action by the white birch can be 3.5 g of sulfuric acid /kg.

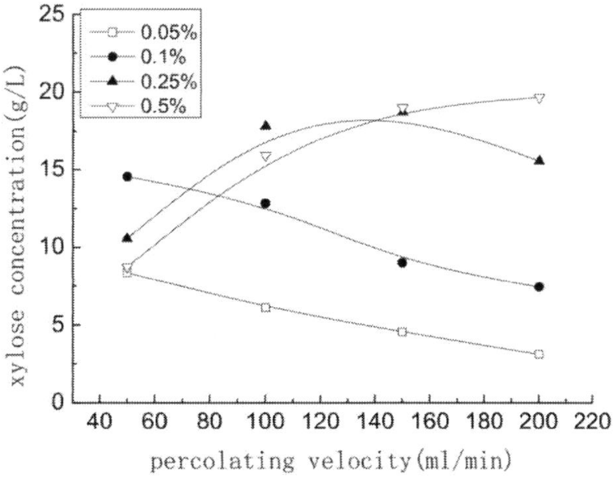

Fig.2. The effect of sulfuric acid concentration on the xylose concentration

When the acid concentration rise to 0.25%, it is evident from observing Fig 2 that decreasing the liquid percolation rate will increase the xylose concentration initially and then decline. The results can be explained as percolation rate decreases, hemicellulose is hydrolyzed more completely, and resulting in xylose in the hydrolysate is accumulated and then reaches the peak. But, a prolonged reaction time also results in a decrease in xylose concentration for its degradation.

The increasing acid concentration dramatically increases the hemicellulose hydrolysis rate and also speeds up the xylose degradation rate. Hydrolysis under the acid concentration of 0.5% did not realize much increase in the xylose concentration. Furthermore, high acid concentration means that operation cost increases and equipment corrosion speeds up. So, the acid concentration of 0.25% was determined as the condition for the highest xylose yield.

Effect of Hydrolysis Temperature

The experiments was proceed within the temperature range of 160-180℃ in the presence of acid with concentration of 0.25%. The results are listed in Fig 3.

At 160□ the rate of xylose generation is slower to reach its peak than those at other temperatures; however sugar preservation is sustained for a longer period of time due to a low level of sugar decomposition. The maximum xylose yields occurred at the temperature of 170℃ with acid concentration of 0.25%.

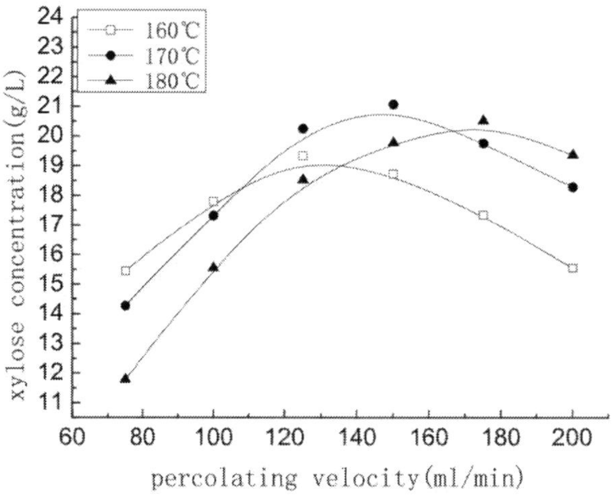

Fig.3. The xylose concentration varys with the temperature and the percolating velocity

Effects of Acidic Pretreatment on Enzymatic Hydrolysis

Fig 4 gives the cellulose contents in the residues after the acid pretreatment. Only a small portion of cellulose is converted when hydrolysis proceeds at 160℃ or 170℃, both with acid concentration of 0.25%. But when temperature is raised to 180℃, the cellulose lost in the residue is noticeable for the cellulose hydrolysis at the pretreatment step. With the hemicellulose and cellulose hydrolysis, the crystalline structure of the cellulose is destructed which favors for the enzymatic hydrolysis (see Fig 5).

Residue of corn stalk after pretreatment operated under various conditions was subsequently hydrolyzed enzymatically. Here, the degree of enzymatic hydrolysis is defined as follows: the amount of glucose so formed is divided by the amount of glucose that is theoretically calculated from the cellulose content in the residue.

Fig 6 indicates that the more severe conditions for the acid hydrolysis can lead to the higher degree of enzymatic hydrolysis. In the runs of tests, the highest degree of enzymatic hydrolysis was over 90%, which implies that a small fraction of cellulose is still unconverted, for the cellulose contained in the residue is not identical.

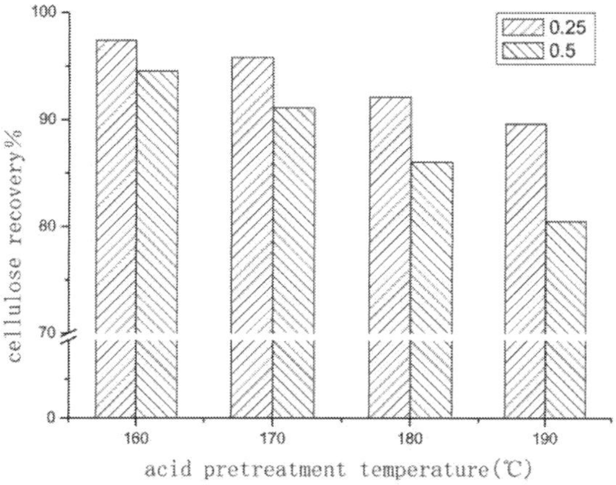

Fig. 4. Cellulose recovery in the water-insoluble solid at different pretreatment conditions, expressed as a percentage of the initial content in the raw material.

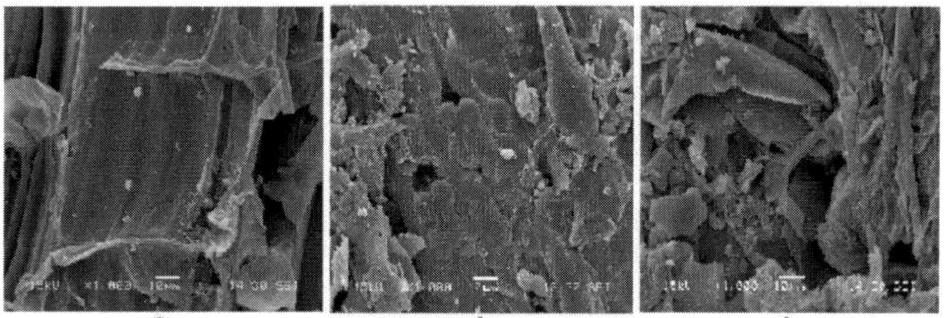

a b c

a: corn stalk; b: the residue treated at the pretreatment condition of 170℃; c: the residue treated at the pretreatment condition of 180℃

Fig. 5 SEM images of corn stalk (a) and acid pretreatment residues (b and c)

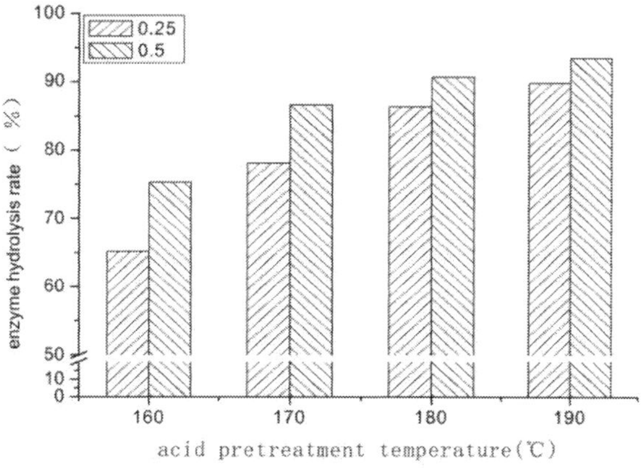

Fig. 6. Enzymatic hydrolysis yield at different pretreatment

Combination of Acid-enzyme Hydrolysis

Concentrations of the total fermentable sugar obtained from both steps of the acid pretreatment and enzymatic hydrolysis are illustrated in Fig 7. Two columns consist of one group with the temperature of pretreatment as parameter; the column in front indicates the concentration of the total fermentable sugar obtained at acid concentration 0.25% as well as the shares of xylose and glucose while the column behind indicates the data obtained at acid concentration of 0.5%.

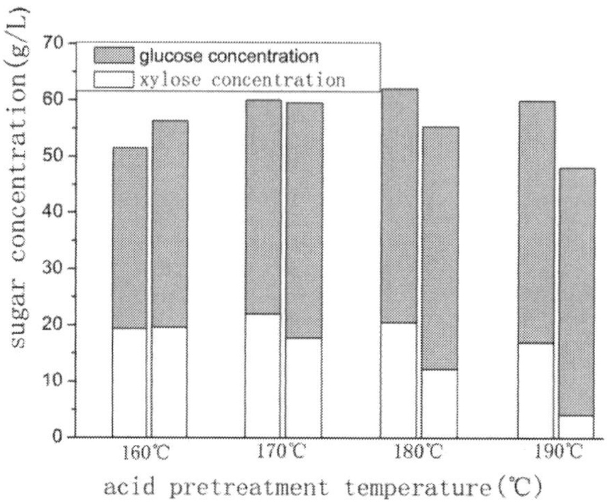

Fig. 7. Overall sugar concentration at different pretreatment temperature

It is known from the Fig 7 that the concentration of total fermentable sugar is the highest when the pretreatment proceeds at 180℃ and acid concentration 0.25%. In case that the xylose concentration is taken as the desired product, the preferential pretreatment conditions are at 170℃ with a sulfuric acid of 0.25%. Compared with the lower temperature (170℃) experiments, there was a 3.1g/L of glucose increased. The results show that the pretreatment conditions for the highest xylose yield did not coincide with that of for the highest total fermentable sugar yield. Therefore, it is essential to control the pretreatment severity when a process of acid-enzyme coupling hydrolysis shall be employed for the lignocellulosic biomass.

Conclusions

In this paper, conclusions are made for the acid-enzyme coupling hydrolysis of corn stalk with the concentration of total fermentable sugar as the target.

(1) The preferential conditions for the pretreatment by the dilute acid are as follows: reaction temperature, 170℃; concentration of sulfuric acid 0.25%; liquid to solid ratio 10:1.

(2) The preferential conditions for acid-enzyme coupling hydrolysis of corn stalk are as follows: pretreatment temperature 180℃; acid concentration 0.25% when enzymatic hydrolysis proceeds.

(3) The highest xylose yield did not coincide with the highest total fermentable sugar yield. The severity of acid pretreatment affects the enzymatic hydrolysis results.

Acknowledgement

The study was supported by National Key Technology Support Program (2011BAD22B01, 2012BAD32B07) and Chinese National High-tech R&D Program (863 Program) (2012AA101806) are greatly appreciated.

References

[1] Brethauer S., Wyman C. E., **(2010)** *Bioresour. Technol.*, 101, 4862–4874.
[2] Qureshi N., Ezeji T. C., Ebener J., Dien B. S., **(2008)** *Bioresour. Technol.*, 99, 5915–5922.
[3] Bai F.W., Anderson W. A., Young M. M., **(2008)** *Biotechnol. Adv.*, 26, 89– 105.
[4] Talebnia F., Karakashev D., Angelidaki I., **(2010)** *Bioresour. Technol.*, 101, 4744–4753.
[5] Zhang S. P., Marechal F., Garssner M., **(2009)** *Energy Fuels*, 23(3), 1759-1765.
[6] Ewanick S. M., Bura R., Saddler J. N., **(2007)** *Biotechnol. Bioeng.*, 98 (4), 736-746.
[7] Kumar P.; Barrett D. M., Delwiche M. J., Stroeve P. **(2009)**, *Ind. Eng. Chem. Res.*, 48 (8), 3713–3729.
[8] Saitoh S., Hasunuma T., Tanaka T., **(2010)** *Appl. Microbiol. Biotechnol.*, 87, 1975-1982.
[9] Rubin E. M., Nat. **2008,** 454, 841-845.
[10] Ruiz C. E., Oliva J. M., Saez F., Castro E., **(2008)** *Bioresour. Technol.*, 99(6), 1869-1876.
[11] Gírio F. M., Fonseca C., Carvalheiro F., **(2010)** Bioresour. Technol., 101, 4775–4800.
[12] Juan E. Morinelly, Jill R. Jensen, Matthew Browne, et al. **(2009)** *Ind. Eng. Chem. Res.*, 28, 9877-9884.
[13] Cara C., Ruiz E., Oliva J. M., Saez F., Castro E., **(2008)** *Bioresour. Technol.*, 99, 1869-1876.
[14] Jensen J. R., Morinelly J. E., Gossen K. R., Brodeur-Campbell M. J., Shonnard D. R., **(2010)** *Bioresour. Technol.*, 101, 2317–2325.
[15] Hsu D. D., **(1995)** Biomass Chemical Analysis and Testing Laboratory Analytical Procedures. NREL,.
[16] Maloney M. T., Chapman T. W., Baker A. J. **(1985)** *Biotechnol. Bioeng.*, XXVII, 355-361.

(675b) Hybrid Thermochemical Processing: Fermentation of Pyrolytic Substrates

Tao Jin[1], Yi Liang[2], Donovan S. Layton[1], Mark M. Deaton[3], Zhanyou Chi[4], Robert C. Brown[5], Zhiyou Wen[2] and **Laura R. Jarboe**[6], (1)Chemical and Biological Engineering, Iowa State University, Ames, IA, (2)Food Science and Human Nutrition, Iowa State University, Ames, IA, (3)Chemical & Biological Engineering, Iowa State University, Ames, IA, (4)Department of Biological Systems Engineering, Washington State University, Pullman, WA, (5)Mechanical Engineering, Iowa State University, Ames, IA, (6)Department of Chemical and Biological Engineering, Iowa State University, Ames, IA

As a temporary storage unit of sunlight-derived energy and atmospheric carbon, biomass is an excellent source of carbon and energy for the production of biorenewable fuels and chemicals. However, the cost- and energy-efficient release of fermentable sugars from this biomass is challenging, largely due to the complex structure of lignocellulose. Thermochemical processing of biomass by fast pyrolysis provides a non-enzymatic route for depolymerization of biomass into fermentable substrates that can be used for the biological production of fuels and chemicals. Fermentative utilization of these pyrolytic substrates faces two formidable challenges. First is the fact that most bio-oil-associated sugars are present in the anhydrous form. Metabolic engineering has enabled utilization of the main anhydrosugar, levoglucosan, in workhorse biocatalysts. The second challenge is the fact that bio-oil is rich in microbial inhibitors. Biomass processing, collection of the pyrolytic substrates, detoxification of the substrates prior to fermentation and increased robustness of the biocatalyst are effective methods for addressing this inhibition. Here we describe metabolic engineering efforts to enable utilization of these pyrolytic substrates.

(675c) Influence of Impeller Design On Power Input and Performance During High Solid Enzymatic Hydrolysis of Pretreated Lignocellulose

Benny Palmqvist and Gunnar Lidén, Department of Chemical Engineering, Lund University, Lund, Sweden

Reaching high ethanol titers after the fermentation step is crucial in cellulosic ethanol production in order to make the process cost competitive and commercially available. To achieve this, a high amount of water insoluble solids (WIS), i.e. lignocellulose fibers, is needed in the hydrolysis step. However, a high WIS content in the hydrolysis step is likely to give a high viscosity. The power requirements needed to operate the impeller at high viscosities can be quite substantial, and high WIS processes therefore may need to be operated with a poor mass- and/or heat transfer. We have previously shown that stirrer speed (or more accurately – mixing-power) strongly affects the hydrolysis of certain lignocellulosic materials, in particular pretreated spruce. An almost 100 % yield increase was achieved by increasing the stirrer speed from 75 to 500 rpm. In contrast, the power input seems to have only a very marginal impact on the hydrolysis performance when using for example the energy crop Arundo Donax (Giant Reed). Another difference between these particular materials is their relative change in rheology due to enzymatic hydrolysis. For example, Arundo Donax quickly (within 2-4 hours) losses its network strength, leaving no significant difference in needed power input between different WIS contents (within the range of 10-20 % WIS). Spruce on the other hand retains its strength, i.e. high viscosity, during a much longer time which results in significantly higher power inputs during a longer time. For the same stirrer rate, approximately, four times as much energy is needed for mixing of spruce compared to Arundo Donax at 20 % WIS.

In the present study, enzymatic hydrolysis at high solid loadings (<20% WIS) was investigated using different impeller types. In addition to sugar yields, impeller torques (and hence power input) were monitored throughout the hydrolysis to compare the suitability of the impeller type for this process. Furthermore, it was possible to control the power input at a fixed set-point value throughout the hydrolysis period by adjusting the stirrer speed. This enables a proper comparison between the different impellers based on an energy input basis. The factors affecting high solid hydrolysis will be discussed.

Enzymatic recovery of polyhydroxybutyrate (PHB) from *Bukerholderia cepacia* by pancreatin

Yuanzhen Wang, Shijie Liu*

Department of Paper and Bioprocess Engineering, SUNY College of Environmental Science and Forestry, 1 Forestry Drive, Syracuse, NY. 13210

*Author for correspondence (E-mail: sliu@esf.edu)

1 Abstract

Enzymatic digestion by pancreatin has been developed as an effective and cost efficient method to recover polydroxybutyrate (PHB) from fermentation culture of *Bukerholdria cepacia.* Pretreatment methods including heat shock, surfactant digestion, and oxidizing agent digestion were screened, and heat pretreatment at 95 °C for 45 min was chosen for high efficiency and no introduction of chemicals. Transmission electron microscopy (TEM) indicated the efficient cell lysis and PHB granules release by pancreatin. Optimized enzymatic digestion condition was found to be 50 °C in a pH 8.0 Tris-HCl buffer, with the pancreatin/biomass ratio fixed as 0.02. PHB recovery and purity reached up to 93.5% and 90.6%, respectively. Successful protein removal was achieved. Polymer degradation was negligible because no significant molecular weight reduction was observed.

Keywords: Polyhydroxybutyrate; Enzymatic recovery; Pancreatin; *Bukerholdria cepacia*

(675e) Screening Wood-Decay Fungi for Demethylation of Kraft Lignin Using Two Novel Assays with Applications in Producing Bio-Methanol and Formaldehyde-Based Polymers

Andrew Gibson, Biotechnology, Lakehead University, Thunder Bay, ON, Canada

Screening Wood-decay Fungi for Demethylation of Kraft Lignin using Two Novel Assays with Applications in Producing Bio-Methanol and Formaldehyde-based Polymers

Andrew Gibson [1,2], Balaji Venkatesagowda [1], Iryna Kandybovich [1],

Aneli M. Barbosa [1], Brian Ross [2], Lada Malek [2], Robert F.H. Dekker [1]

[1] Biorefining Research Institute, [2] Department of Biology, [3] Northern Ontario School of Medicine, Lakehead University, Thunder Bay, Ontario, Canada P7B 5E1

Lignin, the most abundant aromatic polymer present in the biosphere, is found in the cell walls of plants rendering the plant resistant to microbial attack and degradation. This biopolymer has potential applications, yet is currently considered a waste product (black liquor) in Kraft pulp and paper mills. By enzymatically modifying Kraft lignin, two immediate products can be obtained: a modified lignin with a higher phenolic content, and bio-methanol. The first can be used in formaldehyde-based polymer applications, while the second can be used in industrial processes and as biofuel. We have developed two specific assay procedures to screen for enzymatic activity that would produce these products; selected ion flow tube-mass spectrometry (SIFT-MS) and Ti(III)-nitriloacetic acid colorimetry. SIFT-MS is a highly sensitive technique capable of measuring trace amounts of gases including methanol (ppb levels) in the headspace of flask-cultivated fungi, while the Ti(III)-NTA assay measures the changes in vicinal hydroxyl groups during the modification of lignin by various fungal species. The Mycological Herbarium at Lakehead University in conjunction with the Biorefining Research Institute has a collection of >450 fungal isolates from the Boreal Forest in northern Canada; 20 of which were screened using these techniques. Fungal cultures were grown under stationary conditions on a Kraft lignin-minimum salts (Vogel) nutrient medium for 3 weeks at room temperature. The fungal isolates showed a reduction of up to 45% of *O*-methyl content using the Ti(III)-NTA assay, and a 10-fold increase in methanol concentration as measured by SIFT-MS compared to controls. The research presents evidence of *new* enzyme activities that specifically cleave *O*-methyl groups from the lignin molecule. Further work will continue and confirm the existence of these enzyme activities through their isolation and purification.

Supported by funds from NSERC-CRD (CRDPJ 38079-09: Dekker) Canada

(675f) Biological and Enzymatic Demethylation of Kraft Lignin and Lignin-Like Model Compounds by Boreal Forest Fungi Using a Novel Technique: Selected Ion Flow Tube-Mass Spectrometry (SIFT-MS)

Balaji Venkatesagowda[1], Andrew Gibson[2], Aneli M. Barbosa[1], Brian Ross[3], Lada Malek[4] and Robert F.H. Dekker[1], (1)Biorefining Research Institute, Lakehead University, Thunder Bay, ON, Canada, (2)Biorefining Research Institute, Department of Biology, Lakehead University, Thunder Bay, ON, Canada, (3)Northern Ontario School of Medicine, Lakehead University, Thunder Bay, ON, Canada, (4)Department of Biology, Lakehead University, Thunder Bay, ON, Canada

Selected Ion Flow Tube-Mass Spectrometry (SIFT-MS) is a rapid and highly sensitive analytical method for detecting volatile organic compounds by chemical ionization using selected ions. The objective of this work was to detect methanol released from Kraft lignin (KL) and lignin-like model compounds (LMCs) by the action of enzymes produced by ligninolytic fungi that specifically cleave-off *O*-methyl groups. Lignin demethylation is considered a *new* enzyme activity, and such enzymes can be described specifically as lignin demethylases. The culture collection at Lakehead University harboring some 450 wood-decay fungi collected from the Boreal Forest in northern Canada were grown on minimum salts (Vogel) nutrient medium containing Kraft lignin and different low MW LMCs under stationary conditions at room temperature for 4 weeks. Growth on these substrates induced the formation of enzymes that specifically liberated methanol as determined by SIFT-MS. The most promising fungi examined producing methanol from KL were isolates LU-86 (84 ppb), LU-122 (125 ppb), BRI-269 (112 ppb) and BRI-270 (180 ppb), and with LMCs as substrates: BRI-122 (30 ppb), BRI-269 (140 ppb) and BRI-270 (171 ppb). The low levels of methanol produced are attributable to these isolates also producing an enzyme (methanol oxidase) that metabolizes the methanol released. Confirmation of demethylation of lignin was supported by an increase in phenolic (vicinal hydroxyl) content in the treated lignin. In another series of experiments, the same fungal isolates (6) were cultivated on media containing KL for 4 weeks. Mycelia were removed by centrifugation, and the supernatants recovered (ECF) and used as the source of enzyme. Enzymatic demethylation of KL and different LMCs was determined by adding ECF and measuring the amount of methanol liberated by SIFT-MS. Different lignin substrates, such as lignosulfonate (Sigma), periodate-oxidized lignin, Kraft lignin, and lignins extracted from softwood and hardwood black liquors, were also examined for the release of methanol by enzymatic demethylation. The release of methanol from these ligninolytic substrates as measured by SIFT-MS demonstrated unequivocally that *O*-methyl groups were cleaved from these compounds by the fungal and enzyme systems studied. The results obtained will be discussed in detail.

Supported by funds from NSERC-CRD (CRDPJ 38079-09: Dekker) Canada

Tethered immunolipoplex nanoparticle (tILN) device for detection of circulating tumor cells (CTCs) for lung cancer diagnosis

Junyu Ma[1], Kwang Joo Kwak[1], Yun Wu[1], Daniel Gallego-Perez[1], Xinmei Wang [1], Hongyen He[1], Yicheng Mao[1], L. James Lee[1,2,3]

[1]Nanoscale Science and Engineering Center for Affordable Nanoengineering of Polymeric Biomedical Devices, The Ohio State University, 174 West 18th Avenue, Room 1012, Columbus, Ohio 43210, United States
[2]Division of Pharmaceutics, College of Pharmacy, The Ohio State University, 500 West 12th Avenue, Columbus, Ohio, USA)
[3]William G. Lowrie Department of Chemical and Biomolecular Engineering, The Ohio State University, 140 West 19th Avenue,Room 125A, Columbus, Ohio 43210, United States

Lung cancer is the leading cause of cancer deaths worldwide with a disappointing 15% overall 5-year survival rate. Circulating tumor cells (CTCs) which may constitute origin of metastasis have been identified in the peripheral blood of cancer patient. Detection of CTCs have showed great potential in cancer diagnosis and surveillance. However, because CTCs are rare, CTCs detection technique was limited by tremendous technical challenges including a lack of interlaboratory reproducibility, low numbers of isolated CTCs and loss of cell viability during processing which limiting further biological assessment.

We have recently developed a novel and simple 'tethered ImmunoLipoplex Nanoparticle (tILN)' method to detect CTCs in the blood of lung cancer patients. In this study, microRNA-21 (miR-21) and lung specific TTF-1 mRNA were selected as the biomarkers for lung cancer early detection. Briefly, intracellular molecular probes, such as molecular beacons (MBs) corresponding to miR-21 and TTF-1 mRNA respectively and cell surface marker receptors, are encapsulated and inserted respectively in liposomal nanoparticles tethered on a pretreated surface.

The tLN device as first test with non-small cell lung cancer cells (A549) .The results demonstrated A549 with high miR-21and TTF1 expression could be detected by our tLN device, suggesting possibility of CTC detection in whole peripheral blood. To further test our tLN device, prestained A549 cells were spike into health people blood sample. A549 cells on tLN showed stronger green and red fluorescence signals which are corresponding to miR-21 and TTF1 expression respectively compared to peripheral blood mononuclear cells(PBMC), strongly indicating A549 cells were detectable in health people blood. Nanoparticles uptake is enhanced and optimized by utilizing the combined antibody and lipid interactions. Compared to conventional method, tILNs have advantages of higher molecular beacons uptake and lower antibody and molecular beacons consumption.

Our method represents an innovative approach to characterizing the peripheral blood lung cancer. tILNs showed great potential to be an alternative CTCs detection and isolation device which may contribute to cancer diagnosis and understanding of metastasis mechanism.

(676b) Imaging Primo Vascular System, a Newly Found Vascular System

Kyung A. Kang, Chemical Engineering, University of Louisville, Louisville, KY and Kwang-Sup Soh, Nano Primo Research Center, Seoul National University, Suwon, South Korea

Primo Vascular System (PVS) is a newly found organ in the body, which is composed of nodes (Primo Nodes; PNs) and thin vessels (Primo Vessels; PVs) branching out of PNs. The liquid inside the PVS (Primo Fluid) flows at an average velocity of 0.3 mm/s. PVs are mechanically elastic but rather fragile, requiring extensive care during harvesting. This system has been observed in most organs, forming an extensive network throughout the body, including the heart, intestine, liver, adipose tissue, brain, even in some blood and lymphatic vessels.

PVS is extremely difficult to image to the people who are not trained for microsurgery, due to its semi-transparent optical property and small size. The diameter of rabbit PVs is in the range of only 20-50 micrometers and a PV is composed of a bundle of sub-vessels running in parallel. The size of the PN is in the range of 100-1000 micrometers. A few unique characteristics of PVS that could be used for imaging PVS are: that its outer vessel walls are more porous than those of blood or lymphatic vessels and that the nuclei of the PV endothelial cells are rod-shaped instead of circular shape. The first property can used to visualize it with the help of contrast agents because the agents are preferentially infiltrated to PVS, not to blood or lymph vessels. The latter can be used for imaging using the nucleus specific dye (e.g., DAPI) so that the vessels with rod shaped nuclei can be differentiated from that with circular neclei (e.g., blood or lymphatic vessels).

Due to the short history of discovering this system, only limited information on the PVS roles and functions has been revealed, via various imaging techniques. Some of the important properties already reported are:

* The abundant presence of PVS in the adipose tissue may be an important in the study of obesity;

* Some cells transported via PVS present stem cell biomarkers, which may be important for the relationship between PVS and regeneration;

* These stem-cell-like cells are in the PNs at a very high concentration, potentially implying that PNs are for storing these cells;

* A PV runs from the brain via the spinal canal, indicating some type of communication between the brain and rest of body. Also this PV may be used as a special conduit for drug delivery avoiding the known difficulties in the delivery through the blood-brain barrier;

* According to the study results using xenografs of several cancer types (lung, ovarian, skin, gastric and breast cancers, leukemia, etc.), as the cancer is formed PVS is also formed at a high density around the tumor. In addition, the cancer cells were observed to be transported via PVS from the primary to the secondary tumors more actively than via lymph vessels. Optical imaging of cancer cell transport via PVS has been performed using quantum dot-electroporated cancer cells, which was another unique imaging technique.

With the findings listed above alone, PVS appeared to have fundamental roles in human life and image techniques with high resolution that can provide three dimensional PVS maps in the body (e.g., CT and MRI with gold and iron oxide nanoparticles, respectively) may change the paradigm of current medical practice. We feel that it is extremely important to elucidate the roles and functions of PVS as soon as possible, so that these findings can be used in human health management.

(676c) Stabilized Proteolipobead Microenvironments for the Functional Analysis and Drug Screening of the Cancer and Alzheimer's Disease Drug Target Gamma-Secretase

M. Lane Gilchrist[1], Lina Zhong[2], Ji Yuen Hur[3], Jesse Martin[4], Kwangwook Ahn[5] and Yueming Li[5], (1)Chemical Engineering, City College and the Graduate Center of the City University of New York, New York, NY, (2)Department of Chemical Engineering, City College of New York (Graduate Center, CUNY), New York, NY, (3)Molecular Pharmacology and Chemistry Program, Memorial Sloan-Kettering Cancer Center, New York, NY, New York, NY, (4)Biomedical Engineering, City College of New York of CUNY, New York, NY, (5)Molecular Pharmacology and Chemistry Program, Memorial Sloan-Kettering Cancer Center, New York, NY

The intramembrane protease gamma-secretase is a current target of therapeutic intervention, with pivotal pathological functions within Alzheimer's disease and cancer. Our primary objective is to expand methods for *in vitro* studies of the intramembrane proteases (IMPs) with the development of a microsphere-supported biomembrane (proteolipobead) platform. Despite extensive studies, understanding of IMP regulation and catalysis has been hampered by membrane-associated enzymology. The precise reaction mechanisms of the intramembrane proteases remain to be elucidated and furthermore, rigorous analysis of the kinetics of interfacial catalysis in these systems has not yet been undertaken.

High throughput of screening of drug candidates has been carried out in bulk assay systems using cell membrane fragments, solubilized enzymes and is underway in proteoliposomes. A critical barrier to further progress in the study and HTPS of gamma-secretase is that such bulk systems do not allow for the direct in situ quantification of enzyme, substrates, or inhibitors or their relative distributions within the structures under assay. Due to these factors and the instability of these systems there are limitations in the scope of assays possible using solubilized systems and proteoliposomes in bulk solution. We have expanded in vitro models of gamma-secretase to include proteolipobeads (microsphere-supported biomembranes). Using confocal imaging in tandem with flow cytometry, we will present findings from i) direct measurements of substrate and gamma-secretase loading within supported biomembranes, ii) the localization and quantification of enzyme:substrate and enzyme:inhibitor complexes within supported biomembranes, and iii) comparisons of stability of tether-supported assays to existing systems. We have also initiated studies of fused biomembrane systems make new in vitro models of gamma secretase catalysis and regulation.

(676d) Imaging Echogenic Microparticles in the Gastrointestinal Tract: Demonstration of a New Class of Ultrasound Contrast Agents for Noninvasive Diagnosis of Eosinophil Esophagitis

Hedieh Saffari[1], Anne Kennedy[2], Kathryn Peterson[3], Gerald Gleich[3,4] and Leonard F. Pease III[5], (1)Chemical Engineering, University of Utah, Salt Lake City, UT, (2)Clinical Radiology, University of Utah, Salt Lake City, UT, (3)Department of Internal Medicine, Division of Gastroenterology, University of Utah, Salt Lake City, UT, (4)Department of Dermatology, University of Utah, Salt Lake City, UT, (5)Chemical Engineering and Pharmaceutics & Pharmaceutical Chemistry, University of Utah, Salt Lake City, UT

Imaging the transport of microparticles in GI tract is important to using them as ultrasound contrast agents. This is particularly critical in the esophagus where peristalsis driven flow (2-4 cm/s) drags contrast agents through the 20-35 cm long adult esophagus in 5-20 s, yet image contrast agents are essential to understand the spatial distribution of molecular markers of esophageal diseases. For example, eosinophilic esophagitis (EoE) is characterized by patchy invasion by eosinophils, inflammatory leukocytes not found in the healthy esophagus. At the molecular level, eosinophils release granule proteins, such as the cytotoxic major basic protein (MBP-1), upon activation. State-of-the-art diagnosis of EoE necessitates endoscopy with biopsy that is invasive, and biopsy specimens characterize less than 2% of the esophagus even though the inflammation is patchy, suggesting the need for numerous biopsies to ensure that the involved area has been adequately sampled.

Here we report a new ecogenic contrast agent comprised of insulin particles approximately 0.1-3.0 mm in diameter that can be used as ultrasound contrast enhancement agents labeled with eosinophil specific antibodies to facilitate the diagnosis of EoE. We prepared the first eosinophil specific ultrasound contrast agents from insulin particles, tethering EoE specific antibodies to their surfaces, and experimentally modeled the transport and binding of these agents to eosinophil proteins on *ex vivo* monkey esophagi. The results show that insulin particles can be observed by ultrasound and bind to the granule proteins. From serial ultrasound images, we determine the characteristics and concentration profile of the particle slurry in the esophagi, track changes in the binding layer thickness over time, and evaluated surface binding kinetics. Our results show that insulin particles provide strong ultrasound contrast and bind to the granule proteins even at air-water interfaces that are challenging for other classes of agent. These results indicate the potential of this new class of contrast agents to diagnose a broad range of lung and gastrointestinal diseases.

(676e) Dendrimer Based Nanoprobes for Super-Resolution Fluorescence Microscopy (STORM)

Younghoon Kim, Department of Chemical and Biomolecular Engineering, University of Illinois at Urbana-Champaign, Urbana, IL, Sung Hoon Kim, Chemistry, University of Illinois, Urbana-Champaign, Urbana, IL, John A. Katzenellenbogen, Department of Chemistry, University of Illinois at Urbana-Champaign, Urbana, IL and Charles M. Schroeder, Chemical and Biomolecular Engineering, University of Illinois at Urbana-Champaign, Urbana, IL

Recent advances in imaging and single molecule fluorescence microscopy have enabled the direct observation of biological processes at the molecular level. Super-resolution imaging techniques such as photoactivated localization microscopy (PALM) and stochastic optical reconstruction microscopy (STORM) have been developed using photoswitchable fluorescent proteins and organic dyes. However, there is a strong need for development of bright and photostable nanoscale probes for biological imaging to further push the limits of high resolution microscopy.

In this work, we demonstrate the use of dye-conjugated dendrimers as nanoscale imaging probes for high-resolution fluorescence microscopy, which overcomes the diffraction limited optical resolution and enables imaging at a higher resolution compared to standard organic dyes. Generation-5 polyamidoamine (PAMAM) dendrimers were modified to contain multiple fluorescent dyes, and the photophysical properties of these nanoprobes were characterized using single molecule fluorescence microscopy, total internal reflection microscopy (TIRF-M). For dendrimer-based imaging probes, we observed enhanced photophysical properties including highly extended photobleaching lifetimes and enhanced localization precision compared to single organic dyes such as Cy5 or Alexa flour 657. Using this new class of chemical imaging probes, we were able to achieve a 10 nm spatial resolution for reconstructed images from *d*STORM, which surpasses the spatial resolution of single organic probes (25 nm).

(676f) Pinpointed Diagnosis of Early Stage Oral Cancer by Optical Coherence Tomography Using Stimuli-Disassembling Gold Nanoclusters

Chang Soo Kim, Chemical Engineering and Materials Science, Beckman Laser Institute, University of California, Irvine, Irvine, CA, Petra Wilder-Smith, Beckman Laser Institute, University of California, Irvine, CA, Zhongping Chen, University of California, Irvine, Irvine, CA; Biomedical Engineering, University of California, Irvine and **Young Jik Kwon**, Pharmaceutical Sciences, Chemical Engineering and Material Science, Biomedical Engineering, University of California, Irvine, Irvine, CA

Early, accurate, and non-invasive (or minimally invasive) diagnosis plays a determining role in achieving an effective treatment and prognosis of cancer. Optical coherence tomography (OCT) offers almost non-invasive and real time monitoring of cancer pathology with micrometer resolution. A pivotal challenge in OCT imaging is to obtain the maximum signals to noise ratio due to optical interference with heterogeneous tissue structures and fluid movements. Gold nanoparticles (Au NPs) have been increasingly employed as optical contrast agents for their superior optical properties and high biocompatibility. However, they are generally not responsive to pathological triggers. In this study, Au NPs were clustered to form gold nanoclusters (Au NCs) via linkers that are cleavable in tumor. Upon disassembly in cancerous microenvironment such as mildly acidic pH, Au NCs are designed to generate two distinctive and quantitative signal changes: Lowered scattering and increased Doppler variance. Au NCs were synthesized by conjugating activated Au NPs via acid-cleavable ketal linkages. Dynamic light scattering (DLS) particle analysis, transmission electron microscopy (TEM), and UV/Vis spectroscopy confirmed acid-triggered disassembly of Au NCs in addition to successful synthesis. The stimuli-triggered Au NC disassembly was simulated efficient in a phantom tissue under an acidic condition. The novel concept developed in this study was finally demonstrated using a hamster oral cancer model. Significantly diminished OCT signals along with distinctively increased Doppler variance OCT signals after administering stimuli-disassembling Au NCs clearly pinpointed early stage cancer, while no such changes were observed after administering non-disassembling Au NCs. This talk will exemplify the crucial and indispensable roles of combining interdisciplinary principles of engineering, physics, and medicine in developing technologies for an accurate detection of early stage cancer.

(676g) Fluorogenic Peptide Linked to Anti-CD41 Detects Platelet-Localized Thrombin Activity within Blood Clots Formed Under Flow in Microfluidic Assays or in Mouse

John Welsh[1], Thomas Colace[2], Ryan Muthard[3], Timothy J. Stalker[4], Lawrence F. Brass[4] and Scott L Diamond[2], (1)Department of Biochemistry and Molecular Biophysics, University of Pennsylvania, Philadelphia, PA, (2)Chemical and Biomolecular Engineering, University of Pennsylvania, Philadelphia, PA, (3)Department of Chemical Engineering, Penn State University, University Park, University Park, PA, (4)School of Medicine, University of Pennsylvania, Philadelphia, PA

Background: Thrombin is subject to advection and diffusion upon its production during coagulation. In order to monitor thrombin activity on platelets we developed a thrombin-sensing molecule capable of localizing to growing thrombi. *Methods and Results:* A thrombin sensitive peptide (ThS-P), with a fluorescent reporter, was chemically linked to an anti-CD61 antibody using click chemistry to generate a thrombin sensitive platelet specific sensor (ThS-Ab) (Figure 1A). The kinetics of thrombin cleavage of ThS-P by thrombin was determined (K_m = 22 μM, k_{cat} = 1 s^{-1}). Flow cytometry determined that platelets specifically showed a ~20 fold increase in fluorescence signal in response to thrombin generation in whole blood (Figure 1B). ThS-Ab signal showed a dose response to various thrombin environments created in a microfluidic model of thrombosis, and was capable of identifying spatially distinct regions of thrombin generation in a thrombus (Figure 1C). A separate microfluidic model capable of generating varied pressure gradients was utilized to study thrombin localization within intraluminal thrombi. Altered pressure gradients influenced the localization of thrombin and were visualized using ThS-Ab signal. Thrombin generation *in vivo* was monitored using a mouse laser injury model and infused ThS-Ab whose signal temporally co-localized with fibrin deposition. *Conclusion:* We generated a platelet surface specific thrombin sensor, which provides real time information on the localization of thrombin within a thrombus.

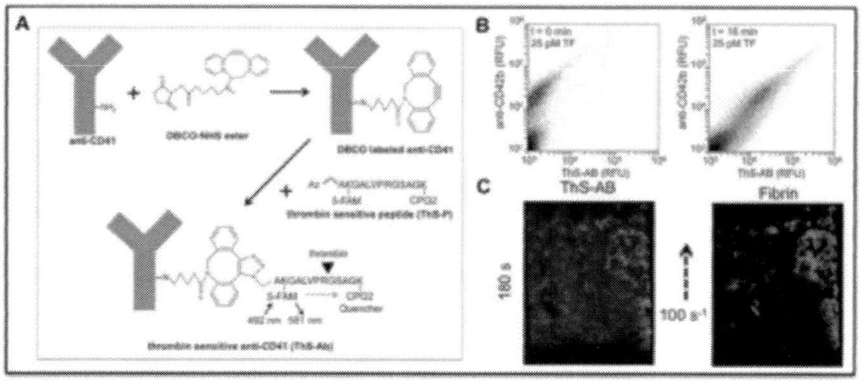

(676h) Fluorinated Nanoparticle Immunolabels for Imaging Specific Membrane Proteins in Parallel with Cell Membrane Lipids Using High-Resolution Secondary Ion Mass Spectrometry

Mary L. Kraft, Department of Chemical & Biomolecular Engineering, University of Illinois at Urbana-Champaign, Urbana, IL, Robert L. Wilson, Chemistry, University of Illinois, Urbana, IL, Jessica F. Frisz, Chemistry, University of Illinois at Urbana-Champaign, Urbana, IL, Kevin J. Carpenter, Glenn T. Seaborg Institute , Lawrence Livermore National Laboratory, Livermore, CA and Peter K. Weber, Lawrence Livermore National Laboratory, Livermore, CA

The cellular plasma membrane is organized into compositionally and functionally distinct domains. Many tools have been developed to visualize protein organization within cellular membranes. Though most biological processes are executed by proteins, the local abundances of specific lipid species near a membrane protein may also influence protein activity. To characterize the relationship between protein activity and the local lipid composition, the protein of interest must be imaged in parallel with the specific lipid species. A combination of high-resolution secondary ion mass spectrometry (SIMS) performed on a Cameca NanoSIMS and metabolic labeling enables imaging the distributions of isotope-labeled lipids in cellular plasma membranes with ~100 nm lateral resolution. To permit imaging specific proteins in parallel with the isotope-labeled lipids using this technique, we have developed fluorine-functionalized gold nanoparticles immunolabels to tag distinct proteins. As a proof of concept, we demonstrate imaging the distribution of influenza hemagglutinin protein in parallel with lipids on the cell surface. This approach may enable evaluating whether distinct proteins are co-localized with specific lipid species in the plasma membrane.

(681a) Preparation of Sensitive Immuno-Sorbent and Immuno-Sensors by Solid-Phase Refolding of PS-Tag-Fused Scfvs

Yoichi Kumada, Bio-molecular Engineering, Kyoto Institute of Technology, Kyoto, Japan

Preparation of sensitive immuno-sorbent and immuno-sensors by solid-phase refolding of PS-tag-fused scFvs

Yoichi Kumada, Kyoto Institute of Technology

Abstract

In this study, we successfully developed a method of site-specific immobilization and activation of single-chain Fv antibodies (scFv) onto polystyrene supports mediated by the novel polystyrene binding peptides (PS-tags: RIIIRRIRR). PS-tag-fused scFvs (scFv-PS) over-expressed in the insoluble fraction of *E. coli* cells were denatured and site-specifically immobilized onto a hydrophilic PS plate in the presence of 0.5 ~ 2 M urea. Their antigen-binding activities were efficiently recovered by washing the PS plate with PBST. The solid-phase refolding mediated by PS-tag was applicable to a number of scFvs such as mouse anti-CRP, RNase, TSH, CEA, CA19-9, AAT, huIgG and huIgA antibody. In comparison with Maxisorp^{TM} immobilized with whole monoclonal antibody (Mab), more than 10 times higher density of PS-tag-fused scFvs were attained and consequently, 10 ~ 100 times higher sensitivity was detected in sandwich ELISA. Furthermore, the activation of a single-chain Fv antibody on the surface of sensor chip for QCM as well as SPR sensors was also investigated in order to develop economical and sensitive immuno-sensor systems for use in clinical diagnosis. On the bare gold surface of sensor chips, approximately 20 ~ 60 nm of hydrophilic polystyrene (phi-PS) thin film that specifically binds the polystyrene-binding peptide (PS-tag) was prepared by spin-coating and O_2-plasma irradiation. When the adsorption of scFv-PS onto the phi-PS surface was directly monitored, the maximum density of scFv-PS attained was 1.56 mg/cm^2. The specific antigen-binding activity of scFv-PS after solid-phase refolding increased with the density of immobilized scFv-PS, and, consequently, activity 1.7 times higher than that of scFv was retained. The scFv-PS-immobilized sensor chips rapidly allowed the detection of clear signals for antigen with high sensitivity, while no signal was detectable for BSA as a negative control. The scFv-PS-immobilized sensors developed in the present study will therefore be very useful for the rapid and highly sensitive detection of biomarkers, and should be applicable to clinical diagnosis.

(681b) High Throughput, Mutliplexed Detection of miRNA Analogs Using Micellar Electrokinetic Chromatography

Johnathan M. Goldman and James W. Schneider, Chemical Engineering, Carnegie Mellon University, Pittsburgh, PA

Short, non-coding micro-RNAs (miRNA) have been implicated in an increasing number of cancers, and miRNA detection has emerged as a promising means of early-stage cancer diagnosis. miRNA are present in extremely low (attomole) levels in cell extracts, making them difficult and time-consuming to detect by conventional means. miRNA are only 22 nucleotides in length, and therefore make marginally stable duplexes with complementary oligonucleotide probes. Complicating matters even further is the observation that many miRNA differ by only a single base in sequence. Available methods, such as microchip arrays and bead-based assays, suffer from poor reproducibility and false-positive readings. Most implementations also require a PCR-type amplification step that presents inefficiencies given the short template length.

Here, we present a fast, specific, high-sensitivity miRNA detection method operating in a capillary electrophoresis (CE) format. Detection relies on the hybridization-based linkage of a surfactant micelle and a dsDNA oligomer pre-loaded with a flurorescent intercalating dye (YOYO-1) in the presence of target miRNA. In so doing, the electrophoretic mobility of the YOYO-1 DNA is shifted into a detection zone of the CE electropherogram. High-affinity, high-selectivity miRNA binding is ensured by use of synthetic DNA analogues (γ-substituted PNAs). The YOYO-1 DNA may also be made highly fluorescent by simply extending its length as required.

We will discuss our recent efforts to implement the method for detection of the let-7 series of miRNA. Preliminary results show that up to 13 targets are reliably identified in a 5-minute run with nanomolar sensitivity. We find that high selectivity, over a wide range of single-base mismatches, is realized for runs performed at slightly elevated temperature (40-45°C). We also discuss recent efforts to increase sensitivity by orders of magnitude using stacking and sweeping methods in inhomogeneous buffer systems.

(681c) Nanoparticle-Based, PCR-Less Detection and Isolation of Molecular Biomarkers

Kalpesh D. Mahajan[1], Greg Vieira[2], Gang Ruan[3], Maryam Lustberg[4], Jeffrey J. Chalmers[3], R. Sooryakumar[5] and **Jessica O. Winter**[6,7], (1)William G. Lowrie Dept. of Chemical and Biomolecular engineering, The Ohio State University, Columbus, OH, (2)Department of Physics, Ohio State University, Columbus, OH, (3)William G. Lowrie Department of Chemical and Biomolecular Engineering, The Ohio State University, Columbus, OH, (4)Division of Medical Oncology, The Ohio State University, Columbus, OH, (5)Department of Physics, The Ohio State University, Columbus, OH, (6)William G. Lowrie Department of Chemical and Biomolecular Engineering, the Ohio State University, Columbus, OH, (7)Department of Biomedical Engineering, The Ohio State University, Columbus, OH

The molecular detection of biomarkers that are physiologically present at very low concentration is important not only for diagnostics, but also for monitoring disease progression and efficacy of therapy. The ability of small capture probes to interact with individual molecules has lead to ultrasensitive nanotechnology-based detection assays; however, these assays are unable to also isolate the molecules of interest for further analysis. Here we present an ultrasensitive, small volume, PCR-less assay for the detection and simultaneous isolation of soluble biomolecules (proteins, ssDNA/RNA) or surface expressed proteins. This assay consists of magnetic-fluorescent nanoparticles (i.e., MagDots) conjugated to molecules of interest, which are then manipulated by mobile magnetic traps on a magnetic microarray using external electromagnets. Using this assay, we have isolated leukocytes based on CD45 receptor expression, with *in situ* receptor quantification using quantum dot fluorescence. We have also detected and isolated avidin and short p53 ssDNA from 10^{-10} M solutions in a 5 μl sample volume. A maximum sensitivity of the system of 10^{-16} M has been demonstrated. Additionally simultaneous detection of protein and DNA targets has also been demonstrated. This technology will eventually be combined with a lab-on-chip platform incorporating mixing, detection, quantification and isolation of biomarkers. This technology has potential applications in diagnostics, nanoscale synthesis, and chromatography.

(681d) Application of a Novel Tablet PCR Platform for Detection of Influenza Subtypes From Clinical Samples

Stephanie Angione[1], Zintis Inde[2], Christina Beck[1], Steve M. Opal[3], Andrew W. Artenstein[3] and Anubhav Tripathi[1], (1)School of Engineering, Brown University, Providence, RI, (2)Department of Biology and Medicine, Brown University, Providence, RI, (3)Brown Medical School, Memorial Hospital of RI, Pawtucket, RI

Developing diagnostics which can quickly and effectively identify viral infections ranging from HIV to hepatitis is crucially important for addressing the global health impact of these diseases. This is particularly true in the case of influenza, a disease which not only spreads quickly, but also mutates rapidly. The gold standard approach for influenza subtyping is RT-PCR (qRT-PCR) which allows for detection of the target sequence and quantification the number of viral particles present. Because RT-PCR and qRT-PCR require thermal cycling in order to amplify the clinical sample for detection, a typical apparatus for these techniques is costly, bulky, and requires a large sample volume. Thus, a microfluidic approach may prove advantageous in addressing these issues. Thus, we present our work on a droplet platform for the differentiation of influenza strains using qRT-PCR using a novel microdroplet device. Previous data have demonstrated this platform's ability to detect both lambda phage DNA and synthetic H3 influenza RNA in a dose-dependent fashion. We now further demonstrate that the platform can effectively differentiate between H1 and H3 strains of influenza, as well as between swine-origin and seasonal strains of H1. Detection using this platform is highly sensitive, amplifying clinical samples as dilute as 10^2 virions/mL. The tablet platform is described and its efficacy demonstrated elsewhere. Primers were designed using consensus sequences from the online NCBI Influenza Virus Sequence Database to differentiate H3 and H1 swine and seasonal subtypes. SYBR Green I dye was used for real-time fluorescence detection. Each RT-PCR reaction consisted of 50ul of RT-PCR mix from the Superscript III RT-PCR kit. Both reverse transcription and PCR were carried out in the same tube or on the droplet platform. This included 1X Taq buffer, 0.2mM dNTPs, 1.5mM $MgCl_2$ and 0.2uM of both the forward and reverse primers. Cycling consisted of a 30 minute RT step at 50°C, followed by 40 cycles of 94°C for 15 seconds, the primer-specific annealing temperature for 15 seconds, and 68°C for 30 seconds. An initial denature was done following the RT step for 15 minutes for off-chip controls and 2 minutes for platform droplet amplification. We first determined primer efficacy with blinded spiked heat inactivated influenza virus particles in transport media as a serial dilution for the H3N2 and H1N1 (seasonal) subtypes. We effectively identified the sample subtypes which are differentiated by amplicon length. Samples A-H (except C) were positive for the H3 subtype, displayed by the amplicon of 233 bp. Sample C is a negative control, while samples J and K are H1 seasonal positive with an amplicon of 183 bp. The lowest concentration detected was sample E, at 10^2 virions/ml. We also tested each sample against the other primer sets and found no false positives. We subsequently performed qRT-PCR on the serially diluted samples A-H for the H3 subtype to develop a standard curve for efficiency utilizing our novel tablet platform. We then tested our PCR protocol against clinical samples of unknown subtype and concentration to identify seasonal H1N1, swine H1N1 or seasonal H3N2. The platform is robust, using a simple apparatus and producing results which are both replicable for influenza and applicable to other diseases for which diagnostics currently rely on PCR. Thus, the development of this platform represents an important step towards improved diagnostic technologies for influenza and other infectious diseases.

(681e) Microfluidic Platform for Antibiotic Susceptibility Screening

Ritika Mohan, Arnab Mukherjee, Jaebum Lee, Emre Sevgen, Charles M. Schroeder and Paul J. A. Kenis, Chemical and Biomolecular Engineering, University of Illinois at Urbana-Champaign, Urbana, IL

In this work, we report a multiplexed microfluidic platform for rapid characterization of antibiotic susceptibilities in microbial pathogens, for healthcare applications. Injudicious use of antibiotics over the last few decades has resulted in an escalating emergence of multidrug resistant pathogens that respond poorly to conventional antibiotic therapy [1]. As resistance to single antibiotics evolves very rapidly, multidrug resistant pathogens are usually treated with combinations of antibiotics. However, existing techniques to elucidate the right therapeutic combinations are expensive, time-consuming, laborious, and require substantial volumes of patient body fluids like sputum and blood. Consequently, there is a pressing need for low cost diagnostic platforms, which enable expedited analyis. To address this, we designed and fabricated a 4x6-array microfluidic chip capable of confining cells and reagents in nanoliter volume compartments. Leak-free isolation of contiguous compartments is achieved with elastomeric valves that are closed at rest [2]. Actuation of the normally closed valves with negative pressure enables controlled exposure of cells to antibiotics that are loaded in adjacent compartments. The miniaturized device design is highly convenient for biological assays as it facilitates portability, uses microliter volumes of samples and reagents for analysis, and is readily amenable to automation. Most importantly, the microfluidic diagnostic platform significantly accelerates assay time due to increased detection sensitivity since it enables analysis at single cell resolution.

We employed the microfluidic platform to investigate susceptibilities of *E. coli* and *P.aeruginosa* cells to several commonly used antibiotics. Specifically, we assessed the effects of ampicillin, tetracycline, chloramphenicol, cephalexin, ciprofloxacin, and combinations thereof on *E.coli* MG1655 and *P.aeruginosa* PAO1 cells. Furthermore, we quantified synergistic effects of several antibiotic combinations. As an example, although ampicillin at 10 μg/mL and tetracycline at 1 μg/mL are ineffective individually, they exhibit significant bactericidal activity against *E.coli* when administered in combination. Finally, we extended the platform to characterize antibiotic resistance in a mixed microbial community comprising *E.coli* and *P.aeruginosa* cells. Overall, these results emphasize the utility of the microfluidic platform for rapidly characterizing antibiotic susceptibilities over a wide range of antibiotic concentrations. The platform improves significantly on existing diagnostic methods (disc diffusion and broth dilution assays) in terms of several key advantages including rapid turn-around time, low cost of analysis, ease-of-use, detection sensitivity, and reduced sample volumes required for diagnosis [3].

References:

[1] K Churski *et al.*, *Lab Chip*, **2012**,12.

[2] R Mohan *et al.*, *Sensors and Actuators B*, **2011**, 1, 1216-1223.

[3] JP Torella *et al.*, *PLoS Comput Biol*, **2010**, 6.

(681f) Identifying the Differentiation Stages of Individual Hematopoietic Cells by Multivariate Analysis of Secondary Ion Mass Spectra

Mary L. Kraft, Department of Chemical & Biomolecular Engineering, University of Illinois at Urbana-Champaign, Urbana, IL, Jessica F. Frisz, Chemistry, University of Illinois at Urbana-Champaign, Urbana, IL, Ji Sun Choi, Chemical and Biomolecular Engineering, University of Illinois at Urbana-Champaign, Urbana, IL, Robert L. Wilson, Chemistry, University of Illinois, Urbana, IL and Brendan A.C. Harley, Chemical & Biomolecular Engineering, University of Illinois at Urbana-Champaign, Urbana, IL

Identification of the mechanical and biochemical properties that induce hematopoietic stem cells (HSCs) to self-renew or differentiate into various types of blood and immune cells is an important goal in tissue engineering. Combinatorial biomaterial substrates permit screening the effects of multiple microenvironments on stem cell fate while using a minimum number of cells. To utilize these substrates, robust and quantitative methods to identify the differentiation stages of individual cells with location specificity are required. For this purpose, we have used time-of-flight secondary ion mass spectrometry (TOF-SIMS) to collect spectra that encodes for the surface chemistries of individual cells from distinct hematopoietic cell populations. We then used a multivariate analysis technique, partial least-squares-discriminant analysis (PLS-DA) to construct a PLS-DA model that could be used to identify the differentiation stages of individual hematopoietic cells that were isolated from mouse bone marrow according to their mass spectra. This approach may facilitate correlating HSC fates with the properties that control them.

(681g) Detailed Cellular Phenotype Measurements Using Peptide-Guided Surface Enhanced Raman Scattering (pg-SERS)

Ryan S. Senger, Biological Systems Engineering Department, Virginia Tech, Blacksburg, VA

A cellular phenotype is determined by its metabolic network and the response of that network to environmental conditions. This means that both genetic and environmental perturbations can be responsible for altering the chemical composition of a cell. This has profound implications with microorganisms as we seek to understand the response of cultures to fermentive products (e.g., tolerance to biofuels) and inhibitory substrates. This is also of considerable interest to understand genetic manipulations designed to engineering higher cellulose content in plants. Thus, cellular composition is almost never static. In fact, cellular composition determination involving mass spectroscopy is time consuming, does not consider spatial variation, and is destructive to dynamically responding cultures. Here, the power of Raman spectroscopy and surface-enhanced Raman spectroscopy (SERS) are demonstrated and interfaced with genome-scale modeling for accurate phenotype predictions. We also demonstrate a new technique of peptide-guided surface enhanced Raman scattering (pg-SERS). This technique has enabled targeting of specific cellular locations (e.g., the inner cell wall) to study the spatial confined and compartment-specific response of microbial cells to stress. Specific examples of the response of *Escherichia coli* K12 to 4-carbon alcohols has revealed distinct responses that are directly related to the length of the hydrophobic chain of the alcohol molecule. We demonstrate methods of global Raman analysis and more specific functional group targeting using a web-based database made available on our research group website.

(681h) Electronic Platform Used As a Proxy to Quantify Cellular Toxicity of Anticancer Drug

Reem Eldawud, Chemical Engineering, West Virginia University, morgantown, WV

Digitoxin is a naturally occurring class of cardiac glycosides known for its efficacy in treatment of heart failure. Recent studies showed that digitoxin exhibits strong cytotoxic effects against several types of cancer from breast to prostate and lung cancer. However, the cellular mechanisms associated with digitoxin exposure and its selectivity to cancer cells when compared to normal cells, are still subject to debate. In this research we are using an electronic platform as a proxy to assess toxicity of digitoxin when incubated with non-small lung cancer cells (NCI-H460) in real time. The approach is based on measuring the behavior of NCI-H460 upon exposure to different concentrations of digitoxin using the electrical cell impedance sensor (ECIS). This non-invasive technique relies on real time impedance measurements of cells seeded on gold electrodes to assess changes in cellular morphology and cell-cell interactions in real time. Such measurements are further correlated with cellular functions to derive mechanisms associated with cytotoxic and apoptotic events and further confirmed by cellular and optical assays. Our results provide novel means to investigate drug kinetics and metabolization in real time using a high throughput electronic platform and promises to extend such cellular-based interface studies to other analysts, all with high sensitivity and in real time.

(682a) Ligation-Independent Cloning with Self-Cleaving Intein As a Tool for High-Throughput Protein Purification

Tiana D. Warren, Michael J. Coolbaugh and David W. Wood, Chemical and Biomolecular Engineering, The Ohio State University, Columbus, OH

Ligation Independent Cloning (LIC) is being incorporated into the ΔI-CM self-cleaving intein system to create a more streamlined method for the cloning, expression and purification of proteins. In comparison to conventional molecular cloning, LIC relies on extended segments of complementary DNA for facilitating highly efficient *in vivo* ligation, thus eliminating the need for *in vitro* restriction digestion and ligation. The self-cleaving intein used for this project, ΔI-CM, was designed by the Wood lab as an alternate method for protein purification. The intein is a 168 amino acid sequence that is expressed along with the target protein and an purification tag to create a tag-intein-target protein fusion. With a mild pH or temperature shift, the intein will self-cleave from the target protein, resulting in a pure protein product, thus obviating the use of expensive proteases to release the protein from the purification tag. Through this newly developed system, different protein genes can be incorporated into various expression vectors and their encoded proteins purified in a high-throughput manner. Two vectors were constructed with the Chiting Binding Domain (CBD) and Elastin-like polypeptide (ELP) purification tags to be used for LIC and intein-mediated protein purification. With these vectors, Maltose Binding Protein and β-lactamase proteins were successfully cloned, expressed and purified using both affinity chromatography and non-chromatographic purification.

Combination of Spectroscopic Methods for Inline Monitoring and Control of Mammalian Cell Cultivations

Stefan Buziol[1], Mark Roth[1], Christine Jung[1], Pedro Felizardo[2], Joana Mendes[2] and Jose Menezes[2]
(1) Roche Diagnostics GmbH, Penzberg, Germany,
(2) 4Tune Engineering Ltd - Technology Services & Products, Lisbon, Portugal

Abstract

The integration of strengths of different spectroscopic methods such as Near Infrared spectroscopy (NIRS), Mid-Infrared spectroscopy (MIRS), 2-Dimensional Fluorescence Spectroscopy (2DFS) and Dielectric Spectroscopy (DS) allowed monitoring and control of different culture attributes to be readily available (inline) instead of being determined in the laboratory (offline). This is an advantage over analytical laboratory methods, since spectroscopic sensors can enable the intensification of in-process analytics.

Moreover, since these spectroscopies are focusing on information of different quality attributes (chemical, physical, physiological) extractable from a biotechnological cultivation process, the combination of those has the potential to enhance the process understanding significantly. An overview of our experience in the integration of above techniques - to establish process knowledge in a reduced number of cultivations for different media, clones and products - will be described.

Introduction

Bioprocesses are complex systems and the need of their monitoring has long been an integral part of industrial processes since it allows increased automation and introducing of advanced control schemes.

In a conventional approach for batch fermentation monitoring, off-line samples are taken and analyzed by laboratory methods to ensure process monitoring of critical analytes (glucose, lactate, ammonium, product, etc.) and cell parameters (for instance, cell density, viability, etc.). Furthermore, on-line measurements of process parameters (pH, temperature, oxygen, carbon dioxide, agitation, feed rates, etc.) are made for open and closed-loop control and monitoring of fermentation runs (Ündey *et al,* 2010). However, this conventional approach presents drawbacks: the control of the critical analytes and cell parameters fails for the low measurement frequency and their time delay, whereas the monitoring of process parameters underperforms for considering the individual analysis of each process parameter instead of their correlation with each other and with critical control analytes and cell parameters.

To overcome these drawbacks the use of Process Analytical Technology tools (PAT) combined with statistical and mathematical methods (chemometrics) has been proposed. Since recent years, applying PAT tools and chemometrics to real-time multivariate data collected by inline sensors have received significant attention (Junker and Wang, 2006; Molony and Undey, 2009) because it provides efficient means of identifying/reducing variation, managing process risks, relating process information to critical analytes and cell parameters and determining process improvement opportunities such as detecting contaminations, increasing yield or reducing impurities variability (Ündey *et al,* 2010).

An overview of our experience integrating different spectroscopic sensors (such as, NIRS, MIRS, 2DFS and DS) for establishing inline monitoring and control and their benefits is presented.

Spectroscopy-based methods: Principles and Equipment Used

Spectroscopy sensors hold great promise for bioprocessing applications because they generally support noninvasive, continuous and multivariable monitoring of animal cell cultures fermentations. A brief overview of spectroscopy sensors used in this work is presented below.

Near Infrared Spectroscopy (NIRS)

Near-infrared spectroscopy (NIRS) is a spectroscopic method that uses the near-infrared region of the electromagnetic spectrum from about 800 nm to 2500 nm (12500 to 4000 cm^{-1}). NIRS is essentially suitable to detect components that contain a heteroatom such as, aliphatic C-H, aromatic or alkenes C-H, amine N-H and O-H bounds. The resulting absorption spectra reflect the molecular overtones and combinations of fundamental vibrations of those functional groups, resulting in a signal of difficult interpretation. These characteristics of NIR spectra require the use of chemometric tools to relate spectral information with a specific compound (Pons et al., 2004). Furthermore, the low adsorption coefficient of near infrared radiation allows high penetration into a sample, increasing the range of matrices that can be submitted to this technique (e.g., fermentation broths). In addition, the influence of both physical and chemical properties of the sample in the spectra signal can be positively used to extract chemical and physical information from one single measurement, which make the use of these sensors even more interesting (Reich, 2005). In the present work, spectra were collected using a Matrix-FT spectrometer (Bruker Optics GmbH) equipped with an immersion fiber optic transflection probe (Solvias) adjusted to 1 mm optical path-length. Spectra were recorded using the spectrometer software OPUS (Bruker Optics GmbH). For each sample, an average of 512 scans at a data spacing of 8 cm^{-1} in the NIR wavenumber range 12800 – 4000 cm^{-1} was taken. In each fermentation run, spectra were recorded automatically at 2 min intervals.

Mid-Infrared Spectroscopy (MIRS)

Mid-infrared spectroscopy (MIR) is an extremely reliable and widely recognized fingerprinting technique. Many substances can be characterized, identified and also quantified by this method, since it is in the mid-infrared spectral region where most of the fundamental structural information is produced.

Within the mid-IR region from 2.5 – 25 μm (4,000 – 400cm^{-1}), IR absorption bands mainly arise from the fundamental vibrational modes and can be assigned to specific functional groups. Moreover, absorption bands associated with individual components in a mixture are frequently isolated from other bands and can be used for quantification purposes. Other advantage of MIRS is the increased sensitivity and fast spectra acquisition, which is particularly advantageous in production facilities.

Data for fermentation monitoring were collected with a Matrix-MF spectrometer (Bruker Optics GmbH) equipped with an ATR diamond probe. The software used to record the spectra was OPUS (Bruker Optics GmbH). An average of 128 scans at a resolution of 8 cm^{-1} was taken for each sample in the wavenumber range of 3500 – 560 cm^{-1}. For each fermentation run, spectra were taken automatically at 2 min intervals.

2D-Flurescence Spectroscopy (2DFS)

Fluorescence spectroscopy is a type of electromagnetic spectroscopy, which analyzes the emission of light subsequent to absorption of ultraviolet or visible light of a fluorescence molecule or substructure, called fluorophore. The fluorophore will absorb energy in a form of light at a given wavelength and will liberate energy in the form of emission of light at a specific higher wavelength.

Since emission and excitation spectra can be obtained from 2D fluorescence spectrometers it facilitates more appropriate and more precise analysis of fluorophores present in complex media. Thus, the technique can provide information about fluorescent molecules in a wide variety of biological samples with a very high sensitivity (100x up to 1000x more than other spectroscopic methods).

Fluorescence spectra were collected using BioView (Delta Light & Optics). Unlike other probes, the sensor's probe is not in direct contact with the cell suspension, the probe is connected to an optical well containing a bottom surface of glass. The BioView system was set up to excite with wavelengths in the range from 260 to 560 nm in steps of 20 nm, and subsequently measure fluorescence light emitted from the solutions and fermentation broth in the range from 300 to 600 nm every 20 nm as well. As described for vibrational spectroscopies, the system was set to collect a full spectrum every 2 minutes.

Dielectric Spectroscopy (DS)

Among other methods, dielectric spectroscopy (DS) is highly recommended for on-line monitoring of biomass. This method relies on the principle that each cell acts as a mini-capacitor, this means that when cells presented in a ionic solution are submitted to an alternating electric field, a polarization is developed at the poles of the cell. This occurs because ion movement within the highly conducting cellular cytoplasm is restricted by the non-conducting plasma membrane. This way cells with intact plasma membranes (viable cells) are polarized which does not happen to gas bubbles, micro carriers, cell debris and other particle in suspension. Therefore, while the electric field is released, the relaxation of ions induces a capacitance that is measured and can be correlated to the membrane enclosed volume fraction of cells in suspension, *i.e.,* the biovolume.

In this project permittivity data was collected using a Biomass System that contains a capacitance probe (Fogale Nanotech). The capacitance data was recorded using Fogale's data acquisition software. The data was collected by performing a frequency scan measuring 17 frequency values from 0.3 to 10 MHz, every minute. The dual-frequency permittivity increment (permittivity of the cell suspension) was defined as the difference in permittivity measurements at these two frequencies (1000 Hz and 10 MHz). The dielectric spectra were also collected.

Fermentation Monitoring with Spectroscopy-based methods

The use of spectroscopy sensors to monitor processes generates a large amount of data (spectra) that contains very useful information about process conditions/state and specific analytes. This spectral information is often not easily interpreted and the use of mathematical and statistical multivariate tools for extracting relevant information from these complex systems (i.e., chemometrics) is common. In this field, the use of spectroscopic measurements for quantification purposes is of special interest and it can be achieved through the development of multivariate calibration methods that establish relationships between responses (e.g. analytes concentration) and variables (spectra). After calibrations for critical analytes and cell parameters are developed, these sensors can be used for process monitoring, although this task is not easy to accomplish as several aspects from samples variability and process variability have to be taken into account (not described is this work). A few examples of possible applications for the spectroscopy sensors described are presented below.

Monitoring specific analytes with Spectroscopy Sensors

Viable cell density (VCD) is one of most important parameters to monitor along the course of fermentations. VCD, stands as the condition for adding feeds and/or other solutions in fed-batch processes. As such, its monitoring is very important. The on-line measurement of culture turbidity or scattering of NIR light, fluorescence, or capacitance properties has been used to estimate the biomass concentration (Kiviharju *et al,* 2008). The biomass concentration influences the scattering properties of the fermentation broth and thus the overall absorption. Several studies have shown that an increased biomass concentration increases the scattering of the fermentation broth thus decreasing the light permeability.

In the present work, DS presented the best results, showing more robustness and better prediction capability, when compared to NIRS and 2DFS. This was expected since DS, on the other hand, offers several distinct advantages over other measurement principles such as NIRS light scattering, which are unable to differentiate viable and nonviable cells and 2DFS that can only capture biomass indirectly by correlating it with fluorescence signal from fluorophores present in the fermentation broth and intracellular cofactors such as NAD(P)H and FAD.

Figure 1A shows predicted VCD using the DS spectra during a fermentation run (black line) against off-line laboratory analytics for the same fermentation (determined by Cedex, Roche Diagnostics GmbH). As it can be seen the predictions are in agreement with the laboratory analytics and the increased frequency of data acquisition (every 2 minutes as opposed to once a day) allows the establishment of an accurate VCD profile where the different phases of the fermentation can be clearly seen. As a result from this higher frequency of sampling acquisition it is possible to extract knowledge from each fermentation run, especially the calculation of inline specific growth rates (Figure 1B), impossible to determine by conventional off-line monitoring of fermentations.

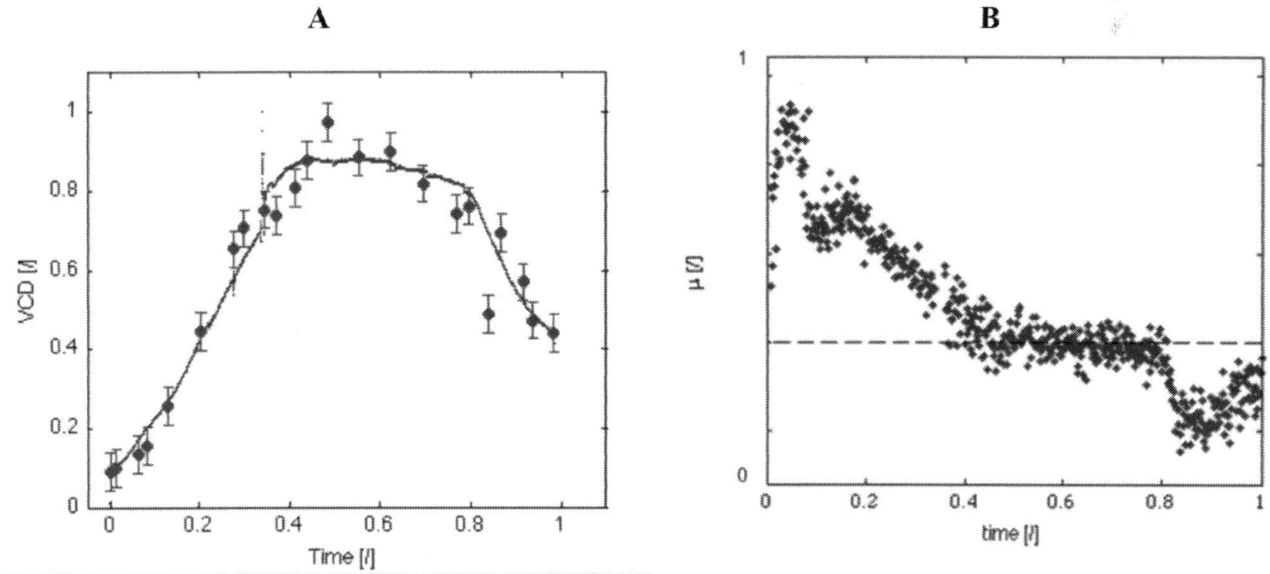

Figure 1 – **(A)** Predicted VCD by DS (black solid line) and off-line cedex analytics (•) for fermentation run; **(B)** Calculate specific growth (•) based on DS predictions and mass balance for a fermentation run.

Monitoring of glucose, lactate, glutamine, glutamate, ammonia and asparagine in fermentations using vibrational spectroscopy (NIRS and MIRS) has been reported in several studies (Arnold et al, 2003; Hagman and Sivertsson, 1998; Roychoudhury et al., 2007; Riley et al., 2001; Riley et al., 1999; RhieL et al., 1999). Even though both NIRS and MIRS offers simultaneous, multi-analyte information within minutes, these two techniques clearly diverge in the areas of sensitivity and accuracy and in terms of ease of use *in situ.* NIRS presents low sensitivity and selectivity caused by the broad and overlapping bands from the overtones and combination transitions that occur in the NIRS region. However, the low adsorption coefficient of near infrared radiation allows high penetration into a sample, increasing the range of matrices that can be submitted to this technique. It is also known that NIRS also captures physical and chemical information of a process and thus its usage is of great interest in fermentations. On the other hand, MIRS offers the ability to quantify analytes present at limiting levels and to distinguish between molecules of very similar structure. This was also proved in this work, especially for glucose concentration monitoring, where the accuracy of MIRS surpassed the performance of NIRS. An example of a fermentation run monitored with MIRS is presented in Figure 2. As it can be seen MIRS can predict missing data with high accuracy, which is not captured by the conventional off-line analytics. This is very important in a fed-batch fermentation run as it allows the calculation of specific substrate uptake rate (combined with the fermenter mass balance). The calculation of the specific substrate uptake rate can be used to establish feeding control based substrate requirement.

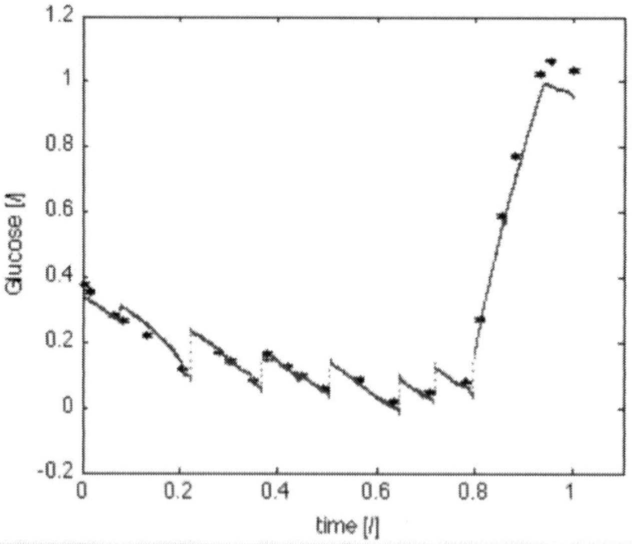

Figure 2 - Predicted Glucose concentration by MIRS (black solid line) and off-line glucose analytics (*) for a fermentation run.

Monitoring mammalian cell bioprocesses using fluorescence is less well covered in the literature, probably due to the complexity of media formulations. Mammalian cells are not able to synthesize any of the fluorescent amino acids, which have to be provided to the initial culture medium or by feeds. During culture time, they are taken up by cells and incorporated into the recombinant protein produced and into intracellular proteins (Teixeira *et al,* 2009a). Therefore, these proteins can be monitored using 2D fluorescence as few studies already proved (Teixeira *et al,* 2009b). Figure 3A shows the performance of 2DFS for predicting recombinant protein yield (solid black line) and its off-line analytics (•) while Figure 3B shows the predicting performance for lactate dehydrogenase (LDH).

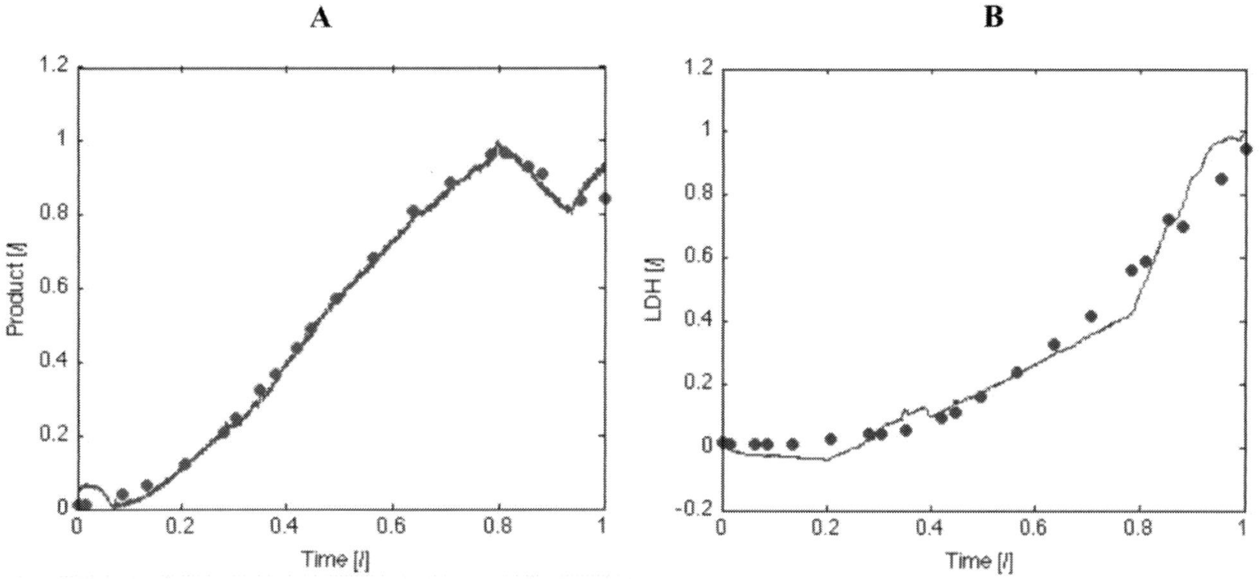

Figure 3 – **(A)** Predicted Product concentration by 2DFS (black solid line) and off-line product analytics (•) for a fermentation run; **(A)** Predicted LDH concentration by 2DFS (black solid line) and off-line LDH analytics (•) for a fermentation run.

Furthermore, the use of 2DFS to monitor the concentration of fluorophores should also be possible. This would allow the determination of specific uptake rate of amino acids and vitamins that are part of the feeding solutions.

Monitoring Process Dynamics and Process Trajectories

Spectroscopic techniques, such as those presented in this work can be very valuable tools to monitor fermentation runs. As it can be seen throughout this article these tools together with chemometrics enable the quantification of several analytes based on their spectra. However, the use of these spectroscopic techniques is not limited to the development of calibrations for monitoring analytes. In fact, monitoring process trajectories is an interesting area where these tools can play an important role. The great advantage of spectroscopy measurements is the ability to increase sampling frequency and with that capture much more information about a specific fermentation than reference analytics with low frequency samples acquisition. Another opportunity to use these sensors is using its raw data to monitor process dynamics, since these spectroscopies can capture much more than the analytes profiles, and as such, they can be very useful for process supervision or guided sampling procedures, even without quantification methods. Two examples of this would be the information contained in the raw spectra that is not directly related with known parameters, for instance in NIRS (Figure 4) or in 2DFS and the extraction of process profiles based on factor decomposition methods (example PCA). Although today's process monitoring is restricted to known parameters, the use of inline spectroscopic tools allow for profound insight into processes waiting for correlation and attribution, which will enable process understanding and building process knowledge.

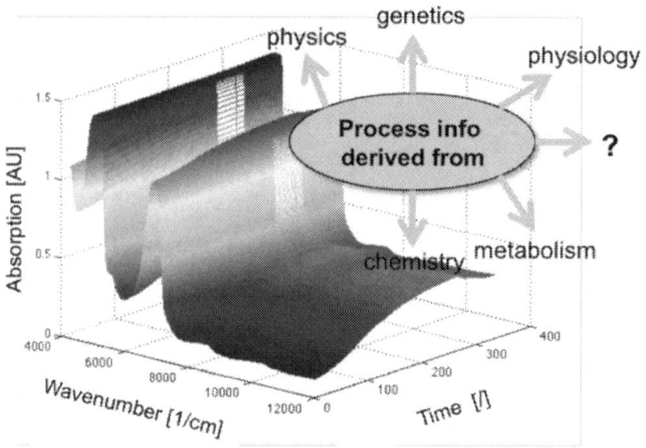

Figure 4 – NIRS and 2DFS spectrum along process time carries a lot of process information waiting for correlation and attribution

References

1. Junker BH and Wang HY (2006). *Bioprocess monitoring and computer control: Key roots of the current PAT initiative*. Biotechnology and Bioengineering, 95: 226–261.
2. Molony M and Undey C (2009). *PAT Tools for Biologics: Considerations and Challenges*. In Quality by Design for Biopharmaceuticals, edited by Rathore A and Mhatre R, John Wiley & Sons, Hoboken, New Jersey, USA.
3. Undey C, Ertunç S, Mistretta T and Pathak M (2010). *Applied Advanced Process Analytics in Biopharmaceutical Manufacturing: Challenges and Prospects in Real-time Monitoring and Control*. Journal of Process Control. 20: 1009–1018.
4. Pons MN, Le Bonté S and Potier O (2004). *Spectral analysis and fingerprinting for biomedia characterisation*. Journal of Biotechnology, 113: 211–230.
5. Reich G (2005). *Near-infrared spectroscopy and imaging: Basic principles and pharmaceutical applications*. Advanced Drug Delivery Reviews 57: 1109–1143.
6. Kiviharju K, Salonen K, Moilanen U and Eerikäinen T (2008). *Biomass measurement on- line: the performance of in situ measurements and software sensors*. Journal of Industrial Microbiology and Biotechnology, 35: 657–665.
7. Arnold SA, Crowley J, Woods N, Harvey LM and McNeil B (2003). *In-situ near infrared spectroscopy to monitor key analytes in mammalian cell cultivation*. Biotechnology and Bioengineering, 84: 13–19.
8. Hagman A and Sivertsson P (1998). *The use of NIR spectroscopy in monitoring and controlling bioprocesses*. Process Control and Quality, 11: 125–128.
9. Roychoudhury P, O'Kennedy R, McNeil B and Harvey LM (2007). *Multiplexing fibre optic near infrared (NIR) spectroscopy as an emerging technology to monitor industrial bioprocess*. Analytical Chimica Acta, 590: 110–117.
10. Riley MR, Crider HM, Nite ME, García RA, Woo J and Wegge RM (2001). *Simultaneous measurement of 19 components in serum containing animal cell culture media by Fourier transform near-infrared spectroscopy*. Biotechnology Progress, 17: 376–378.

11. Riley MR, Okeson CD and Frazier BL (1999). *Rapid calibration of near-infrared spectroscopic measurements of mammalian cell cultivations*. Biotechnology Progress, 15: 1133–1141.
12. Rhiel M, Ziegler T, Ducommun P, von Stockar U and Marison IW (1999). *Mid-Infrared spectroscopic monitoring of animal cell culture bioreactor processes*. In Animal Cell Technology: Products from Cells, Cells as Products. Edited by A. Bernard et al., Kluwer Academic Publishers, Netherlands.
13. Teixeira AP, Oliveira R, Alves PM and Carrondo MJ (2009a). *Advances in on-line monitoring and control of mammalian cell cultures: Supporting the PAT initiative*. Biotechnology Advances, 27: 726–732.
14. Teixeira AP, Portugal CA, Carinhas N, Dias JM, Crespo JP, Alves PM, Carrondo MJ and Oliveira R (2009b). *In situ 2D fluorometry and chemometric monitoring of mammalian cell cultures*. Biotechnology and Bioengineering, 102: 1098–1106.

(682c) A Framework for Using a Well Characterized Performance Variability In a Virus Filtration Process towards a Highly Consistent, Economical Unit Operation

Willem Kools, Biomanufacturing Science Network, EMD Millipore corporation, Billerica, MA

Virus filtration is the most common orthogonal step to clear virus in downstream processes for monoclonal antibody production in mammalian cell cultures. Virus clearance is required by regulation to ensure safety of the biological drug. As parvovirus (20 nm) needs to be removed yet monoclonal antibody (150 kd) needs to be recovered in the filtrate, several types of parvovirus filters are commercially available that achieve this objective. 2 main classes of parvovirus filters exist: low flux filters that are often sized based on flux and high performance filters that are sized based on a capacity and flux component. Flux for the second type of filter is often higher.

As virus filtration needs to be implemented in manufacturing, a robust process needs to be implemented that takes into account various possible sources of variability. QbD approach or other process characterization can be used to define performance differences due to changes within your process window of feed, process and membrane device. This paper will describe the impact of these changes from average performance from a process development perspective. Additionally, this paper will address how a partial pore blockage model in combination with range of input variables can be used to predict the manufacturability of a virus filtration step from a quality as well as from a economic perspective. More advanced models or actual filtration data can be used through a similar frame work.

The learnings of both process development and manufacturing view will finally be translated into some direction towards development of robust virus filtration steps.

(682d) Towards a Next-Generation Harvest Strategy for Monoclonal Antibody Processes: Cost Considerations and New Technology Evaluation

Michael Felo[1], Nripen Singh[2] and Neil Soice[2], (1)Biomanufacturing Sciences Network, EMD Millipore Corp., Billerica, MA, (2)EMD Millipore Corp., Bedford, MA

Throughout the history of biologics processing, continuously increasing cell densities in the bioreactor and decreasing cell viabilities at harvest have posed a significant challenge to cell culture clarification operations. Separation technologies have evolved to meet this challenge, moving from normal flow filtration to microporous tangential flow filtration to centrifugation and depth filtration. Today, two strategies dominate harvest process: two-stage depth filtration for lower volume applications and centrifugation followed by secondary depth filtration for larger volume applications. However, continued difficulty in clarifying high cell titer cultures as well as a need for fully disposable, portable processes are driving improvement and innovation in clarification technologies.

Cost models have been developed to compare current cell culture clarification technologies for a monoclonal antibody (MAb) process. Among model parameters evaluated, the volumetric capacity of the primary depth filter in the two-stage depth filtration process was one of the key factors impacting the relative cost of these two technologies. Pretreatment methods such as acid precipitation and flocculation enhance the efficiency of the clarification process by increasing the average particle size in the cell culture broth. Use of a novel charged polymer for cell culture flocculation and a novel depth filter optimized for the larger particle size and increased solids content of the flocculated feed has achieved primary depth filter capacities in excess of 150 L/m^2. In many cases, no secondary clarification is required prior to sterilizing-grade filtration. This improved clarification technology has been demonstrated to work robustly for multiple cell culture streams of varying cell densities and viabilities, including high density cell cultures.

Analysis of these cell culture clarification technologies in the cost model indicates that a harvest strategy of cell culture pretreatment with the novel charged flocculating polymer followed by clarification with the optimized depth filter is cost advantageous to centrifugation followed by depth filtration at bioreactor volumes of up to 12,000 L. Detailed cost analyses for processes using these various cell culture harvest strategies will be presented. Performance data for the novel cell culture flocculant and optimized depth filter will be shown. Finally, the opportunities and challenges of a large volume, non-centrifuge MAb process will be outlined and discussed.

(682e) Role of Intermolecular Interactions On Ultrafiltration of Pegylated Proteins

Krisada Ruanjaikaen, Chemical Engineering, Pennsylvania State University, University Park, PA and Andrew L. Zydney, Chemical Engineering, The Pennsylvania State University, University Park, PA

PEGylated proteins are attractive as biotherapeutics due to the increased in vivo efficacy associated with the covalent attachment of polyethylene glycol (PEG) to the base protein. Previous studies have demonstrated the feasibility of using ultrafiltration for removal of residual unreacted proteins and PEG (Molek and Zydney, 2006) as well as more highly PEGylated species (Ruanjaikaen and Zydney, 2011) from the desired singly PEGylated product. In addition, ultrafiltration/ diafiltration (UF/DF) can be used for the final formulation of PEGylated proteins to achieve the desired final product concentration and buffer conditions. The objective of this study was to evaluate the effects of intermolecular interactions, mediated primarily by the PEG, on the performance of these ultrafiltration systems.

Experimental studies were performed with α-lactalbumin PEGylated with a 20 kDa activated PEG. Ultrafiltration data were obtained over a range of PEG concentration, solution conditions, and filtrate flux using Ultracel composite regenerated cellulose membranes (both as is and chemically-modified to provide a negatively-charged membrane). Data were analyzed using a recently developed theoretical model that accounts for the increase in effective size associated with the attached PEG along with ion exclusion from the PEG layer surrounding the charged protein (Molek et al., 2010).

The transmission of the PEGylated protein was significantly greater in the presence of high PEG concentrations due to the unfavorable intermolecular interactions between the free PEG and the PEGylated protein. The magnitude of this effect was consistent with previous thermodynamic models of PEG-PEG interactions. These intermolecular interactions also reduced the selectivity in ultrafiltration processes used for purification of the singly-PEGylated product. These results provide new insights into the effects of intermolecular interactions on the design and performance of ultrafiltration processes for the concentration and purification of PEGylated proteins.

References

Molek JR, Ruanjaikaen K, Zydney AL. 2010. Effect of electrostatic interactions on transmission of PEGylated proteins through charged ultrafiltration membranes. J Membr Sci 353:60–69.

Molek JR, Zydney AL. 2007. Separation of PEGylated α-lactalbumin from unreacted precursors and byproducts using ultrafiltration. Biotechnol Prog 23:1417–1424.

Ruanjaikaen K, Zydney AL. 2011. Purification of singly PEGylated α-lactalbumin using charged ultrafiltration membranes. Biotechnol Bioeng 108:822–829.

(682f) Challenges of UF Optimization for High Concentration Therapeutic Proteins

Sunitha Kandula, Protein Purification Development, Merck &Co, Union , NJ, Pragathi Peddi, NJIT, Newark, NJ, Edward Glowacki, Protein Purification Development, Merck & Co., Union, NJ and Thomas Linden, Protein Purification Development, Merck & Co., Inc., Rahway, NJ

High concentration monoclonal antibodies (mabs) are being increasingly used in therapeutic products for injection. Product concentrations greater than 100mg/ml often result in higher viscosity solutions and pose a challenge to the ultrafiltration process. The recovery and stability of the molecule becomes challenging at high concentrations due to several reasons; most importantly viscosity increase and absorption onto the membrane. An optimized and well designed ultrafiltration/diafiltration (UF/DF) process is warranted for these processes, which can yield greater than ninety five percent recovery.

In this study UF/DF process was optimized for two mabs at concentrations between 100 to 200g/L. Several process parameters were studied which includes different membrane types, feed flow rate, transmembrane pressure, control of retentate vs. permeate flow and shear effect by different pumps. The stability of the product was monitored using various analytical tools including Size exclusion chromatography, DLS and DSC. Critical observations from this study include (a) at concentrations higher than 100mg/mL the stability of the molecule was highly molecule specific compared to processing impact (b)The viscosity of the solution impacted flux rate and mass transfer coefficient however the bulk concentration and concentration at the membrane remained same (c) SEC was the best indicator for stability and DLS also provided a relative difference.

(692a) Molecular Simulation Studies of Quantitative Correlation Between the Stability of Drug Nanocrystals Suspension System and Certain Properties of Both the Drug API and the Stabilizer

Wusheng Zhu, Frank Romanski and M. Silvina Tomassone, Chemical and Biochemical Engineering, Rutgers University, Piscataway, NJ

Molecular Simulation Studies of Quantitative Correlation between the Stability of Drug Nanocrystals Suspension System and Certain Properties of Both the Drug API and the Stabilizer

Wusheng Zhu, Frank Romanski, and M. Silvina Tomassone

Department of Chemical and Biochemical Engineering, Rutgers University, Piscataway, New Jersey 08854

Molecular simulations have been applied in the suspension system of poorly water-soluble (BCS Class II) drug nanocrystals in presence of a polymer or surfactant stabilizer. Since the binding energy between the drug crystal surfaces and the stabilizer is believed to link the inhibitory effect of the stabilizer that reflects the relative stability of the drug suspension system, comparison of the simulated drug-stabilizer binding energy, which is evaluated as a statistical average over different cleavage drug crystal faces, is utilized to gauge the relative stabilization capability of the suspension system. To efficiently screen an effective stabilizer for a given drug API that will result in good suspension stability, our research work is to find out the correlation between the drug-stabilizer binding energy and certain properties of the API and the stabilizer. If such a correlation is mapped out, an effective stabilizer can be easily chosen for a given API achieving good suspension stability without doing numerous real screening tests. In previous studies, we have built up quantitative correlations for two cases: i) correlation between the drug-stabilizer binding energy and certain properties such as supersaturation ratio and the molar water solubility of the drug API in presence of the same stabilizer; ii) correlation between the drug-stabilizer binding energy and certain properties such as hydrophobicity and molecular area of the stabilizer absorption on the same drug API nanocrystals. The current ongoing work suggests that a hierarchical high dimensional representation model can be employed to build up the full correlation between the binding energy and the properties of both the drug API and the stabilizer. The correlations obtained for the same stabilizer or the same drug API act as the first order approximation in the high dimensional representation model. The second order correlation can be fitted by simulated binding energies between various stabilizers adsorption on various drug APIs.

(692b) The Properties of Various Coating Polymers and Their Effects On Surface Frictional Behavior As Coatings On Pharmaceutical Tablets

Padma Narayan[1], Karen Balwinski[2], Debora Holbrook[2], Madhusudhan Kodam[1], Meaghan Blake[3] and Karl Jacob[1], (1)Engineering Sciences- Solids Processing, The Dow Chemical Company, Midland, MI, (2)Dow Wolff Cellulosics, The Dow Chemical Company, Midland, MI, (3)Analytical Sciences, The Dow Chemical Company, Midland, MI

Pharmaceutical tablets are typically coated for cosmetic and functional purposes, and the properties of coatings are important to ensure good performance and marketability. For immediate release applications, coatings provide a good barrier for taste masking, ease of swallowing tablets, and aesthetics. Smooth coatings are usually desirable for appearance and 'feel', hence the type of polymer and processing method used can influence the texture of the coating surface. Friction between tablets can also affect their performance in downstream manufacturing steps, such as during conveying, gravity filling, and packaging. Tablets with low values of frictional resistance may have improved throughput and minimal defect formation during conveying operations, especially when extremely large batch sizes (hundreds of kilograms) are processed. Assessment of in-process roughness and tackiness can also help during a coating operation, to prevent tablets from sticking and inducing further defects on the surface.

This study investigates the inter-relationship between physical properties of various coating materials and formulations, such as puncture and adhesion forces, surface roughness, and tablet-tablet sliding friction. A Schulze ring shear tester and Aero-Flow avalanche tester, both typically used for measurement of powders, were utilized for testing the frictional behavior of tablets. It was found that coatings with higher adhesion and low puncture forces (more hard and elastic materials) had lower values of friction than softer and plastically deforming materials. The coating surface roughness parameters did not show correlations with tablet sliding frictional behavior; however had a strong relationship with its pull-off (adhesion) force to the tablet surface. Based on this experimental study, it can be concluded that parameters relating to the elasticity and strength of the coatings influence their frictional behavior, agreeing with previous studies conducted on other material surfaces (Bellemare, et al., 2008 and Hancock, et al., 2010). Future work would include measuring the values of Young's modulus, yield strength, and strain hardening exponent of the films as well as understanding the effects of coating process parameters on the morphology and frictional characteristics of the tablet surfaces.

(692c) Surface Modification of Pollen Shape Carriers for Improved Dry Powder Inhalation Efficiency

Hung Loong Giam, Thi Quynh Ngoc Nguyen and **Raymond Lau**, School of Chemical and Biomedical Engineering, Nanyang Technological University, Singapore, Singapore

Pollen-shape particles were found to have great potential as drug carriers for dry powder inhalation. However, the relatively strong interparticle force on the carrier surface limits the detachment of drug particles after the drug-carrier mixture is released from inhalers. In this study, surface etching technique was employed to modify the carrier particle surface. Four surface etching temperature were examined. A low surface etching temperature were not able to modify the particle surface while a high surface etching temperature destroyed the pollen-shape morphology. Two of the surface etching conditions were then utilized and the physical properties of the resulting particles were characterized. Budesonide was used as the model drug and Rotahaler® is used as the model inhaler. The drug delivery efficiency using the two type of carrier particles are also characterized by Andersen Cascade Impactor (ACI) using inhalation rates of 30 liter/min and 60 liter/min. While there is a reduction in the emitted dose (ED), the fine particle fraction (FPF) of the drug was increased when pollen-shape carriers that were processed by surface etching were used.

ENCAPSULATION OF LACCASE IN CHITOSAN MICRO-CARRIERS BY TWO-FLUID AND THREE-FLUID NOZZLE

Ondrej Kaspar, Frantisek Stepanek, Institute of Chemical Technology Prague, Czech Republic

Introduction

Spray drying is a widely used particle formation process by which aqueous or organic solutions, emulsions and suspensions can be converted into a dry powder with a relatively narrow particle size distribution. Thanks to short contact times, spray drying is well suited even for heat-sensitive materials (enzymes, drugs) [1]. One of the most commonly used compounds for microencapsulation purposes is chitosan thanks to its availability and excellent properties, such as being nontoxic, bioadhesive, and biodegradable [2]. Spray dried microspheres made of pure chitosan cannot be kept suspended in aqueous media for a long time because of their swelling and dissolution. Therefore, non cross-linked chitosan microspheres prepared by spray-drying technique are unsuitable for the purpose of enzyme immobilization. Chitosan has been cross-linked by polyvalent tripolyphosphate (TPP) anions.

The effect of composition (in particular, the TPP/chitosan ratio) and cross-linking method (*ex-situ* vs. *in-situ*) on the properties of chitosan microspheres formed by spray drying by 2-fluid/3-fluid nozzle has been recently described [3]. The particle size distribution, particle morphology, zeta potential and swelling behavior have been systematically investigated. A novel cross-linking method, based on in-situ contact of TPP and chitosan solutions directly at the nozzle orifice (3-fluid nozzle), has been developed and compared with a more conventional approach of spray-drying a pre-mixed TPP/chitosan colloid solution using a 2-fluid nozzle (Fig. 1).

The main objective of this work is encapsulation of enzyme laccase by *ex-situ* and *in-situ* methodology. Laccase (EC 1.10.3.2) is an enzyme belonging to the family of copper-containing oxidases, which are widely distributed in fungi or higher plants [4]. It can oxidize ortho- and para-diphenols, polyphenols, anilines and a number of inorganic ions with the concomitant reduction of molecular oxygen to water. Particles of immobilized laccase could be used in a lot of industrial applications (decolorisation of waste water, biobleaching, antioxidant detectors) [5-7].

Materials and methods

Materials

Low molecular weight chitosan (75-85 % deacetylated), 2,2'-azino-bis(3-ethylbenzthiazoline-6-sulfonic acid) diammonium salt (ABTS) and STPP - sodium tripolyphosphate (>98 % purity) were purchased from Sigma-Aldrich (Germany), acetic acid from Penta (>99 % purity), sodium acetate anhydrous (>99%) was purchased from Fluka, sodium phosphate dibasic dodecahydrate (>99 %) and potassium phosphate monobasic (>99.5 %) from Fluka. Laccase obtained from wood rotten fungi *Trametes hirsute* was a kind gift from TU Graz.

Preparation of chitosan microparticles by spray drying

Chitosan solution (0.5% w/v) has been prepared by dissolving the required amount of chitosan in 25 ml of aqueous solution of acetic acid (2% v/v) at room temperature. After complete dissolution of the polymer required amount of laccase was added. In a case of 2-fluid nozzle cross-linking has been achieved by adding different volumes (0-2 ml) of a 1% (w/v) TPP solution drop-wise with a syringe (type 25Gx1") to the chitosan solution under constant stirring and then further stirred at 650 rpm for 15 min to ensure complete cross-linking of chitosan - "*in-situ* cross-linking". After addition of TPP the solution changes from transparent to opalescent. Opalescence occurs due to the formation of chitosan-TPP nanoparticles. The colloidal suspension of cross-linked chitosan nanoparticles was then spray-dried by a 2-fluid or to form microspheres with uniformly dispersed laccase. The volume of added laccase was 25 up to 200uL. In a case of 3-fluid nozzle was chitosan solution containing laccase and TPP solution pumped simultaneously into nozzle and spray-dried - "*in-situ* cross-linking". The laboratory spray-drier Buchi B-290 has been equipped with a high performance cyclone in open mode configuration. The values of operation parameters that ensured stable and robust process for the entire range of compositions for 2-fluid and 3-fluid nozzle: aspirator rate 90%, liquid flow rate 6.4 ml/min (skin-forming phase) and 3ml (inner phase - for 3 fluid nozzle), and atomizing gas flow rate 414 normlitres/hr (2-fluid nozzle) and 670 normlitres/hr (3-fluid nozzle). Inlet temperature was varied in a range 100 - 190°C. The outlet temperature was typically 40°C (for inlet temperature 100°C) to 85°C (inlet temperature 190°C).

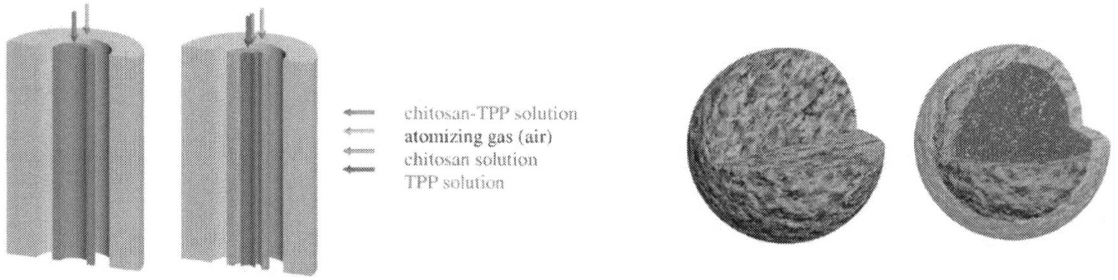

a) b)

Fig. 1 a) Scheme of a 2-fluid and a 3-fluid nozzle b) Schematic droplet structures produced by 2-fluid (left) and 3-fluid (right) nozzle.

Particle size and morphology characterization

Surface morphology of microparticles was examined by Scanning Electron Microscope (SEM) Jeol JCM-5700. Before measurement the samples of prepared microparticles were sputter-coated (Emitech K550X) by a thin layer of gold (5 nm). SEM micrographs were taken at an acceleration voltage of 20 kV. The internal structure of particles produced by the 3-fluid nozzle was determined by a Laser Scanning Confocal Microscope (Olympus Fluoview FV1000 confocal system with Olympus IX81 inverted microscope). The particle size distribution was measured by the static light scattering method (Horiba Partica LA-950 S2) in the wet mode. Chitosan particles were dispersed in absolute ethanol (in which they are not soluble) and placed to an ultrasound bath for 5 minutes. The suspension was then added to the measurement cuvette filled with ethanol and measured five times.

Release and enzymatic studies

Specific enzymatic activity of laccase stock solution was determined by a reaction with ABTS in a 0.1M acetate buffer at pH 5.0. Absorbance of the product (ABTS radical) was measured at 420 nm (characteristic absorbance peak for ABTS radical, ε_{420nm} = 36000 M^{-1} cm^{-1}) at 25°C and 50°C. The value of laccase specific activity was determined by a calibration measurement (0.51 ± 0.02 nmol.min^{-}

[1].uL[-1]). To measure enzymatic activity of laccase immobilized in chitosan powder, a given amount of the sample was placed inside an envelope made of filtrate paper which keeps chitosan microparticles but is fully permeable for ABTS substrate. The envelope was held in the cuvette by a plastic holder with an opening for the beam as shown in Fig. 2. This arrangement was chosen to prevent unwanted disruptions of the signal inside the cuvette by freely floating chitosan microparticles. The measuring cell was filled with 2.8 ml of acetate buffer in each experiment.

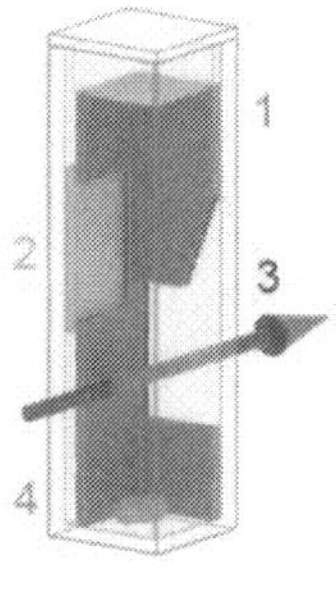

Fig. 2: Spectrophotometric cuvette with a custom attachment for the measurement of release kinetics (1 – plastic holder; 2 – sample; 3 – light beam; 4 – stirrer).

Results

This section is structured as follows. The dependence of inlet parameters (inlet temperature and TPP/chitosan mass ratio) on the laccase activity for two and three-fluid nozzle is discussed. The results will be presented in the form of parametric studies aiming at identifying general trends that can be used for particle design.

Influence of drying temperature (2 and 3-fluid nozzle)

Influence of drying temperature (100 to 190°C) on laccase activity and sample properties has been investigated. At higher temperatures (160 to 190°C) laccase has been significantly deactivated (air outlet temperature > 70°C). Suitable inlet temperature had been chosen 130°C (dry, nonstick product). Lower temperature leads to sticky and wet product. The results of the experiment are showed on Table 1.

Table 1. Influence of drying temperature on laccase activity.

Inlet temperature [°C]	Outlet temperature [°C]	Deactivation with respect to original enzyme [%]	Powder character
190	85	70	overheated
160	75	24	overheated
130	55	16	dry
100	40	14	sticky

Morphology and structure of particles

SEM images of chitosan particles produced by 2-fluid and 3-fluid nozzle are showed at Fig. 3a, b and c. Microparticles of pure chitosan are typically brain-like structured of spherical shape. Cross-linked (same TPP/chitosan ratio) particles produced by 2-fluid nozzle are of irregular collapsed shape and particles produced by 3-fluid nozzle are spherical with smoother surface due to different cross-linking process. These micro-carriers are much more stable [3] and apparently have better flow properties to compare with commonly prepared chitosan-TPP particles of irregular shape. Theoretical core-shell structure has been proven by laser scanning confocal microscopy (Fig. 3d).

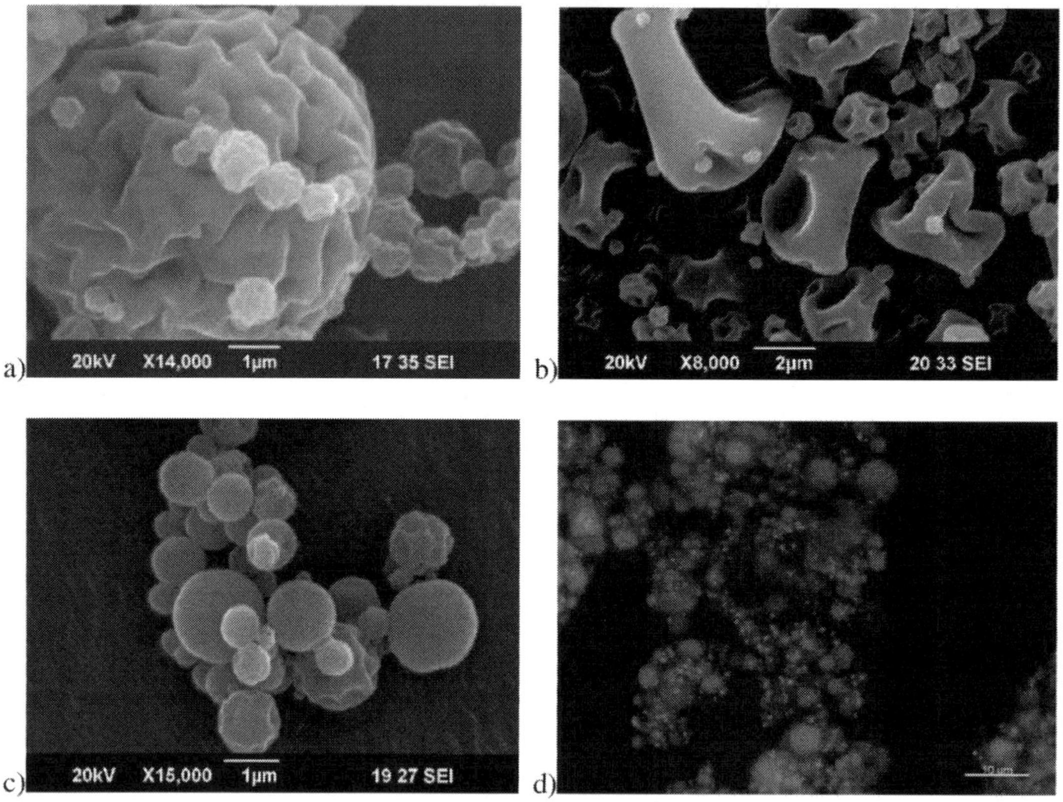

Fig. 3: a) SEM images of uncross-linked b) *ex-situ* cross-linked and c) *in-situ* cross-linked chitosan particles d) LSCM image of core shell structure (curcumin - core, FITC - shell).

Influence of laccase amount, reaction temperature and TPP content (2-fluid nozzle)

Firstly the activity of chitosan powder with different immobilized laccase content at 25 and 50°C has been investigated (Fig. 4a). It can be seen that the reaction temperature has a significant influence on enzymatic activity. In fact, it has been reported in the literature the optimum temperature for laccase *T. hirsute* is around 50°C [4], which is consistent with our findings. Higher TPP content (TPP/chitosan ratio) leads to denser cross-linked chitosan matrix without tendency to swell rapidly in aqueous medium. Increase in TPP/chitosan ratio generally leads to slower substrate diffusion through cross-linked shell to immobilized enzyme (Fig. 4b). Encapsulation efficiency for stable cross-linked samples (TPP/chitosan > 0.08) has been higher than 98%.

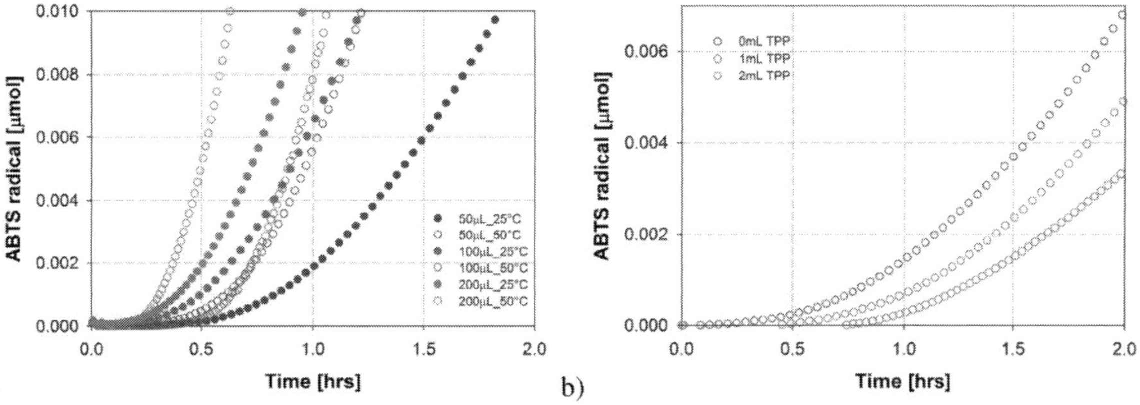

a) b)

Fig. 4: a) Influence of laccase amount, temperature and b) TPP content on enzymatic activity TPP/chitosan ratio a) 0.08 b) 0 to 0.16.

Influence of TPP/chitosan ratio (varied by chitosan and TPP concentration) for 3-fluid nozzle

Similarly to 2-fluid nozzle higher TPP concentration leads to denser cross-linked matrix and slower substrate diffusion in and out of particle (Fig. 4a). Higher chitosan concentration in outer phase caused lower TPP/chitosan ratio and stronger swelling activity in aqueous medium (Fig. 4b). Flow rates of inner (TPP) and outer (chitosan and laccase) fluid has been 3 and 6.4 mL/min. This set-up has been chosen to achieve higher cross-linked fraction without nozzle clogging and deposition of waste material on inner surface of drying chamber. Encapsulation efficiency for stable cross-linked samples (TPP/chitosan ratio > 0.05) has been 91 to 95%.

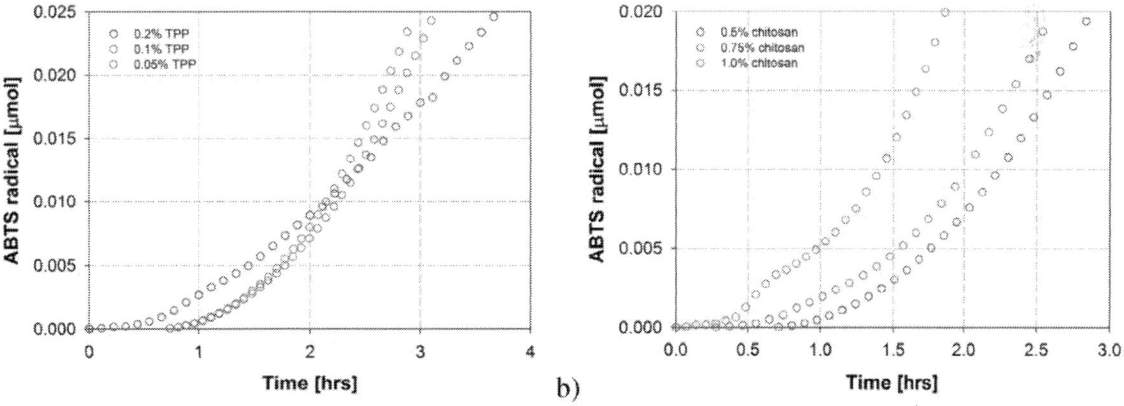

a) b)

Fig. 5: a) Influence of TPP/chitosan ratio on enzymatic activity TPP/chitosan ratio a) 0.05 to 0.19 b) 0.05 to 0.09.

Comparison of particle size distribution and specific enzymatic activity for 2 and 3-fluid nozzle

Particles produced by 3-luid nozzle have been smaller (mean size approx. 2.7 μm) than particles produced by 2-fluid nozzle (3.6 μm) of same composition due to higher flow of atomizing gas (wider nozzle orifice) to avoid deposition of spray-dried material on inner surface of the drying chamber. Typical particle size distributions for both types of nozzles are showed in Fig. 5a. Comparison of enzymatic activity for particles with same composition is showed in Fig. 5b. Enzymatic activity for 3-fluid nozzle product has been higher to compare with 2-fluid nozzle. Possible explanation is difference between cross-linking methods. In a case of 2-fluid nozzle is an enzyme immobilized into uniformly cross-linked chitosan matrix. In the second case chitosan droplets are cross-linked from inside (core) to outside (shell) during very rapid spray drying process (1-2 seconds) and density of cross-linked chitosan strains of shell and core is different. Diffusion of substrate through hydrated shell is faster and particles are much more stable due positively charged chitosan chains on the surface without tendency to form stable agglomerates. On the other hand encapsulation efficiency for 3-fluid nozzle (91 to 95%) is slightly lower than for commonly used 2-fluid nozzle (>98%).

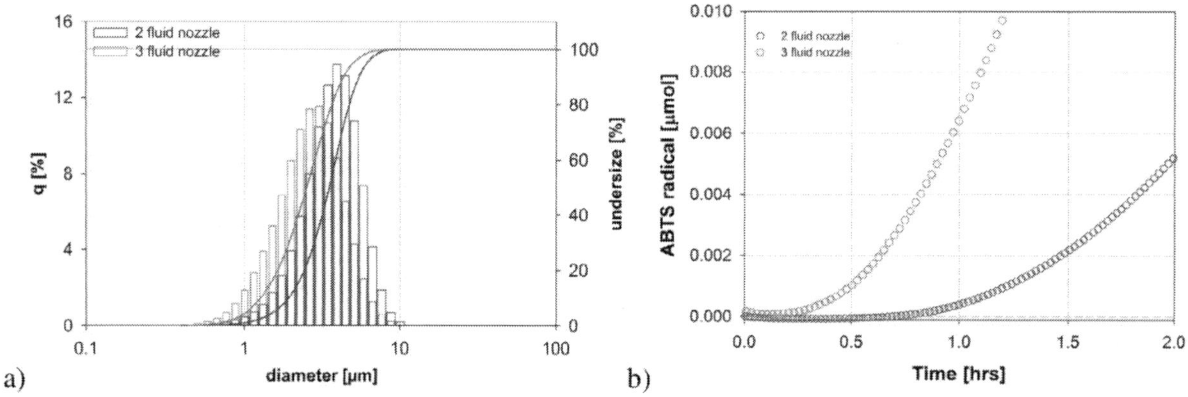

Fig. 6: Particle size distribution and enzymatic activity for samples produced by both types of nozzle (same composition – TPP/chitosan ratio 0.08).

Conclusion

Laccase has been successfully encapsulated by both types of nozzles without significant loss of enzymatic activity – 16% loss for spray-drying at 130°C respect to original. Encapsulation efficiency for 2 and 3-fluid nozzle has been higher than 98% and 91% respectively. Encapsulation efficiency of 3-fluid nozzle could be improved by optimalization of flow ratio of inner and outer fluids to minimize uncross-linked fraction. Particles produced by 3-fluid nozzle are more stable than particle produced by 2-fluid nozzle (higher zeta potential) and could be used in many industrial applications (e.g., decolorization of dyes, biobleaching, food processing, biosensors).

Acknowledgments

This work has been supported by the European Research Council (200580-Chobotix), by the Specific University Research (MSMT 21/2012) and by the Specific Science Cooperation Program AKTION (MEB061108).

References

1. Vehring, R., Foss, W.R.Lechuga-Ballesteros, D., Particle formation in spray drying. Journal of Aerosol Science, 2007. 38(7): pp. 728-746.

2. Agnihotri, S.A., Mallikarjuna, N.N.Aminabhavi, T.M., Recent advances on chitosan-based micro- and nanoparticles in drug delivery. Journal of Controlled Release, 2004. 100(1): pp. 5-28.

3. Kaspar, O., Jakubec, M., Stepanek, F., Characterization of spray dried chitosan–TPP microparticles formed by two- and three-fluid nozzles. Powder Technology, 2012 (in-press).

4. Arora, D.S.Sharma, R.K., Ligninolytic Fungal Laccases and Their Biotechnological Applications. Applied Biochemistry and Biotechnology, 2010. 160(6): pp. 1760-1788.

5. D'Annibale, A., Stazi, S.R., Vinciguerra, V., Di Mattia, E.Sermanni, G.G., Characterization of immobilized laccase from Lentinula edodes and its use in olive-mill wastewater treatment. Process Biochemistry, 1999. 34(6-7): pp. 697-706.

6. Peralta-Zamora, P., Pereira, C.M., Tiburtius, E.R.L., Moraes, S.G., Rosa, M.A., Minussi, R.C.Duran, N., Decolorization of reactive dyes by immobilized laccase. Applied Catalysis B-Environmental, 2003. 42(2): pp. 131-144.

7. Fernandes, S.C., de Oliveira, I.R.W.Z., Fatibello, O., Spinelli, A.Vieira, I.C., Biosensor based on laccase immobilized on microspheres of chitosan crosslinked with tripolyphosphate. Sensors and Actuators B-Chemical, 2008. 133(1): pp. 202-207.

(692e) Engineered Biodegradable Janus Particles for Drug Delivery

Jennifer Winkler, Chemical Engineering, Rutgers University, Piscataway, NJ, Frank Romanski, BASF, Iselin, NJ and Maria. S Tomassone, Department of Chemical and Biochemical Engineering, Rutgers University, Piscataway, NJ

Recent advances in the field of nanotechnology have generated interest in the creation of "Janus" particles, whose dual surface chemistries and unique properties set them apart from spherical isotropic particles. Janus particles present two distinct faces with different shapes, compositions, chemistries, and polarities. Polymeric and hybrid polymer-lipid Janus particles have applications as compartmentalized drug carriers, biological sensors, surfactants, or display materials for electronics. However, biodegradable Janus particles have yet to make their debut in the pharmaceutical industry because current synthesis methods rely on microfluidic devices capable of producing just one single particle at a time. Furthermore, lipid-based materials do not meet the low viscosity requirements of microfluidic devices. Here, we describe a modified emulsion-solvent evaporation template for the large-scale production of biodegradable Janus micro- and nanoparticles from GRAS materials. Biodegradable anisotropic particles are produced through emulsification of the oil phase containing the two GRAS components and the aqueous phase containing stabilizers, followed b homogenization and solvent evaporation. During the slow 6-hour evaporation process, the two immiscible components segregate to an entropically favorable state of separation. However, the immiscible polymer and/or lipid components are forced to co-precipitate within the same oil droplet "microreactor" during evaporation, leading to the formation of an interpenetrating network connecting the two components. Polymeric Janus particles have been formed from 50/50 combinations of GRAS polymers such as PLGA, polycaprolactone, polyvinylpyrrolidone, and poly(4-vinylpyrrolidone). Polymer-lipid "ice cream cone" particles have been formed from 75/25 combinations of GRAS polymers listed above and GRAS lipids such as precirol and compritol. Drug loading is achieved by co-dissolving the API with the polymer and/or lipid components in the dispersed phase prior to emulsification. In some cases, drug loading tremendously alters the phase separation and self-assembly of polymeric Janus particles. Spherical particles with surface pox marks were produced when 5% w/w griseofulvin, omeprazole, or acetaminophen was co-dissolved with the 50/50 polymer mixture in the dispersed phase. However, a Janus-like morphology is still obtained when 5% w/w curcumin or quercetin is introduced into the system. It is hypothesized that certain APIs form interactions with both polymers, thus acting as block copolymer compatibilizers that reduce the interfacial tension between two immiscible polymers. No APIs were found to disrupt the phase separation behavior of the polymer-lipid hybrid particles, presumably because the polymer and lipid are highly dissimilar in structure and chemistry.

(703a) Single-Cell Analysis of Quality Control in S. Cerevisiae: How 'Low' Can You Go?

Carissa L. Young[1,2], Theresa Yuraszeck[3], David Raden[4], Jeffrey Caplan[1], Francis J. Doyle III[5], Kirk J. Czymmek[1] and Anne S. Robinson[6], (1)UD Bio-Imaging Center, Delaware Biotechnology Institute, Newark, DE, (2)Department of Chemical and Biomolecular Engineering, University of Delaware, Newark, DE, (3)Chemical Engineering, University of California, Santa Barbara, CA, (4)Department of Biology, University of Delaware, Newark, DE, (5)Chemical Engineering, University of California Santa Barbara, Santa Barbara, CA, (6)Department of Chemical and Biomolecular Engineering, Tulane University, New Orleans, LA

Within the secretory pathway of eukaryotic cells, the endoplasmic reticulum (ER) is responsible for maintaining the fidelity of protein synthesis and maturation. A variety of disturbances including nutrient deprivation, pathogenic infection, and chemical treatment, collectively termed 'ER stress', induce quality control mechanisms to facilitate the recovery of cell homeostasis. ER-associated degradation (ERAD), unfolded protein response (UPR), and autophagy are quality control pathways that occur on various timescales, encompass variations in the spatial organization of multiple organelles, and alter select protein concentrations and intracellular localization. Surprisingly, all three pathways are activated in several neurodegenerative and hereditary diseases. In all cases, as a result of ER stress, there is evidence of atypical, intracellular protein distribution during disease manifestation. Yet, how the accumulation of disease-specific proteins is involved in compromising quality control remains elusive.

By implementing DNA recombination strategies combined with high-resolution imaging techniques, we have determined that protein redistribution, resultant spatial effects, and organelle modifications are a consequence of the cell's response to ER stress in yeast, *S. cerevisiae*. In pursuit of a thorough analysis of protein redistribution at the subcellular level, multiple yeast expression cassettes [1] have been created to test the effects of codon-optimized fluorescent protein (FP) variants, small epitope tags (reviewed in [2]), polylinker length for N- and C-terminal tags, and the inclusion of essential retrieval sequences for ER luminal chaperones and foldases [3]. Consequently, our approach enables one to monitor the trafficking effects of ER chaperones and foldases by incorporating H/KDEL retrieval sequences fused to FP variants, hence evaluate the spatiotemporal effects of chaperone/co-chaperone interactions [4]. To investigate discrete subpopulations of tagged proteins using live-cell imaging methods and super-resolution techniques (e.g. Fluorescence-Photoactivation Localization Microscopy, F-PALM and Structured Illumination Microscopy, SIM), photoconvertible GFP variants and FlAsH-based technology were implemented.

Utilizing FP variants as probes, proteins were recombinantly expressed from their native promoter in order to monitor protein trafficking, analyze localization effects of proteins involved in ERAD, and examine organelle dynamics and morphology as a consequence of local perturbations to the system [5]. Furthermore, we have confirmed the existence of cellular variability following UPR activation at the level of single-cell analysis and evaluated cytoskeleton modifications. It is now evident that endogenous proteins redistribute in the cell, specifically the ER, in order to perform essential functions that maintain cell homeostasis. Using novel imagining techniques such as Focused Ion Beam (FIB) microscopy [6], entire three-dimensional organelle reconstructions of yeast cells were developed at electron microscope resolution. Our results confirm organelle connectivity during inheritance (e.g. lipid droplets derived from perinuclear ER) and following periods of prolonged stress (e.g. peripheral ER and mitochondria junctions). High-resolution techniques (e.g. SIM and F-PALM) have led to extraordinary biological insights involving organelle biogenesis and intracellular communication [7].

Due to the intrinsic complexity of biological systems, the integration of experimental and computational approaches is crucial when investigating the UPR of *S. cerevisiae*. To establish that the UPR is a global response to localized perturbations, we evaluated the transcriptional effects of UPR activation [8]. Interestingly, microarray analysis and q-PCR validation have confirmed the novel repression of 206 genes – highly enriched in protein synthesis and metabolic biological functions – following ER stress. We have expanded the characterization of the UPR and reaffirmed diverse, global consequences for the cell. UPR activation was assessed at the molecular level by a systematic analysis of yeast deletion strains. Collectively, our experimental approaches have led to a better understanding of ER homeostasis and cell regulation.

References

1. C. L. Young, D. Raden, J. Caplan, K. Czymmek, A. S. Robinson *Optimized Cassettes for Live-Cell Imaging of Proteins and High Resolution Techniques in Yeast*, Yeast, 2012 doi:10.1002/yea.2895. [Epub 2012 Apr 4]

2. C. L. Young, Z. T. Britton, A. S. Robinson *Recombinant Protein Expression and Purification: A Comprehensive Review of Affinity Tags and Microbial Applications*, Biotechnology Journal, 7(4), Jan 10 2012 doi:10.1002/biot.201100155. [Epub ahead of print]

3. C. L. Young, D. L. Raden, A. S. Robinson, *Analysis of Endoplasmic Reticulum Resident Proteins in S. cerevisiae: Implementation of H/KDEL Retrieval Sequences,* 2012 (submitted).

4. M. Griesemer, C. Young, A. Robinson, L. Petzold *Spatial Localization of Chaperone Distribution in the Endoplasmic Reticulum of Yeast.* IET Systems Biology, 2012 doi:10.1049/iet-syb.2011.0006.

5. C. L. Young, D. L. Raden, J. Caplan, B. Chung, K. Czymmek, A. S. Robinson *Dynamics of Endoplasmic Reticulum Resident Proteins and Organelle Morphology in S. cerevisiae,* 2012 (in preparation).

6. D. Wei, S. Jacobs, S. Modla, S. Zhang, C. L. Young, R. Cirino, J. Caplan, K. Czymmek *High-resolution three-dimensional reconstruction of a whole yeast cell using focused-ion beam scanning electron microscopy,* BioTechniques, 2012 (in press).

7. C. L. Young, D. Raden, J. Caplan, K. Czymmek, A. S. Robinson *Spatiotemporal Resolution of Protein Distribution at the Sub-Organelle Level in S. cerevisiae during Cell Division,* 2012 (in preparation).

8. T. Yuraszeck , C. Young , P. Xu, C. A. Gelmi, F. J. Doyle III, A. S. Robinson *Novel down-regulation pathways in the Unfolded Protein Response from S. cerevisiae provide evidence of a complex regulatory response to ER stress,* 2012 (under revision) co-authorship.

(703b) Inter-Kingdom Signaling and Chemotaxis of E. Coli towards the Human Hormone Norepinephrine

Sasi K. Pasupuleti, Chemical Engineering, Texas A&M university, College Station, TX and Arul Jayaraman, Department of Chemical Engineering, Texas A&M University, College Station, TX

Inter-kingdom Signaling and Chemotaxis of *E. coli* towards the human hormone norepinephrine

Sasi Kiran Pasupuleti[a], Mathew Sears[b], Michael D. Manson[b], Arul Jayaraman[a]

Department of Chemical Engineering[a] and Department of Biology[b], Texas A&M University, College Station, Texas 77840.

Abstract

The co-existence of ~10^{14}non-pathogenic commensal bacteria and human mucosa cells (intestinal epithelial cells, dendritic cells, macrophages) in the human gastrointestinal (GI) tract creates an environment rich in molecules produced by both the bacteria (e.g., autoinducer-2, indole) and the host (e.g., norepinephrine, dopamine). The close-proximity of different signals and cells is thought to lead to inter-kingdom signaling where bacteria and human cells recognize and respond to signals produced by the other kingdom. Prior work from our lab [1] has shown that pathogenic bacteria are attracted towards higher concentrations of host-produced hormones (i.e., chemotaxis) and this migration correlates with an increase in pathogen virulence and invasion of host cells [2]. However, the mechanisms through which NE is sensed by pathogenic bacteria are poorly understood.

In this study, we investigated the mechanisms underlying the inter-kingdom recognition and chemotaxis towards norepinephrine using *E. coli* as the model bacterium. Using isogenic mutant strains and a microfluidic chemotaxis model (µFlow) developed in our laboratory [2], we identified that chemotaxis towards norepinephrine requires the serine chemoreceptor (Tsr), and that Tsr is both necessary and sufficient for norepinephrine chemotaxis. The shape of the concentration gradient was important, as chemotaxis was observed only in the presence of a sharp concentration gradient, and not in shallow concentration gradients. We further identified that the serine majority binding site (T156), but not the minority binding site (R69), is required for norepinephrine chemotaxis. We also observed that a robust chemotactic response was observed only when bacteria were pre-exposed to norepinephrine, as untreated cells did not move towards higher concentrations in a norepinephrine concentration gradient. Therefore, we hypothesized that priming bacteria with norepinephrine results in *de novo* synthesis of protein(s) required for chemotaxis towards norepinephrine. Pre-exposure to norepinephrine in the presence of a protein synthesis inhibitor abolished the chemotaxis in a norepinephrine gradient and confirmed the need for *de novo* protein synthesis. Preliminary qRT-PCR studies identified that the enzyme monoamine oxidase A (*maoA*), which converts norepinephrine to 3,4-dihydroxyphenylglycolaldehyde in the GI tract, is up-regulated in *E. coli* upon norepinephrine priming. Our results suggest a novel chemotaxis mechanism in which the invading pathogen metabolizes an abundant molecule such as norepinephrine that is present in the microenvironment to generate a metabolite that is then used as a potent chemoattractant.

References

[1] Bansal, T., et al., *Differential effects of epinephrine, norepinephrine, and indole on Escherichia coli O157:H7 chemotaxis, colonization, and gene expression*. Infection and immunity 75:4597–4607. [2] Englert, DL., et al., *Investigation of bacterial chemotaxis in flow-based microfluidic devices*. Nature Protocols 5, -864 - 872 (2010). [2] Englert, DL., et al., *Investigation of bacterial chemotaxis in flow-based microfluidic devices*. Nature Protocols 5, -864 - 872 (2010).

(703c) Constructing a Synthetic Gene Network to Model and Understand Signaling Interactions

Ashley Jermusyk and Gregory T. Reeves, Chemical and Biomolecular Engineering, North Carolina State University, Raleigh, NC

In multicellular organisms, cellular signaling events are crucial for patterning tissues, as well as for maintaining healthy adult tissues, while improper signaling can lead to disease states, such as cancer. Therefore, cellular signaling process must be tightly regulated. A complex system of gene regulatory circuits controls this signaling process and acts to buffer this system against noise, thereby minimizing mistakes in gene expression and preventing patterning defects or disease states. Despite their importance to patterning and development, hypotheses regarding these gene regulatory circuits have been difficult to test experimentally due to their complexity and high connectivity. Therefore to better understand the fundamental processes involved, we created a synthetic gene network in the fruit fly *Drosophila melanogaster*. This network utilizes genes from yeast and *E. coli*, namely, *gal4*, *gal80*, and *lacZ*. We expressed *gal4* in a graded fashion along the anterior-posterior axis of the embryo, mimicking the intracellular diffusion of an endogenous transcription factor, Bicoid. The *gal4* activates expression of *UAS*-linked *gal80* and *lacZ*. *gal80* inhibits *gal4* activation, creating a negative feedback loop in our system. These genes were chosen since they are not endogenous to *Drosophila*, so all interactions in this network will be fully understood. Our goal is to measure the variability in location of the *lacZ* domain both with and without the negative feedback loop. This system provides a direct experimental test of whether negative feedback loops in cellular signaling events can provide robustness to noisy, diffusive systems. It is this robustness that is important for combating diseases and defects in development and maintenance of expression in the organism.

(703d) A Polyvalent Cell Engineering Strategy to Enhance Intracellular Clearance

Wensi Song, Kiri Kilpatrick and **Laura Segatori**, Chemical and Biomolecular Engineering and Bioengineering, Rice University, Houston, TX

Lysosomes are membrane-enclosed organelles rich in lytic enzymes capable of breaking down a variety of biomolecules, such as proteins, lipids and carbohydrates. They function as the cell's waste disposal system, as they provide a site of degradation for material taken up from the extracellular space or intracellular waste products. Impairment of lysosomal degradation often leads to the accumulation of storage material and to the development of devastating human diseases collectively referred to as lysosomal storage disorders (LSDs). LSDs are clinically highly diverse and can affect most organs, often as part of a multisystem disorder. The most severe forms affect the brain, causing progressive neurodegeneration and early death, and are presently incurable. Several experimental treatments are currently being tested in animal models of LSDs, which address the substitution of the single gene or protein defective in the specific LSD. However, these approaches have severe intrinsic limitations to brain targeting due to the presence of the blood-brain barrier (BBB). The BBB actually separates brain tissues from the body's bloodstream, thus limiting the permeability to molecules, proteins, and DNA-containing vectors that may be used as therapeutic agents. In addition, these approaches are disease-specific, which often means that the knowledge gained in the development of a treatment for a given LSD is usually not directly transferrable to other LSDs. We describe here a polyvalent therapy against multiple neurodegenerative LSDs. We designed a strategy to modulate endogenous cellular pathways involved in lysosomal biogenesis and function, including the folding and processing of lysosomal proteins and autophagic processes. Particularly, we used chemical and biological approaches to activate the transcription factor EB, a master regulator of lysosomal biogenesis. We observed modulation of the transcription factor EB dependent network in association with rescue of common features of LSDs such as the impairment of lysosomal function and the accumulation of undegraded storage material. We also demonstrate that this strategy can be used to promote clearance of cytoplasmic proteinaceous aggregates through chaperone-mediated autophagy, thus suggesting that this strategy could potentially impact a number of clinically distinct human diseases ranging from early childhood lysosomal LSD to late-onset neurodegenerative diseases, such as Alzheimer's, Parkinson's and Huntington's diseases.

(703e) Adherens Junctions Formation Prevents Lentiviral Entry

Roshan Padmashali[1], **Hui You**[1] and Stelios T. Andreadis[1,2,3], (1)Chemical and Biological Engineering, University at Buffalo, State University of New York, Amherst, NY, (2)Center of Excellence in Bioinformatics and Life Sciences, Buffalo, NY, (3)Department of Biomedical Engineering, University at Buffalo, State University of New York, Buffalo, NY

Adherens junctions (AJs) in epithelial cells involve the inter-cellular interactions and are well known to play a role in number of processes such as proliferation, polarity, differentiation, inflammation etc. Here we explore the role of AJs in mediating lentivirus entry into epithelial cells by controlling junctional stability using calcium (Ca2+) switch, shRNA knockdown or dominant negative overexpression of AJ complex proteins. Epithelial cells form colonies with mature AJs in presence of high Ca2+ (1~1.8mM) and grow as single cells under low Ca2+ conditions (0.06 mM). AJ formation upon increasing Ca2+ concentrations resulted in significant reduction of gene transfer in these cells. Interestingly, cell colonies growing in high Ca2+ conditions exhibited higher transduction preferentially towards the periphery which can be attributed to AJ-free edges of peripheral cells. By shRNA knockdown of junction complex protein α-catenin and overexpression of dominant negative mutant E-Cadherin we were able to prevent AJs formation. Owing to loss of AJs, lentiviral entry lost dependence on Ca2+ and transduction efficiency remained unchanged at varying Ca2+ concentrations. With a tet-ON E-Cadherin dominant negative expression system we weakened AJs in a doxycycline dose dependent manner even in the presence of high Ca2+ concentration. Once again weaker AJ led to higher levels of virus entry and transduction efficiency. Interestingly, we also noticed rapid movement of cells upon lentivirus exposure, eventually weakening AJs and cells spreading out. This suggests a role of actin cytoskeleton in mediating this response to virus entry. To confirm this we disrupted cellular stress fibers using the ROCK inhibitor Y27632 and thereby weakened AJs. Gene delivery was significantly increased upon Y27632 treatment and showed dose dependence. With the use of shRNA and dominant negative forms of ROCK we were able to further delineate the role of ROCK in AJ controlled infection. Our work strongly suggests that AJ formation prevents lentiviral entry and disruption of AJs significantly enhances gene transfer to epithelial cells.

(703f) Effect of Microtubule Motors On Microtubule Mechanics in Living Cells

Nandini Shekhar, Jun Wu, Anthony J.C. Ladd, Richard Dickinson and **Tanmay Lele**, Chemical Engineering, University of Florida, Gainesville, FL

Effect of microtubule motors on microtubule mechanics in living cells

Introduction: The mechanical properties of microtubules are important in enabling its various cellular functions during division, migration and intracellular transport. Microtubules under thermal forces have a persistence length on the order of millimeters yet *in vivo* microtubules exhibit bends on micron length scales. This suggests that within a living cell, microtubules are exposed to large non-thermal forces. We show with femtosecond laser ablation that microtubules are under tension due to pulling forces generated by dynein motors[1]. These pulling forces focus bends in microtubules to the periphery and also center the centrosome. We also answer the question: can microtubule motors cause fluctuations in the trajectories of the tips of growing microtubules?

Materials and Methods: Laser ablation experiments were performed with a femtosecond laser on an inverted laser scanning confocal microscope using a 63X, 1.4-NA Plan-Approchromatic oil immersion lens. The growing tips of microtubules in EGFP-EB1 expressing NIH-3T3 cells were imaged on a Leica SP5 DM6000 confocal microscope using a 63X objective. DsRed-CC1 expression was used to inhibit dynein. The growing tips were detected and tracked over time using the software plusTipTracker[2]. Further analysis was done using programs developed in MATLAB.

Results and Discussions: On severing a single microtubule in a living cell using femtosecond laser ablation, we found that the minus-ended microtubule increased in bending in a dynein-dependent manner suggesting that the microtubule is under tension (not compression) along its length due to dynein motor activity(Figure 1 A). Experimental results on single microtubule severing and centrosome centering by dynein pulling forces could be explained with a computational model. We also found that microtubule growth trajectories in control cells spread out considerably at early times compared to trajectories in dynein-inhibited cells (Figure 1 B). Dynein activity thus contributes significantly to the bending of growing microtubules. We have also performed preliminary experiments on understanding the role of kinesin and myosin II in bending growing microtubule tips.

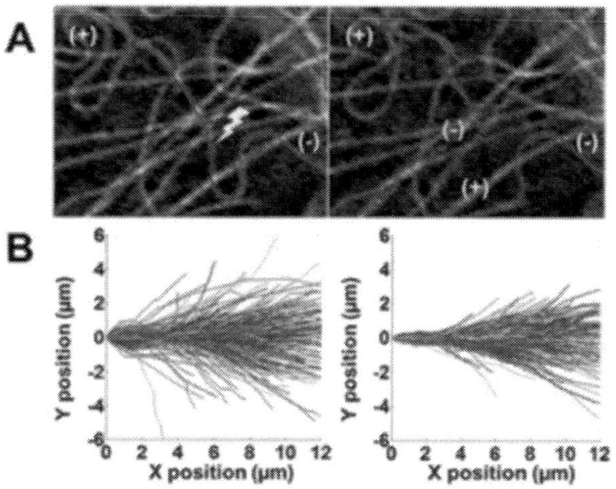

Figure 1. (A) Images highlighting changes in shape after laser-based severing of a single microtubule in a living cell. The newly generated minus end microtubule increases in bending after laser severing. The position of the cut is indicated by the flash and the severed microtubule is highlighted in red. New plus and minus ends created upon severing are indicated by the yellow symbols; the white symbols indicate the ends of the original

microtubule[3]. (B) Trajectories of growing microtubules were reconstructed using plusTipTracker. The plots show that fluctuations in growing microtubules are smaller in dynein-inhibited cells (right) compared to control cells (left).

Conclusions: Dynein pulls along the length of microtubules in living cells and also frictionally resists any lateral motion. Compared to these forces, compressive stresses in the microtubule are negligible (except very close to the periphery). In the absence of dynein, microtubules grow straight while in the presence of dynein, there is significant variance in the growth trajectories. Collectively, our results shed new light on microtubule mechanics in living cells.

References: 1. Wu J. et al. MOL BIOL CELL, 2011, 22, 4834-4841

2. Applegate K. T. *et al.* J STRUCT BIOL, 2011, 176, 168-184

3. Wu J. *et al.* INTEGR BIOL, 2012, DOI: 10.1039/c2ib20015e

(703g) A Mechanism for Adaptative Remodeling in the Bacterial Flagellar Motor

Pushkar Lele and Howard C. Berg, Department of Molecular and Cellular Biology, Harvard University, Cambridge, MA

Flagellar motors of *Escherichia coli* switch stochastically between clockwise (CW) and counter-clockwise (CCW) rotational states, enabling the cell to perform chemotaxis. CW rotation is promoted by the binding of an activated cytoplasmic response regulator, CheY-P, to a rotor component, FliM. Yuan et al, Nature (2012) recently discovered that the rotor complex undergoes remodeling in response to persistent decline in the levels of CheY-P, a mechanism which may be responsible for countering cell to cell variations in concentrations of chemotactic proteins, as well as aiding in navigating rapidly changing environments. However, the feedback mechanism which controls this remodeling remains a mystery. I will present our latest results which establish a link between adaptative remodeling and the dynamics of the motor, and propose a model mechanism for this adaptation. These results suggest a regulatory role for protein exchange in multi-protein complexes, which could be a tactic employed in other biological machines as well.

(703h) Mathematical Modeling of Intracellular Transport in the Squid Giant Axon

Jennifer Anne Pascal[1], Michael Loewenberg[2], Arnaud Chauviere[1], Pamela Seamster[1], Elaine Bearer[1] and Vittorio Cristini[1], (1)Pathology, University of New Mexico, Albuquerque, NM, (2)Chemical Engineering, Yale University, New Haven, CT

Intracellular transport of cargo, including macromolecules, vesicles and organelles, through the attachment to microtubules via molecular motors, such as kinesin and dynein, is a complex process that plays a significant role in neuronal function. Disruption of this transport has been linked to neurodegenerative diseases, such as Alzheimer's and Parkinson's diseases. Thus, studying the interactions among different types of cargo and molecular motors can lead to a better understanding of the complicated processes involved during intracellular transport.

Here, we present a mathematical model based on traffic-like partial differential equations to describe coupled cargo transport within the squid giant. The model is informed using direct microscopic measurements of nano-bead transport within the squid giant axon which allows for meaningful validation of the model framework. An analytical solution of the model equations is obtained under conditions characterized by an excess of molecular motors within the cytoplasm. The solution exhibits wave-like hyperbolic behavior at short times and a convective/diffusive behavior at longer times. A new model is presented for conditions characterized by a deficiency of molecular motors. Under these conditions, transport of different cargo species is nonlinearly coupled. As a first step towards understanding this phenomenon, we analyze experiments in which a protein (APP-C) is co-injected with protein-coated nano-beads in the squid giant axon leading to competition for motors.

(726a) Cyclic Strain Versus Endothelial Cell Presence On MSC Osteogenesis

Mariah S. Hahn, Department of Chemical Engineering, Texas A&M University, College Station, TX

Mesenchymal stem cells (MSCs) are increasingly recognized as a viable cell source for bone regeneration applications. In addition to soluble factors, cyclic stretch has been shown to have a profound effect on mesenchymal stem cell (MSC) osteogenesis. Similarly, endothelial cell (EC) presence has been demonstrated to enhance MSC osteogenic differentiation. However, potential synergistic interactions between mechanical stimulation and EC presence remain to be elucidated. The aim of the present manuscript was therefore to examine the simultaneous influence of cyclic stretch and EC paracrine signaling on MSC osteogenesis in the context of scaffolds with "osteogenic" moduli.

To accomplish this, 10T½ multipotent stem cells were encapsulated in poly(ethylene glycol) diacrylate [PEGDA] hydrogels with moduli within the "osteogenic" range. Half of the constructs were fabricated with a luminal EC layer. EC^+ and EC^- constructs were then subjected to continuous cyclic stretch. Following 10 days of culture, both EC^+ and EC^- constructs were associated with significantly elevated levels of chondrogenic transcription factor sox9 relative to initial (day 0) expression levels. By day 22 of culture, however, sox9 levels in both the EC^+ and EC^- constructs had returned to or fallen below day 0 levels. In contrast, osteocalcin expression was significantly higher in Day 22 EC^+ constructs relative to EC^- constructs and relative to day 0. Similarly, osteopontin and alkaline phosphatase levels were elevated in Day 22 EC^+ constructs relative to EC^- constructs. Cumulatively, the present results suggest that EC paracrine signaling enhances MSC osteogenesis in the presence of cyclic stretch. In addition, the observed transition from chondrogenic to osteogenic protein expression associated with the EC^+ constructs would be consistent with the neovascularization-dependent transition from cartilage matrix to osteoid matrix associated with endochondral bone formation.

(726b) Co-Electrospun Scaffolds with Gradients in Fiber Alignment and Chemistry for Regeneration of the Ligament-Bone Interface

Satyavrata Samavedi, Prudvi Gaddam, Abby Whittington and Aaron Goldstein, Chemical Engineering, Virginia Tech, Blacksburg, VA

Co-electrospun scaffolds with gradients in fiber alignment and chemistry for regeneration of the ligament-bone interface

Introduction: Grafts currently used for the repair of ligament injuries integrate poorly with bone due to a significant mismatch of mechanical and chemical properties between graft and bone. This mismatch stems from the grafts' inability to regenerate gradients of mineral content, collagen fiber alignment and cell phenotype that exist at natural ligament-bone transitions. We envision that an electrospun scaffold, possessing gradients of architecture, mechanical and biochemical properties, will help establish spatial gradients of cell phenotype in bone marrow stromal cells (BMSCs) and thus aid the regeneration and osseointegration of injured ligaments. As a step towards the larger goal of regenerating complex tissue transitions, we describe herein the fabrication of a scaffold possessing gradients of mineral content and fiber alignment, obtained by co-electrospinning two polymer solutions, with different chemistries, from offset spinnerets onto a dual-drum collector. The dual-drum collector used in this study allows the creation of graded scaffolds with transition regions at physiologically relevant length-scales.

Materials and Methods: Solutions of poly-lactic-co-glycolic acid (PLGA) doped with an amorphous calcium phosphate (ACP) phase and polycaprolactone (PCL) were co-electrospun from offset spinnerets onto a custom-designed slowly rotating dual-drum collector (**Figure 1a**). Post-electrospinning, the drums were stretched apart to result in strain-induced fiber alignment in the gap region between the drums. This fabrication procedure resulted in the formation of randomly oriented ACP doped-PLGA fibers on one drum, aligned PCL fibers in the gap region between the drums and a transition region consisting of both types of fibers at the interface. Angular orientation of the fibers from the three regions was characterized from micrographs of the scaffold. Finally, rat BMSC alignment, cell metabolic activity, deposition of Collagen-I (Col-I) and phenotypic makers of cell differentiation were investigated on all regions of the graded scaffolds.

Results and Discussion: Co-electrospinning onto the custom-designed collector resulted in the formation of a graded scaffold consisting of three different regions: a randomly oriented region possessing fibers that were electrospun from one spinneret (**Figure 1b**), an aligned region possessing fibers that were electrospun from the other spinneret (**Figure 1d**) and a transition region possessing a mixture of both types of fibers (**Figure 1 c**). After 1 and 3 days of culture, rat BMSCs were found to be aligned in the aligned region of the scaffold (**Figure 1e**), and randomly oriented on the other regions. Cells were also found to be metabolically active on all regions of the scaffolds after 1 and 7 days of culture. Finally, a gradient of cell phenotype and collagen-I (Col-I) orientation was observed on the graded scaffolds.

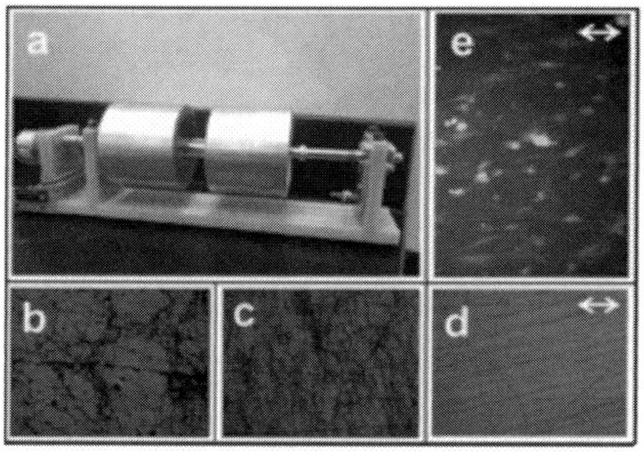

Figure 1:

(a) Slowly-rotating dual-drum set-up with gap region in between the drums.

(b), (c) and (d): Microscopy images of samples collected from a scaffold fabricated with two PCL solutions from offset spinnerets (proof-of concept). White arrow in (d) indicates general direction of fiber alignment.

e) Rat bone marrow stromal cells seeded on the aligned PCL section scaffold and imaged with Calein AM after 1 day of seeding. White arrow indicates general direction of cell alignment.

Conclusions: This study demonstrates that scaffolds with complex architectures and gradients in mechano-chemical properties at physiologically relevant length-scales, can be fabricated by co-electrospinning appropriate polymer solutions from offset spinnerets onto a specially designed collector. The study also indicates that these scaffolds have the potential to be used for the regeneration of graded tissue transitions such as those present at the ligament-bone interface.

(726c) Engineered Arterial Mimics (EAMs) to Quantify Smooth Muscle Cell Contribution to Atherosclerosis

William Herrick, Chemical Engineering, University of Massachusetts Amherst, Amherst, MA and Shelly R. Peyton, Chemical Engineering, University of Massachussetts, Amherst, Amherst, MA

For over a century, coronary heart disease (CHD) has been the leading cause of death in the Western world, surpassing all forms of cancer combined. The primary cause of CHD, atherosclerosis, is an extensively studied, yet incompletely understood disease. We hypothesize that one reason this disease is poorly understood, as well as why so few drug targets discovered *in vitro* are successful *in vivo*, is that the conditions in which cells are cultured and studied *in vitro* do not capture the complex *in vivo* microenvironment. To address this issue, we have created Engineered Arterial Mimics (EAMs) from PEG-PC gels, a hydrogel formed from a PEGDMA (poly(ethylene) glycol dimethacrylate) backbone with zwitterionic phosphorylcholine (PC) pendant groups, and which include integrin-binding extracellular matrix (ECM) proteins found in healthy and atherosclerotic arteries.

PEG-PC gels offer key advantages over PEG hydrogels. First, the Young's modulus can be tuned over a wider range by allowing gel formation at very low PEGDMA concentrations (~2 kPa, 5 kPa, 50 kPa, 200 kPa and 500 kPa at 0.5, 1, 3, 6 and 10% PEGDMA and 20% PC). Second, they do not become opaque at higher Young's moduli, and finally, the zwitterionic PC group makes these gels more hydrophilic than PEGDMA alone and they can, therefore, more efficiently block non-specific protein adsorption. We are exploiting these features to create EAMs that represent the chemical and physical changes that occur during the progression of atherosclerosis. We have EAMs that consists of collagen III, collagen IV and laminin (70/15/15 weight % at 10 ug/cm^2 total protein), which are predominantly found in healthy arteries, and EAMs that consist of fibronectin and collagen I (50/50 weight % at 10 ug/cm^2 total protein), which dominate the atherosclerotic artery. In order to capture the reported mechanical heterogeneity of diseased arteries, EAMs are made over a wide range of Young's moduli: ~5, 50 and 200 kPa (1, 3, 6% PEGDMA and 20% PC, respectively).

Preliminary migration experiments on EAMs varying only integrin-binding proteins found that human aortic smooth muscle cells (SMCs) migrate faster on the diseased ECM than on the healthy ECM (20 μm/hr vs. 15 μm/hr), indicating that the adhesive ligands available to the cells are able to tip the 'phenotypic balance.' We are currently performing migration experiments on PEG-PC gels to inform us of how the integrin-binding and stiffness cues integrate to affect this particular phenotype indicator. Preliminary results indicate that calponin, an early SMC differentiation marker, has a biphasic dependence on substrate stiffness (relative expression on 5, 50 and 200 kPa PEG-PC gels: 1.0, 1.6, 1.2), whereas phosphorylation of the 20 kDa myosin light chain (pMLC
calponin (relative expression on 5,50 and 200 kPa PEG-PC gels: 1.0, 0.69, 0.94). We quantified SMC proliferation over the course of a week on the EAMs in either serum-free DMEM, DMEM containing smooth muscle growth supplement (SMGS; 5% FBS, 2 ng/mL basic fibroblast growth factor (bFGF), 0.5 ng/mL epidermal growth factor, 5 ng/mL heparin, 5 ug/mL insulin) or a defined media condition (serum-free, 25 ng/mL platelet-derived growth factor, and 10 ng/mL bFGF). We found that proliferation increases with stiffness in the no-serum and defined media conditions, proliferation is higher on the diseased EAM than the healthy EAM, and proliferation in the SMGS condition is high on all EAM and stiffness conditions. The latter observation suggests that serum and/or various mitogens can override the effect of ECM ligands and stiffness on phenotype maintenance.

We are continuing to investigate the effect of substrate stiffness and ECM ligands on proliferation, migration and SMC marker expression. In particular, we are interested in the interplay between SMC mechanosensing (integrin binding, cytoskeleton assembly and acto-myosin contractility), growth factor sensitivity and SMC phenotype. If we can better understand how these factors are related in a system, which closely resembles the *in vivo* environment, there is a greater potential to make discoveries that will lead to novel drug targets and treatments.

(726d) Development of Biomimetic Environments with Appropriate Chemical and Mechanical Cues for Cells in Bioengineered Vascular Grafts

Mao-Shih Liang, Chemical and Biological Engineering, State University of New York at Buffalo, Amherst, NY, Maxwell T. Koobatian, Physiology, State University of New York at Buffalo, Amherst, NY, Daniel D Swartz, Pediatrics, Women and Childrens Hospital of Buffalo, University at Buffalo, The State University of New York, Buffalo, NY and Stelios T. Andreadis, Department of Chemical and Biological Engineering, University at Buffalo, the State University of New York, Amherst, NY

Tissue engineered vascular constructs are promising for next generation vascular substitutes. Currently, one major challenge facing the field is developing biomimetic environments capable of providing appropriate biochemical and biomechanical signals to cells within bioengineered vascular grafts. Previously, we showed that TGF-β1 that was conjugated within fibrin hydrogels through the action of factor XIII prolonged phosphorylation of Smad2 and enhanced contractility of vascular grafts as compared to soluble growth factor. Vascular contractility increased significantly with the concentration of immobilized TGF-β1 in a biphasic manner. Here we examined the hypothesis that mechanical force could further enhance the function of TGF-b1 decorated vascular grafts. To this end, we developed a high throughput biomechanical plate to investigate the effects of different mechano-stimulation protocols on control or TGF-b1 decorated tissue engineered vascular constructs (TEVs). We found that mechanical stimulation induced circumferential alignment of MSCs, increased the collagen content and the Young's modulus of TEVs. Notably, pulsation at period of 2s and 5% distention significantly increased vascular reactivity through receptor and non-receptor mediated pathways suggesting increased myogenic function. Interestingly, mechanical stimulation of TGF-β1 conjugated TEVs further increased the vasoreactivity and mechanical properties. These results suggest that the optimal combination of mechanical stimulation and growth factor delivery yields more functional and mechanically robust vascular constructs.

(726f) Sustained Release Systems to Locally Expand Regulatory T Cell Populations and Suppress Inflammation

Stephen C. Balmert[1,2], Siddharth Jhunjhunwala[1,2], Giorgio Raimondi[3,4], John R. Vu[5], Louis D. Falo[2,5], Angus W Thomson[3,4,6] and Steven R. Little[1,2,6,7], (1)Bioengineering, University of Pittsburgh, Pittsburgh, PA, (2)McGowan Institute for Regenerative Medicine, Pittsburgh, PA, (3)Surgery, University of Pittsburgh, Pittsburgh, PA, (4)Starzl Transplantation Institute, Pittsburgh, PA, (5)Dermatology, University of Pittsburgh, Pittsburgh, PA, (6)Immunology, University of Pittsburgh, Pittsburgh, PA, (7)Chemical Engineering, University of Pittsburgh, Pittsburgh, PA

Aberrant inflammation leading to tissue destruction is responsible for transplant rejection, autoimmunity, and inflammatory diseases. Traditional therapies involve harsh regimens of systemic immunosuppression, which leave patients vulnerable to opportunistic infections. Consequently, there is considerable interest in developing novel methods to localize suppression to inflamed tissue.[1] One such therapeutic approach involves increasing the presence of suppressive regulatory T cells (Treg) at local sites,[2] which typically requires systemic infusion of *ex vivo* cultured cells.[3] As inspiration for an alternative method to increase Treg locally, we consider a naturally occurring mechanism employed by the immune system. Specifically, tolerogenic dendritic cells induce the differentiation of naive T cells to Treg by establishing a local immunosuppressive milieu, comprised of secreted cytokines (IL-2 and TGF-β).[4] In an inflammatory microenvironment, these factors may be insufficient to prevent differentiation of naive T cells to pro-inflammatory effector T cells; however, the immunosuppressant drug rapamycin (rapa) has been shown to preferentially suppress generation and proliferation of effector T cells, but not Treg.[5] With this rationale, we have developed and tested biomimetic sustained release microparticle (MP) formulations of various factors, including IL-2, TGF-β, and rapa (collectively FactorMP), which can dramatically increase populations of Treg *in vitro* and suppress local inflammation in a murine model of contact hypersensitivity.

To fabricate sustained release cytokine/drug formulations, IL-2, TGF-β, and rapamycin (rapa) were encapsulated in poly(lactic-co-glycolic acid) (PLGA) using emulsion-evaporation techniques, and the resulting FactorMP were characterized, as described elsewhere.[6] Empty PLGA MP (BlankMP) were used as controls. For *in vitro* Treg induction assays, na•ve CD4 T cells were cultured in the presence of the soluble or encapsulated cytokines and drugs. At day 4, cell phenotypes were analyzed by flow cytometry, and suppressive function was assessed by a standard mixed leukocyte reaction. For the contact hypersensitivity suppression study, either FactorMP or BlankMP were injected into the footpads of mice at days -11 and -6. On day -5, mice were sensitized with dinitrofluorobenzene (DNFB). A contact hypersensitivity response was elicited on day 1 by painting one footpad with DNFB, and measurements of footpad thickness (indicative of inflammation) were taken at 24, 48, and 72 hours post-elicitation. Subsequently, draining lymph nodes were isolated and T cell populations analyzed by flow cytometry.

IL-2, TGF-β, and rapa were successfully encapsulated in PLGA MP, with sizes ranging from 16.7±6.3 to 25.5±7.5μm (large enough to remain at an injection site). Each MP formulation provided sustained release of the encapsulated factor for at least 3 weeks *in vitro*, creating a sustained immunosuppressive, local microenvironment. Similar *in vitro* Treg induction efficiencies were observed for naive CD4 T cells cultured in the presence of FactorMP or soluble factors. Notably, the presence of soluble rapa or rapaMP, significantly ($p<0.05$) enhanced induction efficiency from 57±18% (with IL-2 and TGF-β alone) to 71±15%. Importantly, induced Treg expressed canonical Treg surface markers (CD25 and GITR), as well as FoxP3, a characteristic Treg transcription factor. Furthermore, FactorMP-induced Treg were functionally suppressive in a mixed leukocyte reaction. Finally, data from a contact hypersensitivity model (Figure 1) suggests that establishment of a local immunosuppressive milieu by FactorMP can

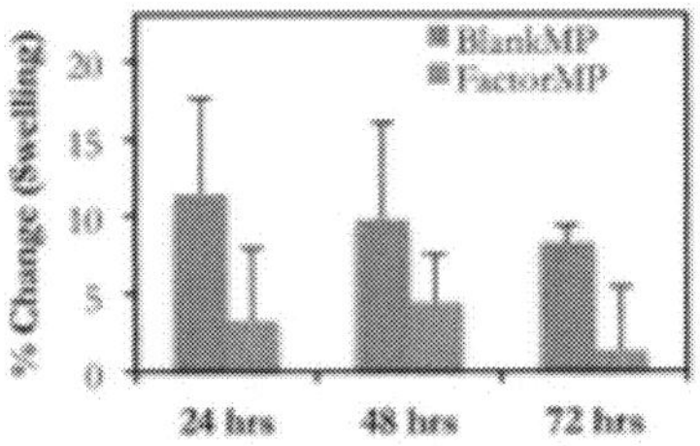

Figure 1. Percent increase in footpad thickness (swelling from inflammation) 24, 48, & 72 hrs post-elicitation with DNFB. Values reported relative to thickness of feet not painted with DNFB. Bars represent mean ± SE (n≥4).

decrease footpad swelling (local inflammation), presumably by increasing local Treg numbers.

In conclusion, we have successfully fabricated MP that provide sustained release of Treg-inducing factors. These FactorMP significantly increase numbers of functional Treg *in vitro*. Preliminary results also suggest that they can suppress inflammation *in vivo*. Ultimately, this FactorMP system may have the potential to treat a variety of inflammatory and autoimmune diseases, as well as transplant rejection.

[1] Thomson AW (ed). *Immunology and Medicine: Therapeutic Immunosuppression* (2001) Springer; 29:1-512.

[2] Sakaguchi S et al. *Immunol Rev* (2001) 182:18-32.

[3] Raimondi G et al. *J Immunol* (2010) 184:624-36.

[4] Horwitz DA et al. *Eur J Immunol* (2008) 38:901-937.

[5] Battaglia M et al. *Blood* (2005) 105:4743-4748.

[6] Jhunjhunwala S et al. *J Control Release* (2012) 159:78-84.

(726e) Wnt5a Conjugated Poly(ethylene glycol) - Gelatin Composite for Vascularized Tissue Engineering

Alpesh Patel[1,2], Akhilesh K. Gaharwar[3,4], Pinar Zorlutuna[1,2], Elif Karaca[2], Lina Schukar[2] and **Ali Khademhosseini**[1,2,4], (1)Harvard-MIT Division of Health Sciences and Technology, Massachusetts Institute of Technology, Cambridge, MA, (2)Center for Biomedical Engineering, Brigham and Women's Hospital, Harvard medical School,, Boston, MA, (3)David H. Koch Institute for Integrative Cancer Research, Massachusetts Institute of Technology, Cambridge, MA, (4)Wyss Institute for Biologically Inspired Engineering, Harvard University, Boston, MA

Embryonic stem cell (ESC) differentiation holds enormous potential for generating a renewable source of cells for regenerative medicine. However, directed in vitro differentiation of ESC has proven to be a challenging task. In early embryonic development, a tightly controlled series of cell-cell, cell-soluble factor and cell-matrix interactions create a microenvironment that provides a coordinated set of cues to developing cells. Wnt signaling molecules are among the most important cues that control key developmental events. Recently we have used microscale technologies to demonstrate the role of noncanonical Wnt11 in cardiogenesis and Wnt5a in directing endothelial cell differentiation of ESCs. Here, we propose to fabricate Wnt5a conjugated poly(ethylene glycol) vinyl sulfone (PEGVS)-gelatin composite to engineer vascular tissues by controlling ESCs differentiation. We fabricated photo cross-linked PEGVS-gelatin hydrogels with tunable mechanical and degradation properties by controlling the amount of gelatin. We then conjugated Wnt5A protein to 4-arm PEGVS via Michael-type addition utilizing the cysteine groups of the Wnt5a protein and analyzed the conjugation using matrix-assisted laser desorption/ionization (MALDI). The results indicated that one or two Wnt5a molecules were conjugated on the PEGVS arms. We further confirmed the conjugation by labeling the Wnt5a with a florescent molecule and observing the release kinetics. The biofunctionality of the Wnt5a conjugated hydrogels was evaluated by monitoring protein and gene expression of the encapsulated EBs. The immunocytochemistry results indicated that conjugation of Wnt5a to copolymer network results in higher expression of CD31 and higher sprouting compared to the control (without Wnt5a). Additionally, it was observed that CD31 was evenly distributed at the cell-cell junction and the expression pattern was similar to the fully mature endothelial cells. Furthermore, the endothelial differentiated of EBs was determined by monitoring the gene expression using quantitative polymerase chain reaction. The results indicated that conjugation of Wnt5a to PEGVS-Gelatin hydrogels up-regulated Wnt5a and CD31 gene expression compared to the control. Overall, we were able to successfully design Wnt5a conjugated PEGVS-gelatin hydrogels and able to modulate ESC signaling microenvironment based on noncanonical Wnt5a pathway for potential application in vascularized tissue engineering.

(744a) Modeling Stress-Induced Hormone Effects On Glucose-Insulin Dynamics in Critically Ill Patients

Thang Ho[1], Gilles Clermont[2], Balaji Yegneswaran[2] and Robert S. Parker[1], (1)Chemical and Petroleum Engineering, University of Pittsburgh, Pittsburgh, PA, (2)Department of Critical Care Medicine, University of Pittsburgh, Pittsburgh, PA

In the U.S., 5 million patients are admitted annually to intensive care units (ICUs) [1]. Critical care patients commonly suffer from stress hyperglycemia, resulting from a decrease in insulin dependent glucose uptake or/and relative lack of insulin secretion. Clinical evidence has shown that targeted glucose control could help reduce morbidity and mortality rate in ICUs [2]. However, the complex interaction of factors impacting patient glucose dynamics is challenging, where stress-induced hormone levels, e.g., epinephrine, cortisol, and glucagon, play pivotal roles in maintaining glucose homeostasis in ICU patients. Thus in order to efficiently and successfully control ICU patient's glucose level, it is necessary to exam the impact of these hormones on glucose-insulin dynamics. The focus of this work is to synthesize a patient-tailorable ICU model of glucose-insulin interactions including stress hormones. This involves the integration of epinephrine and cortisol models into the extended minimal model [3] to address changes in glucose-insulin dynamics resulting from trauma, stress, and inflammation. As an inflammation surrogate, we use lipopolysaccharide (LPS) challenge.

We synthesized a composite model starting from the extended minimal model (EMM) of Roy [3], which is more relevant to diabetic patients than ICU patients. The model's main structure includes the dynamics of free fatty acid and elevated metabolism, in addition to glucose-insulin interactions [3]. To capture the dynamics of pancreatic insulin secretion, a key characteristic of non-diabetic ICU patients, a three-state pancreas model was constructed from the biological mechanism of pancreatic insulin release [4,5]. The non-diabetic EMM successfully captures both glucose and insulin data from intravenous glucose tolerance tests. Two-compartment pharmacokinetic models are incorporated to capture the dynamics of epinephrine and cortisol in the body upon exogenous infusion [6, 7]. The steady state responses from the simulation are in agreement with 9 different epinephrine infusion studies; dynamic fit is also consistent with transient response data. Endogenous epinephrine and cortisol productions are calibrated against LPS challenge.

Elevated plasma epinephrine and cortisol concentrations decrease pancreatic insulin secretion and increase insulin resistance (decrease in insulin dependent glucose uptake). Epinephrine also increases hepatic glucose production. By incorporating these stress-induced hormonal effects, the resulting patient model provides a superior characterization of the glycemic response of trauma in ICU patients. Furthermore, this composite model can be used within a model-based control system to provide decision support to clinicians working to maintain targeted glucose control without hypoglycemia.

Reference:

1. Angus, D.C., Kelly, M.A., Schmitz, R.J., White, A., Popovich, J. (2000) "Current and Projected Workforce Requirements for Care of the Critically Ill and Patients With Pulmonary Disease: Can We Meet the Requirements of an Aging Population?" *Journal of the American Medical Association.* 284(21):2762-2770
2. Van den Berghe, G. et al (2001) "Intensive Insulin Therapy in Critically Ill Patients." *New England Journal of Medicine*.345: 1359-1367
3. Roy, A. (2008). Dynamic Modeling of Free Fatty Acid, Glucose, and Insulin During Rest and Exercise in Insulin Dependent Diabetes Mellitus Patients. Ph.D dissertation, University of Pittsburgh.
4. Bergman, R.N., Phillips, L.S., Cobelli, C. (1981) "Physiological Evaluation of Factors controlling Glucose Tolerance in Man". *J. Clin. Invest.* 68:1456-1467
5. Pedersen, M.G., Toffolo, G.M., Cobelli, C. (2010). "Cellular Modeling: Insight into Oral Minimal Models of Insulin Secretion". *Am. J. Phys. Endocrinol Metab.* 298: E597-E601

6. Rosen S.G., Linares O.A., Sanfield J.A., Zech L.A., Lizzio V.P., Halter J.B.(1989) "Epinephrinekinetics in humans: Radiotracer methodology". *J Clin Endocrinol Metab.* 69(4):753-61
7. Mager, M.E, Jusko, W.J.(2002) "Quantitative Structure-Pharmacokinetic/ Pharmacodynamic Relationships of Corticosteroids in Man". *J. of Pharmaceutical Science*. 91(11):2441-2451

(744b) Bayesian Inference Based Aggregated Gaussian Process Models for Identification of Ovarian Cancer Subtypes and Prediction of Survival Rates of Cancer Therapies

Mudassir Rashid and Jie Yu, Department of Chemical Engineering, McMaster University, Hamilton, ON, Canada

Cancer genome research effort aims to provide a comprehensive molecular characterization to accelerate the understanding of cancer biology and the discovery of new therapeutic targets. The immense wealth of multidimensional genomic data provides a new paradigm for developing new therapeutic solutions to improve the lives of patients.

The standard treatment for high-grade serous ovarian cancer is aggressive surgery followed by platinum-taxane chemotherapy. After therapy, platinum resistant cancer can occasionally reoccur resulting in a low overall five-year survival rate. The comprehensive measurements of genomic and epigenomic abnormalities on high-grade serous ovarian cancer samples are essential to identify the molecular abnormalities that may affect the outcome of different treatments. Moreover, accurate identification of the ovarian cancer subtypes can provide an opportunity to improve specialized cancer therapies. To that end, a genomic identification method based on Bayesian inference and Gaussian process classification is developed by using the gene expression data to identify the subtypes of cancer and predict the survival rates of cancer therapies. The proposed approach uses aggregate models developed through the integration of local Gaussian process classification models using the posterior probabilities estimated from Bayesian inference. Unlike the current standard analysis that involves separate clustering of different genomic data types followed by a manual integration of the cluster assignments, the new approach results in various localized models that can be amalgamated within the Bayesian framework. The application of the proposed approach to ovarian cancer patient samples demonstrates the significantly improved prediction accuracy and identification performance over the traditional techniques.

(744c) A NEW Framework for Online Optimization of Recombinant Protein Production in FED-Batch Fermentation Processes

Zheng Li, Department of Chemical Engineering, Texas Tech University, Lubbock, TX and M. Nazmul Karim, Chemical Engineering Department, Texas Tech University, Lubbock, TX

Online optimization of protein production is a challenging problem in the pharmaceutical industry. A new framework for online optimization of protein production in fed-batch fermentation processes is proposed in this study. This framework consists of trajectory planning and nonlinear model predictive control (NMPC) to realize online optimization of fed-batch fermentation. To avoid the effects of measurement noise and model-process mismatch, the proposed framework also includes dual Unscented Kalman Filter (UKF) to estimate both process states and model parameters, simultaneously. An online optimization of tissue-plasminogen activator (tPA) protein expression in mammalian cell culture is studied by using the proposed framework. In this case study, a modified Monod type kinetic model is applied to precisely predict the states of bioprocess including the nutrients consumption rates, cell numbers in all state of the cells (live, dead. lysed), and product formation rates. With the fed-batch mammalian cell culture model, the trajectory states are determined by solving a nonlinear optimization problem for maximizing the protein productivity. The calculated trajectory states are taken as reference states in NMPC. To avoid control degradation due to model uncertainties and perturbation during the real-time fermentation, a dual estimation strategy based on Unscented Kalman Filter is developed to estimate the actual process states and model parameters at each sampling time. To study the influence of the updated model parameters on the protein productivity in the fed-batch fermentation, variance-based global sensitivity analysis is performed to estimate the importance of the model parameters. The updated model parameters with their total-effect indices are used to evaluate the extent of the process-model mismatch. If the value of inner products of the vector of model parameters and the vector of their corresponding total-effect indices is over a threshold, the trajectory planning is executed again to update the trajectory states of the fermentation. Through the trajectory planning and NMPC algorithm, the optimal feeding rates and concentrations are determined for maximizing the productivity of tPA-I and tPA-II proteins during fermentation. The simulation result of protein production shows 12.64% improvement compared with the batch fermentation. The preliminary results show that under conditions of noisy measurement and structured plant-model mismatch, the dual estimation algorithm can precisely estimate the actual states.

DESIGN OF A MODULAR SAFETY SYSTEM FOR THE ARTIFICIAL PANCREAS: THE HEALTH MONITORING SYSTEM

Rebecca A. Harvey[1,2], *Eyal Dassau*[1,2], *Howard Zisser*[1,2], *Dale E. Seborg*[1,2], *Lois Jovanovič*[1,2], *and Francis J. Doyle III*[1,2]

[1]*University of California Santa Barbara, Santa Barbara, CA*
[2]*Sansum Diabetes Research Institute, Santa Barbara, CA*

Introduction

Type 1 diabetes mellitus (T1DM) is a disease in which blood glucose (BG) cannot be regulated by the body due to the auto-immune destruction of insulin-producing beta cells in the pancreas. Control of this disease may be achieved in a variety of ways, but is essential, because poor control can result in serious complications, both long- and short-term. Automatic control has become realizable recently, due to vast improvements in technology for sensing of BG and administration of insulin.[1,2] The ultimate goal in automatic control of T1DM is the artificial pancreas device system (APDS), a combination of continuous glucose monitors (CGMs), insulin delivery devices, and the control system that executes insulin delivery and monitors the system for safety. The APDS will be a high-risk device, due to its automatic delivery of a crucial hormone that is both necessary for life and dangerous in high doses.[3,4] Like any other high-risk system, the APDS must contain multiple safety layers to ensure the health of the user and the proper condition of the device.

The primary safety concern in the APDS is hypoglycemia (BG<70 mg/dL), or low BG, which can lead to coma or death of severe and prolonged. Hypoglycemia is the most dangerous short-term complication of intensive treatment of T1DM. Other major health concerns include hyperglycemia (BG>180 mg/dL) and missed meals. In addition, the system performance must be monitored, with such issues as pump or sensor error and communication problems, both among devices and in telemedicine.

There are several differences in the design of the safety system when compared to a traditional system, due to the constraints of a biomedical system. The lack of redundant sensors due to space constraints is a major concern, particularly due to the frequent noise encountered. In addition, only glucose can easily be measured in situ; insulin levels can only be measured by assay. One of the most notable disturbances is that the system itself has a human fully incorporated, who may intervene in an unforeseen manner.

Methods

The Health Monitoring System (HMS) has been designed as a process monitoring and alert module that can be executed in real time in parallel to any controller for the APDS (see Figure 1).

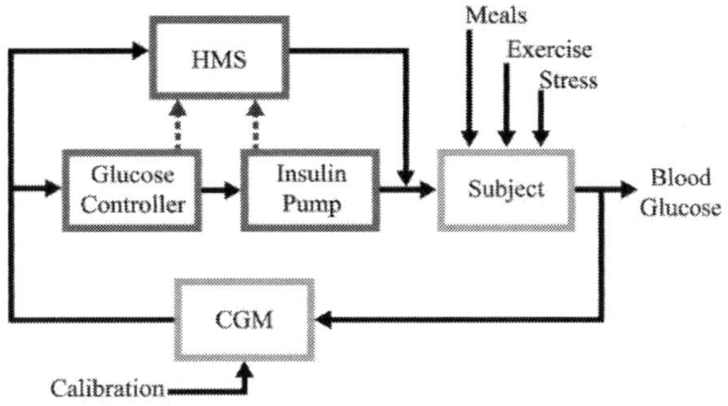

Figure 1: *APDS structure with parallel controller and safety system. Gray dashed lines are optional lines of communication.*

The HMS has been designed to easily accommodate several safety modules, including hypoglycemia, hyperglycemia, and missed meal detection; pump and sensor error detection; and communication monitoring. The primary module of the HMS, the Low Glucose Predictor (LGP), was designed first to predict and prevent hypoglycemia and is presented here.

LGP Design

The LGP was designed using retrospective, ambulatory CGM data from a large patient cohort. It was then tested in silico using the FDA-accepted UVA/Padova metabolic simulator.[5] There are three major modules in the LGP. The first module is a pre-processing module to prepare CGM data for prediction by filtering the data for noise and drift, and detecting sensor calibration. The handling of missing data comprises the second module. The third module is the core algorithm, where predictions of hypoglycemia are made by extrapolating an estimation of the rate of change of the CGM data. See Figure 2 for a flowchart of the LGP module.

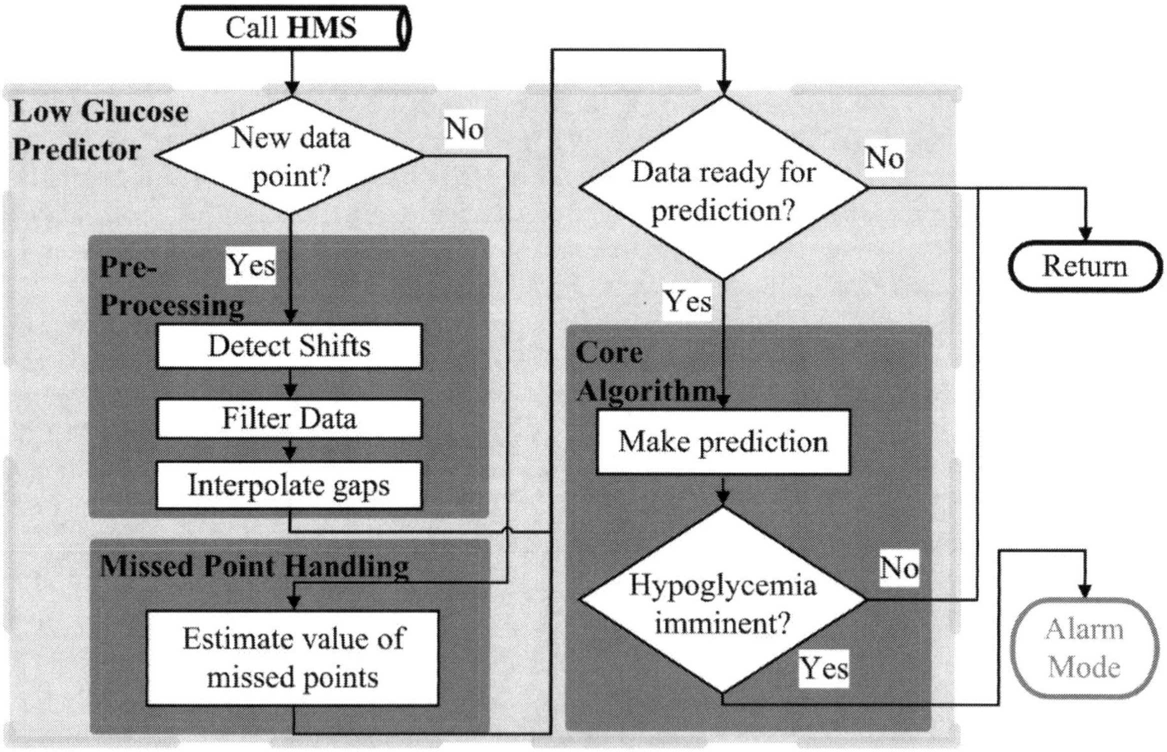

Figure 2: *LGP algorithm flowchart*

The major sub-modules in the LGP are the filter and the prediction. The noisy CGM data are filtered using physiologically-based parameters to ensure that data reflect the glucose value accurately. The algorithm filters the data using a noise spike filter,[6]

$$G_{F,NS}(k) = \begin{cases} G_m(k) & if\,|G_m(k) - G_F(k-1)| \le \Delta G \\ G_F(k-1) - \Delta G & if\,(G_F(k-1) - G_m(k)) > \Delta G \\ G_F(k-1) + \Delta G & if\,(G_m(k) - G_F(k-1)) > \Delta G \end{cases} \qquad (1)$$

where k is the sampling instant, $G_F(k\text{-}1)$ is the previous filtered value, $G_{F,NS}(k)$ is the filtered value resulting from the noise-spike filter, $G_m(k\text{-}1)$ is the measurement, and ΔG is the maximum allowable change from one instant to the next.[6] The high frequency noise is dampened using a low pass filter[6]:

$$G_F(k) = \frac{\Delta t}{\tau_F + \Delta t} G_{F,NS}(k) + \left(1 - \frac{\Delta t}{\tau_F + \Delta t}\right) G_F(k-1) \qquad (2)$$

where $G_F(k)$ is the filtered value, Δt is the sampling period, and τ_F is the filter time constant.[6] An estimate of the rate of change is calculated in the core algorithm section, using filtered data in the following equation, the derivative of the Lagrange interpolation polynomial:

$$\begin{aligned} G'_F(j) \cong & \frac{t(j)-t(j-1)}{\left(t(j-2)-t(j-1)\right)\left(t(j-2)-t(j)\right)} G_F(j-2) + \\ & \frac{t(j)-t(j-2)}{\left(t(j-1)-t(j-2)\right)\left(t(j-1)-t(j)\right)} G_F(j-1) + \\ & \frac{2t(j)-t(j-2)-t(j-1)}{\left(t(j)-t(j-1)\right)\left(t(j)-t(j-2)\right)} G_F(j) \end{aligned} \tag{3}$$

where $G'_F(j)$ is the estimated rate of change at point j, t is time, G_F is the filtered glucose value, $j=k-SAR+1:k$, and SAR is the number of successive alarms required for an alert to be engaged.[7] The rate of change calculation is extrapolated linearly through the hypoglycemia threshold, TH, to decide if hypoglycemia will occur within the prediction horizon, PH.

The LGP outputs the prediction and current measurement to the HMS alarm module, which determines if an alarm must be issued (see *Figure 3*). The alarm mode module issues audible and visual alerts and sends warning short and multimedia messages (SMS and MMS), seen in *Figure 4*. The automatic, redundant alert system of SMS and MMS messages is the key safety feature of the HMS. These messages are sent to the physician in charge or other primary contact with a profile of the current trend and prediction for the upcoming 15 minutes.[8] The HMS alarms are acted upon by administration of 16g of rescue carbohydrates.

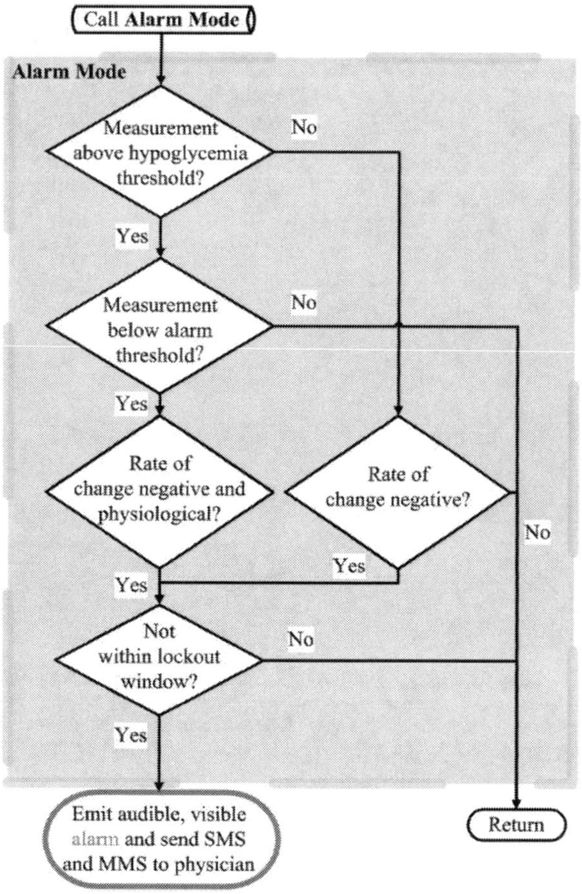

Figure 3: *Alarm mode module flow chart.*

Clinical Evaluation

The HMS was evaluated clinically in parallel to the zone-MPC controller[9,10] via the Artificial Pancreas System (APS©). The zone-MPC controller was designed at UCSB/SDRI to maintain BG in a predefined range, increasing or decreasing insulin delivery accordingly, when the CGM predictions by the internal model violate the boundaries of the zone. The zone-MPC is an updated implementation of MPC for diabetes control; use of MPC in this context was first published by Parker et al.[11,12] The Dexcom® SEVEN® PLUS (San Diego, CA) CGM was used for subcutaneous glucose sensing, and insulin delivery was effected using the Animas® OneTouch® Ping® (West Chester, PA). Reference IV blood samples were taken and assayed using the YSI 2300 STAT Plus (Yellow Springs, OH) approximately every 30 minutes. The protocol included automatic closed-loop control from approximately 5PM followed by an unannounced 50g CHO dinner meal, overnight control, an unannounced 40g CHO breakfast meal, unannounced 16g CHO snack (pre-exercise) if the reference glucose was < 120 mg/dl, 30 minute mild exercise and 16g CHO snack 3h post exercise.

Results and Discussion

The HMS was recently validated in clinic in conjunction with a zone-model predictive control (zone-MPC) controller via the Artificial Pancreas System (APS©), both designed at UCSB.[9,10,13,14] The ability of the HMS to alarm and emit SMS and MMS push notifications was validated (see Figure 4 for a screenshot from the attending physician's phone).

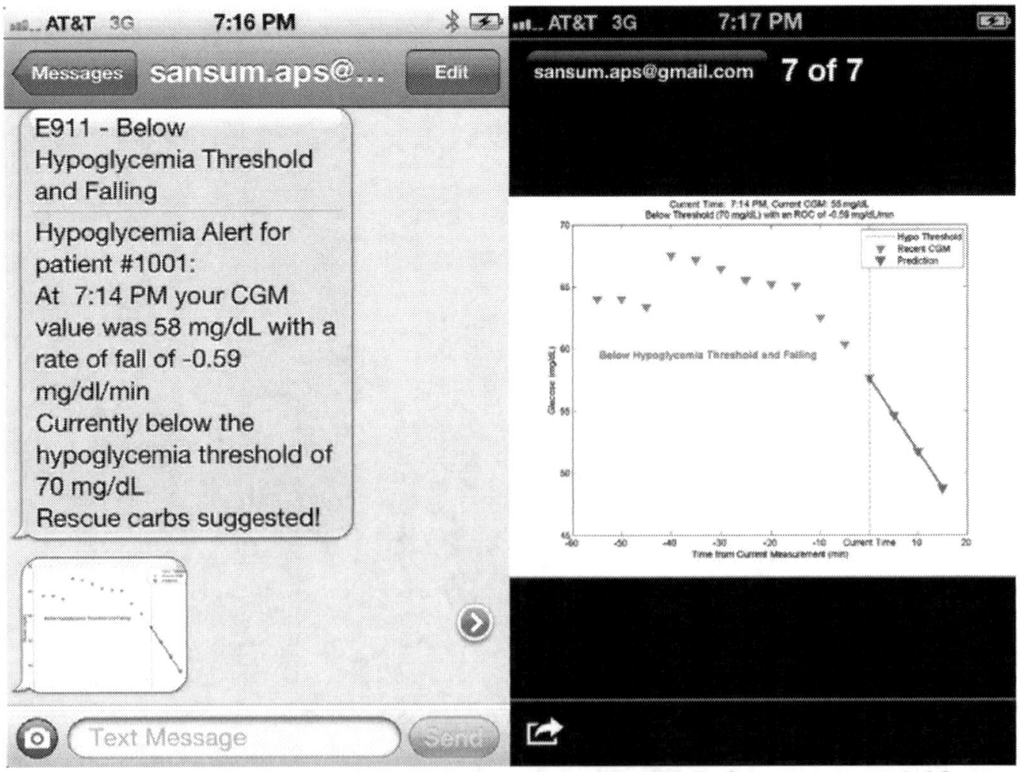

Figure 4: *Screenshots of HMS-issued SMS (left top) and MMS (left bottom), with blow-up of MMS (right) from a clinical trial.*

During the first clinical trial, the subject overdosed ~3U of insulin prior to her arrival to the closed-loop session (not per protocol). Even with insulin reduction by the controller and four carbohydrate supplements after HMS alerts, hypoglycemia was unavoidable but manageable due to the HMS, confirming that a safety system to detect adverse events is an essential part of the APDS.

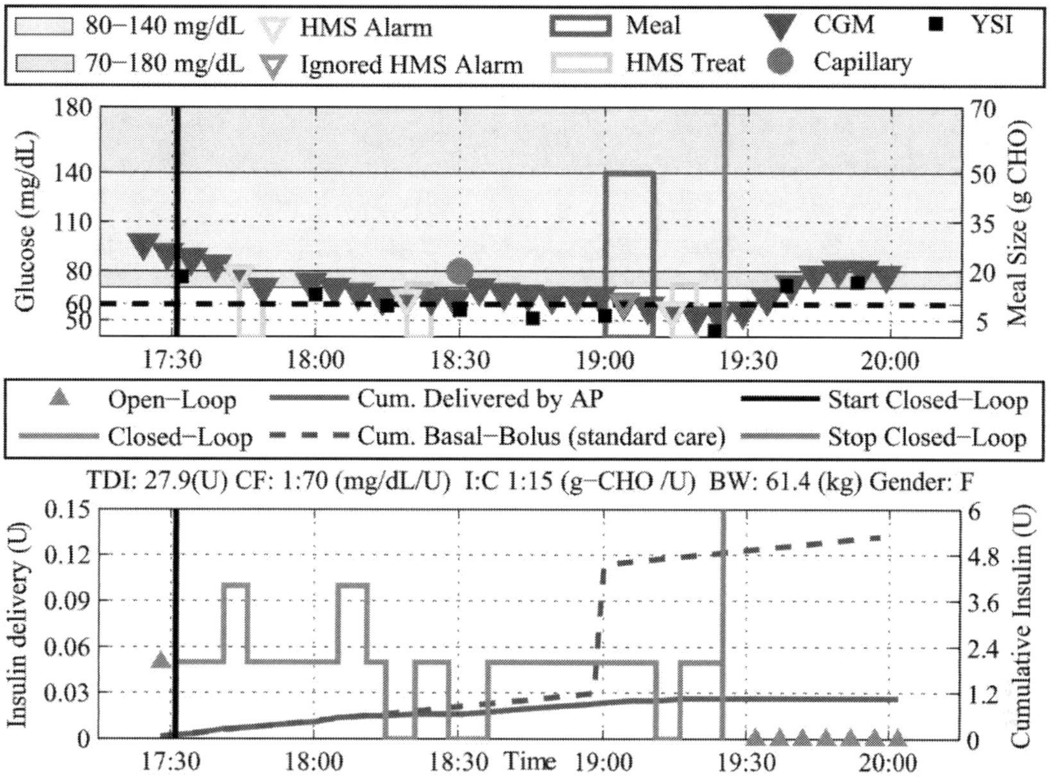

Figure 5: *Clinical case study using the HMS in which the trial was terminated early due to low blood glucose (per protocol). The start and end of closed loop are shown as black and red bars. On the top plot, CGM, YSI, and fingerstick values are shown as blue triangles, black squares, and red dots, respectively. The HMS alarms are plotted on top of the CGM values as orange triangles for ignored alarms and light blue for accepted alarms. Meals and HMS treatments are shown in purple and green, respectively. On the bottom plot, insulin delivery is shown as a green line (automatic) and blue triangles (manual) with Cumulative delivery in blue and the standard care in purple (dotted).*

A total of 12 full-length closed-loop sessions were conducted with the HMS/zone-MPC system (mean time in closed-loop 23.5 ± 0.7 hours). During closed loop, the time spent under 70mg/dL (considered hypoglycemia by the American Diabetes Association)[15] was only 2% by CGM and 1.4% by YSI. This result is a significant improvement over that found in ambulatory data (approximately 10-14%).[7,16] A total of 46 HMS alarms were sounded (mean 4 ± 2 alarms per subject), at an average CGM value of 82 ± 11 mg/dL, with a recovery of 10 mg/dL within 30 minutes on average. The CGM values, aligned from the time of the HMS alarm, are given in Figure 6, showing that alarms were sounded as CGM values were nearing the hypoglycemia region, causing recovery into the normoglycemia zone (70-180 mg/dL). The low standard deviation at the onset of the alarm suggests a high specificity of the safety system and confirmation of the necessity of the alarm. The recovery period of about 0.3-0.4 hours is in agreement with the clinical expectation of recovery in 15 minutes with 15-20g of rescue carbohydrates, depending on the rate of fall and amount of insulin on board.[17]

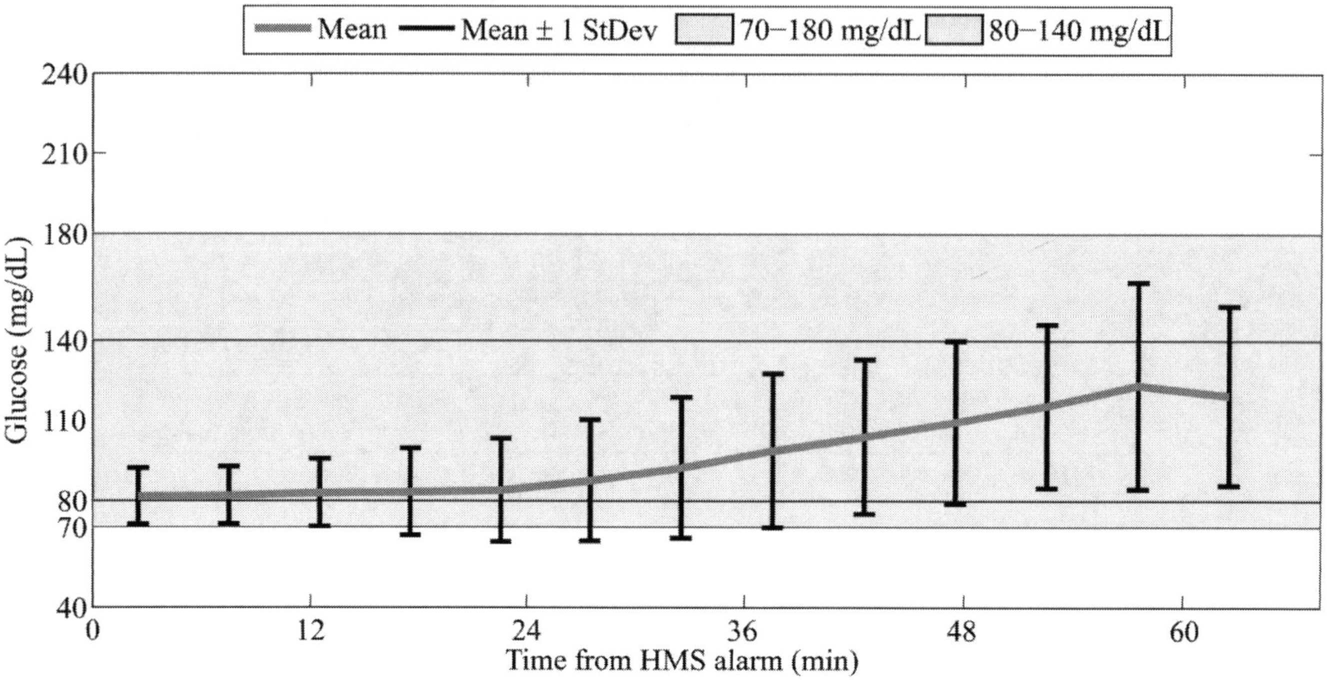

Figure 6: *CGM data from 12 clinical subjects with 46 HMS alarms, aligned from the time the alarm sounded. The average CGM value is in red, with standard deviation given in black brackets. The zone used for the zone-MPC controller is in the green filled area, with normoglycemia in the blue area.*

Conclusions

The HMS has been designed as a safety system in parallel to APDS that alerts for and mitigates adverse events. This design ensures the safety of the user and adds robustness to the overall system without interfering with the primary controller. The ability of the HMS to be an effective alert system that provides a safety layer to the APDS controller has been clearly demonstrated in clinic, detecting events that were unavoidable even with controller braking. With use of the HMS, adverse events that are not handled by the controller are readily detected and mitigated. The HMS has the ability to ensure safety through the implementation of escalating alarms; alarming locally initially, increasing the scope by alarming globally (SMS/MMS) to an emergency contact, followed by a health care provider or call center if no response is detected.

References

1. Zisser, H.C., T.S. Bailey, S. Schwartz, R.E. Ratner, J. Wise, "Accuracy of the SEVEN(R) Continuous Glucose Monitoring System: Comparison with Frequently Sampled Venous Glucose Measurements", *J Diabetes Sci Technol*, 3 (5) pp. 1146-54 (2009).
2. Keenan, D.B., B. Grosman, H.W. Clark, A. Roy, S.A. Weinzimer, R.V. Shah, J.J. Mastrototaro, "Continuous glucose monitoring considerations for the development of a closed-loop artificial pancreas system", *J Diabetes Sci Technol*, 5 (6) pp. 1327-36 (2011).
3. Tanenberg, R.J., C.A. Newton, A.J. Drake, "Confirmation of hypoglycemia in the "dead-in-bed" syndrome, as captured by a retrospective continuous glucose monitoring system", *Endocr. Pract.*, 16 (2) pp. 244-8 (2010).

4. Klonoff, D.C., C.L. Zimliki, L.A. Stevens, P. Beaston, A. Pinkos, S.Y. Choe, G. Arreaza-Rubin, W. Heetderks, "Innovations in technology for the treatment of diabetes: clinical development of the artificial pancreas (an autonomous system)", *J Diabetes Sci Technol*, 5 (3) pp. 804-26 (2011).
5. Kovatchev, B.P., C. Breton, C. Dalla Man, C. Cobelli, "In Silico Preclinical Trials: A Proof of Concept in Closed-Loop Control of Type 1 Diabetes", *J Diabetes Sci Technol*, 3 pp. 44-55 (2009).
6. Seborg , D.E., T.F. Edgar, D.A. Mellichamp, F.J. Doyle III, *Process Dynamics and Control*, 3rd ed., John Wiley & Sons, Hoboken, NJ, 2011.
7. Harvey, R.A., E. Dassau, H.C. Zisser, W. Bevier, D.E. Seborg, L. Jovanovic, F.J. Doyle III, "Clinically relevant hypoglycemia prediction metrics for event mitigation", *Diabetes Technol. Ther.*, 14 (8) pp. 719-27 (2012).
8. Dassau, E., L. Jovanovič, F.J. Doyle III, H. Zisser, "Enhanced 911/GPS Wizard: a Telemedicine application for the Prevention of Severe Hypoglycemia - Monitor, Alert and Locate", *J Diabetes Sci Technol*, 3 pp. 1501-06 (2009).
9. Zisser, H.C., E. Dassau, W. Bevier, R.A. Harvey, L. Jovanovič, F.J. Doyle III, Clinical Evaluation of a Fully-Automated Artificial Pancreas using Zone-Model Predictive Control with Health Monitoring System, in: *American Diabetes Association 72nd Scientific Sessions*, Philadelphia, PA, 2012.
10. Grosman, B., E. Dassau, H.C. Zisser, L. Jovanovič, F.J. Doyle III, "Zone model predictive control: a strategy to minimize hyper- and hypoglycemic events", *J Diabetes Sci Technol*, 4 (4) pp. 961-75 (2010).
11. Parker, R.S., F.J. Doyle III, J.E. Harting, N.A. Peppas, Model predictive control for infusion pump insulin delivery, in: *18th Annual International Conference of the IEEE Engineering in Medicine and Biology Society*, 1996, pp. 1822-3.
12. Parker, R.S., F.J. Doyle III, N.A. Peppas, "A model-based algorithm for blood glucose control in Type 1 diabetic patients", *IEEE Trans. Biomed. Eng.*, 46 (2) pp. 148-57 (1999).
13. Dassau, E., H. Zisser, C.C. Palerm, B.A. Buckingham, L. Jovanovič, F.J. Doyle III, "Modular Artificial β-Cell System: A Prototype for Clinical Research ", *J Diabetes Sci Technol*, 2 (5) pp. 863-72 (2008).
14. van Heusden, K., E. Dassau, H.C. Zisser, D.E. Seborg, F.J. Doyle III, "Control-relevant models for glucose control using a priori patient characteristics", *IEEE Trans. Biomed. Eng.*, 59 (7) pp. 1839-49 (2012).
15. Workgroup on Hypoglycemia, A.D.A., "Defining and reporting hypoglycemia in diabetes: a report from the American Diabetes Association Workgroup on Hypoglycemia", *Diabetes Care*, 28 (5) pp. 1245-9 (2005).
16. Harvey, R.A., E. Dassau, H. Zisser, D.E. Seborg , L. Jovanovič, F.J. Doyle III, "Design of the Health Monitoring System for the Artificial Pancreas: Low Glucose Prediction Module", *J Diabetes Sci Technol*, (2012).
17. American Diabetes Association, "Standards of Medical Care in Diabetes—2012", *Diabetes Care*, 35 (Supplement 1) pp. S11-S63 (2012).

(744e) Optimal Control for Predicting Drug Dosage in Superovulation Stage of in Vitro Fertilization

Kirti Maheshkumar Yenkie, Bioengineering, University of Illinois, Chicago, Chicago, IL, Urmila Diwekar, Vishwamitra Research Institute, Center for Uncertain Systems: Tools for Optimization and Management, Clarendon Hills, IL and Vibha Bhalerao, Jijamata Hospital, Nanded, India, Nanded, India

In vitro fertilization (IVF) is one of the most highly pursued assisted reproductive technology worldwide. The IVF procedure is divided into four stages: Superovulation, Egg-retrieval, Insemination/Fertilization, Embryo transfer. Superovulation is the most crucial stage in IVF, since it involves external injection of hormones to stimulate development and maturation of multiple oocytes. The maximum amount of effort and money for IVF procedure goes into superovulation. Although numerous advancements have been made in IVF procedures, medication quality, etc little attention has been given to modifying the protocols based on a predictive model.

A model for the follicle growth dynamics and number as a function of the injected hormones and patient characteristics is developed. The modeling basics are adapted from batch crystallization moment model, since moments are representatives of specific properties like number, shape and size of the particles under consideration. Based on this model, the dosage of the hormones to stimulate multiple ovulation or follicle growth is predicted by using the theory of optimal control. The objective of successful superovulation is to obtain maximum number of mature oocytes/follicles within a particular size range. Using the mathematical model involving follicle growth dynamics and the optimal control theory, optimal dose and frequency of medication is predicted for obtaining the desired result. The model will be modified to consider the sources of uncertainty due to patient's age, previous medical history, suitability of medicine and protocol used. The optimal drug delivery regime predicted in the presence of uncertainty will be compared to the current dosage regime predicted.

Thus, a phenomenon currently based on trial and error will get a supportive basis to start with. This will aid as a predictive tool for medical professionals and provide them with a specific dosage strategy for a patient. This will bring down the probability of failure, decrease cost of complex monitoring and excess medication. Thus, it will decrease the overall cost of IVF treatment for the patient as well as the physician.

(744f) Dynamic Treatment Strategies for Methanol Intoxication Using Ethanol As Alcohol Dehydrogenase Inhibitor in Human Beings

Ruben D. Vargas, Angel A. Galvis, Jonathan Moreno, Jorge M. Gomez and Watson L. Vargas, Department of Chemical Engineering, Universidad de los Andes, Bogotá, Colombia

Methanol poisoning is a fatal public health problem that affects population in both developed and developing countries particularly as alcohol consumption increases in most countries around the world. Several sources of poisoning are related to this issue (e.g. adulterated beverages, inappropriate distillation processes, methylated spirits, accidental exposure to this organic solvent, suicide attempt cases, among others). The inlet routes to the systemic metabolism are usually ingestion, inhalation or absorption through the skin. Several isolated cases of mass intoxications have taken place worldwide in recent years, Nicaragua (788 cases, 2006); Colombia (31 cases, 1989 and 88 cases, 2004); Jordan (17 cases 2006), Estonia (154 cases, 2001) [1-6]. Nevertheless, the total of cases per year in each country might be higher: Colombia (136 cases, 2007 and 250 cases, 2008); United States (2283 possible cases, 2007) [7].

Methanol toxicity arises due to two main accepted mechanisms: i) central nervous system depression and ii) biotransformation of methanol into formaldehyde and then to formic acid. Therefore, current treatments approaches are related to alcohol dehydrogenase (ADH) inhibition, since this enzyme is the one responsible for methanol metabolism to formaldehyde (the toxic metabolite). In emergency settings and with the purpose of treating patients for Methanol poisoning, inhibition is carried out using intravenous injection of "pure" ethanol or 4-methylpirazole (fomepizole) because these compounds may compete for the active sites on the ADH enzyme.

Several *in-silico* models for alcohols and glycols have been developed and the physiologically based pharmacokinetic (PBPK) ones are the most common because these enhance the fitting with experimental data keeping the physiological meaning of each differential equation [7-8]. Nevertheless, in the state of art there are few models that include non-reactive elimination pathways or dynamic control strategies in order to provide the most accurate doses for the inhibition kinetics of ADH in a physiological media.

In this study a multi-compartment physiologically based pharmacokinetic model including non-reactive elimination pathways, and a dynamic optimization technique is implemented in order to minimize the formaldehyde production on the alcohol dehydrogenase ADH enzyme using intra-venous ethanol as inhibitor. This technique is contrasted with a PID control strategy designed to stabilize the vein formaldehyde concentration under critical levels, the controller parameters are tuned up via nonlinear optimization techniques.

The proposed PBPK model includes separation between venous and arterial blood pools, and the main organs for methanol, ethanol and its metabolites biodistribution have been modeled in detail (i.e. Liver, lungs, kidneys, muscles and fat, stomach, gastrointestinal tract and skin), and the reactive unit (liver) is modeled as series of CSTR reactors with a reversible Michaelis-Menten kinetics. Additionally, the parameters for this model have been adjusted and validated using experimental data from the literature.

Keywords: Methanol intoxication, enzymatic inhibition, dynamic optimization, PID control.

References:

[1] Pan American Health Organization, Methanol Poisoning in Leon, Nicaragua, Report, September, (2006).

[2] Bennett I.L., Cary F.H., Mitchell G.L., et al., Acute methyl alcohol poisoning: A review based on experiences in an outbreak of 323 cases. Medicine 32:431-463, (1953).

[3] Kane R.L., Talbert W., Harlan J., et al., A methanol poisoning outbreak in Kentucky. A clinical epidemiologic study In Archives of Environmental Health, 17(1):119-29. (1968).

[4] Paasma R., Hovda K.E., Tikkerberi A., Jacobsen D., Methanol mass poisoning in Estonia: outbreak in 154 patients. Clin Toxicol (Phila). 45(2):152-7 (2007).

[5] Balaguer J., Intoxicación por Metanol. Toxicología Clínica. Bataller Editorial, Published in Spain p. 60-64 (2004).

[6] Colombian National Health Institute, Methanol monitoring and poisoning control (2010). (*Original language:* Protocolo de vigilancia y control de intoxicaciones por metanol, Instituto nacional de salud de Colombia (2010)).

[7] BrentJ. Fomepizole for Ethylene Glycol and Methanol Poisoning. *The new england journal of medicine* 360; 21 may 21, 2009

(744g) Revisiting the Foundations of Cybernetic Modeling Using Systems Engineering Methods

Aravinda Mandli and Jayant Modak, Department of Chemical Engineering, Indian Institute of Science, Bangalore, India

The cybernetic modeling framework for the growth of microorganisms provides an elegant methodology to account for unknown regulatory interactions using principles of optimal control theory[1, 2]. In this work, we revisit the various assumptions used in the cybernetic modeling framework. We first propose a generalized Monod model incorporating control variables, formulate various optimal control problems involving the generalized Monod model and solve them analytically to show that the long-term objective of maximizing cellular concentration at a fixed final time is equivalent to maximizing instantaneous growth rate during the growth of microorganisms in a single/two substrate batch environment. We then show that reformulation of the state and adjoint equations of the above optimal control problems result in generalized governing dynamic equations of the cybernetic model. We illustrate the influence of various constraints in the cybernetic model on the optimal growth behavior of microorganisms by solving several dynamic optimization problems using genetic algorithms. Finally, we will present a methodology for inferring the values of control variables for enzyme induction u_i of cybernetic models from experimental data.

References:

1. Ramkrishna D, Song HS: Dynamic models of metabolism: Review of the cybernetic approach. *AIChE J* 2012, 58(4):986-997.

2. Young JD, Ramkrishna D: On the matching and proportional laws of cybernetic models. *Biotechnol Prog* 2007, 23(1):83-99.

(752a) Bioremediation of Oil Spill From Oil Exploration Site Using Microbial Consortia

Bina Singh Sr., EIBD, Teri, New Delhi, India, Priyangshu M. Sarma Sr., EIBD, The Energy Resources Institute, New Delhi, India, Ajoy K. Mandal, Energy Resources Institute, The Energy Resources Institute, New Delhi, India and Banwari Lal, Microbilogy Department, The Energy Resources Institute, New Delhi, India

A field study was carried out in Ahmedabad area (four different sites), India, to bio-remediate land contaminated with an oily sludge, using a carrier based hydrocarbon- degrading bacterial consortium and nutrient mixture. A feasibility study carried out at the site showed that the introduced bacterial consortium effectively adapted to the local environment of the soil at bioremediation site. The soil samples were collected from the contaminated sites (plot A, B, C, D and E) after treatment with bacterial consortium at different time intervals, and total petroleum hydrocarbon was calculated gravimetrically after soxhlet extraction. At time zero (just before initiation of bioremediation), the concentration of oil (TPH) in the soil (25-cm horizon) of plot A, B, C and D was 39.90., 18.90, 25.80 and 29.90 % respectively, after 360 days it was reduced to 0.97% , 1.1% , 1.0%, and 1.1 % respectively with microbial consortia, where as in control plot it was reduced from 40%-35% where microbial consortia was not applied. There was only 5% degradation (40 to 35%) after 365 days in the control plot (E) where microbial consortium was not applied.

(752b) Transcriptomic Analysis Reveals Global Regulation of Lignocellulolytic Enzymes within Anaerobic Fungi

Michelle A. O'Malley[1], Diego Borges-Rivera[2], Dawn A. Thompson[2], Michael K. Theodorou[3,4], Chris A. Kaiser[5] and Aviv Regev[2,5], (1)Department of Chemical Engineering, University of California, Santa Barbara, Santa Barbara, CA, (2)Broad Institute of MIT and Harvard, Cambridge, MA, (3)Centre for Process Innovation (CPI), Redcar, United Kingdom, (4)Biological and Biomedical Sciences, Durham University, Durham, United Kingdom, (5)Department of Biology, MIT, Cambridge, MA

Anaerobic gut fungi are among the most active and efficient plant-degrading microbes found in nature. Increased insight into the enzymatic mechanisms responsible for fungal hydrolysis would better inform sustainable chemical production strategies that rely on lignocellulosic depolymerization. Despite the powerful degradation capacity of gut fungi, remarkably little genetic or enzymatic sequence information exists for these microbes. Though fungal hydrolytic activity has been shown to depend on the identity of the substrate, it is unknown how relative cellulase expression levels are coordinated during growth on these substrates. These issues stem from difficulties associated with the isolation and cultivation of gut fungi under controlled conditions, as well as genetic sequencing challenges associated with their A/T-rich genome.

In order to enumerate novel biomass-degrading enzymes and characterize their coordinated expression, we have implemented methods to sustain an anaerobic fungus in batch culture and analyze its transcriptome via RNAseq under a number of growth conditions. A new species of gut fungus was isolated from the digestive tract of a horse, and its proliferation was monitored via a pressure transducer method that coupled growth to the accumulation of fermentation gases. The fungi exhibited high enzymatic reactivity against model cellulosic and lignocellulosic substrates (filter paper, Avicel, reed canary grass), which was repressed in the presence of simple sugars (glucose, fructose). Through strand-specific RNAseq and use of the TRINITY bioinformatic platform, we were able to assemble novel protein coding regions de novo without the need for genetic sequence information. We will discuss the global regulation patterns observed for important cellulolytic enzyme families in the presence of lignocellulose and glucose, and connect these regulation patterns to physical degradation behavior.

Application of biofilter system for removal of ethyl acetate: Column and kinetic studies

Smita Raghuvanshi[a*], Suresh Gupta[a], B. V. Babu[b]

[a]Assistant Professor, Department of Chemical Engineering,
Birla Institute of Technology and Science (BITS), PILANI – 333 031 (Rajasthan) India
Phone: +91-1596-245073 ext. 215, Fax: +91-1596-244183.
Email: smita@bits-pilani.ac.in; smitaraghuvanshi@gmail.com

[a]Assistant Professor, Department of Chemical Engineering,
Birla Institute of Technology and Science (BITS), PILANI – 333 031 (Rajasthan) India
Phone: +91-1596-515454, Fax: +91-1596-244183.
Email: sureshg@bits-pilani.ac.in; sureshg25@gmail.com

[c]Director, Institute of Engineering and Technology (IET), JK Lakshmipat University (JKLU)
JAIPUR - 302026 (Rajasthan) India
Phone: +91-141-2168387 / 2259504 (O); +91-141-2259550 (R); Mobile: +91-9414014762
E-mail: director.iet@jklu.edu.in; profbvbabu@gmail.com
*Corresponding Author

Abstract

Ethyl acetate (EA) is a volatile organic compound (VOC) that is one of the high-priority toxic chemicals and large volumes of EA are emitted into the atmosphere from various paints and solvent industries. In this study, the performance of biofilter column is evaluated for the removal of EA using a mixture of matured compost and coal in the ratio of 2:1. The effect of various operational parameters such as inlet load, column bed height, packing material and shock loading conditions on the performance of biofilter column is investigated experimentally. The experimental results are validated with the Ottengraf-vanden Oever and Michaelis-menten kinetic models for explaining the mechanism of biofiltration. The maximum elimination capacity (EC) achieved is 32.66 g m-3 h-1 at an inlet load of 36 g m-3 h-1. The value of microbial concentration at the end of 15th day is

3.64x106 CFU g-1, which increased to a value of 5.12x108 CFU g-1 of packing material, with an increase in the operation time from 15 to 45 days. The reasonably good value of microbial concentration obtained is responsible for attaining the removal efficiency of around 92% during the shock loading study.

Keywords: Biofiltration; Ethyl acetate; Acclimated mixed culture; Elimination capacity; Shock loading.

1. Introduction

Emissions of VOCs, ethyl acetate (EA) in particular, cause serious environmental and public health concerns. Ethyl acetate is considered to be a toxic chemical and is released into the atmosphere from various paints and solvent industries (Chan & Su, 2008). The inhalation of vapors of ethyl acetate can cause severe irritation of mucous membranes and upper respiratory tract. Exposures to high concentrations of EA may cause liver and kidney damage. Thus streams polluted with the vapors of EA needs to be treated before being released into the atmosphere. The choice of treatment technique depends on the total cost of treatment method, nature of pollutant being treated, concentration and flow rate of the pollutant, and the mode of emissions of the gaseous waste streams & effluent streams (Devinny et al., 1999). Thus bio-based techniques, especially biofiltration is considered to be one of the better options for the removal of low to medium concentration of organic pollutants (Wright, 2005; Raghuvanshi & Babu, 2009, 2010). Biofiltration incorporates all the basic processes such as adsorption, absorption, degradation and desorption of gaseous phase contaminants. The process of biofiltration has the advantage of metabolic and physiological flexibility, as well as the high adaptability of the populations of microbial species to consume the organic pollutants as a carbon source for its growth (Deshusses, 2007; Rene & Swaminathan, 2007).

The removal of EA by using biofiltration has been studied by few researchers. Deshusses and Johnson (1999) investigated the performance of biofiltration for high inlet loads of EA (inlet loads of around 400 g m^3h^{-1}). The work highlighted the operational problems encountered during the handling of high loads of VOCs and their mixtures. Liu et al., (2002) carried out the bench scale biofiltration for the treatment of air streams containing mixtures of ethyl acetate and toluene. The study focused on the inhibition effect of toluene in the presence of ethyl acetate. The elimination capacity of around 400 g $m^{-3}h^{-1}$for ethyl acetate was achieved and 50 g $m^{-3}h^{-1}$ for pure toluene. Huwang et al. (2003) developed the different combinations of microbial cultures using the shake flask study for the removal of ethyl acetate and toluene. Chan and Su, (2008) in their very recent work investigated the biodegradation kinetics for the removal of ethyl acetate and amyl acetate using a composite bead biofilter. Their work was focused on the study of the microbial kinetics related to ethyl acetate and amyl acetate. It was concluded that the biochemical reaction rate of ethyl acetate was greater than that of amyl acetate in the inlet concentration range of 100–400 ppm, and ethyl acetate was degraded by microbial culture easily than the amyl acetate.

The previous works focused on the mixtures of ethyl acetate with another pollutant (mostly toluene). The focus was on the study of inhibition effect of one pollutant in the presence of another pollutant. The present work focuses on the removal of ethyl acetate using the acclimated mixed culture in the biofilter packed with the matured compost and coal. The studies are also carried out under shock loading conditions and the kinetics of microbial degradation are obtained for the removal of EA. Hence the experimental results are analyzed based on the following aspects: start up time for the biofiltration; performance of biofilter column based on the elimination capacity; bed height vs normalized concentration; and shock loading conditions. The obtained experimental data are validated with the Michaelis-menten kinetic model and Ottengraf – van den Oever model to understand the mechanism of biodegradation of pollutant with the acclimated mixed culture.

2. Experimental methods

2.1. Shake flask study

The shake flask study was carried out for obtaining the acclimated mixed culture from the activated sludge from the secondary clarifier section of Sewage Treatment Plant of BITS Pilani, India. The details on experimental methodology of shake flask study for culture growth and batch studies are available in our earlier study (Raghuvanshi & Babu, 2009).

2.2. Experimental set-up and procedure

The experimental set-up of the biofilter column was made up of perspex tube having a 5 cm internal diameter and a length of 1 m (Fig. 1). A total of six sampling ports were placed at a spacing of 10 cm along the 70 cm packed portion of the column. The packing material was supported on a stainless steel mesh installed at the bottom of the column to maintain a homogeneous radial distribution of the EA vapors. Compressed air was passed into the EA bottle to generate saturated vapors of EA. The saturated vapors and air stream from the sampling bottles were then passed separately into the mixing bottle. The vapor of ethyl acetate was then allowed to enter into the biofilter column from the inlet port, in an up flow mode of operation (Fig. 1). The moisture content in the bed was maintained periodically by sprinkling fresh minimal salt medium (MSM) from the top of the biofilter column as shown in Fig 1. The composition of MSM is given in Table 1. The leachate was collected at the bottom of the column. The continuous column experiments were carried out by varying the flow rates of air and EA to cover a wide range of initial concentration of EA and residence time.

The performance of biofilter was observed in terms of elimination capacity for the removal of EA by performing the experiments at different inlet loads during 45 days of biofilter operation. The total duration of biofilter operation was divided into 4 phases (Phase I to Phase IV). Phase I was considered as acclimation phase. The operating conditions for the biofilter operation at different phases are reported in Table 2.

2.3. Preparation of filter material

The filter material used in the present work was comprised of matured compost and coal. The matured compost was prepared from the vegetable waste, which was brought from the JRD Tata Foundation for Research in Yoga, Naturopathy & Ayurvedic Sciences, Chitrakut, UP (India). The coal was obtained from the local market and was sieved through 8 – 10 mm mesh screen. The final coal size obtained was 2.36 mm. The coal was washed with normal water and distilled water. The coal was then kept in the oven (maintained at 100^0 C) for 24 hrs for removing the moisture content. The coal and matured compost was then thoroughly mixed in the ratio of 2:1 (v: v). This material was then mixed with the acclimated mixed culture obtained from the shake flask study, and this mixture is used as the packing material in the column studies. The total amount of acclimated culture (seeding culture) used for the biofiltration study was 700 mL which corresponds to 0.35 g of microbial culture. The leachate collected at the bottom was simultaneously transferred to the biofilter column during the start up period in order to have the best possible interaction between microbial culture and the packing material. The moisture content of the packing material was 63% on wet weight basis.

2.4. Method of analysis

Concentration of ethyl acetate was determined using gas chromatography (Model 5700 series, Nucon Engineers, India) using a poropak column (2m length, 1/8 in i.d.) and flame ionization detector. The samples were collected from different sampling ports using gas tight syringes at different intervals of time during the 45 days of biofilter operation. At each time, 5-6 readings were taken and the average value was used for calculations. The GC-FID was operated at an injection temperature of 130 °C, the detector temperature of 130 °C, and the oven temperature of 200 °C. Nitrogen was used as the carrier gas at a flow rate of 24 mL min^{-1} and hydrogen was used as the fuel. The microbial concentration was measured after phase I (15 days acclimation period for EA) and after the end of phase IV as reported in the literature (Rene et al., 2005).

3. Results and Discussion

3.1. Start-up of the biofilter column

The process was started after the biofilter was packed with the packing material. The column was operated with an air flow rate (Q) of 0.18 $m^3\,h^{-1}$ corresponding to an empty bed residence time (EBRT) of 27.4s with an inlet load ranging from 6.54 - 8.25 g $m^{-3}h^{-1}$ in the initial 15 days (acclimation period or phase I) of the biofilter operation. The removal efficiency (RE) was fluctuating during the starting days and it reached to a value of 75% after the 10th day of biofilter operation. The removal efficiency increased with the operation time and the maximum value of removal efficiency was obtained as 90.12% on 15th day of biofilter operation as given in Table 2. A near linear relationship (Fig. 2) was obtained between inlet load (IL) and elimination capacity (EC) during the acclimation phase. The value of microbial concentration obtained at the end of 15th day was $3.64x10^6$ CFU g^{-1} which represents a reasonably good growth of microorganisms. The result was consistent with the reported acclimation time of several weeks to several months depending on the duration of biofilter column as reported in the literature (Jorio et al., 2000; Juneson et al., 2001).

3.2. Biofilter performance in terms of elimination capacities

The relationship between EC and inlet load of ethyl acetate is shown in Fig. 2. After the acclimation period (phase I), the inlet load was increased gradually. The inlet load of EA was increased in the range of 28.59- 31.65 g $m^{-3}h^{-1}$ for the next 10 days (16th-25th day of biofilter opearation) with an EBRT maintained as 16.5 s. The increased inlet load yielded the value of EC in the range of 19-27.5 g $m^{-3}h^{-1}$. The removal efficiency was decreased to around 68.23% and reached to a steady state value of 86– 87% at the end of this phase (Table 2). The sudden decrease in the RE is attributed to the sudden increase in the inlet load of EA.

During phase III, the EC was found to increase to a maximum value of 33.7 g $m^{-3}h^{-1}$(Fig. 2) for the inlet load of EA in the range of 35 - 37 g $m^{-3}h^{-1}$. The EC obtained was reasonably good during this phase which showed that the acclimated mixed culture was able to degrade by changing inlet loads (Fig. 2). This may be due to the increased value of EBRT (23.5 s) which gave microorganisms sufficient time to interact with EA and degrade it at a faster rate. The high value of EC was also strengthened by achieving better RE of around 92% (Table 2).

Fig.2 also shows that when the inlet load was decreased to a value in the range of 31-34 g $m^{-3}h^{-1}$ in phase IV, the elimination capacity also decreased in the range of 25.2-31.7 g $m^{-3}h^{-1}$. The results of elimination capacity obtained in the present study confirm that even on changing inlet loads, the microorganisms were able to degrade the lower to intermittent inlet loads of EA (29 - 34 g $m^{-3}h^{-1}$). Such behavior is also reported in other studies (Groenestijn & Messelink, 1993; Devinny et al., 1999; Liu et al., 2002). The removal efficiency was also found to increase from 76.1% to 94.11% with an increased operating time from 36th day to 45th day of biofilter operation (Table 2). This may be due to the ability of microorganisms to degrade the lower loads of EA. The increased RE was also strengthened by obtaining the increased microbial concentration as $5.12X10^8$ CFU g^{-1} of packing material with an increase in the operation time from 15 to 45 days.

3.3 Bed height vs normalized concentration

Fig. 3 shows the profiles of normalized concentration versus height of biofilter column for the removal of EA at different operating conditions. The initial EA concentrations were maintained at 0.063, 0.145, 0.24, and 0.193 g m^{-3} and their corresponding air flow rates were maintained at 0.18, 0.3, 0.21, and 0.24 m^3 h^{-1}. It was found that the value of normalized concentration reached to 0.22, 0.57, 0.52, and 0.53 for 0.063, 0.145, 0.24, and 0.193 g m^{-3} of initial EA concentration respectively within the bottom 20 cm height of biofilter column. The upflow mode of operation could be one of the reasons for the maximum removal of ethyl acetate within first 20 cm of the biofilter column. The obtained value of normalized concentration was higher for 0.145 g m^{-3} of initial EA concentration which may be due to the lower value of EBRT (16.5 s).

3.4. Biofilter response to shock loading conditions

The practical problem associated with the waste gas treatment is the sudden fluctuations in inlet load of the polluted streams. The biomass developed in the biofilter over the period must be able to withstand sudden loads (Kim er al., 2008). The high microbial concentration ($5.12X10^8$ CFU g^{-1} of packing material) which was obtained after 45 days of biofilter operation was one of the reasons for carrying out the studies under shock loading conditions. The biofilter column was subjected to different loads varying from 23 to 53 g $m^{-3}h^{-1}$ (Fig. 4). The operating conditions of shock loading study are given in Table 3. The inlet load was maintained in the range of 38.19 – 41.25 g $m^{-3}h^{-1}$ for first three days and the elimination capacity was found in the range of 32.51 – 36.31 g $m^{-3}h^{-1}$. The removal efficiency was obtained in the range of 85.11 to 88.02% (Table 3). The inlet load was increased for the next three days between 48.89 – 52.4 g m^{-3} $h^{-1}$by maintaining EBRT as 20.6 s. The EC was obtained in the range of 40 – 44 g m^{-3} h^{-1}.The value of removal efficiency was slightly decreased in the range of 82.11 – 85.12% which shows that there was no significant effect of changing the operating conditions on the performance of biofilter column. During the last four days of biofilter operation, the inlet load was decreased to the range of 23 - 25 g m^{-3} h^{-1} (corresponding to an EBRT of 27.4 s). A reasonably good value of EC was obtained in the range of 20 - 22 g m^{-3} h^{-1}. This can also be established by achieving a high value of RE (91.11%). The inference that could be drawn with this result is that the biofilter column responded well to the sudden changes in the inlet load once the microbial culture was well developed. This also proves that the biofilter was stable and biomass was well distributed in the column. Kim et al., (2008) have also shown the similar results for the removal of H_2S were reported under different shock load conditions, where the inlet load was varied from 0.1 to 6 g $m^{-3}h^{-1}$ to achieve the removal efficiency in the range of 64 – 99% for 18 days of shock loading study.

3.5. Results of kinetic study on the EA removal

3.5.1. Estimation of Michaelis–Menten kinetic constants

The kinetics of pollutant degradation is often understood using Michaelis-Menten equation as given in Eq. (1). The equation involves the estimation of two kinetic constants maximum degradation rate per unit filter volume, r_{max} (g $m^{-3}h^{-1}$) and saturation constant, K_S (g m^{-3}).

$$\frac{V/Q}{C_{gi} - C_{g0}} = \frac{K_S}{r_{max}} \frac{1}{C_{ln}} + \frac{1}{r_{max}} \qquad (1)$$

where C_{gi} and C_{g0} are the inlet and outlet concentrations of the pollutant in g m^{-3}. V is the volume of the filter bed (m^3), r_{max} is the maximum degradation rate per unit filter volume (g m^{-3} h^{-1}), K_S is the saturation constant (g m^{-3}) in gas phase, and C_{ln} is the log mean concentration difference. These constants were evaluated for the gas phase by plotting $[(V/Q) / (C_{gi} - C_{g0})]$ versus $(1/ C_{ln})$ from Michaelis-Menten equation Eq. (1) which gave a straight line as shown in Fig 5.

The values of r_{max} and K_S obtained were 0.092 g m^{-3} h^{-1} and 1.061 g m^{-3} respectively for EA. The coefficients of determination (R^2) obtained for EA was 0.991 and hence show the applicability of Michaelis-Menten model for the biodegradation of EA. The Michaelis-Menten kinetics is valid for two cases: (i) zero-order kinetics (ii) first-order kinetics. Zero-order kinetics holds true when the inlet concentration of EA is higher than the value of saturation constant and in such situations the rate of biodegradation is approximately equal to r_{max} and does not change with the concentration of the pollutant. When the inlet concentration of pollutant is much below the saturation constant, the rate of biodegradation is proportional to the pollutant concentration and it is said to be following the first-order kinetics. So in the present work, the value of K_S obtained was 1.061 g m^{-3} which was comparable with the value of inlet concentration of EA (0.05 – 0.24 g m^{-3}). So the obtained results show that the biodegradation of EA can neither be explained by the first-order kinetics nor by zero-order kinetics. Hence in such cases, biodegradation of pollutants can be better explained by the Ottengraf-Van den Oever model.

3.5.2 Validation with Ottengraf-Van den Oever model

There have been numerous models developed in order to understand the actual mechanism of biofiltration (Shareefdeen & Baltzis, 1994). The modeling of biofiltration process is considered to be a tedious task as it incorporates all the basic mechanisms of pollutant removal such as adsorption, absorption and microbial degradation of the pollutants. Thus to visualize and to understand the actual mechanism of biofiltration is difficult. These limitations are being overcome by carrying out the modeling studies, and the first study in this direction was carried out by Ottengraf and van der Oever (1983). The model is widely used to discuss the behavior of biofilters. The model describes basically two different approaches: (i) zero-order with diffusion limitation, and (ii) zero-order with reaction limitation. In both the approaches, the elimination capacity (EC) is correlated to the inlet load (IL) as given in Eq. (2).

(i) Zero-order with diffusion limitation:

$$\mathrm{EC} = \mathrm{IL}\left(1-\left(1-K_1\sqrt{\frac{\mathrm{EBRT}}{\mathrm{IL}}}\right)^2\right) \quad (2)$$

$$K_1 = \sqrt{\frac{K_0 D_e a}{2m\delta}} \quad (3)$$

where K_0 is the zero-order kinetic constant, D_e is the effective diffusion coefficient of EA in the biofilm ($m^2\ h^{-1}$), m is the Henry`s coefficient for EA in water, a is the interfacial area per unit volume (m^{-1}), δ is the biofilm thickness (μm). The gas phase concentration at the outlet of the biofilter column is given by Eq. (4)

$$\frac{C_{g0}}{C_{gi}} = \left[1-\mathrm{EBRT}\frac{K_1}{\sqrt{C_{gi}}}\right]^2 \quad (4)$$

The Ottengraf Van den Oever model predicted the performance of the biofilter of this study for phase II, phase III and phase IV of EA removal with the standard deviation values obtained as 2.137, 2.37, and 2.209 respectively (Fig. 6).

(ii) Zero-order with reaction limitation:

The EC was directly calculated by obtaining the constant K_0 which corresponds to the maximum EC in that particular phase. The values of the obtained constants K_0 and K_1 for different phases are presented in Table 4.The C_{critical} and IL_{critical} were obtained at different phases for different operating conditions as described in the previous study [5] and are presented in Table 4. It was observed that the obtained values of critical inlet concentration were higher than the inlet concentration of EA for all the phases.

Based on the above analysis, we can conclude that the biodegradation of ethyl acetate follows the zero-order diffusion limited kinetic model. It is in line with our observation in our earlier study for biodegradation of MIBK which also followed zero-order diffusion limited kinetics. The predicted values of the elimination capacity are presented in Table 4 for different inlet loads of ethyl acetate.

4. Conclusions

- Acclimation time of 15 days was obtained with an inlet load 6.54 - 8.25 g $m^{-3}h^{-1}$ because of the use of well acclimated mixed culture obtained from the shake flask study.
- A reasonably good EC (25.2 – 31.7 g $m^{-3}h^{-1}$) was obtained for low to medium inlet loads of EA when the inlet load was decreased to a value of 31-34 g $m^{-3}h^{-1}$ in last phase.
- The EC was obtained in the range of 20 – 22 g m^{-3} h^{-1} for the lower values of inlet loads under shock loading conditions thus showing its application in the actual working conditions.
- The kinetic analysis carried out using the Michaelis–Menten model yielded the kinetic constants as 0.092 g m^{-3} h^{-1} and 1.061 g m^3 with a coefficient of determination (R^2) value of 0.99.
- The obtained experimental results were found to fit the Ottengraf-Van den Oever model for zero-order diffusion control for phases II to IV and significant constants such as critical inlet concentration and critical inlet load were also estimated.

Acknowledgements

Authors thank the University Grant Commission (UGC), New Delhi, India for their financial support in carrying out the experimental work.

References

[1] Chan WC, Su MQ. Biofiltration of ethyl acetate and amyl acetate using a composite bead biofilter. Bioresour Technol 2008;(99):8016-8021.

[2] Devinny JS, Deshusses, MA, Webster TS. Biofiltration for air pollution control. Lewis Publishers, CRC Press. 1999.

[3] Wright WF. Transient response of vapor-phase biofilters. Chem Eng J 2005;(113) :161-173.

[4] Raghuvanshi S, Babu BV. Experimental studies and kinetic modeling for removal of methyl ethyl ketone using biofiltration. Bioresour Technol 2009;(100):3855-3861.

[5] Raghuvanshi S, Babu BV. Biofiltration for removal of methyl iso butyl ketone (MIBK): experimental studies and kinetic modeling. Environ Technol 2010;(31): 29-40.

[6] Deshusses MA. Biodegradation of mixtures of ketone vapours in biofilters for the treatment of waste air. Ph D Thesis, Swiss Federal Institute of Technology. Zurich, Switzerland, 1994.

[7] Rene ER, Swaminathan T. Biological treatment systems for VOC removal. Chem Week 2007:195-202.

[8] Deshusses MA, Johnson CT. Biofiltration of High Loads of Ethyl Acetate in the Presence of Toluene. J Air & Waste Manage Assoc 1999;(49):973-979.

[9] Liu Y, Quan Y, Sun J, Chen D, Xue J, Chung K. Simultaneous removal of ethyl acetate and toluene in air streams using compost-based biofilters. J Hazard Mater 2002;(B95):199-213.

[10] Hwang SCJ, Lee CM, Lee HC, Pua H.F. Biofiltration of waste gases containing both ethyl acetate and toluene using different combinations of bacterial cultures. J Biotechnol 2003;(105):83-94.

[11] Rene ER, Murthy DVS, Swaminathan T, Performance evaluation of a compost biofilter treating toluene vapours. Process Biochem 2005;(40):2771–2779.

[12] Jorio H, Bibeau L, Heitz M. Biofiltration of air contaminated by styrene: effect of nitrogen supply gas flowrate, and inlet concentration. Environ Sci Technol 2000;(34):1764–1771.

[13] Juneson C, Ward OP, Singh A. Microbial treatment of a styrene-contaminated air stream in a biofilter with highelimination capacities. J Ind Microbiol Biotechnol 2001;(26):196–202.

[14] Groenestijn JW, Messelink PGM. Biotechniques for air pollution Control. Biodegradation 1993;(4):283–301.

[15] Kim JH, Rene ER, Park H. Biological oxidation of hydrogen sulfide under steady and transient state conditions in an immobilized cell biofilter. Bioresour Technol 2008;(99):583-588.

[16] Shareefdeen Z, Baltzis BC. Biofiltration of toluene vapor under steady state and transient conditions: theory and experimental results. Chem Eng Sci 1994;(49): 4347-4360.

[17] Hodge D, Devinny J. Modeling removal of air contaminants by biofiltration, J Environ Eng 1995;(121):21-32.

[18] Ottengraf SPP, van den Oever AHC. Kinetics of organic compound removal from waste gas with a biological filter. Biotechnol Bioeng 1983;(25):3089-3102.

Table 1

S No	Constituents	Composition for ethyl acetate in 1 L of distilled water (g/L)
1.	KH_2PO_4	0.4
2.	$CaSO_4.2H_2O$	0.05
3.	$MgSO_4.7H_2O$	0.5
4.	$(NH_4)_2SO_4$	1.0
5.	$FeSO_4.7\ H_2O$	0.01
6.	KNO_3	0.3
7.	$NaHCO_3$	0.02

Table 2

Phase	Operating time (days)	Air flow rate Q, ($m^3\ h^{-1}$)	EBRT (s)	Inlet loading ($g\ m^{-3}\ h^{-1}$)	Removal efficiency (%)
I (Acclimation)	15	0.18	27.4	6.54-8.25	50.12 - 90.12
II	10	0.30	16.5	28.59-31.65	68.23 – 87
III	10	0.21	23.5	35.29-36.67	73 – 92
IV	10	0.24	20.6	31.43-33.70	76.1 – 94.11

Table 3

Days of biofilter operation	Air flow rate, Q ($m^3\ h^{-1}$)	EBRT (s)	Inlet loading ($g\ m^{-3}\ h^{-1}$)	Removal efficiency (%)
1 - 3	0.21	23.5	38.19 – 41.25	85.11 - 88.02
4 – 6	0.24	20.6	48.89 – 52.4	82.11 – 85.12
7 – 10	0.18	27.4	23.57 – 25.14	86.5 – 91.11

Table 4

Phase	C_{gi}, (g m^{-3})	IL, (g m^{-3} h^{-1})	K_0, (g m^{-3} h^{-1})	K_1, (g m^{-3} h^{-1})	$C_{critical}$ (g m^{-3})	$IL_{critical}$, (g m^{-3} h^{-1})	s.d.
II	0.131 –0.145	28 – 32	27.54	44.64	0.169	36.81	2.137
III	0.231 –0.240	35 – 37	33.77	44.19	0.277	42.44	2.37
IV	0.18– 0.193	31 – 34	31.72	47.26	0.221	38.63	2.209

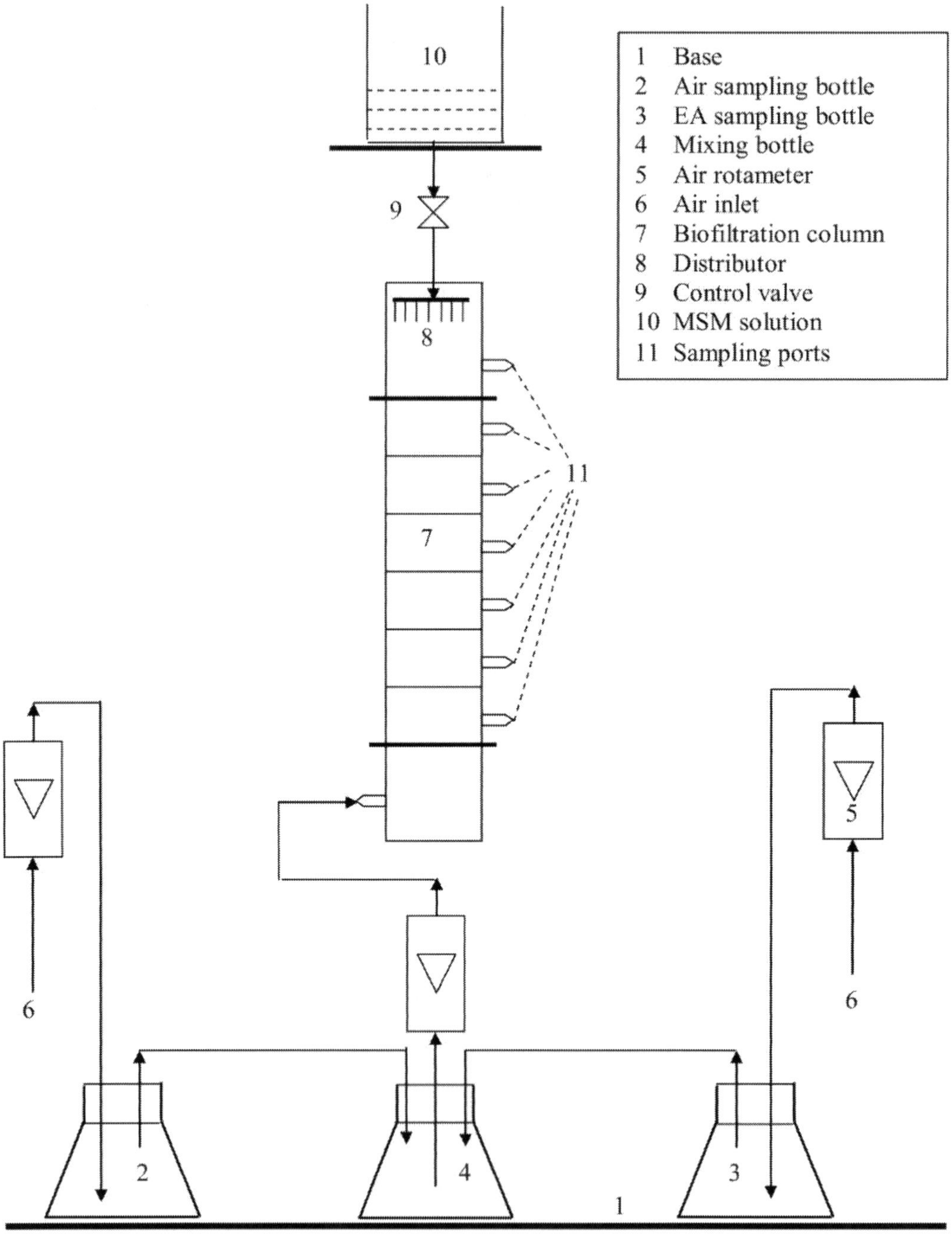

10
9
8
7
11
1 Base
2 Air sampling bottle
3 EA sampling bottle
4 Mixing bottle
5 Air rotameter
6 Air inlet
7 Biofiltration column
8 Distributor
9 Control valve
10 MSM solution
11 Sampling ports
6
5
6
2
4
3
1

Figure 1.

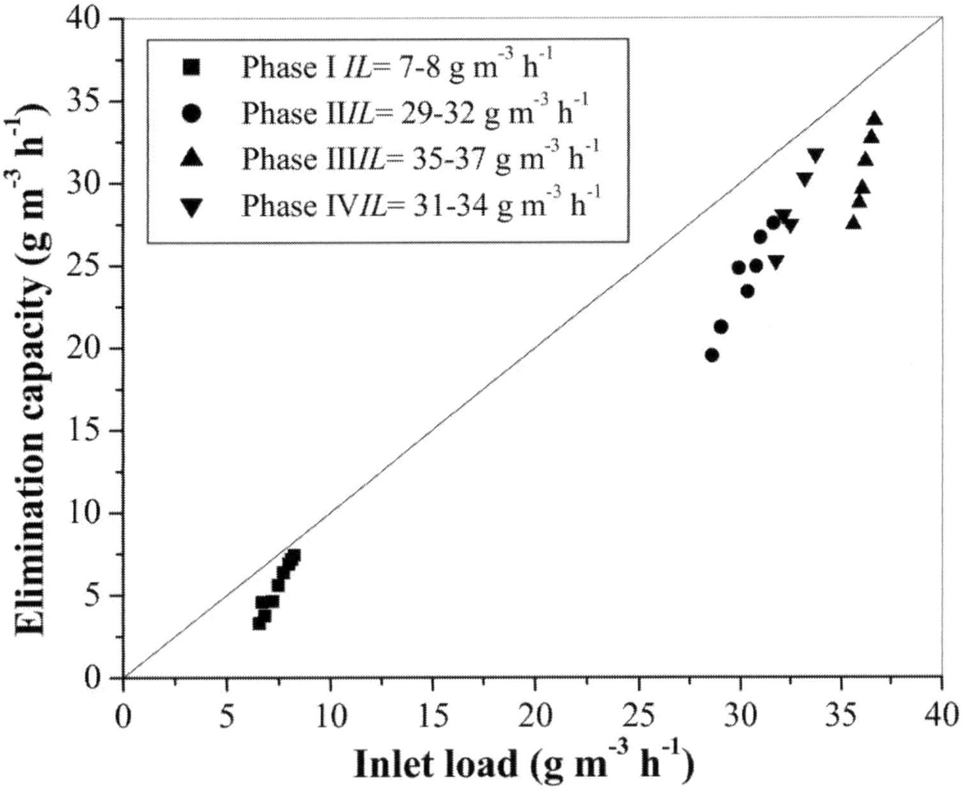

Phase I IL= 7-8 g m-3 h-1
Phase II IL= 29-32 g m-3 h-1
Phase III IL= 35-37 g m-3 h-1
Phase IV IL= 31-34 g m-3 h-1
Elimination capacity (g m-3 h-1)
Inlet load (g m-3 h-1)
0
5
10
15
20
25
30
35
40

Figure 2.

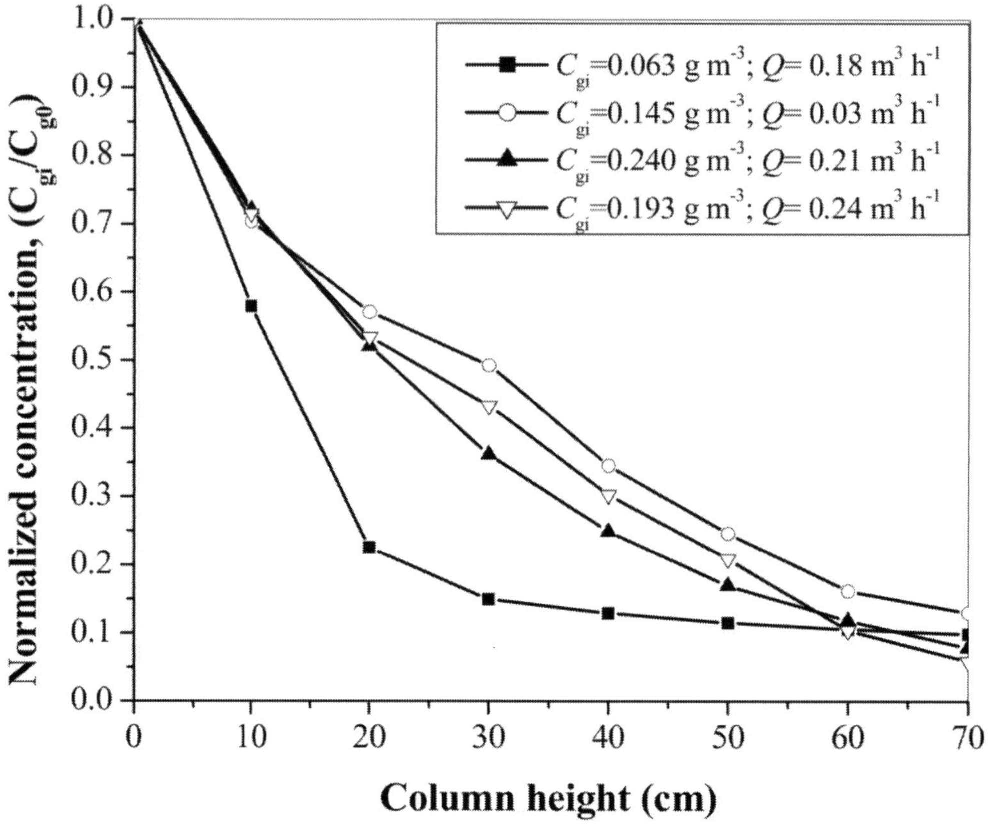

1.0
0.9
0.8
0.7
0.6
0.5
0.4
0.3
0.2
0.1
0.0
Normalized concentration, (C_gi/C_g0)
0
10
20
30
40
50
60
70
Column height (cm)
C_gi=0.063 g m^-3; Q= 0.18 m^3 h^-1
C_gi=0.145 g m^-3; Q= 0.03 m^3 h^-1
C_gi=0.240 g m^-3; Q= 0.21 m^3 h^-1
C_gi=0.193 g m^-3; Q= 0.24 m^3 h^-1

Figure. 3

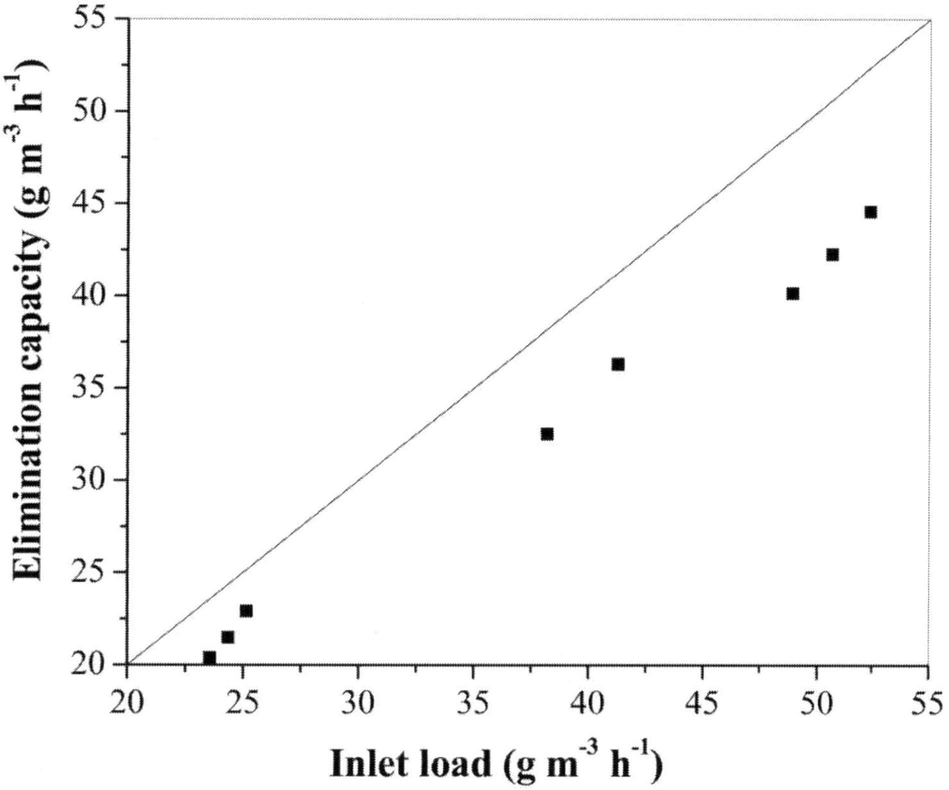

Elimination capacity (g m-3 h-1)
Inlet load (g m-3 h-1)
20
25
30
35
40
45
50
55

Figure 4

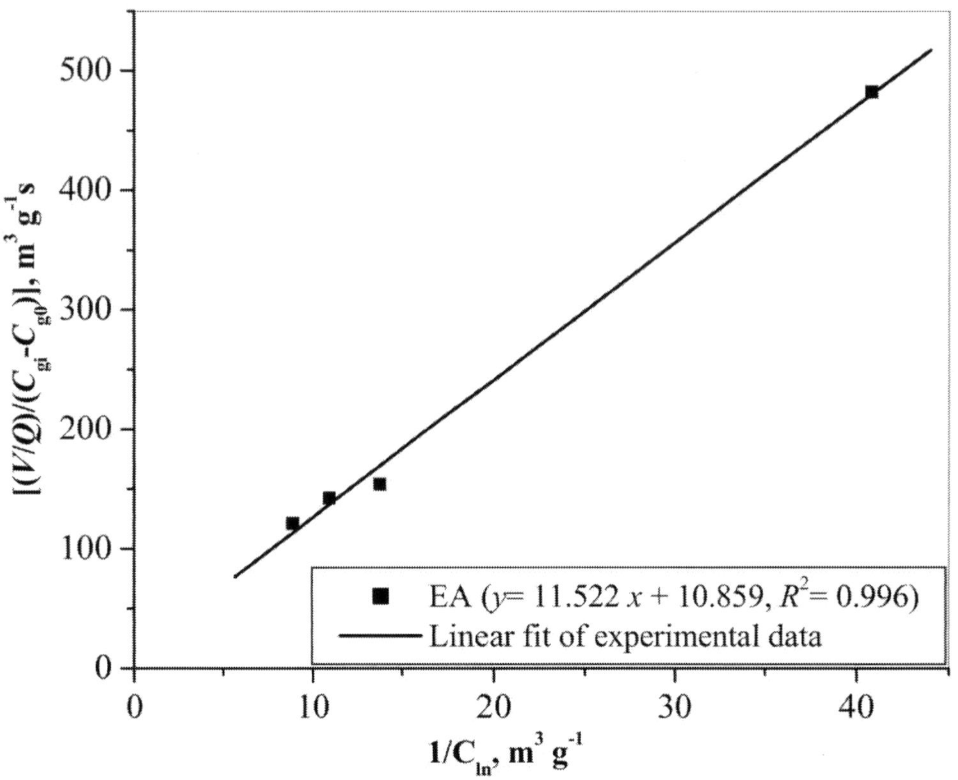
[(V/Q)/(C_gi-C_g0)], m^3 g^-1 s
1/C_ln, m^3 g^-1
0
100
200
300
400
500
0
10
20
30
40
EA (y= 11.522 x + 10.859, R^2= 0.996)
Linear fit of experimental data

Figure 5

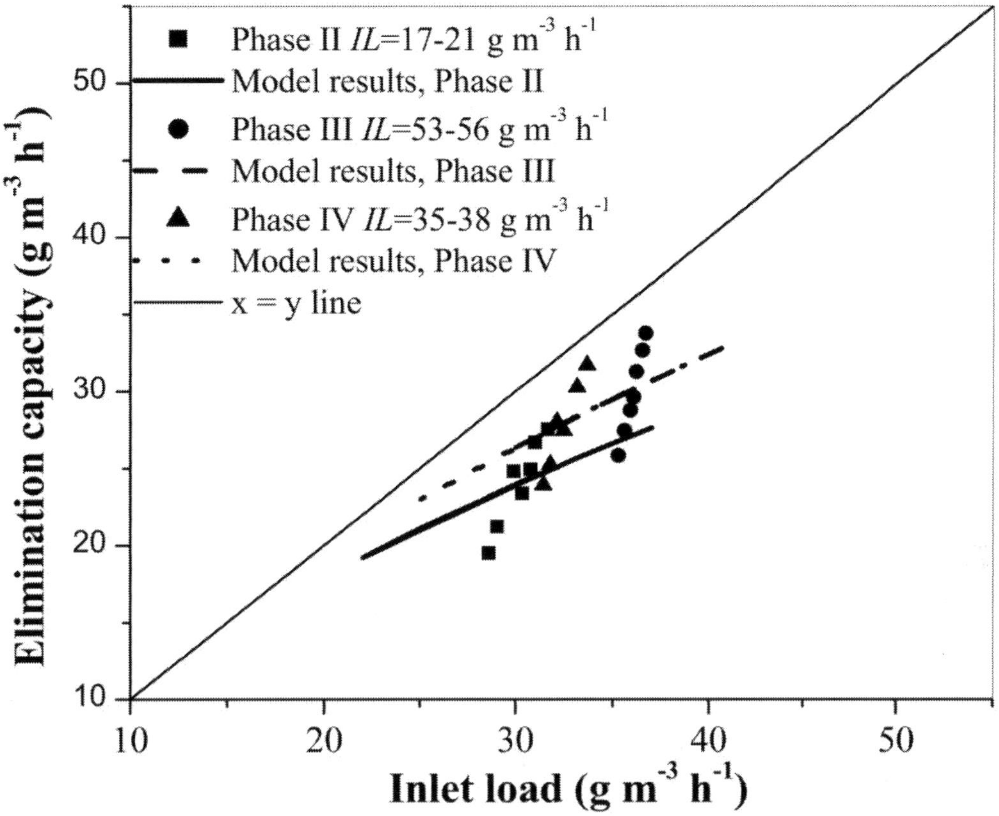

Phase II IL=17-21 g m-3 h-1
Model results, Phase II
Phase III IL=53-56 g m-3 h-1
Model results, Phase III
Phase IV IL=35-38 g m-3 h-1
Model results, Phase IV
x = y line
Elimination capacity (g m-3 h-1)
Inlet load (g m-3 h-1)
10
20
30
40
50

Figure 6

(752d) A Novel up-Flow Inner-Cycle Anoxic Bioreactor (UIAB) System for the Treatment of Sulfide Wastewater and Purification of Biogas

Jianmin Xing, National Key Lab of Biochemical Engineering, Institute of Process Engineering, Chinese Academy of Sciences, Beijing, China and Ziyu Song, Institute of Process Engineering, Chinese Academy of Sciences, Beijing, China

Anaerobic digestion is one of common treatment processes for the organic wastewater disposal, and usually produces methane-rich biogas. Biogas from anaerobic digestion always contains hydrogen sulfide, which is highly toxic, corrosion and malodorous. Biogas desulfurization is essential for its purification processes before utilization. Biological desulfurization process is a method of eliminating sulfide in aqueous by autotrophic colorless sulfide-oxidizing bacteria. Hydrogen sulfide is adsorbed by alkaline solution, and is chemically transformed into sulfide.

In this paper, an up-flow inner-cycle anoxic bioreactor with a novel three phase separator was designed and implemented for the treatment of sulfide wastewater. The sulfide in wastewater could be converted to element sulfur by sulfide oxidizing bacteria, and recovered by simple precipitation. When the oxidation-reduction potential (ORP) was controlled at -100mV, 91.3% of sulfide could be oxidized to element sulfur. To achieve high sulfide removal rate and sulfur conversion yield, the pH of influent should be controlled at the range from 7.0 to 8.0. The optimal desulfurization process was carried out at 400 $mmol \cdot L^{-1} \cdot d^{-1}$ sulfide loading rate and 120 min hydraulic retention time (HRT). The sulfide removal rate was approximately 95.2% and element sulfur conversion rate was above 90.3%. These results demonstrated that the novel up-flow inner cycle anoxic bioreactor system could be regarded as not only the enhancement of sulfide oxidation process but also one important purified stage for biogas, which would be an efficient and economic supplement method for the treatment of sulfide wastewater from gas purification.

(752e) Bioleaching of Metals From Spent Refinery Catalysts Using Acidithiobacillus Thiooxidans and Chelating Agent

Ankit Lal Jr.[1], Bina Singh Sr.[2], Priyangshu M. Sarma Sr.[3] and Banwari Lal Sr.[2], (1)EIBD, Teri, New delhi, India, (2)EIBD, Teri, New Delhi, India, (3)EIBD, The Energy Resources Institute, New Delhi, India

Bioleaching of metals from hazardous spent hydro-processing catalyst was attempted in the second stage after growing the bacteria with sulfur in the first stage. The first stage involved transformation of elemental sulfur particles to sulfuric acid through an oxidation process by acidophilic bacteria (*Acidithiobacillus thiooxidans).* In the second stage, the acidic medium was utilized for the leaching process. Nickel and Molybdenum containing spent catalyst were chelated with EDTA and leached from the solid materials to liquid medium by the action of sulfuric acid that was produced by acidophilic leaching bacteria and compared with chemical leaching. Involved microorganisms were sulfur oxidizing bacteria. Experiments were conducted varying the reaction time, amount of spent catalysts, amount of elemental sulfur and temperature. At 50 g/L spent catalyst concentration and 20 g/L elemental sulfur, 84.3% Ni, 46.3% Mo, were recovered after 7 hrs of heating at 100^0C. Chemical leaching with commercial sulfuric acid of the similar amount that produced by bacteria was compared. Thermodynamic parameters were calculated and the nature of reaction was found to be exothermic. Leaching kinetics of the metals was represented by different reaction kinetic equations, however, only diffusion controlled model showed the best correlation here. During the whole process Mo showed low dissolution because of substantiate precipitation with leach residues as MoO3. Bioleach residues were characterized by EDX and XRD.

(752f) Biodegradation of Carbon Tetrachloride in Laboratory Flow Channels

Sathishkumar Santharam, BG Consultants, Inc., Manhattan, KS, Larry Davis, Biochemistry, Kansas State University, Manhattan, KS and Larry Erickson, Chemical Engineering, Kansas State University, Manhattan, KS

Carbon tetrachloride (CT) is a solvent used in several applications including degreasing and fumigation. CT is a common groundwater pollutant and a suspected carcinogen. Remediation feasibility studies were conducted by mesocosm experiments; a chamber was divided into six channels and filled with soil, and plants were grown on the soil surface. Each channel was fed with CT contaminated water near the bottom and collected at the outlet, simulating groundwater flow conditions. CT was introduced at a concentration of about 2 mg/L (~13 μmoles/L) in three channels, two of them with alfalfa plants and the other with grass. The other three channels were used for a different study. After the system had attained steady state, the concentrations of CT at inlet and outlet were monitored and the amount of CT disappearing in the saturated zone was studied. Since no degradation products were found at the outlet after about 100 days, anaerobic conditions were created in one channel (with alfalfa) by adding one liter of 0.2% glucose solution. The glucose solution was fed once every month starting from day 111 and continued until day 328. From day 203 to day 445, one liter of 0.1% emulsified soy oil methyl esters (SOME) was fed once a month to another channel (with alfalfa). The monthly SOME addition dates were the same as for glucose. Degradation compounds of CT were detected in the outlet liquid of the channels fed with SOME and glucose; however, the extent of degradation in the SOME fed channel was not as great as that of the glucose fed channel. No degradation compounds were observed in the outlet solution of the channel (grass grown on top) in which no carbon and energy supplements were added. The total chlorinated methanes (CMes) decreased to less than 1 μmoles/L in channels 5 and channel 6; however, the CMes remained low for a longer time in the glucose fed channel. In addition to glucose, corn starch and cheese whey were also added to channel 5, as growth substrates. Supplements such as glucose, corn starch, cheese whey and SOME stimulated the indigenous microbes and helped in the degradation of CT. In both glucose and SOME fed channels, the degradation continued several weeks after stopping the feeding of supplements. The soil matrix and the microorganisms were able to store the supplements/degradation products of supplements and provide a long-term source of carbon and hydrogen. This finding is very valuable in the design of remediation systems in field sites since it determines the frequency of supplement addition.

(752g) Microbial Mechanism On Upflow Microaerobic Sludge Blanket Reactor for Municipal Wastewater Treatment

Shaokui Zheng, School of Environment, Beijing Normal University, Beijing, China

In this study, 16S rDNA-based DGGE and clone library were used to investigate the microbial mechanism on the upflow microaerobic sludge blanket reactor (UMSB) for domestic wastewater with an aerobic reactor (AR) as control, with particular emphasis on nitrification. The oxygen half saturation constant (Ko) of activated sludge was measuured to compare the oxygen affinity of sludge obtained from these two reactors at different experimental phases. Like the AR, the UMSB achieved a 60% energy saving, and a COD and NH3-N removals of 98% and 90% respectivity, at DO=0.7-0.9 mg l-1. The majority of Nox-N in the effluent was NO3-N for the AR but NO2-N for the UMSB. The activated sludge was dominated by nitrite oxidation bacteria (NOB) in the UMSB and Ammonia-Oxidizing Bacteria (AOB) in AR, respectively. The low DO in the UMSB favors the growth of filamentous bacteria, e.g., Sphaerotilus, Planctomycetes, Chloroflexi, however, they did not lead to filamentous bulking in the UMSB. Generally, the activated sludge in the UMSB has low Ko values than that in the AR.

(760a) Microrheology of VEGF-Stimulated Nuclear Reorganization in Endothelial Cells

Stephen T. Spagnol[1], James S. Weltz[1] and Kris Noel Dahl[2], (1)Department of Chemical Engineering, Carnegie Mellon University, Pittsburgh, PA, (2)Chemical Engineering, Carnegie Mellon University, Pittsburgh, PA

Introduction: Genome regulation requires delicately balanced temporal control for proper cellular homeostasis and adaptation. Although DNA is a linear genetic code, recent studies have illuminated that genome organization within the nucleus aids in regulation[1], and there is evidence that gene translocation within the nucleus leads to differential expression[2]. The correlation of gene expression with gene position and movement in the mechanically stiff nucleus[3] suggests there is bio-rheological coupling impacting gene expression. We examine the effects of extracellular biochemical stimulation on nuclear organization. These signals are typically examined as chemical signaling cascades that activate nuclear transcription factors leading to gene expression. However, stimulation by certain factors simultaneously reorganizes cellular structures via motor proteins, causing mechanical force propagation to the nucleus. We quantify the relationship between the chemical factors and genome reorganization to compare these rheological changes with the accompanying changes in gene expression. This exciting idea that intracellular force generation correlates with gene expression as manifested by rheological changes allows us to answer fundamental cell biology questions using microrheology.

Materials and Methods: We examine vascular endothelial growth factor (VEGF) stimulated angiogenesis of human umbilical endothelial cells (HUVECs). Cells were transfected with GFP-tagged sub-nuclear markers and were treated with up to 50 ng/mL of VEGF. Using the fiducial GFP particles, nuclear movements were tracked in live cells. Particle trajectories were determined (using custom software[5] to remove translocation and rotation) to calculate the mean square displacement (MSD) versus time, which was fit to a power-law model ($MSD(\tau)=D_{eff}*\tau^{\beta}$), consistent with cell rheological modeling[3].

Results and Discussion: Our results indicate that VEGF stimulation of HUVECs results in an effective intranuclear softening corresponding to increased MSD magnitudes at all time points ($p < 0.001$ from unstimulated control). This effective softening corresponds to chromatin decondensation and reorganization that decreases the resistance to motion. This decondensation is coupled with increased motor protein-driven cytoskeletal stress on the nucleus. This stress has been suggested to increase nucleoplasmic agitation to enhance DNA and protein diffusion and, thus, collision frequency to drive the favorable binding events necessary for transcriptional activation. In addition, the VEGF stimulated response is time dependent, with significant attenuation in the response after 2.5 hours. These results are consistent with a recent DNA microarray study showing a more than two-fold upregulation of 139 genes following VEGF stimulation, with 53 genes directly induced by VEGF within the first 2 hours of exposure[4].

Conclusions: Understanding genome organization within the nucleus and the mechanisms underlying stimulated transitions in gene expression are vital to elucidating genome function. Our results indicate that the large turnover in gene expression from VEGF stimulation corresponds to an effective nuclear softening from global chromatin agitation, rather than a purely localized and gene-specific effect. Thus, our work fortifies the dynamic view of the genome, suggesting stimulated gene repositioning that likely involves motor protein-driven cytoskeletal stress by correlating known VEGF-stimulated changes in gene expression with global nuclear reorganization through microrheology.

[1]T. Misteli, *Cell.* **128**, 787 (2007). [2]S.M. Gasser, *Science.* **296**, 1412 (2002). [3]K.N. Dahl *et al.*, *Biophys. J.* **89**, 2855 (2005). [4]M. Abe *et al.*, *Angiogenesis.* **4**, 289 (2001). [5]G. Yang *et al.*, *Journal of Cell Biology.* **182**, 631-639 (2008).

(760b) Role of ERK Activity in Epithelial-to-Mesenchymal Transition in Lung Cancer

Janine Buonato and Matthew J. Lazzara, Chemical and Biomolecular Engineering, University of Pennsylvania, Philadelphia, PA

Epithelial-to-mesenchymal transition (EMT) is a cellular program critical for tumor invasion and metastasis. During EMT, cells with initially epithelial characteristics (e.g., expression of markers such as E-cadherin and a high degree of cell spreading) become less proliferative and increasingly migratory as they take on mesenchymal characteristics (e.g., expression of markers such as vimentin and reduced cell-cell contacts). Compared to their epithelial counterparts, cells with mesenchymal characteristics also tend to be more resistant to a variety of targeted therapeutics used in the treatment of cancer (e.g., inhibitors of specific kinases). Thus, developing a quantitative understanding of the factors which control EMT may provide a new framework for improving tumor response to targeted therapeutics. In this presentation, we will discuss our recent work demonstrating that the activity of the extracellular regulated kinase (ERK) pathway is of critical importance in the determination of mesenchymal characteristics in non-small cell lung carcinoma (NSCLC). Working with a panel of human NSCLC cell lines, we have demonstrated that ERK activity is required for EMT initiated by exogenous transforming growth factor beta (TGF-beta), one of the most well-established mediators of EMT. Moreover, we have demonstrated and quantitatively determined the extent to which ERK activity levels govern an epithelial versus mesenchymal state even in the absence of TGF-beta. More specifically, we have demonstrated that three distinct cellular perturbations affecting ERK activity levels, each of which may be observed in NSCLC, can alter the balance of epithelial and mesenchymal characteristics of NSCLC cells. These perturbations include the mutational status of the cellular kinase KRAS, amplification of ERK itself, and altered rates of endocytosis of the epidermal growth factor receptor. Finally, we have also connected this control of epithelial/mesenchymal marker expression by ERK to associated changes in cellular phenotypes including cellular migration speed in scratch closure assays, proliferation, and cell death response to targeted chemotherapeutics. Importantly, our results demonstrate that the observed cell phenotypic changes occur with a time scale which corresponds to that with which epithelial and mesenchymal marker expression changes in response to ERK activity perturbation. This time scale is much larger than that for changes in ERK activity itself, reflecting the need for downstream ERK-dependent processes to effect phenotypic changes. Overall, our results suggest a number of opportunities to rationally control EMT processes in cancer to engineer improved clinical outcomes for patients with NSCLC.

(760c) Depletion of SIRT1, but Not SIRT2, Inhibits PMA-Stimulated Megakaryocytic Differentiation of the K562 Cell Line

Mark T. Duncan[1], Zachary Mays[2], Nitin Kini[2] and William M. Miller[3], (1)Chemical and Biological Engineering, Northwestern University, Evanston, IL, (2)Northwestern University, (3)Chemical & Biological Engineering, Northwestern University, Evanston, IL

Depletion of SIRT1, but not SIRT2, inhibits PMA-stimulated megakaryocytic differentiation of the K562 cell line

Mark T. Duncan*, Zachary Mays, Nitin Kini, William M. Miller

Department of Chemical and Biological Engineering, Northwestern University, Evanston, IL, United States of America

*email: mtduncan@northwestern.edu

Sirtuins (SIRT1-7 in humans) are class III NAD-dependent histone/protein deacetylases that have been implicated in modulating the differentiation of many cell types including adipocytes and endothelial, neural, and skeletal muscle cells. Previously, we have demonstrated that treatment with the pan-SIRT inhibitor Nicotinamide (NIC) greatly enhances the polyploidization of human megakaryocytes (MKs) derived from CD34+ cells in culture. This effect was found to be, at least in part, due to inhibition of SIRT1/2 deacetylases, as similar effects were recapitulated by the SIRT1/2 specific inhibitor cambinol. We are currently using the cell line K562 to further investigate the roles of SIRT1/2 in MK differentiation. Addition of phorbol 12-myristate 13-acetate (PMA) induces K562 cells to express the MK antigen CD41 and exhibit morphological characteristics of MK maturation, including enlargement of cell size, polyploidization, and the appearance of cytoplasmic vacuoles. Similar to primary cells, NIC treatment of differentiating K562 cells led to an increased fraction of polyploid cells. In addition, NIC treatment reduced cell expansion and appeared to increase/accelerate apoptosis. Surprisingly, silencing of SIRT1 with lentiviral shRNAs strongly inhibited PMA-mediated differentiation. We consistently observed a moderate reduction in CD41 expression, and a >50% reduction in polyploidization in SIRT1-silenced cells. SIRT2 silencing did not appear to affect differentiation. Thus, we hypothesized that SIRT1 may have an important role in PMA-induced cell signaling, and confirmed its high expression in K562 cells via immunostaining. Unexpectedly, SIRT1 appeared to be predominantly localized to the cytosol (both pre- and post-PMA stimulation). We are currently investigating a possible role for cytosolic SIRT1 in PMA-induced MAP kinase signaling, as a previous report suggests SIRT1 can enhance ERK1/2 activation.[1] In addition, we have begun to examine the effects of SIRT1 silencing on the differentiation of human MKs derived from CD34+ cells in culture. In an initial experiment, SIRT1 silencing reduced MK expansion, again suggesting SIRT1's importance in regulating the cell cycle (though whether this is through a similar or distinct mechanism from the K562 model cell line remains to be determined). We are also currently exploring knockdown of SIRT7, which is upregulated during primary Mk differentiation.

<u>References</u>

1 Huang, J. *et al.* SIRT1 Overexpression Antagonizes Cellular Senescence with Activated ERK/S6k1 Signaling in Human Diploid Fibroblasts. *PLoS ONE* **3**, e1710, doi:10.1371/journal.pone.0001710 (2008).

(760d) Exploring Structural Variations of Cytoskeleton to Understand hESC Differentiation to Insulin Producing Phenotype

Joseph E. Candiello[1], Li Ang Zhang[2], Prashant Kumta[1] and Ipsita Banerjee[3], (1)Bioengineering, University of Pittsburgh, Pittsburgh, PA, (2)Carnegie Mellon University, (3)Chemical and Petroleum Engineering, University of Pittsburgh, Pittsburgh, PA

Introduction:

Embryonic stem cells (ESCs) have the ability to differentiate into cells of all three germ layer designations: mesoderm, endoderm, and ectoderm. As ESCs differentiate along these various lineages they undergo changes in cell phenotype, both structurally and chemically. There are various methods used to categorize the changes in cell behavior apparent in the protein expression, gene expression, and cell morphology. Currently many of the tools available to study changes in ESC morphology and structure, especially that of the cytoskeletal components, are purely qualitative. Evidence has suggested that as a cell differentiates it undergoes changes in structure and expression of these cytoskeletal components. Also, cell shape and cytoskeletal tension have a direct effect on stem cell lineage fate [1]. The cell cytoskeleton components contribute to various cellular functions, including cellular adhesion, movement, intracellular transport and signaling, and cell mechanics so it makes sense that differentiation to a target cell phenotype would include mimicking its cytoskeletal traits.

Our current focus is harnessing the pluripotent nature of human ESCs to produce a therapeutic tool in controlling the disease pathogenesis associated with the loss of function of insulin producing pancreatic Beta cells due to type I diabetes. The actin cytoskeleton plays a role in proper insulin exocytosis. It has been demonstrated that proper insulin production in response to glucose levels in beta cells is linked to cytoskeletal organization, specifically F-actin [2, 3]. In this study we use a previously established model to quantify the features of induced stem cells at different stages along the lineage towards a mature insulin producing cell, by fluorescently labeling components of the cytoskeleton and imaging at various stages of differentiation towards mature insulin producing cells.

Methods:

Human embryonic stem cells grown in culture were differentiated by chemical signaling in media with a well-established protocol first towards definitive endoderm phenotype, later to a pancreatic precursor phenotype, and finally to a mature insulin producing cell. After each stage of differentiation a population of cells were fixed and stained for the cytoskeletal components F-actin using rhodamine phalloidin and B-tubulin using anti-tubulin AlexaFlour 488. Additionally, the differentiated cells were costained for a marker indicative of a cell in the desired phenotype. For undifferentiated embryonic stem cells we stained for the pluripotent marker OCT4, definitive endoderm cells for SOX17, pancreatic precursor cells for PDX1, and finally mature insulin producing cell for C-Peptide. Cells were chosen for analysis based on a positive signal from the marker labeling. After fluorescently imaging the cytoskeletal components we used an established FiberScore algorithm to quantify the structural properties of the cells at all three stages of development [4, 5].

Results:

We were successfully able to locate and image both the actin and beta-tubulin cellular structure for cell phenotypes from the pluripotent hESCs, definitive endoderm, pancreatic precursor, and mature insulin producing cell. Qualitatively we noted an increase in overall cell size, most notably when compared to the pluripotent ESC stage. We also noted a trend regarding the formation of stress fibers, specifically during the definitive endoderm and pancreatic precursor stage. The cells were analyzed by the FiberScore algorithm and structural properties such as fiber intensity, polarity, average width, and total fiber length were found for each group. For instance, we also found a change in the quantitative measurements of the differentiating cells, most notably an increase in beta-tubulin intensity and width between the definitive endoderm and pancreatic precursor stage and fiber intensity in the later stages of differentiation for the actin components.

Conclusion:

This study demonstrates the feasibility of utilizing an established technique to better quantify structural changes that occur during the differentiation of hESCS towards an insulin producing phenotype. While there are various established methods to differentiate towards such a cell type, evidence has suggested that variances in the cytoskeleton and its downstream effects of exocytosis of insulin in response to glucose sitimuli are linked to the cytoskeleton. Gaining a better understanding of how the cytoskeleton changes during differentiation can be used to possibly elucidate a more beneficial pathway to attain the overall goal of a glucose responsive insulin producing cell. This knowledge would also be useful in designing and fine tuning materials substrates for inducing desired differentiation of hESCs towards an insulin producing phenotype.

1. McBeath, R., et al., *Cell shape, cytoskeletal tension, and RhoA regulate stem cell lineage commitment.* Dev Cell, 2004. **6**(4): p. 483-95.

2. Hammar, E., et al., *Role of the Rho-ROCK (Rho-associated kinase) signaling pathway in the regulation of pancreatic beta-cell function.* Endocrinology, 2009. **150**(5): p. 2072-9.

3. Tomas, A., et al., *Regulation of pancreatic beta-cell insulin secretion by actin cytoskeleton remodelling: role of gelsolin and cooperation with the MAPK signalling pathway.* J Cell Sci, 2006. **119**(Pt 10): p. 2156-67.

4. Lichtenstein, N., B. Geiger, and Z. Kam, *Quantitative analysis of cytoskeletal organization by digital fluorescent microscopy.* Cytometry A, 2003. **54**(1): p. 8-18.

5. Shah, S.A., P. Santago, and B.K. Rubin, *Quantification of biopolymer filament structure.* Ultramicroscopy, 2005. **104**(3-4): p. 244-54.

(760e) Hydroxylated Flavones Reduce Amyloid-β Induced Calcium Influx

J. Will Reed, Kayla Pate, John Clegg, McCall Rogers and Melissa A. Moss, Department of Chemical Engineering, University of South Carolina, Columbia, SC

Alzheimer's disease (AD) is the most common form of dementia, characterized by brains that have developed extracellular plaques and intracellular tangles. The formation of plaques, which are composed of aggregated amyloid-β protein (Aβ), is largely believed to be the underlying cause of the disease. Mounting evidence suggests that the plaques themselves do not induce neurodegeneration; it is instead induced by Aβ oligomers, which are formed in an intermediate step of plaque formation. Healthy individuals have low levels of Aβ in blood stream and cerebrospinal fluid that may be neurotrophic; therefore, complete elimination of the protein is not likely a viable method of AD prevention. Rather, the current therapeutic goal is to prevent nascent Aβ monomers from aggregating into toxic oligomers. Unfortunately, the current diagnostic criteria for AD often do not present in early stages of the disease when it is believed intervention can be successful. As a result, naturally occurring compounds that are capable of inhibiting Aβ aggregation would be the ideal therapeutic as they can be safely incorporated into a population's diet in a strategy similar to iodized salt.

Although old age is generally considered the biggest risk factor for AD, the disease is markedly more prevalent in developed countries. By taking advantage of differences in diets between cultures, epidemiological studies have demonstrated a correlation between the increased intake of polyphenols and reduced incidence of AD. The failure of the Western diet to incorporate foods containing polyphenols may be a contributing factor to the AD epidemic. Many fruits, vegetables, and cereals contain flavonoids, which are the most common polyphenolic compounds found in the human diet. The most basic flavonoid is the three aromatic ring structure, flavone. While many flavone derivatives exist, this study focuses on the base structure and two derivatives with hydroxyl functionalized aromatic rings: 3',4' dihydroxyflavone (DHF) and 5,7,3',4'5' pentahydroxyflavone (PHF). We explored the therapeutic potential of these three compounds by examining their ability to inhibit Aβ aggregation using both solution and *in vitro* assays.

$A\beta_{1\ 42}$ oligomers were prepared in solution prior to cellular exposure. Synthetic monomer was solubilized in dimethyl sulfoxide before the addition of phosphate buffered saline (PBS), which initiates oligomerization at room temperature. Flavone, DHF, or PHF was added prior to PBS in order to allow any inhibition to occur. The resulting products formed in solution were stabilized in sodium dodecyl sulfate (SDS). These oligomers were then resolved by SDS-polyacrylamide gel electrophoresis (SDS-PAGE) and electroblotted to analyze size distribution. Western blot analysis demonstrates the presence of DHF and PHF in the oligomerization reactions reduces the maximum size of detectable oligomers. In parallel experiments, cultures of the neuroblastoma cell line SH SY5Y were challenged with the reaction products immediately after formation. Intracellular Ca^{2+} ion levels were measured with confocal microscopy via Fluo-3 fluorescence. Cells exposed to $A\beta_{1\ 42}$ oligomers exhibit Ca^{2+} influx, which is an early indicator of cell death commonly associated with deleterious proteins. Oligomers formed with DHF induce little Ca^{2+} influx similar to untreated controls whereas oligomers formed with PHF induce the same level of Ca^{2+} influx as oligomers alone.

Polyphenol structure influences the effect upon both Aβ oligomerization and physiological activity. PHF greatly reduces the size of oligomers but does not prevent Ca^{2+} influx, suggesting it keeps the oligomers in their most toxic state. Alternatively, the ability of DHF to reduce both the formation of larger $A\beta_{1\ 42}$ oligomers as well as the toxic effects of Ca^{2+} influx suggests bioavailability of DHF could reduce AD symptoms.

(760f) Calcium Stimulated Metabolism Promotes Oxidative Stress in Hepatic Lipotoxicity

Robert Egnatchik and Jamey D. Young, Chemical and Biomolecular Engineering, Vanderbilt University, Nashville, TN

The increasing prevalence of obesity in Western society has resulted in higher rates of associated diseases such as type-2 diabetes, atherosclerosis, and nonalcoholic fatty liver disease (NAFLD). NAFLD is a chronic liver disease resulting from excess lipid accumulation (i.e., steatosis) and affects up to 30% of the US population. Although simple steatosis does not always lead to complications, around 10% of NAFLD patients are at increased risk of developing more serious liver injuries such as nonalcoholic steatohepatitis (NASH) and hepatocellular carcinoma. Free fatty acid (FFA) levels are present in higher concentrations in the plasma of these individuals, suggesting that *in vivo* alterations in FFA metabolism are linked to corresponding changes in disease severity. *In vitro* experiments have demonstrated that saturated fatty acid (SFA) overexposure stimulates acute apoptosis in a diverse range of cell types, although the mechanism of SFA-induced apoptosis may vary depending upon the tissue of origin. In particular, SFA lipotoxicity in hepatic cells is associated with increased reactive oxygen species (ROS) and endoplasmic reticulum (ER) stress but is independent of ceramide synthesis. Furthermore, the response to SFA treatment is altogether different from that of monounsaturated fatty acid (MUFA) treatment, which induces steatotic triglyceride formation without initiating ROS accumulation or apoptosis.

To identify mechanisms of palmitate lipotoxicity in hepatic cells, H4IIEC3 cells were treated with free fatty acids in combination with interventions intended to block specific events leading to palmitate-induced metabolic dysfunction and apoptosis. These experiments indicate that palmitate lipoapoptosis is marked by increased oxygen consumption and accumulation of reactive oxygen species (ROS). Supplementing culture media with the antioxidant N-acetyl-cysteine (NAC) reduced caspase activation and partially restored cell viability, suggesting that ROS accumulation is a critical factor necessary for activating lipoapoptosis. However, NAC did not normalize oxygen uptake rates of palmitate-treated cells, indicating that elevated ROS are not required for palmitate-induced alterations to oxidative metabolism. On the other hand, addition of the mitochondrial antagonist phenformin to palmitate-treated cells eliminated abnormal ROS accumulation, prevented the appearance of apoptotic markers, and normalized oxygen consumption rate. Taken together, these results indicate that palmitate-induced deregulation of mitochondrial metabolism is the primary cause of ROS accumulation and apoptosis in H4IIEC3 cells. Furthermore, chelation of intracellular calcium using BAPTA-AM attenuated ROS accumulation, oxygen consumption, caspase activation, and cell death in palmitate-treated cells. Based on these results, we propose a novel mechanism of palmitate lipotoxicity that is dependent on calcium-stimulated overactivation of oxidative mitochondrial metabolism, which promotes ROS accumulation and apoptosis.

Our presentation will discuss our experiments using modulators of calcium trafficking (BAPTA) and mitochondrial function (Phenformin, Rotenone) which alter ROS production and demonstrate a clear role for calcium activated mitochondrial metabolism in the context of palmitate lipotoxicity. Additionally, we will highlight how these alterations are promising targets in modulating NAFLD and lipoapoptosis.

(760g) Integrated Transcriptomic and Lipidomic Study of Macrophage Response to Liver X Receptor Ligand 25-Hydroxy-Cholesterol

Shakti Gupta[1], Ashok Reddy Dinasarapu[2], **Mano R. Maurya**[2], Eoin Fahy[2], Manish Sud[2] and Shankar Subramaniam[3], (1)Department of Bioengineering, University of California, San Diego, CA, (2)Department of Bioengineering, University of California San Diego, La Jolla, CA, (3)San Diego Supercomputer Center, Department of Bioengineering, Department of Chemistry & Biochemistry, University of California San Diego, La Jolla, CA

Macrophage inflammation and resulting increase in the levels of oxysterols plays an important role in the progression of cardiovascular diseases such as atherosclerosis. Upon stimulation with pathogen markers such as lipopolysaccharide (LPS), primary macrophages such as bone-marrow derived macrophages (BMDMs) as well as RAW 264.7 cells (a leukemia-virus transformed macrophage cell line) show substantial lipidomic and transcriptomic changes. While lipids from several categories such as eicosanoids, other fatty acids, sphingolipids and sterols show changes, our focus here is on the changes in sterol levels. We have found that 25-hydroxy-cholesterol (25-OH-Ch) accumulates inside the cell even though the levels of its precursor, cholesterol, do not change substantially. Since 25-OH-Ch is a ligand for liver X receptors (LXRs), its increase results in up-regulation of its target genes such as ATP-binding cassette transporter (ABC) A1 (Abca1) and Abcg1 and down-regulation of some of the cholesterol biosynthesis genes. Abca1 and Abcg1 contribute to cholesterol-efflux. This results in negative feedback regulation of cholesterol homeostasis. Motivated by this observation, new experiments carried out where BMDMs are treated with 25-OH-Ch show substantial up-regulation of Abca1 and Abcg1 and down-regulation of lanosterol synthase (Lss). Sterol (*de novo*) biosynthetic genes such as HMG-CoA reductase (Hmgcr) and Squalene synthase (Sqle) are also down-regulated, albeit through sterol regulatory element (SRE; 10bp sequence) -binding protein (SREBP) 2, a LXR-independent mechanism. The overall negative feedback regulation of sterol biosynthesis and increased efflux of cholesterol results in relatively reduced levels of cholesterol in the cells. Further, cholesterol efflux results in reverse-cholesterol transport to the liver. Given that high-levels of oxysterols, often resulting from inflammation, contribute to atherosclerosis, LXR and SREBP2 and associated molecules are being investigated as putative drug targets. In this work, we will present a combined analysis of lipidomic and transcriptomic data from 25-OH-Ch stimulated BMDMs as it relates to the regulation of various targets of LXRs and SREBP and remodeling and homeostasis of various types of lipid and other pathways.

Acknowledgement: The experimental data used in this work was provided by the Lipid Metabolite and Pathway Strategy (LIPID MAPS) Consortium experimental laboratories and is available on the LIPID MAPS web site (www.lipidmaps.org). LIPID MAPS is funded by the National Institute of General Medical Sciences.

(760h) Role of TRBP and PACT in Asymmetry Sensing

Phillip Angart and S. Patrick Walton, Chemical Engineering and Materials Science, Michigan State University, East Lansing, MI

Short interfering RNAs (siRNAs) were recognized for their potential as therapeutics soon after their discovery as a method of targeted gene knockdown using the endogenous RNA interference (RNAi) pathway. Mimicking the natural structure of endogenous miRNAs and siRNAs, siRNAs should have 19bp with 2nt 3' overhangs. siRNAs are designed to be fully complementary with their target RNA; still, sequence selection is not necessarily straightforward. This is due to the number of critical inter- and intramolecular interactions that determine the eventual activity of the siRNA in silencing its target. Prominent among these are the interactions of the siRNA with the RNAi pathway proteins that result in the selection of one strand as the guide strand and the subsequent destruction of the complementary passenger strand. The preferential selection of one strand is referred to as asymmetry, and the asymmetry information is contained in the sequence of the siRNA. Thus, an understanding of the mechanism of achieving asymmetry is essential for siRNA design.

Our work focuses on defining characteristics of an effective siRNA, in this case identifying the siRNA characteristics that result in preferential selection of one siRNA strand over the other. siRNA strand selection is believed to occur in a ribonucleoprotein complex that forms before the active complex, called the RNA Induced Silencing Complex (RISC) Loading Complex (RLC). Strand selection then occurs based upon directional binding of the siRNA by the RLC. In this presentation, we will describe our work investigating the roles of various RLC and related proteins on determining siRNA asymmetry and how asymmetry defines eventual silencing efficiency of these siRNAs.

(776a) Isolation of Rare Circulating Tumor Cells and in Situ Culturing

Zhuo Zhang, Meggie M G Grafton and Sunitha Nagrath, Department of Chemical Engineering, University of Michigan, Ann Arbor, MI

Circulating tumor cells (CTCs) shed by primary and metastatic tumors are shown to carry important genetic and phenotypic information about tumor progression. Since late 1990s, massive microfluidic research effort has been devoted to isolate and characterize CTCs as a non-invasive method for cancer diagnosis and prognosis. One of the most effective methods to separate CTCs from blood is the immunoaffinity based CTC-chip with high capture efficiency and purity. However, the ability to use the captured CTCs to perform cellular characterization and drug screening is limited due to the fact that CTCs are rare. In order to overcome this challenge, we present a novel method to capture cancer cells from blood and conduct in situ cell culturing in a microfluidic platform. Low numbers of cancer cells spiked into whole blood are captured in a CTC-chip. Consequently, the captured cells are cultured on the same device. The growth of the cells is monitored and recorded for 72hrs using an on-stage culture chamber. This CTC isolation and culturing system can test the sensitivity of drugs and assist in therapy determination therefore extend the diagnostic potential of current CTC isolation technologies.

(776b) Engineered Microenvironments to Analyze Host-Tumor Cell Interactions

Eline Boghaert[1], Jason P. Gleghorn[1], KangAe Lee[1], Derek C. Radisky[2] and Celeste M. Nelson[1], (1)Princeton University, Princeton, NJ, (2)Mayo Clinic Cancer Center, Jacksonville, FL

The development of normal tissues, both natural and engineered, requires concerted interactions between cells and their local microenvironment. Similarly, breast tumor development is modulated by cues from the surrounding microenvironment, including interactions with neighboring non-tumor cells as well as the extracellular matrix (ECM). In addition to chemical signals, studies using homogeneous populations of breast cancer cell lines cultured in three-dimensional (3D) ECM have shown that the physical properties of the microenvironment, namely increased ECM stiffness, can stimulate the invasion of tumor cells. However, at early stages of breast cancer development, malignant cells are surrounded by normal epithelial cells, which have been shown to have a tumor-suppressive effect on co-cultured cancer cells. Here we used tissue engineering strategies to explore how the biophysical characteristics of the host microenvironment affect the proliferative and invasive tumor phenotype of the earliest stages of tumor development. We used a 3D microfabrication-based approach to engineer ducts comprised of normal mammary epithelial cells that contained a single tumor cell. We found that tumor cell phenotype was dictated by position in the duct, as tumor cell proliferation and invasion were enhanced at the ends, while these effects were blocked when the tumor cell was located elsewhere. Regions of invasion correspond with regions of high endogenous mechanical stress, as shown by finite element modeling and bead displacement experiments, and modulating the contractility of the host epithelium controlled the subsequent invasion of tumor cells. Combining micro-computed tomographic analysis with finite element modeling confirmed that regions of high mechanical stress correspond with regions of tumor formation *in vivo*. This work suggests that the mechanical tone of the non-tumorigenic host epithelium directs the phenotype of tumor cells, and provides additional insight into the instructive role of the mechanical tumor microenvironment. Our study also suggests that tissue engineering strategies can be used to unveil novel aspects of tumor development.

(776c) Elucidation of Mast Cell Localization Using a Microfluidic Device That Generates a Controllable Diffusion-Driven SCF Gradient

Meghaan M. Smith, Amir Shamloo, Maheswaran Mani, Milan Manchandia, Kenneth Weinberg and Sarah C. Heilshorn, Stanford University, Stanford, CA

Mast cells, granular cells that reside in tissue, have become a well-known modulator of wound healing. Investigators have proven that the absence of mast cells greatly impairs healing of skin wounds created in mast cell-deficient murine models. Subsequent reconstitution with bone marrow-derived mast cells (BMMCs) reversed this effect. These studies also demonstrated that the extent of mast cell degranulation, which promotes healing by increasing vascular permeability, neutrophil recruitment, and wound closure, correlates inversely with distance from the lesion. Thus, regulation of mast cell migration and localization is critical for wound healing.

In addition to its chemotactic properties, stem cell factor (SCF, aka kit ligand) is required for mast cell growth, differentiation, and survival; thus, SCF is key in regulating mast cell localization. To shed light on the mechanism behind SCF-dependent localization, we developed a gradient-generating microfluidic device, which offers several advantages over traditional chemotaxis assays, including establishment of a temporally stable gradient, minimal use of expensive reagents and cells, and the ability to directly visualize the migration of single cells. The devices employed for these studies generated a stable linear SCF gradient across a central channel containing BMMCs without exposing the cells to convective flow.The SCF concentration profile is dictated by Fickian diffusion across the cell culture chamber, which enables direct control over the gradient steepness and average concentration. A concentration gradient of 8 ng/ml/mm was determined to be sufficient to induce BMMC chemotaxis. Analysis of the time-lapse microscopic videos enabled tracking of the migratory paths of hundreds of individual cells, which were then quantified to determine speed, persistence time, and chemotactic index (CI) as a function of SCF gradient and concentration. Interestingly, we found that within the same 8 ng/ml/mm gradient, BMMCs exposed to the highest average SCF concentration (10 ng/ml) displayed greater migration efficiency (CI = 0.32), than cells exposed to lower concentrations (6 ng/ml, CI = 0.10, $p<0.05$), while cell speed was largely unaffected. As expected, BMMCs not subjected to an SCF gradient displayed random migration within the devices. These data reinforce the concept that chemokinesis, i.e. the effect of a chemokine on random cell motility and speed, is a distinct process from chemotaxis, which requires asymmetric reorganization of the cytoskeleton. These results further suggest that efficient cell polarization requires both a minimum gradient steepness as well as a minimum SCF concentration. Further studies of BMMC response to SCF may help elucidate the mechanism behind mast cell migration and localization in response to a lesion. The knowledge acquired through these studies will aid in the development of chemokine-delivery therapies for dermal wound healing of chronic lesions.

(776d) Mechanical Stretching Induced Mesenchymal Stem Cell Orientation

Chun Liu[1], Seungik Baek[2] and Christina Chan[1], (1)Chemical Engineering and Materials Science, Michigan State University, East Lansing, MI, (2)Mechanical Engineering, Michigan State University, East Lansing, MI

Tissues and cells in the body are exposed to a complex mechanical microenvironment. Therefore, mechanical stimulation is important for cell proliferation and differentiation. Recent interest has grown on the impact of mechanical stimulation, such as stiffness, surface topography, and mechanical stretching, on cell fate decision. In this study, we created an anisotropic mechanical environment by stretching the substrate membrane in an uniaxial direction, and recorded mesenchymal stem cell orientation response accordingly.

We investigated the alignment behavior of cells on an anisotropic surface. The experimental results agreed with our "cell active sensing" model, and suggest that the cells can sense and respond to the surface anisotropy by orienting in the direction of maximal effective stiffness. We apply this model to a uniaxial pre-stretched surface to predict cell orientation. After stretching a substrate unidirectionally, we applied a thin poly-L-lysine coating (~0.8nm) and culture the MSCs, whereby the cells do not experience the actual stretching. By culturing cells on a pre-stretched PDMS substrate membrane, the effective stiffness the cells sense in the stretched direction is larger than in the unstretched direction. This resulted in the cells aligning in the stretched direction, and contributed to the regulation of the cell focal adhesion, and further modulated cell signaling and differentiation.

Currently, most of the prior studies that focused on mechanical stress attempted to measure the cytoskeletal changes after active stretching, whereby the cells are seeded prior to applying the external force or stretch. However, in our experiment we seed the cells after the membrane is stretched, such that the cells do not experience the stretch due to external changes in force. Nevertheless the cells are able to sense the differential effective stiffness in the two directions and changed their cytoskeleton accordingly to orient in the direction of maximum stiffness. These results support our model of "active mechanosensing".

(776e) Controlling Embryonic Cell Sheet Migration Using Microfluidics

Melis Hazar[1], YongTae Kim[2], Jiho Song[1], Philip R. LeDuc[1], William Messner[1] and Lance A Davidson[3], (1)Mechanical Engineering, Carnegie Mellon University, Pittsburgh, PA, (2)Koch Institute for Integrative Cancer Research Massachsetts Institute of Technology, Cambridge, MA, (3)Bioengineering, University of Pittsburgh, Pittsburgh, PA

Embryonic development consists of a complex series of cell signaling, cell migration and cell differentiation events whereas morphogenesis is the process that controls the organized spatiotemporal distributions of these events. Cell sheet migration is central to embryonic development, to homeostasis of complex organs, and to disease pathology therefore studying cell migration within sheets is important. Embryos from the African Claw-toed frog, *Xenopus laevis*, are used to elucidate genes important in moving cells. However, little is known about the underlying mechanism by which cells in epithelial sheets coordinate their responses to growth factors, directional signals, and motility cues to direct sheet movement *in vivo*. One reason for this knowledge gap is the lack of technologies to control of the chemical microenvironment surrounding multicellular tissues. Here, we manipulate the chemical microenvironment with precise spatiotemporal delivery of cytoskeletal inhibitors and contraction-stimulating compounds using laminar flow interfaces with microfluidics. We find that microfluidics can pattern cell responses and control cell motility of multicellular sheets and may allow engineers the ability to initiate and control the outcome of synthetic morphogenetic programs. We use this approach to study motility of the cells within the sheets to these localized environments and understand how their coordinated behavior works with spatiotemporal control.

Microfluidics provides an opportunity to control chemical stimulation of biological systems using laminar flow interfaces with well-known flow modulation methods. To deliver precise chemical stimulation to a multicellular tissue, we have developed a system for controlling the inlet pressures to a microfluidic device by modulating fluidic resistance and capacitance. We employed this system to deliver chemicals over tissue explants from *Xenopus* embryos with a spatiotemporal control to study mechanical patterning and local control of cell sheet migration during epibolic-type morphogenetic movements. This approach enabled spatiotemporally controlled inhibition of cell sheet migration in an embryonic tissue to investigate the dynamic responses to localized inputs.

We observed that applying blebbistatin (BBS), a myosin inhibitor, to half of the animal cap explant, we could affect cell migration. When we stimulated the animal cap explant locally with BBS locally with our microfluidic system *in vivo*, sheet migration altered during epibolic-type morphogenetic movements. Using time-lapse microscopy, we were able to observe localized response, analyze the dynamics with kymographs, and construct spatiotemporal strain maps.

The development of new technologies to control the form of growing multicellular tissues advance the goals of regeneration and tissue engineering and produce new applications in bioengineering and medicine. We believe that patterning cell mechanics and controlling cell motility provide a means to initiate synthetic morphogenetic programs. In addition, the ability to control the form of multicellular tissues potentially has high impact in tissue engineering and regeneration applications in bioengineering and medicine.

(776f) Mechano-Transduction Pathway Interference with BMP-2 Signaling Cascade

Laure Fourel[1], **Jorge Almodovar**[1], Corinne Albiges-Rizo[2] and Catherine Picart[1], (1)Grenoble Institute of Technology, Grenoble, France, (2)Institut Albert Bonniot, Grenoble, France

The bone morphogenetic proteins (BMP) growth factors are essential signaling molecules in embryogenesis, tissue regeneration or cancer metastasis. Until now, the cellular effects of these proteins have mostly been investigated by adding them in solution in the cell culture medium even though growth factors are naturally not free to diffuse and are rather bound to ECM components including proteins and polysaccharides. In this work, we design a new biomimetic nanoassembly based on polyelectrolyte multilayer film made of a polypeptide and of hyaluronan which combines both matrix-bound presentation of growth factor bone morphogenetic protein 2 (BMP2) and modulation of the material's mechanical properties. We investigated the interplay between the stiffness of a biomaterial and the matrix-bound presentation of the growth factor BMP2 and its impact on cellular processes. Whereas C2C12 myoblast cells are round and poorly spread on soft films, cells displayed an increase of spreading and migration. By using a combined set of techniques including SiRNA strategy and pharmacological approach, we study how the mechano-transduction pathway can potentially interfere with the BMP-2 signalling cascade. We demonstrated a crosstalk between specific integrin and BMP2 receptors. Our results pointed out that signaling can be enhanced by the spatial association of integrin and BMP2 receptors.

(776g) Monocyte Chemoattractant Protein-1 Static Concentration Gradient in a 3D Collagen Matrix and Its Haptotactic Effect On Monocyte Migration

Neda Ghousifam and Heather Gappa Fahlenkamp, School of Chemical Engineering, Oklahoma State University, Stillwater, OK

Atherosclerosis is known as an inflammatory disease which is initiated by the accumulation of lipid substances in the subendothelial layer of major arteries, followed by adhesion of monocytes and lymphocytes to endothelial cells and their subsequent migration across the endothelial layer to the extracellular matrix (ECM). This adhesion and transmigration involves several steps, which are mediated by bioactive molecules named chemotactic cytokines (chemokines). Monocyte Chemoattractant Protein-1 (MCP-1) is one chemokine that plays an important role in monocyte trafficking across the endothelial layer. Many studies investigate the effect of soluble concentration gradients of chemokines, such as MCP-1, on monocyte migration by using a 2D experimental system. Studies have shown that soluble concentration gradients of chemokines drive, and possibly control cell migration. Recently, it was hypothesized that MCP-1 can bind with collagen, which is a major protein in the extracellular matrix (ECM). A 2D system may not adequately predict *in vivo* cell behavior due to the lack of the third dimension that is important for the creation of MCP-1 binding reaction in the ECM. A 3D tissue model consisting of a collagen matrix would be a better experimental model for studying the formation of MCP-1 static concentration gradient within the ECM. The goal of this research is to confirm the formation of an MCP-1 static concentration gradient within a simplified 3D vascular tissue model and examine its haptotactic effect on monocyte migration. The experimental results show that the 3D tissue model pre-treated with MCP-1 had a significant increase in the number of migrated monocytes in comparison with the control group lacking the pre-treatment. Also, when an MCP-1 antibody was used to block the MCP-1 within the collagen matrix, there was no significant difference between the number of migrated monocytes compared to the control group that was not treated with MCP-1. Furthermore, it is shown that the haptotactic (static) gradient of MCP-1 had the same effect and potency on monocyte migration as the chemotactic (soluble + static) gradient of MCP-1. The results suggest that the effect of the static MCP-1 concentration gradient on monocyte transmigration is stronger than the soluble MCP-1 concentration gradient in the collagen matrix. The results of this research will provide new information about the relationship between MCP-1 concentration gradients and monocyte transendothelial migration, and will lead to the development of an improved model to study transendothelial monocyte migration associated with inflammation.

(776h) STEP Enabled Long Time Culture of Primary Hepatocytes in Multiple Layers

Kevin Sheets, School of Biomedical Engineering and Sciences, Virginia Tech, Blacksburg, VA, Ji Wang, Engineering Science and Mechanics, Virginia Tech, Blacksburg, VA and Amrinder S. Nain, Mechanical Engineering, Virginia Tech, Blacksburg, VA

Introduction: Chronic diseases of the liver including hepatitis and cirrhosis account for a large portion of the 16,000 – 17,000 person-long waiting list for a liver transplantation. Only 5,000-6,000 liver donations are available each year, meaning only 63% of people needing a transplant are unable to receive one. To alleviate this deficit, bioartifical liver assist devices (BLADs) have been developed which perform the functions of a normal liver outside of the body. The design and implementation of more successful assist devices is driven by the need to maintain hepatocyte functionality over longer periods of time. In this work, hepatocytes are seeded onto an ECM-mimicking, highly aligned nanofibrous network. As single cells attach and spread, they pull against these fibers and cause deflections which allow migratory forces to be probed. We believe that an understanding of the magnitude and nature of these forces will allow improvements to be made to BLAD construction to create more viable assist devices for the many in need of them.

Figure 1. Hepatic arrangement around STEP Scaffolds

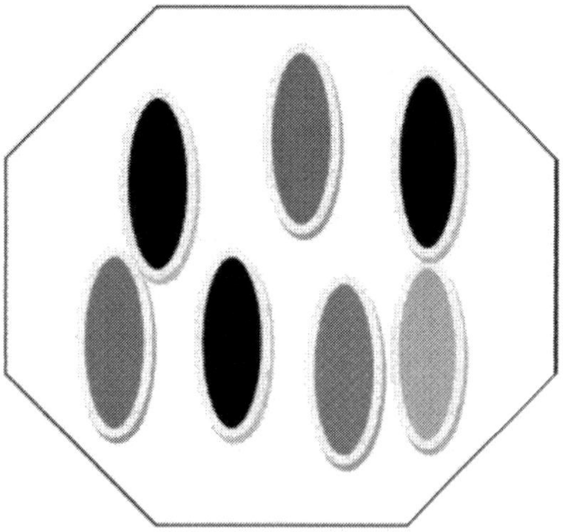

Figure 1. Hepatic arrangement around STEP Scaffolds

Materials and Methods: Using the previously-reported Spinneret-based Tunable Engineered Parameters (STEP) technique, nanofibrous scaffolds were created with fused intersections. Using xylene as a solvent, suspended polystyrene and polyurethane fibers of 500 nm and 300 nm diameter, respectively, were spun in criss-cross layers. Primary rat hepatocytes were seeded onto these constructs, which were then imaged with an incubating microscope. Migration behavior was analyzed via timelapse videos and fluorescence imaging. Cells were fixed and stained for nucleus (DAPI), actin (phalloidin) and focal adhesion (paxillin) locations.

Results and Discussion: After examining over 20 cells who had deflected nearby fibers, it was found that the average hepatocyte generated a force of approximately 1.5 nN.

Conclusions: We have shown that nanofibrous scaffolds carefully arranged for fiber spacing and layouts can cause cells to preferentially attach in specific shapes. According to focal adhesion immunohistochemistry, these configurations inherently display different attachment and migration forces which may prove useful in directed differentiation in the future.

AUTHOR INDEX

AUTHOR INDEX

AUTHOR INDEX

AUTHOR INDEX

AUTHOR INDEX

AUTHOR INDEX

AUTHOR INDEX

AUTHOR INDEX

AUTHOR INDEX

AUTHOR INDEX

AUTHOR INDEX

AUTHOR INDEX

AUTHOR INDEX

AUTHOR INDEX

AUTHOR INDEX

AUTHOR INDEX

AUTHOR INDEX

AUTHOR INDEX

AUTHOR INDEX

AUTHOR INDEX

AUTHOR INDEX